MicroRNA in Human Malignancies

MicroRNA in Human Malignancies

Edited by

Massimo Negrini
Department of Translational Medicine, The University of Ferrara, Ferrara, Italy

George A. Calin
Department of Translational Molecular Pathology, The University of Texas MD Anderson Cancer Center, Houston, TX, United States

Carlo M. Croce
Wexner Medical Center, The Ohio State University, Columbus, OH, United States

ELSEVIER

ACADEMIC PRESS
An imprint of Elsevier

Academic Press is an imprint of Elsevier
125 London Wall, London EC2Y 5AS, United Kingdom
525 B Street, Suite 1650, San Diego, CA 92101, United States
50 Hampshire Street, 5th Floor, Cambridge, MA 02139, United States
The Boulevard, Langford Lane, Kidlington, Oxford OX5 1GB, United Kingdom

Notices
Knowledge and best practice in this field are constantly changing. As new research and experience broaden our understanding, changes in research methods, professional practices, or medical treatment may become necessary.

Practitioners and researchers must always rely on their own experience and knowledge in evaluating and using any information, methods, compounds, or experiments described herein. In using such information or methods they should be mindful of their own safety and the safety of others, including parties for whom they have a professional responsibility.

To the fullest extent of the law, neither the Publisher nor the authors, contributors, or editors, assume any liability for any injury and/or damage to persons or property as a matter of products liability, negligence or otherwise, or from any use or operation of any methods, products, instructions, or ideas contained in the material herein.

Library of Congress Cataloging-in-Publication Data
A catalog record for this book is available from the Library of Congress

British Library Cataloguing-in-Publication Data
A catalogue record for this book is available from the British Library

ISBN 978-0-12-822287-4

For information on all Academic Press publications
visit our website at https://www.elsevier.com/books-and-journals

Publisher: Andre G. Wolff
Editorial Project Manager: Andrea Dulberger
Production Project Manager: Swapna Srinivasan
Cover Designer: Vicky Pearson

Typeset by STRAIVE, India

Contents

Contributors

Numbers in parenthesis indicate the pages on which the authors' contributions begin.

Samuel Akanksha (63), Department of Translational Medicine, The University of Ferrara, Ferrara, Italy

Alessandra Allione (255), Department of Medical Sciences, University of Turin, Turin, Italy

Juan Carlos Álvarez-Perez (153), Department of Biochemistry and Molecular Biology I, Faculty of Sciences, University of Granada; GENYO, Centre for Genomics and Oncological Research: Pfizer/University of Granada/Andalusian Regional Government; Institute of Biomedical Research of Granada (ibs.Granada), Granada, Spain

Alvaro Andrades (153), Department of Biochemistry and Molecular Biology I, Faculty of Sciences, University of Granada; GENYO, Centre for Genomics and Oncological Research: Pfizer/University of Granada/Andalusian Regional Government; Institute of Biomedical Research of Granada (ibs.Granada), Granada, Spain

Alberto M. Arenas (153), Department of Biochemistry and Molecular Biology I, Faculty of Sciences, University of Granada; GENYO, Centre for Genomics and Oncological Research: Pfizer/University of Granada/Andalusian Regional Government; Institute of Biomedical Research of Granada (ibs.Granada), Granada, Spain

Carlos Baliñas-Gavira (153), GENYO, Centre for Genomics and Oncological Research: Pfizer/University of Granada/Andalusian Regional Government; Institute of Biomedical Research of Granada (ibs.Granada), Granada, Spain

Dominik A. Barth (271), Research Unit of Non-Coding RNAs and Genome Editing in Cancer, Division of Clinical Oncology; Department of Medicine; Comprehensive Cancer Center Graz, Medical University of Graz, Graz, Austria

Anna Bielowski (47), Molecular Oncology Group, Biodonostia Research Institute, San Sebastián, Spain; Julius-Maximilians-University Wuerzburg, Würzburg, Germany

Roopa Biswas (77), Department of Anatomy, Physiology and Genetics, School of Medicine, Uniformed Services University of the Health Sciences, Bethesda, MD, United States

Mattia Boeri (245), Tumor Genomics Unit, Department of Research, Fondazione IRCCS Istituto Nazionale dei Tumori, Milan, Italy

Elisabetta Broseghini (339), Department of Experimental, Diagnostic, and Specialty Medicine, University of Bologna, Bologna, Italy

George A. Calin (21), Department of Translational Molecular Pathology, The University of Texas MD Anderson Cancer Center, Houston, TX, United States

Enrica Calura (141), Department of Biology, University of Padova, Padova, Italy

Elisabetta Casalone (255), Department of Medical Sciences, University of Turin, Turin, Italy

Vera Constâncio (291), Cancer Biology and Epigenetics Group, IPO Porto Research Center, Portuguese Oncology Institute of Porto & Porto Comprehensive Cancer Center, Doctoral Programme in Biomedical Sciences, School of Medicine & Biomedical Sciences, University of Porto, Porto, Portugal

Giulia Cosentino (195), Fondazione IRCCS Istituto Nazionale dei Tumori, Milan, Italy

Marina C. Costa (93,109), Instituto de Medicina Molecular João Lobo Antunes; Cardiomics Unit, Centro Cardiovascular da Universidade de Lisboa, Faculdade de Medicina, Universidade de Lisboa, Lisboa, Portugal

Carlo M. Croce (29,167), Wexner Medical Center, The Ohio State University, Columbus, OH, United States

Marta Cuadros (153), GENYO, Centre for Genomics and Oncological Research: Pfizer/University of Granada/Andalusian Regional Government; Department of Biochemistry and Molecular Biology III and Immunology, Faculty of Medicine, University of Granada; Institute of Biomedical Research of Granada (ibs.Granada), Granada, Spain

Christine E. Cutucache (187), STEM Teaching, Research, and Inquiry-Based Learning Center; Department of Biology, University of Nebraska at Omaha; Fred and Pamela Buffett Cancer Center, University of Nebraska Medical Center, Omaha, NE, United States

Ben Davidson (317), Department of Pathology, Oslo University Hospital, Norwegian Radium Hospital; University of Oslo, Faculty of Medicine, Institute of Clinical Medicine, Oslo, Norway

Emi Dika (339), Department of Experimental, Diagnostic, and Specialty Medicine, University of Bologna, Bologna, Italy

Sayra Dilmac (281), Department of Histology and Embryology, School of Medicine, Akdeniz University, Antalya, Turkey

Mihnea P. Dragomir (385), Department of Pathology, Charité-Universitätsmedizin Berlin, Corporate Member of Freie Universität Berlin, Humboldt-Universität zu Berlin and Berlin Institute of Health, Berlin, Germany

Rares Drula (21), Department of Translational Molecular Pathology, The University of Texas MD Anderson Cancer Center, Houston, TX, United States; Research Center for Functional Genomics Biomedicine and Translational Medicine, "Iuliu Hațieganu" University of Medicine and Pharmacy, Cluj-Napoca, Romania

Francisco J. Enguita (93,109), Instituto de Medicina Molecular João Lobo Antunes; Cardiomics Unit, Centro Cardiovascular da Universidade de Lisboa, Faculdade de Medicina, Universidade de Lisboa, Lisboa, Portugal

Zhaohui Feng (37), Department of Radiation Oncology, Rutgers Cancer Institute of New Jersey, Rutgers-State University of New Jersey, New Brunswick, NJ, United States

Manuela Ferracin (339), Department of Experimental, Diagnostic, and Specialty Medicine, University of Bologna, Bologna, Italy

Francesca Fornari (301), Department for Life Quality Studies, University of Bologna, Rimini; Centre for Applied Biomedical Research - CRBA, University of Bologna, St. Orsola Hospital, Bologna, Italy

Orazio Fortunato (245), Tumor Genomics Unit, Department of Research, Fondazione IRCCS Istituto Nazionale dei Tumori, Milan, Italy

André F. Gabriel (93,109), Instituto de Medicina Molecular João Lobo Antunes; Cardiomics Unit, Centro Cardiovascular da Universidade de Lisboa, Faculdade de Medicina, Universidade de Lisboa, Lisboa, Portugal

Daniel J. García (153), GENYO, Centre for Genomics and Oncological Research: Pfizer/University of Granada/Andalusian Regional Government; Department of Biochemistry and Molecular Biology III and Immunology, Faculty of Medicine, University of Granada; Institute of Biomedical Research of Granada (ibs.Granada), Granada, Spain

Laura Gramantieri (301), Centre for Applied Biomedical Research - CRBA, University of Bologna, St. Orsola Hospital; Division of Internal Medicine, Hepatobiliary and Immunoallergic Diseases, IRCCS Azienda Ospedaliero-Universitaria di Bologna, Bologna, Italy

Rui Henrique (291), Cancer Biology and Epigenetics Group, IPO Porto Research Center, Portuguese Oncology Institute of Porto & Porto Comprehensive Cancer Center; Department of Pathology, Portuguese Oncology Institute of Porto; Department of Pathology and Molecular Immunology, Institute of Biomedical Sciences Abel Salazar, University of Porto, Porto, Portugal

Wenwei Hu (37), Department of Radiation Oncology, Rutgers Cancer Institute of New Jersey, Rutgers-State University of New Jersey, New Brunswick, NJ, United States

Marilena V. Iorio (195), Fondazione IRCCS Istituto Nazionale dei Tumori, Milan, Italy

Carmen Jerónimo (291), Cancer Biology and Epigenetics Group, IPO Porto Research Center, Portuguese Oncology Institute of Porto & Porto Comprehensive Cancer Center; Department of Pathology and Molecular Immunology, Institute of Biomedical Sciences Abel Salazar, University of Porto, Porto, Portugal

Sakshi Kamboj (125), Virology Unit and Bioinformatics Centre, Institute of Microbial Technology, Council of Scientific and Industrial Research (CSIR), Chandigarh; Academy of Scientific and Innovative Research, Ghaziabad, India

Manoj Kumar (125), Virology Unit and Bioinformatics Centre, Institute of Microbial Technology, Council of Scientific and Industrial Research (CSIR), Chandigarh; Academy of Scientific and Innovative Research, Ghaziabad, India

Charles Henderson Lawrie (47), Molecular Oncology Group, Biodonostia Research Institute, San Sebastián; IKERBASQUE, Basque Foundation for Science, Bilbao, Spain; Radcliffe Department of Medicine, University of Oxford, Oxford, United Kingdom

Ana Lúcia Leitão (93,109), Faculdade de Ciências e Tecnologia, Universidade Nova de Lisboa, Caparica, Portugal; MEtRICs, Faculdade de Ciências e Tecnologia, Universidade Nova de Lisboa, Caparica, Portugal

Juan Liu (37), Department of Radiation Oncology, Rutgers Cancer Institute of New Jersey, Rutgers-State University of New Jersey, New Brunswick, NJ, United States

Francesca Lovat (167), Department of Cancer Biology and Genetics and Comprehensive Cancer Center, The Ohio State University, Columbus, OH, United States

Laura Masatti (63), Department of Translational Medicine, The University of Ferrara, Ferrara, Italy

Giuseppe Matullo (255,323), Department of Medical Sciences, University of Turin; Medical Genetics Unit, AOU Città della Salute e della Scienza, Turin, Italy

Pedro P. Medina (153), Department of Biochemistry and Molecular Biology I, Faculty of Sciences, University of Granada; GENYO, Centre for Genomics and Oncological Research: Pfizer/University of Granada/Andalusian Regional Government; Institute of Biomedical Research of Granada (ibs.Granada), Granada, Spain

Swati Mohapatra (21), Department of Translational Molecular Pathology, The University of Texas MD Anderson Cancer Center; The University of Texas MD Anderson Cancer Center UTHealth Graduate School of Biomedical Sciences, Houston, TX, United States

Vlad Moisoiu (385), Faculty of Physics, Babeş-Bolyai University, Cluj-Napoca, Romania

Massimo Negrini (63), Department of Translational Medicine, The University of Ferrara, Ferrara, Italy

Vu Hong Loan Nguyen (3), Department of Biology, York University, Toronto, Canada

Jacob Anderson O'Brien (3), Department of Biology, York University, Toronto, Canada

Bulent Ozpolat (281), Department of Experimental Therapeutics, The University of Texas MD Anderson Cancer Center, Houston, TX, United States

Barbara Pardini (323), Italian Institute for Genomic Medicine; Candiolo Cancer Institute, FPO-IRCCS, Candiolo, Italy

Juan Rodrigo Patiño-Mercau (153), Department of Biochemistry and Molecular Biology I, Faculty of Sciences, University of Granada; GENYO, Centre for Genomics and Oncological Research: Pfizer/University of Granada/Andalusian Regional Government; Institute of Biomedical Research of Granada (ibs.Granada), Granada, Spain

Laura Pazzaglia (395), Experimental Oncology Laboratory, IRCCS Istituto Ortopedico Rizzoli, Bologna, Italy

Paola Peinado (153), Department of Biochemistry and Molecular Biology I, Faculty of Sciences, University of Granada; GENYO, Centre for Genomics and Oncological Research: Pfizer/University of Granada/Andalusian Regional Government; Institute of Biomedical Research of Granada (ibs.Granada), Granada, Spain

Yuri Pekarsky (29), Department of Cancer Biology and Genetics, The Wexner Medical Center, Columbus, OH, United States

Chun Peng (3), Department of Biology; Centre for Research in Biomolecular Interactions, York University, Toronto, Canada

George E.D. Petrescu (385), Department of Neurosurgery, Bagdasar-Arseni Clinical Emergency Hospital, Bucharest, Romania

Martin Pichler (271), Research Unit of Non-Coding RNAs and Genome Editing in Cancer, Division of Clinical Oncology; Department of Medicine; Comprehensive Cancer Center Graz; Medical University of Graz, Graz, Austria

Ilaria Plantamura (195), Fondazione IRCCS Istituto Nazionale dei Tumori, Milan, Italy

Reuven Reich (317), Institute of Drug Research, School of Pharmacy, Faculty of Medicine, The Hebrew University of Jerusalem, Jerusalem, Israel

Maria Isabel Rodriguez (153), GENYO, Centre for Genomics and Oncological Research: Pfizer/University of Granada/Andalusian Regional Government; Department of Biochemistry and Molecular Biology III and Immunology, Faculty of Medicine, University of Granada; Institute of Biomedical Research of Granada (ibs.Granada), Granada, Spain

Chiara Romualdi (141), Department of Biology, University of Padova, Padova, Italy

Juan Sanjuan-Hidalgo (153), GENYO, Centre for Genomics and Oncological Research: Pfizer/University of Granada/Andalusian Regional Government, Granada, Spain

Katia Scotlandi (395), Experimental Oncology Laboratory, IRCCS Istituto Ortopedico Rizzoli, Bologna, Italy

Ram C. Shankaraiah (63), Department of Translational Medicine, The University of Ferrara, Ferrara, Italy

Ondrej Slaby (203), Centre for Molecular Medicine, Central European Institute of Technology; Department of Biology, Theoretical Departments, Faculty of Medicine, Masaryk University, Brno, Czech Republic

Carla Solé (47), Molecular Oncology Group, Biodonostia Research Institute, San Sebastián, Spain

Dharmendra Kumar Soni (77), Department of Anatomy, Physiology and Genetics, School of Medicine, Uniformed Services University of the Health Sciences, Bethesda, MD, United States

Gabriella Sozzi (245), Tumor Genomics Unit, Department of Research, Fondazione IRCCS Istituto Nazionale dei Tumori, Milan, Italy

Anamika Thakur (125), Virology Unit and Bioinformatics Centre, Institute of Microbial Technology, Council of Scientific and Industrial Research (CSIR), Chandigarh; Academy of Scientific and Innovative Research, Ghaziabad, India

Angelo Veronese (179), Department of Pharmacy and Drug Sciences, University of Bari Aldo Moro, Bari, Italy

Rosa Visone (179), Department of Medical, Oral and Biotechnological Sciences; Center for Advanced Studies and Technology (CAST), G. d'Annunzio University, Chieti, Italy

Petra Vychytilova-Faltejskova (203), Centre for Molecular Medicine, Central European Institute of Technology, Masaryk University, Brno, Czech Republic

Cen Zhang (37), Department of Radiation Oncology, Rutgers Cancer Institute of New Jersey, Rutgers-State University of New Jersey, New Brunswick, NJ, United States

Section A

miRNAs involvement in basic mechanisms of cancer development

Chapter 1

Overview on miRNA classification, biogenesis, and functions

Jacob Anderson O'Brien[a], Vu Hong Loan Nguyen[a], and Chun Peng[a,b]
[a]*Department of Biology, York University, Toronto, Canada,* [b]*Centre for Research in Biomolecular Interactions, York University, Toronto, Canada*

Introduction

Since the discovery of the first microRNA (miRNA), *lin-4*, in 1993 by the Ambros and Ruvkun groups in *Caenorhabditis elegans* (Lee, Feinbaum, & Ambros, 1993; Wightman, Ha, & Ruvkun, 1993), there have been more than 120,000 publications on the topic of miRNAs. Moreover, more than 1000 human miRNA genes have been validated while many hundreds more potential miRNA genes require further study (de Rie et al., 2017; Kozomara & Griffiths-Jones, 2011). MicroRNA are small noncoding RNA involved in the regulation of gene expression facilitated by the miRNA-induced silencing complex (miRISC). This complex, guided mostly by sequence complementarity between nucleotides 2–8 of miRNA and target RNA, allows for the highly specific regulation of gene expression. It is now well established that miRNAs play critical roles in normal developmental and physiological processes (Hayder et al., 2018; Rajman & Schratt, 2017) and their aberrant expression is associated with many human diseases, including cancer (Paul et al., 2018; Tufekci et al., 2014). In addition to cell-intrinsic gene expression regulation, miRNAs can also be secreted into extracellular fluids and play a role in mediating cell–cell communication (O'Brien et al., 2018). Importantly, extracellular miRNAs have been under intensive investigation for their potential to be used as predictive or prognostic biomarkers for human diseases (Hayes, Peruzzi, & Lawler, 2014; Huang, 2017; Wang, Chen, & Sen, 2016). It is no surprise then that miRNAs have become such a crucial element in our understanding of the cell and why miRNA as a field is so intensely studied.

In this chapter, we will start with an introduction of the complex network of gene regulation, followed by a short overview of miRNA biogenesis and nomenclature and a more in-depth discussion of miRNA functions in regulating gene expression and the underlying mechanisms. Each section will cover multiple independent but related topics that on their own provide a static view, a snapshot, of miRNA biology. Nonetheless, the aim of this chapter is not only to introduce fundamental miRNA theory but also to generate a deeper understanding of the potential reality of miRNA within mammalian cells. As such, in each section, the key concepts will be described assuming a framework of a dynamical nonlinear system to place the abstracted snapshot back into the living cells.

MicroRNA and the complex biological system of gene expression

A small noncoding RNA is typically defined as being between 18 and 31 nucleotides (nt) in length (Dozmorov et al., 2013), although this is not a strict definition but a commonly observed categorization; miRNAs are on average 22 nt in length. An RNA molecule that cannot be translated into a protein is termed noncoding RNA (ncRNA) as opposed to an RNA that contains a protein-coding sequence (CDS), such as messenger RNA (mRNA). The areas of gene regulation that miRNAs are involved in range from transcriptional/posttranscriptional to translational, meaning that miRNA can regulate the amount of mRNA in a cell as well as how much mRNA is translated into protein.

Both these processes can then be further differentiated theoretically and mechanistically. First, both transcription and translation are at least dynamical steady-state equilibrium processes. A dynamical system is a system that changes over time (Janson, 2012). If the amount of mRNA existing within any given cell is a dynamical steady-state process, this implies that the amount of mRNA both changes (regulated creation and destruction) and is maintained over time. There are many ways cells regulate the abundance of mRNA, generally termed mRNA metabolism, and they can be grouped into transcription, maturation, and decay/turnover.

Messenger RNA are transcribed by RNA polymerase II (POLII) stochastically (Cao & Grima, 2020; Elowitz et al., 2002), analogous to randomly. In other words, the average rate of transcription of any given gene may remain the same at

MicroRNA in Human Malignancies. https://doi.org/10.1016/B978-0-12-822287-4.00027-X

steady state (e.g., 3 transcripts/min) but the time interval between each transcription event tends to vary within some probability distribution. Maturation of the primary transcript involves co/posttranscriptional splicing, 3′ poly-adenylation, m^7G capping, and nuclear export to the cytoplasm (Ben-Yishay & Shav-Tal, 2019). Within the cell, mRNAs do not exist alone as naked RNAs but instead associate with a variety of proteins involved in the regulation of their stability (i.e., change in mRNA abundance over time) and translation potential (how likely the mRNA is to be translated into protein); protein-bound mRNAs are termed messenger ribonucleoproteins (mRNPs) (Borbolis & Syntichaki, 2015). It should be generally assumed that when discussing mRNAs, mRNPs are implied.

In the cytoplasm, various processes regulate mRNA turnover. Stability can be modulated by subcellular localization of the mRNA to areas that promote or suppress mRNA decay or as a result of sequence errors detected during translation that induce mRNA degradation (Borbolis & Syntichaki, 2015). Although the cause of degradation can vary, this usually involves poly-deadenylation, m7G cap removal (decapping), and/or endonucleolytic cleavage followed by 5′– 3′ or 3′– 5′ exonucleolytic decay.

Taken together, mRNA metabolism is at a steady state when there is a balanced rate of transcription, maturation, and turnover. If any one of these equilibriums are shifted, such as by increasing transcription, the entire cellular system equilibrates to its new steady state, which may or may not be an increase in mRNA abundance. Moreover, the complexity of the system allows even a small change in gene expression to result in significant and generally nonlinear effects on the overall cellular system (Janson, 2012). There are thousands of genes expressed at any given time, called the transcriptome (Mele et al., 2015), all of which in principle exert some degree of direct or indirect regulation on each other. Consequently, when we talk about gene regulation in general, it is important to keep in mind the interconnectedness of the cellular system.

Lastly, how are we defining a system? In our first example, the system was the entire body, but we could easily define our system as a tissue, single cell, subcellular compartment, signaling pathway, etc. What's important is that elements (nodes) of the system, for example a single gene, are described relative to all other elements in the system. With the application of rules that define the consequence of all pairwise interactions of nodes, the system can be modeled given some initial state. The models of biological behaviors that scientists have generated are described as complex systems, in some cases through networks (Yan et al., 2018) and others through analytical solutions (Janson, 2012). All the interesting biological behaviors that we have come to understand well describe a complex system, one that is made of many heterogeneous elements, filled with nonlinear outputs, stochasticity, dynamics, massive interconnectedness, and feedbacks (Ma'ayan, 2017). A complex system is one where the same change in one parameter can lead to an entirely different set of outcomes with a simple change of the initial state.

MicroRNA biogenesis, nomenclature, and classification

Origins of the canonical primary miRNA transcript

The production of miRNA is a multistep, hierarchical process that involves the processing of RNA transcripts with a characteristic stemloop secondary structure to that of the single-stranded mature miRNA loaded into the Argonaute (AGO) family of proteins at which point they can elicit their gene regulatory potential. An important factor in understanding the first step of miRNA biogenesis, the synthesis of the primary miRNA (pri-miRNA) by POLII, is in the distribution of miRNA genes throughout the genome (Ha & Kim, 2014). Intragenic miRNA genes are located within the body of another protein-coding or noncoding gene called the host gene, and they make up approximately half of all identified miRNA genes (de Rie et al., 2017; Kim & Kim, 2007). Intragenic miRNA can also be further categorized depending on the position they occupy within the host gene, mainly exonic or intronic, with the majority of miRNA genes being intronic (Chang et al., 2015; de Rie et al., 2017). Intergenic miRNAs, as the name suggests, exist independent of a host gene and are transcriptionally regulated by their own promoters (de Rie et al., 2017).

What exactly then is the pri-miRNA? For miRNAs whose expression is coupled to their host gene, the pri-miRNA would be the pre-mRNA of the host gene. Conversely, intergenic miRNAs are transcribed into their own 5′ m^7G capped and 3′ polyadenylated pri-miRNA transcript like any other POLII transcripts (Cai, Hagedorn, & Cullen, 2004; Lee et al., 2004). Unfortunately, this tidy categorization did not hold over time as new discoveries were made. For example, multiple intragenic miRNAs are regulated by their own promoters and can also be expressed independent of their host gene (de Rie et al., 2017; Kim & Kim, 2007; Marsico et al., 2013). Moreover, intragenic miRNA on the antisense strand of the host gene can be produced through antisense transcription (Georgakilas et al., 2014). Intergenic pri-miRNA transcripts can also be coupled to a proximal host gene through either divergent transcription or transcription that carries on past the end of the gene (readthrough transcription) (Georgakilas et al., 2014).

From canonical primary miRNA to mature miRNA

Regardless of the source of pri-miRNA, a common characteristic shared by canonical pri-miRNA is a stemloop (hairpin) structure with a central mismatch embedded within the transcript. The single strands of the stemloop are what will ultimately make up the mature miRNA. The stemloop is cleaved from the pri-miRNA near the basal junction (the base/end of the stemloop structure) into the ~70nt in length precursor miRNA (pre-miRNA) (Denli et al., 2004). The cleavage process typically results in a 2nt 3′ overhang, which plays a role in strand selection and pre-miRNA cleavage in subsequent biogenesis processes (Han et al., 2004; Tian et al., 2014). The cleavage event is accomplished through the ~365kDa microprocessor complex consisting of an RNA binding protein DiGeorge Syndrome Critical Region 8 (DGCR8) dimer (Senturia et al., 2010) and the ribonuclease III enzyme DROSHA (Denli et al., 2004). In addition, there are various motifs, structures, and RNA modifications associated with the pri-miRNA stemloop that promote pri-miRNA processing into pre-miRNA. This is mainly accomplished by promoting DROSHA or DGCR8 binding and/or correct positioning to facilitate efficient cleavage of the stemloop (Auyeung et al., 2013; Chaulk et al., 2011; Conrad et al., 2014; Fang & Bartel, 2015; Roden et al., 2017) (Fig. 1.1).

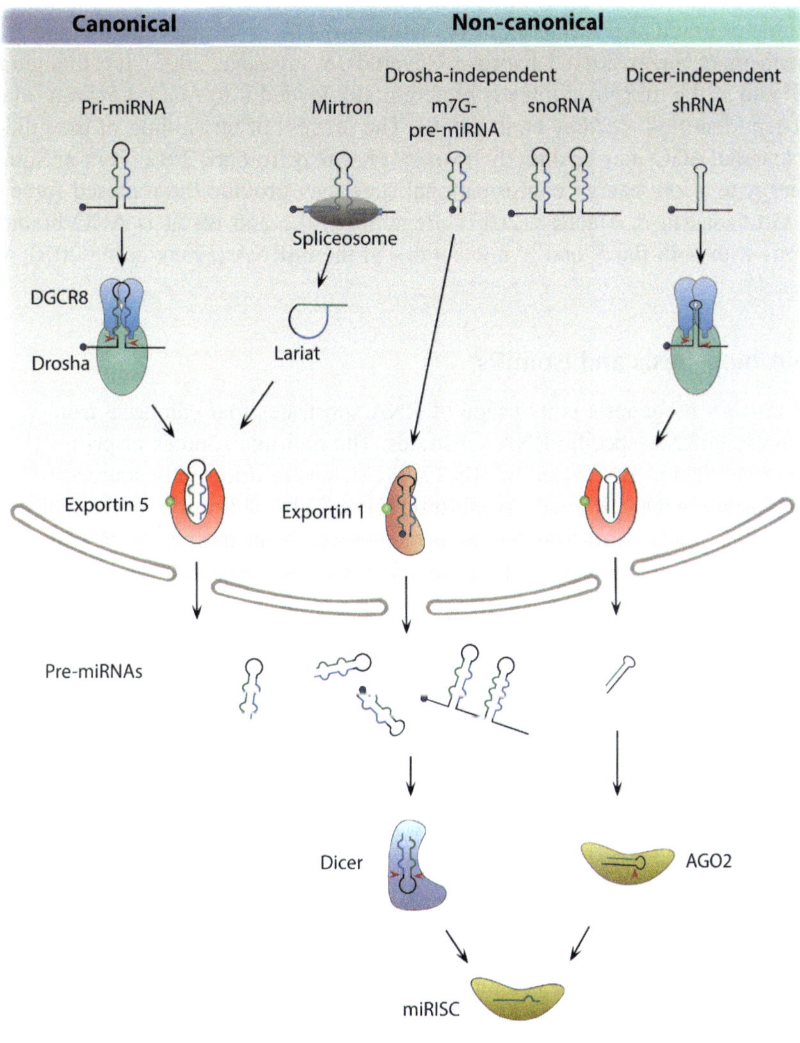

FIG. 1.1 Canonical and noncanonical microRNA biogenesis. Like mRNA, most miRNA genes are transcribed by RNA polymerase II in the nucleus to produce long transcripts termed pri-miRNA. In the canonical pathway, the microprocessor complex, which comprises Drosha and DGCR8, recognizes target pri-miRNA and cleaves both strands of pri-miRNA near the base of the stemloop structure, generating a ~70nt hairpin RNA precursor termed pre-miRNA. Upon the processing event, pre-miRNAs are transported to the cytoplasm by the Exportin 5/Ran-GTP complex, where the final processing event occurs. In the cytoplasm, RNase III endonuclease Dicer recognizes and cleaves pre-miRNA near the apical loop, giving rise to a small miRNA duplex. Subsequently, either strand (-5p or -3p) of the miRNA duplex is preferentially loaded into an AGO protein to form an miRNA-induced silencing complex (miRISC). In addition to the canonical pathway, functional miRNAs can be produced from noncanonical pathways. Some of the examples of the noncanonical pathway include mirtrons, m7G-pre-miRNA, snoRNA, and shRNA. While the production of miRNA from mirtron, m7G-pre-miRNA, and snoRNA is independent of the DROSHA/DGCR8 complex, shRNA has been shown to be processed independently of DICER.

The pre-miRNA is then exported to the cytoplasm via the Exportin-5 (XPO5)/RanGTP complex, where it is processed by the RNase III endonuclease Dicer (Denli et al., 2004; Okada et al., 2009) in cooperation with double-stranded RNA binding protein (dsRBP) trans-activation-responsive RNA-binding protein (TRBP) (Chendrimada et al., 2005). The 2 nt 3′ overhang of the pre-miRNA initially binds to DICER; this interaction is stabilized with the help of TRBP bound to both the pre-miRNA and DICER (Tian et al., 2014; Wilson et al., 2015). The apical loop of the pre-miRNA is then cleaved, leaving the mature miRNA duplex bound to DICER (Lau et al., 2012; Zhang et al., 2004) (Fig. 1.1).

The final step in canonical miRNA biogenesis is the loading of one of the miRNA duplex strands into the Argonaute family of proteins (AGO1-4) to form the miRNA-induced silencing complex (miRISC). This is accomplished through the TRBP-stabilized interaction of AGO and DICER to form the RISC loading complex (Chendrimada et al., 2005; Fareh et al., 2016; Wang et al., 2009). The exact mechanism of loading is not fully understood, but the prevailing theory suggests that there is a direct handoff of the miRNA from DICER to AGO (Nakanishi, 2016). The strand that is loaded into AGO is termed the guide strand and the unloaded is called the passenger strand. Depending on the cellular environment, there can be a strand bias toward either the 5p or 3p strands, ranging from near 50:50 proportions to predominantly one or the other (Medley, Panzade, & Zinovyeva, 2020; Meijer, Smith, & Bushell, 2014). Two identified mechanisms that mediate the strand bias are a 5′ adenine or uracil at position 1 of the mature miRNA or the thermodynamic stability at the 5′ end of the miRNA (Frank, Sonenberg, & Nagar, 2010; Khvorova, Reynolds, & Jayasena, 2003). It is thought that the strand with the weakest pairing at the 5′ end of the miRNA duplex is preferentially loaded into AGO; however, many factors modulating 5p/3p abundance have been identified (Medley et al., 2020). The process of unwinding of the guide strand from the passenger strand is still unclear, but AGO may be directly involved (Kwak & Tomari, 2012; Park & Shin, 2015). The transition of AGO from a high-energy to a low-energy conformational state may provide the required force to unwind the strands (Gu et al., 2012; Sheu-Gruttadauria & MacRae, 2017). Regardless, the end result is AGO bound to a mature miRNA through stable interactions with both the 5′ and 3′ nucleotides of the miRNA (Frank et al., 2010; Ma, Ye, & Patel, 2004; Wang et al., 2008) (Fig. 1.1).

Noncanonical miRNA biogenesis and isomiRs

An intriguing aspect of miRNA biogenesis is its usage of RNA substrates that can arise from various disparate cellular processes rather than entirely miRNA-specific RNA substrates. The multiple sources of pri-miRNA are the first example of substrate dependence rather than miRNA-specific RNA sources; sets of differently sourced RNA transcripts can all act as substrates for nuclear DROSHA/DGCR8 and cytoplasmic DICER. By definition, all canonical miRNAs require both the microprocessor complex and DICER to complete their maturation. Noncanonical miRNAs are currently classified by their independence of either the microprocessor complex or DICER to complete their maturation (Fig. 1.1). Mirtrons, pre-miRNA produced from small introns as a result of splicing, resemble DICER substrates that are not processed by DROSHA and are thus classified as DROSHA-independent (Babiarz et al., 2008; Rorbach, Unold, & Konopka, 2018; Ruby, Jan, & Bartel, 2007). Mirtron pre-miRNAs are then exported to the cytoplasm and processed like canonical pre-miRNA. In addition, 7-methylguanosine (m^7G)-capped pre-miRNA (Xie et al., 2013) and small nucleolar RNA (snoRNA) (Ender et al., 2008) can act as DICER substrates without processing by DROSHA. m^7G-capped pre-miRNA are shuttled to the cytoplasm by exportin 1 (XPO1) and cleaved by DICER. A strong 3p strand bias has been observed in this case, likely due to the cap interfering with the 5′ nucleotide AGO-binding of the 5p strand (Xie et al., 2013). The DICER-independent endogenous short hairpin RNA (shRNA) are another identified source of noncanonical miRNA (Yang et al., 2010). These transcripts are cleaved into pre-miRNA by the microprocessor and transported to the cytoplasm by XPO5 but are too small (dsRNA length of ~18 bp) to act as DICER substrates (Yang et al., 2010). Instead, the entire miRNA duplex is loaded into AGO2 and undergoes AGO2-dependent slicing of the 3p strand followed by 3′-5′ trimming of the remaining 3p fragment to complete AGO2 loading (Cheloufi et al., 2010).

The positioning of the microprocessor complex on pri-miRNA and DICER on pre-miRNA is critical to the final mature miRNA sequence (Neilsen, Goodall, & Bracken, 2012). Differential positioning of these nucleases can lead to pri/pre-miRNAs that undergo alternative cleavage, producing 5′ and 3′ sequence variants called isomiRs (Vermeulen et al., 2005; Wu et al., 2009). Variations to the 5′ seed region have been shown to affect miRNA targeting (Manzano et al., 2015), whereas 3′ variations typically affect miRNA abundance and activity (Yu et al., 2017a). In some cases, single miRNA genes can lead to dozens of expressed 5′/3′ isomiRs (Parafioriti et al., 2020). Nontemplated nucleotide addition (NTA) of adenine or uracil at 3′ ends has been shown to directly affect miRNA stability (Neilsen et al., 2012). Terminal nucleotidyltransferase 2 (TENT2, also known as GLD2) has been shown to catalyze 3′ adenylation and promote miRNA stability (D'Ambrogio et al., 2012; Katoh et al., 2009), although destabilization has also been observed (Boele et al., 2014). Uridylation of 3′ miRNA termina by multiple terminal uridylyltransferase (TUT) family members has been associated with reduced activity and stability (Gutierrez-Vazquez et al., 2017; Jones et al., 2009; Kim et al., 2015). Internal variations in miRNA sequence are termed polymorphic isomiRs. These miRNAs undergo RNA editing, typically adenosine to inosine via adenosine deaminase RNA specific (ADAR) (Nishikura, 2016) or less frequently cytidine to uracil by apolipoprotein B mRNA editing enzyme catalytic (APOBEC) family members (Salter, Bennett, & Smith, 2016). Depending on the position of sequence variation, RNA-edited miRNA can show differences in RNA targets (Kawahara et al., 2007) or miRNA abundance (Kawahara et al., 2008; Vesely et al., 2014).

Nomenclature and classification

Given the relatively complex process of miRNA biogenesis, various naming conventions have been developed to aid in inferring the source of the miRNA, stage of biogenesis, and which arm of the pri-miRNA the mature miRNA is embedded within (Budak et al., 2016). Moreover, various other miRNA features, such as sequence homology, have been used in naming groups of miRNA genes to infer a relation between them (Seal et al., 2020). We will discuss in the next section why sequence homology is a defining feature of the mechanism of miRNA function and why this would infer a relation between miRNAs. It should be noted that naming conventions are still evolving as our understanding of miRNA biology is incomplete.

As discussed previously, miRNA biogenesis involves the processing of primary miRNAs to precursor and finally to mature miRNA. The primary miRNA is transcribed either with a host gene or from its own gene. Regardless if the miRNA is intra- or intergenic, each miRNA is given a gene symbol. In humans, miRNA is shortened to MIR and a number is assigned to it, such as *MIR25*, while in mice only the "M" is capitalized, such as *Mir25*. The number is assigned by chronological order of discovery. There are exceptions, like with the "let" family of miRNA, but this is an example of a historical naming convention.

The pri/pre-miRNA adopts a different naming structure. Using *MIR25* as an example, it would be written hsa-mir-25 for humans or mmu-mir-25 for mice. The species should always be added to the miRNA name within the abstract, the first time it is used in the text, or as needed to avoid confusion if the species cannot be inferred. This will be ever more important in the future as more computer-based natural language processing is applied to miRNA studies. Unfortunately, this naming convention is not well followed.

When referring to the mature, single-stranded miRNA, we capitalize the "R," hsa-miR-25, and identify which arm of the pre-miRNA the mature miRNA was embedded within, hsa-miR-25-5p or hsa-miR-25-3p (Fig. 1.2). A previous naming practice employed the use of the asterisk (*) to denote a passenger strand that was predominantly degraded *in place of* the strand identifier (i.e., 5p or 3p) but this has been largely deprecated. A common example was with the hypoxamir (an miRNA that is induced by hypoxia) hsa-miR-210 and its degraded passenger strand hsa-miR-210* (Huang, Le, & Giaccia, 2010). But this raises the question, which exact miRNA are we referring to when we say miR-210? Without the RNA sequence or a priori knowledge, it would remain ambiguous. As such, we should never refer to a mature miRNA without first identifying the arm it was embedded within, like with the hypoxamir miR-210-3p. It is also common practice to drop the strand identifier when referring to the 5p but for the same reasons above, this should be avoided. Evolving on past nomenclature, when referring to a common and predominant strand bias, the addition of the asterisk can be used in conjunction with the strand identifier (Budak et al., 2016). In the case of miR-210, we could refer to the 5p strand as miR-210-5p* as it is predominantly the passenger strand. Likewise, miRNA* would refer to the mature miRNAs that are predominantly passenger strands.

In some cases, multiple proximal miRNA genes are transcribed within a single RNA molecule. These miRNAs are referred to as clusters (Desvignes et al., 2015). A well-studied example is the mir-19/72 cluster (*MIR17HG*, "HG" meaning host gene), consisting of members mir-17, mir-18a, mir-19a, mir-20a, mir-19b-1, mir-92a-1, also known as oncomiR-1 (Mogilyansky & Rigoutsos, 2013), with oncomiR being the miRNA equivalent of an oncogene. Cluster members also tend to share similar expression profiles but depending on the cellular context, can be differentially expressed (Fang et al., 2017). In some cases, cluster members may also cooperate in the regulation of similar pathways within the cell, as with the mir-17/92 cluster (Mogilyansky & Rigoutsos, 2013) (Fig. 1.3A).

Depending on sequence homology, miRNAs can also be grouped into families (Desvignes et al., 2015). MicroRNA genes that share a 100% sequence identity between mature miRNA sequences have an added incrementing number suffix while maintaining the same gene name. For example, hsa-mir-218-**1** (chr4: *MIR218-1*) and hsa-mir-218-**2** (chr5: *MIR218-2*) contain a 5p strand with identical sequences (Fig. 1.3B). Because the mature miRNAs share a 100% sequence identity, we can refer to the pool of miR-218-1-5p and miR-218-2-5p within the cell as just miR-218-5p. Likewise, exogenous sources of miRNA mimics for the 5p strand would also be referred to as miR-218-5p mimics. The 3p strands of *MIR218-1/-2* are not identical and thus, they are referred to as miR-218-1-3p and miR-218-2-3p. Lastly, miRNAs that share a sequence homology

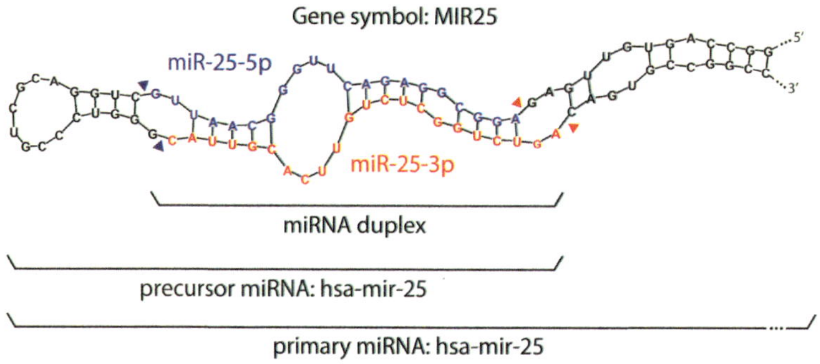

FIG. 1.2 Structure and naming of miRNA. After transcription of the primary (pri-) miRNA, the pri-miRNA is cleaved by nuclear DROSHA (*red arrows*) to produce the precursor (pre-) miRNA. Cleavage by cytoplasmic DICER (*blue arrows*) produces the miRNA duplex. Either the 5p or 3p mature strand is then loaded into the AGO family of proteins, blue and red strands, respectively. *MIR25* pri/pre miRNA are labeled as mir-25 whereas mature miRNAs include the strand identifier (5p or 3p) and a capital R, miR-25-5p.

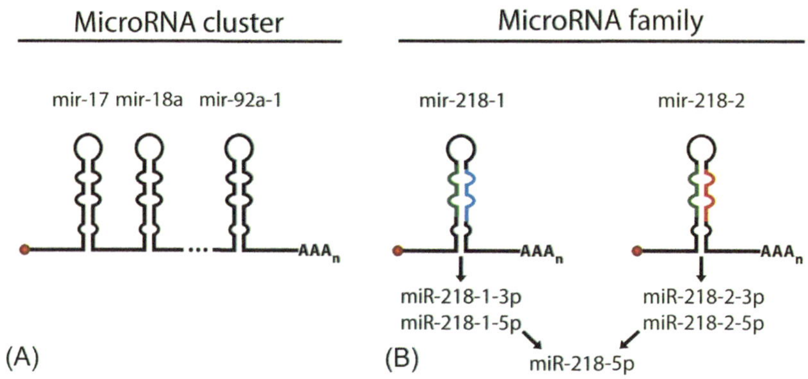

FIG. 1.3 MicroRNA clusters and families. (A) MicroRNA clusters are single RNA molecules containing multiple precursor miRNA, such as with the mir-19/72 cluster. (B) MicroRNA families are classified by shared sequence identities; similar colors indicate 100% sequence identity. MicroRNA genes that share at least one identical mature sequence have a numerical suffix added to the miRNA gene symbol (*MIR218-1*/*MIR218-2*). In this example, the 5p strands of miR-218-1/2 share 100% sequence identity. In addition, because both miR-218-1/2-5p strands are identical, the family suffix can be dropped when referring to the mature miRNA, that is, miR-218-5p refers to both miR-218-1-5p and miR-218-2-5p. However, the mature 3p strands are not identical and must then be referred to as miR-218-1-3p and miR-218-2-3p.

between positions 2–7 but do not share a 100% sequence identity are given an added incrementing letter suffix instead to denote their membership in an miRNA family. For example, hsa-miR-19**a**-5p (5′ **AGUUUUGCA**UAG**UUGCA**CUACA) and hsa-miR-19**b**-1-5p (5′ **AGUUUUGCA**GG**UUUGCA**UCCAGC) share a sequence homology only at positions 1–9 and 13–17 (Fig. 1.3A). It should be noted that miRNA family classification when dealing with evolutionary relation is more complex (Zou et al., 2014) but the nomenclature is the same.

Mechanisms of miRNA-mediated gene regulation

Posttranscriptional regulation of mRNA stability and translation

Once the mature miRNA is loaded into miRISC, this forms a minimally competent complex in which the miRNA acts as a guide to facilitate mRNA target recognition while AGO acts as a mediator of the regulatory potential of the complex (Huntzinger & Izaurralde, 2011; Ipsaro & Joshua-Tor, 2015; Kawamata & Tomari, 2010). Target specificity is a result of sequence complementarity between the miRNA and mRNA target (Jo et al., 2015). The site of interaction on target mRNA is called the miRNA response element (MRE) (Nair et al., 2020). Generally, nucleotides 2–8 of the miRNA are sufficient in conveying robust target specificity (Bartel, 2009; Ellwanger et al., 2011; Xu et al., 2014). Given its importance, it was termed the 5′ seed region of the miRNA and was sufficient for the in silico prediction of many MRE (Quillet et al., 2020). Additional pairing throughout the remaining 3′ region of the miRNA has been shown to increase specificity and stability (Sheu-Gruttadauria et al., 2019a). This would be expected given our current understanding of the miRNA:MRE interaction dynamic. When a miRNA is bound to AGO, the seed region of the miRNA is oriented such that nucleotides 2–4/5 can rapidly bind complementary RNA (Chandradoss et al., 2015; Salomon et al., 2015; Schirle, Sheu-Gruttadauria, & MacRae, 2014).

This interaction is metastable and miRISC will quickly dissociate if the remaining seed region nucleotides are not complementary and cannot stabilize the interaction (Chandradoss et al., 2015). This promotes efficient scanning of potential target sites and more stable, long-lived interactions with true MREs (Chandradoss et al., 2015).

miRISC associations with target MREs are indiscriminate, and have the potential to interact with an MRE anywhere within the RNA molecule (Plotnikova, Baranova, & Skoblov, 2019; Xu et al., 2014). This assumes that the MRE is available to interact with miRISC. However, this does not mean that any given MRE will have equivalent regulatory potential. MRE that are complementary to a 5′ seed region of 7 nucleotides and contain an adenine opposite to position 1 (termed t1A) of the miRNA (8-mer) are generally the highest affinity MRE (Agarwal et al., 2015) where 7/6-mer MRE, lacking the position 1 adenine or with mismatches to the seed region, have a lower affinity and as a result, the miRISC:MRE interaction is less stable (Agarwal et al., 2015; Friedman et al., 2009; Nielsen et al., 2007). RNA secondary structures and RNA-binding proteins can inhibit or promote miRISC interactions to potential MREs (Kedde et al., 2010; Min et al., 2017; Nam et al., 2014). Typically, MREs within the 3′ untranslated region (UTR) of target mRNA lead to translation initiation inhibition and mRNA decay, thereby resulting in a decrease in mRNA and protein levels of the target genes (Brkic et al., 2018; Huntzinger & Izaurralde, 2011; Ipsaro & Joshua-Tor, 2015). Functional MREs within the 5′ UTR and CDS have also been reported but they are uncommon (Forman, Legesse-Miller, & Coller, 2008; Zhang et al., 2018a). The current model suggests that translating ribosomes actively lead to the efficient dissociation of miRISC, preventing miRISC activity (Gu et al., 2009). However, high affinity 3' miRNA:MRE interactions within the CDS have been shown to induce ribosome stalling and mRNA degradation hypothesized via a nonstop decay mechanism (Zhang et al., 2018b).

Of the four human AGO family members (AGO1-4), only AGO2 is known to have endonuclease activity. The degree of complementarity between miRNA and MRE determines if target mRNA can be cleaved. A fully complementary miRNA:MRE interaction induces AGO2 endonuclease activity (Fig. 1.4A) while any central mismatches (positions ~ 9–12) in this interaction prevent mRNA cleavage (Jo et al., 2015). The overwhelming majority of identified human MREs contain central mismatches with their cognate miRNA and as such, mRNA cleavage in humans is generally rare (Hansen et al., 2011; Jonas & Izaurralde, 2015). For clarification, cognate miRNAs are the set of miRNAs that are capable of interacting with a given MRE, whereas cognate MRE are the set of MRE capable of interacting with a given miRNA.

Most miRNA:mRNA interactions lead to target repression (Salem et al., 2018; Valinezhad Orang, Safaralizadeh, & Kazemzadeh-Bavili, 2014; Wang et al., 2019). In order for miRISC to affect mRNA stability without directly cleaving the target mRNA, other factors are recruited to miRISC to inhibit translation initiation and/or to promote mRNA decay (Fig. 1.4B). Inhibition of translation initiation has been shown to involve the dissociation of eukaryotic translation initiation factor 4A1 (EIF4A1)/EIF4A2, preventing formation of the EIF4F translation initiation complex (Behm-Ansmant et al., 2006; Fukao et al., 2014; Meijer et al., 2013). The current model suggests that mRNA decay is accomplished by localizing the poly(A) tail of mRNA to miRISC, followed by the recruitment of multiple deadenylases that can ultimately promote degradation of the mRNA (Christie et al., 2013; Jonas & Izaurralde, 2015). The crucial adapter protein that facilitates poly(A) localization to miRISC is the trinucleotide repeat containing adaptor 6A (TNRC6A), also called GW182 (Liu et al., 2005a). The N-terminal domain of TNRC6A interacts with AGO while the C-terminal domain interacts directly with poly(A)-binding protein C (PABPC) (Huntzinger et al., 2013). The C-terminal also recruits poly(A) nuclease deadenylation complex subunit 2 (PAN2)/PAN3 and the carbon catabolite repressor protein 4 (CCR4):NOT complex (Braun et al., 2011; Chen et al., 2009). PAN2/3 initiates the deadenylation of target mRNA and CCR4:NOT completes the reaction (Jonas & Izaurralde, 2015). Deadenylation promotes the decapping of target mRNA by the mRNA-decapping enzyme subunit 1 (DCP1)/DCP2 complex (Behm-Ansmant et al., 2006). Decapped mRNA can then be degraded by 5′–3′ exoribonuclease 1 (XRN1) (Fig. 1.4B) (Braun et al., 2012).

Models of miRNA-mediated gene regulation within the nucleus

The role of miRNAs within the nucleus is an emerging area of study. MicroRNAs have been detected in the nucleus and several potential functions of miRNAs in the nucleus have been suggested. Perhaps the most striking is their potential to interact with DNA and regulate transcription. However, the nucleus also contains large amounts of RNA ranging from pre-mRNA to ribosomal RNA, raising the possibility that miRNAs may interact with these RNA molecules to regulate their levels.

Depending on cell type, the ratio of nuclear to cytoplasmic miRNA varies, but a significant population of nuclear miRNA has been observed in many cell types ranging from embryonic to neural cells in mice and humans (Catalanotto, Cogoni, & Zardo, 2016; Fan et al., 2020; Sarshad et al., 2018; Turunen et al., 2019; Zhang, Shen, & Cui, 2019). Changes in cell state have been shown to affect miRNA distributions within the cell with senescence increasing nuclear miRNA abundance (Benhamed et al., 2012) and serum starvation depleting it (Kucherenko & Shcherbata, 2018). These large-scale

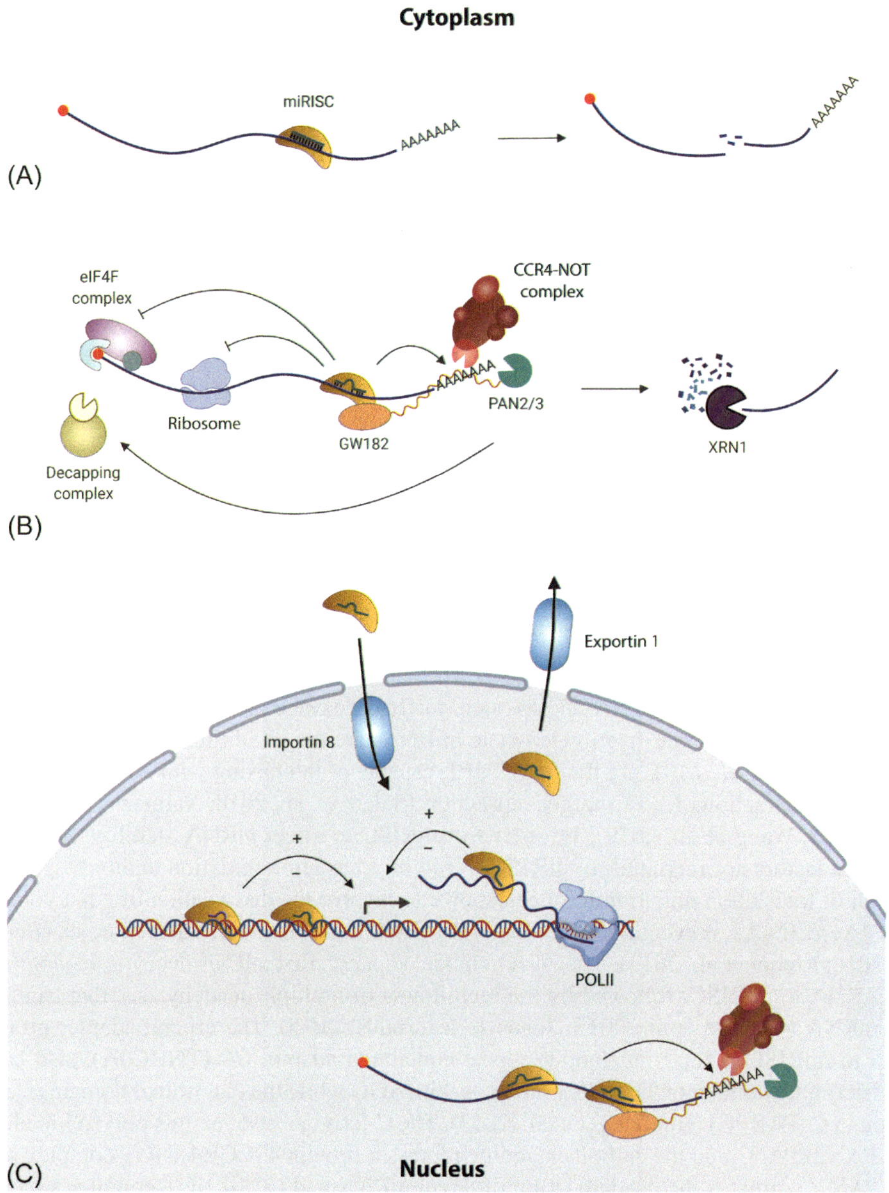

FIG. 1.4 Cytoplasmic and nuclear functions of miRNA. The repressive effect of miRNA at the posttranscriptional level is facilitated through the promotion of either mRNA cleavage or mRNA decay and repression of translation initiation. The sequence of the loaded miRNA gives the miRISC complex target specificity via complementary miRNA:MRE interaction. While perfect complementarity of miRNA and MRE of the target mRNA results in mRNA cleavage (A), a majority of miRNA:MRE interactions have at least a central or 3′ mismatch in mammalian cells. Most miRISC binding in mammalian cells leads to translation initiation inhibition and mRNA decay through the recruitment of other factors (B). Mechanically, translation initiation inhibition is a consequence of eukaryotic translation initiation factor 4A1 (EIF4A1)/EIF4A2 dissociation. In addition, the binding of protein TNRC6A to AGO of the miRISC complex provides a binding platform for poly(A) binding proteins, including PAN2/3 and CCR4/NOT deadenylase complexes, which initiates and completes poly-A tail deadenylation of the target mRNA, respectively. The target mRNA is further decapped by decapping protein 1/2 (DCP1/2) complexes, and then degraded by 5′–3′ exoribonuclease 1 (XRN1). In addition, miRISC can be transported into the nucleus via the Importin 8/RAN-GTP complex and functions in regulating transcription at the target loci (C).

changes in miRNA localization tend to involve interactions between miRISC and cellular proteins whose localization is affected by the change in cell state. For example, unliganded estrogen receptor β (ERβ) led to the nuclear shuttling of ERβ:miRISC complexes (Tarallo et al., 2017) and the inactivation of canonical Wnt signaling increased the nuclear localization of miRNA (Di Mauro et al., 2019). In the case of hypoxia, the differential expression of miRNA was observed between the cytoplasm and nucleus, notably leading to significant increases of hsa-miR-210-3p within the nucleus during hypoxia (Turunen et al., 2019).

Multiple studies have attempted to identify various miRNA sequence motifs regulating steady-state nuclear or cytoplasmic localization, but these have been met with conflicting results. The current model involves a constant two-way shuttling of free miRISC:TNRC6A complexes (miRISC not bound to target RNA) between the nucleus (importin-8; IPO8) and cytoplasm (XPO1) (Nishi et al., 2013; Nishi et al., 2015; Pitchiaya et al., 2017). Instead of a sequence motif mediating localization (Hwang, Wentzel, & Mendell, 2007), the abundance of target RNA within the nucleus is likely to be the most significant miRNA-specific factor modulating miRISC retention within the nucleus, that is, an increase in target RNA within the nucleus is sufficient to mediate cognate miRNA nuclear retention. In the nucleus, free-miRISC diffuses more quickly than bound-miRISC, and as such, miRISC binding to more nuclear RNA targets shifts the steady-state distribution toward the nucleus (Pitchiaya et al., 2017). The current model may ultimately explain the previous conflict of trying to identify a nuclear localization signal (NLS) within miRNA. An enrichment of MRE within the nucleus can promote the increased retention of cognate miRNA. This results in an increase of miRNA within the nucleus containing the sequence complementary to the enriched MRE. Consequently, when applying statistical tools to identify overrepresented nuclear miRNA sequence motifs, such a sequence may be found. But the error then may lie in the assumption that the sequence is an NLS. This may explain why predicted miRNA NLS motifs may not promote nuclear localization in different cell types or why miRNA with similar seed regions tend to localize to the same subcellular compartments (Jeffries, Fried, & Perkins, 2011; Khudayberdiev et al., 2013).

In theory, mature miRNA alone should be insufficient to mediate any of its gene regulatory potential as AGO recruits effector proteins such as TNRC6A and CCR4:NOT. Within the nucleus, miRISC has been found in at least two major forms, a low molecular weight complex of just miRNA and AGO and a high molecular weight complex of miRNA, AGO, TNRC6A, and multiple poly(A) deadenylases (Sarshad et al., 2018). The degree to which either the low or high molecular weight complex is found within the nucleus may depend on cell types in which more stem-like cells contain higher proportions of high molecular weight miRISC. As a result, cell types with the high molecular weight miRISC have been shown to facilitate mRNA decay within the nucleus (Sarshad et al., 2018). Moreover, MRE within the CDS of target mRNA have been shown to elicit mRNA decay (Sarshad et al., 2018), in contrast to cytoplasmic targets, which are predominantly located within the 3′ UTR. Nuclear miRNAs have also been reported to affect alternative splicing (Allo et al., 2009) and target rRNA (Wang et al., 2015).

Nuclear miRISC has also been shown to regulate transcriptional activity both positively and negatively (Benhamed et al., 2012; Dharap et al., 2013; Miao et al., 2016; Nishi et al., 2013; Pitchiaya et al., 2017; Truesdell et al., 2012) (Fig. 1.4C). Genome-wide studies have found that miRISC tends to associate with actively transcribing loci (Xun et al., 2020) and can alter histone modifications to promote or inhibit euchromatin formation (Benhamed et al., 2012; Roberts, 2014). However, the way in which miRISC is known to interact with chromatin is still incomplete (Stavast & Erkeland, 2019). The most plausible model of interaction involves miRISC identifying MREs in nascent RNA cotranscriptionally, enriching chromatin modifiers to sites of active transcription by direct interaction with AGO and adapter proteins (Matsui et al., 2013; Pisignano et al., 2017; Schwartz et al., 2008). Nevertheless, there is growing evidence that miRISC may also interact directly with DNA through Watson-Crick base pairing (Younger, Pertsemlidis, & Corey, 2009) and triplex forming Hoogsteen base pairing (Paugh et al., 2016). The deletion of DNA-associated MRE, determined by sequence complementarity, can reduce the miRNA regulatory effect (Miao et al., 2016). Beyond enriching chromatin remodelers at miRNA target sites on the DNA, miRISC was also shown to sterically hinder the binding of transcription factors, leading to decreased transcription (Miao et al., 2016). Taken together, miRISC may function similarly to transcriptional activators and suppressors, guided to target loci by miRNA:DNA and miRNA:RNA sequence complementarity.

The dynamics of miRNA-mediated gene regulation

Perhaps one of the most exciting shifts in biology over recent decades has been the increased resolution and granularity with which we are able to observe the cell (de Rie et al., 2017; Yan et al., 2018; Zheng et al., 2017). This experimental and technical innovation has also correlated with a significant increase in the amount of information we are able to collect, and as such, the inferences we can draw can be substantially more accurate, encompassing, and complex. MicroRNA research has of course benefited from these advances and we are now capable of observing changes in entire populations of RNA molecules, including their spatiotemporal interactions and localizations. Spatio- refers to the physical locations within two-dimensional (2D) and/or three-dimensional (3D) space and temporal referring to time and changes over time.

The modeling of nonlinear dynamics in complex systems, such as with miRNA biology, is by its nature computationally complex (Walpole, Papin, & Peirce, 2013). In cases where miRNAs are expressed at unphysiologically high levels, a simple systems approach can be sufficient. That is, if we have a target mRNA with a predicted MRE and if the prediction is accurate, overexpression of the cognate miRNA will typically downregulate the target. But this can be a somewhat artificial set-up (Jin et al., 2015), analogous to naively overexpressing a constitutively active kinase, and misses the complexities of the system.

Excluding the spatiotemporal regulation of miRNA and cognate mRNA, a key factor in the overall output of the system is the concept of MRE load and affinity (Denzler et al., 2016). Any given miRNA will typically have many interactable MRE distributed among numerous RNA molecules, that is, one miRNA can target many mRNAs (Baek et al., 2008). Reciprocally, a single mRNA molecule can be targeted by multiple miRNAs (Chou et al., 2018), in some cases synergistically (Denzler et al., 2016; Elkayam et al., 2017; Lemus-Diaz et al., 2017; Li et al., 2014; Pons-Espinal et al., 2017; Wu et al., 2010; Yao et al., 2014). Moreover, MREs for a single miRNA can vary by sequence, such as ACACGAA vs. ACACGAU, leading to cognate MRE with different affinities (Agarwal et al., 2015; Denzler et al., 2016). As affinities are highly important in determining more stable, long-lived interactions between miRISC and target mRNA, affinities can directly modulate the occupancy of any given MRE (Bosson, Zamudio, & Sharp, 2014). In addition, mRNA 3' UTR isoforms (Nam et al., 2014) and RNA binding proteins can all alter the available pool of interactable MRE (Bottini et al., 2017; Kedde et al., 2010).

At steady state, assuming miRNA and MRE are concomitantly and homogeneously spatiotemporally regulated, cognate miRNA would be distributed such that MRE with the highest affinity would have the highest relative miRNA occupancy, that is, the cognate MRE with the highest affinity toward a cognate miRNA are, at any given time, most likely to be bound to that miRNA (Bosson et al., 2014; Denzler et al., 2016). Depending on the different affinities between cognate MRE, and the cognate miRNA is not in excess, this can drastically affect MRE occupancy such that low-affinity MRE can be predominantly unbound even though low-affinity cognate MRE are typically much more abundant relative to high-affinity MRE (Bosson et al., 2014). As such, higher-affinity MRE typically show the greatest sensitivity to miRISC regulation (Baek et al., 2008; Denzler et al., 2016; Molotski & Soen, 2012). Combined with relatively short interaction times (low affinity $\approx$ 1 s and high affinity $\gg$ 5 min (Chandradoss et al., 2015)), miRISC can efficiently sample many targets and respond quickly to stochastic fluctuations of transcription or dynamic changes in gene expression to change their MRE distribution profiles (Wee et al., 2012).

The distribution of miRNA among its MRE pool is at the heart of the competitive endogenous RNA (ceRNA) model of gene regulation. The primary aspect of the theory proposes that a sufficient increase of cognate MRE load within the cell will decrease the occupancy of any given MRE enough to result in a derepression of the target genes, assuming the miRNA was acting to downregulate its target genes (Bosson et al., 2014; Kartha & Subramanian, 2014) (Fig. 1.5).

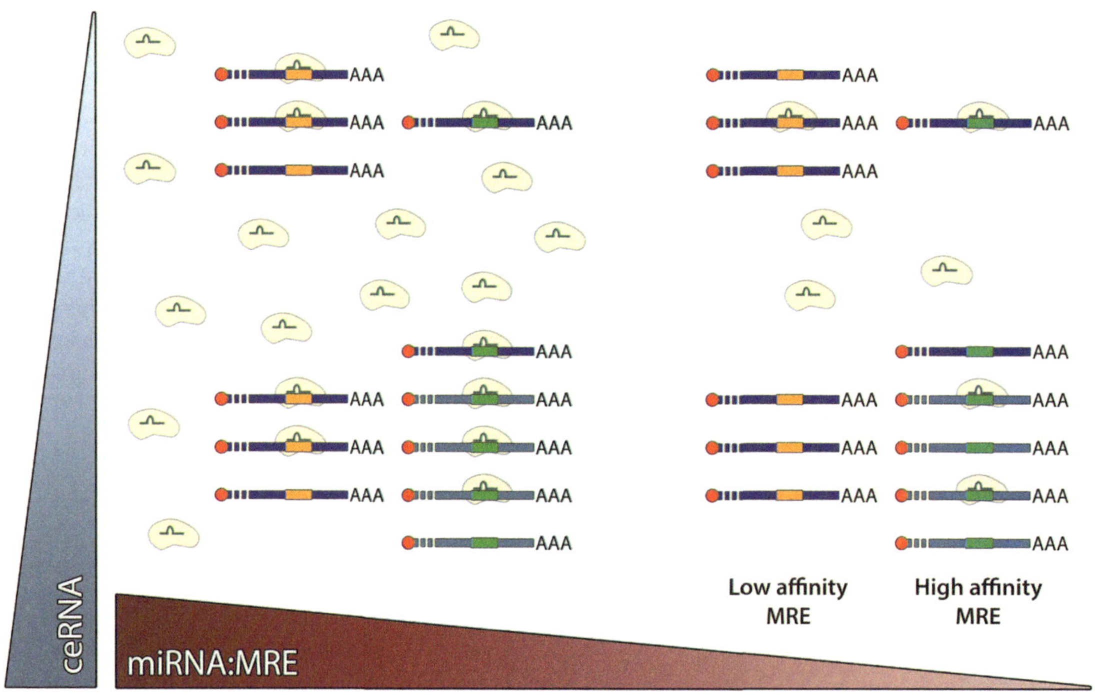

FIG. 1.5 Competitive endogenous RNA (ceRNA) model. The ceRNA model predicts that the significant overexpression of RNAs containing high-affinity MRE results in the sequestering of miRNA from other endogenous targets leading to target derepression. When miRNA abundance is in excess to their MRE load (a high miRNA:MRE ratio; top left), the upregulation of ceRNA (shown as steel-blue RNA molecules) would not produce any appreciable derepression of endogenous targets (bottom left) as endogenous and ceRNA MRE are both maximally occupied. As miRNA:MRE ratios and miRNA abundance decrease, miRNA targets become more susceptible to ceRNA derepression. Before ceRNA upregulation, miRNA targets are still efficiently regulated (top right); however, an increase in ceRNA high-affinity MREs can sequester a limited pool of miRNA away from other targets, leading to target derepression (bottom right).

Typically, this mechanism is attributed to the upregulation of an mRNA, long noncoding RNA, circular RNA, or pseudogene that is the source of the MRE (Barrett & Salzman, 2016; Liu et al., 2019; Militello et al., 2017; Tay, Rinn, & Pandolfi, 2014; Yu et al., 2017b). RNA molecules with repeated high-affinity MRE are termed miRNA sponges and can be potent ceRNA (Militello et al., 2017). This theory, however, has been viewed with controversy due to the potentially unphysiological levels of ceRNA required for the derepression of miRNA targets. Mathematical models have suggested that at least 50% of total cognate MRE from a single ceRNA would be needed in order to sequester enough miRNAs from other targets to promote their derepression (Denzler et al., 2016). Nevertheless, to suggest whether an RNA molecule is acting as a ceRNA depends on the ratio of miRNA to MRE and the overall abundance of miRNA within the cell (Bosson et al., 2014; Liu et al., 2019). MicroRNA that are in great excess relative to their MRE pool (a high miRNA:MRE ratio) and have high copy numbers per cell (in the tens of thousands) tend to be robust to ceRNA derepression. However, as the miRNA copy number and miRNA:MRE ratio decrease, the miRNA can be more susceptible to ceRNA derepression. Although single genes acting as potent ceRNAs may be relatively infrequent, most reported miRNA:MRE ratios tend to be much less than one (Bosson et al., 2014), suggesting a potential wide-spread steady-state ceRNA crosstalk involving complex gene regulatory networks.

Beyond miRNA and MRE abundance and differential MRE affinities, the subcellular localization of miRISC is integral to its regulatory potential, both in terms of control and efficiency. As was discussed earlier, the nuclear localization of miRISC is necessary to regulate nuclear RNA. This process is at least in part dependent on nuclear cognate RNA abundance where increased target abundance increases nuclear miRNA retention. Outside the nucleus, miRISC has been detected within the cytosol, microtubule networks, processing (P)-bodies, stress granules, lysosomes, rough endoplasmic reticulum (rER), polysomes, Golgi apparatus, endosomes, and mitochondria. miRISC can be transported along microtubule networks within the cytosol, which can lead to the nucleation of P-bodies enriched with miRNA, cognate mRNA, deadenylases, and ribonucleases, promoting the efficient degradation of target mRNA (Carbonaro, O'Brate, & Giannakakou, 2011; Eulalio et al., 2007; Liu et al., 2005b; Rajgor et al., 2014). In response to cell stress, miRNA and target mRNA can be exported from the cell or shuttled to stress granules that lack miRISC effector proteins and reduce miRNA posttranscriptional regulation (Detzer et al., 2011; Kucherenko & Shcherbata, 2018; Wang et al., 2010). Stress granules are transient and their components, including miRISC, can return to the cytosol and in some cases be degraded within lysosomes (Leung, Calabrese, & Sharp, 2006; Protter & Parker, 2016; Rajgor et al., 2014). Localization of miRISC:polysome complexes to the rER has been associated with increased silencing of targets, where MRE within the mRNA of polysomes were observed to reach up to ~50% occupancy compared to 5%–20% for a typical MRE (Barman & Bhattacharyya, 2015; Bosson et al., 2014; Molotski & Soen, 2012; Pillai et al., 2005). miRISC within the Golgi and endosomes have been associated with miRISC activity and either miRNA export from the cell or recycling back into the cytosol (Bose et al., 2017). Endosomes are also important players in miRNA import from extracellular fluids, potentially regulating which imported miRNA are cycled back into the cytosol or deposited into lysosomes (Bose et al., 2017; Gibbings et al., 2009). Within the mitochondria, miRISC has been shown to promote translation (Barrey et al., 2011; Zhang et al., 2014).

MicroRNA abundance and turnover

The abundance of miRNA within the cell is dependent on at least the cell state and cell type (Trapnell, 2015), with state referring to the current set of programs (gene expression networks) running within the cell, and cell type referring to stable morphological and/or phenotypically distinct populations of cells. Put another way, any given cell type can run many different cell states and the set of potential cell states is limited by the cell type.

There are substantial differences in miRNA expression profiles within and between different cell types. A large-scale study by the Functional Annotation of the Mammalian Genome (FANTOM5) consortium found that for any given cell type, the top five expressed miRNA represent on average ~50% of the total miRNA pool within the cell (de Rie et al., 2017). Of the total set of expressed human miRNA, half were cell type enriched, a quarter broadly expressed, and the remaining identified miRNA were expressed at low levels regardless of cell type.

Generally, miRNA abundance is strongly correlated with pri-miRNA expression (Kingston & Bartel, 2019). Moreover, the expression rate of an miRNA is considered high if an individual miRNA molecule can be generated at > 100 molecules per minute per cell (Kingston & Bartel, 2019). The overall lifespan of miRNA is also long lived; in some cases, half-lives of more than 100h have been observed. Depending on the cell type, the median half-life can vary drastically, ranging from less than 10h to more than 30h. This is in contrast to median mRNA half-life, which is commonly observed between 2 and 10h (Kingston & Bartel, 2019). These differences in expression rate and half-life result in typically more than 1000 molecules per miRNA per cell vs. ~18 molecules per mRNA per cell (Denzler et al., 2016; Schwanhausser et al., 2011). Although in some cases, more than 100,000 molecules per miRNA per cell have been observed (Song et al., 2017).

Various mechanisms of specific miRNA turnover have been identified, whereas factors affecting global miRNA turnover rates still remain poorly understood (Gebert & MacRae, 2019). Global turnover rates may correlate with the speed at which cells need to remodel their gene expression. This is exemplified in mouse embryonic stem cells (mESC) in which median global miRNA turnover rates are less than 6h versus >20h in other differentiated cell types (Kingston & Bartel, 2019). Long miRNA half-lives would in theory act to help maintain the cell type/state, rendering the system more robust to perturbations and changes to cell state (Kingston & Bartel, 2019; Siciliano et al., 2013).

Although global miRNA turnover rates may be more stable, individual miRNAs are susceptible to target RNA-directed miRNA degradation (TDMD), a consequence of miRNA interacting with MRE that have significant pairing in the 3′ region of the miRNA (Ameres et al., 2010; Guo et al., 2015; Sheu-Gruttadauria et al., 2019b). This pairing is hypothesized to dissociate the 3′ end of the miRNA from AGO, which allows 3' NTA (tailing) and trimming followed by 3′–5′ exonuclease degradation (Ameres et al., 2010). However, what factors are necessary to induce TDMD are uncertain as it is difficult to predict (Haas et al., 2016). Regardless, the 5′ seed region may be an important factor in TDMD, even though it is not always necessary (Park, Shin, & Shin, 2017), as it stabilizes miRISC association with cognate MRE, which then allows more stable 3′ pairing to occur. It can also be the case that 5′ seed interactions with MRE stabilize cognate miRNA levels (Pitchiaya et al., 2017). Nevertheless, the miRNA sequence has been shown to be critical in determining miRNA stability (Guo et al., 2015). For example, miRNA:MRE interactions with 100% complementarity generally lead to unloading of the miRNA from AGO and promote the miRNA's degradation (Ameres et al., 2010; De et al., 2013). This is the mechanism utilized to achieve exogenous antisense miRNA knockdown (Krutzfeldt et al., 2005).

Concluding remarks

The topic of miRNA is an immensely rewarding avenue of research, which is in part due to the strongly dynamical and nonlinear nature of the complex miRNA system. A nonlinear system is one where the sum of the inputs does not necessarily equal the sum of its outputs, in contrast to a linear response where a change in A equates to a proportional change in B. A dynamical system is one that evolves (changes) over time. And lastly, a systems approach takes into consideration a network of connections that describes how each node relates to all the other nodes in the network. In biology, the node could represent a molecule, a set of molecules, a single gene, a protein, a cellular function, or even an entire cell or tissue. The key idea is that when thinking about miRNA or any biological system, keeping in mind this complex interplay of connections is crucial. By thinking of miRNA biology through this framework, it will offer an understanding that cannot be inferred by a more linear approach alone.

MicroRNA, a subset of small noncoding RNA, arise from various RNA sources but the end result is the loading of mature single-stranded miRNA into the Argonaute family of proteins forming the miRNA-induced silencing complex. This complex is shuttled within the cell between the cytoplasm and nucleus, existing both diffusely within these compartments and localized to specific subcellular regions. The recognition of miRNA response elements by the 5′ seed region of the miRNA (nucleotides 2–8) and supplemental 3′ miRNA complementarity within target RNA and/or DNA are sufficient to elicit miRISC regulatory potential. Depending on the effector proteins complexed with miRISC and the subcellular localization, translation initiation inhibition, ribosome stalling, poly(A) deadenylation and mRNA decay, translation activation, and transcriptional suppression and promotion have all been observed. Although miRNAs are most often observed to negatively correlate with cognate mRNA levels, our current understanding of the mechanistic and regulatory potentials of miRNA has so greatly expanded over the last 20 years that miRISC has become insufficiently named to encompass the wide breadth of roles it plays within the cell.

The spatiotemporal steady-state levels of miRNA globally and relative to their mRNA targets are of crucial importance when trying to understand the dynamics of miRNA:mRNA interactions and their regulatory potential. In a system where the number of miRNA molecules is in a 1:1 proportion to target mRNA, what effect would that have on target mRNA stability? Is it sufficient to promote mRNA decay? What are the interaction dynamics like at that ratio? At any given time, would each target mRNA molecule be bound by at least one cognate miRNA? In other words, would there be a 100% occupancy of target mRNA by miRISC at any given time? Does miRNA interaction with MRE have any influence on the miRNA itself? What if cognate miRNA and mRNA are independently spatially regulated, what effect would that have on interaction dynamics and the models we generate to predict those effects? Although we have no definitive answer and no single model to fully explain miRNA biology, there has been truly impressive work in these areas that has greatly impacted and expanded our view of miRNA biology.

There may be hundreds of miRNAs expressed at any given time within any given cells, but only a few miRNAs are highly expressed in any given cell types. These highly expressed miRNA tend to exist in great excess compared to their MRE load and function to broadly suppress the majority of their targets by maximally occupying cognate MRE and as a result, are robust to changes competitive endogenous RNA expression. However, when miRNAs are expressed at lower absolute abundances and are not in excess relative to their MRE load, these miRNA regulatory networks are more susceptible to changes in MRE load. The combination of low miRNA abundance and less than or equal miRNA:MRE proportions leads to quickly changing MRE occupancy distributions following the upregulation of high-affinity targets, which may result in the derepression of previously occupied MRE.

Taken together, miRNAs have evolved to dynamically and efficiently regulate a wide range of targets to maintain a steady-state equilibration of stochastic gene expression networks in an ever-changing environment. By considering global populations of gene expression agents, miRNA being of particular noteworthiness, the study of disease-perturbed and normal gene expression networks alike has led to models with ever greater fidelity, aiding researchers around the world to more completely and accurately understand the complex biological system that is our existence.

References

Agarwal, V., et al. (2015). Predicting effective microRNA target sites in mammalian mRNAs. *eLife*, *4*, e05005.

Allo, M., et al. (2009). Control of alternative splicing through siRNA-mediated transcriptional gene silencing. *Nature Structural & Molecular Biology*, *16*(7), 717–U43.

Ameres, S. L., et al. (2010). Target RNA-directed trimming and tailing of small silencing RNAs. *Science*, *328*(5985), 1534–1539.

Auyeung, V. C., et al. (2013). Beyond secondary structure: Primary-sequence determinants license Pri-miRNA hairpins for processing. *Cell*, *152*(4), 844–858.

Babiarz, J. E., et al. (2008). Mouse ES cells express endogenous shRNAs, siRNAs, and other microprocessor-independent, dicer-dependent small RNAs. *Genes & Development*, *22*(20), 2773–2785.

Baek, D., et al. (2008). The impact of microRNAs on protein output. *Nature*, *455*(7209), 64–U38.

Barman, B., & Bhattacharyya, S. N. (2015). mRNA targeting to endoplasmic reticulum precedes ago protein interaction and microRNA (miRNA)-mediated translation repression in mammalian cells. *The Journal of Biological Chemistry*, *290*(41), 24650–24656.

Barrett, S. P., & Salzman, J. (2016). Circular RNAs: Analysis, expression and potential functions. *Development*, *143*(11), 1838–1847.

Barrey, E., et al. (2011). Pre-microRNA and mature microRNA in human mitochondria. *PLoS ONE*, *6*(5), e20220.

Bartel, D. P. (2009). MicroRNAs: Target recognition and regulatory functions. *Cell*, *136*(2), 215–233.

Behm-Ansmant, I., et al. (2006). mRNA degradation by miRNAs and GW182 requires both CCR4:NOT deadenylase and DCP1:DCP2 decapping complexes. *Genes & Development*, *20*(14), 1885–1898.

Benhamed, M., et al. (2012). Senescence is an endogenous trigger for microRNA-directed transcriptional gene silencing in human cells. *Nature Cell Biology*, *14*(3), 266–275.

Ben-Yishay, R., & Shav-Tal, Y. (2019). The dynamic lifecycle of mRNA in the nucleus. *Current Opinion in Cell Biology*, *58*, 69–75.

Boele, J., et al. (2014). PAPD5-mediated 3' adenylation and subsequent degradation of miR-21 is disrupted in proliferative disease. *Proceedings of the National Academy of Sciences of the United States of America*, *111*(31), 11467–11472.

Borbolis, F., & Syntichaki, P. (2015). Cytoplasmic mRNA turnover and ageing. *Mechanisms of Ageing and Development*, *152*, 32–42.

Bose, M., et al. (2017). Spatiotemporal uncoupling of microRNA-mediated translational repression and target RNA degradation controls microRNP recycling in mammalian cells. *Molecular and Cellular Biology*, *37*(4), e00464–e00516.

Bosson, A. D., Zamudio, J. R., & Sharp, P. A. (2014). Endogenous miRNA and target concentrations determine susceptibility to potential ceRNA competition. *Molecular Cell*, *56*(3), 347–359.

Bottini, S., et al. (2017). Post-transcriptional gene silencing mediated by microRNAs is controlled by nucleoplasmic Sfpq. *Nature Communications*, *8*(1), 1189.

Braun, J. E., et al. (2011). GW182 proteins directly recruit cytoplasmic deadenylase complexes to miRNA targets. *Molecular Cell*, *44*(1), 120–133.

Braun, J. E., et al. (2012). A direct interaction between DCP1 and XRN1 couples mRNA decapping to 5′ exonucleolytic degradation. *Nature Structural & Molecular Biology*, *19*(12), 1324–1331.

Brkic, J., et al. (2018). MicroRNA-218-5p promotes endovascular trophoblast differentiation and spiral artery remodeling. *Molecular Therapy*, *26*(9), 2189–2205.

Budak, H., et al. (2016). MicroRNA nomenclature and the need for a revised naming prescription. *Briefings in Functional Genomics*, *15*(1), 65–71.

Cai, X. Z., Hagedorn, C. H., & Cullen, B. R. (2004). Human microRNAs are processed from capped, polyadenylated transcripts that can also function as mRNAs. *RNA*, *10*(12), 1957–1966.

Cao, Z. X., & Grima, R. (2020). Analytical distributions for detailed models of stochastic gene expression in eukaryotic cells. *Proceedings of the National Academy of Sciences of the United States of America*, *117*(9), 4682–4692.

Carbonaro, M., O'Brate, A., & Giannakakou, P. (2011). Microtubule disruption targets HIF-1alpha mRNA to cytoplasmic P-bodies for translational repression. *The Journal of Cell Biology*, *192*(1), 83–99.

Catalanotto, C., Cogoni, C., & Zardo, G. (2016). MicroRNA in control of gene expression: An overview of nuclear functions. *International Journal of Molecular Sciences*, *17*(10), 1712.

Chandradoss, S. D., et al. (2015). A dynamic search process underlies microRNA targeting. *Cell*, *162*(1), 96–107.

Chang, T. C., et al. (2015). Genome-wide annotation of microRNA primary transcript structures reveals novel regulatory mechanisms. *Genome Research*, *25*(9), 1401–1409.

Chaulk, S. G., et al. (2011). Role of pri-miRNA tertiary structure in miR-17 similar to 92 miRNA biogenesis. *RNA Biology*, *8*(6), 1105–1114.

Cheloufi, S., et al. (2010). A dicer-independent miRNA biogenesis pathway that requires Ago catalysis. *Nature*, *465*(7298), 584–589.

Chen, C. Y. A., et al. (2009). Ago-TNRC6 triggers microRNA-mediated decay by promoting two deadenylation steps. *Nature Structural & Molecular Biology*, *16*(11), 1160–U6.

Chendrimada, T. P., et al. (2005). TRBP recruits the dicer complex to Ago2 for microRNA processing and gene silencing. *Nature*, *436*(7051), 740–744.

Chou, C. H., et al. (2018). miRTarBase update 2018: A resource for experimentally validated microRNA-target interactions. *Nucleic Acids Research*, *46*(D1), D296–D302.

Christie, M., et al. (2013). Structure of the PAN3 pseudokinase reveals the basis for interactions with the PAN2 deadenylase and the GW182 proteins. *Molecular Cell*, *51*(3), 360–373.

Conrad, T., et al. (2014). Microprocessor activity controls differential miRNA biogenesis in vivo. *Cell Reports*, *9*(2), 542–554.

D'Ambrogio, A., et al. (2012). Specific miRNA stabilization by Gld2-catalyzed monoadenylation. *Cell Reports*, *2*(6), 1537–1545.

de Rie, D., et al. (2017). An integrated expression atlas of miRNAs and their promoters in human and mouse. *Nature Biotechnology*, *35*(9), 872–878.

De, N., et al. (2013). Highly complementary target RNAs promote release of guide RNAs from human Argonaute2. *Molecular Cell*, *50*(3), 344–355.

Denli, A. M., et al. (2004). Processing of primary microRNAs by the microprocessor complex. *Nature*, *432*(7014), 231–235.

Denzler, R., et al. (2016). Impact of microRNA levels, target-site complementarity, and cooperativity on competing endogenous RNA-regulated gene expression. *Molecular Cell*, *64*(3), 565–579.

Desvignes, T., et al. (2015). miRNA nomenclature: A view incorporating genetic origins, biosynthetic pathways, and sequence variants. *Trends in Genetics*, *31*(11), 613–626.

Detzer, A., et al. (2011). Cell stress is related to re-localization of Argonaute 2 and to decreased RNA interference in human cells. *Nucleic Acids Research*, *39*(7), 2727–2741.

Dharap, A., et al. (2013). MicroRNA miR-324-3p induces promoter-mediated expression of RelA gene. *PLoS ONE*, *8*(11), e79467.

Di Mauro, V., et al. (2019). Wnt signalling mediates miR-133a nuclear re-localization for the transcriptional control of Dnmt3b in cardiac cells. *Scientific Reports*, *9*(1), 9320.

Dozmorov, M. G., et al. (2013). Systematic classification of non-coding RNAs by epigenomic similarity. *BMC Bioinformatics*, *14*, S2.

Elkayam, E., et al. (2017). Multivalent recruitment of human Argonaute by GW182. *Molecular Cell*, *67*(4), 646–658 (e3).

Ellwanger, D. C., et al. (2011). The sufficient minimal set of miRNA seed types. *Bioinformatics*, *27*(10), 1346–1350.

Elowitz, M. B., et al. (2002). Stochastic gene expression in a single cell. *Science*, *297*(5584), 1183–1186.

Ender, C., et al. (2008). A human snoRNA with microRNA-like functions. *Molecular Cell*, *32*(4), 519–528.

Eulalio, A., et al. (2007). P-body formation is a consequence, not the cause, of RNA-mediated gene silencing. *Molecular and Cellular Biology*, *27*(11), 3970–3981.

Fan, J. H., et al. (2020). Nuclear miR-665 aggravates heart failure via suppressing phosphatase and tensin homolog transcription. *Science China. Life Sciences*, *63*(5), 724–736.

Fang, W. W., & Bartel, D. P. (2015). The menu of features that define primary microRNAs and enable De novo design of microRNA genes. *Molecular Cell*, *60*(1), 131–145.

Fang, L. L., et al. (2017). Expression, regulation and mechanism of action of the miR-17-92 cluster in tumor cells. *International Journal of Molecular Medicine*, *40*(6), 1624–1630.

Fareh, M., et al. (2016). TRBP ensures efficient dicer processing of precursor microRNA in RNA-crowded environments. *Nature Communications*, *7*, 13694.

Forman, J. J., Legesse-Miller, A., & Coller, H. A. (2008). A search for conserved sequences in coding regions reveals that the let-7 microRNA targets dicer within its coding sequence. *Proceedings of the National Academy of Sciences of the United States of America*, *105*(39), 14879–14884.

Frank, F., Sonenberg, N., & Nagar, B. (2010). Structural basis for 5'-nucleotide base-specific recognition of guide RNA by human AGO2. *Nature*, *465*(7299), 818–822.

Friedman, R. C., et al. (2009). Most mammalian mRNAs are conserved targets of microRNAs. *Genome Research*, *19*(1), 92–105.

Fukao, A., et al. (2014). MicroRNAs trigger dissociation of eIF4AI and eIF4AII from target mRNAs in humans. *Molecular Cell*, *56*(1), 79–89.

Gebert, L. F. R., & MacRae, I. J. (2019). Regulation of microRNA function in animals. *Nature Reviews Molecular Cell Biology*, *20*(1), 21–37.

Georgakilas, G., et al. (2014). microTSS: Accurate microRNA transcription start site identification reveals a significant number of divergent pri-miRNAs. *Nature Communications*, *5*, 5700.

Gibbings, D. J., et al. (2009). Multivesicular bodies associate with components of miRNA effector complexes and modulate miRNA activity. *Nature Cell Biology*, *11*(9), 1143–1149.

Gu, S., et al. (2009). Biological basis for restriction of microRNA targets to the 3' untranslated region in mammalian mRNAs. *Nature Structural & Molecular Biology*, *16*(2), 144–150.

Gu, S., et al. (2012). Slicing-independent RISC activation requires the Argonaute PAZ domain. *Current Biology*, *22*(16), 1536–1542.

Guo, Y. W., et al. (2015). Characterization of the mammalian miRNA turnover landscape. *Nucleic Acids Research*, *43*(4), 2326–2341.

Gutierrez-Vazquez, C., et al. (2017). 3' Uridylation controls mature microRNA turnover during CD4 T-cell activation. *RNA*, *23*(6), 882–891.

Ha, M., & Kim, V. N. (2014). Regulation of microRNA biogenesis. *Nature Reviews. Molecular Cell Biology*, *15*(8), 509–524.

Haas, G., et al. (2016). Identification of factors involved in target RNA-directed microRNA degradation. *Nucleic Acids Research*, *44*(6), 2873–2887.

Han, J., et al. (2004). The Drosha-DGCR8 complex in primary microRNA processing. *Genes & Development*, *18*(24), 3016–3027.

Hansen, T. B., et al. (2011). miRNA-dependent gene silencing involving Ago2-mediated cleavage of a circular antisense RNA. *EMBO Journal*, *30*(21), 4414–4422.

Hayder, H., et al. (2018). MicroRNAs: Crucial regulators of placental development. *Reproduction*, *155*(6), R259–R271.

Hayes, J., Peruzzi, P. P., & Lawler, S. (2014). MicroRNAs in cancer: Biomarkers, functions and therapy. *Trends in Molecular Medicine*, *20*(8), 460–469.

Huang, W. (2017). MicroRNAs: Biomarkers, diagnostics, and therapeutics. *Methods in Molecular Biology*, *1617*, 57–67.

Huang, X., Le, Q. T., & Giaccia, A. J. (2010). MiR-210–micromanager of the hypoxia pathway. *Trends in Molecular Medicine*, *16*(5), 230–237.

Huntzinger, E., & Izaurralde, E. (2011). Gene silencing by microRNAs: Contributions of translational repression and mRNA decay. *Nature Reviews. Genetics*, *12*(2), 99–110.

Huntzinger, E., et al. (2013). The interactions of GW182 proteins with PABP and deadenylases are required for both translational repression and degradation of miRNA targets. *Nucleic Acids Research*, *41*(2), 978–994.

Hwang, H. W., Wentzel, E. A., & Mendell, J. T. (2007). A hexanucleotide element directs microRNA nuclear import. *Science*, *315*(5808), 97–100.

Ipsaro, J. J., & Joshua-Tor, L. (2015). From guide to target: Molecular insights into eukaryotic RNA-interference machinery. *Nature Structural & Molecular Biology*, *22*(1), 20–28.

Janson, N. B. (2012). Non-linear dynamics of biological systems. *Contemporary Physics*, *53*(2), 137–168.

Jeffries, C. D., Fried, H. M., & Perkins, D. O. (2011). Nuclear and cytoplasmic localization of neural stem cell microRNAs. *RNA*, *17*(4), 675–686.

Jin, H. Y., et al. (2015). Transfection of microRNA mimics should be used with caution. *Frontiers in Genetics*, *6*, 340.

Jo, M. H., et al. (2015). Human Argonaute 2 has diverse reaction pathways on target RNAs. *Molecular Cell*, *59*(1), 117–124.

Jonas, S., & Izaurralde, E. (2015). Towards a molecular understanding of microRNA-mediated gene silencing. *Nature Reviews. Genetics*, *16*(7), 421–433.

Jones, M. R., et al. (2009). Zcchc11-dependent uridylation of microRNA directs cytokine expression. *Nature Cell Biology*, *11*(9), 1157–U258.

Kartha, R. V., & Subramanian, S. (2014). Competing endogenous RNAs (ceRNAs): New entrants to the intricacies of gene regulation. *Frontiers in Genetics*, *5*, 8.

Katoh, T., et al. (2009). Selective stabilization of mammalian microRNAs by 3' adenylation mediated by the cytoplasmic poly(A) polymerase GLD-2. *Genes & Development*, *23*(4), 433–438.

Kawahara, Y., et al. (2007). Redirection of silencing targets by adenosine-to-inosine editing of miRNAs. *Science*, *315*(5815), 1137–1140.

Kawahara, Y., et al. (2008). Frequency and fate of microRNA editing in human brain. *Nucleic Acids Research*, *36*(16), 5270–5280.

Kawamata, T., & Tomari, Y. (2010). Making RISC. *Trends in Biochemical Sciences*, *35*(7), 368–376.

Kedde, M., et al. (2010). A Pumilio-induced RNA structure switch in p27-3' UTR controls miR-221 and miR-222 accessibility. *Nature Cell Biology*, *12*(10), 1014–1020.

Khudayberdiev, S. A., et al. (2013). A comprehensive characterization of the nuclear microRNA repertoire of post-mitotic neurons. *Frontiers in Molecular Neuroscience*, *6*, 43.

Khvorova, A., Reynolds, A., & Jayasena, S. D. (2003). Functional siRNAs and miRNAs exhibit strand bias. *Cell*, *115*(2), 209–216.

Kim, Y. K., & Kim, V. N. (2007). Processing of intronic microRNAs. *The EMBO Journal*, *26*(3), 775–783.

Kim, B., et al. (2015). TUT7 controls the fate of precursor microRNAs by using three different uridylation mechanisms. *EMBO Journal*, *34*(13), 1801–1815.

Kingston, E. R., & Bartel, D. P. (2019). Global analyses of the dynamics of mammalian microRNA metabolism. *Genome Research*, *29*(11), 1777–1790.

Kozomara, A., & Griffiths-Jones, S. (2011). miRBase: Integrating microRNA annotation and deep-sequencing data. *Nucleic Acids Research*, *39*, D152–D157.

Krutzfeldt, J., et al. (2005). Silencing of microRNAs in vivo with 'antagomirs'. *Nature*, *438*(7068), 685–689.

Kucherenko, M. M., & Shcherbata, H. R. (2018). miRNA targeting and alternative splicing in the stress response—events hosted by membrane-less compartments. *Journal of Cell Science*, *131*(4), jcs202002.

Kwak, P. B., & Tomari, Y. (2012). The N domain of Argonaute drives duplex unwinding during RISC assembly. *Nature Structural & Molecular Biology*, *19*(2), 145–151.

Lau, P. W., et al. (2012). The molecular architecture of human dicer. *Nature Structural & Molecular Biology*, *19*(4), 436–440.

Lee, R. C., Feinbaum, R. L., & Ambros, V. (1993). The *C. elegans* heterochronic gene lin-4 encodes small RNAs with antisense complementarity to lin-14. *Cell*, *75*(5), 843–854.

Lee, Y., et al. (2004). MicroRNA genes are transcribed by RNA polymerase II. *EMBO Journal*, *23*(20), 4051–4060.

Lemus-Diaz, N., et al. (2017). Dissecting miRNA gene repression on single cell level with an advanced fluorescent reporter system. *Scientific Reports*, *7*, 45197.

Leung, A. K., Calabrese, J. M., & Sharp, P. A. (2006). Quantitative analysis of Argonaute protein reveals microRNA-dependent localization to stress granules. *Proceedings of the National Academy of Sciences of the United States of America*, *103*(48), 18125–18130.

Li, Y., et al. (2014). Mirsynergy: Detecting synergistic miRNA regulatory modules by overlapping neighbourhood expansion. *Bioinformatics*, *30*(18), 2627–2635.

Liu, J. D., et al. (2005a). A role for the P-body component GW182 in microRNA function. *Nature Cell Biology*, *7*(12), 1261–1266.

Liu, J., et al. (2005b). MicroRNA-dependent localization of targeted mRNAs to mammalian P-bodies. *Nature Cell Biology*, *7*(7), 719–723.

Liu, Y. W., et al. (2019). Competitive endogenous RNA is an intrinsic component of EMT regulatory circuits and modulates EMT. *Nature Communications*, *10*(1), 1637.

Ma, J. B., Ye, K. Q., & Patel, D. J. (2004). Structural basis for overhang-specific small interfering RNA recognition by the PAZ domain. *Nature*, *429*(6989), 318–322.

Ma'ayan, A. (2017). Complex systems biology. *Journal of the Royal Society Interface*, *14*(134), 20170391.

Manzano, M., et al. (2015). Divergent target recognition by coexpressed 5'-isomiRs of miR-142-3p and selective viral mimicry. *RNA*, *21*(9), 1606–1620.

Marsico, A., et al. (2013). PROmiRNA: A new miRNA promoter recognition method uncovers the complex regulation of intronic miRNAs. *Genome Biology*, *14*(8), R84.

Matsui, M., et al. (2013). Promoter RNA links transcriptional regulation of inflammatory pathway genes. *Nucleic Acids Research*, *41*(22), 10086–10109.

Meijer, H. A., Smith, E. M., & Bushell, M. (2014). Regulation of miRNA strand selection: Follow the leader? *Biochemical Society Transactions*, *42*(4), 1135–1140.

Medley, J. C., Panzade, G., & Zinovyeva, A. Y. (2020). microRNA strand selection: Unwinding the rules. *Wiley Interdisciplinary Reviews. RNA*, *12*(3), e1627.

Meijer, H. A., et al. (2013). Translational repression and eIF4A2 activity are critical for microRNA-mediated gene regulation. *Science*, *340*(6128), 82–85.

Mele, M., et al. (2015). Human genomics. The human transcriptome across tissues and individuals. *Science*, *348*(6235), 660–665.

Miao, L., et al. (2016). A dual inhibition: microRNA-552 suppresses both transcription and translation of cytochrome P450 2E1. *Biochimica et Biophysica Acta*, *1859*(4), 650–662.

Militello, G., et al. (2017). Screening and validation of lncRNAs and circRNAs as miRNA sponges. *Briefings in Bioinformatics*, *18*(5), 780–788.

Min, K. W., et al. (2017). AUF1 facilitates microRNA-mediated gene silencing. *Nucleic Acids Research*, *45*(10), 6064–6073.

Mogilyansky, E., & Rigoutsos, I. (2013). The miR-17/92 cluster: A comprehensive update on its genomics, genetics, functions and increasingly important and numerous roles in health and disease. *Cell Death and Differentiation*, *20*(12), 1603–1614.

Molotski, N., & Soen, Y. (2012). Differential association of microRNAs with polysomes reflects distinct strengths of interactions with their mRNA targets. *RNA*, *18*(9), 1612–1623.

Nair, A. A., et al. (2020). Frequency of MicroRNA response elements identifies pathologically relevant signaling pathways in triple-negative breast cancer. *Iscience*, *23*(6), 101249.

Nakanishi, K. (2016). Anatomy of RISC: How do small RNAs and chaperones activate Argonaute proteins? *Wiley Interdisciplinary Reviews. RNA*, *7*(5), 637–660.

Nam, J. W., et al. (2014). Global analyses of the effect of different cellular contexts on microRNA targeting. *Molecular Cell*, *53*(6), 1031–1043.

Neilsen, C. T., Goodall, G. J., & Bracken, C. P. (2012). IsomiRs—The overlooked repertoire in the dynamic microRNAome. *Trends in Genetics*, *28*(11), 544–549.

Nielsen, C. B., et al. (2007). Determinants of targeting by endogenous and exogenous microRNAs and siRNAs. *RNA*, *13*(11), 1894–1910.

Nishi, K., et al. (2013). Human TNRC6A is an Argonaute-navigator protein for microRNA-mediated gene silencing in the nucleus. *RNA*, *19*(1), 17–35.

Nishi, K., et al. (2015). Control of the localization and function of a miRNA silencing component TNRC6A by Argonaute protein. *Nucleic Acids Research*, *43*(20), 9856–9873.

Nishikura, K. (2016). A-to-I editing of coding and non-coding RNAs by ADARs. *Nature Reviews Molecular Cell Biology*, *17*(2), 83–96.

O'Brien, J., et al. (2018). Overview of MicroRNA biogenesis, mechanisms of actions, and circulation. *Frontiers in Endocrinology*, *9*, 402.

Okada, C., et al. (2009). A high-resolution structure of the pre-microRNA nuclear export machinery. *Science*, *326*(5957), 1275–1279.

Parafioriti, A., et al. (2020). Expression profiling of microRNAs and isomiRs in conventional central chondrosarcoma. *Cell Death Discovery*, *6*(1), 46.

Park, J. H., & Shin, C. (2015). Slicer-independent mechanism drives small-RNA strand separation during human RISC assembly. *Nucleic Acids Research*, *43*(19), 9418–9433.

Park, J. H., Shin, S. Y., & Shin, C. (2017). Non-canonical targets destabilize microRNAs in human Argonautes. *Nucleic Acids Research*, *45*(4), 1569–1583.

Paugh, S. W., et al. (2016). MicroRNAs form triplexes with double stranded DNA at sequence-specific binding sites; a eukaryotic mechanism via which microRNAs could directly alter gene expression. *PLoS Computational Biology*, *12*(2), e1004744.

Paul, P., et al. (2018). Interplay between miRNAs and human diseases. *Journal of Cellular Physiology*, *233*(3), 2007–2018.

Pillai, R. S., et al. (2005). Inhibition of translational initiation by Let-7 MicroRNA in human cells. *Science*, *309*(5740), 1573–1576.

Pisignano, G., et al. (2017). A promoter-proximal transcript targeted by genetic polymorphism controls E-cadherin silencing in human cancers. *Nature Communications*, *8*, 15622.

Pitchiaya, S., et al. (2017). Resolving subcellular miRNA trafficking and turnover at single-molecule resolution. *Cell Reports*, *19*(3), 630–642.

Plotnikova, O., Baranova, A., & Skoblov, M. (2019). Comprehensive analysis of human microRNA-mRNA interactome. *Frontiers in Genetics*, *10*, 933.

Pons-Espinal, M., et al. (2017). Synergic functions of miRNAs determine neuronal fate of adult neural stem cells. *Stem Cell Reports*, *8*(4), 1046–1061.

Protter, D. S., & Parker, R. (2016). Principles and properties of stress granules. *Trends in Cell Biology*, *26*(9), 668–679.

Quillet, A., et al. (2020). Improving bioinformatics prediction of microRNA targets by ranks aggregation. *Frontiers in Genetics*, *10*, 1330.

Rajgor, D., et al. (2014). Mammalian microtubule P-body dynamics are mediated by nesprin-1. *The Journal of Cell Biology*, *205*(4), 457–475.

Rajman, M., & Schratt, G. (2017). MicroRNAs in neural development: From master regulators to fine-tuners. *Development*, *144*(13), 2310–2322.

Roberts, T. C. (2014). The microRNA biology of the mammalian nucleus. *Molecular Therapy- -Nucleic Acids*, *3*(8), e188.

Roden, C., et al. (2017). Novel determinants of mammalian primary microRNA processing revealed by systematic evaluation of hairpin-containing transcripts and human genetic variation. *Genome Research*, *27*(3), 374–384.

Rorbach, G., Unold, O., & Konopka, B. M. (2018). Distinguishing mirtrons from canonical miRNAs with data exploration and machine learning methods. *Scientific Reports*, *8*.

Ruby, J. G., Jan, C. H., & Bartel, D. P. (2007). Intronic microRNA precursors that bypass Drosha processing. *Nature*, *448*(7149), 83–86.

Salem, M., et al. (2018). miR-590-3p promotes ovarian cancer growth and metastasis via a novel FOXA2-Versican pathway. *Cancer Research*, *78*(15), 4175–4190.

Salomon, W. E., et al. (2015). Single-molecule imaging reveals that Argonaute reshapes the binding properties of its nucleic acid guides. *Cell*, *162*(1), 84–95.

Salter, J. D., Bennett, R. P., & Smith, H. C. (2016). The APOBEC protein family: United by structure, divergent in function. *Trends in Biochemical Sciences*, *41*(7), 578–594.

Sarshad, A. A., et al. (2018). Argonaute-miRNA complexes silence target mRNAs in the nucleus of mammalian stem cells. *Molecular Cell*, *71*(6), 1040.

Schirle, N. T., Sheu-Gruttadauria, J., & MacRae, I. J. (2014). Structural basis for microRNA targeting. *Science*, *346*(6209), 608–613.

Schwanhausser, B., et al. (2011). Global quantification of mammalian gene expression control. *Nature*, *473*(7347), 337–342.

Schwartz, J. C., et al. (2008). Antisense transcripts are targets for activating small RNAs. *Nature Structural & Molecular Biology*, *15*(8), 842–848.

Seal, R. L., et al. (2020). A guide to naming human non-coding RNA genes. *The EMBO Journal*, *39*(6), e103777.

Senturia, R., et al. (2010). Structure of the dimerization domain of DiGeorge critical region 8. *Protein Science*, *19*(7), 1354–1365.

Sheu-Gruttadauria, J., & MacRae, I. J. (2017). Structural foundations of RNA silencing by Argonaute. *Journal of Molecular Biology*, *429*(17), 2619–2639.

Sheu-Gruttadauria, J., et al. (2019a). Beyond the seed: Structural basis for supplementary microRNA targeting by human Argonaute2. *EMBO Journal*, 38(13).

Sheu-Gruttadauria, J., et al. (2019b). Structural basis for target-directed microRNA degradation. *Molecular Cell*, *75*(6), 1243.

Siciliano, V., et al. (2013). MiRNAs confer phenotypic robustness to gene networks by suppressing biological noise. *Nature Communications*, *4*, 2364.

Song, Y. K., et al. (2017). Determination of absolute expression profiles using multiplexed miRNA analysis. *PLoS ONE*, *12*(7), e0180988.

Stavast, C. J., & Erkeland, S. J. (2019). The non-canonical aspects of microRNAs: Many roads to gene regulation. *Cells*, *8*(11), 1465.

Tarallo, R., et al. (2017). The nuclear receptor ER beta engages AGO2 in regulation of gene transcription, RNA splicing and RISC loading. *Genome Biology*, *18*(1), 189.

Tay, Y., Rinn, J., & Pandolfi, P. P. (2014). The multilayered complexity of ceRNA crosstalk and competition. *Nature*, *505*(7483), 344–352.

Tian, Y., et al. (2014). A phosphate-binding pocket within the platform-PAZ-connector helix cassette of human dicer. *Molecular Cell*, *53*(4), 606–616.

Trapnell, C. (2015). Defining cell types and states with single-cell genomics. *Genome Research*, *25*(10), 1491–1498.

Truesdell, S. S., et al. (2012). MicroRNA-mediated mRNA translation activation in quiescent cells and oocytes involves recruitment of a nuclear microRNP. *Scientific Reports*, *2*, 842.

Tufekci, K. U., et al. (2014). The role of microRNAs in human diseases. *Methods in Molecular Biology*, *1107*, 33–50.

Turunen, T. A., et al. (2019). Changes in nuclear and cytoplasmic microRNA distribution in response to hypoxic stress. *Scientific Reports*, *9*(1), 10332.

Valinezhad Orang, A., Safaralizadeh, R., & Kazemzadeh-Bavili, M. (2014). Mechanisms of miRNA-mediated gene regulation from common downregulation to mRNA-specific upregulation. *International Journal of Genomics*, *2014*, 970607.

Vermeulen, A., et al. (2005). The contributions of dsRNA structure to dicer specificity and efficiency. *RNA*, *11*(5), 674–682.

Vesely, C., et al. (2014). ADAR2 induces reproducible changes in sequence and abundance of mature microRNAs in the mouse brain. *Nucleic Acids Research*, *42*(19), 12155–12168.

Walpole, J., Papin, J. A., & Peirce, S. M. (2013). Multiscale computational models of complex biological systems. *Annual Review of Biomedical Engineering*, *15*, 137–154.

Wang, J., Chen, J., & Sen, S. (2016). MicroRNA as biomarkers and diagnostics. *Journal of Cellular Physiology*, *231*(1), 25–30.

Wang, Y. L., et al. (2008). Structure of the guide-strand-containing argonaute silencing complex. *Nature*, *456*(7219), 209–U34.

Wang, H. W., et al. (2009). Structural insights into RNA processing by the human RISC-loading complex. *Nature Structural & Molecular Biology*, *16*(11), 1148–U4.

Wang, K., et al. (2010). Export of microRNAs and microRNA-protective protein by mammalian cells. *Nucleic Acids Research*, *38*(20), 7248–7259.

Wang, X. Y., et al. (2015). Down-regulation of 5S rRNA by miR-150 and miR-383 enhances c-Myc-rpL11 interaction and inhibits proliferation of esophageal squamous carcinoma cells. *FEBS Letters*, *589*(24), 3989–3997.

Wang, N. Y., et al. (2019). Single-cell microRNA-mRNA co-sequencing reveals non-genetic heterogeneity and mechanisms of microRNA regulation. *Nature Communications*, *10*.

Wee, L. M., et al. (2012). Argonaute divides its RNA guide into domains with distinct functions and RNA-binding properties. *Cell*, *151*(5), 1055–1067.

Wightman, B., Ha, I., & Ruvkun, G. (1993). Posttranscriptional regulation of the heterochronic gene lin-14 by lin-4 mediates temporal pattern formation in *C. elegans*. *Cell*, *75*(5), 855–862.

Wilson, R. C., et al. (2015). Dicer-TRBP complex formation ensures accurate mammalian microRNA biogenesis. *Molecular Cell*, *57*(3), 397–407.

Wu, H. Q., et al. (2009). Alternative processing of primary microRNA transcripts by Drosha generates 5 ' end variation of mature microRNA. *PLoS ONE*, *4*(10), e7566.

Wu, S., et al. (2010). Multiple microRNAs modulate p21Cip1/Waf1 expression by directly targeting its 3′ untranslated region. *Oncogene*, *29*(15), 2302–2308.

Xie, M., et al. (2013). Mammalian 5′-capped microRNA precursors that generate a single microRNA. *Cell*, *155*(7), 1568–1580.

Xu, W., et al. (2014). Identifying microRNA targets in different gene regions. *BMC Bioinformatics*, *15*(Suppl. 7), S4.

Xun, Y., et al. (2020). Purification and identification of miRNA target sites in genome using DNA affinity precipitation. *Frontiers in Genetics*, *10*, 778.

Yan, J. W., et al. (2018). Network approaches to systems biology analysis of complex disease: Integrative methods for multi-omics data. *Briefings in Bioinformatics*, *19*(6), 1370–1381.

Yang, J. S., et al. (2010). Conserved vertebrate mir-451 provides a platform for dicer-independent, Ago2-mediated microRNA biogenesis. *Proceedings of the National Academy of Sciences of the United States of America*, *107*(34), 15163–15168.

Yao, W., et al. (2014). The application of multiple miRNA response elements enables oncolytic adenoviruses to possess specificity to glioma cells. *Virology*, *458–459*, 69–82.

Younger, S. T., Pertsemlidis, A., & Corey, D. R. (2009). Predicting potential miRNA target sites within gene promoters. *Bioorganic & Medicinal Chemistry Letters*, *19*(14), 3791–3794.

Yu, F., et al. (2017a). Naturally existing isoforms of miR-222 have distinct functions. *Nucleic Acids Research*, *45*(19), 11371–11385.

Yu, C. Y., et al. (2017b). The circular RNA circBIRC6 participates in the molecular circuitry controlling human pluripotency. *Nature Communications*, *8*(1), 1149.

Zhang, H., et al. (2004). Single processing center models for human Dicer and bacterial RNase III. *Cell*, *118*(1), 57–68.

Zhang, X., et al. (2014). MicroRNA directly enhances mitochondrial translation during muscle differentiation. *Cell*, *158*(3), 607–619.

Zhang, J., et al. (2018a). Oncogenic role of microRNA-532-5p in human colorectal cancer via targeting of the 5'UTR of RUNX3. *Oncology Letters*, *15*(5), 7215–7220.

Zhang, K., et al. (2018b). A novel class of microRNA-recognition elements that function only within open reading frames. *Nature Structural & Molecular Biology*, *25*(11), 1019.

Zhang, X. Y., Shen, B., & Cui, Y. L. (2019). Ago HITS-CLIP expands microRNA-mRNA interactions in nucleus and cytoplasm of gastric cancer cells. *BMC Cancer*, *19*(1), 29.

Zheng, G. X., et al. (2017). Massively parallel digital transcriptional profiling of single cells. *Nature Communications*, *8*, 14049.

Zou, Q., et al. (2014). miRClassify: An advanced web server for miRNA family classification and annotation. *Computers in Biology and Medicine*, *45*, 157–160.

Chapter 2

microRNA in cancer: An overview

Rares Drula[a,b,*], Swati Mohapatra[a,c,*], and George A. Calin[a]

[a]*Department of Translational Molecular Pathology, The University of Texas MD Anderson Cancer Center, Houston, TX, United States,* [b]*Research Center for Functional Genomics Biomedicine and Translational Medicine, "Iuliu Haţieganu" University of Medicine and Pharmacy, Cluj-Napoca, Romania,* [c]*The University of Texas MD Anderson Cancer Center UTHealth Graduate School of Biomedical Sciences, Houston, TX, United States*

The extent of the dogmatic shift in molecular biology caused by the discovery of noncoding RNAs was initially difficult to assess, as the rather restrictive implications and orthologues of the first observed microRNA (miRNA) were yet to be explored in humans, even less so in pathological contexts such as cancer. Therefore, the mapping of two microRNA to the frequently deleted region in chronic lymphocytic leukemia patients came as a great surprise, as it gave a new purpose to previously disregarded regions of the human genome. The extension of the noncoding genome paradigm came as more roles were attributed to these transcripts in different pathological conditions, laying the foundations for a completely new area of research in oncology. As such, the aim of this overview is not only to mention the groundbreaking studies that have shaped cancer miRNA research, but also to emphasize their influence on cancer research to this day. Additionally, we will approach the crucial terms in miRNA biology with a steady introduction into the more mechanistic and cancer-specific processes, which will be detailed in the later chapters of this book along with the evolution of investigation methods and their future perspectives in diagnostics and therapy.

The dawn of miRNAs in human biology

The first validated miRNA in humans was the let-7 RNA. At that time, it was the only known heterochronic regulatory RNA responsible for the transition from late larval to adult cell fates in *C. elegans* (Reinhart et al., 2000). What drew attention was the sequence conservation in both vertebrate and several invertebrate clades, indicative of the functional evolutionary relevance of these short RNAs. Sustaining this fact, the study of let-7 in lower vertebrates (*Danio rerio*) indicated that, like in the case of *C. elegans*, the expression of let-7 was also temporarily regulated. Referring specifically to the expression levels of let-7 in human tissues, Pasquinelli et al. reported that immature hematopoietic bone marrow cells showed the lowest expression levels of let-7 in comparison to the other investigated tissues. This indicated a programmed upregulation in later developmental and differentiation pathways, further highlighting the conserved heterochronic nature of this RNA (Pasquinelli et al., 2000). The first extended confirmation regarding miRNA conservation in mammals came from the study of Lagos-Quintana et al. in 2002. While the study mainly focused on the characterization of specific miRNA tissue signatures in mice, the group also approached and assessed the sequence conservation of 34 miRNAs between humans, mice, and other nonmammalian vertebrates (Lagos-Quintana et al., 2002). The identification of the mammalian orthologues of lin-4, miR-125a, and miR-125b variants in brain tissues and the identification of miR-155 as a lymphoid-specific transcript are worth mentioning. This is in concordance with the expression pattern of the previously described BIC (Tam, 2001), the long noncoding transcript from which miR-155's sequence was originally mapped. This would foreshadow their future implications in human biology.

First reported miRNA cancer-associated alterations

These studies can be considered the dawn of miRNA in human biology, which opened new research avenues focused on mapping similar or paralog processes and possible implications in the pathophysiology of human diseases. The study that laid the foundation of a completely new area of research in oncology was the identification of the miR-15/16 cluster on chromosome 13q14 in 2002 by Calin et al. from the Croce laboratory, at that time located at the Kimmel Cancer Center,

* These authors contributed equally to this work.

MicroRNA in Human Malignancies. https://doi.org/10.1016/B978-0-12-822287-4.00014-1

Philadelphia, United States. Calin et al. were studying the frequent deletions in B-cell chronic lymphocytic leukemia patients and were looking for tumor suppressor genes harbored in the 13q14 chromosomal region. Surprisingly, they found that this region sheltered two important miRNA genes, namely miR-15 and miR-16 (Calin et al., 2002). This was a major discovery that opened a wide field of research to the community by linking miRNA and cancer. For the first time in the history of miRNAs, two candidates—miRNA-15 and miRNA-16—were seen to be involved in cancer where they were either deleted or downregulated, acting as tumor suppressors. Based on this indication, the group later investigated the other mapped miRNA loci at the time, 186 specifically, to confirm that more than half of them were located in regions associated with frequent deletions (Calin et al., 2004).

The indication that a significant number of miRNAs were harbored in fragile chromosomal regions had initially raised many more questions than it has answered, as at that time the extent of the regulatory activity of the assumed lost and/or downregulated miRNAs was still widely unknown. Mainly, it was not clear whether the loss of these miRNAs offered any kind of evolutionary advantage or it acted as a bystander with the already reported cancer-specific chromosomal alterations . Yet again, this was the driving force for investigating the lost sequences in the first place, as the previous extensive loss of heterozygosity (LOH), mutation, and expression studies failed to demonstrate the consistent involvement of any of the coding genes located in the deleted regions. This conundrum did not last for long, as the functional consequence of the 13q14 harbored miRNA loss came a year later in 2005, when the Croce lab highlighted the interaction between the downregulated miR-15/16 and the BCL2 oncogene (Cimmino et al., 2005). A few months earlier, Johnson et al. revealed that the ultra-conserved miRNA, let-7, could target 3′UTRs of the RAS mRNA. This was the first reported bona fide posttranscriptional regulation of an oncogene by an miRNA in human models, after observing a similar interaction with let-60, the *C. elegans* paralog of RAS (Johnson et al., 2005).

The age of discovery: Microarrays and miRNA cancer signatures

The first indications of a distinguishable cancer-specific miRNA signature was reported in colon cancer when Michael et al. characterized the expression patterns of a set of 28 miRNAs. From this subset, miR-143 and miR-145 were highlighted as the main downregulated transcripts in precancerous and neoplastic colorectal tissue (Michael, O'Connor, van Holst Pellekaan, Young, & James, 2003). As the functional aspects were still under investigation, the potential biomarker utility of these transcripts was in a very uncertain light, as the methods were still rudimentary. The gold standard was northern blotting at that time. Although it was reliable, it was also laborious and included the use of radioactive isotopes; it was not suitable for high-throughput screening and multiplexing. Thus, the demand for newer and more accessible technologies was increasing proportionally with the interest in these transcripts. In 2004, multiple groups reported the development of adapted miRNA microarrays, making the profiling of the new "miRNome" a much less tedious task (Babak, Zhang, Morris, Blencowe, & Hughes, 2004; Calin et al., 2004; Miska et al., 2004; Nelson et al., 2004; Sun et al., 2004; Thomson, Parker, Perou, & Hammond, 2004). These studies gave a start to a new exploratory era in human cancer biology, as miRNA just became a new variable in an already complex system. As researchers all over the world started looking into miRNAs for their pathology of interest, the number of studies started growing exponentially, contouring a common consensus of what the main goals of this new area of research were: to establish clear miRNA cancer-specific signatures, to map their interactions with oncogenes, and to assess their clinical value. One of the pioneering uses of miRNA microarrays came from the Croce lab in 2005, where the reported breast cancer signature consisting of mir-125b, mir-145, mir-21, and mir-155 was able to differentiate not only between tumor vs. normal tissues but also between different molecular subtypes. Other hybrid-based methods were to be developed soon, including the bead-based flow cytometry method developed by Lu et al. in 2005, which offered one of the most extensive multicancer profiling studies to this day (Lu et al., 2005). Nevertheless, while these profiling technologies are still in use today, the advent of next-generation sequencing provided newer methods that offer significantly higher throughput and sensibility in miRNA detection. In addition, the development of fluorescent qPCR probes greatly facilitated the methodology in studying smaller miRNA subsets.

Functional classification of miRNA: The line between tumor suppressor and oncomiRs

The increased capacity for multiplexing offered by the array technologies facilitated the discovery of the dysregulated miRNAs in most, if not all, cancer types. This provided a plethora of investigative opportunities, both for functional and biomarker studies. Referring strictly to the functional aspects of cancer-associated miRNAs, the functional versatility given by the numerous transcripts they can target, makes classification a relative endeavor, as it is highly dependent on the biological context and regulated targets. There is an expanding number of miRNAs predominantly displaying tumor-promoting or suppressor roles, leading to their popular referral as ***oncomiRs*** or tumor ***suppressor miRs***, respectively.

Referring to the transcripts in the OncomIRs category, these are generally reported as overexpressed in cancer cells, either as a result of chromosomal or epigenetic alterations. They have high potential in developing tumorigenesis and invasive programs by targeting important tumor suppressor genes and/or upregulating oncogenes. One of the first and most extensively studied example oncomirs belongs to the appropriately named "*oncomir-1*" cluster, consisting of a single polycistronic precursor for seven microRNAs, namely miR-17-5p, miR-17-3p, miR-18, miR-19a, miR-20, miR-19b-1, and miR-92-1. The mir-17-92 cluster was reported as amplified in several types of B-cell lymphomas by He et al. They further investigated its oncogenic role in conjunction with the *c-Myc* oncogene in *E*μ-*myc*-induced lymphoma murine models, hinting toward a functional interaction between the two (He et al., 2005). The functional validation came shortly after in the same year when O'Donnell et al. confirmed that c-Myc modulated the expression of the oncomiR-1 cluster by directly binding its promoter (O'Donnell, Wentzel, Zeller, Dang, & Mendell, 2005). Multiple groups were reporting similar expression patterns of the cluster members in different cancers, including Hayashita et al. in lung cancer (Hayashita et al., 2005). Another relevant study came from the Croce lab from 2006, and they were one of the first groups to coin the term "miRnome." The profiling study included an extensive number of samples derived from lung, breast, stomach, prostate, colon, and pancreatic tumors, intending to establish an extensive pan-cancer signature (Volinia et al., 2006). The highlighted enriched miRNA signature consisted of miR-17-5p, miR-20a, miR-21, miR-92, miR-106a, and miR-155, several of which were also validated as targeting tumor suppressors (Volinia et al., 2006). Several other groundbreaking studies emerged around the same time with similar approaches and goals. This not only led to the consolidation of the research area as a whole, but it also created a consensus regarding the role of specific transcripts and their presumptive role as either an oncomiR or tumor suppressor. For example, miR-21 was one of the transcripts recurrently identified as overexpressed in multiple types of cancer and is now considered a bona fide oncomiR. Although miR-21 was first reported in glioma (Chan, Krichevsky, & Kosik, 2005), subsequent profiling studies, including previously mentioned ones, confirmed its overexpressed status in most cancer types (Meng et al., 2007). This made miR-21 a very interesting transcript for investigation, being attributed as both a potential pan-cancer biomarker (Lampis et al., 2021; Mohapatra & Calin, 2021) and a targetable oncomiR, as functional studies emerged showcasing important tumor suppressor genes it was targeting, such as PTEN (Meng et al., 2007). More so, miR-21 was one of the first transcripts in any type of cancer to be associated with an "OncomiR addiction," meaning that miR-21 was crucial to the development of the cancer phenotype, in this case from pre-B-cell lymphoma, and its inactivation resulted in complete tumor regression (Medina, Nolde, & Slack, 2010). MiR-155 is another example of an established oncomiR, often investigated in leukemia pathological contexts (Cui et al., 2014; Ferrajoli et al., 2013) but with major implications in other solid tumors, such as lung (Van Roosbroeck et al., 2017). As described previously in this chapter, miR-155 (or BIC) was initially reported as enriched in several types of B cell lymphomas by Eis et al. in 2005 and was proposed as a potential prognostic indicator (Eis et al., 2005). There were several in vivo models that were designed to capture the genetic events, which were not possible in the artificial milieu of in vitro experiments. The Croce laboratory in 2009 reported the first mice model developing a leukemic phenotype by the overexpression of the miR-155 transgene (Costinean et al., 2006). In 2010, the Slack group reported that miR-21 was enough for the development of precursor B-cell lymphoma. The authors also showed that tumor size was significantly reduced and better survival was found upon the shutdown of miR-21 overexpression (Medina et al., 2010). This showed that the tumors could become addicted to miR-21, similar to the oncogene addiction phenomenon.

In 2007, the Li Ma team showed for the first time the involvement of miRNAs in metastasis and invasion (Ma, Teruya-Feldstein, & Weinberg, 2007). This study came from Robert Weinberg's laboratory, where they decrypted a mechanism of miR-10b regulating cell migration and invasion positively. Also, the initiation of robust metastasis and invasion happens upon the overexpression of miR-10b in nonmetastatic breast tumors. MiR-10b was found to be induced by Twist by binding to its promoter, thereby inhibiting the translational machinery of the mRNA-encoding homeobox D10, leading to the overexpression of the prometastatic gene RHOC (Fig. 2.1).

In a direct antithesis to the functional class of the oncomiRs, tumor suppressor miRs are generally downregulated as a result of either cancer-associated genomic alterations or epigenetic silencing, indirectly providing aids to sustain the proliferative programs. It was clear at this point that although chromosomal abnormalities were responsible for the alterations in the case of at least a subset of miRNA, it was highly unlikely that this was the case for all the altered miRNA identified up to that point. The first insights into miRNAs subjected to epigenetic silencing came from the study of Saito et al. in 2006. After profiling the expression levels following treatment with chromatin-modifiers, the group noticed a synchronous upregulation of a subset of miRNA. Further validations highlighted miR-127, which mapped to a CpG island and the transcription of which was highly induced following the treatment-induced demethylation. The reported functional inactivation observed in cancer cells correlated with the increased expression levels of one of its targets, the proto-oncogene *BCL6*, providing a clear example of a tumor-suppressing miRNAs subjected to epigenetic silencing (Saito et al., 2006). Some of the most important tumor suppressor miRNAs belong to the miR-34 family, represented by the miR-34a, b, and c variants. The initial reporting came from the miRNA profiling comparison of wild-type mice with *P53−/−* models, which indicated clearly that the miR-34 family expression levels

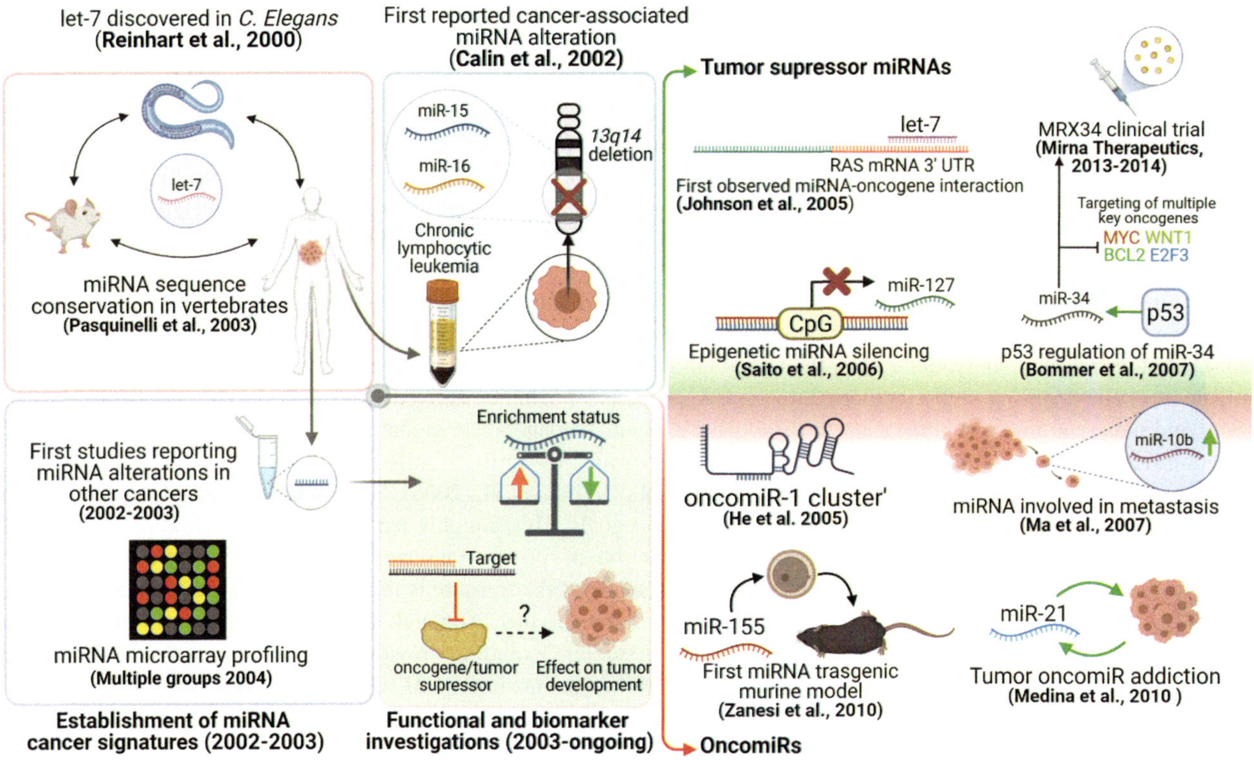

FIG. 2.1 General overview highlighting several important events and discoveries in cancer miRNA research. Created with BioRender.com.

were positively correlated with P53 status, as p53 seems to regulate their expression level (He et al., 2007). This was an especially important event, as the functional loss of TP53 is a well-known event in many types of cancer (Hermeking, 2007). Further studies confirmed its downregulated status in most cancers, associated with both loss of its harboring 1p36 chromosomal locus but also with methylation-associated silencing, especially associated with metastatic cancers (Lujambio et al., 2008).

Development of miRNA as biomarkers

Lawrie et al. in 2008 employed miRNAs (miR-155, miR-210, miR-21) to detect diffuse large B-cell lymphoma (DLBCL) in the serum of patients; this was the first report to utilize miRNAs as biomarkers. They compared the expression levels of these miRNAs and elucidated a correlation between disease-free survival to their up/downregulation, thereby showing miRNAs to have promising roles as noninvasive diagnostic markers for several cancers, including DLBCL (Lawrie et al., 2008).

MiR-21 (upregulated in cancer) and let-7 (downregulated in cancer) are known to be fairly common ones associated with patient outcomes. The overexpression of miR-21 in many cancers such as hepatocellular carcinoma (Huang et al., 2015), colorectal cancer (Lampis et al., 2021; Yu, Zhu, Wang, Yu, & Guo, 2018), stomach cancer (Li, Zhou, Li, Lin, & Tomuleasa, 2014), glioblastoma (Yang et al., 2014), breast cancer (Yan et al., 2008), and prostate cancer (Kumar et al., 2018) targets mostly tumor-suppressor genes, thereby lending a protumorigenic environment with functional roles such as invasion, proliferation, and apoptosis.

Coming to the utilization of miRNAs for disease classification, Lu et al. in 2005 showed how miRNA expression signatures are correlated with the origin of the tumor subtype (Lu et al., 2005). This further led to the acceleration of research proposing tumor classifications based on differential miRNA expressions. In 2008, Rosenfeld et al. reported tumor classification according to the tissue of origin based on an independent blinded test with more than 90% accuracy (Rosenfeld et al., 2008). Later, specific miRNA expression patterns were found to have a significant correlation with the disease subtype, including acute myeloid leukemia (Garzon et al., 2008), prostate cancer (Liu et al., 2012), and breast cancer (Farazi et al., 2014); this is discussed in detail in later chapters. miRNA signatures were shown to have an association with different events of metastases, especially in assisting the endothelial to mesenchymal transitioning that has been shown in breast cancer (Andorfer, Necela, Thompson, & Perez, 2011) and endometrial cancer (Diaz-Martin et al., 2014). The identification of such miRNA signatures in different types of cancer would establish a network that may help in the better assessment of disease prognosis along with understanding the mRNA/protein alterations.

miRNA polymorphisms have also been explored and found to have significant roles in predisposing cancers (Ziebarth, Bhattacharya, & Cui, 2012). Such single nucleotide polymorphisms are often associated with the risk of a poor outcome in patients or in developing resistance to therapy (Chin et al., 2008; Ding, Li, Wang, Li, & Guo, 2013; Wu et al., 2011).

The transition of miRNA into clinics

The application of miRNA therapeutics has not only expanded in cancer but also in other diseases that have risen in the past decade. In 2004, the potential clinical utility of *let-7* in human lung cancers was reported for the first time. The Takamizawa team identified a poor survival rate in lung cancer patients having reduced *let-7* expression, especially after curative resection. In 2005, the Calin team reported 13 miRNA signatures that helped in the better stratification of CLL patients by establishing a link between disease prognosis and progression (Calin et al., 2005). In 2008, Chin et al. identified a single nucleotide variant in the KRAS 3' UTR that affected the *let-7* binding and eventually revealed that patients carrying this mutant allele are at a higher risk of developing nonsmall cell lung cancer (NSCLC) at a later stage. In 2010, Jason Wiggins et al. reported that ectopic miR-34a blocked tumor expansion in a patient-derived xenograft model of lung cancer (Wiggins et al., 2010). This was the first study to provide proof of principle by successfully delivering miRNAs utilizing a synthetic mimic and a lipid vesicle. Further, many follow-up studies backed up the idea, including one from the Slack group in 2011 where they utilized neutral lipid emulsion to deliver miR-34a in a mice model of NSCLC driven by *KRASG12D* that showed a significant reduction in the tumor burden. In 2012, another report from the same laboratory determined that lentiviral-delivered miR034a helped in preventing both initiation and progression of cancer in aggressive mice model of NSCLC. It was not a surprise when miR-34a became the first miRNA therapeutic that reached Phase I clinical trials (ClinicalTrials.gov Identifier: NCT01829971) in liver cancer patients; that trial was sponsored by Mirna Therapeutics (MRX34). The clinical trial for MRX34, a liposomal formulation of miR-34, took place between 2013 and 2014 and included patients with varying types of solid tumors, mainly liver and lung. The MRX34 administration showed acceptable tolerance and evidence of antitumor activity, yet the study had to be terminated at phase I due to unexpected fatal immune adverse effects. While similar reactions did not occur in preclinical toxicological models, the study indicated that the effective delivery of the miRNA constructs that resulted in systemic immune activation was the main limitation (Hong et al., 2020). While at this point there are no additional clinical studies focused on miR-34, future studies will benefit from the insight regarding dosing, tolerance, and other important aspects provided from this trial.

Recently, the initiation of the first-in-human miRNA therapeutics was a remarkable achievement in the field. The Phase I clinical trial is completed, and they have moved into Phase II for advanced tumors for some of the therapeutics, including Cobomarsen (locked nucleic acid-based oligonucleotide inhibitor of miR-155) in T-cell leukemia/lymphoma (Foss et al., 2018); TargomiR (miR-16 mimic-based therapies) in mesothelioma (van Zandwijk et al., 2017); and Miravirsen (anti-miR-122) in individuals infected with the hepatitis C virus (HCV) (Janssen et al., 2013). miRNA mimics would act as replacements for downregulated endogenous miRNAs. For example, the miR-138 mimic can specifically target and bind to the mRNA of PDL1 and CTLA4 and suppress their expressions, mimicking the roles of anti-PD-1 and anti-CTLA-4 antibodies (Smolle, Calin, Pichler, & Calin, 2017).

Conclusion and future perspectives

The still-expanding field of miRNA biology has contributed substantially to the deciphering of many mechanistic aspects in oncology, including pathophysiology, oncogene regulation, and their association with genomic and epigenetic alterations. An impressive number of miRNA networks have been characterized since the discovery of the miR-15a and miR-16-1 deletions in CLL, and the number is very likely to grow as the advent of modern deep-sequencing techniques will both expand and consolidate the field. Along with this, the functional investigations focused on assessing the importance of miRNA regulation of important gene targets will define the involvement of these transcripts in oncogenesis, invasion, and metastasis. Additionally, extensive validations could provide reliable miRNA biomarker candidates with important utility in patient stratification, therapeutic efficiency, and prognosis assessment. Lastly, while their therapeutic potential is still commencing, important steps have been taken in that direction. Given the number of emerging clinical studies, there are still important lessons to be learned before their clinical implementation, especially regarding delivery systems and off-target effects.

All the previously mentioned aspects will be covered in extenso in the following chapters of this book. The first part will focus on the fundamentals of miRNA cancer biology, followed by an extensive description of the state-of-the-art methodologies utilized in current miRNA research. Further, the focus will shift toward a more translational approach, highlighting the potential and implications of miRNA therapeutics. The final part of the book will consist of the current aspects regarding the pathophysiology of miRNAs in several cancer types, with a focus on mechanisms and key pathology-specific processes.

References

Andorfer, C. A., Necela, B. M., Thompson, E. A., & Perez, E. A. (2011). MicroRNA signatures: Clinical biomarkers for the diagnosis and treatment of breast cancer. *Trends in Molecular Medicine*, *17*(6), 313–319. https://doi.org/10.1016/j.molmed.2011.01.006.

Babak, T., Zhang, W., Morris, Q., Blencowe, B. J., & Hughes, T. R. (2004). Probing microRNAs with microarrays: Tissue specificity and functional inference. *RNA*, *10*(11), 1813–1819. https://doi.org/10.1261/rna.7119904.

Calin, G. A., Dumitru, C. D., Shimizu, M., Bichi, R., Zupo, S., Noch, E., et al. (2002). Frequent deletions and down-regulation of micro- RNA genes miR15 and miR16 at 13q14 in chronic lymphocytic leukemia. *Proceedings of the National Academy of Sciences of the United States of America*, *99*(24), 15524–15529. https://doi.org/10.1073/pnas.242606799.

Calin, G. A., Ferracin, M., Cimmino, A., Di Leva, G., Shimizu, M., Wojcik, S. E., et al. (2005). A MicroRNA signature associated with prognosis and progression in chronic lymphocytic leukemia. *The New England Journal of Medicine*, *353*(17), 1793–1801. https://doi.org/10.1056/NEJMoa050995.

Calin, G. A., Liu, C. G., Sevignani, C., Ferracin, M., Felli, N., Dumitru, C. D., et al. (2004). MicroRNA profiling reveals distinct signatures in B cell chronic lymphocytic leukemias. *Proceedings of the National Academy of Sciences of the United States of America*, *101*(32), 11755–11760. https://doi.org/10.1073/pnas.0404432101.

Calin, G. A., Sevignani, C., Dumitru, C. D., Hyslop, T., Noch, E., Yendamuri, S., et al. (2004). Human microRNA genes are frequently located at fragile sites and genomic regions involved in cancers. *Proceedings of the National Academy of Sciences of the United States of America*, *101*(9), 2999–3004. https://doi.org/10.1073/pnas.0307323101.

Chan, J. A., Krichevsky, A. M., & Kosik, K. S. (2005). MicroRNA-21 is an antiapoptotic factor in human glioblastoma cells. *Cancer Research*, *65*(14), 6029–6033. https://doi.org/10.1158/0008-5472.CAN-05-0137.

Chin, L. J., Ratner, E., Leng, S., Zhai, R., Nallur, S., Babar, I., et al. (2008). A SNP in a let-7 microRNA complementary site in the KRAS 3′ untranslated region increases non-small cell lung cancer risk. *Cancer Research*, *68*(20), 8535–8540. https://doi.org/10.1158/0008-5472.CAN-08-2129.

Cimmino, A., Calin, G. A., Fabbri, M., Iorio, M. V., Ferracin, M., Shimizu, M., et al. (2005). miR-15 and miR-16 induce apoptosis by targeting BCL2. *Proceedings of the National Academy of Sciences of the United States of America*, *102*(39), 13944–13949. https://doi.org/10.1073/pnas.0506654102.

Costinean, S., Zanesi, N., Pekarsky, Y., Tili, E., Volinia, S., Heerema, N., et al. (2006). Pre-B cell proliferation and lymphoblastic leukemia/high-grade lymphoma in E(mu)-miR155 transgenic mice. *Proceedings of the National Academy of Sciences of the United States of America*, *103*(18), 7024–7029. https://doi.org/10.1073/pnas.0602266103.

Cui, B., Chen, L., Zhang, S., Mraz, M., Fecteau, J. F., Yu, J., et al. (2014). MicroRNA-155 influences B-cell receptor signaling and associates with aggressive disease in chronic lymphocytic leukemia. *Blood*, *124*(4), 546–554. https://doi.org/10.1182/blood-2014-03-559690.

Diaz-Martin, J., Diaz-Lopez, A., Moreno-Bueno, G., Castilla, M. A., Rosa-Rosa, J. M., Cano, A., et al. (2014). A core microRNA signature associated with inducers of the epithelial-to-mesenchymal transition. *The Journal of Pathology*, *232*(3), 319–329. https://doi.org/10.1002/path.4289.

Ding, C., Li, C., Wang, H., Li, B., & Guo, Z. (2013). A miR-SNP of the XPO5 gene is associated with advanced non-small-cell lung cancer. *Oncotargets and Therapy*, *6*, 877–881. https://doi.org/10.2147/OTT.S48284.

Eis, P. S., Tam, W., Sun, L., Chadburn, A., Li, Z., Gomez, M. F., et al. (2005). Accumulation of miR-155 and BIC RNA in human B cell lymphomas. *Proceedings of the National Academy of Sciences of the United States of America*, *102*(10), 3627–3632. https://doi.org/10.1073/pnas.0500613102.

Farazi, T. A., Ten Hoeve, J. J., Brown, M., Mihailovic, A., Horlings, H. M., van de Vijver, M. J., et al. (2014). Identification of distinct miRNA target regulation between breast cancer molecular subtypes using AGO2-PAR-CLIP and patient datasets. *Genome Biology*, *15*(1), R9. https://doi.org/10.1186/gb-2014-15-1-r9.

Ferrajoli, A., Shanafelt, T. D., Ivan, C., Shimizu, M., Rabe, K. G., Nouraee, N., et al. (2013). Prognostic value of miR-155 in individuals with monoclonal B-cell lymphocytosis and patients with B chronic lymphocytic leukemia. *Blood*, *122*(11), 1891–1899. https://doi.org/10.1182/blood-2013-01-478222.

Foss, F. M., Querfeld, C., Kim, Y. H., Pinter-Brown, L. C., William, B. M., Porcu, P., et al. (2018). Ph 1 study of MRG-106, an inhibitor of miR-155, in CTCL. *Journal of Clinical Oncology*, *36*(15_suppl), 2511. https://doi.org/10.1200/JCO.2018.36.15_suppl.2511.

Garzon, R., Volinia, S., Liu, C. G., Fernandez-Cymering, C., Palumbo, T., Pichiorri, F., et al. (2008). MicroRNA signatures associated with cytogenetics and prognosis in acute myeloid leukemia. *Blood*, *111*(6), 3183–3189. https://doi.org/10.1182/blood-2007-07-098749.

Hayashita, Y., Osada, H., Tatematsu, Y., Yamada, H., Yanagisawa, K., Tomida, S., et al. (2005). A polycistronic microRNA cluster, miR-17-92, is overexpressed in human lung cancers and enhances cell proliferation. *Cancer Research*, *65*(21), 9628–9632. https://doi.org/10.1158/0008-5472.CAN-05-2352.

He, L., He, X., Lim, L. P., de Stanchina, E., Xuan, Z., Liang, Y., et al. (2007). A microRNA component of the p53 tumour suppressor network. *Nature*, *447*(7148), 1130–1134. https://doi.org/10.1038/nature05939.

He, L., Thomson, J. M., Hemann, M. T., Hernando-Monge, E., Mu, D., Goodson, S., et al. (2005). A microRNA polycistron as a potential human oncogene. *Nature*, *435*(7043), 828–833. https://doi.org/10.1038/nature03552.

Hermeking, H. (2007). p53 enters the microRNA world. *Cancer Cell*, *12*(5), 414–418. https://doi.org/10.1016/j.ccr.2007.10.028.

Hong, D. S., Kang, Y. K., Borad, M., Sachdev, J., Ejadi, S., Lim, H. Y., et al. (2020). Phase 1 study of MRX34, a liposomal miR-34a mimic, in patients with advanced solid tumours. *British Journal of Cancer*, *122*(11), 1630–1637. https://doi.org/10.1038/s41416-020-0802-1.

Huang, C. S., Yu, W., Cui, H., Wang, Y. J., Zhang, L., Han, F., et al. (2015). Increased expression of miR-21 predicts poor prognosis in patients with hepatocellular carcinoma. *International Journal of Clinical and Experimental Pathology*, *8*(6), 7234–7238. Retrieved from https://www.ncbi.nlm.nih.gov/pubmed/26261620.

Janssen, H. L., Reesink, H. W., Lawitz, E. J., Zeuzem, S., Rodriguez-Torres, M., Patel, K., et al. (2013). Treatment of HCV infection by targeting microRNA. *The New England Journal of Medicine*, *368*(18), 1685–1694. https://doi.org/10.1056/NEJMoa1209026.

Johnson, S. M., Grosshans, H., Shingara, J., Byrom, M., Jarvis, R., Cheng, A., et al. (2005). RAS is regulated by the let-7 microRNA family. *Cell, 120*(5), 635–647. https://doi.org/10.1016/j.cell.2005.01.014.

Kumar, B., Rosenberg, A. Z., Choi, S. M., Fox-Talbot, K., De Marzo, A. M., Nonn, L., et al. (2018). Cell-type specific expression of oncogenic and tumor suppressive microRNAs in the human prostate and prostate cancer. *Scientific Reports, 8*(1), 7189. https://doi.org/10.1038/s41598-018-25320-z.

Lagos-Quintana, M., Rauhut, R., Yalcin, A., Meyer, J., Lendeckel, W., & Tuschl, T. (2002). Identification of tissue-specific microRNAs from mouse. *Current Biology, 12*(9), 735–739. https://doi.org/10.1016/s0960-9822(02)00809-6.

Lampis, A., Hahne, J. C., Gasparini, P., Cascione, L., Hedayat, S., Vlachogiannis, G., … Valeri, N. (2021). MIR21-induced loss of junctional adhesion molecule A promotes activation of oncogenic pathways, progression and metastasis in colorectal cancer. *Cell Death & Differentiation*. https://doi.org/10.1038/s41418-021-00820-0.

Lawrie, C. H., Gal, S., Dunlop, H. M., Pushkaran, B., Liggins, A. P., Pulford, K., et al. (2008). Detection of elevated levels of tumour-associated microRNAs in serum of patients with diffuse large B-cell lymphoma. *British Journal of Haematology, 141*(5), 672–675. https://doi.org/10.1111/j.1365-2141.2008.07077.x.

Li, L., Zhou, L., Li, Y., Lin, S., & Tomuleasa, C. (2014). MicroRNA-21 stimulates gastric cancer growth and invasion by inhibiting the tumor suppressor effects of programmed cell death protein 4 and phosphatase and tensin homolog. *Journal of BUON, 19*(1), 228–236. Retrieved from https://www.ncbi.nlm.nih.gov/pubmed/24659669.

Liu, C., Kelnar, K., Vlassov, A. V., Brown, D., Wang, J., & Tang, D. G. (2012). Distinct microRNA expression profiles in prostate cancer stem/progenitor cells and tumor-suppressive functions of let-7. *Cancer Research, 72*(13), 3393–3404. https://doi.org/10.1158/0008-5472.CAN-11-3864.

Lu, J., Getz, G., Miska, E. A., Alvarez-Saavedra, E., Lamb, J., Peck, D., et al. (2005). MicroRNA expression profiles classify human cancers. *Nature, 435*(7043), 834–838. https://doi.org/10.1038/nature03702.

Lujambio, A., Calin, G. A., Villanueva, A., Ropero, S., Sanchez-Cespedes, M., Blanco, D., et al. (2008). A microRNA DNA methylation signature for human cancer metastasis. *Proceedings of the National Academy of Sciences of the United States of America, 105*(36), 13556–13561. https://doi.org/10.1073/pnas.0803055105.

Ma, L., Teruya-Feldstein, J., & Weinberg, R. A. (2007). Tumour invasion and metastasis initiated by microRNA-10b in breast cancer. *Nature, 449*(7163), 682–688. https://doi.org/10.1038/nature06174.

Medina, P. P., Nolde, M., & Slack, F. J. (2010). OncomiR addiction in an in vivo model of microRNA-21-induced pre-B-cell lymphoma. *Nature, 467*(7311), 86–90. https://doi.org/10.1038/nature09284.

Meng, F., Henson, R., Wehbe-Janek, H., Ghoshal, K., Jacob, S. T., & Patel, T. (2007). MicroRNA-21 regulates expression of the PTEN tumor suppressor gene in human hepatocellular cancer. *Gastroenterology, 133*(2), 647–658. https://doi.org/10.1053/j.gastro.2007.05.022.

Michael, M. Z., O'Connor, S. M., van Holst Pellekaan, N. G., Young, G. P., & James, R. J. (2003). Reduced accumulation of specific microRNAs in colorectal neoplasia. *Molecular Cancer Research, 1*(12), 882–891. Retrieved from https://www.ncbi.nlm.nih.gov/pubmed/14573789.

Miska, E. A., Alvarez-Saavedra, E., Townsend, M., Yoshii, A., Sestan, N., Rakic, P., et al. (2004). Microarray analysis of microRNA expression in the developing mammalian brain. *Genome Biology, 5*(9), R68. https://doi.org/10.1186/gb-2004-5-9-r68.

Mohapatra, S., & Calin, G. (2021). JAM-ming miR-21. *Cell Death & Differentiation*. https://doi.org/10.1038/s41418-021-00825-9.

Nelson, P. T., Baldwin, D. A., Scearce, L. M., Oberholtzer, J. C., Tobias, J. W., & Mourelatos, Z. (2004). Microarray-based, high-throughput gene expression profiling of microRNAs. *Nature Methods, 1*(2), 155–161. https://doi.org/10.1038/nmeth717.

O'Donnell, K. A., Wentzel, E. A., Zeller, K. I., Dang, C. V., & Mendell, J. T. (2005). c-Myc-regulated microRNAs modulate E2F1 expression. *Nature, 435*(7043), 839–843. https://doi.org/10.1038/nature03677.

Pasquinelli, A. E., Reinhart, B. J., Slack, F., Martindale, M. Q., Kuroda, M. I., Maller, B., et al. (2000). Conservation of the sequence and temporal expression of let-7 heterochronic regulatory RNA. *Nature, 408*(6808), 86–89. https://doi.org/10.1038/35040556

Reinhart, B. J., Slack, F. J., Basson, M., Pasquinelli, A. E., Bettinger, J. C., Rougvie, A. E., et al. (2000). The 21-nucleotide let-7 RNA regulates developmental timing in *Caenorhabditis elegans*. *Nature, 403*(6772), 901–906. https://doi.org/10.1038/35002607.

Rosenfeld, N., Aharonov, R., Meiri, E., Rosenwald, S., Spector, Y., Zepeniuk, M., et al. (2008). MicroRNAs accurately identify cancer tissue origin. *Nature Biotechnology, 26*(4), 462–469. https://doi.org/10.1038/nbt1392.

Saito, Y., Liang, G., Egger, G., Friedman, J. M., Chuang, J. C., Coetzee, G. A., et al. (2006). Specific activation of microRNA-127 with downregulation of the proto-oncogene BCL6 by chromatin-modifying drugs in human cancer cells. *Cancer Cell, 9*(6), 435–443. https://doi.org/10.1016/j.ccr.2006.04.020.

Smolle, M. A., Calin, H. N., Pichler, M., & Calin, G. A. (2017). Noncoding RNAs and immune checkpoints-clinical implications as cancer therapeutics. *The FEBS Journal, 284*(13), 1952–1966. https://doi.org/10.1111/febs.14030.

Sun, Y., Koo, S., White, N., Peralta, E., Esau, C., Dean, N. M., et al. (2004). Development of a micro-array to detect human and mouse microRNAs and characterization of expression in human organs. *Nucleic Acids Research, 32*(22). https://doi.org/10.1093/nar/gnh186, e188.

Tam, W. (2001). Identification and characterization of human BIC, a gene on chromosome 21 that encodes a noncoding RNA. *Gene, 274*(1–2), 157–167. https://doi.org/10.1016/s0378-1119(01)00612-6.

Thomson, J. M., Parker, J., Perou, C. M., & Hammond, S. M. (2004). A custom microarray platform for analysis of microRNA gene expression. *Nature Methods, 1*(1), 47–53. https://doi.org/10.1038/nmeth704.

Van Roosbroeck, K., Fanini, F., Setoyama, T., Ivan, C., Rodriguez-Aguayo, C., Fuentes-Mattei, E., et al. (2017). Combining anti-Mir-155 with chemotherapy for the treatment of lung cancers. *Clinical Cancer Research, 23*(11), 2891–2904. https://doi.org/10.1158/1078-0432.CCR-16-1025.

van Zandwijk, N., Pavlakis, N., Kao, S. C., Linton, A., Boyer, M. J., Clarke, S., et al. (2017). Safety and activity of microRNA-loaded minicells in patients with recurrent malignant pleural mesothelioma: A first-in-man, phase 1, open-label, dose-escalation study. *The Lancet Oncology, 18*(10), 1386–1396. https://doi.org/10.1016/S1470-2045(17)30621-6.

Volinia, S., Calin, G. A., Liu, C. G., Ambs, S., Cimmino, A., Petrocca, F., et al. (2006). A microRNA expression signature of human solid tumors defines cancer gene targets. *Proceedings of the National Academy of Sciences of the United States of America*, *103*(7), 2257–2261. https://doi.org/10.1073/pnas.0510565103.

Wiggins, J. F., Ruffino, L., Kelnar, K., Omotola, M., Patrawala, L., Brown, D., et al. (2010). Development of a lung cancer therapeutic based on the tumor suppressor microRNA-34. *Cancer Research*, *70*(14), 5923–5930. https://doi.org/10.1158/0008-5472.CAN-10-0655.

Wu, Y., Xiao, Y., Ding, X., Zhuo, Y., Ren, P., Zhou, C., et al. (2011). A miR-200b/200c/429-binding site polymorphism in the 3′ untranslated region of the AP-2alpha gene is associated with cisplatin resistance. *PLoS ONE*, *6*(12). https://doi.org/10.1371/journal.pone.0029043, e29043.

Yan, L. X., Huang, X. F., Shao, Q., Huang, M. Y., Deng, L., Wu, Q. L., et al. (2008). MicroRNA miR-21 overexpression in human breast cancer is associated with advanced clinical stage, lymph node metastasis and patient poor prognosis. *RNA*, *14*(11), 2348–2360. https://doi.org/10.1261/rna.1034808.

Yang, C. H., Yue, J., Pfeffer, S. R., Fan, M., Paulus, E., Hosni-Ahmed, A., et al. (2014). MicroRNA-21 promotes glioblastoma tumorigenesis by down-regulating insulin-like growth factor-binding protein-3 (IGFBP3). *The Journal of Biological Chemistry*, *289*(36), 25079–25087. https://doi.org/10.1074/jbc.M114.593863.

Yu, W., Zhu, K., Wang, Y., Yu, H., & Guo, J. (2018). Overexpression of miR-21-5p promotes proliferation and invasion of colon adenocarcinoma cells through targeting CHL1. *Molecular Medicine*, *24*(1), 36. https://doi.org/10.1186/s10020-018-0034-5.

Ziebarth, J. D., Bhattacharya, A., & Cui, Y. (2012). Integrative analysis of somatic mutations altering microRNA targeting in cancer genomes. *PLoS ONE*, *7*(10). https://doi.org/10.1371/journal.pone.0047137, e47137.

Chapter 3

miR-15/16 in human malignancies

Yuri Pekarsky[a,b] and Carlo M. Croce[c]

[a]Department of Cancer Biology and Genetics, The Wexner Medical Center, Columbus, OH, United States, [b]The Ohio State University, Comprehensive Cancer Center, Columbus, OH, United States, [c]Wexner Medical Center, The Ohio State University, Columbus, OH, United States

13q deletions in CLL and *miR-15/16*

Chronic lymphocytic leukemia (CLL) is the most frequent leukemia in humans. About 12,000 CLL cases are diagnosed every year, accounting for more than 30% of all leukemia cases in the United States (Sgambati, Linet, & Devesa, 2001). CLL leukemic cells are morphologically mature B-cells that cannot be easily grown in vitro (Bullrich & Croce, 2001; Sgambati et al., 2001). On the other hand, in tissue culture conditions these cells do not die for several weeks but do not proliferate either, and have a morphologically mature B-cell appearance (Bullrich & Croce, 2001; Pekarsky, Balatti, & Croce, 2018; Sgambati et al., 2001). CLL lymphocytes are the population of CD5 positive mature B-cells. Normally, CD5 is expressed on T-cells and very rarely on B-cells, thus CLL lymphocytes represent an expansion of a rare population of CD5 positive B-cells. CLL can have a very variable clinical course, as some patients can have relatively mild symptoms and live without treatment for several years (Sgambati et al., 2001). However, quite often newly diagnosed CLL patients have a progressive disease and need treatments right away (Pekarsky et al., 2018; Sgambati et al., 2001). CLL has a number of prognostic markers. Generally, patients expressing low levels of ZAP-70 (zeta-chain-associated protein kinase 70) and mutated IgH V_H usually have indolent disease and can survive for extended periods of time without treatment. On the other hand, CLL patients expressing high levels of ZAP-70 and displaying unmutated IgH V_H have a more aggressive form of CLL with immediate treatment often necessary. This general trend has many exceptions though (Bullrich & Croce, 2001; Pekarsky et al., 2018; Sgambati et al., 2001).

CLL has many subgroups because more than 80% of CLL cases show genomic deletions or gains. In many cases, several rearrangements occur at the same time. The most common chromosomal abnormalities in CLL are 13q deletions, which are detected in more than 60% of cases. The other abnormalities include 11q deletions, 17p deletions, and trisomy 12 (Dohner et al., 2000; Edelmann et al., 2012). We previously showed that in aggressive CLL, trisomy 12 is associated with *NOTCH1* (notch homolog 1), activating mutations (Balatti et al., 2012). *TP53* and ATM (ataxia-telangiectasia mutated) genes are associated with 17p and 11q deletions, respectively. In addition, 17p deletions in CLL are associated with the loss of *miR-3676*, a negative regulator of the *TCL1* oncogene, a critical molecule in aggressive CLL (Balatti et al., 2015; Gaudio et al., 2012; Palamarchuk et al., 2012). Generally, 17p and 11q deletions are predictors of clinically aggressive disease. On the other hand, CLLs showing 13q deletions as a sole abnormality are mostly indolent (Pekarsky et al., 2018).

Twenty years ago, we investigated 13q deletions in CLL, trying to identify gene(s) downregulated by this most common deletion in CLL. We hypothesized that the 13q deleted region contains a commonly inactivated tumor-suppressor gene involved in CLL development (Calin et al., 2002). At that time, we also knew that besides CLL, the 13q14 chromosomal region is also rearranged or deleted in almost half of mantle cell lymphomas (Stilgenbauer et al., 1998) and in half of prostate cancers (Dong, Boyd, & Frierson Jr., 2001).

The deleted 13q14 genomic region was extensively studied by us and other laboratories using LOH and positional cloning approaches. At that time, we sequenced a large 1 Mb region located within the 13q14 deleted region to try to identify specific targets of these deletions (Bullrich et al., 2001; Migliazza et al., 2001).

Despite extensive studies, including identifying and sequencing all protein coding genes located in the 13q14 deleted region as well as searching for mutations or downregulation of these genes in CLL, we and others were not able to identify any protein coding genes affected by 13q deletions in CLL (Bullrich et al., 2001; Mertens et al., 2002; Migliazza et al., 2001; Rondeau et al., 2001).

We previously reviewed the discovery of *miR-15/16* as a target of 13q deletions in several articles. Briefly, in 2001–02, we studied several CLL cases showing deletions and translocations at 13q14. Using somatic cell hybrids, we identified a small (~30 kb) deletion within the *DLEU2* gene. We also identified a translocation breakpoint located in the same region (Calin et al., 2002; Pekarsky, Calin, & Aqeilan, 2005). *DLEU2* was previously studied and found not to be affected by 13q

MicroRNA in Human Malignancies. https://doi.org/10.1016/B978-0-12-822287-4.00011-6

deletions (Calin et al., 2002; Pekarsky et al., 2005). There were no other protein coding genes within this 30 kb deleted region. At that time, genomic databases were not fully developed, but in one of the updates we found that two microRNAs, *miR-15a* and *miR-16-1 (miR-15/16)*, were located in a cluster within this 30 kb region (Calin et al., 2002). *miR-15/16* were the only genes in this region. To determine if *miR-15/16* are targets of 13q deletions in CLL, we studied its expression in 13q deleted CLL samples by northern blot and determined that *miR-15/16* expression was downregulated in most CLLs compared to normal controls (Calin et al., 2002). As mentioned above, all protein coding genes within the 13q deleted region were already excluded (Bullrich et al., 2001; Migliazza et al., 2001; Pekarsky et al., 2005). Thus, the only possible conclusion was that 13q deletions in CLL target *miR-15/16* expression (Calin et al., 2002). The discovery of *miR-15/16* involvement in CLL was the first example of the involvement of small noncoding RNAs in cancer.

Because *miR-15/16* expression is downregulated in most CLLs, we investigated whether there are important oncogenes in CLL that *miR-15/16* may target. By using bioinformatics, we identified a list of predicted targets for *miR-15/16*. Interestingly, we found that *BCL2*, an important oncogene in CLL, was one of the top predicted targets. *BCL2* is a well-known oncogene overexpressed in many blood malignancies as well as in solid tumors. Bcl2 is mostly localized in mitochondria. It increases cell survival and inhibits apoptosis by inhibiting the release of cytochrome C from mitochondria into the cytoplasm (Grabow et al., 2012; Sanchez-Beato, Sanchez-Aguilera, & Piris, 2003). We analyzed the homology between *miR-15/16* and the *BCL2* mRNA and determined that *miR-15/16* target a single site in the 3′ untranslated region (3′ UTR) of *BCL2*; using luciferase experiments, we confirmed this targeting (Cimmino et al., 2005). We then studied the expression of *miR-15a*, *miR-16*, and *BCL2* in CLL. We found the inverse correlation in the expression of *BCL2* and *miR-15a*, and *BCL2* and *miR-16*. Thus, we concluded that the 13q deletions in CLL result in the downregulation of *miR-15/16*, and that causes *BCL2* overexpression in CLL (Cimmino et al., 2005; Pekarsky et al., 2018). Because *miR-15/16* are downregulated in most CLLs, it would be logical to use it as a drug to treat this disease. However, to date there are no efficient methods to deliver microRNAs into cancer cells. The overexpression of Bcl2 is a hallmark of CLL, thus it is likely that the inhibition of Bcl2 could work the same way as the introduction of *miR-15/16* into CLL cells. A few years ago, Abbott was able to do exactly that, as they developed a drug called Venetoclax that inhibited Bcl2 by inhibiting its important interactions, causing increased cell survival (Souers et al., 2013). Currently, Venetoclax is FDA approved to treat CLL patients. It was proven to be very effective in CLL. Even in previously treated relapsed CLL patients, Venetoclax had an 80% response rate (Croce & Reed, 2016).

Other mechanisms of *miR-15/16* dysregulation

Many tumor suppressors are silenced by several different mechanisms. Our initial report concluded that *miR-15/16* are silenced in CLL by 13q deletions. However, in many instances *miR-15/16* expression was downregulated in CLLs with no 13q deletions observed, suggesting that additional mechanisms could be responsible for *miR-15/16* silencing. One report demonstrated a link between histone deacetylases and *miR-15/16*, showing that the overexpression of histone deacetylases in CLL results in increases in aberrant epigenetic silencing of *miR-15/16* (Allegra et al., 2014; Sampath et al., 2012). Another report studied CLL cases lacking *miR-15/16* expression and determined that in several such cases, DROSHA does not correctly process *miR-15/16* precursors. That resulted in the downregulation of *miR-15/16* expression by defective processing (Allegra et al., 2014). Thus, *miR-15/16* expression can also be shut down by aberrant epigenetic modifications or flawed processing (Pekarsky et al., 2018).

Many tumor suppressors are inactivated by loss of function mutations, and the identification of such mutations is a very important proof of the tumor-suppressor function of any gene. To determine if *miR-15/16* can be inactivated by mutations, we investigated and sequenced 75 CLL cases and 160 normal controls, including the genomic area of *miR-15/16* (Calin et al., 2005). We observed one germline change in the area very close to the *miR-15/16* sequence in two CLL patients while no such changes were observed in the controls. This germline change was only seven base pairs downstream of the *miR-16-1* precursor (Calin et al., 2005). Interestingly, this region is strongly conserved among all primates, indicating that this single nucleotide change could be important for the regulation of *miR-15/16* expression in these patients. We then investigated *miR-15/16* expression in both patients and confirmed the downregulation of *miR-15/16* expression in both cases (Calin et al., 2005). Using in vitro transfection experiments, we showed that this germline change causes a significant decrease of *miR-15/16* expression; thus, this change is classified as a loss of function mutation (Calin et al., 2005). In conclusion, in both patients we observed a classic scenario for tumor-suppressor gene inactivation: one *miR-15/16* allele was inactivated by deletion, and the second allele was mutated (Calin et al., 2005; Pekarsky et al., 2018).

For any tumor-suppressor gene, functional proof of such activity is required. To carry out these experiments, we chose MEG-01 cells, which lack *miR-15/16* expression. We transfected *miR-15/16* into MEG-01 cells and tested the tumorigenicity of these cells in nude mice. Several weeks after injection, we observed no tumors in mice injected with cells expressing *miR-15/16*; at the same time, we observed large tumors in mice injected with cells transfected with the control vector (Calin et al., 2008). This was the first evidence of the tumor-suppressor function of *miR-15/16* in vivo in MEG-01 cells.

To investigate additional mechanisms of *miR-15/16* dysregulation, we analyzed the genomic region surrounding *miR-15/16* to find possible transcription factor binding sites. Interestingly, we observed several p53 binding sites just upstream of *miR-15/16* (Fabbri et al., 2011). We then proceeded to prove that p53 can actually bind to these sites by using the chromatin immunoprecipitation approach. These experiments demonstrated that p53 indeed can be bound to these sites upstream of *miR-15/16*. We further determined that p53 can activate the expression of *miR-15/16* (Fabbri et al., 2011). We then analyzed the 3′ UTR of *TP53* and found putative binding sites for *miR-15/16*. Using standard luciferase experiments, we found that *miR-15/16* can bind these sites and regulate *TP53* expression. Thus, we observed the feedback loop between *miR-15/16* and *TP53*. P53 can activate *miR-15/16* expression while at the same time, increased *miR-15/16* expression results in a decrease of p53 expression (Fabbri et al., 2011; Pekarsky et al., 2018).

Interplay between ROR1, Bcl2, and *miR-15/16* in CLL

Receptor kinase-like orphan receptor 1 (ROR1) is an oncoembryonic antigen and a very important molecule in CLL because it expresses on most CLL B-cells but not on normal B-cells (Baskar et al., 2008; Broome, Rassenti, Wang, Meyer, & Kipps, 2011). In addition, antibodies against ROR1 can inhibit the growth of CLL cells, particularly human CLL cells of an aggressive phenotype. (Widhopf II et al., 2014). Because both *ROR1* and *miR-15/16* are important in CLL pathogenesis, we investigated whether *miR-15/16* target *ROR1* expression. By analyzing the 3′ UTR of *ROR1,* we found a single *miR-15/16* binding site. To prove that *miR-15/16* target *ROR1* expression, we used luciferase experiments and determined that *miR-15/16* expression decreases the luciferase activity of the *ROR1* luciferase construct, thus regulating *ROR1* expression (Pekarsky & Croce, 2019; Rassenti et al., 2017). Although most CLLs abundantly express ROR1 (Fukuda et al., 2008), a small proportion of CLLs (~10%) are almost ROR1 negative, expressing very low levels of ROR1. Thus, we studied *miR-15/16* expression in CLLs with high and low levels of ROR1 (Rassenti et al., 2017). We found that CLL samples expressing high levels of ROR1 showed low *miR-15/16* expression, and CLL samples expressing low levels of ROR1 showed high *miR-15/16* expression (Rassenti et al., 2017), confirming that *miR-15/16* target ROR1 expression in CLL. Because *miR-15/16* regulate the expression of both ROR1 and Bcl2 in CLL, we investigated the expression of ROR1 and Bcl2 by western blotting. As expected, we found a direct correlation between Bcl2 and ROR1 expression; the CLL cells either expressed high levels of both Bcl2 and ROR1, or low levels of both Bcl2 and ROR1 (Rassenti et al., 2017). In addition, we determined that the microRNA signature of 17 microRNAs can distinguish between CLLs expressing high and low levels of ROR1 (Pekarsky & Croce, 2019; Rassenti et al., 2017).

Our results showed that in addition to *BCL2, miR-15/16* also targeting *ROR1* can serve as a rationale for the development of new CLL treatments by interfering with different biological mechanisms. Most CLL cases can be treated with anti-ROR1 antibodies because ROR1 is expressed only on CLL cells. On the other hand, some CLL cells can escape this treatment due to possible mutations in *ROR1* or other reasons. In this case, the combination of Cirmtuzumab (anti-ROR1 antibodies) and Venetoclax (Bcl2 inhibiting drug) should be able to eliminate all CLL cells (Rassenti et al., 2017). In fact, our results confirmed that Cirmtuzumab and Venetoclax have a synergistic effect in killing CLL cells in vitro (Pekarsky & Croce, 2019; Rassenti et al., 2017) (Fig. 3.1).

miR-15/16 in mice

The development of tumor phenotypes in knockout mouse models is very important evidence for the tumor-suppressor activity of any gene. Such models for *miR-15/16* were generated by Klein et al.; two knockout strains were generated (Klein et al., 2010). In one case, the minimal deleted region (*MDR*) was deleted. MDR containing *miR-15/16* and the *Dleu2* gene was located nearby. This deletion mimicked a minimum deleted region in CLL. The second knockout strain had only *miR-15/16* deleted (Klein et al., 2010). In both cases, aging mice developed a CLL-like disease. 42% of *MDR* mice and 26% of *miR-15/16* deleted mice developed B-cell malignancies or proliferations at 18 months (Klein et al., 2010). In MDR mice, CD5 positive B-cell proliferations were already detectable at 12 months (Klein et al., 2010). Interestingly, *miR-15/16* deleted mice developed a milder phenotype than *MDR* knockouts, suggesting that some regulatory elements or sequences in MDR also contribute to the CLL phenotype (Klein et al., 2010; Pekarsky et al., 2018). These results confirmed that *miR-15/16* has a tumor-suppressor function in CLL.

Among multiple mouse strains, the New Zealand Black (NZB) mice developed CD5+ B-cell accumulations at 18–24 months (Raveche, 1990). Dr. Raveche and colleagues reported a genome-wide mapping study to isolate the genomic regions responsible for CD5+ B-cell proliferation in NZB mice (Raveche et al., 2007). Several genomic *loci* were identified in this study located on mouse chromosomes 14, 18, and 19. Interestingly, the region on mouse chromosome 14 corresponds to the *miR-15/16* region at 13q14 in humans deleted in CLL (Raveche et al., 2007). Further investigations revealed that NZB mice express reduced levels of *miR-15/16* due to a single nucleotide change near mouse *miR-15/16*, located in the position very close to a point mutation detected in human CLL samples (Calin et al., 2005; Pekarsky et al., 2018; Raveche et al., 2007).

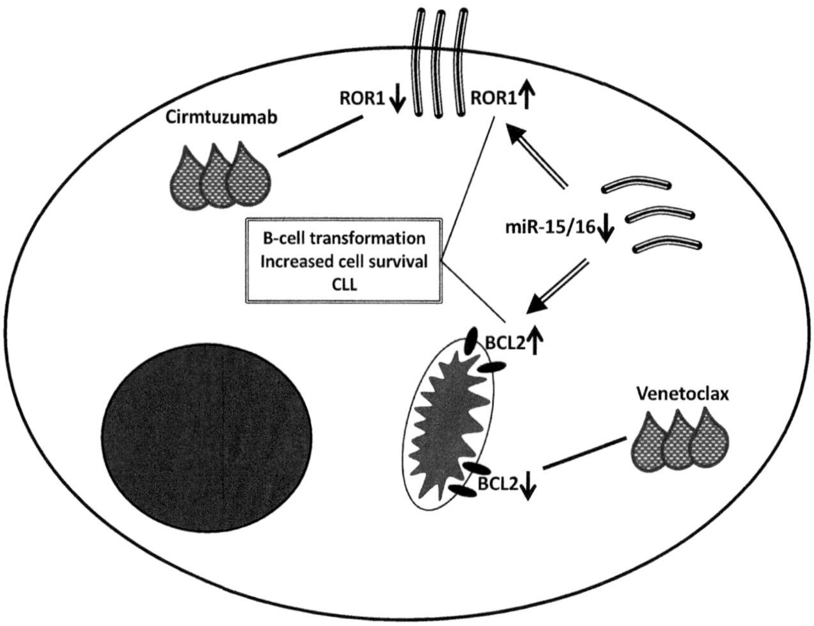

FIG. 3.1 Interplay between ROR1, Bcl2, and *miR-15/16* in CLL.

The human genome contains two *miR-15/16* clusters. We referred to the *miR-15a/16-1* cluster as *miR-15/16*. The second *miR-15/16* cluster (*miR-15b/16-2*) is located at 3q25 (Lovat et al., 2015). Both clusters are almost identical and all four microRNAs have the same seed region and target the same genes. Because *miR-15a/16-1* function as a tumor suppressor in CLL, we developed an *miR-15b/16-2* knockout mouse model and investigated the phenotype of these mice (Lovat et al., 2015). Similarly to *miR-15a/16-1* knockouts, *miR-15b/16-2* knockout mice developed proliferations of CD5+ B-cells, and some developed B-cell lymphomas. These malignancies were developed in animals that were 15–18 months old with a penetrance rate of more than 50% (Lovat et al., 2015; Pekarsky et al., 2018). In an additional study, we crossed *miR-15a/16-1* and *miR-15b/16-2* knockouts, generating mice lacking the expression of both *miR-15/16* clusters (DKO) (Lovat et al., 2018). The lack of expression of any *miR-15/16* in these mice was confirmed by northern blot.

DKO mice also developed leukemias and lymphomas. Overall, only 35% of DKO mice survived to 12 months. All DKO mice developed significant splenomegaly; however, these malignancies mostly were not B-cell malignancies (Lovat et al., 2018). Surprisingly, histological and FACS analysis revealed that 77% of DKO mice developed malignancies closely resembling human acute myeloid leukemia (AML) while only 23% of DKO mice developed malignancies of B-cell origin. The AML phenotype was confirmed by observing 60%–70% of CD11b/Gr1 cells in the bone marrow of DKO mice (Lovat et al., 2018; Pekarsky & Croce, 2019).

miR-15/16 in human AML and MDS

The development of AML by DKO mice may suggest that *miR-15/16* is also involved in the pathogenesis of human AML. Very recently, we carried out a study on the role of *miR-15/16* in human AML and myelodysplastic syndrome (MDS). Because MDS can often be further transformed into AML, we chose and analyzed three different groups of patient samples: a group that had MDS only, a group of MDS patients later transformed to AML, and a group of AML patients who previously had MDS (Lovat et al., 2020). In addition, we tested a larger group of AML patient samples and AML cell lines for the expression of *miR-15a*, *miR-15b*, *miR-16-1*, and *miR-16-2* (*miR-16-1* and *miR-16-2* are identical, so we could only assess their combined expression). We found that the expression levels of *miR-15a, miR-15b*, and *miR-16* in the MDS group was significantly higher than in the other two groups, and also significantly higher in the MDS group in comparison with the general AML group. Our results indicated that all three microRNAs can predict the progression of MDS to AML (Lovat et al., 2020). In addition, we determined that all three microRNAs can serve as prognostic predictors in AML as patients with a higher expression of *miR-15a, miR-15b,* and *miR-16* had a more favorable prognosis compared to patients expressing the lower levels. These results were confirmed by studying AML cell lines, as we found that 40% of AML cell lines lack *miR-15a and miR-15b* expression, and most of these cell lines overexpressed Bcl2 and ROR1, targets of *miR-15/16*. To identify the causes of decreased *miR-15/16* expression in AML, we studied the methylation status of sequences immediately upstream of *miR-15/16* and determined that the CpG methylation of these sequences is partially responsible for the downregulation of *miR-15/16* expression in AML. Thus, we determined that the loss of *miR-15/16* expression, including the loss of expression of one or both *miR-15/16* clusters, contributes to the pathogenesis of human myeloid malignancies, MDS, and AML (Lovat et al., 2020).

miR-15/16 in solid cancers

Neuroblastoma (NB) is a common human malignancy with a poor survival rate. The *MYCN* oncogene plays an important role in the pathogenesis of NB. *MYCN* is commonly upregulated and amplified in NB, and its overexpression is also associated with the recurrence of NB (Chava et al., 2020). A recent report studied the regulation of *MYCN* in NB. They found that in tumor xenografts expressing high levels of *MYCN*, *miR-15a*, *miR-15b,* and *miR-16* were downregulated (Chava et al., 2020). Given the TargetScan prediction, the authors tested the targeting of *MYCN* by *miR-15a*, *miR-15b,* and *miR-16* and found that all three microRNAs bind the 3′ UTR of *MYCN*. In transfection experiments, specific microRNA inhibitors significantly increased *MYCN* mRNA and protein expression, and overexpression of all three microRNAs resulted in a significant reduction of *MYCN* expression. In addition, the overexpression of *miR-15a*, *miR-15b,* and *miR-16* in NB cells significantly reduced the tumorigenicity of these cells in mice. The interaction of all three microRNAs with *MYCN* mRNA was also confirmed in vivo by Ago2 immunoprecipitation assays. Moreover *miR-15a*, *miR-15b,* and *miR-16* showed tumor-suppressor activity in NB cell lines by reducing invasion and migration. These data suggest that both *miR-15/16* clusters function as tumor suppressors by targeting *MYCN* in NB (Chava et al., 2020).

The loss of heterozygosity (LOH) at 13q14 in prostate cancer is well documented and associated with progression and metastatic disease (Dong et al., 2001; Jin et al., 2018). Because the TGF-β signaling pathway is important in prostate cancer progression, Jin et al. recently investigated the tumor-suppressor role of *miR-15/16* in this pathway (Jin et al., 2018). By using TargetScan and other prediction software, the authors identified several potential *miR-15/16* targets within the TGF-β pathway. The authors tested the targeting of these candidates by *miR-15/16* and found that these microRNAs bind the 3′ UTRs of the *SMAD3* and *ACVR2A* genes (Jin et al., 2018). In transfection experiments, the overexpression of *miR-15/16* resulted in a significant reduction of Smad3 and ACVR2A protein expression in prostate cancer cells, confirming the targeting of these two genes by *miR-15/16*. In addition, exogenous *miR-15/16* expression resulted in the reduction of invasion of LNCaP prostate cancer cells (Jin et al., 2018). Interestingly, *miR-15/16* overexpression also caused the overexpression of epithelial-mesenchymal transition (EMT) promoting factors Snail and Twist. Thus, the inactivation of *miR-15/16* plays an important role in metastatic prostate cancer by promoting EMT through the TGF-β pathway (Jin et al., 2018).

The B-lymphoma Moloney murine leukemia virus insertion region-1 protein (BMI1) is an important oncogene in a number of solid tumors. Its overexpression is associated with tumor progression, and in breast cancer it enhances metastatic disease. BMI1 has several functions, including the inhibition of apoptosis and the induction of EMT (Patel et al., 2016, 2017). In addition, it participates in DNA repair by increasing histone H2A ubiquitination. Because *BMI1* is a predicted *miR-15/16* target, two reports investigated the relationship between *BMI1* and *miR-15/16* in breast cancer (Patel et al., 2016, 2017). Luciferase experiments revealed a specific binding of *miR-15a* and *miR-16* at the 3′ UTR of *BMI1*. Transfection experiments in breast cancer cells revealed the significant downregulation of BMI1 expression by *miR-15a* and *miR-16* in breast cancer cells. In addition, exogenous *miR-15a* and *miR-16* expression triggered mitochondrial mediated apoptosis. Moreover, increased *miR-15/16* expression impeded DNA repair by downregulating the expression of BMI1 (Patel et al., 2016, 2017). Thus, *miR-15/16* functions as a tumor suppressor in breast cancer by targeting *BMI1* and inhibiting oncogenic pathways downstream of *BMI1*.

The 40S ribosomal protein S6 Kinase (S6K1) encoded by the *RPS6KB1* gene is a downstream target of mTOR and is overexpressed in breast cancer (Ramaiah et al., 2014). Because *RPS6KB1* is a predicted *miR-15/16* target, one study investigated the targeting of *RPS6KB1* by *miR-15/16*. The authors were able to prove that *miR-15/16* were able to specifically target the 3’ UTR of *RPS6KB1* and inhibit S6K1 expression in breast cancer cells. This inhibition resulted in the inhibition of cell viability, an increase in the G1 phase, and increased apoptosis (Ramaiah et al., 2014).

Fatty acid synthase (FASN) is overexpressed in many cancers, including breast cancer (Wang et al., 2016). In breast cancer, FASN works upstream of HER2 signaling and its overexpression leads to increased cell proliferation and decreased apoptosis. Because *FASN* is a predicted *miR-15/16* target, a recent study investigated a relationship between *FASN* and *miR-15/16* in breast cancer cells. Luciferase experiments confirmed that *miR-15/16* specifically target the 3′ UTR of *FASN* (Wang et al., 2016). In addition, the exogeneous expression of *miR-15/16* caused a reduction of FASN levels in breast cancer cells and reduced cell proliferation (Wang et al., 2016). In summary, *miR-15/16* plays a crucial tumor-suppressor role in breast cancer by targeting multiple oncogenes.

To assess the role of *miR-15/16* in lung cancer, Bandi et al. investigated the expression of *miR-15/16* in nonsmall cell lung cancer (NSCLC) (Bandi et al., 2009). Using microdissected lung adenocarcinomas, squamous cell carcinomas, and matched normal lung tissues, they determined that 82% of adenocarcinomas and 67% of squamous cell carcinomas have reduced *miR-16* expression (Bandi et al., 2009). This reduced expression was associated with deletions of *miR-15/16* in most cases. Because *CCND1*, *CCND2*, and *CCNE1* (encoding cyclins D1, D2, and E1, respectively) are predicted *miR-15/16* targets, the authors investigated whether *miR-15/16* target these genes (Bandi et al., 2009). Luciferase experiments confirmed that *miR-15/16* specifically target the 3′ UTRs of all three genes. In addition, *miR-15/16* expression was inversely correlated with cyclin D1 levels in NSCLC samples. Moreover, the exogenous expression of *miR-15/16* expression induced cell cycle arrest in NSCLC cells (Bandi et al., 2009).

Another report also studied *miR-15/16* expression in NSCLC samples and confirmed the downregulation of *miR-15/16* in NSCLC (Ke, Zhao, Xiong, & Cao, 2013). They focused on another predicted *miR-15/16* oncogenic target, hepatoma-derived growth factor (*HDGF*). The authors confirmed that HDGF is directly targeted by *miR-15/16* in NSCLC cells. The exogenous expression of *miR-16* suppressed cell growth, migration, and invasion, possibly due to the downregulation of *HDGF* expression (Ke et al., 2013).

MAPK1 kinase 1 (MEK1) is a member of the MAPK signaling pathway activated in many tumor types. MEK1 phosphorylates ERK, activating the MAPK signaling pathway (Chen et al., 2019). Because *miR-15/16* are downregulated in lung cancer, and *MEK1* is a predicted target of *miR-15/16*, a recent report investigated the relationship between *miR-16* and MEK1 expression. Luciferase experiments showed that *miR-16* specifically target MEK1 expression in lung cancer cells. In addition, increased *miR-16* expression caused a decrease of MEK1 expression levels (Chen et al., 2019).

Cyclin B1 (encoded by the *CCNB1* gene) plays an important oncogenic role in colorectal cancer (Dai et al., 2012). Because *CCNB1* is a predicted *miR-15/16* target, another report investigated the expression of *miR-15/16* and cyclin B1 in colorectal cancer cells. They found an inverse correlation between *miR-15/16* and *CCNB1* expression levels and a specific interaction between *miR-15/16* and the 3′ UTR of *CCNB1*. This suggested that *miR-15/16* targets *CCNB1* in colorectal cancer cells (Dai et al., 2012).

The investigation of *miR-16* expression in oral squamous cell carcinoma (OSCC) in 131 tumor samples and 131 adjacent normal tissues revealed the significant downregulation of *miR-16* expression in OSCC (Hu et al., 2018). In addition, higher *miR-16* expression in OSCC samples correlated with the higher survival rate of OSCC patients. Tousled-like kinase 1 (TLK1) is involved in chromatin assembly and DNA repair (Hu et al., 2018). TLK1 expression was found to be upregulated in most OSCC tumor samples (vs adjacent normal tissues) and cell lines. Moreover, the authors found a significant reverse correlation between *miR-16* and TLK1 expression (Hu et al., 2018). Because *TLK1* is a predicted target of *miR-15/16*, luciferase experiments were carried out, and they proved that *miR-16* directly and specifically target the 3′ UTR of *TLK1*. In addition, the exogenous expression of *miR-16* in oral cancer cells decreased TLK1 expression and inhibited cell growth by increasing G2/M arrest (Hu et al., 2018). Thus, *miR-16* is downregulated in OSCC cells, and functions as a tumor suppressor by inhibiting TLK1 expression (Hu et al., 2018) (Fig. 3.2).

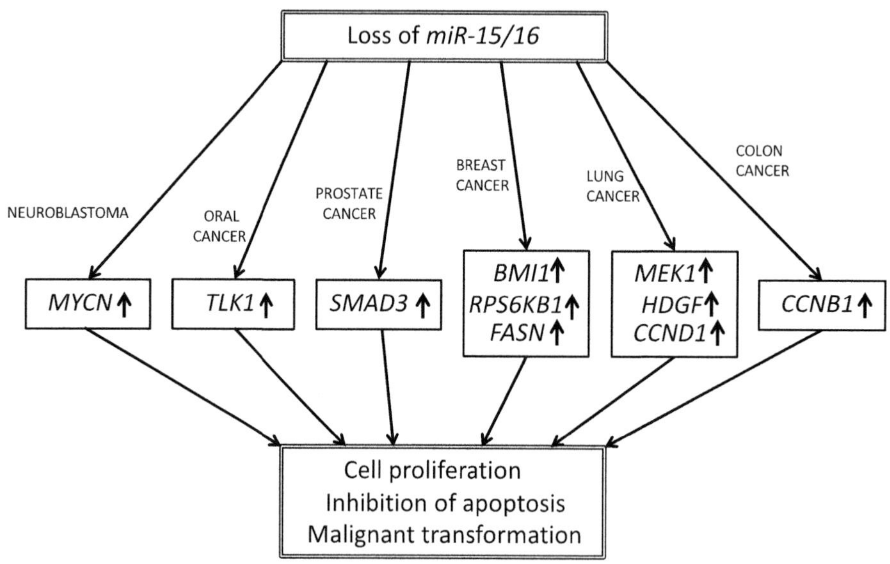

FIG. 3.2 *miR-15/16* in solid cancers.

References

Allegra, D., Bilan, V., Garding, A., Dohner, H., Stilgenbauer, S., Kuchenbauer, F., et al. (2014). Defective DROSHA processing contributes to downregulation of MiR-15/-16 in chronic lymphocytic leukemia. *Leukemia*, *28*(1), 98–107.

Balatti, V., Bottoni, A., Palamarchuk, A., Alder, H., Rassenti, L. Z., Kipps, T. J., et al. (2012). NOTCH1 mutations in CLL associated with trisomy 12. *Blood*, *119*(2), 329–331.

Balatti, V., Rizzotto, L., Miller, C., Palamarchuk, A., Fadda, P., Pandolfo, R., et al. (2015). TCL1 targeting miR-3676 is codeleted with tumor protein p53 in chronic lymphocytic leukemia. *Proceedings of the National Academy of Sciences of the United States of America*, *112*(7), 2169–2174.

Bandi, N., Zbinden, S., Gugger, M., Arnold, M., Kocher, V., Hasan, L., et al. (2009). miR-15a and miR-16 are implicated in cell cycle regulation in a Rb-dependent manner and are frequently deleted or down-regulated in non-small cell lung cancer. *Cancer Research*, *69*(13), 5553–5559.

Baskar, S., Kwong, K. Y., Hofer, T., Levy, J. M., Kennedy, M. G., Lee, E., et al. (2008). Unique cell surface expression of receptor tyrosine kinase ROR1 in human B-cell chronic lymphocytic leukemia. *Clinical Cancer Research: An Official Journal of the American Association for Cancer Research*, *14*(2), 396–404.

Broome, H. E., Rassenti, L. Z., Wang, H. Y., Meyer, L. M., & Kipps, T. J. (2011). ROR1 is expressed on hematogones (non-neoplastic human B-lymphocyte precursors) and a minority of precursor-B acute lymphoblastic leukemia. *Leukemia Research*, *35*(10), 1390–1394.

Bullrich, F., & Croce, C. (2001). Molecular biology of chronic lymphocytic leukemia. In B. Cheson (Ed.), *Chronic lymphocytic leukemias* (2nd ed., pp. 9–32). New York: Marcel Dekker, Inc. Revised and expanded.

Bullrich, F., Fujii, H., Calin, G., Mabuchi, H., Negrini, M., Pekarsky, Y., et al. (2001). Characterization of the 13q14 tumor suppressor locus in CLL: Identification of ALT1, an alternative splice variant of the LEU2 gene. *Cancer Research*, *61*(18), 6640–6648.

Calin, G. A., Cimmino, A., Fabbri, M., Ferracin, M., Wojcik, S. E., Shimizu, M., et al. (2008). MiR-15a and miR-16-1 cluster functions in human leukemia. *Proceedings of the National Academy of Sciences of the United States of America*, *105*(13), 5166–5171.

Calin, G. A., Dumitru, C. D., Shimizu, M., Bichi, R., Zupo, S., Noch, E., et al. (2002). Frequent deletions and down-regulation of micro- RNA genes miR15 and miR16 at 13q14 in chronic lymphocytic leukemia. *Proceedings of the National Academy of Sciences of the United States of America*, *99*(24), 15524–15529.

Calin, G. A., Ferracin, M., Cimmino, A., Di Leva, G., Shimizu, M., Wojcik, S. E., et al. (2005). A MicroRNA signature associated with prognosis and progression in chronic lymphocytic leukemia. *The New England Journal of Medicine*, *353*(17), 1793–1801.

Chava, S., Reynolds, C. P., Pathania, A. S., Gorantla, S., Poluektova, L. Y., Coulter, D. W., et al. (2020). miR-15a-5p, miR-15b-5p, and miR-16-5p inhibit tumor progression by directly targeting MYCN in neuroblastoma. *Molecular Oncology*, *14*(1), 180–196.

Chen, T., Xiao, Q., Wang, X., Wang, Z., Hu, J., Zhang, Z., et al. (2019). miR-16 regulates proliferation and invasion of lung cancer cells via the ERK/MAPK signaling pathway by targeted inhibition of MAPK kinase 1 (MEK1). *The Journal of International Medical Research*, *47*(10), 5194–5204.

Cimmino, A., Calin, G. A., Fabbri, M., Iorio, M. V., Ferracin, M., Shimizu, M., et al. (2005). miR-15 and miR-16 induce apoptosis by targeting BCL2. *Proceedings of the National Academy of Sciences of the United States of America*, *102*(39), 13944–13949.

Croce, C. M., & Reed, J. C. (2016). Finally, an apoptosis-targeting therapeutic for cancer. *Cancer Research*, *76*(20), 5914–5920.

Dai, L., Wang, W., Zhang, S., Jiang, Q., Wang, R., Dai, L., et al. (2012). Vector-based miR-15a/16-1 plasmid inhibits colon cancer growth in vivo. *Cell Biology International*, *36*(8), 765–770.

Dohner, H., Stilgenbauer, S., Benner, A., Leupolt, E., Krober, A., Bullinger, L., et al. (2000). Genomic aberrations and survival in chronic lymphocytic leukemia. *The New England Journal of Medicine*, *343*(26), 1910–1916.

Dong, J. T., Boyd, J. C., & Frierson, H. F., Jr. (2001). Loss of heterozygosity at 13q14 and 13q21 in high grade, high stage prostate cancer. *The Prostate*, *49*(3), 166–171.

Edelmann, J., Holzmann, K., Miller, F., Winkler, D., Buhler, A., Zenz, T., et al. (2012). High resolution genomic profiling of chronic lymphocytic leukemia reveals new recurrent genomic alterations. *Blood*, *120*(24), 4783–4794.

Fabbri, M., Bottoni, A., Shimizu, M., Spizzo, R., Nicoloso, M. S., Rossi, S., et al. (2011). Association of a microRNA/TP53 feedback circuitry with pathogenesis and outcome of B-cell chronic lymphocytic leukemia. *JAMA: The Journal of the American Medical Association*, *305*(1), 59–67.

Fukuda, T., Chen, L., Endo, T., Tang, L., Lu, D., Castro, J. E., et al. (2008). Antisera induced by infusions of autologous Ad-CD154-leukemia B cells identify ROR1 as an oncofetal antigen and receptor for Wnt5a. *Proceedings of the National Academy of Sciences of the United States of America*, *105*(8), 3047–3052.

Gaudio, E., Spizzo, R., Paduano, F., Luo, Z., Efanov, A., Palamarchuk, A., et al. (2012). Tcl1 interacts with Atm and enhances NF-kappaB activation in hematologic malignancies. *Blood*, *119*(1), 180–187.

Grabow, S., Waring, P., Happo, L., Cook, M., Mason, K. D., Kelly, P. N., et al. (2012). Pharmacological blockade of Bcl-2, Bcl-x(L) and Bcl-w by the BH3 mimetic ABT-737 has only minor impact on tumour development in p53-deficient mice. *Cell Death and Differentiation*, *19*(4), 623–632.

Hu, S., Wang, H., Yan, D., Lu, W., Gao, P., Lou, W., et al. (2018). Loss of miR-16 contributes to tumor progression by activation of tousled-like kinase 1 in oral squamous cell carcinoma. *Cell Cycle*, *17*(18), 2284–2295.

Jin, W., Chen, F., Wang, K., Song, Y., Fei, X., & Wu, B. (2018). miR-15a/miR-16 cluster inhibits invasion of prostate cancer cells by suppressing TGF-beta signaling pathway. *Biomedicine & Pharmacotherapy*, *104*, 637–644.

Ke, Y., Zhao, W., Xiong, J., & Cao, R. (2013). Downregulation of miR-16 promotes growth and motility by targeting HDGF in non-small cell lung cancer cells. *FEBS Letters*, *587*(18), 3153–3157.

Klein, U., Lia, M., Crespo, M., Siegel, R., Shen, Q., Mo, T., et al. (2010). The DLEU2/miR-15a/16-1 cluster controls B cell proliferation and its deletion leads to chronic lymphocytic leukemia. *Cancer Cell*, *17*(1), 28–40.

Lovat, F., Fassan, M., Gasparini, P., Rizzotto, L., Cascione, L., Pizzi, M., et al. (2015). miR-15b/16-2 deletion promotes B-cell malignancies. *Proceedings of the National Academy of Sciences of the United States of America*, *112*(37), 11636–11641.

Lovat, F., Fassan, M., Sacchi, D., Ranganathan, P., Palamarchuk, A., Bill, M., et al. (2018). Knockout of both miR-15/16 loci induces acute myeloid leukemia. *Proceedings of the National Academy of Sciences of the United States of America*, *115*(51), 13069–13074.

Lovat, F., Nigita, G., Distefano, R., Nakamura, T., Gasparini, P., Tomasello, L., et al. (2020). Combined loss of function of two different loci of miR-15/16 drives the pathogenesis of acute myeloid leukemia. *Proceedings of the National Academy of Sciences of the United States of America*, *117*(22), 12332–12340.

Mertens, D., Wolf, S., Schroeter, P., Schaffner, C., Dohner, H., Stilgenbauer, S., et al. (2002). Down-regulation of candidate tumor suppressor genes within chromosome band 13q14.3 is independent of the DNA methylation pattern in B-cell chronic lymphocytic leukemia. *Blood*, *99*(11), 4116–4121.

Migliazza, A., Bosch, F., Komatsu, H., Cayanis, E., Martinotti, S., Toniato, E., et al. (2001). Nucleotide sequence, transcription map, and mutation analysis of the 13q14 chromosomal region deleted in B-cell chronic lymphocytic leukemia. *Blood*, *97*(7), 2098–2104.

Palamarchuk, A., Yan, P. S., Zanesi, N., Wang, L., Rodrigues, B., Murphy, M., et al. (2012). Tcl1 protein functions as an inhibitor of de novo DNA methylation in B-cell chronic lymphocytic leukemia (CLL). *Proceedings of the National Academy of Sciences of the United States of America*, *109*(7), 2555–2560.

Patel, N., Garikapati, K. R., Pandita, R. K., Singh, D. K., Pandita, T. K., Bhadra, U., et al. (2017). miR-15a/miR-16 down-regulates BMI1, impacting Ub-H2A mediated DNA repair and breast cancer cell sensitivity to doxorubicin. *Scientific Reports*, *7*(1), 4263.

Patel, N., Garikapati, K. R., Ramaiah, M. J., Polavarapu, K. K., Bhadra, U., & Bhadra, M. P. (2016). miR-15a/miR-16 induces mitochondrial dependent apoptosis in breast cancer cells by suppressing oncogene BMI1. *Life Sciences*, *164*, 60–70.

Pekarsky, Y., Balatti, V., & Croce, C. M. (2018). BCL2 and miR-15/16: From gene discovery to treatment. *Cell Death and Differentiation*, *25*(1), 21–26.

Pekarsky, Y., Calin, G. A., & Aqeilan, R. (2005). Chronic lymphocytic leukemia: Molecular genetics and animal models. *Current Topics in Microbiology and Immunology*, *294*, 51–70.

Pekarsky, Y., & Croce, C. M. (2019). Noncoding RNA genes in cancer pathogenesis. *Advances in Biological Regulation*, *71*, 219–223.

Ramaiah, M. J., Lavanya, A., Honarpisheh, M., Zarea, M., Bhadra, U., & Bhadra, M. P. (2014). miR-15/16 complex targets p70S6 kinase1 and controls cell proliferation in MDA-MB-231 breast cancer cells. *Gene*, *552*(2), 255–264.

Rassenti, L. Z., Balatti, V., Ghia, E. M., Palamarchuk, A., Tomasello, L., Fadda, P., et al. (2017). MicroRNA dysregulation to identify therapeutic target combinations for chronic lymphocytic leukemia. *Proceedings of the National Academy of Sciences of the United States of America*, *114*(40), 10731–10736.

Raveche, E. S. (1990). Possible immunoregulatory role for CD5 + B cells. *Clinical Immunology and Immunopathology*, *56*(2), 135–150.

Raveche, E. S., Salerno, E., Scaglione, B. J., Manohar, V., Abbasi, F., Lin, Y. C., et al. (2007). Abnormal microRNA-16 locus with synteny to human 13q14 linked to CLL in NZB mice. *Blood*, *109*(12), 5079–5086.

Rondeau, G., Moreau, I., Bezieau, S., Petit, J. L., Heilig, R., Fernandez, S., et al. (2001). Comprehensive analysis of a large genomic sequence at the putative B-cell chronic lymphocytic leukaemia (B-CLL) tumour suppresser gene locus. *Mutation Research*, *458*(3–4), 55–70.

Sampath, D., Liu, C., Vasan, K., Sulda, M., Puduvalli, V. K., Wierda, W. G., et al. (2012). Histone deacetylases mediate the silencing of miR-15a, miR-16, and miR-29b in chronic lymphocytic leukemia. *Blood*, *119*(5), 1162–1172.

Sanchez-Beato, M., Sanchez-Aguilera, A., & Piris, M. A. (2003). Cell cycle deregulation in B-cell lymphomas. *Blood*, *101*(4), 1220–1235.

Sgambati, M., Linet, M., & Devesa, S. (2001). Chronic lymphocytic leukemia, epidemiological, familial, and genetic aspects. In B. Cheson (Ed.), *Chronic lymphocytic leukemias* (2nd ed., pp. 33–62). New York: Marcel Dekker, Inc. Revised and expanded.

Souers, A. J., Leverson, J. D., Boghaert, E. R., Ackler, S. L., Catron, N. D., Chen, J., et al. (2013). ABT-199, a potent and selective BCL-2 inhibitor, achieves antitumor activity while sparing platelets. *Nature Medicine*, *19*(2), 202–208.

Stilgenbauer, S., Nickolenko, J., Wilhelm, J., Wolf, S., Weitz, S., Dohner, K., et al. (1998). Expressed sequences as candidates for a novel tumor suppressor gene at band 13q14 in B-cell chronic lymphocytic leukemia and mantle cell lymphoma. *Oncogene*, *16*(14), 1891–1897.

Wang, J., Zhang, X., Shi, J., Cao, P., Wan, M., Zhang, Q., et al. (2016). Fatty acid synthase is a primary target of MiR-15a and MiR-16-1 in breast cancer. *Oncotarget*, *7*(48), 78566–78576.

Widhopf, G. F., II, Cui, B., Ghia, E. M., Chen, L., Messer, K., Shen, Z., et al. (2014). ROR1 can interact with TCL1 and enhance leukemogenesis in Emu-TCL1 transgenic mice. *Proceedings of the National Academy of Sciences of the United States of America*, *111*(2), 793–798.

Chapter 4

microRNAs and tumor suppressor p53 regulation

Zhaohui Feng, Juan Liu, Cen Zhang, and Wenwei Hu

Department of Radiation Oncology, Rutgers Cancer Institute of New Jersey, Rutgers-State University of New Jersey, New Brunswick, NJ, United States

p53 and its signaling pathway

The tumor suppressor p53 plays a key role in tumor suppression, and loss or attenuation of the p53 function is crucial for the initiation and/or progression of many human cancers (Donehower et al., 2019; Levine, 2019; Levine, Hu, & Feng, 2006; Muller & Vousden, 2013; Vousden & Prives, 2009; Zhang et al., 2020). The key role of p53 in tumor suppression has been established by extensive studies in cells, mouse models, and humans. For instance, germline heterozygous p53 mutations cause Li-Fraumeni syndrome, a rare autosomal-dominant disorder that predisposes patients to tumors. Of Li-Fraumeni syndrome patients, 50% develop one to multiple tumors, including sarcomas, brain tumors, and breast cancers, before the age of 30. Mice with p53 deletion are highly susceptible to tumors; the majority of them develop tumors, mainly lymphomas and sarcomas, before 6 month of age whereas wild-type mice usually never develop tumors before 9 month of age. The p53 gene has been shown to be the most frequently mutated gene in human cancers; p53 is mutated in more than 50% of all human cancers and in almost every type of cancer. Furthermore, p53 function is frequently impaired in human cancers without p53 mutations. In addition to DNA mutations, different mechanisms have been reported to attenuate the tumor-suppressive function of p53 in cancer, which in turn initiates and/or promotes tumorigenesis. p53 can be inactivated by many tumor virus-encoded proteins, such as the large T-antigen of simian virus 40 (SV40) and the E6 oncoprotein of human papilloma virus (HPV). p53 function can also be impaired by the overexpression of its negative regulators (e.g., MDM2, MDM4, and PPM1D) or downregulation of its positive regulators in cancers (e.g., p19ARF). MDM2, MDM4 (also known as MDMX), and PPM1D (also known as Wip1) are often amplified and/or overexpressed in different types of cancers, and their overexpression is often mutually exclusive with p53 mutations in cancers, supporting their important roles in tumorigenesis through the negative regulation of p53. p19ARF, a tumor suppressor that inhibits MDM2 activity to activate p53, is frequently deleted or silenced in different types of cancers, which in turn activates MDM2 to impair p53 function (Donehower et al., 2019; Levine, 2019; Levine et al., 2006; Muller & Vousden, 2013; Vousden & Prives, 2009; Zhang et al., 2020).

p53 is a transcription factor that forms a homotetramer in cells and binds to p53-responsive elements in its target genes to induce (for the majority of target genes) or repress the expression of these genes. The p53-responsive element is a degenerate DNA sequence of PuPuPuC(A/T) (A/T) GPyPyPy repeated twice with a variable spacer (usually 0–20), where Pu is a purine and Py a pyrimidine. More than 200 protein-coding genes have been identified to be direct p53 target genes. Under nonstressed conditions, the p53 protein has a very short half-life (usually a few minutes) and is maintained at a low level in normal cells and tissues. In response to a wide variety of intrinsic and extrinsic stress signals such as DNA damage, nutritional deprivation, impaired ribosome biogenesis, hypoxia, activation of oncogenes, etc., the half-life and levels of p53 protein are drastically increased in cells, leading to the activation of p53 function. Through transcriptional regulation of select target genes in a cell/tissue type- and stress signal-specific manner, p53 regulates a wide variety of cellular processes, including apoptosis, cell cycle arrest, senescence, DNA repair, autophagy, ferroptosis, metabolism, antioxidant defense, cell stemness, immune response, etc., to exert its function in tumor suppression (Fig. 4.1; Donehower et al., 2019; Levine, 2019; Levine et al., 2006; Muller & Vousden, 2013; Vousden & Prives, 2009; Zhang et al., 2020). In addition to tumor suppression, p53 has also been demonstrated to play an important role in many other fundamental biological (e.g., antiinfection, reproduction, and development) and pathological processes (e.g., metabolic diseases, neurodegenerative diseases, and aging) (Checler & Alves Da Costa, 2014; Hu, Feng, Teresky, & Levine, 2007; Levine, 2019; Liu, Tavana, & Gu, 2019; Liu, Zhang, Hu, & Feng, 2019; Vousden & Prives, 2009; Zhang et al., 2020; Zhao, 2018).

MicroRNA in Human Malignancies. https://doi.org/10.1016/B978-0-12-822287-4.00001-3

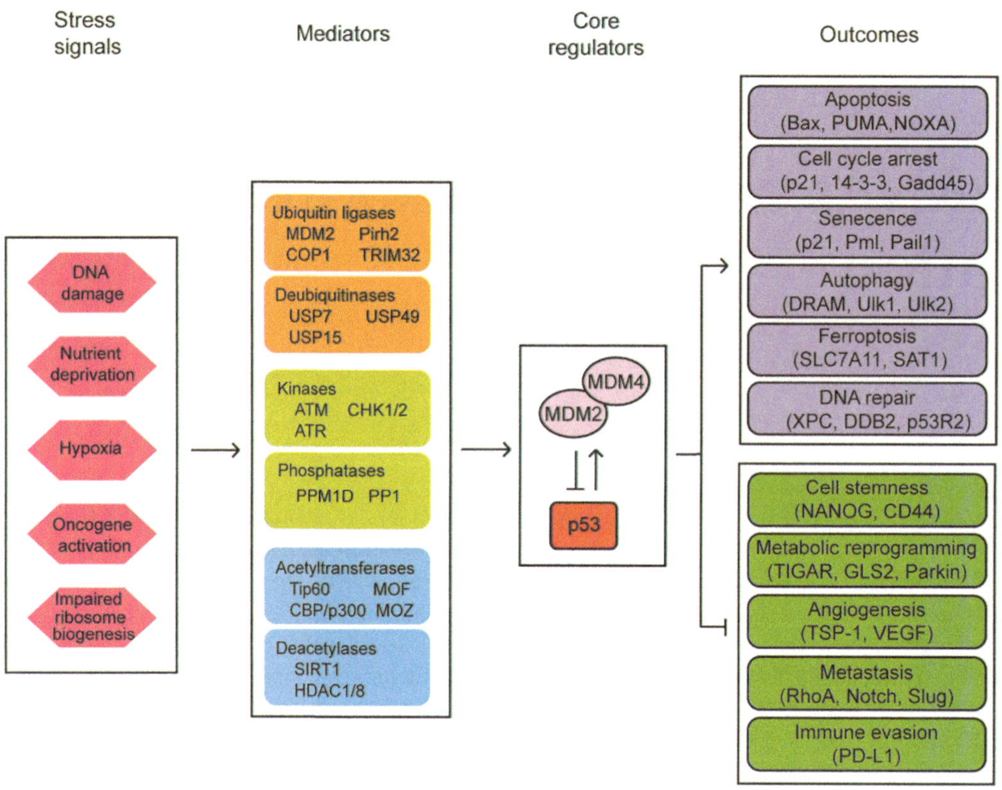

FIG. 4.1 The p53 signaling pathway. In response to various stress signals, p53 is activated and binds to its target genes to regulate their expression in a context-dependent manner. Thus, p53 plays a critical role in the regulation of a wide variety of cellular processes, contributing to the tumor-suppressive function of p53. To maintain its proper function, p53 is regulated by many different mechanisms, including the posttranslational modifications, which are the most critical and efficient mechanism for p53 regulation. Many proteins, including ubiquitin ligases, deubiquitinases, kinases, phosphatases, acetyltransferases, deacetylases, etc., are involved in the posttranslational modifications of p53.

To ensure that p53 can properly exert its critical roles in tumor suppression and other fundamental biological processes as described above, p53 protein levels and activity are tightly and exquisitely regulated by many different mechanisms in cells. The posttranslational modifications, including ubiquitination, phosphorylation, and acetylation, are the most critical and efficient mechanism to regulate p53 levels and activity. Many proteins are involved to positively or negatively regulate p53, including E3 ubiquitin ligases, deubiquitinases, kinases, phosphatases, acetyltransferases, deacetylases, etc. (Levine, 2019; Levine, Hu, & Feng, 2006; Liu, Tavana, & Gu, 2019; Meek, 2015; Fig. 4.1). Among these regulators, E3 ubiquitin ligase MDM2 is the most critical negative regulator of the p53 protein; MDM2 directly binds to the p53 protein, leading to the ubiquitination and proteasomal degradation of p53. Because MDM2 is also a p53 target gene and is transcriptionally induced by p53, MDM2 forms a negative feedback loop with p53 to tightly regulate p53 in cells. Additionally, MDM2 also promotes p53 nuclear export to suppress p53 transcriptional regulation of its target genes in the nucleus. Many other proteins have been shown to modulate p53 through their regulation of MDM2 levels and activity, which underscores the critical role of MDM2 in p53 regulation. For instance, tumor-suppressor p19ARF binds to MDM2 and inhibits MDM2-mediated ubiquitination and degradation of p53, which in turn upregulates p53 levels and activity. MDM4, a structural homolog of MDM2 without detectable E3 ubiquitin ligase activity, binds to MDM2 to promote MDM2-mediated ubiquitination and degradation of p53. Further, MDM4 also directly inhibits p53 transcriptional activity through its binding to the p53 protein. Gankyrin, an ankyrin repeat oncoprotein that is frequently overexpressed in cancers, directly interacts with MDM2 to facilitate MDM2-p53 binding, promoting MDM2-mediated ubiquitination and degradation of p53. PP2A phosphatase dephosphorylates MDM2 to enhance MDM2 activity, leading to the increased p53 ubiquitination and degradation. The cell cycle regulator Cyclin G1 encoded by CCNG1 (a p53 target gene), binds to PP2A and forms a negative feedback loop with p53. In response to ribosomal biogenesis stress, some ribosomal proteins (e.g., RPL11, RPL5, and RPL23) have been reported to bind to MDM2 and block MDM2-mediated ubiquitination and degradation of p53.

Besides MDM2, many additional E3 ubiquitin ligases have been reported to directly bind to p53 and ubiquitinate p53 for proteasomal degradation, including Cop1, Pirh2, CHIP, and some TRIM family proteins (e.g., TRIM 23, TRIM 24, TRIM 32, TRIM 39, etc.) (Hafner, Bulyk, Jambhekar, & Lahav, 2019; Levine, 2019; Liu, Tavana, et al., 2019; Liu, Zhang, et al., 2019; Liu, Zhang, Wang, Hu, & Feng, 2020). Some of these E3 ubiquitin ligases, such as Cop1, Pirh2, and TRIM32, are also p53 target genes, and thus, they form negative feedback loops with p53 to regulate p53. The p53 protein ubiquitination is elegantly regulated by E3 ubuiquitin ligases and deubiquitinases, which can bind to p53 and deubiquitinate p53 to stabilize and activate p53. For instance, deubiquitinases USP7 (also known as HAUSP), USP15, and USP49 deubiquitinate p53 to stabilize it. In addition, phosphorylation also plays a critical role in regulating p53 levels and activity. The ATM/ATR and CHK1/2 kinases phosphorylate p53 to inhibit MDM2-mediated p53 ubiquitination to stabilize and activate p53 in response to DNA damage, which is critical for the function of p53 in DNA repair. In contrast, phosphatases PPM1D and PP1 (protein phosphatase 1) dephosphorylate p53 to inactivate p53. Further, acetylation modification on p53 has been shown to regulate p53 transcriptional activity. The transcriptional coactivators CBP/p300 and some histone acetyltransferases (HATs), such as Tip60, MOF, and MOZ, acetylate p53 to enhance p53 transcriptional activity. In contrast, NAD-dependent protein deacetylase SIRT1 and some histone deacetylases (HDACs), such as HDAC1 and HDAC8, deacetylate p53 to reduce p53 transcriptional activity (Hafner et al., 2019; Levine, 2019; Liu, Tavana, et al., 2019).

Thus, p53 and its signaling are regulated by many different mechanisms involving different positive and negative regulators and various positive and negative feedback loops in a highly dynamic and context-dependent manner to ensure its proper functions in tumor suppression and the regulation of many other biological processes. Disruption of this exquisite regulation will lead to pathological conditions and diseases. For instance, the attenuation or loss of p53 function results in cancer development and maternal reproduction defects, whereas the inappropriate activation of p53 promotes neurodegenerative diseases and aging (Checler & Alves Da Costa, 2014; Hu et al., 2007; Levine, 2019; Zhang et al., 2020; Zhao, 2018).

GOF mutant p53 in cancer

In addition to the loss of the tumor-suppressive function of wild-type p53, many mutant p53 (mutp53) display gain-of-function (GOF) activities to promote cancer progression (Donehower et al., 2019; Freed-Pastor and Prives, 2012; Levine et al., 2006; Mantovani, Collavin, & Del Sal, 2019; Muller & Vousden, 2013; Yue et al., 2017; Zhang et al., 2020). Unlike many other well-known tumor suppressor genes, such as RB, BRCA1, APC, or PTEN, which are usually inactivated by deletions or truncating mutations in cancers, the majority of p53 mutations in human cancers are missense mutations, which produce full-length mutp53 proteins with only one amino acid difference from wild-type p53. Notably, mutp53 proteins frequently accumulate to very high levels in cancer cells. Around 90% of p53 mutations occur in the DNA-binding domain of p53, including six mutational hotspots (e.g., R175, R245, R248, R249, R273, and R282) that account for around one-third of p53 mutations in cancers. Many missense mutp53 proteins, including the hotspot mutp53, have been shown to display GOF activities to promote cancer progression independently of wild-type p53. For example, compared with p53 knockout mice, R172H and R270H (equivalent to human R175H and R273H, respectively) mutp53 knock-in mice develop more malignant tumors and these tumors are often more metastatic; humanized R248Q mutp53 knock-in mice have an earlier tumor onset. In Li-Fraumeni syndrome patients, germline p53 missense mutations are associated with an earlier cancer onset (~9 years) compared with germline p53 deletion mutations. Many GOF activities of mutp53 have been reported, including promoting cancer cell proliferation, migration/metastasis, stemness, metabolic reprogramming (e.g., enhanced glycolysis and nucleotide synthesis, and activated mevalonate pathway), resistance to therapies, and immune evasion (Donehower et al., 2019; Levine, 2019; Levine et al., 2006; Muller & Vousden, 2013; Vousden & Prives, 2009; Zhang et al., 2020). Although GOF mutp53 usually loses the ability to bind to the p53-responsive elements and transcriptionally regulate target genes of wild-type p53 due to the mutations in the DNA-binding domain of the p53 protein, mutp53 has been shown to exert its GOF activities to promote tumorigenesis through many different mechanisms. For instance, GOF mutp53 can bind to different transcription factors and cofactors, such as p63/p73 (two p53 family member proteins), NF-Y, VDR, SREBPs, ETS1/2, and Sp1, and is recruited to their binding sites to affect the expression of their target genes. In addition, GOF mutp53 can also specifically interact with many other proteins, including oncoproteins and tumor suppressors, to affect their levels, activities, and/or functions. In addition, mutp53 binds to chromatin regulatory genes MLL1/2 and MOZ and upregulates their expression, leading to the genome-wide increases of histone methylation and acetylation to regulate gene expression (Donehower et al., 2019; Levine, 2019; Levine et al., 2006; Muller & Vousden, 2013; Vousden & Prives, 2009; Zhang et al., 2020).

miRNAs regulated by wild-type p53

In addition to protein-coding genes, p53 also regulates the levels of many noncoding genes, such as miRNAs. At the same time, many miRNAs regulate p53 protein levels and activity. Thus, miRNAs not only function as new downstream target genes that mediate various functions of p53, but also act as a new class of p53 regulators in cells (Dangelmaier,

Lazar, & Lal, 2019; Feng, Zhang, Wu, & Hu, 2011; Javad, Tahereh, & Atieh, 2020; Liao, Cao, Zhou, & Lu, 2014; Liu, Zhang, Zhao, & Feng, 2017; Luo, Cui, Tili, & Croce, 2018). Many miRNAs have been reported to be p53 target genes, and mediate p53 functions (Fig. 4.2). The miR-34 gene family members miR-34a/b/c were the first reported miRNAs that are directly regulated by p53. miR-34 family members directly repress the expression of a group of targets, including BCL2, SIRT1, Survivin, CDK4/6, cyclin D1, Cyclin E2, c-Myc, N-Myc, E2F3, and Met, to regulate apoptosis and cell cycle arrest, contributing to the tumor-suppressive function of p53. miR-34 family members also target and downregulate some cancer stem cell (CSC) markers, such as CD133, CD44, and BMI1, and regulators for induced pluripotent stem cells (iPSC), such as NANOG, SOX2, N-Myc and SIRT1, contributing to the role of p53 in regulating cell stemness. miR-34 family members target and suppress many metabolic enzymes, including HK1/2, PDK1, and LDHA involved in glucose metabolism; ACSL1 and ACSL4 involved in lipid synthesis; and IMPDH involved in nucleotide synthesis, contributing to the function of p53 in maintaining metabolic homeostasis in normal cells. miR-34 family members also target and repress the regulators of the epithelial-mesenchymal transition (EMT), including Snail, Slug, ZFN281 as well as Notch1 and its ligand Jagged1, to inhibit EMT and cancer metastasis. Furthermore, miR-34a targets and represses PD-L1, diacylglycerol kinase ζ (DGK ζ), and CCL22 to promote immune surveillance and inhibit tumor immune evasion (Dangelmaier et al., 2019; Feng et al., 2011; Javad et al., 2020; Liu et al., 2017; Luo et al., 2018; Navarro & Lieberman, 2015).

In addition to the miR-34 family, p53 transcriptionally induces the expression of many other miRNAs, including miR-145, miR-107, miR-200 family members (miR-200a, miR-200b, miR200c, miR-141, and miR-429), miR-192, miR-194, miR-215, miR-15a, miR-16-1, miR-27b-3p, miR-445-3p, and miR-30c (Fig. 4.2; Dangelmaier et al., 2019; Feng et al., 2011; Javad et al., 2020; Liao et al., 2014; Liu et al., 2017; Luo et al., 2018). Many of these miRNAs display a tumor-suppressive function, and their induction by p53 contributes to the tumor suppressive function of p53. For instance, miR-145 directly represses c-Myc, E2F3, and CDK4/6 to induce cell cycle arrest; represses OCT4, SOX2, and KLF4 to inhibit cell stemness; represses MUC1, FSCN1, and NEDD9 to inhibit metastasis and represses VEGF and N-Ras to inhibit angiogenesis. miR-107 mediates p53 function in inducing cell cycle arrest through repressing cell cycle regulators CDK6/8 and RBL2 (RB-related 2). Furthermore, miR-107 targets HIF-1β to inhibit angiogenesis and cancer metabolic reprogramming, and represses NF1 to inhibit cell migration and invasion. p53 induces the expression of the miR-200 family members, which

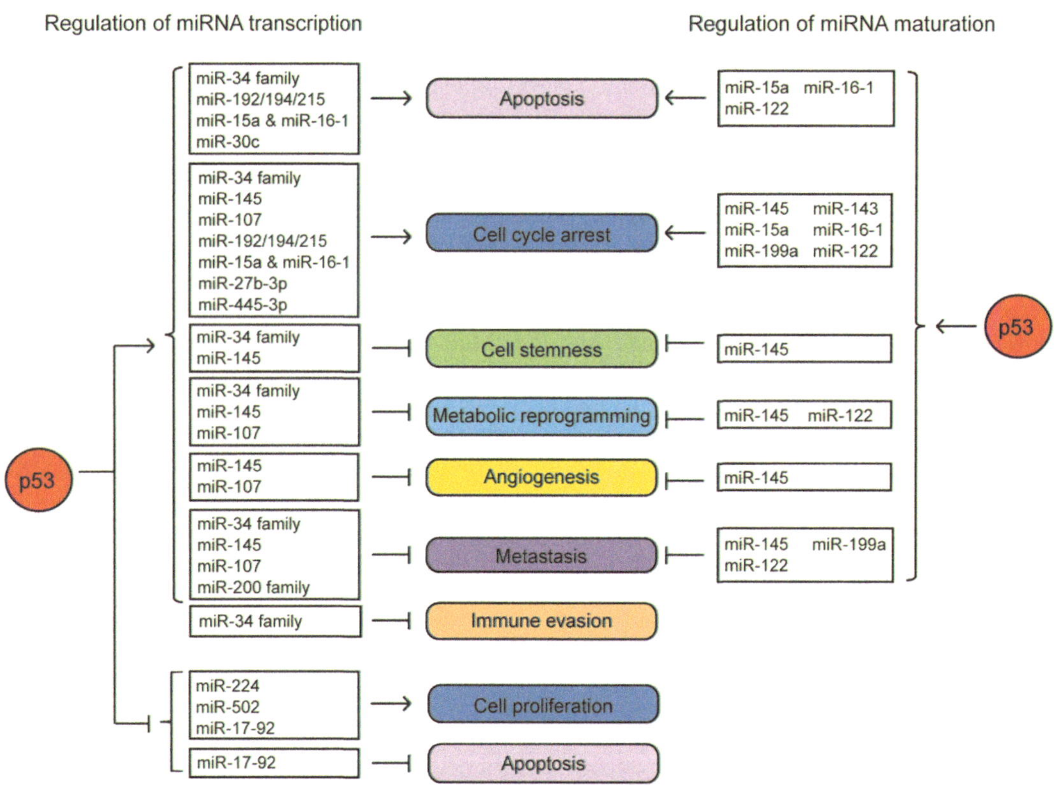

FIG. 4.2 p53 regulates the transcription and maturation of miRNAs involved in a wide variety of cellular processes. p53 transcriptionally induces or represses the expression of many miRNAs, and also promotes the maturation of specific miRNAs. The regulation of these miRNAs contributes to the role of p53 in the regulation of different cellular processes and the function of p53 in tumor suppression.

suppress EMT by directly repressing ZEB1 and ZEB2, two important EMT transcription factors. MiR-192, miR-194, miR-215, miR-15a, and miR-16-1 are transcriptionally upregulated by p53 to mediate p53 function in inducing apoptosis and cell cycle arrest. p53 also induces the expression of miR-27b-3p and miR-445-3p, which target cell cycle regulators CKS1B and CAC1, respectively, leading to the reduction of p27 polyubiquitination to induce quiescence. Additionally, p53 activates the transcription of miR-30c, which in turn reduces the expression of DNA repair protein FANCF and translesion synthesis DNA polymerase REV1 to promote apoptosis.

Notably, in addition to transcriptionally inducing the expression of many miRNAs, p53 has also been reported to transcriptionally repress the expression of some miRNAs, such as miR-224, miR-502, and the miR-17-92 cluster (Dangelmaier et al., 2019; Feng et al., 2011; Javad et al., 2020; Liao et al., 2014; Liu et al., 2017; Luo et al., 2018). miR-224 promotes tumor progression in nonsmall cell lung cancer through directly targeting TNFAIP1 and SMAD4. p53 represses miR-224 expression to suppress cell proliferation. In addition, p53 represses miR-17-92 expression to induce apoptosis in response to hypoxia and DNA damage.

Besides the transcriptional regulation of miRNAs, p53 also modulates miRNA maturation and target selection (Fig. 4.2). p53 interacts with DEAD-box RNA helicase p68 (also known as DDX5), a member of the Drosha-processing complex, to facilitate the processing of primary (pri)-miRNA into pre-miRNAs. Thus, p53 promotes the maturation of miR-15a, miR-16-1, miR-143, miR-145, miR-199a, and miR-122 to inhibit cell proliferation and metastasis (Suzuki et al., 2009). RBM38 (RNA-binding-motif protein 38) binds to 3′-UTR of mRNAs and blocks the access of miRNAs to target mRNAs. RBM38 can effectively counteract the effects of miR-17/106b on the p21 3′-UTR, miR-153 on the DDIT4 3′-UTR, and miR-372/373 on the LATS2 3′-UTR. p53 induces the expression of RBM38 and blocks the function of these miRNAs, which in turn enhances the expression of p21, DDIT4, and LAST2, leading to cell cycle arrest and DNA repair (Leveille, 2011).

Taken together, wild-type p53 directly or indirectly regulates a group of miRNAs, which in turn modulate the levels of their targets to mediate the function of p53 in inducing apoptosis and cell cycle arrest, and the inhibition of cell stemness, metabolic reprogramming, angiogenesis, metastasis, and immune evasion. Thus, these miRNAs regulated by p53 act as a new class of downstream targets of p53, contributing to the function of p53 in tumor suppression (Fig. 4.2).

miRNAs that directly regulate p53

Many miRNAs have been identified as direct negative regulators of p53 through directly binding to the 3′-UTR of p53 mRNA to decrease the p53 protein levels (Fig. 4.3; Dangelmaier et al., 2019; Feng et al., 2011; Javad et al., 2020; Liao

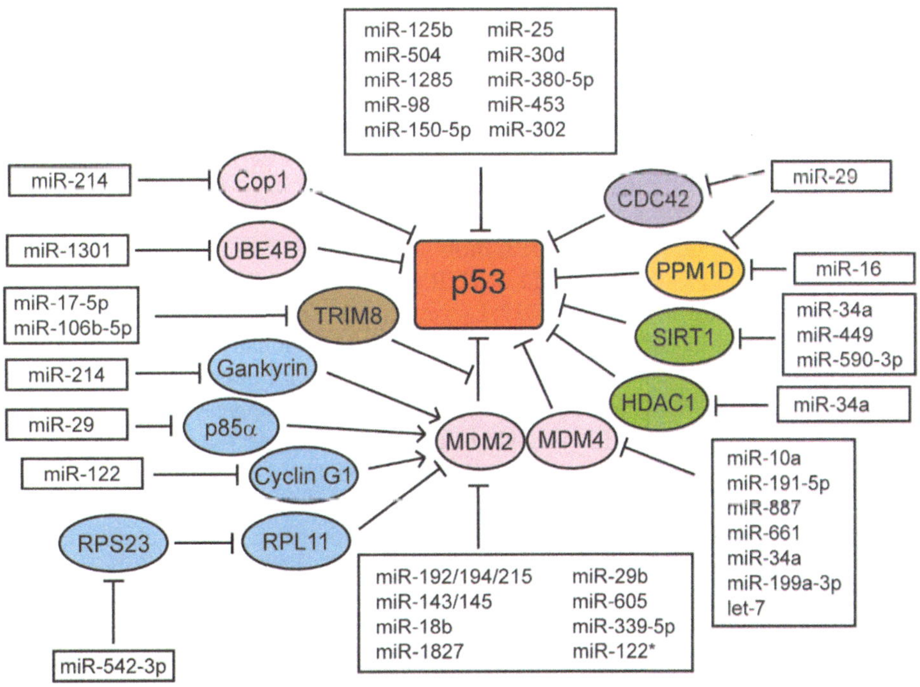

FIG. 4.3 miRNAs regulate p53 directly or indirectly through regulating its positive and negative regulators. While many miRNAs directly target the 3′-UTR of p53 mRNA to downregulate p53 protein levels and function, many miRNAs can regulate p53 protein levels and function indirectly through targeting different positive or negative regulators of p53.

et al., 2014; Liu et al., 2017; Luo et al., 2018). miR-125b and miR-504 were the first two miRNAs identified to directly target p53 to downregulate p53 levels and functions (Hu et al., 2010; Le et al., 2009). miR-125b is a brain-enriched miRNA. It decreases the p53 levels and activity, inhibiting apoptosis in both human cells and zebrafish brains (Le et al., 2009). It was reported that the expression of miR-125b is elevated in colorectal cancer (CRC), and its overexpression is associated with the poor prognosis and decreased survival of CRC patients. Furthermore, miR-125b is a potential marker for hepatocellular carcinoma (HCC) because its expression levels are higher in the serum of hepatitis B virus (HBV)-positive HCC patients compared with patients with chronic HBV infection. miR-125b also directly represses the expression of a group of p53 target genes involved in apoptosis and cell cycle regulation to regulate tissue stem cell homeostasis. These findings are consistent with the function of miR-125b as a negative regulator of p53 signaling. miR-504 can also reduce the p53 protein levels and inhibit p53-mediated apoptosis and cell cycle arrest in cancer cells. The overexpression of miR-504 promotes the growth of colorectal xenograft tumors in vivo in a largely p53-dependent manner (Hu et al., 2010). miR-504 also promotes cisplatin resistance through the negative regulation of p53 in osteosarcoma cells. Interestingly, miR-504 resides within the FGF13 (fibroblast growth factor 13) gene, which is overexpressed in various types of cancers. Furthermore, miR-504 and its host gene FGF13 are transcriptionally repressed by p53. Thus p53 and miR-504 form a negative feedback loop, adding a new layer of p53 regulation to the p53 network (Bublik et al., 2017).

Following the identification of miR-125b and miR-504, a group of miRNAs has been identified as miRNAs that directly target p53 and negatively regulate p53 levels and function (Dangelmaier et al., 2019; Feng et al., 2011; Javad et al., 2020; Liao et al., 2014; Liu et al., 2017; Luo et al., 2018). For instance, miR-1285 represses p53 and inhibits p53-mediated cell cycle arrest. miR-380-5p targets p53 and decreases p53 levels in neuroblastomas. The inhibition of miR-380-5p induces apoptosis in neuroblastoma cells and inhibits tumor formation in an orthotropic neuroblastoma mouse model through p53 activation. In multiple myeloma, miR-25 and miR-30d directly target and downregulate p53 and impair the function of p53 in inducing cell cycle arrest, apoptosis, and senescence. In lung cancer cells, miR-98 and miR-453 downregulate p53 expression. Cisplatin treatment decreases the expression of miR-98 and induces the p53 protein in cells, suggesting that the reduction of miR-98 by cisplatin may contribute to p53 activation. In CRC cells, miR-150-5p targets p53 to decrease its levels, and miR-150-5p expression levels are significantly elevated in clinical CRC specimens. The inhibition of miR-150-5p increases p53, leading to cell cycle arrest, apoptosis, and suppression of proliferation of CRC cells. Thus, the miR-150-5p-p53 pathway plays an important role in CRC, which is a potential target for therapy. In hematopoietic stem cells (HSC), miR-33 promotes the repopulation capacity of HSCs by targeting p53 to reduce its levels. In addition, miR-302 directly targets p53 to reduce p53 levels in differentiation-incompetent Dgcr8 −/− embryonic stem cells (ESCs). p53 serves as a barrier restricting neural differentiation. miR-302 expression is sufficient to enable neural differentiation of Dgcr8 −/− ESCs.

miRNAs that indirectly regulate p53

In addition to directly targeting p53 to reduce its levels, some miRNAs can indirectly regulate p53 levels and functions through targeting and repressing p53 regulators, such as MDM2, MDM4, PPM1D, Cop1, HDACs, SIRT1, Gankyrin, Cyclin G1, etc. (Dangelmaier et al., 2019; Feng et al., 2011; Javad et al., 2020; Liao et al., 2014; Liu et al., 2017; Luo et al., 2018; Fig. 4.3). As the major negative regulator of p53, MDM2 has been reported to be directly targeted and regulated by a group of miRNAs. For instance, miR-192/194/215, miR-143/145, miR-29b, miR-605, miR-18b, miR-339-5p, and miR-1827 have been reported to directly target MDM2 3′-UTR and decrease MDM2 levels, which in turn activates p53 (Agostino, 2020; Dangelmaier et al., 2019; Liu et al., 2017; Luo et al., 2018; Zhang et al., 2014, 2016). Interestingly, among these miRNAs, miR-192/194/215, miR-143/145, and miR-605 are transcriptionally induced by p53, and therefore, they form positive feedback loops with p53. Notably, the liver-specific miR-122*, the complementary strand of miR-122 that is induced by p53, targets and represses MDM2 to enhance p53 levels, leading to suppression of growth of human HCC cells, which suggests that some p53 regulation by miRNAs could be cell/tissue type-specific (Simerzin et al., 2016).

Some miRNAs have been shown to indirectly regulate MDM2 to modulate p53. For instance, several miRNAs, including miR-10a, miR-191-5p, miR-887, miR-661, miR-34a, miR-199a-3p, and let-7, can upregulate p53 levels and function through targeting and repressing MDM4, which activates MDM2 to ubiquitinate p53 (Dangelmaier et al., 2019; Feng et al., 2011; Javad et al., 2020; Liao et al., 2014; Liu et al., 2017; Luo et al., 2018). miR-542-3p directly targets and suppresses the ribosomal protein RPS23 to upregulate RPL11, which in turn stabilizes p53 through sequestrating MDM2 and inhibiting the MDM2-mediated degradation of p53. miR-122, a p53-inducible miRNA, targets and downregulates Cyclin G1, which inhibits the recruitment of the PP2A phosphatase to MDM2 to inhibit MDM2 activity, thereby indirectly blocking MDM2 to increase p53 levels and activity. miR-214 targets and represses Gankyrin, which enhances MDM2-p53 interaction and MDM2 activity toward p53, to activate p53 in human myeloma cell lines. Therefore, it functions as a tumor suppressor in multiple myeloma through p53 activation. miR-29 downregulates MDM2 levels to activate p53 by targeting and repressing p85α, a regulatory subunit of the PI3K that phosphorylates and activates MDM2 to downregulate p53 levels (Park, Lee, Ha, Nam, & Kim, 2009). Interestingly, the TRIM family protein TRIM8 binds to p53 to inhibit its interaction with MDM2, which in turn stabilizes p53 and leads to cell cycle arrest. miR-17-5p and miR-106b-5p, two miRNAs induced by oncogene N-Myc, target TRIM8 and reduce its expression, which in turn downregulates p53 levels to lead to the chemotherapeutic resistance of cancer cells.

In addition to regulating p53 through modulating MDM2 directly or indirectly, some miRNAs have been reported to target other p53 regulators to regulate p53 (Dangelmaier et al., 2019; Feng et al., 2011; Javad et al., 2020; Liao et al., 2014; Liu et al., 2017; Luo et al., 2018). For instance, miR-214 targets Cop1, an E3 ubiquitin ligase of p53, to increase p53 levels. Cop1 is overexpressed in canine hemangiosarcoma (HSA) and miR-214 overexpression induces p53-mediated apoptosis in HSA cell lines. The p53-inducible miRNA miR-16 targets PPM1D and reduces its levels, which in turn enhances p53-mediated apoptosis and cell cycle arrest. The level of miR-16 is induced by DNA damage, which postpones the accumulation of the PPM1D protein and thus allows cells to initiate p53-dependent cell cycle checkpoints in the early stage of DNA damage response. miR-449 and miR-590-3p target SIRT1 to activate p53 and induce apoptosis. miR-34a induces p53 transcriptional activity and protein levels by targeting multiple p53 negative regulators, including SIRT1, HDAC1, and MDM4. As a p53 target gene, miR-34a forms a positive feedback loop with p53. miR-29 targets and represses CDC42, a Rho GTPase, to activate p53. How CDC42 negatively regulates p53 is still unclear (Park et al., 2009). Further, miR-29 also targets and suppresses PPM1D to activate p53. Because miR-29 is a p53-inducible miRNA, miR-29 and p53 form a positive feedback loop. The expression levels of the miR-29 are decreased in gastric cancer and HCC, and the low miR-29 expression levels are associated with poor survival in HCC. In addition, miR-1301 stabilizes and activates p53 by directly suppressing UBE4B, an E3 and E4 ubiquitin ligase. The ectopic expression of miR-1301 inhibits the dissemination and metastasis of tumor cells in a p53-dependent manner.

Collectively, these findings have demonstrated that p53 levels and activity can be regulated by the network of miRNAs (Fig. 4.3), and the dysregulation of these miRNAs in cancer can contribute to the attenuation of p53 function, promoting cancer progression.

miRNAs regulated by GOF mutp53

Studies have also shown that mutp53 can regulate the expression of different miRNAs to display its GOF activities (Fig. 4.4; Agostino, 2020; Dangelmaier et al., 2019; Li, Jones, Subramanian, & Lal, 2014; Liu et al., 2017; Luo et al., 2018). For instance, mutp53 R175H induces the expression of miR-128-2, which targets and represses E2F5 to inhibit apoptosis, leading to reduced apoptosis and enhanced resistance to chemotherapeutic agents such as cisplatin, doxorubicin, and 5-fluorouracyl in lung cancer cells. It is known that mutp53 binds to p63 to suppress the transcription activity of p63 as a GOF mechanism. Mutp53 binds to p63 and inhibits its transcriptional activity, which in turn induces the expression of miR-155, an miRNA repressed by p63, in breast cancer cells. miR-155 directly represses the expression of ZNF652, a zinc-finger transcriptional

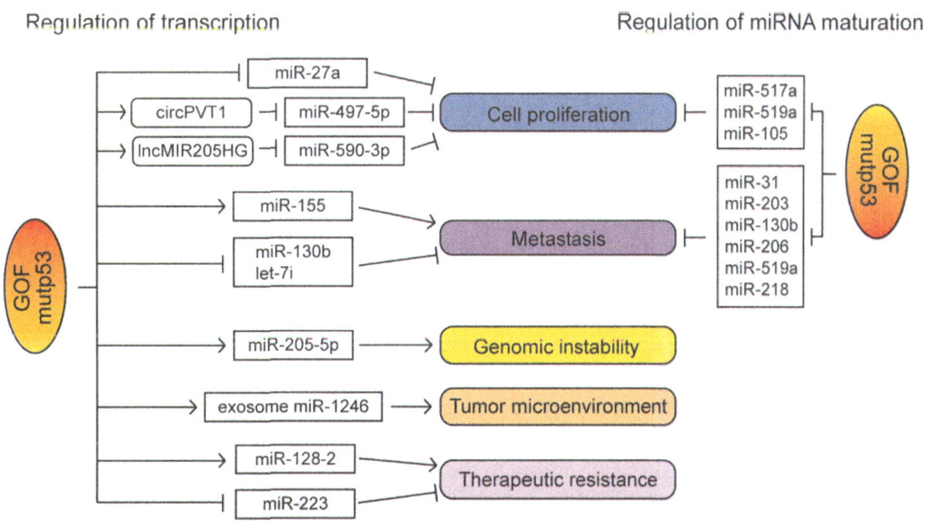

FIG. 4.4 Mutp53 regulates the transcription or maturation of miRNAs as a GOF mechanism to promote tumor progression. GOF mutp53 transcriptionally induces or represses some miRNAs and inhibits the maturation of some miRNAs to promote cancer cell proliferation, migration/metastasis, therapeutic resistance, tumor microenvironment reshaping, and immune evasion.

repressor that reduces the expression of some important drivers of cancer cell invasion and metastasis, such as TGFB1/2, TGFBR2, EGFR, SMAD2, and VIM, promoting breast cancer metastasis. miR-205-5p, whose expression is upregulated in head and neck squamous cell carcinoma (HNSCC) with p53 mutations, can directly repress the expression of BRCA1 and RAD17, two pivotal DNA repair proteins. Thus, miR-205-5p contributes to mutp53 GOF in tumorigenesis through suppressing DNA repair and enhancing chromosomal instability. Further, CRC cells expressing R248W mutp53 can selectively secrete miR-1246-enriched exosomes, which reprogram neighboring tumor-associated macrophages to a tumor-supportive and antiinflammatory state, leading to cancer progression and metastasis.

Mutp53 can repress the expression of some miRNAs (Fig. 4.4; Agostino, 2020; Dangelmaier et al., 2019; Li et al., 2014; Liu et al., 2017; Luo et al., 2018). Mutp53 R273H suppresses miR-27a expression, which in turn increases the levels of EGFR, a miR-27a target, and activates EGFR/ERK signaling to promote cell proliferation. Mutp53 transcriptionally represses miR-130b in endometrial cancer, which in turn upregulates miR-130b target ZEB1, a key EMT-related transcription factor, to promote EMT and cancer metastasis. Mutp53 reduces the expression of miR-223 in breast and CRC cells to enhance resistance to chemotherapeutic agents via enhancing the expression of miR-223 target STMN-1, an oncoprotein that can confer chemotherapeutic resistance. miR-223 directly targets and represses mutp53 in lung cancer cells, and thus, miR-223 forms a negative feedback loop with mutp53. Mutp53 also promotes tumorigenesis by binding to p63 and p73 to repress the transcription of let-7i, the miRNA known to inhibit migration, invasion, and metastasis by repressing the expression of E2F5, Lin28b, Myc, and N-Ras.

Mutp53 also regulates miRNA function through the regulation of other noncoding RNAs. For instance, mutp53 interacts with YAP to enhance the expression of circular RNA (circRNAs) circPVT1, an inhibitory miRNA sponge of miR-497-5p, leading to the enhanced expression of cell cycle regulators such as Aurka, Ki67, Bub1, Mcm7, Cdc20, and Foxm1 to promote cell proliferation in HNSCC. In HNSCC, mutp53 not only induces the expression of miR-205-5p, but also promotes the transcription of a long noncoding RNA (lncRNA) in the same host gene, lncMIR205HG, through interacting with NF-Y and E2F1. LncMIR205HG depletes endogenous miR-590-3p, leading to the increased expression of Cyclin B, CDK1, and YAP, all of which are miR-590-3p targets, to promote cell proliferation (Agostino, 2020; Dangelmaier et al., 2019).

Interestingly, mutp53 is also involved in regulating the maturation of some miRNAs. Mutp53 represses the maturation of some miRNAs, including miR-31, miR-203, miR-130b, and miR-206, through reducing the expression of Dicer by both p63-dependent and -independent depletion of Dicer to promote the invasion and metastasis of cancer cells. Mutp53 also binds and sequesters p72/82 from the microprocessor complex, which interferes with Drosha-pri-miRNAs association and leads to the inhibition of the maturation of miR-517a, miR-519a, miR-218, and miR-105, all of which display tumor-suppressive functions by promoting apoptosis, inhibiting proliferation, or suppressing the migration of cancer cells (Agostino, 2020; Li et al., 2014; Liu et al., 2017; Luo et al., 2018).

Conclusion

As summarized above, studies have demonstrated a close and complex interplay between miRNAs and p53 signaling. p53 regulates the expression of many miRNAs, which mediate different functions of p53. At the same time, many miRNAs regulate p53 levels and activity through targeting p53 or p53 regulators. Further, some of these miRNAs form positive or negative feedback loops with p53. Thus, miRNAs have become a new class of p53-regulated targets and p53 regulators as well, indicating the critical role of miRNAs in the p53 network. Despite the constant progress made in the miRNAs and p53 fields, many important and urgent questions remain to be addressed. Usually one miRNA targets multiple genes, and one gene is regulated by multiple miRNAs. Further, the expression of miRNAs often shows cell/tissue specificity, and some miRNAs can even display opposite effects in different cells and tissues. Currently, the biological functions and target genes of many of these miRNAs involved in the p53 signaling pathway are still unclear. It is also unclear how these miRNAs work coordinately to dynamically regulate the p53 signaling in a tissue-type- and stress-specific manner in different biological and pathological processes. While ~50% of human cancers contain p53 mutations, it was estimated that the majority of the remaining human cancers without p53 mutations have the attenuated p53 function. Dysregulation of these miRNAs that regulate the p53 level and activity and these miRNAs regulated by p53 to mediate p53 function may contribute to the attenuation of p53 function in cancer. Because p53 exerts its tumor-suppressive function mainly through regulating its downstream protein-coding target genes, the dysregulation of miRNAs regulating these p53 target genes may also constitute another important mechanism contributing to the attenuation of p53 function in cancer. In addition to cancer, the dysregulation of p53 signaling is also involved in many other diseases, such as neurodegenerative diseases, metabolic diseases, and the aging process. For instance, inappropriate p53 activation contributes to neurodegenerative diseases by inducing p53-dependent neuronal cell death (e.g., apoptosis, autophagy, and ferroptosis). It is unclear how the dysregulation of miRNAs in neuronal cells contributes to the inappropriate p53 activation, and mediates the p53 activation-induced neuronal cell death. Another urgent question to be addressed is how miRNAs can be used to regulate p53 signaling for treatments of cancer and other diseases, such as neurodegenerative diseases. The introduction of miRNAs or their antagonists to restore the proper levels and functions of p53 is an attractive therapeutic strategy for cancer and other diseases. For instance, based on the scenario of the low expression of miR-34 in cancers with p53 mutations, the delivery of miR-34 has been tested to recover tumor-suppressive activities and induce cell apoptosis (Agostino, 2020; Rokavec, Li, Jiang, & Hermeking, 2014; Slabakova, 2017). However, no clinical trials of miRNA-based therapies are under way to date (Hanna, Hossain, & Kocerha, 2019; Seo et al., 2019). The systematic identification of all miRNAs involved in the p53 signaling regulation and a dynamic analysis of their roles under different physiological and pathological conditions in different cells/tissues will be a future challenge. This will be crucial to reveal the mechanism and impact of the interplay between p53 and miRNAs upon different biological and pathological processes. A better understanding of this interplay will yield new therapeutic avenues for cancer and other diseases.

References

Agostino, D. (2020). The impact of mutant p53 in the noncoding RNA world. *Biomolecules*, *10*(3), 472.

Bublik, D. R., Bursać, S., Sheffer, M., Oršolić, I., Shalit, T., Tarcic, O., et al. (2017). Regulatory module involving FGF13, miR-504, and p53 regulates ribosomal biogenesis and supports cancer cell survival. *Proceedings of the National Academy of Sciences of the United States of America*, *114*(4), E496–E505. https://doi.org/10.1073/pnas.1614876114.

Checler, F., & Alves Da Costa, C. (2014). P53 in neurodegenerative diseases and brain cancers. *Pharmacology and Therapeutics*, *142*(1), 99–113. https://doi.org/10.1016/j.pharmthera.2013.11.009.

Dangelmaier, E., Lazar, S. B., & Lal, A. (2019). Long noncoding RNAs: P53's secret weapon in the fight against cancer? *PLoS Biology*, *17*(2). https://doi.org/10.1371/journal.pbio.3000143.

Donehower, L. A., Soussi, T., Korkut, A., Liu, Y., Schultz, A., Cardenas, M., et al. (2019). Integrated analysis of TP53 gene and pathway alterations in the cancer genome atlas. *Cell Reports*, *28*(5), 1370–1384.e5. https://doi.org/10.1016/j.celrep.2019.07.001.

Feng, Z., Zhang, C., Wu, R., & Hu, W. (2011). Tumor suppressor p53 meets microRNAs. *Journal of Molecular Cell Biology*, *3*(1), 44–50. https://doi.org/10.1093/jmcb/mjq040.

Freed-Pastor, W. A., & Prives, C. (2012). Mutant p53: One name, many proteins. *Genes and Development*, *26*(12), 1268–1286. https://doi.org/10.1101/gad.190678.112.

Hafner, A., Bulyk, M. L., Jambhekar, A., & Lahav, G. (2019). The multiple mechanisms that regulate p53 activity and cell fate. *Nature Reviews Molecular Cell Biology*, *20*(4), 199–210. https://doi.org/10.1038/s41580-019-0110-x.

Hanna, J., Hossain, G. S., & Kocerha, J. (2019). The potential for microRNA therapeutics and clinical research. *Frontiers in Genetics*, *10*, 478.

Hu, W., Chan, C. S., Wu, R., Zhang, C., Sun, Y., Song, J. S., et al. (2010). Negative regulation of tumor suppressor p53 by microRNA miR-504. *Molecular Cell*, *38*(5), 689–699. https://doi.org/10.1016/j.molcel.2010.05.027.

Hu, W., Feng, Z., Teresky, A. K., & Levine, A. J. (2007). p53 regulates maternal reproduction through LIF. *Nature*, *450*(7170), 721–724. https://doi.org/10.1038/nature05993.

Javad, S., Tahereh, E., & Atieh, A. (2020). The P53/microRNA network: A potential tumor suppressor with a role in anticancer therapy. *Pharmacological Research*. https://doi.org/10.1016/j.phrs.2020.105179. 105179.

Le, M. T. N., Teh, C., Shyh-Chang, N., Xie, H., Zhou, B., Korzh, V., et al. (2009). MicroRNA-125b is a novel negative regulator of p53. *Genes and Development*, *23*(7), 862–876. https://doi.org/10.1101/gad.1767609.

Leveille, N. (2011). Selective inhibition of microRNA accessibility by RBM38 is required for p53 activity. *Nature Communications*, *2*, 513.

Levine, A. J. (2019). The many faces of p53: Something for everyone. *Journal of Molecular Cell Biology*, *11*(7), 524–530. https://doi.org/10.1093/jmcb/mjz026.

Levine, A. J., Hu, W., & Feng, Z. (2006). The P53 pathway: What questions remain to be explored? *Cell Death and Differentiation*, *13*(6), 1027–1036. https://doi.org/10.1038/sj.cdd.4401910.

Li, X. L., Jones, M. F., Subramanian, M., & Lal, A. (2014). Mutant p53 exerts oncogenic effects through microRNAs and their target gene networks. *FEBS Letters*, *588*(16), 2610–2615. https://doi.org/10.1016/j.febslet.2014.03.054.

Liao, J. M., Cao, B., Zhou, X., & Lu, H. (2014). New insights into p53 functions through its target microRNAs. *Journal of Molecular Cell Biology*, *6*(3), 206–213. https://doi.org/10.1093/jmcb/mju018.

Liu, Y., Tavana, O., & Gu, W. (2019). P53 modifications: Exquisite decorations of the powerful guardian. *Journal of Molecular Cell Biology*, *11*(7), 564–577. https://doi.org/10.1093/jmcb/mjz060.

Liu, J., Zhang, C., Hu, W., & Feng, Z. (2019). Tumor suppressor p53 and metabolism. *Journal of Molecular Cell Biology*, *11*(4), 284–292. https://doi.org/10.1093/jmcb/mjy070.

Liu, J., Zhang, C., Wang, X., Hu, W., & Feng, Z. (2020). Tumor suppressor p53 cross-talks with TRIM family proteins. *Genes and Diseases*, *8*(4), 463–474. https://doi.org/10.1016/j.gendis.2020.07.003.

Liu, J., Zhang, C., Zhao, Y., & Feng, Z. (2017). MicroRNA control of p53. *Journal of Cellular Biochemistry*, *118*(1), 7–14. https://doi.org/10.1002/jcb.25609.

Luo, Z., Cui, R., Tili, E., & Croce, C. (2018). Friend or foe: MicroRNAs in the p53 network. *Cancer Letters*, *419*, 96–102. https://doi.org/10.1016/j.canlet.2018.01.013.

Mantovani, F., Collavin, L., & Del Sal, G. (2019). Mutant p53 as a guardian of the cancer cell. *Cell Death and Differentiation*, *26*(2), 199–212. https://doi.org/10.1038/s41418-018-0246-9.

Meek, D. W. (2015). Regulation of the p53 response and its relationship to cancer. *Biochemical Journal*, *469*(3), 325–346. https://doi.org/10.1042/BJ20150517.

Muller, P. A. J., & Vousden, K. H. (2013). P53 mutations in cancer. *Nature Cell Biology*, *15*(1), 2–8. https://doi.org/10.1038/ncb2641.

Navarro, F., & Lieberman, J. (2015). miR-34 and p53: New insights into a complex functional relationship. *PLoS One*, *10*(7), e0132767.

Park, S. Y., Lee, J. H., Ha, M., Nam, J. W., & Kim, V. N. (2009). miR-29 miRNAs activate p53 by targeting p85α and CDC42. *Nature Structural and Molecular Biology*, *16*(1), 23–29. https://doi.org/10.1038/nsmb.1533.

Rokavec, M., Li, H., Jiang, L., & Hermeking, H. (2014). The p53/miR-34 axis in development and disease. *Journal of Molecular Cell Biology*, *6*(3), 214–230. https://doi.org/10.1093/jmcb/mju003.

Seo, H. A., Moeng, S., Sim, S., Kuh, H. J., Choi, S. Y., & Park, J. K. (2019). MicroRNA-based combinatorial cancer therapy: Effects of microRNAs on the efficacy of anti-cancer therapies. *Cell*, *9*(1). https://doi.org/10.3390/cells9010029.

Simerzin, A., Zorde-Khvalevsky, E., Rivkin, M., Adar, R., Zucman-Rossi, J., Couchy, G., et al. (2016). The liver-specific microRNA-122*, the complementary strand of microRNA-122, acts as a tumor suppressor by modulating the p53/mouse double minute 2 homolog circuitry. *Hepatology*, *64*(5), 1623–1636. https://doi.org/10.1002/hep.28679.

Slabakova, E. (2017). Alternative mechanisms of miR-34a regulation in cancer. *Cell Death & Disease*, *8*(10), e3100.

Suzuki, H. I., Yamagata, K., Sugimoto, K., Iwamoto, T., Kato, S., & Miyazono, K. (2009). Modulation of microRNA processing by p53. *Nature*, *460*(7254), 529–533. https://doi.org/10.1038/nature08199.

Vousden, K. H., & Prives, C. (2009). Blinded by the light: The growing complexity of p53. *Cell*, *137*(3), 413–431. https://doi.org/10.1016/j.cell.2009.04.037.

Yue, X., Zhao, Y., Xu, Y., Zheng, M., Feng, Z., & Hu, W. (2017). Mutant p53 in cancer: Accumulation, gain-of-function, and therapy. *Journal of Molecular Biology*, *429*(11), 1595–1606. https://doi.org/10.1016/j.jmb.2017.03.030.

Zhang, C., Liu, J., Tan, C., Yue, X., Zhao, Y., Peng, J., et al. (2016). microRNA-1827 represses MDM2 to positively regulate tumor suppressor p53 and suppress tumorigenesis. *Oncotarget*, *7*(8), 8783–8796. https://doi.org/10.18632/oncotarget.7088.

Zhang, C., Liu, J., Wang, X., Wu, R., Lin, M., Laddha, S. V., et al. (2014). MicroRNA-339-5p inhibits colorectal tumorigenesis through regulation of the MDM2/p53 signaling. *Oncotarget*, *5*(19), 9106–9117. https://doi.org/10.18632/oncotarget.2379.

Zhang, C., Liu, J., Xu, D., Zhang, T., Hu, W., & Feng, Z. (2020). Gain-of-function mutant p53 in cancer progression and therapy. *Journal of Molecular Cell Biology*, *12*(9), 674–687. https://doi.org/10.1093/jmcb/mjaa040.

Zhao, Y. (2018). A polymorphism in the tumor suppressor p53 affects aging and longevity in mouse models. *eLife*, *7*, e34701.

Chapter 5

MicroRNA involvement in invasion and metastasis

Carla Solé[a], Anna Bielowski[a,b], and Charles Henderson Lawrie[a,c,d]

[a]Molecular Oncology Group, Biodonostia Research Institute, San Sebastián, Spain, [b]Julius-Maximilians-University Wuerzburg, Würzburg, Germany, [c]IKERBASQUE, Basque Foundation for Science, Bilbao, Spain, [d]Radcliffe Department of Medicine, University of Oxford, Oxford, United Kingdom

Introduction—The process of metastasis

Tumor metastasis, the spread of tumor material and its accumulation in organs distant from the primary tumor, is not only a major feature of cancer diseases, but also the main cause of cancer mortality. Half of all cancer patients show metastatic growth at the time of diagnosis, which underlines the necessity of better understanding the process of metastasis and finding ways to therapeutically target it (Robert, 2013; Sole et al., 2019). Metastasis starts with malignant cells of the primary tumor invading and escaping the surrounding tissues, followed by the dissemination of circulating tumor cells (CTC) through the lymphatic and blood systems, a process whereby only 0.2% of CTCs survive, with an even lower rate of cells successfully extravasating and settling at the metastatic site (Chambers, Groom, & MacDonald, 2002; Solé & Lawrie, 2019). Tumor invasion involves multiple interactions between tumor cells and the tumor microenvironment (TME) (Fig. 5.1) (Robert, 2013; Steeg, 2016). The TME includes a versatile extracellular matrix (ECM), a network of proteins, proteoglycans, hyaluronic acid, cytokines, and growth factors (GFs). The major cellular components of the TME are stromal fibroblasts that differentiate into myofibroblast-like cancer associated fibroblasts (CAFs) during tumor progression, a process regulated by transforming growth factor-β (TGFβ). In addition, fibronectin splice variants containing extra domains A and B (ED-A and B) are highly expressed in the tumor stroma and act to release and activate bound TGFβ from within the ECM (Eble & Niland, 2019; Kalluri, 2016). Laminin-332, which also acts to enhance CAF differentiation and its expression in tumor stroma, is associated with increased tumor invasiveness and poorer prognosis (Kim et al., 2011a). When formed, CAFs modify the TME by facilitating angiogenesis through the expression of vascular endothelial growth factor-A (*VEGF-A*) and platelet-derived growth factor (*PDGF*), and enhance cross-linkage of the ECM, driven by tumor cells (Eble & Niland, 2019; Kalluri, 2016). CAFs can also influence immune cell recruitment and activation (immune editing) by expressing a number of cytokines, chemokines, and other factors, including interleukin IL-1β, IL-4, IL-6, chemokine CXC motif ligand 14 (CXCL14), forkhead box 03 (FOX03), and indoleamine 2,3-dioxygenase (IDO). IL-6 induces T-cell anergy and mast cell recruitment while IDO is involved in IL-6-dependent myeloid-derived suppressor cell (MDSC)-driven immune escape and CXCL14 affects macrophage recruitment to the tumor, generating an immune-suppressive TME (Becker et al., 2013; Kalluri, 2016). These metabolic and cellular changes are a prerequisite for tumor cell intravasation and dissemination in order to establish metastasis. The ECM also acts to connect the tumor to blood vessels and nerves and the transport systems through the body during dissemination (Eble & Niland, 2019).

The process of tumor invasion into the basement membrane, and consequent passage to surrounding tissues and intravasation into the blood stream, requires epithelial-mesenchymal transition (EMT) (Brabletz et al., 2018; Kalluri & Weinberg, 2009). EMT is characterized by a shift from an epithelial and immotile phenotype to a motile, spindle-shaped, mesenchymal phenotype. This change includes the loss of epithelial cell polarity and cell-cell adhesions and occurs under physiological conditions during embryonic development. This process is tightly regulated by a range of EMT transcription factors (EMT-TFs), including *SNAI1* (Snail), *SNAI2* (Slug), Twist-related protein 1 (*TWIST1*), and zinc-finger *E*-box-binding homeobox 1 and 2 (*ZEB1* and *ZEB2*), which are also associated with the induction of stemness and cell survival, further promoting the metastatic growth of tumor cells (Kalluri & Weinberg, 2009; Stemmler et al., 2019). In addition to EMT-TFs, epithelial markers such as cytokeratin and *E*-cadherin, major components of cellular adherent junctions, play key roles in metastasis. The inhibition of E-cadherin expression induces β-catenin nuclear translocation and consequently the upregulation of the pro-oncogenic factors, *c-MYC* and *CD44*. In contrast, markers of a fibroblastic, mesenchymal

MicroRNA in Human Malignancies. https://doi.org/10.1016/B978-0-12-822287-4.00005-0

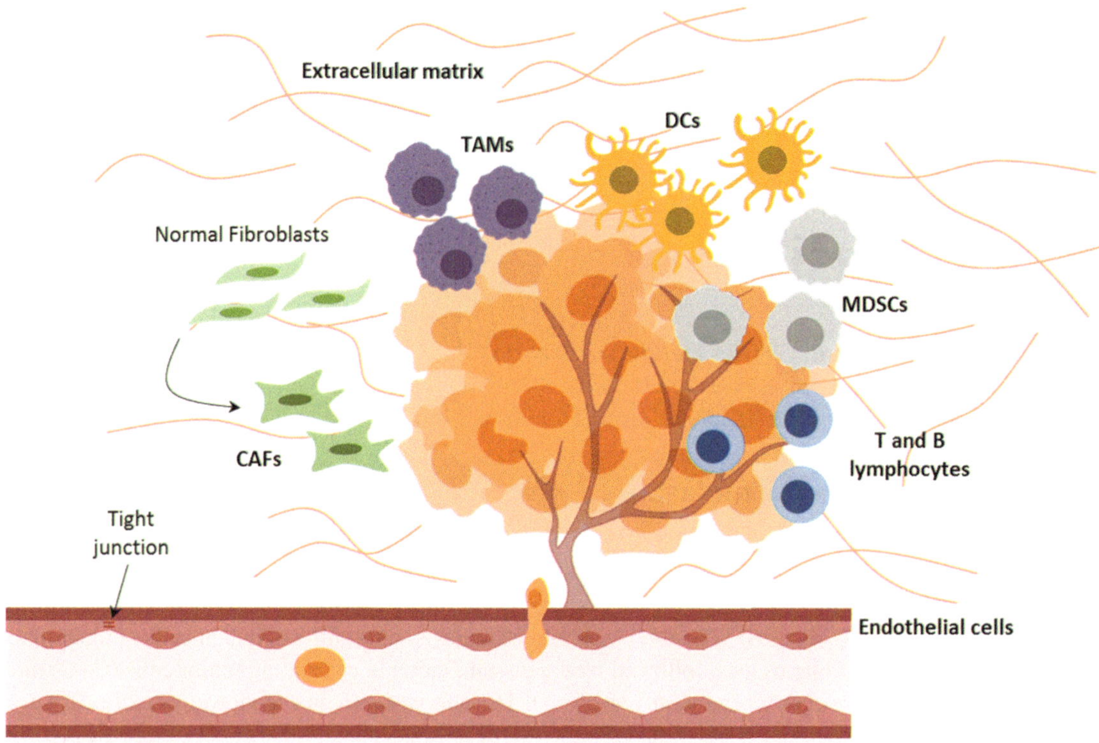

FIG. 5.1 Schematic representation of tumor microenvironment and the cells that form it, which includes an extracellular matrix, cancer-associated fibroblasts (CAFs), tumor-associated macrophages (TAMs), dendritic cells (DCs), myeloid-derived suppressor cells (MDSCs), T and B lymphocytes, and endothelial cells, which help cancer cells to survive, grow, and spread to form metastasis.

phenotype including fibroblast-specific protein 1 (FSP1), vimentin (VIM), and α-SMA are induced by the EMT process. Secreted cytokine TGFβ is a key regulator in the EMT process (Kalluri & Weinberg, 2009; Song et al., 2014).

After intravasation, CTCs continue circulating until the microenvironment of their secondary site signals them to extravasate and colonize with the help of reversed EMT, the mesenchymal-epithelial transition (MET). This procolonization signaling is regulated by a wide range of chemokines, such as chemokine (C-C motif) receptor 9 (CCR9) and CXCL12, that act to promote the infiltration of immune cells in inflammation and regulate homeostasis under physiological conditions. Local primary tumor cells also help to form a secondary tumor site with the expression of soluble factors, such as VEGF, that stimulates the migration of myeloid-derived suppressor cells (MDSCs), which in turn upregulate inflammatory cytokines and fibronectin disposition on the extracellular matrix, leading to the attraction of CTCs (Eble & Niland, 2019; Steeg, 2016). In addition to this role, VEGF is also a key regulator of angiogenesis and the formation of new blood vessels from preexisting vessels through sprouting, thereby ensuring the supply of nutrients and oxygen for propagating tumor cells. This is a key process for tumor progression, as malignant tissues often suffer from deficient oxygenation (hypoxia) as a result of aberrant growth. Hypoxia leads to the induction of Lysyl oxidase (LOX), which crosslinks collagens and elastin in the extracellular matrix (ECM) as well as supporting tumor colonization by mobilizing bone marrow cells toward the potential secondary site (Steeg, 2016; Walsh et al., 2014).

Many of the above-described steps and factors that constitute metastasis are regulated directly and indirectly by multiple microRNAs (miRNAs) (Fig. 5.2) (Ma, Teruya-Feldstein, & Weinberg, 2007; Ma & Weinberg, 2008; Sole et al., 2019; Yu et al., 2007). In this chapter, we will describe the role of the most important, or rather most studied, miRNAs in the metastatic process, including in the regulation of EMT and angiogenesis regulation, their role and potential as cancer biomarkers, and their potential importance as novel therapeutics.

Direct regulation of metastasis

Metastasis-promoting miRNAs

Many miRNAs have been found to promote metastasis either directly or indirectly through the activation of associated pathways such as migration and/or invasion in many different cancer types (Fig. 5.2), including breast cancer, glioblastoma

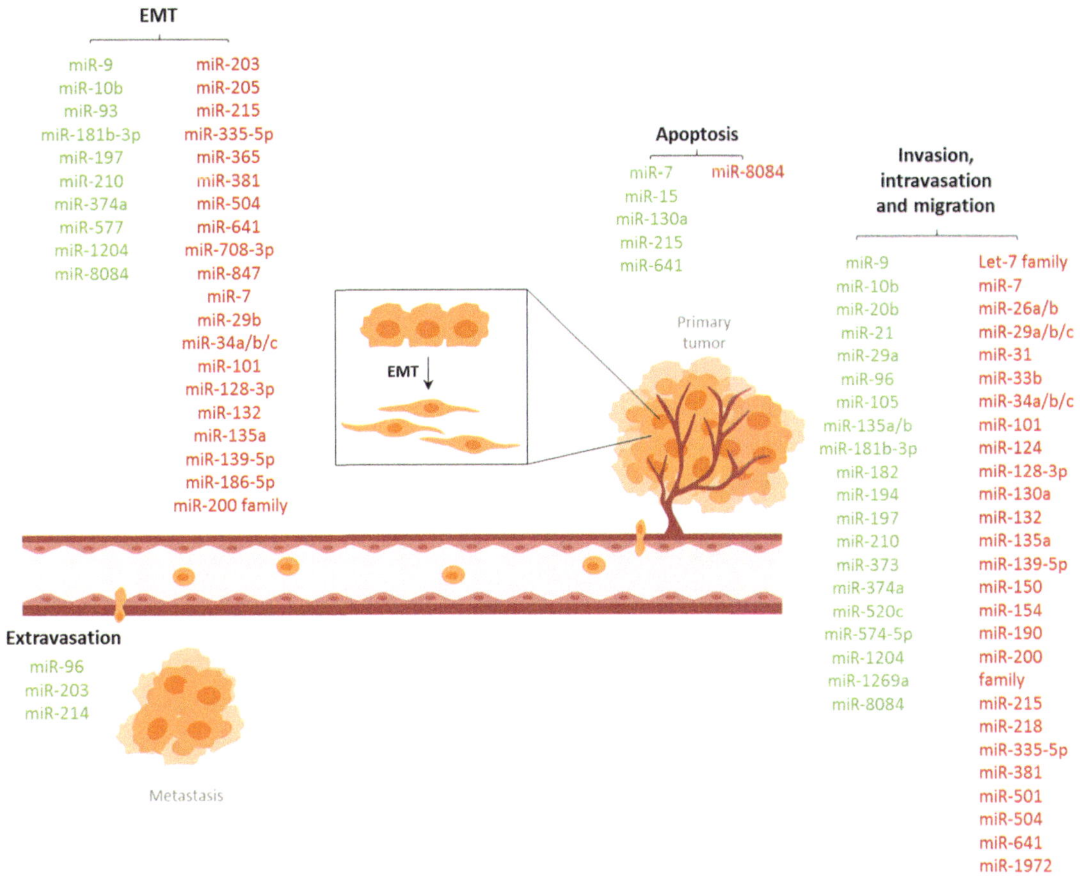

FIG. 5.2 Schematic depiction of the process of metastasis from the primary to the secondary tissue, including various involved miRNAs enhancing or inhibiting apoptosis, epithelial-mesenchymal transition (EMT), angiogenesis, invasion, intravasation, and extravasation; specific targets and impaired signaling pathways are further described in the following chapter.

multiforme (GBM), pancreatic carcinomas, gastric cancer (GC), and testicular germ cell tumors (Table 5.1) (Li et al., 2016; Ma & Weinberg, 2008).

The miRNA *miR-10b* is a key regulator of metastasis in multiple cancer types including breast cancer, colorectal cancer (CRC), and hepatocellular carcinoma (HCC) (Li et al., 2018a; Ma et al., 2007; Xie et al., 2019). In breast cancer, increased levels of *miR-10b* have been associated with the presence of brain metastases in a patient study cohort (Ahmad et al., 2014). This miRNA was shown to be upregulated by TWIST1 binding to an upstream promoter region, and conversely to influence the EMT process through TWIST1 and E-cadherin regulation (Ahmad et al., 2014; Liu et al., 2012a; Ma et al., 2007). *miR-10b* has also been demonstrated to inhibit translation of Homeobox D10 (*HOXD10*) mRNA, leading to increased expression of the prometastatic *RHOC* gene as well as targeting T-box transcription factor 5 (*TBX5*), dual specificity tyrosine-phosphorylation-regulated kinase 1A (*DYRK1A*), and Krüppel-like factor 4 (*KLF4*) mRNA (Ahmad et al., 2014; Kim et al., 2016; Ma et al., 2007). In HCC, *miR-10b* was found to target cell adhesion molecule 2 (*CADM2*), leading to increased levels of focal adhesion kinase (*FAK*) and overactivation of the AKT pathway, driving EMT, tumor invasion, and cell proliferation (Jiang et al., 2020; Li et al., 2018a; Xie et al., 2019).

The key hypoxia-associated miRNA *miR-210* also plays an important role in the metastatic process of multiple cancer types, including HCC as well as breast and prostate cancers (Eminaga et al., 2018; Tang et al., 2018a; Ying et al., 2011). This miRNA directly targets *E*-cadherin, thereby promoting cell invasion and migration (Tang et al., 2018a). In addition to *miR-210*, *E*-cadherin is indirectly regulated by *miR-181b-3p* through downregulation of the *YWHAG* gene. This leads to Snail stabilization, which in turn regulates E-cadherin expression (Yoo et al., 2016).

Like *miR-210*, *miR-96* is also commonly overexpressed in metastases of multiple cancers. For example, *miR-96* enhancement is detected in prostate cancer bone metastases, along with TGFβ, which is also a positive regulator of *miR-96* in the bone matrix (Voss et al., 2020). In addition, *miR-96* was shown to correlate with increased levels of EpCAM and E-cadherin, both of which are involved in cell-cell adhesion (Voss et al., 2020). In HCC, *miR-96* was found to promote

TABLE 5.1 Metastasis promoting miRNAs.

miRNA	Downregulated in	Target	References
miR-9	EOC	*E-Cadherin*	Song et al. (2014)
miR-10b	Breast cancer	*HOXD10; TBX5*	Kim et al. (2016), Ma et al. (2007)
	CRC	*HOXD10; KLF4*	Liu et al. (2012b), Xie et al. (2019)
	NSCLC	*KLF4*	Liu et al. (2013)
	HCC	*HOXD10*	Liao et al. (2014)
	GBM	*E-Cadherin, Apaf1, PTEN*	Ma et al. (2017)
	HCC	*CADM2*	Li et al. (2018a)
miR-10b-3p	ESCC	*TSGA20*	Zhang et al. (2019a)
miR-20b	EOC	*PTEN*	Wang, Yang, and Xiao (2016)
miR-21	VCT	*PDCD4; TPM1*	Zhu et al. (2008)
miR-29a	BCSCs	*SUV420H2*	Wu et al. (2019)
miR-93	Endometrial carcinoma	*FOXA1*	Chen et al. (2016)
miR-96	Breast cancer	*PTPN9, FOXO1, FOXO3a*	Hong et al. (2016)
	EOC	*RECK*	Xia et al. (2014)
	HCC	*EphrinA5*	Wang et al. (2016)
	NSCLC	*SAMD9*	Wu et al. (2016)
	Prostate cancer	*E-Cadherin, EpCAM*	Voss et al. (2020)
miR-96-3p	PTC	*SDHB*	Zhao, Li, and Zhou (2019)
miR-105	Breast cancer	*ZO-1*	Zhou et al. (2014)
miR-135a	Breast cancer	*HOXA10*	Chen et al. (2012)
	cHL	*JAK2*	Navarro et al. (2009)
	HCC	*FOXO1*	Zeng et al. (2016)
miR-135a-5p	Lung cancer	*LOXL4*	Zhang et al. (2019b)
miR-135b	Colon cancer	*APC, CaSR, TGFβ*	Fetahu et al. (2016)
	HNSCC	*FIH*	Zhang et al. (2013)
miR-194	Ovarian cancer	*PTPN12*	Liang et al. (2016)
miR-197	HCC	*Axin-2, DKK2, NKD1*	Hu et al. (2018)
miR-210	BCSC	*E-Cadherin*	Eminaga et al. (2018)
	Prostate cancer	*E-Cadherin*	Tang et al. (2018a)
	HCC	*VMP1*	Ying et al. (2011)
miR-210-5p	Osteosarcoma	*PIK3R5*	Liu et al. (2020)
miR-373	Breast cancer	*CD44*	Huang et al. (2008)
	Testicular germ cell tumors	*LATS2*	Voorhoeve et al. (2006)
miR-374a	Breast cancer	*PTEN, WNT5a, WIF*	Cai et al. (2013)
miR-520c	GC	*IRF2*	Li et al. (2016)
miR-574-5p	NSCLC	*PTPRU*	Zhou et al. (2016)
miR-577	GC	*SDPR*	Luo et al. (2019)
miR-1204	Breast cancer	*VDR*	Liu et al. (2018a)
miR-1269a	CRC	*HOXD10, SMAD7*	Bu et al. (2015)
miR-8084	Breast cancer	*ING2*	Gao et al. (2018)

BCSCs, breast cancer stem cells; *cHL*, classic Hodgkin lymphoma; *CRC*, colorectal carcinoma; *EOC*, esophageal cancer; *ESCC*, esophageal squamos cel carcinoma; *GC*, gastric cancer; *HCC*, hepatocellular carcinoma; *HNSCC*, head and neck squamous cell carcinoma; *NSCLC*, nonsmall cell lung cancer; *PTC*, papillary thyroid cancer; *TNBC*, triple negative breast cancer.

metastasis through inhibition of Ephrin A5 expression (Wang et al., 2016). *MiR-96* can also promote cellular proliferation and invasion through targeting the protein tyrosine phosphatase and tumor suppressor nonreceptor type 9 (*PTPN9*) gene as well as the *FOXO1* and *FOXO3a* genes in metastatic breast cancer (Hong et al., 2016). Other validated targets for *miR-96* include *RECK* (reversion inducing cysteine rich protein) in esophageal cancer (EOC) and *SMAD9* (sterile alpha motif domain containing 9) in nonsmall cell lung cancer (NSCLC) (Hong et al., 2016; Wu et al., 2016; Xia et al., 2014).

In addition to *miR-96, FOXO1* is also targeted by *miR-135a.* In HCC, downregulation of *FOXO1* by *miR-135a* leads to upregulation of the *Snail* and matrix metalloproteinase-2 (*MMP-2*) genes, in turn causing AKT phosphorylation and activation thereby promoting metastasis through the PI3K/AKT pathway (Zeng et al., 2016). In breast cancer, *miR-135a* was also demonstrated to promote tumor migration and invasion in vitro by targeting *HOXA10* (Chen et al., 2012). In classic Hodgkin lymphoma (cHL), *miR-135a* mediates Janus kinase 2 (*JAK2*) downregulation and decreases both the mRNA and protein levels of the antiapoptotic gene Bcl-xL (BCL2L1), leading to apoptosis in vitro (Navarro et al., 2009). In breast cancer, *miR-135b* overexpression was associated with the presence of the aggressive triple-negative breast cancer (TNBC) subtype. Furthermore, overexpression of this miRNA was demonstrated to significantly contribute to the invasive phenotype in vitro and in vivo by the suppression of APC, a Wnt pathway suppressor (Lv et al., 2019). Other negative regulators of the Wnt/β-catenin signaling pathway, such as *WIF1* (Wnt inhibitory factor 1), *PTEN* (phosphatase and tensin homolog), and *WNT5a*, are targeted by *miR-374a*, leading to pathway hyperactivation, and overexpression of this miRNA is associated with poor survival in metastatic breast cancer patients (Cai et al., 2013).

Metastasis—Suppressing miRNAs

In addition to miRNAs that promote metastasis, miRNAs have also been identified that act in a contrary manner, that is, as metastasis inhibitors. Consequently, these miRNAs are generally found to be downregulated in many metastatic cancers including in breast cancer, colorectal carcinoma (CRC), and pancreatic cancer (Xi et al., 2015; Yu et al., 2007) (Fig. 5.2 and Table 5.2).

TABLE 5.2 Overview over metastasis inhibiting miRNAs.

miRNA	Downregulated in	Target	References
let-7	Pancreatic cancer	*HMGA2, RAS*	Yu et al. (2007)
miR-7	Breast cancer	*EGFR; FAK, PAK 1; RELA*	Kong et al. (2012), Li et al. (2020), Reddy et al. (2008)
	CRC	*TRIP6*	Ling et al. (2019)
	GBM	*EGFR, IRS-1, IRS-2*	Kefas et al. (2008), Pan et al. (2020)
miR-15a	HCC	*BCL2*	Budhu et al. (2008)
miR-26a/b	HNSCC	*LOXL2*	Fukumoto et al. (2016)
miR-29a/b/c	ccRCC	*LOXL2*	Nishikawa et al. (2015)
	HNSCC	*ITGA6, LAMC2, LOXL2*	Fukumoto et al. (2016)
miR-29b	Prostate cancer	*Snail*	Lamouille et al. (2013)
miR-29c	Lung cancer	*Integrin 1, MMP2*	Wang et al. (2013)
miR-31	Breast cancer	*β1 & 3 Integrin α-subunits, WAVE 3*	Augoff et al. (2011), Sossey-Alaoui et al. (2011)
miR-33b	Breast cancer	*HMGA2, SALL4, Twist1*	Lin et al. (2015)
miR-34a/b/c	CRC	*Snail*	Siemens et al. (2011)
miR-34a	Breast cancer	*CXCL10*	Xu et al. (2018)
	Prostate CSCs	*CD44*	Liu et al. (2011)
miR-34c	Breast cancer	*GIT1*	Tao et al. (2016)
miR-101	Ovarian carcinoma	*ZEB1* and *ZEB2*	Guo et al. (2014)
miR-124	Breast cancer	*ROCK1*	Xi et al. (2015)
miR-128-3p	ESCC	*ZEB1*	Zhao et al. (2018)

Continued

TABLE 5.2 Overview over metastasis inhibiting miRNAs—cont'd

miRNA	Downregulated in	Target	References
miR-130a	CLL	*ATG2B, DICER1*	Kovaleva et al. (2012)
	TNBC	*FOSL1*	Chen et al. (2018)
miR-132	CRC NSCLC	*ZEB2*	You et al. (2014), Zheng et al. (2014)
miR-135a	GC	*ROCK1*	Shin et al. (2014)
miR-139-5p	GBM	*ZEB1* and *ZEB2*	Yue et al. (2015)
miR-141	Breast cancer	*HIPK1*	Liu et al. (2018b)
	Prostate CSC	*EZH2, CD44, Rho-GTPases*	Liu et al. (2017)
miR-150	TNBC	*HMGA2*	Tang et al. (2018b)
miR-154	Breast cancer	*E2F5*	Xu et al. (2016)
miR-186-5p	CRC	*ZEB1*	Li et al. (2018b)
miR-190	Breast cancer	*STC2*	Sun, Liu, and Han (2019)
miR-200 family	Breast cancer ccRCC NSCLC Ovarian cancer Prostate cancer	*ZEB1, ZEB2*	Damiano et al. (2017), Korpal et al. (2008), Krebs et al. (2017), Pang et al. (2018), Wang et al. (2020a)
miR-200b	TNBC	*MYC, ZEB1, ZEB2*	Pang et al. (2018)
miR-200c	Breast cancer	*HIPK1*	Liu et al. (2018b)
miR-203	Breast cancer	*TGFβ, Slug*	Ding et al. (2013)
	CRC	*EIF5A2*	Deng et al. (2016)
	GC	*Annexin A5*	Li et al. (2019b)
	HNSCC	*LASP1, NUAK1; SPARC*	Benaich et al. (2014)
	Melanoma	*Slug*	Lohcharoenkal et al. (2018)
	NSCLC	*RGS17*	Chi et al. (2017)
	Ovarian cancer	*BIRC5*	Wang et al. (2018a)
	Prostate cancer	*Slug*	Tian et al. (2018)
miR-205	Breast cancer	*ZEB1, ZEB2*	Lamouille et al. (2013)
miR-215	NSCLC	*ZEB2*	Hou et al. (2015)
miR-218	HNSCC	*LOXL2*	Fukumoto et al. (2016)
miR-335-5p	Osteosarcoma	*ROCK1*	Wang et al. (2018b)
miR-365	Human lung adenocarcinoma	*HMGA2*	Lamouille et al. (2013)
miR-381	Breast cancer	*ROCK1*	Mohammadi-Yeganeh, Hosseini, and Paryan (2019)
miR-429	Breast cancer	*ZEB1*	Bracken et al. (2008)
miR-504	NSCLC	*LOXL2*	Ye et al. (2018)
miR-641	Cervical cancer	*ZEB1*	Yao et al. (2018)
miR-708-3p	Breast cancer	*CDH2, VIM, ZEB1*	Lee et al. (2018)
miR-874	NSCLC	*AQP3*	Wang et al. (2020b)
miR-1972	Osteosarcoma	*ROCK1*	Wang et al. (2018b)
miR-4319	Breast CSCs	*E2F2*	Chu et al. (2018)

ccRCC, clear cell renal cell carcinoma; *CLL*, chronic lymphatic leukemia; *CRC*, colorectal cancer; *CSC*, cancer stem cell; *ESCC*, esophageal squamous-cell cancer; *GC*, gastric cancer; *GBM*, glioblastoma multiforme; *HCC*, hepatocellular carcinoma; *HNSCC*, head and neck squamous cell carcinoma; *NSCLC*, nonsmall cell lung cancer; *TNBC*, triple negative breast cancer; *VCT*, various cancer types.

For example, *miR-7*, which is frequently downregulated in breast cancer, targets the *FAK* gene, a cell survival and EMT-mediating enzyme (Kong et al., 2012). *MiR-7* levels were positively correlated with *E*-cadherin levels, and negatively with VIM, fibronectin, and Snail expression, supporting an epithelial, noninvading phenotype. Decreased expression of *miR-7* was frequently observed in metastatic breast cancer tissue. Moreover, *miR-7* has been demonstrated to directly target a variety of additional oncogenes, including *PAK1* activated kinase 1 (*PAK1*), *EGFR*, and *Raf1* in breast cancer (Kong et al., 2012; Reddy et al., 2008; Webster et al., 2009). *MiR-7* downregulation has also been observed in other cancer types, including NSCLC and cervical cancer (Li et al., 2019a). In GBM, *miR-7* was shown to target the EMT regulator T-box 2 gene (*TBX2*) and upstream AKT pathway regulators, namely Insulin receptor substrate 1 & 2 (IRS-1 and IRS-2), and in that way to decrease AKT phosphorylation and activation. Loss of *miR-7* therefore enhances EMT and AKT pathway signaling and thus an antiapoptotic, proliferative, and invasive phenotype (Kefas et al., 2008; Pan et al., 2020).

Members of the *miR-200* family (i.e., *miR-200a*, *miR-200b*, *miR-200c*, *miR-141*, and *miR-429*) as well as *miR-205* have been shown to inhibit the expression of EMT-TFs ZEB1 and ZEB2, thereby enhancing *E*-cadherin expression and inhibiting the EMT process. For example, ZEB1 depletion decreased tumor grade, invasion, and distant metastasis in a pancreatic cancer KPC mouse model (Krebs et al., 2017). Studies of HCC cell lines and of patient material in vitro demonstrated that increasing *miR-200* levels prevented EMT development through targeting of p53 expression (Kim et al., 2011b). In contrast, it has been shown that during EMT, ZEB1 can inhibit *miR-200* expression via reciprocal negative feedback, thereby enhancing EMT progression (Bracken et al., 2008; Korpal et al., 2008; Krebs et al., 2017; Lamouille et al., 2013). Members of the *miR-200* family have also been shown to be downregulated in various cancer types, including NSCLC, breast and ovarian cancer, and clear cell renal carcinoma (ccRCC) (Damiano et al., 2017; Wang et al., 2020a). In TNBC for example, *miR-200b* downregulation has been associated with increased expression of the proto-oncogene *MYC* and DNA methyltransferase 3A (*DNMT3A*) (Pang et al., 2018). Moreover, *MYC* is important for *DNMT3A* recruitment to the *miR-200b* promoter, resulting in proximal CpG island hypermethylation and subsequent epigenetic *miR-200b* repression, which in turn directly inhibits *DNMT3A* expression (Pang et al., 2018). In prostate adenocarcinoma, a similar situation was found, whereby *miR-200b* directly targets Slug, which conversely represses the expression of *miR-200b* (Lamouille et al., 2013). ZEB1 and Slug have also been found downregulated by *miR-34a* as well as various stemness factors such as CD44, CD133, and c-MYC (Lamouille et al., 2013; Siemens et al., 2011; Stemmler et al., 2019). In addition to targeting EMT and stemness factors, members of the *miR-200* family can inhibit angiogenesis as well. *MiR-141*, for example, does so by targeting a wide range of angiogenesis driving targets, including *CXCL12*β, *GAB1, GATA6,* and *TGF*β*2*. Hypoxia-induced *miR-429* targets HIF1A (hypoxia-inducible factor 1A) that leads to decreased VEGFA levels. *MiR-200b* is negatively regulated by HIF1 and hypoxia, and targets Ets-1 (v-ets erythroblastosis virus E26 oncogene homolog 1), an important angiogenesis-related transcription factor (Bartoszewska et al., 2015; Chan et al., 2011; Dong et al., 2019).

Another frequently encountered metastasis-suppressor miRNA is *miR-203* that negatively regulates invasiveness in several cancer types, including breast cancer, CRC, HCC, leukemia, and esophageal cancer (Deng et al., 2016; He et al., 2019; Swellam et al., 2018; Wei et al., 2013). In CRC, downregulation of *miR-203* was found to correlate with the presence of lymph node metastasis, advanced metastasis stage, and poor survival. This study further showed that this miRNA regulates invasion and migration through targeting EIF5A (eukaryotic initiation factor 5A2) in vitro (Deng et al., 2016). However, in contrast Wang et al. reported that *miR-203*-associated EMT inhibition enhances a MET cellular state supporting lung infiltration (Wang et al., 2018a). In breast cancer, *miR-203* regulates EMT via a double negative feedback loop targeting EMT enhancer TGFß and Slug (Ding et al., 2013). In melanoma, *miR-203* also negatively correlates with poor overall survival and also targets Slug, whereas in GC it negatively regulates EMT via Annexin A4 (Li et al., 2019b; Lohcharoenkal et al., 2018).

The *miR-29* family (i.e., *miR-29a/b/c*) is downregulated in many cancer types, including metastatic lung cancer, head and neck squamous cell carcinoma (HNSCC), and breast cancer (Rostas et al., 2014; Wang et al., 2013). Under physiological conditions, these miRNAs inhibit proliferation, migration, and invasion through targeting α6 integrin (*ITGA6*) and laminin γ2 (*LAMC2*) genes as part of the focal adhesion laminin-integrin pathway (Kinoshita et al., 2013). In ccRCC, HNSCC and NSCLC *miR-29* suppress invasion and migration by targeting LOXL2 (LOX-like 2), which is also targeted by multiple miRNAs, including *miR-26*, *miR-218*, and *miR-501* and which overexpression was found to be connected to disease progression (Fukumoto et al., 2016; Nishikawa et al., 2015; Ye et al., 2018). In addition, *miR-29b* suppresses EMT by targeting Snail and LOXL4 and is often downregulated in breast and prostate cancer tissues (Lamouille et al., 2013; Yin et al., 2020). In contrast, *miR-29c* acts as an adhesion, invasion, and metastasis suppressor through targeting integrin 1 and MMP2 (Wang et al., 2013).

Several miRNAs, including *miR-124*, *miR-135a*, and *miR-381*, have been found to directly target Rho-associated protein kinase 1 (ROCK1) expression, thereby inhibiting EMT and cell proliferation, cell migration, and invasion. *MiR-124*, for example, is downregulated in metastatic CRC and other cancer types. In GC, *miR-135a* levels were found to be frequently downregulated and correlated with advanced stage disease and higher lymph node metastatic rates. *MiR-381* is also involved in the regulation of additional WNT signaling pathway molecules, including RhoA and c-MYC, and is downregulated in TNBC (Mohammadi-Yeganeh et al., 2019; Shin et al., 2014; Xi et al., 2015).

Indirect regulation of metastasis

Regulation of miRNA biosynthesis

Aside from the direct targeting of metastasis-associated genes by miRNAs, the factors that regulate levels of miRNA can also indirectly change the metastatic phenotype (Fig. 5.3). For example, DICER and DROSHA are components of the miRNA biosynthetic pathway (Bartel, 2004; Kwon et al., 2016) that are frequently deregulated in multiple cancer types, including ovarian cancer where low levels of *DICER* are associated with less successful surgical cytoreduction and advanced tumor stages (Merritt et al., 2008). In NSCLC, high expression of *DROSHA* is associated with a poorer prognostic outcome (Díaz-García et al., 2013). Moreover, *DROSHA* was found to be upregulated in diffuse large B-cell lymphoma (DLBCL) cancer patients (Lawrie et al., 2009).

Under hypoxic conditions, oxygen-dependent H3K27me3 demethylases KDM6A/B are inhibited, leading to methylation of the *DICER* promoter, which in turn leads to decreased levels of *miR-200* and upregulation, enhanced stemness, and EMT induction in breast cancer (van den Beucken et al., 2014). TAp63 is part of the tumor suppressor *P53* family that binds to the *DICER* promoter and is critical for tumor invasion (Su et al., 2010). Under physiological conditions, Argonaute 2 (AGO2) binds to DICER and HIV-1 transactivating response RNA-binding protein (TRBP) to form the RISC-loading complex (Cikaluk et al., 1999; Diederichs & Haber, 2007; Shen et al., 2013) and is also the major protein partner of circulating miRNAs and therefore crucial for their localization (Arroyo et al., 2011; Turchinovich et al., 2011). AGO2 phosphorylation mediates EGFR-associated stemness and invasiveness and supports the prometastatic phenotype and correlates with poorer survival in breast cancer patients (Cui et al., 2019; Shen et al., 2013). In addition to indirect regulation of miRNA levels by regulation of components of the miRNA biosynthetic pathway, those components themselves can be targeted by miRNAs, resulting in a negative feedback loop. For example, in breast cancer high levels of *miR-103/107* that target DICER are

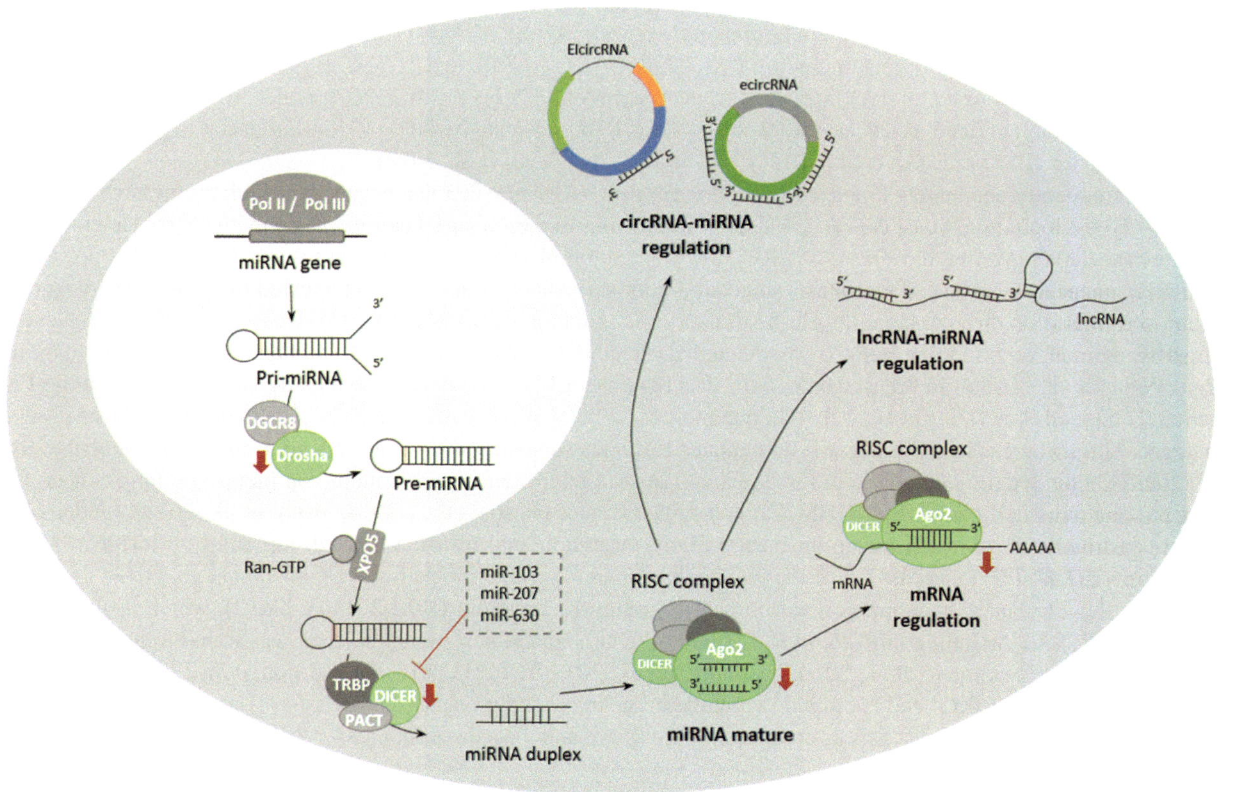

FIG. 5.3 Indirect regulation of metastasis-involved miRNAs. A schematic representation of components of miRNA biogenesis that have been described as downregulated in different cancers, leading to a decrease in certain miRNAs associated with metastasis. Moreover, miRNAs can also be downregulated by interaction with other ncRNAs, such as circRNAs and lncRNAs, which decreased its availability to regulate genes related with metastasis. *XPO5*, exportin 5; *lncRNA*, long noncoding RNA; *ElcircRNA*, exon-intron circular RNA; *ecircRNA*, exonic circRNA.

associated with a poor clinical outcome and also *miR-630* that is upregulated under hypoxic conditions is associated with the presence of metastasis (Martello et al., 2010; Rupaimoole et al., 2016).

RNA-RNA interaction-based miRNA regulation

In addition to direct and indirect regulation of metastasis-associated genes by miRNAs, long ncRNAs (lncRNA) can also regulate metastasis through their ability to absorb or "sponge" miRNAs, thereby decreasing available levels (Fig. 5.2). For example, *H19* modulates *miR-29b-3p* expression in bladder cancer as well as *miR-200b/c* and *let-7b* expression in breast cancer, promoting a metastatic phenotype (Lv et al., 2017; Zhou et al., 2017). The lncRNA *DANCR* can decoy *miR-335-5p* and *miR-1972*, reversing ROCK1 inhibition promoting metastasis and tumor cell proliferation in osteosarcoma. It has also been demonstrated that individual microRNAs can be targeted by multiple lncRNAs, such as is the case for *miR-200*, which can be targeted by both *MALAT1* and *H19* (Expósito-Villén, Aránega, & Franco, 2018; Wang et al., 2018b). Circular RNAs (circRNAs), a subgroup of lncRNAs, are characterized by covalent circular linkage of their 5′ and 3′-moieties. Similar to other lncRNAs, they can compete with miRNAs for the miRNA response elements (MRE) of the 3-UTR end of mRNAs (Fig. 5.2). For example, *circPTPRA* suppresses EMT and metastasis in NSCLC by sponging *miR-96*, whereas *circ_0021977* sponges *miR-10b-5p* in CRC, leading to decreased cancer cell invasion and migration (Lu et al., 2020; Pardini et al., 2019; Wei et al., 2019). In addition to these examples, there are many circRNAs that have been reported to have an effect on migration, EMT, and other metastasis-related processes in different cancer types, including HCC and prostate cancer (reviewed in Zhang et al., 2021). Conversely miRNAs can also target lncRNAs, leading to negative-negative feedback loops that warrant further investigation (Expósito-Villén et al., 2018; Wang et al., 2018b).

Translational application of metastasis-associated miRNAs

Noninvasive biomarkers

MiRNAs have been shown to be present in many bodily fluids, including blood, saliva, cerebral spinal fluid, tears, semen, ocular fluid, etc. In part due to their size, circulating miRNAs are remarkably stable and resistant to RNAse activity in a manner that mRNA and longer RNAs such as lncRNAs are not. This makes them eminently suitable as noninvasive biomarkers of cancer. MiRNAs are found in biological fluids packaged as exosomes (Caby et al., 2005; Stoorvogel, 2012), into lipid-based carriers as HDK or LDL (Vickers & Remaley, 2012), or as cell-free miRNA bound to proteins, such as Argonaute 2 (Arroyo et al., 2011; Turchinovich et al., 2011). As well as being passively transported, miRNAs can function at a distance as a form of communication between cells in near or far sites (Cortez et al., 2011; Pardini & Calin, 2019). Many different circulating miRNAs have been identified from the blood of metastatic cancer patients (Mitchell et al., 2008; Zhang et al., 2017).

One of the most studied metastasis-associated circulating miRNAs is *miR-21*, which has been found to be upregulated in multiple cancers. For example, high levels of this miRNA in either serum or associated with exosomes has been linked with lymph node metastasis in pancreatic ductal carcinoma, NSCLC, GC, ESCC, and cervical and breast cancer (Abue et al., 2015; Kim et al., 2013; Tanaka et al., 2013; Wang et al., 2011; Yadav et al., 2016; Zhang et al., 2016). High levels of *miR-21* were also found in serum exosomes from pancreatic ductal carcinoma and plasma exosomes and serum from CRC patients, associated with liver metastasis (Abue et al., 2015; Tsukamoto et al., 2017; Yin et al., 2014). Moreover, expression of this miRNA was associated with distant metastasis in ESCC, CRC, and breast cancer (Tanaka et al., 2013; Toiyama et al., 2013; Yadav et al., 2016). Another frequently reported metastasis-associated circulating miRNA is *miR-141*, which has been found to be upregulated in plasma, serum exosomes, and serum of metastatic patients with CRC, GC, and prostate cancer, respectively (Brase et al., 2011; Huang et al., 2014; Maierthaler et al., 2017). In breast cancer patients, *miR-141* was highly expressed in the plasma of metastatic patients compared with localized cancer patients while high levels in serum were associated with brain metastasis (Debeb et al., 2016; Zhang et al., 2017). Additionally, *miR-141* has been observed to be upregulated in serum from CRC patients with liver metastases (Yin et al., 2014).

In breast cancer, it was demonstrated that serum levels of *miR-155* significantly discriminated M0-patients from healthy women, whereas *miR-10b*, *miR-34a*, and *miR-155* could discriminate patients with metastasis (M1) from healthy controls (Roth et al., 2010). On the other hand, plasma levels of *miR-331* and *miR-195* have been found to distinguish metastatic from local breast cancer, with an area under the curve (AUC) of 0.902 (McAnena et al., 2019). Levels of plasma *miR-200a*, *miR-200b*, *miR-200c*, *miR-210*, *miR-215*, and *miR-486-5p* were significantly correlated with the onset of metastases up to 2 years before clinical diagnosis (Madhavan et al., 2016). Moreover, an algorithm based on the levels of a combination of eight circulating miRNAs was demonstrated to predict the onset of metastasis (AUC 0.79) in breast cancer after eribulin treatment (Satomi-Tsushita et al., 2019).

miRNAs as therapeutics

Interest in the therapeutic targeting of miRNAs has been increasing rapidly over the last few years, and currently there are >7000 US and >5000 EU patents and >45 registered clinical trials underway that use miRNAs are therapeutics, the majority of which target cancer and metastatic cancer in particular (Akao, Nakagawa, & Naoe, 2006; Bonci et al., 2008; Garzon et al., 2009; He et al., 2005; Klein et al., 2010; Rupaimoole & Slack, 2017; Trang et al., 2010; Yu et al., 2007; Zhao et al., 2010). There are two major strategies to modulate miRNA expression, direct strategies including oligonucleotides or virus-based constructs to block the expression of onco-miRNAs or to reestablish the expression of tumor-suppressor miRNAs; and indirect strategies, including the use of drugs that modulate expression of miRNAs, targeting miRNA processing and transcription (Li & Rana, 2014). Particularly promising is the use of CRISPR/Cas9 construct technology to reduce the expression of miRNAs in vitro and in vivo (Chang et al., 2016).

The modulation of miRNAs is most commonly carried out by viral vectors, including lentiviruses, adenoviruses, and adeno-associated viruses. However, persistent safety concerns have led to great interest in the use of nonviral methods for miRNA delivery such as lipid and polymeric nanoparticles, which protect miRNAs from degradation and increase their half-life in circulation (Bai et al., 2019; Dasgupta & Chatterjee, 2021). For example, one strategy used was to combine *let-7g* with an aptamer that binds and antagonizes the oncogenic receptor tyrosine kinase Axl in the lung cancer cell line (A549) and the breast cell line (MCF7) (Esposito et al., 2014).

Further studies have shown that miRNA treatments can reduce the number of metastases or reduce vascularization. For example, although the systemic treatment of a murine breast cancer model with *miR-10b* antagomirs did not reduce the primary tumor size, it did decrease the number of pulmonary metastases by 83% and displayed minimal toxic effects (Ma et al., 2010). More recently, it has been described that therapeutic silencing of *miR-214*, with anti-miRNA or an miRNA sponge, in malignant melanomas, triple negative breast cancer, and neuroendocrine pancreatic tumors, reduced metastasis in vivo without toxicity effects (Dettori et al., 2018). In CRC, the overexpression of *miR-27b* by intratumoral injection of a mimic repressed tumor growth and decreased capillary blood vessels in vivo (Ye et al., 2013). Exosomal *miR-126* derived from breast cancer cells was demonstrated to suppress lung cancer proliferation and lung metastasis in vivo (Nie et al., 2020). Although considerable advances have been made in these therapeutic techniques, there is still much work to be done with major problems including the efficient delivery of drugs into target tissue and drug stability, and perhaps most challenging how to control for unwanted off-target effects (Li & Rana, 2014).

Conclusions and perspectives

As demonstrated above, miRNAs are intimately involved in the regulation of many components of the metastatic process. As such, they represent a rich source of novel biomarkers and therapeutics for this devastating phenomenon. MiRNAs show perhaps their greatest, and certainly most immediate, potential as biomarkers of metastatic cancer. We provide many examples of their usefulness as diagnostic, prognostic, and predictive indicators. A particularly attractive characteristic of miRNAs as biomarkers is their presence and detection in multiple bodily fluids, which has allowed the development of noninvasive biomarkers, potentially allowing for routine management of cancer patients for the first time (Sole et al., 2019). However, caution should be taken when considering the current state of evidence and efforts made to systematically evaluate biomarker studies as many use different protocols for sample collection as well as RNA extraction and detection. This fact is reflected by many contradictory and nonoverlapping reports in the literature, and represents a clear barrier in the routine application of miRNAs as biomarkers in the clinic (Fernandez-Mercado et al., 2015). Similar drawbacks are even more evident in miRNA therapeutic studies, which are almost all single center and carried out in underpowered patient cohorts.

Nevertheless, metastasis-related miRNAs offer remarkable advantages for monitoring and modulating metastatic tumor spread over other molecular classes and as such represent a clear candidate for continuing research to the eventual benefit of cancer patients.

Acknowledgment

C.S. is recipient of a Sara Borrell postdoctoral fellowship from the Instituto de Salud Carlos III, Spain (CD19/00196).

References

Abue, M., et al. (2015). Circulating miR-483-3p and miR-21 is highly expressed in plasma of pancreatic cancer. *International Journal of Oncology*, *46*(2), 539–547.

Ahmad, A., et al. (2014). Up-regulation of microRNA-10b is associated with the development of breast cancer brain metastasis. *American Journal of Translational Research*, *6*(4), 384–390.

Akao, Y., Nakagawa, Y., & Naoe, T. (2006). let-7 microRNA functions as a potential growth suppressor in human colon cancer cells. *Biological and Pharmaceutical Bulletin*, *29*(5), 903–906.

Arroyo, J. D., et al. (2011). Argonaute2 complexes carry a population of circulating microRNAs independent of vesicles in human plasma. *Proceedings of the National Academy of Sciences of the United States of America*, *108*(12), 5003–5008.

Augoff, K., et al. (2011). miR-31 is a broad regulator of beta1-integrin expression and function in cancer cells. *Molecular Cancer Research*, *9*(11), 1500–1508.

Bai, Z., et al. (2019). Non-viral nanocarriers for intracellular delivery of microRNA therapeutics. *Journal of Materials Chemistry B*, *7*(8), 1209–1225.

Bartel, D. P. (2004). MicroRNAs: Genomics, biogenesis, mechanism, and function. *Cell*, *116*(2), 281–297.

Bartoszewska, S., et al. (2015). The hypoxia-inducible miR-429 regulates hypoxia-inducible factor-1α expression in human endothelial cells through a negative feedback loop. *FASEB Journal*, *29*(4), 1467–1479.

Becker, J. C., et al. (2013). Immune-suppressive properties of the tumor microenvironment. *Cancer Immunology, Immunotherapy*, *62*(7), 1137–1148.

Benaich, N., et al. (2014). Rewiring of an epithelial differentiation factor, miR-203, to inhibit human squamous cell carcinoma metastasis. *Cell Reports*, *9*(1), 104–117.

Bonci, D., et al. (2008). The miR-15a-miR-16-1 cluster controls prostate cancer by targeting multiple oncogenic activities. *Nature Medicine*, *14*(11), 1271–1277.

Brabletz, T., et al. (2018). EMT in cancer. *Nature Reviews Cancer*, *18*(2), 128–134.

Bracken, C. P., et al. (2008). A double-negative feedback loop between ZEB1-SIP1 and the microRNA-200 family regulates epithelial-mesenchymal transition. *Cancer Research*, *68*(19), 7846–7854.

Brase, J. C., et al. (2011). Circulating miRNAs are correlated with tumor progression in prostate cancer. *International Journal of Cancer*, *128*(3), 608–616.

Bu, P., et al. (2015). miR-1269 promotes metastasis and forms a positive feedback loop with TGF-beta. *Nature Communications*, *6*, 6879.

Budhu, A., et al. (2008). Identification of metastasis-related microRNAs in hepatocellular carcinoma. *Hepatology*, *47*(3), 897–907.

Caby, M. P., et al. (2005). Exosomal-like vesicles are present in human blood plasma. *International Immunology*, *17*(7), 879–887.

Cai, J., et al. (2013). MicroRNA-374a activates Wnt/beta-catenin signaling to promote breast cancer metastasis. *Journal of Clinical Investigation*, *123*(2), 566–579.

Chambers, A. F., Groom, A. C., & MacDonald, I. C. (2002). Dissemination and growth of cancer cells in metastatic sites. *Nature Reviews Cancer*, *2*(8), 563–572.

Chan, Y. C., et al. (2011). miR-200b targets Ets-1 and is down-regulated by hypoxia to induce angiogenic response of endothelial cells. *Journal of Biological Chemistry*, *286*(3), 2047–2056.

Chang, H., et al. (2016). CRISPR/cas9, a novel genomic tool to knock down microRNA in vitro and in vivo. *Scientific Reports*, *6*, 22312.

Chen, Y., et al. (2012). miRNA-135a promotes breast cancer cell migration and invasion by targeting HOXA10. *BMC Cancer*, *12*, 111.

Chen, S., et al. (2016). MicroRNA-93 promotes epithelial-mesenchymal transition of endometrial carcinoma cells. *PLoS One*, *11*(11), e0165776.

Chen, X., et al. (2018). microRNA-130a suppresses breast cancer cell migration and invasion by targeting FOSL1 and upregulating ZO-1. *Journal of Cellular Biochemistry*, *119*(6), 4945–4956.

Chi, Y., et al. (2017). miR-203 inhibits cell proliferation, invasion, and migration of non-small-cell lung cancer by downregulating RGS17. *Journal of Cellular and Molecular Medicine*, *108*(12), 2366–2372.

Chu, J., et al. (2018). MiR-4319 suppress the malignancy of triple-negative breast cancer by regulating self-renewal and tumorigenesis of stem cells. *Cellular Physiology and Biochemistry*, *48*(2), 593–604.

Cikaluk, D. E., et al. (1999). GERp95, a membrane-associated protein that belongs to a family of proteins involved in stem cell differentiation. *Molecular Biology of the Cell*, *10*(10), 3357–3372.

Cortez, M. A., et al. (2011). MicroRNAs in body fluids—The mix of hormones and biomarkers. *Nature Reviews Clinical Oncology*, *8*(8), 467–477.

Cui, M., et al. (2019). Circulating microRNAs in cancer: Potential and challenge. *Frontiers in Genetics*, *10*, 626.

Damiano, V., et al. (2017). Epigenetic silencing of miR-200c in breast cancer is associated with aggressiveness and is modulated by ZEB1. *Genes, Chromosomes and Cancer*, *56*(2), 147–158.

Dasgupta, I., & Chatterjee, A. (2021). Recent advances in miRNA delivery systems. *Methods and Protocols*, *4*(1), 10.

Debeb, B. G., et al. (2016). miR-141-mediated regulation of brain metastasis from breast cancer. *Journal of the National Cancer Institute*, *108*(8), djw026.

Deng, B., et al. (2016). MiRNA-203 suppresses cell proliferation, migration and invasion in colorectal cancer via targeting of EIF5A2. *Scientific Reports*, *6*, 28301.

Dettori, D., et al. (2018). Therapeutic silencing of miR-214 inhibits tumor progression in multiple mouse models. *Molecular Therapy*, *26*(8), 2008–2018.

Díaz-García, C. V., et al. (2013). DICER1, DROSHA and miRNAs in patients with non-small cell lung cancer: Implications for outcomes and histologic classification. *Carcinogenesis*, *34*(5), 1031–1038.

Diederichs, S., & Haber, D. A. (2007). Dual role for Argonautes in microRNA processing and posttranscriptional regulation of microRNA expression. *Cell*, *131*(6), 1097–1108.

Ding, X., et al. (2013). Signaling between transforming growth factor beta (TGF-beta) and transcription factor SNAI2 represses expression of microRNA miR-203 to promote epithelial-mesenchymal transition and tumor metastasis. *Journal of Biological Chemistry*, *288*(15), 10241–10253.

Dong, H., et al. (2019). The regulatory network of miR-141 in the inhibition of angiogenesis. *Angiogenesis*, *22*(2), 251–262.

Eble, J. A., & Niland, S. (2019). The extracellular matrix in tumor progression and metastasis. *Clinical and Experimental Metastasis*, *36*(3), 171–198.

Eminaga, O., et al. (2018). The upregulation of hypoxia-related miRNA 210 in primary tumor of lymphogenic metastatic prostate cancer. *Epigenomics*, *10*(10), 1347–1359.

Esposito, C. L., et al. (2014). Multifunctional aptamer-miRNA conjugates for targeted cancer therapy. *Molecular Therapy*, *22*(6), 1151–1163.

Expósito-Villén, A., Aránega, A. E., & Franco, D. (2018). Functional role of non-coding RNAs during epithelial-to-mesenchymal transition. *Noncoding RNA*, *4*(2), 14.

Fernandez-Mercado, M., et al. (2015). The circulating transcriptome as a source of non-invasive cancer biomarkers: Concepts and controversies of non-coding and coding RNA in body fluids. *Journal of Cellular and Molecular Medicine*, *19*(10), 2307–2323.

Fetahu, I. S., et al. (2016). miR-135b- and miR-146b-dependent silencing of calcium-sensing receptor expression in colorectal tumors. *International Journal of Cancer*, *138*(1), 137–145.

Fukumoto, I., et al. (2016). Tumor-suppressive microRNAs (miR-26a/b, miR-29a/b/c and miR-218) concertedly suppressed metastasis-promoting LOXL2 in head and neck squamous cell carcinoma. *Journal of Human Genetics*, *61*(2), 109–118.

Gao, Y., et al. (2018). Tumor-promoting properties of miR-8084 in breast cancer through enhancing proliferation, suppressing apoptosis and inducing epithelial-mesenchymal transition. *Journal of Translational Medicine*, *16*(1), 38.

Garzon, R., et al. (2009). MicroRNA-29b induces global DNA hypomethylation and tumor suppressor gene reexpression in acute myeloid leukemia by targeting directly DNMT3A and 3B and indirectly DNMT1. *Blood*, *113*(25), 6411–6418.

Guo, F., et al. (2014). MiR-101 suppresses the epithelial-to-mesenchymal transition by targeting ZEB1 and ZEB2 in ovarian carcinoma. *Oncology Reports*, *31*(5), 2021–2028.

He, L., et al. (2005). A microRNA polycistron as a potential human oncogene. *Nature*, *435*(7043), 828–833.

He, R., et al. (2019). Reduced miR-203 predicts metastasis and poor survival in esophageal carcinoma. *Aging (Albany NY)*, *11*(24), 12114–12130.

Hong, Y., et al. (2016). miR-96 promotes cell proliferation, migration and invasion by targeting PTPN9 in breast cancer. *Scientific Reports*, *6*, 37421.

Hou, Y., et al. (2015). miR-215 functions as a tumor suppressor and directly targets ZEB2 in human non-small cell lung cancer. *Oncology Letters*, *10*(4), 1985–1992.

Hu, Z., et al. (2018). MicroRNA-197 promotes metastasis of hepatocellular carcinoma by activating Wnt/beta-catenin signaling. *Cellular Physiology and Biochemistry*, *51*(1), 470–486.

Huang, Q., et al. (2008). The microRNAs miR-373 and miR-520c promote tumour invasion and metastasis. *Nature Cell Biology*, *10*(2), 202–210.

Huang, D., et al. (2014). miRNA27a is a biomarker for predicting chemosensitivity and prognosis in metastatic or recurrent gastric cancer. *Journal of Cellular Biochemistry*, *115*(3), 549–556.

Jiang, N., et al. (2020). Role of PI3K/AKT pathway in cancer: The framework of malignant behavior. *Molecular Biology Reports*, *47*(6), 4587–4629.

Kalluri, R. (2016). The biology and function of fibroblasts in cancer. *Nature Reviews Cancer*, *16*(9), 582–598.

Kalluri, R., & Weinberg, R. A. (2009). The basics of epithelial-mesenchymal transition. *Journal of Clinical Investigation*, *119*(6), 1420–1428.

Kefas, B., et al. (2008). microRNA-7 inhibits the epidermal growth factor receptor and the Akt pathway and is down-regulated in glioblastoma. *Cancer Research*, *68*(10), 3566–3572.

Kim, B. G., et al. (2011a). Laminin-332-rich tumor microenvironment for tumor invasion in the interface zone of breast cancer. *American Journal of Pathology*, *178*(1), 373–381.

Kim, T., et al. (2011b). p53 regulates epithelial-mesenchymal transition through microRNAs targeting ZEB1 and ZEB2. *Journal of Experimental Medicine*, *208*(5), 875–883.

Kim, S. Y., et al. (2013). Validation of circulating miRNA biomarkers for predicting lymph node metastasis in gastric cancer. *Journal of Molecular Diagnostics*, *15*(5), 661–669.

Kim, J., et al. (2016). Ablation of miR-10b suppresses oncogene-induced mammary tumorigenesis and metastasis and reactivates tumor-suppressive pathways. *Cancer Research*, *76*(21), 6424–6435.

Kinoshita, T., et al. (2013). Tumour-suppressive microRNA-29s inhibit cancer cell migration and invasion by targeting laminin-integrin signalling in head and neck squamous cell carcinoma. *British Journal of Cancer*, *109*(10), 2636–2645.

Klein, U., et al. (2010). The DLEU2/miR-15a/16-1 cluster controls B cell proliferation and its deletion leads to chronic lymphocytic leukemia. *Cancer Cell*, *17*(1), 28–40.

Kong, X., et al. (2012). MicroRNA-7 inhibits epithelial-to-mesenchymal transition and metastasis of breast cancer cells via targeting FAK expression. *PLoS One*, *7*(8), e41523.

Korpal, M., et al. (2008). The miR-200 family inhibits epithelial-mesenchymal transition and cancer cell migration by direct targeting of E-cadherin transcriptional repressors ZEB1 and ZEB2. *Journal of Biological Chemistry*, *283*(22), 14910–14914.

Kovaleva, V., et al. (2012). miRNA-130a targets ATG2B and DICER1 to inhibit autophagy and trigger killing of chronic lymphocytic leukemia cells. *Cancer Research*, *72*(7), 1763–1772.

Krebs, A. M., et al. (2017). The EMT-activator Zeb1 is a key factor for cell plasticity and promotes metastasis in pancreatic cancer. *Nature Cell Biology*, *19*(5), 518–529.

Kwon, S. C., et al. (2016). Structure of human DROSHA. *Cell*, *164*(1–2), 81–90.

Lamouille, S., et al. (2013). Regulation of epithelial-mesenchymal and mesenchymal-epithelial transitions by microRNAs. *Current Opinion in Cell Biology*, *25*(2), 200–207.

Lawrie, C. H., et al. (2009). Aberrant expression of microRNA biosynthetic pathway components is a common feature of haematological malignancy. *British Journal of Haematology*, *145*(4), 545–548.

Lee, J. W., et al. (2018). MicroRNA-708-3p mediates metastasis and chemoresistance through inhibition of epithelial-to-mesenchymal transition in breast cancer. *Cancer Science*, *109*(5), 1404–1413.

Li, Z., & Rana, T. M. (2014). Therapeutic targeting of microRNAs: Current status and future challenges. *Nature Reviews Drug Discovery*, *13*(8), 622–638.

Li, Y. R., et al. (2016). MicroRNA-520c enhances cell proliferation, migration, and invasion by suppressing IRF2 in gastric cancer. *FEBS Open Bio*, *6*(12), 1257–1266.

Li, D., et al. (2018a). CADM2, as a new target of miR-10b, promotes tumor metastasis through FAK/AKT pathway in hepatocellular carcinoma. *Journal of Experimental and Clinical Cancer Research*, *37*(1), 46.

Li, J., et al. (2018b). MiR-186-5p upregulation inhibits proliferation, metastasis and epithelial-to-mesenchymal transition of colorectal cancer cell by targeting ZEB1. *Archives of Biochemistry and Biophysics*, *640*, 53–60.

Li, Q., et al. (2019a). MicroRNA-7-5p induces cell growth inhibition, cell cycle arrest and apoptosis by targeting PAK2 in non-small cell lung cancer. *FEBS Open Bio*, *9*(11), 1983–1993.

Li, J., et al. (2019b). miR-203 inhibits the invasion and EMT of gastric cancer cells by directly targeting Annexin A4. *Oncology Research*, *27*(7), 789–799.

Li, M., et al. (2020). miR-7 reduces breast cancer stem cell metastasis via inhibiting RELA to decrease ESAM expression. *Molecular Therapy Oncolytics*, *18*, 70–82.

Liang, T., et al. (2016). MicroRNA-194 promotes the growth, migration, and invasion of ovarian carcinoma cells by targeting protein tyrosine phosphatase nonreceptor type 12. *Oncotargets and Therapy*, *9*, 4307–4315.

Liao, C. G., et al. (2014). miR-10b is overexpressed in hepatocellular carcinoma and promotes cell proliferation, migration and invasion through RhoC, uPAR and MMPs. *Journal of Translational Medicine*, *12*, 234.

Lin, Y., et al. (2015). MicroRNA-33b inhibits breast cancer metastasis by targeting HMGA2, SALL4 and Twist1. *Scientific Reports*, *5*, 9995.

Ling, Y., et al. (2019). TRIP6, as a target of miR-7, regulates the proliferation and metastasis of colorectal cancer cells. *Biochemical and Biophysical Research Communications*, *514*(1), 231–238.

Liu, C., et al. (2011). The microRNA miR-34a inhibits prostate cancer stem cells and metastasis by directly repressing CD44. *Nature Medicine*, *17*(2), 211–215.

Liu, Y., et al. (2012a). MicroRNA-10b targets E-cadherin and modulates breast cancer metastasis. *Medical Science Monitor*, *18*(8), BR299–BR308.

Liu, Z., et al. (2012b). miR-10b promotes cell invasion through RhoC-AKT signaling pathway by targeting HOXD10 in gastric cancer. *International Journal of Oncology*, *40*(5), 1553–1560.

Liu, Y., et al. (2013). MicroRNA-10b overexpression promotes non-small cell lung cancer cell proliferation and invasion. *European Journal of Medical Research*, *18*(1), 41.

Liu, C., et al. (2017). MicroRNA-141 suppresses prostate cancer stem cells and metastasis by targeting a cohort of pro-metastasis genes. *Nature Communications*, *8*, 14270.

Liu, X., et al. (2018a). miR-1204 targets VDR to promotes epithelial-mesenchymal transition and metastasis in breast cancer. *Oncogene*, *37*(25), 3426–3439.

Liu, B., et al. (2018b). miR-200c/141 regulates breast cancer stem cell heterogeneity via targeting HIPK1/beta-catenin axis. *Theranostics*, *8*(21), 5801–5813.

Liu, W., et al. (2020). miR-210-5p promotes epithelial-mesenchymal transition by inhibiting PIK3R5 thereby activating oncogenic autophagy in osteosarcoma cells. *Cell Death and Disease*, *11*(2), 93.

Lohcharoenkal, W., et al. (2018). Genome-wide screen for sicroRNAs reveals a role for miR-203 in melanoma metastasis. *Journal of Investigative Dermatology*, *138*(4), 882–892.

Lu, C., et al. (2020). The circ_0021977/miR-10b-5p/P21 and P53 regulatory axis suppresses proliferation, migration, and invasion in colorectal cancer. *Journal of Cellular Physiology*, *235*(3), 2273–2285.

Luo, Y., et al. (2019). miR-577 regulates TGF-beta induced cancer progression through a SDPR-modulated positive-feedback loop with ERK-NF-kappaB in gastric cancer. *Molecular Therapy*, *27*(6), 1166–1182.

Lv, M., et al. (2017). lncRNA H19 regulates epithelial-mesenchymal transition and metastasis of bladder cancer by miR-29b-3p as competing endogenous RNA. *Biochimica et Biophysica Acta, Molecular Cell Research*, *1864*(10), 1887–1899.

Lv, Z. D., et al. (2019). miR-135b promotes proliferation and metastasis by targeting APC in triple-negative breast cancer. *Journal of Cellular Physiology*, *234*(7), 10819–10826.

Ma, L., Teruya-Feldstein, J., & Weinberg, R. A. (2007). Tumour invasion and metastasis initiated by microRNA-10b in breast cancer. *Nature*, *449*(7163), 682–688.

Ma, L., & Weinberg, R. A. (2008). Micromanagers of malignancy: Role of microRNAs in regulating metastasis. *Trends in Genetics*, *24*(9), 448–456.

Ma, L., et al. (2010). Therapeutic silencing of miR-10b inhibits metastasis in a mouse mammary tumor model. *Nature Biotechnology*, *28*(4), 341–347.

Ma, C., et al. (2017). MicroRNA-10b mediates TGF-β1-regulated glioblastoma proliferation, migration and epithelial-mesenchymal transition. *International Journal of Oncology*, *50*(5), 1739–1748.

Madhavan, D., et al. (2016). Circulating miRNAs with prognostic value in metastatic breast cancer and for early detection of metastasis. *Carcinogenesis*, *37*(5), 461–470.

Maierthaler, M., et al. (2017). Plasma miR-122 and miR-200 family are prognostic markers in colorectal cancer. *International Journal of Cancer*, *140*(1), 176–187.

Martello, G., et al. (2010). A microRNA targeting dicer for metastasis control. *Cell*, *141*(7), 1195–1207.

McAnena, P., et al. (2019). Circulating microRNAs miR-331 and miR-195 differentiate local luminal a from metastatic breast cancer. *BMC Cancer*, *19*(1), 436.

Merritt, W. M., et al. (2008). Dicer, Drosha, and outcomes in patients with ovarian cancer. *New England Journal of Medicine*, *359*(25), 2641–2650.

Mitchell, P. S., et al. (2008). Circulating microRNAs as stable blood-based markers for cancer detection. *Proceedings of the National Academy of Sciences of the United States of America*, *105*(30), 10513–10518.

Mohammadi-Yeganeh, S., Hosseini, V., & Paryan, M. (2019). Wnt pathway targeting reduces triple-negative breast cancer aggressiveness through miRNA regulation in vitro and in vivo. *Journal of Cellular Physiology*, *234*(10), 18317–18328.

Navarro, A., et al. (2009). Regulation of JAK2 by miR-135a: Prognostic impact in classic Hodgkin lymphoma. *Blood, 114*(14), 2945–2951.

Nie, H., et al. (2020). Use of lung-specific exosomes for miRNA-126 delivery in non-small cell lung cancer. *Nanoscale, 12*(2), 877–887.

Nishikawa, R., et al. (2015). Tumour-suppressive microRNA-29s directly regulate LOXL2 expression and inhibit cancer cell migration and invasion in renal cell carcinoma. *FEBS Letters, 589*(16), 2136–2145.

Pan, C. M., et al. (2020). MicroRNA-7 targets T-Box 2 to inhibit epithelial-mesenchymal transition and invasiveness in glioblastoma multiforme. *Cancer Letters, 493*, 133–142.

Pang, Y., et al. (2018). MYC and DNMT3A-mediated DNA methylation represses microRNA-200b in triple negative breast cancer. *Journal of Cellular and Molecular Medicine, 22*(12), 6262–6274.

Pardini, B., et al. (2019). Noncoding RNAs in extracellular fluids as cancer biomarkers: The new frontier of liquid biopsies. *Cancers (Basel), 11*(8), 1170.

Pardini, B., & Calin, G. A. (2019). MicroRNAs and long non-coding RNAs and their hormone-like activities in cancer. *Cancers (Basel), 11*(3), 378.

Reddy, S. D., et al. (2008). MicroRNA-7, a homeobox D10 target, inhibits p21-activated kinase 1 and regulates its functions. *Cancer Research, 68*(20), 8195–8200.

Robert, J. (2013). Biology of cancer metastasis. *Bulletin du Cancer, 100*(4), 333–342.

Rostas, J. W., 3rd, et al. (2014). microRNA-29 negatively regulates EMT regulator N-myc interactor in breast cancer. *Molecular Cancer, 13*, 200.

Roth, C., et al. (2010). Circulating microRNAs as blood-based markers for patients with primary and metastatic breast cancer. *Breast Cancer Research, 12*(6), R90.

Rupaimoole, R., & Slack, F. J. (2017). MicroRNA therapeutics: Towards a new era for the management of cancer and other diseases. *Nature Reviews Drugs Discovery, 16*(3), 203–222.

Rupaimoole, R., et al. (2016). Hypoxia-upregulated microRNA-630 targets Dicer, leading to increased tumor progression. *Oncogene, 35*(33), 4312–4320.

Satomi-Tsushita, N., et al. (2019). Serum microRNA-based prediction of responsiveness to eribulin in metastatic breast cancer. *PLoS One, 14*(9), e0222024.

Shen, J., et al. (2013). EGFR modulates microRNA maturation in response to hypoxia through phosphorylation of AGO2. *Nature, 497*(7449), 383–387.

Shin, J. Y., et al. (2014). MicroRNA 135a suppresses lymph node metastasis through down-regulation of ROCK1 in early gastric cancer. *PLoS One, 9*(1), e85205.

Siemens, H., et al. (2011). miR-34 and SNAIL form a double-negative feedback loop to regulate epithelial-mesenchymal transitions. *Cell Cycle, 10*(24), 4256–4271.

Sole, C., et al. (2019). The circulating transcriptome as a source of cancer liquid biopsy biomarkers. *Seminars in Cancer Biology, 58*, 100–108.

Solé, C., & Lawrie, C. H. (2019). MicroRNAs and metastasis. *Cancers (Basel), 12*(1), 96.

Song, Y., et al. (2014). MicroRNA-9 promotes tumor metastasis via repressing E-cadherin in esophageal squamous cell carcinoma. *Oncotarget, 5*(22), 11669–11680.

Sossey-Alaoui, K., et al. (2011). WAVE3, an actin remodeling protein, is regulated by the metastasis suppressor microRNA, miR-31, during the invasion-metastasis cascade. *International Journal of Cancer, 129*(6), 1331–1343.

Steeg, P. S. (2016). Targeting metastasis. *Nature Reviews Cancer, 16*(4), 201–218.

Stemmler, M. P., et al. (2019). Non-redundant functions of EMT transcription factors. *Nature Cell Biology, 21*(1), 102–112.

Stoorvogel, W. (2012). Functional transfer of microRNA by exosomes. *Blood, 119*(3), 646–648.

Su, X., et al. (2010). TAp63 suppresses metastasis through coordinate regulation of Dicer and miRNAs. *Nature, 467*(7318), 986–990.

Sun, G., Liu, M., & Han, H. (2019). Overexpression of microRNA-190 inhibits migration, invasion, epithelial-mesenchymal transition, and angiogenesis through suppression of protein kinase B-extracellular signal-regulated kinase signaling pathway via binding to stanniocalicin 2 in breast cancer. *Journal of Cellular Physiology, 234*(10), 17824–17838.

Swellam, M., et al. (2018). Aberrant expression of some circulating miRNAs in childhood acute lymphoblastic leukemia. *Biochemical Genetics, 56*(4), 283–294.

Tanaka, Y., et al. (2013). Clinical impact of serum exosomal microRNA-21 as a clinical biomarker in human esophageal squamous cell carcinoma. *Cancer, 119*(6), 1159–1167.

Tang, T., et al. (2018a). Up-regulation of miR-210 induced by a hypoxic microenvironment promotes breast cancer stem cells metastasis, proliferation, and self-renewal by targeting E-cadherin. *FASEB Journal, 32*(12), 6965–6981.

Tang, W., et al. (2018b). MicroRNA-150 suppresses triple-negative breast cancer metastasis through targeting HMGA2. *Oncotargets and Therapy, 11*, 2319–2332.

Tao, W. Y., et al. (2016). MicroRNA-34c suppresses breast cancer migration and invasion by targeting GIT1. *Journal of Cancer, 7*(12), 1653–1662.

Tian, X., et al. (2018). The miR-203/SNAI2 axis regulates prostate tumor growth, migration, angiogenesis and stemness potentially by modulating GSK-3β/β-CATENIN signal pathway. *Cancer Science, 70*(3), 224–236.

Toiyama, Y., et al. (2013). Serum miR-21 as a diagnostic and prognostic biomarker in colorectal cancer. *Journal of the National Cancer Institute, 105*(12), 849–859.

Trang, P., et al. (2010). Regression of murine lung tumors by the let-7 microRNA. *Oncogene, 29*(11), 1580–1587.

Tsukamoto, M., et al. (2017). Circulating exosomal microRNA-21 as a biomarker in each tumor stage of colorectal cancer. *Oncology, 92*(6), 360–370.

Turchinovich, A., et al. (2011). Characterization of extracellular circulating microRNA. *Nucleic Acids Research, 39*(16), 7223–7233.

van den Beucken, T., et al. (2014). Hypoxia promotes stem cell phenotypes and poor prognosis through epigenetic regulation of DICER. *Nature Communications, 5*, 5203.

Vickers, K. C., & Remaley, A. T. (2012). Lipid-based carriers of microRNAs and intercellular communication. *Current Opinion in Lipidology, 23*(2), 91–97.

Voorhoeve, P. M., et al. (2006). A genetic screen implicates miRNA-372 and miRNA-373 as oncogenes in testicular germ cell tumors. *Cell, 124*(6), 1169–1181.

Voss, G., et al. (2020). Regulation of cell-cell adhesion in prostate cancer cells by microRNA-96 through upregulation of E-cadherin and EpCAM. *Carcinogenesis*, *41*(7), 865–874.

Walsh, J. C., et al. (2014). The clinical importance of assessing tumor hypoxia: Relationship of tumor hypoxia to prognosis and therapeutic opportunities. *Antioxidants and Redox Signaling*, *21*(10), 1516–1554.

Wang, B., Yang, J., & Xiao, B. (2016). MicroRNA-20b (miR-20b) promotes the proliferation, migration, invasion, and tumorigenicity in esophageal cancer cells via the regulation of phosphatase and tensin homologue expression. *PLoS One*, *11*(10), e0164105.

Wang, Z. X., et al. (2011). Prognostic significance of serum miRNA-21 expression in human non-small cell lung cancer. *Journal of Surgical Oncology*, *104*(7), 847–851.

Wang, H., et al. (2013). miRNA-29c suppresses lung cancer cell adhesion to extracellular matrix and metastasis by targeting integrin beta1 and matrix metalloproteinase2 (MMP2). *PLoS One*, *8*(8), e70192.

Wang, T. H., et al. (2016). OncomiR miR-96 and miR-182 promote cell proliferation and invasion through targeting ephrinA5 in hepatocellular carcinoma. *Molecular Carcinogenesis*, *55*(4), 366–375.

Wang, B., et al. (2018a). miR-203 inhibits ovarian tumor metastasis by targeting BIRC5 and attenuating the TGFbeta pathway. *Journal of Experimental and Clinical Cancer Research*, *37*(1), 235.

Wang, Y., et al. (2018b). Long noncoding RNA DANCR, working as a competitive endogenous RNA, promotes ROCK1-mediated proliferation and metastasis via decoying of miR-335-5p and miR-1972 in osteosarcoma. *Molecular Cancer*, *17*(1), 89.

Wang, J., et al. (2020a). miR-429-CRKL axis regulates clear cell renal cell carcinoma malignant progression through SOS1/MEK/ERK/MMP2/MMP9 pathway. *Biomedicine and Pharmacotherapy*, *127*, 110215.

Wang, S., et al. (2020b). miR-874 directly targets AQP3 to inhibit cell proliferation, mobility and EMT in non-small cell lung cancer. *Thoracic Cancer*, *11*(6), 1550–1558.

Webster, R. J., et al. (2009). Regulation of epidermal growth factor receptor signaling in human cancer cells by microRNA-7. *Journal of Biological Chemistry*, *284*(9), 5731–5741.

Wei, W., et al. (2013). miR-203 inhibits proliferation of HCC cells by targeting survivin. *Cell Biochemistry and Function*, *31*(1), 82–85.

Wei, S., et al. (2019). The circRNA circPTPRA suppresses epithelial-mesenchymal transitioning and metastasis of NSCLC cells by sponging miR-96-5p. *EBioMedicine*, *44*, 182–193.

Wu, L., et al. (2016). miR-96 induces cisplatin chemoresistance in non-small cell lung cancer cells by downregulating SAMD9. *Oncology Letters*, *11*(2), 945–952.

Wu, Y., et al. (2019). miR-29a contributes to breast cancer cells epithelial-mesenchymal transition, migration, and invasion via down-regulating histone H4K20 trimethylation through directly targeting SUV420H2. *Cell Death and Disease*, *10*(3), 176.

Xi, Z. W., et al. (2015). Downregulation of rho-associated protein kinase 1 by miR-124 in colorectal cancer. *World Journal of Gastroenterology*, *21*(18), 5454–5464.

Xia, H., et al. (2014). MiR-96 promotes proliferation and chemo- or radioresistance by down-regulating RECK in esophageal cancer. *Biomedicine and Pharmacotherapy*, *68*(8), 951–958.

Xie, Y., et al. (2019). MicroRNA-10b controls the metastasis and proliferation of colorectal cancer cells by regulating Krüppel-like factor 4. *Artificial Cells, Nanomedicine, and Biotechnology*, *47*(1), 1722–1729.

Xu, H., et al. (2016). MicroRNA-154 inhibits growth and invasion of breast cancer cells through targeting E2F5. *American Journal of Translational Research*, *8*(6), 2620–2630.

Xu, M., et al. (2018). MicroRNA-34a inhibition of the TLR signaling pathway via CXCL10 suppresses breast cancer cell invasion and migration. *Cellular Physiology and Biochemistry*, *46*(3), 1286–1304.

Yadav, P., et al. (2016). Serum microRNA-21 expression as a prognostic and therapeutic biomarker for breast cancer patients. *Tumour Biology*, *37*(11), 15275–15282.

Yao, R., et al. (2018). miRNA-641 inhibits the proliferation, migration, and invasion and induces apoptosis of cervical cancer cells by directly targeting ZEB1. *Oncotargets and Therapy*, *11*, 8965–8976.

Ye, J., et al. (2013). miRNA-27b targets vascular endothelial growth factor C to inhibit tumor progression and angiogenesis in colorectal cancer. *PLoS One*, *8*(4), e60687.

Ye, M. F., et al. (2018). MiR-504 inhibits cell proliferation and invasion by targeting LOXL2 in non small cell lung cancer. *Biomedicine and Pharmacotherapy*, *97*, 1289–1295.

Yin, J., et al. (2014). Differential expression of serum miR-126, miR-141 and miR-21 as novel biomarkers for early detection of liver metastasis in colorectal cancer. *Chinese Journal of Cancer Research*, *26*(1), 95–103.

Yin, H., et al. (2020). EZH2-mediated epigenetic silencing of miR-29/miR-30 targets LOXL4 and contributes to tumorigenesis, metastasis, and immune microenvironment remodeling in breast cancer. *Theranostics*, *10*(19), 8494–8512.

Ying, Q., et al. (2011). Hypoxia-inducible microRNA-210 augments the metastatic potential of tumor cells by targeting vacuole membrane protein 1 in hepatocellular carcinoma. *Hepatology*, *54*(6), 2064–2075.

Yoo, J. O., et al. (2016). miR-181b-3p promotes epithelial-mesenchymal transition in breast cancer cells through Snail stabilization by directly targeting YWHAG. *Biochimica et Biophysica Acta*, *1863*(7 Pt A), 1601–1611.

You, J., et al. (2014). MiR-132 suppresses the migration and invasion of lung cancer cells via targeting the EMT regulator ZEB2. *PLoS One*, *9*(3), e91827.

Yu, F., et al. (2007). let-7 regulates self renewal and tumorigenicity of breast cancer cells. *Cell*, *131*(6), 1109–1123.

Yue, S., et al. (2015). miR-139-5p suppresses cancer cell migration and invasion through targeting ZEB1 and ZEB2 in GBM. *Tumour Biology*, *36*(9), 6741–6749.

Zeng, Y. B., et al. (2016). miRNA-135a promotes hepatocellular carcinoma cell migration and invasion by targeting forkhead box O1. *Cancer Cell International*, *16*, 63.

Zhang, L., et al. (2013). MicroRNA-135b acts as a tumor promoter by targeting the hypoxia-inducible factor pathway in genetically defined mouse model of head and neck squamous cell carcinoma. *Cancer Letters*, *331*(2), 230–238.

Zhang, L., et al. (2016). Circulating microRNA-21 is involved in lymph node metastasis in cervical cancer by targeting RASA1. *International Journal of Gynecological Cancer*, *26*(5), 810–816.

Zhang, C., et al. (2021). Circular RNA in tumor metastasis. *Molecular Therapy—Nucleic Acids*, *23*, 1243–1257.

Zhang, G., et al. (2017). MicroRNA-200c and microRNA-141 are regulated by a FOXP3-KAT2B axis and associated with tumor metastasis in breast cancer. *Breast Cancer Research*, *19*(1), 73.

Zhang, Q., et al. (2019a). Hypoxia-induced microRNA-10b-3p promotes esophageal squamous cell carcinoma growth and metastasis by targeting TSGA10. *Aging (Albany NY)*, *11*(22), 10374–10384.

Zhang, Y., et al. (2019b). Downregulation of lysyl oxidase-like 4 LOXL4 by miR-135a-5p promotes lung cancer progression in vitro and in vivo. *Journal of Cellular Physiology*, *234*(10), 18679–18687.

Zhao, X., Li, Y., & Zhou, Y. (2019). MicroRNA-96-3p promotes metastasis of papillary thyroid cancer through targeting SDHB. *Cancer Cell International*, *19*, 287.

Zhao, J. J., et al. (2010). microRNA expression profile and identification of miR-29 as a prognostic marker and pathogenetic factor by targeting CDK6 in mantle cell lymphoma. *Blood*, *115*(13), 2630–2639.

Zhao, L., et al. (2018). Tumor suppressor miR-128-3p inhibits metastasis and epithelial-mesenchymal transition by targeting ZEB1 in esophageal squamous-cell cancer. *Acta Biochimica et Biophysica Sinica Shanghai*, *50*(2), 171–180.

Zheng, Y. B., et al. (2014). miR-132 inhibits colorectal cancer invasion and metastasis via directly targeting ZEB2. *World Journal of Gastroenterology*, *20*(21), 6515–6522.

Zhou, W., et al. (2014). Cancer-secreted miR-105 destroys vascular endothelial barriers to promote metastasis. *Cancer Cell*, *25*(4), 501–515.

Zhou, R., et al. (2016). MicroRNA-574-5p promotes metastasis of non-small cell lung cancer by targeting PTPRU. *Scientific Reports*, *6*, 35714.

Zhou, W., et al. (2017). The lncRNA H19 mediates breast cancer cell plasticity during EMT and MET plasticity by differentially sponging miR-200b/c and let-7b. *Science Signaling*, *10*(483), eaak9557.

Zhu, S., et al. (2008). MicroRNA-21 targets tumor suppressor genes in invasion and metastasis. *Cancer Cell International*, *18*(3), 350–359.

Chapter 6

microRNAs and metabolism

Ram C. Shankaraiah, Laura Masatti, Samuel Akanksha, and Massimo Negrini
Department of Translational Medicine, The University of Ferrara, Ferrara, Italy

Introduction

Cancer cells adapt their metabolism to meet the needs associated with increased biomass. Because metabolic alterations are commonly described in various types of cancer cells, reprogramming of metabolism is considered a hallmark of cancer (Hanahan & Weinberg, 2011; Pavlova & Thompson, 2016)

The best known metabolic phenotype reprogrammed in cancer cells is known as the Warburg effect. It is characterized by high rates of glycolysis, which leads to the conversion of pyruvate into lactic acid even in the presence of oxygen (Warburg, Wind, & Negelein, 1927) . However, this does not necessarily imply that the major source of ATP for cancer cells is from glycolysis. Numerous studies have indeed indicated that while glycolysis is upregulated in most cancer cells, the mitochondrial respiration (*tricarboxylic acid cycle* (TCA) cycle followed by oxidative phosphorylation (HOXPHOS)) continues to function in proportion to oxygen supply. In fact, other pathways, mainly deriving from the catabolism of proteins and fatty acids, can provide the intermediates necessary for the functioning of the mitochondrial respiration (Fig. 6.1).

As regards the upregulation of glycolysis, it is instead emerging that the main advantage provided to cancer cells is the availability of precursors of anabolic pathways for the synthesis of macromolecules (De Berardinis & Chandel, 2016). Due to a general greater need for macromolecules of cancer cells to cope with the increase in mass compared to normal cells, the production of anabolic or intermediate molecules is functional to tumor growth. These products include nicotinamide adenine dinucleotide phosphate (NADPH) and ribonucleotides via the pentose phosphate (PPP) pathway; glycogen via the hexosamine pathway; amino acids via the serine/glycine biosynthesis pathway; and the Acetyl CoA used for the biosynthesis of fatty acids (Fig. 6.2). Moreover, the evidence that a significant amount of glycolytic intermediates is diverted toward the PPP pathway for the production of NADPH also has implications for the way cancer cells cope with and regulate the generation of reactive oxygen species (ROS) (Fig. 6.2).

Hence, the changes involved in the reprogramming of glucose metabolism represent characteristic and functionally relevant factors for the survival and growth of neoplastic cells. The knowledge of the governing elements could provide important indications for the design of potential antitumor strategies through targeting the molecular mechanisms linked to the reprogramming of glucose metabolism (Hamanaka & Chandel, 2012). This dependency makes the glucose metabolism an element of potential vulnerability of neoplastic cells. In this context, for example, various molecules capable of interfering with the synthesis of nucleotides are used in the clinic as chemotherapeutic agents against different types of cancer.

Several mechanisms have been described that cause the activation of glycolysis and promote cell proliferation. Examples include the activation of oncogenes (e.g., RAS [viral oncogene homolog of rat sarcoma] or MYC [viral oncogene homolog of avian myelocytomatosis]), the inactivation of tumor suppressor genes (e.g., tumor protein TP53 or PTEN [homolog of phosphatase and tensin]), growth factor-mediated signal transduction (e.g., IGF1R [Insulin-like growth factor receptor] or EGFR [epidermal growth factor receptor] or HER2 [epidermal growth factor receptor 2]) or hypoxia-mediated signaling (HIF1 [hypoxia-inducible factor-1α]) (Denko, 2008; Ward & Thompson, 2012).

In this chapter, we will discuss how noncoding RNAs (ncRNAs) are involved in the regulation of glycolysis. The dysregulation of miRNAs and lncRNAs in cancer cells is responsible for their functioning as oncogenes or tumor suppressors and can promote various cancer-associated processes, including the reprogramming of the metabolism. In fact, an increasing number of studies have shown that ncRNAs can alter glucose metabolism either directly by modulating the traffic and consumption of glucose in the glycolytic path or indirectly by modulating the signaling pathways associated with cancer.

MicroRNA in Human Malignancies. https://doi.org/10.1016/B978-0-12-822287-4.00006-2

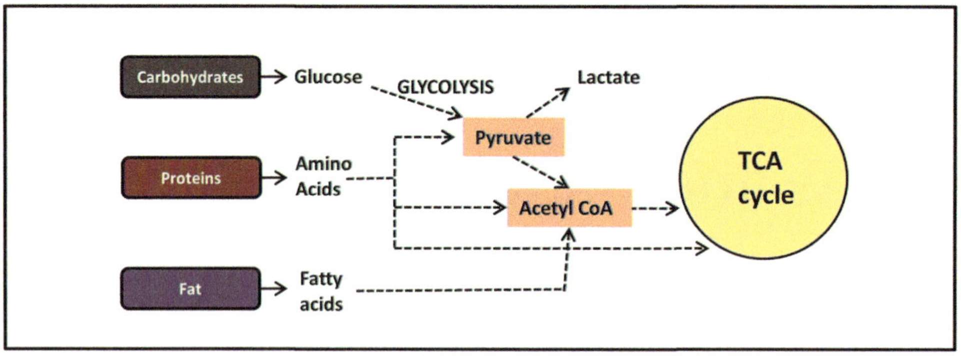

FIG. 6.1 Fuels of mitochondrial respiration. Catabolism of sugars, proteins, and fatty acids provides products used in the cycle of tricarboxylic acids (TCA) for the production of NADH and FADH2, subsequently required for the production of ATP during oxidative phosphorylation.

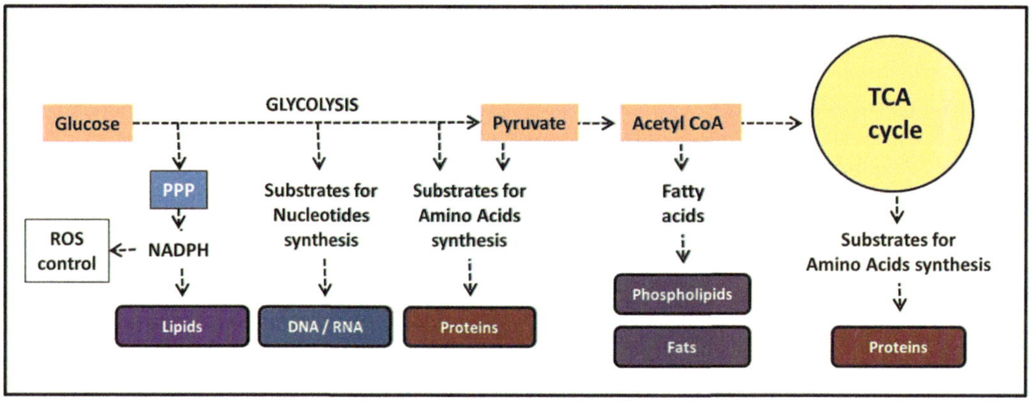

FIG. 6.2 Anabolic pathways originating from glycolysis intermediates. Various products of glycolysis (shunt PPP, pyruvate, and Acetyl Coa) provide intermediate molecules used by the cell for the anabolism of nucleotides, amino acids, and fatty acids necessary for the biosynthesis of cellular macromolecules and to control the levels of reactive oxygen species (ROS) of the cell.

Trafficking and consumption of glucose

NcRNAs regulate glucose trafficking in cancer cells by altering GLUT levels

As previously mentioned, tumor cells exhibit increased glycolysis, which allows for an increase in the amount of intermediate products of anabolic pathways and, to some degree, a compensation for inefficiency in the production of adenosine triphosphate (ATP) (two ATPs generated by glycolysis versus ~38 from OXPHOS). This is achieved initially by upregulating the expression of glucose transporters (GLUTs) capable of increasing glucose uptake (De Berardinis & Chandel, 2016), a property exploited, for example, for diagnostic imaging in positron emission tomography by use of a radio-labeled glucose analogue as a reporter (2-deoxy-2-fluoro-18fluoro-D-glucose) (Avril et al., 2000).

The transport of glucose across the cell membrane along a concentration gradient is an energy-independent process. Passive glucose transport is facilitated by the solute transporter 2A (SLC2A) and by proteins of the GLUT family. Among the 14 GLUT isoforms (Mueckler & Thorens, 2013), GLUT1 is known to be overexpressed in cancer. One mechanism leading to GLUT1 overexpression involves miR-1291. It is in fact described that the GLUT1 gene is the target of miR-1291: mimics can reduce the expression of the GLUT1 protein (Yamasaki et al., 2013). However, such miRNA is downregulated in tumor tissues compared to adjacent noncancerous tissue. Therefore, the increased level of these proteins, in particular of the GLUT1 protein, facilitates the transport of glucose into the cytoplasm of cancer cells (Fig. 6.3). These results suggest that GLUT1 may represent a possible target with some degree of specificity for cancer cells.

NcRNAs regulate key glycolytic enzymes

Once the glucose has entered the cell, there are three key steps that commit the cell toward the advancement of the glycolytic process: (i) the phosphorylation of glucose by hexokinases (HKs) to generate glucose-6-phosphate (G6P); (ii) the conversion of fructose-6-phosphate (F6P) into fructose-1,6-diphosphate (F1.6P) catalyzed by the enzyme phosphofructokinase (PFK1); and (iii) the conversion of phosphoenolpyruvate (PEP) to pyruvate (PYR) catalyzed by pyruvate kinase (PKs) (Fig. 6.3).

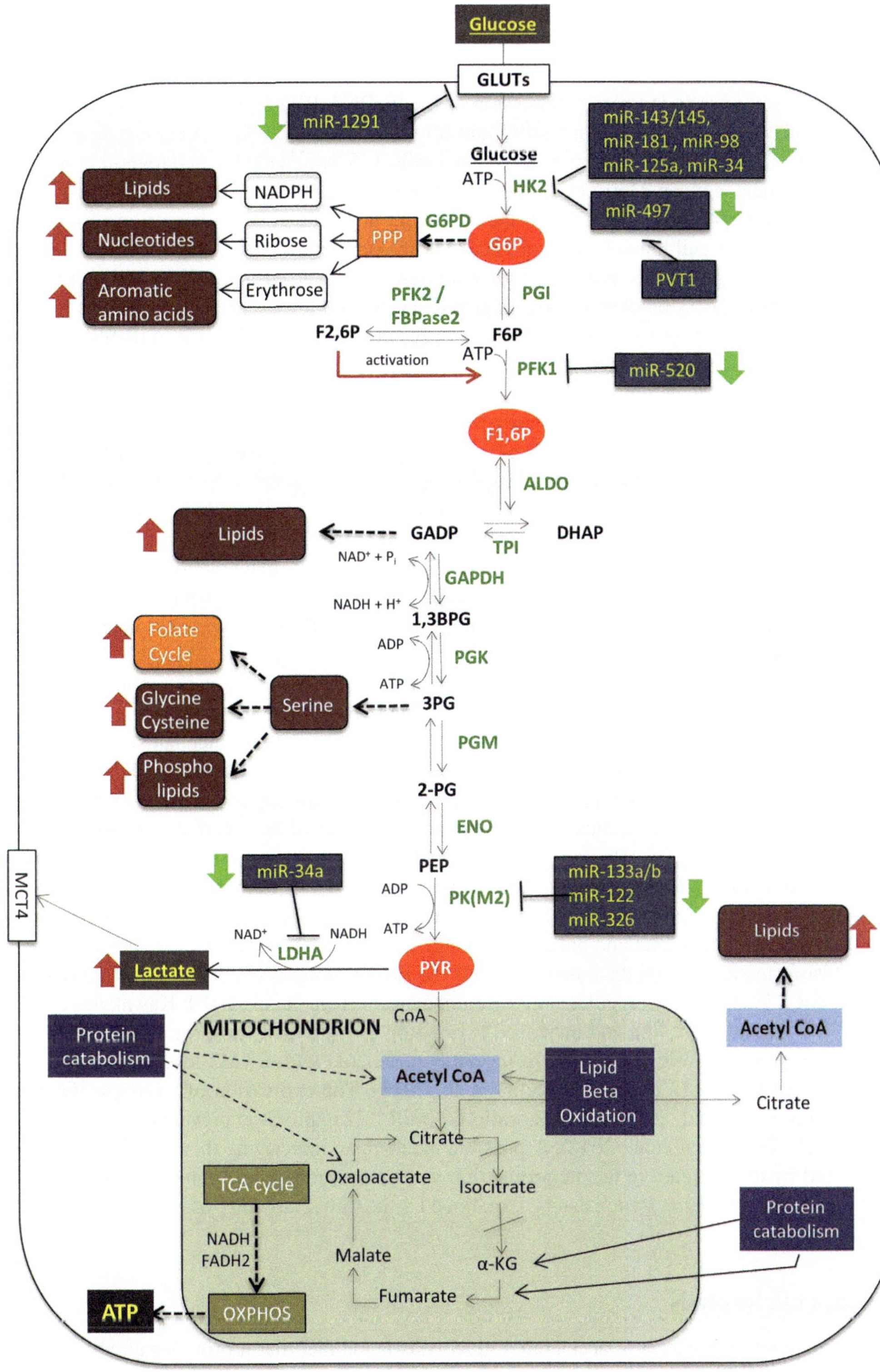

FIG. 6.3 Aerobic glycolysis and microRNAs that exert control on key components. The *green* and *red* arrows indicate the regulation of the indicated molecules in cancer cells. Abbreviations are described in the corresponding paragraph of the chapter text and in Table 6.2.

As mentioned above, the first key step is glucose phosphorylation by hexokinases (HKs) to generate glucose-6-phosphate (G6P). This phase allows maintaining a low glucose concentration in the cytoplasm, thus promoting the continuous transport of new glucose from the outside to the inside of the cell and, additionally, once phosphorylated glucose can no longer escape and remains trapped within the cell to undergo subsequent glycolysis steps or be used in the pentose phosphate pathway (PPP). In addition to the HK1 enzyme, which is constitutively expressed, tumor cells overexpress HK2.

It is described that the reduction of HK2, but not HK1, can restore OXPHOS and, with the increase in oxygen consumption, also the susceptibility to cell death mediated by radiation and chemotherapy (Mueckler & Thorens, 2013). The 3' untranslated region (UTR) of HK2 mRNA presents a binding site for miR-497. In osteosarcoma, HK2 is overexpressed while miR-497 is downregulated, possibly due to increased expression of the lncRNA PVT1 (plasmacytoma variant translocation 1), which acts as a molecular sponge by trapping miR-497 and thus contributing to the upregulation of expression of HK2 (Song et al., 2017). Additional HK2 regulators are miR-143 and miR-145, which are located together on chromosome 5q32 and are downregulated in many cancers (Iorio et al., 2005; Spizzo et al., 2010; Volinia et al., 2006; Zhang et al., 2013). The restoration of mature miR-143 and -145 through the use of mimics causes the suppression of the expression of the HK2 protein in various cancerous cell lines (Fang et al., 2012; Peschiaroli et al., 2013; Yoshino et al., 2013), making miR-143/145 important regulators of glycolysis in cancer. Other studies have revealed that HK2 is a direct target of miR-181b (Tao et al., 2017). The ectopic expression of miR-181b in gastric cancer cell lines represses glycolysis through HK2 targeting and has a negative effect on cell proliferation and viability (Li et al., 2016). Likewise, miR-98 suppresses glycolysis and cell proliferation in colon cancer cell lines by directly targeting HK2 (Zhu et al., 2017). In summary, the previously mentioned miRNAs are generally downregulated in cancer cells, thus favoring the overexpression of the enzyme HK2 and the consequent use of glucose-6-phosphate in the glycolytic pathway or in the PPP shunt by the cell.

The second key step of glycolysis is the conversion of F6P into F1.6P catalyzed by the enzyme phosphofructokinase (PFK1). It is a key step as the addition of the second phosphate group makes possible the subsequent phases of glycolysis. The presence of two charged groups not only makes the sugar to become energetically less stable but also prevents the subsequent triose phosphate products of glycolysis to escape from cell. TARDBP (TAT-activating regulatory DNA-binding protein), highly expressed in hepatocellular carcinoma (HCC), is reported to regulate glycolysis by inducing expression of the platelet isoform of PFK1 through the suppression of the miR-520a/b/e expression, which prevents their binding to the PFK1 3'-UTR and boosts the expression of the PFK1 protein (Park et al., 2013). F6P is also converted to F2,6P by PFK2, which allosterically activates PFK1.

The third key step in glycolysis is catalyzed by pyruvate kinase (PK), which converts phosphoenolpyruvate (PEP) into pyruvate. As described above, cancer cells employ various strategies to increase glycolysis. One of these, apparently puzzling, is linked to the attenuation of PK functionality. One of the most studied mechanisms leading to reduced PK functionality is given by the expression of a low affinity dimeric isoform, PKM2, instead of PKM1 (Spoden et al., 2009; Wong et al., 2008). PKM2 overexpression is particularly known in pancreatic cancers. Given its low enzymatic activity, PKM2 produces a slowdown in glycolysis, thus favoring the accumulation of intermediate products, which can be used for the synthesis of nucleotides and lipids (Dayton, Jacks, & Vander Heiden, 2016). A number of dysregulated miRNAs in cancer have been found to modulate PKM2 expression. For example, the PKM2 transcript has binding sites for miR-133a and -133b. In vitro studies have shown that miR-133a/b is able to target PKM2 and block protein expression. In particular, miR-133b can suppress the Warburg effect and induce autophagy in colon adenocarcinoma and gastric cancer cell lines by inhibiting the PKM2 isoform and shifting the balance toward the PKM1 isoform (Sugiyama et al., 2016; Taniguchi et al., 2016). Interestingly, miR-133a/b are downregulated in different cancers (Kano et al., 2010; Kawakami et al., 2012; Kojima et al., 2012), leading to a loss of inhibition and increased expression of the PKM2 isoform. MiR-326 also directly targets PKM2 and inhibits its expression. miR-326 levels are inversely correlated with PKM2 levels in glioma cells, suggesting an endogenous mechanism for PKM2 regulation (Kefas et al., 2010). The expression of liver-specific miR-122 is often reduced in HCC (Thakral & Ghoshal, 2015). The restoration of miR-122 expression prevents the development of HCC in mouse models (Hsu et al., 2012; Tsai et al., 2012). A possible mechanism underlying this effect is the reversal of aerobic glycolysis, as suggested by the decrease in lactate production and the increase in oxygen consumption observed in HCC cell lines. The reversal of the glycolytic switch can be explained by the direct targeting of PKM2 by miR-122 (Fong et al., 2015; Liu et al., 2014).

Branched pathways of glycolysis

As previously indicated, a relevant aspect of accelerated glycolysis is the fact that it provides an abundance of intermediates that can be used by anabolic pathways that branch out from the glycolytic pathway. In cancer cells, the increased functionality of HK and PFK and the reduced activity of PKM2 lead to the accumulation of glycolytic intermediates that favor anabolic processes, such as the pentose phosphate pathway (PPP) and serine biosynthesis, which stimulate synthesis of lipid, nucleotides, and amino acids to support the rapid growth of cancer cells (Balsa-Martinez & Puigserver, 2015; Koczula et al., 2016).

Moreover, the conversion of pyruvate to lactate is advantageous to cancer cells for a number of reasons. It is used to generate NAD+ necessary for maintaining the flow of energy in the branched pathways. The accumulation of lactate causes acidification of the tumor microenvironment, which facilitates tumor invasion processes. The secreted lactate can also be absorbed and recycled by stromal cells adjacent to the tumor to generate pyruvate that can be used for energy production (Feng et al., 2018; Fukumura et al., 2001; Hirschhaeuser, Sattler, & Mueller-Klieser, 2011).

The conversion of pyruvate to lactate is catalyzed by lactate dehydrogenase (LDH). One of the subunits that forms LDH isoenzymes, LDHA, plays an important role in cancer cells. LDHA was one of the first MYC-induced glycolytic genes to be identified (Dang, Le, & Gao, 2009; He et al., 2015). LDH expression levels correlate with patient survival in various tumor types (Petrelli et al., 2015; Wulaningsih et al., 2015). LDHA is known to be a direct target of miR-34a. The repression of LDHA by miR-34a inhibits glycolysis (Xiao et al., 2016). However, miR-34a is a known effector of the p53 protein and is therefore downregulated in tumors with dysfunctional p53 (Navarro & Lieberman, 2015). Thus, the loss of miR-34a and the parallel upregulation of LDHA can divert the use of pyruvate toward the production of lactate at the intersection of molecular pathways involved in cancer.

NcRNAs can influence glucose metabolism by regulating cancer-associated signaling pathways

The involvement of glucose metabolism reprogramming in tumorigenesis would suggest a wide presence of genetic alterations affecting genes coding for metabolic enzymes in cancer cells. Indeed, a number of mutations in these genes have been identified, such as in succinate dehydrogenase (SDH) or fumarate hydratase (FH) (Kaelin & Ratcliffe, 2008); in isocitrate dehydrogenase 1 or 2 (IDH1 or IDH2), which are common in gliomas, acute myelogenous leukemias (AMLs), and other types of cancer (Mardis et al., 2009; Yan et al., 2009); or the amplification and overexpression of PHGDH in a fraction of breast cancer and melanoma (Locasale et al., 2011; Possemato et al., 2011). These data on one side support the relevance of these metabolic enzymes in tumorigenesis, but on the other, overall mutations in these genes are infrequent. Conversely, it has been reported that the loss of tumor suppressors and the activation of oncogenes not only can promote cell proliferation but simultaneously cause activation of glycolysis. Examples include the activation of oncogenes (e.g., RAS or MYC), the inactivation of tumor suppressor genes (e.g., tumor protein TP53 or PTEN), or growth factor-mediated signal transduction (e.g., IGF1R or EGFR, or HER2 or HIF1) (Denko, 2008; Ward & Thompson, 2012), thus relieving the needs of genetic alterations targeting metabolic enzymes. Furthermore, several studies have indicated that the aforementioned alterations affect metabolites that are used by molecular pathways involved in the epigenetic regulation of gene expression, causing defects in the cell differentiation process rather than having effects on energy metabolism (Kaelin and McKnight, 2013).

PI3K/Akt/mTOR and HIF-1 signaling pathway

The PI3K/Akt/mTOR signaling pathway is one of the most frequently altered pathways in cancer. Physiologically, PI3K is activated by extracellular stimuli such as growth factors that activate receptor tyrosine kinases (RTKs) (Fig. 6.4).

In tumor cells, the pathway is constitutively activated through mutations affecting the catalytic subunit of PI3K, or PTEN, or activating mutations in RTKs or in RAS and mTOR proteins. Once activated, PI3K phosphorylates phosphatidylinositol-4,5-bisphosphate (PIP2) to produce phosphatidylinositol-3,4,5-triphosphate (PIP3), a key lipid messenger of the pathway. PTEN lipid phosphatase negatively regulates this pathway by converting PIP3 to PIP2 (Cantley & Neel, 1999). PIP3 can associate with the pleckstrin-homologous domain of AKT. Once attached to the plasma membrane, the AKT protein kinase is activated by its phosphorylation at Thr-308 and complete activation occurs by phosphorylation at Ser-472, executed by the mTOR (mTORC) 2 complex (Alessi et al., 1997; Saci, Cantley, & Carpenter, 2011). The oncogenic activation of PI3K/Akt/mTOR signaling contributes to the Warburg effect through (i) increased expression of GLUT; and (ii) increased expression of the glycolytic enzymes HK2 and PKM2.

Many miRNAs have been found to modulate PI3K/Akt/mTOR signaling (Lien, Lyssiotis, & Cantley, 2016) (Lien et al., 2016). These include, for example, miR-451, miR-199-3p, and miR-125a (Callegari et al., 2018; Guo et al., 2016; Jin et al., 2017; Tang, Li, Liang, Zhou, & Wang, 2015), often downregulated in cancer cells. MiRNAs of the Let-7 family are also downregulated in cancer and function as tumor suppressors by negatively regulating the RAS members (Roncarati, Lupini, Shankaraiah, & Negrini, 2019), linked to the PI3K, AKT, and mTOR pathways. In addition, numerous miRNAs overexpressed in tumor cells can inhibit the expression of PTEN, the negative regulator of the PI3K pathway. Examples include miR-17, -21, -214, -19, and -26 (Tay, Song, & Pandolfi, 2013).

Oncoproteins and tumor suppressors contribute to the Warburg effect through mechanisms that overlap the cellular response mediated by hypoxia. The increased activity of the PI3K/AKT pathway leads in turn to an increased expression of HIF-1α (Fig. 6.4). The HIF1 transcriptional complex consists of the oxygen-regulated HIF-1α subunit and the constitutively expressed HIF-1β subunit. The microenvironment of solid tumors is known to be oxygen-deficient. Under hypoxic

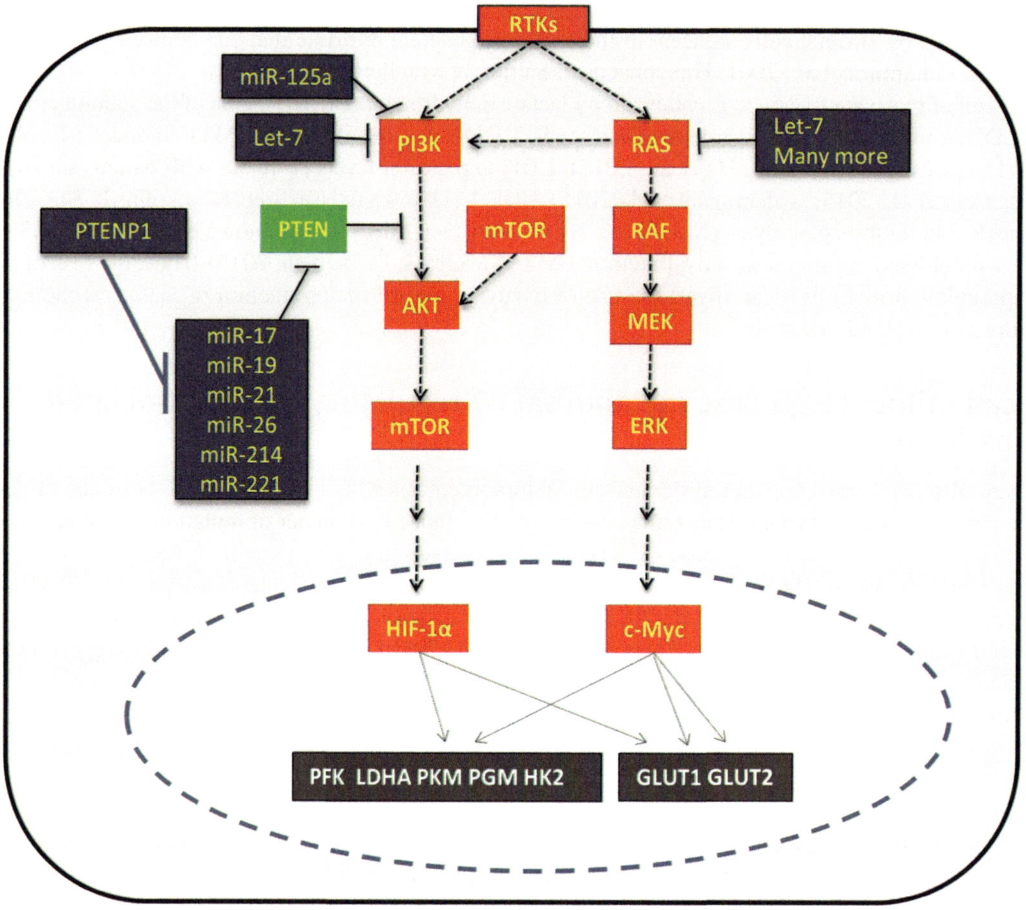

FIG. 6.4 miRNAs involved in the control of the MAPK and PI3K pathways. The final effectors of PI3K and MAPK pathways, HIF-1 and MYC, can control the expression of a number of glycolytic enzymes. Abbreviations are described in Table 6.2.

stress, HIF-1α dimerizes with HIF-1β to form the active HIF-1 complex, which regulates the transcription of various genes by binding to hypoxia response elements (Maxwell, Pugh, & Ratcliffe, 2001). The activation of the PI3K pathway can similarly activate HIF-1 activity. In turn, HIF-1 stimulates glycolysis thanks to (i) the upregulation of GLUT1 (Hayashi et al., 2004); (ii) the upregulation of glycolytic enzymes to increase glucose consumption (Iyer et al., 1998); and (iii) the upregulation of LDHA, which creates an acidic tumor microenvironment (Ryan, Lo, & Johnson, 1998). Under hypoxic stress, glucose uptake and lactate production are therefore increased following the upregulation of GLUT1 and LDHA.

p53 signaling pathway

The dysfunction of the tumor suppressor gene TP53 also promotes the Warburg effect. The best characterized functions of the p53 protein are related to cell cycle control and cell death. However, p53 is also implicated in the Warburg effect as it can modulate the transition between glycolysis and OXPHOS (Fig. 6.5).

MiR-34a, inducible by p53, has been shown to target lactate dehydrogenase (LDHA), which uses pyruvate to produce lactate. The dysfunctionality of p53 in the majority of neoplastic cells eliminates this control mechanism and leads to an overexpression of the LDHA enzyme, diverting pyruvate from its normal use for the production of acetyl CoA, used in the TCA cycle, toward the production of lactate. miR-34a is also capable of inhibiting HK2 expression. In the absence of this control, G6P is produced, which not only proceeds in the glycolytic pathway but is also the starting point of the PPP pathway, the importance of which has been discussed previously (Kim, Roe, Lee, Cho, & Youn, 2013). P53 induces the expression of miR-145, which promotes cell death activity (Spizzo et al., 2010). However, the expression of miR-483-3p is increased in conditions of elevated intracellular glucose and by the activity of the enzyme β-*N*-acetylglucosamine transferase (Lupini et al., 2016; Pepe et al., 2017). Thus, in the presence of elevated glucose, miR-483-3p is upregulated and cells are protected from the proapoptotic action of the p53/miR-145 axis.

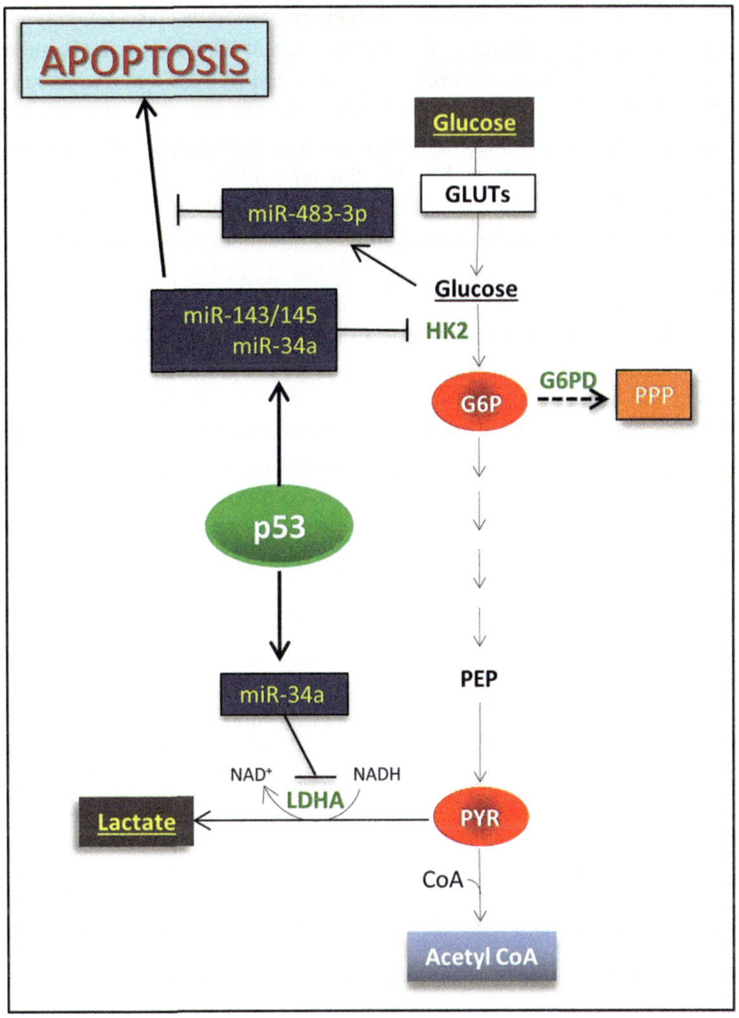

FIG. 6.5 p53 controls miRNAs involved in the modulation of the glycolytic metabolism *No Permission Required* Abbreviations are described in Table 6.2.

Conclusion and future perspectives

The high rate of aerobic glycolysis is beneficial to the proliferation of cancer cells. However, current knowledge in the field of cancer metabolism has not yet been fully translated into clinically useful applications. The high albeit inefficient rate of ATP production by glycolysis is not the main advantage for cancer cells. Instead, cells benefit from the high levels of intermediates, which are diverted to side pathways employed to generate nucleotides, amino acids, lipids, and NADPH (De Berardinis, Lum, Hatzivassiliou, & Thompson, 2008; Heiden, Matthew, Cantley, & Thompson, 2009). Identifying critical nodes in the network of pathways that regulate aerobic glycolysis may reveal new targets for anticancer therapies.

Among the various enzymes involved in the glycolytic cascade, GLUT-1, HK2, LDHA, and PKM2, which are known to be overexpressed in cancer, can represent potential targets (Bluemlein et al., 2011; Chan et al., 2011; Le et al., 2010; Wolf et al., 2011). However, many challenges remain to this approach. First, the critical pathways to be targeted must be ascertained. This can be difficult to achieve given the phenotypic and functional heterogeneity not only between different tumors, but also in individual tumors. Furthermore, the plasticity of cancer cells can cause them to modify their metabolic profile to adapt to new conditions. Second, even though enzyme inhibitors have high specificity, cells can recruit other glycolytic enzyme isoforms in response to inhibition of the cancer-associated isoform. Third, the toxic effects of the inhibition of glycolytic enzymes in cells of normal tissues must be clarified. In fact, the effectiveness could be limited by the fact that, in addition to cancer cells, aerobic glycolysis is also a key process in immune and stem cells.

Precision treatment strategies adapted to individual patients are becoming a reality. A personalized view of the mechanisms underlying metabolic processes in a patient can provide an additional layer of useful information. Advances in system biology, including the reconstruction of genomic-scale metabolic models (GEMs), may allow the systematic evaluation of metabolic profiles (Agren et al., 2014; Folger et al., 2011; Frezza et al., 2011). GEMs integrate omics data to provide a better understanding of metabolic aberrations, also revealing potential biomarkers and specific metabolites in different types of cancer (Björnson et al., 2015).

The role of ncRNAs in the regulation of cellular metabolism is well established, albeit incomplete (Table 6.1).

The aerobic glycolysis-related ncRNAs mentioned in this chapter could potentially represent diagnostic biomarkers of reprogrammed glucose metabolism in cancer. For example, lincRNA-p21 has been described as an independent prognostic marker for both time to relapse and cancer-specific survival in lung adenocarcinoma (Castellano et al., 2016), or mir-135b in fecal samples could be a strong diagnostic marker for CRC (Wu et al., 2014).

Targeting the molecules involved in the Warburg effect may be an effective therapeutic strategy for cancer treatment (Hamanaka & Chandel, 2012; Tennant, Durán, & Gottlieb, 2010). miRNA and lncRNA have multiple physiological targets, which include enzymes and signaling pathway components associated with aerobic glycolysis. Several preclinical studies have demonstrated the effects of targeting glucose metabolism reprogramming on the survival and proliferation of cancer cells. With technological and computational advances, miRNA mimics, anti-miRNA oligonucleotides that modulate glycolysis, and improvements in methods for in vivo delivery, ncRNAs can be used to make cancer cells more susceptible to existing anticancer therapies. A combination of precision medicine able to target activated oncogenes together with the targeting of the Warburg effect could represent a potentially effective anticancer strategy. See Table 6.2.

TABLE 6.1 miRNAs known to regulate glucose metabolism.

	Target	ncRNA	Reference
Glucose uptake	GLUT1	miR-1291	(Yamasaki et al., 2013)
Glycolytic enzymes	HK2	miR-497/lncRNA PVT1	(Song et al., 2017).
		miR-143-145 cluster	(Fang et al., 2012; Peschiaroli et al., 2013; Yoshino et al., 2013)
		miR-181b	(Li et al., 2016; Tao et al., 2017)
		miR-98	(Zhu et al., 2017)
	PFK	miR-520s	(Park et al., 2013)
	PKM2	miR-133a/b	(Sugiyama et al., 2016; Taniguchi et al., 2016)
		miR-326	(Kefas et al., 2010)
		miR-122	(Fong et al., 2015; Liu et al., 2014)
	LDHA	miR-34a	(Xiao et al., 2016)
MAPK and PI3K pathways	RAS	let-7	(Roncarati et al., 2019)
		several others	
	PI3K	miR-125a	(Jin et al., 2017; Tang et al., 2015),
		let-7	(Fang et al., 2012)
		miR-451	(Guo et al., 2016)
	PTEN	miR-21, miR-221	(Roncarati et al., 2019)
		miR-17, miR-21, miR-214, miR-19, miR-26	(Tay et al., 2013)
	MTOR	miR-199	(Callegari et al., 2018)
p53 pathway	TP53	miR-34a	(Spizzo et al., 2010)

TABLE 6.2 Abbreviations.

1,3BPG	1,3-Bisphosphoglycerate
18F-FDG	2-deoxy-2-fluorine-18fluoro-D-glucose
2PG	2-Phosphoglycerate
3PG	3-Phosphoglycerate
AKT	Ak strain transforming
ALDO	Fructose-bisphosphate aldolase
ATP	Adenosine triphosphate
BC	Bladder cancer
CAFs	Cancer-associated fibroblasts
CoA	Coenzyme A
COX	Cytochrome c oxidase
CT	Computed tomography
DHAP	Dihydroxyacetone phosphate
EGFR	Epidermal growth factor receptor
ENO	Enolase
ERK	Extracellular-signal-regulated kinase
F1,6P	Fructose 1,6-phosphate
F2,6P	Fructose 2,6-phosphate
F6P	Fructose 6-phosphate
G6P	α-D-Glucose-6-phosphate
G6PD	Glucose-6-phosphate dehydrogenase
GADP	Glyceraldehyde 3-phosphate
GAPDH	Glyceraldehyde phosphate dehydrogenase
GBM	Glioblastoma multiforme
GLUT	Glucose transporter
HCC	Hepatocellular carcinoma
HER2	Human epidermal growth factor receptor 2
HIF1	Hypoxia-inducible factor 1
HK	Hexokinase
HMIT	H+/myo-inositol transporter
HOTAIR	HOX transcript antisense RNA
IGFR	Insulin-like growth factor receptor
LDH	Lactate dehydrogenase
LDHA	Lactate dehydrogenase alpha subunit
lncRNA	Long noncoding RNA
MEG3	Maternally expressed gene 3
MEK	MAPK/ERK kinase
miRNA	MicroRNA

Continued

TABLE 6.2 Abbreviations—cont'd

mRNA	Messenger RNA
mTOR	Mammalian target of rapamycin
mTORC1	Mammalian target of rapamycin complex 1
mTORC2	Mammalian target of rapamycin complex 2
MYC	V-Myc avian myelocytomatosis viral oncogene homolog
NADPH	Nicotinamide adenine dinucleotide phosphate
ncRNA	Noncoding RNA
OS	Osteosarcoma
OXPHOS	Oxidative phosphorylation
PDPK1	Phosphoinositide-dependent protein kinase 1
PEP	Phosphoenolpyruvate
PET	Positron emission tomography
PFK1	Phosphofructokinase 1
PFK-2/FBPase-2	Phosphofructokinase 2/fructose-2,6-bisphosphatase
PFKFB	Phosphofructokinase 2-F2,6Bpase
PGI	Phosphoglucose isomerase
PGK	Phosphoglycerate kinase
PGM	Phosphoglycerate mutase
PI3K	Phosphoinositide 3-kinase
PI3K	Phospho inositide 3-phosphate kinase
PIP2	Phosphatidylinositol-4,5-biphosphate
PIP3	Phosphatidylinositol-3,4,5-triphosphate
piRNA	PIWI-interacting RNA
PK	Pyruvate kinase
PKM1	Pyruvate kinase M1
PKM2	Pyruvate kinase M2
PPP	Pentose phosphate pathway
PTEN	Phosphatase and tensin homolog
PTENP1	PTEN pseudogene
PVT1	Pvt1 oncogene (Nonprotein coding)
PYR	Pyruvate
RAF	Rapidly accelerated fibrosarcoma
RAS	Rat sarcoma viral oncogene
RCC	Renal cell carcinoma
ROS	Reactive oxygen species
RTK	Receptor tyrosine kinase
SCO2	Cytochrome c oxidase 2
siRNA	Small interfering RNA
snoRNA	Small nucleolar RNA

TABLE 6.2 Abbreviations—cont'd

STAT3	Signal transducer and activator pf transcription 3
TCA cycle	Tricarboxylic acid cycle
TIGAR	TP53-induced glycolysis and apoptosis regulator gene
tiRNA	Transcription initiation RNA
TP53	Tumor protein P53
TPI	Triosephosphate isomerase
UCA1/CUDR	Urothelial cancer-associated 1
WRAP53	WD repeat containing antisense of p53

References

Agren, R., Mardinoglu, A., Asplund, A., Kampf, C., Uhlen, M., & Nielsen, J. (2014). Identification of anticancer drugs for hepatocellular carcinoma through personalized genome-scale metabolic modeling. *Molecular Systems Biology*, *10*(3). https://doi.org/10.1002/msb.145122.

Alessi, D. R., James, S. R., Downes, C. P., Holmes, A. B., Gaffney, P. R. J., Reese, C. B., et al. (1997). Characterization of a 3-phosphoinositide-dependent protein kinase which phosphorylates and activates protein kinase Bα. *Current Biology*, *7*(4), 261–269. https://doi.org/10.1016/s0960-9822(06)00122-9.

Avril, N., Rosé, C. A., Schelling, M., Dose, J., Kuhn, W., Bense, S., et al. (2000). Breast imaging with positron emission tomography and fluorine-18 fluorodeoxyglucose: Use and limitations. *Journal of Clinical Oncology*, *18*(20), 3495–3502. https://doi.org/10.1200/JCO.2000.18.20.3495.

Balsa-Martinez, E., & Puigserver, P. (2015). Cancer cells hijack gluconeogenic enzymes to fuel cell growth. *Molecular Cell*, *60*(4), 509–511. https://doi.org/10.1016/j.molcel.2015.11.005.

Björnson, E., Mukhopadhyay, B., Asplund, A., Pristovsek, N., Cinar, R., Romeo, S., et al. (2015). Stratification of hepatocellular carcinoma patients based on acetate utilization. *Cell Reports*, *13*(9), 2014–2026. https://doi.org/10.1016/j.celrep.2015.10.045.

Bluemlein, K., Grüning, N. M., Feichtinger, R. G., Lehrach, H., Kofler, B., & Ralser, M. (2011). No evidence for a shift in pyruvate kinase PKM1 to PKM2 expression during tumorigenesis. *Oncotarget*, *2*(5), 393–400. https://doi.org/10.18632/oncotarget.278.

Callegari, E., D'Abundo, L., Guerriero, P., Simioni, C., Elamin, B. K., Russo, M., et al. (2018). miR-199a-3p modulates MTOR and PAK4 pathways and inhibits tumor growth in a hepatocellular carcinoma transgenic mouse model. *Molecular Therapy - Nucleic Acids*, *11*, 485–493. https://doi.org/10.1016/j.omtn.2018.04.002.

Cantley, L. C., & Neel, B. G. (1999). New insights into tumor suppression: PTEN suppresses tumor formation by restraining the phosphoinositide 3-kinase/AKT pathway. *Proceedings of the National Academy of Sciences of the United States of America*, *96*(8), 4240–4245. https://doi.org/10.1073/pnas.96.8.4240.

Castellano, J. J., Navarro, A., Viñolas, N., Marrades, R. M., Moises, J., Cordeiro, A., et al. (2016). LincRNA-p21 impacts prognosis in resected non-small cell lung cancer patients through angiogenesis regulation. *Journal of Thoracic Oncology*, *11*(12), 2173–2182. https://doi.org/10.1016/j.jtho.2016.07.015.

Chan, D. A., Sutphin, P. D., Nguyen, P., Turcotte, S., Lai, E. W., Banh, A., et al. (2011). Targeting GLUT1 and the Warburg effect in renal cell carcinoma by chemical synthetic lethality. *Science Translational Medicine*, *3*(94). https://doi.org/10.1126/scitranslmed.3002394, 94ra70.

Dang, C. V., Le, A., & Gao, P. (2009). MYC-induced cancer cell energy metabolism and therapeutic opportunities. *Clinical Cancer Research: An Official Journal of the American Association for Cancer Research*, *15*(21), 6479–6483. https://doi.org/10.1158/1078-0432.CCR-09-0889.

Dayton, T. L., Jacks, T., & Vander Heiden, M. G. (2016). PKM2, cancer metabolism, and the road ahead. *EMBO Reports*, *17*(12), 1721–1730. https://doi.org/10.15252/embr.201643300.

De Berardinis, R. J., & Chandel, N. S. (2016). Fundamentals of cancer metabolism. *Science Advances*, *2*(5). https://doi.org/10.1126/sciadv.1600200, e1600200.

De Berardinis, R. J., Lum, J. J., Hatzivassiliou, G., & Thompson, C. B. (2008). The biology of cancer: Metabolic reprogramming fuels cell growth and proliferation. *Cell Metabolism*, *7*(1), 11–20. https://doi.org/10.1016/j.cmet.2007.10.002.

Denko, N. C. (2008). Hypoxia, HIF1 and glucose metabolism in the solid tumour. *Nature Reviews Cancer*, *8*(9), 705–713. https://doi.org/10.1038/nrc2468.

Fang, R., Xiao, T., Fang, Z., Sun, Y., Li, F., Gao, Y., et al. (2012). MicroRNA-143 (miR-143) regulates cancer glycolysis via targeting hexokinase 2 gene. *Journal of Biological Chemistry*, *287*(27), 23227–23235. https://doi.org/10.1074/jbc.M112.373084.

Feng, Y., Xiong, Y., Qiao, T., Li, X., Jia, L., & Han, Y. (2018). Lactate dehydrogenase A: A key player in carcinogenesis and potential target in cancer therapy. *Cancer Medicine*, *7*(12), 6124–6136. https://doi.org/10.1002/cam4.1820.

Folger, O., Jerby, L., Frezza, C., Gottlieb, E., Ruppin, E., & Shlomi, T. (2011). Predicting selective drug targets in cancer through metabolic networks. *Molecular Systems Biology*, *7*. https://doi.org/10.1038/msb.2011.35.

Fong, M. Y., Zhou, W., Liu, L., Alontaga, A. Y., Chandra, M., Ashby, J., et al. (2015). Breast-cancer-secreted miR-122 reprograms glucose metabolism in premetastatic niche to promote metastasis. *Nature Cell Biology*, *17*(2), 183–194. https://doi.org/10.1038/ncb3094.

Frezza, C., Zheng, L., Folger, O., Rajagopalan, K. N., MacKenzie, E. D., Jerby, L., et al. (2011). Haem oxygenase is synthetically lethal with the tumour suppressor fumarate hydratase. *Nature*, *477*(7363), 225–228. https://doi.org/10.1038/nature10363.

Fukumura, D., Xu, L., Chen, Y., Gohongi, T., Seed, B., & Jain, R. K. (2001). Hypoxia and acidosis independently up-regulate vascular endothelial growth factor transcription in brain tumors in vivo. *Cancer Research*, *61*(16), 6020–6024.

Guo, H., Nan, Y., Zhen, Y., Zhang, Y., Guo, L., Yu, K., et al. (2016). miRNA-451 inhibits glioma cell proliferation and invasion by downregulating glucose transporter 1. *Tumor Biology*, *37*(10), 13751–13761. https://doi.org/10.1007/s13277-016-5219-3.

Hamanaka, R. B., & Chandel, N. S. (2012). Targeting glucose metabolism for cancer therapy. *Journal of Experimental Medicine*, *209*(2), 211–215. https://doi.org/10.1084/jem.20120162.

Hanahan, D., & Weinberg, R. A. (2011). Hallmarks of cancer: The next generation. *Cell*, 646–674. https://doi.org/10.1016/j.cell.2011.02.013.

Hayashi, M., Sakata, M., Takeda, T., Yamamoto, T., Okamoto, Y., Sawada, K., et al. (2004). Induction of glucose transporter 1 expression through hypoxia-inducible factor 1α under hypoxic conditions in trophoblast-derived cells. *Journal of Endocrinology*, *183*(1), 145–154. https://doi.org/10.1677/joe.1.05599.

He, T. L., Zhang, Y. J., Jiang, H., Li, X. H., Zhu, H., & Zheng, K. L. (2015). The c-Myc-LDHA axis positively regulates aerobic glycolysis and promotes tumor progression in pancreatic cancer. *Medical Oncology*, *32*.

Heiden, V., Matthew, G., Cantley, L. C., & Thompson, C. B. (2009). Understanding the Warburg effect: The metabolic requirements of cell proliferation. *Science (New York, N.Y.)*, *324*(5930), 1029–1033. https://doi.org/10.1126/science.1160809.

Hirschhaeuser, F., Sattler, U. G. A., & Mueller-Klieser, W. (2011). Lactate: A metabolic key player in cancer. *Cancer Research*, *71*(22), 6921–6925. https://doi.org/10.1158/0008-5472.CAN-11-1457.

Hsu, S. H., Wang, B., Kota, J., Yu, J., Costinean, S., Kutay, H., et al. (2012). Essential metabolic, anti-inflammatory, and anti-tumorigenic functions of miR-122 in liver. *Journal of Clinical Investigation*, *122*(8), 2871–2883. https://doi.org/10.1172/JCI63539.

Iorio, M. V., Ferracin, M., Liu, C.-G., Veronese, A., Spizzo, R., Sabbioni, S., et al. (2005). MicroRNA gene expression deregulation in human breast cancer. *Cancer Research*, *65*(16), 7065–7070.

Iyer, N. V., Kotch, L. E., Agani, F., Leung, S. W., Laughner, E., Wenger, R. H., et al. (1998). Cellular and developmental control of O2 homeostasis by hypoxia-inducible factor 1 alpha. *Genes & Development*, *12*(2), 149–162.

Jin, F., Wang, Y., Zhu, Y., Li, S., Liu, Y., Chen, C., et al. (2017). The MIR-125a/HK2 axis regulates cancer cell energy metabolism reprogramming in hepatocellular carcinoma. *Scientific Reports*, *7*(1). https://doi.org/10.1038/s41598-017-03407-3.

Kaelin, W. G., & McKnight, S. L. (2013). Influence of metabolism on epigenetics and disease. *Cell*, *153*(1), 56–69. https://doi.org/10.1016/j.cell.2013.03.004.

Kaelin, W. G., & Ratcliffe, P. J. (2008). Oxygen sensing by metazoans: The central role of the HIF hydroxylase pathway. *Molecular Cell*, *30*(4), 393–402. https://doi.org/10.1016/j.molcel.2008.04.009.

Kano, M., Seki, N., Kikkawa, N., Fujimura, L., Hoshino, I., Akutsu, Y., et al. (2010). MiR-145, miR-133a and miR-133b: Tumor-suppressive miRNAs target FSCN1 in esophageal squamous cell carcinoma. *International Journal of Cancer*, *127*(12), 2804–2814. https://doi.org/10.1002/ijc.25284.

Kawakami, K., Enokida, H., Chiyomaru, T., Tatarano, S., Yoshino, H., Kagara, I., et al. (2012). The functional significance of miR-1 and miR-133a in renal cell carcinoma. *European Journal of Cancer*, *48*(6), 827–836. https://doi.org/10.1016/j.ejca.2011.06.030.

Kefas, B., Comeau, L., Erdle, N., Montgomery, E., Amos, S., & Purow, B. (2010). Pyruvate kinase M2 is a target of the tumorsuppressive microRNA-326 and regulates the survival of glioma cells. *Neuro-Oncology*, *12*(11), 1102–1112. https://doi.org/10.1093/neuonc/noq080.

Kim, H. R., Roe, J. S., Lee, J. E., Cho, E. J., & Youn, H. D. (2013). P53 regulates glucose metabolism by miR-34a. *Biochemical and Biophysical Research Communications*, *437*(2), 225–231. https://doi.org/10.1016/j.bbrc.2013.06.043.

Koczula, K. M., Ludwig, C., Hayden, R., Cronin, L., Pratt, G., Parry, H., et al. (2016). Metabolic plasticity in CLL: Adaptation to the hypoxic niche. *Leukemia*, *30*(1), 65–73. https://doi.org/10.1038/leu.2015.187.

Kojima, S., Chiyomaru, T., Kawakami, K., Yoshino, H., Enokida, H., Nohata, N., et al. (2012). Tumour suppressors miR-1 and miR-133a target the oncogenic function of purine nucleoside phosphorylase (PNP) in prostate cancer. *British Journal of Cancer*, *106*(2), 405–413. https://doi.org/10.1038/bjc.2011.462.

Le, A., Cooper, C. R., Gouw, A. M., Dinavahi, R., Maitra, A., Deck, L. M., et al. (2010). Inhibition of lactate dehydrogenase A induces oxidative stress and inhibits tumor progression. *Proceedings of the National Academy of Sciences of the United States of America*, *107*(5), 2037–2042. https://doi.org/10.1073/pnas.0914433107.

Li, L. Q., Yang, Y., Chen, H., Zhang, L., Pan, D., & Xie, W. J. (2016). MicroRNA-181b inhibits glycolysis in gastric cancer cells via targeting hexokinase 2 gene. *Cancer Biomarkers*, *17*(1), 75–81. https://doi.org/10.3233/CBM-160619.

Lien, E. C., Lyssiotis, C. A., & Cantley, L. C. (2016). Metabolic reprogramming by the PI3K-Akt-mTOR pathway in cancer. *Recent Results in Cancer Research. Fortschritte Der Krebsforschung. Progres Dans Les Recherches Sur Le Cancer*, *207*, 39–72. https://doi.org/10.1007/978-3-319-42118-6_3.

Liu, A. M., Xu, Z., Shek, F. H., Wong, K. F., Lee, N. P., Poon, R. T., et al. (2014). MiR-122 targets pyruvate kinase M2 and affects metabolism of hepatocellular carcinoma. *PLoS One*, *9*(1). https://doi.org/10.1371/journal.pone.0086872.

Locasale, J. W., Grassian, A. R., Melman, T., Lyssiotis, C. A., Mattaini, K. R., Bass, A. J., et al. (2011). Phosphoglycerate dehydrogenase diverts glycolytic flux and contributes to oncogenesis. *Nature Genetics*, *43*(9), 869–874. https://doi.org/10.1038/ng.890.

Lupini, L., Pepe, F., Ferracin, M., Braconi, C., Callegari, E., Pagotto, S., et al. (2016). Over-expression of the miR-483-3p overcomes the miR-145/TP53 pro-apoptotic loop in hepatocellular carcinoma. *Oncotarget*, *7*(21), 31361–31371. https://doi.org/10.18632/oncotarget.8913.

Mardis, E. R., Ding, L., Dooling, D. J., Larson, D. E., McLellan, M. D., Chen, K., et al. (2009). Recurring mutations found by sequencing an acute myeloid leukemia genome. *New England Journal of Medicine*, *361*(11), 1058–1066. https://doi.org/10.1056/NEJMoa0903840.

Maxwell, P. H., Pugh, C. W., & Ratcliffe, P. J. (2001). Activation of the HIF pathway in cancer. *Current Opinion in Genetics and Development*, *11*(3), 293–299. https://doi.org/10.1016/S0959-437X(00)00193-3.

Mueckler, M., & Thorens, B. (2013). The SLC2 (GLUT) family of membrane transporters. *Molecular Aspects of Medicine*, *34*(2–3), 121–138. https://doi.org/10.1016/j.mam.2012.07.001.

Navarro, F., & Lieberman, J. (2015). miR-34 and p53: New insights into a complex functional relationship. *PLoS One*, *10*(7). https://doi.org/10.1371/journal.pone.0132767.

Park, Y. Y., Kim, S. B., Han, H. D., Sohn, B. H., Kim, J. H., Liang, J., et al. (2013). Tat-activating regulatory DNA-binding protein regulates glycolysis in hepatocellular carcinoma by regulating the platelet isoform of phosphofructokinase through microRNA 520. *Hepatology*, *58*(1), 182–191. https://doi.org/10.1002/hep.26310.

Pavlova, N. N., & Thompson, C. B. (2016). The emerging hallmarks of cancer metabolism. *Cell Metabolism*, *23*(1), 27–47. https://doi.org/10.1016/j.cmet.2015.12.006.

Pepe, F., Pagotto, S., Soliman, S., Rossi, C., Lanuti, P., Braconi, C., et al. (2017). Regulation of miR-483-3p by the O-linked N-acetylglucosamine transferase links chemosensitivity to glucose metabolism in liver cancer cells. *Oncogenesis*, *6*(5). https://doi.org/10.1038/oncsis.2017.35.

Peschiaroli, A., Giacobbe, A., Formosa, A., Markert, K. E., Bongiorno-Borbone, L., Levine, A. J., et al. (2013). miR-143 regulates hexokinase 2 expression in cancer cells. *Oncogene*, *32*, 797–802. https://doi.org/10.1038/onc.2012.100.

Petrelli, F., Cabiddu, M., Coinu, A., Borgonovo, K., Ghilardi, M., Lonati, V., et al. (2015). Prognostic role of lactate dehydrogenase in solid tumors: A systematic review and meta-analysis of 76 studies. *Acta Oncologica*, *54*(7), 961–970. https://doi.org/10.3109/0284186X.2015.1043026.

Possemato, R., Marks, K. M., Shaul, Y. D., Pacold, M. E., Kim, D., Birsoy, K., et al. (2011). Functional genomics reveal that the serine synthesis pathway is essential in breast cancer. *Nature*, *476*(7360), 346–350. https://doi.org/10.1038/nature10350.

Roncarati, R., Lupini, L., Shankaraiah, R. C., & Negrini, M. (2019). The importance of microRNAs in RAS oncogenic activation in human cancer. *Frontiers in Oncology*, *9*, 988. https://doi.org/10.3389/fonc.2019.00988.

Ryan, H. E., Lo, J., & Johnson, R. S. (1998). HIF-1α is required for solid tumor formation and embryonic vascularization. *EMBO Journal*, *17*(11), 3005–3015. https://doi.org/10.1093/emboj/17.11.3005.

Saci, A., Cantley, L. C., & Carpenter, C. L. (2011). Rac1 regulates the activity of mTORC1 and mTORC2 and controls cellular size. *Molecular Cell*, *42*(1), 50–61. https://doi.org/10.1016/j.molcel.2011.03.017.

Song, J., Wu, X., Liu, F., Li, M., Sun, Y., Wang, Y., et al. (2017). Long non-coding RNA PVT1 promotes glycolysis and tumor progression by regulating miR-497/HK2 axis in osteosarcoma. *Biochemical and Biophysical Research Communications*, *490*(2), 217–224. https://doi.org/10.1016/j.bbrc.2017.06.024.

Spizzo, R., Nicoloso, M. S., Lupini, L., Lu, Y., Fogarty, J., Rossi, S., et al. (2010). MiR-145 participates with TP53 in a death-promoting regulatory loop and targets estrogen receptor-α in human breast cancer cells. *Cell Death and Differentiation*, *17*(2), 246–254. https://doi.org/10.1038/cdd.2009.117.

Spoden, G. A., Rostek, U., Lechner, S., Mitterberger, M., Mazurek, S., & Zwerschke, W. (2009). Pyruvate kinase isoenzyme M2 is a glycolytic sensor differentially regulating cell proliferation, cell size and apoptotic cell death dependent on glucose supply. *Experimental Cell Research*, *315*(16), 2765–2774. https://doi.org/10.1016/j.yexcr.2009.06.024.

Sugiyama, T., Taniguchi, K., Matsuhashi, N., Tajirika, T., Futamura, M., Takai, T., et al. (2016). MiR-133b inhibits growth of human gastric cancer cells by silencing pyruvate kinase muscle-splicer polypyrimidine tract-binding protein 1. *Cancer Science*, *107*(12), 1767–1775. https://doi.org/10.1111/cas.13091.

Tang, H., Li, R. P., Liang, P., Zhou, Y. L., & Wang, G. W. (2015). miR-125a inhibits the migration and invasion of liver cancer cells via suppression of the PI3K/AKT/mTOR signaling pathway. *Oncology Letters*, *10*(2), 681–686. https://doi.org/10.3892/ol.2015.3264.

Taniguchi, K., Sakai, M., Sugito, N., Kumazaki, M., Shinohara, H., Yamada, N., et al. (2016). PTBP1-associated microRNA-1 and -133b suppress the Warburg effect in colorectal tumors. *Oncotarget*, *7*(14), 18940–18952. https://doi.org/10.18632/oncotarget.8005.

Tao, T., Chen, M., Jiang, R., Guan, H., Huang, Y., Su, H., et al. (2017). Involvement of EZH2 in aerobic glycolysis of prostate cancer through miR-181b/HK2 axis. *Oncology Reports*, *37*(3), 1430–1436. https://doi.org/10.3892/or.2017.5430.

Tay, Y., Song, S. J., & Pandolfi, P. P. (2013). The Lilliputians and the giant: An emerging oncogenic microRNA network that suppresses the PTEN tumor suppressor in vivo. *MicroRNA (Shariqah, United Arab Emirates)*, *2*(2), 127–136. https://doi.org/10.2174/22115366113029990017.

Tennant, D. A., Durán, R. V., & Gottlieb, E. (2010). Targeting metabolic transformation for cancer therapy. *Nature Reviews Cancer*, *10*(4), 267–277. https://doi.org/10.1038/nrc2817.

Thakral, S., & Ghoshal, K. (2015). miR-122 is a unique molecule with great potential in diagnosis, prognosis of liver disease, and therapy both as miRNA mimic and antimir. *Current Gene Therapy*, *15*(2), 142–150. https://doi.org/10.2174/1566523214666141224095610.

Tsai, W. C., Hsu, S. D., Hsu, C. S., Lai, T. C., Chen, S. J., Shen, R., et al. (2012). MicroRNA-122 plays a critical role in liver homeostasis and hepatocarcinogenesis. *Journal of Clinical Investigation*, *122*(8), 2884–2897. https://doi.org/10.1172/JCI63455.

Volinia, S., Calin, G. A., Liu, C. G., Ambs, S., Cimmino, A., Petrocca, F., et al. (2006). A microRNA expression signature of human solid tumors defines cancer gene targets. *Proceedings of the National Academy of Sciences of the United States of America*, *103*(7), 2257–2261. https://doi.org/10.1073/pnas.0510565103.

Warburg, O., Wind, F., & Negelein, E. (1927). The metabolism of tumors in the body. *Journal of General Physiology*, *8*(6), 519–530. https://doi.org/10.1085/jgp.8.6.519.

Ward, P. S., & Thompson, C. B. (2012). Signaling in control of cell growth and metabolism. *Cold Spring Harbor Perspectives in Biology*, *4*(7), 1–15. https://doi.org/10.1101/cshperspect.a006783.

Wolf, A., Agnihotri, S., Micallef, J., Mukherjee, J., Sabha, N., Cairns, R., et al. (2011). Hexokinase 2 is a key mediator of aerobic glycolysis and promotes tumor growth in human glioblastoma multiforme. *Journal of Experimental Medicine*, *208*(2), 313–326. https://doi.org/10.1084/jem.20101470.

Wong, T. S., Liu, X. B., Ho, A. C. W., Yuen, A. P. W., Ng, R. W. M., & Wei, W. I. (2008). Identification of pyruvate kinase type M2 as potential oncoprotein in squamous cell carcinoma of tongue through microRNA profiling. *International Journal of Cancer*, *123*(2), 251–257. https://doi.org/10.1002/ijc.23583.

Wu, C. W., Ng, S. C., Dong, Y., Tian, L., Ng, S. S. M., Leung, W. W., et al. (2014). Identification of microrna-135b in stool as a potential noninvasive biomarker for colorectal cancer and adenoma. *Clinical Cancer Research*, *20*(11), 2994–3002. https://doi.org/10.1158/1078-0432.CCR-13-1750.

Wulaningsih, W., Holmberg, L., Garmo, H., Malmstrom, H., Lambe, M., Hammar, N., et al. (2015). Serum lactate dehydrogenase and survival following cancer diagnosis. *British Journal of Cancer*, *113*(9), 1389–1396. https://doi.org/10.1038/bjc.2015.361.

Xiao, X., Huang, X., Ye, F., Chen, B., Song, C., Wen, J., et al. (2016). The MIR-34a-LDHA axis regulates glucose metabolism and tumor growth in breast cancer. *Scientific Reports*, *6*. https://doi.org/10.1038/srep21735.

Yamasaki, T., Seki, N., Yoshino, H., Itesako, T., Yamada, Y., Tatarano, S., et al. (2013). Tumor-suppressive microRNA-1291 directly regulates glucose transporter 1 in renal cell carcinoma. *Cancer Science*, *104*(11), 1411–1419. https://doi.org/10.1111/cas.12240.

Yan, H., Parsons, D. W., Jin, G., McLendon, R., Rasheed, B. A., Yuan, W., et al. (2009). IDH1 and IDH2 mutations in gliomas. *New England Journal of Medicine*, *360*(8), 765–773. https://doi.org/10.1056/NEJMoa0808710.

Yoshino, H., Enokida, H., Itesako, T., Kojima, S., Kinoshita, T., Tatarano, S., et al. (2013). Tumor-suppressive microRNA-143/145 cluster targets hexokinase-2 in renal cell carcinoma. *Cancer Science*, *104*(12), 1567–1574. https://doi.org/10.1111/cas.12280.

Zhang, J., Sun, Q., Zhang, Z., Ge, S., Han, Z. G., & Chen, W. T. (2013). Loss of microRNA-143/145 disturbs cellular growth and apoptosis of human epithelial cancers by impairing the MDM2-p53 feedback loop. *Oncogene*, *32*(1), 61–69. https://doi.org/10.1038/onc.2012.28.

Zhu, W., Huang, Y., Pan, Q., Xiang, P., Xie, N., & Yu, H. (2017). MicroRNA-98 suppress Warburg effect by targeting HK2 in colon cancer cells. *Digestive Diseases and Sciences*, *62*(3), 660–668. https://doi.org/10.1007/s10620-016-4418-5.

Chapter 7

microRNAs in inflammation processes

Dharmendra Kumar Soni and Roopa Biswas

Department of Anatomy, Physiology and Genetics, School of Medicine, Uniformed Services University of the Health Sciences, Bethesda, MD, United States

Introduction

Since their discovery in 1993, microRNAs (miRNAs, miRs) have gained notable prominence due to their decisive regulatory function in gene expression at the posttranscriptional level. Lee and colleagues discovered a small RNA, lin-4 (the first identified miRNA) and reported that at the posttranscriptional level, lin-4 downregulated the expression of LIN-14, which has a role in the progression of larval stages of *Caenorhabditis elegans*. Later in 2000, let-7 (the second identified miRNA) was identified in *C. elegans*, which also has a role in the progression of larval stages. Subsequently, numerous studies on miRNA-based posttranscriptional regulation mechanism have been reported in many other organisms, including humans.

MiRNAs are small (~ 19–22 nucleotides) single-stranded noncoding RNAs that are highly conserved across species and present in nearly all biological tissues. The association of miRNA with target messenger RNA (mRNA) results in mRNA degradation or translation inhibition. In addition, one single miRNA can interact with hundreds of mRNAs and regulate their expression. In the human genome, 239 different miRNA families have been identified that express approximately 2654 miRNAs. Moreover, miRNAs have been suggested in the regulation of more than 50% of the protein-coding genome. MiRNAs play pivotal roles in a wide range of biological processes, including cell death, differentiation, proliferation, metabolism, apoptosis, stress response, immune response, and inflammatory response. Dysregulation of specific miRNAs is concomitant with immune and inflammatory responses, leading to various human diseases including inflammatory diseases.

Inflammatory diseases mainly arise because of dysregulated innate and adoptive immune responses and inflammation. Usually, inflammation is a highly regulated and sequential signaling cascade of host defense responses to infections and injuries, which is self-limiting and takes a few hours to days in maintaining homeostasis. However, in some circumstances, inflammation persists for a longer period of time and the host defense system senses a consistent state of alert, which misguides the immune response. As a result, the immune system treats its own cells or tissues as harmful substances and damages them, which leads to the initiation, progression, and development of various inflammatory diseases such as allergic, respiratory, cardiovascular, autoimmune, cancer, and other diseases.

The increasing evidence of the decisive role of miRNA in inflammatory responses opens a new paradigm for the regulation of genes and pathways implicated in inflammatory diseases. In this chapter, we first provide an overview of miRNA biogenesis and functions and then discuss briefly inflammation and the mechanism of activation. Further, we present some of the most-studied inflammatory diseases that are associated with dysregulated miRNAs with discussion of using miRNAs as biomarkers in diagnosis and targets for the prevention or treatment of inflammatory diseases.

MiRNA biogenesis and functions

MiRNAs are present throughout the human genome through intronic, exonic, or intergenic regions. The biogenesis and maturation of miRNA are multistep processes, as illustrated in Fig. 7.1. Fig. 7.1 Briefly, *step 1: transcription of primary miRNA*—the synthesis of miRNA starts with the transcription of primary miRNA (pri-miRNA) from the DNA sequence (gene encoding miRNA) by RNA polymerase II (Pol II) within the nucleus. It has a stemloop- or hairpin-like structure and the lengths of transcripts vary from several hundred to a few thousand nucleotides. *Step 2: synthesis of precursor miRNA*—consequently, microprocessor complex (Drosha ribonuclease III (RNase III Drosha) and its binding partner, Di George syndrome Critical Region 8 (DGCR8)), bind to the hairpin structures in pri-miRNA and cleave it into about 60–70 nucleotides known as precursor miRNA (pre-miRNA). *Step 3: export of pre-miRNA into the cytoplasm*—after nuclear processing, the pre-miRNA is translocated to the cytoplasm from the nucleus through the GTP-binding nuclear protein Ran (RanGTP)/exportin-5 (XPO-5) complex. *Step 4: production of miRNA duplex*—after translocation to the cytoplasm, the pre-miRNA is

MicroRNA in Human Malignancies. https://doi.org/10.1016/B978-0-12-822287-4.00019-0

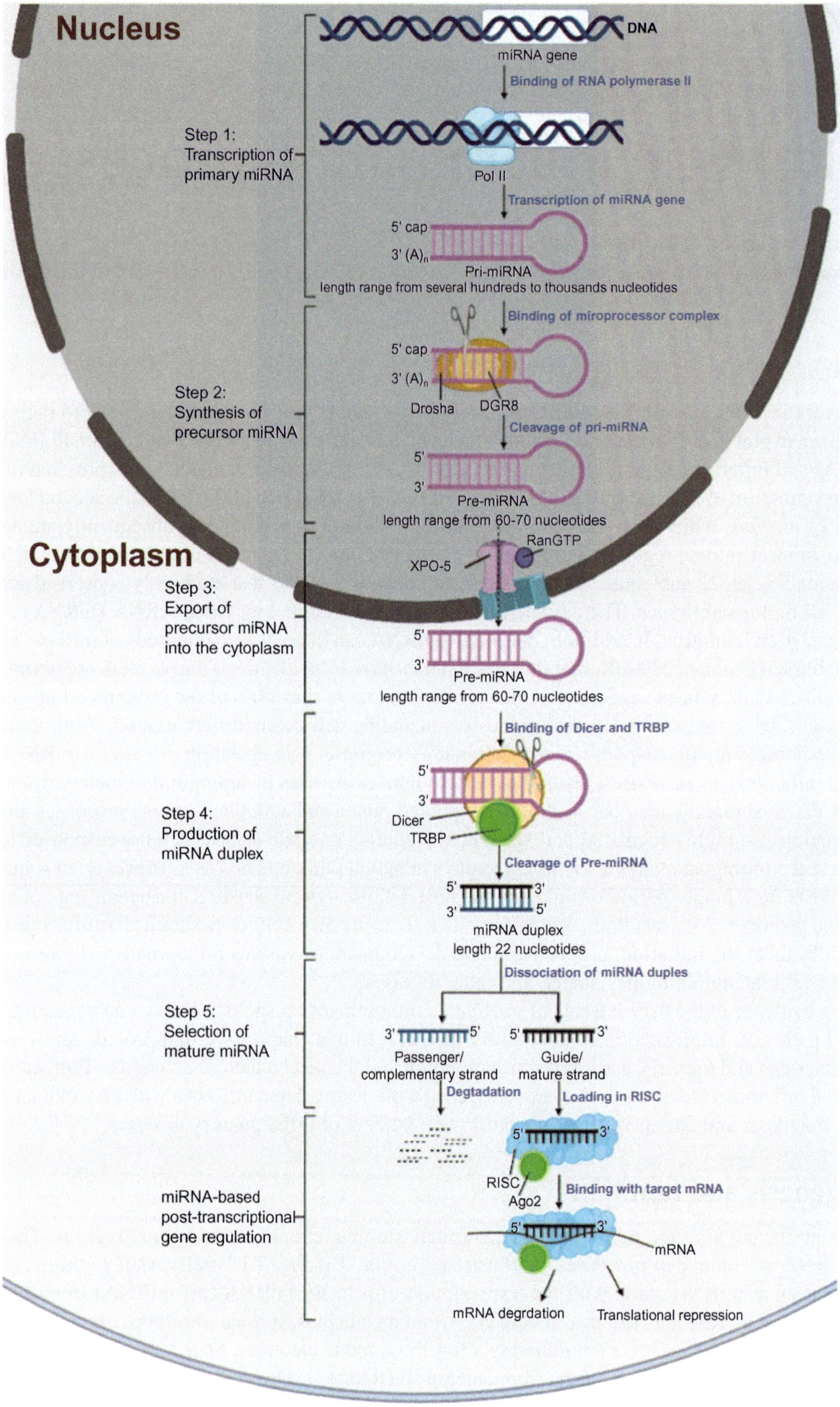

FIG. 7.1 MicroRNA biogenesis. The biogenesis of microRNAs is a multistep process.

further cleaved by another ribonuclease III enzyme, RNase III Dicer and its cofactor transactivation response RNA-binding protein (TRBP), to generate ~22-bp-long double-stranded miRNA (ds-miRNA). *Step 5: selection of guide strand/mature miRNA strand*—subsequently, strand selection takes place, where one strand (passenger/complementary strand) of the ds-miRNA is degraded while the other strand (guide/mature strand) of the ds-miRNA binds to the RNA-induced silencing complex (RISC), which contains Argonaute 2 (Ago2) proteins.

The regulation of posttranscriptional gene expression by miRNA is sequence-specific. RISC directs the guide/mature miRNA strand to bind with target messenger RNA (mRNA) through sequence complementarity. Predominantly, the seed region (nucleotides 2–8 from the 5′ end of the sequence) of the miRNA interacts with the miRNA response element (MRE, present in the 3′ untranslated region (3′ UTR)) of the target mRNA. However, miRNAs can also interact through either the seed region or noncanonical seed regions (other regions of miRNAs sequence) with MREs and MREs can also be present in either 3′ UTR, 5′ UTR, or the coding region of the target mRNA. The complete sequence complementarity between an miRNA and target mRNA induces the degradation of target mRNA, whereas incomplete sequence complementarity results in the repression of mRNA translation. Thus, in most circumstances, they act as negative regulators of gene expression. Conversely, in some circumstances, miRNA can also enhance RNA stability and upregulate gene expression by increasing the rate of transcription and translation of target mRNA. Moreover, multiple miRNAs can interact with a single mRNA and regulate its expression. On the other hand, one single miRNA can interact with multiple mRNAs and regulate their expression. Multiple lines of evidence indicate that miRNAs are able to modulate various cellular and developmental processes, including cell death, differentiation, proliferation, metabolism, apoptosis, stress response, immune response, and inflammation, which are associated with many diseases such as allergic, respiratory, cardiovascular, autoimmune, cancer, and other diseases. Therefore, miRNAs can play pivotal roles in the regulation of various pathological conditions and may be used as biomarkers and therapeutic agents in different diseases.

Inflammation and mechanism of activation

Inflammation is a highly regulated and sequential biological response from our body to protect and heal itself by fighting against potentially harmful internal or external stimuli such as irritants, toxins, injuries, damaged cells, or infections. However, on the basis of severity and duration, inflammation is mainly categorized into two types: acute and chronic (Fig. 7.2). Fig. 7.2 Acute inflammation is the redness, warmth, swelling, and pain around tissues and joints. It is a normal and healthy response from the host defense system. Briefly, inflammatory responses are initiated by cell or tissue damage that generates molecular patterns. These molecular patterns are recognized by the host immune system *via* pattern recognition receptors (PRRs). PRRs such as Toll-like receptors (TLRs) promote the production of proinflammatory cytokines and chemokines to circumvent unwarranted inflammation and repair damage. Thus, it is an important process for our body's healing system and normally self-limiting, taking a few hours to days in maintaining homeostasis.

In some instances, dysregulation of the PRR-mediated signaling cascades and immune responses can prolong inflammation for a long period of time. Due to unresolved immune responses, the body senses a consistent state of alert and the immune system continues to fight for an indefinite period. As a result, the immune system misguidedly perceives the body's own cells or tissues as harmful substances and leads to dysregulated inflammation with persistent pain, redness, swelling, and stiffness, which is known as chronic inflammation. Chronic inflammation leads to a wide array of inflammatory diseases, such as allergic, respiratory, cardiovascular, autoimmune, cancer, and other diseases, as presented in Fig. 7.2.

miRNAs in inflammatory disease

In recent years, several studies on miRNA-based posttranscriptional regulatory mechanisms have been concomitant with the regulation of immune and inflammatory responses. Based on their functions with different target mRNAs, miRNAs can not only initiate and/or increase the inflammatory processes but also decrease and/or terminate the inflammatory processes. Hence, in order to ascertain the appropriate action of specific miRNAs associated with immune and inflammatory processes, miRNA biogenesis and the action mechanism are very well timed and well regulated at the transcriptional as well as posttranscriptional level through the alternation of transcription and the processing or stabilization of mature or precursor miRNA transcripts. Further, the involvement of multiple miRNAs in the coordinated regulation of numerous genes and proteins of immune and inflammatory processes make these processes even more regulated. However, in some circumstances, the aberrant expression of miRNAs can lead to dysregulated immune and inflammatory responses that may result in dysregulated/chronic inflammation and eventually lead to inflammatory diseases. Here, we are discussing some well-studied inflammatory diseases (such as respiratory diseases, cardiovascular diseases, autoimmune diseases, and cancer) that are associated with aberrantly expressed miRNAs.

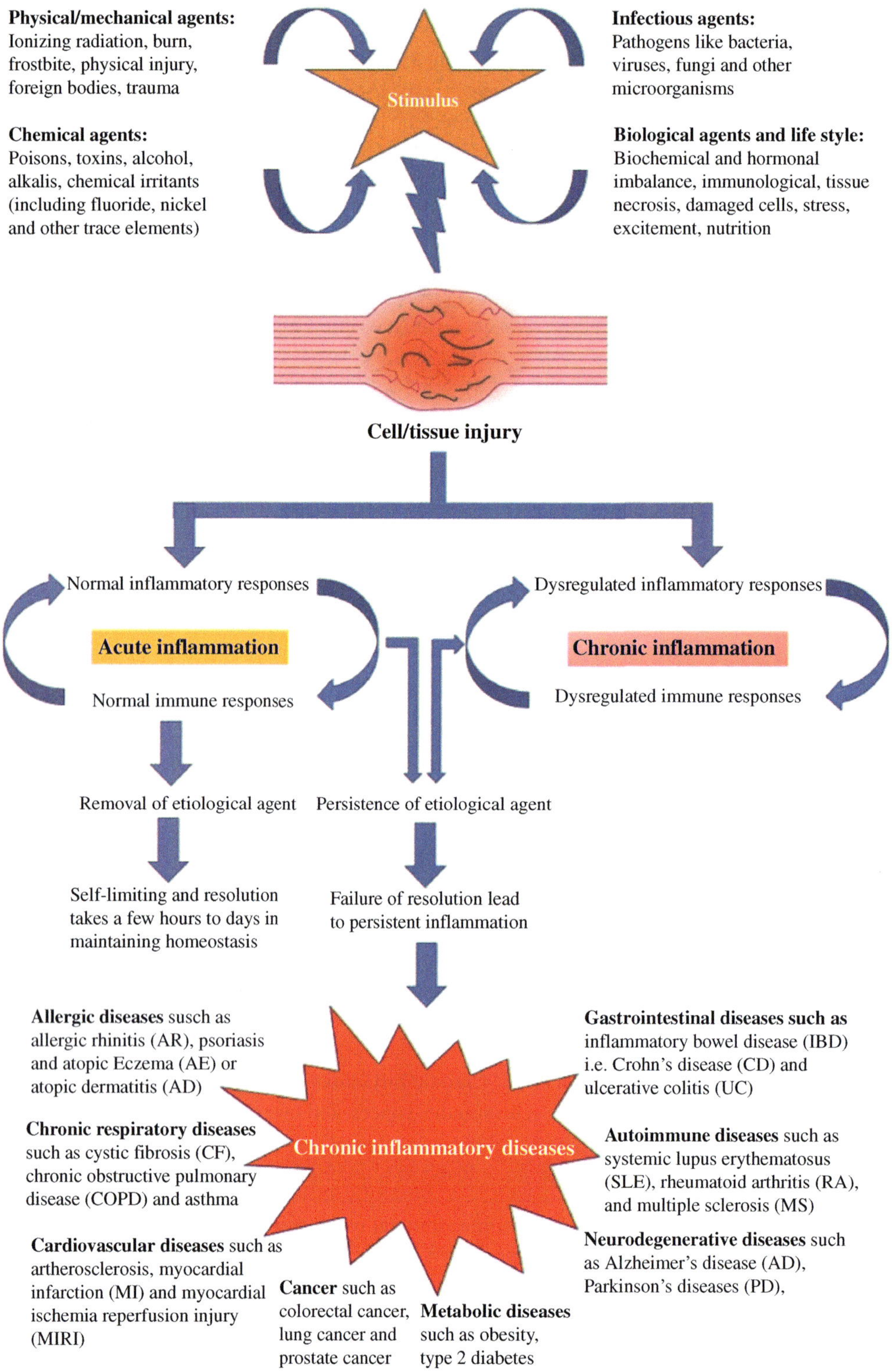

FIG. 7.2 Activation of inflammation. The severity and duration of inflammation are mainly categorized into two types: acute and chronic.

Chronic respiratory diseases

Chronic respiratory diseases are complex multifactorial disorders in which the impairment and inflammation of airways and lungs lead to irreversible respiratory obstruction. These diseases have an effect on several million people globally, and thereby have a major impact on human health. MiRNAs are abundantly present in lung tissue and changes in miRNA expressions and functions have been widely associated with the development and progression of chronic respiratory diseases such as cystic fibrosis (CF), chronic obstructive pulmonary disease (COPD), and asthma.

Cystic fibrosis (CF)

Excessive bronchial inflammation is one of the leading factors in the development of CF lung disease. This is mainly due to the mutations in the CF transmembrane conductance regulator (*CFTR*) gene that encodes the chloride ion channel. This defect in chloride ion transport causes dehydration of the epithelial layer and accumulation of thick mucus in the airway, which affects innate immune defenses and fosters bacterial colonization and infection/inflammation. As a result, neutrophil-dominated bronchial inflammation leads to airway remodeling and bronchiectasis. Oglesby et al. were the first to report the association of miRNAs in CF (Oglesby et al., 2010). The authors found that in the lung epithelium of CF patients, miR-126 downregulation results in the elevated expression of Target of Myb1 (TOM1) and the repressed expression of nuclear factor (NF)-kB-induced cytokines, such as interleukin 8 (IL-8). TOM1 regulates the intracellular trafficking of ubiquitinated proteins and also is a negative regulator of innate immune responses. The proinflammatory IL-8 cytokine is a chemotactic agent for neutrophils and the hypersecretion of IL-8 in the lung epithelium is an inflammatory hallmark of CF. Thus, it was suggested that the increased expression of TOM1 may lend itself to indemnifying the increased proinflammatory responses in the lung epithelium of CF patients (Oglesby et al., 2010). Further, the elevated expression of miR-155 was linked with the hyperexpression of IL-8 in CF epithelial cells (Bhattacharyya et al., 2011). Besides, the direct involvement of miR-17 was also established in the suppression of IL-8 expression (Oglesby et al., 2015).

In addition, miRNAs have also been shown in the regulation of *CFTR* biosynthesis, expression, and function. The binding of miR-138 with SIN3 homolog A (SIN3A, a transcriptional regulatory protein implicated in CTCF-mediated gene regulation) was shown to promote increased expression and biosynthesis of F508del-CFTR in the airway epithelial cells (Ramachandran et al., 2012). The involvement of miR-16 was also reported in the redemption of F508del-CFTR trafficking defects, possibly by the repression of Heat shock protein 90 (HSP90), a chaperone implicated in protein-folding activity (Kumar et al., 2015). Beside these reports, several studies have also identified miRNAs such as miR-101, miR-144, miR-145, miR-223, and miR-494, which directly target *CFTR* (Gillen et al., 2011; Megiorni et al., 2011; Hassan et al., 2012; Oglesby et al., 2013; Ramachandran et al., 2013). The elevated expression of miR-145 and miR-494 has been shown to reduce the expression of *CFTR* in the primary lung epithelial cells obtained from CF patients (Gillen et al., 2011; Megiorni et al., 2011; Oglesby et al., 2013; Ramachandran et al., 2013).

Chronic obstructive pulmonary disease (COPD)

COPD is a chronic heterogeneous disease of the airways and lungs. It is characterized by excessive and persistent inflammation in the airways and lungs, progressive and irreversible respiratory obstruction, destruction of lung parenchyma and epithelium, and impaired lung function. In the development of these COPD pathological conditions, cigarette smoking is considered a predominant risk factor. Several investigations conducted in COPD patients revealed differentially expressed miRNA between nonsmokers and smokers. The downregulation of miR-181c, which is known to induce neutrophil infiltration, the generation of reactive oxygen species, and inflammatory cytokines, was reported in COPD patients who were cigarette smokers (Du et al., 2017). Further, the downregulation of miR-126, which is involved in the activation of the ataxia telangiectasia-mutated (ATM) kinase of the DNA damage pathway, was reported in COPD patients who were smokers (Paschalaki et al., 2018). The authors suggest that the downregulation of miR-126 could accelerate lung aging and COPD pathogenesis. Conickx et al. reported the downregulation of miR-218-5p in COPD patients who were smokers without airflow limitation, and suggest its protective role in cigarette smoke (CS)-induced inflammation in COPD patients (Conickx et al., 2017). In addition, the roles of upregulated miRNAs such as miR-101, miR-144, and miR-223 were also shown in smoke-exposed COPD patients (Hassan et al., 2012; Leuenberger et al., 2016). However, the development of COPD pathological conditions (as mentioned above) was only found in 20% of smokers (Pauwels & Rabe, 2004; Hogg & Timens, 2009), which indicates that a combination of environmental, genetic, and epigenetic factors have critical roles in the initiation and progression of COPD.

Among various epigenetic regulators, miRNAs have emerged as predominant regulators and are implicated in many cellular and molecular pathways such as the transforming growth factor-β (TGF-β) signaling pathway that regulates the pathogenesis of COPD. Multiple lines of evidence showed that miRNAs regulate several TGF-β signaling molecules such as SMAD7, SMAD3, and SERPINE1. As with COPD patients, Ezzie et al. showed that the elevated expression of miR-15b results in the decreased expression of SMAD7 (Ezzie et al., 2012). Francis et al. showed that reduced expressions of miR-34c result in decreased expression of SERPINE1 (Savarimuthu Francis et al., 2014). O'Leary et al. showed that the elevated expression of miR-145 results in the increased expression of SMAD3 (O'Leary et al., 2016). O'Leary et al. also reported that the elevated expression of miR-145 repressed the release of IL-6 and CXCL8. Further, it has been demonstrated that miR-96 has a predominant regulatory role in the p53/hypoxia signaling pathway, which is also associated with the pathogenesis of COPD (Kusko et al., 2016). These findings suggest the interrelationship between the regulation of miRNA and signaling pathways that play critical roles in the pathogenesis of COPD.

Asthma

Asthma is a disease of the lungs that is characterized by chronic inflammation, airflow limitation, hyperresponsiveness, and reversible obstruction of airways and lungs. It is a heterogeneous disorder with the predominant phenotype of eosinophilic airway inflammation concomitant with persistent infiltration of T helper type 2 (Th2) cells and mast cells. Combinations of genetic and environmental factors have critical roles in the initiation and progression of asthma. Recently, an increasing number of reports established the involvement of miRNAs in eosinophilic airway inflammation and asthma by using allergen (such as ovalbumin or house dust mite)-induced mouse models of asthma. In the ovalbumin-induced mouse model of asthma, the elevated expression of miR-21 was found to be associated with IL-12 expression and Th2 response (Lu et al., 2009; Lu et al., 2011). In asthmatic mice, let-7 administration has been linked with reduced IL-13 expression and Th2 response (Kumar et al., 2011). In the house dust mite-induced mouse model, the suppression of miR-126 has been implicated in reduced airway hyperresponsiveness and impaired Th2 response (Mattes et al., 2009). Collison et al. reported that the suppression of miR-145 in the house dust mite-induced mouse model resulted in reduced IL-5 and IL-13 expression (Collison et al., 2011). The elevated expression of miR-155 was found in the house dust mite-induced mouse model (Okoye et al., 2014). Likewise, Malmhäll et al. reported decreased IL-5 and IL-13 expression and Th2 cell numbers in the ovalbumin-induced miR-155 null mice (Malmhall et al., 2014).

Several studies have explored the role of miRNAs in asthmatic patients and identified a number of differentially expressed miRNAs in the airway epithelial cells of patients compared to healthy controls (Jardim et al., 2012; Solberg et al., 2012). Patients with asthma had significant upregulation of miR-221 that regulates inflammation and abnormal growth (Liu et al., 2012), upregulation of miR-19a that regulates IL-13 cytokine production (Simpson et al., 2014), upregulation of miR-126 and miR-21 that regulates Th2 response (Wu et al., 2014), and upregulation of miR-146a that regulates the expression of cyclooxygenase-2 (COX-2) and RNA binding protein human antigen R (HuR) (Comer et al., 2014). Further, it was reported that the expression of miR-155 and COX-2 is positively correlated in the airway epithelial cells of asthmatic patients (Comer et al., 2015).

Cardiovascular diseases

Cardiovascular diseases are complex multifaceted disorders that affect the structures or functions of the heart and blood vessels and can lead to serious complications such as heart attacks or heart failure. Dysregulated inflammation, for example in the myocardium and vasculature, is one of the predominant characteristics that contributes to the development and progression of various cardiovascular diseases, including atherosclerosis, myocardial infarction (MI), and myocardial ischemia reperfusion injury (MIRI). Recently, a number of studies clearly showed the regulatory role of miRNAs in these cardiac inflammatory diseases.

Atherosclerosis

Atherosclerosis is a chronic cardiac inflammatory disease that is characterized by the accumulation of lipids and fibrous elements in blood vessels that leads to complete obstruction in blood flow through the formation of plaque. The development of this disease usually takes several decades with the involvement of various complex cellular and molecular mechanisms that eventually result in chronic inflammation and the development and progression of plaque. Among various risk factors, dyslipidemia, a cholesterol imbalance, is an important player that has a role in the chronic inflammatory response and disease progression to atherosclerosis. Cholesterol homeostasis is achieved by the delivery of low-density lipoprotein (LDL) cholesterol and the removal of high-density lipoprotein (HDL) cholesterol from cells and tissues. Excess LDL cholesterol level initiates an inflammatory reaction by provoking endothelial cells to release proinflammatory mediators (such as cytokines and adhesion molecules), recruiting monocytes and elevating macrophage differentiation, which ultimately cause atherosclerotic lesion formation. Several reports indicate the involvement of miRNAs in cholesterol homeostasis to attain the proper ratio of atherogenic LDL and atheroprotective HDL levels and function. The suppression of miR-122 in mice and African green monkeys caused a decrease in plasma cholesterol levels (Esau et al., 2006; Elmen et al., 2008). Moreover, the suppression of miR-122 also reduced cholesterol biosynthesis (Esau et al., 2006). Additionally, miR-27, miR-128-1, miR-148a, miR-130b, and miR-301b have been shown to modulate the level of LDL cholesterol (Wagschal et al., 2015; Xie et al., 2016). Further, it has been demonstrated that the level of plasma HDL cholesterol is increased by the suppression of miR-33, which leads to atherosclerotic plaque regression in mice (Rayner et al., 2011). Interestingly, miR-223 has been found to be associated with various processes of atherosclerotic disease progression such as the elevation HDL cholesterol and total cholesterol levels, the regulation of endothelial cell apoptosis, and the development of inflammatory responses (Johnnidis et al., 2008; Pan et al., 2014). The suppression of miR-223 was shown in reduced monocyte differentiation (Ismail et al., 2013). Further, HDL-transferred miR-223 promotes the repression of intercellular adhesion molecule-1 (ICAM-1) expression in endothelial cells and the regulation of vascular inflammation (Tabet et al., 2014). Moreover, miR-223 regulates several inflammatory signaling cascades such as NLRP3 (NACHT, LRR, and PYD domains containing protein 3), IKK-α (I-kappa-B kinase subunit alpha), and STAT3 (signal transducer and activator of transcription 3) (Li et al., 2010; Bauernfeind et al., 2012; Chen & Juo, 2012; Haneklaus et al., 2012; Haneklaus et al., 2013).

The involvement of miR-126 was shown in endothelial dysfunction, where miR-126 downregulates the sprout-related protein 1 (Spred-1) and contributes to the build up of angiogenesis (Wang et al., 2008). In addition, the regulatory role of miR-126 was also shown in the PI3K/Akt signaling pathway and the suppression of vascular endothelial cell apoptosis (Chen et al., 2016). Interestingly, in apolipoprotein E-deficient mouse models, the systemic addition of miR-181b mimics resulted in the reduced activity of NFkB that significantly suppressed atherosclerotic lesion formation (Sun et al., 2014). The functions of miR-181a-5p, miR-181a-3p, and miR-181b have been shown in the inhibition of endothelial inflammatory responses and adhesion molecule expression including VCAM-1, ICAM-1 and E-selectin (Su et al., 2019). Further, in apolipoprotein E-deficient mouse models, the rescue of these miRNAs leads to a significant reduction of atherosclerotic lesion formation (Su et al., 2019).

Myocardial infarction (MI) and ischemia reperfusion injury (MIRI)

MIs arise due to a decreased and/or discontinuous supply of blood flow to the heart, mainly by coronary atherosclerosis that causes a blockage in the coronary vessel and leads to cardiac remodeling and heart failure. Reperfusion therapy is the only way to treat this conditions, but the associated myocardial injury that is known as myocardial ischemia reperfusion injury (MIRI) leads to more cardiac complications, injury, and subsequent heart failure. Cardiac inflammation is one of the major factors that contributes to the initiation and progression of MI, MIRI, and cardiac remodeling. MiRNAs have been shown to be key players in the regulation of cardiac inflammation . The reduced level of miRNA let-7i is reported in angiotensin II-based cardiac inflammatory-stressed mice. The modulation of let-7i leads to the suppression of cardiac fibrosis by targeting IL-6 and collagen type 1 alpha 2 (Col1α2) (Wang et al., 2015). MiR-223 is elevated by MIRI. In miR-223 KO mice, increased post-MIRI cardiomyocyte necrosis and TNF-α and IL-1β levels were observed, suggesting its role as a cardioprotective and antiinflammatory molecule (Qin et al., 2016). Additionally, it was found that MI induced increased levels of miR-155. Consistently in miR-155 KO mice, the levels of inflammatory cytokines are decreased after MI, suggesting its role as a proinflammatory molecule (Wang et al., 2017). Another miRNA, miR-21, has been reported as a cardioprotective miR in MIRI (Tu et al., 2013). Further, Yang et al. reported elevated inflammation in miR-21 KO mice after MI, suggesting its role as an antiinflammatory molecule (Yang et al., 2018).

Autoimmune diseases

Autoimmune diseases arise due to dysregulation of immune responses, leading to the unwarranted production of autoantibodies, and inflammatory cytokines that target host cells, tissues, and/or organs. These assaults lead to chronic inflammation and damage and/or destruction of host tissues, which eventually results in autoimmune diseases. Multiple lines of evidence have shown that the aberrant expression of miRNAs leads to the excessive production of autoimmune responses, autoantibodies, and inflammatory cytokines. Consequently, this leads to the initiation and development of a variety of autoimmune diseases such as systemic lupus erythematosus (SLE), rheumatoid arthritis (RA), and multiple sclerosis (MS).

Systemic lupus erythematosus (SLE)

SLE is commonly regarded as an archetypal autoimmune disease where the gratuitous production of autoantibodies and autoreactive lymphocytes provokes systemic inflammation in connective tissues and stimulates numerous proinflammatory pathways that lead to self-tissue damage in a variety of organs and systems. The pathophysiology of SLE is extremely intricate with the involvement of a wide range of clinical manifestations that differ among affected individuals and therefore, cause challenges in the diagnosis and treatment of this disease. Emerging evidence has demonstrated that aberrant miRNA expressions are implicated in the overactivation of B cells, T cells, inflammasome, and inflammatory pathways, resulting in the overproduction of autoreactive autoantibodies and inflammatory cytokines that contributes to the progression and development of SLE. For example, increased expression of miR-21 in SLE patients elevates IL-10 expression by targeting programmed cell death 4 (PDCD4) and plays an important role in B and T cell hyperactivity (Stagakis et al., 2011). Similarly, increased expression of miR-30a in SLE patients stimulates B cell hyperactivity by directly targeting LYN (Liu et al., 2013).

The decreased expression of miR-155 was found in SLE patients and the overexpression of miR-155 was shown to repress the expression of the suppressor of cytokine signaling 1 (SOCS1), resulting in the elevated expression of IL-21 that has an imperative role in the proliferation and differentiation of B cells and the generation of autoantibodies (Rasmussen et al., 2015). Furthermore, the decreased expression of miR-17-5p in SLE patients induces the IFN pathway by targeting E2F1, resulting in the elevated expression of inflammatory cytokines (Sarhan et al., 2015). Recently, it has been observed that decreased expression of miR-633 in SLE patients stimulates the AKT/mTOR inflammatory pathway by targeting AKT1 and elevates the expression of various inflammatory cytokines such as IL-4, IL-17, and IFN-γ (Chen et al., 2019).

Rheumatoid arthritis (RA)

RA is an inflammatory autoimmune disease that is a consequence of synovial tissue inflammation. This alteration of the synovial environment into the inflammatory environment favors the production of inflammatory cytokines and chemokines and stimulates the secretion of proteases, which leads to bone and cartilage erosion and eventually results in joint damage and functional disability. In the past few decades, a number of studies have established the association of miRNAs with RA. For example, miR-17, miR-19, and miR-20 have been demonstrated as antiinflammatories in rheumatoid arthritis synovial fibroblasts (RASFs), where these miRNAs modulate the TNFα pathway by regulating the expressions of Toll-like receptor 2 (TLR) or mitogen-activated protein kinase (MAPK) (Philippe et al., 2012; Philippe et al., 2013; Akhtar et al., 2016). Furthermore, the role of miR-20 has also been shown in the modulation of the NLRP3-inflammasome pathway in RASFs by regulating the expressions of TXNIP (Li et al., 2016). Besides, the proinflammatory roles of several miRNAs such as miR-18 and miR-125b have also been demonstrated, where these miRNAs directly target and regulate the expression of various components involved in the NF-κB pathway (Trenkmann et al., 2013; Zhang et al., 2017).

Multiple sclerosis (MS)

MS is a serious central nervous system (CNS) disorder, characterized by the formation of demyelinating plaques. It is an immune-mediated chronic inflammatory disease in which dysregulated immune responses, mainly the production of cytokines by autoreactive T lymphocytes, result in chronic oxidative damage and inflammation leading to axonal demyelination and degeneration. Recent studies in the literature clearly show the association between dysregulation in miRNA function and the pathogenesis of MS. The elevated expression of miR-128 and miR-27b in naïve cells and the elevated expression of miR-340 in $CD4^+$ T cells of MS patients suppress the development of Th2 cells and activate the differentiation of Th1 cells (Guerau-de-Arellano et al., 2011). The elevated expression of miR-155 was observed in $CD4^+$ T cells during experimental autoimmune encephalomyelitis (EAE) and the deletion of miR-155 in mice showed reduced Th1 and Th17 responses in the CNS and peripheral lymphoid organs, respectively (Murugaiyan et al., 2011). Further, it was revealed that miR-155 regulates the expression of IL-23R in Th17 cells by directly targeting Ets1; miR-$155^{-/-}$ Th17 cells are hugely lacking in the ability to cause EAE (Hu et al., 2013). The elevated expression of miR-301a, miR-21, and miR-155 was reported in $CD4^+$ T cells in response to the myelin oligodendrocyte protein (MOG) antigen (Mycko et al., 2012). In the same study, it was shown that the overexpression of miR-301a in T cells infiltrating the CNS in animals with EAE contributes to Th17 differentiation and the pathogenesis of autoimmune demyelination by inhibiting its target PIAS3. The reduced expression of miR-320a and the increased expression of its target mRNA matrix metallopeptidase-9 (MMP-9) that disrupts the blood-brain barrier were observed in B cells of MS patients (Aung et al., 2015).

Cancer

Cancer is a group of diseases characterized by the uncontrolled division of abnormal cells; it can arise in various parts of the body. Among numerous factors, dysregulation of the cell cycle and apoptotic genes as well as a tumor microenvironment abundant in growth/survival factor and inflammatory cells are the most predominant participants in tumor initiation, promotion, development, and metastasis. Chronic inflammation has been widely accepted in the pathophysiology and development of cancer. In recent years, the role of inflammation-related miRNAs has been explored in different types of chronic inflammation-induced cancer such as colorectal, lung, and prostate cancer.

Colorectal cancer (CRC)

Cancer in the colon and/or rectum is known as CRC. The abnormal growths in the inner lining of the colon or rectum, called polyps, sometimes become cancerous and develop CRC. Multiple lines of evidence established the role of miRNAs in the regulation of CRC development by modulation of inflammatory responses. The upregulation of miR-21 was reported in various cancers, including CRC (Volinia et al., 2006). The activation of miR-21 and miR-181b-1 through IL-6-induced STAT3 leads to the activation of the NF-κB pathway that induces a stable transformed state by targeting PTEN (phosphatase and tensin homolog) and CYLD (cylindromatosis) expression (Iliopoulos et al., 2010). The role of miR-126 was found on tumor suppression in CRC cells, which partially relies on the regulation of CXC chemokine receptor 4 (CXCR4) expression via AKT and extracellular signal-regulated kinase 1/2 (ERK1/2) signaling pathways (Liu et al., 2014). Further, it was reported that miR-126 inhibits the activation of the PI3K/AKT signaling pathway by reducing Sphingosine 1 phosphate receptor 2 (S1PR2) expression in inflammatory bowel disease (IBD) (Chen et al., 2017). Besides, it has been shown that inflammatory inducers regulate the miRNA expression. In colon epithelial cells treated with TNF-α, reduced miR-192 expression was observed (Wu et al., 2008). In intestinal myofibroblasts, elevated miR-155 expression was induced by TNF-α and lipopolysaccharide (LPS) (Pathak et al., 2015). In intestinal epithelial cells, elevated let-7i, miR-21, and miR-449b expression was induced by IL-1β, TNF-α, and INF-γ, respectively (Ben-Shachar et al., 2016). Furthermore, in IBD CD4+T cells, elevated miR-301a expression was induced by TNF-α (He et al., 2016). In the same study, it was also found that transduction with miR-301a lentivirus in IBD CD4+T cells resulted in increased TNF-α production and Th17 cell differentiation by reducing the expression of Smad nuclear interacting protein 1 (SNIP1).

Lung cancer (LC)

Cancer in the lung is known as LC. Abnormal cell growth, mainly in the lining of the bronchi and bronchioles or alveoli, becomes cancerous, developing into LC. Chronic airway inflammation is considered one of the most important risk factors in LC development. Because of their role in the undesirable remodeling of bronchial epithelium, the proliferation of quiescent bronchoalveolar stem cells and the alteration of the lung microenvironment lead to pulmonary carcinogenesis. Recent studies demonstrated the regulatory function of miRNAs in lung inflammatory mechanisms and LC. In lung cancer tissues, miR-101 expression is repressed by IL-1β, which promotes the activation of HIF1α and NF-κB via the COX-2-HIF1α pathway and the NF-κB/COX-2 pathway, respectively (Wang et al., 2014; Lv et al., 2015). COX-2 negatively regulates miR-101 expression in nonsmall cell lung cancer (NSCLC) (Lv et al., 2015). These mechanisms lead to the upregulation of *Lin28B,* which establishes the connection between inflammation and lung cancer (Wang et al., 2014). In lung cancer tissues, the expression of miR-153, an inhibitor of AKT, is reduced, which suggests its antitumor function in lung cancer (Yuan et al., 2015). The increased expression of miR-301a, which promotes the activation of NF-κB and Stat3, is reported in lung cancer. Its deletion resulted in the suppression of lung tumorigenesis in mice, which suggests its role in creating a proinflammatory microenvironment that assists tumorigenesis (Ma et al., 2015). Yin et al. reported that miR-451 reduced the expression of PSMB8/NOS2 inflammatory factors and suppressed cell proliferation and migration in lung cancer (Yin et al., 2015).

Prostate cancer (PC)

Cancer in the prostate (a small walnut-shaped gland found only in the pelvis of males) is known as PC. The abnormal cell growths in the prostate gland become cancerous and develop PC. Emerging reports indicate that among various risk factors, miRNAs associated with inflammation play a crucial role in creating a tumor microenvironment and promoting the initiation and progression of PC. For example, miR-101 directly targets COX-2 and inhibits its expression, and the overexpression of miR-101 suppresses the growth of PC cells (Hao et al., 2011). Another miRNA, let-7, has been found to associate with the regulation of IL-6 and NF-B expression in several cancers. In PC, the reduced expression of let-7 was reported and the overexpression of let-7 showed decreased growth of PC cells in human PC xenografts (Liu et al., 2012; Nadiminty et al., 2012). The involvement of miR-146 has been reported in the NF-kB and proinflammatory cytokine signaling pathways. In PC, the reduced expression of mir-146 was reported and the overexpression of miR-146 showed decreased growth of PC cells (Sun et al., 2014). Further, a positive correlation between the expression of miR-146 and its target genes involved in inflammation has been reported (Ding et al., 2016).

Conclusion

As briefly illustrated in this chapter, miRNA-based posttranscriptional regulatory mechanisms and their role in inflammation, inflammatory pathways, and diseases have gained substantial attention in the last few decades because of the high severity, frequency, and economic burden of inflammation-associated diseases. One of the striking features of miRNA is binding with its target mRNA, which leads to the degradation or translation inhibition of mRNA and modulates gene and protein expression. Additionally, one single miRNA can also interact with hundreds of mRNAs and modulate their expression. Together with this, the decisive roles of miRNAs in nearly all the biological processes, including inflammation, ascertain miRNAs as crucial modulators of inflammatory pathways that regulate immune and inflammatory responses.

The aberrant miRNA expressions concomitant with immune and inflammatory responses trigger and develop various inflammatory diseases such as allergic, respiratory, cardiovascular, autoimmune, cancer, and other diseases. Thus, miRNAs can be used as biomarkers in diagnosis and targets for the prevention or treatment of inflammatory diseases. For example, some miRNAs that function as crucial inhibitors either by directly inhibiting or targeting activators of the inflammatory system can be administrated to downregulate target genes that are upregulated during the disease condition. Blocking miRNA inhibitory functions can be exploited to upregulate target genes that are downregulated during the disease condition. In addition, other miRNAs can act as enhancers by targeting the inhibitors of the inflammatory system and can be administrated to upregulate target genes that are downregulated during the disease condition. Hence, the identification of miRNAs and their targets concomitant with inflammatory pathways could open a new paradigm for the regulation of genes and pathways implicated in inflammatory diseases. Subsequently, this regulatory information may offer miRNAs as potential targets and can be utilized for the development of miRNA based-therapeutic strategies, either by delivery of miRNA inhibitors or mimics.

Moreover, miRNA-based therapeutic strategies have several benefits over the approach of targeting genes and proteins. The synthesis of miRNAs is easier and they can target multiple genes, which can be more rewarding than targeting multiple genes individually. However, further research is needed to attain complete knowledge of the molecular mechanism and pathways for miRNAs and targets, and the implementation of this strategy for therapeutic management. Thus, miRNAs can be considered potential candidates in the diagnosis, prevention, and treatment of inflammatory diseases.

References

Akhtar, N., Singh, A. K., & Ahmed, S. (2016). MicroRNA-17 suppresses TNF-alpha signaling by interfering with TRAF2 and cIAP2 association in rheumatoid arthritis synovial fibroblasts. *Journal of Immunology*, *197*, 2219–2228. https://doi.org/10.4049/jimmunol.1600360.

Aung, L. L., Mouradian, M. M., Dhib-Jalbut, S., & Balashov, K. E. (2015). MMP-9 expression is increased in B lymphocytes during multiple sclerosis exacerbation and is regulated by microRNA-320a. *Journal of Neuroimmunology*, *278*, 185–189. https://doi.org/10.1016/j.jneuroim.2014.11.004.

Bauernfeind, F., Rieger, A., Schildberg, F. A., Knolle, P. A., Schmid-Burgk, J. L., & Hornung, V. (2012). NLRP3 inflammasome activity is negatively controlled by miR-223. *Journal of Immunology*, *189*, 4175–4181. https://doi.org/10.4049/jimmunol.1201516.

Ben-Shachar, S., Yanai, H., Sherman Horev, H., Elad, H., Baram, L., Issakov, O., et al. (2016). MicroRNAs expression in the ileal pouch of patients with ulcerative colitis is robustly up-regulated and correlates with disease phenotypes. *PLoS One*, *11*. https://doi.org/10.1371/journal.pone.0159956, e0159956.

Bhattacharyya, S., Balakathiresan, N. S., Dalgard, C., Gutti, U., Armistead, D., Jozwik, C., et al. (2011). Elevated miR-155 promotes inflammation in cystic fibrosis by driving hyperexpression of interleukin-8. *The Journal of Biological Chemistry*, *286*, 11604–11615. https://doi.org/10.1074/jbc.M110.198390.

Chen, K. C., & Juo, S. H. (2012). MicroRNAs in atherosclerosis. *The Kaohsiung Journal of Medical Sciences*, *28*, 631–640. https://doi.org/10.1016/j.kjms.2012.04.001.

Chen, S., Wang, Y., Qin, H., Lin, J., Xie, L., Chen, S., et al. (2019). Downregulation of miR-633 activated AKT/mTOR pathway by targeting AKT1 in lupus CD4(+) T cells. *Lupus*, *28*, 510–519. https://doi.org/10.1177/0961203319829853.

Chen, L., Wang, J., Wang, B., Yang, J., Gong, Z., Zhao, X., et al. (2016). MiR-126 inhibits vascular endothelial cell apoptosis through targeting PI3K/Akt signaling. *Annals of Hematology*, *95*, 365–374. https://doi.org/10.1007/s00277-015-2567-9.

Chen, T., Xue, H., Lin, R., & Huang, Z. (2017). MiR-126 impairs the intestinal barrier function via inhibiting S1PR2 mediated activation of PI3K/AKT signaling pathway. *Biochemical and Biophysical Research Communications*, *494*, 427–432. https://doi.org/10.1016/j.bbrc.2017.03.043.

Collison, A., Mattes, J., Plank, M., & Foster, P. S. (2011). Inhibition of house dust mite-induced allergic airways disease by antagonism of microRNA-145 is comparable to glucocorticoid treatment. *The Journal of Allergy and Clinical Immunology*, *128*, 160–167. e164 https://doi.org/10.1016/j.jaci.2011.04.005.

Comer, B. S., Camoretti-Mercado, B., Kogut, P. C., Halayko, A. J., Solway, J., & Gerthoffer, W. T. (2014). MicroRNA-146a and microRNA-146b expression and anti-inflammatory function in human airway smooth muscle. *American Journal of Physiology. Lung Cellular and Molecular Physiology*, *307*, L727–L734. https://doi.org/10.1152/ajplung.00174.2014.

Comer, B. S., Camoretti-Mercado, B., Kogut, P. C., Halayko, A. J., Solway, J., & Gerthoffer, W. T. (2015). Cyclooxygenase-2 and microRNA-155 expression are elevated in asthmatic airway smooth muscle cells. *American Journal of Respiratory Cell and Molecular Biology*, *52*, 438–447. https://doi.org/10.1165/rcmb.2014-0129OC.

Conickx, G., Mestdagh, P., Avila Cobos, F., Verhamme, F. M., Maes, T., Vanaudenaerde, B. M., et al. (2017). MicroRNA profiling reveals a role for MicroRNA-218-5p in the pathogenesis of chronic obstructive pulmonary disease. *American Journal of Respiratory and Critical Care Medicine*, *195*, 43–56. https://doi.org/10.1164/rccm.201506-1182OC.

Ding, Y., Wu, H., Warden, C., Steele, L., Liu, X., Iterson, M. V., et al. (2016). Gene expression differences in prostate cancers between young and old men. *PLoS Genetics*, *12*. https://doi.org/10.1371/journal.pgen.1006477, e1006477.

Du, Y., Ding, Y., Chen, X., Mei, Z., Ding, H., Wu, Y., et al. (2017). MicroRNA-181c inhibits cigarette smoke-induced chronic obstructive pulmonary disease by regulating CCN1 expression. *Respiratory Research*, *18*, 155. https://doi.org/10.1186/s12931-017-0639-1.

Elmen, J., Lindow, M., Schutz, S., Lawrence, M., Petri, A., Obad, S., et al. (2008). LNA-mediated microRNA silencing in non-human primates. *Nature*, *452*, 896–899. https://doi.org/10.1038/nature06783.

Esau, C., Davis, S., Murray, S. F., Yu, X. X., Pandey, S. K., Pear, M., et al. (2006). miR-122 regulation of lipid metabolism revealed by in vivo antisense targeting. *Cell Metabolism*, *3*, 87–98. https://doi.org/10.1016/j.cmet.2006.01.005.

Ezzie, M. E., Crawford, M., Cho, J. H., Orellana, R., Zhang, S., Gelinas, R., et al. (2012). Gene expression networks in COPD: microRNA and mRNA regulation. *Thorax*, *67*, 122–131. https://doi.org/10.1136/thoraxjnl-2011-200089.

Gillen, A. E., Gosalia, N., Leir, S. H., & Harris, A. (2011). MicroRNA regulation of expression of the cystic fibrosis transmembrane conductance regulator gene. *The Biochemical Journal*, *438*, 25–32. https://doi.org/10.1042/BJ20110672.

Guerau-de-Arellano, M., Smith, K. M., Godlewski, J., Liu, Y., Winger, R., Lawler, S. E., et al. (2011). Micro-RNA dysregulation in multiple sclerosis favours pro-inflammatory T-cell-mediated autoimmunity. *Brain*, *134*, 3578–3589. https://doi.org/10.1093/brain/awr262.

Haneklaus, M., Gerlic, M., Kurowska-Stolarska, M., Rainey, A. A., Pich, D., McInnes, I. B., et al. (2012). Cutting edge: miR-223 and EBV miR-BART15 regulate the NLRP3 inflammasome and IL-1beta production. *Journal of Immunology*, *189*, 3795–3799. https://doi.org/10.4049/jimmunol.1200312.

Haneklaus, M., Gerlic, M., O'Neill, L. A., & Masters, S. L. (2013). miR-223: Infection, inflammation and cancer. *Journal of Internal Medicine*, *274*, 215–226. https://doi.org/10.1111/joim.12099.

Hao, Y., Gu, X., Zhao, Y., Greene, S., Sha, W., Smoot, D. T., et al. (2011). Enforced expression of miR-101 inhibits prostate cancer cell growth by modulating the COX-2 pathway in vivo. *Cancer Prevention Research (Philadelphia, Pa.)*, *4*, 1073–1083. https://doi.org/10.1158/1940-6207.CAPR-10-0333.

Hassan, F., Nuovo, G. J., Crawford, M., Boyaka, P. N., Kirkby, S., Nana-Sinkam, S. P., et al. (2012). MiR-101 and miR-144 regulate the expression of the CFTR chloride channel in the lung. *PLoS One*, *7*. https://doi.org/10.1371/journal.pone.0050837, e50837.

He, C., Shi, Y., Wu, R., Sun, M., Fang, L., Wu, W., et al. (2016). miR-301a promotes intestinal mucosal inflammation through induction of IL-17A and TNF-alpha in IBD. *Gut*, *65*, 1938–1950. https://doi.org/10.1136/gutjnl-2015-309389.

Hogg, J. C., & Timens, W. (2009). The pathology of chronic obstructive pulmonary disease. *Annual Review of Pathology*, *4*, 435–459. https://doi.org/10.1146/annurev.pathol.4.110807.092145.

Hu, R., Huffaker, T. B., Kagele, D. A., Runtsch, M. C., Bake, E., Chaudhuri, A. A., et al. (2013). MicroRNA-155 confers encephalogenic potential to Th17 cells by promoting effector gene expression. *Journal of Immunology*, *190*, 5972–5980. https://doi.org/10.4049/jimmunol.1300351.

Iliopoulos, D., Jaeger, S. A., Hirsch, H. A., Bulyk, M. L., & Struhl, K. (2010). STAT3 activation of miR-21 and miR-181b-1 via PTEN and CYLD are part of the epigenetic switch linking inflammation to cancer. *Molecular Cell*, *39*, 493–506. https://doi.org/10.1016/j.molcel.2010.07.023.

Ismail, N., Wang, Y., Dakhlallah, D., Moldovan, L., Agarwal, K., Batte, K., et al. (2013). Macrophage microvesicles induce macrophage differentiation and miR-223 transfer. *Blood*, *121*, 984–995. https://doi.org/10.1182/blood-2011-08-374793.

Jardim, M. J., Dailey, L., Silbajoris, R., & Diaz-Sanchez, D. (2012). Distinct microRNA expression in human airway cells of asthmatic donors identifies a novel asthma-associated gene. *American Journal of Respiratory Cell and Molecular Biology*, *47*, 536–542. https://doi.org/10.1165/rcmb.2011-0160OC.

Johnnidis, J. B., Harris, M. H., Wheeler, R. T., Stehling-Sun, S., Lam, M. H., Kirak, O., et al. (2008). Regulation of progenitor cell proliferation and granulocyte function by microRNA-223. *Nature*, *451*, 1125–1129. https://doi.org/10.1038/nature06607.

Kumar, M., Ahmad, T., Sharma, A., Mabalirajan, U., Kulshreshtha, A., Agrawal, A., et al. (2011). Let-7 microRNA-mediated regulation of IL-13 and allergic airway inflammation. *The Journal of Allergy and Clinical Immunology*, *128*, 1077–1085. e1071-1010 https://doi.org/10.1016/j.jaci.2011.04.034.

Kumar, P., Bhattacharyya, S., Peters, K. W., Glover, M. L., Sen, A., Cox, R. T., et al. (2015). miR-16 rescues F508del-CFTR function in native cystic fibrosis epithelial cells. *Gene Therapy*, *22*, 908–916. https://doi.org/10.1038/gt.2015.56.

Kusko, R. L., Brothers, J. F., 2nd, Tedrow, J., Pandit, K., Huleihel, L., Perdomo, C., et al. (2016). Integrated genomics reveals convergent transcriptomic networks underlying chronic obstructive pulmonary disease and idiopathic pulmonary fibrosis. *American Journal of Respiratory and Critical Care Medicine*, *194*, 948–960. https://doi.org/10.1164/rccm.201510-2026OC.

Leuenberger, C., Schuoler, C., Bye, H., Mignan, C., Rechsteiner, T., Hillinger, S., et al. (2016). MicroRNA-223 controls the expression of histone deacetylase 2: A novel axis in COPD. *Journal of Molecular Medicine (Berlin, Germany)*, *94*, 725–734. https://doi.org/10.1007/s00109-016-1388-1.

Li, T., Morgan, M. J., Choksi, S., Zhang, Y., Kim, Y. S., & Liu, Z. G. (2010). MicroRNAs modulate the noncanonical transcription factor NF-kappaB pathway by regulating expression of the kinase IKKalpha during macrophage differentiation. *Nature Immunology*, *11*, 799–805. https://doi.org/10.1038/ni.1918.

Li, X. F., Shen, W. W., Sun, Y. Y., Li, W. X., Sun, Z. H., Liu, Y. H., et al. (2016). MicroRNA-20a negatively regulates expression of NLRP3-inflammasome by targeting TXNIP in adjuvant-induced arthritis fibroblast-like synoviocytes. *Joint, Bone, Spine*, *83*, 695–700. https://doi.org/10.1016/j.jbspin.2015.10.007.

Liu, Y., Dong, J., Mu, R., Gao, Y., Tan, X., Li, Y., et al. (2013). MicroRNA-30a promotes B cell hyperactivity in patients with systemic lupus erythematosus by direct interaction with Lyn. *Arthritis and Rheumatism*, *65*, 1603–1611. https://doi.org/10.1002/art.37912.

Liu, C., Kelnar, K., Vlassov, A. V., Brown, D., Wang, J., & Tang, D. G. (2012). Distinct microRNA expression profiles in prostate cancer stem/progenitor cells and tumor-suppressive functions of let-7. *Cancer Research*, *72*, 3393–3404. https://doi.org/10.1158/0008-5472.CAN-11-3864.

Liu, F., Qin, H. B., Xu, B., Zhou, H., & Zhao, D. Y. (2012). Profiling of miRNAs in pediatric asthma: Upregulation of miRNA-221 and miRNA-485-3p. *Molecular Medicine Reports*, *6*, 1178–1182. https://doi.org/10.3892/mmr.2012.1030.

Liu, Y., Zhou, Y., Feng, X., An, P., Quan, X., Wang, H., et al. (2014). MicroRNA-126 functions as a tumor suppressor in colorectal cancer cells by targeting CXCR4 via the AKT and ERK1/2 signaling pathways. *International Journal of Oncology*, *44*, 203–210. https://doi.org/10.3892/ijo.2013.2168.

Lu, T. X., Hartner, J., Lim, E. J., Fabry, V., Mingler, M. K., Cole, E. T., et al. (2011). MicroRNA-21 limits in vivo immune response-mediated activation of the IL-12/IFN-gamma pathway, Th1 polarization, and the severity of delayed-type hypersensitivity. *Journal of Immunology*, *187*, 3362–3373. https://doi.org/10.4049/jimmunol.1101235.

Lu, T. X., Munitz, A., & Rothenberg, M. E. (2009). MicroRNA-21 is up-regulated in allergic airway inflammation and regulates IL-12p35 expression. *Journal of Immunology*, *182*, 4994–5002. https://doi.org/10.4049/jimmunol.0803560.

Lv, P., Zhang, P., Li, X., & Chen, Y. (2015). Micro ribonucleic acid (RNA)-101 inhibits cell proliferation and invasion of lung cancer by regulating cyclooxygenase-2. *Thoracic Cancer*, *6*, 778–784. https://doi.org/10.1111/1759-7714.12283.

Ma, X., Yan, F., Deng, Q., Li, F., Lu, Z., Liu, M., et al. (2015). Modulation of tumorigenesis by the pro-inflammatory microRNA miR-301a in mouse models of lung cancer and colorectal cancer. *Cell Discovery*, *1*, 15005. https://doi.org/10.1038/celldisc.2015.5.

Malmhall, C., Alawieh, S., Lu, Y., Sjostrand, M., Bossios, A., Eldh, M., et al. (2014). MicroRNA-155 is essential for T(H)2-mediated allergen-induced eosinophilic inflammation in the lung. *The Journal of Allergy and Clinical Immunology*, *133*, 1429–1438. 1438.e1421-1427 https://doi.org/10.1016/j.jaci.2013.11.008.

Mattes, J., Collison, A., Plank, M., Phipps, S., & Foster, P. S. (2009). Antagonism of microRNA-126 suppresses the effector function of TH2 cells and the development of allergic airways disease. *Proceedings of the National Academy of Sciences of the United States of America*, *106*, 18704–18709. https://doi.org/10.1073/pnas.0905063106.

Megiorni, F., Cialfi, S., Dominici, C., Quattrucci, S., & Pizzuti, A. (2011). Synergistic post-transcriptional regulation of the Cystic Fibrosis Transmembrane conductance Regulator (CFTR) by miR-101 and miR-494 specific binding. *PLoS One*, *6*. https://doi.org/10.1371/journal.pone.0026601, e26601.

Murugaiyan, G., Beynon, V., Mittal, A., Joller, N., & Weiner, H. L. (2011). Silencing microRNA-155 ameliorates experimental autoimmune encephalomyelitis. *Journal of Immunology*, *187*, 2213–2221. https://doi.org/10.4049/jimmunol.1003952.

Mycko, M. P., Cichalewska, M., Machlanska, A., Cwiklinska, H., Mariasiewicz, M., & Selmaj, K. W. (2012). MicroRNA-301a regulation of a T-helper 17 immune response controls autoimmune demyelination. *Proceedings of the National Academy of Sciences of the United States of America*, *109*, E1248–E1257. https://doi.org/10.1073/pnas.1114325109.

Nadiminty, N., Tummala, R., Lou, W., Zhu, Y., Zhang, J., Chen, X., et al. (2012). MicroRNA let-7c suppresses androgen receptor expression and activity via regulation of Myc expression in prostate cancer cells. *The Journal of Biological Chemistry*, *287*, 1527–1537. https://doi.org/10.1074/jbc.M111.278705.

Oglesby, I. K., Bray, I. M., Chotirmall, S. H., Stallings, R. L., O'Neill, S. J., McElvaney, N. G., et al. (2010). miR-126 is downregulated in cystic fibrosis airway epithelial cells and regulates TOM1 expression. *Journal of Immunology*, *184*, 1702–1709. https://doi.org/10.4049/jimmunol.0902669.

Oglesby, I. K., Chotirmall, S. H., McElvaney, N. G., & Greene, C. M. (2013). Regulation of cystic fibrosis transmembrane conductance regulator by microRNA-145, -223, and -494 is altered in DeltaF508 cystic fibrosis airway epithelium. *Journal of Immunology*, *190*, 3354–3362. https://doi.org/10.4049/jimmunol.1202960.

Oglesby, I. K., Vencken, S. F., Agrawal, R., Gaughan, K., Molloy, K., Higgins, G., et al. (2015). miR-17 overexpression in cystic fibrosis airway epithelial cells decreases interleukin-8 production. *The European Respiratory Journal*, *46*, 1350–1360. https://doi.org/10.1183/09031936.00163414.

Okoye, I. S., Czieso, S., Ktistaki, E., Roderick, K., Coomes, S. M., Pelly, V. S., et al. (2014). Transcriptomics identified a critical role for Th2 cell-intrinsic miR-155 in mediating allergy and antihelminth immunity. *Proceedings of the National Academy of Sciences of the United States of America*, *111*, E3081–E3090. https://doi.org/10.1073/pnas.1406322111.

O'Leary, L., Sevinc, K., Papazoglou, I. M., Tildy, B., Detillieux, K., Halayko, A. J., et al. (2016). Airway smooth muscle inflammation is regulated by microRNA-145 in COPD. *FEBS Letters*, *590*, 1324–1334. https://doi.org/10.1002/1873-3468.12168.

Pan, Y., Liang, H., Liu, H., Li, D., Chen, X., Li, L., et al. (2014). Platelet-secreted microRNA-223 promotes endothelial cell apoptosis induced by advanced glycation end products via targeting the insulin-like growth factor 1 receptor. *Journal of Immunology*, *192*, 437–446. https://doi.org/10.4049/jimmunol.1301790.

Paschalaki, K. E., Zampetaki, A., Baker, J. R., Birrell, M. A., Starke, R. D., Belvisi, M. G., et al. (2018). Downregulation of MicroRNA-126 augments DNA damage response in cigarette smokers and patients with chronic obstructive pulmonary disease. *American Journal of Respiratory and Critical Care Medicine*, *197*, 665–668. https://doi.org/10.1164/rccm.201706-1304LE.

Pathak, S., Grillo, A. R., Scarpa, M., Brun, P., D'Inca, R., Nai, L., et al. (2015). MiR-155 modulates the inflammatory phenotype of intestinal myofibroblasts by targeting SOCS1 in ulcerative colitis. *Experimental & Molecular Medicine*, *47*. https://doi.org/10.1038/emm.2015.21, e164.

Pauwels, R. A., & Rabe, K. F. (2004). Burden and clinical features of chronic obstructive pulmonary disease (COPD). *Lancet*, *364*, 613–620. https://doi.org/10.1016/S0140-6736(04)16855-4.

Philippe, L., Alsaleh, G., Pichot, A., Ostermann, E., Zuber, G., Frisch, B., et al. (2013). MiR-20a regulates ASK1 expression and TLR4-dependent cytokine release in rheumatoid fibroblast-like synoviocytes. *Annals of the Rheumatic Diseases*, *72*, 1071–1079. https://doi.org/10.1136/annrheumdis-2012-201654.

Philippe, L., Alsaleh, G., Suffert, G., Meyer, A., Georgel, P., Sibilia, J., et al. (2012). TLR2 expression is regulated by microRNA miR-19 in rheumatoid fibroblast-like synoviocytes. *Journal of Immunology*, *188*, 454–461. https://doi.org/10.4049/jimmunol.1102348.

Qin, D., Wang, X., Li, Y., Yang, L., Wang, R., Peng, J., et al. (2016). MicroRNA-223-5p and -3p Cooperatively Suppress Necroptosis in Ischemic/Reperfused Hearts. *The Journal of Biological Chemistry*, *291*, 20247–20259. https://doi.org/10.1074/jbc.M116.732735.

Ramachandran, S., Karp, P. H., Jiang, P., Ostedgaard, L. S., Walz, A. E., Fisher, J. T., et al. (2012). A microRNA network regulates expression and biosynthesis of wild-type and DeltaF508 mutant cystic fibrosis transmembrane conductance regulator. *Proceedings of the National Academy of Sciences of the United States of America*, *109*, 13362–13367. https://doi.org/10.1073/pnas.1210906109.

Ramachandran, S., Karp, P. H., Osterhaus, S. R., Jiang, P., Wohlford-Lenane, C., Lennox, K. A., et al. (2013). Post-transcriptional regulation of cystic fibrosis transmembrane conductance regulator expression and function by microRNAs. *American Journal of Respiratory Cell and Molecular Biology*, *49*, 544–551. https://doi.org/10.1165/rcmb.2012-0430OC.

Rasmussen, T. K., Andersen, T., Bak, R. O., Yiu, G., Sorensen, C. M., Stengaard-Pedersen, K., et al. (2015). Overexpression of microRNA-155 increases IL-21 mediated STAT3 signaling and IL-21 production in systemic lupus erythematosus. *Arthritis Research & Therapy*, *17*, 154. https://doi.org/10.1186/s13075-015-0660-z.

Rayner, K. J., Sheedy, F. J., Esau, C. C., Hussain, F. N., Temel, R. E., Parathath, S., et al. (2011). Antagonism of miR-33 in mice promotes reverse cholesterol transport and regression of atherosclerosis. *The Journal of Clinical Investigation*, *121*, 2921–2931. https://doi.org/10.1172/JCI57275.

Sarhan, R. A., Aboelenein, H. R., Sourour, S. K., Fawzy, I. O., Salah, S., & Abdelaziz, A. I. (2015). Targeting E2F1 and c-Myc expression by microRNA-17-5p represses interferon-stimulated gene MxA in peripheral blood mononuclear cells of pediatric systemic lupus erythematosus patients. *Discovery Medicine*, *19*, 419–425.

Savarimuthu Francis, S. M., Davidson, M. R., Tan, M. E., Wright, C. M., Clarke, B. E., Duhig, E. E., et al. (2014). MicroRNA-34c is associated with emphysema severity and modulates SERPINE1 expression. *BMC Genomics*, *15*, 88. https://doi.org/10.1186/1471-2164-15-88.

Simpson, L. J., Patel, S., Bhakta, N. R., Choy, D. F., Brightbill, H. D., Ren, X., et al. (2014). A microRNA upregulated in asthma airway T cells promotes TH2 cytokine production. *Nature Immunology*, *15*, 1162–1170. https://doi.org/10.1038/ni.3026.

Solberg, O. D., Ostrin, E. J., Love, M. I., Peng, J. C., Bhakta, N. R., Hou, L., et al. (2012). Airway epithelial miRNA expression is altered in asthma. *American Journal of Respiratory and Critical Care Medicine*, *186*, 965–974. https://doi.org/10.1164/rccm.201201-0027OC.

Stagakis, E., Bertsias, G., Verginis, P., Nakou, M., Hatziapostolou, M., Kritikos, H., et al. (2011). Identification of novel microRNA signatures linked to human lupus disease activity and pathogenesis: miR-21 regulates aberrant T cell responses through regulation of PDCD4 expression. *Annals of the Rheumatic Diseases*, *70*, 1496–1506. https://doi.org/10.1136/ard.2010.139857.

Su, Y., Yuan, J., Zhang, F., Lei, Q., Zhang, T., Li, K., et al. (2019). MicroRNA-181a-5p and microRNA-181a-3p cooperatively restrict vascular inflammation and atherosclerosis. *Cell Death & Disease*, *10*, 365. https://doi.org/10.1038/s41419-019-1599-9.

Sun, X., He, S., Wara, A. K. M., Icli, B., Shvartz, E., Tesmenitsky, Y., et al. (2014). Systemic delivery of microRNA-181b inhibits nuclear factor-kappaB activation, vascular inflammation, and atherosclerosis in apolipoprotein E-deficient mice. *Circulation Research*, *114*, 32–40. https://doi.org/10.1161/CIRCRESAHA.113.302089.

Sun, Q., Zhao, X., Liu, X., Wang, Y., Huang, J., Jiang, B., et al. (2014). miR-146a functions as a tumor suppressor in prostate cancer by targeting Rac1. *Prostate*, *74*, 1613–1621. https://doi.org/10.1002/pros.22878.

Tabet, F., Vickers, K. C., Cuesta Torres, L. F., Wiese, C. B., Shoucri, B. M., Lambert, G., et al. (2014). HDL-transferred microRNA-223 regulates ICAM-1 expression in endothelial cells. *Nature Communications*, *5*, 3292. https://doi.org/10.1038/ncomms4292.

Trenkmann, M., Brock, M., Gay, R. E., Michel, B. A., Gay, S., & Huber, L. C. (2013). Tumor necrosis factor alpha-induced microRNA-18a activates rheumatoid arthritis synovial fibroblasts through a feedback loop in NF-kappaB signaling. *Arthritis and Rheumatism*, *65*, 916–927. https://doi.org/10.1002/art.37834.

Tu, Y., Wan, L., Fan, Y., Wang, K., Bu, L., Huang, T., et al. (2013). Ischemic postconditioning-mediated miRNA-21 protects against cardiac ischemia/reperfusion injury via PTEN/Akt pathway. *PLoS One*, *8*. https://doi.org/10.1371/journal.pone.0075872, e75872.

Volinia, S., Calin, G. A., Liu, C. G., Ambs, S., Cimmino, A., Petrocca, F., et al. (2006). A microRNA expression signature of human solid tumors defines cancer gene targets. *Proceedings of the National Academy of Sciences of the United States of America*, *103*, 2257–2261. https://doi.org/10.1073/pnas.0510565103.

Wagschal, A., Najafi-Shoushtari, S. H., Wang, L., Goedeke, L., Sinha, S., de Lemos, A. S., et al. (2015). Genome-wide identification of microRNAs regulating cholesterol and triglyceride homeostasis. *Nature Medicine*, *21*, 1290–1297. https://doi.org/10.1038/nm.3980.

Wang, S., Aurora, A. B., Johnson, B. A., Qi, X., McAnally, J., Hill, J. A., et al. (2008). The endothelial-specific microRNA miR-126 governs vascular integrity and angiogenesis. *Developmental Cell*, *15*, 261–271. https://doi.org/10.1016/j.devcel.2008.07.002.

Wang, X., Wang, H. X., Li, Y. L., Zhang, C. C., Zhou, C. Y., Wang, L., et al. (2015). MicroRNA Let-7i negatively regulates cardiac inflammation and fibrosis. *Hypertension*, *66*, 776–785. https://doi.org/10.1161/HYPERTENSIONAHA.115.05548.

Wang, C., Zhang, C., Liu, L., Chen, X. A. B., Li, Y., & Du, J. (2017). Macrophage-derived mir-155-containing exosomes suppress fibroblast proliferation and promote fibroblast inflammation during cardiac injury. *Molecular Therapy*, *25*, 192–204. https://doi.org/10.1016/j.ymthe.2016.09.001.

Wang, L., Zhang, L. F., Wu, J., Xu, S. J., Xu, Y. Y., Li, D., et al. (2014). IL-1beta-mediated repression of microRNA-101 is crucial for inflammation-promoted lung tumorigenesis. *Cancer Research*, *74*, 4720–4730. https://doi.org/10.1158/0008 5472.CAN-14-0960.

Wu, X. B., Wang, M. Y., Zhu, H. Y., Tang, S. Q., You, Y. D., & Xie, Y. Q. (2014). Overexpression of microRNA-21 and microRNA-126 in the patients of bronchial asthma. *International Journal of Clinical and Experimental Medicine*, *7*, 1307–1312.

Wu, F., Zikusoka, M., Trindade, A., Dassopoulos, T., Harris, M. L., Bayless, T. M., et al. (2008). MicroRNAs are differentially expressed in ulcerative colitis and alter expression of macrophage inflammatory peptide-2 alpha. *Gastroenterology*, *135*, 1624–1635. e1624 https://doi.org/10.1053/j.gastro.2008.07.068.

Xie, W., Li, L., Zhang, M., Cheng, H. P., Gong, D., Lv, Y. C., et al. (2016). MicroRNA-27 prevents atherosclerosis by suppressing lipoprotein lipase-induced lipid accumulation and inflammatory response in Apolipoprotein E knockout mice. *PLoS One*, *11*. https://doi.org/10.1371/journal.pone.0157085, e0157085.

Yang, L., Wang, B., Zhou, Q., Wang, Y., Liu, X., Liu, Z., et al. (2018). MicroRNA-21 prevents excessive inflammation and cardiac dysfunction after myocardial infarction through targeting KBTBD7. *Cell Death & Disease*, *9*, 769. https://doi.org/10.1038/s41419-018-0805-5.

Yin, P., Peng, R., Peng, H., Yao, L., Sun, Y., Wen, L., et al. (2015). MiR-451 suppresses cell proliferation and metastasis in A549 lung cancer cells. *Molecular Biotechnology*, *57*, 1–11. https://doi.org/10.1007/s12033-014-9796-3.

Yuan, Y., Du, W., Wang, Y., Xu, C., Wang, J., Zhang, Y., et al. (2015). Suppression of AKT expression by miR-153 produced anti-tumor activity in lung cancer. *International Journal of Cancer*, *136*, 1333–1340. https://doi.org/10.1002/ijc.29103.

Zhang, B., Wang, L. S., & Zhou, Y. H. (2017). Elevated microRNA-125b promotes inflammation in rheumatoid arthritis by activation of NF-kappaB pathway. *Biomedicine & Pharmacotherapy*, *93*, 1151–1157. https://doi.org/10.1016/j.biopha.2017.07.042.

Section B

miRNAs methodologies

Chapter 8

Wet-lab methods for miRNA analysis

Marina C. Costa[a,b], Ana Lúcia Leitão[c,d], André F. Gabriel[a,b], and Francisco J. Enguita[a,b]

[a]*Instituto de Medicina Molecular João Lobo Antunes, Faculdade de Medicina, Universidade de Lisboa, Lisboa, Portugal,* [b]*Cardiomics Unit, Centro Cardiovascular da Universidade de Lisboa, Faculdade de Medicina, Universidade de Lisboa, Lisboa, Portugal,* [c]*Faculdade de Ciências e Tecnologia, Universidade Nova de Lisboa, Caparica, Portugal,* [d]*MEtRICs, Faculdade de Ciências e Tecnologia, Universidade Nova de Lisboa, Caparica, Portugal*

Introduction

MicroRNAs (miRNAs) are a class of small noncoding RNAs of about 19–25 nucleotides acting as negative posttranscriptional regulators of gene expression. The miRNA regulatory action is exerted through their association with an RNA-induced silencing complex (RISC), which will be guided by the sequence complementarity between the miRNA and its cognate mRNA target (Bartel, 2004; Hausser & Zavolan, 2014). The first evidence of the existence of miRNAs was described by Lee and coworkers in 1993 in *Caenorhabditis elegans* (Lee, Feinbaum, & Ambros, 1993), and since then they have been described in many eukaryotic organisms. In humans, they are responsible for the regulation of more than 60% of the protein coding genes (Hausser & Zavolan, 2014). Following their discovery, miRNAs have been identified across all clades of life, including viruses, bacteria, plants, and metazoans. In mammals, it is estimated that approximately 1%–3% of the genome codes for miRNA genes. Also, studies suggest that target sequences for miRNAs, called miRNA response elements (MREs), are encoded in the sequences of the majority of the coding transcripts (Bartel, 2004; Friedman, Farh, Burge, & Bartel, 2009). As a result, miRNAs have been shown to modulate important cellular and biological processes, such as cell fate determination (Cayirlioglu et al., 2008) and stem cell maintenance (Shcherbata et al., 2007), among others. Additionally, miRNAs have been correlated to the onset and progression of a vast number of human pathological conditions (Li & Kowdley, 2012; Soifer, Rossi, & Sætrom, 2007), including several types of cancer (Rupaimoole & Slack, 2017).

Despite their diagnostic and biomedical potential, there are several technical challenges that hinder the study of miRNAs. First, they have very short sequences that exhibit a high level of homology among them, particularly between those belonging to the same family (Roush & Slack, 2008). Therefore, a very high level of specificity is required to selectively analyze each miRNA. Also, miRNAs only represent around 0.01% of total RNA mass in a given transcriptome (Dong et al., 2013) and usually may need to be studied in a highly complex cellular or molecular environment. Moreover, the regulatory effect of miRNAs is exerted by an imperfect base pairing between the miRNA and its targets, which is prone to promiscuity; in complex organisms, a single miRNA is able to regulate hundreds of mRNAs (Breda, Rzepiela, Gumienny, Van Nimwegen, & Zavolan, 2015). In consequence, the laboratory techniques designed for the analysis of miRNAs should offer specificity and sensitivity to detect low-abundance species and flexibility to characterize the complex regulatory networks where miRNAs are involved.

Methods for miRNA discovery and detection

Robust and accurate miRNA detection has led to the identification of numerous miRNAs implicated in the regulation of a variety of key cellular and molecular processes (Gulyaeva & Kushlinskiy, 2016; Liu & Olson, 2010; Paul et al., 2018; Russo et al., 2018). The development of highly specific and sensitive detection techniques has also played an important role in the discovery of miRNA-based biomarkers in cancer, cardiovascular, autoimmune, and other diseases (Detassis et al., 2019; Hagiwara, Kantharidis, & Cooper, 2014; Wang, Chen, & Sen, 2016). Accurate techniques to quantify and detect aberrant miRNA expression are paramount for research and for early clinical diagnosis. However, due to their small sequence size as well as the often low abundance and high sequence homology among family members, correct miRNA detection and quantification prove to be challenging tasks (Table 8.1).

At a cellular level, the presence of miRNA can only be accurately demonstrated by the existence of the mature miRNA form. Over the last decades, several methods have been used and new ones developed to study and evaluate the abundance

MicroRNA in Human Malignancies. https://doi.org/10.1016/B978-0-12-822287-4.00008-6

TABLE 8.1 Conventional miRNA discovery and detection methods.

Detection method	Application	Advantages	Disadvantages
Northern blotting	Discovery detection	High specificity Established technique that does not require specific laboratory equipment	Low-throughput and low sensitivity Laborious and very time consuming, requires large amounts of RNA
Microarray analysis	Detection high-throughput screening quantification	Low-cost and high-throughput considering the number of samples that can be processed per day Capacity to multiplexing	Semiquantitative, because it is suitable for comparing the relative abundance of specific miRNAs between two states; Low sensitivity Long turnaround time Poor degree of automation Does not identify new miRNAs
qPCR-based methods	High-throughput screening quantification	High sensitivity and specificity Can be used for absolute quantification Easy to incorporate into the workflow for laboratories that are used to qPCR Highly reproducible and suitable for routine measurements in laboratory settings	Does not allow the identification of novel miRNAs Costly
Next-generation sequencing (NGS)	Discovery detection quantification	Widest range of applications Very high sensitivity and accuracy in distinguishing different forms of miRNAs very similar in sequence	Generates large amounts of information that require substantial computational support for data analysis Costly technique, not yet suitable for high-throughput laboratory settings

of miRNAs across various cell types, tissues, and organisms. Northern blotting is frequently used to identify and determine the expression of individual miRNAs while microarray analysis, real-time qPCR, or next-generation sequencing (NGS) are able to evaluate the levels of several miRNAs in parallel.

Northern blot

Northern blotting is a classical method extensively used for the detection of mature miRNAs as well as its precursors and it does not require specialized equipment (Berezikov, Cuppen, & Plasterk, 2006; Lagos-Quintana, Rauhut, Lendeckel, & Tuschl, 2001). In northern blotting, total RNA preparations are separated by gel electrophoresis, transferred to a nylon membrane, and further hybridized with complementary probes to detect miRNAs (Pall, Codony-Servat, Byrne, Ritchie, & Hamilton, 2007; Torres, Fabani, Vigorito, & Gait, 2011).

Although Northern blotting is a reliable and specific method, it has some limitations, including being semiquantitative, time-consuming, requiring large amounts of RNA, and showing limited analytical output. Additionally, because of the small size and low abundance of miRNA molecules, the use of probes with increased sensitivity is often critical to detect miRNAs. Since the first application of Northern blotting for miRNA identification, and considering the limitations mentioned, numerous improvements have been described to increase the sensitivity and specificity of this method (Koscianska et al., 2011; Koscianska, Starega-Roslan, Czubala, & Krzyzosiak, 2011; Pall et al., 2007; Pall & Hamilton, 2008; Ramkissoon, Mainwaring, Sloand, Young, & Kajigaya, 2006; Válóczi et al., 2004; Várallyay, Burgyán, & Havelda, 2007). Most of the advances were concentrated on the labeling and design of the probes used to identify the miRNAs. In general, probe labeling in Northern blotting is based on the incorporation of radio isotopes, namely 32-phosphorus (^{32}P). However, isotope labeling is commonly hazardous, inconvenient, and limited by some facilities. Alternatively, digoxigenin (DIG) labeling is a safer system with several advantages, namely a higher sensitivity, increased safety, and a shorter exposure time. In fact, the use of DIG labeled probes for the detection of miRNAs has proved to be more sensitive than ^{32}P labeled probes (Kim et al., 2010;

Ramkissoon et al., 2006; Wu et al., 2014). Important progress was also achieved in probe design strategies, as traditional DNA oligonucleotide probes have been progressively replaced by locked nucleic acid (LNA) oligonucleotide probes that significantly improve the sensitivity in detecting miRNAs (Várallyay et al., 2007). LNA-modified oligonucleotide probes were first applied for the detection of miRNAs by Northern blotting analysis in plants in 2004 (Válóczi et al., 2004). LNAs are nucleic acid analogs that contain a methylene bridge connecting the 2′ oxygen and 4′ carbon. This bridge locks the ribose in the 3′-endo conformation, restricting the flexibility of the ribofuranose ring and locking the structure into a rigid bicyclic formation. This structural conformation promotes enhanced thermal stability, increasing the melting temperature and hybridization specificity of a primer or a probe. Designing acid nucleic probes with LNA can significantly improve the thermal stability, hybridization efficiency, and sensitivity in miRNA determination (Várallyay et al., 2007).

Hybridization microarrays

Like Northern blotting, microarray RNA analysis depends on the sensitive and specific hybridization of the target RNA to a complementary oligonucleotide probe. Microarrays were developed in the early 1990s for whole-genome DNA analysis and later were applied as one of the first high-throughput methods for the analysis of miRNA expression (Pritchard, Cheng, & Tewari, 2012; Thomson, Parker, Perou, & Hammond, 2004). Microarray assays allow the parallel analysis of large numbers of miRNAs and can be used to detect their presence and relative expression (Li & Kowdley, 2012; Soifer et al., 2007). Due to the nature of the detection protocol based on nucleic acid hybridization, microarray results always require an additional validation step, usually performed by quantitative PCR (Dangwal, Bang, & Thum, 2012).

The first step in miRNA microarray profiling is the purification of miRNAs from cells. For this purpose, several protocols have been designed for the extraction of high-quality RNA using commercial kits and reagents. Even though it is possible to use total RNA for microarray analysis, miRNA enrichment protocols can increase sensitivity because small RNAs represent around 0.01% of all varying significantly across different cells and tissues (Gaur et al., 2007; Liang et al., 2005). MiRNA labeling is usually achieved by dephosphorylation of the 5′ end of the miRNA followed by ligation of a labeled oligonucleotide with T4 ligase. The dephosphorylation of the 5′ end is a crucial step to avoid self-circularization of the miRNA and/or adapter sequences (Berezikov et al., 2006). The miRNA labeling can be carried out either directly on the miRNA, by binding a fluorescent dye (Babak, Zhang, Morris, Blencowe, & Hughes, 2004; Thomson et al., 2004) or a biotin residue (Gaur et al., 2007; Liang et al., 2005), or indirectly following a reverse-transcription step (Gulyaeva & Kushlinskiy, 2016; Liu & Olson, 2010; Paul et al., 2018; Russo et al., 2018) and/or a PCR amplification (Baskerville & Bartel, 2005; Miska et al., 2004). After labeling, the miRNA is then directly hybridized with the appropriate capture probes specific for mature miRNAs immobilized on a glass slide on the array surface, typically followed by two-channel fluorescent imaging.

Due to the small sequence size and low abundance of microarrays, their use in studying and quantifying miRNAs is challenging. In gene expression microarrays, capture probes consist of either synthetic oligonucleotides or cDNA fragments, designed to have a high specificity and affinity for individual transcripts. As miRNA are small, probe design is difficult, and considering the miRNA sequence, their melting temperature (*Tm*) can vary significantly (from 45°C to 74°C). In a microarray experiment, if the probes' *Tm* diverges significantly, the binding affinities of the miRNAs in analysis will differ accordingly. Therefore, the method cannot be considered quantitative. Therefore, the size and sequence of the probe should be carefully considered in the function of the physicochemical characteristics of the target miRNA. This will enable a greater balance between melting temperatures of the miRNA probes and make the affinity comparable for all probes, providing a more precise analysis (Detassis et al., 2019; Hagiwara et al., 2014; Wang et al., 2016). To improve probe design and miRNA binding affinities, several modifications to microarray-based protocols have been developed (Lee et al., 1993). In early reports, DNA-based microarrays were described for the detection of miRNAs (Liu, Calin, Volinia, & Croce, 2008). However, DNA-based oligonucleotide array platforms presented a major disadvantage due to the difficulty in normalizing the *Tm* of the probe set, necessary for genome-wide miRNA profiling. Again, the introduction of LNA modified probes in the miRNA microarray methodology allowed for *Tm* normalization of the array capture probes, increasing the *Tm* to 5–8°C and promoting the binding affinity of the capture probe (Castoldi, Schmidt, Benes, Hentze, & Muckenthaler, 2008). The incorporation of LNA favored discrimination between single nucleotide differences, improving the identification between closely related miRNA family members. Also, the higher sensitivity prevented the need for RNA size selection and/or amplification (Castoldi et al., 2006). Other strategies to standardize the assay *Tm* included the use of 2′-*O*-(2-methoxyethyl)-modified oligoribonucleotides (Beuvink et al., 2007) and adjusting the length of probes according to the physicochemical traits of the target miRNA (Wang, Ach, & Curry, 2007). The addition of the hairpin structure onto the 5′ end of the probe allows distinguishing the targeted miRNA from larger RNAs, such as miRNA precursors (Liu, Calin, et al., 2008).

The detection of a positive hybridization signal in microarrays depends on the proper sample labeling. Despite the initial use of radioactive isotopes for miRNA sample labeling (Krichevsky, King, Donahue, Khrapko, & Kosik, 2003), the preferred labeling protocols for miRNA detection by microarrays are based on the use of fluorescent probes (Neely et al., 2006; Nelson et al., 2004). Because of the particularly small size of miRNA molecules, direct labeling might be advantageous and can be accomplished by enzymatic or chemical labeling. Enzymatic labeling consists of the use of T4 RNA ligase to attach a fluorophore-conjugated nucleotide or a short oligonucleotide to the 3′ end of miRNA, directly or after the 3′-tailing of miRNA (Pritchard et al., 2012). In both cases, the enzyme may favor certain sequences over forming sequence artifacts, which is more evident in degraded RNA samples. Also, such biases tend to be more serious in degraded samples (Baker, 2010; Sato, Tsuchiya, Terasawa, & Tsujimoto, 2009). Another limitation of the enzymatic labeling techniques is that they are unable to detect 5′ end mismatches due to their 3′ activity. Alternatively, miRNA labeling can be performed by chemical procedures comprising alkylation-based labeling of the miRNA and protocols based on platinum coordination chemistry with nucleic acids (Pritchard et al., 2012). Still, chemical labeling methods do not distinguish 3′ end modifications, and can introduce bias by labeling some nucleotides with higher efficiency (Wark, Lee, & Corn, 2008). In both enzymatic and chemical methods, coexisting pre-miRNAs can also be labeled, resulting in a background signal and cross-hybridization (Duan et al., 2011).

Although DNA microarrays are relatively easy to use and analyze, there are some limitations in their application to miRNA detection and profiling (Hunt, Broyles, Head, & Deo, 2015). MiRNA microarrays are only semiquantitative due to the lack of a calibration curve in the experimental workflow. Therefore, microarrays are best suited to determine the relative difference of miRNA expression levels between two states (i.e., control vs. disease). On the other hand, to guarantee specificity, microarrays often require additional validation. Microarrays also have a smaller dynamic range than both qRT-PCR and next-generation sequencing. Also, although being a high-throughput and highly multiplexed technique, classic miRNA microarrays show a relatively low sensitivity, with detection limits around the nanomolar range.

Quantitative PCR

In both research and clinical diagnosis, qRT-PCR has become a regular and reliable technique for detecting miRNA expression due to its high sensitivity and specificity and its broad functional dynamic range (Duan et al., 2011). The principle of the qRT-PCR method entails two main steps: cDNA synthesis using reverse transcription followed by detection of the amplified products using a conventional qPCR with either an intercalating dye or a TaqMan probe. In the first stage of the qRT-PCR procedure, RNA is converted to cDNA by reverse transcription. However, the limited length of the mature miRNA (~22 nt), the fact that the mature miRNA sequence is also present in the pri- and pre-miRNA transcript, and the lack of a common sequence feature such as a poly(A) tail, prevent the use of a conventional reverse-transcription protocol for miRNA analysis. There are two different strategies for the reverse-transcription of miRNAs: the first approach uses specific primers for each miRNA, whereas in the second method, polyadenylation of all miRNAs by Escherichia coli poly(A) polymerase is followed by reverse transcription using universal primers with an oligo (dT) sequence on its 5′ end (Reichenstein, Aizenberg, Goshen, Bentwich, & Avni, 2010; Ro, Park, Jin, Sanders, & Yan, 2006).

For the reverse transcription using miRNA-specific primers, the prepolyadenylation step is avoided and miRNAs can be directly reverse transcribed into cDNA with stemloop primers (Chen et al., 2005; Varkonyi-Gasic & Hellens, 2011), linear primers (Raymond, Roberts, Garrett-Engele, Lim, & Johnson, 2005), or two-tailed primers (Androvic, Valihrach, Elling, Sjoback, & Kubista, 2017). For several years, the miRNA-specific reverse transcription, with stemloop primers, was a popular method to study mature miRNAs (Beuvink et al., 2007; Krichevsky et al., 2003; Liu, Spizzo, Calin, & Croce, 2008). Stemloop primers are designed with a short single-stranded region complementary to the sequence on the 3′ end of the miRNA, a double-stranded part (the stem), and a loop containing the universal primer-binding sequence. The cDNA generated is then used as a template for qPCR using an miRNA-specific forward primer and a universal reverse primer. To detect amplification, an miRNA-specific TaqMan probe is used and the fluorescence is measured to quantify mature miRNA levels. Modified versions of the TaqMan assay that allow for multiplexing (Lao et al., 2006), the use of a universal TaqMan probe (Jung, Jiang, Kaufmann, & Patzel, 2013), or SYBR green detection (Tong, Xue, Xiong, Xiao, & Zhou, 2015) have also been reported.

In the miRNA reverse transcription approaches based on polyadenylation, all 3′ ends of miRNAs are first tailed to have a common sequence, such as poly(A or U) by using poly(A or U) polymerase and employing a poly(T) adapter to generate a cDNA (Mei et al., 2012; Shi & Chiang, 2005). For PCR amplification, an miRNA-specific forward primer and a reverse primer that binds to a region in the poly(T) adapter are used, and quantification is measured through SYBR green fluorescence. Polyadenylation-based methods tend to be more suitable for detecting several miRNAs from a very small amount of starting material, such as biofluids (Baskerville & Bartel, 2005; Liu et al., 2004). However, they do not accurately differentiate between pre- and pri-miRNAs and mature miRNAs because all RNA molecules will be polyadenylated (Benes & Castoldi, 2010). Because optimal annealing conditions may vary substantially among miRNAs due to their sequence variability and size, performing parallel qRT-PCR for the detection of multiple miRNAs may prove to be challenging. To overcome this obstacle and standardize optimal miRNA primer annealing conditions, LNA nucleotides can be incorporated in the primer sequence (Pritchard et al., 2012). Over the last decade, several biochemistry companies, including Applied Biosystems, Exiqon, Qiagen, Fluidigm, Thermo-Fisher, Genecopeia, Eurogentec, and SA-Biosystems, have offered qRT-PCR panels that can evaluate the expression of hundreds of miRNAs in parallel, with some offering customizable assays (Nelson et al., 2004). The development of qRT-PCR methods for miRNA detection and quantification has improved the sensitivity down to a few nanograms of total RNA (Dangwal et al., 2012). However, parameters such as RNA integrity control, cDNA synthesis, primer design, amplicon detection, and normalization must be taken into account to obtain significant and reproducible results.

An important part of the qRT-PCR experiment is analyzing the data after generation. As miRNAs represents only ~0.01% of the total RNA extracted from a sample, the effective amount present may vary considerably depending on the sample preparation, extraction yield, and integrity. Therefore, the normalization of miRNA amplification data is of extreme importance (Vandesompele et al., 2002). Widespread normalization methods rely on the expression of genes for small RNAs that might not be transcribed by the same polymerases and therefore are less related to miRNA regulation (Baker, 2010). MiRNAs found to have higher stability than U6 snoRNA, across several human tissues, could be used for qRT-PCR normalization, namely miR-191, miR-93, miR-106a, miR-17–5p, and miR-25 (Peltier & Latham, 2008). The use of external controls, such as exogenous "spike-in" RNAs, is significant for data normalization, particularly for the quantification of miRNAs in body fluids such as plasma, where the level of miRNAs is low and inconsistent (Dangwal et al., 2012; Hagiwara et al., 2014). Spike-in RNAs controls are also frequently used, mainly to account for variability in RNA extractions. Several authors have reviewed and proposed different strategies for the normalization of miRNA expression data, in particular for the multiplex analysis of miRNAs (Mestdagh et al., 2009; Mohammadian et al., 2013; Qureshi & Sacan, 2013; Schwarzenbach, Da Silva, Calin, & Pantel, 2015; Wylie, Shelton, Choudhary, & Adai, 2011). When a large number of miRNAs is being evaluated in parallel, the geometric mean miRNA expression present in each sample can be used to normalize qRT-PCR data (Mestdagh et al., 2009; Mohammadian et al., 2013; Qureshi & Sacan, 2013; Schwarzenbach et al., 2015; Wylie et al., 2011).

Next-generation sequencing

Over the last two decades, technological advances have revolutionized the way that genetic and transcriptomic studies are undertaken. The development of large-scale massive parallel sequencing, or next-generation sequencing (NGS) technologies, was prompted by the need for higher-throughput sequencing of large genomes at a lower cost, and led to the development of many second-generation sequencing technologies using an array of original methodologies (Bar et al., 2008; Juzenas et al., 2017). Throughout the last 15 years, high-throughput RNA sequencing (RNA-seq) has rapidly evolved and new techniques are continually being developed and commercialized. RNA-Seq has clear advantages over existing approaches, providing new means for both mapping and quantification, and is expected to transform how transcriptomes are studied.

The development of NGS platforms has promoted a third major approach for miRNA detection and expression profiling. With the continuing decrease of the sequencing cost, this technology is rapidly becoming a leading method applied to the analysis of miRNAs. Although the technologies and procedures vary substantially depending on the specific platform, several steps are common in the laboratory part of miRNA sequencing. The first step involves the extraction of good quality RNA. The quality of the RNA needs to be verified before sequencing to ensure that all the biologically relevant fragments are sequenced and degradation products do not influence the results. Next, total RNA is fractioned and miRNAs are size separated by gel electrophoresis, where the band corresponding to the miRNAs is selected for further processing. This process eliminates the bigger fragments, including mRNAs and rRNAs, from the samples. After miRNA preparation, cDNA libraries are prepared with a reverse transcription in a process similar to RT-PCR. Next, adaptors are ligated to both 5′ and 3′ ends of the cDNA fragments, and the resulting products are attached to either a planar or bead-based immobilization surface. Following library preparation, millions of individual cDNA fragments are ready to be sequenced in parallel. The application of NGS methodologies for miRNA research brings great advantages as it is suitably sensitive and cost-effective due to the high capacity for sample multiplexing. In addition, it allows for the discovery of novel miRNA species (Friedländer et al., 2008) and enables the detection of miRNA isoforms as a result of the NGS ability for single-nucleotide resolution measurements (Neilsen, Goodall, & Bracken, 2012).

For miRNA deep sequencing, there are important considerations associated with small RNA library preparation, such as the need to size-select the small RNA species, biased adapter ligation, formation of adapter dimers, and the need to adapt for very low RNA input protocols, in particular when biofluids (e.g., plasma) are analyzed (Alon et al., 2011; Baker, 2010; Hafner et al., 2011; Linsen et al., 2009). In fact, some adapter-miRNA sequences may be favored over others during ligation and PCR amplification, and thereby result in over- or underrepresentation of miRNAs in the sequencing library preparation (Fuchs, Sun, Zhuang, & Robb, 2015; Raabe, Tang, Brosius, & Rozhdestvensky, 2014). On the other hand, the formation of adapter dimers or ineffective size fractioning may hinder the enrichment of miRNAs over other RNA species and therefore reduce the number of reads (Raabe et al., 2014). Additionally, the use of NGS techniques in the development of miRNA-based biomarkers in biofluids is still complex due to the fact that there is very low RNA present. One more hindrance that prevents the universal use of NGS technology for miRNA research, and in the clinical setting, is the fact that it requires computational analysis and validation to distinguish significant data from the noise.

Nevertheless, studies with RNA-seq data (mainly from Illumina or SOLiD platforms) have shown the large potential of this technology for identifying the complete profiles of miRNA variability in several cells and tissues (Bar et al., 2008; Juzenas et al., 2017; Morin et al., 2008, 2010). A comparison of the miRNA detection performance between a microarray (Illumina) and two NGS (SOLiD and Illumina) platforms and the NanoString nCounter system reported a strong correlation between methods and a better detection rate of NGS, including the analyses of formalin-fixed, paraffin-embedded tissues (Tam, De Borja, Tsao, & Mcpherson, 2014). The available data have demonstrated that NGS, as qRT-PCR, exhibits a wide dynamic range and does not experience fold-change compression, as with the microarray assays.

Functional miRNA analysis

MiRNAs exert their regulatory action by sharing a spatial-temporal background with their cognate targets. This background is conditioned by the target complementarity, but also by the presence of additional regulatory factors. The demonstration of an miRNA functional effect over an mRNA target requires considering all these elements. Ideally, the miRNA functional analysis should be performed in the original cell or tissue, but sometimes system models are required, especially in validation studies.

Classical target validation methods: Reporter genes and RISC immunoprecipitation

After collecting evidence about the putative regulation of an miRNA exerted over an mRNA transcript, the next step is to demonstrate their functional relationship. Over the last two decades, the golden standard to validate miRNA targets was the use of gene reporters (Fig. 8.1). A typical validation set-up includes cloning the target sequence (typically the 3′-UTR of the mRNA) downstream to a reporter gene, and the use of a parallel cellular system to transfect the reporter vector and the

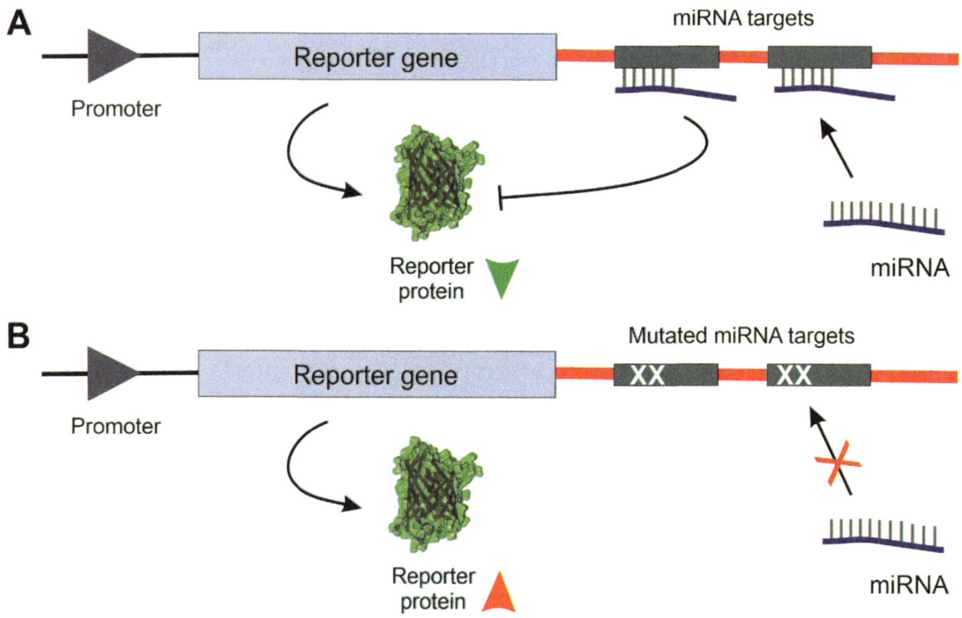

FIG. 8.1 miRNA target validation by reporter assays. (A) the 3′-UTR of the putative miRNA target (*red line*) is cloned downstream to the reporter gene, which is controlled by a strong promoter. Cotransfection of the reporter cassette together with the tested miRNA will result in an miRNA binding to the target sites in the cloned 3′-UTR and a subsequent decrease on the reported protein. (B) Targeted mutations of the miRNA binding sites that abolish miRNA binding will result in an increase of expression of the reporter gene.

selected miRNA (Nicolas, 2011). A regulatory effect of the miRNA over its target will induce a decrease in the expression of the reporter gene that can be quantified by the detection of the encoded protein. The generation of site-directed point mutations in the miRNA-binding sites will prevent the regulatory effect of the miRNA, resulting in an increase of the reporter levels (Alvarez, 2014). A reporter gene should encode a stable and easy-to-detect protein. The most widely used reporters for miRNA target validation are the firefly luciferase (Nicolas, 2011) and less frequently the fluorescent proteins GFP, YFP, or mCherry (Turk et al., 2018).

Despite their wide use, gene reporters are not optimal systems for miRNA target validation because they present important technical problems. One of the most important drawbacks of gene reporters, especially observed in the luciferase system, is related to the global normalization of the signal across all the analyzed samples. A possible solution relies on the use of a quantitative dual luciferase reporter system, where firefly luciferase is used to report the microRNA-target sequence interaction and Renilla luciferase is used as an internal standard to normalize expression between replicates (Moyle et al., 2017). Moreover, due to their experimental characteristics, gene reporter experiments are limited in the sample size, which prevents their use in high-throughput validations. To increase the sample size, Wolter and coworkers introduced a medium-throughput assay based on luciferase and designated it as 3′-LIFE (luminescent identification of functional elements in 3′-UTRs) (Wolter, Kotagama, Pierre-Bez, Firago, & Mangone, 2014). The system is based on the use of a precloned library of 3′-UTR fragments that are cotransfected together with an miRNA in cell growth in a microtiter plate to perform luciferase reactions. The 3′-LIFE method demonstrated its feasibility using a library composed of 275 human 3′-UTRs and two cancer-related miRNAs, let-7c and miR-10b, in HEK293T cells (Wolter et al., 2014).

The use of reporter genes is a flexible technology that is easy to implement but time consuming and limited in its output. Another characteristic to take into account is that the manipulation of the reporter involves transfection into model cell systems, meaning that the detection and quantification of the miRNA regulatory effect over the target is determined in a cellular system that could be totally unrelated to the original source (Tomasello, Cluts, & Croce, 2019). The readout of the reporter systems is also prone to produce false positive results, due to the transient high levels of the reporter mRNA and the transfected miRNA (Moyle et al., 2017). Due to these technical limitations, the use of reporter genes for miRNA target validation is currently limited to confirmation experiments.

The limitations of the use of gene reporters can be partially circumvented by the isolation of RISC-associated RNAs and their quantification by qPCR or other methods (Dzikiewicz-Krawczyk et al., 2018). This approach has a clear advantage over the reporters because it allows a direct characterization of the miRNAs and targets engaged in the RISC complex in the studied cells, avoiding artificial manipulations of the biological system (Teteloshvili et al., 2017). A typical protocol for the isolation of RISC-associated RNAs is based on the RNA immunoprecipitation method (RIP), where the precipitating antibodies are normally targeted to Ago2, the main component of the RISC complex. To ensure the stability of the RNA-protein complexes during the immunoprecipitation, it would be advisable to perform a chemical or UV crosslinking reaction prior to treatment with the antibodies. The protein fraction of the immunoprecipitated RNA-protein complexes is eliminated by proteinase K digestion and the remaining RNAs are used for the specific detection of miRNAs and mRNA transcripts. This protocol has been successfully employed for the characterization of mRNA targets of the oncogenic miR-21 in a cell line model by the quantification of the relative enrichment levels of specific mRNAs in the RISC-associated RNA fraction (Teteloshvili et al., 2017). The method has enough flexibility to be adapted to different biological and clinical samples. For instance, Dzikiewicz-Krawczyk and coworkers described the use of Ago2-RNA immunoprecipitation for the characterization of different primary Burkitt lymphoma tumors and normal B cells (Dzikiewicz-Krawczyk et al., 2018). The analysis of the immunoprecipitated RISC-associated RNAs allowed determining that in contrast to normal B cells, Ago2-associated mRNAs in primary Burkitt lymphoma were enriched for targets of the oncogenic miR-17 to 92 cluster, including genes involved in the cell cycle and regulation of apoptosis (Dzikiewicz-Krawczyk et al., 2018). The RISC-associated RNAs can also be isolated by taking advantage of other RISC components and different capturing approaches. For instance, an overexpressed GFP-tagged version of the GW182 protein has been used to identify miRNA-target associations in *Caenorhabditis elegans* by capturing the RISC complexes with an anti-GFP antibody (Zhang et al., 2007).

The major drawbacks of the described capture methods are related to the abundance of false positive results (Steinkraus, Toegel, & Fulga, 2016) and also with the limited relevance of the RISC enrichment events in specific biological circumstances (Zhang et al., 2007). The number of false positive results can be minimized by the stabilization of RNA-protein complexes prior to capture, either by chemical or radiation-induced crosslinking (Wang, Wilfred, Hu, Stromberg, & Nelson, 2010). The relevance of the results can be enhanced by a careful experimental design, which could be facilitated by the selection of proper cell models and testing protocols. The application of external factors to the system such as treatment with drugs or other stimuli needs to be compared with proper controls to determine the differences in the enrichment patterns of RISC-associated RNAs (Zhang et al., 2007). In conclusion, RISC capture protocols are an interesting approach for the study of RISC-associated RNAs in a wide range of biological systems, but they do not provide a direct demonstration of the miRNA-mRNA targeting, and need to be complemented with other methods.

High-throughput methods for target identification

Eukaryotic cells are extremely redundant in their metabolic pathways and regulatory events, having many different alternatives to control the same biological process. At the organism level, the redundancy in control and regulatory functions increases organism complexity (Sugimoto et al., 2009). Translating these facts to the miRNA regulatory functions, in simpler organisms such as worms or flies it is more likely to find miRNAs that are solely involved in the regulation of important cellular processes such as proliferation or apoptosis (Chen et al., 2005). However, in complex eukaryotic organisms, the existence of entangled regulatory networks where several miRNAs are simultaneously involved in the regulation of a specific metabolic process is expected (Panni, Lovering, Porras, & Orchard, 2020; Vandevenne, Delmarcelle, & Galleni, 2019). The study of miRNA-centered regulatory networks in high eukaryotic organisms requires the application of postulates of systems biology (Gabriel, Costa, & Enguita, 2020). In consequence, the identification of the functional relationships between miRNAs and their cognate targets should be performed by using high-throughput methods. The transcriptome-wide analysis of miRNA-target interactions is performed by an initial capture of the RISC-engaged RNAs, followed by an analysis by next-generation sequencing (Fig. 8.2).

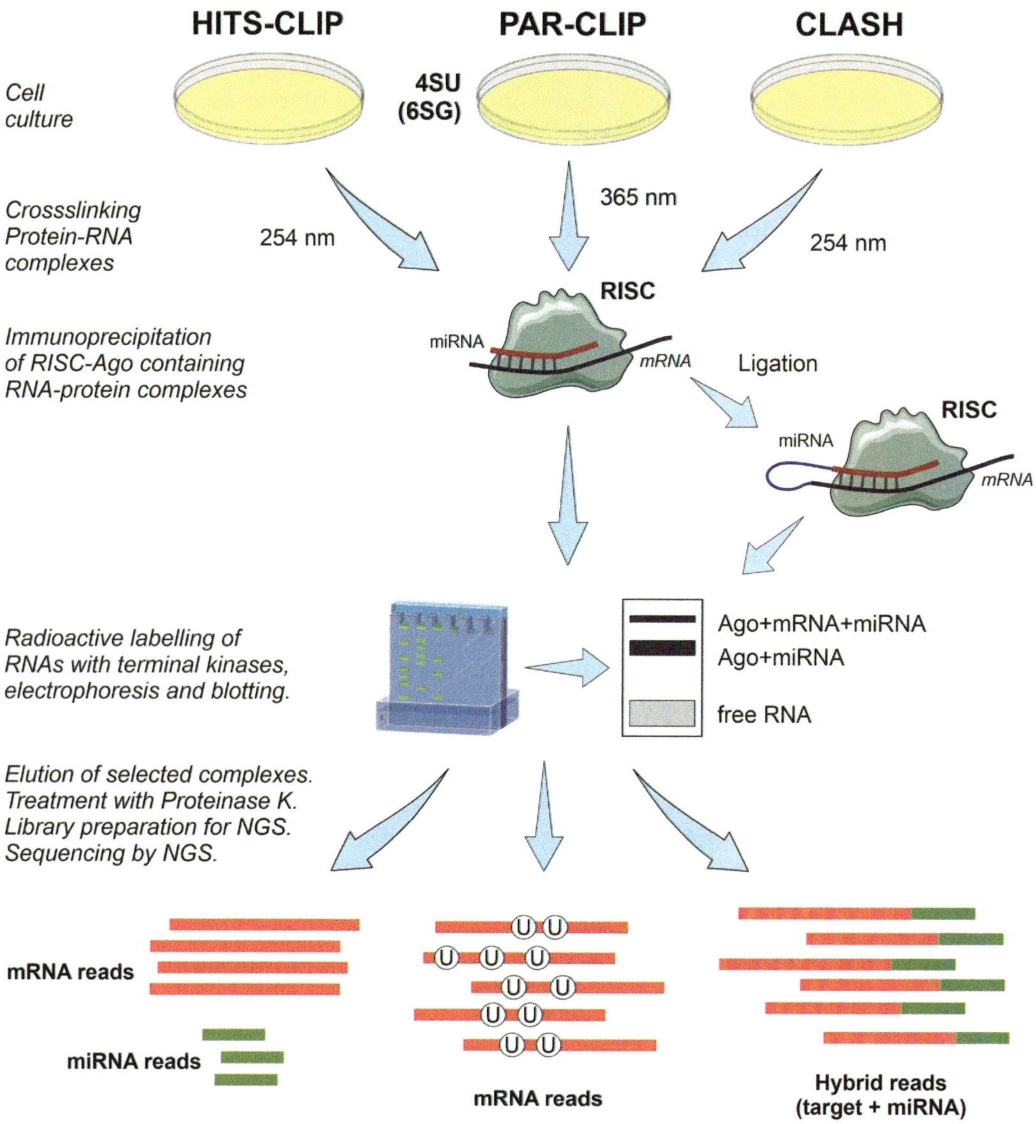

FIG. 8.2 High-throughput methods for the identification of miRNA targets based on the application of crosslinking, the immunoprecipitation of protein-RNA complexes, and next-generation sequencing. In the HITS-CLIP and CLASH methods, the RNA-protein complexes that contain miRNAs and their cognate targets are first crosslinked by UV radiation, and immunoprecipitated by anti-Ago2 specific antibodies. The PAR-CLIP method uses 4-thiouridine in the cell culture medium, which will facilitate a more specific and less noisy crosslinking of protein-RNA complexes by low-energy UV light irradiation.

The initial platforms for high-throughput identification of miRNA targets were modifications of the crosslinking and immunoprecipitation method (CLIP). The high-throughput sequencing of the Ago2-associated RNA (HITS-CLIP) method was originally designed for the study of RNA-binding proteins, but lately modified for the analysis of miRNA-mRNA interactions (Grozdanov & Macdonald, 2014). In the HITS-CLIP protocol, the RISC-associated RNAs are first crosslinked to their interacting proteins by UV irradiation at 254 nm. After crosslinking and cell disruption, a treatment with the RNAse enzyme will trim the RNAs, which are not protected by the RISC proteins, and the Ago2-containing complexes will be further immunoprecipitated with anti-Ago2 antibodies (Spengler et al., 2016). Probably the most technically demanding step of HITS-CLIP is the separation of the Ago2-miRNA-mRNA complexes from the Ago2-miRNA ones. The separation is performed by initial RNA radio-labeling, followed by SDS-PAGE electrophoresis and blotting onto nitrocellulose membranes. After the electrophoretic separation, the higher molecular weight bands will correspond to the ternary complexes formed by the RISC-associated miRNAs acting over their cognate mRNA targets. The classical HITS-CLIP protocol would continue by an in situ treatment of membranes with proteinase K to eliminate proteins from the complexes, and the RNA isolation for sequencing. After library preparation and next-generation sequencing, an HITS-CLIP experiment will generate two different datasets, one for the miRNAs and the other for their targets (Zhang & Darnell, 2011). The functional relationships between miRNAs and their targets are established on the basis of their presence in the HITS-CLIP results, complemented by previous information collected from miRNA target prediction algorithms (Chi, Zang, Mele, & Darnell, 2009). The HITS-CLIP method has been extremely useful not only for the characterization of miRNA-mRNA interactomes, but also to pinpoint new rules in the definition of miRNA targets (Chi, Hannon, & Darnell, 2012). The HITS-CLIP protocol has been successfully employed for the characterization and mapping of the miRNA-binding sites in the human myocardium, showing the advantages of the application of the method over nonproliferative tissue (Spengler et al., 2016). In a clinical context, Ago2 HITS-CLIP data showed the production of miRNAs from the herpes virus associated with Kaposi's sarcoma and their functional targets in the infected cells (Haecker et al., 2012). The establishment of complex miRNA-centered regulatory networks in gastric cancer has also been characterized by HITS-CLIP, showing the importance of a specific miRNA fingerprint in cancer progression and metastasis (Zhang, Shen, & Cui, 2019).

However, the HITS-CLIP method uses a harsh crosslinking protocol based on high-energy UV radiation, being prone to the production of false positives and a relatively noisy background in the next-generation sequencing data (Grozdanov & Macdonald, 2014). The crosslinking efficiency in CLIP experiments was improved by the development of the PAR-CLIP method (photoactivatable ribonucleoside enhanced-CLIP). In this protocol, cell cultures are grown in the presence of 4-thiouridine (4SU) or 6-thioguanosine (6SG), which are incorporated in the synthesized RNAs in a random fashion (Spitzer et al., 2014). These nucleotides are very sensitive to photoactivation at longer wavelengths (365–370 nm), allowing a more selective crosslinking of RNA-protein complexes. The incorporation of 4SU in RNA molecules will induce the transition of thymidine to cytidine in the preparation of the sequencing cDNA libraries (Hafner et al., 2010). This conversion occurs at higher frequencies in crosslinked sites, and in consequence, in the location of miRNA targets; it constitutes an elegant strategy to filter true signals from noise (Fig. 8.2.). The analysis of PAR-CLIP data often results in the construction of more detailed molecular maps of the miRNA-mRNA interactions, when compared with the classical HITS-CLIP protocol (Erhard, Dölken, Jaskiewicz, & Zimmer, 2013; Jens, 2016). The PAR-CLIP method has been successfully applied for the construction of miRNA regulatory networks in several contexts (Farazi et al., 2014; Haecker & Renne, 2014).

HITS-CLIP and PAR-CLIP were originally designed for the transcriptome mapping of RNA-binding proteins, and adapted to the analysis of miRNA targets by selective immunoprecipitation with anti-Ago2 antibodies. Both protocols are able to map miRNA-binding sites, but they are limited in the establishment of functional relationships between a specific miRNA and its mRNA targets (Haecker & Renne, 2014). A new method named CLASH (crosslinking, ligation, and sequencing of hybrids) was developed with the objective of finding direct connections between miRNAs and targets in a genome-wide approach (Helwak, Kudla, Dudnakova, & Tollervey, 2013). The CLASH protocol is essentially an HITS-CLIP that includes an additional ligation step after Ago2 immunoprecipitation (Helwak & Tollervey, 2016). The ligation step will produce a RNA hybrid by the covalent connection of every miRNA and its target when they are blocked within the RISC complex. In consequence, the protocol will establish a direct functional connection between miRNAs and targets (Li & Hu, 2019) that will be detected in the form of chimeric RNA reads composed by miRNAs and their mRNA targets (Li & Hu, 2019). The CLASH protocol allows determining RNA interaction maps that are extremely detailed, and the flexibility of the protocol allows its application to several biological scenarios not necessarily reduced to cell cultures or simple models (Helwak et al., 2013; Helwak & Tollervey, 2016). CLASH interaction maps have been extremely useful not only to characterize regulatory networks, but also to rewrite the rules that govern the target pairing between miRNAs and mRNAs (Lu & Leslie, 2016).

Functional validation

Phenotypic consequences of miRNA regulation can be detected at different levels, ranging from the cells to the whole organisms. The external manipulation of the miRNA levels is a useful method to study their regulatory actions, complementing the information obtained from the target analysis and validation. The laboratory methods used for miRNA level manipulation can act at the transcriptomic or genomic levels (Steinkraus et al., 2016).

When an isolated miRNA or a group of miRNAs is upregulated in a particular process and functionally related with the observation of a phenotype, a decrease in miRNA levels will be correlated with the observed characteristics. To decrease the miRNA levels, the most widely used protocol is based on the inhibition by complementary base pairing using miRNA antagonists or "antagomirs" (Li et al., 2010). Antagomirs are short complementary oligonucleotides that have a reverse complement sequence to a specific miRNA. Their chemistry can be diverse, but the RNA-based antagomirs are preferred due to their favorable characteristics for generating stable hybrids with miRNAs (Lu et al., 2016). The administration of antagomirs can be performed in cellular systems by transfection but they are also suitable to be incorporated into animal models (Zhou et al., 2016). There are several examples in the literature describing the use of antagomirs in cellular models. For instance, the transfection of human corneal epithelial keratinocytes with the anti-miR-205 antagomir reduced the internal levels of miR-205, and enhanced the adhesion ability of these cells by reverting the regulatory effect of miR-205 (Li et al., 2010). In more complex systems such as animal models, antagomirs administered by intravenous injection have shown interesting properties as therapeutic strategies for those conditions involving a dysregulation of miRNA levels. In a rodent model of spinal cord injury, the intravenous administration of antagonists against miR-223 promoted an increase of local angiogenesis and a functional recovery of the injured animals (Liu et al., 2015). Similar approaches were used in other recent examples of animal models of disease, such as in the treatment of rat models of ocular cataracts by the application of an miR-326 antagomir (Ren et al., 2018), or the amelioration of ischemic brain injury in mice after an induced stroke by the treatment with an miR-15a/16-1 antagomir (Yang et al., 2017). Besides the use of miRNA antagonists to reduce their expression levels, other alternative strategies mimic the natural phenomenon of the complementary endogenous RNAs (ceRNAs) (Liang et al., 2016). The transfection of linear or circular synthetic RNAs that contain several miRNA-binding sites allows them to act as miRNA sponges, reducing the local concentration of one or several miRNAs simultaneously (Kluiver et al., 2012). This strategy has been applied to block the action of oncogenic miRNAs such as miR-10b in MCF-7 breast cancer cells (Liang et al., 2016). However, the toxic effects and the erratic inhibition observed in some short antagomirs and miRNA sponges prompted the design of new alternatives for miRNA inhibition. Recently, new miRNA inhibitors based on a stable RNA secondary structure were developed (Hooykaas et al., 2018). The "tough decoys" are short synthetic RNAs that are able to fold into very stable secondary structures and harbor complementary sequences to a specific miRNA. They are extremely efficient in decreasing the levels of miRNAs, and have been used in cellular and animal models (Hrdlicka et al., 2021; Jeong et al., 2018).

The protocols for increasing miRNA levels are less diverse, and usually centered on the direct transfection of miRNA mimics (Xiao, Chen, Betenbaugh, Martin, & Shiloach, 2015) or the construction of expression vectors that resemble the structure of a pri-miRNA (Thakral & Ghoshal, 2015). In practical matters, they are less efficient for the modification of miRNA concentration when compared with the use of inhibitors or sponges.

Final remarks

Wet-lab protocols for the study of miRNAs are essential tools for the detailed analysis of the regulatory action of these noncoding RNAs. In the last two decades, the protocols for the characterization of miRNA regulation have rapidly evolved from the use of Northern blotting to next-generation sequencing. Due to the specific nature of the miRNA posttranscriptional regulation, the application of bioinformatic analysis together with the laboratory protocols is essential. In fact, the field of miRNA research is probably one of the better-established paradigms of collaboration between bioinformatics and wet-lab scientists. The evolution of wet-lab methods has been accompanied by a parallel development of specific computer software for data processing and analysis.

The specific nature of miRNA regulation facilitates the application of the principles of systems biology for the analysis of their functions in biological systems. These principles rely on the experimental observations that define miRNAs as a complementary regulatory layer located on the top of the transcriptional control mechanisms. For the global characterization of these complex miRNA-centered regulatory networks, the selection of proper laboratory methods is essential. Scientists working in the miRNA field need a deep knowledge of all the available wet-lab techniques with their advantages, drawbacks, and specific applications. A careful choice of the experimental protocols is essential for obtaining reliable output and deriving functional consequences.

Acknowledgments

This work is supported by COST (European Cooperation in Science and Technology) Action EU-ardioRNA CA17129 and the Portuguese Foundation for Science and Technology (FCT) under the framework of the research grant PTDC-MED-GEN-29389-2017. The authors would like to acknowledge Francisco Enguita Jr. for excellent technical assistance and support in the design and conceptualization of the manuscript.

References

Alon, S., Vigneault, F., Eminaga, S., Christodoulou, D. C., Seidman, J. G., Church, G. M., et al. (2011). Barcoding bias in high-throughput multiplex sequencing of miRNA. *Genome Research*, *21*(9), 1506–1511. https://doi.org/10.1101/gr.121715.111.

Alvarez, M. L. (2014). Faster experimental validation of microRNA targets using cold fusion cloning and a dual firefly-renilla luciferase reporter assay. *Methods in Molecular Biology*, *1182*, 227–243. https://doi.org/10.1007/978-1-4939-1062-5_21.

Androvic, P., Valihrach, L., Elling, J., Sjoback, R., & Kubista, M. (2017). Two-tailed RT-qPCR: A novel method for highly accurate miRNA quantification. *Nucleic Acids Research*, *45*(15). https://doi.org/10.1093/nar/gkx588, e144.

Babak, T., Zhang, W., Morris, Q., Blencowe, B. J., & Hughes, T. R. (2004). Probing microRNAs with microarrays: Tissue specificity and functional inference. *RNA*, *10*(11), 1813–1819. https://doi.org/10.1261/rna.7119904.

Baker, M. (2010). MicroRNA profiling: Separating signal from noise. *Nature Methods*, *7*(9), 687–692. https://doi.org/10.1038/nmeth0910-687.

Bar, M., Wyman, S. K., Fritz, B. R., Qi, J., Garg, K. S., Parkin, R. K., et al. (2008). MicroRNA discovery and profiling in human embryonic stem cells by deep sequencing of small RNA libraries. *Stem Cells*, *26*(10), 2496–2505. https://doi.org/10.1634/stemcells.2008-0356.

Bartel, D. P. (2004). MicroRNAs: Genomics, biogenesis, mechanism, and function. *Cell*, *116*(2), 281–297. https://doi.org/10.1016/S0092-8674(04)00045-5.

Baskerville, S., & Bartel, D. P. (2005). Microarray profiling of microRNAs reveals frequent coexpression with neighboring miRNAs and host genes. *RNA*, *11*(3), 241–247. https://doi.org/10.1261/rna.7240905.

Benes, V., & Castoldi, M. (2010). Expression profiling of microRNA using real-time quantitative PCR, how to use it and what is available. *Methods*, *50*(4), 244–249. https://doi.org/10.1016/j.ymeth.2010.01.026.

Berezikov, E., Cuppen, E., & Plasterk, R. H. A. (2006). Approaches to microrna discovery. *Nature Genetics*, *38*(6S), S2. https://doi.org/10.1038/ng1794.

Beuvink, I., Kolb, F. A., Budach, W., Garnier, A., Lange, J., Natt, F., et al. (2007). A novel microarray approach reveals new tissue-specific signatures of known and predicted mammalian microRNAs. *Nucleic Acids Research*, *35*(7). https://doi.org/10.1093/nar/gkl1118.

Breda, J., Rzepiela, A. J., Gumienny, R., Van Nimwegen, E., & Zavolan, M. (2015). Quantifying the strength of miRNA-target interactions. *Methods*, *85*, 90–99. https://doi.org/10.1016/j.ymeth.2015.04.012.

Castoldi, M., Schmidt, S., Benes, V., Hentze, M. W., & Muckenthaler, M. U. (2008). miChip: An array-based method for microRNA expression profiling using locked nucleic acid capture probes. *Nature Protocols*, *3*(2), 321–329. https://doi.org/10.1038/nprot.2008.4.

Castoldi, M., Schmidt, S., Benes, V., Noerholm, M., Kulozik, A. E., Hentze, M. W., et al. (2006). A sensitive array for microRNA expression profiling (miChip) based on locked nucleic acids (LNA). *RNA*, *12*(5), 913–920. https://doi.org/10.1261/rna.2332406.

Cayirlioglu, P., Kadow, I. G., Zhan, X., Okamura, K., Suh, G. S. B., Gunning, D., et al. (2008). Hybrid neurons in a microRNA mutant are putative evolutionary intermediates in insect CO2 sensory systems. *Science*, *319*(5867), 1256–1260. https://doi.org/10.1126/science.1149483.

Chen, C., Ridzon, D. A., Broomer, A. J., Zhou, Z., Lee, D. H., Nguyen, J. T., et al. (2005). Real-time quantification of microRNAs by stem-loop RT-PCR. *Nucleic Acids Research*, *33*(20). https://doi.org/10.1093/nar/gni178, e179.9.

Chi, S. W., Hannon, G. J., & Darnell, R. B. (2012). An alternative mode of microRNA target recognition. *Nature Structural and Molecular Biology*, *19*(3), 321–327. https://doi.org/10.1038/nsmb.2230.

Chi, S. W., Zang, J. B., Mele, A., & Darnell, R. B. (2009). Argonaute HITS-CLIP decodes microRNA-mRNA interaction maps. *Nature*, *460*(7254), 479–486. https://doi.org/10.1038/nature08170.

Dangwal, S., Bang, C., & Thum, T. (2012). Novel techniques and targets in cardiovascular microRNA research. *Cardiovascular Research*, *93*(4), 545–554. https://doi.org/10.1093/cvr/cvr297.

Detassis, S., Grasso, M., Tabraue-Chávez, M., Marín-Romero, A., López-Longarela, B., Ilyine, H., et al. (2019). New platform for the direct profiling of microRNAs in biofluids. *Analytical Chemistry*, *91*(9), 5874–5880. https://doi.org/10.1021/acs.analchem.9b00213.

Dong, H., Lei, J., Ding, L., Wen, Y., Ju, H., & Zhang, X. (2013). MicroRNA: Function, detection, and bioanalysis. *Chemical Reviews*, *113*(8), 6207–6233. https://doi.org/10.1021/cr300362f.

Duan, D., Zheng, K. X., Shen, Y., Cao, R., Jiang, L., Lu, Z., et al. (2011). Label-free high-throughput microRNA expression profiling from total RNA. *Nucleic Acids Research*, *39*(22). https://doi.org/10.1093/nar/gkr774, e154.

Dzikiewicz-Krawczyk, A., Diepstra, A., Rutgers, B., Kortman, G., de Jong, D., Koerts, J., et al. (2018). Argonaute 2 RNA immunoprecipitation reveals distinct miRNA targetomes of primary Burkitt lymphoma tumors and normal B cells. *American Journal of Pathology*, *188*(5), 1289–1299. https://doi.org/10.1016/j.ajpath.2018.01.018.

Erhard, F., Dölken, L., Jaskiewicz, L., & Zimmer, R. (2013). PARma: Identification of microRNA target sites in AGO-PAR-CLIP data. *Genome Biology*, *14*(7). https://doi.org/10.1186/gb-2013-14-7-r79.

Farazi, T. A., ten Hoeve, J. J., Brown, M., Mihailovic, A., Horlings, H. M., van de Vijver, M. J., et al. (2014). Identification of distinct miRNA target regulation between breast cancer molecular subtypes using AGO2-PAR-CLIP and patient datasets. *Genome Biology*, *15*(1). https://doi.org/10.1186/gb-2014-15-1-r9.

Friedländer, M. R., Chen, W., Adamidi, C., Maaskola, J., Einspanier, R., Knespel, S., et al. (2008). Discovering microRNAs from deep sequencing data using miRDeep. *Nature Biotechnology*, *26*(4), 407–415. https://doi.org/10.1038/nbt1394.

Friedman, R. C., Farh, K. K. H., Burge, C. B., & Bartel, D. P. (2009). Most mammalian mRNAs are conserved targets of microRNAs. *Genome Research, 19*(1), 92–105. https://doi.org/10.1101/gr.082701.108.

Fuchs, R. T., Sun, Z., Zhuang, F., & Robb, G. B. (2015). Bias in ligation-based small RNA sequencing library construction is determined by adaptor and RNA structure. *PLoS One, 10*(5). https://doi.org/10.1371/journal.pone.0126049.

Gabriel, A. F., Costa, M. C., & Enguita, F. J. (2020). Circular RNA-centered regulatory networks in the physiopathology of cardiovascular diseases. *International Journal of Molecular Sciences, 21*(2). https://doi.org/10.3390/ijms21020456.

Gaur, A., Jewell, D. A., Liang, Y., Ridzon, D., Moore, J. H., Chen, C., et al. (2007). Characterization of microRNA expression levels and their biological correlates in human cancer cell lines. *Cancer Research, 67*(6), 2456–2468. https://doi.org/10.1158/0008-5472.CAN-06-2698.

Grozdanov, P. N., & Macdonald, C. C. (2014). High-throughput sequencing of RNA isolated by cross-linking and immunoprecipitation (HITS-CLIP) to determine sites of binding of CstF-64 on nascent RNAs. *Methods in Molecular Biology, 1125*, 187–208. https://doi.org/10.1007/978-1-62703-971-0_17.

Gulyaeva, L. F., & Kushlinskiy, N. E. (2016). Regulatory mechanisms of microRNA expression. *Journal of Translational Medicine, 14*(1), 143. https://doi.org/10.1186/s12967-016-0893-x.

Haecker, I., Gay, L. A., Yang, Y., Hu, J., Morse, A. M., McIntyre, L. M., et al. (2012). Ago HITS-CLIP expands understanding of Kaposi's Sarcoma-associated Herpesvirus miRNA function in primary effusion lymphomas. *PLoS Pathogens, 8*(8). https://doi.org/10.1371/journal.ppat.1002884.

Haecker, I., & Renne, R. (2014). HITS-CLIP and PAR-CLIP advance viral miRNA targetome analysis. *Critical Reviews in Eukaryotic Gene Expression, 24*(2), 101–116. https://doi.org/10.1615/CritRevEukaryotGeneExpr.2014006367.

Hafner, M., Landthaler, M., Burger, L., Khorshid, M., Hausser, J., Berninger, P., et al. (2010). Transcriptome-wide identification of RNA-binding protein and microRNA target sites by PAR-CLIP. *Cell, 141*(1), 129–141. https://doi.org/10.1016/j.cell.2010.03.009.

Hafner, M., Renwick, N., Brown, M., Mihailović, A., Holoch, D., Lin, C., et al. (2011). RNA-ligase-dependent biases in miRNA representation in deep-sequenced small RNA cDNA libraries. *RNA, 17*(9), 1697–1712. https://doi.org/10.1261/rna.2799511.

Hagiwara, S., Kantharidis, P., & Cooper, M. E. (2014). microRNA as biomarkers and regulator of cardiovascular development and disease. *Current Pharmaceutical Design, 20*(14), 2347–2370. https://doi.org/10.2174/13816128113199990495.

Hausser, J., & Zavolan, M. (2014). Identification and consequences of miRNA-target interactions-beyond repression of gene expression. *Nature Reviews Genetics, 15*(9), 599–612. https://doi.org/10.1038/nrg3765.

Helwak, A., Kudla, G., Dudnakova, T., & Tollervey, D. (2013). Mapping the human miRNA interactome by CLASH reveals frequent noncanonical binding. *Cell, 153*(3), 654–665. https://doi.org/10.1016/j.cell.2013.03.043.

Helwak, A., & Tollervey, D. (2016). Identification of miRNA-target RNA interactions using clash. In *Vol. 1358. Methods in molecular biology* (pp. 229–251). Humana Press Inc. https://doi.org/10.1007/978-1-4939-3067-8_14.

Hooykaas, M. J. G., Soppe, J. A., De Buhr, H. M., Kruse, E., Wiertz, E. J. H. J., & Lebbink, R. J. (2018). RNA accessibility impacts potency of Tough Decoy microRNA inhibitors. *RNA Biology, 15*(11), 1410–1419. https://doi.org/10.1080/15476286.2018.1537746.

Hrdlicka, H. C., Pereira, R. C., Shin, B., Yee, S. P., Deymier, A. C., Lee, S. K., et al. (2021). Inhibition of miR-29-3p isoforms via tough decoy suppresses osteoblast function in homeostasis but promotes intermittent parathyroid hormone-induced bone anabolism. *Bone, 143*. https://doi.org/10.1016/j.bone.2020.115779.

Hunt, E. A., Broyles, D., Head, T., & Deo, S. K. (2015). MicroRNA detection: Current technology and research strategies. *Annual Review of Analytical Chemistry, 8*, 217–237. https://doi.org/10.1146/annurev-anchem-071114-040343.

Jens, M. (2016). A pipeline for PAR-CLIP data analysis. In *Vol. 1358. Methods in molecular biology* (pp. 197–207). Humana Press Inc. https://doi.org/10.1007/978-1-4939-3067-8_12.

Jeong, D., Yoo, J., Lee, P., Kepreotis, S. V., Lee, A., Wahlquist, C., et al. (2018). miR-25 Tough decoy enhances cardiac function in heart failure. *Molecular Therapy, 26*(3), 718–729. https://doi.org/10.1016/j.ymthe.2017.11.014.

Jung, U., Jiang, X., Kaufmann, S. H. E., & Patzel, V. (2013). A universal TaqMan-based RT-PCR protocol for cost-efficient detection of small noncoding RNA. *RNA, 19*(12), 1864–1873. https://doi.org/10.1261/rna.040501.113.

Juzenas, S., Venkatesh, G., Hübenthal, M., Hoeppner, M. P., Du, Z. G., Paulsen, M., et al. (2017). A comprehensive, cell specific microRNA catalogue of human peripheral blood. *Nucleic Acids Research, 45*(16), 9290–9301. https://doi.org/10.1093/nar/gkx706.

Kim, S. W., Li, Z., Moore, P. S., Monaghan, A. P., Chang, Y., Nichols, M., et al. (2010). A sensitive non-radioactive northern blot method to detect small RNAs. *Nucleic Acids Research, 38*(7). https://doi.org/10.1093/nar/gkp1235, e98.

Kluiver, J., Slezak-Prochazka, I., Smigielska-Czepiel, K., Halsema, N., Kroesen, B. J., & van den Berg, A. (2012). Generation of miRNA sponge constructs. *Methods, 58*(2), 113–117. https://doi.org/10.1016/j.ymeth.2012.07.019.

Koscianska, E., Starega-Roslan, J., Czubala, K., & Krzyzosiak, W. J. (2011). High-resolution northern blot for a reliable analysis of microRNAs and their precursors. *TheScientificWorldJOURNAL, 11*, 102–117. https://doi.org/10.1100/tsw.2011.11.

Koscianska, E., Starega-Roslan, J., Sznajder, L. J., Olejniczak, M., Galka-Marciniak, P., & Krzyzosiak, W. J. (2011). Northern blotting analysis of microRNAs, their precursors and RNA interference triggers. *BMC Molecular Biology, 12*. https://doi.org/10.1186/1471-2199-12-14.

Krichevsky, A. M., King, K. S., Donahue, C. P., Khrapko, K., & Kosik, K. S. (2003). A microRNA array reveals extensive regulation of microRNAs during brain development. *RNA, 9*(10), 1274–1281. https://doi.org/10.1261/rna.5980303.

Lagos-Quintana, M., Rauhut, R., Lendeckel, W., & Tuschl, T. (2001). Identification of novel genes coding for small expressed RNAs. *Science, 294*(5543), 853–858. https://doi.org/10.1126/science.1064921.

Lao, K., Xu, N. L., Yeung, V., Chen, C., Livak, K. J., & Straus, N. A. (2006). Multiplexing RT-PCR for the detection of multiple miRNA species in small samples. *Biochemical and Biophysical Research Communications, 343*(1), 85–89. https://doi.org/10.1016/j.bbrc.2006.02.106.

Lee, R. C., Feinbaum, R. L., & Ambros, V. (1993). The *C. elegans* heterochronic gene lin-4 encodes small RNAs with antisense complementarity to lin-14. *Cell, 75*(5), 843–854. https://doi.org/10.1016/0092-8674(93)90529-Y.

Li, J., Bai, H., Zhu, Y., Wang, X. Y., Wang, F., Zhang, J. W., et al. (2010). Antagomir dependent microRNA-205 reduction enhances adhesion ability of human corneal epithelial keratinocytes. *Chinese Medical Sciences Journal*, *25*(2), 65–70. https://doi.org/10.1016/S1001-9294(10)60024-7.

Li, X., & Hu, H. (2019). Improving miRNA target prediction using CLASH data. In *Vol. 1970. Methods in molecular biology* (pp. 75–83). Humana Press Inc. https://doi.org/10.1007/978-1-4939-9207-2_6.

Li, Y., & Kowdley, K. V. (2012). MicroRNAs in common human diseases. *Genomics, Proteomics & Bioinformatics*, *10*(5), 246–253. https://doi.org/10.1016/j.gpb.2012.07.005.

Liang, R. Q., Li, W., Li, Y., Tan, C.y., Li, J. X., Jin, Y. X., et al. (2005). An oligonucleotide microarray for microRNA expression analysis based on labeling RNA with quantum dot and nanogold probe. *Nucleic Acids Research*, *33*(2). https://doi.org/10.1093/nar/gni019, e17.

Liang, A. L., Zhang, T. T., Zhou, N., Yun Wu, C., Lin, M. H., & Liu, Y. J. (2016). MiRNA-10b sponge: An anti-breast cancer study in vitro. *Oncology Reports*, *35*(4), 1950–1958. https://doi.org/10.3892/or.2016.4596.

Linsen, S. E. V., de Wit, E., Janssens, G., Heater, S., Chapman, L., Parkin, R. K., et al. (2009). Limitations and possibilities of small RNA digital gene expression profiling. *Nature Methods*, *6*(7), 474–476. https://doi.org/10.1038/nmeth0709-474.

Liu, C. G., Calin, G. A., Meloon, B., Gamliel, N., Sevignani, C., Ferracin, M., et al. (2004). An oligonucleotide microchip for genome-wide microRNA profiling in human and mouse tissues. *Proceedings of the National Academy of Sciences of the United States of America*, *101*(26), 9740–9744. https://doi.org/10.1073/pnas.0403293101.

Liu, C. G., Calin, G. A., Volinia, S., & Croce, C. M. (2008). MicroRNA expression profiling using microarrays. *Nature Protocols*, *3*(4), 563–578. https://doi.org/10.1038/nprot.2008.14.

Liu, D., Huang, Y., Jia, C., Li, Y., Liang, F., & Fu, Q. (2015). Administration of Antagomir-223 inhibits apoptosis, promotes angiogenesis and functional recovery in rats with spinal cord injury. *Cellular and Molecular Neurobiology*, *35*(4), 483–491. https://doi.org/10.1007/s10571-014-0142-x.

Liu, N., & Olson, E. N. (2010). MicroRNA regulatory networks in cardiovascular development. *Developmental Cell*, *18*(4), 510–525. https://doi.org/10.1016/j.devcel.2010.03.010.

Liu, C. G., Spizzo, R., Calin, G. A., & Croce, C. M. (2008). Expression profiling of microRNA using oligo DNA arrays. *Methods*, *44*(1), 22–30. https://doi.org/10.1016/j.ymeth.2007.10.010.

Lu, Y., Hippen, K. L., Lemire, A. L., Gu, J., Wang, W., Ni, X., et al. (2016). MiR-146b antagomir-treated human Tregs acquire increased GVHD inhibitory potency. *Blood*, *128*(10), 1424–1435. https://doi.org/10.1182/blood-2016-05-714535.

Lu, Y., & Leslie, C. S. (2016). Learning to predict miRNA-mRNA interactions from AGO CLIP sequencing and CLASH data. *PLoS Computational Biology*, *12*(7). https://doi.org/10.1371/journal.pcbi.1005026.

Mei, Q., Li, X., Meng, Y., Wu, Z., Guo, M., Zhao, Y., et al. (2012). A facile and specific assay for quantifying microRNA by an optimized RT-qPCR approach. *PLoS One*, *7*(10).

Mestdagh, P., Vlierberghe, P., Van Weer, D., Muth, A., Westermann, D., Speleman, F., et al. (2009). A novel and universal method for microRNA RT-qPCR data normalization. *Genome Biology*, *10*(6).

Miska, E. A., Alvarez-Saavedra, E., Townsend, M., Yoshii, A., Sestan, N., Rakic, P., et al. (2004). Microarray analysis of microRNA expression in the developing mammalian brain. *Genome Biology*, *5*(9), R68.

Mohammadian, A., Mowla, S. J., Elahi, E., Tavallaei, M., Nourani, M. R., & Liang, Y. (2013). Normalization of miRNA qPCR high-throughput data: A comparison of methods. *Biotechnology Letters*, *35*(6), 843–851. https://doi.org/10.1007/s10529-013-1150-5.

Morin, R. D., O'Connor, M. D., Griffith, M., Kuchenbauer, F., Delaney, A., Prabhu, A. L., et al. (2008). Application of massively parallel sequencing to microRNA profiling and discovery in human embryonic stem cells. *Genome Research*, *18*(4), 610–621. https://doi.org/10.1101/gr.7179508.

Morin, R. D., Zhao, Y., Prabhu, A. L., Dhalla, N., McDonald, H., Pandoh, P., et al. (2010). Preparation and analysis of microRNA libraries using the illumina massively parallel sequencing technology. *Methods in Molecular Biology*, *650*, 173–199. https://doi.org/10.1007/978-1-60761-769-3_14.

Moyle, R. L., Carvalhais, L. C., Pretorius, L. S., Nowak, E., Subramaniam, G., Dalton-Morgan, J., et al. (2017). An optimized transient dual luciferase assay for quantifying microRNA directed repression of targeted sequences. *Frontiers in Plant Science*, *8*. https://doi.org/10.3389/fpls.2017.01631.

Neely, L. A., Patel, S., Garver, J., Gallo, M., Hackett, M., McLaughlin, S., et al. (2006). A single-molecule method for the quantitation of microRNA gene expression. *Nature Methods*, *3*(1), 41–46. https://doi.org/10.1038/nmeth825.

Neilsen, C. T., Goodall, G. J., & Bracken, C. P. (2012). IsomiRs—The overlooked repertoire in the dynamic microRNAome. *Trends in Genetics*, *28*(11), 544–549. https://doi.org/10.1016/j.tig.2012.07.005.

Nelson, P. T., Baldwin, D. A., Scearce, L. M., Oberholtzer, J. C., Tobias, J. W., & Mourelatos, Z. (2004). Microarray-based, high-throughput gene expression profiling of microRNAs. *Nature Methods*, *1*(2), 155–161. https://doi.org/10.1038/nmeth717.

Nicolas, F. E. (2011). Experimental validation of microRNA targets using a luciferase reporter system. *Methods in Molecular Biology (Clifton, N.J.)*, *732*, 139–152. https://doi.org/10.1007/978-1-61779-083-6_11.

Pall, G. S., Codony-Servat, C., Byrne, J., Ritchie, L., & Hamilton, A. (2007). Carbodiimide-mediated cross-linking of RNA to nylon membranes improves the detection of siRNA, miRNA and piRNA by northern blot. *Nucleic Acids Research*, *35*(8). https://doi.org/10.1093/nar/gkm112.

Pall, G. S., & Hamilton, A. J. (2008). Improved northern blot method for enhanced detection of small RNA. *Nature Protocols*, *3*(6), 1077–1084. https://doi.org/10.1038/nprot.2008.67.

Panni, S., Lovering, R. C., Porras, P., & Orchard, S. (2020). Non-coding RNA regulatory networks. *Biochimica et Biophysica Acta, Gene Regulatory Mechanisms*, *1863*(6). https://doi.org/10.1016/j.bbagrm.2019.194417.

Paul, P., Chakraborty, A., Sarkar, D., Langthasa, M., Rahman, M., Bari, M., et al. (2018). Interplay between miRNAs and human diseases. *Journal of Cellular Physiology*, *233*(3), 2007–2018. https://doi.org/10.1002/jcp.25854.

Peltier, H. J., & Latham, G. J. (2008). Normalization of microRNA expression levels in quantitative RT-PCR assays: Identification of suitable reference RNA targets in normal and cancerous human solid tissues. *RNA*, *14*(5), 844–852. https://doi.org/10.1261/rna.939908.

Pritchard, C. C., Cheng, H. H., & Tewari, M. (2012). MicroRNA profiling: Approaches and considerations. *Nature Reviews Genetics*, *13*(5), 358–369. https://doi.org/10.1038/nrg3198.

Qureshi, R., & Sacan, A. (2013). A novel method for the normalization of microRNA RT-PCR data. In *Vol. 6. BMC medical genomics*. https://doi.org/10.1186/1755-8794-6-S1-S14. Issue 1.

Raabe, C. A., Tang, T. H., Brosius, J., & Rozhdestvensky, T. S. (2014). Biases in small RNA deep sequencing data. *Nucleic Acids Research*, *42*(3), 1414–1426. https://doi.org/10.1093/nar/gkt1021.

Ramkissoon, S. H., Mainwaring, L. A., Sloand, E. M., Young, N. S., & Kajigaya, S. (2006). Nonisotopic detection of microRNA using digoxigenin labeled RNA probes. *Molecular and Cellular Probes*, *20*(1), 1–4. https://doi.org/10.1016/j.mcp.2005.07.004.

Raymond, C. K., Roberts, B. S., Garrett-Engele, P., Lim, L. P., & Johnson, J. M. (2005). Simple, quantitative primer-extension PCR assay for direct monitoring of microRNAs and short-interfering RNAs. *RNA*, *11*(11), 1737–1744. https://doi.org/10.1261/rna.2148705.

Reichenstein, I., Aizenberg, N., Goshen, M., Bentwich, Z., & Avni, Y. S. (2010). A novel qPCR assay for viral encoded microRNAs. *Journal of Virological Methods*, *163*(2), 323–328. https://doi.org/10.1016/j.jviromet.2009.10.018.

Ren, H., Tao, H., Gao, Q., Shen, W., Niu, Z., Zhang, J., et al. (2018). MiR-326 antagomir delays the progression of age-related cataract by upregulating FGF1-mediated expression of betaB2-crystallin. *Biochemical and Biophysical Research Communications*, *505*(2), 505–510. https://doi.org/10.1016/j.bbrc.2018.09.049.

Ro, S., Park, C., Jin, J., Sanders, K. M., & Yan, W. (2006). A PCR-based method for detection and quantification of small RNAs. *Biochemical and Biophysical Research Communications*, *351*(3), 756–763. https://doi.org/10.1016/j.bbrc.2006.10.105.

Roush, S., & Slack, F. J. (2008). The let-7 family of microRNAs. *Trends in Cell Biology*, *18*(10), 505–516. https://doi.org/10.1016/j.tcb.2008.07.007.

Rupaimoole, R., & Slack, F. J. (2017). MicroRNA therapeutics: Towards a new era for the management of cancer and other diseases. *Nature Reviews Drug Discovery*, *16*(3), 203–221. https://doi.org/10.1038/nrd.2016.246.

Russo, F., Fiscon, G., Conte, F., Rizzo, M., Paci, P., & Pellegrini, M. (2018). Interplay between long noncoding RNAs and microRNAs in cancer. In *Vol. 1819. Methods in molecular biology* (pp. 75–92). Humana Press Inc. https://doi.org/10.1007/978-1-4939-8618-7_4.

Sato, F., Tsuchiya, S., Terasawa, K., & Tsujimoto, G. (2009). Intra-platform repeatability and inter-platform comparability of microRNA microarray technology. *PLoS One*, *4*(5). https://doi.org/10.1371/journal.pone.0005540.

Schwarzenbach, H., Da Silva, A. M., Calin, G., & Pantel, K. (2015). Data normalization strategies for microRNA quantification. *Clinical Chemistry*, *61*(11), 1333–1342. https://doi.org/10.1373/clinchem.2015.239459.

Shcherbata, H. R., Ward, E. J., Fischer, K. A., Yu, J. Y., Reynolds, S. H., Chen, C. H., et al. (2007). Stage-specific differences in the requirements for germline stem cell maintenance in the Drosophila Ovary. *Cell Stem Cell*, *1*(6), 698–709. https://doi.org/10.1016/j.stem.2007.11.007.

Shi, R., & Chiang, V. L. (2005). Facile means for quantifying microRNA expression by real-time PCR. *BioTechniques*, *39*(4), 519–524. https://doi.org/10.2144/000112010.

Soifer, H. S., Rossi, J. J., & Sætrom, P. (2007). MicroRNAs in disease and potential therapeutic applications. *Molecular Therapy*, *15*(12), 2070–2079. https://doi.org/10.1038/sj.mt.6300311.

Spengler, R. M., Zhang, X., Cheng, C., McLendon, J. M., Skeie, J. M., Johnson, F. L., et al. (2016). Elucidation of transcriptome-wide microRNA binding sites in human cardiac tissues by Ago2 HITS-CLIP. *Nucleic Acids Research*, *44*(15), 7120–7131. https://doi.org/10.1093/nar/gkw640.

Spitzer, J., Hafner, M., Landthaler, M., Ascano, M., Farazi, T., Wardle, G., et al. (2014). PAR-CLIP (Photoactivatable Ribonucleoside-Enhanced Crosslinking and Immunoprecipitation): A step-by-step protocol to the transcriptome-wide identification of binding sites of RNA-binding proteins. In *Vol. 539. Methods in enzymology* (pp. 113–161). Academic Press Inc. https://doi.org/10.1016/B978-0-12-420120-0.00008-6.

Steinkraus, B. R., Toegel, M., & Fulga, T. A. (2016). Tiny giants of gene regulation: Experimental strategies for microRNA functional studies. *Wiley Interdisciplinary Reviews: Developmental Biology*, *5*(3), 311–362. https://doi.org/10.1002/wdev.223.

Sugimoto, N., Yoshida, K., Tatsumi, Y., Yugawa, T., Narisawa-Saito, M., Waga, S., et al. (2009). Redundant and differential regulation of multiple licensing factors ensures prevention of re-replication in normal human cells. *Journal of Cell Science*, *122*(8), 1184–1191. https://doi.org/10.1242/jcs.041889.

Tam, S., De Borja, R., Tsao, M. S., & Mcpherson, J. D. (2014). Robust global microRNA expression profiling using next-generation sequencing technologies. *Laboratory Investigation*, *94*(3), 350–358. https://doi.org/10.1038/labinvest.2013.157.

Teteloshvili, N., Smigielska-Czepiel, K., Yuan, Y., Seitz, A., de Jong, D., Rutgers, B., et al. (2017). Argonaute 2 immunoprecipitation revealed large tumor suppressor kinase 1 as a novel proapoptotic target of miR-21 in T cells. *FEBS Journal*, *284*(4), 555–567. https://doi.org/10.1111/febs.14011.

Thakral, S., & Ghoshal, K. (2015). miR-122 is a unique molecule with great potential in diagnosis, prognosis of liver disease, and therapy both as miRNA mimic and antimir. *Current Gene Therapy*, *15*(2), 142–150. https://doi.org/10.2174/1566523214666141224095610.

Thomson, J. M., Parker, J., Perou, C. M., & Hammond, S. M. (2004). A custom microarray platform for analysis of microRNA gene expression. *Nature Methods*, *1*(1), 47–53. https://doi.org/10.1038/nmeth704.

Tomasello, L., Cluts, L., & Croce, C. M. (2019). Experimental validation of MicroRNA targets: Luciferase reporter assay. In *Vol. 1970. Methods in molecular biology* (pp. 315–330). Humana Press Inc. https://doi.org/10.1007/978-1-4939-9207-2_17.

Tong, L., Xue, H., Xiong, L., Xiao, J., & Zhou, Y. (2015). Improved RT-PCR assay to quantitate the pri-, pre-, and mature microRNAs with higher efficiency and accuracy. *Molecular Biotechnology*, *57*(10), 939–946. https://doi.org/10.1007/s12033-015-9885-y.

Torres, A. G., Fabani, M. M., Vigorito, E., & Gait, M. J. (2011). MicroRNA fate upon targeting with anti-miRNA oligonucleotides as revealed by an improved Northern-blot-based method for miRNA detection. *RNA*, *17*(5), 933–943. https://doi.org/10.1261/rna.2533811.

Turk, M. A., Chung, C. Z., Manni, E., Zukowski, S. A., Engineer, A., Badakhshi, Y., et al. (2018). MiRAR-miRNA activity reporter for living cells. *Genes*, *9*(6). https://doi.org/10.3390/genes9060305.

Válóczi, A., Hornyik, C., Varga, N., Burgyán, J., Kauppinen, S., & Havelda, Z. (2004). Sensitive and specific detection of microRNAs by northern blot analysis using LNA-modified oligonucleotide probes. *Nucleic Acids Research*, *32*(22). https://doi.org/10.1093/nar/gnh171, e175.

Vandesompele, J., De Preter, K., Pattyn, F., Poppe, B., Van Roy, N., De Paepe, A., et al. (2002). Accurate normalization of real-time quantitative RT-PCR data by geometric averaging of multiple internal control genes. *Genome Biology*, *3*(7). RESEARCH0034.

Vandevenne, M., Delmarcelle, M., & Galleni, M. (2019). RNA regulatory networks as a control of stochasticity in biological systems. *Frontiers in Genetics*, *10*(May). https://doi.org/10.3389/fgene.2019.00403.

Várallyay, E., Burgyán, J., & Havelda, Z. (2007). Detection of microRNAs by Northern blot analyses using LNA probes. *Methods*, *43*(2), 140–145. https://doi.org/10.1016/j.ymeth.2007.04.004.

Varkonyi-Gasic, E., & Hellens, R. P. (2011). Quantitative stem-loop RT-PCR for detection of microRNAs. *Methods in Molecular Biology (Clifton, N.J.)*, *744*, 145–157. https://doi.org/10.1007/978-1-61779-123-9_10.

Wang, H., Ach, R. A., & Curry, B. O. (2007). Direct and sensitive miRNA profiling from low-input total RNA. *RNA*, *13*(1), 151–159. https://doi.org/10.1261/rna.234507.

Wang, J., Chen, J., & Sen, S. (2016). MicroRNA as biomarkers and diagnostics. *Journal of Cellular Physiology*, *231*(1), 25–30. https://doi.org/10.1002/jcp.25056.

Wang, W. X., Wilfred, B. R., Hu, Y., Stromberg, A. J., & Nelson, P. T. (2010). Anti-Argonaute RIP-Chip shows that miRNA transfections alter global patterns of mRNA recruitment to microribonucleoprotein complexes. *RNA*, *16*(2), 394–404. https://doi.org/10.1261/rna.1905910.

Wark, A. W., Lee, H. J., & Corn, R. M. (2008). Multiplexed detection methods for profiling microRNA expression in biological samples. *Angewandte Chemie, International Edition*, *47*(4), 644–652. https://doi.org/10.1002/anie.200702450.

Wolter, J. M., Kotagama, K., Pierre-Bez, A. C., Firago, M., & Mangone, M. (2014). 3'LIFE: A functional assay to detect miRNA targets in high-throughput. *Nucleic Acids Research*, *42*(17). https://doi.org/10.1093/nar/gku626.

Wu, W., Gong, P., Li, J., Yang, J., Zhang, G., Li, H., et al. (2014). Simple and nonradioactive detection of microRNAs using digoxigenin (DIG)-labeled probes with high sensitivity. *RNA*, *20*(4), 580–584. https://doi.org/10.1261/rna.042150.113.

Wylie, D., Shelton, J., Choudhary, A., & Adai, A. T. (2011). A novel mean-centering method for normalizing microRNA expression from high-throughput RT-qPCR data. *BMC Research Notes*, *4*.

Xiao, S., Chen, Y. C., Betenbaugh, M. J., Martin, S. E., & Shiloach, J. (2015). MiRNA mimic screen for improved expression of functional neurotensin receptor from HEK 293 cells. *Biotechnology and Bioengineering*, *112*(8), 1632–1643. https://doi.org/10.1002/bit.25567.

Yang, X., Tang, X., Sun, P., Shi, Y., Liu, K., Hassan, S. H., et al. (2017). MicroRNA-15a/16-1 antagomir ameliorates ischemic brain injury in experimental stroke. *Stroke*, *48*(7), 1941–1947. https://doi.org/10.1161/STROKEAHA.117.017284.

Zhang, C., & Darnell, R. B. (2011). Mapping in vivo protein-RNA interactions at single-nucleotide resolution from HITS-CLIP data. *Nature Biotechnology*, *29*(7), 607–614. https://doi.org/10.1038/nbt.1873.

Zhang, L., Ding, L., Cheung, T. H., Dong, M. Q., Chen, J., Sewell, A. K., et al. (2007). Systematic identification of *C. elegans* miRISC proteins, miRNAs, and mRNA targets by their interactions with GW182 proteins AIN-1 and AIN-2. *Molecular Cell*, *28*(4), 598–613. https://doi.org/10.1016/j.molcel.2007.09.014.

Zhang, X., Shen, B., & Cui, Y. (2019). Ago HITS-CLIP expands microRNA-mRNA interactions in nucleus and cytoplasm of gastric cancer cells. *BMC Cancer*, *19*(1), 29. https://doi.org/10.1186/s12885-018-5246-0.

Zhou, S., Wang, Y., Meng, Y., Xiao, C., Liu, Z., Brohawn, P., et al. (2016). In vivo therapeutic success of MicroRNA-155 antagomir in a mouse model of lupus alveolar hemorrhage. *Arthritis and Rheumatology*, *68*(4), 953–964. https://doi.org/10.1002/art.39485.

Chapter 9

Bioinformatics utilities, web resources and integrative strategies for the analysis of miRNA regulatory networks

Ana Lúcia Leitão[a,b], Marina C. Costa[c,d], André F. Gabriel[c,d], and Francisco J. Enguita[c,d]

[a]*Faculdade de Ciências e Tecnologia, Universidade Nova de Lisboa, Caparica, Portugal,* [b]*MEtRICs, Faculdade de Ciências e Tecnologia, Universidade Nova de Lisboa, Caparica, Portugal,* [c]*Instituto de Medicina Molecular João Lobo Antunes, Faculdade de Medicina, Universidade de Lisboa, Lisboa, Portugal,* [d]*Cardiomics Unit, Centro Cardiovascular da Universidade de Lisboa, Faculdade de Medicina, Universidade de Lisboa, Lisboa, Portugal*

Introduction

In 1993, the research groups from Victor Ambros and Gary Ruvkun published two seminal papers describing the existence of a small noncoding RNA designated as lin-4 that was able to act as a negative posttranscriptional regulator of specific mRNA transcripts (Lee, Feinbaum, & Ambros, 1993; Wightman, Ha, & Ruvkun, 1993). Further investigations discovered the widespread presence of such small noncoding RNAs (ncRNAs), which were designated as microRNAs (miRNAs), in other eukaryotic organisms, including humans (Bar et al., 2008; Berezikov, Cuppen, & Plasterk, 2006). MiRNAs were connected early with the homeostasis of cellular processes, where they act as modulators of the genomic output, being involved in processes such as the cell cycle (He, Luo, Wan, & Jiang, 2011; Malhas, Saunders, & Vaux, 2010), embryogenesis (Avital, França, & Yanai, 2018; Siddiqui, Abbas, Ansari, & Khan, 2019), and organ development (Lu et al., 2012; Sun et al., 2010). In all cases, the regulatory action of miRNAs can be compared with a buffer that is able to keep a biochemical parameter between a homeostatic interval (Bartel, 2009; Bartel & Chen, 2004).

These tiny regulators are generated from specific genomic information units, initially transcribed by RNA polymerase II to form primary miRNAs (pri-miRNAs). The pri-miRNA transcripts always contain a segment that folds, adopting a hairpin-loop structure that will be recognized and processed by a tandem of nuclear proteins, Drosha and DGCR8 (Jin, Wang, Liu, Wang, & Xu, 2020), to generate the precursor miRNA or pre-miRNA by endonuclease cleavage (Han et al., 2004). Pre-miRNAs are short RNA stemloops with a length around 75 nucleotides that are further exported to the cytoplasm with the help of the ATP-dependent Exportin-5 protein (Yi, Qin, Macara, & Cullen, 2003). In the cytoplasm, the pre-miRNAs are cleaved again by the Dicer endonuclease to form an approximately 22-nucleotide double-stranded RNA molecule with short 3′ overhangs (Lee et al., 1993; Wightman et al., 1993). These RNA duplexes are transferred by Dicer to AGO2, which together with other helper proteins constitutes the RNA-induced silencing complex, or RISC (Bartel, 2009). In the transfer process, one of the strands of the duplex (designated as the "passenger strand") may be discarded and the other one will be selected for regulatory action in a process called "strand selection" (Meijer, Smith, & Bushell, 2014). Recent data demonstrated that the thermodynamic features of the RNA duplex appear to play an important role in this decision; the strand showing the weakest complementarity at its 5′ end is more likely to become the guide strand (Hu et al., 2009; Medley, Panzade, & Zinovyeva, 2020). However, on many occasions the strand selection is not an absolute process and the same pre-miRNA can be the source of two different mature miRNAs, generated either by the 5′ or 3′ ends (O'Brien, Hayder, Zayed, & Peng, 2018; Zhou et al., 2018). The relative abundance of each mature miRNA (designated as 5p or 3p miRNAs) generated from the same precursor depends on the type of cell and its physiological state. The human genome produces up to 2,588 mature miRNAs generated from 1,881 precursors, which are distributed across all chromosomes. In higher eukaryotes, genomic *loci* containing miRNA transcriptional units are frequently clustered and in the human genome, there are 153 clusters that include 468 different miRNA precursors (Backes et al., 2018).

Mature miRNAs are single-stranded RNAs of 19–23 nucleotides in length, which generally function as posttranscriptional silencers over mRNA transcripts. The specificity of miRNA action is ensured by its base complementarity with the targeted mRNA transcripts, which generally guides the RISC complex to the 3′-UTR of mRNAs. The core component of the

MicroRNA in Human Malignancies. https://doi.org/10.1016/B978-0-12-822287-4.00009-8

cytoplasmic RISC complex is AGO2, a protein belonging to the Argonaute family composed of four members, AGO1-4 (Sasaki, Shiohama, Minoshima, & Shimizu, 2003; Wang et al., 2009). The Argonaute family integrates an evolutionary conserved group of dsRNA-binding proteins that contains four structural domains (Li et al., 2007). Two of these AGO2 domains, the PIWI and MID domains, form a molecular pocket that can accommodate a short segment of a dsRNA molecule comprising nucleotides 2–7 of the miRNAs, also known as the "seed sequence" (Fig. 9.1), and the corresponding paired segment in the target mRNA (Tomasello et al., 2020). In plants, the complementarity between the miRNAs and their cognate targets is almost complete, and the functional consequence is the degradation of the targeted mRNA by the slicing activity of the AGO2 protein (Jones-Rhoades & Bartel, 2004; Parizotto, Dunoyer, Rahm, Himber, & Voinnet, 2004). On the other hand, animal miRNAs are only partially complementary to their targets, which catalyzes an RISC-mediated

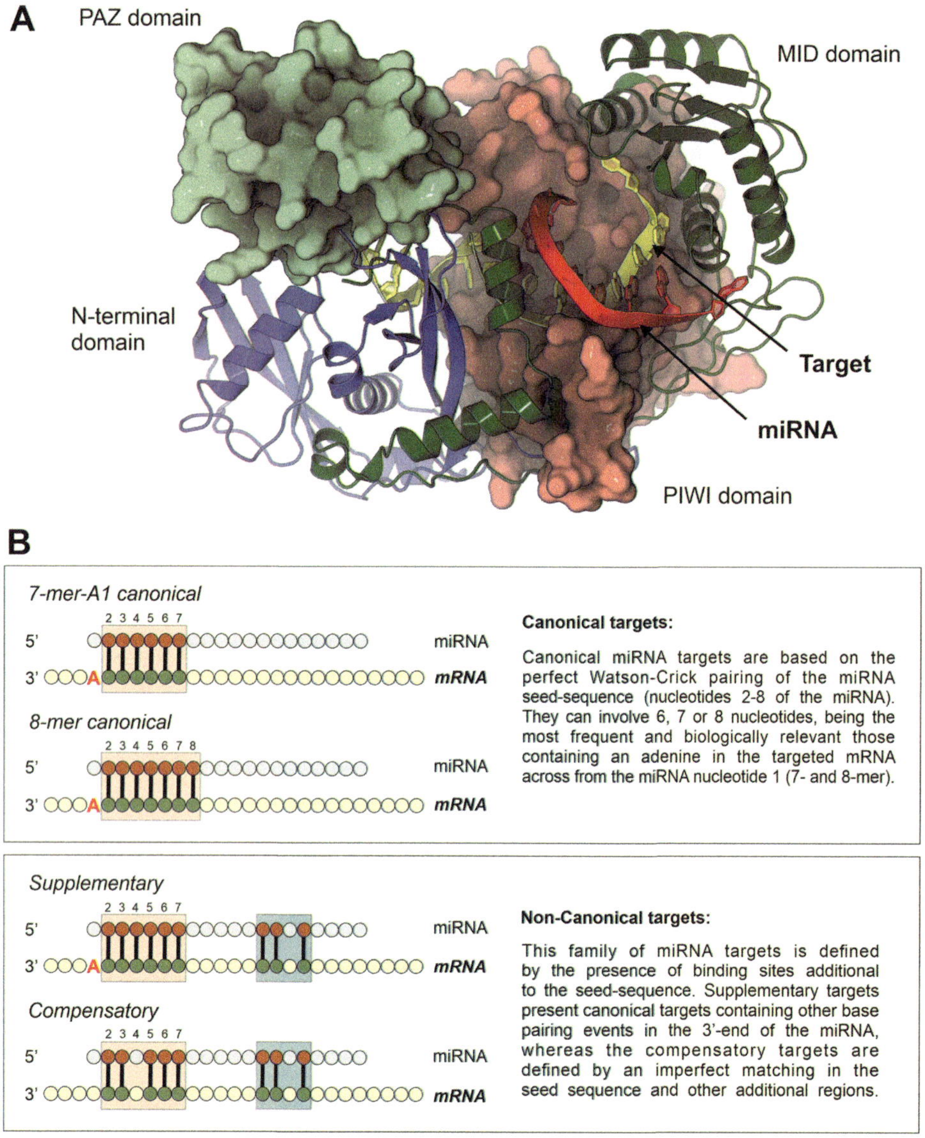

FIG. 9.1 General rules governing the miRNA-target interactions in animal cells involving the Ago2 protein. (A) The structure of the Ago2 protein is composed of four domains. The PIWI and MID domains are involved in the stabilization of the miRNA-mRNA complex by forming a molecular pocket that includes a short segment of the dsRNA hybrid. The figure was prepared with the atomic coordinates of the crystal structure of human Ago2 in a complex with a synthetic miRNA-mRNA hybrid (PDB code: 4W5Q), and rendered with Protein Imager software (Tomasello, Armenia, & Molla, 2020). (B) The first 2–8 nucleotides in the miRNA constitute the so-called seed sequence, which drives mRNA target recognition by a base pairing mechanism. Canonical miRNA targets are only based on the Watson-Crick pairing of the seed sequence and the cognate mRNA target. In noncanonical targets, additional binding sites are observed in the 3′ region of the miRNA. Experimental evidence showed that noncanonical targets are abundant in complex eukaryotes, and that the 3′ half of the miRNA is involved in controlling the target specificity and also the stability of the miRNA itself (Chipman & Pasquinelli, 2019).

translational repression of the target that could evolve to an mRNA destabilization when the miRNA action continues for prolonged times (Bartel, 2018). The pairing of the miRNA seed sequence to its target site has been commonly considered as the minimal regulatory element necessary for proper target recognition and regulation (Bartel, 2018). However, recent evidence showed that AGO2 is able to use a more sophisticated mechanism to engage miRNAs with their targets. The paired interactions between miRNAs and mRNAs catalyzed by the AGO2 protein use a sequential recognition model where the seed sequence pairs first and then nucleotides in the 3′ end of the miRNA can also contribute to the establishment of the miRNA-mRNA hybrid (Schirle, Sheu-Gruttadauria, & MacRae, 2014). This mechanism of extended miRNA pairing allows additional interactions with the target and has been implicated in the miRNA specificity, regulatory action, and miRNA stability (Lee et al., 1993; Wightman et al., 1993).

The detailed principles of miRNA target recognition have been extensively reviewed elsewhere (Agarwal, Bell, Nam, & Bartel, 2015; Bartel, 2009; Bartel & Chen, 2004). Based on experimental data, Bartel and coworkers proposed a structural classification of miRNA targets into two families of miRNA targets that differ on the degree of complementarity between the miRNA and its cognate target (Bartel, 2009). Canonical targets are defined by a perfect Watson-Crick complementarity between the miRNA seed sequence and the mRNA target (Fig. 9.1). The most prevalent canonical targets are the 7-mer-A1, where the seed sequence is constituted by nucleotides 2–7 of the miRNA 5′ end, and the 8-mer, defined by an miRNA seed sequence comprising nucleotides 2–8. In both canonical targeting mechanisms, it is common to find an adenine nucleotide in the targeted mRNA in the first nucleotide of the 3′ region (Agarwal et al., 2015; Agarwal, Subtelny, Thiru, Ulitsky, & Bartel, 2018; Bartel, 2009). Recent experimental data validations by ligation-dependent protocols associated with next-generation sequencing allowed characterizing other canonical sites, spanning a number of seed nucleotides of the miRNA from 6 to 10 (Li et al., 2007). The denomination as "canonical" refers to the specific mechanism for the miRNA-target interaction but it could not represent the most abundant regulatory mechanism for a miRNA in a specific cell or tissue (Castellano & Stebbing, 2013). The second family of miRNA targets is designated as "noncanonical" because it is defined by the presence of additional binding sites to the classical seed sequence at the 3′ end of the miRNA (Fig. 9.1). In "supplementary" target recognition, a classical match in the miRNA seed sequence is accompanied by additional sequence matches at the miRNA 3′ end. In the "compensatory" subclass, an imperfect match in the seed sequence is complemented by a short complementary sequence at the miRNA 3′ end.

Data extracted from transcriptome-wide studies allowed determining that the binding of most miRNAs includes the 5′ seed region, but around 60% of seed interactions are noncanonical, containing bulged or mismatched nucleotides (Helwak, Kudla, Dudnakova, & Tollervey, 2013). Moreover, seed interactions are generally accompanied by a specific nonseed base pairing, and 18% of miRNA-mRNA interactions involve the miRNA 3′ end, with little evidence for 5′ contacts. Analyses of miRNA:mRNA base pairing showed that miRNA species systematically differ in their target RNA interactions, and strongly overrepresented motifs were found in the interaction sites of several miRNAs (Castellano & Stebbing, 2013; Helwak et al., 2013).

Target prediction algorithms

The intrinsic promiscuity of miRNA action is related with the extremely large number of potential target sites of any given miRNA, especially in animals where the complementarity between miRNA and its cognate targets is partial (Akhtar, Micolucci, Islam, Olivieri, & Procopio, 2019; Chen, Zhang, & Feng, 2019). Bioinformatic methods are essential tools to narrow down and select the number of potential targets of a given miRNA, and a complementary approach to facilitate the wet-lab biological validation. However, the prediction of miRNA targets is a challenging enterprise due to the small size of miRNAs and their limited complementarity to the target, which prevents the application of classical sequence alignment protocols (Mount, 2007; Zhang, 2003). The first miRNA target prediction algorithms were developed by using experimental data obtained from simple organisms such as *Caenorhabditis elegans* (Lewis, Shih, Jones-Rhoades, Bartel, & Burge, 2003), and further refined by including data from other complex organisms (Sualp & Can, 2011; Witkos, Koscianska, & Krzyzosiak, 2011; Xia, Cao, & Shao, 2009). Current available algorithms often combine computational approaches with information derived from advanced techniques for the genome-wide characterization of the miRNA-target interactome such as HITS-CLIP (Chi, Zang, Mele, & Darnell, 2009; Licatalosi et al., 2008) or CLASH (Helwak & Tollervey, 2016; Li & Hu, 2019).

MiRNA target prediction algorithms use different features to filter the possible solutions and to avoid false positives. The most extended features include the presence of a seed match and the free energy of the established RNA-RNA hybrid. The seed match follows the previously defined rules of miRNA target recognition, where nucleotides 2–8 of the miRNA 5′ end are essential for its function (Bartel & Chen, 2004). Different degrees of miRNA-mRNA complementarity can contribute to the prediction score, depending on the number of established Watson-Crick pairs (6mer, 7mer or 8mer) between the miRNA seed sequence and the target (Bartel, 2009). The presence of noncanonical miRNA binding and the compensatory action that the additional pairing at the 3′ end of the miRNA exerts on target recognition could also be used to refine the ranking of predicted targets (Chipman & Pasquinelli, 2019; Sheu-Gruttadauria, Xiao, Gebert, & MacRae, 2019). Some target prediction tools often consider the stability of the miRNA-mRNA hybrid by calculating the theoretical values of the Gibbs function and include this value in the ranking score of the putative targets (Yue, Liu, & Huang, 2009). To complement these general features, miRNA target prediction strategies can also include two other general characteristics: site conservation and physical accessibility. Conservation is referred to as the presence of a particular target or miRNA sequence in different species (Paraskevopoulou et al., 2013). Conservation analysis of miRNA targets across different species provides additional evidence that a particular target may be functional because it has been selected by evolution, and it is incorporated into several target prediction algorithms (Maragkakis et al., 2011). The physical accessibility parameter refers to the structural constraints that may prevent the binding of an miRNA to its target. During transcription, mRNAs adopt secondary and tertiary structures due to intrachain base pairing that could interfere with the binding of regulatory molecules, including proteins and miRNAs (Pan & Sosnick, 2006). If an miRNA target is located in a highly structured mRNA region, the miRNA binding will only happen if there is favorable stability toward the formation of the miRNA-mRNA hybrid. The presence of structured regions in targeted mRNAs can be included as an additional constraint for miRNA recognition in target prediction software with the help of associated RNA folding algorithms (Kertesz, Iovino, Unnerstall, Gaul, & Segal, 2007; Ullah, Sahoo, Steinhöfel, & Albrecht, 2012).

Recently, a new group of target prediction algorithms has been developed with the help of artificial intelligence (AI) methods and by using the increasing amount of available information on miRNA regulatory networks from genome-wide studies (Parveen, Mustafa, Yadav, & Kumar, 2019; Sedaghat, Fathy, Modarressi, & Shojaie, 2018). AI applications are computer algorithms able to show intelligent behavior for the analysis of experimental data. All the different flavors of AI algorithms need preliminary training using test datasets where a specific variable is present or absent (Vougas, Almpanis, & Gorgoulis, 2020). During training, the AI algorithm is able to extract the features that characterize the positive and negative cohorts of data, which can be further used for the analysis of new data. In machine learning, the AI algorithm uses a predefined set of features or descriptors to characterize data (supervised learning), whereas in deep learning the AI algorithm is able to automatically extract these features (unsupervised learning) and to apply them to the characterization of data (Wang et al., 2019). Supervised and unsupervised learning systems have been used for the prediction of miRNA targets by applying AI algorithms previously trained on databases of validated miRNA targets (Pla, Zhong, & Rayner, 2018; Sedaghat et al., 2018).

Single predictors

Those target prediction algorithms that use an individual protocol to infer miRNA-mRNA relationships are designated as "single predictors." Thanks to the development of computer software, web interfaces, and the availability of experimental data and database processing algorithms, the number of single predictors for miRNA target analysis has increased exponentially in the last decade. We have compiled the most relevant ones with their characteristics in Table 9.1. The first applications designed after the discovery of miRNAs were miRanda (Enright et al., 2003) and TargetScan (Lewis et al., 2005). Both have a web user interface and use the complementarity of the miRNA and the free energy of the hybrid as features for positive scoring. TargetScan also considers the sequence conservation across species and the presence of additional binding sites in the 3′ end of the miRNA. Interestingly, miRanda has a standalone version for local installation, which improves its flexibility for the analysis of nonmodel transcriptomes and miRNAs. Other classical algorithms for miRNA target prediction are RNA22 (Loher & Rigoutsos, 2012), PicTar (Krek et al., 2005), and PITA (Kertesz et al., 2007). RNA22 and PicTar use an approach based on the analysis of base pairing probabilities and Gibbs function for the miRNA-mRNA hybrid. PITA introduces an additional analysis feature considering the predicted secondary structure of the target binding site, which could impose a potential restriction for miRNA binding. In a general overview, these classical algorithms have low efficiency in the prediction of noncanonical miRNA targets, and the tendency to overestimate the number of predicted targets.

Despite their intrinsic utility in some specific applications, these classical single predictors were overtaken by a second generation of algorithms represented by Diana-microT (Maragkakis et al., 2009; Maragkakis et al., 2011) and MirWalk (Sticht et al., 2018). Both predictors are based on two main pillars: a clear, easy, customizable, and versatile web interface, and a frequent schedule of updates. Diana-microT and MirWalk are in fact web portals for the analysis of miRNA function, including submodules for target prediction, access to databases of validated targets, and a functional module that allows the user to predict the regulatory actions of miRNAs over specific metabolic pathways or their involvement in diseases (Vlachos et al., 2012; Vlachos et al., 2015). Regarding the descriptors used by both applications, Diana-microT is a more

TABLE 9.1 Single miRNA target predictors.

Algorithm	Parameters for score calculation	Description	Web server	Artificial Intelligence	References
miRanda	Complementarity and free energy	Algorithm for genomic target finding, useful with nonmodel genomes. A standalone version of the software for local installation is available	YES	NO	(Enright et al., 2003)
miRWalk	Complementarity, seed match, and base pairing probability	Integrated miRNA prediction portal including its own algorithm and cross-talk with other prediction platforms. Two versions, 2.0 and 3.0, and web interfaces are available	YES	NO	(Dweep & Gretz, 2015, Dweep, Gretz, & Sticht, 2014, Sticht, De La Torre, Parveen, & Gretz, 2018)
miSTAR	Complementarity, base pairing probability, and genomic content	miSTAR uses logistic regression and random forest models to combine information about potential miRNA binding sites present in a target mRNA and the genomic context of each individual site	YES	NO	(Van Peer et al., 2017)
PicTar	Free energy, seed match, and species conservation	PicTar is a classical algorithm for target prediction, specially focused on the influence of target conservation across species	YES	NO	(Krek et al., 2005)
TargetScan	Seed match, 3′ complementarity for additional targeting, position distribution, and species conservation	TargetScan is a classical algorithm, regularly updated with new information. It contains information for mammalian genomes, fish, and worms. It is particularly efficient in the prediction of noncanonical targets	YES	NO	(Agarwal et al., 2015, Agarwal et al., 2018, Lewis, Burge, & Bartel, 2005)
Diana-microT	Free energy and complementarity of the miRNA	Integrated portal for miRNA target prediction and analysis. The prediction module uses neural networks	YES	YES	(Maragkakis et al., 2009, Maragkakis et al., 2011, Paraskevopoulou et al., 2013)
PITA	Target site accessibility and binding energy	Target prediction considers the secondary structure of the mRNA target	YES	NO	(Kertesz et al., 2007, Ullah et al., 2012)
RNA22	Seed sequence complementarity and free energy of the miRNA-mRNA hybrid	Classic algorithm for target prediction. Can be customized for user sequences	YES	NO	(Loher & Rigoutsos, 2012)
DeepMirTar	Free energy, base pairing probability, complementarity, and experimentally validated data	Standalone application implementing a machine learning approach for miRNA target prediction.	NO	YES	(Wen, Cong, Zhang, Lu, & Li, 2018)
miRAW	Free energy, base pairing probability, complementarity, and experimentally validated data	Standalone application implementing a deep learning approach for miRNA target prediction. Training dataset is constructed from CLIP, CLASH, and iPAR-CLIP data.	NO	YES	(Pla et al., 2018)

advanced algorithm because its prediction module is based on neural networks, whereas MirWalk uses the classical features of complementarity, seed match, and pairing probabilities between the miRNA and its targets.

The third generation of miRNA target predictors is based on AI algorithms, taking advantage of the increasing availability of validated interactions between miRNAs and mRNAs in different biological contexts (Chi et al., 2009; Licatalosi et al., 2008). In Table 9.1, we highlighted two AI-based predictors, DeepMirTar (Wen et al., 2018) and miRAW (Pla et al., 2018). Both use AI for extracting specific descriptors from validated miRNA targets and applying them to the prediction of targets in a problem dataset. DeepMirTar uses a supervised approach based on the principles of machine learning, whereas miRAW works in a totally automated manner using a deep-learning algorithm. AI predictors are typically standalone applications that need to be locally installed in a suitable computer, are operated in a command-line basis, and sometimes require specific training with experimental datasets, making these applications less accessible to the general public (Parveen et al., 2019; Sedaghat et al., 2018). However, for trained bioinformaticians, AI predictors have the advantage of being extremely flexible, reliable, customizable, and adaptable for miRNA target prediction in any transcriptome (Joung, Hwang, Nam, Kim, & Zhang, 2007; Qi, Fiori, Turecki, & Trakadis, 2020).

Multiple predictors

MiRNA target prediction algorithms are prone to produce false positives, mainly due to the intrinsic nature of the miRNA-target recognition mechanisms, which involve partial complementarity between the miRNA sequence and its cognate mRNA target. Despite the availability of many target prediction tools that use different strategies and parameters to search for miRNA targets, there is a lack of consensus about the best tool. A possible strategy to circumvent these limitations is the use of multiple prediction algorithms, and a further accumulative analysis of predictions (Fig. 9.2). Multiple miRNA predictors are web-based resources that interrogate individual target prediction algorithms and combine their results to generate either a matrix with the list of positive target matches across all the applications (Dweep et al., 2014; Dweep, Sticht, Pandey, & Gretz, 2011), or a calculated consensus score among all the predictions (Oliveira et al., 2017; Tokar et al., 2018). The first resources grouping several miRNA-target prediction algorithms were created in the second half of the 2000s, and further developed until now (Xiao et al., 2009). The successful utilization of these resources is strictly dependent on their proper updates considering the general miRNA annotation databases and the results of the different target prediction algorithms (Witkos et al., 2011).

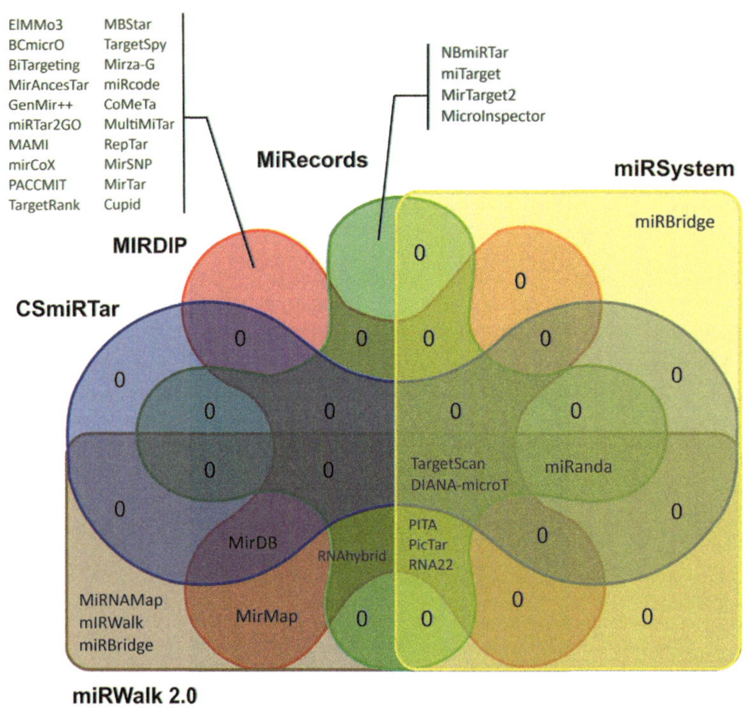

FIG. 9.2 Overview of web-based platforms and resources combining multiple miRNA target prediction algorithms discussed in the text, showing all the individual algorithms used by each portal. With the exception of MIRDIP, all the depicted multiple predictors use an accumulative strategy. MIRDIP combines target prediction algorithms by calculating a consensus score among them, allowing a quantitation of the efficiency of all the protocols.

The first attempt to compile and structure miRNA target prediction algorithms together in a web resource was the development of the miRecords platform (Xiao et al., 2009). This web tool is divided into two sections devoted to predicted and validated miRNA targets. The miRecords target prediction module integrates 11 different algorithms across nine species of model organisms, including humans, mammals, flies, and worms, and can be searched by miRNA or targeted gene. The primary output of any search in the prediction's module is composed of a binary output table indicating which target prediction algorithm gives a positive hit. Further details of each positive prediction can be obtained by direct navigation in the results table. Unfortunately, the miRecords platform is heavily out of date, as its last revision dates from 2013. Using the same skeleton of miRecords, the MiRWalk system was designed in 2011 with the objective of integrating different analysis tools for miRNAs, including not only target prediction algorithms but also applications to assess the functional roles of these regulators within metabolic pathways and diseases (Dweep et al., 2011). MiRWalk versions 2.0 and 3.0 were developed as web tools with improved user interfaces and include information for human, mouse, rat, dog, cow, and zebrafish genomes, constituting one of the most complete resources for miRNA target analysis (Dweep & Gretz, 2015; Sticht et al., 2018). In the transition from MirWalk 2.0 to 3.0, the number of target prediction algorithms was substantially reduced due to the lack of updates in some of the databases. As a main difference, the latest MirWalk 3.0 version uses the new TarPmiR prediction algorithm, which is based on a random forest approach analysis, and also includes an expanded module of databases containing information about experimentally validated miRNA targets (Ding, Li, & Hu, 2016).

Other data integration portals such as CSmiRTar (Wu, Tu, Chen, Hou, & Tseng, 2017) and miRSystem (Lu et al., 2012; Sun et al., 2010) are designed for the functional analysis of miRNAs by combining the predictions of several algorithms. CSmiRTar collects computationally predicted targets of 2588 human miRNAs and 1945 mouse miRNAs from four existing databases (miRDB, TargetScan, miRanda, and DIANA-microT), allowing the use of combined filters for the selection of specific target prediction databases, the tissue expression of miRNA targets, and their involvement in human diseases (Wu et al., 2017). The MiRSystem portal has a different approach, combining seven different target predictors (DIANA-microT, miRanda, miRBridge, PicTar, PITA, RNA22, and TargetScan) with functional analysis based on pathway enrichment and gene ontology. This approach allows the analysis of miRNA function following the postulates of systems biology by the construction of miRNA regulatory networks (Lu, Lee, et al., 2012).

MIRDIP, the multiple predictor that includes a larger number of algorithms, was designed with a different strategy (Shirdel, Xie, Mak, & Jurisica, 2011). Based on the previous data showing that the accumulative use of several target prediction strategies enhances the predictive power (Oliveira et al., 2017), MIRDIP calculates a consensus score by using a combination of all the target prediction algorithms (Tokar et al., 2018). It has a user-friendly interface with the possibility of bidirectional searching between miRNAs and their targets, and a way to export results for analysis with other software. MIRDIP is exclusively designed for human miRNA targets, and includes a total of 2586 mature miRNAs and predictions for more than 15 million targeting events.

In silico functional assessment of miRNAs: Tips and tricks

The regulatory networks established by miRNAs and their cognate mRNA targets are more complex in higher organisms, where it is very unlikely to find a single miRNA capable of regulating by itself a very relevant biological process. In higher eukaryotes such as mammals and primates, the redundancy observed in all the cellular processes and the intrinsic promiscuity of miRNAs are related to the presence of groups of miRNAs that act together in the regulation of biological tasks (Davis, Saunders, Mann, & Backofen, 2017; Russo & Giordano, 2009). MiRNAs that are involved in the regulation of a specific physiological or pathological event have a tendency to regulate genes related to the same biochemical pathways (Bian, Wang, Liu, Xia, & Long, 2015; Wang et al., 2014). Taking into account these facts, practical questions include how we can determine which miRNAs are more relevant for a particular study, how to select them, and how to infer their biological roles. The in silico functional assessment of miRNA function is an intelligent way to start answering these questions (Stępień, Costa, & Enguita, 2018).

General principles to take into consideration for the in silico analysis of miRNA function include (Fig. 9.3):

1. *Sources of information*. The datasets used for the functional assessment of miRNA functions in a specific process are frequently derived from screening experiments. Transcriptomic profiling experiments can be performed by using high-throughput techniques, including next-generation sequencing, hybridization microarrays, or quantitative PCR (Rai, Rai, Saeidian, & Rai, 2016). Starting datasets from transcriptomics often include a quantification of miRNAs and other RNAs, such as coding mRNAs, and their differential representation among experimental conditions (Pantaleão & Ozanne, 2018).
2. *Establishment of miRNA-mRNA connections*. This step is essential to construct the putative miRNA-centered regulatory networks. Usually, the predicted or validated targets of the miRNAs that are differentially represented in the study need

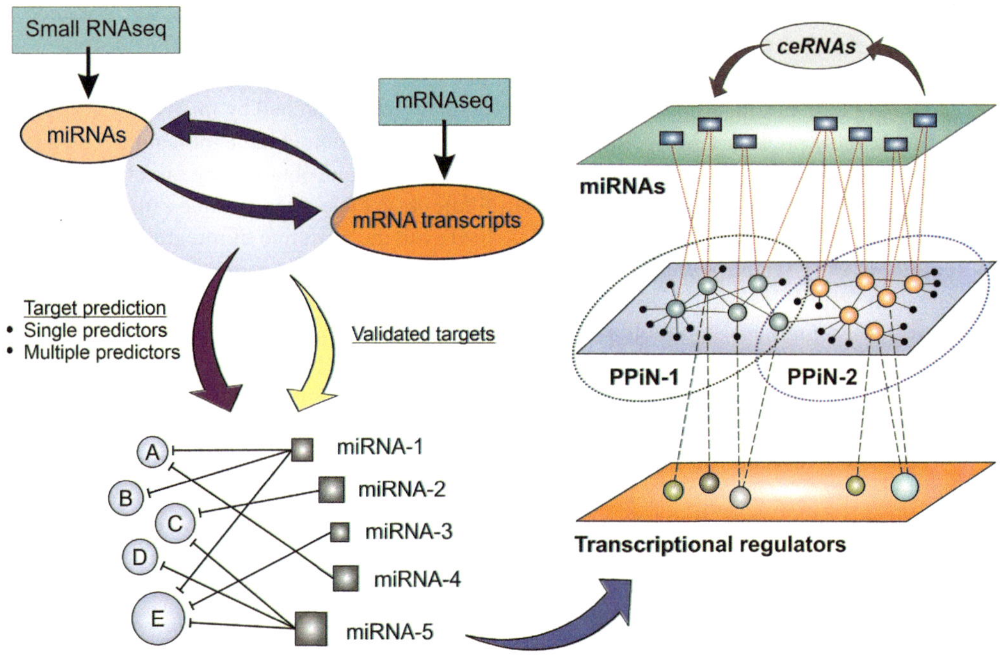

FIG. 9.3 Conceptual framework for the analysis of miRNA function using experimental data and bioinformatics applications. Data extracted from RNA sequencing experiments are frequent sources of information to characterize the functions of miRNAs. Target prediction algorithms can be applied to postulate the functional relationships between mRNAs and their regulatory miRNAs. This information can also be enriched with validated targets and other relevant data. Once the landscape of miRNA-mRNA interactions is built, the analyzed biological process can be complemented by including additional regulatory layers. This approach, founded in the principles of systems biology, is useful to dissect and analyze the complexity of each biological process. Regulatory layers might include the presence of complementary endogenous RNAs (ceRNAs) that could act as buffers of miRNA levels; transcription factors and other regulatory proteins that could modify the expression levels of mRNAs; and protein-protein interaction networks (PPiNs). Other relevant strategies that could be used to group the target mRNAs include enrichment techniques based on descriptors such as the Gene-ontology terms, or the functional enrichment by biochemical pathways. The knowledge of complex regulatory networks combining several layers of information is also an essential tool for the design of wet-lab experiments that will contribute to the overall understanding of a particular biological process.

to be considered. In the case of predicted targets, we strongly recommend the use of multiple predictors where a combination of individual algorithms increases the specificity of the prediction, eliminating the confounding false positives (Kyrollos, Reid, Dick, & Green, 2020; Tokar et al., 2018). Ideally, the multiple predictor should be able to integrate the results from the individual algorithms by calculating a combination score (Tokar et al., 2018).

3. *Graphical representation*. The analysis of miRNA-mRNA interaction data using graphical methods has clear advantages because it provides a conceptual framework that could be filtered, structured, and complemented with other information. Following this idea, the predicted functional connections determined from target prediction algorithms can be graphically represented by considering miRNAs and their mRNA targets as network nodes that will be connected by edges that are depictions of their targeting relationships. To construct the interaction networks, specialized software is recommended. The most popular software for the construction of regulatory networks is the Cytoscape platform (Smoot, Ono, Ruscheinski, Wang, & Ideker, 2011), a=complete system with numerous plugins that allows a very detailed analysis of biological networks. Other open-access alternatives are Gephi (Kauffman, Kittas, Bennett, & Tsoka, 2014) and NAViGaTOR (Brown et al., 2009).
4. *Network simplification and customization*. Due to the promiscuity of the miRNA regulatory action, the interpretation of miRNA-mRNA interaction networks could be complicated, even using a graphical representation. In our experience, networks could be simplified by eliminating those nodes establishing a small number of interactions, following the principle that the most important nodes will be the ones involved in more functional connections (Stępień et al., 2018; Sualp & Can, 2011). This simplification is an iterative and empirical process that depends on the number of miRNAs and predicted targets (Leitão, Costa, & Enguita, 2014). To improve their interpretation, the networks can also be completed with the incorporation of graphical tricks such as color and size codes for the nodes to depict the number of established interactions.
5. *Network enrichment by information overlay*. Functional dissection of the miRNA-centered network will also require the enrichment of the graphical representation with additional information. Depending on the objectives of the study, the

interaction network could be complemented with information about protein-protein interaction networks (PPiN), regulatory proteins such as transcription factors, and interactions with other RNA molecules considering the establishment of complementary endogenous RNA (ceRNA) loops (Fig. 9.3). Ideally, the overlaid information should be represented in planes containing a family of regulatory events. To improve the readability of the network, it would also be advisable to group the mRNA targets by the functional analysis of their source genes. This grouping could be achieved by applying gene enrichment analysis techniques based either on the consideration of the Gene-ontology terms that characterize every gene (GO-enrichment analysis) or on the participation of the gene products in biochemical pathways (pathway enrichment analysis). Recently developed web portals such as miRNet allow an online automated analysis of validated miRNA targets by functional enrichment using GO terms, biochemical pathways, and involvement in other RNA regulatory networks (Chang, Zhou, Soufan, & Xia, 2020; Fan & Xia, 2018). We have previously applied this strategy for the analysis of miRNA regulatory networks in the context of human disease, allowing us to pinpoint specific functional groups of genes under the control of differentially expressed miRNAs and involved in the onset and progression of some chronic diseases (Cortez-Dias et al., 2016).

6. *Selection of network nodes for wet-lab experiments*. The enriched regulatory network could allow not only a systemic analysis of the miRNA regulatory network, but also a rational selection of miRNAs and gene targets that will deserve further laboratory investigation. We recommend selecting those miRNAs and target mRNAs involved in more functional connections for laboratory validation. The analyzed functional connections in targeted mRNAs should consider whether they belong to specific metabolic pathways related to the studied biological event and also their involvement in functional interactions with other partners (PPiN, transcription factors, etc.).

Functional analysis of miRNAs associated with survival time in lung adenocarcinoma: A test case

Lung adenocarcinoma is one of the most prevalent cancers in western countries. It belongs to the subgroup of nonsmall cell lung carcinomas (NSCLC) and is associated with a high risk of mortality. Despite the advances in its treatment, there is a current medical need for diagnosis and prognosis biomarkers in NSCLC. Yekurala-Sathipati and Ho proposed the use of a complex miRNA signature as a biomarker for the determination of the survival rate in patients with lung adenocarcinoma (Sathipati & Ho, 2017). A dataset of 521 lung adenocarcinoma patients selected from the TCGA database allowed characterizing 17 different miRNAs whose expression in lung biopsies is correlated with the survival rate: hsa-let-7f-5p, hsa-let-7g-5p, hsa-miR-141-3p, hsa-miR-155-5p, hsa-miR-16-5p, hsa-miR-18a-5p, hsa-miR-192-5p, hsa-miR-193b-3p, hsa-miR-219a-5p, hsa-miR-3136-5p, hsa-miR-342-3p, hsa-miR-3653-3p, hsa-miR-3934-5p, hsa-miR-217, hsa-miR-3199, hsa-miR-1254, and hsa-miR-1291 (Sathipati & Ho, 2017).

As a working example, we can postulate that the presence of a specific pattern of miRNA expression could be related to the regulatory functions over lung cells. The final regulatory effect exerted on cancer cells will be a consequence of the combined actions of all miRNAs (Chandra, Singh, & Bhattacharya, 2019). To analyze the combined effect of all human miRNAs, we recommend MIRDIP as a starting application because it allows combining a high number of target prediction algorithms and calculating a consensus score (Tokar et al., 2018). Note that MIRDIP is a human-exclusive portal that cannot be used for the prediction of miRNA targets from other species.

In the MIRDIP web portal (http://ophid.utoronto.ca/mirDIP), the user can introduce a list of human genes or miRNAs to perform a target prediction analysis with multiple algorithms. The user can also select the level of confidence for the predictions quantified by the consensus score, and also the type of output (Fig. 9.4). Considering the 17 different miRNAs included in the lung adenocarcinoma, the MIRDIP algorithm was able to predict a total number of 15488 targets regulated by 14 miRNAs, with a "very high" confidence level (Fig. 9.4). In practical terms, the handling and analysis of such a high number of targets are not feasible and require rational filtering. Considering that a group of miRNAs involved in a biological process might act together, we proposed using a functional filtering of the predicted targets by selecting those mRNAs simultaneously targeted by different miRNAs. For the running example, we only considered those targets regulated by five or more miRNAs, reducing the number of mRNAs to 198 (Fig. 9.5). We strongly recommend performing the filtering processing by using a graphical representation of the miRNA-mRNA network. In our case study, all the graphical representations of regulatory networks connecting miRNAs and their targets were built by using the NAViGaTOR software (Brown et al., 2009). Independently of the representation software, the interaction networks are built from the data exported from MIRDIP or other target prediction algorithms where mRNA and miRNAs are defined as network nodes (Fig. 9.5). It would also be interesting to load quantitative information to the edges connecting miRNAs and mRNAs such as the consensus score calculated by MIRDIP. For visual purposes, a

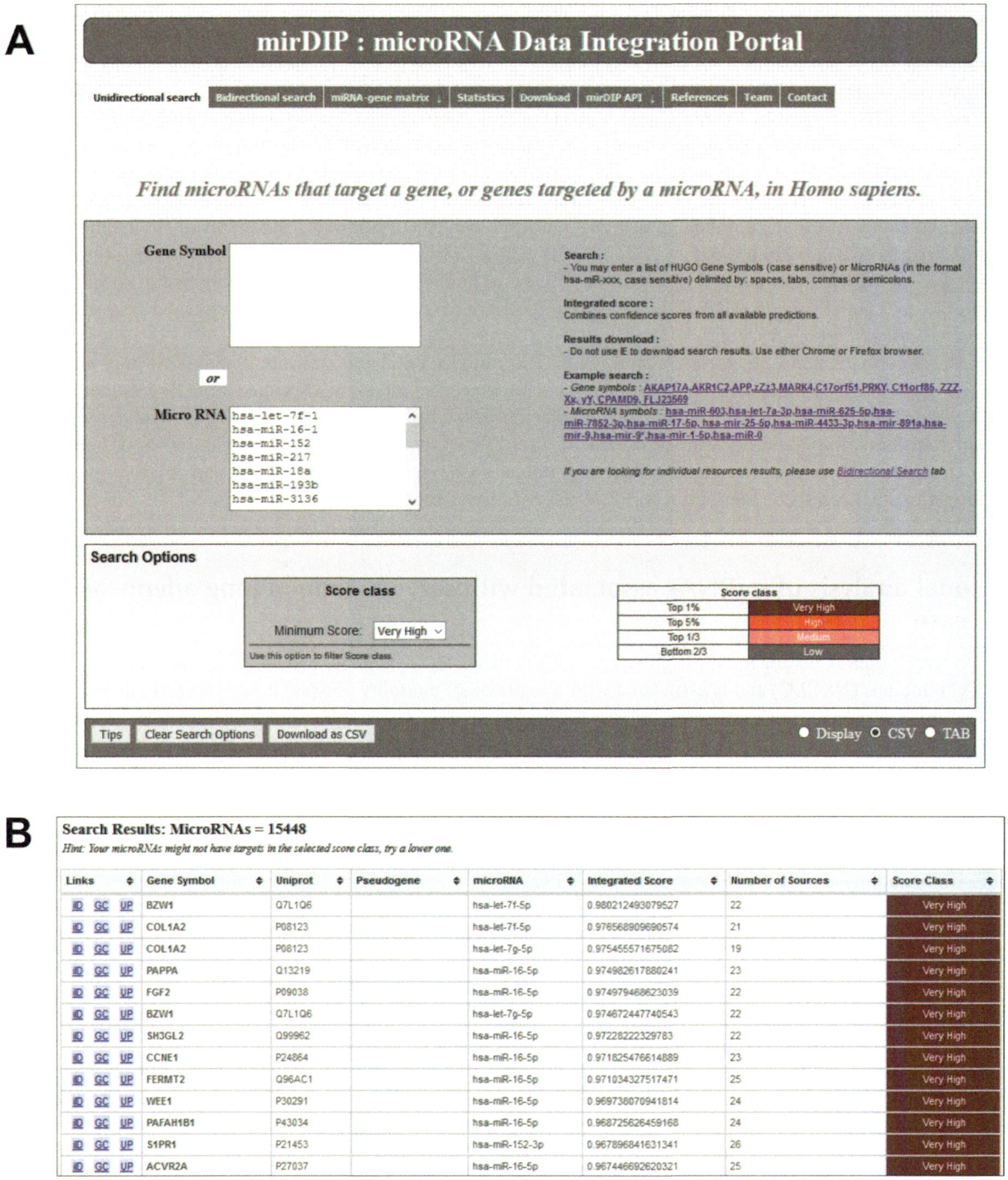

FIG. 9.4 MIRDIP interface and results for the example test case. (A) Initial input screen of the MIRDIP portal where the user can introduce either a list of target genes or a list of miRNAs for the analysis. In this screen the user can also select the "score class," a parameter derived from the calculated consensus score that classifies the miRNA targets into quartiles (very high, high, medium, and low) corresponding to the distribution of the solutions and the respective confidence of the prediction. (B) Output MIRDIP results obtained from the analysis of the miRNA signature determined in the work from Yekurala-Sathipati and Ho (Sathipati & Ho, 2017), filtered by a "very high" score class. The output table shows the correspondence between each miRNA and its cognate targets, together with the integrated MIRDIP score and the number of predictors giving a positive result.

circular layered representation is recommended, where the mRNA targets are ranked in concentric circles. Moreover, the sizes of every node can be adjusted proportionally to the number of interactions established, depicting the relevance of miRNAs and their targets (Fig. 9.5). As we proposed previously, those miRNAs regulating more targets and those mRNAs being regulated simultaneously by a higher number of miRNAs could be more relevant in a particular biological process (Stępień et al., 2018).

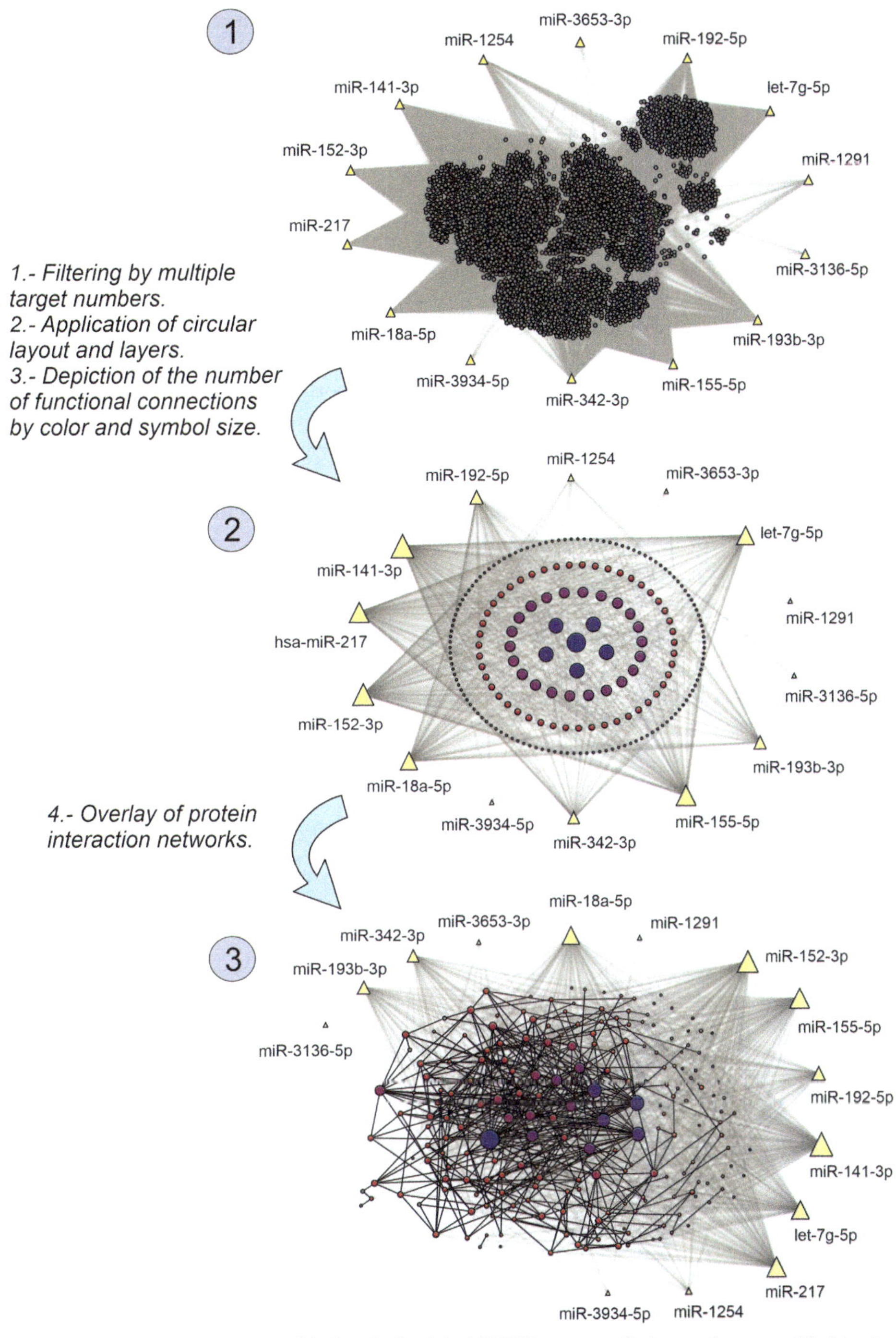

FIG. 9.5 Proposed flowchart for the filtering and graphical analysis of the MIRDIP target prediction results, exemplified by the miRNA signature in lung adenocarcinoma (Sathipati & Ho, 2017). All the graphical representations were prepared with the NAViGaTOR software (Brown et al., 2009). In the first step, (1) the miRNA-target interaction network was constructed, including all the targets predicted by MIRDIP. This initial network was further filtered and simplified by eliminating the nodes establishing less than four interactions, and represented in a concentric circular layout with the mRNA targets ordered by the number of regulating miRNAs. The outer circular layer (filled circles) represents the mRNAs being targeted simultaneously by five of the selected miRNAs, and the successive concentric layers are composed by the mRNAs targeted by six, seven, eight, and nine miRNAs, respectively (2). The symbols representing miRNAs (filled triangles) and mRNAs (filled circles) were resized considering the number of established interactions. In the last step (3), the miRNA-mRNA interaction landscape was completed by overlaying data from the STRING database (Szklarczyk et al., 2015) for the inclusion of the protein interaction networks (PPiN).

In eukaryotic cells, the regulatory action of miRNAs is frequently exerted in a pathway-dependent manner. In this context, the protein-protein interaction networks (PPiNs) are very relevant factors to understand how a particular miRNA is able to regulate a cellular pathway (Chang et al., 2020; Davis et al., 2017). PPiNs can be extracted and incorporated into the graphical representations by importing data from databases such as STRING (Szklarczyk et al., 2015). Software for network analysis such as NAViGaTOR or Cytoscape is able to handle superpositions of networks with different parameters, as we depicted in Fig. 9.5. Using this representation, the interaction between proteins shows the existence of functional relationships among all the targeted mRNAs in the context of lung adenocarcinoma. However, the interpretation of regulatory networks with overlaid information, which in this particular example is constituted by the miRNA-mRNA connections and PPiNs, could be difficult. Several strategies to simplify and organize such overlaid networks have been described in the literature (Jacomy, Venturini, Heymann, & Bastian, 2014; Kuchaiev, Stevanović, Hayes, & Pržulj, 2011; Ou-Yang, Yan, & Zhang, 2017). One of the most powerful methods consists of the application of functional clustering over the targeted mRNAs, either by including information on gene ontology (GO-term enrichment) or biochemical pathways (Pathway enrichment) (Cortez-Dias et al., 2016). Using the functional clustering algorithm included in NAViGaTOR, the interaction network can be organized and simplified, as depicted in Fig. 9.6. Other algorithms such as Genclip (Wang et al., 2019) or ToppGene (Chen, Bardes, Aronow, & Jegga, 2009) can also be applied for functional clustering of the miRNA targets. In the case study of lung adenocarcinoma, the regulatory network represented in Fig. 9.6 give us a clear picture of the most relevant elements classified by functional families. The lung adenocarcinoma molecular signature is composed of miRNAs that could target mRNA transcripts of genes that are mainly included into two functional categories: genome maintenance and cellular fate and organization (Fig. 9.6). The regulatory network also pinpoints the relative importance of mRNAs and miRNAs by the number of established interactions, and paves the way for the design of wet-lab validation experiments. The functional category related to genome maintenance shows at least four genes with an increased number of interactions demonstrated either by PPiNs or by miRNA-mRNA targeting: PTEN, DICER1, YY1, and IGF1. In the enriched category focused on cellular fate, the genes KRAS, NCOA3, MECP2, SP1, KMT2C, and CBL should be considered for further

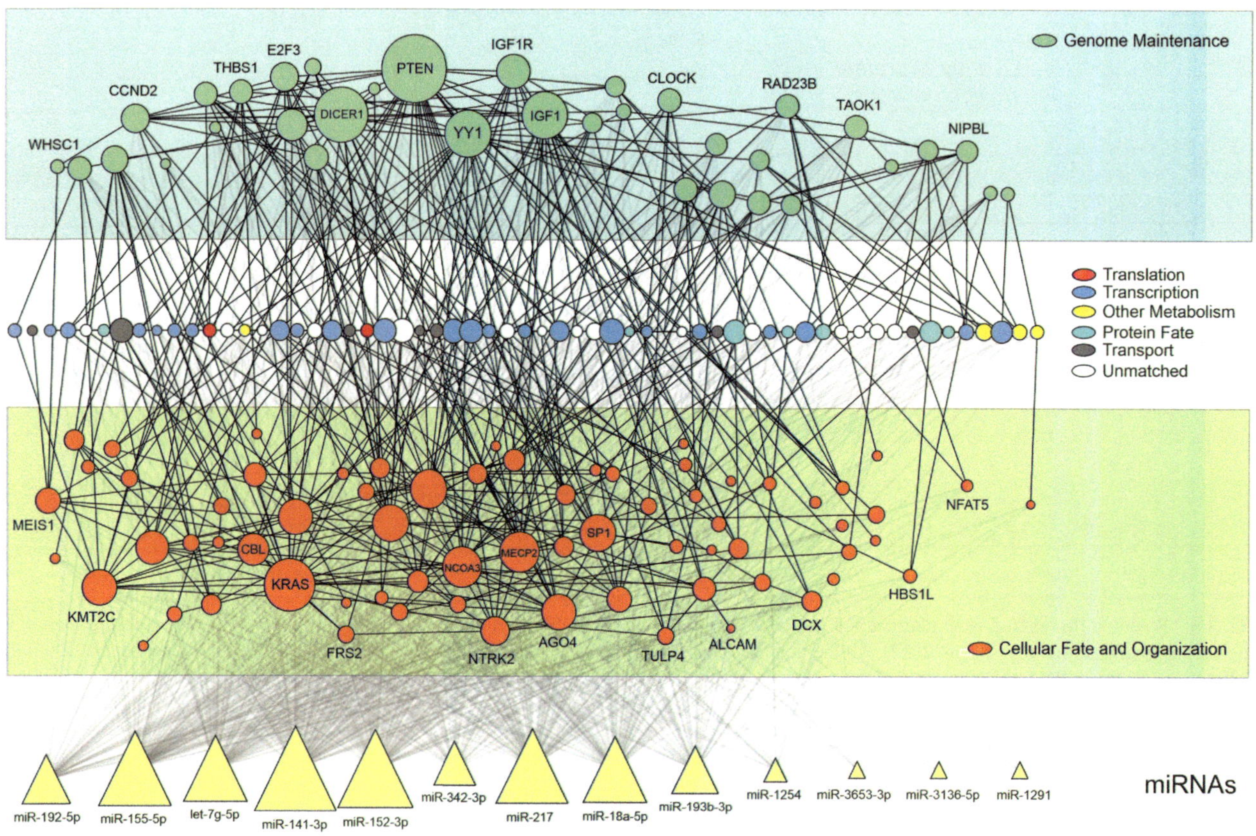

FIG. 9.6 Regulatory landscape of the miRNA lung adenocarcinoma signature including different functional layers: miRNA-mRNA interactions (*grey connecting lines*), protein-protein interaction networks (*black connecting lines*), and functional clustering of the mRNA targets (*symbol color code*). The size of the symbols, triangles for miRNAs and circles for regulated targets, is proportional to the number of functional interactions. The interactions between miRNA target transcripts were functionally clustered into categories by using the NAViGaTOR software (Brown et al., 2009).

validation. On the other hand, the miRNAs regulating a higher number of targets should be more relevant for the progression of lung adenocarcinoma, and those include hsa-miR-155-5p, hsa-miR-141-3p, hsa-miR-152-5p, and hsa-miR-217.

Conclusions

MiRNAs are important posttranscriptional regulators involved in cell homeostasis and pathological processes. Understanding the miRNA action in complex processes such as human diseases requires a paradigm switch to apply the principles and rules of systems biology to determine which miRNAs and cognate targets are more relevant within the studied biological context. To infer possible functional relationships between miRNAs and their targets, we proposed an analysis method based on the construction of graphical representations that are overlaid with different data. The depicted information as a functional interaction network can be simplified and processed by using enrichment methods to generate an overall landscape of a particular biological process. This method can be applied to the study of complex systems where a high number of miRNAs and targets are considered because it allows dynamic filtering with different criteria.

Acknowledgments

This work is supported by COST (European Cooperation in Science and Technology) Action EU-CardioRNA CA17129 and the Portuguese Foundation for Science and Technology (FCT) under the framework of the research grant PTDC-MED-GEN-29389-2017. The authors would like to acknowledge Francisco Enguita Jr. for excellent technical assistance and support in the design and conceptualization of the manuscript.

References

Agarwal, V., Bell, G. W., Nam, J.-W., & Bartel, D. P. (2015). Predicting effective microRNA target sites in mammalian mRNAs. *eLife*, *4*. https://doi.org/10.7554/eLife.05005.

Agarwal, V., Subtelny, A. O., Thiru, P., Ulitsky, I., & Bartel, D. P. (2018). Predicting microRNA targeting efficacy in Drosophila. *Genome Biology*, *19*(1), 152. https://doi.org/10.1186/s13059-018-1504-3.

Akhtar, M. M., Micolucci, L., Islam, M. S., Olivieri, F., & Procopio, A. D. (2019). A practical guide to miRNA target prediction. In *Vol. 1970. Methods in molecular biology* (pp. 1–13). Humana Press Inc. https://doi.org/10.1007/978-1-4939-9207-2_1.

Avital, G., França, G. S., & Yanai, I. (2018). Bimodal evolutionary developmental miRNA program in animal embryogenesis. *Molecular Biology and Evolution*, *35*(3), 646–654. https://doi.org/10.1093/molbev/msx316.

Backes, C., Fehlmann, T., Kern, F., Kehl, T., Lenhof, H. P., Meese, E., et al. (2018). MiRCarta: A central repository for collecting miRNA candidates. *Nucleic Acids Research*, *46*(1), D160–D167. https://doi.org/10.1093/nar/gkx851.

Bar, M., Wyman, S. K., Fritz, B. R., Qi, J., Garg, K. S., Parkin, R. K., et al. (2008). MicroRNA discovery and profiling in human embryonic stem cells by deep sequencing of small RNA libraries. *Stem Cells*, *26*(10), 2496–2505. https://doi.org/10.1634/stemcells.2008-0356.

Bartel, D. P. (2009). MicroRNAs: Target recognition and regulatory functions. *Cell*, *136*(2), 215–233. https://doi.org/10.1016/j.cell.2009.01.002.

Bartel, D. P. (2018). Metazoan MicroRNAs. *Cell*, *173*(1), 20–51. https://doi.org/10.1016/j.cell.2018.03.006.

Bartel, D. P., & Chen, C. Z. (2004). Micromanagers of gene expression: The potentially widespread influence of metazoan microRNAs. *Nature Reviews Genetics*, *5*(5), 396–400. https://doi.org/10.1038/nrg1328.

Berezikov, E., Cuppen, E., & Plasterk, R. H. A. (2006). Approaches to microrna discovery. *Nature Genetics*, *38*(6S), S2. https://doi.org/10.1038/ng1794.

Bian, L., Wang, Y., Liu, Q., Xia, J., & Long, J. E. (2015). Prediction of signaling pathways involved in enterovirus 71 infection by algorithm analysis based on miRNA profiles and their target genes. *Archives of Virology*, *160*(1), 173–182. https://doi.org/10.1007/s00705-014-2249-2.

Brown, K. R., Otasek, D., Ali, M., McGuffin, M. J., Xie, W., Devani, B., et al. (2009). NAViGaTOR: Network analysis, visualization and graphing Toronto. *Bioinformatics*, *25*(24), 3327–3329. https://doi.org/10.1093/bioinformatics/btp595.

Castellano, L., & Stebbing, J. (2013). Deep sequencing of small RNAs identifies canonical and non-canonical miRNA and endogenous siRNAs in mammalian somatic tissues. *Nucleic Acids Research*, *41*(5), 3339–3351. https://doi.org/10.1093/nar/gks1474.

Chandra, N. K., Singh, R., & Bhattacharya, S. (2019). A novel bayesian multiple testing approach to deregulated miRNA discovery harnessing positional clustering. *Biometrics*, *75*(1), 202–209. https://doi.org/10.1111/biom.12967.

Chang, L., Zhou, G., Soufan, O., & Xia, J. (2020). miRNet 2.0: Network-based visual analytics for miRNA functional analysis and systems biology. *Nucleic Acids Research*, *48*(1), W244–W251. https://doi.org/10.1093/nar/gkaa467.

Chen, H., Zhang, Z., & Feng, D. (2019). Prediction and interpretation of miRNA-disease associations based on miRNA target genes using canonical correlation analysis. *BMC Bioinformatics*, *20*(1). https://doi.org/10.1186/s12859-019-2998-8.

Chen, J., Bardes, E. E., Aronow, B. J., & Jegga, A. G. (2009). ToppGene Suite for gene list enrichment analysis and candidate gene prioritization. *Nucleic Acids Research*, *37*(2), W305–W311. https://doi.org/10.1093/nar/gkp427.

Chi, S. W., Zang, J. B., Mele, A., & Darnell, R. B. (2009). Argonaute HITS-CLIP decodes microRNA-mRNA interaction maps. *Nature*, *460*(7254), 479–486. https://doi.org/10.1038/nature08170.

Chipman, L. B., & Pasquinelli, A. E. (2019). miRNA targeting: Growing beyond the seed. *Trends in Genetics*, *35*(3), 215–222. https://doi.org/10.1016/j.tig.2018.12.005.

Cortez-Dias, N., Costa, M. C., de Sousa, J., Fiúza, M., Gallego, J., Nobre, Â., et al. (2016). Expanding the functional role of miRNAs in the establishment of permanent atrial fibrillation. *International Journal of Cardiology*, *222*, 340–341. https://doi.org/10.1016/j.ijcard.2016.07.296.

Davis, J. A., Saunders, S. J., Mann, M., & Backofen, R. (2017). Combinatorial ensemble miRNA target prediction of co-regulation networks with non-prediction data. *Nucleic Acids Research*, *45*(15), 8745–8757. https://doi.org/10.1093/nar/gkx605.

Ding, J., Li, X., & Hu, H. (2016). TarPmiR: A new approach for microRNA target site prediction. *Bioinformatics*, *32*(18), 2768–2775. https://doi.org/10.1093/bioinformatics/btw318.

Dweep, H., & Gretz, N. (2015). a comprehensive atlas of microRNA-target interactions. *Nature Methods*, *12*(8).

Dweep, H., Gretz, N., & Sticht, C. (2014). MiRWalk database for miRNA-target interactions. *Methods in Molecular Biology*, *1182*, 289–305. https://doi.org/10.1007/978-1-4939-1062-5_25.

Dweep, H., Sticht, C., Pandey, P., & Gretz, N. (2011). MiRWalk-Database: Prediction of possible miRNA binding sites by\walking\the genes of three genomes. *Journal of Biomedical Informatics*, *44*(5), 839–847. https://doi.org/10.1016/j.jbi.2011.05.002.

Enright, A. J., John, B., Gaul, U., Tuschl, T., Sander, C., & Marks, D. S. (2003). MicroRNA targets in Drosophila. *Genome Biology*, *5*(1).

Fan, Y., & Xia, J. (2018). miRNet—Functional analysis and visual exploration of miRNA–target interactions in a network context. In *Vol. 1819. Methods in molecular biology* (pp. 215–233). Humana Press Inc. https://doi.org/10.1007/978-1-4939-8618-7_10.

Han, J., Lee, Y., Yeom, K. H., Kim, Y. K., Jin, H., & Kim, V. N. (2004). The Drosha-DGCR8 complex in primary microRNA processing. *Genes and Development*, *18*(24), 3016–3027. https://doi.org/10.1101/gad.1262504.

He, J. F., Luo, Y. M., Wan, X. H., & Jiang, D. (2011). Biogenesis of MiRNA-195 and its role in biogenesis, the cell cycle, and apoptosis. *Journal of Biochemical and Molecular Toxicology*, *25*(6), 404–408. https://doi.org/10.1002/jbt.20396.

Helwak, A., Kudla, G., Dudnakova, T., & Tollervey, D. (2013). Mapping the human miRNA interactome by CLASH reveals frequent noncanonical binding. *Cell*, *153*(3), 654–665. https://doi.org/10.1016/j.cell.2013.03.043.

Helwak, A., & Tollervey, D. (2016). Identification of miRNA-target RNA interactions using clash. In *Vol. 1358. Methods in molecular biology* (pp. 229–251). Humana Press Inc. https://doi.org/10.1007/978-1-4939-3067-8_14.

Hu, H. Y., Yan, Z., Xu, Y., Hu, H., Menzel, C., Zhou, Y. H., et al. (2009). Sequence features associated with microRNA strand selection in humans and flies. *BMC Genomics*, *10*, 413. https://doi.org/10.1186/1471-2164-10-413.

Jacomy, M., Venturini, T., Heymann, S., & Bastian, M. (2014). ForceAtlas2, a continuous graph layout algorithm for handy network visualization designed for the Gephi software. *PLoS One*, *9*(6). https://doi.org/10.1371/journal.pone.0098679.

Jin, W., Wang, J., Liu, C. P., Wang, H. W., & Xu, R. M. (2020). Structural basis for pri-miRNA recognition by Drosha. *Molecular Cell*, *78*(3), 423–433.e5. https://doi.org/10.1016/j.molcel.2020.02.024.

Jones-Rhoades, M. W., & Bartel, D. P. (2004). Computational identification of plant MicroRNAs and their targets, including a stress-induced miRNA. *Molecular Cell*, *14*(6), 787–799. https://doi.org/10.1016/j.molcel.2004.05.027.

Joung, J. G., Hwang, K. B., Nam, J. W., Kim, S. J., & Zhang, B. T. (2007). Discovery of microRNA-mRNA modules via population-based probabilistic learning. *Bioinformatics*, *23*(9), 1141–1147. https://doi.org/10.1093/bioinformatics/btm045.

Kauffman, J., Kittas, A., Bennett, L., & Tsoka, S. (2014). DyCoNet: A Gephi plugin for community detection in dynamic complex networks. *PLoS One*, *9*(7).

Kertesz, M., Iovino, N., Unnerstall, U., Gaul, U., & Segal, E. (2007). The role of site accessibility in microRNA target recognition. *Nature Genetics*, *39*(10), 1278–1284. https://doi.org/10.1038/ng2135.

Krek, A., Grün, D., Poy, M. N., Wolf, R., Rosenberg, L., Epstein, E. J., et al. (2005). Combinatorial microRNA target predictions. *Nature Genetics*, *37*(5), 495–500. https://doi.org/10.1038/ng1536.

Kuchaiev, O., Stevanović, A., Hayes, W., & Pržulj, N. (2011). GraphCrunch 2: Software tool for network modeling, alignment and clustering. *BMC Bioinformatics*, *12*. https://doi.org/10.1186/1471-2105-12-24.

Kyrollos, D. G., Reid, B., Dick, K., & Green, J. R. (2020). RPmirDIP: Reciprocal perspective improves miRNA targeting prediction. *Scientific Reports*, *10*(1). https://doi.org/10.1038/s41598-020-68251-4.

Lee, R. C., Feinbaum, R. L., & Ambros, V. (1993). The *C. elegans* heterochronic gene lin-4 encodes small RNAs with antisense complementarity to lin-14. *Cell*, *75*(5), 843–854. https://doi.org/10.1016/0092-8674(93)90529-Y.

Leitão, A. L., Costa, M. C., & Enguita, F. J. (2014). A guide for miRNA target prediction and analysis using web-based applications. *Methods in Molecular Biology*, *1182*, 265–277. https://doi.org/10.1007/978-1-4939-1062-5_23.

Lewis, B. P., Burge, C. B., & Bartel, D. P. (2005). Conserved seed pairing, often flanked by adenosines, indicates that thousands of human genes are microRNA targets. *Cell*, *120*(1), 15–20. https://doi.org/10.1016/j.cell.2004.12.035.

Lewis, B. P., Shih, I. H., Jones-Rhoades, M. W., Bartel, D. P., & Burge, C. B. (2003). Prediction of mammalian microRNA targets. *Cell*, *115*(7), 787–798. https://doi.org/10.1016/S0092-8674(03)01018-3.

Li, X., & Hu, H. (2019). Improving miRNA target prediction using CLASH data. In *Vol. 1970. Methods in molecular biology* (pp. 75–83). Humana Press Inc. https://doi.org/10.1007/978-1-4939-9207-2_6.

Li, F., Pallan, P. S., Maier, M. A., Rajeev, K. G., Mathieu, S. I., Kreutz, C., et al. (2007). Crystal structure, stability and in vitro RNAi activity of oligoribonucleotides containing the ribo-difluorotoluyl nucleotide: Insights into substrate requirements by the human RISC Ago2 enzyme. *Nucleic Acids Research*, *35*(19), 6424–6438. https://doi.org/10.1093/nar/gkm664.

Licatalosi, D. D., Mele, A., Fak, J. J., Ule, J., Kayikci, M., Chi, S. W., et al. (2008). HITS-CLIP yields genome-wide insights into brain alternative RNA processing. *Nature*, *456*(7221), 464–469. https://doi.org/10.1038/nature07488.

Loher, P., & Rigoutsos, I. (2012). Interactive exploration of RNA22 microRNA target predictions. *Bioinformatics*, *28*(24), 3322–3323. https://doi.org/10.1093/bioinformatics/bts615.

Lu, T. P., Lee, C. Y., Tsai, M. H., Chiu, Y. C., Hsiao, C. K., Lai, L. C., et al. (2012). MiRSystem: An integrated system for characterizing enriched functions and pathways of microRNA targets. *PLoS One*, *7*(8). https://doi.org/10.1371/journal.pone.0042390.

Lu, L., Zhou, L., Chen, E. Z., Sun, K., Jiang, P., Wang, L., et al. (2012). A Novel YY1-miR-1 regulatory circuit in skeletal myogenesis revealed by genome-wide prediction of YY1-miRNA network. *PLoS One*, *7*(2).

Malhas, A., Saunders, N. J., & Vaux, D. J. (2010). The nuclear envelope can control gene expression and cell cycle progression via miRNA regulation. *Cell Cycle*, *9*(3), 531–539. https://doi.org/10.4161/cc.9.3.10511.

Maragkakis, M., Reczko, M., Simossis, V. A., Alexiou, P., Papadopoulos, G. L., Dalamagas, T., et al. (2009). DIANA-microT web server: Elucidating microRNA functions through target prediction. *Nucleic Acids Research*, *37*(2), W273–W276. https://doi.org/10.1093/nar/gkp292.

Maragkakis, M., Vergoulis, T., Alexiou, P., Reczko, M., Plomaritou, K., Gousis, M., et al. (2011). DIANA-microT Web server upgrade supports Fly and Worm miRNA target prediction and bibliographic miRNA to disease association. *Nucleic Acids Research*, *39*(2), W145–W148. https://doi.org/10.1093/nar/gkr294.

Medley, J. C., Panzade, G., & Zinovyeva, A. Y. (2020). microRNA strand selection: Unwinding the rules. *Wiley Interdisciplinary Reviews: RNA*. https://doi.org/10.1002/wrna.1627.

Meijer, H. A., Smith, E. M., & Bushell, M. (2014). Regulation of miRNA strand selection: Follow the leader? In *Vol. 42*. *Biochemical society transactions* (pp. 1135–1140). Portland Press Ltd. https://doi.org/10.1042/BST20140142. Issue 4.

Mount, D. W. (2007). Steps used by the BLAST algorithm. *CSH Protocols*, *41*.

O'Brien, J., Hayder, H., Zayed, Y., & Peng, C. (2018). Overview of microRNA biogenesis, mechanisms of actions, and circulation. *Frontiers in Endocrinology*, *9*. https://doi.org/10.3389/fendo.2018.00402.

Oliveira, A. C., Bovolenta, L. A., Nachtigall, P. G., Herkenhoff, M. E., Lemke, N., & Pinhal, D. (2017). Combining results from distinct microRNA target prediction tools enhances the performance of analyses. *Frontiers in Genetics*, *8*(May). https://doi.org/10.3389/fgene.2017.00059.

Ou-Yang, L., Yan, H., & Zhang, X. F. (2017). A multi-network clustering method for detecting protein complexes from multiple heterogeneous networks. *BMC Bioinformatics*, *18*. https://doi.org/10.1186/s12859-017-1877-4.

Pan, T., & Sosnick, T. (2006). RNA folding during transcription. *Annual Review of Biophysics and Biomolecular Structure*, *35*, 161–175. https://doi.org/10.1146/annurev.biophys.35.040405.102053.

Pantaleão, L. C., & Ozanne, S. E. (2018). Small RNA sequencing: A technique for miRNA profiling. In *Vol. 1735*. *Methods in molecular biology* (pp. 321–330). Humana Press Inc. https://doi.org/10.1007/978-1-4939-7614-0_21.

Paraskevopoulou, M. D., Georgakilas, G., Kostoulas, N., Vlachos, I. S., Vergoulis, T., Reczko, M., et al. (2013). DIANA-microT web server v5.0: Service integration into miRNA functional analysis workflows. *Nucleic Acids Research*, *41*, W169–W173. https://doi.org/10.1093/nar/gkt393.

Parizotto, E. A., Dunoyer, P., Rahm, N., Himber, C., & Voinnet, O. (2004). In vivo investigation of the transcription, processing, endonucleolytic activity, and functional relevance of the spatial distribution of a plant miRNA. *Genes and Development*, *18*(18), 2237–2242. https://doi.org/10.1101/gad.307804.

Parveen, A., Mustafa, S. H., Yadav, P., & Kumar, A. (2019). Applications of machine learning in miRNA discovery and target prediction. *Current Genomics*, *20*(8), 537–544. https://doi.org/10.2174/1389202921666200106111813.

Pla, A., Zhong, X., & Rayner, S. (2018). miRAW: A deep learning-based approach to predict microRNA targets by analyzing whole microRNA transcripts. *PLoS Computational Biology*, *14*(7). https://doi.org/10.1371/journal.pcbi.1006185.

Qi, B., Fiori, L. M., Turecki, G., & Trakadis, Y. J. (2020). Machine learning analysis of blood microRNA data in major depression: A case-control study for biomarker discovery. *The International Journal of Neuropsychopharmacology*, *23*(8), 505–510. https://doi.org/10.1093/ijnp/pyaa029.

Rai, G., Rai, R., Saeidian, A. H., & Rai, M. (2016). Microarray to deep sequencing: Transcriptome and miRNA profiling to elucidate molecular pathways in systemic lupus erythematosus. *Immunologic Research*, *64*(1), 14–24. https://doi.org/10.1007/s12026-015-8672-y.

Russo, G., & Giordano, A. (2009). miRNAs: From biogenesis to networks. *Methods in Molecular Biology (Clifton, N.J.)*, *563*, 303–352. https://doi.org/10.1007/978-1-60761-175-2_17.

Sasaki, T., Shiohama, A., Minoshima, S., & Shimizu, N. (2003). Identification of eight members of the Argonaute family in the human genome. *Genomics*, *82*(3), 323–330. https://doi.org/10.1016/S0888-7543(03)00129-0.

Sathipati, Y., & Ho, S. Y. (2017). Identifying the miRNA signature associated with survival time in patients with lung adenocarcinoma using miRNA expression profiles. *Scientific Reports*, *7*(1).

Schirle, N. T., Sheu-Gruttadauria, J., & MacRae, I. J. (2014). Structural basis for microRNA targeting. *Science*, *346*(6209), 608–613. https://doi.org/10.1126/science.1258040.

Sedaghat, N., Fathy, M., Modarressi, M. H., & Shojaie, A. (2018). Combining supervised and unsupervised learning for improved miRNA target prediction. *IEEE/ACM Transactions on Computational Biology and Bioinformatics*, *15*(5), 1594–1604. https://doi.org/10.1109/TCBB.2017.2727042.

Sheu-Gruttadauria, J., Xiao, Y., Gebert, L. F. R., & MacRae, I. J. (2019). Beyond the seed: Structural basis for supplementary microRNA targeting by human Argonaute2. *EMBO Journal*, *38*(13). https://doi.org/10.15252/embj.2018101153.

Shirdel, E. A., Xie, W., Mak, T. W., & Jurisica, I. (2011). NAViGaTing the micronome—Using multiple microRNA prediction databases to identify signalling pathway-associated microRNAs. *PLoS One*, *6*(2).

Siddiqui, Z. H., Abbas, Z. K., Ansari, M. W., & Khan, M. N. (2019). The role of miRNA in somatic embryogenesis. *Genomics*, *111*(5), 1026–1033. https://doi.org/10.1016/j.ygeno.2018.11.022.

Smoot, M. E., Ono, K., Ruscheinski, J., Wang, P. L., & Ideker, T. (2011). Cytoscape 2.8: New features for data integration and network visualization. *Bioinformatics*, *27*(3), 431–432. https://doi.org/10.1093/bioinformatics/btq675.

Stępień, E., Costa, M. C., & Enguita, F. J. (2018). miRNAtools: Advanced training using the miRNA web of knowledge. *Non-Coding RNA*, *4*(1). https://doi.org/10.3390/ncrna4010005.

Sticht, C., De La Torre, C., Parveen, A., & Gretz, N. (2018). Mirwalk: An online resource for prediction of microrna binding sites. *PLoS One*, *13*(10). https://doi.org/10.1371/journal.pone.0206239.

Sualp, M., & Can, T. (2011). Using network context as a filter for miRNA target prediction. *Biosystems*, *105*(3), 201–209. https://doi.org/10.1016/j.biosystems.2011.04.002.

Sun, Y., Ge, Y., Drnevich, J., Zhao, Y., Band, M., & Chen, J. (2010). Mammalian target of rapamycin regulates miRNA-1 and follistatin in skeletal myogenesis. *Journal of Cell Biology*, *189*(7), 1157–1169. https://doi.org/10.1083/jcb.200912093.

Szklarczyk, D., Franceschini, A., Wyder, S., Forslund, K., Heller, D., Huerta-Cepas, J., et al. (2015). STRING v10: Protein-protein interaction networks, integrated over the tree of life. *Nucleic Acids Research*, *43*(1), D447–D452. https://doi.org/10.1093/nar/gku1003.

Tokar, T., Pastrello, C., Rossos, A. E. M., Abovsky, M., Hauschild, A. C., Tsay, M., et al. (2018). integrative database of human microRNA target predictions. *Nucleic Acids Research*, *46*(D1).

Tomasello, G., Armenia, I., & Molla, G. (2020). The Protein Imager: A full-featured online molecular viewer interface with server-side HQ-rendering capabilities. *Bioinformatics*, *36*(9), 2909–2911. https://doi.org/10.1093/bioinformatics/btaa009.

Ullah, A. Z. M. D., Sahoo, S., Steinhöfel, K., & Albrecht, A. A. (2012). Derivative scores from site accessibility and ranking of miRNA target predictions. *International Journal of Bioinformatics Research and Applications*, *8*(3–4), 171–191. https://doi.org/10.1504/IJBRA.2012.048966.

Van Peer, G., De Paepe, A., Stock, M., Anckaert, J., Volders, P. J., Vandesompele, J., … Waegeman, W. (2017). miSTAR: miRNA target prediction through modeling quantitative and qualitative miRNA binding site information in a stacked model structure. *Nucleic Acids Research*, *45*(7), e51.

Vlachos, I. S., Kostoulas, N., Vergoulis, T., Georgakilas, G., Reczko, M., Maragkakis, M., et al. (2012). DIANA miRPath v.2.0: Investigating the combinatorial effect of microRNAs in pathways. *Nucleic Acids Research*, *40*(1), W498–W504. https://doi.org/10.1093/nar/gks494.

Vlachos, I. S., Zagganas, K., Paraskevopoulou, M. D., Georgakilas, G., Karagkouni, D., Vergoulis, T., et al. (2015). DIANA-miRPath v3.0: Deciphering microRNA function with experimental support. *Nucleic Acids Research*, *43*(1), W460–W466. https://doi.org/10.1093/nar/gkv403.

Vougas, K., Almpanis, S., & Gorgoulis, V. (2020). Deep learning: Shaping the medicine of tomorrow. *Molecular & Cellular Oncology*, *7*(3). https://doi.org/10.1080/23723556.2020.1723462.

Wang, F., Casalino, L. P., & Khullar, D. (2019). Deep learning in medicine—Promise, progress, and challenges. *JAMA Internal Medicine*, *179*(3), 293–294. https://doi.org/10.1001/jamainternmed.2018.7117.

Wang, F., Chan, L. W. C., Law, H. K. W., Cho, W. C. S., Tang, P., Yu, J., et al. (2014). Exploring microRNA-mediated alteration of EGFR signaling pathway in non-small cell lung cancer using an mRNA: MiRNA regression model supported by target prediction databases. *Genomics*, *104*(6), 504–511. https://doi.org/10.1016/j.ygeno.2014.09.004.

Wang, B., Li, S., Qi, H. H., Chowdhury, D., Shi, Y., & Novina, C. D. (2009). Distinct passenger strand and mRNA cleavage activities of human Argonaute proteins. *Nature Structural and Molecular Biology*, *16*(12), 1259–1266. https://doi.org/10.1038/nsmb.1712.

Wang, J.-H., Zhao, L.-F., Wang, H.-F., Wen, Y.-T., Jiang, K.-K., Mao, X.-M., et al. (2019). GenCLiP 3: Mining human genes' functions and regulatory networks from PubMed based on co-occurrences and natural language processing. *Bioinformatics*, *36*(6), 1973–1975. https://doi.org/10.1093/bioinformatics/btz807.

Wen, M., Cong, P., Zhang, Z., Lu, H., & Li, T. (2018). DeepMirTar: A deep-learning approach for predicting human miRNA targets. *Bioinformatics*, *34*(22), 3781–3787. https://doi.org/10.1093/bioinformatics/bty424.

Wightman, B., Ha, I., & Ruvkun, G. (1993). Posttranscriptional regulation of the heterochronic gene lin-14 by lin-4 mediates temporal pattern formation in *C. elegans*. *Cell*, *75*(5), 855–862. https://doi.org/10.1016/0092-8674(93)90530-4.

Witkos, T. M., Koscianska, E., & Krzyzosiak, W. J. (2011). Practical aspects of microRNA target prediction. *Current Molecular Medicine*, *11*(2), 93–109. https://doi.org/10.2174/156652411794859250.

Wu, W. S., Tu, B. W., Chen, T. T., Hou, S. W., & Tseng, J. T. (2017). CSmiRTar: Condition-specific microRNA targets database. *PLoS One*, *12*(7). https://doi.org/10.1371/journal.pone.0181231.

Xia, W., Cao, G. J., & Shao, N. S. (2009). Progress in miRNA target prediction and identification. *Science in China, Series C: Life Sciences*, *52*(12), 1123–1130. https://doi.org/10.1007/s11427-009-0159-4.

Xiao, F., Zuo, Z., Cai, G., Kang, S., Gao, X., & Li, T. (2009). miRecords: An integrated resource for microRNA-target interactions. *Nucleic Acids Research*, *37*(1), D105–D110. https://doi.org/10.1093/nar/gkn851.

Yi, R., Qin, Y., Macara, I. G., & Cullen, B. R. (2003). Exportin-5 mediates the nuclear export of pre-microRNAs and short hairpin RNAs. *Genes and Development*, *17*(24), 3011–3016. https://doi.org/10.1101/gad.1158803.

Yue, D., Liu, H., & Huang, Y. (2009). Survey of computational algorithms for MicroRNA target prediction. *Current Genomics*, *10*(7), 478–492. https://doi.org/10.2174/138920209789208219.

Zhang, H. (2003). Alignment of BLAST high-scoring segment pairs based on the longest increasing subsequence algorithm. *Bioinformatics*, *19*(11), 1391–1396. https://doi.org/10.1093/bioinformatics/btg168.

Zhou, L., Lim, M. Y. T., Kaur, P., Saj, A., Bortolamiol-Becet, D., Gopal, V., et al. (2018). Importance of miRNA stability and alternative primary miRNA isoforms in gene regulation during Drosophila development. *eLife*, *7*. https://doi.org/10.7554/eLife.38389.

Chapter 10

Computational resources for analysis of miRNA targetome

Anamika Thakur[a,b], Sakshi Kamboj[a,b], and Manoj Kumar[a,b]

[a]*Virology Unit and Bioinformatics Centre, Institute of Microbial Technology, Council of Scientific and Industrial Research (CSIR), Chandigarh, India,* [b]*Academy of Scientific and Innovative Research, Ghaziabad, India*

Introduction

MicroRNAs (miRNAs) are double-stranded, short noncoding RNA (ncRNA) molecules, having 21–23 nucleotides (nts) that are produced endogenously in cells. They are considered to bind to complementary target sequences and can cause cleavage of target mRNA or translational repression (Rupaimoole & Slack, 2017). RNA-polymerase II/III enzyme is used to form primary miRNA (pri-miRNA) transcripts from the miRNA genes inside the nucleus. The pri-miRNA is further processed to pre-miRNA using a drosha enzyme (Ha & Kim, 2014). Then, the exportin-5 protein shifts the pre-miRNA from the nucleus to the cytoplasm, where the dicer enzyme cleaves the hairpin loop forming the miRNA duplex and loads onto the RISC complex, forming mature miRNA. The mature miRNA binds to the target mRNA that further results in translational repression or mRNA degradation (Rupaimoole & Slack, 2017).

In 1993, the Ambros group discovered the first miRNA, that is, lin-4, in *Caenorhabditis elegans (C. elegans)*. Lin-4 is involved in the developmental stage of the nematode *C. elegans* (Lee, Feinbaum, & Ambros, 1993). Then the second miRNA, *let-7*, was also discovered in *C. elegans* (Reinhart et al., 2000). Since then, numerous miRNAs have been discovered from different species, including humans, mice, rats, flies, etc. (Kim et al., 2004; Lagos-Quintana et al., 2002). Plant miRNAs have also been identified in different species such as *Arabidopsis thaliana* (Xu et al., 2018), *Oryza sativa* (rice) (Wang, Zhou, Chen, Luo, & Qu, 2004), *Zea mays* (maize) (Huang et al., 2016), etc. Similarly, miRNAs were also discovered for viruses such as the Epstein-Barr virus (EBV) in 2004 (Pfeffer et al., 2004), the human cytomegalovirus (HCMV) (Pfeffer et al., 2005), Kaposi's sarcoma-associated herpes virus (KSHV) (Lin, Li, Liang, & Lan, 2012), etc.

The main function of miRNA is gene expression regulation by binding to the 3′ UTR region of the target mRNA. One of the important component in a mature miRNA sequence is its seed region encompassing the 2–8 nucleotide positions which binds to the target mRNA. Also, isomiRs were considered important in the miRNA sequence, and are defined as the miRNA sequence variation as well as nucleotide addition/deletion at the 5′/3′ end (Guo & Chen, 2014). Another important feature of miRNA is the cooperativity and multiplicity (Qureshi, Thakur, Monga, Thakur, & Kumar, 2014). Cooperativity is a phenomenon where multiple miRNAs bind to a single target gene. For example, the PPP1R12A gene is targeted by many miRNAs, including hsa-miR-30a, hsa-miR-30b, hsa-miR-30c, and hsa-miR-30d. In the case of multiplicity, a single miRNA targets more than one gene. For example, ebv-miR-bart2-3p miRNA targets three genes: TMED4, IER3, and SBNO1 (Gottwein et al., 2011).

Various computational resources have been developed for experimentally validated miRNAs such as miRBase (Kozomara, Birgaoanu, & Griffiths-Jones, 2019), PMRD (Zhenhai et al., 2010), miRNEST (Szczesniak & Makalowska, 2014), bloodmiRs (Juzenas et al., 2017), miRmine (Panwar, Omenn, & Guan, 2017), VIRmiRNA (Qureshi et al., 2014), etc. The miRBase database covers information about the miRNA sequence and its annotation for many organisms. The latest version of the miRBase22 database has 38,589 miRNAs for 271 organisms. Currently, there are 1917 pre-miRNA sequences for the human genome GRCh38. PMRD is a plant miRNA database that includes data for the majority of plants such as *Arabidopsis*, rice, wheat, etc. VIRmiRNA is a dedicated resource developed by our group for experimental miRNAs and their targets. Currently, this resource contains 1308 viral miRNAs for 44 viruses (Qureshi et al., 2014). There are several reports for experimental as well as predicted mRNA target sequences. Repositories have also been developed for miRNA target sequences such as TarBase (Sethupathy, Corda, & Hatzigeorgiou, 2006), miRDB (Wang et al., 2004), miRTarBase (Hsu, Lin, et al., 2011), PMTED (Sun et al., 2013), etc. Also, tools have been developed for miRNA target

MicroRNA in Human Malignancies. https://doi.org/10.1016/B978-0-12-822287-4.00020-7

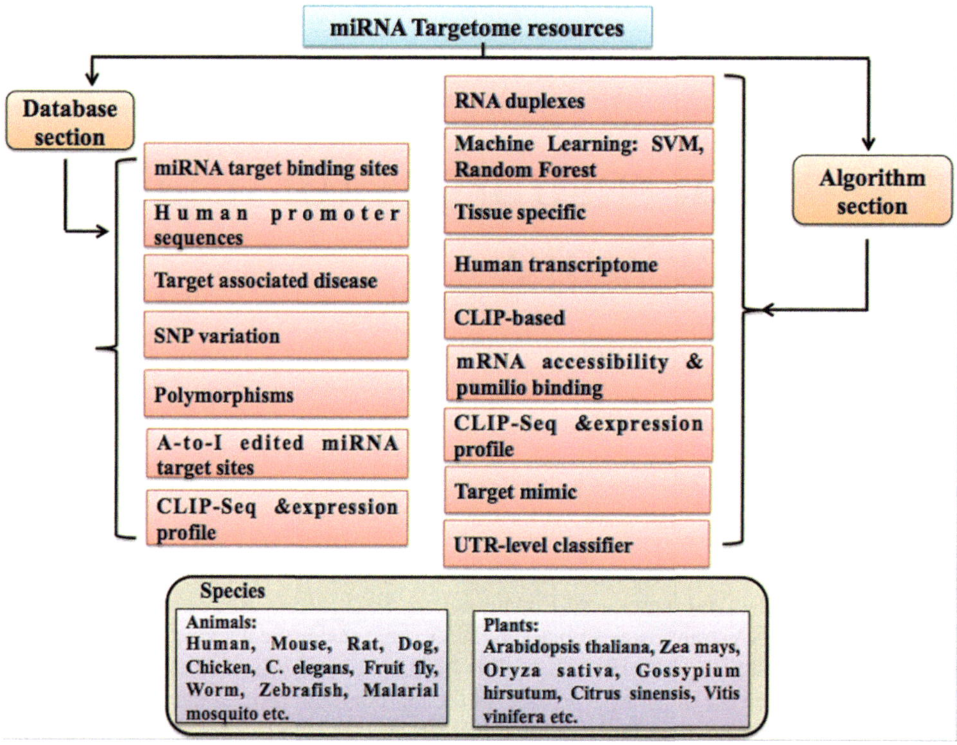

FIG. 10.1 Schematic representation of miRNA targetome resources. Figure depicts two sections of miRNA targetome resources—databases and algorithms for different organisms.

prediction such as miTarget (Kim et al., 2004), MicroTar (Thadani & Tammi, 2006), ViTa (Hsu, Lin, et al., 2011), miRWalk (Sticht, De La Torre, Parveen, & Gretz, 2018), miRAW (Pla, Zhong, & Rayner, 2018), etc.

In the last decade, there has been a tremendous increase in the development of computational resources for miRNAs. Efforts have been made to summarize the list of tools and databases developed for miRNAs in different fields (Monga & Kumar, 2019). Still, there is a need to update the miRNA targets in a broad perspective. In this chapter, we mainly focus on tools and databases developed for the miRNA target. It is divided into two sections-miRNA target databases and miRNA target algorithms, which address the databases/tools developed for human and other animal and plant species. A schematic representation of the miRNA targetome resources is shown in Fig. 10.1.

miRNA target databases

The identification and characterization of targets for different miRNAs are of utmost importance for understanding different disease conditions and drug development. miRNA targets are identified and validated by various experimental techniques such as microarray, CLIP-Seq, degradome sequencing, qRT-PCR, western blotting, luciferase reporter assay, pSILAC, etc. (Singh, 2017). Other than experimental techniques, different machine learning-based prediction methods are also used to predict the potential targets for miRNAs. Various research groups in the past 15 years have predicted and validated targets for miRNAs in different organisms and developed repositories for miRNA targets (Riolo, Cantara, Marzocchi, & Ricci, 2020). In this section, we described databases of miRNA targets developed for different organisms.

miRNA target databases for human

Many databases for miRNA targets have been developed to address miRNA targets and their roles in the regulation of different physiological processes and pathological conditions in humans. The *HOCTAR (Host Gene Oppositely Correlated Targets) database* (Gennarino et al., 2011) is a database of predicted target genes for intragenic miRNAs in humans and contains miRNA target genes for 290 annotated intragenic miRNAs. It analyzes the miRNA target genes predicted by the PicTar (Krek et al., 2005), TargetScan (Friedman, Farh, Burge, & Bartel, 2009), and miRanda (Enright et al., 2003) prediction algorithms. *microPIR (microRNA-Promoter Interaction Resource)* is a database containing predicted potential target sites for miRNAs in the promoter regions of human genes (Piriyapongsa, Bootchai, Ngamphiw, & Tongsima, 2012). This resource contains > 15 million target sites identified for all human gene promoter regions. The updated version of this database, *microPIR2* (Piriyapongsa, Bootchai, Ngamphiw, & Tongsima, 2014), has about 80 million human and 40 million mice target sites. miRdSNP (Bruno et al., 2012) is a comprehensive data source of predicted miRNA target sites and annotated disease-associated single nucleotide polymorphisms (dSNPs) on the 3′ UTR of human genes. It contains 786 dSNP-disease associations for 630 unique dSNPs and 204 disease types.

miRNAs are important regulators of various cellular processes and can target multiple genes. Many resources have been developed to provide insight into the single nucleotide polymorphisms (SNPs) in miRNA targets, gene expression, and functional and genome annotations in miRNA target sites. *miRGator* (Nam, Kim, Shin, & Lee, 2008) is a database that contains a list of predicted miRNA target sites as well as miRNA-mRNA expression profiles, functional annotations, and their disease associations. It provides miRNA, miRNA target genes, and binding sites for both humans and mice. This database is also updated with the incorporation of expression profiles of miRNA in experimental conditions, in miRNA perturbation, and paired miRNA-mRNA profiles (Cho et al., 2011). Further, this resource also compiled the deep sequencing data for miRNA target analysis (Cho et al., 2013). Many other SNP-related resources were developed for miRNA such as *MSDD* and *dbMTS*. The *MiRNA SNP Disease Database* (*MSDD*) (Yue et al., 2018) is a manually curated database that provides information on the experimentally supported SNPs in the miRNA and their association with human diseases. *dbMTS* (Li et al., 2020) is a database for single nucleotide variations (SRVs) in miRNA target sites with their functional annotations. The *Polymorphism in microRNA Target Site (PolymiRTS) database* compiles polymorphisms or variations in the sequences of miRNAs and their targets (Bao et al., 2007). It provides information about the effects of SNPs and insertions/deletions on miRNA target sites, seed regions, and miRNA-mRNA interactions (Bhattacharya, Ziebarth, & Cui, 2014; Ziebarth, Bhattacharya, Chen, & Cui, 2012). A complete list of miRNA target resources for humans is shown in Table 10.1.

miRNA target databases for other species

Apart from humans, many resources have been developed to provide information about miRNA targets and their related aspects for species such as mice, rats, chickens, *Drosophila*, *C. elegans*, zebrafish, rice, maize, cotton, *Arabidopsis thaliana*, viruses, etc. *miRNAMap* is a database that stores known and putative miRNA genes and miRNA targets for humans, dogs, rats, and mice (Hsu et al., 2006). In the updated version of this database, *miRNAMap 2.0* (Hsu et al., 2008) experimentally verified and predicted miRNA genes and targets for other species such as *C. elegans*, mosquitoes, chickens, fruit flies, zebrafish, fugu, frogs (*Xenopus tropicalis*), and opossums have also been added along with the expression profiles of miRNAs in 18 major human tissues. *TarBase* (Sethupathy et al., 2006) compiles a manually curated collection of experimentally tested miRNA targets for animal species such as humans, mice, *Drosophila*, *C. elegans*, and zebrafish. The extended versions of this database incorporate a significant number of new miRNA targets derived from experimental techniques, including high-throughput techniques as well as miRNA-gene interactions for > 65,000 targets (Papadopoulos, Reczko, Simossis, Sethupathy, & Hatzigeorgiou, 2009; Vergoulis et al., 2012). It is also linked to other useful external databases such as the UCSC genome browser, gene ontology (GO), HUGO, Enscmbl, and SwissProt. The latest version of this resource, *DIANA-TarBase v8.0* (Karagkouni et al., 2018), provides > 1 million experimentally supported miRNA-mRNA interactions with about 670,000 unique miRNA target pairs for about 600 cell types from 18 species.

miRDB is an online resource for predicted mature miRNAs and their targets and functional annotations of miRNAs for animal species—humans, mice, rats, dogs, and chickens. This resource uses an in-house developed prediction algorithm called MirTarget2 (Wang & El Naqa, 2008) for miRNA target prediction. In its first version, 1437 miRNAs, 389,729 gene targets, and 47,946 unique gene targets were compiled (Wang, 2008). In the updated versions of miRDB, there has been an increase in the number of predicted miRNAs and targets to 7000 and 3.5 million, respectively, that also incorporate transcriptome data-RNA-seq and CLIP-seq of about 1000 cell lines for miRNA expression profiling (Chen & Wang, 2020; Nathan & Xiaowei, 2015). *miRGate* (Andrés-León, Peña, Gómez-López, & Pisano, 2015) is an open resource for predicted miRNA-mRNA pairs for humans, rats, and mice, which uses miRNA and target data from databases such as miRBase 20 (Kozomara & Griffiths-Jones, 2014), Tarbase (Sethupathy et al., 2006), miRTarbase (Hsu, Lin, et al., 2011), miRecords (Xiao et al., 2009), and OncomirDB (Wang, Gu, Wang, & Ding, 2014) as well as prediction algorithms such as miRanda (Enright et al., 2003), RNAHybrid (Krüger & Rehmsmeier, 2006), MicroTar (Thadani & Tammi, 2006), and TargetScan (Friedman et al., 2009).

There are many other resources for miRNA targets that are freely available such as miRNA_Targets (Kumar, Wong, Tizard, Moore, & Lefèvre, 2012), miRSel (Naeem, Küffner, Csaba, & Zimmer, 2010), etc. Further, resources for miRNA target interactions are also available. *miRecords* is an integrated resource of predicted and validated miRNA targets for seven animal species and miRNA target interactions. It contains 1135 miRNA target interactions records for 301 validated

TABLE 10.1 Computational resources showing miRNA target data for humans.

S. no.	Name	Year	Description	PMID
1	HOCTARdb	2011	Resource for microRNA target prediction	21435384
2	miRGator v2.0	2011	An integrated system for the functional investigation of microRNAs	21062822
3	microPIR	2012	An integrated database of microRNA target sites within human promoter sequences	22439011
4	miRdSNP	2012	Database of disease-associated SNPs and microRNA target sites on 3′ UTRs of human genes	22276777
5	miRNASNP	2012	Genome-wide identification of SNPs in microRNA genes and the SNP effects on microRNA target binding and biogenesis	22045659
6	MirSNP	2012	Database of polymorphisms altering miRNA target sites, identifies miRNA-related SNPs in GWAS SNPs and eQTLs	23173617
7	miRGator v3.0	2013	microRNA portal for deep sequencing, expression profiling, and mRNA targeting	23193297
8	SomamiR	2013	A database for somatic mutations impacting microRNA function in cancer	23180788
9	targetHub	2013	A programmable interface for miRNA-gene interactions	24013925
10	Pharmaco-miR	2014	Linking microRNAs and drug effects	23376192
11	TargetCompare	2014	A web interface to compare simultaneous miRNA targets	25352731
12	ComiRNet	2015	A web-based system for the analysis of miRNA-gene regulatory networks	26051695
13	miRNASNP v2.0	2015	An update of the miRNASNP database for better SNP selection by GWAS data, miRNA expression, and online tools	25877638
14	SomamiR 2.0	2016	Database of cancer somatic mutations altering microRNA-ceRNA interactions	26578591
15	IMOTA	2018	An interactive multiomics tissue atlas for the analysis of human miRNA target interactions	28977416
16	MSDD	2018	Database of experimentally supported associations among miRNAs, SNPs, and human diseases	29106642
17	dbMTS	2020	A comprehensive database of putative human microRNA target site SNVs and their functional predictions	32227657
18	miRNASNP-v3	2021	Comprehensive database for SNPs and disease-related variations in miRNAs and miRNA targets	32990748

miRNA and 902 target genes (Xiao et al., 2009). The *miRTarBase* (Hsu, Lin, et al., 2011) database provides experimentally validated miRNA target interaction data. The first version curated 3576 miRNA target interactions between 2297 target genes and 657 miRNAs for 17 species including humans, mice, rats, chickens, sheep, zebrafish, fruit flies, silkworms, *C. elegans*, plants, and viruses. So far, four updated versions of this database have been launched (Chou et al., 2016, 2018; Hsu et al., 2014; Huang et al., 2020) with significant increases and improvements from the initial release. In the latest version, miRNA and target gene expression profiling; their association with diseases; and high-throughput data from CLIP-seq, RNA-seq, reporter assays, and text mining has been made available for users. *miRWalk* (Dweep, Gretz, & Sticht, 2014; Dweep, Sticht, Pandey, & Gretz, 2011) is a database that provides a collection of experimentally verified and predicted miRNA binding sites for all coding regions, upstream flanking regions, and mitochondrial genes for humans, rats, and mice. *miRNASNP* (Gong et al., 2012) is a database that compiles the SNPs in pre-miRNAs, miRNA target sites, and miRNA flanking regions in 9 animal species: humans, chimpanzees, mice, dogs, horses, rats, cows, zebrafish, and chickens. In addition, this group has also explored the effect of SNPs on miRNA target interactions in association with diseases and drugs, especially in cancer (Jing et al., 2015; Liu et al., 2021).

miRNA target resources for model organisms such as *Drosophila* have also been developed. The *Dme-Hsa Disease Database (DHDD)* (Wei et al., 2018) is a database for miRNAs and their targets related to human diseases and orthologs in *D. melanogaster*. VIRmiRNA (Qureshi et al., 2014) is an online resource that contains 1308 experimentally validated miRNAs for 44 viruses and 7283 target sites. A few databases for miRNA targets for plants have also been developed. The *Plant MiRNA Target Expression Database (PMTED)* is a database that provides miRNAs, their targets, gene ontology, and expression profiles for many plant species, including *A. thaliana*, citrus, cotton, rice, sugarcane, tomato, wheat, grape, and maize. It also provides miRNA target prediction section as well as miRNA target expression profiles from the manually curated microarray data (Sun et al., 2013). The *Plant Noncoding RNA Database (PNRD)* is an online resource that provides a collection of noncoding RNAs, their targets, and expression profiles. It contains 25,739 entries for 11 ncRNA types for 150 plant species, including *A. thaliana*, rice, maize, soybean, wheat, potato, cotton, mustard, tobacco, papaya, tomato, etc., as well as 178,138 miRNA target pairs for 46 plant species and 35 miRNA expression profiles (Yi, Zhang, Ling, Xu, & Su, 2015). The degradome-based plant miRNA target interaction and network database (DPMIND) is a database that compiles degradome sequencing data-based miRNA target interactions. It contains 3794 miRNAs, 34,514 predicted targets, 64,195 predicted miRNA target interactions, 28,666 validated miRNA target interactions, and 69 degradomes from 10 plant species (Fei et al., 2018). A list of computational resources for miRNA targets for various species other than humans is given in Table 10.2.

TABLE 10.2 Computational resources showing miRNA target data for other species.

S. no.	Name	Year	Description	Species	PMID
1	TargetFinder	1999	Searching annotated sequence databases for target genes of transcription factors	NA	10089203
2	miRNAMap	2006	Genomic maps of microRNA genes and their target genes in mammalian genomes	Human, mice, rat, and dog	16381831
3	TarBase	2006	Comprehensive database of experimentally supported animal microRNA targets	Human, mice, fruit fly, worm, and zebrafish	16373484
4	PolymiRTS Database	2007	Database for linking polymorphisms in microRNA target sites with complex traits	Human, mice	17099235
5	miRDB	2008	microRNA target prediction and functional annotation database with a wiki interface	Human, mice, rat, dog, chicken	18426918
6	miRDB	2008	Prediction of both conserved and nonconserved microRNA targets in animals	Human, mice, rat, dog, chicken	18048393
7	miRNAMap 2.0	2008	Genomic maps of microRNAs in metazoan genomes	Human, mice, rat, dog, chicken, fruit fly, worm, zebrafish, fugu, frog (*Xenopus tropicalis*), malarial mosquito, and opossum	18029362
8	miRGator	2008	Integrated system for functional annotation of microRNAs	Human and mice	17942429
9	miRecords	2009	An integrated resource for microRNA target interactions	Animals	18996891
10	miRSel	2010	Automated extraction of associations between microRNAs and genes from the biomedical literature	Human, mice, rat	20233441

Continued

TABLE 10.2 Computational resources showing miRNA target data for other species—cont'd

S. no.	Name	Year	Description	Species	PMID
11	MIR@NT@N	2011	A framework integrating transcription factors, microRNAs, and their targets to identify subnetwork motifs in a metaregulation network model	*C. elegans, Danio rerio, D. melanogaster, Gallus gallus,* human, mice, and *Rattus norvegicus*	21375730
12	miRTarBase	2011	A database of experimentally validated microRNA-target interactions	Human, mice, rat, chicken, sheep, zebrafish, fruit fly, silkworm, nematode, plants, viruses	21071411
13	miRWalk	2011	Database of possible miRNA binding sites by "walking" the genes of three genomes	Human, mice, rat	21605702
14	starBase	2011	Database for exploring microRNA-mRNA interaction maps from Argonaute CLIP-Seq and Degradome-Seq data	Human, mice, *C. elegans, Arabidopsis, Oryza sativa, Vitis vinifera*	21037263
15	RepTar	2011	Database of predicted cellular targets of host and viral miRNAs	Human and mice	21149264
16	miRNA_Targets	2012	A database for miRNA target predictions in coding and noncoding regions of mRNAs	Human, mice, cow, chicken, zebrafish, fruit fly, and *C. elegans*	22940442
17	TarBase 6.0	2012	Capturing the exponential growth of miRNA targets with experimental support	All species	22135297
18	PolymiRTS Database 2.0	2012	Database of linking polymorphisms in microRNA target sites with human diseases	Human, mice	22080514
19	miR-EdiTar	2012	Database of predicted A-to-I edited miRNA target sites	NA	23044546
20	PMTED	2013	Plant microRNA target expression database	*A. thaliana, Glycine* max, *Zea mays, Oryza sativa, Gossypium hirsutum, Citrus sinensis, Medicago truncatula, Populus trichocarpa, Solanum lycopersicum, Saccharum officinarum, Vitis vinifera,* and *Triticum aestivum*	23725466
21	miRTarBase	2014	Resource for experimentally validated miRNA target interactions	Human, mice, rat, chicken, cattle, sheep, frogs, Japanese medaka, zebrafish, fruit fly, silkworm, nematode, thale cress, viruses	24304892
22	miRWalk	2014	miRWalk database for miRNA target interactions	Human, mice, and rat	25055920
23	starBase v2.0	2014	Database decoding miRNA-ceRNA, miRNA-ncRNA, and protein-RNA interaction networks from large-scale CLIP-Seq data	Human, mice and *C. elegans*	24297251
24	multiMiR	2014	The multiMiR R package and database: integration of microRNA-target interactions along with their disease and drug associations	Human, mice	25063298

TABLE 10.2 Computational resources showing miRNA target data for other species—cont'd

S. no.	Name	Year	Description	Species	PMID
25	PolymiRTS Database 3.0	2014	Database for linking polymorphisms in microRNAs and their target sites with human diseases and biological pathways	Human, mice	24163105
26	microPIR2	2014	A comprehensive database for human-mice comparative study of microRNA-promoter interactions	Human and mice	25425035
27	VIRmirTar (VIRmiRNA)	2014	A comprehensive resource for experimentally validated viral miRNAs and their targets	Virus	25380780
28	DIANA-miRPath v3.0	2015	Deciphering microRNA function with experimental support	Humans, *Mus musculus*, *Rattus norvegicus*, *Drosophila melanogaster*, *Caenorhabditis elegans*, *Gallus gallus*, and *Danio rerio*	25977294
29	DIANA-TarBase v7.0	2015	Database indexing more than 500,000 experimentally supported miRNA:mRNA interactions	24 species	25416803
30	miRDB	2015	Resource for microRNA target prediction and functional annotations	Human, mice, rat, dog, chicken	25378301
31	miRGate	2015	A curated database of human, mice, and rat miRNA-mRNA targets	Human, mice, rat	25858286
32	MtiBase	2015	Database for decoding microRNA target sites located within CDS and 5′ UTR regions from CLIP-Seq and expression profile datasets	Human, MICE	26490638
33	PNRD	2015	Plant noncoding RNA database	150 plant species	25398903
34	miRTarBase	2016	Updates to the experimentally validated miRNA target interactions database	NA	26590260
35	STarMirDB	2016	A database of microRNA binding sites	Human, mice, and worm	27144897
36	CSmiRTar	2017	Condition-specific microRNA target database	Human, mice	28704505
37	miRPathDB	2017	New dictionary on microRNAs and target pathways	Human, mice	27742822
38	DIANA-TarBase v8	2018	Collection of experimentally supported miRNA-gene interactions	18 species	29156006
39	miRTarBase	2018	A resource for experimentally validated microRNA-target interactions	NA	29126174
40	Dme-Hsa Disease Database (DHDD)	2018	Conserved human disease-related mIRNA and their targeting genes in *Drosophila melanogaster*	*D. melanogaster*	30200613
41	DPMIND	2018	Degradome-based plant miRNA target interaction and network database	Plants	29280990
42	miRDB	2020	Database for prediction of functional microRNA targets	Human, mice, rat, dog, chicken	31504780
43	miRTarBase	2020	Updates to the experimentally validated microRNA-target interaction database	NA	31647101
44	miRPathDB 2.0	2020	miRNA pathway dictionary database	Human, mice	31691816

miRNA algorithms

Because it is incredibly time consuming to validate a potential target mRNA for a given miRNA, in silico methods are used to identify a specific target for the validation. Different approaches were implemented to identify miRNA targets from the computational models. Each algorithm has a distinct strategy for miRNA target prediction. Tools have been developed based on their binding sites such as 3′/5′ UTR or the evolutionarily conservation of miRNA regions (ReyesHerrera & Ficarra, 2012). These methods also identify targets using features such as seed match, expression, target inhibition, conservation, free energy, composition of nucleotides, etc. (Peterson et al., 2014; Srivastava, Moturu, Pandey, Baldwin, & Pandey, 2014). Also, the targets can be identified using machine learning techniques (MLTs) such as support vector machine (SVM), random forest (RF), artificial neural network (ANN), etc. Therefore, different tools have different aspects, and the user can decide which tool to use. We discuss the resources developed for the in silico prediction of miRNA target sites for different species in this section.

miRNA target prediction for humans

There are several computational resources that predict miRNA binding sites for humans. In the last 15 years, algorithms have been developed for miRNA target prediction in humans on different aspects. The *miTarget* tool was developed using an SVM classifier (Kim, Nam, Rhee, Lee, & Zhang, 2006). The algorithm uses structural, thermodynamic, and position-based features. Further, it also performs gene ontology (GO) analysis as well as the feature selection experiment. *MTar* is a novel computational algorithm for the human transcriptome (Chandra, Girijadevi, Nair, Pillai, & Pillai, 2010). This tool uses 16 positional, thermodynamic, and structural parameters in miRNA:mRNA pairs to identify the potential targets for miRNA. *miRTar* is an integrated tool that identifies the relationship between miRNAs and their targets in 3′/5′ UTR coding regions and also the genes found in the KEGG pathway (Hsu, Chiu, et al., 2011). *miRTar Hunter* is also used to predict target sites in the presence/absence of a seed match. This approach utilizes a dynamic programming algorithm that uses sequence-specific features as well as those features that show complementarities between miRNAs and their targets (Park & Kim, 2013). The *HomoTarget* algorithm is also a miRNA target-based predictor in humans. The algorithm is developed using a pattern recognition neural network (PRNN) and principal component analysis (PCA) architecture. The model used in this predictor has also incorporated 12 structural, thermodynamic, and positional features of miRNA:mRNA binding sites to select the potential target sites (Ahmadi et al., 2013). *MREdictor* (Incarnato, Neri, Diamanti, & Oliviero, 2013) is a computational method that predicts the target containing pumilio recognition elements (PRE) in the miRNA binding sites in the 3′ UTR. The tool was built using a training dataset of humans and mice. The *RFMirTarget* (Mendoza et al., 2013) predictor was developed for humans using a random forest classifier using structural, thermodynamics, alignment, seed, and position-based features. *mirMark* (Menor, Ching, Zhu, Garmire, & Garmire, 2014) tool uses WEKA software for the prediction of miRNA target sites at the UTR level. This method uses experimentally verified miRNA targets from miRecords (Xiao et al., 2009) and mirTarBase (Hsu, Lin, et al., 2011) for training and considers around 700 features. *miRAW* (Pla et al., 2018) was developed to identify miRNA target sites in the 3′ UTR region. This tool uses deep artificial neural networks (ANN) by analyzing whole mature miRNA transcripts. *maTE* (Yousef, Abdallah, & Allmer, 2019) is also a machine learning (ML)-based approach that uses random forest (RF) software. This classifier provides a group of genes targeted by a particular miRNA. *miRNATIP* (Fiannaca, La Rosa, La Paglia, Rizzo, & Urso, 2016) is an miRNA target interaction prediction algorithm for humans and *C. elegans*. This approach uses a self-organizing map (SOM) method trained with the miRNA seed sequence and then projected into the mRNA sequence to identify the potential target.

miRNA target prediction for other species

In addition to humans, algorithms have been developed for species such as mice, rats, chickens, *D. melanogaster*, *C. elegans*, zebrafish, etc. *miRanda* (Enright et al., 2003) is a computational method used to predict miRNA targets in *D. melanogaster*, *D. pseudoobscura*, and *Anopheles gambiae*. The target genes are selected based on their sequence complementarity, free energies of RNA-RNA duplexes, and conservation sites in the genome. *MovingTargets* (Burgler & Macdonald, 2005) software performs a genome-wide search to predict miRNA targets in *Drosophila* using five different biological constraints. *MicroInspector* (Rusinov, Baev, Minkov, & Tabler, 2005) is a web tool that searches miRNA binding sites in the RNA sequence. The tool was developed for different classes of organisms such as nematodes, plants, viruses, arthropods, and vertebrates. *TargetBoost* (Sætrom, Snøve Jr., & Sætrom, 2005) uses a machine-learning based method for miRNA identification in *C. elegans* and *D. melanogaster* using the sequence complementarity steps. The *MicroTar* (Thadani & Tammi, 2006) algorithm predicts miRNA targets from RNA duplexes. The tool is developed using experimentally verified miRNA target data of *C. elegans*, *Drosophila*, and mice. *MiRTif* (Yang, Wang, & Li, 2008) software is a postprocessing filter that uses predicted miRNA target interactions using the SVM classifier. The software was developed using the interaction data of *Drosophila*, *C. elegans*, humans, mice, rats, and zebrafish. The *TargetMiner* (Bandyopadhyay & Mitra, 2009) predictor is an SVM-based classifier that was developed using miRNA-transcript pairs as positive and negative examples. *TargetSpy* (Sturm, Hackenberg, Langenberger, & Frishman, 2010) is a computational method that was trained on mice and also shows good performance in humans and *Drosophila*. This method was developed using the MultiBoost technique and was also compared to other methods such as SVM, naive Bayes, C4.5, AdaBoost, and WEKA. *MiRSystem* (Lu et al., 2012) is a user-friendly tool developed for predicting the miRNA target genes along with the enriched functions and their pathwaysin humans and mice. The *STarMir* (Kanoria et al., 2016) web server also predicts miRNA binding sites in humans, mice, and *C. elegans*. The process uses high-throughput miRNA binding data obtained from cross-inking immunoprecipitation (CLIP) studies. miRWalk (Sticht et al., 2018) predicts miRNA target sites using random-forest-based TarPmiR software. Multiple features have been implicated in this resource such as region complementarity, binding site conservation, and target site accessibility. Further, ViTa tool was developed to predict the host miRNA targets for humans, mice, rats, and chickens on the viruses (Hsu, Lin, Hsu, Hsu, & Huang, 2007). A complete list of resources is given in Table 10.3.

TABLE 10.3 Computational resources for miRNA target prediction in animal species.

S. no.	Name	Year	Description	Species	PMID
1	miRanda	2003	MicroRNA targets in *Drosophila*	*D. melanogaster*, *D. pseudoobscura*, and *Anopheles gambiae*	14709173
2	MovingTargets	2005	Prediction and verification of miRNA targets	*Drosophila*	15943864
3	MicroInspector	2005	Web tool for detection of miRNA binding sites in an RNA sequence	*C. elegans*, *Caenorhabditis briggsae*, *D. melanogaster*, *D. pseudopscura*, *Oryza sativa*, *Zea mays*, *A. thaliana*, human, mice, rat, *Gallus*, zebrafish	15980566
4	PicTar	2005	Combinatorial microRNA target predictions	Vertebrates	15806104
5	Targetboost	2005	Weighted sequence motifs as an improved seeding step in microRNA target prediction algorithms	*D. melanogaster* and *C. elegans*	15928346
6	TargetFinder	2005	Based on MAST and secondary structures of target mRNA	NA	15598838
7	MicroTar	2006	Predicting miRNA targets from RNA duplexes	*C. elegans*, *Drosophila*, and mice	17254305
8	miTarget	2006	Prediction using a support vector machine	Human	16978421
9	RNAhybrid	2006	microRNA target prediction	*C. elegans*	16845047
10	ViTa	2007	microRNAs targets on viruses	Human, mice, rat, and chicken	17148483
11	MiRTif	2008	SVM-based miRNA target interaction filter	*Drosophila*, *C. elegans*, human, mice, rat, and zebrafish	19091027
12	miRGator	2008	Integrated system for functional annotation of microRNAs	Human and mice	17942429
13	TargetMiner	2009	miRNA target prediction of tissue-specific negative examples	Human, mice, and *Drosophila*	19692556
14	TargetSpy	2009	Supervised machine learning approach	Human, chimp, mice, rat, chicken, fly, *D. melanogaster*, *D. yakuba*, *D. ananassae*, and *D. pseudoobscura*	20509939

Continued

TABLE 10.3 Computational resources for miRNA target prediction in animal species—cont'd

S. no.	Name	Year	Description	Species	PMID
15	MTar	2010	Human transcriptome	Human	20122191
16	Targetfinder.org	2010	Resource for systematic discovery of transcription factor target genes	NA	20460454
17	miRTar	2011	An integrated system for identifying miRNA target interactions in humans	Human	21791068
18	DIANA-microT web server	2011	Supports fly and worm miRNA target prediction and bibliographic miRNA to disease association	Human, mice, *D. melanogaster*, and *C. elegans*	21551220
19	miRGator v2.0	2011	Integrated system for functional investigation of microRNAs	Human	21062822
20	MirSOM	2011	Prediction of microRNA targets in *C. elegans* using a self-organizing map	*C. elegans*	21422073
21	MiRSystem	2012	Integrated system for characterizing enriched functions and pathways of microRNA targets	Human and mice	22870325
22	RNA22	2012	Interactive exploration of RNA22 microRNA target predictions	Human, mice, *D. melanogaster*, and *C. elegans*	23074262
23	HomoTarget	2013	Prediction of microRNA targets	Human	23174671
24	miRTar Hunter	2013	microRNA target sites	Human	23475422
25	Sfold	2013	CLIP-based prediction	Mammals	23703212
26	MREdictor	2013	Two-step dynamic interaction model that accounts for mRNA accessibility and Pumilio binding accurately predicts microRNA targets	Human and mice	23863844
27	RFMirTarget	2013	Random forest classifier	Human	23922946
28	DIANA-microT web server	2013	Service integration into miRNA functional analysis workflows	Human, mice, *D. melanogaster*, and *C. elegans*	23680784
29	miRGator v3.0	2013	microRNA portal for deep sequencing, expression profiling, and mRNA targeting	Human	23193297
30	mirMark	2014	A site-level and UTR-level classifier	Human	25344330
31	STarMir	2014	microRNA binding sites	Human, mice, and *C. elegans*	24803672
32	STarMir	2016	microRNA binding sites	Human, mice, and *C. elegans*	27665594
33	MiRNATIP	2016	SOM-based miRNA target interaction predictor	Human and *C. elegans*	28185545
34	DIANA-mirExTra v2.0	2016	Uncovering microRNAs and transcription factors with crucial roles in NGS expression data	Human, mice, *D. melanogaster*, *C. elegans*	27207881
35	miRWalk	2018	Prediction of microRNA binding sites	Human, mice, rat, dog, and cow	30335862
36	miRAW	2018	Deep learning-based approach to predict microRNA targets by analyzing whole microRNA transcripts	Human	30005074
37	STarMir	2019	Sfold tools	Human, mice, and *C. elegans*	30963486
38	maTE	2019	Discovering expressed interactions between microRNAs and their targets	Human	30895309

miRNA target prediction for plants

miRNA target tools have also been developed for plant species such as *A. thaliana*, rice, tomato, etc. For example, *miRcheck* (Jones-Rhoades & Bartel, 2004) has been developed for *A. thaliana* and rice to identify genomic approaches in miRNA and their targets. The *miRU* (Zhang, 2005) server has been used to predict miRNA targets for plants that results in the higher sensitivity and selectivity. The *p-TAREF* (Jha & Shankar, 2011) algorithm was developed using an SVM regression-based approach for identifying plant miRNA targets by utilizing position-specific dinucleotide density variation information around the target sites. The algorithm was developed using plant species such as *Arabidopsis*, *Medicago*, rice, and tomato. Further, comTAR (Chorostecki & Palatnik, 2014) tool predicts targets in different plant species such as *A. thaliana*, poplar, and rice. psRNATarget (Dai & Zhao, 2011; Dai, Zhuang, & Zhao, 2018)() predicts plant targets by matching small RNA with the target mRNA sequence and by target site accessibility. The PAREameters (Thody, Moulton, & Mohorianu, 2020) computational tool was built using a sequencing dataset for *A. thaliana*. The *TAPIR* (Bonnet, He, Billiau, & van de Peer, 2010) web server is a general predictor for plants that was developed using two algorithms, that are, FASTA local alignment program and RNAhybrid. *Target-align* (Xie & Zhang, 2010) is a general plant miRNA target-based predictor developed using Smith-Waterman-like alignment tool. *PsRobot* (Wu, Ma, Chen, Wang, & Wang, 2012) is also a web-based tool for plants to identify small RNA and their target genes by performing metaanalysis. A list of miRNA target algorithms is given in Table 10.4.

TABLE 10.4 Computational resources for miRNA target prediction in plants.

S. no.	Name	Year	Description	Species	PMID
1	miRcheck	2004	microRNAs and their target identification	*Arabidopsis thaliana* and rice (*Oryza sativa*)	15200956
2	miRU	2005	An automated plant miRNA target prediction server	Plants	15980567
3	miRU	2005	An automated plant miRNA target prediction server	Plants	15980567
4	TAPIR	2010	microRNA targets, including target mimics	Plants	20430753
5	Target-align	2010	Tool for plant microRNA target identification	Plants	20934992
6	miRTour	2011	Plant miRNA and target prediction tool	*Arabidopsis* and Orysa protein sequences	21887016
7	p-TAREF	2011	Machine learning for miRNA target identification	*Arabidopsis*, *Medicago*, rice, and tomato	22206472
8	C-mii	2012	Plant miRNA and target identification	*Arabidopsis*	23281648
9	psRobot	2012	Small RNA metaanalysis toolbox	Plants	22693224
10	comTAR	2014	Web tool for the prediction of conserved microRNA targets in plants	*A. thaliana*, poplar, and rice	24632500
11	NA	2018	Integrative mRNA-miRNA interaction analysis based on transcriptome database	Sea cucumber (*Apostichopus japonicus*)	29054825
12	psRNATarget	2018, 2011	Plant small RNA target analysis server (2017 release)	Plants	29718424, 21622958
13	PAREameters	2020	Tool for computational inference of plant miRNA-mRNA targeting rules using small RNA and degradome sequencing data	*A. thaliana*	31943065
14	findMiRNA	2005	Computational prediction of miRNAs in *Arabidopsis thaliana*	*A. thaliana*	15632092

Conclusion

In conclusion, we have compiled a list of miRNA-target databases as well as the tools developed to address different aspects of miRNA. We have provided the list of miRNA target databases developed for different species. Also, we have incorporated the list of algorithms that predict miRNA targets using various computational methods for many species. This chapter would be helpful for biologists to select the relevant tool required for their miRNA research work.

References

Ahmadi, H., Ahmadi, A., Azimzadeh-Jamalkandi, S., Shoorehdeli, M. A., Salehzadeh-Yazdi, A., Bidkhori, G., & Masoudi-Nejad, A. (2013). HomoTarget: A new algorithm for prediction of microRNA targets in Homo sapiens. *Genomics*, *101*(2), 94–100. https://doi.org/10.1016/j.ygeno.2012.11.005.

Andrés-León, E., Peña, D. G., Gómez-López, G., & Pisano, D. G. (2015). MiRGate: A curated database of human, mouse and rat miRNA-mRNA targets. *Database*. https://doi.org/10.1093/database/bav035.

Bandyopadhyay, S., & Mitra, R. (2009). TargetMiner: MicroRNA target prediction with systematic identification of tissue-specific negative examples. *Bioinformatics*, *25*(20), 2625–2631. https://doi.org/10.1093/bioinformatics/btp503.

Bao, L., Zhou, M., Wu, L., Lu, L., Goldowitz, D., Williams, R. W., & Cui, Y. (2007). PolymiRTS database: Linking polymorphisms in microRNA target sites with complex traits. *Nucleic Acids Research*, *35*(1), D51–D54. https://doi.org/10.1093/nar/gkl797.

Bhattacharya, A., Ziebarth, J. D., & Cui, Y. (2014). PolymiRTS database 3.0: Linking polymorphisms in microRNAs and their target sites with human diseases and biological pathways. *Nucleic Acids Research*, *42*(1), D86–D91. https://doi.org/10.1093/nar/gkt1028.

Bonnet, E., He, Y., Billiau, K., & van de Peer, Y. (2010). TAPIR, a web server for the prediction of plant microRNA targets, including target mimics. *Bioinformatics*, *26*(12), 1566–1568. https://doi.org/10.1093/bioinformatics/btq233.

Bruno, A. E., Li, L., Kalabus, J. L., Pan, Y., Yu, A., & Hu, Z. (2012). miRdSNP: A database of disease-associated SNPs and microRNA target sites on 3'UTRs of human genes. *BMC Genomics*, *13*(1). https://doi.org/10.1186/1471-2164-13-44.

Burgler, C., & Macdonald, P. M. (2005). Prediction and verification of microRNA targets by moving targets, a highly adaptable prediction method. *BMC Genomics*, *6*. https://doi.org/10.1186/1471-2164-6-88.

Chandra, V., Girijadevi, R., Nair, A. S., Pillai, S. S., & Pillai, R. M. (2010). MTar: A computational microRNA target prediction architecture for human transcriptome. *BMC Bioinformatics*, *11*(1). https://doi.org/10.1186/1471-2105-11-S1-S2.

Chen, Y., & Wang, X. (2020). miRDB: An online database for prediction of functional microRNA targets. *Nucleic Acids Research*, *48*(D1), D127–D131. https://doi.org/10.1093/nar/gkz757. 31504780.

Cho, S., Jang, I., Jun, Y., Yoon, S., Ko, M., Kwon, Y., … Lee, S. (2013). MiRGator v3.0: A microRNA portal for deep sequencing, expression profiling and mRNA targeting. *Nucleic Acids Research*, *41*(1), D252–D257. https://doi.org/10.1093/nar/gks1168.

Cho, S., Jun, Y., Lee, S., Choi, H. S., Jung, S., Jang, Y., … Kim, W. (2011). miRGator v2.0: An integrated system for functional investigation of microRNAs. *Nucleic Acids Research*, *39*, 158–162.

Chorostecki, U., & Palatnik, J. F. (2014). ComTAR: A web tool for the prediction and characterization of conserved microRNA targets in plants. *Bioinformatics*, *30*(14), 2066–2067. https://doi.org/10.1093/bioinformatics/btu147.

Chou, C.-H., Chang, N.-W., Shrestha, S., Hsu, S.-D., Lin, Y.-L., Lee, W.-H., … Huang, H.-D. (2016). miRTarBase 2016: Updates to the experimentally validated miRNA-target interactions database. *Nucleic Acids Research*, *44*(D1), D239–D247. https://doi.org/10.1093/nar/gkv1258. 26590260.

Chou, C.-H., Shrestha, S., Yang, C.-D., Chang, N.-W., Lin, Y.-L., Liao, K.-W., … Huang, H.-D. (2018). miRTarBase update 2018: A resource for experimentally validated microRNA-target interactions. *Nucleic Acids Research*, *46*(D1), D296–D302. https://doi.org/10.1093/nar/gkx1067. 29126174.

Dai, X., & Zhao, P. X. (2011). psRNATarget: A plant small RNA target analysis server. *Nucleic Acids Research*, *39*(Web Server issue), W155–W159. https://doi.org/10.1093/nar/gkr319. 21622958.

Dai, X., Zhuang, Z., & Zhao, P. X. (2018). psRNATarget: A plant small RNA target analysis server (2017 release). *Nucleic Acids Research*, *46*(W1), W49–W54. https://doi.org/10.1093/nar/gky316. 29718424.

Dweep, H., Gretz, N., & Sticht, C. (2014). MiRWalk database for miRNA target interactions. *Methods in Molecular Biology*, *1182*, 289–305. https://doi.org/10.1007/978-1-4939-1062-5_25.

Dweep, H., Sticht, C., Pandey, P., & Gretz, N. (2011). MiRWalk—Database: Prediction of possible miRNA binding sites by\walking\the genes of three genomes. *Journal of Biomedical Informatics*, *44*(5), 839–847. https://doi.org/10.1016/j.jbi.2011.05.002.

Enright, A. J., John, B., Gaul, U., Tuschl, T., Sander, C., & Marks, D. S. (2003). MicroRNA targets in Drosophila. *Genome Biology*, *5*(R1), 1–14. https://doi.org/10.1186/gb-2003-5-1-r1.

Fei, Y., Wang, R., Li, H., Liu, S., Zhang, H., & Huang, J. (2018). DPMIND: Degradome-based plant miRNA target interaction and network database. *Bioinformatics*, *34*(9), 1618–1620. https://doi.org/10.1093/bioinformatics/btx824.

Fiannaca, A., La Rosa, M., La Paglia, L., Rizzo, R., & Urso, A. (2016). MiRNATIP: A SOM-based miRNA target interactions predictor. *BMC Bioinformatics*, *17*. https://doi.org/10.1186/s12859-016-1171-x.

Friedman, R. C., Farh, K. K. H., Burge, C. B., & Bartel, D. P. (2009). Most mammalian mRNAs are conserved targets of microRNAs. *Genome Research*, *19*(1), 92–105. https://doi.org/10.1101/gr.082701.108.

Gennarino, V. A., Sardiello, M., Mutarelli, M., Dharmalingam, G., Maselli, V., Lago, G., & Banfi, S. (2011). HOCTAR database: A unique resource for microRNA target prediction. *Gene*, *480*(1–2), 51–58. https://doi.org/10.1016/j.gene.2011.03.005.

Gong, J., Tong, Y., Zhang, H. M., Wang, K., Hu, T., Shan, G., … Guo, A. Y. (2012). Genome-wide identification of SNPs in MicroRNA genes and the SNP effects on MicroRNA target binding and biogenesis. *Human Mutation*, *33*(1), 254–263. https://doi.org/10.1002/humu.21641.

Gottwein, E., Corcoran, D. L., Mukherjee, N., Skalsky, R. L., Hafner, M., Nusbaum, J. D., … Cullen, B. R. (2011). Viral microRNA targetome of KSHV-infected primary effusion lymphoma cell lines. *Cell Host & Microbe*, *10*(5), 515–526. https://doi.org/10.1016/j.chom.2011.09.012.

Guo, L., & Chen, F. (2014). A challenge for miRNA: Multiple isomiRs in miRNAomics. *Gene*, *544*(1), 1–7. https://doi.org/10.1016/j.gene.2014.04.039.

Ha, M., & Kim, V. N. (2014). Regulation of microRNA biogenesis. *Nature Reviews Molecular Cell Biology*, *15*(8), 509–524. https://doi.org/10.1038/nrm3838.

Hsu, J. B., Chiu, C. M., Hsu, S. D., Huang, W. Y., Chien, C. H., Lee, T. Y., & Huang, H. D. (2011). MiRTar: An integrated system for identifying miRNA target interactions in human. *BMC Bioinformatics*, *12*. https://doi.org/10.1186/1471-2105-12-300.

Hsu, S. D., Chu, C. H., Tsou, A. P., Chen, S. J., Chen, H. C., Hsu, P. W. C., … Huang, H. D. (2008). miRNAMap 2.0: Genomic maps of microRNAs in metazoan genomes. *Nucleic Acids Research*, *36*(1), D165–D169. https://doi.org/10.1093/nar/gkm1012.

Hsu, P. W. C., Huang, H. D., Hsu, S. D., Lin, L. Z., Tsou, A. P., Tseng, C. P., … Hofacker, I. L. (2006). miRNAMap: Genomic maps of microRNA genes and their target genes in mammalian genomes. *Nucleic Acids Research*, *34*, D135–D139. https://doi.org/10.1093/nar/gkj135.

Hsu, P. W. C., Lin, L. Z., Hsu, S. D., Hsu, J. B. K., & Huang, H. D. (2007). ViTa: Prediction of host microRNAs targets on viruses. *Nucleic Acids Research*, *35*(1), D381–D385. https://doi.org/10.1093/nar/gkl1009.

Hsu, S. D., Lin, F. M., Wu, W. Y., Liang, C., Huang, W. C., Chan, W. L., … Huang, H. D. (2011). MiRTarBase: A database curates experimentally validated microRNA-target interactions. *Nucleic Acids Research*, *39*(1), D163–D169. https://doi.org/10.1093/nar/gkq1107.

Hsu, S. D., Tseng, Y. T., Shrestha, S., Lin, Y. L., Khaleel, A., Chou, C. H., … Huang, H. D. (2014). MiRTarBase update 2014: An information resource for experimentally validated miRNA target interactions. *Nucleic Acids Research*, *42*(1), D78–D85. https://doi.org/10.1093/nar/gkt1266.

Huang, H., Long, J., Zheng, L., Li, Y., Hu, Y., Yu, G., … Huang, Y. (2016). Identification and characterization of microRNAs in maize endosperm response to exogenous sucrose using small RNA sequencing. *Genomics*, *108*(5–6), 216–223. https://doi.org/10.1016/j.ygeno.2016.10.007.

Huang, H.-Y., Lin, Y.-C.-D., Li, J., Huang, K.-Y., Shrestha, S., Hong, H.-C., … Huang, H.-D. (2020). miRTarBase 2020: Updates to the experimentally validated microRNA-target interaction database. *Nucleic Acids Research*, *48*(D1), D148–D154. https://doi.org/10.1093/nar/gkz896. 31647101.

Incarnato, D., Neri, F., Diamanti, D., & Oliviero, S. (2013). MREdictor: A two-step dynamic interaction model that accounts for mRNA accessibility and Pumilio binding accurately predicts microRNA targets. *Nucleic Acids Research*, *41*(18), 8421–8433. https://doi.org/10.1093/nar/gkt629.

Jha, A., & Shankar, R. (2011). Employing machine learning for reliable miRNA target identification in plants. *BMC Genomics*, *12*. https://doi.org/10.1186/1471-2164-12-636.

Jing, G., Chunjie, L., Wei, L., Yuliang, W., Zhaowu, M., Hu, C., & An-Yuan, G. (2015). An update of miRNASNP database for better SNP selection by GWAS data, miRNA expression and online tools. *Database*. https://doi.org/10.1093/database/bav029.

Jones-Rhoades, M. W., & Bartel, D. P. (2004). Computational identification of plant microRNAs and their targets, including a stress-induced miRNA. *Molecular Cell*, *14*(6), 787–799. https://doi.org/10.1016/j.molcel.2004.05.027.

Juzenas, S., Venkatesh, G., Hübenthal, M., Hoeppner, M. P., Du, Z. G., Paulsen, M., … Hemmrich-Stanisak, G. (2017). A comprehensive, cell specific microRNA catalogue of human peripheral blood. *Nucleic Acids Research*, *45*(16), 9290–9301. https://doi.org/10.1093/nar/gkx706.

Kanoria, S., Rennie, W., Liu, C., Carmack, C. S., Lu, J., & Ding, Y. (2016). STarMir tools for prediction of microRNA binding sites. In *Vol. 1490. Methods in Molecular Biology* (pp. 73–82). Humana Press. https://doi.org/10.1007/978-1-4939-6433-8_6.

Karagkouni, D., Paraskevopoulou, M. D., Chatzopoulos, S., Vlachos, I. S., Tastsoglou, S., Kanellos, I., … Hatzigeorgiou, A. G. (2018). DIANA-TarBase v8: A decade-long collection of experimentally supported miRNA-gene interactions. *Nucleic Acids Research*, *46*(1), D239–D245. https://doi.org/10.1093/nar/gkx1141.

Kim, J., Krichevsky, A., Grad, Y., Hayes, G. D., Kosik, K. S., Church, G. M., & Ruvkun, G. (2004). Identification of many microRNAs that copurify with polyribosomes in mammalian neurons. *Proceedings of the National Academy of Sciences of the United States of America*, *101*(1), 360–365. https://doi.org/10.1073/pnas.2333854100.

Kim, S. K., Nam, J. W., Rhee, J. K., Lee, W. J., & Zhang, B. T. (2006). miTarget: MicroRNA target gene prediction using a support vector machine. *BMC Bioinformatics*, *7*. https://doi.org/10.1186/1471-2105-7-411.

Kozomara, A., Birgaoanu, M., & Griffiths-Jones, S. (2019). MiRBase: From microRNA sequences to function. *Nucleic Acids Research*, *47*(1), D155–D162. https://doi.org/10.1093/nar/gky1141.

Kozomara, A., & Griffiths-Jones, S. (2014). MiRBase: Annotating high confidence microRNAs using deep sequencing data. *Nucleic Acids Research*, *42*(1), D68–D73. https://doi.org/10.1093/nar/gkt1181.

Krek, A., Grün, D., Poy, M. N., Wolf, R., Rosenberg, L., Epstein, E. J., … Rajewsky, N. (2005). Combinatorial microRNA target predictions. *Nature Genetics*, *37*(5), 495–500. https://doi.org/10.1038/ng1536.

Krüger, J., & Rehmsmeier, M. (2006). RNAhybrid: MicroRNA target prediction easy, fast and flexible. *Nucleic Acids Research*, *34*, W451–W454. https://doi.org/10.1093/nar/gkl243.

Kumar, A., Wong, A. K. L., Tizard, M. L., Moore, R. J., & Lefèvre, C. (2012). MiRNA_Targets: A database for miRNA target predictions in coding and non-coding regions of mRNAs. *Genomics*, *100*(6), 352–356. https://doi.org/10.1016/j.ygeno.2012.08.006.

Lagos-Quintana, M., Rauhut, R., Yalcin, A., Meyer, J., Lendeckel, W., & Tuschl, T. (2002). Identification of tissue-specific microRNAs from mouse. *Current Biology*, *12*(9), 735–739. https://doi.org/10.1016/S0960-9822(02)00809-6.

Lee, R. C., Feinbaum, R. L., & Ambros, V. (1993). The *C. elegans* heterochronic gene lin-4 encodes small RNAs with antisense complementarity to lin-14. *Cell*, *75*(5), 843–854. https://doi.org/10.1016/0092-8674(93)90529-Y.

Li, C., Mou, C., Swartz, M. D., Yu, B., Bai, Y., Tu, Y., & Liu, X. (2020). dbMTS: A comprehensive database of putative human microRNA target site SNVs and their functional predictions. *Human Mutation*, *41*, 1123–1130.

Lin, X., Li, X., Liang, D., & Lan, K. (2012). MicroRNAs and unusual small RNAs discovered in Kaposi's sarcoma-associated herpesvirus virions. *Journal of Virology*, *86*(23), 12717–12730. https://doi.org/10.1128/JVI.01473-12.

Liu, C. J., Fu, X., Xia, M., Zhang, Q., Gu, Z., & Guo, A. Y. (2021). MiRNASNP-v3: A comprehensive database for SNPs and disease-related variations in miRNAs and miRNA targets. *Nucleic Acids Research*, *49*(1), D1276–D1281. https://doi.org/10.1093/nar/gkaa783.

Lu, T. P., Lee, C. Y., Tsai, M. H., Chiu, Y. C., Hsiao, C. K., Lai, L. C., & Chuang, E. Y. (2012). MiRSystem: An integrated system for characterizing enriched functions and pathways of microRNA targets. *PLoS One*, *7*(8). https://doi.org/10.1371/journal.pone.0042390.

Mendoza, M. R., da Fonseca, G. C., Loss-Morais, G., Alves, R., Margis, R., & Bazzan, A. L. C. (2013). RFMirTarget: Predicting human microRNA target genes with a random forest classifier. *PLoS One*, *8*(7). https://doi.org/10.1371/journal.pone.0070153.

Menor, M., Ching, T., Zhu, X., Garmire, D., & Garmire, L. X. (2014). mirMark: A site-level and UTR-level classifier for miRNA target prediction. *Genome Biology*, *15*(10), 500. https://doi.org/10.1186/s13059-014-0500-5.

Monga, I., & Kumar, M. (2019). Computational resources for prediction and analysis of functional miRNA and their targetome. In *Vol. 1912*. *Methods in molecular biology* (pp. 215–250). Humana Press. https://doi.org/10.1007/978-1-4939-8982-9_9.

Naeem, H., Küffner, R., Csaba, G., & Zimmer, R. (2010). MiRSel: Automated extraction of associations between microRNAs and genes from the biomedical literature. *BMC Bioinformatics*, *11*. https://doi.org/10.1186/1471-2105-11-135.

Nam, S., Kim, B., Shin, S., & Lee, S. (2008). miRGator: An integrated system for functional annotation of microRNAs. *Nucleic Acids Research*, *36*(1), D159–D164. https://doi.org/10.1093/nar/gkm829.

Nathan, W., & Xiaowei, W. (2015). miRDB: An online resource for microRNA target prediction and functional annotations. *Nucleic Acids Research*, D146–D152. https://doi.org/10.1093/nar/gku1104.

Panwar, B., Omenn, G. S., & Guan, Y. (2017). MiRmine: A database of human miRNA expression profiles. *Bioinformatics*, *33*(10), 1554–1560. https://doi.org/10.1093/bioinformatics/btx019.

Papadopoulos, G. L., Reczko, M., Simossis, V. A., Sethupathy, P., & Hatzigeorgiou, A. G. (2009). The database of experimentally supported targets: A functional update of TarBase. *Nucleic Acids Research*, *37*(1), D155–D158. https://doi.org/10.1093/nar/gkn809.

Park, K., & Kim, K. B. (2013). MiRTar hunter: A prediction system for identifying human microRNA target sites. *Molecules and Cells*, *35*(3), 195–201. https://doi.org/10.1007/s10059-013-2165-4.

Peterson, S. M., Thompson, J. A., Ufkin, M. L., Sathyanarayana, P., Liaw, L., & Congdon, C. B. (2014). Common features of microRNA target prediction tools. *Frontiers in Genetics*, *5*. https://doi.org/10.3389/fgene.2014.00023.

Pfeffer, S., Sewer, A., Lagos-Quintana, M., Sheridan, R., Sander, C., Grässer, F. A., … Tuschl, T. (2005). Identification of microRNAs of the herpesvirus family. *Nature Methods*, *2*(4), 269–276. https://doi.org/10.1038/nmeth746.

Pfeffer, S., Zavolan, M., Grässer, F. A., Chien, H., Russo, J. J., Ju, J., … Tuschl, T. (2004). Identification of virus-encoded microRNAs. *Science*, *304*(5671), 734–736. https://doi.org/10.1126/science.1096781.

Piriyapongsa, J., Bootchai, C., Ngamphiw, C., & Tongsima, S. (2012). MicroPIR: An integrated database of microRNA target sites within human promoter sequences. *PLoS One*, *7*(3). https://doi.org/10.1371/journal.pone.0033888.

Piriyapongsa, J., Bootchai, C., Ngamphiw, C., & Tongsima, S. (2014). MicroPIR2: A comprehensive database for human-mouse comparative study of microRNA-promoter interactions. *Database*, *2014*. https://doi.org/10.1093/database/bau115.

Pla, A., Zhong, X., & Rayner, S. (2018). miRAW: A deep learning-based approach to predict microRNA targets by analyzing whole microRNA transcripts. *PLoS Computational Biology*, *14*(7). https://doi.org/10.1371/journal.pcbi.1006185.

Qureshi, A., Thakur, N., Monga, I., Thakur, A., & Kumar, M. (2014). VIRmiRNA: A comprehensive resource for experimentally validated viral miRNAs and their targets. *Database: The Journal of Biological Databases and Curation*, *2014*. https://doi.org/10.1093/database/bau103.

Reinhart, B. J., Slack, F. J., Basson, M., Pasquienelll, A. E., Bettlnger, J. C., Rougvle, A. E., … Ruvkun, G. (2000). The 21-nucleotide let-7 RNA regulates developmental timing in Caenorhabditis elegans. *Nature*, *403*(6772), 901–906. https://doi.org/10.1038/35002607.

ReyesHerrera, P. H., & Ficarra, E. (2012). One decade of development and evolution of microRNA target prediction algorithms. *Genomics, Proteomics & Bioinformatics*, *10*(5), 254–263. https://doi.org/10.1016/j.gpb.2012.10.001.

Riolo, G., Cantara, S., Marzocchi, C., & Ricci, C. (2020). miRNA targets: From prediction tools to experimental validation. *Methods and Protocols*, *4*(1), 1–20. https://doi.org/10.3390/mps4010001.

Rupaimoole, R., & Slack, F. J. (2017). MicroRNA therapeutics: Towards a new era for the management of cancer and other diseases. *Nature Reviews Drug Discovery*, *16*(3), 203–221. https://doi.org/10.1038/nrd.2016.246.

Rusinov, V., Baev, V., Minkov, I. N., & Tabler, M. (2005). MicroInspector: A web tool for detection of miRNA binding sites in an RNA sequence. *Nucleic Acids Research*, *33*(2), W696–W700. https://doi.org/10.1093/nar/gki364.

Sætrom, O. L. A., Snøve, O. L. A., Jr., & Sætrom, P.Å. L. (2005). Weighted sequence motifs as an improved seeding step in microRNA target prediction algorithms. *RNA*, 995–1003. https://doi.org/10.1261/rna.7290705.

Sethupathy, P., Corda, B., & Hatzigeorgiou, A. G. (2006). TarBase: A comprehensive database of experimentally supported animal microRNA targets. *RNA*, *12*(2), 192–197. https://doi.org/10.1261/rna.2239606.

Singh, N. K. (2017). miRNAs target databases: Developmental methods and target identification techniques with functional annotations. *Cellular and Molecular Life Sciences*, *74*(12), 2239–2261. https://doi.org/10.1007/s00018-017-2469-1.

Srivastava, P. K., Moturu, T. R., Pandey, P., Baldwin, I. T., & Pandey, S. P. (2014). A comparison of performance of plant miRNA target prediction tools and the characterization of features for genome-wide target prediction. *BMC Genomics*, *15*(348), 1–15. https://doi.org/10.1186/1471-2164-15-348.

Sticht, C., De La Torre, C., Parveen, A., & Gretz, N. (2018). Mirwalk: An online resource for prediction of microrna binding sites. *PLoS One*, *13*(10). https://doi.org/10.1371/journal.pone.0206239.

Sturm, M., Hackenberg, M., Langenberger, D., & Frishman, D. (2010). TargetSpy: A supervised machine learning approach for microRNA target prediction. *BMC Bioinformatics*, *11*. https://doi.org/10.1186/1471-2105-11-292.

Sun, X., Dong, B., Yin, L., Zhang, R., Du, W., Liu, D., … Mao, L. (2013). PMTED: A plant microRNA target expression database. *BMC Bioinformatics*, *14*(1). https://doi.org/10.1186/1471-2105-14-174.

Szczesniak, M. W., & Makalowska, I. (2014). miRNEST 2.0: A database of plant and animal microRNAs. *Nucleic Acids Research*, *42*, 74–77.

Thadani, R., & Tammi, M. T. (2006). MicroTar: Predicting microRNA targets from RNA duplexes. *BMC Bioinformatics*, *7*(5). https://doi.org/10.1186/1471-2105-7-S5-S20.

Thody, J., Moulton, V., & Mohorianu, I. (2020). PAREameters: A tool for computational inference of plant miRNA-mRNA targeting rules using small RNA and degradome sequencing data. *Nucleic Acids Research*, *48*(5), 2258–2270. https://doi.org/10.1093/nar/gkz1234.

Vergoulis, T., Vlachos, I. S., Alexiou, P., Georgakilas, G., Maragkakis, M., Reczko, M., … Hatzigeorgiou, A. G. (2012). TarBase 6.0: Capturing the exponential growth of miRNA targets with experimental support. *Nucleic Acids Research*, *40*(1), D222–D229. https://doi.org/10.1093/nar/gkr1161.

Wang, X. (2008). miRDB: A microRNA target prediction and functional annotation database with a wiki interface. *RNA*, *14*(6), 1012–1017. https://doi.org/10.1261/rna.965408.

Wang, X., & El Naqa, I. M. (2008). Prediction of both conserved and nonconserved microRNA targets in animals. *Bioinformatics*, *24*(3), 325–332. https://doi.org/10.1093/bioinformatics/btm595.

Wang, D., Gu, J., Wang, T., & Ding, Z. (2014). OncomiRDB: A database for the experimentally verified oncogenic and tumor-suppressive microRNAs. *Bioinformatics*, *30*(15), 2237–2238. https://doi.org/10.1093/bioinformatics/btu155.

Wang, J. F., Zhou, H., Chen, Y. Q., Luo, Q. J., & Qu, L. H. (2004). Identification of 20 microRNAs from Oryza sativa. *Nucleic Acids Research*, *32*(5), 1688–1695. https://doi.org/10.1093/nar/gkh332.

Wei, G., Sun, L., Qin, S., Li, R., Chen, L., Jin, P., & Ma, F. (2018). Dme-Hsa disease database (DHDD): Conserved human disease-related miRNA and their targeting genes in Drosophila melanogaster. *International Journal of Molecular Sciences*, *19*(9). https://doi.org/10.3390/ijms19092642.

Wu, H. J., Ma, Y. K., Chen, T., Wang, M., & Wang, X. J. (2012). PsRobot: A web-based plant small RNA meta-analysis toolbox. *Nucleic Acids Research*, *40*(1), W22–W28. https://doi.org/10.1093/nar/gks554.

Xiao, F., Zuo, Z., Cai, G., Kang, S., Gao, X., & Li, T. (2009). miRecords: An integrated resource for microRNA-target interactions. *Nucleic Acids Research*, *37*(1), D105–D110. https://doi.org/10.1093/nar/gkn851.

Xie, F., & Zhang, B. (2010). Target-align: A tool for plant microrna target identification. *Bioinformatics*, *26*(23), 3002–3003. https://doi.org/10.1093/bioinformatics/btq568.

Xu, L., Hu, Y., Cao, Y., Li, J., Ma, L., Li, Y., & Qi, Y. (2018). An expression atlas of miRNAs in Arabidopsis thaliana. *Science China. Life Sciences*, *61*(2), 178–189. https://doi.org/10.1007/s11427-017-9199-1.

Yang, Y., Wang, Y. P., & Li, K. B. (2008). MiRTif: A support vector machine-based microRNA target interaction filter. *BMC Bioinformatics*, *9*(12). https://doi.org/10.1186/1471-2105-9-S12-S4.

Yi, X., Zhang, Z., Ling, Y., Xu, W., & Su, Z. (2015). PNRD: A plant non-coding RNA database. *Nucleic Acids Research*, *43*(1), D982–D989. https://doi.org/10.1093/nar/gku1162.

Yousef, M., Abdallah, L., & Allmer, J. (2019). maTE: Discovering expressed interactions between microRNAs and their targets. *Bioinformatics*, *35*, 4020–4028.

Yue, M., Zhang, Y., Zhou, D., Zhi, H., Wang, P., Zhang, Y., … Li, X. (2018). MSDD: A manually curated database of experimentally supported associations among miRNAs, SNPs and human diseases. *Nucleic Acids Research*, *46*(1), D181–D185. https://doi.org/10.1093/nar/gkx1035.

Zhang, Y. (2005). miRU: An automated plant miRNA target prediction server. *Nucleic Acids Research*, *33*(Web Server issue), W701–W704. https://doi.org/10.1093/nar/gki383. 15980567.

Zhenhai, Z., Jingyin, Y., Daofeng, L., Zuyong, Z., Fengxia, L., Xin, Z., … Zhen, S. (2010). PMRD: Plant microRNA database. *Nucleic Acids Research*, D806–D813. https://doi.org/10.1093/nar/gkp818.

Ziebarth, J. D., Bhattacharya, A., Chen, A., & Cui, Y. (2012). PolymiRTS database 2.0: Linking polymorphisms in microRNA target sites with human diseases and complex traits. *Nucleic Acids Research*, *40*(1), D216–D221. https://doi.org/10.1093/nar/gkr1026.

Chapter 11

miRNA bioinformatics and pathway analysis

Enrica Calura and Chiara Romualdi
Department of Biology, University of Padova, Padova, Italy

The microRNA synergy

The microRNA (miRNA) mode of action is cooperative in many different ways, leading to an intricate network of coregulations, interactions, and competitive effects on gene target regulations. The first indication of their synergistic activity was highlighted by an analysis of the chromosomal distribution of miRNAs in relation to the studied pathology: clustered miRNAs (multiple miRNAs that share the same promoter because they were originated by a long primary transcript) are not only coexpressed but also act in concerted ways. The fraction of disease-associated miRNAs within a cluster are 1.4 times on average higher than the background frequency, while polycistronic loci are on average 3.5 times more disease-associated than expected by chance (Ruepp et al., 2010). Malumbres asserted that the deregulation of miRNA clusters occurs systematically in human diseases, and that not just one miRNA acts as the causative agent but instead multiple miRNAs synergistically act to regulate cellular processes (Malumbres, 2012). Additionally, it is widely known that cells express multiple miRNAs at the same time, and a single gene can be targeted by multiple miRNAs. Clearly, to characterize the miRNA functions, this intrinsic layer of complexity cannot be ignored (Krek et al., 2005; Lewis, Burge, & Bartel, 2005; Peter, 2010).

It has been estimated that generally, the target downregulation yielded by a single miRNA is quantitatively moderated at the protein level (Baek et al., 2008), suggesting that multiple miRNA contributions determine whether a gene is expressed. The single contribution can be weak but the sum of several weak contributions can have a significant impact on the cell system. This leads to a combinatorial diversity yielding to a fine tuning of expression regulations and to a synergy in biological effects that allow concerted action on cellular systems (Peter, 2010).

Poliseno et al., Tay et al., and Salmena et al. also demonstrated the complementary implications of the miRNA-target regulation, that is, that different types of mRNAs competitively act as decoys for miRNAs (Poliseno et al., 2010; Salmena, Poliseno, Tay, Kats, & Pandolfi, 2011; Tay et al., 2011). In his seminal paper, Poliseno studied PTEN and its pseudogene, which share the same 3′ UTR and the seeds for a set of miRNAs. They demonstrated that an upregulation of the pseudogene induces an increase of PTEN mRNA abundance, due to the competitive recruitment of PTEN miRNAs acting as a miRNA sponge and mitigating miRNA inhibitory action. Today we can generalize this theory and can assert that all the coding and noncoding as well as long and short transcripts in the cell are able to regulate each other's expression posttranscriptionally, forming a large-scale regulatory network crossing all cellular processes.

In addition to this fascinating layer of posttranscriptional regulations, we should not forget the regular one: the expression of coding and noncoding genes is influenced by transcription factors that regulate their abundances to solve all the cellular functions and that, consequently, the gene expressions are a proxy of active cell processes.

Role of microRNAs in biological signaling pathways

Especially for their rapid action and multigene regulatory capacity, miRNAs are the best candidates to play a pivotal role in signal transduction modulation in time and space. Because multiple signals fluctuate in the cell system and a stimulus doesn't always elicit an unambiguous on/off situation, often the cell has to distinguish between real signals and inputs that are too weak or transient. In this context, miRNAs are perfect for this role, and we have examples of their involvement in both signal transduction mechanisms (Inui, Martello, & Piccolo, 2010). In fact, although miRNAs downregulate gene expressions, their function in signaling pathways is not only repressive but is dependent on the contextual topology of the signaling. Due to their intrinsic nature, miRNAs are able to function as a signal rheostat, influencing the amount of signal perceived by the cell. miRNAs in signaling pathways have the role of setting up the threshold of sensitivity of the system,

MicroRNA in Human Malignancies. https://doi.org/10.1016/B978-0-12-822287-4.00021-9

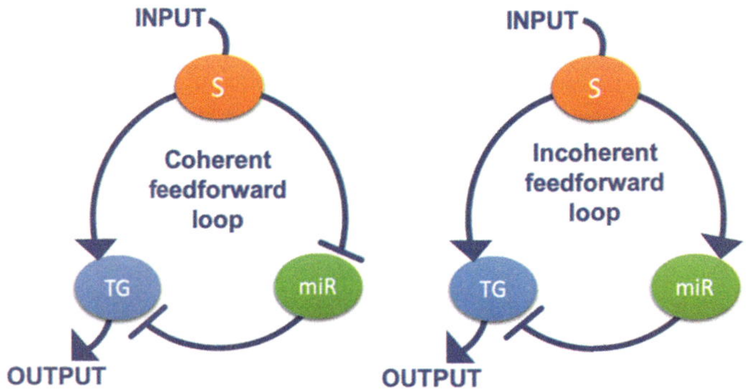

FIG. 11.1 Schema of coherent and incoherent feed-forward loops. A stimulus (S) regulates both the target gene (TG) and the miRNA (miR).

thus helping in discriminating signals that can or cannot overcome that threshold. In network biology, these systems are called coherent and incoherent feed-forward loops, as drawn in Fig. 11.1. While the coherent feed-forward loop represents the classical way of action of miRNAs, the incoherent feed-forward loop has a fundamental advantage: it prevents undesired pathway activation by random signal fluctuations, placing the threshold of the system sensitivity at a higher level (Herranz & Cohen, 2010). Moreover, the temporal difference to produce miRNAs and proteins (miRNA is faster processed) allows miRNA to affect gene expression more rapidly than what is done by transcription factors, triggering also a faster decay of bound transcripts (Hausser & Zavolan, 2014).

Another fundamental consideration is that the miRNA presence is strictly dependent on the cell type in which the reaction occurs, determining in this way the modulation of the signal that depends on the context. The unique miRNA milieu of each cell type gives great plasticity to the system, suggesting why different cell types with such different functions can share the same cascades of the signal.

Pathways: The manually curated maps of the cell interactome

One of the most challenging goals of modern biology is to decipher and describe the complexity of cell systems. Using the analogy between biological pathways and electronic circuits, Lazebnik in 2002 noted the importance of using standard maps through which even a biologist–without any specific knowledge–could fix a radio. And, without the integration of knowledge coming from different fields of science, the efforts of reverse engineering the cell are destined to fail (Beltrame et al., 2011).

Since the beginning of the *omic* era, a variety of databases containing information on cell signaling pathways have been developed in conjunction with efforts to define common languages for sharing information, developing tools and models, running effective simulations, and visualizing, analyzing, and integrating high-throughput data (Bauer-Mehren, Furlong, & Sanz, 2009). Moreover, in spite of the commitment of communities dedicated to the development of standard formats, different databases are still characterized by different graphical and computational annotations. To fill this gap, Pathway Commons is a resource that aggregates data from publicly available biological pathway and molecular interaction databases using BioPax standard format and SBGN graphical notation. Although these initiatives are successful results of the joint efforts of wider communities in many fields of biology, they struggle to keep pace with biology discoveries. Pathway maps require curated manual annotation, and this by definition results in maps always being a step back from the latest discoveries, thus incomplete. Moreover only a part of the annotated interactions are contained in all the repositories. This gap is particularly large when we look at miRNAs.

Many miRNA target genes have been validated with direct experimental methods; these can be found in the literature and are collected in public databases. However, despite the increase of experimentally validated miRNA targets, the majority of them remain unknown, and the in silico predictions are the only solutions to preinvestigate the data in a rapid way. Reactome (Jassal et al., 2020) and KEGG (Kanehisa, Furumichi, Tanabe, Sato, & Morishima, 2016), among the most famous pathway databases, include few miRNAs in their pathways. At the time of writing, Reactome has 47 miRNAs in 40 pathways for a total of 272 reactions while KEGG has 106 miRNAs in only four pathways. All miRNAs are included in the "microRNAs in cancer" pathway while the "proteoglycans in cancer" has three miRNAs. "Fluid shear stress and atherosclerosis" and "axon regeneration" have one miRNA each. Considering the number of known miRNAs (38,589 in Mirbase version 22.1) and the number of validated miRNA-target interactions with luciferase assays (7676 miRTarBase release 8.0; (Huang et al., 2020)), we can assert that in pathway repositories, miRNAs are virtually absent.

Thus, even if we are intensively studying miRNAs by decades and despite the fact that we have a large variety of data about the paramount importance of miRNAs in cell signaling, virtually all pathway databases lack solid and exhaustive miRNA annotations.

The miRNA interactome

The intricate miRNA and pathway relations can be represented using networks composed of miRNA-gene and gene–gene interactions, such as the one in Fig. 11.2.

It is increasingly clear that we cannot continue studying an miRNA and its target genes as isolated entities, separately outside the context of their network interactions.

The analysis of this network gives us a panoramic view about the state of the art of miRNA-target research. As expected, the addition of miRNAs increases the size of almost all pathway databases (Table 11.1), with slight differences across map annotations and pathways (Table 11.2). Table 11.2 shows the top 10 pathways ordered by the ratio between the number of miRNAs and the original size of the pathway (with only genes). From these statistics, it is also clear that specific fields of investigation are more deeply impacted by miRNA research (Table 11.2). Cancer and disease-related pathways triple their original size including miRNAs, and this is almost due to the amount of miRNA studies in these fields. Accordingly, the

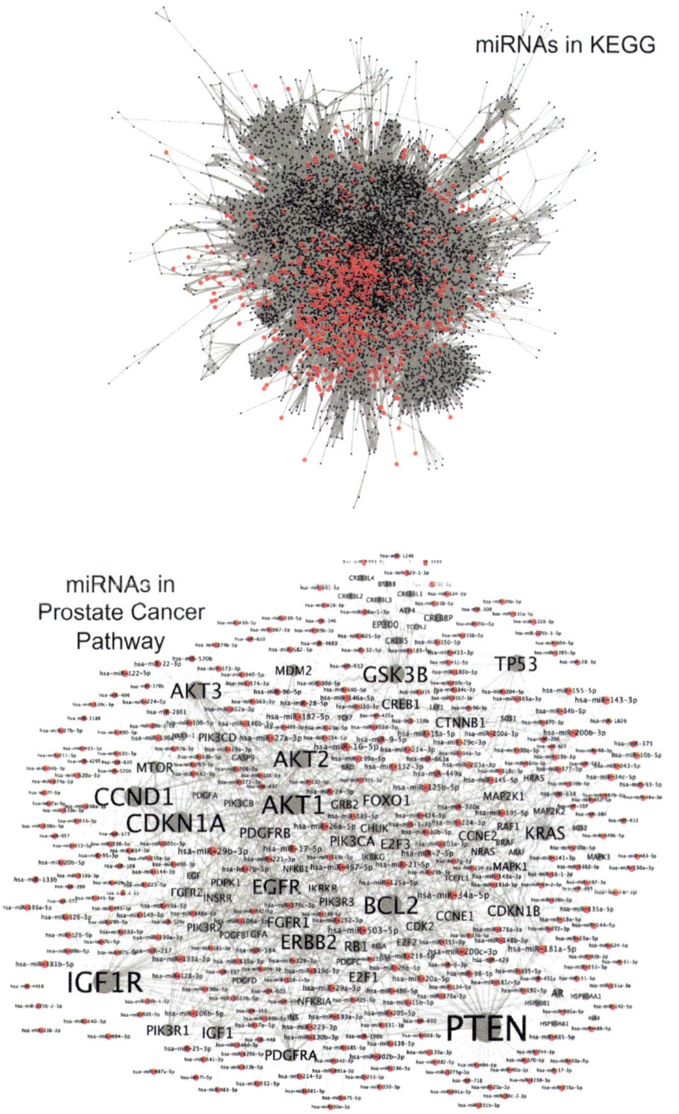

FIG. 11.2 miRNA pathways. The KEGG network and the prostate cancer pathway with miRNAs.

TABLE 11.1 Statistics of miRNAs targeting genes/proteins (only interactions validated with luciferase assay) in different pathway databases.

Pathway db	Number of validated miRNAs targeting pathway genes	Number of genes in db maps	miRNA/ genes
Biocarta	500	1035	0.48
Kegg	650	6358	0.1
NCI	603	2388	0.25
Panther	457	1165	0.39
Pathbank	441	1311	0.34
Reactome	684	10,075	0.07

TABLE 11.2 Statistics of miRNAs in KEGG pathways (only miRNA-target interactions validated with luciferase assays).

Pathway name	Number of validated miRNAs targeting pathway genes	Number of pathway genes	miRNA/genes
Prostate cancer	847	85	9.96
Endometrial cancer	515	52	9.9
Melanoma	675	70	9.64
Platinum drug resistance	368	39	9.44
Colorectal cancer	695	74	9.39
Chronic myeloid leukemia	704	76	9.26
EGFR tyrosine kinase inhibitor resistance	730	79	9.24
Glioma	657	75	8.76
Endocrine resistance	792	93	8.52
Nonsmall cell lung cancer	587	71	8.27

Top 10 KEGG pathways ordered by decreasing ratio (miRNAs/genes) are shown.

ranking of target genes by the number of experimentally validated miRNAs identifies the most famous cancer genes, as shown in Table 11.3.

In this perspective, pathway and network analysis are the biggest challenges in integrative biology. The combination of different kinds of data, such as gene and miRNA expression, needs the development of new statistical methods combined with easy and enjoyable algorithm implementations for their large use by the biological community.

miRNA pathway analysis methods

Gene set analysis (GSA) is one of the main computational approaches for the analysis and interpretation of transcriptomic data. GSA's goal is to contextualize transcriptomic alterations moving from individual genes to groups of genes (called *gene sets*) defined by their functional relationships, highlighting the cellular processes involved in the biological problem at hand.

TABLE 11.3 Top 10 miRNAs in KEGG pathways ordered by the decreasing number of interactors.

miRNA	Number of target genes in pathways	Target genes in the pathways	Number of miRNAs targeting the gene
hsa-miR-155-5p	101	PTEN	75
hsa-miR-34a-5p	89	VEGFA	55
hsa-miR-21-5p	82	BCL2	47
hsa-miR-145-5p	81	CCND1	45
hsa-miR-125b-5p	73	CDKN1A	45
hsa-miR-29a-3p	64	IGF1R	45
hsa-miR-200c-3p	60	CDK6	36
hsa-miR-124-3p	59	TP53	34
hsa-miR-17-5p	58	MYC	32
hsa-miR-29b-3p	56	SMAD4	30

Genes in KEGG pathways ordered by the decreasing number of miRNAs targeting them.

Pathway analysis (PA) is a GSA method where predefined gene sets are represented by biological pathways (such as those available in public repositories such as KEGG and Reactome).

Several PA methods have been developed, both in the univariate and multivariate context. Following (Khatri, Sirota, & Butte, 2012), they can be divided into two broad categories: (i) overrepresentation analysis (ORA) based on the statistical evaluation of the fraction of genes in a particular pathway found among the set of differentially expressed genes and (ii) functional class scoring (FCS) approaches based on the idea that not only large alterations, but also weaker coordinated changes in sets of functionally related genes (i.e., pathways) can have significant effects. While ORA approaches are generally all based on the same statistical test (Fisher exact test and its modifications), the FCS category includes a wide range of different approaches, among which the most representative is gene set enrichment analysis (GSEA) proposed by the Broad Institute in 2005 (Subramanian et al., 2005). Both FCS and ORA can be further divided into topological and nontopological methods according to their ability to deal with the explicit gene connectivity derived from pathway annotation (for a comprehensive review see (Khatri et al., 2012)).

With the increasingly evident role of miRNAs in the modulation of all the cellular processes, many pathway analysis tests, originally developed for mRNA, have been adapted to deal with miRNAs. However, each miRNA has the potential to target hundreds of different genes, meaning that the perturbation of a single miRNA may affect many biological functions. Pathway annotations are still poorly characterized in terms of miRNA presence, so the interpretation of miRNA differential expression is not straightforward and the miRNA role in cellular pathways is still poorly understood. In the following paragraphs, we will describe some of the widely used approaches to functionally characterize miRNA alterations along with the bioinformatic tools available for the analyses.

ORA on miRNA targets

The most common method of miRNA functional enrichment analysis is based on the application of the ORA methods on the list of the in silico predicted miRNA targets. This method consists of three steps: (i) finding genes that are predicted targets of selected miRNAs, (ii) annotating target genes for their participation in pathways, and (iii) statistical testing for overrepresentation of a biological process in the set of targeted genes.

For the first and second steps, public databases are used to retrieve putative target predictions and pathway annotations while for the final step, the hypergeometric distribution, or equivalently the Fisher's exact test (or in some cases its approximation with the chi-squared test) is used to test for enrichment. The test is performed for each gene set to estimate the probability of observing a number of features in a given set by chance: the lower the probability, the higher the enrichment of the gene set.

Although widely used, these enrichment methods have been developed under the assumption that target genes are selected uniformly at random from a finite population, an assumption that is not generally true for target genes selected based on the query miRNAs (Bleazard, Lamb, & Griffiths-Jones, 2015). Indeed, as we showed in Tables 11.1–11.3, miRNA targets significantly enriched specific pathways mainly related to cancer and diseases. Therefore, these pathways have more chances to be selected if this imbalance is not taken into account. On the other hand, there are less investigated biological processes with few known miRNA targets that have less chance to be identified.

The violation of the uniformity assumption leads to biased (Bleazard et al., 2015) and highly unspecific outcomes (Godard & Van Eyll, 2015). To address this issue, both Bleazard et al. (2015) and Godard and Van Eyll (2015) proposed alternative strategies to the classical ORA methods that were able to eliminate the bias, providing more realistic results.

FCS on miRNA targets

Most FCS methods are based on a general framework that consists of the following three steps. First, a gene-level statistic is computed using the molecular measurements available for the experiment. The choice of the statistic is up to the user; it could be the simple log fold change or test statistics suitable for the type of data used. Second, the gene-level statistics for all genes in a pathway are aggregated into a single pathway-level statistic. This aggregation statistic can be multivariate accounting for gene correlations, or it can be univariate. Finally, to test the significance of a given pathway/gene set, competitive or full-contained approaches are proposed (Khatri et al., 2012). The competitive method compares the set of genes in a pathway with the set of genes not in the pathway, while the full-contained approach ignores the genes that are not in the pathway. Clearly, the two approaches are based on completely different null hypotheses and their comparison needs to be grounded on the biological question that the user has to address.

The first FCS competitive approach, the GSEA, was proposed for coding genes by Subramanian et al. (2005). Here, the pathway-level statistic was based on the Kolmogorov–Smirnov statistic modified in order to weigh genes according to their differential expression. The GSEA approach can be applied to miRNAs without any modification, providing that the miRNAs were functionally annotated using the approach described by Godard and Van Eyll (2015) in which a miRNA inherits the functional annotations of all its target genes.

Topological approaches of pathway analysis with miRNAs

As presented above, pathway maps can be converted into networks following biologically driven rules (Sales, Calura, Cavalieri, & Romualdi, 2012; Sales, Calura, & Romualdi, 2019) where nodes represent genes and edges represent different biological interactions. The update of pathway annotation (namely the inclusion of a new gene or of a new interaction among already present molecules) is slow compared to the increase of the new available knowledge of specific cellular processes, then many currently available pathway maps lack of important molecular information. However, pathway graphs still represent valuable information to take into account during the analysis.

Among the pathway analysis methods developed to study gene expression and considering the pathway topology, we found: (i) ORA approaches that simply map the differentially expressed genes on the graph structure and try to identify the connected components of the graph (called modules) enriched by DEGs, and (ii) FCS approaches based on a multivariate setting that exploit the theory of graphical models to infer pathway alterations (Martini, Chiogna, Calura, & Romualdi, 2019; Martini, Sales, Calura, et al., 2014; Martini, Sales, Massa, Chiogna, & Romualdi, 2012; Salviato, Djordjilović, Chiogna, & Romualdi, 2019). In the latter class, providing a gene expression matrix and using the graph structure as a backbone, two different parameters can be estimated: (i) the multivariate mean vector, that is, the vector of the means of the genes belonging to the pathway, and (ii) the covariance matrix, that is, the matrix that defines the level of correlation between pairs of connected pathway nodes. These two quantities can be used to compare groups of samples identifying significant altered pathways or portions of pathways (often called the subpath). As a general statement, most of these pathway analysis approaches proposed for gene expression can be applied even in the presence of miRNAs without any modifications. However, we have to take into account that they are limited by the presence/absence of target genes in pathway annotations and that the increased dimensions of pathways augmented with miRNAs could limit the parameter estimates of the tests, if a larger sample size is not available.

Computational tools for miRNA pathway analyses

The vast majority of tools available for miRNA functional analyses are based on ORA methods applied to the list of targets of one or more miRNAs; only few of them are based on FCS methods and even fewer on topological approaches. In general, to show the multi-to-multi relationships in data (many miRNAs target many targets and many genes are targets of many miRNAs), they often provide network-based visualizations where miRNA and targets are connected according to their validated or predicted interactions retrieved from several public resources. In the following, we briefly overview the most recent and updated tools implementing pathway analyses with ORA, FCS, and topological approaches.

mirNET2.0 (http://www.mirnet.ca) (Chang, Zhou, Soufan, & Xia, 2020) is a web-based tool offering many modules of analyses. The tool allows starting the analyses from a list of miRNAs (or an expression matrix) or from a list (or matrix) of targets. Starting from these inputs, the tool identifies the significant features, selects their predicted or validated targets, identifies the enriched functional categories, and visualizes the results in an efficient and customizable network-based environment. Ten different species are supported and 11 different databases are used to annotate miRNAs. Of particular interest are the modules: (i) for the analysis and exploration of single nucleotide polymorphisms, (ii) to include the transcription factors and evaluating their cooperation with miRNAs in gene expression tuning through feedback loops, and (iii) to select binding predictions on noncoding elements such as long noncoding RNAs and circular RNAs. mirNET2.0 applies two types of enrichment tests, the hypergeometric test with both the classical null distribution and the empirical sampling distribution for bias correction.

DIANA-miRPath 3.0 (http://www.microrna.gr/miRPathv3) (Vlachos et al., 2015) annotates miRNAs from seven species according to their association with KEGG pathways and GO categories and tests their functional association using ORA approaches with bias correction. As output, they provide KEGG maps updated with the inclusion of the miRNAs whose targets belong to the pathway.

Similarly, **miEAA 2.0** (https://www.ccb.uni-saarland.de/mieaa2) (Kern et al., 2020) provides many different miRNA functional annotations for 10 species tested with both a classical ORA approach (without bias correction) and an FCS approach (GSEA methods, see paragraph above). Word clouds and heat maps represent efficiently the significant categories.

mirPathDB 2.0 (https://mpd.bioinf.uni-sb.de/) (Kehl et al., 2020) is a database focused on the miRNA effect on pathways. Along with GO categories, mirPathDB is the only tool that gathers pathway annotations not only from KEGG but also from WikiPathway and Reactome. It provides these modules of analyses: (i) an miRNA-centric view, where for each miRNA a list of significantly targeted pathways is reported; (ii) a pathway-centric view, where for each pathway a list of miRNAs that significantly target the pathway is reported; (iii) a multi-miRNA visualization, where given an miRNA list, the heat map of the significantly targeted pathways by at least one miRNA is reported. The peculiarities of this tool include: (i) the miRNA similarity analysis that defines groups of similar miRNAs based on their seed sequence, mature sequence, target genes, and target pathways, and (ii) the maximum coverage analysis that finds a smaller number of miRNAs that are sufficient to regulate a given pathway.

Among the topological ORA methods, we found **Subpathway-Gmir** (Feng et al., 2015) available as an R package. **Subpathway-Gmir** is a generalization of the previous Subpathway-GM approach proposed for coding genes (Li et al., 2013). The method maps differentially expressed genes and miRNAs on pathway graphs augmented with miRNAs. Then, using the shortest path among differentially expressed elements of the graph (lenient distance similarity), it identifies a subpathway for each pathway. These subpathways are supposed to identify key regions that might be representative of the corresponding entire pathways. For each subpathways identified, a Fisher exact test is used to evaluate its enrichment in terms of differentially genes and miRNAs.

A hybrid topological approach, proposed for coding genes, is **SPIA** (Tarca et al., 2009). SPIA combines the ORA of differentially expressed genes in a pathway with a topological FCS approach based on the estimation of a perturbation probability measured by propagating expression changes through the pathway topology. Exploiting this approach, Alaimo, Marceca, Ferro, and Pulvirenti (2017) proposed **MITHrIL,** a generalization of the SPIA model suitable for pathways in which miRNA are included based on the presence of a predicted/validated target within the pathway.

Micrographite (Calura et al., 2014) was the first graph-based method proposed to analyze pathways augmented with miRNAs. Micrographite augments pathway annotations with validated and predicted miRNAs; an miRNA is added to the pathway when at least one target gene is a pathway gene (Fig. 11.3). Given the large amount of in silico miRNA-target predicted interactions, only those with a significant correlation are selected. Then, using a modified version of the CliPPER approach (Martini, Chiogna, et al., 2019; Martini, Sales, Calura, et al., 2014; Martini, Sales, Massa, et al., 2012; Salviato et al., 2019), it identifies significantly altered pathways according both to the mean and to the covariance matrix. Then, with a recursive strategy it decomposes the graph into smaller connected components (Fig. 11.3, steps 2–3) with the aim to: (i) identify the portion of the pathways responsible for the differential expression, and (ii) reduce pathway redundancy. The final outcome is a regulatory circuit composed of miRNAs and genes showing significant alterations between two groups of samples (Fig. 11.3, steps 4–5). The micrographite approach has been successfully applied to study

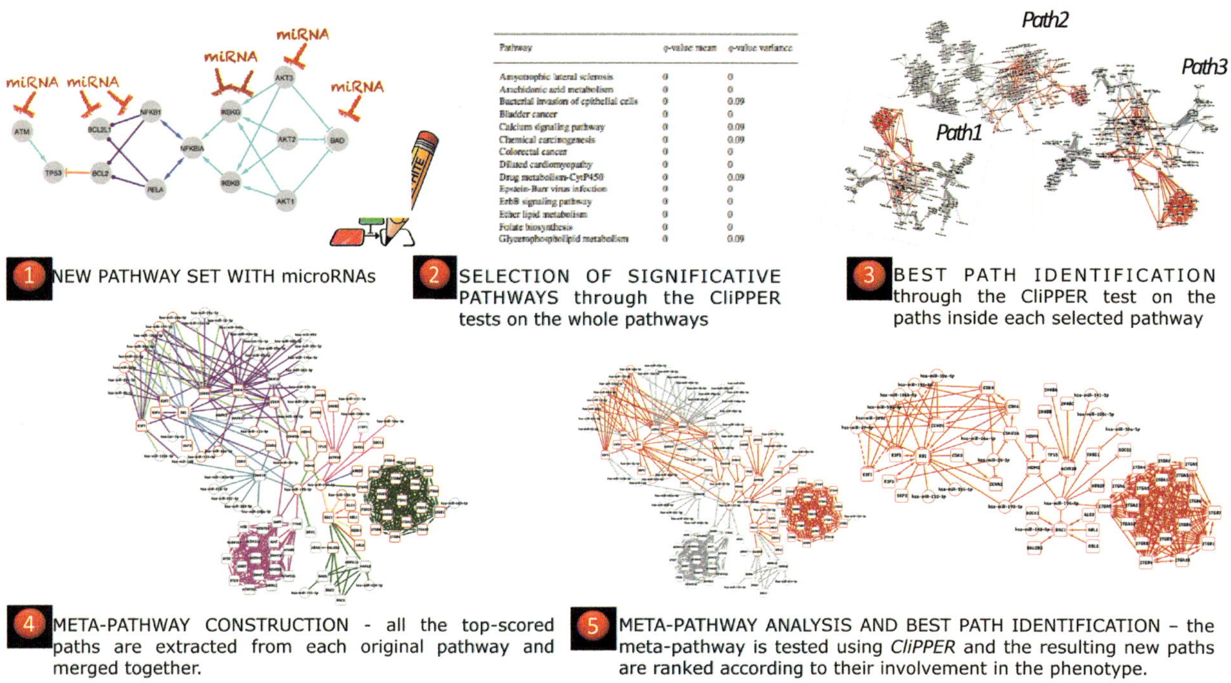

FIG. 11.3 Micrographite. Graphical outline of the micrographite computational approach. (From Calura, E., Martini, P., Sales, G., Beltrame, L., Chiorino, G., D'Incalci, M., Marchini, S., & Romualdi, C. (2014). Wiring miRNAs to pathways: A topological approach to integrate miRNA and mRNA expression profiles. *Nucleic Acids Research, 42*(11), e96. https://doi.org/10.1093/nar/gku354.)

ovarian cancer aspects and myeloproliferative diseases (Calura et al., 2013, 2019; Calura, Paracchini, et al., 2016; Calura, Pizzini, et al., 2016).

Conclusions

miRNAs are fundamental elements of signaling pathways conferring temporal as well as quantitative precision. However, the complex network of interactions in which miRNAs are involved is the cause of objective difficulties in categorizing an miRNA as an oncogene or an oncosuppressor. Data interpretations are often intricate. Tahe results are sometimes debated and re dependent on the sampled tissue, cell type, and disease. Our final message is that even in the case of a single miRNA study, many interactors, coregulators, and promoters have to be taken into account to characterize the miRNA function, considering multiple main and side-effects. Such a high level of connectivity can only be accounted for and managed using bioinformatics tools in a system biology perspective.

In this chapter, we presented the state of the art of the available bioinformatic tools to deal with the complicated world of the network biology of miRNAs. We hope that this chapter will prove useful for researchers and students who wish to learn more about this challenging area of work and maybe entice someone to join us on the journey.

Funding

This work was supported by the My First AIRC grant to Enrica Calura (MFAG 2019, Grant 23522) and by the IG grant to Chiara Romualdi (Grant IG 21837), both provided by the Italian Association for Cancer Research.

References

Alaimo, S., Marceca, G., Ferro, A., & Pulvirenti, A. (2017). Detecting disease specific pathway substructures through an integrated systems biology approach. *Non-Coding RNA*, *3*(2), 20. https://doi.org/10.3390/ncrna3020020.

Baek, D., Villén, J., Shin, C., Camargo, F. D., Gygi, S. P., & Bartel, D. P. (2008). The impact of microRNAs on protein output. *Nature*, *455*(7209), 64–71. https://doi.org/10.1038/nature07242. Epub.

Bauer-Mehren, A., Furlong, L. I., & Sanz, F. (2009). Pathway databases and tools for their exploitation: Benefits, current limitations and challenges. *Molecular Systems Biology*, *5*. https://doi.org/10.1038/msb.2009.47.

Beltrame, L., Calura, E., Popovici, R. R., Rizzetto, L., Guedez, D. R., Donato, M., et al. (2011). The biological connection markup language: A SBGN-compliant format for visualization, filtering and analysis of biological pathways. *Bioinformatics*, *27*(15), 2127–2133. https://doi.org/10.1093/bioinformatics/btr339.

Bleazard, T., Lamb, J. A., & Griffiths-Jones, S. (2015). Bias in microRNA functional enrichment analysis. *Bioinformatics*, *31*(10), 1592–1598. https://doi.org/10.1093/bioinformatics/btv023.

Calura, E., Ciciani, M., Sambugaro, A., Paracchini, L., Benvenuto, G., Milite, S., et al. (2019). Transcriptional characterization of stage I epithelial ovarian cancer: A multicentric study. *Cells*, *8*(12). https://doi.org/10.3390/cells8121554.

Calura, E., Fruscio, R., Paracchini, L., Bignotti, E., Ravaggi, A., Martini, P., et al. (2013). MiRNA landscape in stage i epithelial ovarian cancer defines the histotype specificities. *Clinical Cancer Research*, *19*(15), 4114–4123. https://doi.org/10.1158/1078-0432.CCR-13-0360.

Calura, E., Martini, P., Sales, G., Beltrame, L., Chiorino, G., D'Incalci, M., et al. (2014). Wiring miRNAs to pathways: A topological approach to integrate miRNA and mRNA expression profiles. *Nucleic Acids Research*, *42*(11). https://doi.org/10.1093/nar/gku354, e96.

Calura, E., Paracchini, L., Fruscio, R., DiFeo, A., Ravaggi, A., Peronne, J., et al. (2016). A prognostic regulatory pathway in stage I epithelial ovarian cancer: New hints for the poor prognosis assessment. *Annals of Oncology*, *27*(8), 1511–1519. https://doi.org/10.1093/annonc/mdw210.

Calura, E., Pizzini, S., Bisognin, A., Coppe, A., Sales, G., Gaffo, E., et al. (2016). A data-driven network model of primary myelofibrosis: Transcriptional and post-transcriptional alterations in CD34+ cells. *Blood Cancer Journal*, *6*(6). https://doi.org/10.1038/bcj.2016.47.

Chang, L., Zhou, G., Soufan, O., & Xia, J. (2020). miRNet 2.0: Network-based visual analytics for miRNA functional analysis and systems biology. *Nucleic Acids Research*, *48*(W1), W244–W251. https://doi.org/10.1093/nar/gkaa467.

Feng, L., Xu, Y., Zhang, Y., Sun, Z., Han, J., Zhang, C., et al. (2015). Subpathway-GMir: Identifying miRNA-mediated metabolic subpathways by integrating condition-specific genes, microRNAs, and pathway topologies. *Oncotarget*, *6*(36), 39151–39164. https://doi.org/10.18632/oncotarget.5341.

Godard, P., & Van Eyll, J. (2015). Pathway analysis from lists of microRNAs: Common pitfalls and alternative strategy. *Nucleic Acids Research*, *43*(7), 3490–3497. https://doi.org/10.1093/nar/gkv249.

Hausser, J., & Zavolan, M. (2014). Identification and consequences of miRNA-target interactions-beyond repression of gene expression. *Nature Reviews Genetics*, *15*(9), 599–612. https://doi.org/10.1038/nrg3765.

Herranz, H., & Cohen, S. M. (2010). MicroRNAs and gene regulatory networks: Managing the impact of noise in biological systems. *Genes and Development*, *24*(13), 1339–1344. https://doi.org/10.1101/gad.1937010.

Huang, H. Y., Lin, Y. C. D., Li, J., Huang, K. Y., Shrestha, S., Hong, H. C., et al. (2020). MiRTarBase 2020: Updates to the experimentally validated microRNA-target interaction database. *Nucleic Acids Research*, *48*(1), D148–D154. https://doi.org/10.1093/nar/gkz896.

Inui, M., Martello, G., & Piccolo, S. (2010). MicroRNA control of signal transduction. *Nature Reviews Molecular Cell Biology*, *11*(4), 252–263. https://doi.org/10.1038/nrm2868.

Jassal, B., Matthews, L., Viteri, G., Gong, C., Lorente, P., Fabregat, A., et al. (2020). The reactome pathway knowledgebase. *Nucleic Acids Research*, *48*(D1), D498–D503. https://doi.org/10.1093/nar/gkz1031.

Kanehisa, M., Furumichi, M., Tanabe, M., Sato, Y., & Morishima, K. (2016). KEGG: New perspectives on genomes, pathways, diseases and drugs. *Nucleic Acids Research*, *45*(D1), D353–D361. https://doi.org/10.1093/nar/gkw1092. Epub.

Kehl, T., Kern, F., Backes, C., Fehlmann, T., Stöckel, D., Meese, E., et al. (2020). miRPathDB 2.0: A novel release of the miRNA pathway dictionary database. *Nucleic Acids Research*, *48*(D1), D142–D147. https://doi.org/10.1093/nar/gkz1022.

Kern, F., Fehlmann, T., Solomon, J., Schwed, L., Grammes, N., Backes, C., et al. (2020). miEAA 2.0: Integrating multi-species microRNA enrichment analysis and workflow management systems. *Nucleic Acids Research*, *48*(W1), W521–W528. https://doi.org/10.1093/nar/gkaa309.

Khatri, P., Sirota, M., & Butte, A. J. (2012). Ten years of pathway analysis: Current approaches and outstanding challenges. *PLoS Computational Biology*, *8*(2). https://doi.org/10.1371/journal.pcbi.1002375.

Krek, A., Grün, D., Poy, M. N., Wolf, R., Rosenberg, L., Epstein, E. J., et al. (2005). Combinatorial microRNA target predictions. *Nature Genetics*, *37*(5), 495–500. https://doi.org/10.1038/ng1536.

Lewis, B. P., Burge, C. B., & Bartel, D. P. (2005). Conserved seed pairing, often flanked by adenosines, indicates that thousands of human genes are microRNA targets. *Cell*, *120*(1), 15–20. https://doi.org/10.1016/j.cell.2004.12.035.

Li, C., Han, J., Yao, Q., Zou, C., Xu, Y., Zhang, C., et al. (2013). Subpathway-GM: Identification of metabolic subpathways via joint power of interesting genes and metabolites and their topologies within pathways. *Nucleic Acids Research*, *41*(9), e101. https://doi.org/10.1093/nar/gkt161.

Malumbres, M. (2012). MiRNAs versus oncogenes: The power of social networking. *Molecular Systems Biology*, *8*. https://doi.org/10.1038/msb.2012.2.

Martini, P., Chiogna, M., Calura, E., & Romualdi, C. (2019). MOSClip: Multi-omic and survival pathway analysis for the identification of survival associated gene and modules. *Nucleic Acids Research*, *47*(14). https://doi.org/10.1093/nar/gkz324, e80.

Martini, P., Sales, G., Calura, E., Cagnin, S., Chiogna, M., & Romualdi, C. (2014). TimeClip: Pathway analysis for time course data without replicates. *BMC Bioinformatics*, *15*. https://doi.org/10.1186/1471-2105-15-S5-S3.

Martini, P., Sales, G., Massa, M. S., Chiogna, M., & Romualdi, C. (2012). Along signal paths: An empirical gene set approach exploiting pathway topology. *Nucleic Acids Research*, *41*(1), e19. https://doi.org/10.1093/nar/gks866. Epub.

Peter, M. E. (2010). Targeting of mRNAs by multiple miRNAs: The next step. *Oncogene*, *29*(15), 2161–2164. https://doi.org/10.1038/onc.2010.59.

Poliseno, L., Salmena, L., Zhang, J., Carver, B., Haveman, W., & Pandolfi, P. P. (2010). A coding-independent function of gene and pseudogene mRNAs regulates tumour biology. *Nature*, *465*(7301). https://doi.org/10.1038/nature09144. 20577206.

Ruepp, A., Kowarsch, A., Schmidl, D., Buggenthin, F., Brauner, B., Dunger, I., et al. (2010). PhenomiR: A knowledgebase for microRNA expression in diseases and biological processes. *Genome Biology*, *11*(1). https://doi.org/10.1186/gb-2010-11-1-r6.

Sales, G., Calura, E., Cavalieri, D., & Romualdi, C. (2012). Graphite—A Bioconductor package to convert pathway topology to gene network. *BMC Bioinformatics*, *13*(1). https://doi.org/10.1186/1471-2105-13-20.

Sales, G., Calura, E., & Romualdi, C. (2019). metaGraphite—A new layer of pathway annotation to get metabolite networks. *Bioinformatics*, *35*(7). https://doi.org/10.1093/bioinformatics/bty719. 30184047.

Salmena, L., Poliseno, L., Tay, Y., Kats, L., & Pandolfi, P. P. (2011). A ceRNA hypothesis: The Rosetta stone of a hidden RNA language? *Cell*, *146*(3), 353–358. https://doi.org/10.1016/j.cell.2011.07.014.

Salviato, E., Djordjilović, V., Chiogna, M., & Romualdi, C. (2019). SourceSet: A graphical model approach to identify primary genes in perturbed biological pathways. *PLoS Computational Biology*, *15*(10). https://doi.org/10.1371/journal.pcbi.1007357.

Subramanian, A., Tamayo, P., Mootha, V. K., Mukherjee, S., Ebert, B. L., Gillette, M. A., et al. (2005). Gene set enrichment analysis: A knowledge-based approach for interpreting genome-wide expression profiles. *Proceedings of the National Academy of Sciences of the United States of America*, *102*(43), 15545–15550. https://doi.org/10.1073/pnas.0506580102.

Tarca, A. L., Draghici, S., Khatri, P., Hassan, S. S., Mittal, P., Kim, J., et al. (2009). A novel signaling pathway impact analysis. *Bioinformatics*, *25*(1), 75–82. https://doi.org/10.1093/bioinformatics/btn577.

Tay, Y., Kats, L., Salmena, L., Weiss, D., Tan, S. M., Ala, U., et al. (2011). Coding-independent regulation of the tumor suppressor PTEN by competing endogenous mRNAs. *Cell*, *147*(2), 344–357. https://doi.org/10.1016/j.cell.2011.09.029.

Vlachos, I. S., Zagganas, K., Paraskevopoulou, M. D., Georgakilas, G., Karagkouni, D., Vergoulis, T., et al. (2015). DIANA-miRPath v3.0: Deciphering microRNA function with experimental support. *Nucleic Acids Research*, *43*(W1), W460–W466. https://doi.org/10.1093/nar/gkv403.

Section C

miRNAs involvement in therapeutics/theranostics

Chapter 12

Opportunities of miRNAs in cancer therapeutics

Alberto M. Arenas[a,b,d,*], Alvaro Andrades[a,b,d,*], Juan Rodrigo Patiño-Mercau[a,b,d], Juan Sanjuan-Hidalgo[b], Marta Cuadros[b,c,d], Daniel J. García[b,c,d], Paola Peinado[a,b,d], Maria Isabel Rodriguez[b,c,d], Carlos Baliñas-Gavira[b,d], Juan Carlos Álvarez-Perez[a,b,d], and Pedro P. Medina[a,b,d]

[a]*Department of Biochemistry and Molecular Biology I, Faculty of Sciences, University of Granada, Granada, Spain,* [b]*GENYO, Centre for Genomics and Oncological Research: Pfizer/University of Granada/Andalusian Regional Government, Granada, Spain,* [c]*Department of Biochemistry and Molecular Biology III and Immunology, Faculty of Medicine, University of Granada, Granada, Spain,* [d]*Institute of Biomedical Research of Granada (ibs.Granada), Granada, Spain*

Introduction

MicroRNAs (miRNAs) are small (~ 18–25 nucleotides), highly conserved noncoding RNAs that regulate gene expression (Gebert & MacRae, 2019). According to miRBase (http://www.mirbase.org, release 22.1), 1917 miRNAs have been identified in humans. Many miRNAs are involved in cancer pathways, a fact that makes miRNAs attractive targets for cancer therapy. In this chapter, we present an overview of the current status and the future perspectives of miRNA-based cancer therapies, which we define as cancer therapies in which miRNAs or miRNA-inspired molecules are either a drug or a therapeutic target.

MicroRNAs in cancer. OncomiRs and tumor-suppressor miRNAs

MicroRNA-based cancer therapies rely on the unique genomic features, biogenesis, and mechanisms of action of miRNAs. MicroRNA loci can be intergenic or intronic, and their primary transcripts can be monocistronic or polycistronic (Gebert & MacRae, 2019). MicroRNA loci that encode polycistronic transcripts are termed miRNA clusters. In addition, some miRNAs constitute families: groups of ancestrally related miRNAs that may be scattered across multiple chromosomes. When an miRNA gene is transcribed, its primary transcript is processed by the Drosha ribonuclease, and then exported to the cytosol. In the cytosol, the miRNA precursor is cleaved by the Dicer ribonuclease, generating a double stranded mature miRNA. Then, the guide strand of the miRNA binds to Argonaute proteins, forming the miRNA-induced silencing complex (miRISC), whereas the complementary passenger strand is discarded. In the miRISC, the miRNA guides the complex to complementary sequences in the 3′ untranslated regions of target mRNAs. When the miRISC binds to its target mRNAs, it triggers mechanisms that prevent their translation, such as degradation of the mRNAs or repression of their translation (Gebert & MacRae, 2019). Thus, each miRNA silences multiple mRNAs in a specific manner by sequence complementarity, and each mRNA may be silenced by multiple miRNAs, causing miRNAs to have central roles in networks of gene expression regulation.

Because miRNAs play key roles in the regulation of gene expression, they are often altered in cancer (Coira et al., 2015). Just as protein-coding genes may act as oncogenes or as tumor-suppressor genes in cancer, miRNAs in cancer may act as "oncomiRs" (oncogenic miRNAs) or as tumor-suppressor miRNAs (Medina & Slack, 2008). OncomiRs preferentially target tumor-suppressor genes, so that when an oncomiR is overexpressed it excessively represses tumor-suppressor genes. Some clinically interesting oncomiRs are miR-21 and miR-155 (Medina, Nolde, & Slack, 2010; Rupaimoole & Slack, 2017). In mice, the overexpression of either miR-21 or miR-155 alone generates lymphoid tumors, which regress completely when the expression of the miRNA is reverted to its normal levels (Cheng et al., 2015; Medina, Nolde, & Slack, 2010). Conversely, tumor suppressor miRNAs preferentially target oncogenes. Thus, a loss of function of tumor suppressor miRNAs leads to the increased activation of oncogenes. One such example is the miR-34 family, which targets antiapoptotic proteins, cell cycle regulators, and metastasis-related proteins, among others (Rupaimoole & Slack, 2017). Other

[*]First author-shared.

MicroRNA in Human Malignancies. https://doi.org/10.1016/B978-0-12-822287-4.00015-3

tumor suppressor miRNAs with high therapeutic potential in cancer include the let-7 family and the miR-15/16 cluster (Gilles & Slack, 2018; Rupaimoole & Slack, 2017; Trang et al., 2010), which are being studied as therapeutic targets, as we will discuss below.

Alterations in miRNAs and their appeal for cancer therapy

Genes can promote cancer if they harbor driver mutations, carry epigenetic alterations, or display aberrant expression. However, cancer driver mutations in miRNAs are extremely rare (Rheinbay et al., 2020). On the other hand, epigenetic alterations in miRNAs seem to be common in cancer, but further research is required to explore possible therapeutic applications (Piletič & Kunej, 2016). In contrast, aberrant miRNA expression is pervasive across multiple cancers and it has been explored extensively (Rupaimoole & Slack, 2017). As a consequence, all miRNA-based therapeutics that are currently under development aim to revert aberrant miRNA expression.

MicroRNAs share several features that make them attractive targets for cancer therapy. First, miRNAs have a relatively simple mechanism of action based on sequence complementarity, which helps researchers design miRNA-based oligonucleotide drugs and predict their targets and off-targets reliably (Bartel, 2009). Second, miRNAs can be strong tumor drivers and play central roles in networks of gene expression regulation (Cheng et al., 2015; Garzon, Marcucci, & Croce, 2010; Medina, Nolde, & Slack, 2010). As a consequence, when the altered expression of a miRNA drives tumor formation, reverting the expression of the miRNA to its normal levels can be enough to induce tumor regression (Cheng et al., 2015; Medina, Nolde, & Slack, 2010). Third, because each miRNA targets multiple genes in vivo, a single miRNA-based drug can modulate multiple genes of interest (Garzon, Marcucci, & Croce, 2010). Interestingly, a single miRNA or miRNA cluster can target multiple genes from the same signaling pathway (Mestdagh et al., 2010; Uhlmann et al., 2012). Therefore, a single miRNA-based drug may modulate a whole pathway at multiple levels. Finally, miRNAs often display highly tumor-specific expression (Lu et al., 2005), and thus miRNA-based drugs may also be tumor-specific. Overall, miRNA-based therapeutics have emerged as a promising approach for cancer treatment. However, they also pose their own challenges, as will be detailed in later sections of this chapter.

MicroRNA-based drug design

Designing miRNA-based drugs is a multistep process that must ensure the optimal safety and efficacy of the drug. Here, we will address the three key choices that must be made: (i) the therapeutic strategy; (ii) chemical modifications of the RNA-based drugs; and (iii) the method of delivery to the target tissue.

Choosing the therapeutic strategy: miRNA mimics, antagomiRs, and more

Different types of miRNA-based therapeutic molecules have been devised, and the choice depends on the characteristics of the target. The most common types of miRNA-based therapeutic molecules are synthetic small RNA drugs. Such drugs may either bind to an oncomiR, blocking its function (antagomiRs), or bind to the targets of a tumor-suppressor miRNA, mimicking its function (miRNA mimics) (Bonneau, Neveu, Kostantin, Tsongalis, & De Guire, 2019; Rupaimoole & Slack, 2017). While antagomiRs are single-stranded, miRNA mimics are usually double-stranded because they must resemble the endogenous miRNA to be loaded into the miRISC effectively. Currently, all miRNA-based drugs under clinical trials for cancer treatment are either antagomiRs or miRNA mimics.

Besides antagomiRs and miRNA mimics, small molecules that affect miRNA activity are undergoing clinical trials, albeit not for cancer treatment. Small molecules are an attractive option because they can be delivered to the tumor more easily than RNA drugs. Small molecules may either increase the levels of an endogenous tumor-suppressor miRNA or block the biogenesis and function of an oncomiR. An example of the former is ABX464, which specifically increases the levels of miR-124, an antiinflammatory and antiviral miRNA (Vautrin et al., 2019). As of December 2020, ABX464 is under phase 2 clinical trials for the treatment of inflammatory bowel disease (ClinicalTrials.gov Identifiers: NCT03368118, NCT03760003, NCT03905109, and NCT04023396), rheumatoid arthritis (NCT04049448), and HIV (NCT02990325), and under phase 2/3 clinical trials for treating complications of COVID-19 (NCT04393038). On the other hand, small molecule inhibitors of oncomiRs are being extensively researched at the preclinical level (Monroig, Chen, Zhang, & Calin, 2015). In most cases, they act at a posttranscriptional level by binding to a precursor of the oncomiR and preventing its processing. A recent example is Targapremir-210, which selectively binds the Dicer processing site of miR-210, an oncomiR involved in the response to hypoxia (Costales et al., 2017). Targapremir-210 has an antitumoral effect in mouse xenograft models of hypoxic triple negative breast cancer (Costales et al., 2017).

Other miRNA-based therapeutic approaches are being studied at a preclinical stage. Overall, to our knowledge, none of these approaches has progressed beyond preclinical studies, but they may prove to have a promising future:

- **MicroRNA sponges**: RNA molecules that contain multiple binding sites for one or several oncomiRs, blocking their function (Miroshnichenko & Patutina, 2019; Tay, Lim, Zhu, Hin, & Wang, 2015).
- **MicroRNA masks**: Small RNA molecules that are complementary to the binding sites of an miRNA, so that they prevent the miRNA from binding to those sites (Miroshnichenko & Patutina, 2019).
- **Small RNA zippers**: Small RNA molecules that contain two binding sites for a target oncomiR, one for each half of the target (Meng et al., 2017). As a result, in tumor cells, oncomiR and zipper molecules bind to each other end to end, forming long multimers that trap the oncomiR molecules.
- **CRISPR/Cas strategies**: CRISPR/Cas9 mutagenesis can be used to suppress oncomiRs, usually by mutating the precursor genes in essential regions for miRNA biogenesis (Aquino-Jarquin, 2017). Although CRISPR/Cas systems have been extensively used in basic miRNA research, they have not yet progressed toward clinical studies because of concerns about their off-target effects and the in vivo safety and efficacy of the delivery methods (Aquino-Jarquin, 2017).
- **Artificial ribonucleases**: Synthetic RNA-cleaving peptides covalently linked to RNA molecules that bind to a target oncomiR (Miroshnichenko & Patutina, 2019; Patutina et al., 2017). In this way, Patutina et al. designed an "miRNAse" that specifically suppressed miR-21 in lymphosarcoma cells (Patutina et al., 2017).

MicroRNA-based drugs may be more effective if administered along with other drugs than if administered alone (reviewed in (Miroshnichenko & Patutina, 2019)). One possibility involves mixing multiple antagomiRs or miRNA mimics into a single cocktail. Alternatively, miRNA-based drugs can be combined with standard chemotherapeutic agents. In theory, when multiple drugs are combined, they may have synergies; for example, certain miRNA-based drugs could increase the sensitivity of tumor cells to chemotherapy. However, in practice, the combined effects are often less than additive, and synergies are rare (Miroshnichenko & Patutina, 2019). Therefore, future research should explore further combinations of drugs to identify clinically useful synergies.

Chemical modifications of RNA-based drugs

Most miRNA-based drugs are RNA molecules. However, RNA is readily degraded by ribonucleases because its ribose units have free 2′-OH chemical groups that hydrolyze the phosphodiester bonds. Therefore, replacing the 2′-OH or the phosphate groups may increase RNA stability. As an added benefit, such modified RNA drugs often bind to their targets more effectively than unmodified RNA drugs. RNA drugs can also be modified to improve their delivery. Several chemical modifications of RNA drugs have been proposed (Adams, Parsons, Walker, Zhang, & Slack, 2017; Segal & Slack, 2020):

- **Substituting the 2′-OH by other chemical groups**, such as 2′-*O*-methyl, 2′-*O*-(2-methoxyethyl), or 2-fluoro. Another option is using DNA nucleotides whose deoxyribose backbone lacks the 2′-OH groups.
- **Locked nucleic acids (LNAs)**. Ribose units in LNAs contain a 2′-O,4′-C methylene bridge that prevents the RNA from being degraded by ribonucleases and greatly increases the binding affinity of the RNA to its targets.
- **Modifying the phosphodiester bonds** may also protect the RNA from nucleases. The most common modification is the phosphorothioate bond.
- **Peptide nucleic acids (PNAs)** have a peptide-like backbone that is highly stable and has a neutral net charge, which facilitates delivery into cells.
- **Conjugation to other biomolecules**. RNA drugs are often conjugated to other molecules, such as aptamers, that improve their delivery. Delivery of RNA drugs is further discussed in the next section.

To increase their efficacy, RNA drugs often combine different chemical modifications. For example, Cobomarsen (MRG-106, PubChem ID: 145994613) is an anti-miR-155 whose nucleotides are either LNAs (9 of 14 nucleotides) or deoxyribonucleotides (5 of 14 nucleotides). In addition, all nucleotides are linked by phosphorothioate bonds. Cobomarsen is under clinical trials for the treatment of various lymphoid malignancies (Seto et al., 2018).

Optimizing the chemical modifications of miRNA mimics is especially challenging because the modified mimics must be recognized by the endogenous RISC. For this reason, miRNA mimics are often double-stranded and bear most modifications on their passenger strand, which is not loaded into RISC (Adams et al., 2017). To protect miRNA mimics from degradation, 2′-*O*-methyl modifications are often enough, and further modifications such as LNAs may interfere with the function of the mimic (Adams et al., 2017).

Delivery of RNA-based drugs

Much like any other gene therapy, miRNA-based drugs face one serious challenge for their application in clinics, which is finding a delivery system that ensures the integrity of the easily degradable miRNAs as well as the specificity to target tissue or cells (Table 12.1). Traditionally, delivery systems for gene therapy are classified as viral and nonviral vectors. Although viral vectors (lentivirus, adenovirus, adeno-associated virus) display a higher delivery efficiency overall, the limited packaging capacity of viruses and their side effects, such as immunogenicity and mutagenicity, limit their in vivo applications. In contrast, nonviral vectors such as liposomes, polymers, and nanoparticles are less immunogenic and display more packaging capacity than viral vectors, but they lack the transfection efficiency of viruses (McErlean, McCrudden, & McCarthy, 2016).

MiRNA-based drugs can be administered either locally or systemically. Local delivery such as topical administration or intratumoral injection shows better bioavailability and reduced toxicity. However, this is only applicable to certain subtypes of tumors where the tumor mass is readily accessible, such as melanoma, breast, or cervical cancer. Systemic delivery via blood circulation allows the miRNA-based drugs to reach more inaccessible tumor sites, at the cost of increased toxicity, miRNA degradation, and difficulties to specifically reach the target tissue (Chen et al., 2015).

Chemical modifications

As previously explained, chemical modifications of miRNAs increase their stability and may also help with delivery. For example, conjugating the miRNAs with *N*-acetylgalactosamine (GalNAc) greatly improves hepatic delivery thanks to the overexpression of GalNAc receptors in hepatocytes. In addition, miRNAs can also be bound to aptamers, short peptides or oligonucleotides that recognize and bind to other tissue-specific molecules, allowing targeted delivery to a concrete cell type (Esposito, Catuogno, & de Franciscis, 2016).

Liposomes, micelles, and exosomes

Due to the lipophilic nature of cell membranes, encapsulating miRNA-based drugs into a single (micelle) or double (liposome and exosome) layer of lipids protects them from degradation and facilitates cellular uptake. Exosomes are small vesicles released by several cell types, and contrary to artificially made liposomes, they have a complex protein composition in their membranes, which mimics the structure of cell membranes better, thus making them more biocompatible (Antimisiaris, Mourtas, & Marazioti, 2018).

Liposomal phospholipids can be functionalized with other chemical groups for specific delivery to certain tissues. In parallel, exosome engineering allows the customization of the exosome load and the membrane elements, which can be modified at will by properly choosing the exosome-producing cell (e.g., brain endothelial cells produce exosomes that are able to bypass the blood-brain barrier), and by infecting the exosome-producing cells with viral particles containing the desired miRNA or surface membrane receptors (Gilligan & Dwyer, 2017).

Nanoparticles

Nanomaterials have opened multiple research horizons in terms of nanomedicine and drug delivery. Several types of nanoparticle (NP) constructs are available:

TABLE 12.1 Examples of successfully used local or systemic delivery systems for miRNA in vivo.

Delivery method	Examples	References
Cholesterol conjugation	miR-375	Chen, Gao, and Huang (2015)
Liposomes	miR-143	Akao et al. (2010)
Gold nanoparticles	miR-182	Kouri et al. (2015)
PEI-linked poly-arginine complex	miR-145	Zhang, Xue, He, and Hsieh (2015)
Lentiviral vector	let-7, miR-15a/16	Chen et al. (2015)
Adeno-associated viruses	miR-26a	Chen et al. (2015)
EDV nanocells	miR-16 (TargomiRs)	van Zandwijk et al. (2017)

- *Inorganic nanomaterials*: They are easy to functionalize, nontoxic, and biocompatible. Gold NPs stand out for their capacity to bind directly to nucleic acids, and they have been used to deliver miRNAs in brain, breast, and prostate cancer (Lee et al., 2019). Other inorganic nanomaterials are graphene-derived materials (Assali et al., 2018), quantum dots (Liang et al., 2016), and iron magnetic NPs (Lee et al., 2019).
- *Carbon nanotubes (CNTs)*: The inertness and small size of CNTs make them potential tools for efficient in vivo delivery (Masotti et al., 2016). However, they must be first functionalized to increase their water solubility.
- *Peptidic NPs*: Proteins are fully biocompatible and biodegradable, so their use for nucleic acid delivery is attractive. Several proteins have been studied as skeletons for NP delivery, such as gelatin, elastin, spider silk, albumin, zein (a corn protein), and poly-aspartic acid (Riley & Vermerris, 2017).
- *Lipid-based NPs*: Lipids are biocompatible and can be conjugated with other functionalizing groups to create lipidoid nanoparticles (LNPs). LNPs have been successfully used to deliver siRNAs against an antiapoptotic protein in non-Hodgkin's lymphoma cell lines (Knapp, He, Lister, & Whitehead, 2016).
- *Polymer-based NPs*: Polymers are formed by linear or branched repetitions of monomers, and they have different properties depending on their chemical composition. Polymers are broadly used as coating agents for other nanocomplexes, as this coating functionalizes the complex, protects it from degradation, and helps it escape the endosome by acting as a "proton sponge": polymeric matrices capture protons, which increase osmotic pressure inside the vesicles, making them burst and liberating their content to the cytoplasm. Fairly used polymers for designing NPs are polyethylene glycol (PEG), a nontoxic, biodegradable polymer, and polyethyleneimine (PEI), which acts as a proton sponge allowing endosome escape (Riley & Vermerris, 2017).

Viral vectors

Viruses naturally inject their genetic material into host cells to replicate themselves and further spread. This fact can be exploited to our benefit by packaging the miRNAs into a viral capsid and infecting target cells with them. There are several types of viral vectors (Herrera-Carrillo, Liu, & Berkhout, 2017):

- *Lentiviruses (LVs)* are retroviruses derived from HIV with several modifications so that they cannot produce de novo infectious viral particles, making them safer. LVs allow the packaging of relatively large DNA sequences (up to ~8 kb), they are able to infect nondividing cells, and they integrate their sequences into random locations at the target cell's genome in order to achieve stable integration and expression.
- *Adenoviruses (AVs)* are naked viruses with double-stranded DNA (dsDNA). Unlike LVs, they do not integrate into the genome of host cells. Therefore, they are safer than LVs and they are preferred in terms of in vivo applications. In addition, AVs can be directed to target specific tissues or cell types.
- *Adeno-associated viruses (AAVs)* are single-stranded DNA (ssDNA) viruses dependent on coinfection with other viruses. They are replication-deficient, so they are safer, and because of their ssDNA they have an improved efficiency over AVs, being the preferred option for in vivo clinical trials. Due to their small size, AAVs have a limited packaging capacity (~4 kb), but this should not be a problem for miRNA-based therapy.

Bacteria-based vectors (EDV nanocells)

EDV nanocells (technology developed by EnGeneIC, https://engeneic.com/edv-technology/) are formed when genetically modified bacteria divide at one of their poles at the same time as centrally. The EDV nanocells are formed by the smaller portions obtained after the division, which do not contain a bacterial chromosome. Conventional or nucleic acid-based drugs can then be loaded into the nanocells, and specific antibodies can also be attached to the surface of the EDV nanocells for tissue-specific delivery (MacDiarmid & Brahmbhatt, 2011).

MicroRNA-based therapeutics in the current clinic

Historically, research on small RNA-based therapies has focused on RNA interference (RNAi). RNAi silences target mRNAs using exogenous small RNAs, and thus it shares many similarities with miRNA-based therapies. While RNAi therapies attracted the attention of some pharmaceutical companies shortly after the discovery of RNAi in the late 1990s and the first successful preclinical trials in the early 2000s, miRNA-based therapies lagged behind by almost a decade (Eisenstein, 2019). Therefore, the journey of RNAi therapeutics and the lessons learned from them can also be informative for miRNA-based therapeutics.

TABLE 12.2 MicroRNA-based drugs against cancer in clinical or preclinical studies.

Drug (company)	miRNA (antagomiR/ miRNA mimic)	Disease	Stage	Source
MRX34 (Mirna Therapeutics)	miR-34a (miRNA mimic)	Advanced solid tumors	Phase I (terminated)	ClinicalTrials. gov Identifier: NCT0182997
Cobomarsen (miRagen Therapeutics)	miR-155 (antagomiR)	Cutaneous T-cell lymphoma/mycosis fungoides	Phase II (terminated early due to "business reasons")	ClinicalTrials. gov Identifier: NCT03837457
TargomiRs (Asbestos Diseases Research Foundation & EnGeneIC Limited)	miR-16a (miRNA mimic)	Malignant pleural mesothelioma/nonsmall cell lung cancer	Phase I	ClinicalTrials. gov Identifier: NCT02369198
RGLS5579 (Regulus)	miR-10b (antagomir)	Glioblastoma multiforme	Preclinical	Regulus Therapeutics Inc. (http://regulusrx.com)
INT-1B3 (InteRNA)	miR-193a-3p (miRNA mimic)	Triple negative breast cancer	Preclinical	InteRNA Technologies B.V. (https://interna-technologies.com/)

Although RNAi therapies were initially met with great enthusiasm by pharmaceutical companies and investors, this initial enthusiasm faded in the early 2010s (Eisenstein, 2019). In 2018, one of the few companies that pushed on, Alnylam Pharmaceuticals, obtained the first FDA approval of patisiran, an RNAi drug targeting the transthyretin (TTR) gene, for the treatment of TTR-mediated amyloidosis, a rare hereditary disease (Garber, 2018). Since then, one more RNAi drug, givosiran, has been approved in the United States and Europe (Second RNAi drug approved, 2020). The approval of patisiran and givosiran has paved the way for future approval of RNA-based therapeutics, including miRNA-based drugs. Pharmaceutical companies are investing again in RNA-based therapeutics (Eisenstein, 2019) and several miRNA-based drugs have entered clinical trials in recent years.

The first successful clinical trial for an miRNA-based drug started in the late 2000s and early 2010s (Janssen et al., 2013). The drug, Miravirsen, is a miR-122 antagonist that is still under development for the treatment of hepatitis C. Since then, several other miRNA-based drugs have entered clinical trials with varying degrees of success. Although we will focus on cancer treatment, miRNA-based drugs are also being developed for the treatment of other diseases, such as hepatitis C, irritable bowel disease, Alport syndrome, nonalcoholic steatohepatitis, fibrosis, and amyotrophic lateral sclerosis (Bonneau et al., 2019). As of December 2020, no miRNA-based anticancer drug has progressed to phase 3 clinical trials (Table 12.2). Below we summarize the current progress of the most promising miRNA-based anticancer drugs.

MRX34

MRX34 was an miRNA-based drug that reached phase I clinical trials for patients with advanced solid tumors (ClinicalTrials.gov Identifier: NCT01829971). MRX34 is a double-stranded miRNA mimic of miR-34a, a tumor-suppressor miRNA that downregulates several oncogenes and immune evasion-related genes, such as MYC and PD-L1. The miR-34a mimic is encapsulated into an anionic liposome to form the MRX34 drug, which is then intravenously administered to the patients (Beg et al., 2017).

In this clinical trial, two administration schedules were tested. Initially, patients were administered MRX34 intravenously twice weekly (BIW schedule). After observing adverse effects upon administration, a new schedule was tested (QDx5) where MRX34 was infused for 2h for 5 consecutive days in 1 week, followed by 2 weeks off. The QDx5 schedule was supposed to increase tolerance to the treatment, but adverse effects were still observed, and the study had to be halted due to the deaths of four patients because of immune-related toxicities. Overall, MRX34 had a 4% response rate that did not outweigh the risk of serious immune-related adverse effects (Hong et al., 2020).

Cobomarsen

Cobomarsen (MRG-106) is an oligonucleotide inhibitor of miR-155 especially designed for delivery to CD4+ T cells and mycosis fungoides (MF) cells, a subtype of cutaneous T-cell lymphoma (CTCL) (Seto et al., 2018). The oncomiR miR-155 was the first miRNA found to be overexpressed in cancer (Eis et al., 2005), and it exerts its function by downregulating important tumor-suppressor genes and DNA-repair factors, such as TP53INP and MLH1/MSH6, respectively (Witten & Slack, 2020). Under normal physiological conditions, miR-155 is expressed in the hematopoietic system and it has a role in immune response and inflammation. Thus, miR-155 is mainly found altered in hematological malignancies, such as acute myeloid leukemia (AML), acute and chronic lymphocytic leukemia, diffuse large B-cell lymphoma (DLBCL), and mantle cell lymphoma (Witten & Slack, 2020).

As of December 2020, the two current clinical studies (ClinicalTrials.gov Identifiers: NCT03713320 & NCT03837457) are in phase II. However, both have been terminated early "for business reasons," and no results are available. In these studies, recruited patients with CTCL were treated with cobomarsen and its efficacy was compared to those treated with vorinostat, a drug approved for the treatment of CTCL in several countries around the world. Over the time of the study, the patients were examined for the severity of their skin lesions, the presence of any disease-associated symptoms, the time that the disease stayed stable or improved, and the overall quality of life.

TargomiRs

TargomiRs are miR-16a mimics encapsulated into EDV minicells engineered to target EGFR-expressing tumor cells. Both miR-15 and miR-16 compose a family of tumor-suppressor miRNAs targeting the oncogene BCL2. The miR-15/16 family is downregulated in malignant pleural mesothelioma (MPM), an aggressive tumor type caused by the inhalation of asbestos, and restoration of miR-16 in MPM cell lines impaired tumor growth both in vitro and in vivo (Reid et al., 2013).

A targomiRs clinical study (ClinicalTrials.gov Identifier: NCT02369198) aimed to treat 26 patients suffering from MPM or nonsmall cell lung cancer (NSCLC) using escalating doses of targomiRs. After the maximum tolerated dose was defined, mild secondary effects of targomiR treatment were observed in the majority of patients, mainly transient lymphopenia and hypophosphatemia. Also, 5 of the 26 patients experienced different cardiac events. Out of the 22 patients who were assessed for a response by computerized tomography (CT), only one of them showed a response to the treatment (5% response rate) (van Zandwijk et al., 2017). The study concluded that patients show an acceptable tolerance to targomiR therapy, as lymphopenia and hypophosphatemia are easily correctable symptoms after EDV perfusion. Further studies in a larger cohort, including biopsies to check for the inhibition of miR-16a targets and combination with other therapies, will be necessary to bring targomiRs to the clinic (van Zandwijk et al., 2017).

Remlarsen

Remlarsen (MRG-201) is a mimic of miR-29 that has reached phase II clinical trials for the treatment of keloidal scars, as miR-29 downregulates collagen and other molecules involved in scar formation (ClinicalTrials.gov Identifier: NCT03601052). Although no results of this study have been confirmed yet, because the miR-29 family has been reported as a tumor-suppressor miRNA downregulated in several human cancers (Qi et al., 2017), Miroshnichenko and Patutina (2019) hypothesized that remlarsen could be also applied in cancer therapy in future clinical studies if the clinical study for keloidal scars is successful (Miroshnichenko & Patutina, 2019).

Other preclinical studies

Several miRNA-based drugs are currently in preclinical trials for the treatment of multiple diseases such as hepatitis C, amyotrophic lateral sclerosis, and inflammatory bowel disease. In the case of cancer, in addition to the aforementioned examples of miRNA-based drugs in clinical trials, two companies (Regulus and InteRNA) have miRNA-based drugs in development in their preclinical pipelines: RGLS5579 for glioblastoma multiforme, and INT-1B3 for triple negative breast cancer, respectively (Bonneau et al., 2019; Table 12.2).

The future of miRNAs in the clinic

Current challenges in the use of miRNAs

MicroRNA-based drugs face two great difficulties before their clinical application:

- *Delivery*: RNA molecules are biologically fragile and easily degraded. They must remain intact until they reach their action site inside the cytoplasm of the tumor cells they are targeting. Besides, specific delivery to the tumor site is essential for targeted miRNA-based gene therapies.
- *Efficacy and specificity*. MicroRNA-based drugs, as with any other antineoplastic drug, must be able to impede tumor cell proliferation and they must be safe, including watching out for off-target effects and unwanted nontumor cell death.

Thus, the current challenges for the administration of miRNA-based drugs in clinics are:

Quick degradation in the bloodstream

Nucleic acids that enter systemic circulation are quickly degraded and cleared out before they can even reach their target tissues. There are two main causes for their quick disappearance from blood circulation. First, the degradation by A-type RNase enzymes that are found in the serum, which unspecifically recognize the naked 2′-OH ribose ends in miRNAs and degrade them (Segal & Slack, 2020) (Fig. 12.1A). Secondly, particles smaller than 200 nm or 5000 Da in molecular weight are filtered at the kidney and thus, their half-life in systemic circulation is very short. Encapsulating miRNAs into lipidic or polymeric particles protects them from RNase degradation and renal filtering. However, particles larger than the renal threshold tend to accumulate in the liver, lungs, or spleen, where they are phagocytized and degraded by the mononuclear phagocytic system (McErlean, McCrudden, & McCarthy, 2016).

Extrahepatic delivery to tumor site

Hepatic delivery is easy to perform as miRNA-based drugs tend to naturally accumulate in the liver, and GalNAc modification enhances liver uptake (Fig. 12.1B). However, in any extrahepatic cancer, miRNA-based drugs need to arrive at the specific tumor site.

Tumor cells tend to promote the formation of new blood vessels (angiogenesis) to provide themselves with nutrients and oxygen. Besides, they generate a tumor microenvironment with special characteristics, such as an acidic pH, hypoxia, or the presence of cancer-associated biomolecules. This, however, can also be exploited to our benefit. Because the newly formed blood vessels are leakier than the normal ones, a prolonged circulation of miRNA-associated particles would increase their extravasation and accumulation preferentially into the tumor site. This phenomenon is called the enhanced permeability and retention (EPR) effect. The EPR effect has been observed in many cancer drug delivery studies in vivo. However, its translation to humans is controversial and yet to be demonstrated (Nichols & Bae, 2014).

Furthermore, it could be possible to enhance reaching the tumor-site by encapsulating miRNAs in pH- or tumor matrix metalloproteinase-sensitive particles. Thus, thanks to the special tumor microenvironment conditions, the miRNAs could be preferentially released into the tumor site (Liu et al., 2015).

Limited tissue penetration

Delivering miRNA-based drugs to certain target tissues may pose additional difficulties. The most notable case is brain-targeted miRNAs, which must bypass the blood-brain barrier (BBB) (Fig. 12.1C). The BBB prevents 95% of molecules in blood from entering the central nervous system (CNS), thus hindering any drug therapy of CNS tumors, mainly gliomas. Scientists have researched ways of improving drug delivery to the brain for more than 30 years, but no clear solution has been devised yet: receptor/transporter-mediated transcytosis studies show conflicting results, and strategies based on temporarily opening the BBB prior to drug treatment to enhance delivery need further research on BBB leakage and permeability (Dong, 2018).

Endocytosis and endosome entrapment

Due to the anionic nature of naked nucleic acids such as miRNAs, direct assimilation by the anionic bilayer of lipids that conforms the nuclear membrane is not possible. Instead, miRNAs must be encapsulated into cationic vehicles (i.e., poly-L-lysine) that allow interaction with the cell membrane and their internalization by endocytosis (McErlean, McCrudden, & McCarthy, 2016). Another option is coupling cell-penetrating proteins (CPPs) to the miRNA vector. CPPs form a pore in the surface of the cell membrane, allowing the miRNA vector to enter the cell directly without the need for a specific receptor (Tripathi, Arami, Banga, Gupta, & Gandhi, 2018).

Furthermore, upon entering the cells, cellular uptake is enclosed within endosomes, which can be then returned to the membrane, or fused with the lysosomes and degraded (Fig. 12.1D). In order to reach their goal at the cytoplasm, miRNA-based drugs must escape from the endosomes before being degraded. This can be achieved by coupling the delivery vehicle with cationic proteins that facilitate the endosome escape. For example, the RALA peptide is an amphipathic protein whose

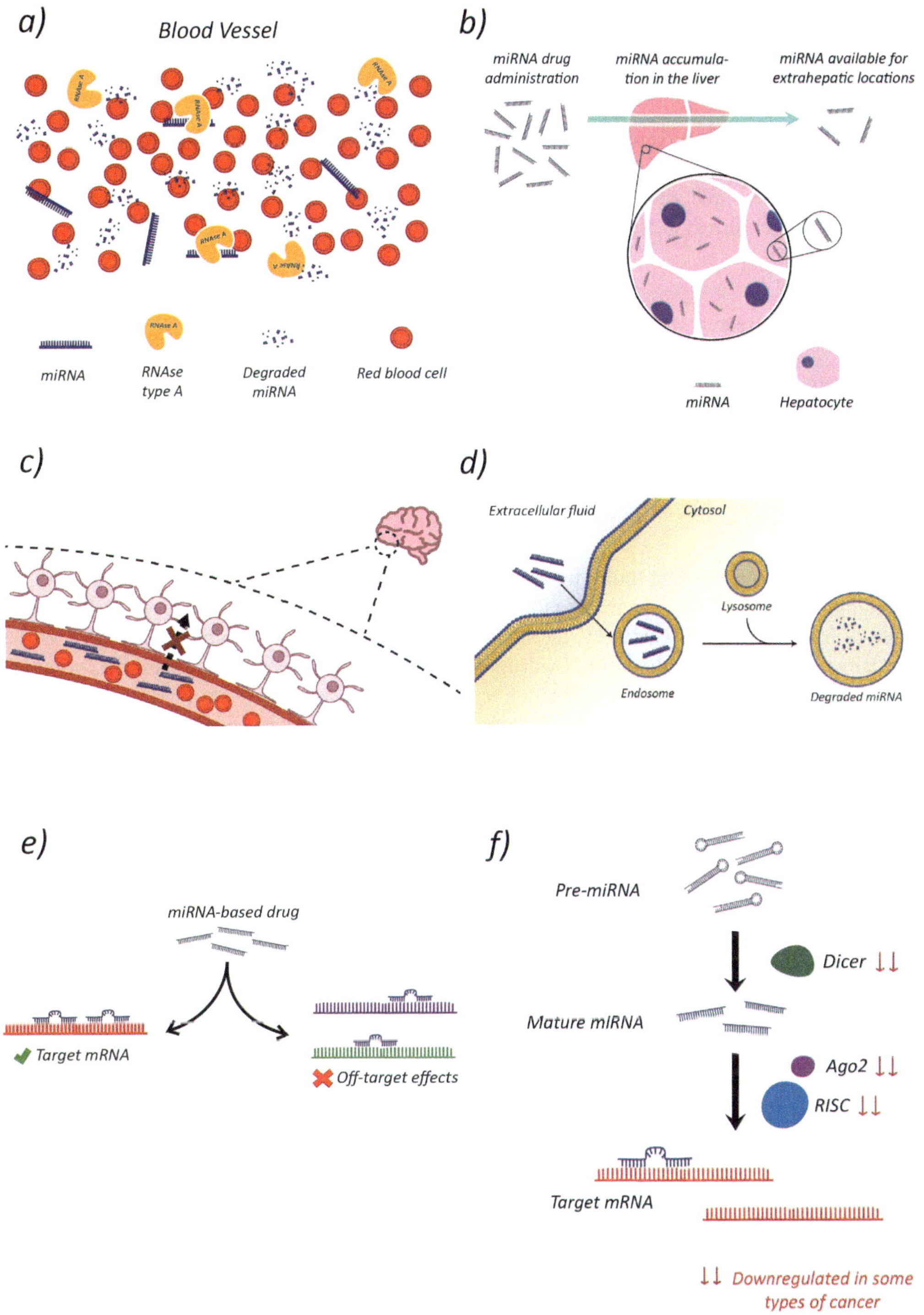

FIG. 12.1 Current challenges in the clinical use of miRNA-based drugs. Quick degradation by blood nucleases (A), extrahepatic delivery (B), limited tissue penetration (C), endocytosis and endosome entrapment (D), off-target effects (E), and dependency on functional miRNA-processing machinery (F).

positively charged arginine residues interact with the anionic lipids in the endosome membrane, generating pores that disrupt the membrane and allow the leakage from the interior of the endosome (McCarthy et al., 2014).

Off-targets and side effects

Although miRNA regulation is very specific, off-target binding and silencing of unwanted genes might cause severe side effects (Fig. 12.1E). In order to avoid that, it is proposed to administer low doses of a combination of miRNA-based drugs targeting the same gene. For example, cotransfection of the miRNAs miR-34a and miR-15a/16, which both target CCNE1 and CCND3, induced cell cycle arrest in NSCLC (Bandi & Vassella, 2011).

Other challenges

Other challenges still under research arise when trying to apply miRNA-based drugs to the clinics. For instance, we still lack a lot of information about the immune response to miRNA-based drugs that might pose a threat to some patients. In addition, miRNA therapy is dependent on properly-working intracellular miRNA processing machinery (DICER, Ago2, RISC). However, some of these effectors are downregulated in some subtypes of cancer (Fig. 12.1F). Using mature miRNAs or siRNAs skips these processing steps, but it requires more precise dosing control to avoid off-target effects (Chen et al., 2015).

Future perspectives

MicroRNA-based drugs hold a promising future for specific cancer treatment. The basic design of an miRNA-based drug is as simple as finding complementary sequences to the target RNA, and chemical modifications of RNA to increase their stability and efficacy are well studied. Currently, the main hurdle that slows the development of miRNA-based drugs is optimizing their delivery and safety. Although RNA drug delivery is being extensively researched, in many tumor types there are no convincing solutions for tumor-specific delivery. Still, the recent successes of RNAi drugs prove that these hurdles can be overcome and that RNA-based drugs can be clinically useful. Despite early failures, several clinical trials involving miRNA-based drugs are under way and it may be a matter of time before the first successful miRNA-based drugs are commercialized.

References

Adams, B. D., Parsons, C., Walker, L., Zhang, W. C., & Slack, F. J. (2017). Targeting noncoding RNAs in disease. The Journal of Clinical Investigation, 127, 761–771. https://doi.org/10.1172/jci84424.

Akao, Y., Nakagawa, Y., Hirata, I., Iio, A., Itoh, T., Kojima, K., et al. (2010). Role of anti-oncomirs miR-143 and -145 in human colorectal tumors. Cancer Gene Therapy, 17(6), 398–408. https://doi.org/10.1038/cgt.2009.88.

Antimisiaris, S. G., Mourtas, S., & Marazioti, A. (2018). Exosomes and exosome-inspired vesicles for targeted drug delivery. Pharmaceutics, 10. https://doi.org/10.3390/pharmaceutics10040218.

Aquino-Jarquin, G. (2017). Emerging role of CRISPR/Cas9 technology for microRNAs editing in cancer research. Cancer Research, 77, 6812–6817. https://doi.org/10.1158/0008-5472.can-17-2142.

Assali, A., Akhavan, O., Mottaghitalab, F., Adeli, M., Dinarvand, R., Razzazan, S., et al. (2018). Cationic graphene oxide nanoplatform mediates miR-101 delivery to promote apoptosis by regulating autophagy and stress. International Journal of Nanomedicine, 13, 5865–5886. https://doi.org/10.2147/IJN.S162647.

Bandi, N., & Vassella, E. (2011). miR-34a and miR-15a/16 are co-regulated in non-small cell lung cancer and control cell cycle progression in a synergistic and Rb-dependent manner. Molecular Cancer, 10, 55. https://doi.org/10.1186/1476-4598-10-55.

Bartel, D. P. (2009). MicroRNAs: Target recognition and regulatory functions. Cell, 136, 215–233. https://doi.org/10.1016/j.cell.2009.01.002.

Beg, M. S., Brenner, A. J., Sachdev, J., Borad, M., Kang, Y. K., Stoudemire, J., et al. (2017). Phase I study of MRX34, a liposomal miR-34a mimic, administered twice weekly in patients with advanced solid tumors. Investigational New Drugs, 35, 180–188. https://doi.org/10.1007/s10637-016-0407-y.

Bonneau, E., Neveu, B., Kostantin, E., Tsongalis, G. J., & De Guire, V. (2019). How close are miRNAs from clinical practice? A perspective on the diagnostic and therapeutic market. EJIFCC, 30, 114–127. https://www.ncbi.nlm.nih.gov/pmc/articles/PMC6599191/pdf/ejifcc-30-114.pdf.

Chen, Y., Gao, D. Y., & Huang, L. (2015). In vivo delivery of miRNAs for cancer therapy: Challenges and strategies. Advanced Drug Delivery Reviews, 81, 128–141. https://doi.org/10.1016/j.addr.2014.05.009.

Cheng, C. J., Bahal, R., Babar, I. A., Pincus, Z., Barrera, F., Liu, C., et al. (2015). MicroRNA silencing for cancer therapy targeted to the tumour microenvironment. Nature, 518, 107–110. https://doi.org/10.1038/nature13905.

Coira, I. F., Rufino-Palomares, E. E., Romero, O. A., Peinado, P., Metheetrairut, C., Boyero-Corral, L., et al. (2015). Expression inactivation of SMARCA4 by microRNAs in lung tumors. Human Molecular Genetics, 24(5), 1400–1409. https://doi.org/10.1093/hmg/ddu554.

Costales, M. G., Haga, C. L., Velagapudi, S. P., Childs-Disney, J. L., Phinney, D. G., & Disney, M. D. (2017). Small molecule inhibition of microRNA-210 reprograms an oncogenic hypoxic circuit. *Journal of the American Chemical Society*, *139*(9), 3446–3455. https://doi.org/10.1021/jacs.6b11273.

Dong, X. (2018). Current strategies for brain drug delivery. *Theranostics*, *8*(6), 1481–1493. https://doi.org/10.7150/thno.21254.

Eis, P. S., Tam, W., Sun, L., Chadburn, A., Li, Z., Gomez, M. F., … Dahlberg, J. E. (2005). Accumulation of miR-155 and BIC RNA in human B cell lymphomas. *Proceedings of the National Academy of Sciences of the United States of America*, *102*(10), 3627–3632. https://doi.org/10.1073/pnas.0500613102.

Eisenstein, M. (2019). Pharma's roller-coaster relationship with RNA therapies. *Nature*, *574*, S4–S6. https://doi.org/10.1038/d41586-019-03069-3.

Esposito, C. L., Catuogno, S., & de Franciscis, V. (2016). Aptamer-MiRNA conjugates for cancer cell-targeted delivery. *Methods in Molecular Biology*, *1364*, 197–208. https://doi.org/10.1007/978-1-4939-3112-5_16.

Garber, K. (2018). Alnylam launches era of RNAi drugs. *Nature Biotechnology*, *36*(9), 777–778. https://doi.org/10.1038/nbt0918-777.

Garzon, R., Marcucci, G., & Croce, C. M. (2010). Targeting microRNAs in cancer: Rationale, strategies and challenges. *Nature Reviews Drug Discovery*, *9*(10), 775–789. https://doi.org/10.1038/nrd3179.

Gebert, L. F. R., & MacRae, I. J. (2019). Regulation of microRNA function in animals. *Nature Reviews Molecular Cell Biology*, *20*(1), 21–37. https://doi.org/10.1038/s41580-018-0045-7.

Gilles, M.-E., & Slack, F. J. (2018). Let-7 microRNA as a potential therapeutic target with implications for immunotherapy. *Expert Opinion on Therapeutic Targets*, *22*(11), 929–939. https://doi.org/10.1080/14728222.2018.1535594.

Gilligan, K. E., & Dwyer, R. M. (2017). Engineering exosomes for cancer therapy. *International Journal of Molecular Sciences*, *18*(6). https://doi.org/10.3390/ijms18061122.

Herrera-Carrillo, E., Liu, Y. P., & Berkhout, B. (2017). Improving miRNA delivery by optimizing miRNA expression cassettes in diverse virus vectors. *Human Gene Therapy Methods*, *28*(4), 177–190. https://doi.org/10.1089/hgtb.2017.036.

Hong, D. S., Kang, Y.-K., Borad, M., Sachdev, J., Ejadi, S., Lim, H. Y., … Beg, M. S. (2020). Phase 1 study of MRX34, a liposomal miR-34a mimic, in patients with advanced solid tumours. *British Journal of Cancer*, *122*(11), 1630–1637. https://doi.org/10.1038/s41416-020-0802-1.

Janssen, H. L. A., Reesink, H. W., Lawitz, E. J., Zeuzem, S., Rodriguez-Torres, M., Patel, K., … Hodges, M. R. (2013). Treatment of HCV infection by targeting microRNA. *The New England Journal of Medicine*, *368*(18), 1685–1694. https://doi.org/10.1056/NEJMoa1209026.

Knapp, C. M., He, J., Lister, J., & Whitehead, K. A. (2016). Lipidoid nanoparticle mediated silencing of Mcl-1 induces apoptosis in mantle cell lymphoma. *Experimental Biology and Medicine (Maywood)*, *241*(9), 1007–1013. https://doi.org/10.1177/1535370216640944.

Kouri, F. M., Hurley, L. A., Daniel, W. L., Day, E. S., Hua, Y., Hao, L., et al. (2015). miR-182 integrates apoptosis, growth, and differentiation programs in glioblastoma. Genes & Development, 29(7), 732–745. https://doi.org/10.1101/gad.257394.114.

Lee, S. W. L., Paoletti, C., Campisi, M., Osaki, T., Adriani, G., Kamm, R. D., … Chiono, V. (2019). MicroRNA delivery through nanoparticles. *Journal of Controlled Release*, *313*, 80–95. https://doi.org/10.1016/j.jconrel.2019.10.007.

Liang, G., Li, Y., Feng, W., Wang, X., Jing, A., Li, J., & Ma, K. (2016). Polyethyleneimine-coated quantum dots for miRNA delivery and its enhanced suppression in HepG2 cells. *International Journal of Nanomedicine*, *11*, 6079–6088. https://doi.org/10.2147/IJN.S120828.

Liu, Y., Zhang, D., Qiao, Z.-Y., Qi, G.-B., Liang, X.-J., Chen, X.-G., & Wang, H. (2015). A peptide-network weaved nanoplatform with tumor microenvironment responsiveness and deep tissue penetration capability for cancer therapy. *Advanced Materials*, *27*(34), 5034–5042. https://doi.org/10.1002/adma.201501502.

Lu, J., Getz, G., Miska, E. A., Alvarez-Saavedra, E., Lamb, J., Peck, D., … Golub, T. R. (2005). MicroRNA expression profiles classify human cancers. *Nature*, *435*(7043), 834–838. https://doi.org/10.1038/nature03702.

MacDiarmid, J. A., & Brahmbhatt, H. (2011). Minicells: Versatile vectors for targeted drug or si/shRNA cancer therapy. *Current Opinion in Biotechnology*, *22*(6), 909–916. https://doi.org/10.1016/j.copbio.2011.04.008.

Masotti, A., Miller, M. R., Celluzzi, A., Rose, L., Micciulla, F., Hadoke, P. W. F., … Caporali, A. (2016). Regulation of angiogenesis through the efficient delivery of microRNAs into endothelial cells using polyamine-coated carbon nanotubes. *Nanomedicine*, *12*(6), 1511–1522. https://doi.org/10.1016/j.nano.2016.02.017.

McCarthy, H. O., McCaffrey, J., McCrudden, C. M., Zholobenko, A., Ali, A. A., McBride, J. W., … Robson, T. (2014). Development and characterization of self-assembling nanoparticles using a bio-inspired amphipathic peptide for gene delivery. *Journal of Controlled Release*, *189*, 141–149. https://doi.org/10.1016/j.jconrel.2014.06.048.

McErlean, E. M., McCrudden, C. M., & McCarthy, H. O. (2016). Delivery of nucleic acids for cancer gene therapy: Overcoming extra- and intra-cellular barriers. *Therapeutic Delivery*, *7*(9), 619–637. https://doi.org/10.4155/tde-2016-0049.

Medina, P. P., Nolde, M., & Slack, F. J. (2010). OncomiR addiction in an in vivo model of microRNA-21-induced pre-B-cell lymphoma. *Nature*, *467*(7311), 86–90. https://doi.org/10.1038/nature09284.

Medina, P. P., & Slack, F. J. (2008). MicroRNAs and cancer: An overview. *Cell Cycle*, *7*(16), 2485–2492. https://doi.org/10.4161/cc.7.16.6453.

Meng, L., Liu, C., Lü, J., Zhao, Q., Deng, S., Wang, G., … Yu, Z. (2017). Small RNA zippers lock miRNA molecules and block miRNA function in mammalian cells. *Nature Communications*, *8*, 13964. https://doi.org/10.1038/ncomms13964.

Mestdagh, P., Boström, A.-K., Impens, F., Fredlund, E., Van Peer, G., De Antonellis, P., … Vandesompele, J. (2010). The miR-17-92 microRNA cluster regulates multiple components of the TGF-β pathway in neuroblastoma. *Molecular Cell*, *40*(5), 762–773. https://doi.org/10.1016/j.molcel.2010.11.038.

Miroshnichenko, S., & Patutina, O. (2019). Enhanced inhibition of tumorigenesis using combinations of miRNA-targeted therapeutics. *Frontiers in Pharmacology*, *10*, 488. https://doi.org/10.3389/fphar.2019.00488.

Monroig, P. D. C., Chen, L., Zhang, S., & Calin, G. A. (2015). Small molecule compounds targeting miRNAs for cancer therapy. *Advanced Drug Delivery Reviews*, *81*, 104–116. https://doi.org/10.1016/j.addr.2014.09.002.

Nichols, J. W., & Bae, Y. H. (2014). EPR: Evidence and fallacy. *Journal of Controlled Release*, *190*, 451–464. https://doi.org/10.1016/j.jconrel.2014.03.057.

Patutina, O. A., Bichenkova, E. V., Miroshnichenko, S. K., Mironova, N. L., Trivoluzzi, L. T., Burusco, K. K., … Zenkova, M. A. (2017). miRNases: Novel peptide-oligonucleotide bioconjugates that silence miR-21 in lymphosarcoma cells. *Biomaterials*, *122*, 163–178. https://doi.org/10.1016/j.biomaterials.2017.01.018.

Piletič, K., & Kunej, T. (2016). MicroRNA epigenetic signatures in human disease. *Archives of Toxicology*, *90*(10), 2405–2419. https://doi.org/10.1007/s00204-016-1815-7.

Qi, Y., Huang, Y., Pang, L., Gu, W., Wang, N., Hu, J., … Li, F. (2017). Prognostic value of the microRNA-29 family in multiple human cancers: A meta-analysis and systematic review. *Clinical and Experimental Pharmacology and Physiology*, *44*(4), 441–454. https://doi.org/10.1111/1440-1681.12726.

Reid, G., Pel, M. E., Kirschner, M. B., Cheng, Y. Y., Mugridge, N., Weiss, J., … van Zandwijk, N. (2013). Restoring expression of miR-16: A novel approach to therapy for malignant pleural mesothelioma. *Annals of Oncology*, *24*(12), 3128–3135. https://doi.org/10.1093/annonc/mdt412.

Rheinbay, E., Nielsen, M. M., Abascal, F., Wala, J. A., Shapira, O., Tiao, G., … PCAWG Consortium. (2020). Analyses of non-coding somatic drivers in 2,658 cancer whole genomes. *Nature*, *578*(7793), 102–111. https://doi.org/10.1038/s41586-020-1965-x.

Riley, M. K., & Vermerris, W. (2017). Recent advances in nanomaterials for gene delivery—A review. *Nanomaterials (Basel)*, *7*(5). https://doi.org/10.3390/nano7050094.

Rupaimoole, R., & Slack, F. J. (2017). MicroRNA therapeutics: Towards a new era for the management of cancer and other diseases. *Nature Reviews Drug Discovery*, *16*(3), 203–222. https://doi.org/10.1038/nrd.2016.246.

(2020). Second RNAi drug approved. *Nature Biotechnology*, *38*(4), 385. https://doi.org/10.1038/s41587-020-0494-3.

Segal, M., & Slack, F. J. (2020). Challenges identifying efficacious miRNA therapeutics for cancer. *Expert Opinion on Drug Discovery*, *15*(9), 987–992. https://doi.org/10.1080/17460441.2020.1765770.

Seto, A. G., Beatty, X., Lynch, J. M., Hermreck, M., Tetzlaff, M., Duvic, M., & Jackson, A. L. (2018). Cobomarsen, an oligonucleotide inhibitor of miR-155, co-ordinately regulates multiple survival pathways to reduce cellular proliferation and survival in cutaneous T-cell lymphoma. *British Journal of Haematology*, *183*(3), 428–444. https://doi.org/10.1111/bjh.15547.

Tay, F. C., Lim, J. K., Zhu, H., Hin, L. C., & Wang, S. (2015). Using artificial microRNA sponges to achieve microRNA loss-of-function in cancer cells. *Advanced Drug Delivery Reviews*, *81*, 117–127. https://doi.org/10.1016/j.addr.2014.05.010.

Trang, P., Medina, P. P., Wiggins, J. F., Ruffino, L., Kelnar, K., Omotola, M., … Slack, F. J. (2010). Regression of murine lung tumors by the let-7 microRNA. *Oncogene*, *29*(11), 1580–1587. https://doi.org/10.1038/onc.2009.445.

Tripathi, P. P., Arami, H., Banga, I., Gupta, J., & Gandhi, S. (2018). Cell penetrating peptides in preclinical and clinical cancer diagnosis and therapy. *Oncotarget*, *9*(98), 37252–37267. https://doi.org/10.18632/oncotarget.26442.

Uhlmann, S., Mannsperger, H., Zhang, J. D., Horvat, E.-Á., Schmidt, C., Küblbeck, M., … Sahin, O. (2012). Global microRNA level regulation of EGFR-driven cell-cycle protein network in breast cancer. *Molecular Systems Biology*, *8*, 570. https://doi.org/10.1038/msb.2011.100.

van Zandwijk, N., Pavlakis, N., Kao, S. C., Linton, A., Boyer, M. J., Clarke, S., et al. (2017). Safety and activity of microRNA-loaded minicells in patients with recurrent malignant pleural mesothelioma: a first-in-man, phase 1, open-label, dose-escalation study. Lancet Oncology, 18, 1386–1396. https://doi.org/10.1016/S1470-2045(17)30621-6.

Vautrin, A., Manchon, L., Garcel, A., Campos, N., Lapasset, L., Laaref, A. M., … Tazi, J. (2019). Both anti-inflammatory and antiviral properties of novel drug candidate ABX464 are mediated by modulation of RNA splicing. *Scientific Reports*, *9*(1), 792. https://doi.org/10.1038/s41598-018-37813-y.

Witten, L., & Slack, F. J. (2020). miR-155 as a novel clinical target for hematological malignancies. *Carcinogenesis*, *41*(1), 2–7. https://doi.org/10.1093/carcin/bgz183.

Zhang, T., Xue, X., He, D., & Hsieh, J. T. (2015). A prostate cancer-targeted polyarginine-disulfide linked PEI nanocarrier for delivery of microRNA. Cancer Letters, 365(2), 156–165. https://doi.org/10.1016/j.canlet.2015.05.003.

Section D

miRNAs involvement in human cancer: Pathophysiology and translational opportunities

Chapter 13

Pathophysiology roles and translational opportunities of miRNAs in acute leukemias

Francesca Lovat[a] and Carlo M. Croce[b]

[a]*Department of Cancer Biology and Genetics and Comprehensive Cancer Center, The Ohio State University, Columbus, OH, United States,* [b]*Wexner Medical Center, The Ohio State University, Columbus, OH, United States*

Introduction

Acute myeloid leukemia (AML) originates from the abnormal proliferation of clonal transformed immature hematopoietic cells, commonly in the bone marrow, following epigenetic and genetic changes in hematopoietic stem cells and progenitor cells (Döhner, Weisdorf, & Bloomfield, 2015; Papaemmanuil et al., 2016). This malignancy represents the most common type of acute leukemia in adult patients. In the majority of cases can occur as a de novo disease in healthy individuals or as a secondary event (De Kouchkovsky & Abdul-Hay, 2016; Wang, Hoshino, Redner, Kajigaya, & Liu, 1998). These myeloid diseases are caused by acquired well-characterized chromosomal translocations resulting in fusion genes that encode aberrant proteins with oncogenic capabilities. Some examples of frequently observed cytogenetic alterations include t(8;21)(q22;q22) [RUNX1/RUNX1T1] (De Kouchkovsky & Abdul-Hay, 2016; Wang et al., 1998), t(15;17)(q24;q21) [PML/RARA] (Melnick & Licht, 1999; Zelent, Guidez, Melnick, Waxman, & Licht, 2001), inv(16)(p13q22) [CBFB/MYH11] (Kogan et al., 1998; Liu et al., 1993) and t(9;11)(p22;q23) [MLLT3/MLL] (Bloomfield et al., 2002; De Braekeleer, Morel, Le Bris, Herry, & Douet-Guilbert, 2005; Ruault, Brun, Ventura, Roizès, & De Sario, 2002). In addition to chromosomal rearrangements, more subtle gene alterations have been involved in the development of hematopoietic malignancies. These somatically acquired genetic alterations are important for diagnostic and prognostic stratification and targeted therapeutic strategies. The most common recurrent genetic abnormalities in AML-involved genes are FLT3, NPM1, CEBPα, IDH1/IDH2, DNMT3A, RUNX1, WT1, and MLL (Bacher, Haferlach, Kern, Haferlach, & Schnittger, 2008; Barjesteh van Waalwijk van Doorn-Khosrovani et al., 2003; Colombo, Alcalay, & Pelicci, 2011; Figueroa et al., 2010; Gombart et al., 2002; Mardis et al., 2009; Schnittger et al., 2002; Thiede et al., 2006). These genes are linked to treatment response and AML progression and represent attractive targets for direct therapies. Another class of regulators of hematopoietic stem cell stemness or cell fate and lineage commitment is represented by microRNAs (miRNAs) (Ambros, 2004). Playing a role in the normal differentiation of hematopoietic cells, miRNA dysregulation may promote the development of hematopoietic malignancies such as AML. It has been well documented that miRNAs can act as oncomiRs or tumor-suppressors in different subtypes of AML, altering leukemic processes such as differentiation, survival, proliferation, stemness, DNA damage, and resistance to chemotherapy. Different mechanisms can promote miRNA dysregulation in AML development, such as miRNA overexpression or downregulation caused by translocations, deletions, epigenetic alteration, copy number variations, dysregulated aberrant fusion proteins or transcription factors and altered miRNA processing. In this chapter, we will focus our discussion on miRNA functional changes that occur in AML development, explaining their role as biomarkers and presenting potential applications as therapeutic targets.

MiRNAs involved in the pathogenesis of AML

The first descriptions of the important role of miRNAs in tumorigenesis and progression involved the miR-15/16 family. The deletion of the miR-15a/16-1 cluster located on chromosome 13q14 represents a hallmark of more than 70% of B-cell chronic lymphocytic leukemia (CLL) (Calin et al., 2002, 2005). Mouse models have been useful to demonstrate the role of the miR-15/16 cluster in the development of CLL. In fact, the miR-15a/16-1 knockout mouse model develops an indolent form of CLL late in life (Klein et al., 2010), similarly to human patients. Belonging to the miR-15/16 cluster

MicroRNA in Human Malignancies. https://doi.org/10.1016/B978-0-12-822287-4.00016-5

is also the miR-15b/16-2 located on chromosome 3. It has been shown that mice lacking this second miR-15b/16-2 locus develop mainly CLL and a few diffuse large B cell lymphomas (DLBCL) (Lovat et al., 2015). The combined loss of both miR-15a/16-1 and miR-15b/16-2 loci leads to the development of AML in 77% of analyzed double knockout mice and a much shorter lifespan than the CLLs of the miR-15a/16-1 and miR-15b/16-2 in single knockout mice (Lovat et al., 2018). The loss of miR-15/16 expression in AML is caused by a combination of genetic loss and silencing by epigenetic alteration such as methylation (Lovat et al., 2020). Previously, Garzon et al. (Garzon, Volinia, et al., 2008) reported that miR-15a as well as 13 other miRs, including miR-34b and miR-29a/b/c, were downregulated in patients with t(11q23) versus all other AML patients, correlating miRNA expression with genetic deletions. One of the most studied miRNA families in AML is the miR-29 cluster. The miR-29 family is composed of three members, miR-29a, miR-29b, and miR-29c. The cluster known as miR-29a/b-1 is located on chromosome 7q32.3, a region frequently deleted in myelodysplastic syndromes (MDSs) and therapy-related AML (Le Beau et al., 1996), and the miR-29b-2/-29c cluster is found on chromosome 1q32.2. MiR-29 has been shown to play a role as a tumor suppressor or an oncogene in myeloproliferative disorders. It has been described that the miR-29 family expression is downregulated in primary AML patients with t(6;11) and t(9;11) translocations (Garzon et al., 2009) and in primary cytogenetically normal (CN) AML patients with the wild-type nucleophosmin gene (NPM1) compared to patients with NPM1 mutations (Garzon, Volinia, et al., 2008). In particular, miR-29b is downregulated in AML patients with alteration of chromosome 7 (monosomy or deletion), with CEBPA mutations or with the loss of CEBPA due to t(8;21) or inv(16) chromosome alterations (Eyholzer, Schmid, Wilkens, Mueller, & Pabst, 2010), and in *KIT* mutated patients (Kogan et al., 1998; Liu et al., 2010). A lower level of miR-29a in core binding factor AML patients is due to a polymorphism in the miR-29b-1/miR-29a cluster (Ngankeu et al., 2018). The miR-29 family is known to regulate different mechanisms such as epigenetic regulators, proliferation, apoptosis, and hematopoietic differentiation. The downregulation of miR-29b results in the upregulation of its direct targets DNMT3A and DNMT3B and the indirect target DNMT1, promoting DNA hypermethylation in acute myeloid leukemia. These DNA methyltransferases (DNMTs) usually repress gene transcription through CpG island methylation, whereas the TET family activates DNA demethylation. In this scenario, the downregulation of miR-29 family, which normally target and repress TET expression, can promote DNA demethylation, increasing hematopoietic stem cell self-renewal and myeloid transformation. The miR-29 family is also involved in the regulation of cell proliferation and apoptosis by targeting CDK6 and CCND2 (Gong et al., 2014). In AML, miR-155 expression is upregulated by the activation of the oncogenic protein (FLT3-ITD) and by MLL rearrangements through HoxA9 and Meis1 (Gerloff et al., 2015; Schneider et al., 2016) or methylation in pediatric AML (Narayan et al., 2017). Several studies reported that miR-155 expression is upregulated in adult and pediatric AML (Danen-van Oorschot et al., 2012; Garzon, Garofalo, et al., 2008). However, the ectopic overexpression of miR-155 in AML cell lines induced cell cycle arrest, apoptosis, and myelomonocytic differentiation (Forrest et al., 2010; Palma et al., 2014). Narayan et al. (Narayan, Bracken, & Ekert, 2018) speculated about this dual role of miR-155 in AML based on its expression and its targeting and consequently downstream gene regulation. In fact, different miR-155 expression levels, called "high" or "intermediate," regulate different genes. For example: genes associated with loss of function in AML and myeloid differentiation such as PU.1 (Gerloff et al., 2015), C/EBPbeta (Alemdehy et al., 2016), and Tle2 (Narayan et al., 2017), are regulated by "intermediate" miR-155 expression, whereas genes normally required for AML maintenance or activated in AML such as MYB (Basova et al., 2014) and c-Kit (Boissel et al., 2006), are regulated by "high" miR-155 expression. Therefore, the differential miR-155 expression not only proposes a new role of the function of the miR expression level inside the same disease (rather than different cellular contexts), but triggers different miR-transcription factor interactions that in turn control different downstream targets. miR-9 has been shown to be overexpressed in MLL-rearranged AML (Chen et al., 2013) and in AML patients with adverse or intermediate cytogenetic risk (Marcucci, Mrózek, Radmacher, Garzon, & Bloomfield, 2011) and downregulated in AML patients with more favorable cytogenetic findings such as t(8;21), inv(16), and t(15;17) (Emmrich et al., 2014; Marcucci et al., 2011). Different mechanisms can regulate miR-9 expression, such as: (i) the MLL fusion protein directly targets miR-9, causing its upregulation, (ii) increased activity or expression of HDACs and subsequently decreased myocyte enhancer factor 2C (Mef2C) transcription factor activity reduces the activity of the miR-9 promoter (Khosravi, Alizadeh, Jalili, Shirzad, & Saki, 2018), (iii) the overexpression of CREB, a leucine zipper transcription factor that promotes the proliferation and survival of myeloid cells and induces miR-9 promoter expression (Laneve et al., 2010), (iv) the ectopic viral integration site 1 (EVI1), upregulated in AML, inhibits miR-9 expression by causing the hypermethylation of the miR-9 promoter (Senyuk et al., 2013). miR-9 targets genes involved in the development of hematologic malignancies such as: HMGA2 and LIN28B. miR-9 in cooperation with let-7 inhibits the oncogenic LIN28B/HMGA2 axis, repressing leukemic proliferation and inducing monocytic differentiation of t(8;21) AML cells (Emmrich et al., 2014). miR-9 overexpression in MLL-rearranged AML reduces RYB and RHOH expression (Chen et al., 2013) increasing proliferation and cell survival. Downregulation of miR-9 promotes

HES1 expression preventing cell proliferation and survival (Tian et al., 2016). The downregulation of miR-9 promotes HES1 expression, preventing cell proliferation and survival. The polycistronic miR-17 family has been described to be highly expressed in AML containing MLL rearrangements (Li et al., 2008). MLL fusion proteins can activate and regulate miR-17 cluster expression (Schreiner et al., 2001) that normally contributes to maintaining leukemia stem cell (LSC) self-renewal. A downstream target of the miR-17 family is p21 (Wong et al., 2010), a regulator of cell cycle progression through the inhibition of cyclin-CDK2 or CDK4 complexes. Wong et al. proposed an additional level of p21 deregulation by miR-17 family targeting, given that MLL fusion proteins downregulate p53-mediated p21 induction in response to DNA damage (Wiederschain, Kawai, Shilatifard, & Yuan, 2005) (Fig. 13.1).

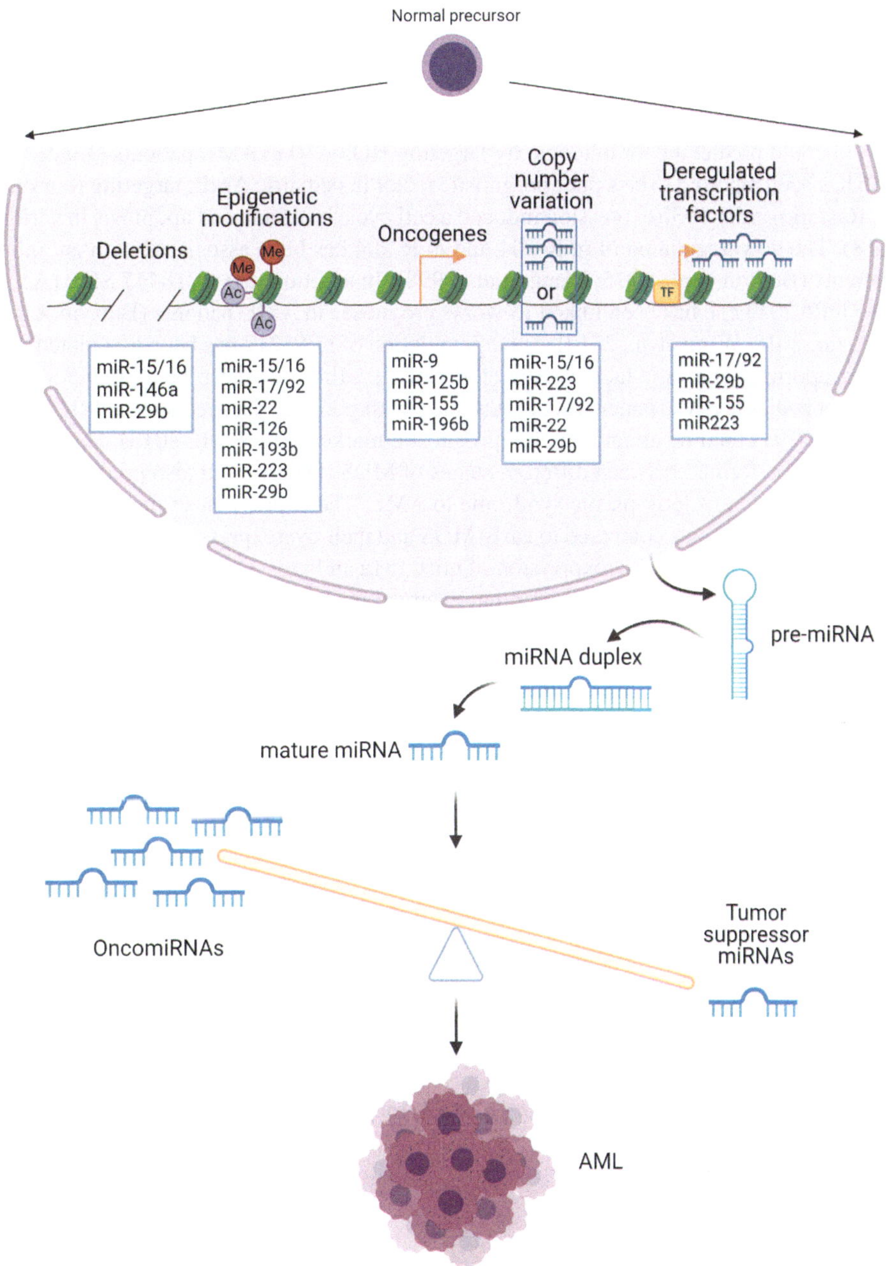

FIG. 13.1 Schematic representation of mechanisms involved in the deregulation of miRNAs in AML. Different mechanisms can contribute to compromise the balance during the generation and processing of mature miRNAs and consequently lead to the development of AML. These mechanisms include: (i) deletions, (ii) epigenetic modifications, (iii) proximity to an oncogenic region, (iv) copy number alteration, and (v) miRNA promoter region targeted by deregulated transcription factors.

miRNAs as biomarkers

A highly promising area for miRNA research involves the potential role of miRNAs as disease biomarkers. The presence of highly conserved sequences in various tissue and fluids, the high stability in blood and other body fluids, and the sensitive and easy detection methods can make miRNAs good biomarkers (Grasedieck et al., 2013). Mi et al. (Mi et al., 2007) identified four miRNAs (miR-223, miR-128a, miR-128b, and let-7b) able to discriminate between acute lymphoblastic leukemia (ALL) and AML with an accuracy greater than 95%. Molecular rearrangements, specific genetic subtypes, and chromosomal aberrations have been correlated with miRNA expression signatures, suggesting that miRNAs could be useful in the classification of AML. Mutations of NPM1, a favorable prognostic factor, have been associated with a particular signature that involves the upregulation of miR-10a, miR-10b, miR-196a, and miR-196b, which target different HOX genes (Becker et al., 2010). Moreover, mutated CEBPA and/or FLT3-ITD associate with the overexpression of miR-181 and identify a subset of patients with a favorable clinical outcome within the CN-AML group (Marcucci et al., 2008). Noteworthy, miRNA expression profiles may also help to provide prognostic information and predict survival in AML patients. Díaz-Beyá et al. (Díaz-Beyá et al., 2014) identified a signature of four miRNAs (miR-196b, miR-644, miR-135a, and miR-409-3p) as independent prognostic factors in intermediate-risk cytogenetic AML (IR-AML) patients. An additional study demonstrated that low expression of miR-135a could predict a poor outcome by targeting HOXA10 in AML patients (Xu & Wen, 2018). Low expression of miR-193b has been reported to be a poor prognostic factor in pediatric AML, targeting four different genes within the MAPK pathway. Restoring miR-193b expression induced a cell cycle blockade and apoptosis in various AML subgroups (Bhayadia et al., 2018). The downregulation of miR-204 and miR-304 has been associated with an unfavorable prognostic outcome in AML patients (Butrym et al., 2015; Wang et al., 1998). In addition, the rs718447 SNP (A > G) in the upstream flanking region of the miR-204 gene has been linked to worse prognoses in AML patients (Butrym, Łacina, Kuliczkowski, Bogunia-Kubik, & Mazur, 2018; Wang et al., 2018). The upregulation of miR-212 has been associated with a better survival independently of other reported prognostic factors (Sun et al., 2013). Mir-29 can also be a predictor biomarker. The lower expression of miR-29a in pediatric AML patients was associated with poor relapse-free survival (RFS) and overall survival (OS), suggesting that miR-29a could be an unfavorable prognostic marker (Zhu et al., 2013). Differential miRNA expression profiles have also been performed between different stages of MDS, which could also predict the involvement of these miRNAs in the progression from myelodysplastic syndrome to AML. The expression of the miR-181 family and miR-222 increased in later stages of MDS patients compared to early MDS and their overexpression increased further in AML patients (Pons et al., 2009; Sokol et al., 2011). The overexpression of miR-181a and miR-181b contributes to a negative regulation of hematopoiesis and granulocyte and macrophage differentiation targeting (Su et al., 2015). Recently, it has been demonstrated that during the progression from MDS to AML, the expression of the miR-15/16 cluster was significantly reduced and could stratify MDS patients who will potentially transform in AML (Lovat et al., 2020). The majority of miRNAs is localized in the intracellular compartment, however there are cell-free circulating miRNAs. Their stability is preserved because they are either vesicle-associated (as cargo of exosomes, microvesicles, and apoptotic bodies) or associated with ribonucleoprotein such as Ago2 or NPM1 or high-density lipoprotein (Cui et al., 2019). Several researchers have suggested circulating miRNA screening as a potential early, useful, noninvasive, and sensitive approach to detect miRNA expression in the peripheral blood serum. Downregulation of miR-150 and miR-342 in the plasma of AML patients at diagnosis compared to healthy donors and to patients in complete remission has been described as a candidate biomarker and a prognostic tool for relapse detection (Fayyad-Kazan et al., 2013). Another serum miRNA expression analysis has identified a set of six miRNAs (miR-10a-5p, miR-93-5p, miR-129-5p, miR-155-5p, miR-181b-5p, and miR-320d) that was significantly differently expressed in 140 AML patients compared to the control serum from healthy donors. In particular, miR-181b-5p has also been recognized as a biomarker for overall survival in AML patients (Zhi et al., 2013). Koutova et al. (Koutova et al., 2015) identified a plasma signature of six miRNAs (miR-199b-5p, miR-301b, miR-326, miR-361-5p, miR-625, and miR-655) that seemed to be a potential tool for monitoring AML patients before and after chemotherapy. These early promising discoveries make circulating miRNAs promising biomarkers for early detection as well as for prognosis and treatment of AML patients. However, these studies have also produced discordant results suggesting that several issues need to be addressed to validate these potential circulating biomarkers.

miRNAs in chemoresistance

The combined therapy based on cytarabine and anthracycline has been the standard treatment for AML in the past 40 years. Up to 40% of AML patients fail to respond the initial therapy and, even when this aggressive chemotherapy regimen is effective, many patients relapse within 5 years. The cause of this poor long-term survival is due to drug resistance. In fact, current therapies can target dividing blast cell population but lack the ability to eradicate LCS, therapy-resistant AML clones, or to battle the new mutations secondary to the treatments. Nanbakhsh et al. (Nanbakhsh et al., 2015) demonstrated that the acquisition of leukemia cell resistance to daunorubicin (DNR), the most common anthracycline used to treat AML, is associated with miR-181a downregulation and the consequent upregulation of the BCL2 and MCL1 family and MAP3K10 and MAP2K1 tyrosine kinases. Furthermore, the inhibition of histone deacetylases (HDACs), a class of chromatin-modulating proteins, induced overexpression of miR-182 and subsequent targeted Rad51. Downregulation of Rad51 compromised the capability of AML cells to perform homologous recombination repair and therefore sensitized cells to sapacitabine, a double-strand break-inducing drug in AML (Lai et al., 2016). Downregulation of miR-331-5p and miR-27a has been linked to doxorubicin resistance in AML cells and to the process of leukemia relapse, considering that the expression of these miRNAs was lower in relapse patients compared to primary patients at diagnosis (Feng et al., 2011). High miR-335 expression in bone marrow and serum samples from adult patients with AML was correlated with poor chemotherapy response and worse clinical outcomes after treatment (Yingchun, Rong, Kun, Ying, & Zhuogang, 2015). Blum et al. (2010) identified a possible relationship between miR-29b expression and clinical response to decitabine and bortezomib treatment. In fact, miR-29 can modulate the sensitivity of AML cells to chemotherapeutic drugs. In particular, bortezomib increased miR-29b expression levels by directly targeting DNMT3a and DNMT3b and indirectly DNMT1 through Sp1, thus increasing sensitivity to decitabine. Recently, Vandewalle et al. (Vandewalle et al., 2021) demonstrated that overexpression of miR-15a and miR-21-5p in a subgroup of AML, cytogenetically normal AML patients with wild-type NPM1 without FLT3-IDT, reduced apoptosis induced by cytarabine and daunorubicin treatment. The chemoresistance provoked by these miRNAs was due to the alteration of three proapoptotic genes PDCD4, ARL2, and BTG2. Correction of the altered expression of miRNAs might be useful as an important addition to therapeutic strategies aiming to counteract cancer cell resistance.

miRNAs as therapeutic targets

Considering that miRNAs are endogenous noncoding RNAs able to modulate multiple targets and different biological processes, miRNA-based therapeutic approaches are fascinating and promising. Thus, targeting oncogenes or tumor suppressors by one miRNA-based treatment can actually affect many pathways at the same time. One example of a miRNA-based therapeutic approach is represented by miR-22. Downregulation of miR-22 was found in the majority of AML patient samples and its enforced expression resulted in the repression of critical oncogenic targets such as CRCT1, FLT3, and MYCBP and the consequent deregulation of CREB and MYC signaling pathways. Delivery of miR-22 expression via polyamidoamine dendrimer nanoparticles led to a delay in AML progression and prolonged survival of leukemic mice. This study highlights the possibility of a combination of miRNA-based therapies with standard chemotherapeutic drugs or small inhibitors against MYC or CREB pathways to increase AML responsiveness and obtain an antileukemic effect with minimal side effects (Jiang et al., 2016). Another example of miRNA-based therapy was represented by miR-29b, a tumor suppressor in AML. A nanoparticle-based system was designed to deliver miR-29b into the AML blast, causing downregulation of its direct or indirect targets such as DNMTs, CDK6, SP1, FLT3, and KIT. This treatment resulted in decreased leukemia cell growth and improved survival in an AML mouse model (Huang et al., 2013). Shortly after the previous referenced study was published, another group confirmed the results, introducing in vivo miR-29 via lentiviral transduction and observing a reduction of AML engraftment in the mouse spleen and bone marrow associated with decreased cell proliferation and increased apoptotic activity in the spleen (Gong et al., 2014). Dorrence et al. (Dorrance et al., 2015) demonstrated that mir-126 expression is higher in cytogenetically normal AML patients and is associated with worse outcomes. In particular, miR-126 overexpression leads to LSC self-renewal and long-term maintenance. AntagomiR-126 delivery via transferrin or an antibody (anti-CD45.2) conjugated in nanoparticles caused a reduction of functional LSC activity in vivo. Another study showed that higher miR-126 expression was found in inv(16) AML CD34+ and CD34− blasts, contributing to AML initiation and reduction of apoptosis in LSC maintaining their quiescence status. A CpG-anti-miR-126 oligonucleotide inhibitor treatment, named miRsten, previously tested in chronic myeloid leukemia (Zhang et al., 2018), reduced AML initiating ability and in combination with other chemotherapeutic regimens, reduced the AML burden and prolonged survival. Velu et al. (Velu et al., 2014) studied the therapeutic inhibition of miR-21 and miR-196b in a murine model of MLL-AF9 leukemia. They pinpointed that co-treatment with antagomiR delivered via an implanted osmotic pump improved the outcome of MLL-AF9 leukemia without any apparent toxicity. Moreover, they suggested that naked antagomiR for both miR-21 and miR-196 had less efficacy in the in vivo model compared to ones with cholesterol modification, probably due to the interaction between lipoprotein particles and the lipoprotein receptors expressed on myeloid cells. Several studies have reported that the modification of miRNA expression levels can modulate the sensitivity to chemotherapeutic drugs. In this context, Lu et al. (Lu et al., 2014) demonstrated that the overexpression of miR-181 increased the sensitivity of leukemic cells to cytarabine (Ara-C) or doxorubicin (DOX) in vitro, inducing apoptosis through the inhibition of HMGB1 and Mcl-1. Gao et al. (Gao et al., 2011) reported that the upregulation of miR-193a expression for c-kit-positive AML in combination with 5-azacytidine (5-aza) treatment led to increased apoptosis and granulocytic differentiation. Moreover, it was demonstrated that miR-10a inhibition affected the sensitivity of MDM2 inhibitor Nutlin-3a and cytarabine in vitro and in vivo AML models, inducing apoptosis through p53 activation (Vu et al., 2020). Recently, it was reported that the reduced expression of the miR-15/16 cluster stratified AML patients and affected the expression of two direct targets of miR-15/16, Bcl-2 and ROR-1. Reduced expression of this cluster rendered leukemic cells more sensitive to venetoclax, a highly selective inhibitor of Bcl-2 (Lovat et al., 2020). Moreover, emerging evidence revealed that available drugs can influence and modulate miRNA expression levels by affecting the pathways of miRNA regulation. Lenalidomide induced the endogenous expression of miR-181a in AML patients through the activation of C/EBPα-p30 translation and sensitized AML cells to Ara-C therapy via miR-181a (Hickey et al., 2013). The drug bortezomib, a proteasome inhibitor, has been shown to induce miR-29b expression by affecting both Sp1/NFκB and Sp1/NFκB/HDAC complexes and consequently causing Sp1 downregulation and inhibition of the KIT. These events in combination with decitabine treatment resulted in an inhibition of leukemic cell growth (Blum et al., 2012). MLN4924, known as Pevonedistat, is an inhibitor of the NEDD8-activating enzyme and is therefore able to reduce the activity of the NF-κB pathway. It has been demonstrated to downregulate the miR-155 expression level in FLT3-IDT AML cell lines. These results led to apoptosis and monocytic differentiation in vitro and in vivo by the upregulation of SHIP1 and PU.1 (Khalife et al., 2015). Despite the significant progress represented by the above-mentioned miRNA-based therapeutic approaches, the application of these methods is still far from clinical use, especially due to the lack of specific delivery systems. The discovery of new techniques for a more efficient delivery of miRNA mimics or antagomiRNAs and combined treatments with multiple miRNAs and drugs are the challenges that need progress for a potential miRNA-based therapy in AML.

Role of miR-15/16 clusters in AML pathogenesis

One of the most common genetic aberrations in CLL is represented by the deletion at chromosome 13q14, where the miR-15a/16-1 cluster is located (Calin et al., 2002, 2005). Moreover, Calin et al. (Calin et al., 2005) discovered that rare mutations are present at seven nucleotides in the 3′ direction of pre-miR-16-1 in CLL. This mutation can affect the miRNA transcription process by compromising the recognition of the pri-miRNA transcript and the consequent cleavage by the Drosha complex, thus resulting in a downregulation of the mature form of miR-15a/16-1. Analyzing the lymphoid tissue from the New Zealand Black (NZB) mouse, a spontaneous model for studying human CLL, a point mutation in the flanking region 3′ of the pre-miR-16-1 was found, causing the alteration of miRNA maturation processing (Raveche et al., 2007). Klein et al. (Klein et al., 2010) demonstrated that the loss of miR-15a/16-1 promotes the development of CLL late in life (around 18 months) with a penetrance of nearly 40% in a mouse model. These discoveries highlighted that the loss of miR-15/16 expression at chromosome 13q14 can cause the development of CLL in humans and in mouse models. Belonging to the miR-15/16 family, a second locus is well-conserved both in humans and in mice, represented by miR-15b/16-2 at chromosome 3q25. The generation of an miR-15b/16-2 knockout mouse model demonstrated that this locus also has a role in the development of CLL with higher penetrance and earlier appearance of the disease than miR-15a/16-1 knockout mice (Lovat et al., 2015). Crossbreeding the miR-15a/16-1 and miR-15b/16-2 knockout mice, the combined double knockout mice surprisingly developed AML earlier than the CLLs evidenced in the single miR-15a or miR-15b knockout mice (Lovat et al., 2018). Moreover, a study of miR-15/16 expression levels was conducted in three subgroups of MDS patients and in a large cohort of de novo AML patients. Patients with MDS that will transform into AML and patients with AML developing from MDS showed progressive downregulation of miR-15a, miR-15b, and miR-16 expression levels compared to patients with MDS, suggesting a role for these miRNAs during the progression of the disease. In addition, the expression of these miR-15/16 clusters in AML patients was reduced compared to MDS patients, in particular a fraction of AML patients (nearly 22%) presented a combined loss of expression of both miR-15/16 loci. The researchers demonstrated that this loss was due to genetic deletion or/and epigenetic mechanisms such as methylation. Downregulation of miR-15/16 clusters affected the expression of direct and indirect targets of these miRNAs such as Bcl-2, ROR1, Cyclin D1, Mcl-1, mTOR, and PI3K p85 (Lovat et al., 2020). In particular, Bcl-2 promotes cell survival inhibiting apoptosis and is overexpressed in most CLLs (Pekarsky, Balatti, & Croce, 2018), whereas ROR1, a tyrosine-protein kinase transmembrane receptor, can promote cell proliferation and survival of leukemia cells and is expressed on the surface of CLL cells (Rassenti et al., 2017). Recently, a selective drug for Bcl-2, venetoclax, was developed and tested first in patients with CLL (Souers et al., 2013). It is now approved by the US Food and Drug Administration (FDA) as a treatment for CLL. Moreover, it has been demonstrated that cirmtuzumab, a humanized anti-ROR1 monoclonal antibody, in combination with venetoclax enhanced its cytotoxic activity on CLL cells (Rassenti et al., 2017). Considering that, the loss of expression of miR-15/16 clusters helps to identify a fraction of AML patients that can be treated with a combined therapy based on venetoclax and cirmtuzumab (Fig. 13.2).

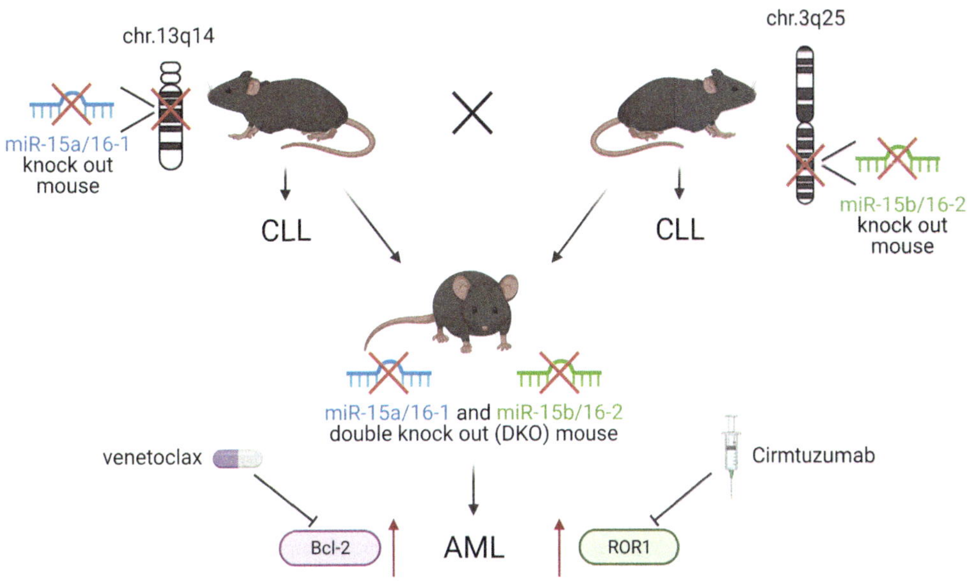

FIG. 13.2 Schematic representation of miR-15/16 clusters in the pathogenesis of AML. MiR-15a/16-1 at chromosome 13q14 and miR-15b/16-2 at chromosome 3q25 knockout mouse model develop CLL late in life. A double knockout mouse model (DKO) of both miR-15/16 loci developed acute myeloid leukemia. Downregulation of miR-15/16 clusters affected the expression of Bcl-2 and ROR1, targets of venetoclax and cirmtuzumab, respectively.

Conclusions

Cellular and circulating miRNAs and their mRNA targets are important regulators of AML pathogenesis and can potentially be employed as disease biomarkers. Specific miRNA expression profiles can be useful as markers for better stratifying diagnosis, disease progression, treatment response and clinical outcome. Despite the progress concerning the mechanisms of how miRNAs function in AML, many questions remain to be answered. Thus, it is important to identify the critical pathways that lead to leukemia to design future therapeutic treatments. Addressing issues regarding miRNA delivery by using mimics or inhibitors as therapeutic targets will provide important opportunities for AML treatment. Moreover, the combination of miRNA-based therapy with current chemotherapeutic drugs could potentially target to eliminate AML LSCs. Overall, these considerations highlight the need to continue exploring emerging concepts of miRNA roles in AML biology. This might produce a better understanding of the mechanisms that govern the pathogenesis of AML, which ultimately may lead to an improved disease outcome.

References

Alemdehy, M. F., de Looper, H. W., Kavelaars, F. G., Sanders, M. A., Hoogenboezem, R., Löwenberg, B., et al. (2016). MicroRNA-155 induces AML in combination with the loss of C/EBPA in mice. *Leukemia*, *30*, 2238–2241. https://doi.org/10.1038/leu.2016.171.

Ambros, V. (2004). The functions of animal microRNAs. *Nature*, *431*, 350–355. https://doi.org/10.1038/nature02871.

Bacher, U., Haferlach, C., Kern, W., Haferlach, T., & Schnittger, S. (2008). Prognostic relevance of FLT3-TKD mutations in AML: The combination matters—An analysis of 3082 patients. *Blood*, *111*, 2527–2537. https://doi.org/10.1182/blood-2007-05-091215.

Barjesteh van Waalwijk van Doorn-Khosrovani, S., Erpelinck, C., van Putten, W. L., Valk, P. J., van der Poel-van de Luytgaarde, S., Hack, R., et al. (2003). High EVI1 expression predicts poor survival in acute myeloid leukemia: A study of 319 de novo AML patients. *Blood*, *101*, 837–845. https://doi.org/10.1182/blood-2002-05-1459.

Basova, P., Pospisil, V., Savvulidi, F., Burda, P., Vargova, K., Stanek, L., et al. (2014). Aggressive acute myeloid leukemia in PU.1/p53 double-mutant mice. *Oncogene*, *33*, 4735–4745. https://doi.org/10.1038/onc.2013.414.

Becker, H., Marcucci, G., Maharry, K., Radmacher, M. D., Mrózek, K., Margeson, D., et al. (2010). Favorable prognostic impact of NPM1 mutations in older patients with cytogenetically normal de novo acute myeloid leukemia and associated gene- and microRNA-expression signatures: A cancer and leukemia group B study. *Journal of Clinical Oncology*, *28*, 596–604. https://doi.org/10.1200/jco.2009.25.1496.

Bhayadia, R., Krowiorz, K., Haetscher, N., Jammal, R., Emmrich, S., Obulkasim, A., et al. (2018). Endogenous tumor suppressor microRNA-193b: Therapeutic and prognostic value in acute myeloid leukemia. *Journal of Clinical Oncology*, *36*, 1007–1016. https://doi.org/10.1200/jco.2017.75.2204.

Bloomfield, C. D., Archer, K. J., Mrózek, K., Lillington, D. M., Kaneko, Y., Head, D. R., et al. (2002). 11q23 balanced chromosome aberrations in treatment-related myelodysplastic syndromes and acute leukemia: Report from an international workshop. *Genes, Chromosomes & Cancer*.

Blum, W., Garzon, R., Klisovic, R. B., Schwind, S., Walker, A., Geyer, S., et al. (2010). Clinical response and *miR-29b* predictive significance in older AML patients treated with a 10-day schedule of decitabine. *Proceedings of the National Academy of Sciences*, *107*, 7473–7478. https://doi.org/10.1073/pnas.1002650107.

Blum, W., Schwind, S., Tarighat, S. S., Geyer, S., Eisfeld, A. K., Whitman, S., et al. (2012). Clinical and pharmacodynamic activity of bortezomib and decitabine in acute myeloid leukemia. *Blood*, *119*, 6025–6031. https://doi.org/10.1182/blood-2012-03-413898.

Boissel, N., Leroy, H., Brethon, B., Philippe, N., de Botton, S., Auvrignon, A., et al. (2006). Incidence and prognostic impact of c-Kit, FLT3, and Ras gene mutations in core binding factor acute myeloid leukemia (CBF-AML). *Leukemia*, *20*, 965–970. https://doi.org/10.1038/sj.leu.2404188.

Butrym, A., Łacina, P., Kuliczkowski, K., Bogunia-Kubik, K., & Mazur, G. (2018). Genetic variation of the gene coding for microRNA-204 (miR-204) is a risk factor in acute myeloid leukaemia. *BMC Cancer*, *18*, 107. https://doi.org/10.1186/s12885-018-4045-y.

Butrym, A., Rybka, J., Baczyńska, D., Tukiendorf, A., Kuliczkowski, K., & Mazur, G. (2015). Low expression of microRNA-204 (miR-204) is associated with poor clinical outcome of acute myeloid leukemia (AML) patients. *Journal of Experimental & Clinical Cancer Research*, *34*, 68. https://doi.org/10.1186/s13046-015-0184-z.

Calin, G. A., Dumitru, C. D., Shimizu, M., Bichi, R., Zupo, S., Noch, E., et al. (2002). Frequent deletions and down-regulation of micro- RNA genes miR15 and miR16 at 13q14 in chronic lymphocytic leukemia. *Proceedings of the National Academy of Sciences of the United States of America*, *99*, 15524–15529. https://doi.org/10.1073/pnas.242606799.

Calin, G. A., Ferracin, M., Cimmino, A., Di Leva, G., Shimizu, M., Wojcik, S. E., et al. (2005). A MicroRNA signature associated with prognosis and progression in chronic lymphocytic leukemia. *The New England Journal of Medicine*, *353*, 1793–1801. https://doi.org/10.1056/NEJMoa050995.

Chen, P., Price, C., Li, Z., Li, Y., Cao, D., Wiley, A., et al. (2013). miR-9 is an essential oncogenic microRNA specifically overexpressed in *mixed lineage leukemia*–rearranged leukemia. *Proceedings of the National Academy of Sciences*. https://doi.org/10.1073/pnas.1310144110, 201310144.

Colombo, E., Alcalay, M., & Pelicci, P. G. (2011). Nucleophosmin and its complex network: A possible therapeutic target in hematological diseases. *Oncogene*, *30*, 2595–2609. https://doi.org/10.1038/onc.2010.646.

Cui, M., Wang, H., Yao, X., Zhang, D., Xie, Y., Cui, R., et al. (2019). Circulating microRNAs in cancer: Potential and challenge. *Frontiers in Genetics*, *10*, 626. https://doi.org/10.3389/fgene.2019.00626.

Danen-van Oorschot, A. A., Kuipers, J. E., Arentsen-Peters, S., Schotte, D., de Haas, V., Trka, J., et al. (2012). Differentially expressed miRNAs in cytogenetic and molecular subtypes of pediatric acute myeloid leukemia. *Pediatric Blood & Cancer*, *58*, 715–721. https://doi.org/10.1002/pbc.23279.

De Braekeleer, M., Morel, F., Le Bris, M. J., Herry, A., & Douet-Guilbert, N. (2005). The MLL gene and translocations involving chromosomal band 11q23 in acute leukemia. *Anticancer Research*, *25*, 1931–1944.

De Kouchkovsky, I., & Abdul-Hay, M. (2016). Acute myeloid leukemia: A comprehensive review and 2016 update. *Blood Cancer Journal*, *6*. https://doi.org/10.1038/bcj.2016.50, e441.

Díaz-Beyá, M., Brunet, S., Nomdedéu, J., Tejero, R., Díaz, T., Pratcorona, M., et al. (2014). MicroRNA expression at diagnosis adds relevant prognostic information to molecular categorization in patients with intermediate-risk cytogenetic acute myeloid leukemia. *Leukemia*, *28*, 804–812. https://doi.org/10.1038/leu.2013.281.

Döhner, H., Weisdorf, D. J., & Bloomfield, C. D. (2015). Acute myeloid leukemia. *The New England Journal of Medicine*, *373*, 1136–1152. https://doi.org/10.1056/NEJMra1406184.

Dorrance, A. M., Neviani, P., Ferenchak, G. J., Huang, X., Nicolet, D., Maharry, K. S., et al. (2015). Targeting leukemia stem cells in vivo with antagomiR-126 nanoparticles in acute myeloid leukemia. *Leukemia*, *29*, 2143–2153. https://doi.org/10.1038/leu.2015.139.

Emmrich, S., Katsman-Kuipers, J. E., Henke, K., Khatib, M. E., Jammal, R., Engeland, F., et al. (2014). miR-9 is a tumor suppressor in pediatric AML with t(8;21). *Leukemia*, *28*, 1022–1032. https://doi.org/10.1038/leu.2013.357.

Eyholzer, M., Schmid, S., Wilkens, L., Mueller, B. U., & Pabst, T. (2010). The tumour-suppressive miR-29a/b1 cluster is regulated by CEBPA and blocked in human AML. *British Journal of Cancer*, *103*, 275–284. https://doi.org/10.1038/sj.bjc.6605751.

Fayyad-Kazan, H., Bitar, N., Najar, M., Lewalle, P., Fayyad-Kazan, M., Badran, R., et al. (2013). Circulating miR-150 and miR-342 in plasma are novel potential biomarkers for acute myeloid leukemia. *Journal of Translational Medicine*, *11*, 31. https://doi.org/10.1186/1479-5876-11-31.

Feng, D. D., Zhang, H., Zhang, P., Zheng, Y. S., Zhang, X. J., Han, B. W., et al. (2011). Down-regulated miR-331-5p and miR-27a are associated with chemotherapy resistance and relapse in leukaemia. *Journal of Cellular and Molecular Medicine*, *15*, 2164–2175. https://doi.org/10.1111/j.1582-4934.2010.01213.x.

Figueroa, M. E., Abdel-Wahab, O., Lu, C., Ward, P. S., Patel, J., Shih, A., et al. (2010). Leukemic IDH1 and IDH2 mutations result in a hypermethylation phenotype, disrupt TET2 function, and impair hematopoietic differentiation. *Cancer Cell*, *18*, 553–567. https://doi.org/10.1016/j.ccr.2010.11.015.

Forrest, A. R., Kanamori-Katayama, M., Tomaru, Y., Lassmann, T., Ninomiya, N., Takahashi, Y., et al. (2010). Induction of microRNAs, mir-155, mir-222, mir-424 and mir-503, promotes monocytic differentiation through combinatorial regulation. *Leukemia*, *24*, 460–466. https://doi.org/10.1038/leu.2009.246.

Gao, X. N., Lin, J., Li, Y. H., Gao, L., Wang, X. R., Wang, W., et al. (2011). MicroRNA-193a represses c-kit expression and functions as a methylation-silenced tumor suppressor in acute myeloid leukemia. *Oncogene*, *30*, 3416–3428. https://doi.org/10.1038/onc.2011.62.

Garzon, R., Garofalo, M., Martelli, M. P., Briesewitz, R., Wang, L., Fernandez-Cymering, C., et al. (2008). Distinctive microRNA signature of acute myeloid leukemia bearing cytoplasmic mutated nucleophosmin. *Proceedings of the National Academy of Sciences*, *105*, 3945–3950. https://doi.org/10.1073/pnas.0800135105.

Garzon, R., Heaphy, C. E., Havelange, V., Fabbri, M., Volinia, S., Tsao, T., et al. (2009). MicroRNA 29b functions in acute myeloid leukemia. *Blood*, *114*, 5331–5341. https://doi.org/10.1182/blood-2009-03-211938.

Garzon, R., Volinia, S., Liu, C. G., Fernandez-Cymering, C., Palumbo, T., Pichiorri, F., et al. (2008). MicroRNA signatures associated with cytogenetics and prognosis in acute myeloid leukemia. *Blood*, *111*, 3183–3189. https://doi.org/10.1182/blood-2007-07-098749.

Gerloff, D., Grundler, R., Wurm, A. A., Bräuer-Hartmann, D., Katzerke, C., Hartmann, J. U., et al. (2015). NF-κB/STAT5/miR-155 network targets PU.1 in FLT3-ITD-driven acute myeloid leukemia. *Leukemia*, *29*, 535–547. https://doi.org/10.1038/leu.2014.231.

Gombart, A. F., Hofmann, W. K., Kawano, S., Takeuchi, S., Krug, U., Kwok, S. H., et al. (2002). Mutations in the gene encoding the transcription factor CCAAT/enhancer binding protein alpha in myelodysplastic syndromes and acute myeloid leukemias. *Blood*, *99*, 1332–1340. https://doi.org/10.1182/blood.v99.4.1332.

Gong, J. N., Yu, J., Lin, H. S., Zhang, X. H., Yin, X. L., Xiao, Z., et al. (2014). The role, mechanism and potentially therapeutic application of microRNA-29 family in acute myeloid leukemia. *Cell Death and Differentiation*, *21*, 100–112. https://doi.org/10.1038/cdd.2013.133.

Grasedieck, S., Sorrentino, A., Langer, C., Buske, C., Döhner, H., Mertens, D., et al. (2013). Circulating microRNAs in hematological diseases: Principles, challenges, and perspectives. *Blood*, *121*, 4977–4984. https://doi.org/10.1182/blood-2013-01-480079.

Hickey, C. J., Schwind, S., Radomska, H. S., Dorrance, A. M., Santhanam, R., Mishra, A., et al. (2013). Lenalidomide-mediated enhanced translation of C/EBPα-p30 protein up-regulates expression of the antileukemic microRNA-181a in acute myeloid leukemia. *Blood*, *121*, 159–169. https://doi.org/10.1182/blood-2012-05-428573.

Huang, X., Schwind, S., Yu, B., Santhanam, R., Wang, H., Hoellerbauer, P., et al. (2013). Targeted delivery of *microRNA-29b* by transferrin-conjugated anionic lipopolyplex nanoparticles: A novel therapeutic strategy in acute myeloid leukemia. *Clinical Cancer Research*, *19*, 2355–2367. https://doi.org/10.1158/1078-0432.ccr-12-3191.

Jiang, X., Hu, C., Arnovitz, S., Bugno, J., Yu, M., Zuo, Z., et al. (2016). miR-22 has a potent anti-tumour role with therapeutic potential in acute myeloid leukaemia. *Nature Communications*, *7*. https://doi.org/10.1038/ncomms11452, 11452.

Khalife, J., Radomska, H. S., Santhanam, R., Huang, X., Neviani, P., Saultz, J., et al. (2015). Pharmacological targeting of miR-155 via the NEDD8-activating enzyme inhibitor MLN4924 (Pevonedistat) in FLT3-ITD acute myeloid leukemia. *Leukemia*, *29*, 1981–1992. https://doi.org/10.1038/leu.2015.106.

Khosravi, A., Alizadeh, S., Jalili, A., Shirzad, R., & Saki, N. (2018). The impact of Mir-9 regulation in normal and malignant hematopoiesis. *Oncology Reviews*, *12*, 348. https://doi.org/10.4081/oncol.2018.348.

Klein, U., Lia, M., Crespo, M., Siegel, R., Shen, Q., Mo, T., et al. (2010). The DLEU2/miR-15a/16-1 cluster controls B cell proliferation and its deletion leads to chronic lymphocytic leukemia. *Cancer Cell*, *17*, 28–40. https://doi.org/10.1016/j.ccr.2009.11.019.

Kogan, S. C., Lagasse, E., Atwater, S., Bae, S. C., Weissman, I., Ito, Y., et al. (1998). The PEBP2betaMYH11 fusion created by Inv(16)(p13;q22) in myeloid leukemia impairs neutrophil maturation and contributes to granulocytic dysplasia. *Proceedings of the National Academy of Sciences of the United States of America*.

Koutova, L., Sterbova, M., Pazourkova, E., Pospisilova, S., Svobodova, I., Horinek, A., et al. (2015). The impact of standard chemotherapy on miRNA signature in plasma in AML patients. *Leukemia Research*, *39*, 1389–1395. https://doi.org/10.1016/j.leukres.2015.08.009.

Lai, T. H., Ewald, B., Zecevic, A., Liu, C., Sulda, M., Papaioannou, D., et al. (2016). HDAC inhibition induces microRNA-182, which targets RAD51 and impairs HR repair to sensitize cells to sapacitabine in acute myelogenous leukemia. *Clinical Cancer Research*, *22*, 3537–3549. https://doi.org/10.1158/1078-0432.ccr-15-1063.

Laneve, P., Gioia, U., Andriotto, A., Moretti, F., Bozzoni, I., & Caffarelli, E. (2010). A minicircuitry involving REST and CREB controls miR-9-2 expression during human neuronal differentiation. *Nucleic Acids Research*, *38*, 6895–6905. https://doi.org/10.1093/nar/gkq604.

Le Beau, M. M., Espinosa, R., 3rd, Davis, E. M., Eisenbart, J. D., Larson, R. A., & Green, E. D. (1996). Cytogenetic and molecular delineation of a region of chromosome 7 commonly deleted in malignant myeloid diseases. *Blood*, *88*, 1930–1935.

Li, Z., Lu, J., Sun, M., Mi, S., Zhang, H., Luo, R. T., et al. (2008). Distinct microRNA expression profiles in acute myeloid leukemia with common translocations. *Proceedings of the National Academy of Sciences of the United States of America*, *105*, 15535–15540. https://doi.org/10.1073/pnas.0808266105.

Liu, P., Tarlé, S. A., Hajra, A., Claxton, D. F., Marlton, P., Freedman, M., et al. (1993). Fusion between transcription factor CBF beta/PEBP2 beta and a myosin heavy chain in acute myeloid leukemia. *Science*, *261*, 1041–1044. https://doi.org/10.1126/science.8351518.

Liu, S., Wu, L. C., Pang, J., Santhanam, R., Schwind, S., Wu, Y. Z., et al. (2010). Sp1/NFkappaB/HDAC/miR-29b regulatory network in KIT-driven myeloid leukemia. *Cancer Cell*, *17*, 333–347. https://doi.org/10.1016/j.ccr.2010.03.008.

Lovat, F., Fassan, M., Gasparini, P., Rizzotto, L., Cascione, L., Pizzi, M., et al. (2015). miR-15b/16-2 deletion promotes B-cell malignancies. *Proceedings of the National Academy of Sciences of the United States of America*, *112*, 11636–11641. https://doi.org/10.1073/pnas.1514954112.

Lovat, F., Fassan, M., Sacchi, D., Ranganathan, P., Palamarchuk, A., Bill, M., et al. (2018). Knockout of both miR-15/16 loci induces acute myeloid leukemia. *Proceedings of the National Academy of Sciences of the United States of America*, *115*, 13069–13074. https://doi.org/10.1073/pnas.1814980115.

Lovat, F., Nigita, G., Distefano, R., Nakamura, T., Gasparini, P., Tomasello, L., et al. (2020). Combined loss of function of two different loci of miR-15/16 drives the pathogenesis of acute myeloid leukemia. *Proceedings of the National Academy of Sciences of the United States of America*, *117*, 12332–12340. https://doi.org/10.1073/pnas.2003597117.

Lu, F., Zhang, J., Ji, M., Li, P., Du, Y., Wang, H., et al. (2014). miR-181b increases drug sensitivity in acute myeloid leukemia via targeting HMGB1 and Mcl-1. *International Journal of Oncology*, *45*, 383–392. https://doi.org/10.3892/ijo.2014.2390.

Marcucci, G., Maharry, K., Radmacher, M. D., Mrózek, K., Vukosavljevic, T., Paschka, P., et al. (2008). Prognostic significance of, and gene and microRNA expression signatures associated with, CEBPA mutations in cytogenetically normal acute myeloid leukemia with high-risk molecular features: A Cancer and Leukemia Group B Study. *Journal of Clinical Oncology*, *26*, 5078–5087. https://doi.org/10.1200/jco.2008.17.5554.

Marcucci, G., Mrózek, K., Radmacher, M. D., Garzon, R., & Bloomfield, C. D. (2011). The prognostic and functional role of microRNAs in acute myeloid leukemia. *Blood*, *117*, 1121–1129. https://doi.org/10.1182/blood-2010-09-191312.

Mardis, E. R., Ding, L., Dooling, D. J., Larson, D. E., McLellan, M. D., Chen, K., et al. (2009). Recurring mutations found by sequencing an acute myeloid leukemia genome. *The New England Journal of Medicine*, *361*, 1058–1066. https://doi.org/10.1056/NEJMoa0903840.

Melnick, A., & Licht, J. D. (1999). Deconstructing a disease: RARalpha, its fusion partners, and their roles in the pathogenesis of acute promyelocytic leukemia. *Blood*, *93*, 3167–3215.

Mi, S., Lu, J., Sun, M., Li, Z., Zhang, H., Neilly, M. B., et al. (2007). MicroRNA expression signatures accurately discriminate acute lymphoblastic leukemia from acute myeloid leukemia. *Proceedings of the National Academy of Sciences*, *104*, 19971–19976. https://doi.org/10.1073/pnas.0709313104.

Nanbakhsh, A., Visentin, G., Olive, D., Janji, B., Mussard, E., Dessen, P., et al. (2015). miR-181a modulates acute myeloid leukemia susceptibility to natural killer cells. *OncoImmunology*, *4*. https://doi.org/10.1080/2162402X.2014.996475, e996475.

Narayan, N., Bracken, C. P., & Ekert, P. G. (2018). MicroRNA-155 expression and function in AML: An evolving paradigm. *Experimental Hematology*, *62*, 1–6. https://doi.org/10.1016/j.exphem.2018.03.007.

Narayan, N., Morenos, L., Phipson, B., Willis, S. N., Brumatti, G., Eggers, S., et al. (2017). Functionally distinct roles for different miR-155 expression levels through contrasting effects on gene expression, in acute myeloid leukaemia. *Leukemia*, *31*, 808–820. https://doi.org/10.1038/leu.2016.279.

Ngankeu, A., Ranganathan, P., Havelange, V., Nicolet, D., Volinia, S., Powell, B. L., et al. (2018). Discovery and functional implications of a miR-29b-1/miR-29a cluster polymorphism in acute myeloid leukemia. *Oncotarget*, *9*, 4354–4365. https://doi.org/10.18632/oncotarget.23150.

Palma, C. A., Al Sheikha, D., Lim, T. K., Bryant, A., Vu, T. T., Jayaswal, V., et al. (2014). MicroRNA-155 as an inducer of apoptosis and cell differentiation in acute myeloid leukaemia. *Molecular Cancer*, *13*, 79. https://doi.org/10.1186/1476-4598-13-79.

Papaemmanuil, E., Gerstung, M., Bullinger, L., Gaidzik, V. I., Paschka, P., Roberts, N. D., et al. (2016). Genomic classification and prognosis in acute myeloid leukemia. *New England Journal of Medicine*, *374*(23), 2209–2221. https://doi.org/10.1056/NEJMoa1516192.

Pekarsky, Y., Balatti, V., & Croce, C. M. (2018). BCL2 and miR-15/16: From gene discovery to treatment. *Cell Death and Differentiation*, *25*, 21–26. https://doi.org/10.1038/cdd.2017.159.

Pons, A., Nomdedeu, B., Navarro, A., Gaya, A., Gel, B., Diaz, T., et al. (2009). Hematopoiesis-related microRNA expression in myelodysplastic syndromes. *Leukemia & Lymphoma*, *50*, 1854–1859. https://doi.org/10.3109/10428190903147645.

Rassenti, L. Z., Balatti, V., Ghia, E. M., Palamarchuk, A., Tomasello, L., Fadda, P., et al. (2017). *MicroRNA* dysregulation to identify therapeutic target combinations for chronic lymphocytic leukemia. *Proceedings of the National Academy of Sciences*, *114*, 10731–10736. https://doi.org/10.1073/pnas.1708264114.

Raveche, E. S., Salerno, E., Scaglione, B. J., Manohar, V., Abbasi, F., Lin, Y. C., et al. (2007). Abnormal microRNA-16 locus with synteny to human 13q14 linked to CLL in NZB mice. *Blood*, *109*, 5079–5086. https://doi.org/10.1182/blood-2007-02-071225.

Ruault, M., Brun, M. E., Ventura, M., Roizès, G., & De Sario, A. (2002). MLL3, a new human member of the TRX/MLL gene family, maps to 7q36, a chromosome region frequently deleted in myeloid leukaemia. *Gene*, *284*, 73–81. https://doi.org/10.1016/s0378-1119(02)00392-x.

Schneider, E., Staffas, A., Röhner, L., Krowiorz, K., Heuser, M., Döhner, K., et al. (2016). MicroRNA-155 is upregulated in MLL-rearranged AML but its absence does not affect leukemia development. *Experimental Hematology*, *44*, 1166–1171. https://doi.org/10.1016/j.exphem.2016.08.012.

Schnittger, S., Schoch, C., Dugas, M., Kern, W., Staib, P., Wuchter, C., et al. (2002). Analysis of FLT3 length mutations in 1003 patients with acute myeloid leukemia: Correlation to cytogenetics, FAB subtype, and prognosis in the AMLCG study and usefulness as a marker for the detection of minimal residual disease. *Blood*, *100*, 59–66. https://doi.org/10.1182/blood.v100.1.59.

Schreiner, S., Birke, M., García-Cuéllar, M.-P., Zilles, O., Greil, J., & Slany, R. K. (2001). MLL-ENL causes a reversible and *myc*-dependent block of myelomonocytic cell differentiation. *Cancer Research*, *61*, 6480–6486. https://cancerres.aacrjournals.org/content/canres/61/17/6480.full.pdf.

Senyuk, V., Zhang, Y., Liu, Y., Ming, M., Premanand, K., Zhou, L., et al. (2013). Critical role of miR-9 in myelopoiesis and EVI1-induced leukemogenesis. *Proceedings of the National Academy of Sciences*, *110*, 5594–5599. https://doi.org/10.1073/pnas.1302645110.

Sokol, L., Caceres, G., Volinia, S., Alder, H., Nuovo, G. J., Liu, C. G., et al. (2011). Identification of a risk dependent microRNA expression signature in myelodysplastic syndromes. *British Journal of Haematology*, *153*, 24–32. https://doi.org/10.1111/j.1365-2141.2011.08581.x.

Souers, A. J., Leverson, J. D., Boghaert, E. R., Ackler, S. L., Catron, N. D., Chen, J., et al. (2013). ABT-199, a potent and selective BCL-2 inhibitor, achieves antitumor activity while sparing platelets. *Nature Medicine*, *19*, 202–208. https://doi.org/10.1038/nm.3048.

Su, R., Lin, H. S., Zhang, X. H., Yin, X. L., Ning, H. M., Liu, B., et al. (2015). MiR-181 family: Regulators of myeloid differentiation and acute myeloid leukemia as well as potential therapeutic targets. *Oncogene*, *34*, 3226–3239. https://doi.org/10.1038/onc.2014.274.

Sun, S. M., Rockova, V., Bullinger, L., Dijkstra, M. K., Döhner, H., Löwenberg, B., et al. (2013). The prognostic relevance of miR-212 expression with survival in cytogenetically and molecularly heterogeneous AML. *Leukemia*, *27*, 100–106. https://doi.org/10.1038/leu.2012.158.

Thiede, C., Koch, S., Creutzig, E., Steudel, C., Illmer, T., Schaich, M., et al. (2006). Prevalence and prognostic impact of NPM1 mutations in 1485 adult patients with acute myeloid leukemia (AML). *Blood*, *107*, 4011–4020. https://doi.org/10.1182/blood-2005-08-3167.

Tian, C., You, M. J., Yu, Y., Zhu, L., Zheng, G., & Zhang, Y. (2016). MicroRNA-9 promotes proliferation of leukemia cells in adult CD34-positive acute myeloid leukemia with normal karyotype by downregulation of Hes1. *Tumour Biology*, *37*, 7461–7471. https://doi.org/10.1007/s13277-015-4581-x.

Vandewalle, V., Essaghir, A., Bollaert, E., Lenglez, S., Graux, C., Schoemans, H., et al. (2021). miR-15a-5p and miR-21-5p contribute to chemoresistance in cytogenetically normal acute myeloid leukaemia by targeting PDCD4, ARL2 and BTG2. *Journal of Cellular and Molecular Medicine*, *25*, 575–585. https://doi.org/10.1111/jcmm.16110.

Velu, C. S., Chaubey, A., Phelan, J. D., Horman, S. R., Wunderlich, M., Guzman, M. L., et al. (2014). Therapeutic antagonists of microRNAs deplete leukemia-initiating cell activity. *The Journal of Clinical Investigation*, *124*, 222–236. https://doi.org/10.1172/jci66005.

Vu, T. T., Stölzel, F., Wang, K. W., Röllig, C., Tursky, M. L., Molloy, T. J., et al. (2020). miR-10a as a therapeutic target and predictive biomarker for MDM2 inhibition in acute myeloid leukemia. *Leukemia*. https://doi.org/10.1038/s41375-020-01095-z.

Wang, J., Hoshino, T., Redner, R. L., Kajigaya, S., & Liu, J. M. (1998). ETO, fusion partner in t(8;21) acute myeloid leukemia, represses transcription by interaction with the human N-CoR/mSin3/HDAC1 complex. *Proceedings of the National Academy of Sciences*, *95*, 10860–10865. https://www.pnas.org/content/pnas/95/18/10860.full.pdf.

Wang, Z., Luo, H., Fang, Z., Fan, Y., Liu, X., Zhang, Y., et al. (2018). MiR-204 acts as a potential therapeutic target in acute myeloid leukemia by increasing BIRC6-mediated apoptosis. *BMB Reports*, *51*, 444–449. https://doi.org/10.5483/bmbrep.2018.51.9.036.

Wiederschain, D., Kawai, H., Shilatifard, A., & Yuan, Z. M. (2005). Multiple mixed lineage leukemia (MLL) fusion proteins suppress p53-mediated response to DNA damage. *The Journal of Biological Chemistry*, *280*, 24315–24321. https://doi.org/10.1074/jbc.M412237200.

Wong, P., Iwasaki, M., Somervaille, T. C. P., Ficara, F., Carico, C., Arnold, C., et al. (2010). The *miR-17-92* microRNA polycistron regulates *MLL* leukemia stem cell potential by modulating p21 expression. *Cancer Research*, *70*, 3833–3842. https://doi.org/10.1158/0008-5472.can-09-3268.

Xu, H., & Wen, Q. (2018). Downregulation of miR-135a predicts poor prognosis in acute myeloid leukemia and regulates leukemia progression via modulating HOXA10 expression. *Molecular Medicine Reports*, *18*, 1134–1140. https://doi.org/10.3892/mmr.2018.9066.

Yingchun, L., Rong, Z., Kun, Y., Ying, Y., & Zhuogang, L. (2015). Bone marrow microRNA-335 level predicts the chemotherapy response and prognosis of adult acute myeloid leukemia. *Medicine (Baltimore)*, *94*. https://doi.org/10.1097/md.0000000000000986, e0986.

Zelent, A., Guidez, F., Melnick, A., Waxman, S., & Licht, J. D. (2001). Translocations of the RARalpha gene in acute promyelocytic leukemia. *Oncogene*, *20*, 7186–7203. https://doi.org/10.1038/sj.onc.1204766.

Zhang, B., Nguyen, L. X. T., Li, L., Zhao, D., Kumar, B., Wu, H., et al. (2018). Bone marrow niche trafficking of miR-126 controls the self-renewal of leukemia stem cells in chronic myelogenous leukemia. *Nature Medicine*, *24*, 450–462. https://doi.org/10.1038/nm.4499.

Zhi, F., Cao, X., Xie, X., Wang, B., Dong, W., Gu, W., et al. (2013). Identification of circulating microRNAs as potential biomarkers for detecting acute myeloid leukemia. *PLoS ONE*, *8*. https://doi.org/10.1371/journal.pone.0056718, e56718.

Zhu, C., Wang, Y., Kuai, W., Sun, X., Chen, H., & Hong, Z. (2013). Prognostic value of miR-29a expression in pediatric acute myeloid leukemia. *Clinical Biochemistry*, *46*, 49–53. https://doi.org/10.1016/j.clinbiochem.2012.09.002.

Chapter 14

Pathophysiology roles and translational opportunities of miRNAs in CLL

Angelo Veronese[a] and Rosa Visone[b,c]

[a]Department of Pharmacy and Drug Sciences, University of Bari Aldo Moro, Bari, Italy, [b]Department of Medical, Oral and Biotechnological Sciences, G. d'Annunzio University, Chieti, Italy, [c]Center for Advanced Studies and Technology (CAST), G. d'Annunzio University, Chieti, Italy

MicroRNAs and the cellular origin of CLL cells

The normal cellular counterpart of CLL cells is still debated (Darwiche, Gubler, Marolleau, & Ghamlouch, 2018; Klein et al., 2001; Seifert et al., 2012). The heterogeneous results collected on the cellular origin of CLL depend on the diversity of either the technical approaches or the B-cell cellular subsets used as a reference (Darwiche et al., 2018). MicroRNAs were also used for this aim. Based on global microRNA expression, the antigen-experienced marginal zone-like B cells (MZB) were identified as the populations showing the highest similarities with the CLL cells and, likely, those from which CLL could originate (Negrini et al., 2014). MZB cells are mostly localized in the marginal area of the lymphoid organ and have unique properties compared to the most representative B cells, named the follicular (FO) B cells. Evidence demonstrates that microRNAs are involved in this type of B cell decision fate. Indeed, mice carrying a deletion of Dicer, a protein involved in microRNA processing, display an overrepresentation of transitional and MZB cells together with an impaired generation of B cells (Belver, de Yébenes, & Ramiro, 2010). By comparing the microRNAs deregulated in CLL with those deregulated in Dicer-deficient mice, the miR-185 stands out. It is known to be implied in BCR signaling by downregulating the Bruton tyrosine kinase (Belver et al., 2010), and miR-146a is demonstrated to be involved in MZB cell differentiation (King et al., 2016). These studies highlight the fundamental role of microRNAs in B cell differentiation and CLL pathogenesis. Some genetic studies strengthen the hypothesis that B-CLL cells do not originate from differentiating cells by demonstrating the presence of both typical genes and chromosomal alterations in the early hematopoietic progenitors of these patients (Gahn et al., 1997; Quijada-Álamo et al., 2017). Moreover, analyzing the VDJ immunoglobulin genes, it was demonstrated that the hematopoietic stem cells (HSC) from CLL patients, but not those from healthy donors, lead to monoclonal B lymphocytosis (MBL) after xenogeneic transplantation in mice (Kikushige et al., 2011). All these data highlight the relevance of microRNAs related to B lineage differentiation from HSC to immature B cells in the bone marrow, and from transitional B cells to antigen-experienced B cells during peripheral maturation (Sanz et al., 2019) in the pathophysiology of CLL.

MicroRNAs and CLL genetics

Several studies have demonstrated the importance of chromosome abnormality in CLL (Baliakas et al., 2014; Puiggros, Blanco, & Espinet, 2014). Aneuploidy is a frequent event in cancer and miR-155, highly expressed in CLL, is involved in this process. Indeed, miR-155 targets proteins of the outer plate of the kinetochores, such as the mitotic checkpoint serine/threonine kinase budding uninhibited by benzimidazoles 1 (BUB1), the zw10 kinetochore protein (ZW10), and the centromere protein F (CENP-F), responsible for the correct alignment of the metaphase plate and the separation of chromatids in daughter cells. Reduction of the target proteins induces aneuploidy in the early phase of cancer transformation. Indeed, in the advanced stage, although miR-155 is high in CLL, the binding site to the BUB1 is not available, suggesting that other genes will be the target of this miR, as we will discuss later (Pagotto et al., 2019, 2018). Other microRNAs are located at the fragile sites whose break can induce their loss. This is the case of the miR-15a/16-1 cluster. After the pioneering study that located the microRNA cluster miR-15a/16-1 at the most deleted chromosomal region in CLL patients, del13q14 (Calin et al., 2002) and the functional studies that display the BCL2 antiapoptotic gene as the physiologic target of miR-15a and miR-16-1 (Cimmino et al., 2005), the final incontestable evidence of the involvement of this cluster in CLL pathogenesis

MicroRNA in Human Malignancies. https://doi.org/10.1016/B978-0-12-822287-4.00023-2

was demonstrated by Klein et al. They reported lymphoproliferative disorders such as CLL and MBL in miR-15a/16-1-deficient mice, and in mice carrying the deletion of the cluster only in B lymphocytes (Darwiche et al., 2018; Klein et al., 2010; Seifert et al., 2012; Simonetti, Bertilaccio, Ghia, & Klein, 2014). Fine mapping of the 13q chromosome region in CLL patients identifies the minimal deleted region (MDR) starting about 20 kb upstream to the microRNA cluster encompassing the regulative sequences of the host gene, DLEU2, and of the miR-15a/16-1 cluster itself (Edelmann et al., 2012; Lerner et al., 2009; Veronese et al., 2015), pointing to the relevance of miR-15a/16-1 transcriptional regulation above its complete deletion. The involvement of the miR-15a/16-1 cluster in the development of CLL was evident also by a different murine model. New Zealand Black (NZB) mice are known for their propensity to develop a lymphoproliferative disorder that mimics the human B-CLL disease (Marti, Metcalf, & Raveche, 1995). They display decreased levels of miR-16-1 in lymphoid tissues because of a mutation in the 3′ flanking region of the miR-16-1 that impairs microRNA maturation (Scaglione et al., 2007). Germline nucleotide variations impairing microRNA maturation of the miR-15a/16-1 cluster were also described in CLL patients (Calin et al., 2005; Veronese et al., 2015). Note that the 13q14 deletion is considered a clonal aberration that occurs early during the pathogenesis of the disease (Landau et al., 2013). Therefore, although still debated, it is a common consensus that the 13q deletion is a karyotypic aberration that also recurs in the undifferentiated counterpart of B cell populations of CLL patients (Gahn et al., 1997; Quijada-Álamo et al., 2017). Indeed, some data describe how the miR-15 family strengthens the shift from proliferation to differentiation in pre-B cells by impairing cyclin E1 and cyclin D3 (Lindner et al., 2017). The microRNAs of the miR-15 family share the same seed sequence and comprise the miR-15a/16-1 (13q14), the miR-15b/16-2 (3q25), and the miR-497/195 (17p13). However, while a clear involvement in CLL was demonstrated for the miR-15b/16-2 cluster (Lovat et al., 2015), the miR-497/195 cluster at the 17p13 chromosome region does not display any pathologic role in CLL (Hutter et al., 2020). Generally, CLLs harboring 13q deletion are indolent with a better prognosis compared to other karyotypes, such as the 17p deletion where the tumor-suppressor TP53 gene lies. A molecular model has been proposed to locate the miR-15a/16-1 cluster in the indolent behavior of CLL. It draws a regulative negative feedback loop between the miR-15a/16-1 cluster and its target TP53 gene that, in turn, inhibits the prognostic factor zeta-chain (TCR)-associated protein kinase 70 kDa (ZAP70) by transactivating the transcription of the miR-34b,c, and miR-34a. In the pathologic view, the downregulation of the miR-15a/16-1 cluster sustains a TP53 nuclear activation and consequent transcription of the miR-34s. This results in increased expression of the miR-15a/16-1 targets such as the antiapoptotic proteins BCL2 and MCL1, decreased expression of the ZAP70 protein and the miR-34 target, and increased expression of the TP53 downstream effectors such as the cyclin-dependent kinase inhibitor 1A (p21, Cip1; CDKN1A) and the P53 upregulated modulator of apoptosis (BBC3/PUMA) (Fabbri et al., 2011). The resulting short circuiting flows in an indolent pathologic disequilibrium between the cell cycle and the apoptotic effectors. Moreover, the role of TP53 in the miR-15a/16-1 regulation in 13q deleted CLLs could be different on the basis of which allele is deleted. Indeed, there are findings showing a physiologic allelic imbalance at the 13q14 region (Mertens et al., 2006) that could be partially explained by the transcriptional regulation of the miR-15a/16-1; the miR-15a/16-1 alleles are normally transcribed in B cells, one through the RNA polymerase II (RPII) machinery together with the host gene DLEU2, and the other one by the RNA polymerase III (RPIII) independently of the host gene (Veronese et al., 2015) It is well known that TP53 inhibits RPIII activity (Crighton et al., 2003), whereas RPII activity induces specific DLEU2/miR15a/16-1 expression (Fabbri et al., 2011), increasing the clinical and molecular importance of the TP53/miR-15a/16-1/miR-34s/ZAP70 circuit in CLL. The commitment of miR-15a/16-1 in the pathophysiology of CLL is evident by genetic and functional observations. It connects among the most relevant clinical, molecular and genetic characteristics of the disease: (1) the most deleted chromosome region in CLL (13q); (2) the frequently deleted regions 11q23 and 17p13 where the miR-34b,c, and the TP53 genes respectively lie; (3) the functional downstream targets that count between the most relevant genes in leukemia and cancer as BCL2, MCL1, TP53, CCND1, CCNE1 and several others (Chou et al., 2018); (4) the CLL prognostic factor ZAP70.

microRNAs in proliferation and survival of CLL cells

The proliferation of leukemic cells is due to an abnormal expression of intrinsic and extrinsic factors. It takes place in lymph nodes, peripheral blood, and bone marrow; however, the higher percentage of the newly born cells is in the lymph node compartment (Cui et al., 2014; Herndon et al., 2017) In lymph node several factors sustain proliferation and survival of CLL cells and some microRNAs could be considered the effector of such elements.

microRNAs and BCR signaling

Altered signaling that guarantees the proliferation and survival of CLL cells relies on B cell receptor (BCR) clustering and its constitutive activation. The biological outcomes of BCR signaling may vary depending on the B-cell differentiation stage, the strength of the signals, and costimulatory mediators in the environment (Niiro & Clark, 2002). CLL cases are grouped in subsets based on BCR Ig stereotypy and linked to clinical features and CLL clone reactivity (Jaramillo et al., 2019). Indeed, CLL clones respond differently to Ig stimulation depending on the stereotyped BCRs and their mutational status. For instance, CLL cells in subset #4 are anergic and become responsive to BCR when TLR1/2 receptors are stimulated. TLR1/2 signaling breaks anergy by increasing the expression of microRNAs, namely miR-15a, and the miR-17-92 cluster, which in turn downregulates c-Jun N-terminal kinase (JNK) and mitogen-activated protein kinase kinase 3 (MAP2K3), influencing the immunomodulatory signaling (Ntoufa et al., 2016). This occurs specifically in subset #4 because the TLR1/2 stimulation of other CLL subsets does not affect the expression of the cluster. The miR-17 expression is also sensitive to the TLR-9 agonist CpG unmethylated oligos in IGHV unmutated CLL. In this context, the miR-17 targets the tumor suppressor genes tumor protein P53 inducible nuclear protein 1 (TP53INP1), E2F transcription factor 5 (E2F5), tripartite motif containing 8 (TRIM8), and zinc finger and BTB domain containing 4 (ZBTB4), promoting the survival of CLL in response to TLR-9 signaling (Bomben et al., 2012). Other miRNAs have not been reported to be directly regulated by BCR stimulation; however, they interfere with this cascade leading to cell survival. Indeed, the miR-21, whose high expression is associated with poor prognosis and drug resistance in CLL (Ferracin et al., 2010; Rossi et al., 2010), targets Sprouty 2 (SPRY2), by blocking the activity of three kinases while also weakening B-cell receptor and MAPK-Erk signaling in B cells from healthy donors and CLL patients (Shukla et al., 2016). Large weights as effectors of the BCR signaling in CLL have several miRNAs implicated in the development and differentiation of B cells, such as miR-155, miR-150, and miR-181b. MiR-155, physiologically important to regulate the germinal center response (Li, Wan, Ji, Fang, & Wu, 2013), is highly expressed in circulating CLL cells exiting from the lymphoid compartment and having higher levels of proliferation and antiapoptotic genes. In these cells, miR-155 enhances BCR signaling after ectopic expression or if increased by stimulation with either the CD40 ligand (CD40L) or the B cell-activating factor (BAFF). In this context, its expression is inversely correlated to the SH2-containing inositol-5 phosphatase-1 (SHIP-1) (Cui et al., 2014), which is involved in turning down the BCR signaling. MiR-150 is involved in the transition from proB to preB; its ectopic expression blocks the cells in a proB stage by targeting the transcription factor c-MYB. Deficiency of the miR-150 in murine models induces an increase of the splenic B1 cells CD19+/CD5+ (Xiao et al., 2007). This microRNA is the most abundant in CLL and the low expression has been associated with an unfavorable prognosis. MiR-150 interferes with BCR signaling by targeting GRB2-associated binding protein 1 (GAB1) and forkhead box P1 (FOXP1). FOXP1 is a transcriptional factor required for immunoglobulin rearrangement during the steps of B-cell maturation. The abundance of this gene promotes efficient BCR triggering (Mraz et al., 2014). A similar role has miR-34 in B cell differentiation and BCR signaling. This miRNA, low expressed in CLL, is induced following activation of the DNA damage response (Cerna et al., 2019). MiR-181b is one of the miRNAs that drives the differentiation of hematopoietic stem cells in the B cell lineage and, accordingly, miR-181a1b1-deficient mice have a reduced germinal center and peripheral B cells (Chen, Li, Lodish, & Bartel, 2004; Fragoso et al., 2012). Activation of B cells via BCR ligation induces the downregulation of miR-181b in normal and leukemic B cells; however, its role as a direct effector of BCR signaling in CLL has not been demonstrated yet (Li et al., 2011). Several other microRNAs directly target Burton's tyrosine kinase. Histone deacetylase 1 (HDAC1) is recruited to their promoter, silencing their expression; however, their relevance and pathophysiologic role in CLL have not been established yet (Bottoni et al., 2016).

MicroRNAs and microenvironment

CLL cells recirculate between peripheral blood and lymphoid organs receiving supportive stimuli from the microenvironment. This is very heterogeneous in regard to the components and tissue organization. The microenvironment is composed of cells and soluble factors, such as cytokines, that support CLL cell survival and protect them from apoptosis (ten Hacken & Burger, 2016). The most used systems to study the effect of the tissue microenvironment on CLL are the mesenchymal bone marrow stromal cells (BMSC) and the CD40L, a molecule that is expressed on activated T lymphocytes. Both stromal and T cells sustain the survival of leukemic cells by deregulating microRNAs. MicroRNAs play a critical role in this context. Among them, miR-21 increased while miR-29, miR-34, and miR-181b decreased in CLL grown on BMCS. MiR-21 has a cytoprotective effect on leukemic cells (Carabia et al., 2017), whereas the downregulated miRs fail to target the T-cell leukemia/lymphoma 1 (TCL1), a critical oncogene for the onset of CLL (Sivina et al., 2012). These miRs are also involved in CD40L/CD40 signaling. Stimulation of CLL by CD40L mimics the T cell interaction and is strikingly linked to BCR signaling. Indeed CD40L induces different fates in B cells depending on the BCR binding antigen (foreign or self) and on its status (bound or unbound). Specifically, miR-29 reduced by BCR stimulation promotes CLL cell survival by interfering with the NFKb pathway upon CD40L stimulation (Sharma et al., 2020). The miR-181b that increases in CLL cells when stimulated by activated T cells expressing CD40L leads to the death of leukemic cells. This is due to a concomitant long exposure of CLL cells to T cells CD40L positive and high miR-181b levels, both reducing interleukin-10, an immunosuppressive cytokine (Di Marco et al., 2021). Another microRNA induced by the stimulation of CLL cells with CD40L/interleukin 4 is miR-22. It is overexpressed in the proliferating fraction of the leukemic cells and silences the onco-suppressor PTEN activating the mitogenic AKT/PI3K pathway (Palacios et al., 2015). The CLL supportive microenvironment also includes extracellular vesicles that represent a way of communication between malignant and surrounding cells. In immunodeficient mice, the injection of CLL-derived exosomes together with CLL cells promotes tumor progression by inducing stromal cells to acquire the characteristic of fibroblast cancer cells (Paggetti et al., 2015). MiR-155 is abundant in CLL-derived exosomes and is involved in this process (Yeh et al., 2015). Exosomes carrying miR-155 also control the proliferation and release of cytokines from myeloid-derived suppressor cells, a heterogeneous population of progenitors and mature myeloid cells that suppress T cell functions (Li et al., 2014). Vitamin D treatment reduces miR-155 in CLL cells and exosomes, which breaches myeloid-derived suppressor effect of cells (Bruns et al., 2017).

MicroRNAs and apoptosis

A defect in the apoptotic pathway is a hallmark of cancer cells, especially in B-CLL where leukemic cells in the blood are quiescent and unable to undergo programmed death if needed. This results in an accumulation of tumor cells in the bone marrow, lymph nodes, and peripheral blood. This pathway is strictly correlated with the microenvironment and stimuli (ten Hacken & Burger, 2016), connecting all the clinical, cellular, and molecular aspects discussed here. Indeed, in the indolent CLL harboring the 13q14 deletion, the impaired miR-15a/16-1-P53-miR34s axis results in the upregulation of the antiapoptotic proteins BCL2 and MCL1 by the unsuccessful targeting of miR-15a (Fabbri et al., 2011; Mackus et al., 2005). These proteins belong to the BCL2 gene family, which regulates the mitochondrial pathway of caspase-dependent apoptosis (Billard, 2014; Buggins & Pepper, 2010). In the case of genetic aberrations of the TP53 gene (mutations; 17p deletion), also the proapoptotic downstream target of TP53 will be impaired, causing a more aggressive disease. miR-181b, which is under the CD40/CD40L pathway and reduces IL10 (Di Marco et al., 2021), targets BCL2 and MCL1 as well (Visone et al., 2011), linking the mitochondrial with the extrinsic apoptotic pathways (Fas (CD95)-mediated apoptosis of CD40 activated cells) (Dicker, Kater, Fukuda, & Kipps, 2005). Targeting the antiapoptotic factors that drive B cells to survive in CLL has been a tantalizing idea for a long time (Kolb, Kern, Quiney, Roman, & Billard, 2003; Packham & Stevenson, 2005). It recently became more concrete with Venetoclax, a promising new drug approved by the US Food and Drug Administration for the treatment of poor prognosis CLL. The drug targets BCL2, as the miR-15a/16-1 cluster and miR-181 do (Pekarsky, Balatti, & Croce, 2018; Zhu et al., 2012). So far, many studies have delineated a pathophysiologic role of miRNAs, associating them with a specific pathway in CLL. In Fig. 14.1, we summarize the miRNAs most frequently dysregulated in CLL and the most important signaling at which they are linked up to now.

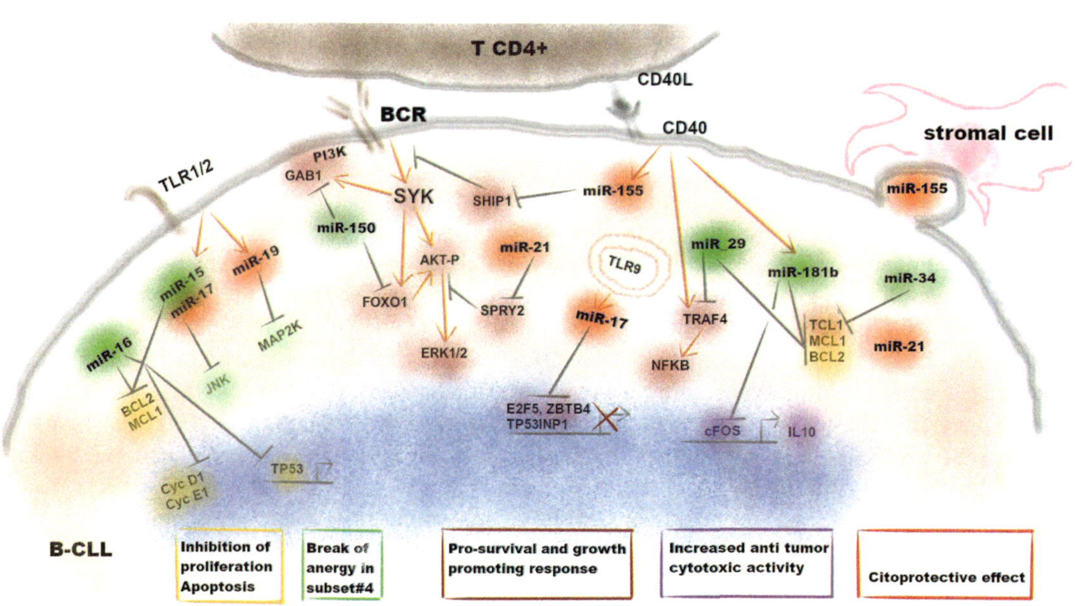

FIG. 14.1 MicroRNA involvement in the most relevant pathways deregulated in CL. *Red and green* microRNAs indicate the upregulated and down-regulated miRNAs in CLL, respectively. The molecule belonging to the different pathophysiologic effect induced by the microRNAs is indicated by as many colors.

Translational opportunities

MicroRNA research defined a new field of potential therapeutic approaches in cancer, especially in leukemias, where the delivery could turn out to be easier than in solid tumors. However, there are concerns and fundamental questions to be addressed: (1) the choice of the right miRNA depends on cancer type; (2) the miRNA clearance in the body; (3) the miRNAs off-target effects and toxicity; (4) the miRNAs spatio-temporal regulation, and (5) efficient and targeted delivery. Note that all these aspects can be specific for every chosen miRNA. Academic and industry researchers are working on these topics nowadays to overcome toxicity and maximize efficacy for a few microRNAs (Chakraborty, Sharma, Sharma, & Lee, 2021; Fortunato & Iorio, 2020; Rupaimoole & Slack, 2017). Regarding CLL, a phase I clinical trial recently ended to ascertain the toxicity of Cobomarsen (MRG-106), an inhibitor of miR-155 (ClinicalTrials.gov Identifier: NCT02580552). As described, miR-155 plays a crucial role in the differentiation and proliferation of normal B cells and B-CLL cells. Inhibition of this microRNA is believed to drive B-CLL cells to a physiologic-like condition and reduce aberrant leukemic cell proliferation. With the Eμ-TCL1 CLL mouse model that recapitulates an aggressive CLL (Bichi et al., 2002; Bresin et al., 2016), the administration of miR-26a by anti-CD38 conjugated nanoparticles significantly increases antileukemic activity by increasing cellular apoptosis (D'Abundo et al., 2017). Using an adoptive model achieved by the transplantation of leukemic cells obtained from mice overexpressing the TCL1 gene with the 3′UTR miR-181b binding site in syngeneic mice, the delivery of miR-181b undermines the proliferative and survival pathways of leukemic cells (Bresin et al., 2015). Therefore, although clinical and preclinical studies using microRNAs as therapeutic tools in CLL are promising, a closer clinical milestone involving these molecules resides in their recruitment as biomarkers to stratify and prognosticate CLL patients. Stratifying CLL patients based on drug susceptibility, predicting the switch from responsive to nonresponsive treatment, or following the course of the disease with a parameter able to identify the minimal residual disease are the most challenging clinical efforts, especially in this era of new drugs for CLL treatment. The efficacies of Ibrutinib (BTK inhibitor), Idelalisib (PI3Kδ inhibitor), and Venetoclax (a BCL-2 inhibitor) have revolutionized the standard treatment of CLL patients, giving much longer progression-free survival than chemoimmunotherapy, especially in patients stratified as high-risk CLL. The growing clinical data based on these new treatments enforce a continuous updating in guidelines all over the world toward the precision medicine model. Treatment failure could drive the disease to a more aggressive state such as the Richter transformation, that could develop early during the first 15 months of Ibrutinib treatment or later with CLL progression. Therefore, there is an urgent need for tools to sustain clinicians in patients management, such as predictive markers for drug response or progression of the disease. MicroRNAs have already stratified CLL patients for clinical and molecular features: microRNA signatures associated with (i) ZAP70 expression and IGVH mutational status (Calin et al., 2004); (ii) disease prognosis (Calin et al., 2005); (iii) CLL-related karyotypes (Visone et al., 2009); (iv) fludarabine resistance (Ferracin et al., 2010); (v) disease risk stratification (Moussay et al., 2011; Stamatopoulos et al., 2009); and (vi) disease progression (Visone et al., 2011; Visone, Veronese, Balatti, & Croce, 2012). These findings did not substitute the traditional clinical practice, but help in the management of CLL patients treated with next-generation drugs. In the future, with more details and better understanding about the microRNA role in the disease, treatment with a personalized activation or inhibition of specific microRNAs could become a reality.

References

Baliakas, P., Iskas, M., Gardiner, A., Davis, Z., Plevova, K., Nguyen-Khac, F., et al. (2014). Chromosomal translocations and karyotype complexity in chronic lymphocytic leukemia: A systematic reappraisal of classic cytogenetic data. *American Journal of Hematology*, *89*(3), 249–255. https://doi.org/10.1002/ajh.23618.

Belver, L., de Yébenes, V. G., & Ramiro, A. R. (2010). MicroRNAs prevent the generation of autoreactive antibodies. *Immunity*, *33*(5), 713–722. https://doi.org/10.1016/j.immuni.2010.11.010.

Bichi, R., Shinton, S. A., Martin, E. S., Koval, A., Calin, G. A., Cesari, R., et al. (2002). Human chronic lymphocytic leukemia modeled in mouse by targeted TCL1 expression. *Proceedings of the National Academy of Sciences*, *99*(10), 6955–6960. https://doi.org/10.1073/pnas.102181599.

Billard, C. (2014). Apoptosis inducers in chronic lymphocytic leukemia. *Oncotarget*, *5*(2), 309–325. https://doi.org/10.18632/oncotarget.1480.

Bomben, R., Gobessi, S., Dal Bo, M., Volinia, S., Marconi, D., Tissino, E., et al. (2012). The miR-17~92 family regulates the response to toll-like receptor 9 triggering of CLL cells with unmutated IGHV genes. *Leukemia*, *26*(7), 1584–1593. https://doi.org/10.1038/leu.2012.44.

Bottoni, A., Rizzotto, L., Lai, T.-H., Liu, C., Smith, L. L., Mantel, R., et al. (2016). Targeting BTK through microRNA in chronic lymphocytic leukemia. *Blood*, *128*(26), 3101–3112. https://doi.org/10.1182/blood-2016-07-727750.

Bresin, A., Callegari, E., D'Abundo, L., Cattani, C., Bassi, C., Zagatti, B., et al. (2015). miR-181b as a therapeutic agent for chronic lymphocytic leukemia in the Eμ-TCL1 mouse model. *Oncotarget*, *6*(23), 19807–19818. https://doi.org/10.18632/oncotarget.4415.

Bresin, A., D'Abundo, L., Narducci, M. G., Fiorenza, M. T., Croce, C. M., Negrini, M., et al. (2016). TCL1 transgenic mouse model as a tool for the study of therapeutic targets and microenvironment in human B-cell chronic lymphocytic leukemia. *Cell Death & Disease*, *7*. https://doi.org/10.1038/cddis.2015.419, e2071.

Bruns, H., Böttcher, M., Qorraj, M., Fabri, M., Jitschin, S., Dindorf, J., et al. (2017). CLL-cell-mediated MDSC induction by exosomal miR-155 transfer is disrupted by vitamin D. *Leukemia*, *31*(4), 985–988. https://doi.org/10.1038/leu.2016.378.

Buggins, A. G. S., & Pepper, C. J. (2010). The role of Bcl-2 family proteins in chronic lymphocytic leukaemia. *Leukemia Research*, *34*(7), 837–842. https://doi.org/10.1016/j.leukres.2010.03.011.

Calin, G. A., Dumitru, C. D., Shimizu, M., Bichi, R., Zupo, S., Noch, E., et al. (2002). Frequent deletions and down-regulation of micro- RNA genes miR15 and miR16 at 13q14 in chronic lymphocytic leukemia. *Proceedings of the National Academy of Sciences of the United States of America*, *99*(24), 15524–15529. https://doi.org/10.1073/pnas.242606799.

Calin, G. A., Ferracin, M., Cimmino, A., Di Leva, G., Shimizu, M., Wojcik, S. E., et al. (2005). A MicroRNA signature associated with prognosis and progression in chronic lymphocytic leukemia. *The New England Journal of Medicine*, *353*(17), 1793–1801. https://doi.org/10.1056/NEJMoa050995.

Calin, G. A., Liu, C.-G., Sevignani, C., Ferracin, M., Felli, N., Dumitru, C. D., et al. (2004). MicroRNA profiling reveals distinct signatures in B cell chronic lymphocytic leukemias. *Proceedings of the National Academy of Sciences of the United States of America*, *101*(32), 11755–11760. https://doi.org/10.1073/pnas.0404432101.

Carabia, J., Carpio, C., Abrisqueta, P., Jiménez, I., Purroy, N., Calpe, E., et al. (2017). Microenvironment regulates the expression of MIR-21 and tumor suppressor genes PTEN, PIAS3 and PDCD4 through ZAP-70 in chronic lymphocytic leukemia. *Scientific Reports*, *7*(1). https://doi.org/10.1038/s41598-017-12135-7.

Cerna, K., Oppelt, J., Chochola, V., Musilova, K., Seda, V., Pavlasova, G., et al. (2019). MicroRNA miR-34a downregulates FOXP1 during DNA damage response to limit BCR signalling in chronic lymphocytic leukaemia B cells. *Leukemia*, *33*(2), 403–414. https://doi.org/10.1038/s41375-018-0230-x.

Chakraborty, C., Sharma, A. R., Sharma, G., & Lee, S.-S. (2021). Therapeutic advances of miRNAs: A preclinical and clinical update. *Journal of Advanced Research*, *28*, 127–138. https://doi.org/10.1016/j.jare.2020.08.012.

Chen, C.-Z., Li, L., Lodish, H. F., & Bartel, D. P. (2004). MicroRNAs modulate hematopoietic lineage differentiation. *Science*, *303*(5654), 83–86. https://doi.org/10.1126/science.1091903.

Chou, C. H., Shrestha, S., Yang, C. D., Chang, N. W., Lin, Y. L., Liao, K. W., et al. (2018). MiRTarBase update 2018: A resource for experimentally validated microRNA-target interactions. *Nucleic Acids Research*, *46*(1), D296–D302. https://doi.org/10.1093/nar/gkx1067.

Cimmino, A., Calin, G. A., Fabbri, M., Iorio, M. V., Ferracin, M., Shimizu, M., et al. (2005). miR-15 and miR-16 induce apoptosis by targeting BCL2. *Proceedings of the National Academy of Sciences of the United States of America*, *102*(39), 13944–13949. https://doi.org/10.1073/pnas.0506654102.

Crighton, D., Woiwode, A., Zhang, C., Mandavia, N., Morton, J. P., Warnock, L. J., et al. (2003). p53 represses RNA polymerase III transcription by targeting TBP and inhibiting promoter occupancy by TFIIIB. *The EMBO Journal*, *22*(11), 2810–2820. https://doi.org/10.1093/emboj/cdg265.

Cui, B., Chen, L., Zhang, S., Mraz, M., Fecteau, J.-F., Yu, J., et al. (2014). MicroRNA-155 influences B-cell receptor signaling and associates with aggressive disease in chronic lymphocytic leukemia. *Blood*, *124*(4), 546–554. https://doi.org/10.1182/blood-2014-03-559690.

D'Abundo, L., Callegari, E., Bresin, A., Chillemi, A., Elamin, B. K., Guerriero, P., et al. (2017). Anti-leukemic activity of microRNA-26a in a chronic lymphocytic leukemia mouse model. *Oncogene*, *36*(47), 6617–6626. https://doi.org/10.1038/onc.2017.269.

Darwiche, W., Gubler, B., Marolleau, J.-P., & Ghamlouch, H. (2018). Chronic lymphocytic leukemia B-cell normal cellular counterpart: Clues from a functional perspective. *Frontiers in Immunology*, *9*. https://doi.org/10.3389/fimmu.2018.00683.

Di Marco, M., Veschi, S., Lanuti, P., Ramassone, A., Pacillo, S., Pagotto, S., et al. (2021). Enhanced expression of miR-181b in B cells of CLL improves the anti-tumor cytotoxic T cell response. *Cancers*, *13*(2), 257. https://doi.org/10.3390/cancers13020257.

Dicker, F., Kater, A. P., Fukuda, T., & Kipps, T. J. (2005). Fas-ligand (CD178) and TRAIL synergistically induce apoptosis of CD40-activated chronic lymphocytic leukemia B cells. *Blood*, *105*(8), 3193–3198. https://doi.org/10.1182/blood-2003-10-3684.

Edelmann, J., Holzmann, K., Miller, F., Winkler, D., Bühler, A., Zenz, T., et al. (2012). High-resolution genomic profiling of chronic lymphocytic leukemia reveals new recurrent genomic alterations. *Blood*, *120*(24), 4783–4794. https://doi.org/10.1182/blood-2012-04-423517.

Fabbri, M., Bottoni, A., Shimizu, M., Spizzo, R., Nicoloso, M. S., Rossi, S., et al. (2011). Association of a MicroRNA/TP53 feedback circuitry with pathogenesis and outcome of B-cell chronic lymphocytic leukemia. *JAMA*, *305*(1), 59–67. https://doi.org/10.1001/jama.2010.1919.

Ferracin, M., Zagatti, B., Rizzotto, L., Cavazzini, F., Veronese, A., Ciccone, M., et al. (2010). MicroRNAs involvement in fludarabine refractory chronic lymphocytic leukemia. *Molecular Cancer*, *9*(1), 123. https://doi.org/10.1186/1476-4598-9-123.

Fortunato, O., & Iorio, M. V. (2020). The therapeutic potential of micrornas in cancer: Illusion or opportunity? *Pharmaceuticals*, *13*(12), 1–12. https://doi.org/10.3390/ph13120438.

Fragoso, R., Mao, T., Wang, S., Schaffert, S., Gong, X., Yue, S., et al. (2012). Modulating the strength and threshold of NOTCH oncogenic signals by mir-181a-1/b-1. *PLoS Genetics*, *8*(8). https://doi.org/10.1371/journal.pgen.1002855, e1002855.

Gahn, B., Schäfer, C., Neef, J., Troff, C., Feuring-Buske, M., Hiddemann, W., et al. (1997). Detection of trisomy 12 and Rb-deletion in CD34+ cells of patients with B-cell chronic lymphocytic leukemia. *Blood*, *89*(12), 4275–4281.

Herndon, T. M., Chen, S.-S., Saba, N. S., Valdez, J., Emson, C., Gatmaitan, M., et al. (2017). Direct in vivo evidence for increased proliferation of CLL cells in lymph nodes compared to bone marrow and peripheral blood. *Leukemia*, *31*(6), 1340–1347. https://doi.org/10.1038/leu.2017.11.

Hutter, K., Rülicke, T., Drach, M., Andersen, L., Villunger, A., & Herzog, S. (2020). Differential roles of miR-15a/16-1 and miR-497/195 clusters in immune cell development and homeostasis. *The FEBS Journal*. https://doi.org/10.1111/febs.15493.

Jaramillo, S., Agathangelidis, A., Schneider, C., Bahlo, J., Robrecht, S., Tausch, E., et al. (2019). Prognostic impact of prevalent chronic lymphocytic leukemia stereotyped subsets: Analysis within prospective clinical trials of the German CLL Study Group (GCLLSG). *Haematologica*. https://doi.org/10.3324/haematol.2019.231027.

Kikushige, Y., Ishikawa, F., Miyamoto, T., Shima, T., Urata, S., Yoshimoto, G., et al. (2011). Self-renewing hematopoietic stem cell is the primary target in pathogenesis of human chronic lymphocytic leukemia. *Cancer Cell*, *20*(2), 246–259. https://doi.org/10.1016/j.ccr.2011.06.029.

King, J. K., Ung, N. M., Paing, M. H., Contreras, J. R., Alberti, M. O., Fernando, T. R., et al. (2016). Regulation of marginal zone B-cell differentiation by MicroRNA-146a. *Frontiers in Immunology*, *7*, 670. https://doi.org/10.3389/fimmu.2016.00670.

Klein, U., Lia, M., Crespo, M., Siegel, R., Shen, Q., Mo, T., et al. (2010). The DLEU2/miR-15a/16-1 cluster controls B cell proliferation and its deletion leads to chronic lymphocytic leukemia. *Cancer Cell*, *17*(1), 28–40. https://doi.org/10.1016/j.ccr.2009.11.019.

Klein, U., Tu, Y., Stolovitzky, G. A., Mattioli, M., Cattoretti, G., Husson, H., et al. (2001). Gene expression profiling of B cell chronic lymphocytic leukemia reveals a homogeneous phenotype related to memory B cells. *The Journal of Experimental Medicine*, *194*(11), 1625–1638. https://doi.org/10.1084/jem.194.11.1625.

Kolb, J. P., Kern, C., Quiney, C., Roman, V., & Billard, C. (2003). Re-establishment of a normal apoptotic process as a therapeutic approach in B-CLL. *Current Drug Targets. Cardiovascular & Haematological Disorders*, *3*(4), 261–286. https://doi.org/10.2174/1568006033481384.

Landau, D. A., Carter, S., Stojanov, P., McKenna, A., Stevenson, K., Lawrence, M. S., … Wu, C. J. (2013). Evolution and impact of subclonal mutations in chronic lymphocytic leukemia. *Cell*, *152*(4), 714–726. https://doi.org/10.1016/j.cell.2013.01.019.

Lerner, M., Harada, M., Lovén, J., Castro, J., Davis, Z., Oscier, D., et al. (2009). DLEU2, frequently deleted in malignancy, functions as a critical host gene of the cell cycle inhibitory microRNAs miR-15a and miR-16-1. *Experimental Cell Research*, *315*(17), 2941–2952. https://doi.org/10.1016/j.yexcr.2009.07.001.

Li, J., Wan, Y., Ji, Q., Fang, Y., & Wu, Y. (2013). The role of microRNAs in B-cell development and function. *Cellular & Molecular Immunology*, *10*(2), 107–112. https://doi.org/10.1038/cmi.2012.62.

Li, L., Zhang, J., Diao, W., Wang, D., Wei, Y., Zhang, C.-Y., et al. (2014). MicroRNA-155 and MicroRNA-21 promote the expansion of functional myeloid-derived suppressor cells. *The Journal of Immunology*, *192*(3), 1034–1043. https://doi.org/10.4049/jimmunol.1301309.

Li, S., Moffett, H. F., Lu, J., Werner, L., Zhang, H., Ritz, J., et al. (2011). MicroRNA expression profiling identifies activated B cell status in chronic lymphocytic leukemia cells. *PLoS ONE*, *6*(3). https://doi.org/10.1371/journal.pone.0016956.

Lindner, S. E., Lohmüller, M., Kotkamp, B., Schuler, F., Knust, Z., Villunger, A., et al. (2017). The miR-15 family reinforces the transition from proliferation to differentiation in pre-B cells. *EMBO Reports*, *18*(9), 1604–1617. https://doi.org/10.15252/embr.201643735.

Lovat, F., Fassan, M., Gasparini, P., Rizzotto, L., Cascione, L., Pizzi, M., et al. (2015). MiR-15b/16-2 deletion promotes B-cell malignancies. *Proceedings of the National Academy of Sciences of the United States of America*, *112*(37), 11636–11641. https://doi.org/10.1073/pnas.1514954112.

Mackus, W. J. M., Kater, A. P., Grummels, A., Evers, L. M., Hooijbrink, B., Kramer, M. H. H., et al. (2005). Chronic lymphocytic leukemia cells display p53-dependent drug-induced Puma upregulation. *Leukemia*, *19*(3), 427–434. https://doi.org/10.1038/sj.leu.2403623.

Marti, G. E., Metcalf, R. A., & Raveche, E. (1995). The natural history of a lymphoproliferative disorder in aged NZB mice. *Current Topics in Microbiology and Immunology*, *194*, 117–126. https://doi.org/10.1007/978-3-642-79275-5_15.

Mertens, D., Wolf, S., Tschuch, C., Mund, C., Kienle, D., Ohl, S., et al. (2006). Allelic silencing at the tumor-suppressor locus 13q14.3 suggests an epigenetic tumor-suppressor mechanism. *Proceedings of the National Academy of Sciences of the United States of America*, *103*(20), 7741–7746. https://doi.org/10.1073/pnas.0600494103.

Moussay, E., Wang, K., Cho, J. H., Van Moer, K., Pierson, S., Paggetti, J., et al. (2011). MicroRNA as biomarkers and regulators in B-cell chronic lymphocytic leukemia. *Proceedings of the National Academy of Sciences of the United States of America*, *108*(16), 6573–6578. https://doi.org/10.1073/pnas.1019557108.

Mraz, M., Chen, L., Rassenti, L. Z., Ghia, E. M., Li, H., Jepsen, K., et al. (2014). miR-150 influences B-cell receptor signaling in chronic lymphocytic leukemia by regulating expression of GAB1 and FOXP1. *Blood*, *124*(1), 84–95. https://doi.org/10.1182/blood-2013-09-527234.

Negrini, M., Cutrona, G., Bassi, C., Fabris, S., Zagatti, B., Colombo, M., et al. (2014). microRNAome expression in chronic lymphocytic leukemia: Comparison with normal B-cell subsets and correlations with prognostic and clinical parameters. *Clinical Cancer Research: An Official Journal of the American Association for Cancer Research*, *20*(15), 4141–4153. https://doi.org/10.1158/1078-0432.CCR-13-2497.

Niiro, H., & Clark, E. A. (2002). Regulation of B-cell fate by antigen-receptor signals. *Nature Reviews Immunology*, *2*(12), 945–956. https://doi.org/10.1038/nri955.

Ntoufa, S., Papakonstantinou, N., Apollonio, B., Gounari, M., Galigalidou, C., Fonte, E., et al. (2016). B cell anergy modulated by TLR1/2 and the miR-17~92 cluster underlies the indolent clinical course of chronic lymphocytic leukemia stereotyped subset #4. *Journal of Immunology (Baltimore, Md. : 1950)*, *196*(10), 4410–4417. https://doi.org/10.4049/jimmunol.1502297.

Packham, G., & Stevenson, F. K. (2005). Bodyguards and assassins: Bcl-2 family proteins and apoptosis control in chronic lymphocytic leukaemia. *Immunology*, *114*(4), 441–449. https://doi.org/10.1111/j.1365-2567.2005.02117.x.

Paggetti, J., Haderk, F., Seiffert, M., Janji, B., Distler, U., Ammerlaan, W., et al. (2015). Exosomes released by chronic lymphocytic leukemia cells induce the transition of stromal cells into cancer-associated fibroblasts. *Blood*, *126*(9), 1106–1117. https://doi.org/10.1182/blood-2014-12-618025.

Pagotto, S., Veronese, A., Soranno, A., Balatti, V., Ramassone, A., Guanciali-Franchi, P. E., et al. (2019). HNRNPL restrains miR-155 targeting of BUB1 to stabilize aberrant karyotypes of transformed cells in chronic lymphocytic leukemia. *Cancers*, *11*(4), 575. https://doi.org/10.3390/cancers11040575.

Pagotto, S., Veronese, A., Soranno, A., Lanuti, P., Di Marco, M., Russo, M. V., et al. (2018). Hsa-miR-155-5p drives aneuploidy at early stages of cellular transformation. *Oncotarget*, *9*(16), 13036–13047. https://doi.org/10.18632/oncotarget.24437.

Palacios, F., Abreu, C., Prieto, D., Morande, P., Ruiz, S., Fernández-Calero, T., et al. (2015). Activation of the PI3K/AKT pathway by microRNA-22 results in CLL B-cell proliferation. *Leukemia*, *29*(1), 115–125. https://doi.org/10.1038/leu.2014.158.

Pekarsky, Y., Balatti, V., & Croce, C. M. (2018). BCL2 and miR-15/16: From gene discovery to treatment. *Cell Death and Differentiation*, *25*(1), 21–26. https://doi.org/10.1038/cdd.2017.159.

Puiggros, A., Blanco, G., & Espinet, B. (2014). Genetic abnormalities in chronic lymphocytic leukemia: Where we are and where we go. *BioMed Research International*. https://doi.org/10.1155/2014/435983.

Quijada-Álamo, M., Hernández-Sánchez, M., Robledo, C., Hernández-Sánchez, J.-M., Benito, R., Montaño, A., et al. (2017). Next-generation sequencing and FISH studies reveal the appearance of gene mutations and chromosomal abnormalities in hematopoietic progenitors in chronic lymphocytic leukemia. *Journal of Hematology & Oncology*, *10*(1), 83. https://doi.org/10.1186/s13045-017-0450-y.

Rossi, S., Shimizu, M., Barbarotto, E., Nicoloso, M. S., Dimitri, F., Sampath, D., et al. (2010). microRNA fingerprinting of CLL patients with chromosome 17p deletion identify a miR-21 score that stratifies early survival. *Blood*, *116*(6), 945–952. https://doi.org/10.1182/blood-2010-01-263889.

Rupaimoole, R., & Slack, F. J. (2017). MicroRNA therapeutics: Towards a new era for the management of cancer and other diseases. *Nature Reviews Drug Discovery*, *16*(3), 203–221. https://doi.org/10.1038/nrd.2016.246.

Sanz, I., Wei, C., Jenks, S. A., Cashman, K. S., Tipton, C., Woodruff, M. C., et al. (2019). Challenges and opportunities for consistent classification of human B cell and plasma cell populations. *Frontiers in Immunology*, *10*. https://doi.org/10.3389/fimmu.2019.02458.

Scaglione, B. J., Salerno, E., Balan, M., Coffman, F., Landgraf, P., Abbasi, F., et al. (2007). Murine models of CLL: Role of microRNA-16 in the NZB mouse model. *British Journal of Haematology*, *139*(5), 645–657. https://doi.org/10.1111/j.1365-2141.2007.06851.x.

Seifert, M., Sellmann, L., Bloehdorn, J., Wein, F., Stilgenbauer, S., Dürig, J., et al. (2012). Cellular origin and pathophysiology of chronic lymphocytic leukemia. *Journal of Experimental Medicine*, *209*(12), 2183–2198. https://doi.org/10.1084/jem.20120833.

Sharma, S., Pavlasova, G., Seda, V., Cerna, K., Vojackova, E., Filip, D., et al. (2020). miR-29 modulates CD40 signaling in chronic lymphocytic leukemia by targeting TRAF4: An axis affected by BCR inhibitors. *Blood*. https://doi.org/10.1182/blood.2020005627, blood.2020005627.

Shukla, A., Rai, K., Shukla, V., Chaturvedi, N. K., Bociek, R. G., Pirruccello, S. J., et al. (2016). Sprouty 2: A novel attenuator of B-cell receptor and MAPK-Erk signaling in CLL. *Blood*, *127*(19), 2310–2321. https://doi.org/10.1182/blood-2015-09-669317.

Simonetti, G., Bertilaccio, M. T. S., Ghia, P., & Klein, U. (2014). Mouse models in the study of chronic lymphocytic leukemia pathogenesis and therapy. *Blood*, *124*(7), 1010–1019. https://doi.org/10.1182/blood-2014-05-577122.

Sivina, M., Hartmann, E., Vasyutina, E., Boucas, J. M., Breuer, A., Keating, M. J., et al. (2012). Stromal cells modulate TCL1 expression, interacting AP-1 components and TCL1-targeting micro-RNAs in chronic lymphocytic leukemia. *Leukemia*, *26*(8), 1812–1820. https://doi.org/10.1038/leu.2012.63.

Stamatopoulos, B., Meuleman, N., Haibe-Kains, B., Saussoy, P., Van Den Neste, E., Michaux, L., et al. (2009). microRNA-29c and microRNA-223 down-regulation has in vivo significance in chronic lymphocytic leukemia and improves disease risk stratification. *Blood*, *113*(21), 5237–5245. https://doi.org/10.1182/blood-2008-11-189407.

ten Hacken, E., & Burger, J. A. (2016). Microenvironment interactions and B-cell receptor signaling in chronic lymphocytic leukemia: Implications for disease pathogenesis and treatment. *Biochimica et Biophysica Acta-Molecular Cell Research*, *1863*(3), 401–413. https://doi.org/10.1016/j.bbamcr.2015.07.009.

Veronese, A., Pepe, F., Chiacchia, J., Pagotto, S., Lanuti, P., Veschi, S., et al. (2015). Allele-specific loss and transcription of the miR-15a/16-1 cluster in chronic lymphocytic leukemia. *Leukemia*, *29*(1), 86–95. https://doi.org/10.1038/leu.2014.139.

Visone, R., Rassenti, L. Z., Veronese, A., Taccioli, C., Costinean, S., Aguda, B. D., et al. (2009). Karyotype-specific microRNA signature in chronic lymphocytic leukemia. *Blood*, *114*(18), 3872–3879. https://doi.org/10.1182/blood-2009-06-229211.

Visone, R., Veronese, A., Balatti, V., & Croce, C. M. (2012). MiR-181b: New perspective to evaluate disease progression in chronic lymphocytic leukemia. *Oncotarget*, *3*(2), 195–202. http://www.impactjournals.com/oncotarget/index.php?journal=oncotarget&page=article&op=download&path%5B%5D=448&path%5B%5D=741.

Visone, R., Veronese, A., Rassenti, L. Z., Balatti, V., Pearl, D. K., Acunzo, M., et al. (2011). miR-181b is a biomarker of disease progression in chronic lymphocytic leukemia. *Blood*, *118*(11), 3072–3079. https://doi.org/10.1182/blood-2011-01-333484.

Xiao, C., Calado, D. P., Galler, G., Thai, T.-H., Patterson, H. C., Wang, J., et al. (2007). MiR-150 controls B cell differentiation by targeting the transcription factor c-Myb. *Cell*, *131*(1), 146–159. https://doi.org/10.1016/j.cell.2007.07.021.

Yeh, Y.-Y., Ozer, H. G., Lehman, A. M., Maddocks, K., Yu, L., Johnson, A. J., et al. (2015). Characterization of CLL exosomes reveals a distinct microRNA signature and enhanced secretion by activation of BCR signaling. *Blood*, *125*(21), 3297–3305. https://doi.org/10.1182/blood-2014-12-618470.

Zhu, D. X., Zhu, W., Fang, C., Fan, L., Zou, Z. J., Wang, Y. H., et al. (2012). MiR-181a/b significantly enhances drug sensitivity in chronic lymphocytic leukemia cells via targeting multiple anti-apoptosis genes. *Carcinogenesis*, *33*(7), 1294–1301. https://doi.org/10.1093/carcin/bgs179.

Chapter 15

Pathophysiology roles and translational opportunities of miRNAs in lymphoma

Christine E. Cutucache[a,b,c]

[a]STEM Teaching, Research, and Inquiry-Based Learning Center, University of Nebraska at Omaha, Omaha, NE, United States, [b]Department of Biology, University of Nebraska at Omaha, Omaha, NE, United States, [c]Fred and Pamela Buffett Cancer Center, University of Nebraska Medical Center, Omaha, NE, United States

Summary

The role of miRNA in lymphoma remains to be fully elucidated. However, as we have observed from many hematological malignancies, the function of miRNA can directly affect tumorigenesis, disease progression, and pathophysiology. With improved understanding of miRNA, we can expand our diagnostic methods to improve the early identification of lymphoma, and ultimately inform prognostication to give patients clarity and a range of personalized treatment opportunities. This chapter highlights the incidence, prevalence, and existing data surrounding miRNA within lymphomas, with a special focus on B-cell lymphomas, diffuse large B-cell lymphoma, and splenic marginal zone lymphoma.

microRNAs (miRNAs) are still considered nuanced regulators, identified only recently. They provide a unique opportunity to realize the contributions to advancement within the field of genetics over the last 20 years. As you have already seen across the chapters within this book, miRNAs have endless possibilities for diagnostics and prognostics while their role in disease pathogenesis remains an ongoing area of research.

Novel contributions from regulatory elements, such as miRNAs, provide the ability to think more broadly about how we are studying hematological malignancies, how we are working to manage them, and how we are tracking things such as pathophysiology to include disease progression and regression.

miRNAs are small RNA molecules of only 21–23 nucleotides in length, and they can bind to partially complementary regions of the 3′ UTR of target transcripts. This binding relationship regulates gene expression at the posttranscriptional level. miRNAs have homeostatic functions that range from regulating development to contributing to tumorigenesis through both brake and acceleration roles.

Because we are focusing on human malignancies, and particularly lymphoma, it makes sense to compare and contrast homeostatic, healthy immune cell development and regulation with that of the changes that take place to drive tumorigenesis. As you will read below, much of these changes take place within the primary and secondary lymphoid tissues, and during cellular differentiation. Consequently, malignancies of these organs with documented miRNA involvement will be elaborated upon to demonstrate similarities and differences by tissue site as well as by incidence, to include diffuse large B-cell lymphoma (DLBCL) and splenic marginal zone lymphoma (SMZL). While DLBCL is one of the more common lymphomas (and the most common adult lymphoma, CLL, is covered elsewhere in this text) and SMZL is one of the rarest, the involvement of miRNA will be discussed, demonstrating the importance of miRNA for clinical diagnosis and ultimately management of the malignancy.

The expression profiles and cytokine profiling within tumors and the tumor microenvironment, especially within B-cell lymphomas, have aided our understanding of these events in both a healthy environment and during tumorigenesis. The identification of the role of miRNAs (and their unique expression profiles) has improved our understanding of regulatory changes that drive tumorigenesis (whether through a brake or accelerator mechanism), and can make a patient refractory to treatment. While much of what we know as a field surrounding lymphoma pathophysiology comes from understanding molecular events involving gene expression, miRNA "ups the ante" and gives us another lens through which we can better understand the uniqueness of progression between individuals, not yet captured simply by gene-based events.

This chapter will focus on the pathophysiological roles of miRNA identified in lymphomas to date. Further, we will discuss translational opportunities for the management of lymphoma, particularly highlighting evidence from B-cell malignancies.

MicroRNA in Human Malignancies. https://doi.org/10.1016/B978-0-12-822287-4.00010-4

miRNAs and pathophysiology

Research articles published on the contributions of miRNA to the onset and progression of malignancies have increased 10-fold over the past 15 years. While the field is expanding to document the relevance and role of miRNA in lymphoma, there are three emergent roles documented to date for miRNA, and these are related to: (1) tumorigenesis, (2) progression with prognostics, and (3) potential therapeutics.

At the close of 2018, there were just more than 500,000 new cases of NHL annually in the world (World Health Organization (WHO), 2020). This places NHL in 11th place for the most common incidences worldwide, trailing only lung, breast, colorectal, prostate, stomach, liver, esophagus, cervix uteri, thyroid, and bladder cancer (Fig. 15.1). Further, NHL remains more common in western and developed countries (Fig. 15.1). Also, NHL is more prevalent in males than females, regardless of geographic location (Figs. 15.2 and 15.3; World Health Organization (WHO), 2020).

miRNAs bind to the 3′ untranslated region (UTR) of a target message and inhibit the translation of such a message as well as initiate its degradation process (Ambros, 2004; Calin & Croce, 2006; Hwang & Mendell, 2006; Jovanovic & Hengartner, 2006; Kloosterman & Plasterk, 2006; Malumbres et al., 2009). Beginning in the early 2000s, miRNA signatures have been identified to function similarly to gene expression signatures to uniquely identify and distinguish cancer subtypes; this will be much of the focus of this chapter.

miRNA coding sequences are often found at genomic regions that are associated with tumorigenesis (Calin & Croce, 2006; Malumbres et al., 2009). It was well reasoned that the inclusion of miRNA regulation during hematopoesis and resultant cellular differentiation also occurred to influence cellular function. Therefore, Malumbres and colleagues investigated the miRNA profiles within healthy/normal peripheral b-cell subpopulations (centroblasts, naïve, and memory B-cells) and T-cells, and observed unique changes in miRNA expression during different stages of healthy B-cell development (Malumbres et al., 2009). Via these investigations of healthy B-cell differentiation, a differential expression signature of miRNAs was observed in germinal center (GC) lymphocytes (essentially, centroblasts). For example, miR-125b downregulates the expression of core genes (namely *PRDM1/BLIMP1 and IRF4*) that signal differentiation from centroblasts to memory and plasma cells (Angelin-Duclos, Cattoretti, Lin, & Calame, 2000; Klein et al., 2006; Kuo et al., 2007; Malumbres et al., 2009). Importantly, these findings served as the underpinnings to demonstrate the functional importance of the regulation of miRNA expression for future investigations in determining the significance and influence of miRNA in leukemia and lymphoma. This is because these data provided foundational evidence for the involvement of miRNAs in the regulation of immune cell differentiation processes.

While the potential for miRNA in the diagnosis, prognosis, treatment, and overall management of human malignancies (particularly lymphoma) is very promising, major challenges exist with miRNAs because of their unpredictable nature. For example, one miRNA can regulate multiple target mRNAs (and vice versa); therefore while we can make classifications and predictions upon miRNA, the presence of paralogs and the unpredictable nature of miRNAs make them a "moving target."

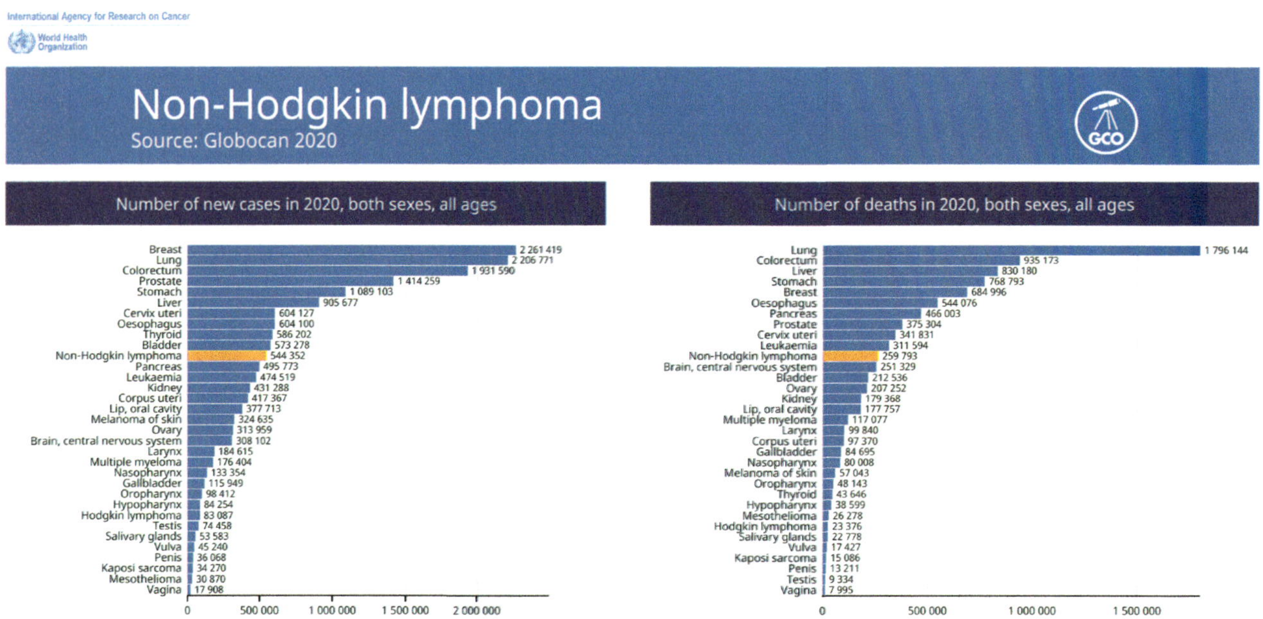

FIG. 15.1 NHL incidence and death rates worldwide. Side-by-side comparison of incidence and deaths for the most common cancers worldwide, including non-Hodgkin lymphomas. (Courtesy of SEER. https://gco.iarc.fr/today/data/factsheets/cancers/34-Non-hodgkin-lymphoma-fact-sheet.pdf (Accessed October 2020).)

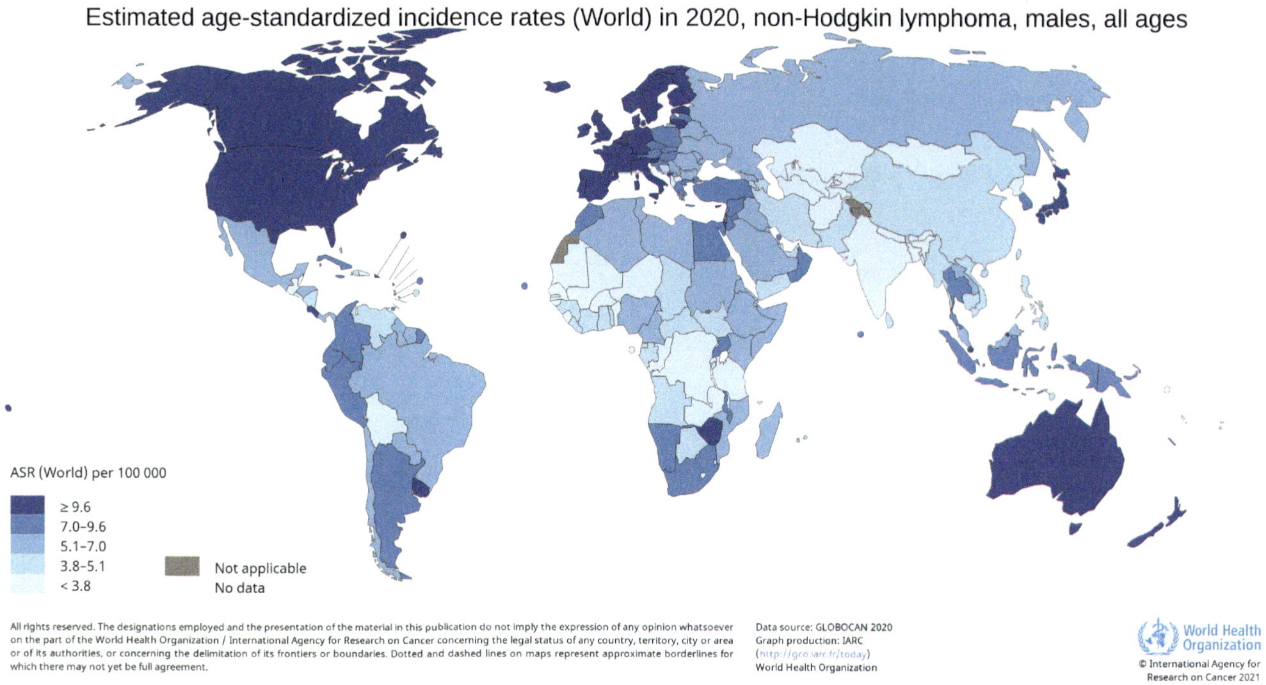

FIG. 15.2 Worldwide incidence of non-Hodgkin lymphoma for males. Worldwide incidence of males affected with non-Hodgkin lymphoma to demonstrate the preponderance for men, especially in western countries. (Courtesy of SEER. https://gco.iarc.fr/today/data/factsheets/cancers/34-Non-hodgkin-lymphoma-fact-sheet.pdf (Accessed October 2020).)

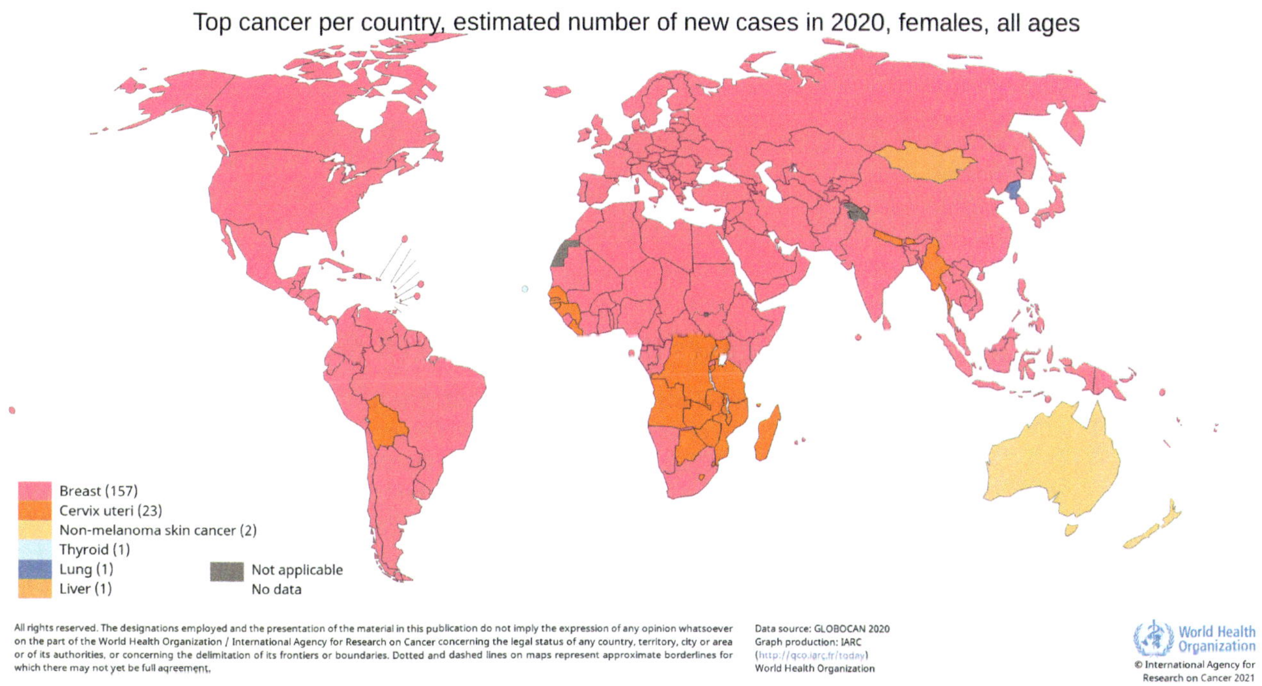

FIG. 15.3 Worldwide incidence of females affected with non-Hodgkin lymphoma. Worldwide incidence of females affected with non-Hodgkin lymphoma to demonstrate the preponderance for men as opposed to women while reemphasizing the prevalence in western countries. (Courtesy of: SEER. https://gco.iarc.fr/today/data/factsheets/cancers/34-Non-hodgkin-lymphoma-fact-sheet.pdf (Accessed October 2020).)

However, there are a few miRNAs that have been identified as having critical roles within lymphoma, including miR-155, miR15a-, and miR-16. The influence of miR-15a and miR16 is discussed in other chapters that focus on chronic lymphocytic leukemia (CLL). In this chapter, we will discuss miRNAs with established roles in tumorigenesis, progression, and/or management of diffuse large B-cell lymphoma (DLBCL), mantle cell lymphoma, follicular lymphoma, and nodal marginal zone lymphomas (NMZL), with elaboration upon the subtypes. The roles of these miRNAs within lymphoma will be highlighted herein.

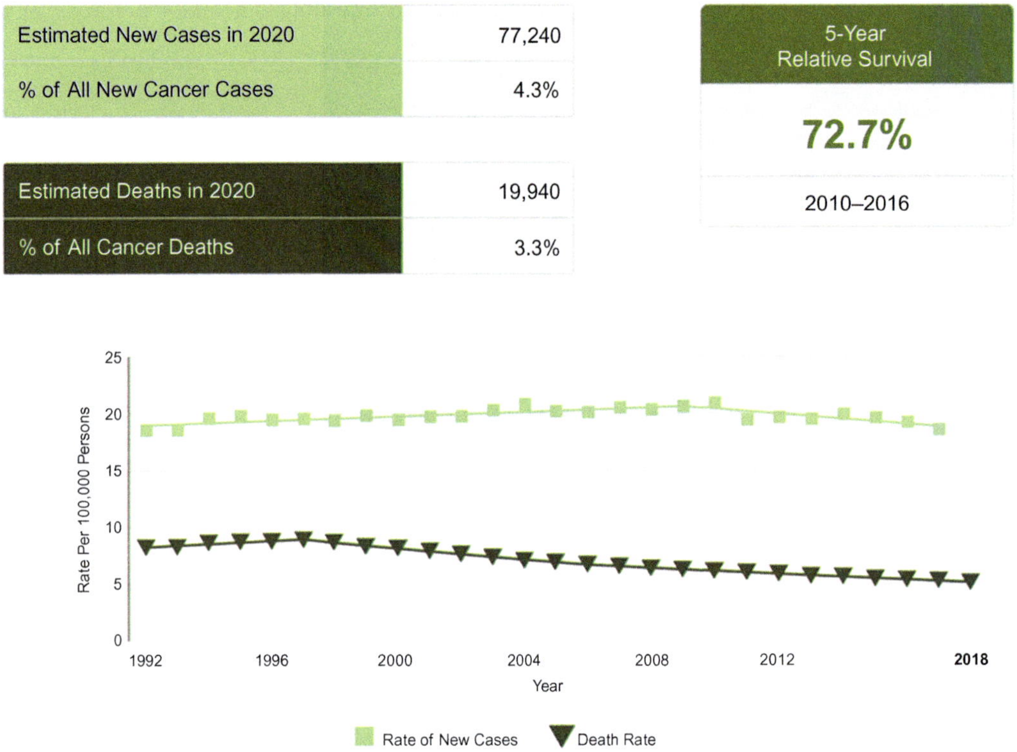

FIG. 15.4 Incidence rate and death rate changes over three decades. The cancer incidence, death rates, and 5-year relative survival data are presented from SEER estimates; https://seer.cancer.gov. Fortunately, the cases of cancer remain mostly stagnant despite fluctuations in the overall worldwide population, and similarly death rates have declined by nearly half over the last 30 years.

Finally, while the incidence and prevalence rates remain high worldwide, the death rate within the United States (a country with high incidence) has been decreasing since the late 1990s. Additionally, in the United States, the rate of new cases has remained largely unchanged from the early 1990s to present. The most recent 5-year survival data for men and women diagnosed with NHL in the United States is from 2010 to 2016, and they show a 73% 5-year survival rate (SEER, 2020; Fig. 15.4).

miRNAs regulate the expression of genes that promote B-cell differentiation

Given that miRNAs are differentially expressed temporally in various stages of B-cell development, it was well reasoned that they similarly influenced the plasticity of these very cells, thus resulting in the promotion of malignancy. Importantly, two miRNAs contribute to the downregulation of *BCL2*, a core molecular in the differentiation of B-cells. These two miRNAs, miR-15a and miR-16, suppress *BCL2*, thereby impacting GC lymphocyte development and actually promoting apoptosis (G. Calin & Croce, 2006; G.A. Calin et al., 2008; Cimmino et al., 2005; Xia et al., 2008).

Lymphomagenesis

While traditionally either chromosomal changes and then gene expression changes have aided cancer diagnosis, with key molecular features such as the upregulation of oncogenes or the suppression or loss of tumor suppressors signaling oncogenesis, miRNAs serve as another unique classifier that aids in further delineating between subtypes of diseases. Essentially, malignant lymphomas come from healthy B-cells at various stages of differentiation.

Because miRNAs are involved in oncogenesis in a variety of cancers, we will instead focus on the role of miRNAs as unique contributors to lymphomagenesis across B-cell lymphomas only in an effort to describe their unique role even among diseases with a singular etiology.

The potential for miRNAs to contribute to lymphomagenesis is well documented (Kapsogeorgou & Tzioufas, 2018).

DLBCL and miRNA

DLBCL is a heterogeneous group of tumors that represents approximately 40% of new cases, making it the most common non-Hodgkin lymphoma (NHL) (Campo et al., 2011). The disease is characterized by three characteristically distinct subtypes, all with divergent response rates to immunotherapy and chemotherapy regimens (Iqbal et al., 2015; Troppan et al., 2014). Currently, DLBCL is considered curable in approximately half of all cases. Due to this heterogeneity in presentation and responsiveness, several groups have tried to further differentiate DLBCL into subtypes for both diagnostic and prognostic classifications.

These subtypes reflect their cells of origin, and include the germinal center B-cell (GCB subtype), the activated B-cell (ABC subtype), and a primary bone marrow (PBML-DLBCL) signature as well as nonclassifiable (NC) cases (Iqbal et al., 2015; Rosenwald et al., 2003; Troppan et al., 2014). While these categories were widely identified as helping to aid the classification of DLBCL, a further narrowing of categories to improve diagnostics was completed with the use of miRNA expression-based classification (Rosenwald et al., 2003).

Malumbres and colleagues demonstrated the stage of differentiation-specific miRNAs for B-cell development, and subsequently tumorigenesis. Importantly, this work aimed to classify miRNA expression by the cellular differentiation stage (i.e., from naïve B-cells to centroblasts, and ultimately memory B-cells) and to align across the corresponding etiology of DLBCL subtypes to identify similarities and differences that may contribute to diagnosis and that may aid in understanding the pathophysiology of DLBCL.

In order to further narrow diagnosis within these DLBCL subtypes, unique miRNA profiles were identified through unsupervised hierarchal clustering using expression profiles. Specifically, Iqbal and colleagues used a Bayesian algorithm to distinguish a unique, 27-miRNA classifier (this unique diagnostic profile had 11 upregulated and six downregulated miRNAs) to further distinguish Burkitt lymphoma from DLBCL. Ultimately, an eight-miRNA_classifier was derived to distinguish DLBCL subgroups with more than 80% probability (Iqbal et al., 2015; Fig. 15.5) to include miR-155 and miR-542-3p upregulated in ABC-DLBCL, and miR-28-3p, miR-28-5p, miR-129-3p, and miR-589 upregulated in GCB-DLBCL (Iqbal et al., 2015). The unique regulation of miRNAs was observed from both patient cases and within patient-derived cell

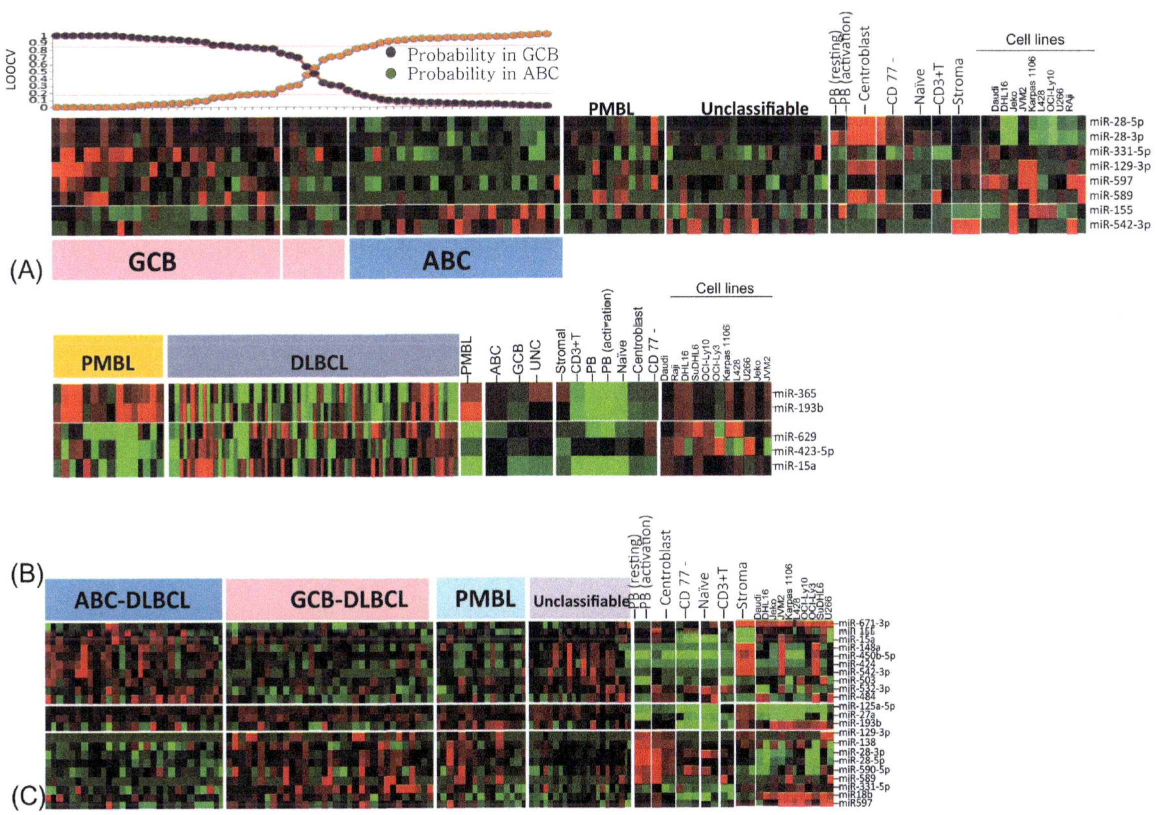

FIG. 15.5 Diffuse large B-cell lymphoma subtype discernment by expression profile. (Reproduced with permission from Dr. J. Iqbal, February 2021. PMID 25498913 Iqbal and colleagues discern subtypes of DLBCL based upon unique expression profiles.)

lines, further emphasizing the uniqueness of these classifiers to aid in diagnosis and the potential they have as therapeutic targets. The ability to distinguish and classify cell lines of ABC-type and GCB-type DLBCL cell lines was shown to have 100% predictability for classification by miRNA expression patterns by Malumbres et al. (2009). These same outcomes have been replicated in clinical cases since. Arguably, the most salient information related to pathophysiology of DLBCL is that these miRNA distinguish between the ABC and GCB subtypes of DLBCL. Furthermore, this contribution underscored that while these tumor types have divergent cells of origin (activated B-cells and germinal center B-cells, respectively), they also likely have different biological contributors to tumorigenesis.

Perhaps even more importantly, miRNA markers have aided in understanding which cases are sensitive versus refractory to particular therapeutics. For instance, miR-155 has been associated with cyclin-dependent kinase inhibitor 1A (CDKN1A or p21) and overall miR-155-expressing cells are more sensitive to AKT IV inhibition when compared with other inhibitors. Most importantly, for DLBCL an miR-155 signature was predictive of treatment *failure* in ABC-DLBCL, but not in GCB-DLBCL (Iqbal et al., 2015).

Upon further expansion of the miRNA signatures to mine for impact on prognostication, Iqbal and colleagues used a semisupervised prediction algorithm in a cohort of 222 DLBCL cases that was treated using an R-CHOP regimen. They determined that six transcripts representing three unique genes (*KLHL5*, *PSIP1*, and *CHD9*) were significantly correlated with event-free survival. This signature of low expression of *KLHL5*, *PSIP1*, and *CHD9* was associated with poor response rates from R-CHOP treatment (Iqbal et al., 2015). These data have ultimately informed a prognostication for DLBCL involving miR-155.

MZL and miRNA

Similarly to the Iqbal study, our group investigated the nodal zone lymphomas to address a similar issue of a challenge of delineation of subtypes, and subsequent classification and stratification (Robinson et al., 2020).

Diagnosis

While the REAL (REAL Classification for Lymphoma, 2020 and WHO (World Health Organization (WHO), 2020) classifications distinguished cellular etiology, it could be argued that it essentially only delineated the *stage* of progression more than the actual lymphoma subtype. Consequently, the decades that have followed focused on expanding the diagnosis (and prognosis, detailed in the next section) for these malignancies. Herein, the focus will be on B-cell lymphomas, but similar detail can be described for T-cells. For example, peripheral T-cell lymphoma—not otherwise specified, a so-called "waste basket grouping of T-cell leukemia/lymphoma," has similarly been further delineated by subcategory (Herek, Shew, Spurgin, & Cutucache, 2015).

The proliferation of gene expression profile data, the expanded banks of chromosomal aberration data, and the use of flow cytometry and additional multiplex staining methods coupled with the worldwide review of the rarer lymphomas largely contributed to the further delineation of subtypes using both histochemical and molecular traits, coupled with cellular and tissue morphology for diagnosis.

Prognosis

While splenic marginal zone lymphoma remains a more rare B-cell malignancy, the challenge of diagnosing and ultimately providing an accurate prognosis remains. SMZL has largely been determined via an exclusionary diagnosis, not unlike the DLBCL scenario early on in investigation. Consequently, we aimed to determine a unique diagnostic signature that would aid in the accuracy of classification and ultimately serve as a pilot for further association studies to determine prognostic markers from the same panel.

Part of the challenge in conveying prognostic details for SMZL is the sheer rarity of the disease, affecting approximately 0.13 in 100,000 persons annually, with overall survival being around a decade. Multiple research teams have identified genes associated with SMZL progression, but they seem to vary and are not associated clearly with prognosis. One of the genes speculated to be involved in caveolin-1 (CAV1) Fig. 15.6. *CAV1* is a common, Janis-faced tumor suppressor and oncogene whose function is determined largely by the surrounding tumor microenvironment (Deb et al., 2014) Fig. 15.6.

A graphical representation of the affected region on chromosome 7q involving a common Janis-faced oncogene and tumor suppressor, CAV1. The complexity of miRNAs that regulate CAV1 is also depicted, namely miR-376, miR-485, and miR-199a.

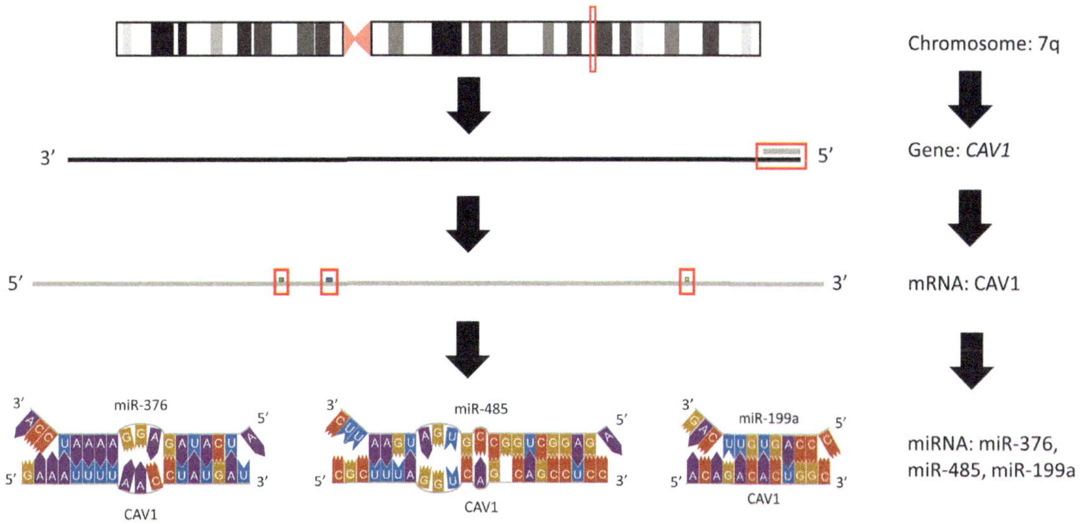

FIG. 15.6 Graphical representation of the affected region on 7q in SMZL, CLL, and other solid tumor malignancies. (From Robinson, J. E., & Cutucache, C. E. (2018). Deciphering splenic marginal zone lymphoma pathogenesis: the proposed role of microRNA. *Oncotarget, 9*(52), 30005–30022. https://doi.org/10.18632/oncotarget.25487.)

We have reported an example of a panel with clinically significant B-cell biological significance, and with early studies indicating prognostic potential (Robinson, Greiner, Bouska, Iqbal, & Cutucache, 2020). This panel, coined the SMZL-specific gene expression signature (SSGES), does not provide a catch-all for SMZL diagnosis and prognosis, further hinting at the involvement of miRNA in this B-cell lymphoma likely to mirror that of DLBCL, CLL, and others defined within this text. Needless to say, while the past 20 years have opened our eyes to the involvement of miRNA in cancer biology, diagnosis, and prognosis, we have substantial investigation ahead of us to discern the particular contributors, unique expression profiles, and prognostic significance that miRNAs warrant in additional lymphomas.

van den Brand and van Krieken's (2013) systematic review of nodal zone lymphoma emphasizes some of the challenges associated with the delineation of lymphoma types that have plagued the field for decades. Specifically, this includes the approach of not integrating clinical and pathological data (and, prior to molecular genomics, the exclusion of tumor expression profiles). Rather, the approach has largely been through a process of exclusion as opposed to identifying unique, independent markers to couple with clinical data for the identification of specific subtypes of malignancies. It is important to recognize the overall context of these diagnostic challenges within lymphoma. Unlike other types of human malignancies, some of the subtypes of leukemias and lymphomas are tremendously rare. For example, NMZL represents just over 1% of all lymphoid malignancies (2020). It is common for patients to only present later in life without viable samples *during* tumorigenesis and progression to study the disease pathophysiology. Finally, murine models do not exist for the majority of B-cell lymphomas. Essentially, only one accepted mouse model exists to model human clinical tumorigenesis, progression, and heterogeneity (the heterogeneity perhaps being the most important concept to model within lymphoma), and that is the Bichi and Croce Emu-*TCL1* model for CLL (Bichi et al., 2002). Consequently, clinicians and researchers alike are limited to using banked clinical samples, active cases, and databases of banked data to study these rare malignancies.

References

Ambros, V. (2004). The functions of animal microRNAs. *Nature, 431*, 350–355.

Angelin-Duclos, C., Cattoretti, G., Lin, K. I., & Calame, K. (2000). Commitment of B lymphocytes to a plasma cell fate is associated with Blimp-1 expression in vivo. *Journal of Immunology, 165*(10), 5462–5471. https://doi.org/10.4049/jimmunol.165.10.5462.

Bichi, R., Shinton, S. A., Martin, E. S., Koval, A., Calin, G. A., Cesari, R., … Croce, C. M. (2002). Human chronic lymphocytic leukemia modeled in mouse by targeted TCL1 expression. *Proceedings of the National Academy of Sciences of the United States of America, 99*(10), 6955–6960. https://doi.org/10.1073/pnas.102181599.

Calin, G., & Croce, C. (2006). MicroRNA signatures in human cancers. *Nature Reviews Cancer, 6*, 857–866.

Calin, G. A., Cimmino, A., Fabbri, M., Ferracin, M., Wojcik, S. E., Shimizu, M., … Croce, C. M. (2008). MiR-15a and miR-16-1 cluster functions in human leukemia. *Proceedings of the National Academy of Sciences of the United States of America, 105*(13). https://doi.org/10.1073/pnas.0800121105. 5166–71.

Campo, E., Swerdlow, S. H., Harris, N. L., Pileri, S., Stein, H., & Jaffe, E. S. (2011). The 2008 WHO classification of lymphoid neoplasms and beyond: Evolving concepts and practical applications. *Blood, 117*(19), 5019–5032. https://doi.org/10.1182/blood-2011-01-293050.

Cimmino, A., Calin, G. A., Fabbri, M., Iorio, M. V., Ferracin, M., Shimizu, M., … Croce, C. M. (2005). miR-15 and miR-16 induce apoptosis by targeting BCL2. *Proceedings of the National Academy of Sciences of the United States of America*, *102*(39), 13944–13949. https://doi.org/10.1073/pnas.0506654102.

Deb, M., Sengupta, D., Kar, S., Rath, S. K., Parbin, S., Shilpi, A., … Patra, S. K. (2014). Elucidation of caveolin 1 both as a tumor suppressor and metastasis promoter in light of epigenetic modulators. *Tumour Biology: The Journal of the International Society for Oncodevelopmental Biology and Medicine*, *35*(12), 12031–12047. https://doi.org/10.1007/s13277-014-2502-z.

Herek, T. A., Shew, T. D., Spurgin, H. N., & Cutucache, C. E. (2015). Conserved molecular underpinnings and characterization of a role for caveolin-1 in the tumor microenvironment of mature T-cell lymphomas. *PLoS One*, *10*(11). https://doi.org/10.1371/journal.pone.0142682.

Hwang, H. W., & Mendell, J. T. (2006). MicroRNAs in cell proliferation, cell death, and tumorigenesis. *British Journal of Cancer*, *94*(6), 776–780. https://doi.org/10.1038/sj.bjc.6603023.

Iqbal, J., Shen, Y., Huang, X., Liu, Y., Wake, L., Liu, C., … Chan, W.C. (2015). Global microRNA expression profiling uncovers molecular markers for classification and prognosis in aggressive B-cell lymphoma. *Blood*, *125*(7), 1137–1145. https://doi.org/10.1182/blood-2014-04-566778.

Jovanovic, M., & Hengartner, M. O. (2006). miRNAs and apoptosis: RNAs to die for. *Oncogene*, *25*(46), 6176–6187. https://doi.org/10.1038/sj.onc.1209912.

Kapsogeorgou, E. K., & Tzioufas, A. G. (2018). Role of miR200b-5p miRNA in lymphomagenesis associated with Sjögren's syndrome (SS). *Mediterranean Journal of Rheumatology*, *29*(1), 56–58. https://doi.org/10.31138/mjr.29.1.56.

Klein, U., Casola, S., Cattoretti, G., Shen, Q., Lia, M., Mo, T., … Dalla-Favera, R. (2006). Transcription factor IRF4 controls plasma cell differentiation and class-switch recombination. *Nature Immunology*, *7*(7), 773–782. https://doi.org/10.1038/ni1357.

Kloosterman, W. P., & Plasterk, R. H. A. (2006). The diverse functions of microRNAs in animal development and disease. *Developmental Cell*, *11*(4), 441–450. https://doi.org/10.1016/j.devcel.2006.09.009.

Kuo, T. C., Shaffer, A. L., Haddad, J., Yong, S. C., Staudt, L. M., & Calame, K. (2007). Repression of BCL-6 is required for the formation of human memory B cells in vitro. *Journal of Experimental Medicine*, *204*(4), 819–830. https://doi.org/10.1084/jem.20062104.

Malumbres, R., Sarosiek, K. A., Cubedo, E., Ruiz, J. W., Jiang, X., Gascoyne, R. D., … Lossos, I. S. (2009). Differentiation stage-specific expression of microRNAs in B lymphocytes and diffuse large B-cell lymphomas. *Blood*, *113*(16), 3754–3764. https://doi.org/10.1182/blood-2008-10-184077.

REAL. (2020). *Classification for lymphoma*. https://training.seer.cancer.gov/lymphoma/abstract-code-stage/morphology/real.html.

Robinson, J. E., Greiner, T. C., Bouska, A. C., Iqbal, J., & Cutucache, C. E. (2020). Identification of a splenic marginal zone lymphoma signature: Preliminary findings with diagnostic potential. *Frontiers in Oncology*, *10*, 640. https://doi.org/10.3389/fonc.2020.00640.

Rosenwald, A., Wright, G., Leroy, K., Yu, X., Gaulard, P., Gascoyne, R. D., … Staudt, L.M. (2003). Molecular diagnosis of primary mediastinal B cell lymphoma identifies a clinically favorable subgroup of diffuse large B cell lymphoma related to Hodgkin lymphoma. *The Journal of Experimental Medicine*, *198*(6), 851–862.

SEER (Surveillance, Epidemiology, and End Results). (2020). *Cancer incidence and prevalence statistics*. https://gco.iarc.fr/today/data/factsheets/cancers/34-Non-hodgkin-lymphoma-fact-sheet.pdf.

Troppan, K., Wenzl, K., Deutsch, A., Ling, H., Neumeister, P., & Pichler, M. (2014). MicroRNAs in diffuse large B-cell lymphoma: Implications for pathogenesis, diagnosis, prognosis and therapy. *Anticancer Research*, *34*(2), 557–564. http://ar.iiarjournals.org/content/34/2/557.full.pdf+html.

van den Brand, M., & van Krieken, J. H. J. M. (2013). Recognizing nodal marginal zone lymphoma: Recent advances and pitfalls. A systematic review. *Haematologica*, *98*(7), 1003–1013.

World Health Organization (WHO). (2020). www.who.int.

Xia, L., Zhang, D., Du, R., Pan, Y., Zhao, L., Sun, S., … Fan, D. (2008). miR-15b and miR-16 modulate multidrug resistance by targeting BCL2 in human gastric cancer cells. *International Journal of Cancer*, *123*(2), 372–379. https://doi.org/10.1002/ijc.23501.

Chapter 16

Pathophysiology roles and translational opportunities of miRNAs in breast cancer

Giulia Cosentino*, Ilaria Plantamura*, and Marilena V. Iorio
Fondazione IRCCS Istituto Nazionale dei Tumori, Milan, Italy

Breast physiology and pathology

Breast physiology

The mammary gland is an exocrine gland characterizing mammalians, including humans, that is mainly embedded in adipose tissue, and located between the skin and the thorax. It is composed of multiple \glandular structures called *lobules*, interconnected to form a *lobe*. A single mammary gland is made of 15–20 lobes. The mammary gland function is the production of milk, which is secreted by the *nipple*, where it is canalized through *lactiferous ducts*. The functional units of the lobule, called *alveoli*, are characterized by luminal epithelium leaning on a basal membrane, where mioepithelial cells are interconnected and contribute to milk secretion. It is thought that mammary stem cells reside in a basal position between these two populations and give rise to progenitor cells and both lineages of fully differentiated cells (Tiede & Kang, 2011). In addition to epithelial cells, mammary glands are composed of a variety of other cell types, including endothelial cells, fibroblasts, and immune cells (Watson & Khaled, 2008). In mice, a bulb-shaped mammary bud can be discerned in the fetus at 18–19weeks of gestation. However, mammary gland development is mainly a postnatal process in humans and mice. Indeed, at birth only rudimentary ducts are present, and in mice the ductal outgrowth takes place between the third and the sixth week of life during puberty, when the ducts extend throughout the fat pad until the periphery of the gland (Richert, Schwertfeger, Ryder, & Anderson, 2000). This invasive growth of the ducts occurs at their distal tips, called terminal end buds (TEBs). The female mouse is considered a mature virgin after 8weeks. Only in concomitance with the gestation and the lactation does the alveolar proliferation and the functional differentiation of alveolar cells occur. After weaning, the mammary gland undergoes the involution process to restore a simple ductal structure.

Breast pathology

Breast cancer is a heterogeneous disease, deriving from the abnormal proliferation of cells from ducts or lobuli. According to the expression detected by immunohistochemistry of the three main receptors, the estrogen receptor (ER), the progesteron receptor (PgR), and the human epidermal growth factor receptor (ErbB2 or HER2), breast cancer is classified as luminal, HER2-positive, or triple negative. With the advent of omic technologies, in particular the evaluation of the gene expression profile, breast cancer has been further subclassified in five "intrinsic" molecular subtypes: normal-like, luminal A and B, HER2-positive, and basal-like (Perou et al., 2000). More recently, gene expression profiles have more deeply dissected the molecular heterogeneity of triple negative breast cancer, identifying six different groups displaying unique ontologies, including two basal-like (BL1 and BL2), an immunomodulatory (IM), a mesenchymal (M), a mesenchymal stem-like (MSL), and a luminal androgen receptor (LAR) subtype (Lehmann et al., 2011). More than 75% of breast tumor cases affect individuals over 50years of age. Juvenile occurrences are less common. In general, breast cancer risk factors can be divided into two main categories: nonmodifiable factors such as genetic predisposition and personal history of breast cancer, and modifiable factors that individuals can control such as tobacco or diet (Torre, Siegel, Ward, & Jemal, 2016). Breast cancers can be noninvasive (in situ) or invasive. Carcinomas in situ (CIS) are highly heterogeneous both histologically as well as in their malignant potential. Ductal and lobular carcinomas in situ (respectively DCIS and LCIS) are the two main CIS subtypes. LCIS is less frequent than DCIS and it is generally not thought to be a precursor of invasive cancer.

* These authors equally contributed to this work.

MicroRNA in Human Malignancies. https://doi.org/10.1016/B978-0-12-822287-4.00026-8

It is also possible to find rare cases with characteristics of both types of carcinomas. Most breast cancers (more than 80%) are invasive and, like in situ lesions, are highly heterogeneous. The major subtypes include infiltrating ductal (70%–80% of all malignancies), invasive lobular, ductal/lobular, mucinous (colloid), tubular, and medullary and papillary carcinomas (Malhotra, Zhao, Band, & Band, 2010). The stage of breast cancer (from I to IV) is defined according to the TNM classification, where T is the size of the disease, N the involvement of lymph nodes, and M the presence of metastases at distant sites. Overall, if the disease is diagnosed and treated at the initial stage, survival at 5years is 98%, even though the percentage of recurrence is extremely variable, and it depends on tumor subtype and heterogeneous responsiveness to anticancer treatments. Overall survival at 5years is 75% if the disease is present in the lymph nodes at the diagnosis, and drops to an average of 2years if distant metastases are present. These data underline the need for an early diagnosis and tools to predict responsiveness to therapies to select patients who will likely benefit from a specific treatment, avoiding otherwise detrimental side effects and saving time in switching to a new treatment schedule.

MicroRNAs in breast physiology

MicroRNAs are regulated in the different stages of mammary gland development. In particular, the modulation of specific temporally coregulated miRNA clusters has been described from mouse puberty to late mammary gland involution (Avril-Sassen et al., 2009). Interestingly, Avril-Sassen and colleagues reported the modulation of miR-205, in particular an increase during pregnancy and lactation and later during involution. Despite the association of miR-205 expression with stem cell progenitors of normal breast tissue, surprisingly the knockdown of this microRNA in mouse models did not cause defective morphogenesis of the mammary gland; instead, it increased ductal outgrowth, the spontaneous formation of hyperplatic lesions, and focal mammary carcinoma in 60% of miR-205 KD mice (Chao et al., 2014; Greene, Herschkowitz, & Rosen, 2010; Plantamura, Cataldo, Cosentino, & Iorio, 2021). This was consistent with the previously reported role as an oncosuppressor in breast cancer (Iorio et al., 2009; Piovan et al., 2012; Wu, Zhu, & Mo, 2009). Another microRNA, often associated with miR-205 for similarities in targeting and functions, and differentially expressed as well in pregnancy and lactation, is miR-200a. MiR-200a is stimulated by lactogenic hormone treatment of mammary epithelial cells, and contributes to E-cadherin and milk β-casein expression (Nagaoka, Zhang, Watanabe, & Taya, 2013). Another oncosuppressive microRNA playing a crucial role in a normal mammary gland is miR-489, which inhibits mammary gland development in early ages with a specific impact on highly proliferative cells (Patel et al., 2019). MicroRNA-212/132 is one of the first microRNA families described as indispensable during the development of mammary glands in mice, in particular for the outgrowth of the epithelial ducts (Ucar et al., 2010). Interestingly, both miRNAs are expressed exclusively in the stroma, and their absence leads to increased MMP9 production, impaired collagen deposition, TGF-β signaling activation, and impairment of ductal outgrowth. However, a few years later Kayo and colleagues obtained contrasting results performing a comparative analysis with their own independently generated miR-212/132 KO mouse model (Kayo et al., 2014). They concluded that these two microRNAs are instead dispensable for mammary gland development under certain genetic and/or physiological conditions, and that the previously observed effect might be due do the deletion of genetic elements near the targeted locus (Kayo et al., 2014). Tanaka and colleagues used an epithelial mammary gland cell line model (HC11) to describe the role of miR-101, which increases during differentiation and involution by directly targeting Cox-2 and affecting the expression of β-casein (Tanaka, Haneda, Imakawa, Sakai, & Nagaoka, 2009). More recently, using the same cell line model, Wang, Aydoğdu, Mukhopadhyay, Helguero, and Williams (2019) described a role of miR-206 in inducing the mammary differentiation process. MiR-126-3p is instead involved in pregnancy, in particular in the transition from pregnancy to lactation (Cui, Li, Feng, & Ding, 2011). By using mouse mammary epithelial cells as cell models in vitro, the authors defined the progesterone receptor (PGR) as a direct miRNA target. Another microRNA involved in lactation is miR-30b. MiR-30b-transgenic mice showed an altered mammary gland structure, with reduced lumen of fat pad acini, impaired lactation capability, and consequently defective pup growth (Guillou et al., 2012). By using a KO mouse model, Silva and colleagues demonstrated a crucial role of miR-424(322)/503 in epithelial involution after weaning (Llobet-Navas et al., 2014). Activated by TGFbeta, a well-characterized regulator of mammary involution, this miRNA cluster regulates antiapoptotic BCL2 and insulin-like growth factor receptor (IGFR1) expression. Recently, it has been shown that during involution, this miRNA cluster is directly regulated by the SOX10 transcription factor (Mertelmeyer et al., 2020). Interestingly, an analysis in human milk revealed an abundant expression of several miRNAs, which are primarily synthesized in the mammary epithelium and likely play significant roles in the development and normal function of the lactating mammary gland (Alsaweed, Lai, Hartmann, Geddes, & Kakulas, 2016). Lactation-related miRNAs have also been detected in microvesicles (MVs) isolated from human umbilical cord blood immediately after delivery (Wang et al., 2016).

MicroRNAs in breast cancer

Oncosuppressive and oncogenic miRNAs

In the last two decades, it has become clear how miRNA expression is not just an epiphenomenon associated with the development of human cancer, but how these molecules play functional roles in the mechanisms of occurrence and/or progression of the disease as well as in responsiveness to specific therapies (Iorio & Croce, 2012; Lee & Dutta, 2009). In breast cancer, several microRNAs have already been described (Eroles, Tormo, Pineda, Espin, & Lluch, 2016; Iorio et al., 2005; Iorio & Croce, 2009; Khoshnaw, Green, Powe, & Ellis, 2009). More recently, greater attention has been focused on the capability of microRNAs to target multiple molecules/pathways, giving rise to a complex network of regulatory interactions. MiR-142-3p, for instance, has been correlated with responsiveness to chemotherapy in triple negative breast cancer by targeting the HIF-1α/LOX/ITGA5/FN1 axis (Saatci et al., 2020). Moreover, microRNAs have been demonstrated to act not only directly on neoplastic cells, but also on components of the tumor microenvironment, contributing to shaping the tumor niche. MiR-9, for instance, is secreted by TNBC cells packaged in exosomes and delivered to human breast fibroblasts, where it contributes to the acquisition of protumoral features by modulating Fibulin3 (EFEMP1) (Baroni et al., 2016; Cosentino, Romero-Cordoba, Plantamura, Cataldo, & Iorio, 2020). Another microRNA secreted in exosomes in TNBC is miR-939, which represses VE-cadherin expression in endothelial cells, thus promoting vessel permeability, transendothelial migration, and likely disease progression (Di Modica et al., 2017). COX-2-induced miR-526b and miR-655 contribute to lymphangiogenesis by inducing VEGF (De Paz Linares, Opperman, Majumder, & Lala, 2021). Exosome-encapsulated miR-370-3p was also found to promote fibroblast activation by downregulating the expression of CYLD and activating the NF-κB signaling pathway (Ren et al., 2021). A very interesting and recent report shows how cancer cells can educate fibroblasts by also rewiring their metabolism during periodic nutrient fluctuations; in particular, cancer-derived exosomal miR-105 and miR-204 target RAGC, a component of Rag GTPases regulating mTORC1 signaling to globally reduce mRNA translation (Fong et al., 2021). Vice versa, exosomal miR-500a-5p secreted by cancer-associated fibroblasts triggers breast cancer cell growth and migration (Chen et al., 2021). MicroRNAs can also interfere in the crosstalk between tumor cells and immune components. Mouse 4T1-derived exosomes, for instance, induce proinflammatory cytokine production in macrophages by delivering miR-183 (Guo et al., 2020). In recent years, the role of other ncRNAs such as lncRNAs (long noncoding RNAs) has clearly emerged. HOTAIR is one of the first lncRNAs described in cancer, and in particular in breast cancer is. It is shown to promote tumor invasiveness by reprogramming chromatin (Gupta et al., 2010). Moreover, a regulatory interaction between microRNAs and lncRNAs has been demonstrated, such as between HOTAIR and the miR-20a-5p/HMGA2 axis, with consequences on cell growth, apoptosis, and metastasis (Zhao, Geng, Li, Chen, & Sun, 2018). Another example is the controversial MALAT1. Originally described as a metastasis-promoter lncRNA, more recently it was reported by Ma L and coworkers to have an opposite effect mediated by the capability to bind and inactivate the prometastatic transcription factor TEAD, impairing the association with its coactivator YAP and target gene promoters (Arun et al., 2016; Kim et al., 2018). As for HOTAIR, also MALAT1 has been associated with several microRNA regulations, such as miR-20a, correlated with poor survival in breast cancer patients and able to boost cell growth and migration (Zhao et al., 2018).

From biology to the clinics: microRNAs as biomarkers (tissue and ct-miRNAs) of diagnosis, prognosis, and resistance to therapy

The search for new clinical approaches to improve cancer prevention or its early detection represents one of the main goals in translational cancer research (Falzone, Salomone, & Libra, 2018). In the last two decades, the study of miRNAs has increased interest in the scientific community (Iorio et al., 2009) by showing that their aberrant expression is associated with the occurrence and progression of human tumors as well as resistance to specific anticancer treatments, suggesting a potential use of these molecules as diagnostic, prognostic, and predictive biomarkers (Iorio & Croce, 2012). Moreover, microRNAs have also been detected circulating in the blood and secreted in other body fluids, such as saliva, urine, or others (Iorio et al., 2009; Piovan et al., 2012; Wu et al., 2009). A liquid biopsy represents a milestone for the diagnosis and monitoring of cancer in a noninvasive manner. Liquid biopsies are less invasive and easier to do than tissue biopsies (Lianidou & Pantel, 2019; (Quirico & Orso, 2020)). Due to their small size, circulating miRNAs are more stable than long RNAs and their detection is more specific and less expensive than classical protein biomarkers (Quirico & Orso, 2020). Moreover, the quantification of extracellular miRNAs is relatively simple and highly sensitive, also thanks to the introduction of novel technological platforms (e.g., NGS) able to improve the sensitivity and accuracy in detecting nucleic acids. Secreted miRNAs can regulate cellular communication triggering "hormone-like" signals, and behave as autocrine, paracrine, and endocrine regulators of cellular functions. As for tissue microRNAs, also extracellular miRNAs have been associated with tumor occurrence as well as disease staging, aggressiveness, and prognosis (Wu et al., 2021). In the literature over the past few years, different microRNAs have been described as biomarkers able to discriminate breast cancer from healthy patients. For instance, Roth C and colleagues showed that miR-155, miR-10b, and miR-34a expression can discriminate metastatic breast cancer from healthy subjects (Roth et al., 2010). Furthermore, miR-34a levels were significantly higher in breast cancer advanced stage than in early tumors. Heneghan et al. analyzed blood samples from patients with breast cancer, prostate cancer, colon cancer, renal cancer, and healthy controls, reporting the differential expression of miR-155, let-7a, and miR-10b (Heneghan, Miller, Kelly, Newell, & Kerin, 2010). More recently, Li and colleagues showed that miR-329 was downregulated in both serum and tumor tissue of BC patients compared to the normal counterpart, and its low expression was associated with metastasis to the lymph node and the tumor node metastases (TNM) stage (Li et al., 2014). Moreover, McAnena's study identified circulating miR-331 and miR-195, upregulated and downmodulated, respectively, in metastatic breast cancer luminal A patients compared to patients with local tumors or healthy individuals (McAnena et al., 2019). Tan et al. showed that miR-106b is more expressed in circulating tumor cells (CTCs) from patients with metastatic breast cancer versus primary tumors or healthy donors (Tan et al., 2019). In addition, they demonstrated that the combination of the three markers–miR-106b, vimentin, and E-cadherin–in CTCs could predict overall survival. MicroRNA expression has also been associated with responsiveness to therapies. Shao and colleagues demonstrated that high levels of miR-200a and miR-210 in the plasma of metastatic BC was related to chemotherapy resistance (Shao et al., 2019). Furthermore, miR-210 correlated with brain, liver, and lung metastasis and the miR-200a expression results linked with the stage. Finally, Di Cosimo described how ct-miRNAs (at the baseline or after treatment) can predict pCR after neoadjuvant anti-HER2 therapies. In particular, levels of plasma miRNAs after 2weeks of neoadjuvant treatment were associated with responsiveness to trastuzumab (Di Cosimo et al., 2019).

From biology to the clinics: Is a miRNA-therapy feasible?

The relevance of miRNA dysregulation in tumor biology is a commonly accepted fact in cancer research. What is still debated is the feasibility of using miRNAs as therapeutic tools. Over the years, both miRNA mimics and inhibitors have been proposed to be exploited in clinics, but very few trials have currently reached phase II (Chakraborty, Sharma, Sharma, & Lee, 2021; Hanna, Hossain, & Kocerha, 2019). Indeed, many are the challenges that researchers have to face to achieve this goal (Segal & Slack, 2020). First of all, naked miRNAs are unstable in the extracellular compartments, and they are degraded within seconds by nucleases. In a biological system, miRNAs are transferred to other cells or districts protected in lipoprotein complexes or packaged into lipid microvesicles, often exosomes (Baroni et al., 2016; Hasan, 2016; Liu, Peng, & Chen, 2019; Yu, Odenthal, & Fries, 2016). Thus, in order to be delivered as drugs in the bloodstream or locally at the tumor site, miRNAs have to be either chemically modified or encapsulated into nanoparticles. The latter strategy could also be used to overcome other problems: the presence of leaky blood vessels in tumors causes poor blood perfusion and, consequently, reduced miRNA payload at the target tissue. Moreover, once miRNAs are internalized into the cells, they are transferred from early to late endosomes and finally to lysosomes, which contain an acidic environment and nucleases (Gibbings, Ciaudo, Erhardt, & Voinnet, 2009). One of the main drawbacks of using miRNA-based therapies is actually one of the strengths of these micromolecules, which is the ability to regulate multiple targets at once. Physiologically, the miRNA mechanism of action is finely tuned and often cell- and tissue-specific. This might not happen once the miRNA is ectopically introduced in the system, potentially leading to toxicities due to undesired gene silencing, the so-called off-target effect (Seok, Lee, Jang, & Chi, 2018). Recently, the administration of low doses of cooperating miRNAs was proposed to overcome such phenomena (Lai, Eberhardt, Schmitz, & Vera, 2019). The delivery of exogenous molecules, being double-stranded RNA or its vehicle, also holds the risk of activating an adverse immune response because they are seen as pathogens (Korangath et al., 2020; Meng & Lu, 2017). Designing carriers able to capitalize or bypass all these features would permit an efficient cargo delivery, lower dosages, and reduced side effects (Orellana et al., 2019; Park et al., 2019; Shi, van der Meel, Chen, & Lammers, 2020). At present, a wide range of nanoparticle formulations has been designed and tested both in vitro and in vivo as therapeutics for multiple diseases (Chakraborty et al., 2021; (O'Neill & Dwyer, 2020)). A recent review by *Silva-Cázares* et al. summarizes some of the delivery systems chosen over the years to test for breast cancer treatment (Silva-Cázares et al., 2020). Because TNBC malignances still lack targeted therapies, many recent options have been proposed for this subtype. *Ramchandani* et al. suggest gold nanoparticles for miRNA mimic delivery as an antimetastatic drug. *Valcourt and Day* encapsulated miR-34a mimics in poly lactic-*co*-glycolic acid (PLGA) nanoparticles that were functionalized with Notch1 antibodies to specifically target tumor cells. Ahir et al. (2020), instead, introduced tailored mesoporous silica nanoparticles for the codelivery of an miR-34a mimic and antisense-miR-10b, coated with an hyaluronic acid-appended PEG-PLGA polymer for specific targeting (Ramchandani et al., 2019; Valcourt & Day, 2020).

Conclusions

The evidence collected so far clearly underlines a crucial role of microRNAs in both physiological and pathological processes of the mammary gland. By affecting multiple pathways, microRNAs are indeed not only differentially expressed in different stages of mammary gland development, from puberty to involution, but they actively modulate the molecular mechanisms underlying these phenomena. Similarly, tissue or circulating microRNA expression can be associated with tumor occurrence and/or progression as well as resistance to therapies, suggesting the use of these molecules as diagnostic, prognostic, or predictive biomarkers. Nevertheless, their functional role in finely tuning the biology of neoplastic cells and the tumor microenvironment has increased interest toward miRNA-based therapy, where the ectopic modulation of miRNA expression might be exploited to restore an equilibrium lost in cancer. Certainly, several issues still need to be faced to concretely translate an miRNA-based therapy into the clinics, such as safety and delivery efficiency. However the evidence reported so far and ongoing clinical trials are promising.

References

Ahir, M., Upadhyay, P., Ghosh, A., Sarker, S., Bhattacharya, S., Gupta, P., et al. (2020). Delivery of dual miRNA through CD44-targeted mesoporous silica nanoparticles for enhanced and effective triple-negative breast cancer therapy. *Biomaterials Science*, *8*, 2939–2954. https://doi.org/10.1039/D0BM00015A.

Alsaweed, M., Lai, C. T., Hartmann, P. E., Geddes, D. T., & Kakulas, F. (2016). Human Milk MiRNAs primarily originate from the mammary gland resulting in unique MiRNA profiles of fractionated Milk. *Scientific Reports*, *6*, 20680. https://doi.org/10.1038/srep20680.

Arun, G., Diermeier, S., Akerman, M., Chang, K.-C., Wilkinson, J. E., Hearn, S., et al. (2016). Differentiation of mammary tumors and reduction in metastasis upon Malat1 LncRNA loss. *Genes & Development*, *30*, 34–51. https://doi.org/10.1101/gad.270959.115.

Avril-Sassen, S., Goldstein, L. D., Stingl, J., Blenkiron, C., Le Quesne, J., Spiteri, I., et al. (2009). Characterisation of microRNA expression in post-natal mouse mammary gland development. *BMC Genomics*, *10*, 548. https://doi.org/10.1186/1471-2164-10-548.

Baroni, S., Romero-Cordoba, S., Plantamura, I., Dugo, M., D'Ippolito, E., Cataldo, A., et al. (2016). Exosome-mediated delivery of MiR-9 induces cancer-associated fibroblast-like properties in human breast fibroblasts. *Cell Death & Disease*, *7*. https://doi.org/10.1038/cddis.2016.224, e2312.

Chakraborty, C., Sharma, A. R., Sharma, G., & Lee, S.-S. (2021). Therapeutic advances of MiRNAs: A preclinical and clinical update. *Journal of Advanced Research*, *28*, 127–138. https://doi.org/10.1016/j.jare.2020.08.012.

Chao, C.-H., Chang, C.-C., Wu, M.-J., Ko, H.-W., Wang, D., Hung, M.-C., et al. (2014). MicroRNA-205 signaling regulates mammary stem cell fate and tumorigenesis. *The Journal of Clinical Investigation*, *124*, 3093–3106. https://doi.org/10.1172/JCI73351.

Chen, B., Sang, Y., Song, X., Zhang, D., Wang, L., Zhao, W., et al. (2021). Exosomal MiR-500a-5p derived from cancer-associated fibroblasts promotes breast cancer cell proliferation and metastasis through targeting USP28. *Theranostics*, *11*, 3932–3947. https://doi.org/10.7150/thno.53412.

Cosentino, G., Romero-Cordoba, S., Plantamura, I., Cataldo, A., & Iorio, M. V. (2020). MiR-9-mediated inhibition of EFEMP1 contributes to the Acquisition of pro-Tumoral Properties in Normal fibroblasts. *Cell*, *9*. https://doi.org/10.3390/cells9092143.

Cui, W., Li, Q., Feng, L., & Ding, W. (2011). MiR-126-3p regulates progesterone receptors and involves development and lactation of mouse mammary gland. *Molecular and Cellular Biochemistry*, *355*, 17–25. https://doi.org/10.1007/s11010-011-0834-1.

De Paz Linares, G. A., Opperman, R. M., Majumder, M., & Lala, P. K. (2021). Prostaglandin E2 receptor 4 (EP4) as a therapeutic target to impede breast cancer-associated angiogenesis and Lymphangiogenesis. *Cancers*, *13*, 942. https://doi.org/10.3390/cancers13050942.

Di Cosimo, S., Appierto, V., Pizzamiglio, S., Tiberio, P., Iorio, M. V., Hilbers, F., et al. (2019). Plasma MiRNA levels for predicting therapeutic response to neoadjuvant treatment in HER2-positive breast cancer: Results from the NeoALTTO trial. *Clinical Cancer Research*, *25*, 3887–3895. https://doi.org/10.1158/1078-0432.CCR-18-2507.

Di Modica, M., Regondi, V., Sandri, M., Iorio, M. V., Zanetti, A., Tagliabue, E., et al. (2017). Breast cancer-secreted MiR-939 downregulates VE-cadherin and destroys the barrier function of endothelial monolayers. *Cancer Letters*, *384*, 94–100. https://doi.org/10.1016/j.canlet.2016.09.013.

Eroles, P., Tormo, E., Pineda, B., Espin, E., & Lluch, A. (2016). MicroRNAs in breast cancer: One more turn in regulation. *Current Drug Targets*, *17*, 1083–1100. https://doi.org/10.2174/1389450116666150213114103.

Falzone, L., Salomone, S., & Libra, M. (2018). Evolution of cancer pharmacological treatments at the turn of the third millennium. *Frontiers in Pharmacology*, *9*, 1300. https://doi.org/10.3389/fphar.2018.01300.

Fong, M. Y., Yan, W., Ghassemian, M., Wu, X., Zhou, X., Cao, M., et al. (**2021**). Cancer-secreted MiRNAs regulate amino-acid-induced MTORC1 signaling and fibroblast protein synthesis. *EMBO Reports*, *22*. https://doi.org/10.15252/embr.202051239, e51239.

Gibbings, D. J., Ciaudo, C., Erhardt, M., & Voinnet, O. (2009). Multivesicular bodies associate with components of miRNA effector complexes and modulate miRNA activity. *Nature Cell Biology*, *11*, 1143–1149. https://doi.org/10.1038/ncb1929.

Greene, S. B., Herschkowitz, J. I., & Rosen, J. M. (2010). The ups and downs of MiR-205: Identifying the roles of MiR-205 in mammary gland development and breast cancer. *RNA Biology*, *7*, 300–304. https://doi.org/10.4161/rna.7.3.11837.

Guillou, S. L., Sdassi, N., Laubier, J., Passet, B., Vilotte, M., Castille, J., et al. (2012). Overexpression of MiR-30b in the developing mouse mammary gland causes a lactation defect and delays involution. *PLoS One*, *7*. https://doi.org/10.1371/journal.pone.0045727, e45727.

Guo, J., Duan, Z., Zhang, C., Wang, W., He, H., Liu, Y., et al. (2020). Mouse 4T1 breast cancer cell-derived exosomes induce Proinflammatory cytokine production in macrophages via MiR-183. *Journal of Immunology*, *205*, 2916–2925. https://doi.org/10.4049/jimmunol.1901104.

Gupta, R. A., Shah, N., Wang, K. C., Kim, J., Horlings, H. M., Wong, D. J., et al. (2010). Long non-coding RNA HOTAIR reprograms chromatin state to promote cancer metastasis. *Nature*, *464*, 1071–1076. https://doi.org/10.1038/nature08975.

Hanna, J., Hossain, G. S., & Kocerha, J. (2019). The potential for microRNA therapeutics and clinical research. *Frontiers in Genetics*, *10*. https://doi.org/10.3389/fgene.2019.00478.

Hasan, S. M. (2016). Extracellular/circulating MicroRNAs: Release mechanisms, functions and challenges. *Achievements in the Life Sciences*, 175–186. https://doi.org/10.1016/j.als.2016.11.007.

Heneghan, H. M., Miller, N., Kelly, R., Newell, J., & Kerin, M. J. (2010). Systemic MiRNA-195 differentiates breast cancer from other malignancies and is a potential biomarker for detecting noninvasive and early stage disease. *The Oncologist*, *15*, 673–682. https://doi.org/10.1634/theoncologist.2010-0103.

Iorio, M. V., Casalini, P., Piovan, C., Di Leva, G., Merlo, A., Triulzi, T., et al. (2009). MicroRNA-205 regulates HER3 in human breast cancer. *Cancer Research*, *69*, 2195–2200. https://doi.org/10.1158/0008-5472.CAN-08-2920.

Iorio, M. V., & Croce, C. M. (2009). MicroRNAs in cancer: Small molecules with a huge impact. *Journal of Clinical Oncology*, *27*, 5848–5856. https://doi.org/10.1200/JCO.2009.24.0317.

Iorio, M. V., & Croce, C. M. (2012). MicroRNA dysregulation in Cancer: Diagnostics, monitoring and therapeutics. A comprehensive review. *EMBO Molecular Medicine*, *4*, 143–159. https://doi.org/10.1002/emmm.201100209.

Iorio, M. V., Ferracin, M., Liu, C.-G., Veronese, A., Spizzo, R., Sabbioni, S., et al. (2005). MicroRNA gene expression deregulation in human breast cancer. *Cancer Research*, *65*, 7065–7070. https://doi.org/10.1158/0008-5472.CAN-05-1783.

Kayo, H., Kiga, K., Fukuda-Yuzawa, Y., Hedlund, S., Murakami, K., De La Rosa-Velazquez, I. A., et al. (2014). MiR-212 and MiR-132 are dispensable for mouse mammary gland development. *Nature Genetics*, *46*, 802–804. https://doi.org/10.1038/ng.2990.

Khoshnaw, S. M., Green, A. R., Powe, D. G., & Ellis, I. O. (2009). MicroRNA involvement in the pathogenesis and management of breast cancer. *Journal of Clinical Pathology*, *62*, 422–428. https://doi.org/10.1136/jcp.2008.060681.

Kim, J., Piao, H.-L., Kim, B.-J., Yao, F., Han, Z., Wang, Y., et al. (2018). Long noncoding RNA MALAT1 suppresses breast cancer metastasis. *Nature Genetics*, *50*, 1705–1715. https://doi.org/10.1038/s41588-018-0252-3.

Korangath, P., Barnett, J. D., Sharma, A., Henderson, E. T., Stewart, J., Yu, S.-H., et al. (**2020**). Nanoparticle interactions with immune cells dominate tumor retention and induce T cell-mediated tumor suppression in models of breast cancer. *Science Advances*, *6*. https://doi.org/10.1126/sciadv.aay1601, eaay1601.

Lai, X., Eberhardt, M., Schmitz, U., & Vera, J. (2019). Systems biology-based investigation of cooperating microRNAs as monotherapy or adjuvant therapy in cancer. *Nucleic Acids Research*, *47*, 7753–7766. https://doi.org/10.1093/nar/gkz638.

Lee, Y. S., & Dutta, A. (2009). MicroRNAs in cancer. *Annual Review of Pathology*, *4*, 199–227. https://doi.org/10.1146/annurev.pathol.4.110807.092222.

Lehmann, B. D., Bauer, J. A., Chen, X., Sanders, M. E., Chakravarthy, A. B., Shyr, Y., et al. (2011). Identification of human triple-negative breast cancer subtypes and preclinical models for selection of targeted therapies. *The Journal of Clinical Investigation*, *121*, 2750–2767. https://doi.org/10.1172/JCI45014.

Li, Q., Liu, M., Ma, F., Luo, Y., Cai, R., Wang, L., et al. (2014). Circulating MiR-19a and MiR-205 in serum may predict the sensitivity of luminal a subtype of breast cancer patients to neoadjuvant chemotherapy with Epirubicin plus paclitaxel. *PLoS One*, *9*. https://doi.org/10.1371/journal.pone.0104870, e104870.

Lianidou, E., & Pantel, K. (2019). Liquid Biopsies. *Genes, Chromosomes and Cancer*, *58*, 219–232. https://doi.org/10.1002/gcc.22695.

Liu, Q., Peng, F., & Chen, J. (2019). The role of exosomal microRNAs in the tumor microenvironment of breast cancer. *International Journal of Molecular Sciences*, *20*. https://doi.org/10.3390/ijms20163884.

Llobet-Navas, D., Rodriguez-Barrueco, R., de la Iglesia-Vicente, J., Olivan, M., Castro, V., Saucedo-Cuevas, L., et al. (2014). The microRNA 424/503 cluster reduces CDC25A expression during cell cycle arrest imposed by transforming growth factor β in mammary epithelial cells. *Molecular and Cellular Biology*, *34*, 4216–4231. https://doi.org/10.1128/MCB.00611-14.

Malhotra, G. K., Zhao, X., Band, H., & Band, V. (2010). Histological, molecular and functional subtypes of breast cancers. *Cancer Biology & Therapy*, *10*, 955–960. https://doi.org/10.4161/cbt.10.10.13879.

McAnena, P., Tanriverdi, K., Curran, C., Gilligan, K., Freedman, J. E., Brown, J. A. L., et al. (2019). Circulating microRNAs MiR-331 and MiR-195 differentiate local luminal a from metastatic breast cancer. *BMC Cancer*, *19*, 436. https://doi.org/10.1186/s12885-019-5636-y.

Meng, Z., & Lu, M. (2017). RNA interference-induced innate immunity, off-target effect, or immune adjuvant? *Frontiers in Immunology*, *8*, 331. https://doi.org/10.3389/fimmu.2017.00331.

Mertelmeyer, S., Weider, M., Baroti, T., Reiprich, S., Fröb, F., Stolt, C. C., et al. (2020). The transcription factor Sox10 is an essential determinant of branching morphogenesis and involution in the mouse mammary gland. *Scientific Reports*, *10*, 17807. https://doi.org/10.1038/s41598-020-74664-y.

Nagaoka, K., Zhang, H., Watanabe, G., & Taya, K. (2013). Epithelial cell differentiation regulated by microRNA-200a in mammary glands. *PLoS One*, *8*. https://doi.org/10.1371/journal.pone.0065127, e65127.

O'Neill, C. P., & Dwyer, R. M. (2020). Nanoparticle-based delivery of tumor suppressor microRNA for cancer therapy. *Cell*, *9*. https://doi.org/10.3390/cells9020521.

Orellana, E. A., Abdelaal, A. M., Rangasamy, L., Tenneti, S., Myoung, S., Low, P. S., et al. (2019). Enhancing microRNA activity through increased endosomal release mediated by Nigericin. *Molecular Therapy- -Nucleic Acids*, *16*, 505–518. https://doi.org/10.1016/j.omtn.2019.04.003.

Park, J., Choi, Y., Chang, H., Um, W., Ryu, J. H., & Kwon, I. C. (2019). Alliance with EPR effect: Combined strategies to improve the EPR effect in the tumor microenvironment. *Theranostics*, *9*, 8073–8090. https://doi.org/10.7150/thno.37198.

Patel, Y., Soni, M., Awgulewitsch, A., Kern, M. J., Liu, S., Shah, N., et al. (2019). Overexpression of MiR-489 derails mammary hierarchy structure and inhibits HER2/Neu-induced tumorigenesis. *Oncogene*, *38*, 445–453. https://doi.org/10.1038/s41388-018-0439-1.

Perou, C. M., Sørlie, T., Eisen, M. B., van de Rijn, M., Jeffrey, S. S., Rees, C. A., et al. (2000). Molecular portraits of human breast tumours. *Nature*, *406*, 747–752. https://doi.org/10.1038/35021093.

Piovan, C., Palmieri, D., Di Leva, G., Braccioli, L., Casalini, P., Nuovo, G., et al. (2012). Oncosuppressive role of P53-induced MiR-205 in triple negative breast cancer. *Molecular Oncology*, *6*, 458–472. https://doi.org/10.1016/j.molonc.2012.03.003.

Plantamura, I., Cataldo, A., Cosentino, G., & Iorio, M. V. (2021). MiR-205 in breast Cancer: State of the art. *International Journal of Molecular Sciences*, *22*, 27. https://doi.org/10.3390/ijms22010027.

Quirico, L., & Orso, F. (2020). The power of microRNAs as diagnostic and prognostic biomarkers in liquid biopsies. *Cancer Drug Resistance*, *3*, 117–139. https://doi.org/10.20517/cdr.2019.103.

Ramchandani, D., Lee, S. K., Yomtoubian, S., Han, M. S., Tung, C.-H., & Mittal, V. (2019). Nanoparticle delivery of MiR-708 mimetic impairs breast cancer metastasis. *Molecular Cancer Therapeutics*, *18*, 579–591. https://doi.org/10.1158/1535-7163.MCT-18-0702.

Ren, Z., Lv, M., Yu, Q., Bao, J., Lou, K., & Li, X. (2021). MicroRNA-370-3p shuttled by breast cancer cell-derived extracellular vesicles induces fibroblast activation through the CYLD/Nf-KB Axis to promote breast cancer progression. *The FASEB Journal*, *35*. https://doi.org/10.1096/fj.202001430RR, e21383.

Richert, M. M., Schwertfeger, K. L., Ryder, J. W., & Anderson, S. M. (2000). An atlas of mouse mammary gland development. *Journal of Mammary Gland Biology and Neoplasia*, *5*, 227–241. https://doi.org/10.1023/a:1026499523505.

Roth, C., Rack, B., Müller, V., Janni, W., Pantel, K., & Schwarzenbach, H. (2010). Circulating microRNAs as blood-based markers for patients with primary and metastatic breast cancer. *Breast Cancer Research*, *12*, R90. https://doi.org/10.1186/bcr2766.

Saatci, O., Kaymak, A., Raza, U., Ersan, P. G., Akbulut, O., Banister, C. E., et al. (2020). Targeting Lysyl oxidase (LOX) overcomes chemotherapy resistance in triple negative breast cancer. *Nature Communications*, *11*, 2416. https://doi.org/10.1038/s41467-020-16199-4.

Segal, M., & Slack, F. J. (2020). Challenges identifying efficacious MiRNA therapeutics for cancer. *Expert Opinion on Drug Discovery*, *15*, 987–991. https://doi.org/10.1080/17460441.2020.1765770.

Seok, H., Lee, H., Jang, E.-S., & Chi, S. W. (2018). Evaluation and control of miRNA-like off-target repression for RNA interference. *Cellular and Molecular Life Sciences*, *75*, 797–814. https://doi.org/10.1007/s00018-017-2656-0.

Shao, B., Wang, X., Zhang, L., Li, D., Liu, X., Song, G., et al. (2019). Plasma MicroRNAs predict chemoresistance in patients with metastatic breast cancer. *Technology in Cancer Research & Treatment*, *18*. https://doi.org/10.1177/1533033819828709.

Shi, Y., van der Meel, R., Chen, X., & Lammers, T. (2020). The EPR effect and beyond: Strategies to improve tumor targeting and cancer nanomedicine treatment efficacy. *Theranostics*, *10*, 7921–7924. https://doi.org/10.7150/thno.49577.

Silva-Cázares, M. B., Saavedra-Leos, M. Z., Jordan-Alejandre, E., Nuñez-Olvera, S. I., Cómpean-Martínez, I., & López-Camarillo, C. (2020). Lipid-based nanoparticles for the therapeutic delivery of non-coding RNAs in breast cancer (Review). *Oncology Reports*, *44*, 2353–2363. https://doi.org/10.3892/or.2020.7791.

Tan, W., Liang, G., Xie, X., Jiang, W., Tan, L., Sanders, A. J., et al. (2019). Incorporating microRNA into molecular phenotypes of circulating tumor cells enhances the prognostic accuracy for patients with metastatic breast cancer. *The Oncologist*, *24*, e1044–e1054. https://doi.org/10.1634/theoncologist.2018-0697.

Tanaka, T., Haneda, S., Imakawa, K., Sakai, S., & Nagaoka, K. (2009). A MicroRNA, MiR-101a, controls mammary gland development by regulating cyclooxygenase-2 expression. *Differentiation*, *77*, 181–187. https://doi.org/10.1016/j.diff.2008.10.001.

Tiede, B., & Kang, Y. (2011). From milk to malignancy: The role of mammary stem cells in development, pregnancy and breast cancer. *Cell Research*, *21*, 245–257. https://doi.org/10.1038/cr.2011.11.

Torre, L. A., Siegel, R. L., Ward, E. M., & Jemal, A. (2016). Global cancer incidence and mortality rates and trends—an update. *Cancer Epidemiology, Biomarkers & Prevention*, *25*, 16–27. https://doi.org/10.1158/1055-9965.EPI-15-0578.

Ucar, A., Vafaizadeh, V., Jarry, H., Fiedler, J., Klemmt, P. A. B., Thum, T., et al. (2010). MiR-212 and MiR-132 are required for epithelial stromal interactions necessary for mouse mammary gland development. *Nature Genetics*, *42*, 1101–1108. https://doi.org/10.1038/ng.709.

Valcourt, D. M., & Day, E. S. (2020). Dual regulation of miR-34a and notch signaling in triple-negative breast cancer by antibody/miRNA nanocarriers. *Molecular Therapy- -Nucleic Acids*, *21*, 290–298. https://doi.org/10.1016/j.omtn.2020.06.003.

Wang, D., Wang, C., Wang, Y., Qiao, H., Fang, L., & Wang, Z. (2016). Lactation-related microRNA expression in microvesicles of human umbilical cord blood. *Medical Science Monitor*, *22*, 4542–4554. https://doi.org/10.12659/MSM.901695.

Wang, J., Aydoğdu, E., Mukhopadhyay, S., Helguero, L. A., & Williams, C. (2019). A MiR-206 regulated gene landscape enhances mammary epithelial differentiation. *Journal of Cellular Physiology*, *234*, 22220–22233. https://doi.org/10.1002/jcp.28789.

Watson, C. J., & Khaled, W. T. (2008). Mammary development in the embryo and adult: A journey of morphogenesis and commitment. *Development*, *135*, 995–1003. https://doi.org/10.1242/dev.005439.

Wu, H., Zhu, S., & Mo, Y.-Y. (2009). Suppression of cell growth and invasion by MiR-205 in breast cancer. *Cell Research*, *19*, 439–448. https://doi.org/10.1038/cr.2009.18.

Wu, Y., Li, Q., Zhang, R., Dai, X., Chen, W., & Xing, D. (2021). Circulating MicroRNAs: Biomarkers of disease. *Clinica Chimica Acta*, *516*, 46–54. https://doi.org/10.1016/j.cca.2021.01.008.

Yu, X., Odenthal, M., & Fries, J. W. U. (2016). Exosomes as miRNA carriers: Formation–function–future. *International Journal of Molecular Sciences*, *17*, 2028. https://doi.org/10.3390/ijms17122028.

Zhao, W., Geng, D., Li, S., Chen, Z., & Sun, M. (2018). LncRNA HOTAIR influences cell growth, migration, invasion, and apoptosis via the MiR-20a-5p/HMGA2 Axis in breast cancer. *Cancer Medicine*, *7*, 842–855. https://doi.org/10.1002/cam4.1353.

Chapter 17

Pathophysiology roles and translational opportunities of miRNAs in colorectal cancer

Petra Vychytilova-Faltejskova[a] and Ondrej Slaby[a,b]

[a]*Centre for Molecular Medicine, Central European Institute of Technology, Masaryk University, Brno, Czech Republic,* [b]*Department of Biology, Theoretical Departments, Faculty of Medicine, Masaryk University, Brno, Czech Republic*

Introduction

Colorectal cancer (CRC) accounts for 10.2% of all cancers. It is the third most common cancer worldwide and a major health problem due to its high mortality rates (Bray et al., 2018). The main reasons for the high mortality are late diagnosis, insufficient clinical stratification, and low level of therapy individualization. In addition, CRC is a disease with a large degree of intra- and intertumoral heterogeneity and the current TNM classification does not include biological characteristics and predictors of tumor behavior (Khaliq et al., 2021). The prognosis of patients depends primarily on the extent of the disease, as the curative surgical intervention is feasible only for patients with localized tumors (Guraya, 2019). However, around 20% of patients have metastatic disease at the time of diagnosis (Biller & Schrag, 2021). Thus, the early detection of CRC together with precancerous lesions is one of the main prerequisites for mortality reduction and more successful treatment.

Currently, a fecal occult blood test (FOBT) together with endoscopic techniques are the predominant screening methods used for the early detection of CRC. Although these approaches significantly reduce both incidence and mortality (Gini et al., 2020), screening effectiveness is limited by high costs and the invasive nature of a colonoscopy as well as by the relatively poor sensitivity and specificity of FOBT (Kościelniak-Merak, Radosavljević, Zając, & Tomasik, 2018). Similarly, conventional cancer biomarkers including carcinoembryonic antigen (CEA), CA19-9, or CA72-4 evince insufficient diagnostic performance and their combined detection is preferentially used to monitor the recurrence of the disease or the presence of metastasis after surgery (Wu, Mo, & Wu, 2020). Thus, the development of a simple noninvasive and accurate test is highly desirable and extended efforts are dedicated to the identification of novel promising biomarkers using genomic, transcriptomic, proteomic, or metabolomic techniques (Di Lena, Travaglio, & Altomare, 2013).

Almost 30 years ago, the first microRNA (miRNA), lin-4, was described as an essential regulator to control the development timing in *Caenorhabditis elegans* (Lee, Feinbaum, & Ambros, 1993; Wightman, Ha, & Ruvkun, 1993). However, it was only in 2000 when the second miRNA, let-7, was identified to control the transition of larval development through binding to the 3′ untranslated region (UTR) of lin-41 (Reinhart et al., 2000). Since then, thousands of miRNAs have been described in humans and other species. Today, miRNAs belong to the well-documented small noncoding RNAs (ncRNAs) that are involved in the pathophysiology of many diseases including CRC (Liu & Li, 2019), and are currently the subject of intensive translational research (To, Tong, Mingxia, & Cho, 2018). In principle, two approaches are applied today to reveal the connection between miRNAs and CRC. First, miRNAs' involvement in the regulation of key tumor suppressor or oncogenic signaling pathways is extensively studied using numerous in vitro and in vivo approaches, as summarized in the first part of this text. Second, hundreds of studies analyzed the expression profiles of tissue and circulating miRNAs to identify novel diagnostic, prognostic, and predictive biomarkers enabling the early noninvasive diagnosis of the disease, the better prediction of therapy response, and improved stratification of CRC patients based on their prognosis. The most promising findings together with potential clinical applications will be discussed in the second part of this chapter.

MicroRNA in Human Malignancies. https://doi.org/10.1016/B978-0-12-822287-4.00007-4

Involvement of miRNAs in the pathophysiology of CRC

The pathogenesis of CRC is a complex process affected by various mechanisms, including both genetic and epigenetic alterations (Nguyen & Duong, 2018). In addition, reprogramming of the energy metabolism, involvement of the immune system, and contribution of the tumor microenvironment (TME) were proved to play substantial roles in tumor progression and metastasis. Importantly, miRNAs have been repeatedly shown to be involved in CRC pathophysiology with an impact on all cancer hallmarks, as defined by Hanahan and Weinberg (Hanahan & Weinberg, 2011). Currently, the connection between miRNAs' deregulation and key signaling pathways and cellular processes associated with CRC is extensively studied.

Wnt/β-catenin signaling pathway

The wingless-related integration site (Wnt)/β-catenin signaling pathway is an evolutionary conserved pathway crucial for tissue development and homeostasis in all animal species (Logan & Nusse, 2004). It plays an essential role in the early phases of CRC development. The progression of this disease from normal colonic epithelial cells to the malignant phenotype is associated with the presence of mutations in important oncogenes as well as tumor-suppressor genes, including adenomatous polyposis coli (APC). This gene is mutated in hereditary colorectal tumors and in more than 80% of all sporadic forms. Importantly, APC mutations impair β-catenin degradation, leading to increased transcription of the target genes responsible for cancer initiation and progression (Nadeem et al., 2021).

Previous studies have proved that the expression levels of miR-135a, miR-135b (Nagel et al., 2008), miR-103a, miR-1827 (Fasihi, Soltani, Atashi, & Nasiri, 2018), miR-494 (Zhang et al., 2018), miR-942 (Fasihi et al., 2019), and miR-501-3p (Wu et al., 2019) are significantly elevated in the tumor tissue of CRC patients compared to adjacent normal mucosa and bind the 3′ UTR of APC, which in turn leads to the activation of the Wnt/β-catenin pathway and disease progression. On the other hand, the enhanced expression of miR-122a (Wang, Lam, Zhang, Jin, & Sung, 2009) but the inhibition of miRNAs from the miR-17-92 cluster were observed after the restoration of APC function in CRC cell lines. Importantly, binding sites for activated β-catenin were found in the promoter region of the miR-17-92 cluster and the expression of miR-19a significantly correlated with β-catenin levels and disease aggressiveness (Li et al., 2016). Similarly, c-MYC as a major downstream target of the Wnt/β-catenin pathway increases the levels of oncogenic miRNAs from the miR-17-92 cluster (Dews et al., 2010) and miR-552 (Kwak, Kim, Kim, Kim, & Kim, 2018) by direct binding to their promoters. On the contrary, c-MYC expression may be inhibited by diverse tumor-suppressive miRNAs, including miR-145 (Yao et al., 2020), miR-320b (Wang et al., 2015), or miR-185 (Dong-Xu, Jia, & Su-Juan, 2015).

Besides APC, levels of free β-catenin in cytoplasm may be regulated also by other molecules including AXIN1, glycogen synthase kinase 3 (GSK3), or E-cadherin (CDH1). Recent evidence indicates the miR-34 family, miR-103, and miR-107 dependent repression of AXIN2 (Chen et al., 2019; Kim et al., 2013) while miR-155-5p targets preferentially AXIN1 (Prossomariti et al., 2018). Recently, Saieva et al. (2020) proved that hypoxia-induced miR-675-5p supports the transport of β-catenin into the nucleus by inhibiting the expression of GSK3β. Direct regulation of this kinase by miRNAs was also previously confirmed in miR-224 (Li et al., 2016) and miR-452. Importantly, T-cell factor/lymphoid enhancer factor (TCF/LEF) transcription factors, which are crucial downstream molecules of the Wnt/β-catenin signaling pathway, were confirmed to be involved in miR-452 transcription (Li et al., 2018), indicating the presence of a positive feedback loop inducing CRC proliferation and migration. Finally, several miRNAs such as miR-150 (He et al., 2020), miR-93 (Tang et al., 2015), and miR-137 (Fasihi et al., 2018) were identified to directly inhibit the expression of β-catenin. Interestingly, miR-21 was found to be responsible for the increased phosphorylation of β-catenin at Ser552 via the phosphatase and tensin homolog (PTEN)/AKT axis and its subsequent nuclear translocation in APC mutated but not in APC-wild type patients (Lin et al., 2014). This oncogenic miRNA is one of the most studied in CRC as well as in other cancers affecting numerous biological processes. Importantly, TCF4 was revealed to enhance the transcription of miR-21 in colon carcinoma cells, contributing to their increased proliferation and invasion (Lan et al., 2012). On the other hand, the attenuation of TCF4 function by miR-155 (Prossomariti et al., 2018) and miR-29b (Subramanian, Rao, Thacker, Chatterjee, & Karunagaran, 2014) was observed. Lastly, WNT1 as an activator of the whole pathway may be regulated through miRNAs, including miR-185 (Zhang et al., 2018) and miR-200b-3p, which are commonly downregulated in CRC (Chen et al., 2018). The involvement of miRNAs in the Wnt/β-catenin signaling pathway is depicted in Fig. 17.1.

EGFR signaling

The epidermal growth factor receptor (EGFR) contributes not only to the initiation but also to the progression of CRC due to its pivotal role in key signaling pathways affecting cell proliferation, differentiation, apoptosis, angiogenesis, migration,

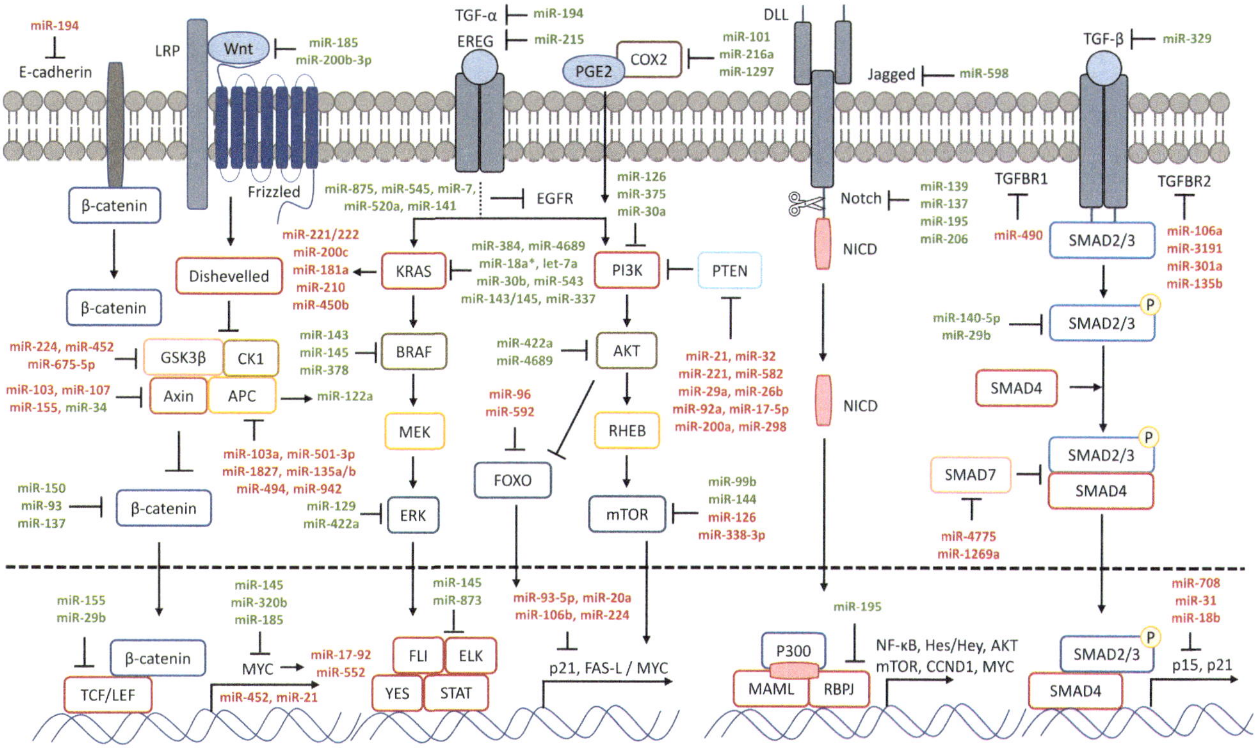

FIG. 17.1 Involvement of miRNAs in key signaling pathways in colorectal cancer. *TCF,* transcription factor 7; *LEF,* lymphoid enhancer binding factor; *LRP,* leucine-responsive regulatory protein; *GSK3β,* glycogen synthase kinase-3 beta; *CK1,* casein kinase 1; *APC,* adenomatous polyposis coli; *TGF-α/β,* transforming growth factor α/β; *EREG,* epiregulin; *EGFR,* epidermal growth factor receptor; *KRAS,* Kirsten rat sarcoma 2 viral oncogene homolog; *BRAF,* v-raf murine sarcoma viral oncogene homolog B1; *MEK,* mitogen-activated protein kinase kinase; *ERK,* mitogen-activated protein kinase 1; *STAT,* signal transducer and activator of transcription; *FLI,* Friend leukemia virus integration 1; *FOXO,* forkhead box O; *PGE2,* prostaglandin E2; *COX2,* cyclooxygenase 2; *PI3K,* phosphatidylinositol-3-kinase; *PTEN,* phosphatase and tensin homolog; *AKT,* protein kinase B (PKB); *RHEB,* Ras homolog enriched in brain; *mTOR,* mechanistic target of rapamycin; *FAS-L,* FAS ligand; *DLL,* distal-less; *NICD,* notch intracellular domain; *MAML,* mastermind-like transcriptional coactivator; *RBPJ,* recombination signal binding protein for immunoglobulin kappa J region; *TGFBR1/2,* transforming growth factor binding receptor 1/2; *SMAD,* mothers against decapentaplegic homolog; *NF-κB,* nuclear factor kappa B; *CCND1,* cyclin D1; green miRNAs function as tumor suppressors, red miRNAs function as oncogenes.

or adhesion. Thus, several molecules involved in the EGFR signaling pathway are considered promising therapeutic targets of anticancer drugs. EGFR is a transmembrane glycoprotein and a member of the human ErbB family of receptor tyrosine kinases that can be selectively activated through the binding of related ligands, including epidermal growth factor (EGF), transforming growth factor α (TGFα), epiregulin (EREG), amphiregulin (AREG), heparin-binding EGF-like growth factor (HB-EGF), connective tissue growth factor (CTGF), betacellulin (BTC), or epigen (Rayego-Mateos et al., 2018). After activation, dimerization of the receptor followed by autophosphorylation of the intracellular domain leads to the activation of two major downstream signaling pathways—KRAS/RAF/ERK and PI3K/AKT (Mlcochova, Faltejskova, Nemecek, Svoboda, & Slaby, 2013).

EGFR is commonly upregulated in CRC and its expression was proved to be negatively regulated through the binding of miR-875-5p (Zhang et al., 2016), miR-545 (Huang & Lu, 2017), miR-520a-3p (Qu et al., 2018), miR-7 (Fan et al., 2019), and miR-141-3p (Xing, Jing, Zhang, Suo, & Qian, 2020). On the contrary, miR-143 and miR-145 expression was found to be downregulated by EGFR in human and murine colonic cells (Zhu et al., 2011). Concerning the EGFR ligands, miR-218 (Lun, Wu, Deng, & Zhi, 2018), miR-133b (Guo et al., 2015), and miR-375 (Alam et al., 2017) were identified as important regulators of cell proliferation, migration, and angiogenesis through their binding to the 3′ UTR of CTGF while miR-194 affects the expression of TGF-α (Cai et al., 2018); miR-215-5p was associated with decreased levels of EREG (Vychytilova-Faltejskova et al., 2017).

Kirsten rat sarcoma 2 viral oncogene homolog (KRAS) is one of the most studied molecules of EGFR signaling. It is responsible for the activation of the mitogen-activated protein kinase (MAPK) cascade and the subsequent regulation of a large number of transcription factors involved in the regulation of cell proliferation, apoptosis, or migration. Thus, KRAS is considered a major driver of CRC progression. Importantly, mutations of this protein were found in more than

40% of CRC tumors and were associated with constitutive activation of the KRAS pathway independently of EGFR and decreased therapy response to anti-EGFR antibodies (Benvenuti et al., 2007). The first evidence of KRAS regulation by miRNAs was published in 2009, when miR-143 (Chen et al., 2009) and miR-18a* (Tsang & Kwok, 2009) were identified to directly target the 3′ UTR of this protein. Interestingly, miR-18a* successfully targeted KRAS, but not the v-Ha-ras Harvey rat sarcoma viral oncogene homolog (HRAS) or the neuroblastoma RAS viral oncogene homolog (NRAS). Later on, direct binding between KRAS and miR-145 (Pagliuca et al., 2013), let-7 (Smits et al., 2011), miR-30b (Liao et al., 2014), miR-4689 (Hiraki et al., 2015), miR-622 (Fang, Sun, Li, Chen, & Xiang, 2016), miR-543 (Fan et al., 2016), miR-384 (Wang et al., 2016), miR-337 (Liu et al., 2017), and miR-19a (Chen, Lin, & Wang, 2018) was described. Interestingly, miR-126 was confirmed to selectively affect KRAS-mutant cells through the repression of multiple genes essential for their survival (Hara et al., 2014). By contrast, KRAS downregulates the levels of tumor suppressor miR-4689 (Hiraki et al., 2015), but increases the levels of oncogenic miR-200c, miR-221, miR-222 (Tsunoda et al., 2011), miR-181a, miR-210 (Ota et al., 2012), and miR-450b-5p (Ye et al., 2016) in a three-dimensional (3D) specific manner in CRC cells. Besides the KRAS, other molecules from the MAPK pathway also may be regulated by miRNAs. Tumor-suppressor miRNAs miR-143, miR-145 (Pagliuca et al., 2013), and miR-378-5p (Wang et al., 2015), which are commonly downregulated in CRC tissue compared to adjacent mucosa, control the expression of B-Raf proto-oncogene, serine/threonine kinase (BRAF), which is mutated in 5%–20% of CRC patients (Di Nicolantonio et al., 2008). Other miRNAs with decreased levels in CRC—miR-422a (Wei et al., 2017) and miR-129 (Wu et al., 2010) were found to inhibit cell proliferation by targeting MAPK1/2 (ERK1/2).

The second central pathway downstream from EGFR and important in CRC development is the phosphatidylinositol 3-kinase (PI3K) pathway, which is associated with cell proliferation, migration, angiogenesis, and apoptosis. Importantly, multiple components of this pathway are mutated in CRC, including phosphatidylinositol-4,5-bisphosphate 3-kinase catalytic subunit α (PIK3CA) or PTEN (Sood et al., 2012). In 2008, Guo et al. (2008) revealed p85β, a regulatory subunit of PI3K, as one of the targets of miR-126. Later, PIK3CA was found to be regulated through miR-375 binding (Wang, Tang, Li, Jiang, & Wang, 2014), while the downregulation of catalytic subunit δ (PIK3CD) was observed in CRC cells after the transfection of miR-30a (Zhong, Bian, & Wu, 2013). Expression of PI3K may be inhibited not only by miRNAs, but also by the well-known tumor-suppressor protein PTEN. Nevertheless, downregulated levels of PTEN in CRC cells have been observed after the introduction of oncogenic miRNAs-miR-32 (Wu et al., 2013), miR-221 (Xue et al., 2013), miR-92a (Zhang et al., 2014), miR-17-5p (Fang et al., 2014), miR-29a (Wang et al., 2016), miR-200a (Li et al., 2016), miR-582 (Song, Long, Liu, Zhang, & Liu, 2017), miR-21 (Wu et al., 2017), miR-26b (Fan et al., 2018), or miR-298 (Arabsorkhi et al., 2020). Recently, lncRNA P73 antisense RNA 1T (TP73-AS1) was identified to function as a competing endogenous RNA (ceRNA) for miR-103. Importantly, PTEN expression was significantly induced due to the binding of this miRNA by TP73-AS1 (Jia et al., 2019). These results indicate the importance of cooperation between miRNAs and lncRNAs in the regulation of signaling pathways in CRC. On the other hand, the activation of PI3K by prostaglandins was described. These molecules are produced by cyclooxygenases (COXs) during the oxidation of arachidonic acid. Nevertheless, the regulation of COX2 overexpression by miR-101 (Strillacci et al., 2009), miR-1297 (Chen, Wang, Pan, & Guo, 2014; Chen, Xi, Wang, & Wei, 2014), or miR-216a-3p (Wang, Li, Zhang, Li, & Yu, 2018) was documented.

Protein kinase B, usually termed AKT, plays a central role in the whole cascade thanks to its pleiotropic effects. Two tumor-suppressor miRNAs with downregulated levels in CRC—miR-422a (Wei et al., 2017) and miR-4689 (Hiraki et al., 2015)—were proved to directly target AKT. Interestingly, miR-4689 targets not only AKT but also KRAS, and thus negatively regulates key components of two major branches in EGFR signaling. These results indicate a potential therapeutic use of this miRNA in CRC. Finally, the mechanistic target of rapamycin (mTOR) kinase acts downstream of the PI3K/AKT pathway to regulate cellular growth, apoptosis, and metabolism; it is frequently deregulated in a variety of cancers (Iwaya et al., 2012). Downregulation of mTOR by miR-99b-5p (Li et al., 2015) or miR-144 (Iwaya et al., 2012) contributed to the suppression of liver metastasis and disease recurrence while miR-126 (Wei et al., 2020) and miR-338-3p induced autophagy and increased the resistance of p53 mutant CRC cells to 5-fluorouracil (5-FU) treatment by reducing the expression of this protein (Han et al., 2017).

Similarly to the Wnt/β-catenin pathway, transcription factors activated by EGFR signaling including ELK1, YES1, JUN, FLI, MYC, signal transducers, and activators of transcription (STATs), or forkhead box O (FOXO) proteins may interact with miRNAs (Oda, Matsuoka, Funahashi, & Kitano, 2005). Interestingly, FLI1 (Zhang et al., 2011), YES, and STAT1 (Gregersen et al., 2010) were found to be regulated by miR-145, which also affects the expression of other proteins from EGFR activated pathways. Proliferation and migration of CRC cells may also be inhibited by miR-873 targeting ELK1 (Fan et al., 2019) or activated through the binding of miR-96 and miR-592 to the 3′ UTR of FOXO3A or FOXO1 (Fu et al., 2016; Gao & Wang, 2015). Taken together, numerous miRNAs significantly contribute to CRC progression by affecting the expression of a large number of molecules involved in the EGFR signaling pathway (Fig. 17.1).

TGF-β signaling pathway

Components of the TGF-β signaling pathway are mutated in 87% of hypermutated tumors and 27% of nonhypermutated CRC tumors and the deregulation of this pathway is a common event during cancerogenesis (Muzny et al., 2012). There is growing evidence demonstrating that miRNAs can target individual components of this pathway such as TGFβ1/β2, TGF-β receptors, regulatory SMAD1-3/5/9, coactivator SMAD4, and inhibitory SMAD6/7 (Fig. 17.1). Importantly, in normal cells and early stage cancers, TGF-β signaling regulates the transcription of several inhibitors of cyclin dependent kinases (CDKs), including $p21^{Cip1}$ and $p15^{Ink4B}$, and thus suppresses tumor formation by inhibiting cell growth and influencing apoptosis. However, due to the acquired mutations, cells become gradually resistant to these antigrowth effects and the whole cascade may evince oncogenic properties, especially through involvement in epithelial-to-mesenchymal transition (EMT). Regulation of TGFBR2 by miRNAs was first described in 2012, when miR-106a was identified to inhibit this receptor and thus increase CRC cell migration and invasion (Feng et al., 2012). Similar effects were observed in miR-3191 (He et al., 2019), miR-301a (Zhang et al., 2014), and miR-135b, which is one of the most upregulated miRNAs in CRC tissues commonly associated with increased metastatic potential (Li et al., 2015; Xu et al., 2012). Interestingly, Akbari et al. (2015) proved that TGFBR1 kinase inhibitor SD-208 evinces its anticancer effects by downregulation of miR-135b in SW48 cells and orthotopically implanted tumors in nude mice. Similarly to TGFBR2, the binding of miRNAs to the 3′ UTR of TGFBR1 was observed, including miR-490-3p (Xu et al., 2015). Besides the receptors, also the growth factors may be regulated through miRNAs. Tumor-suppressive miR-329 was found to target TGF-β1, which resulted in decreased cell proliferation and invasion (Li, Huang, Liu, Wen, & Sun, 2017). On the other hand, TGFβ1 treatment in CRC cells caused a significant decrease in the expression of miR-130b and increased cell migration. Subsequently, the luciferase reporter assay revealed that this miRNA binds to the 3′ UTR of integrin α5, which is involved in cell motility (Yi et al., 2016).

The key components of the whole cascades are transcription factors from the SMAD family. Zhai, Fesler, Ba, Wu, and Ju (2015) observed a progressive loss of miR-140-5p expression from normal colorectal mucosa to primary tumor tissues with further reduction in liver metastasis. This miRNA was proved to regulate cell proliferation, migration, and autophagy through the downregulation of SMAD2. On the contrary, miR-4775 (Zhao et al., 2017) and miR-1269a (Bu et al., 2015) decreased the expression of inhibitory SMAD7, increasing TGF-β signaling and the formation of distant metastasis. Interestingly, levels of miR-1269a may be elevated by TGF-β via SOX4, indicating the existence of a positive feedback loop. Finally, miR-29b may target SMAD3 and regulated TGF-β signaling. Nevertheless, two sponging sites for this miRNA within exons of lncRNA LOC646329-variant D were identified. The overexpression of this lncRNA was followed by increased SMAD3 and p21 levels and downregulated levels of cyclin D1 (CCND1), indicating a tumor-suppressor effect of the LOC646329-variant D in CRC (Javanmard, Dokanehiifard, Bohlooli, & Soltani, 2020).

Notch signaling pathway

The Notch signaling pathway is one of the most important pathways during normal development but also regulates many cellular processes associated with cancerogenesis such as cell proliferation, apoptosis, migration, invasion, and angiogenesis; emerging evidence suggests its frequent deregulation in CRC (Tyagi, Sharma, & Damodaran, 2020). To date, four members of Notch receptors have been identified in humans (Notch-1-4). They may be activated by the binding of five different ligands, including delta-like 1 (DLL-1), DLL-3, DLL-4, Jagged-1 (JAG-1), and JAG-2. Upon activation, Notch is cleaved, releasing the Notch intracellular domain (NICD) that is subsequently translocated into the nucleus for the transcriptional activation of Notch target genes such as the Hes family, the Hey family, nuclear factor-kappa B (NF-κB), vascular endothelial growth factor (VEGF), mTOR, CCND1, c-MYC, p21, p27, and AKT (Wang et al., 2010). Recently, several miRNAs have been shown to crosstalk with the Notch pathway (Fig. 17.1). Notch-1 was found to be overexpressed in 86% of CRC tissues and its elevated levels positively correlated with miR-21 expression, particularly in advanced stages (Xiong et al., 2014). Similarly, miR-137 (Xu et al., 2020) and miR-139-5p (Liu et al., 2016) downregulate this receptor by binding to its 3′ UTR. Interestingly, miR-137 was found to be sponged by lncRNA DSCAM-AS1, which is highly expressed in CRC (Xu et al., 2020), while miR-139-5p increased 5-FU-induced apoptosis (Liu, Yin, et al., 2016) and could serve as a promising therapeutic target in case of multidrug resistance. Decreased stemness and chemoresistance were observed after treatment of CRC cells with miR-195-5p, which binds the Notch signaling protein Notch-2 and the recombination signal binding protein for the immunoglobulin kappa J region (RBPJ) (Jin et al., 2018), a putative cofactor of hypoxia inducible factor 1 subunit alpha (HIF1α) (Díaz-Trelles et al., 2016). Finally, the Notch-3 receptor may be downregulated through the binding of miR-206, and this interaction is associated with the attenuation of cell proliferation and migration (Wang et al., 2015). On the contrary, lncRNA LINC00707 was identified to act as a molecular sponge by competing for miR-206, thus indirectly modulating the Notch signaling pathway (Zhu et al., 2019). Concerning the ligands of Notch receptors, miR-598 inhibits the metastatic potential of CRC cells by suppressing JAG1 and so hindering EMT (Chen et al., 2017).

Cell cycle regulation

Uncontrolled cell replication is a key mechanism of cancerogenesis. The *TP53* gene, encoding the p53 protein, is one of the most important tumor suppressors, and it controls the expression of several genes involved in cell cycle arrest, senescence, DNA repair, apoptosis, and autophagy. In addition, many metabolic pathways may be regulated by this protein under stress conditions. These multiple levels of regulation enable p53 complex regulation of cellular processes (Feroz & Sheikh, 2020). Unfortunately, mutations of this protein are found in more than 60% of CRCs (Nakayama & Oshima, 2019). Concerning cell cycle regulation, p53 induces the transcription of the $p21^{WAF1}$ protein, which is necessary for inhibition of cell cycle progression from the G1 to the S phase and directly inhibits DNA replication during the S phase. Further, the expression of cyclin A1 (CCNA1) is downregulated by p53, providing a secondary checkpoint of cell cycle progression through the S phase (Shaw, 1996). Recently, miRNAs have been recognized as mediators and regulators of p53 signaling. Kumar et al. (2011) revealed that the human *TP53* gene is negatively regulated by miR-25 and miR-30d. Later, the significant downregulation of this protein by miR-125b (Nishida et al., 2011), miR-27a (Maqbool, Lone, & Hussain, 2016), miR-300 (Wang & Yu, 2016), and miR-150-5p (Liu & Di Wang, 2019) was observed. In addition, miR-600 was confirmed to target mutant variants of p53, thus inhibiting the gain of novel oncogenic properties of mutant p53 expressing CRC cell lines (Zhang et al., 2017). On the contrary, miR-214 overexpression enhanced the levels of p53, indicating the therapeutic potential of this miRNA (Subramanian et al., 2014). On the other hand, p53 may induce the expression and maturation of numerous tumor-suppressor miRNAs commonly downregulated in CRC, including miR-30e-5p (Laudato et al., 2017), miR-34a-c, miR-145, miR-23b, miR-192, miR-194, miR-215, miR-15a/16, miR-200a-c, and several other miRNAs, as summarized previously (Rokavec et al., 2014). Importantly, the repression of the oncogenic miR-17-92 cluster by p53 was described under hypoxia conditions (Yan et al., 2009).

Numerous studies proved that miRNAs may influence the cell cycle at multiple levels. Mullany et al. (2018) investigated the expression of mRNAs and miRNAs involved in cell cycle regulation. Together, they identified unique interactions between these molecules comprising 32 mRNAs and 48 miRNAs. The most significant results were found in miRNAs from paralogous clusters miR-17-92, miR-106a-363, and miR-106b-25, as they appear to regulate G1 to S transition through E2F and MYC feedback and feed-forward loops. Similarly, 11 deregulated genes together with 37 deregulated miRNAs were identified in tumor tissues of CRC patients and associated with p53 signaling. Importantly, cell cycle arrest was one of the major cellular responses to this deregulation; however, further laboratory-based validation of these results will be necessary (Slattery et al., 2019). Besides p53, also other molecules including cyclins, cyclin-dependent kinases (CDK), and their inhibitors $p16^{Ink4A}$ (CDKN2A), $p15^{Ink4B}$ (CDKN2B), $p21^{Cip1}$ (CDKN1A), and $p27^{Kip1}$ (CDKN1B) play key roles in the regulation of cell cycle progression. Cyclins control the progression of the cell cycle by activating CDKs. While cyclin A (CCNA1/2), CCND1-3, and CCNE1/2 are essential for G1 to S transition, CCNB1-3 controls the progression from the G2 phase to the M phase of the cell cycle. To date, the binding of miR-93 to CCNB1 was described as inhibiting the early relapse of CRC (Yang et al., 2012) while the MCF2LAS1/miR-874-3p/CCNE1 axis was identified in CRC cell lines (Huang et al., 2021). Further, unconventional cyclin homolog CCNG2, whose high levels are associated with lymph node metastasis and poor overall survival (OS), was confirmed as a direct target of miR-1246 in LOVO and HCT-116 cells (Wang et al., 2016). Finally, the regulation of cyclin D by binding of tumor-suppressor miRNAs was observed in miR-526b-3p that may be sponged by lncRNA HOXD-AS1 (Yan et al., 2020), miR-374a (Chen et al., 2016) (CCND1), miR-1297 (Wang et al., 2017), miR-141 (Ye et al., 2017) (CCND2), and miR-592 (CCND3) (Liu et al., 2015). Importantly, the inhibition of CCND2 by miR-141 led to decreased cell proliferation and stemness of CRC cells, thereby enhancing the response to drug treatment (J. Ye et al., 2017).

Expression of CDK2 was found to be decreased after miR-200c binding (Gao et al., 2020) while levels of CDK4 were affected by tumor-suppressor miR-142-3p, which is commonly downregulated in CRC tissues (Zhu et al., 2018). Similarly, the association between CDK6 and miR-597-5p (Chen, Lang, et al., 2019), miR-539-5p (Gong et al., 2020), and miR-106b (Xu et al., 2017) was confirmed. Interestingly, the expression of these miRNAs may be significantly influenced by circular RNAs (circRNAs) circ_101555 (Chen et al., 2019), circ_000984 (Xu et al., 2017), and lncRNA CASC21, respectively (Gong et al., 2020), indicating the regulation of the cell cycle through ceRNAs. The function of CDKs may be inhibited not only through miRNAs but also by their respective inhibitors. While $p15^{Ink4B}$ and $p16^{Ink4A}$ form a complex with CDK4/6 and prevent the activation of cyclin D, $p21^{Cip1}$ and $p27^{Kip1}$ target CDK2 inhibiting cyclin E and subsequent G1 to S transition. Functional studies proved the binding of miR-708, miR-31 (Lei et al., 2014), and miR-18b to the 3′ UTR of CDKN2B ($p15^{Ink4B}$). The oncogenic properties of these miRNAs have been previously described and associated with the increased proliferation and migration of CRC cells (Yin et al., 2016). Being a major target of p53 activity, $p21^{Cip1}$ belongs to the frequently studied proteins in context with miRNA regulation. Wang et al. (2017) identified miR-93-5p as downregulating this inhibitor and thus increasing the multidrug resistance of CRC cells. Similarly, miR-106b induces cell radio resistance via the $p21^{Cip1}$ inhibition

(Zheng et al., 2015). Further, the direct binding of miR-20a (Sokolova et al., 2015) and miR-224 (Zhang et al., 2014) to this inhibitor was proved while $p27^{Kip1}$ regulation by miR-200b was observed in TGFBR2null CRCs (Fu et al., 2014).

Among the critical regulators of cell cycle belong also the transcription factors from E2F family, which function as downstream effectors of the retinoblastoma protein (Rb). Three of them (E2F1-3a) are activators while six others (E2F3b, E2F4-8) function as suppressors of cell cycle progression and DNA synthesis (Brehm, Miska, Reid, Bannister, & Kouzarides, 1999). Targeting of E2F1 by miR-362-3p, miR-329, miR-328-5p, and miR-610 was described in CRC cells and associated with cell cycle arrest, decreased proliferation, and increased sensitivity to 5-FU treatment (Christensen et al., 2013; Yin et al., 2018; Zhao & Chen, 2020; Zhou et al., 2020). As with other miRNAs, miR-328-5p and miR-610 levels may be sponged by circ_CAMSAP1 (Zhou et al., 2020) and circ_MAT2B (Zhao & Chen, 2020), respectively. Interestingly, miR-34c expression may be promoted by binding of E2F1 to its promoter and this interaction is interrupted by promoter hypermethylation (Yang et al., 2016). Expression of E2F3 may be inhibited by miR-503 (Chang, Yue, Wang, & Zhang, 2015) as well as by miR-34a. Decreased levels of transcription factor were associated with the attenuated resistance of CRC cells to 5-FU-based chemotherapy (Wu et al., 2018). Finally, the inhibition of tumor-suppressor protein Rb by oncogenic miR-675 (Tsang et al., 2010) or by radiation-induced miR-622 (Ma et al., 2015) was observed. The overview of miRNAs involved in cell cycle regulation is summarized in Fig. 17.2.

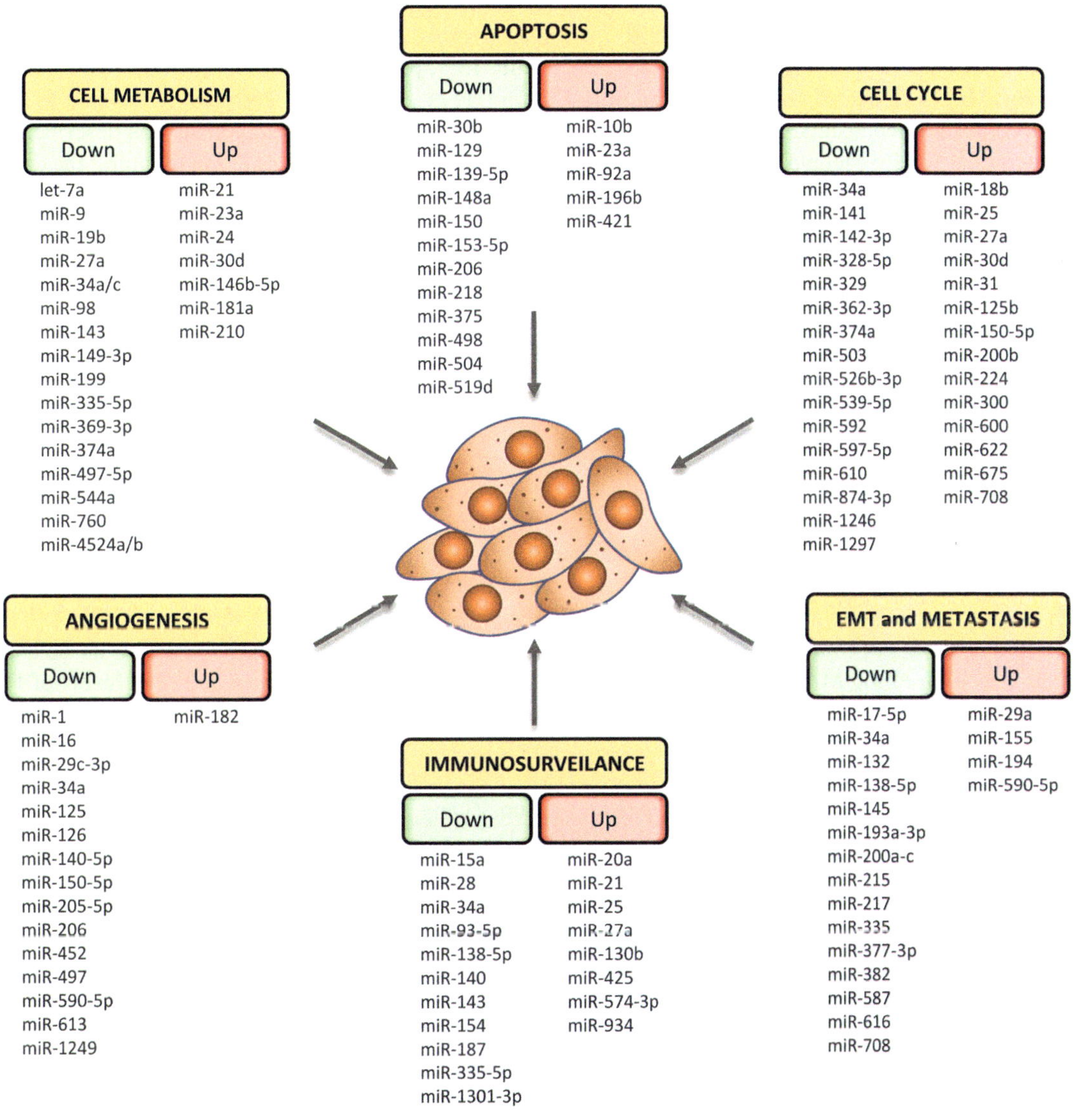

FIG. 17.2 A list of miRNAs with deregulated expression in colorectal cancer and involved in cell metabolism, apoptosis, cell cycle regulation, EMT, metastasis, immunosurveillance, and angiogenesis. Down—miRNAs downregulated in tumor tissue compared to adjacent tissue, Up—miRNAs upregulated in tumor tissue compared to adjacent tissue.

Regulation of apoptosis

Apoptosis is a process of programmed cell death that occurs during normal development, maintaining the balance between the generation and loss of cells. The two main pathways are termed extrinsic and intrinsic. The extrinsic pathway is activated through the binding of ligands to death receptors that are members of the tumor necrosis factor (TNF) receptor gene superfamily while the intrinsic pathway is activated upon stress conditions and is associated with changes in the integrity of the mitochondrial membrane and the release of cytochrome *c* (cyt-c). The primary mediators of apoptosis are caspases in cooperation with proapoptotic proteins BCL2-associated agonist of cell death (BAD) and BCL2-interacting mediator of cell death (BIM/BCL2L11), whereas the inhibition of this process is predominantly performed through the antiapoptotic molecules from the BCL-2 family (Wang, 2020).

Numerous studies proved that miRNAs may contribute to the promotion or inhibition of apoptosis in CRC cells (Fig. 17.2). The miR-17-92 cluster generates a single polycistronic primary transcript that yields six mature oncogenic miRNAs. Performing copy number analyses, miR-92a was shown to be transcribed at the highest levels in both adenomas and carcinomas. Subsequently, the direct targeting of BIM by this miRNA was confirmed, indicating the pivotal role of miR-92a in the development of CRC (Tsuchida et al., 2011). Similarly, the binding of miR-10b to the 3′ UTR of BIM was described and correlated with chemoresistance and worse survival (Nishida et al., 2012). Until now, the binding of miRNAs to BAD in CRC has not yet been described. On the contrary, tumor-suppressor miRNAs were identified to regulate the levels of antiapoptotic protein BCL-2, including miR-148a (Zhang et al., 2011), miR-129 (Karaayvaz, Zhai, & Ju, 2013), miR-30b (Liao et al., 2014), miR-375 (Zaharie et al., 2015), miR-139-5p (Li et al., 2016), miR-206 (Meng & Fu, 2018), miR-498 (Wang, Ma, Li, Ding, & Gao, 2020), and miR-153-5p (He et al., 2020). A correlation between BCL-2 inhibition and enhanced chemotherapeutic sensitivity of CRC cells to 5-FU and oxaliplatin treatment was described (He et al., 2020; Karaayvaz et al., 2013; Meng & Fu, 2018). Concerning the caspases, miR-421 was found to inhibit the expression of caspase 3 (Zhou et al., 2020) while the decreased activation of caspases 3, 7, and 9 was documented after binding of miR-23a to apoptosis-activating factor 1 (APAF-1) (Shang et al., 2014). Finally, cellular FADD-like IL-1β-converting enzyme inhibitory protein (cFLIP), which binds to FADD, caspase 8 and/or caspase 10, and TRAIL receptor 5 (DR5) (Safa, 2012), was found to be regulated through miR-218 (Meng et al., 2018), miR-150, miR-504, and miR-519d in CRC cells (Rong, Yang, Wu, & Wu, 2019), whereas the direct binding of miR-196b to FAS was described (Mo et al., 2015). Taken together, these data indicate a significant involvement of miRNAs in the regulation of apoptosis, especially in connection with response to chemotherapeutic treatments.

Regulation of angiogenesis

Angiogenesis is a coordinated multistep process responsible for fulfilling the nutrient and oxygen demand of normal as well as cancer tissues. The balance between angiogenic and antiangiogenic factors is crucial and may be significantly influenced by miRNA deregulation. Apart from the PI3K/AKT, TGF-β, EGFR, and WNT/β-catenin signaling pathways described previously, multiple signaling molecules such as HIF1α, VEGF and its receptor (VEGFR), thrombospondin-1 (TSP-1), and PTEN (Soheilifar et al., 2020) contribute to CRC angiogenesis. Enhanced expression of VEGF in CRC tissue is associated with the hyperactivation of the MAPK cascade, the PI3K/AKT/mTOR signaling pathway, and high levels of HIF1α and free β-catenin. In addition, decreased levels of tumor-suppressor miRNAs miR-126 (Zhang et al., 2013), miR-34a (Zhang, Zhou, & Dong, 2014), miR-140-5p (Zhang et al., 2015), miR-497 (Qiu et al., 2016), miR-125 (Wu et al., 2018), miR-1 (Zhu et al., 2018), miR-150-5p (Chen et al., 2018), p53 induced miR-1249 (Chen et al., 2019), miR-452 (Mo, Park, Choi, Yun, & Chae, 2019), miR-613 (Li et al., 2020), miR-205-5p (Liu, Li, Sun, Zhang, & Xu, 2020), miR-206 (Zheng et al., 2020), and miR-29c-3p (Chen et al., 2020) targeting the 3′ UTR of VEGF were detected in CRC. Importantly, low expression of miR-613, miR-205-5p, miR-206, and miR-29c-3p was associated with increased levels of circRNAs circ_CCT3 (Li, Xu, et al., 2020), circ_0056618 (Zheng et al., 2020), circ_001971 (Chen, Huang, et al., 2020), and lncRNA NEAT1, respectively (Liu et al., 2020), indicating the critical cooperation between miRNAs and lncRNAs in the regulation of CRC angiogenesis. Further, nuclear factor 90 (NF90), a positive regulator of VEGF, was shown to be a direct target of miR-590-5p (Zhou et al., 2016). Interestingly, the oncogenic cluster miR-17-92 suppresses tumor angiogenesis and progression by targeting HIF1α and VEGF in vivo and in vitro, suggesting the presence of cellular context-dependent pro- and anticancer effects of these clustered miRNAs (Ma et al., 2015). Concerning VEGFR, high blood sugar levels suppressed the expression of miR-16, which resulted in the elevated expression of this receptor in CRC patients (Yang et al., 2016).

Thrombospondin belongs to the most important suppressors of angiogenesis and its expression is under the direct control of p53. Due to the frequent mutations of p53 in various cancers, decreased levels of TSP-1 were detected in CRC tissues and correlated with increased activation of angiogenic factors. Interestingly, tumor-suppressor p53-responsive miR-194 targets TSP-1 during ischemia and promotes angiogenesis, indicating the context-dependent contribution of this miRNA to neoplastic growth (Sundaram et al., 2011). Further, miR-182 was found to target the 3′ UTR of TSP-1. In addition, the transfection of HCT-116 and HT-29 cells with antimiR-182 was associated with increased translocation of EGR-1 and SP-1 into the nucleus and the loading of these transcription factors on the TSP-1 promoter (Amodeo et al., 2013). Finally, exosomal miRNAs can shuttle between CRC cells and the tumor microenvironment, transducing the angiogenesis-regulating signals between different cellular components (Soheilifar et al., 2020). The list of miRNAs involved in the regulation of angiogenesis is summarized in Fig. 17.2.

Regulation of epithelial-to-mesenchymal transition and metastatic capacity

Metastatic disease is the major cause of death in CRC. The formation of metastasis is a multistep process in which malignant cells disseminate from the primary tumor to colonize distant organs (De Krijger, Mekenkamp, Punt, & Nagtegaal, 2011). Apart from the signaling pathways described earlier, EMT together with cytoskeleton reconstructions and changes in cell morphology are significantly involved in the metastatic potential of cancer cells. EMT plays a pivotal role in tumor progression and the formation of distant metastasis in CRC. During this process, cancer cells lose their epithelial phenotype to acquire the mesenchymal one. The whole cascade is controlled by EMT-related transcription factors from five different families–Zinc finger protein SNAI1 (SNAIL), Zinc finger E-box binding (ZEB), Twist-related protein 1 (TWIST), Prospero homeobox protein 1 (PROX1), and Forkhead box proteins (FOX) (Balacescu et al., 2018). In addition, the enhanced expression of matrix metalloproteinases (MMPs) and urokinase plasminogen activator (uPA) is associated with degradation of the extracellular matrix (ECM), the entrance of tumor cells into blood circulation, and facilitated migration (Gershtein, Korotkova, Prorokov, & Kushlinskii, 2013).

Previously, miR-29a was found to promote CRC metastasis by regulating MMP2 and CDH1 through the Krüppel-like factor 4 (KLF4) transcription factor (Tang et al., 2014). Similarly, miR-194 influences EMT markers by downregulating CDH1 and upregulating vimentin and MMP2, thus increasing the cell polarization of SW480 cells (Cai et al., 2017). Vimentin may be further regulated via direct binding of miR-34a (Zhu et al., 2019) or miR-17-5p (Kim et al., 2020). In 2013, miR-155 was described as promoting the invasion of CRC cells through the regulation of claudin-1 (Zhang et al., 2013). Recently, the association between this miRNA and ZEB2 was proved (Yang, Lin, & Mu, 2019), indicating the important involvement of miR-155 in CRC invasiveness. Concerning the ZEB proteins, the targeting of ZEB1 by miR-708 (Sun et al., 2019), miR-200c (Karimi Dermani, Amini, Saidijam, & Najafi, 2018), miR-382, miR-587, miR-616 (Ren et al., 2020), and miR-217 (Bian et al., 2019) was observed while miR-132 (Zheng et al., 2014), miR-335 (Sun et al., 2014), miR-377-3p (Huang et al., 2020), miR-200a-c (Ranković, Zidar, Žlajpah, & Boštjančič, 2019), miR-138-5p (Yan, Bi, Zhang, Song, & Hong, 2020), and miR-215 (Chen, Zhang, et al., 2017) may bind the 3′ UTR of ZEB2. Besides the ZEB transcription factors, TWIST1 is inhibited through tumor-suppressive miR-145 (Shen et al., 2019), whereas the upregulation of TWIST2 is detected in CRC cells with decreased levels of miR-138 (Long et al., 2013). Interestingly, the regulation of miRNA expression by SNAIL during EMT was described in HT29 cells, when SNAIL abundance repressed the levels of miR-192 and miR-194 while the expression of miR-205 and let-7i increased significantly. In addition, the content of extracellular vesicles (EVs) released from cancer cells was impacted (Przygodzka et al., 2019). Finally, the degradation of ECM by MMPs may be facilitated thanks to the hypoxia-induced miR-590-5p directly targeting RECK, a membrane-anchored glycoprotein that inhibits MMP function (Kim et al., 2018). In addition, decreased levels of miR-193a-3p lead to the increased activation of uPA, enhanced ECM degradation, and invasion of CRC cells (Lin et al., 2019).

As in previously described signaling pathways, a close connection between genes involved in EMT, miRNAs, and other ncRNAs was documented. First, the sponging of miR-138 and miR-200a by lncRNA H19 or TUG1 promoted EMT and migration in vitro as well as in vivo due to the increased levels of ZEB1, ZEB2, and vimentin (Liang et al., 2015; Yan, Bi, et al., 2020). Further, lncRNAs UICLM (Chen et al., 2017) and FTX (Yang et al., 2018) function as ceRNAs for tumor-suppressor miR-215 to regulate ZEB2 expression. At last, ZEB1 levels are increased in response to miR-217, miR-382, miR-587, and miR-616 sponging by KCNQ1OT1 (Bian et al., 2019) or circ_0001178 (Ren et al., 2020). These results indicate a complex regulation of numerous processes associated with the formation of distant metastasis by different classes of ncRNAs. An overview of miRNAs deregulated during EMT and metastasis is depicted in Fig. 17.2.

Reprogramming of cellular metabolism

Metabolic reprogramming is one of the important hallmarks in cancers comprising the adaptive changes to sustain the high-energy requirements for rapid cell proliferation. Increasing evidence suggests the involvement of miRNAs in basic metabolic pathways such as glycolysis, oxidative phosphorylation, or metabolism of lactate and fatty acids in

CRC cells (Hon, Abidin, Othman, & Naidu, 2020). The Warburg effect is thought to be an early event in CRC development characterized by increased rates of aerobic glycolysis, enhanced lactate production, and decreased production of ATP from oxidative phosphorylation in mitochondria (Lu, Tan, & Cai, 2015). This metabolic shift enables more rapid production of energy, biosynthesis of macromolecules essential for cell growth, and lower generation of reactive oxygen species (ROS) (Movahed, Rastegari-Pouyani, Hossein Mohammadi, & Mansouri, 2019). The whole process is accompanied by alterations of the glucose transporters (GLUT) and key glycolytic enzymes such as hexokinase (HK), lactate dehydrogenase (LDH), and pyruvate kinase (PK). In this context, miR-181a was identified to be overexpressed in CRC and its high levels were associated with increased glucose uptake and lactate secretion. Further, PTEN was found to be a direct target of this miRNA and its suppression led to increased AKT S473 phosphorylation and miR-181a-induced glycolysis (Wei, Cui, Mei, Liu, & Zhang, 2014). Recently, miR-760 was described to bind GLUT1 and decrease the glycolysis of CRC cells. On the other hand, this miRNA may be sponged by circ_DENND4C, which is overexpressed in CRC and its levels correlates with GLUT1 (Zhang, Zhang, Qin, Wang, & Fu, 2020). Concerning the glycolytic enzymes, lncRNA lncARSR sponges miR-34a-5p to promote CRC invasion and metastasis via HK1-mediated glycolysis (Li et al., 2020). Similarly, levels of HK1 were significantly affected by miR-27a (Barisciano et al., 2020) and miR-143 (Gregersen et al., 2010) binding while the expression of HK2 negatively correlated with miR-9, miR-98, and miR-199 levels (Snezhkina et al., 2018). However, the direct binding to this enzyme needs further validation using a luciferase assay. Several studies have reported increased levels of LDH in CRC, but its exact role has not been established. Wang et al. (2015) investigated miRNA-induced changes in LDHA expression and observed the direct binding of miR-34a, miR-34c, miR-369-3p, miR-374a, and miR-4524a/b to the 3′ UTR of this protein, resulting in glycolysis inhibition and a slower growth rate of CRC cells. The beta subunit of LDH (LDHB) may be downregulated through miR-335-5p binding (Zhang & Yang, 2019). Finally, miR-149-3p was identified as decreasing the chemoresistance of CRC cells by targeting pyruvate dehydrogenase kinase isoform 2 (PDK2), thus affecting the glucose metabolism (Liang et al., 2020).

Once D-glucose is converted into pyruvate, it enters oxidative phosphorylation in the mitochondria. This pathway coupled with the Krebs cycle is the main producer of energy in the form of ATP in mitochondria. Although the tumor cells usually shift to aerobic glycolysis, they are highly dependent on oxidative proliferation and conversion of available nutrients into cellular components to increase their proliferation (Hon et al., 2020). In CRC, miR-142-5p was identified to target succinate dehydrogenase-B (SDHB) to promote cancerogenesis by facilitating the generation of aerobic glycolysis (Liu et al., 2017). Similarly, miR-146b-5p binds the 3′ UTR of pyruvate dehydrogenase-B (PDHB) to promote cell growth, invasion, and glycolysis (Zhu, Wu, Yan, Zhan, & Sun, 2017). Further, miR-23a was shown to decrease the expression of pyruvate dehydrogenase lipoamide kinase isozyme 4 (PDK4) to induce the activation of pyruvate dehydrogenase (PDH) and subsequent oxidative phosphorylation in order to gain a sufficient amount of ATP for increased cell proliferation of CRC cells (Deng et al., 2018). Finally, let-7a could regulate the secretion of EVs and mitochondrial oxidative phosphorylation by targeting synaptosomal-associated protein 23 (SNAP23), an important regulator of vesicle transport and an essential component of membrane fusion machinery (Liu et al., 2021).

Besides the changes in glycolysis and oxidative phosphorylation, the deregulation of the lipid metabolism as well as hypoxia-induced reprogramming of the metabolism have been observed in CRC. Most cancers upregulate the de novo synthesis of fatty acids and cholesterol synthesis pathways to fulfill the high requirements of cells (Daniëls et al., 2014). Thirty-one putative miRNAs were predicted to target the enzymes comprising the acyl-CoA synthetase/stearoylCoA desaturase (ACSL/SCD) network. Further analyses showed miR-544a, miR-142, and miR-19b-1 to be the main regulators of this pathway and decreased levels of miR-19b-1 significantly correlated with a shorter OS of CRC patients (Cruz-Gil et al., 2018). Recently, miR-497-5p was found to mediate starvation-induced death by targeting acyl-CoA synthetase-5 (ACSL5), a key molecule of the lipid metabolism highly expressed in CRC cells, indicating a great potential of fatty acid metabolism modulation in cancer therapy (Gharib, Nasri Nasrabadi, & Reza Zali, 2020). Concerning the hypoxia involvement in metabolism reprogramming, Nijhuis et al. (2017) showed a correlation between an altered amino acid metabolism and hypoxia-responsive miRNAs in CRC cells under hypoxia conditions. Further, the expression of miR-210 was elevated in hypoxic areas while increased levels of miR-21 and miR-30d were detected in 5-FU resistant cells. Interestingly, the miR-23a-24 cluster was found to be one of the most upregulated clusters in CRC cells under hypoxia that functions as a critical regulator of the metabolism switch from oxidative phosphorylation to glycolysis through an miR-24/VHL/HIF1α double-negative feedback loop (Jin et al., 2019). Taken together, these data confirm the significant involvement of miRNAs in the regulation of metabolism in cancer cells. However, many questions are yet to be elucidated. The miRNAs involved in the reprogramming of the CRC metabolism are summarized in Fig. 17.2.

Involvement in tumor immunosurveillance

During recent years, it has become increasingly clear that local immune response is a key factor in the modulation of the multistep process of cancer initiation and progression. The final outcome of pro- versus antitumoral innate immune response is dependent on TME, which evolves dynamically and may be significantly affected by chronic inflammation (Graziella, 2018). Previously, three major immune hallmarks acquired by cancer cells were presented, including the ability to evade immune recognition, suppress immune reactivity, and thrive in a chronically inflamed microenvironment (Cavallo, De Giovanni, Nanni, Forni, & Lollini, 2011). A growing amount of evidence indicates miRNA involvement in the regulation of cancer immune surveillance and escape (Fig. 17.2). Notably, cancer-derived miRNAs may be packed into exosomes or microvesicles to mediate the intercellular communication and affect the expression of a broad range of immunity-associated genes (Yi et al., 2020).

Tumor-suppressor miRNAs miR-138-5p (Zhao et al., 2016), miR-140-3p (Jiang et al., 2019), miR-140 (Naba, Tolay, Erman, & Sayi Yazgan, 2020), and miR-93-5p (Chen, Wang, Lin, & Huang, 2020) are implicated in the control of antitumor immune response through the regulation of immune checkpoint proteins, including programmed cell death 1 (PD-1/CD279) and/or its ligand (PD-L1/B7-H1). Further, doxorubicin was found to inhibit miR-140-3p and upregulate PD-L1 levels (Naba et al., 2020) while circRNA circ_0020397 sponges miR-138, thus contributing to the increased expression of this ligand and CRC progression (Zhang, Xu, & Wang, 2017). Importantly, miR-15a carried by mesenchymal stem cell-derived EVs inhibits the immune evasion of CRC cells by regulating the histone lysine demethylase 4B (KDM4B)/HOXC4/PD-L1 axis, indicating the important role of miRNAs encapsulated within EVs in CRC pathogenesis (Liu et al., 2021). Besides PD-L1, the B7H3 (CD276) inhibitory molecule is targeted by miR-187 (Wang et al., 2016) while the expression of circulating miR-1301-3p, miR-335-5p, and miR-28-5p was negatively correlated with the levels of this molecule, TNM stage, and metastasis (Wang, Zhao, Xu, Zhou, & Yan, 2019).

Apart from the upregulation of immune checkpoint proteins, antigen processing and presentation may be disturbed by miRNAs targeting components of antigen presenting machinery and major histocompatibility complex class I (MHC-I) in CRC cells. Using proteomic screening, miR-27a was identified to repress MHC-I expression through calreticulin, influence the level of CD8+ T-cell infiltration and cytotoxic activity, and contribute to the formation of distant metastasis and poor prognosis of CRC patients (Colangelo et al., 2016). Further, the activating receptor NKG2D and its ligand MICA play essential roles in natural killer (NK) cell-mediated immune response to tumors. Tang, Fu, Xu, and Zhou (2019) described that oncogenic miR-20a regulates the sensitivity of CRC cells to NK cells by downregulating MICA through direct binding to its 3′ UTR and promoting CRC growth. Further, IL-18 promotes the antitumor ability of NK cells via the miR-574-3p/TGF-β1 axis in HCT116 cells (Li, Du, Zhang, & Yang, 2021).

Macrophages are key players at the interface between innate and adaptive immunity. They are involved in antigen presentation, phagocytosis, and inflammation promotion (Graziella, 2018). Recently, tumor-derived exosomal miR-934, miR-25, miR-130b, and miR-425 were identified to induce the polarization of tumor-associated macrophages M2 (TAMs). Further, polarized TAMs induced premetastatic niche formation and promoted CRC liver metastasis by secreting CXCL12/13 and activating the NF-κB signaling pathway and VEGF secretion (Wang et al., 2020; Zhao et al., 2020). In addition, they were found to promote cell proliferation via IL-1β-mediated downregulation of miR-28-3p (Fan, Liao, Zhang, & Zheng, 2020). One of the most studied oncogenic miRNAs, miR-21, is also highly enriched in CRC-derived EVs and is essential for creating a liver proinflammatory phenotype through binding to Toll-like receptor 7 (TLR7) in macrophages and increasing the secretion of IL-6 (Shao et al., 2018). In addition, miR-21 targets chemokine CCL20 in CRC cell lines. Interestingly, high levels of both molecules were found in CRC tissues, indicating they are probably not expressed in the same cell. Further, miR-21 expression was located in stromal cells such as tumor-associated fibroblasts while CCL20 was primarily detected in tumor-infiltrating immune cells, suggesting the coexpression of both molecules in TME (Vicinus et al., 2013). Signaling mediated via the IL-6/IL-6R plays a pivotal role during immune response in cancer. In 2014, IL-6R was proved to be targeted by tumor suppressor miR-34a (Rokavec et al., 2014). A year later, also a soluble variant of this receptor (s-IL-6R) was confirmed as a direct target of this miRNA and upregulation of miR 34a affected both CRC cells with and without IL-6R expression, indicating its essential role in inflammatory bases of the disease (Li, Rokavec, & Hermeking, 2015). Finally, the production of proinflammatory cytokines may be inhibited through targeting of TLRs by miRNAs. Previously, binding of miR-143 (Guo et al., 2013) and miR-154 (Xin, Zhang, & Liu, 2014) to the 3′ UTR of TLR2 was observed while an association between miR-6869-5p and TLR4 is documented (Yan et al., 2018). These results indicate the significant involvement of the immune system as well as miRNAs in the pathophysiology of CRC. However, a better understanding of this complex interplay will be necessary to develop more efficient and optimized cancer immune-based therapies.

The interconnection between miRNAs and epigenetics in CRC

Epigenetics is the study of processes that alter gene activity without changing the DNA sequence and may be heritable. Neoplastic transformation involves alterations that are both genetic and epigenetic. Today, several different epigenetic processes are known, including DNA methylation, histone modifications, and regulation by ncRNAs, especially miRNAs. Genes for miRNAs may be epigenetically regulated by DNA methylation and/or histone modification that results in their overexpression or downregulation in numerous pathologies, including CRC. According to recent studies, almost half the miRNA genes are associated with CpG islands and almost 12% of them were prone to epigenetic inactivation by methylation in 23 different types of cancers (Kunej et al., 2011). Further, methylation was found to be involved in the regulation of expression of miRNA biogenesis genes (Glaich et al., 2019) while multiple miRNAs can be deregulated as a result of the aberrant expression of specific epigenetic regulators. In turn, several miRNAs called epi-miRNAs were recognized to target enzymes involved in epigenetic modulation. It was observed that epigenetic regulators are strongly enriched among the miRNA predicted targets, indicating the importance of these small molecules in pluripotency, development, and cell reprogramming (Gruber & Zavolan, 2013). Recent studies also suggest direct epigenetic functions of miRNAs by recruiting specific protein complexes to genomic DNA (Kim, Sætrom, Snøve, & Rossi, 2008). These observations indicate the existence of a regulatory circuit between miRNAs and epigenetic modulation (Gregorova, Vychytilova-Faltejskova, & Sevcikova, 2021).

So far, DNA hypermethylation was confirmed to be responsible for decreased levels of miR-125a/b (H. Chen & Xu, 2015), miR-124a, miR-34a-c (Deng et al., 2011; Lodygin et al., 2008; Toyota et al., 2008; Vogt et al., 2011), miR-9 (Bandres et al., 2009), miR-137, miR-342 (Balaguer et al., 2010; Chen et al., 2017; Kashani et al., 2019), miR-1, miR-133a (Chen et al., 2012), miR-149 (Wang et al., 2013), miR-212 (Meng et al., 2013), miR-126 (Zhang, Wang, et al., 2013), miR-181b (Zhao et al., 2016), miR-490-3p (Zheng et al., 2016), miR-4500 (Yu et al., 2016), and miR-1247 (Liang et al., 2018) in CRC. Interestingly, the hypermethylation of miR-34b/c and the miR-124a promoter occurs early in CRC cancerogenesis. It is present in all histological types, polyps, and even in the adjacent normal mucosa of CRC patients, but not in colonic mucosa from patients without cancer or with ulcerative colitis (Deng, Kakar, & Kim, 2011). Similarly, the epigenetic silencing of miR-137 was found to be an early event observed in the tumor tissue of CRC patients as well as in tissues of patients with adenomas and different CRC cell lines (Balaguer et al., 2010). Importantly, DNA methyl-CpG-binding protein 2 (MECP2) was identified to be involved in the epigenetic regulation of this miRNA (Chen, Cai, et al., 2017).

Concerning the regulation of key epigenetic enzymes by miRNAs, miR-124 and miR-506 were described to inhibit CRC cell progression by direct targeting of DNA methyltransferase 3B (DNMT3B) and indirect targeting of DNMT1, leading to reduced global DNA methylation and restored expression of CDH1, O^6-methylguanine-DNA-methyltransferase (MGMT), and p16 (Chen et al., 2015). Similarly, miR-203 regulates the expression of the de novo DNA methyltransferase DNMT3B (To, Leung, & Ng, 2017) while the well-known tumor-suppressor miR-143 targets DNMT3A to inhibit the progression of CRC (Ng et al., 2009). On the other hand, oncogenic miR-21 binds the 3′ UTR of 10–11 translocation 1 (TET1), thus preventing the removal of methylation marks by this protein and promoting cell proliferation (Ma et al., 2018). Concerning the histone modifications, miR-761 was found to inhibit CRC progression by targeting histone deacetylase 1 (HDAC1) (Xiong et al., 2019), whereas miR-101 suppresses the levels of the enhancer of the zeste homolog 2 (EZH2) (Zhuoran, Xiaomin, & Jun, 2020), a functional enzymatic component of the Polycomb repressive complex 2 (PRC2), which participates in the methylation of histone H3 at lysine 27 (H3K27met) and thus the inhibition of transcription. Recently, direct regulation of this protein by the cooperative action of miRNAs and other ncRNAs was described, resulting in increased levels of EZH2 due to the sponging of miR-137 by lncRNA XIST (Liu, Cui, & Hua, 2018), miR-26a/b by SNHG6 (Xu et al., 2019), miR-144 by ZNFX-AS1 (Shi et al., 2019), and miR-600 by circular RNA circ_0071589 (Yong et al., 2018). This suggests that EZH2 plays a critical regulatory role in CRC pathogenesis and could serve as a novel promising target of therapy. In conclusion, recent studies uncovered a complex network of interactions among different epigenetic layers, which is involved in the reversible regulation of gene expression.

Translational opportunities of miRNAs in CRC

Expression profiling of miRNAs in CRC patients

Alterations in miRNA expression profiles have been successfully detected in many types of cancers, including CRC. These widespread differential expressions between malignant and normal cells may be explained by the gene location in cancer-associated regions, alterations in the miRNA processing machinery, or epigenetic mechanisms (Garzon, Calin, & Croce, 2009). Nevertheless, lower miRNA levels are detected in tumors compared to adjacent noncancerous tissues and their decreased expression is connected to poor differentiation (Lu et al., 2005). Importantly, the small size and hairpin structure of miRNAs assure the high stability of these molecules in tissues as well as in different body fluids and enable large retrospective studies using formalin-fixed paraffin-embedded (FFPE) samples (Xi et al., 2007). In addition, expression profiling of miRNAs may be used not only for better tumor classification, but also for tracing the tissue of origin of cancers of unknown primary origin (Rosenfeld et al., 2008). In the last 15 years, miRNAs have been highlighted as promising diagnostic, prognostic, and predictive biomarkers in CRC, and technological advances have enabled the development of various platforms for their profiling and subsequent validation.

The first study dealing with expression profiling in CRC was published in 2003 by Michael, O'Connor, Van Holst Pellekaan, Young, & James (2003). Using cloning technology followed by northern blotting, significantly reduced levels of miR-143 and miR-145 were detected in the adenomatous and carcinoma tissues. Three years later, the expression of 156 mature miRNAs was measured in CRC tumors and adjacent nonneoplastic tissues using RT-qPCR. In total, 13 miRNAs were identified to be differentially expressed between these two groups of samples, including upregulated miR-31, miR-96, miR-135b, and miR-183, while levels of miR-145 and miR-133b were significantly decreased (Bandrés et al., 2006). In the same year, Cummins et al. (2006) developed an experimental method called miRNA serial analysis of gene expression (miRAGE) to perform one of the largest analyses of human miRNAs in CRC. Using this approach, they identified 200 known mature miRNAs, 133 novel miRNAs, and 112 previously uncharacterized miRNA forms, suggesting that the human genome contains many more miRNAs than was expected. In another profiling study, downregulated levels of miR-30c and upregulated levels of miR-21, miR-17-5p, miR-191, miR-155, miR-106a, miR-29b, miR-223, and miR-203 were detected in CRC samples by miRNA microarray (Volinia et al., 2006). Using Miltenyi's miRXplore platform and subsequent RT-qPCR validation, miR-21, miR-24, miR-145, miR-150, and miR-378 were found to be significantly deregulated in CRCs as well as colorectal adenomas compared to adjacent mucosa (Gattolliat et al., 2015). Using *mir*Vana miRNA bioarrays, the expression profile of miRNAs was assessed in eight CRC cell lines, 45 CRC samples, and four matched adjacent tissues. In total, 37 miRNAs were found to be differentially expressed in CRC, including tumor-suppressor miR-422a, miR-143, miR-145, miR-30a/c, miR-195, miR-378*, and miR-125a and oncogenic miR-31, miR-21, or miRNAs from the miR-17-92 cluster (Arndt et al., 2009). By using a highly advanced combined approach of bias reduction, a small RNA deep sequencing protocol, and the Illumina small RNA TruSeq method for sample bar coding, Sun et al. (2016) revealed a signature of miRNAs in CRC and validated many previously published results. First, miR-21, which is commonly upregulated in CRC tissue, and a tumor-suppressor miR-143 were the two miRNAs that dominated the miRNA population in colon/rectum tissue and could be the most important miRNAs in CRC. Further, these two miRNAs together with another eight–miR-148a, miR-194, miR-192, miR-200b, miR-200c, miR-10b, miR-26a, and miR-145–constituted the top 10 highly expressed miRNAs in colon/rectum as well as in CRC tissues. Finally, novel molecular characterization of colorectal primary tumors based on the expression of miRNAs was determined using in situ hybridization arrays and FFPE samples to screen 1436 different miRNAs. Importantly, miRNAs associated with diagnosis age, Lynch syndrome, stroma abundance, peritumoral inflammation, mucinous component, tumor location, grade, stage, and progression were assessed, indicating the potential of molecular characterization of tumors based on miRNAs for more accurate patient reclassification and efficient patient management (Conde et al., 2019).

So far, several studies described the metastasis-specific miRNA signature in CRC patients comparing the expression profiles of primary CRCs (pCRCs) with those in liver metastasis (LMs) (Drusco et al., 2014; Hur et al., 2015; Li et al., 2017; Neerincx et al., 2015; Vychytilova-Faltejskova et al., 2016). Using next-generation sequencing (NGS) of 220 fresh-frozen samples, 2245 known unique mature miRNAs and 515 novel candidate miRNAs were revealed. Interestingly, only five known miRNAs were differentially expressed between pCRCs and corresponding LMs, including miR-320b/d, miR-3117, miR-1246, and miR-663b (Neerincx et al., 2015). Similarly, the deregulation of 33 miRNAs in LMs was detected by miRNA Ready to Use PCR Panels (Exiqon, 752 miRNAs) and six of them were subsequently validated by RT-qPCR. Importantly, downregulated levels of miR-10b and upregulated levels of miR-885-5p were observed, which is in agreement with previous studies (Hur et al., 2015; Vychytilova-Faltejskova et al., 2016). Recently, 25 miRNAs were identified to be differentially expressed comparing advanced CRC cases without metastasis, patients with LMs, and patients with metastatic spread to the peritoneum. From those, oncogenic miR-31 was shown to be significantly increased in peritoneal metastasis compared to LMs. However, the highest levels were detected in advanced cases without metastasis (Heublein et al., 2018). These results indicate the possibility to predict the risk of metastatic spread from the molecular profile of the primary tumor.

Further, several studies compared the expression profiles of miRNAs among CRC samples with high microsatellite instability (MSI), samples characterized by microsatellite stability, and those from patients with hereditary nonpolyposis CRC (HNPCC)-associated CRC (Earle et al., 2010; Lanza et al., 2007). Similarly, Slattery et al. (2011) examined 40 colon tumors, 30 rectal tumors, and 30 adjacent tissues to identify different miRNAs by tumor location and subtype. Interestingly,

287 miRNAs were differentially expressed between the colon and rectal tissues, 135 miRNAs were associated with *KRAS* mutations, and 139 miRNAs were deregulated in *TP53*-mutated colon tumors. Concerning rectal cancer, different expressions of miR-31, miR-126, and miR-143 between colon and rectal cancer were detected using a miRCURY LNA Array and subsequent RT-qPCR validation (Li et al., 2012). Recently, miRNA expression signatures associated with *BRAF*- vs. *KRAS*-mutated CRCs were analyzed using the Affymetrix GeneChip miRNA 4.0 Array. In total, 10 differentially expressed miRNAs associated with *BRAF*-mutated CRCs were identified, including miR-31-5p, miR-877-5p, miR-362-5p, and miR-425-3p, and these miRNAs were further associated with an aggressive phenotype (Choi et al., 2016). To identify miRNA signatures connected to *KRAS* status, Mosakhani et al. (2012) used Agilent's miRNA microarray profiling of 40 CRC patients with wt-*KRAS* and 15 patients with mut-*KRAS*. The presence of the KRAS mutation was associated with the upregulation of miR-127-3p, miR-92a, and miR-486-3p while the levels of miR-378 were downregulated. To date, a large number of studies dealing with expression profiling of miRNAs in CRC have been performed. However, a lack of agreement among those studies, especially due to the use of different technologies, is the common drawback of miRNA expression profiling. A list of the 20 most studied miRNAs in CRC using expression profiling and/or functional studies is shown in Table 17.1.

Circulating and fecal miRNAs as promising biomarkers for early CRC detection

The early detection of CRC as well as premalignant stages is one of the main requirements of successful treatment and reduction of mortality from this disease and a major challenge of present day medicine. The discovery of extracellular miRNA abundance in biological fluids has opened a new era of cancer liquid biopsy research. Importantly, they can be found in the circulation alone or in combination with proteins or high-density lipoproteins. In addition, they can be actively secreted into extracellular fluids by cancer and other cells and transported by EVs, especially exosomes (Marcuello et al., 2019). In recent years, numerous reports have demonstrated the diagnostic potential of circulating miRNAs in CRC. However, only some of them have implemented sufficiently large patient cohorts to provide information of relevant statistical power. Thus, only studies with at least 100 CRC patients are discussed in the following paragraphs and summarized in Table 17.2.

The first bigger study was performed in 2009 by Ng, Tsang, et al. (2009), who assessed the expression of 95 miRNAs in plasma samples of CRC patients using the cancer miRNA array QuantiMir system (System Biosciences) and subsequent RT-qPCR validation. They found significantly elevated levels of miR-17-3p and miR-92 in patient samples compared to the control group. In addition, the results indicated the potential of miR-92 to differentiate CRC from gastric cancer, inflammatory bowel disease (IBD), and healthy subjects with high sensitivity and specificity. A year later, these results were confirmed in an independent study performed by Huang et al. (2010). Importantly, the combined detection of miR-92a together with miR-29a revealed an increased diagnostic performance and enabled distinguishing healthy controls from advanced adenomas (AAs). Similarly, combined analyses of this miRNA together with miR-223 in plasma as well as in stool samples yielded a sensitivity of almost 97% and a specificity of 75% for CRC detection (Chang et al., 2016). The elevated levels of miR-92a were subsequently confirmed also in serum samples (Liu et al., 2013; Wang, Tang, et al., 2014), indicating the real potential of this miRNA to serve as a promising biomarker for the early detection of CRC. Besides miR-92a, oncogenic miR-21 belongs to the most studied miRNAs in circulation and its increased levels were detected in both serum and plasma of CRC patients (Guo et al., 2018; Liu et al., 2013; Luo et al., 2013; Wang et al., 2014). In addition, the expression of miR-21 was significantly higher in preoperative samples of patients and dropped in postoperative serum samples after curative surgery. Importantly, the detection of circulating miR-21 also enabled differentiating patients with adenoma from healthy controls (Toiyama et al., 2013). Its increased levels were found in exosomes from CRC serum samples (Zhu, Wu, et al., 2017), indicating the active secretion of this miRNA into circulation by tumor cells.

Due to the high heterogeneity of the disease, single miRNA detection may be inadequate for the precise detection of CRC or adenomas and the diagnostic performance of miRNA-based biomarkers could be substantially magnified by the simultaneous detection of more miRNAs. Assessment of miRNA panels is enabled especially due to the rapid development of modern high-throughput techniques such as microarrays or NGS. Using miRCURY LNA Universal RT microRNA PCR systems (Exiqon), Wang et al. (2012) identified 22 miRNAs to be significantly deregulated in plasma samples of CRC patients compared to healthy controls. Subsequent RT-qPCR validation confirmed downregulated levels of miR-601 and miR-760 in samples from CRC patients and these two miRNAs together enabled the sensitive detection of CRCs as well as AAs. A similar approach was applied in two other studies. First, serum levels of miR-19a-3p, miR-21-5p, and miR-425-5p were significantly higher in CRC patients than in control samples and this three-miRNA serum panel enabled the sensitive detection of the disease. In addition, increased levels of these miRNAs were found in tumor tissues as well as in exosomes, indicating their CRC-specific character (Zhu et al., 2017). Further, a four-stage study including screening, training, testing, and external validation was performed to assess a panel of plasma miRNAs for early CRC detection. In total, seven miRNAs (miR-103a-3p, miR-127-3p, miR-151a-5p, miR-17-5p, miR-181a-5p, miR-18a-5p, and miR-18b-5p) were proved to be significantly

TABLE 17.1 The most studied tissue miRNAs in colorectal cancer using expression profiling and functional studies.

miRNA	Number of studies	Expression profiling studies	Functional studies	Function	Validated targets in CRC[a]
miR-21	80	58	22	O	PDCD4, PTEN, CCL20, TET1, TGFBR2
miR-31	39	30	9	O	SATB2, FIH-1
miR-145	38	20	18	S	NRAS, IRS1, FSCN1, RAD18, PXN, ERG, LASP1, SIP1/ZEB2, MYO6, TUSC3, RHBDD1,c-MYC, IGF1
miR-143	31	18	13	S	DNMT3A, MACC1, TLR2, MMP7, FOSL2, ITGA6, ASAP3, CTNND1
miR-17	27	13	14	O	RND3, PTEN, CYP7B1, PAR4, SIK1, VIM
miR-34a	26	11	15	S	AXL, VEGF, LDHA, HMGB1, SIRT1
miR-200c	26	14	12	S	SOX2, VLDLR, BMI1
miR-92a	23	16	7	O	PTEN, RBM4, KLF4, RECK
miR-124	21	4	17	S	STAT3, PRRX1, ROCK1, DNMT3B, DNMT1, IQGAP1, SGK1, PLCB1, TRAF6, P4HA1, FZD4, PD-L1
miR-141	20	11	9	O/S	SIP1/ZEB2, DLC1, MAP2K4, PDCD4
miR-135b	20	13	7	O	MTSS1, TGFBR2, MTUS1, ST6GALNAC2
miR-182	19	10	9	O/S	SATB2, ST6GALNAC2, FOXO3A, MTDH, DAB2IP
miR-133b	19	10	9	O/S	MET, TBPL1, CTGF, HOXA9
miR-195	19	10	9	S	FGF2, GDPD5
miR-149	19	3	16	S	FOXM1, HOXB8
miR-20a	18	13	5	O	SMAD4, TIMP2, ATG5, MICA
miR-137	17	3	14	S	FMNL2, c-MET, WNT3A, CTNNB1
miR-150	16	9	7	S	VEGFA
miR-375	16	6	11	S	PIK3CA, KLF4, FZD8, YAP1, SP1, TYMS
miR-139-5p	15	4	11	S	IGF1R, RAP1B, NOTCH1, AMFR, BCL2, YAP1

[a] *Validated targets in colorectal cancer based on miRWalk database (http://mirwalk.umm.uni-heidelberg.de/).*
O, oncogene; *S*, suppressor; *CRC*, colorectal cancer; *PDCD4*, programmed cell death protein 4; *PTEN*, phosphatase and tensin homolog; *CCL20*, chemokine C–C motif ligand 20; *TET1*, ten-eleven translocation methylcytosine dioxygenase 1; *TGFBR2*, transforming growth factor beta receptor 2; *SATB2*, special AT-rich sequence-binding protein 2; *FIH-1*, factor inhibiting hypoxia-inducible factor 1; *NRAS*, neuroblastoma RAS viral oncogene homolog; *IRS1*, insulin receptor substrate 1; *FSCN1*, fascin actin-bundling protein 1; *RAD18*, RAD18 E3 ubiquitin protein ligase; *PXN*, paxillin; *ERG*, ETS-related gene; *LASP1*, LIM and SH3 protein 1; *SIP1*, smad interacting protein 1; *ZEB2*, zinc finger E-box-binding homeobox 2; *MYO6*, myosin 6; *TUSC3*, tumor suppressor candidate 3; *RHBDD1*, rhomboid-related protein 4; *IGF1*, insulin-like growth factor 1; *DNMT3A*, DNA methyltransferase 3A; *MACC1*, metastasis-associated in colon cancer protein 1; *TLR2*, Toll-like receptor 2; *MMP7*, matrix metalloproteinase 7; *FOSL2*, FOS-like 2, AP-1 transcription factor subunit; *ITGA6*, integrin subunit alpha 6; *ASAP3*, ArfGAP with SH3 domain, ankyrin repeat and PH domain 3; *CTNND1*, catenin delta 1; *RND3*, Rho family GTPase 3; *CYP7B1*, cytochrome P450 family 7 subfamily B member 1; *PAR4*, protease-activated receptor 4; *SIK1*, salt inducible kinase 1; *VIM*, vimentin; *AXL*, AXL receptor tyrosine kinase; *VEGF*, vascular endothelial growth factor; *LDHA*, lactate dehydrogenase A; *HMGB1*, high-mobility group box 1; *SIRT1*, sirtuin; *SOX2*, SRY (sex determining region Y)-box 2; *VLDLR*, very low density lipoprotein receptor; *BMI1*, polycomb complex protein; *RBM4*, RNA-binding protein 4; *KLF4*, Kruppel-like factor 4; *RECK*, reversion inducing cysteine rich protein with kazal motifs; *STAT3*, signal transducer and activator of transcription 3; *PRRX1*, paired related homeobox 1; *ROCK1*, rho associated coiled-coil containing protein kinase 1; *DNMT3B/1*, DNA methyltransferase 3B/1; *IQGAP1*, IQ motif containing GTPase activating protein 1; *SGK1*, serum/glucocorticoid regulated kinase 1; *PLCB1*, phospholipase C beta 1; *TRAF6*, tumor necrosis factor receptor associated factor 6; *P4HA1*, prolyl 4-hydroxalase subunit alpha-1 precursor; *FZD4*, Frizzled 4; *PD-L1*, programmed death-ligand 1; *DLC1*, deleted in liver cancer 1; *MAP2K4*, mitogen-activated protein kinase kinase 4; *MTSS1*, metastasis suppressor protein 1; *ST6GALNAC2*, alpha-*N*-acetylgalactoaminide alpha-2,6-sialyltransferase 2; *FOXO3A*, forkhead box O3; *MTDH*, metadherin; *DAB2IP*, disabled homolog 2-interacting protein; *TBPL1*, TATA-box binding protein like 1; *CTGF*, connective tissue growth factor; *HOXA9*, homeobox A9; *FGF2*, fibroblast growth factor 2; *GDPD5*, glycerophosphodiester phosphodiesterase domain containing 5; *FOXM1*, forkhead box M1; *HOXB8*, homeobox B8; *SMAD4*, mothers against decapentaplegic homolog 4; *TIMP2*, tissue inhibitor of metalloproteinase 2; *ATG5*, autophagy related 5; *MICA*, MHC class I polypeptide-related sequence A; *FMNL2*, formin-like protein 2; *CTNNB1*, beta catenin; *PIK3CA*, phosphatidylinositol-4;5-bisphosphate 3-kinase catalytic subunit alpha; *FZD8*, Frizzled 8; *YAP1*, yes-associated protein 1; *TYMS*, thymidylate synthase; *IGF1R*, insulin-like growth factor receptor 1; *RAP1B*, ras-related protein Rap-1b precursor; *NOTCH1*, neurogenic locus notch homolog protein 1; *AMFR*, autocrine motility factor receptor; *BCL2*, B-cell lymphoma 2.

TABLE 17.2 Circulating and stool miRNAs with diagnostic potential in colorectal cancer (only studies with at least 100 patients were included).

miRNA/miRNA panel	Cases (N)	Controls (N)	Adenoma/CRC	Sens/ Spec (%)	AUC	Expression	Detection method	References
Blood serum								
miR-21	198	65	CRC	83/91	0.919	up	RT-qPCR	Toiyama et al. (2013)
	43	53	adenoma	77/81	0.813			
miR-21 + miR-92a	200	80	CRC	68/91	0.847	up, up	RT-qPCR	Liu et al. (2013)
	50	80	adenoma	70/70	0.722			
miR-21 + let-7g + miR-31 + miR-92a + miR-181b + miR-203	113	89	CRC	96/88	0.923	up, up, down, down, down, down	RT-qPCR	Wang et al. (2014)
miR-155	146	60	CRC	58/95	0.776	up	RT-qPCR	Lv, Fan, Chen, and Zhao (2015)
miR-145 + miR-106a + miR-17-3p	195	150	CRC	79/83	0.886	down, up, up	TLDA, RT-qPCR	Li et al. (2015)
miR-23a-3p + miR-27a-3p + miR-142-5p + miR-376c-3p	427	276	CRC	89/81	0.917	up, up, up, up	NGS, RT-qPCR	Vychytilova-Faltejskova et al. (2016)
miR-1290	223	69	CRC	70/91	0.830	up	3D-gene, RT-qPCR	Imaoka et al. (2016)
	68	69	adenoma	46/91	0.718			
miR-223	130	24	CRC	n.a.	0.838	up	RT-qPCR	Zekri et al. (2016)
miR-210	268	102	CRC	75/74	0.821	up	RT-qPCR	Wang et al. (2017)
miR-139-3p	117	90	CRC	97/98	0.994	down	RT-qPCR	Ng et al. (2017)
miR-19a-3p + miR-21-5p + miR-425-5p	196	138	CRC	n.a.	0.783	up, up, up	miRCURY arrays, RT-qPCR	Zhu et al. (2017)
miR-1246 + miR-202-3p + miR-21-3p + miR-1229-3p + miR-532-3p	217	190	CRC	92/92	0.960	down, up, up, down, down	Modified three-step RT-qPCR	Guo et al. (2018)
	168	190	adenoma	94/85	0.951			
miR-199a	107	60	CRC	78/83	0.864	down	RT-qPCR	Tan, Zheng, and Liu (2018)
miR-99b-5p[a]	169	155	CRC	32/91	0.628	down	NGS, RT-qPCR	Zhao et al. (2019)
miR-150-5p[a]	169	155	CRC	75/59	0.707	down	NGS, RT-qPCR	Zhao et al. (2019)
miR-592	149	65	CRC	83/78	0.844	up	RT-qPCR	Pan and Miao (2020)
	50	65	adenoma	69/78	0.747			
miR-92a-1	148	68	CRC	82/96	0.914	up	RT-qPCR	Shi and Liu (2020)

miR-203a-3p + miR-145-5p + miR-375-3p + miR-200c-3p	135	135	CRC	81/73	0.893	up, down, down, down	RT-qPCR	Huang et al. (2020)
miR-126 + miR-1290 + miR-23a + miR-940[a]	100	35	CRC	90/89	0.950	up, up, up, up	RT-qPCR	Shi, Zhuang, Zhang, Chen, and Wu (2021)
Blood plasma								
miR-17-3p	120	75	CRC	64/70	0.717	up	QuantiMir, RT-qPCR	Ng et al. (2009)
miR-92	120	75	CRC	89/70	0.885	up	QuantiMir, RT-qPCR	Ng et al. (2009)
miR-92a + miR-29a	100	59	CRC	83/85	0.883	up, up	RT-qPCR	Huang et al. (2010)
	37	59	adenoma	73/80	0.773			
miR-221	103	37	CRC	86/41	0.606	up	RT-qPCR	Pu et al. (2010)
miR-601 + miR-760	100	68	CRC	83/69	0.792	down, down	miRCURY arrays, RT-qPCR	Wang et al. (2012)
	43	68	adenoma	72/62	0.683			
12 miRNA panel[c]	130	194	CRC	n.a.	0.745	up (9/12)[c]	TLDA, RT-qPCR	Luo, Stock, Burwinkel, and Brenner (2013)
miR-409-3p + miR-7 + miR-93	124	117	CRC	82/89	0.897	up, down, down	Agilent arrays, RT-qPCR	Wang et al. (2015)
miR-106a	100	79	CRC	74/44	0.605	up	RT-qPCR	Chen et al. (2015)
miR-20a	100	79	CRC	46/73	0.590	up	RT-qPCR	Chen et al. (2015)
miR-24 + miR-320a + miR-423-5p	111	130	CRC	93/71	0.899	down, down, down	RT-qPCR	Fang et al. (2015)
miR-223 + miR-92a	215	183	CRC	76/69	0.780	up, up	RT-qPCR	Chang et al. (2016)
miR-96	227	57	CRC	65/73	0.740	up	TLDA, RT-qPCR	Sun et al. (2016)
miR-29b	200	400	CRC	61/73	0.743	up	RT-qPCR	Li et al. (2016)
miR-144-3p + miR-425-5p + miR-1260b	101	134	CRC	94/91	0.954	down, down, down	S-poly(T) Plus RT-PCR assay	Tan et al. (2019)
miR-103a-3p + miR-127-3p + miR-151a-5p + miR-17-5p + miR-181a-5p + miR-18a-5p + miR-18b-5p	139	132	CRC	97/53 85/35 77/87[b]	0.762 0.824 0.895	up, up, up, up, up, up, up	miRCURY arrays, RT-qPCR	Zhang et al. (2019)
Stool								
miR-17-92 + miR-135b	197	119	CRC	74/79	n.a.	up, up	RT-qPCR	Koga et al. (2010)
miR-106a	117	107	CRC	34/97	n.a.	up	RT-qPCR	Koga et al. (2013)

Continued

TABLE 17.2 Circulating and stool miRNAs with diagnostic potential in colorectal cancer (only studies with at least 100 patients were included)—cont'd

miRNA/miRNA panel	Cases (N)	Controls (N)	Adenoma/CRC	Sens/ Spec (%)	AUC	Expression	Detection method	References
miR-221 + miR-18a	198	198	CRC	66/75	0.75	up, up	TLDA, RT-qPCR	Yau et al. (2014)
miR-135b	104	109	CRC	78/68	0.79	up	TLDA, RT-qPCR	Wu et al. (2014)
	169	109	adenoma	65/68	0.71			
miR-20a	198	198	CRC	55/82	0.73	up	RT-qPCR	Yau et al. (2016)
miR-223 + miR-92a	138	309	CRC	74/82	0.810	up, up	RT-qPCR	Chang et al. (2016)

[a] *Exosomal miRNAs.*

[b] *Data available only separately for 3 different phases—training, testing, validation.*

[c] *12-miRNa-based panel including miR-18a, miR-20a, miR-21, miR-29a, miR-92a, miR-106b, miR-133a, miR-143, miR-145, miR-181b, miR-342-3p, miR-532-3p-miR-181b, miR-342-3p, miR-532-3p (miR-181b, miR-342-3p, miR-532-3p were not significantly changed between CRC and healthy controls), TLDA—TaqMan Low Density Array, Human Pools A and B v2.1 (Applied Biosystems), NGS—next generation sequencing (Illumina), 3D-Gene Human miRNA Oligo Chips (Toray, Kamakura, Japan), miRCURY LNA Universal RT mikroRNA PCR system (Exiqon), Cancer MicroRNA Array QuantiMir system (System Biosciences), SurePrint Human miRNA Microarrays (Agilent Technologies).*

CRC, colorectal cancer; *sens*, sensitivity; *spec*, specificity; *AUC*, area under the curve; *n.a.*, not available.

increased in CRC samples. The area under the curve (AUC) of this seven-miRNA signature was 0.895 in the validation phase of the study. Additionally, miR-103a-3p, miR-127-3p, miR-17-5p, and miR-18a-5p were also upregulated in tumor tissues while miR-17-5p, miR-181a-5p, miR-18a-5p, and miR-18b-5p were associated with plasma exosomes (Zhang et al., 2019). Using TaqMan Low Density Arrays (Applied Biosystems) and a literature search, a 12-miRNA plasma-based panel was published for the early detection of CRC. Nevertheless, none of the miRNAs showed different expression between AAs and healthy controls (Luo et al., 2013). Later, another plasma miRNA panel consisting of miR-409-3p, miR-7, and miR-93 and enabling CRC detection was assessed using Human microRNA Microarrays from Agilent Technologies. Importantly, a high diagnostic performance of this panel was observed also in discrimination between healthy controls and nonmetastatic CRC stages (Wang, Cao, et al., 2015). In our previous study, we performed high-throughput expression profiling using Illumina small RNA sequencing. We established a four-miRNA signature consisting of miR-23a-3p, miR-27a-3p, miR-142-5p, and miR-376c-3p that enabled us to differentiate between CRC and healthy controls with a sensitivity of 89% and a specificity of 81%. Importantly, the panel exhibited high diagnostic performance when only patients in the early stages of the disease (T1-4N0M0) were included (AUC = 0.877) (Vychytilova-Faltejskova et al., 2016).

Besides the free circulating miRNAs, exosome-associated miRNAs as well as stool miRNAs have emerged as potential new noninvasive biomarkers for CRC with more disease specificity. Using NGS, Zhao et al. (2019) identified 29 serum exosomal miRNAs with > 1.5-fold difference between CRC patients and healthy donors. From those, miR-99b-5p and miR-150-5p were downregulated in patient samples and their expression increased significantly after the surgery. Most recently, a panel of four exosomal miRNAs including miR-126, miR-1290, miR-23a, and miR-940 was assessed, enabling highly sensitive discrimination between healthy controls and CRC patients at TNM stage I (Shi et al., 2021). Interestingly, EV-associated miRNAs were detected not only in serum or plasma, but also in peritoneal lavage. Using TaqMan OpenArray technology, 210 different miRNAs were found to be deregulated between CRC patients and the control group. The top 10 miRNAs with AUC higher than 0.95 included miR-199b-5p, miR-150-5p, miR-29c-5p, miR-218-5p, miR-99a-3p, miR-383-5p, miR-199a-3p, miR-193a-5p, miR-10b-5p, and miR-181c-5p. Thus, peritoneal lavage could serve as an untapped source of novel biomarkers for CRC (Roman-Canal et al., 2019). Concerning stool miRNAs, the expression of the oncogenic miR-17-92 cluster, miR-21, and miR-135 was analyzed in exfoliated colonocytes isolated from feces using RT-qPCR. Interestingly, miR-21 levels did not differ significantly between CRC patients and healthy controls. Nevertheless, increased levels of other miRNAs were observed in CRC samples and their combined expression enabled the detection of CRC patients with high sensitivity and specificity dependent on tumor location (Koga et al., 2010). Elevated levels of stool miR-20a and miR-18a from the miR-17-92 cluster as well as the increased expression of miR-135b in feces were confirmed in later independent studies (Wu et al., 2014; Yau et al., 2014, 2016). Importantly, miR-135b showed a significant increasing trend across the adenoma to cancer sequence and levels of this miRNA in stool decreased significantly upon the removal of CRC or adenoma (Wu et al., 2014). Finally, saliva presents many advantages as a source of noninvasive biomarkers over serum, plasma, or stool. Aberrant expression profiles of salivary miRNAs have been detected in different types of cancers. However, so far only two studies analyzed the expression profiles of salivary miRNAs in CRC. Surprisingly, the estimated diagnostic sensitivity and specificity of miR-21 was 97% and 91%, respectively (Sazanov, Kiselyova, Zakharenko, Romanov, & Zaraysky, 2017). Further, 22 salivary miRNAs were found to be deregulated between CRC patients and healthy controls. From those, five miRNAs were confirmed to be overexpressed in CRC and this five-miRNA signature together with CEA enabled more sensitive detection of the disease than CEA alone (Rapado-González et al., 2019).

Although an increasing number of circulating miRNAs have been identified as potential biomarkers for the early diagnosis of CRC, the lack of consistency between individual studies highlights the major obstacle for the development of robust miRNA biomarkers. Obviously, different experimental designs, procedures, starting materials, inconsistent normalization approaches, ethnic and racial differences as well as low numbers of included samples could contribute to the contradictory results published so far. Thus, in order to achieve true translational relevance and to bring circulating miRNAs into routine diagnostics, multicentric clinical trials with a coordinated and synchronized set of experimental procedures performed in agreement with the conclusions of the Worldwide Strategic Consensus Conference on Biomarker Research are necessary (Hinestrosa et al., 2007).

Prognostic and predictive potential of miRNAs in CRC

An increasing number of studies have shown that miRNAs could serve not only as diagnostic tools, but also as prognostic and predictive factors in the clinical routine. Deregulated expression of these molecules has been associated with different clinicopathological characteristics as well as patient survival, risk of disease recurrence, or response to different therapies. Many recent studies attempted to identify prognostic and predictive miRNAs in tumor tissue or plasma/serum samples as reviewed previously (dos Santos et al., 2021; Gao, Zhao, Wu, Zhang, & Zhang, 2018). However, only studies with at least 100 CRC patients are discussed in the following paragraphs and summarized in Table 17.3 (tissue) and Table 17.4 (serum/plasma).

TABLE 17.3 Tissue miRNAs with prognostic and predictive potential in colorectal cancer (only studies with at least 100 patients were included).

miRNA/miRNA panel	Cases (N)	Expression in TT/NT	Type of biomarker	Significances/outcome	References
miR-106a	110	down	prognostic	Correlation with DFS, OS	Díaz et al. (2008)
miR-21	197	up	prognostic	Correlation with stage, OS	Schetter et al. (2008)
miR-21, miR-155	156	up, up	prognostic	Correlation with venous invasion, LM, stage, DFS, OS	Shibuya, Iinuma, Shimada, Horiuchi, and Watanabe (2011)
miR-150	247	down	prognostic predictive	Correlation with OS response to adjuvant chemotherapy	Yanlei et al. (2012)
miR-181a	162	up	prognostic	Correlation with OS	Nishimura et al. (2012)
miR-148a	273	down	prognostic predictive	Correlation with DFS, OS worse therapeutic response to oxaliplatin and 5-FU	Takahashi et al. (2012)
miR-29a	110	down	prognostic	Correlation with DFS, risk of recurrence in stage II	Weissmann-Brenner et al. (2012)
miR-181b	142	up	prognostic	Correlation with OS in Black patients	Bovell et al. (2013)
miR-182	148	up	prognostic	Correlation with size, LNM, stage, OS	Liu et al. (2013)
miR-16	143	down	prognostic	Correlation with grade, LNM, stage, tumor recurrence, OS	Qian, Jiang, Li, Chen, and Fang (2013)
miR-218	189	down	prognostic	Correlation with stage, LNM, grade, OS	Yu et al. (2013)
miR-15 + miR-16	126	down, down	prognostic	Correlation with stage, grade, LNM, DFS, OS	Xiao et al. (2014)
miR-200a/c, miR-429, miR-141	127	up, up, up	prognostic predictive	Correlation with DFS, OS worse response to fluoropyrimidine treatment	Diaz et al. (2014)
miR-182	138	up	prognostic	Correlation with stage, LNM, distant metastasis, grade, OS	Wang, Yang, Wang, Lin, and Ding (2014)
miR-133a	125	down	prognostic	Correlation with distant metastasis, stage, grade, OS	Wan et al. (2014)
miR-133a	169	down	prognostic	Correlation with depth of invasion, grade, LNM, stage, OS	Wang et al. (2014)
miR-100	138	down	prognostic	Correlation with size, LNM, stage, OS	Chen, Xi, Wang, and Wei (2014)
miR-193a-5p	111	down	prognostic	Correlation with LNM, DFS, OS	Zhang et al. (2014)
miR-196a + miR-96b	126	up, up	prognostic	Correlation with LNM, grade, stage, DFS, OS	Ge, Chen, Li, Lu, and Xiao (2014)
miR-21	306	up	prognostic	Correlation with depth of invasion, LVI, LM, stage, DFS, OS	Fukushima, Iinuma, Tsukamoto, Matsuda, and Hashiguchi (2015)
miR-224	115	up	prognostic	Correlation with OS, short-term relapse	Adamopoulos, Kontos, Rapti, Papadopoulos, & Scorilas (2015)

TABLE 17.3 Tissue miRNAs with prognostic and predictive potential in colorectal cancer (only studies with at least 100 patients were included)—cont'd

miRNA/miRNA panel	Cases (N)	Expression in TT/NT	Type of biomarker	Significances/outcome	References
miR-320e	337	up	prognostic	Correlation with DFS, OS, tumor recurrence (stage II + III)	Perez-Carbonell et al. (2015)
miR-34a-5p	268	down	prognostic	Correlation with DFS, tumor recurrence (stage II + III)	Gao et al. (2015)
miR-29b	245	down	prognostic	Correlation with LNM, T classification, DFS	Inoue et al. (2015)
miR-7	105	down	prognostic predictive	Correlation with OS influence sensitivity to cetuximab treatment	Suto et al. (2014)
miR-93 + miR-195 + let-7b + miR-141 + miR-494 + miR-7	104	down, down, down, up, up, up	prognostic	Enable to distinguish an early relapse from nonearly relapse in postoperative patients (sensitivity 77%, specificity 71%, AUC = 0.834)	Yang et al. (2016)
let-7a-5p	192	down	prognostic	Correlation with size, stage, LNM, DFS, OS	Liu et al. (2016)
miR-422a	102	down	prognostic	Correlation with local invasion, LNM, stage, OS	Zheng et al. (2016)
miR-1260b	120	up	prognostic	Correlation with LNM, venous invasion, OS	Liu, Guan, Gao, Jiang, and Kang (2016)
miR-429	116	up	predictive	Associated with response to 5-FU-based treatment	Dong, Cai, and Li (2016)
miR-10b	246	up	prognostic	Correlation with stage, LNM, LM, venous infiltration, grade, OS	Jiang et al. (2016)
miR-224	306	up	prognostic	Higher in MSS, left colon and rectum, correlation with stage, OS	Ling et al. (2016)
miR-7	196	up	prognostic	Correlation with tumor depth, LVI, LNM, LM, OS	Nagano et al. (2016)
miR-106a	138	up	prognostic	Correlation with size, preoperative CEA levels, stage, OS	Hao et al. (2017)
miR-132	187	down	prognostic	Correlation with metastasis, OS	Mokutani et al. (2016)
miR-215-5p	448	down	prognostic	Correlation with LNM, stage, OS	Vychytilova-Faltejskova et al. (2017)
miR-16	182	up	prognostic	Correlation with DFS, OS	Ma et al. (2016)
miR-24-3p	182	not changed	prognostic	High levels associated with shorter DFS, OS, relapse	Kerimis, Kontos, Christodoulou, Papadopoulos, and Scorilas (2017)
miR-31	157	up	predictive	Higher levels associated with better DFS and OS in patients treated with oxaliplatin-based chemotherapy	Li et al. (2018)

Continued

TABLE 17.3 Tissue miRNAs with prognostic and predictive potential in colorectal cancer (only studies with at least 100 patients were included)—cont'd

miRNA/miRNA panel	Cases (N)	Expression in TT/NT	Type of biomarker	Significances/outcome	References
miR-1256	178	down	prognostic	Correlation with TNM stage, LNM, DFS, OS	Liu et al. (2018)
miR-888	126	up	prognostic	Correlation with stage, grade, DFS, OS	Gao et al. (2018)
miR-200c	109	up	prognostic	Correlation with grade, stage, LVI, KRAS mutation, RFS, OS	Roh et al. (2018)
miR-552	183	up	prognostic	Correlation with grade, LNM, stage, OS	Wang and Liu (2018)
miR-9	357	down	prognostic	Correlation with stage and OS	Kim et al. (2020)

TT, tumor tissue; *NT*, normal adjacent tissue; *DFS*, diseased free survival; *OS*, overall survival; *LM*, liver metastasis; *5-FU*, 5-fluorouracil; *LNM*, lymph node metastasis; *LVI*, lymphovascular invasion; *LM*, liver metastasis; *CRC*, colorectal cancer; *AUC*, area under the curve; *MSS*, microsatellite stable CRC; *CEA*, carcinoembryonic antigen; *RFS*, recurrence-free survival.

TABLE 17.4 Circulating miRNAs with prognostic and predictive potential in colorectal cancer (only studies with at least 100 patients were included).

miRNA/miRNA panel	Cases (N)	Expression in CRC/HC	Type of biomarker	Significances/outcome	References
Blood serum					
miR-21	198	up	prognostic	Correlation with size, distant metastasis, OS	Toiyama et al. (2013)
miR-92a	200	up	prognostic	Correlation with OS	Liu et al. (2013)
miR-21	102	up	prognostic	Correlation with OS, tumor recurrence	Menéndez et al. (2013)
miR-20a+miR-130+miR-145+miR-216+miR-372	253	n.a.	predictive	Prediction of response to chemotherapy	Zhang et al. (2014)
miR-200c	446	up	prognostic	Correlation with LNM, distant metastasis, OS, tumor recurrence	Toiyama et al. (2014)
miR-17-92[a]	299	up	prognostic	Association with tumor recurrence	Matsumura et al. (2015)
miR-17-3p, miR-106a	175	up, up	prognostic predictive	Correlation with DFS, OS, tumor recurrence association with response to chemotherapy	Li et al. (2015)
miR-155	146	up	prognostic	Correlation with PFS, OS	Lv, Fan, Chen, and Zhao (2014)
miR-1290	211	up	prognostic	Correlation with stage, size, LNM, LVI, distant metastasis, DFS, OS, tumor recurrence	Imaoka et al. (2016)
miR-23a-3p+miR-376c-3p	261	up, up	prognostic	Correlation with OS	Vychytilova-Faltejskova, Pesta, et al. (2016)
miR-210	268	up	prognostic	Correlation with DFS, OS	Wang et al. (2017)

TABLE 17.4 Circulating miRNAs with prognostic and predictive potential in colorectal cancer (only studies with at least 100 patients were included)—cont'd

miRNA/miRNA panel	Cases (N)	Expression in CRC/HC	Type of biomarker	Significances/outcome	References
miR-139-5p	250	up	prognostic	Correlation with distant metastasis, OS, disease recurrence	Miyoshi et al. (2017)
miR-203[a]	240	up	prognostic	Correlation with distant metastasis, OS	Takano et al. (2017)
miR-203	330	up	prognostic	Correlation with stage, LNM, distant metastasis, OS	Hur et al. (2017)
miR-638[a]	192	down	prognostic	Correlation with stage, LM, DFS, OS	Yan et al. (2017)
miR-6803-5p[a]	168	up	prognostic	Correlation with stage, LNM, LM, DFS, OS	Yan et al. (2018)
miR-652-3p+miR-342-3p+miR-501-3p+miR-328-3p	322	not changed	prognostic predictive	Correlation with DFS, OS, tumor recurrence associated with poor chemotherapy response	Ji et al. (2018)
miR-122-5p+miR-146b-5p+miR-186-5p+miR-193a-5p	436	up, up, up, down	prognostic	Preoperative prediction of LNM	Qu et al. (2018)
miR-199a	107	down	prognostic	Correlation with stage, LNM, distant metastasis, DFS, OS	Tan et al. (2018)
miR-101	263	down	prognostic	Correlation with stage, OS	He et al. (2018)
miR-548c-5p	108	down	prognostic	Correlation with stage, LM, OS	Peng, Gu, and Yan (2019)
miR-497	110	down	prognostic	Correlation with aggressive disease, OS	Zou, Wang, and Wang (2019)
miR-150-5p	133	down	prognostic	Correlation with grade, stage, LNM, OS	Zou et al. (2019)
miR-30a-5p	138	down	prognostic	Correlation with aggressive disease, OS	Sun et al. (2019)
miR-142-3p	363	down	prognostic	Correlation with size, stage, OS	Gao, Pang, and Yu (2019)
miR-98	115	down	prognostic	Correlation with stage, LNM, grade, DFS, OS	Wang et al. (2019)
miR-210	280	up	prognostic	Correlation with LNM, distant metastasis, grade, DFS, OS, disease recurrence after RA	Zhang, Zhou, Zhang, and Li (2020)
miR-874[a]	125	down	prognostic	Correlation with stage, grade, LNM, distant metastasis, OS	Zhang et al. (2020)
miR-618	104	up	prognostic	Upregulated levels associated with longer OS	Radanova et al. (2021)
Blood plasma					
miR-221	103	up	prognostic	Correlation with OS	Pu et al. (2010)
miR-141	257	up	prognostic	Correlation with OS	Cheng et al. (2011)

Continued

TABLE 17.4 Circulating miRNAs with prognostic and predictive potential in colorectal cancer (only studies with at least 100 patients were included)—cont'd

miRNA/miRNA panel	Cases (N)	Expression in CRC/HC	Type of biomarker	Significances/outcome	References
miR-106a, miR-484, miR-130b	174	up, up, up	prognostic predictive	Correlation with PFS, OS associated with decreased response to oxaliplatin therapy	Kjersem et al. (2014)
miR-183	118	up	prognostic	Correlation with LNM, distant metastasis, stage, DFS, OS, tumor recurrence	Yuan, Li, Zhu, Yan, and Dang (2015)
miR-122, miR-200 family	543	up, down	prognostic	Correlation with LM, OS, disease recurrence	Maierthaler et al. (2017)
miR-31, miR-29a	144	up, up	prognostic	Associated with disease recurrence	Yuan et al. (2017)
miR-21/miR-24	186	ratio	prognostic	Correlation with size, stage, LNM, grade, distant metastasis, OS	(Hao and Ma (2018)
miR-100+miR-92a+miR-16+miR-30e+miR-144-5p+let-7i[a]	210	up, up, down, down, down, down	predictive	Association with response to FOLFOX/XELOX chemotherapy	Han et al. (2020)

[a] *Exosomal miRNAs.*
CRC, colorectal cancer; *HC*, healthy control; *OS*, overall survival; *LNM*, lymph node metastasis; *DFS*, disease-free survival; *PFS*, progression-free survival; *LVI*, lymphovascular invasion; *LM*, liver metastasis; *RFS*, recurrence-free survival; *RA*, radiofrequency ablation.

Besides its diagnostic potential, miR-21 is the most studied miRNA in association with prognosis. Its increased tissue and blood levels correlate with tumor size, TNM stage, venous invasion, presence of distant metastasis, disease recurrence, and shorter survival of CRC patients (Fukushima et al., 2015; Menéndez et al., 2013; Schetter et al., 2008; Shibuya et al., 2011; Toiyama et al., 2013). Similarly, the overexpression of miRNAs from the miR-200 family was found in patients with advanced disease and worse prognosis including shorter disease-free survival (DFS) and OS. In addition, increased levels of miR-200a/c, miR-141, and miR-429 were associated with decreased response to 5-FU-based therapy (Diaz et al., 2014; Dong et al., 2016). Further, prognostic significance of tissue miR-106a (Díaz et al., 2008; Hao et al., 2017), miR-182 (Liu, Zhou, et al., 2013; Wang, Huang, et al., 2014), miR-16 (Diamantopoulos, Kontos, Kerimis, Papadopoulos, & Scorilas, 2017; Qian et al., 2013; Xiao et al., 2014), miR-133a (Wan et al., 2014; Wang, Du, et al., 2014), miR-224 (Adamopoulos et al., 2015; Ling et al., 2016), and miR-7 (Nagano et al., 2016; Suto et al., 2014; Yang, Wu, Sun, et al., 2016) was analyzed in more studies including at least 100 CRC patients (Table 17.3). Contrary to diagnostic miRNAs, most researchers analyze a single miRNA instead of miRNA panels, especially in the case of tissue biomarkers. Nevertheless, Yang et al. (2016) assessed a panel of six miRNAs including miR-7, miR-93, miR-195, miR-141, miR-494, and let-7b that enabled distinguishing early relapsed postoperative CRC patients from nonearly relapsed with a sensitivity of 77% and a specificity of 71%. Importantly, by combining this 6-miRNA-based panel with six clinicopathological factors, the sensitivity and specificity increased to 89%, indicating the high predictive potential of the panel. Today, drug resistance is a major obstacle to effective cancer therapy. Thus, selecting patients profiting from drug treatment would help to promote therapeutic efficacy and improve the survival of patients. In this context, low levels of tissue miR-150 and miR-148a were associated with decreased response to adjuvant chemotherapy (Takahashi et al., 2012; Yanlei et al., 2012) while miR-7 was found to regulate cetuximab sensitivity through EGFR regulation (Suto et al., 2014).

Concerning the circulating miRNAs, miR-21, miR-203, miR-210, miR-200 family, and miRNAs from miR-17-92/106a-363 clusters belong to the most studied prognostic and predictive biomarkers in serum or plasma in large-scale studies and their upregulated levels are associated with advanced disease and shorter survival of CRC patients (Table 17.4). Compared to tissue miRNAs, several research groups introduced their miRNA panels with prognostic or predictive potential. First, Kjersem et al. (2014) assessed a panel of three plasma miRNAs comprising miR-106a, miR-484, and miR-130b and its increased levels were associated with shorter progression-free survival (PFS), OS, and decreased response to oxaliplatin treatment. Similarly, two different studies analyzing the serum miRNAs proved the higher prognostic and predictive potential of miRNA panels compared to individual miRNAs (Ji et al., 2018; Zhang et al., 2014). One of the largest studies involving 436 serum samples from CRC patients proved the high predictive potential of 4-miRNA-based panel (miR-122-5p, miR-146b-5p, miR-186-5p, miR-193a-5p) in connection to the presence of preoperative lymph node metastasis (LNM) (Qu et al., 2018).

Finally, circulating exosomal miRNAs may be associated with the severity of CRC and therapy response, thus serving as noninvasive prognostic and predictive molecules. So far, decreased levels of exosomal miR-638 (Yan et al., 2017) and miR-874 (Zhang, Zhang, Qin, et al., 2020) and increased levels of miR-203 (Takano et al., 2017), miR-6803-5p (Yan, Liu, et al., 2018), and the miR-17-92 cluster (Matsumura et al., 2015) were observed in CRC patients with worse prognosis and shorter survival. In addition, a panel of six exosomal plasma miRNAs containing miR-100, miR-92a, miR-16, miR-30e, miR-144-5p, and let-7i was identified to significantly distinguish between chemo-resistant and chemo-sensitive CRC patients. In conclusion, tissue miRNAs as well as circulating miRNAs are promising molecules of high prognostic and predictive value in CRC.

Perspectives and future challenges

Twenty years ago, miRNAs were first described, and this discovery has substantially changed our view of gene expression regulation. In recent years, an increasing number of studies have identified the deregulated expression of these molecules in CRC and confirmed their involvement in key signaling pathways associated with disease pathogenesis. In addition, expression profiling studies proved the high diagnostic, prognostic, and predictive potential of miRNAs and functional studies indicated their therapeutic capacity. Importantly, circulating, stool, and exosomal miRNAs seem to be promising noninvasive biomarkers for early disease detection, improved cancer classification, identification of patients with poor prognosis, and therapy response prediction. Nevertheless, there are still many issues to be addressed before establishing miRNAs as clinical diagnostic tools. First, the development and application of standard operating procedures and regulating processes related to miRNA analysis are crucial to increase the overlaps among individual studies. Further, larger study cohorts as well as prospective clinical trials are needed to verify current data and prove the real clinical utility of miRNAs. Finally, the assessment of miRNA panels in combination with other types of biomarkers may be necessary to enhance the sensitivity and specificity and finally develop an accurate CRC screening test.

Funding

This work was supported by the Ministry of Health of the Czech Republic, grant NU20-03-00127.

References

Adamopoulos, P. G., Kontos, C. K., Rapti, S. M., Papadopoulos, I. N., & Scorilas, A. (2015). MiR-224 overexpression is a strong and independent prognosticator of short-term relapse and poor overall survival in colorectal adenocarcinoma. *International Journal of Oncology*, *46*(2), 849–859. https://doi.org/10.3892/ijo.2014.2775.

Akbari, A., Ghahremani, M. H., Mobini, G. R., Abastabar, M., Akhtari, J., Bolhassani, M., & Heidari, M. (2015). Down-regulation of miR-135b in colon adenocarcinoma induced by a TGF-β receptor i kinase inhibitor (SD-208). *Iranian Journal of Basic Medical Sciences*, *18*(9), 861–866. http://ijbms.mums.ac.ir/pdf_5206_96ca7ee47e13b1356f33e4e257214c56.html.

Alam, K. J., Mo, J. S., Han, S. H., Park, W. C., Kim, H. S., Yun, K. J., & Chae, S. C. (2017). MicroRNA 375 regulates proliferation and migration of colon cancer cells by suppressing the CTGF-EGFR signaling pathway. *International Journal of Cancer*, *141*(8), 1614–1629. https://doi.org/10.1002/ijc.30861.

Amodeo, V., Bazan, V., Fanale, D., Insalaco, L., Caruso, S., Cicero, G., … Russo, A. (2013). Effects of anti-miR-182 on TSP-1 expression in human colon cancer cells: There is a sense in antisense? *Expert Opinion on Therapeutic Targets*, *17*(11), 1249–1261. https://doi.org/10.1517/14728222.2013.832206.

Arabsorkhi, Z., Gharib, E., Yaghmoorian Khojini, J., Farhadieh, M. E., Nazemalhosseini-Mojarad, E., & Zali, M. R. (2020). miR-298 plays a pivotal role in colon cancer invasiveness by targeting PTEN. *Journal of Cellular Physiology*, *235*(5), 4335–4350. https://doi.org/10.1002/jcp.29310.

Arndt, G. M., Dossey, L., Cullen, L. M., Lai, A., Druker, R., Eisbacher, M., … Raponi, M. (2009). Characterization of global microRNA expression reveals oncogenic potential of miR-145 in metastatic colorectal cancer. *BMC Cancer*, *9*, 374. https://doi.org/10.1186/1471-2407-9-374.

Balacescu, O., Sur, D., Cainap, C., Visan, S., Cruceriu, D., Manzat-Saplacan, R., … Irimie, A. (2018). The impact of miRNA in colorectal cancer progression and its liver metastases. *International Journal of Molecular Sciences*, *19*(12), 3711. https://doi.org/10.3390/ijms19123711.

Balaguer, F., Link, A., Lozano, J. J., Cuatrecasas, M., Nagasaka, T., Boland, C. R., & Goel, A. (2010). Epigenetic silencing of miR-137 is an early event in colorectal carcinogenesis. *Cancer Research*, *70*(16), 6609–6618. https://doi.org/10.1158/0008-5472.CAN-10-0622.

Bandrés, E., Cubedo, E., Agirre, X., Malumbres, R., Zárate, R., Ramirez, N., … García-Foncillas, J. (2006). Identification by real-time PCR of 13 mature microRNAs differentially expressed in colorectal cancer and non-tumoral tissues. *Molecular Cancer*, *5*, 29. https://doi.org/10.1186/1476-4598-5-29.

Bandres, E., Agirre, X., Bitarte, N., Ramirez, N., Zarate, R., Roman-Gomez, J., … Garcia-Foncillas, J. (2009). Epigenetic regulation of microRNA expression in colorectal cancer. *International Journal of Cancer*, *125*(11), 2737–2743. https://doi.org/10.1002/ijc.24638.

Barisciano, G., Colangelo, T., Rosato, V., Muccillo, L., Taddei, M. L., Ippolito, L., … Sabatino, L. (2020). miR-27a is a master regulator of metabolic reprogramming and chemoresistance in colorectal cancer. *British Journal of Cancer*, *122*(9), 1354–1366. https://doi.org/10.1038/s41416-020-0773-2.

Benvenuti, S., Sartore-Bianchi, A., Di Nicolantonio, F., Zanon, C., Moroni, M., Veronese, S., … Bardelli, A. (2007). Oncogenic activation of the RAS/RAF signaling pathway impairs the response of metastatic colorectal cancers to anti-epidermal growth factor receptor antibody therapies. *Cancer Research*, *67*(6), 2643–2648. https://doi.org/10.1158/0008-5472.CAN-06-4158.

Bian, Y., Gao, G., Zhang, Q., Qian, H., Yu, L., Yao, N., … Qian, X. (2019). KCNQ1OT1/miR-217/ZEB1 feedback loop facilitates cell migration and epithelial-mesenchymal transition in colorectal cancer. *Cancer Biology and Therapy*, *20*(6), 886–896. https://doi.org/10.1080/15384047.2019.1579959.

Biller, L. H., & Schrag, D. (2021). Diagnosis and treatment of metastatic colorectal cancer: A review. *JAMA: The Journal of the American Medical Association*, *325*(7), 669–685. https://doi.org/10.1001/jama.2021.0106.

Bovell, L. C., Shanmugam, C., Putcha, B. D. K., Katkoori, V. R., Zhang, B., Bae, S., … Manne, U. (2013). The prognostic value of MicroRNAs varies with patient race/ethnicity and stage of colorectal cancer. *Clinical Cancer Research*, *19*(14), 3955–3965. https://doi.org/10.1158/1078-0432.CCR-12-3302.

Bray, F., Ferlay, J., Soerjomataram, I., Siegel, R. L., Torre, L. A., & Jemal, A. (2018). Global cancer statistics 2018: GLOBOCAN estimates of incidence and mortality worldwide for 36 cancers in 185 countries. *CA: a Cancer Journal for Clinicians*, *68*(6), 394–424. https://doi.org/10.3322/caac.21492.

Brehm, A., Miska, E., Reid, J., Bannister, A., & Kouzarides, T. (1999). The cell cycle regulating transcription factors E2F-RB. *British Journal of Cancer*, *80*(1), 38–41. https://doi.org/10.1038/sj.bjc.6690318.

Bu, P., Wang, L., Chen, K. Y., Rakhilin, N., Sun, J., Closa, A., … Shen, X. (2015). miR-1269 promotes metastasis and forms a positive feedback loop with TGF-β. *Nature Communications*, *6*, 6879. https://doi.org/10.1038/ncomms7879.

Cai, H. K., Chen, X., Tang, Y. H., & Deng, Y. C. (2017). MicroRNA-194 modulates epithelial–mesenchymal transition in human colorectal cancer metastasis. *Oncotargets and Therapy*, *10*, 1269–1278. https://doi.org/10.2147/OTT.S125172.

Cai, Y., Yan, P., Zhang, G., Yang, W., Wang, H., & Cheng, X. (2018). Long non-coding RNA TP73-AS1 sponges miR-194 to promote colorectal cancer cell proliferation, migration and invasion via up-regulating TGF α. *Cancer Biomarkers*, *23*(1), 145–156. https://doi.org/10.3233/CBM-181503.

Cavallo, F., De Giovanni, C., Nanni, P., Forni, G., & Lollini, P. L. (2011). 2011: The immune hallmarks of cancer. *Cancer Immunology, Immunotherapy*, *60*(3), 319–326. https://doi.org/10.1007/s00262-010-0968-0.

Chang, S. W., Yue, J., Wang, B. C., & Zhang, X. L. (2015). MiR-503 inhibits cell proliferation and induces apoptosis in colorectal cancer cells by targeting E2F3. *International Journal of Clinical and Experimental Pathology*, *8*(10), 12853–12860. http://www.ijcep.com/files/ijcep0013974.pdf.

Chang, P. Y., Chen, C. C., Chang, Y. S., Tsai, W. S., You, J. F., Lin, G. P., … Chan, E. C. (2016). MicroRNA-223 and microRNA-92a in stool and plasma samples act as complementary biomarkers to increase colorectal cancer detection. *Oncotarget*, *7*(9), 10663–10675. https://doi.org/10.18632/oncotarget.7119.

Chen, P., Xi, Q., Wang, Q., & Wei, P. (2014). Downregulation of microRNA-100 correlates with tumor progression and poor prognosis in colorectal cancer. *Medical Oncology*, *31*(10), 235. https://doi.org/10.1007/s12032-014-0235-x.

Chen, H., & Xu, Z. (2015). Hypermethylation-associated silencing of miR-125a and miR-125b: A potential marker in colorectal cancer. *Disease Markers*, *2015*. https://doi.org/10.1155/2015/345080.

Chen, X., Guo, X., Zhang, H., Xiang, Y., Chen, J., Yin, Y., … Zhang, C. Y. (2009). Role of miR-143 targeting KRAS in colorectal tumorigenesis. *Oncogene*, *28*(10), 1385–1392. https://doi.org/10.1038/onc.2008.474.

Chen, W. S., Leung, C. M., Pan, H. W., Hu, L. Y., Li, S. C., Ho, M. R., & Tsai, K. W. (2012). Silencing of miR-1-1 and miR-133a-2 cluster expression by DNA hypermethylation in colorectal cancer. *Oncology Reports*, *28*(3), 1069–1076. https://doi.org/10.3892/or.2012.1899.

Chen, P., Wang, B. L., Pan, B. S., & Guo, W. (2014). MiR-1297 regulates the growth, migration and invasion of colorectal cancer cells by targeting cyclooxygenase-2. *Asian Pacific Journal of Cancer Prevention*, *15*(21), 9185–9190. https://doi.org/10.7314/APJCP.2014.15.21.9185.

Chen, P., Xi, Q., Wang, Q., & Wei, P. (2014). Downregulation of microRNA-100 correlates with tumor progression and poor prognosis in colorectal cancer. *Medical Oncology*, *31*(10), 1–6. https://doi.org/10.1007/s12032-014-0235-x.

Chen, W. Y., Zhao, X. J., Yu, Z. F., Hu, F. L., Liu, Y. P., Cui, B. B., … Zhao, Y. S. (2015). The potential of plasma miRNAs for diagnosis and risk estimation of colorectal cancer. *International Journal of Clinical and Experimental Pathology*, *8*(6), 7092–7101. http://www.ijcep.com/files/ijcep0008327.pdf.

Chen, Z., Liu, S., Tian, L., Wu, M., Ai, F., Tang, W., … Tang, A. (2015). miR-124 and miR-506 inhibit colorectal cancer progression by targeting DNMT3B and DNMT1. *Oncotarget*, *6*(35), 38139–38150. https://doi.org/10.18632/oncotarget.5709.

Chen, Y., Jiang, J., Zhao, M., Luo, X., Liang, Z., Zhen, Y., … Fang, W. (2016). MicroRNA-374a suppresses colon cancer progression by directly reducing CCND1 to inactivate the PI3K/AKT pathway. *Oncotarget*, *7*(27), 41306–41319. https://doi.org/10.18632/oncotarget.9320.

Chen, D. L., Lu, Y. X., Zhang, J. X., Wei, X. L., Wang, F., Zeng, Z. L., … Xu, R. H. (2017). Long non-coding RNA UICLM promotes colorectal cancer liver metastasis by acting as a ceRNA for microRNA-215 to regulate ZEB2 expression. *Theranostics*, *7*(19), 4836–4849. https://doi.org/10.7150/thno.20942.

Chen, T., Cai, S. L., Li, J., Qi, Z. P., Li, X. Q., Ye, L. C., … Zhong, Y. S. (2017). Mecp2-mediated epigenetic silencing of miR-137 contributes to colorectal adenoma-carcinoma sequence and tumor progression via relieving the suppression of c-Met. *Scientific Reports*, *7*, 44543. https://doi.org/10.1038/srep44543.

Chen, J., Zhang, H., Chen, Y., Qiao, G., Jiang, W., Ni, P., … Ma, L. (2017). miR-598 inhibits metastasis in colorectal cancer by suppressing JAG1/Notch2 pathway stimulating EMT. *Experimental Cell Research*, *352*(1), 104–112. https://doi.org/10.1016/j.yexcr.2017.01.022.

Chen, L., Wang, X., Zhu, Y., Zhu, J., & Lai, Q. (2018). miR-200b-3p inhibits proliferation and induces apoptosis in colorectal cancer by targeting Wnt1. *Molecular Medicine Reports*, *18*(3), 2571–2580. https://doi.org/10.3892/mmr.2018.9287.

Chen, M., Lin, M., & Wang, X. (2018). Overexpression of miR-19a inhibits colorectal cancer angiogenesis by suppressing KRAS expression. *Oncology Reports*, *39*(2), 619–626. https://doi.org/10.3892/or.2017.6110.

Chen, X., Xu, X., Pan, B., Zeng, K., Xu, M., Liu, X., … Wang, S. (2018). miR-150-5p suppresses tumor progression by targeting VEGFA in colorectal cancer. *Aging*, *10*(11), 3421–3437. https://doi.org/10.18632/aging.101656.

Chen, H. Y., Lang, Y. D., Lin, H. N., Liu, Y. R., Liao, C. C., Nana, A. W., … Chen, R. H. (2019). miR-103/107 prolong Wnt/β-catenin signaling and colorectal cancer stemness by targeting Axin2. *Scientific Reports*, *9*(1). https://doi.org/10.1038/s41598-019-41053-z.

Chen, X., Zeng, K., Xu, M., Liu, X., Hu, X., Xu, T., … Wang, S. (2019). P53-induced miR-1249 inhibits tumor growth, metastasis, and angiogenesis by targeting VEGFA and HMGA2. *Cell Death & Disease*, *10*(2), 131. https://doi.org/10.1038/s41419-018-1188-3.

Chen, H.-Y., Lang, Y.-D., Lin, H.-N., Liu, Y.-R., Liao, C.-C., Nana, A. W., … Chen, R.-H. (2019). miR-103/107 prolong Wnt/β-catenin signaling and colorectal cancer stemness by targeting Axin2. *Scientific Reports*, *9*(1), 9687. https://doi.org/10.1038/s41598-019-41053-z.

Chen, Z., Ren, R., Wan, D., Wang, Y., Xue, X., Jiang, M., … Zhi, Q. (2019). Hsa_circ_101555 functions as a competing endogenous RNA of miR-597-5p to promote colorectal cancer progression. *Oncogene*, *38*(32), 6017–6034. https://doi.org/10.1038/s41388-019-0857-8.

Chen, C., Huang, Z., Mo, X., Song, Y., Li, X., Li, X., & Zhang, M. (2020). The circular RNA 001971/miR-29c-3p axis modulates colorectal cancer growth, metastasis, and angiogenesis through VEGFA. *Journal of Experimental & Clinical Cancer Research*, *39*(1), 91. https://doi.org/10.1186/s13046-020-01594-y.

Chen, Y. L., Wang, G. X., Lin, B. A., & Huang, J. S. (2020). MicroRNA-93-5p expression in tumor tissue and its tumor suppressor function via targeting programmed death ligand-1 in colorectal cancer. *Cell Biology International*, *44*(5), 1224–1236. https://doi.org/10.1002/cbin.11323.

Cheng, H., Zhang, L., Cogdell, D. E., Zheng, H., Schetter, A. J., Nykter, M., … Zhang, W. (2011). Circulating plasma MiR-141 is a novel biomarker for metastatic colon cancer and predicts poor prognosis. *PLoS One*, *6*(3), e17745. https://doi.org/10.1371/journal.pone.0017745.

Choi, Y. W., Song, Y. S., Lee, H., Yi, K., Kim, Y. B., Suh, K. W., & Lee, D. (2016). MicroRNA expression signatures associated with BRAF-mutated versus KRAS-mutated colorectal cancers. *Medicine (United States)*, *95*(15), e3321. https://doi.org/10.1097/MD.0000000000003321.

Christensen, L. L., Tobiasen, H., Holm, A., Schepeler, T., Ostenfeld, M. S., Thorsen, K., … Andersen, C. L. (2013). MiRNA-362-3p induces cell cycle arrest through targeting of E2F1, USF2 and PTPN1 and is associated with recurrence of colorectal cancer. *International Journal of Cancer*, *133*(1), 67–78. https://doi.org/10.1002/ijc.28010.

Colangelo, T., Polcaro, G., Ziccardi, P., Pucci, B., Muccillo, L., Galgani, M., … Colantuoni, V. (2016). Proteomic screening identifies calreticulin as a MIR-27a direct target repressing MHC class I cell surface exposure in colorectal cancer. *Cell Death & Disease*, *7*(2), e2120. https://doi.org/10.1038/cddis.2016.28.

Conde, E., Pascual, A., Prieto-Cuadra, D., Laza, V. F., Molina-Cerrillo, J., Ramos-Muñoz, M. E., … Guillén-Ponce, C. (2019). Novel molecular characterization of colorectal primary tumors based on mirnas. *Cancers*, *11*(3), 346. https://doi.org/10.3390/cancers11030346.

Cruz-Gil, S., Sanchez-Martinez, R., De Cedron, M. G., Martin-Hernandez, R., Vargas, T., Molina, S., … De Molina, A. R. (2018). Targeting the lipid metabolic axis ACSL/SCD in colorectal cancer progression by therapeutic miRNAs: MiR-19b-1 role. *Journal of Lipid Research*, *59*(1), 14–24. https://doi.org/10.1194/jlr.M076752.

Cummins, J. M., He, Y., Leary, R. J., Pagliarini, R., Diaz, L. A., Sjoblom, T., … Velculescu, V. E. (2006). The colorectal microRNAome. *Proceedings of the National Academy of Sciences of the United States of America*, *103*(10), 3687–3692. https://doi.org/10.1073/pnas.0511155103.

Daniëls, V. W., Smans, K., Royaux, I., Chypre, M., Swinnen, J. V., & Zaidi, N. (2014). Cancer cells differentially activate and thrive on de novo lipid synthesis pathways in a low-lipid environment. *PLoS One*, *9*(9), e106913. https://doi.org/10.1371/journal.pone.0106913.

De Krijger, I., Mekenkamp, L. J. M., Punt, C. J. A., & Nagtegaal, I. D. (2011). MicroRNAs in colorectal cancer metastasis. *Journal of Pathology*, *224*(4), 438–447. https://doi.org/10.1002/path.2922.

Deng, G., Kakar, S., & Kim, Y. S. (2011). MicroRNA-124a and microRNA-34b/c are frequently methylated in all histological types of colorectal cancer and polyps, and in the adjacent normal mucosa. *Oncology Letters*, *2*(1), 175–180. https://doi.org/10.3892/ol.2010.222.

Deng, Y. H., Deng, Z. H., Hao, H., Wu, X. L., Gao, H., Tang, S. H., & Tang, H. (2018). MicroRNA-23a promotes colorectal cancer cell survival by targeting PDK4. *Experimental Cell Research*, *373*(1–2), 171–179. https://doi.org/10.1016/j.yexcr.2018.10.010.

Dews, M., Fox, J. L., Hultine, S., Sundaram, P., Wang, W., Liu, Y. Y., … Thomas-Tikhonenko, A. (2010). The Myc-miR-17~92 axis blunts TGFβ signaling and production of multiple TGFβ-dependent antiangiogenic factors. *Cancer Research*, *70*(20), 8233–8246. https://doi.org/10.1158/0008-5472.CAN-10-2412.

Di Lena, M., Travaglio, E., & Altomare, D. F. (2013). New strategies for colorectal cancer screening. *World Journal of Gastroenterology*, *19*(12), 1855–1860. https://doi.org/10.3748/wjg.v19.i12.1855.

Di Nicolantonio, F., Martini, M., Molinari, F., Sartore-Bianchi, A., Arena, S., Saletti, P., … Bardelli, A. (2008). Wild-type BRAF is required for response to panitumumab or cetuximab in metastatic colorectal cancer. *Journal of Clinical Oncology*, *26*(35), 5705–5712. https://doi.org/10.1200/JCO.2008.18.0786.

Diamantopoulos, M. A., Kontos, C. K., Kerimis, D., Papadopoulos, I. N., & Scorilas, A. (2017). Upregulated miR-16 expression is an independent indicator of relapse and poor overall survival of colorectal adenocarcinoma patients. *Clinical Chemistry and Laboratory Medicine*, *55*(5), 737–747. https://doi.org/10.1515/cclm-2016-0756.

Díaz, R., Silva, J., García, J. M., Lorenzo, Y., García, V., Peña, C., … Domínguez, G. (2008). Deregulated expression of miR-106a predicts survival in human colon cancer patients. *Genes, Chromosomes and Cancer*, *47*(9), 794–802. https://doi.org/10.1002/gcc.20580.

Diaz, T., Tejero, R., Moreno, I., Ferrer, G., Cordeiro, A., Artells, R., … Monzo, M. (2014). Role of miR-200 family members in survival of colorectal cancer patients treated with fluoropyrimidines. *Journal of Surgical Oncology*, *109*(7), 676–683. https://doi.org/10.1002/jso.23572.

Díaz-Trelles, R., Scimia, M. C., Bushway, P., Tran, D., Monosov, A., Monosov, E., … Mercola, M. (2016). Notch-independent RBPJ controls angiogenesis in the adult heart. *Nature Communications*, *7*, 12088. https://doi.org/10.1038/ncomms12088.

Dong, S. J., Cai, X. J., & Li, S. J. (2016). The clinical significance of MiR-429 as a predictive biomarker in colorectal cancer patients receiving 5-fluorouracil treatment. *Medical Science Monitor*, *22*, 3352–3361. https://doi.org/10.12659/MSM.900674.

Dong-Xu, W., Jia, L., & Su-Juan, Z. (2015). MicroRNA-185 is a novel tumor suppressor by negatively modulating the Wnt/β-catenin pathway in human colorectal cancer. *Indian Journal of Cancer*, *52*(7), e182–e185. https://doi.org/10.4103/0019-509X.186576.

dos Santos, I. L., Penna, K. G. B. D., dos Santos Carneiro, M. A., Libera, L. S. D., Ramos, J. E. P., & Saddi, V. A. (2021). Tissue micro-RNAs associated with colorectal cancer prognosis: A systematic review. *Molecular Biology Reports*, *48*(2), 1853–1867. https://doi.org/10.1007/s11033-020-06075-1.

Drusco, A., Nuovo, G. J., Zanesi, N., Di Leva, G., Pichiorri, F., Volinia, S., … Croce, C. M. (2014). MicroRNA profiles discriminate among colon cancer metastasis. *PLoS One*, *9*(6), e96670. https://doi.org/10.1371/journal.pone.0096670.

Earle, J. S. L., Luthra, R., Romans, A., Abraham, R., Ensor, J., Yao, H., & Hamilton, S. R. (2010). Association of MicroRNA expression with microsatellite instability status in colorectal adenocarcinoma. *Journal of Molecular Diagnostics*, *12*(4), 433–440. https://doi.org/10.2353/jmoldx.2010.090154.

Fan, C., Lin, Y., Mao, Y., Huang, Z., Liu, A. Y., Ma, H., … Ouyang, G. (2016). MicroRNA-543 suppresses colorectal cancer growth and metastasis by targeting KRAS, MTA1 and HMGA2. *Oncotarget*, *7*(16), 21825–21839. https://doi.org/10.18632/oncotarget.7989.

Fan, D., Lin, X., Zhang, F., Zhong, W., Hu, J., Chen, Y., … Wu, X. (2018). MicroRNA 26b promotes colorectal cancer metastasis by downregulating phosphatase and tensin homolog and wingless-type MMTV integration site family member 5A. *Cancer Science*, *109*(2), 354–362. https://doi.org/10.1111/cas.13451.

Fan, C., Lin, B., Huang, Z., Cui, D., Zhu, M., Ma, Z., … Liu, Y. (2019). MicroRNA-873 inhibits colorectal cancer metastasis by targeting ELK1 and STRN4. *Oncotarget*, *10*(41), 4192–4204. https://doi.org/10.18632/oncotarget.24115.

Fan, X., Liu, M., Tang, H., Leng, D., Hu, S., Lu, R., … Yuan, S. (2019). MicroRNA-7 exerts antiangiogenic effect on colorectal cancer via ERK signaling. *Journal of Surgical Research*, *240*, 48–59. https://doi.org/10.1016/j.jss.2019.02.035.

Fan, H. N., Liao, X. H., Zhang, J., & Zheng, H. M. (2020). Macrophages promote cell proliferation in colorectal cancer via il-1β-mediated downregulation of mir-28-3p. *Journal of Biological Regulators and Homeostatic Agents*, *34*(5), 1657–1668. https://www.biolifesas.org/biolife/category/journals/journal-of-biological-regulators-and-homeostatic-agents/.

Fang, L., Li, H., Wang, L., Hu, J., Jin, T., Wang, J., & Yang, B. B. (2014). MicroRNA-17-5p promotes chemotherapeutic drug resistance and tumour metastasis of colorectal cancer by repressing PTEN expression. *Oncotarget*, *5*(10), 2974–2987. https://doi.org/10.18632/oncotarget.1614.

Fang, Z., Tang, J., Bai, Y., Lin, H., You, H., Jin, H., … Zhang, Z. Y. (2015). Plasma levels of microRNA-24, microRNA-320a, and microRNA-423-5p are potential biomarkers for colorectal carcinoma. *Journal of Experimental & Clinical Cancer Research*, *34*(1), 86. https://doi.org/10.1186/s13046-015-0198-6.

Fang, Y., Sun, B., Li, Z., Chen, Z., & Xiang, J. (2016). MiR-622 inhibited colorectal cancer occurrence and metastasis by suppressing K-Ras. *Molecular Carcinogenesis*, *55*(9), 1369–1377. https://doi.org/10.1002/mc.22380.

Fasihi, A. M., Soltani, B., Atashi, A., & Nasiri, S. (2018). Introduction of hsa-miR-103a and hsa-miR-1827 and hsa-miR-137 as new regulators of Wnt signaling pathway and their relation to colorectal carcinoma. *Journal of Cellular Biochemistry*, *119*(7), 5104–5117. https://doi.org/10.1002/jcb.26357.

Fasihi, A., Soltani, B. M., Ranjbaran, Z. S., Bahonar, S., Norouzi, R., & Nasiri, S. (2019). Hsa-miR-942 fingerprint in colorectal cancer through Wnt signaling pathway. *Gene*, *712*, 143958. https://doi.org/10.1016/j.gene.2019.143958.

Feng, B., Dong, T. T., Wang, L. L., Zhou, H. M., Zhao, H. C., Dong, F., & Zheng, M. H. (2012). Colorectal cancer migration and invasion initiated by microRNA-106a. *PLoS One*, *7*(8), e43452. https://doi.org/10.1371/journal.pone.0043452.

Feroz, W., & Sheikh, A. M. A. (2020). Exploring the multiple roles of guardian of the genome: P53. *Egyptian Journal of Medical Human Genetics*, *21*(1), 1–23. https://doi.org/10.1186/s43042-020-00089-x.

Fu, Y., Liu, X., Zhou, N., Du, L., Sun, Y., Zhang, X., & Ge, Y. (2014). MicroRNA-200b stimulates tumour growth in TGFBR2-null colorectal cancers by negatively regulating p27/kip1. *Journal of Cellular Physiology*, *229*(6), 772–782. https://doi.org/10.1002/jcp.24497.

Fu, Q., Du, Y., Yang, C., Zhang, D., Zhang, N., Liu, X., … Yang, Y. (2016). An oncogenic role of miR-592 in tumorigenesis of human colorectal cancer by targeting Forkhead Box O3A (FoxO3A). *Expert Opinion on Therapeutic Targets*, *20*(7), 771–782. https://doi.org/10.1080/14728222.2016.1181753.

Fukushima, Y., Iinuma, H., Tsukamoto, M., Matsuda, K., & Hashiguchi, Y. (2015). Clinical significance of microRNA-21 as a biomarker in each Dukes' stage of colorectal cancer. *Oncology Reports*, *33*(2), 573–582. https://doi.org/10.3892/or.2014.3614.

Gao, F., & Wang, W. (2015). MicroRNA-96 promotes the proliferation of colorectal cancer cells and targets tumor protein p53 inducible nuclear protein 1, forkhead box protein O1 (FOXO1) and FOXO3a. *Molecular Medicine Reports*, *11*(2), 1200–1206. https://doi.org/10.3892/mmr.2014.2854.

Gao, J., Li, N., Dong, Y., Li, S., Xu, L., Li, X., … Yu, J. (2015). MiR-34a-5p suppresses colorectal cancer metastasis and predicts recurrence in patients with stage II/III colorectal cancer. *Oncogene*, *34*(31), 4142–4152. https://doi.org/10.1038/onc.2014.348.

Gao, S. J., Chen, L., Lu, W., Zhang, L., Wang, L., & Zhu, H. H. (2018). MiR-888 functions as an oncogene and predicts poor prognosis in colorectal cancer. *Oncology Letters*, *15*(6), 9101–9109. https://doi.org/10.3892/ol.2018.8461.

Gao, S., Zhao, Z. Y., Wu, R., Zhang, Y., & Zhang, Z. Y. (2018). Prognostic value of microRNAs in colorectal cancer: A meta-analysis. *Cancer Management and Research*, *10*, 907–929. https://doi.org/10.2147/CMAR.S157493.

Gao, W., Pang, D., & Yu, S. (2019). Serum level of miR-142-3p predicts prognostic outcome for colorectal cancer following curative resection. *Journal of International Medical Research*, *47*(5), 2116–2125. https://doi.org/10.1177/0300060519834815.

Gao, Z., Zhou, H., Wang, Y., Chen, J., & Ou, Y. (2020). Regulatory effects of lncRNA ATB targeting miR-200c on proliferation and apoptosis of colorectal cancer cells. *Journal of Cellular Biochemistry*, *121*(1), 332–343. https://doi.org/10.1002/jcb.29180.

Garzon, R., Calin, G. A., & Croce, C. M. (2009). MicroRNAs in cancer. *Annual Review of Medicine*, *60*, 167–179. https://doi.org/10.1146/annurev.med.59.053006.104707.

Gattolliat, C. H., Uguen, A., Pesson, M., Trillet, K., Simon, B., Doucet, L., … Corcos, L. (2015). MicroRNA and targeted mRNA expression profiling analysis in human colorectal adenomas and adenocarcinomas. *European Journal of Cancer*, *51*(3), 409–420. https://doi.org/10.1016/j.ejca.2014.12.007.

Ge, J., Chen, Z., Li, R., Lu, T., & Xiao, G. (2014). Upregulation of microRNA-196a and microRNA-196b cooperatively correlate with aggressive progression and unfavorable prognosis in patients with colorectal cancer. *Cancer Cell International*, *14*(1), 128. https://doi.org/10.1186/s12935-014-0128-2.

Gershtein, E. S., Korotkova, E. A., Prorokov, V. V., & Kushlinskii, N. E. (2013). Tumor associated proteases—Prognostic markers of colorectal cancer. *Klinichescheskaya Laboratornaya Diagnostika*, *10*, 43–47. http://www.medit.ru.

Gharib, E., Nasri Nasrabadi, P., & Reza Zali, M. (2020). miR-497-5p mediates starvation-induced death in colon cancer cells by targeting acyl-CoA synthetase-5 and modulation of lipid metabolism. *Journal of Cellular Physiology*, *235*(7–8), 5570–5589. https://doi.org/10.1002/jcp.29488.

Gini, A., Jansen, E. E. L., Zielonke, N., Meester, R. G. S., Senore, C., Anttila, A., … van Ballegooijen, M. (2020). Impact of colorectal cancer screening on cancer-specific mortality in Europe: A systematic review. *European Journal of Cancer*, *127*, 224–235. https://doi.org/10.1016/j.ejca.2019.12.014.

Glaich, O., Parikh, S., Bell, R. E., Mekahel, K., Donyo, M., Leader, Y., … Levy, C. (2019). DNA methylation directs microRNA biogenesis in mammalian cells. *Nature Communications*, *10*(1), 5657. https://doi.org/10.1038/s41467-019-13527-1.

Gong, T., Li, Y., Feng, L., Fang, M. Z., Dai, G., Huang, X., … Liu, S. (2020). CASC21, a FOXP1 induced long non-coding RNA, promotes colorectal cancer growth by regulating CDK6. *Aging*, *12*(12), 12086–12106. https://doi.org/10.18632/aging.103376.

Graziella, C. (2018). MiRNAs at the crossroads between innate immunity and cancer: Focus on macrophages. *Cell*, *7*, 12. https://doi.org/10.3390/cells7020012.

Gregersen, L. H., Jacobsen, A. B., Frankel, L. B., Wen, J., Krogh, A., & Lund, A. H. (2010). MicroRNA-145 targets YES and STAT1 in colon cancer cells. *PLoS One*, *5*(1), e8836. https://doi.org/10.1371/journal.pone.0008836.

Gregorova, J., Vychytilova-Faltejskova, P., & Sevcikova, S. (2021). Epigenetic regulation of microrna clusters and families during tumor development. *Cancers*, *13*(6), 1–45. https://doi.org/10.3390/cancers13061333.

Gruber, A. J., & Zavolan, M. (2013). Modulation of epigenetic regulators and cell fate decisions by miRNAs. *Epigenomics*, *5*(6), 671–683. https://doi.org/10.2217/epi.13.65.

Guo, C., Sah, J. F., Beard, L., Willson, J. K. V., Markowitz, S. D., & Guda, K. (2008). The noncoding RNA, miR-126, suppresses the growth of neoplastic cells by targeting phosphatidylinositol 3-kinase signaling and is frequently lost in colon cancers. *Genes, Chromosomes and Cancer*, *47*(11), 939–946. https://doi.org/10.1002/gcc.20596.

Guo, H., Chen, Y., Hu, X., Qian, G., Ge, S., & Zhang, J. (2013). The regulation of toll-like receptor 2 by miR-143 suppresses the invasion and migration of a subset of human colorectal carcinoma cells. *Molecular Cancer*, *12*(1), 77. https://doi.org/10.1186/1476-4598-12-77.

Guo, Y., Li, X., Lin, C., Zhang, Y., Hu, G., Zhou, J., … Deng, H. (2015). MicroRNA-133b inhibits connective tissue growth factor in colorectal cancer and correlates with the clinical stage of the disease. *Molecular Medicine Reports*, *11*(4), 2805–2812. https://doi.org/10.3892/mmr.2014.3075.

Guo, S., Zhang, J., Wang, B., Zhang, B., Wang, X., Huang, L., … Jia, B. (2018). A 5-serum miRNA panel for the early detection of colorectal cancer. *Oncotargets and Therapy*, *11*, 2603–2614. https://doi.org/10.2147/OTT.S153535.

Guraya, S. Y. (2019). Pattern, stage, and time of recurrent colorectal cancer after curative surgery. *Clinical Colorectal Cancer*, *18*(2), e223–e228. https://doi.org/10.1016/j.clcc.2019.01.003.

Han, J., Li, J., Tang, K., Zhang, H., Guo, B., Hou, N., & Huang, C. (2017). miR-338-3p confers 5-fluorouracil resistance in p53 mutant colon cancer cells by targeting the mammalian target of rapamycin. *Experimental Cell Research*, *360*(2), 328–336. https://doi.org/10.1016/j.yexcr.2017.09.023.

Han, J., Sun, W., Liu, R., Zhou, Z., Zhang, H., Chen, X., & Ba, Y. (2020). Plasma exosomal miRNA expression profile as oxaliplatin-based chemoresistant biomarkers in colorectal adenocarcinoma. *Frontiers in Oncology*, *10*, 1495. https://doi.org/10.3389/fonc.2020.01495.

Hanahan, D., & Weinberg, R. A. (2011). Hallmarks of cancer: The next generation. *Cell*, *144*(5), 646–674. https://doi.org/10.1016/j.cell.2011.02.013.

Hao, J. P., & Ma, A. (2018). The ratio of miR-21/miR-24 as a promising diagnostic and poor prognosis biomarker in colorectal cancer. *European Review for Medical and Pharmacological Sciences*, *22*(24), 8649–8656. https://www.europeanreview.org/wp/wp-content/uploads/8649-8656.pdf.

Hao, H., Liu, L., Zhang, D., Wang, C., Xia, G., Zhong, F., & Hu, X. (2017). Diagnostic and prognostic value of miR-106a in colorectal cancer. *Oncotarget*, *8*(3), 5038–5047. https://doi.org/10.18632/oncotarget.13766.

Hara, T., Jones, M. F., Subramanian, M., Li, X. L., Ou, O., Zhu, Y., … Lal, A. (2014). Selective targeting of KRAS-mutant cells by miR-126 through repression of multiple genes essential for the survival of KRAS-mutant cells. *Oncotarget*, *5*(17), 7635–7650. https://doi.org/10.18632/oncotarget.2284.

He, D., Yue, Z., Li, G., Chen, L., Feng, H., & Sun, J. (2018). Low serum levels of miR-101 are associated with poor prognosis of colorectal cancer patients after curative resection. *Medical Science Monitor*, *24*, 7475–7481. https://doi.org/10.12659/MSM.909768.

He, H., Zhao, X., Zhu, Z., Du, L., Chen, E., Liu, S., … Lei, L. (2019). MicroRNA-3191 promotes migration and invasion by downregulating TGFBR2 in colorectal cancer. *Journal of Biochemical and Molecular Toxicology*, *33*(6), e22308. https://doi.org/10.1002/jbt.22308.

He, Y., Zhang, L., Tan, F., Wang, L. F., Liu, D. H., Wang, R. J., & Yin, X. Z. (2020). MiR-153-5p promotes sensibility of colorectal cancer cells to oxaliplatin via targeting Bcl-2-mediated autophagy pathway. *Bioscience, Biotechnology, and Biochemistry*, *84*(8), 1645–1651. https://doi.org/10.1080/09168451.2020.1760784.

He, Z., Dang, J., Song, A., Cui, X., Ma, Z., & Zhang, Y. (2020). The involvement of miR-150/β-catenin axis in colorectal cancer progression. *Biomedicine and Pharmacotherapy*, *121*, 109495. https://doi.org/10.1016/j.biopha.2019.109495.

Heublein, S., Albertsmeier, M., Pfeifer, D., Loehrs, L., Bazhin, A. V., Kirchner, T., … Angele, M. K. (2018). Association of differential miRNA expression with hepatic vs. peritoneal metastatic spread in colorectal cancer. *BMC Cancer*, *18*(1), 201. https://doi.org/10.1186/s12885-018-4043-0.

Hinestrosa, M. C., Dickersin, K., Klein, P., Mayer, M., Noss, K., Slamon, D., … Visco, F. M. (2007). Shaping the future of biomarker research in breast cancer to ensure clinical relevance. *Nature Reviews Cancer*, *7*(4), 309–315. https://doi.org/10.1038/nrc2113.

Hiraki, M., Nishimura, J., Takahashi, H., Wu, X., Takahashi, Y., Miyo, M., … Yamamoto, H. (2015). Concurrent targeting of KRAS and AKT by MiR-4689 is a novel treatment against mutant KRAS colorectal cancer. *Molecular Therapy- -Nucleic Acids*, *4*(3), e231. https://doi.org/10.1038/mtna.2015.5.

Hon, K. W., Abidin, S. A. Z., Othman, I., & Naidu, R. (2020). Insights into the role of micrornas in colorectal cancer (Crc) metabolism. *Cancers*, *12*(9), 1–17. https://doi.org/10.3390/cancers12092462.

Huang, X., & Lu, S. (2017). MicroR-545 mediates colorectal cancer cells proliferation through up-regulating epidermal growth factor receptor expression in HOTAIR long non-coding RNA dependent. *Molecular and Cellular Biochemistry*, *431*(1–2), 45–54. https://doi.org/10.1007/s11010-017-2974-4.

Huang, Z., Huang, D., Ni, S., Peng, Z., Sheng, W., & Du, X. (2010). Plasma microRNAs are promising novel biomarkers for early detection of colorectal cancer. *International Journal of Cancer*, *127*(1), 118–126. https://doi.org/10.1002/ijc.25007.

Huang, L., Liu, Z., Hu, J., Luo, Z., Zhang, C., Wang, L., & Wang, Z. (2020). MiR-377-3p suppresses colorectal cancer through negative regulation on Wnt/β-catenin signaling by targeting XIAP and ZEB2. *Pharmacological Research*, *156*, 104774. https://doi.org/10.1016/j.phrs.2020.104774.

Huang, G., Wei, B., Chen, Z., Wang, J., Zhao, L., Peng, X., … Ni, L. (2020). Identification of a four-microRNA panel in serum as promising biomarker for colorectal carcinoma detection. *Biomarkers in Medicine*, *14*(9), 749–760. https://doi.org/10.2217/bmm-2019-0605.

Huang, F. K., Zheng, C. Y., Huang, L. K., Lin, C. Q., Zhou, J. F., & Wang, J. X. (2021). Long non-coding RNA MCF2L-AS1 promotes the aggressiveness of colorectal cancer by sponging miR-874-3p and thereby up-regulating CCNE1. *The Journal of Gene Medicine*, *23*(1), e3285. https://doi.org/10.1002/jgm.3285.

Hur, K., Toiyama, Y., Schetter, A. J., Okugawa, Y., Harris, C. C., Boland, C. R., & Goel, A. (2015). Identification of a metastasis-specific microRNA signature in human colorectal cancer. *Journal of the National Cancer Institute*, *107*(3). https://doi.org/10.1093/jnci/dju492, dju492.

Hur, K., Toiyama, Y., Okugawa, Y., Ide, S., Imaoka, H., Boland, C. R., & Goel, A. (2017). Circulating microRNA-203 predicts prognosis and metastasis in human colorectal cancer. *Gut*, *66*(4), 654–664. https://doi.org/10.1136/gutjnl-2014-308737.

Imaoka, H., Toiyama, Y., Fujikawa, H., Hiro, J., Saigusa, S., Tanaka, K., … Kusunoki, M. (2016). Circulating microRNA-1290 as a novel diagnostic and prognostic biomarker in human colorectal cancer. *Annals of Oncology*, *27*(10), 1879–1886. https://doi.org/10.1093/annonc/mdw279.

Inoue, A., Yamamoto, H., Uemura, M., Nishimura, J., Hata, T., Takemasa, I., … Mori, M. (2015). MicroRNA-29b is a novel prognostic marker in colorectal cancer. *Annals of Surgical Oncology*, *22*, 1410–1418. https://doi.org/10.1245/s10434-014-4255-8.

Iwaya, T., Yokobori, T., Nishida, N., Kogo, R., Sudo, T., Tanaka, F., … Mimori, K. (2012). Downregulation of miR-144 is associated with colorectal cancer progression via activation of mTOR signaling pathway. *Carcinogenesis*, *33*(12), 2391–2397. https://doi.org/10.1093/carcin/bgs288.

Javanmard, A.-R., Dokanehiifard, S., Bohlooli, M., & Soltani, B. M. (2020). LOC646329 long non-coding RNA sponges miR-29b-1 and regulates TGFβ signaling in colorectal cancer. *Journal of Cancer Research and Clinical Oncology*, *146*(5), 1205–1215. https://doi.org/10.1007/s00432-020-03145-6.

Ji, D., Qiao, M., Yao, Y., Li, M., Chen, H., Dong, Q., … Gu, J. (2018). Serum-based microRNA signature predicts relapse and therapeutic outcome of adjuvant chemotherapy in colorectal cancer patients. *eBioMedicine*, *35*, 189–197. https://doi.org/10.1016/j.ebiom.2018.08.042.

Jia, Z., Peng, J., Yang, Z., Chen, J., Liu, L., Luo, D., & He, P. (2019). Long non-coding RNA TP73-AS1 promotes colorectal cancer proliferation by acting as a ceRNA for miR-103 to regulate PTEN expression. *Gene*, *685*, 222–229. https://doi.org/10.1016/j.gene.2018.11.072.

Jiang, H., Liu, J., Chen, Y., Ma, C., Li, B., & Hao, T. (2016). Up-regulation of mir-10b predicate advanced clinicopathological features and liver metastasis in colorectal cancer. *Cancer Medicine*, *5*(10), 2932–2941. https://doi.org/10.1002/cam4.789.

Jiang, W., Li, T., Wang, J., Jiao, R., Shi, X., Huang, X., & Ji, G. (2019). miR-140-3p suppresses cell growth and induces apoptosis in colorectal cancer by targeting PD-L1. *Oncotargets and Therapy*, *12*, 10275–10285. https://doi.org/10.2147/OTT.S226465.

Jin, Y., Wang, M., Hu, H., Huang, Q., Chen, Y., & Wang, G. (2018). Overcoming stemness and chemoresistance in colorectal cancer through miR-195-5p-modulated inhibition of notch signaling. *International Journal of Biological Macromolecules*, *117*, 445–453. https://doi.org/10.1016/j.ijbiomac.2018.05.151.

Jin, F., Yang, R., Wei, Y., Wang, D., Zhu, Y., Wang, X., … Li, L. (2019). HIF-1α-induced miR-23a~27a~24 cluster promotes colorectal cancer progression via reprogramming metabolism. *Cancer Letters*, *440–441*, 211–222. https://doi.org/10.1016/j.canlet.2018.10.025.

Karaayvaz, M., Zhai, H., & Ju, J. (2013). MiR-129 promotes apoptosis and enhances chemosensitivity to 5-fluorouracil in colorectal cancer. *Cell Death & Disease*, *4*(6). https://doi.org/10.1038/cddis.2013.193.

Karimi Dermani, F., Amini, R., Saidijam, M., & Najafi, R. (2018). miR-200c, a tumor suppressor that modulate the expression of cancer stem cells markers and epithelial-mesenchymal transition in colorectal cancer. *Journal of Cellular Biochemistry*, *119*(7), 6288–6295. https://doi.org/10.1002/jcb.26880.

Kashani, E., Hadizadeh, M., Chaleshi, V., Mirfakhraie, R., Young, C., Savabkar, S., … Bonab, M. A. (2019). The differential DNA hypermethylation patterns of microRNA-137 and microRNA-342 locus in early colorectal lesions and tumours. *Biomolecules*, *9*(10), 519. https://doi.org/10.3390/biom9100519.

Kerimis, D., Kontos, C. K., Christodoulou, S., Papadopoulos, I. N., & Scorilas, A. (2017). Elevated expression of miR-24-3p is a potentially adverse prognostic factor in colorectal adenocarcinoma. *Clinical Biochemistry*, *50*(6), 285–292. https://doi.org/10.1016/j.clinbiochem.2016.11.034.

Khaliq, A. M., Kurt, Z., Grunvald, M. W., Erdogan, C., Turgut, S. S., Rand, T., … Masood, A. (2021). Redefining colorectal cancer classification and clinical stratification through a single-cell atlas. *BioRxiv*. https://doi.org/10.1101/2021.02.02.429256.

Kim, D. H., Sætrom, P., Snøve, O., & Rossi, J. J. (2008). MicroRNA-directed transcriptional gene silencing in mammalian cells. *Proceedings of the National Academy of Sciences*, *105*(42), 16230–16235. https://doi.org/10.1073/pnas.0808830105.

Kim, N. H., Cha, Y. H., Kang, S. E., Lee, Y., Lee, I., Cha, S. Y., … Kim, H. S. (2013). p53 regulates nuclear GSK-3 levels through miR-34-mediated Axin2 suppression in colorectal cancer cells. *Cell Cycle*, *12*(10), 1578–1587. https://doi.org/10.4161/cc.24739.

Kim, C. W., Oh, E. T., Kim, J. M., Park, J. S., Lee, D. H., Lee, J. S., … Park, H. J. (2018). Hypoxia-induced microRNA-590-5p promotes colorectal cancer progression by modulating matrix metalloproteinase activity. *Cancer Letters*, *416*, 31–41. https://doi.org/10.1016/j.canlet.2017.12.018.

Kim, H., Kim, T., Jaygal, G., Woo, J., Kim, C. J., Baek, M. J., & Jeong, D. (2020). Downregulation of miR-9 correlates with poor prognosis in colorectal cancer. *Pathology Research and Practice*, *216*(8). https://doi.org/10.1016/j.prp.2020.153044, 153044.

Kim, T. W., Lee, Y. S., Yun, N. H., Shin, C. H., Hong, H. K., Kim, H. H., & Cho, Y. B. (2020). MicroRNA-17-5p regulates EMT by targeting vimentin in colorectal cancer. *British Journal of Cancer*, *123*(7), 1123–1130. https://doi.org/10.1038/s41416-020-0940-5.

Kjersem, J. B., Ikdahl, T., Lingjaerde, O. C., Guren, T., Tveit, K. M., & Kure, E. H. (2014). Plasma microRNAs predicting clinical outcome in metastatic colorectal cancer patients receiving first-line oxaliplatin-based treatment. *Molecular Oncology*, *8*(1), 59–67. https://doi.org/10.1016/j.molonc.2013.09.001.

Kościelniak-Merak, B., Radosavljević, B., Zając, A., & Tomasik, P. J. (2018). Faecal occult blood point-of-care tests. *Journal of Gastrointestinal Cancer*, *49*(4), 402–405. https://doi.org/10.1007/s12029-018-0169-1.

Koga, Y., Yasunaga, M., Takahashi, A., Kuroda, J., Moriya, Y., Akasu, T., … Matsumura, Y. (2010). MicroRNA expression profiling of exfoliated colonocytes isolated from feces for colorectal cancer screening. *Cancer Prevention Research*, *3*(11), 1435–1442. https://doi.org/10.1158/1940-6207.CAPR-10-0036.

Koga, Y., Yamazaki, N., Yamamoto, Y., Yamamoto, S., Saito, N., Kakugawa, Y., … Matsumura, Y. (2013). Fecal miR-106a is a useful marker for colorectal cancer patients with false-negative results in immunochemical fecal occult blood test. *Cancer Epidemiology, Biomarkers and Prevention*, *22*(10), 1844–1852. https://doi.org/10.1158/1055-9965.EPI-13-0512.

Kumar, M., Lu, Z., Takwi, A. A. L., Chen, W., Callander, N. S., Ramos, K. S., … Li, Y. (2011). Negative regulation of the tumor suppressor p53 gene by microRNAs. *Oncogene*, *30*(7), 843–853. https://doi.org/10.1038/onc.2010.457.

Kunej, T., Godnic, I., Ferdin, J., Horvat, S., Dovc, P., & Calin, G. A. (2011). Epigenetic regulation of microRNAs in cancer: An integrated review of literature. *Mutation Research: Fundamental and Molecular Mechanisms of Mutagenesis*, *717*(1–2), 77–84. https://doi.org/10.1016/j.mrfmmm.2011.03.008.

Kwak, B., Kim, D. U., Kim, T. O., Kim, H. S., & Kim, S. W. (2018). MicroRNA-552 links Wnt signaling to p53 tumor suppressor in colorectal cancer. *International Journal of Oncology*, *53*(4), 1800–1808. https://doi.org/10.3892/ijo.2018.4505.

Lan, F., Yue, X., Han, L., Shi, Z., Yang, Y., Pu, P., … Kang, C. (2012). Genome-wide identification of TCF7L2/TCF4 target miRNAs reveals a role for miR-21 in Wnt-driven epithelial cancer. *International Journal of Oncology*, *40*(2), 519–526. https://doi.org/10.3892/ijo.2011.1215.

Lanza, G., Ferracin, M., Gafà, R., Veronese, A., Spizzo, R., Pichiorri, F., … Negrini, M. (2007). mRNA/microRNA gene expression profile in microsatellite unstable colorectal cancer. *Molecular Cancer*, *6*, 54. https://doi.org/10.1186/1476-4598-6-54.

Laudato, S., Patil, N., Abba, M. L., Leupold, J. H., Benner, A., Gaiser, T., … Allgayer, H. (2017). P53-induced miR-30e-5p inhibits colorectal cancer invasion and metastasis by targeting ITGA6 and ITGB1. *International Journal of Cancer*, *141*(9), 1879–1890. https://doi.org/10.1002/ijc.30854.

Lee, R. C., Feinbaum, R. L., & Ambros, V. (1993). The *C. elegans* heterochronic gene lin-4 encodes small RNAs with antisense complementarity to lin-14. *Cell*, *75*(5), 843–854. https://doi.org/10.1016/0092-8674(93)90529-Y.

Lei, S. L., Zhao, H., Yao, H. L., Chen, Y., Lei, Z. D., Liu, K. J., & Yang, Q. (2014). Regulatory roles of microRNA-708 and microRNA-31 in proliferation, apoptosis and invasion of colorectal cancer cells. *Oncology Letters*, *8*(4), 1768–1774. https://doi.org/10.3892/ol.2014.2328.

Li, X., Zhang, G., Luo, F., Ruan, J., Huang, D., Feng, D., … Wu, W. (2012). Identification of aberrantly expressed miRNAs in rectal cancer. *Oncology Reports*, *28*(1), 77–84. https://doi.org/10.3892/or.2012.1769.

Li, J., Liu, Y., Wang, C., Deng, T., Liang, H., Wang, Y., … Ba, Y. (2015). Serum miRNA expression profile as a prognostic biomarker of stage II/III colorectal adenocarcinoma. *Scientific Reports*, *5*. https://doi.org/10.1038/srep12921, 12921.

Li, J., Liang, H., Bai, M., Ning, T., Wang, C., Fan, Q., … Ba, Y. (2015). miR-135b promotes cancer progression by targeting transforming growth factor beta receptor II (TGFBR2) in colorectal cancer. *PLoS One*, *10*(6), e0130194. https://doi.org/10.1371/journal.pone.0130194.

Li, W., Chang, J., Wang, S., Liu, X., Peng, J., Huang, D., … Li, J. (2015). miRNA-99b-5p suppresses liver metastasis of colorectal cancer by down-regulating mTOR. *Oncotarget*, *6*(27), 24448–24462. https://doi.org/10.18632/oncotarget.4423.

Li, H., Rokavec, M., & Hermeking, H. (2015). Soluble IL6R represents a miR-34a target: Potential implications for the recently identified IL-6R/STAT3/miR-34a feed-back loop. *Oncotarget*, *6*(16), 14026–14032. https://doi.org/10.18632/oncotarget.4334.

Li, Q., Liang, X., Wang, Y., Meng, X., Xu, Y., Cai, S., … Cai, G. (2016). MiR-139-5p inhibits the epithelial-mesenchymal transition and enhances the chemotherapeutic sensitivity of colorectal cancer cells by downregulating BCL2. *Scientific Reports*, *6*, 27157. https://doi.org/10.1038/srep27157.

Li, Y., Sun, J., Cai, Y., Jiang, Y., Wang, X., Huang, X., … Li, H. (2016). MiR-200a acts as an oncogene in colorectal carcinoma by targeting PTEN. *Experimental and Molecular Pathology*, *101*(3), 308–313. https://doi.org/10.1016/j.yexmp.2016.10.006.

Li, T., Lai, Q., Wang, S., Cai, J., Xiao, Z., Deng, D., … Liao, W. (2016). MicroRNA-224 sustains Wnt/β-catenin signaling and promotes aggressive phenotype of colorectal cancer. *Journal of Experimental & Clinical Cancer Research*, *35*(1), 21. https://doi.org/10.1186/s13046-016-0287-1.

Li, L., Guo, Y., Chen, Y., Wang, J., Zhen, L., Guo, X., … Jing, C. (2016). The diagnostic efficacy and biological effects of microRNA-29b for colon cancer. *Technology in Cancer Research & Treatment*, *15*(6), 772–779. https://doi.org/10.1177/1533034615604797.

Li, Y., Lauriola, M., Kim, D., Francesconi, M., D'Uva, G., Shibata, D., … Cheng, J. Q. (2016). Adenomatous polyposis coli (APC) regulates miR17-92 cluster through β-catenin pathway in colorectal cancer. *Oncogene*, *35*(35), 4558–4568. https://doi.org/10.1038/onc.2015.522.

Li, W., Chang, J., Tong, D., Peng, J., Huang, D., Guo, W., … Li, J. (2017). Differential microRNA expression profiling in primary tumors and matched liver metastasis of patients with colorectal cancer. *Oncotarget*, *8*(22), 35783–35791. https://doi.org/10.18632/oncotarget.16206.

Li, B., Huang, M., Liu, M., Wen, S., & Sun, F. (2017). MicroRNA-329 serves a tumor suppressive role in colorectal cancer by directly targeting transforming growth factor beta-1. *Molecular Medicine Reports*, *16*(4), 3825–3832. https://doi.org/10.3892/mmr.2017.7077.

Li, Y., Xin, S., Wu, H., Xing, C., Duan, L., Sun, W., … Zhang, F. (2018). High expression of microRNA-31 and its host gene LOC554202 predict favorable outcomes in patients with colorectal cancer treated with oxaliplatin. *Oncology Reports*, *40*(3), 1706–1724. https://doi.org/10.3892/or.2018.6571.

Li, T., Jian, X., He, H., Lai, Q., Li, X., Deng, D., … Ding, Y. (2018). MiR-452 promotes an aggressive colorectal cancer phenotype by regulating a Wnt/β-catenin positive feedback loop. *Journal of Experimental & Clinical Cancer Research*, *37*(1), 238. https://doi.org/10.1186/s13046-018-0879-z.

Li, S., Zhu, K., Liu, L., Gu, J., Niu, H., & Guo, J. (2020). lncARSR sponges miR-34a-5p to promote colorectal cancer invasion and metastasis via hexokinase-1-mediated glycolysis. *Cancer Science*, *111*(10), 3938–3952. https://doi.org/10.1111/cas.14617.

Li, W., Xu, Y., Wang, X., Cao, G., Bu, W., Wang, X., … Tao, Q. (2020). CircCCT3 modulates vascular endothelial growth factor A and Wnt signaling to enhance colorectal cancer metastasis through sponging miR-613. *DNA and Cell Biology*, *39*(1), 118–125. https://doi.org/10.1089/dna.2019.5139.

Li, Y. P., Du, X. R., Zhang, R., & Yang, Q. (2021). Interleukin-18 promotes the antitumor ability of natural killer cells in colorectal cancer via the miR-574-3p/TGF-β1 axis. *Bioengineered*, *12*(1), 763–778. https://doi.org/10.1080/21655979.2021.1880717.

Liang, W. C., Fu, W. M., Wong, C. W., Wang, Y., Wang, W. M., Hu, G. X., … Waye, M. M. Y. (2015). The LncRNA H19 promotes epithelial to mesenchymal transition by functioning as MiRNA sponges in colorectal cancer. *Oncotarget*, *6*(26), 22513–22525. https://doi.org/10.18632/oncotarget.4154.

Liang, J., Zhou, W., Sakre, N., DeVecchio, J., Ferrandon, S., Ting, A. H., … Kalady, M. F. (2018). Epigenetically regulated miR-1247 functions as a novel tumour suppressor via MYCBP2 in methylator colon cancers. *British Journal of Cancer*, *119*(10), 1267–1277. https://doi.org/10.1038/s41416-018-0249-9.

Liang, Y., Hou, L., Li, L., Li, L., Zhu, L., Wang, Y., … Meng, X. (2020). Dichloroacetate restores colorectal cancer chemosensitivity through the p53/miR-149-3p/PDK2-mediated glucose metabolic pathway. *Oncogene*, *39*(2), 469–485. https://doi.org/10.1038/s41388-019-1035-8.

Liao, W. T., Ye, Y. P., Zhang, N. J., Li, T. T., Wang, S. Y., Cui, Y. M., … Ding, Y. Q. (2014). MicroRNA-30b functions as a tumour suppressor in human colorectal cancer by targeting KRAS, PIK3CD and BCL2. *Journal of Pathology*, *232*(4), 415–427. https://doi.org/10.1002/path.4309.

Lin, P. L., Wu, D. W., Huang, C. C., He, T. Y., Chou, M. C., Sheu, G. T., & Lee, H. (2014). MicroRNA-21 promotes tumour malignancy via increased nuclear translocation of β-catenin and predicts poor outcome in APC-mutated but not in APC-wild-type colorectal cancer. *Carcinogenesis*, *35*(10), 2175–2182. https://doi.org/10.1093/carcin/bgu110.

Lin, M., Zhang, Z., Gao, M., Yu, H., Sheng, H., & Huang, J. (2019). MicroRNA-193a-3p suppresses the colorectal cancer cell proliferation and progression through downregulating the PLAU expression. *Cancer Management and Research*, *11*, 5353–5363. https://doi.org/10.2147/CMAR.S208233.

Ling, H., Pickard, K., Ivan, C., Isella, C., Ikuo, M., Mitter, R., … Nicoloso, M.S. (2016). The clinical and biological significance of MIR-224 expression in colorectal cancer metastasis. *Gut*, *65*(6), 977–989. https://doi.org/10.1136/gutjnl-2015-309372.

Liu, F., & Di Wang, X. (2019). miR-150-5p represses TP53 tumor suppressor gene to promote proliferation of colon adenocarcinoma. *Scientific Reports*, *9*(1), 6740. https://doi.org/10.1038/s41598-019-43231-5.

Liu, G., & Li, B. (2019). Role of miRNA in transformation from normal tissue to colorectal adenoma and cancer. *Journal of Cancer Research and Therapeutics*, *15*(2), 278–285. https://doi.org/10.4103/jcrt.JCRT-135-18.

Liu, H., Du, L., Wen, Z., Yang, Y., Li, J., Wang, L., … Wang, C. (2013). Up-regulation of miR-182 expression in colorectal cancer tissues and its prognostic value. *International Journal of Colorectal Disease*, *28*(5), 697–703. https://doi.org/10.1007/s00384-013-1674-0.

Liu, G. H., Zhou, Z. G., Chen, R., Wang, M. J., Zhou, B., Li, Y., & Sun, X. F. (2013). Serum miR-21 and miR-92a as biomarkers in the diagnosis and prognosis of colorectal cancer. *Tumor Biology*, *34*(4), 2175–2181. https://doi.org/10.1007/s13277-013-0753-8.

Liu, Z., Wu, R., Li, G., Sun, P., Xu, Q., & Liu, Z. (2015). MiR-592 inhibited cell proliferation of human colorectal cancer cells by suppressing of CCND3 expression. *International Journal of Clinical and Experimental Medicine*, *8*(3), 3490–3497. http://www.ijcem.com/files/ijcem0004960.pdf.

Liu, T. P., Huang, C. C., Yeh, K. T., Ke, T. W., Wei, P. L., Yang, J. R., & Cheng, Y. W. (2016). Down-regulation of let-7a-5p predicts lymph node metastasis and prognosis in colorectal cancer: Implications for chemotherapy. *Surgical Oncology*, *25*(4), 429–434. https://doi.org/10.1016/j.suronc.2016.05.016.

Liu, D. R., Guan, Q. L., Gao, M. T., Jiang, L., & Kang, H. X. (2016). miR-1260b is a potential prognostic biomarker in colorectal cancer. *Medical Science Monitor*, *22*, 2417–2423. https://doi.org/10.12659/MSM.898733.

Liu, H., Yin, Y., Hu, Y., Feng, Y., Bian, Z., Yao, S., … Huang, Z. (2016). miR-139-5p sensitizes colorectal cancer cells to 5-fluorouracil by targeting NOTCH-1. *Pathology Research and Practice*, *212*(7), 643–649. https://doi.org/10.1016/j.prp.2016.04.011.

Liu, X., Wang, Y., & Zhao, J. (2017). MicroRNA-337 inhibits colorectal cancer progression by directly targeting KRAS and suppressing the AKT and ERK pathways. *Oncology Reports*, *38*(5), 3187–3196. https://doi.org/10.3892/or.2017.5997.

Liu, S., Xiao, Z., Ai, F., Liu, F., Chen, X., Cao, K., … Zhang, D. (2017). miR-142-5p promotes development of colorectal cancer through targeting SDHB and facilitating generation of aerobic glycolysis. *Biomedicine and Pharmacotherapy*, *92*, 1119–1127. https://doi.org/10.1016/j.biopha.2017.05.134.

Liu, X., Cui, L., & Hua, D. (2018). Long noncoding RNA XIST regulates mir-137-EZH2 axis to promote tumor metastasis in colorectal cancer. *Oncology Research*, *27*(1), 99–106. https://doi.org/10.3727/096504018X15195193936573.

Liu, Z. Y., Yang, L., & Chang, H. Y. (2018). Clinicopathologic and prognostic relevance of miR-1256 in colorectal cancer: A preliminary clinical study. *European Review for Medical and Pharmacological Sciences*, *22*(22), 7704–7709. https://doi.org/10.26355/eurrev-201811-16391.

Liu, H., Li, A., Sun, Z., Zhang, J., & Xu, H. (2020). Long non-coding RNA NEAT1 promotes colorectal cancer progression by regulating miR-205-5p/VEGFA axis. *Human Cell*, *33*(2), 386–396. https://doi.org/10.1007/s13577-019-00301-0.

Liu, Y. D., Zhuang, X. P., Cai, D. L., Cao, C., Gu, Q. S., Liu, X. N., … Yan, D. W. (2021). Let-7a regulates EV secretion and mitochondrial oxidative phosphorylation by targeting SNAP23 in colorectal cancer. *Journal of Experimental & Clinical Cancer Research*, *40*(1), 31. https://doi.org/10.1186/s13046-020-01813-6.

Liu, L., Yu, T., Jin, Y., Mai, W., Zhou, J., & Zhao, C. (2021). MicroRNA-15a carried by mesenchymal stem cell-derived extracellular vesicles inhibits the immune evasion of colorectal cancer cells by regulating the KDM4B/HOXC4/PD-L1 Axis. *Frontiers in Cell and Development Biology*, *9*. https://doi.org/10.3389/fcell.2021.629893, 629893.

Lodygin, D., Tarasov, V., Epanchintsev, A., Berking, C., Knyazeva, T., Körner, H., … Hermeking, H. (2008). Inactivation of miR-34a by aberrant CpG methylation in multiple types of cancer. *Cell Cycle*, *7*(16), 2591–2600. https://doi.org/10.4161/cc.7.16.6533.

Logan, C. Y., & Nusse, R. (2004). The Wnt signaling pathway in development and disease. *Annual Review of Cell and Developmental Biology*, *20*, 781–810. https://doi.org/10.1146/annurev.cellbio.20.010403.113126.

Long, L., Huang, G., Zhu, H., Guo, Y., Liu, Y., & Huo, J. (2013). Down-regulation of miR-138 promotes colorectal cancer metastasis via directly targeting TWIST2. *Journal of Translational Medicine*, *11*(1), 275. https://doi.org/10.1186/1479-5876-11-275.

Lu, J., Getz, G., Miska, E. A., Alvarez-Saavedra, E., Lamb, J., Peck, D., … Golub, T. R. (2005). MicroRNA expression profiles classify human cancers. *Nature*, *435*(7043), 834–838. https://doi.org/10.1038/nature03702.

Lu, J., Tan, M., & Cai, Q. (2015). The Warburg effect in tumor progression: Mitochondrial oxidative metabolism as an anti-metastasis mechanism. *Cancer Letters*, *356*(2), 156–164. https://doi.org/10.1016/j.canlet.2014.04.001.

Lun, W., Wu, X., Deng, Q., & Zhi, F. (2018). MiR-218 regulates epithelial-mesenchymal transition and angiogenesis in colorectal cancer via targeting CTGF. *Cancer Cell International*, *18*(1), 83. https://doi.org/10.1186/s12935-018-0575-2.

Luo, X., Stock, C., Burwinkel, B., & Brenner, H. (2013). Identification and evaluation of plasma MicroRNAs for early detection of colorectal cancer. *PLoS One*, *8*(5), e62880. https://doi.org/10.1371/journal.pone.0062880.

Lv, Z.-c., Fan, Y.-s., Chen, H.-b., & Zhao, D.-w. (2015). Investigation of microRNA-155 as a serum diagnostic and prognostic biomarker for colorectal cancer. *Tumor Biology*, *36*(3), 1619–1625. https://doi.org/10.1007/s13277-014-2760-9.

Lv, Z. C., Fan, Y. S., Chen, H. B., & Zhao, D. W. (2014). Investigation of microRNA-155 as a serum diagnostic and prognostic biomarker for colorectal cancer. *Tumor Biology*, *36*(3), 1619–1625. https://doi.org/10.1007/s13277-014-2760-9.

Ma, W., Yu, J., Qi, X., Liang, L., Zhang, Y., Ding, Y., … Ding, Y. (2015). Radiation-induced microRNA-622 causes radioresistance in colorectal cancer cells by down-regulating Rb. *Oncotarget*, *6*(18), 15984–15994. https://doi.org/10.18632/oncotarget.3762.

Ma, H., Pan, J. S., Jin, L. X., Wu, J., Ren, Y. D., Chen, P., … Han, J. (2016). MicroRNA-17~92 inhibits colorectal cancer progression by targeting angiogenesis. *Cancer Letters*, *376*(2), 293–302. https://doi.org/10.1016/j.canlet.2016.04.011.

Ma, Y., Qin, C., Li, L., Miao, R., Jing, C., & Cui, X. (2018). MicroRNA-21 promotes cell proliferation by targeting tumor suppressor TET1 in colorectal cancer. *International Journal of Clinical and Experimental Pathology*, *11*(3), 1439–1445.

Maierthaler, M., Benner, A., Hoffmeister, M., Surowy, H., Jansen, L., Knebel, P., … Burwinkel, B. (2017). Plasma miR-122 and miR-200 family are prognostic markers in colorectal cancer. *International Journal of Cancer*, *140*(1), 176–187. https://doi.org/10.1002/ijc.30433.

Maqbool, R., Lone, S. N., & Hussain, M. U. (2016). Post-transcriptional regulation of the tumor suppressor p53 by a novel miR-27a, with implications during hypoxia and tumorigenesis. *Biochemical Journal*, *473*(20), 3597–3610. https://doi.org/10.1042/BCJ20160359.

Marcuello, M., Vymetalkova, V., Neves, R. P. L., Duran-Sanchon, S., Vedeld, H. M., Tham, E., … Gironella, M. (2019). Circulating biomarkers for early detection and clinical management of colorectal cancer. *Molecular Aspects of Medicine*, *69*, 107–122. https://doi.org/10.1016/j.mam.2019.06.002.

Matsumura, T., Sugimachi, K., Iinuma, H., Takahashi, Y., Kurashige, J., Sawada, G., … Mimori, K. (2015). Exosomal microRNA in serum is a novel biomarker of recurrence in human colorectal cancer. *British Journal of Cancer*, *113*(2), 275–281. https://doi.org/10.1038/bjc.2015.201.

Menéndez, P., Padilla, D., Villarejo, P., Palomino, T., Nieto, P., Menéndez, J. M., & Rodríguez-Montes, J. A. (2013). Prognostic implications of serum microRNA-21 in colorectal cancer. *Journal of Surgical Oncology*, *108*(6), 369–373. https://doi.org/10.1002/jso.23415.

Meng, X., & Fu, R. (2018). MiR-206 regulates 5-FU resistance by targeting Bcl-2 in colon cancer cells. *Oncotargets and Therapy*, *11*, 1757–1765. https://doi.org/10.2147/OTT.S159093.

Meng, X., Wu, J., Pan, C., Wang, H., Ying, X., Zhou, Y., … Huang, W. (2013). Genetic and epigenetic down-regulation of microRNA-212 promotes colorectal tumor metastasis via dysregulation of MnSOD. *Gastroenterology*, *145*(2), 426–e6. https://doi.org/10.1053/j.gastro.2013.04.004.

Meng, Q., Chen, Y., Lian, B., Shang, Y., & Yang, H. (2018). miR-218 promotes apoptosis of SW1417 human colon cancer cells by targeting c-FLIP. *Oncology Reports*, *40*(2), 916–922. https://doi.org/10.3892/or.2018.6460.

Michael, M. Z., O'Connor, S. M., Van Holst Pellekaan, N. G., Young, G. P., & James, R. J. (2003). Reduced accumulation of specific microRNAs in colorectal neoplasia. *Molecular Cancer Research*, *1*(12), 882–891.

Miyoshi, J., Toden, S., Yoshida, K., Toiyama, Y., Alberts, S. R., Kusunoki, M., … Goel, A. (2017). MiR-139-5p as a novel serum biomarker for recurrence and metastasis in colorectal cancer. *Scientific Reports*, *7*, 43393. https://doi.org/10.1038/srep43393.

Mlcochova, J., Faltejskova, P., Nemecek, R., Svoboda, M., & Slaby, O. (2013). MicroRNAs targeting EGFR signalling pathway in colorectal cancer. *Journal of Cancer Research and Clinical Oncology*, *139*(10), 1615–1624. https://doi.org/10.1007/s00432-013-1470-9.

Mo, J. S., Alam, K. J. A., Kang, I. H., Park, W. C., Seo, G. S., Choi, S. C., … Chae, S. C. (2015). MicroRNA 196B regulates FAS-mediated apoptosis in colorectal cancer cells. *Oncotarget*, *6*(5), 2843–2855. https://doi.org/10.18632/oncotarget.3066.

Mo, J. S., Park, W. C., Choi, S. C., Yun, K. J., & Chae, S. C. (2019). MicroRNA 452 regulates cell proliferation, cell migration, and angiogenesis in colorectal cancer by suppressing VEGFA expression. *Cancers*, *11*(10), 1613. https://doi.org/10.3390/cancers11101613.

Mokutani, Y., Uemura, M., Munakata, K., Okuzaki, D., Haraguchi, N., Takahashi, H., … Yamamoto, H. (2016). Down-regulation of microRNA-132 is associated with poor prognosis of colorectal cancer. *Annals of Surgical Oncology*, *23*, 599–608. https://doi.org/10.1245/s10434-016-5133-3.

Mosakhani, N., Sarhadi, V. K., Borze, I., Karjalainen-Lindsberg, M. L., Sundström, J., Ristamäki, R., … Knuutila, S. (2012). MicroRNA profiling differentiates colorectal cancer according to KRAS status. *Genes, Chromosomes and Cancer*, *51*(1), 1–9. https://doi.org/10.1002/gcc.20925.

Movahed, Z. G., Rastegari-Pouyani, M., Hossein Mohammadi, M., & Mansouri, K. (2019). Cancer cells change their glucose metabolism to overcome increased ROS: One step from cancer cell to cancer stem cell? *Biomedicine and Pharmacotherapy*, *112*, 108690. https://doi.org/10.1016/j.biopha.2019.108690.

Mullany, L. E., Herrick, J. S., Sakoda, L. C., Samowitz, W., Stevens, J. R., Wolff, R. K., & Slattery, M. L. (2018). miRNA involvement in cell cycle regulation in colorectal cancer cases. *Genes & Cancer*, *9*(1–2), 860–872. https://doi.org/10.18632/genesandcancer.167.

Muzny, D. M., Bainbridge, M. N., Chang, K., Dinh, H. H., Drummond, J. A., Fowler, G., … Thomson, E. (2012). Comprehensive molecular characterization of human colon and rectal cancer. *Nature*, *487*(7407), 330–337. https://doi.org/10.1038/nature11252.

Naba, N. M., Tolay, N., Erman, B., & Sayi Yazgan, A. (2020). Doxorubicin inhibits miR-140 expression and upregulates PD-L1 expression in HCT116 cells, opposite to its effects on MDA-MB-231 cells. *Turkish Journal of Biology*, *44*(1), 15–23. https://doi.org/10.3906/biy-1909-12.

Nadeem, A., Aung, K. M., Ray, T., Alam, A., Persson, K., Pal, A., … Wai, S. N. (2021). Suppression of β-catenin signaling in colon carcinoma cells by a bacterial protein. *International Journal of Cancer*, *149*(2), 442–459. https://doi.org/10.1002/ijc.33562.

Nagano, Y., Toiyama, Y., Okugawa, Y., Imaoka, H., Fujikawa, H., Yasuda, H., … Kusunoki, M. (2016). MicroRNA-7 is associated with malignant potential and poor prognosis in human colorectal cancer. *Anticancer Research*, *36*(12), 6521–6526. https://doi.org/10.21873/anticanres.11253.

Nagel, R., Le Sage, C., Diosdado, B., Van Der Waal, M., Oude Vrielink, J. A. F., Bolijn, A., … Agami, R. (2008). Regulation of the adenomatous polyposis coli gene by the miR-135 family in colorectal cancer. *Cancer Research*, *68*(14), 5795–5802. https://doi.org/10.1158/0008-5472.CAN-08-0951.

Nakayama, M., & Oshima, M. (2019). Mutant p53 in colon cancer. *Journal of Molecular Cell Biology*, *11*(4), 267–276. https://doi.org/10.1093/jmcb/mjy075.

Neerincx, M., Sie, D. L. S., Van De Wiel, M. A., Van Grieken, N. C. T., Burggraaf, J. D., Dekker, H., … Verheul, H. M. W. (2015). MiR expression profiles of paired primary colorectal cancer and metastases by next-generation sequencing. *Oncogene*, *4*, e170. https://doi.org/10.1038/oncsis.2015.29.

Ng, E. K. O., Chong, W. W. S., Jin, H., Lam, E. K. Y., Shin, V. Y., Yu, J., … Sung, J. J. Y. (2009). Differential expression of microRNAs in plasma of patients with colorectal cancer: A potential marker for colorectal cancer screening. *Gut*, *58*(10), 1375–1381. https://doi.org/10.1136/gut.2008.167817.

Ng, E. K. O., Tsang, W. P., Ng, S. S. M., Jin, H. C., Yu, J., Li, J. J., … Sung, J. J. Y. (2009). MicroRNA-143 targets DNA methyltransferases 3A in colorectal cancer. *British Journal of Cancer*, *101*(4), 699–706. https://doi.org/10.1038/sj.bjc.6605195.

Ng, L., Wan, T. M. H., Man, J. H. W., Chow, A. K. M., Iyer, D., Chen, G., … Law, W. L. (2017). Identification of serum miR-139-3p as a non-invasive biomarker for colorectal cancer. *Oncotarget*, *8*(16), 27393–27400. https://doi.org/10.18632/oncotarget.16171.

Nguyen, H. T., & Duong, H. Q. (2018). The molecular characteristics of colorectal cancer: Implications for diagnosis and therapy (review). *Oncology Letters*, *16*(1), 9–18. https://doi.org/10.3892/ol.2018.8679.

Nijhuis, A., Thompson, H., Adam, J., Parker, A., Gammon, L., Lewis, A., … Silver, A. (2017). Remodelling of microRNAs in colorectal cancer by hypoxia alters metabolism profiles and 5-fluorouracil resistance. *Human Molecular Genetics*, *26*(8), 1552–1564. https://doi.org/10.1093/hmg/ddx059.

Nishida, N., Yokobori, T., Mimori, K., Sudo, T., Tanaka, F., Shibata, K., … Mori, M. (2011). MicroRNA miR-125b is a prognostic marker in human colorectal cancer. *International Journal of Oncology*, *38*(5), 1437–1443. https://doi.org/10.3892/ijo.2011.969.

Nishida, N., Yamashita, S., Mimori, K., Sudo, T., Tanaka, F., Shibata, K., … Mori, M. (2012). MicroRNA-10b is a prognostic indicator in colorectal cancer and confers resistance to the chemotherapeutic agent 5-fluorouracil in colorectal cancer cells. *Annals of Surgical Oncology*, *19*(9), 3065–3071. https://doi.org/10.1245/s10434-012-2246-1.

Nishimura, J., Handa, R., Yamamoto, H., Tanaka, F., Shibata, K., Mimori, K., … Mori, M. (2012). microRNA-181a is associated with poor prognosis of colorectal cancer. *Oncology Reports*, *28*(6), 2221–2226. https://doi.org/10.3892/or.2012.2059.

Oda, K., Matsuoka, Y., Funahashi, A., & Kitano, H. (2005). A comprehensive pathway map of epidermal growth factor receptor signaling. *Molecular Systems Biology*, *1*. https://doi.org/10.1038/msb4100014, 2005-0010.

Ota, T., Doi, K., Fujimoto, T., Tanaka, Y., Ogawa, M., Matsuzaki, H., … Tsunoda, T. (2012). KRAS up-regulates the expression of miR-181a, miR-200c and miR-210 in a three-dimensional-specific manner in DLD-1 colorectal cancer cells. *Anticancer Research*, *32*(6), 2271–2276. http://ar.iiarjournals.org/content/32/6/2271.full.pdf+html.

Pagliuca, A., Valvo, C., Fabrizi, E., Di Martino, S., Biffoni, M., Runci, D., … Ricci-Vitiani, L. (2013). Analysis of the combined action of miR-143 and miR-145 on oncogenic pathways in colorectal cancer cells reveals a coordinate program of gene repression. *Oncogene*, *32*(40), 4806–4813. https://doi.org/10.1038/onc.2012.495.

Pan, Z., & Miao, L. (2020). Serum microrna-592 serves as a novel potential biomarker for early diagnosis of colorectal cancer. *Oncology Letters*, *20*(2), 1119–1126. https://doi.org/10.3892/ol.2020.11682.

Peng, Z. Y., Gu, R. H., & Yan, B. (2019). Downregulation of exosome-encapsulated miR-548c-5p is associated with poor prognosis in colorectal cancer. *Journal of Cellular Biochemistry*, *120*(2), 1457–1463. https://doi.org/10.1002/jcb.27291.

Perez-Carbonell, L., Sinicrope, F. A., Alberts, S. R., Oberg, A. L., Balaguer, F., Castells, A., … Goel, A. (2015). MiR-320e is a novel prognostic biomarker in colorectal cancer. *British Journal of Cancer*, *113*(1), 83–90. https://doi.org/10.1038/bjc.2015.168.

Prossomariti, A., Piazzi, G., D'Angelo, L., Miccoli, S., Turchetti, D., Alquati, C., … Ricciardiello, L. (2018). MiR-155 is downregulated in familial adenomatous polyposis and modulates Wnt signaling by targeting AXIN1 and TCF4. *Molecular Cancer Research*, *16*(12), 1965–1976. https://doi.org/10.1158/1541-7786.MCR-18-0115.

Przygodzka, P., Papiewska-Pająk, I., Bogusz-Koziarska, H., Sochacka, E., Boncela, J., & Kowalska, M. A. (2019). Regulation of miRNAs by Snail during epithelial-to-mesenchymal transition in HT29 colon cancer cells. *Scientific Reports*, *9*(1). https://doi.org/10.1038/s41598-019-39200-7, 2165.

Pu, X. X., Huang, G. L., Guo, H. Q., Guo, C. C., Li, H., Ye, S., … Lin, T. Y. (2010). Circulating miR-221 directly amplified from plasma is a potential diagnostic and prognostic marker of colorectal cancer and is correlated with p53 expression. *Journal of Gastroenterology and Hepatology (Australia)*, *25*(10), 1674–1680. https://doi.org/10.1111/j.1440-1746.2010.06417.x.

Qian, J., Jiang, B., Li, M., Chen, J., & Fang, M. (2013). Prognostic significance of microRNA-16 expression in human colorectal cancer. *World Journal of Surgery*, *37*(12), 2944–2949. https://doi.org/10.1007/s00268-013-2205-4.

Qiu, Y., Yu, H., Shi, X., Xu, K., Tang, Q., Liang, B., … Yin, P. (2016). microRNA-497 inhibits invasion and metastasis of colorectal cancer cells by targeting vascular endothelial growth factor-A. *Cell Proliferation*, *49*(1), 69–78. https://doi.org/10.1111/cpr.12237.

Qu, A., Yang, Y., Zhang, X., Wang, W., Liu, Y., Zheng, G., … Wang, C. (2018). Development of a preoperative prediction nomogram for lymph node metastasis in colorectal cancer based on a novel serum miRNA signature and CT scans. *eBioMedicine*, *37*, 125–133. https://doi.org/10.1016/j.ebiom.2018.09.052.

Qu, X., Yang, L., Shi, Q., Wang, X., Wang, D., & Wu, G. (2018). Lidocaine inhibits proliferation and induces apoptosis in colorectal cancer cells by upregulating mir-520a-3p and targeting EGFR. *Pathology – Research and Practice*, *214*(12), 1974–1979. https://doi.org/10.1016/j.prp.2018.09.012.

Qu, X., Yang, L., Shi, Q., Wang, X., Wang, D., & Wu, G. (2018). Lidocaine inhibits proliferation and induces apoptosis in colorectal cancer cells by upregulating mir-520a-3p and targeting EGFR. *Pathology Research and Practice*, *214*(12), 1974–1979. https://doi.org/10.1016/j.prp.2018.09.012.

Radanova, M., Mihaylova, G., Mihaylova, Z., Ivanova, D., Tasinov, O., Nazifova-Tasinova, N., … Donev, I. (2021). Circulating mir-618 has prognostic significance in patients with metastatic colon cancer. *Current Oncology*, *28*(2), 1204–1215. https://doi.org/10.3390/curroncol28020116.

Ranković, B., Zidar, N., Žlajpah, M., & Boštjančič, E. (2019). Epithelial-mesenchymal transition-related microRNAs and their target genes in colorectal cancerogenesis. *Journal of Clinical Medicine*, *8*(10), 1603. https://doi.org/10.3390/jcm8101603.

Rapado-González, Ó., Majem, B., Álvarez-Castro, A., Díaz-Peña, R., Abalo, A., Suárez-Cabrera, L., … Suarez-Cunqueiro, M. M. (2019). A novel saliva-based miRNA signature for colorectal cancer diagnosis. *Journal of Clinical Medicine*, *8*(12), 2029. https://doi.org/10.3390/jcm8122029.

Rayego-Mateos, S., Rodrigues-Diez, R., Morgado-Pascual, J. L., Valentijn, F., Valdivielso, J. M., Goldschmeding, R., & Ruiz-Ortega, M. (2018). Role of epidermal growth factor receptor (EGFR) and its ligands in kidney inflammation and damage. *Mediators of Inflammation*, *2018*. https://doi.org/10.1155/2018/8739473, 8739473.

Reinhart, B. J., Slack, F. J., Basson, M., Pasquienelll, A. E., Bettlnger, J. C., Rougvle, A. E., … Ruvkun, G. (2000). The 21-nucleotide let-7 RNA regulates developmental timing in Caenorhabditis elegans. *Nature*, *403*(6772), 901–906. https://doi.org/10.1038/35002607.

Ren, C., Zhang, Z., Wang, S., Zhu, W., Zheng, P., & Wang, W. (2020). Circular RNA hsa_circ_0001178 facilitates the invasion and metastasis of colorectal cancer through upregulating ZEB1 via sponging multiple miRNAs. *Biological Chemistry*, *401*(4), 487–496. https://doi.org/10.1515/hsz-2019-0350.

Roh, M. S., Lee, H. W., Jung, S. B., Kim, K., Lee, E. H., Park, M.i., … Kim, M. S. (2018). Expression of miR-200c and its clinicopathological significance in patients with colorectal cancer. *Pathology Research and Practice*, *214*(3), 350–355. https://doi.org/10.1016/j.prp.2018.01.005.

Rokavec, M., Li, H., Jiang, L., & Hermeking, H. (2014). The p53/microRNA connection in gastrointestinal cancer. *Clinical and Experimental Gastroenterology*, *7*, 395. https://doi.org/10.2147/CEG.S43738.

Rokavec, M., Öner, M. G., Li, H., Jackstadt, R., Jiang, L., Lodygin, D., … Hermeking, H. (2014). IL-6R/STAT3/miR-34a feedback loop promotes EMT-mediated colorectal cancer invasion and metastasis. *Journal of Clinical Investigation*, *124*(4), 1853–1867. https://doi.org/10.1172/JCI73531.

Roman-Canal, B., Tarragona, J., Moiola, C. P., Gatius, S., Bonnin, S., Ruiz-Miró, M., … Colas, E. (2019). EV-associated miRNAs from peritoneal lavage as potential diagnostic biomarkers in colorectal cancer. *Journal of Translational Medicine*, *17*(1), 208. https://doi.org/10.1186/s12967-019-1954-8.

Rong, G., Yang, X., Wu, H., & Wu, Y. (2019). miR-150-504-519d inhibits the growth of human colorectal cancer cell line SW48 and downregulates c-FLIP receptor. *Journal of Cellular Biochemistry*, *120*(5), 7962–7969. https://doi.org/10.1002/jcb.28073.

Rosenfeld, N., Aharonov, R., Meiri, E., Rosenwald, S., Spector, Y., Zepeniuk, M., … Barshack, I. (2008). MicroRNAs accurately identify cancer tissue origin. *Nature Biotechnology*, *26*(4), 462–469. https://doi.org/10.1038/nbt1392.

Safa, A. R. (2012). c-FLIP, a master anti-apoptotic regulator. *Experimental Oncology*, *34*(3), 176–184. http://exp-oncology.com.ua/article/3429/c-flip-a-master-anti-apoptotic-regulator.

Saieva, L., Barreca, M. M., Zichittella, C., Prado, M. G., Tripodi, M., Alessandro, R., & Conigliaro, A. (2020). Hypoxia-induced MiR-675-5p supports β-catenin nuclear localization by regulating GSK3-β activity in colorectal cancer cell lines. *International Journal of Molecular Sciences*, *21*(11), 3832. https://doi.org/10.3390/ijms21113832.

Sazanov, A. A., Kiselyova, E. V., Zakharenko, A. A., Romanov, M. N., & Zaraysky, M. I. (2017). Plasma and saliva miR-21 expression in colorectal cancer patients. *Journal of Applied Genetics*, *58*(2), 231–237. https://doi.org/10.1007/s13353-016-0379-9.

Schetter, A. J., Leung, S. Y., Sohn, J. J., Zanetti, K. A., Bowman, E. D., Yanaihara, N., … Harris, C. C. (2008). MicroRNA expression profiles associated with prognosis and therapeutic outcome in colon adenocarcinoma. *JAMA: The Journal of the American Medical Association*, *299*(4), 425–436. https://doi.org/10.1001/jama.299.4.425.

Shang, J., Yang, F., Wang, Y., Wang, Y., Xue, G., Mei, Q., … Sun, S. (2014). MicroRNA-23a antisense enhances 5-fluorouracil chemosensitivity through apaf-1/caspase-9 apoptotic pathway in colorectal cancer cells. *Journal of Cellular Biochemistry*, *115*(4), 772–784. https://doi.org/10.1002/jcb.24721.

Shao, Y., Chen, T., Zheng, X., Yang, S., Xu, K., Chen, X., … Zheng, S. (2018). Colorectal cancer-derived small extracellular vesicles establish an inflammatory premetastatic niche in liver metastasis. *Carcinogenesis*, *39*(11), 1368–1379. https://doi.org/10.1093/carcin/bgy115.

Shaw, P. H. (1996). The role of p53 in cell cycle regulation. *Pathology Research and Practice*, *192*(7), 669–675. https://doi.org/10.1016/S0344-0338(96)80088-4.

Shen, X., Jiang, H., Chen, Z., Lu, B., Zhu, Y., Mao, J., … Chen, W. (2019). MicroRNA-145 inhibits cell migration and invasion in colorectal cancer by targeting TWIST. *OncoTargets and Therapy*, *12*, 10799–10809. https://doi.org/10.2147/OTT.S216147.

Shi, Y., & Liu, Z. (2020). Serum miR-92a-1 is a novel diagnostic biomarker for colorectal cancer. *Journal of Cellular and Molecular Medicine*, *24*(15), 8363–8367. https://doi.org/10.1111/jcmm.15282.

Shi, L., Hong, X., Ba, L., He, X., Xiong, Y., Ding, Q., … Peng, G. (2019). Long non-coding RNA ZNFX1-AS1 promotes the tumor progression and metastasis of colorectal cancer by acting as a competing endogenous RNA of miR-144 to regulate EZH2 expression. *Cell Death & Disease*, *10*(3), 150. https://doi.org/10.1038/s41419-019-1332-8.

Shi, Y., Zhuang, Y., Zhang, J., Chen, M., & Wu, S. (2021). Four circulating exosomal miRNAs as novel potential biomarkers for the early diagnosis of human colorectal cancer. *Tissue and Cell*, *70*, 101499. https://doi.org/10.1016/j.tice.2021.101499.

Shibuya, H., Iinuma, H., Shimada, R., Horiuchi, A., & Watanabe, T. (2011). Clinicopathological and prognostic value of microRNA-21 and microRNA-155 in colorectal cancer. *Oncology*, *79*(3–4), 313–320. https://doi.org/10.1159/000323283.

Slattery, M. L., Wolff, E., Hoffman, M. D., Pellatt, D. F., Milash, B., & Wolff, R. K. (2011). MicroRNAs and colon and rectal cancer: Differential expression by tumor location and subtype. *Genes, Chromosomes and Cancer*, *50*(3), 196–206. https://doi.org/10.1002/gcc.20844.

Slattery, M. L., Mullany, L. E., Wolff, R. K., Sakoda, L. C., Samowitz, W. S., & Herrick, J. S. (2019). The p53-signaling pathway and colorectal cancer: Interactions between downstream p53 target genes and miRNAs. *Genomics*, *111*(4), 762–771. https://doi.org/10.1016/j.ygeno.2018.05.006.

Smits, K. M., Paranjape, T., Nallur, S., Wouters, K. A. D., Weijenberg, M. P., Schouten, L. J., … Van Engeland, M. (2011). A let-7 microRNA SNP in the KRAS 3′UTR is prognostic in early-stage colorectal cancer. *Clinical Cancer Research*, *17*(24), 7723–7731. https://doi.org/10.1158/1078-0432.CCR-11-0990.

Snezhkina, A. V., Krasnov, G. S., Zhikrivetskaya, S. O., Karpova, I. Y., Fedorova, M. S., Nyushko, K. M., … Kudryavtseva, A. V. (2018). Overexpression of microRNAs miR-9, -98, and -199 correlates with the downregulation of HK2 expression in colorectal cancer. *Molekuliarnaia Biologiia*, *52*(2), 220–230. https://doi.org/10.7868/S0026898418020052.

Soheilifar, M. H., Grusch, M., Neghab, H. K., Amini, R., Maadi, H., Saidijam, M., & Wang, Z. (2020). Angioregulatory microRNAs in colorectal cancer. *Cancers*, *12*(1), 71. https://doi.org/10.3390/cancers12010071.

Sokolova, V., Fiorino, A., Zoni, E., Crippa, E., Reid, J. F., Gariboldi, M., & Pierotti, M. A. (2015). The effects of miR-20a on p21: Two mechanisms blocking growth arrest in TGF-β-responsive colon carcinoma. *Journal of Cellular Physiology*, *230*(12), 3105–3114. https://doi.org/10.1002/jcp.25051.

Song, B., Long, Y., Liu, D., Zhang, W., & Liu, C. (2017). MicroRNA-582 promotes tumorigenesis by targeting phosphatase and tensin homologue in colorectal cancer. *International Journal of Molecular Medicine*, *40*(3), 867–874. https://doi.org/10.3892/ijmm.2017.3059.

Sood, A., McClain, D., Maitra, R., Basu-Mallick, A., Seetharam, R., Kaubisch, A., … Goel, S. (2012). PTEN gene expression and mutations in the PIK3CA gene as predictors of clinical benefit to anti-epidermal growth factor receptor antibody therapy in patients with KRAS wild-type metastatic colorectal cancer. *Clinical Colorectal Cancer*, *11*(2), 143–150. https://doi.org/10.1016/j.clcc.2011.12.001.

Strillacci, A., Griffoni, C., Sansone, P., Paterini, P., Piazzi, G., Lazzarini, G., … Tomasi, V. (2009). MiR-101 downregulation is involved in cyclooxygenase-2 overexpression in human colon cancer cells. *Experimental Cell Research*, *315*(8), 1439–1447. https://doi.org/10.1016/j.yexcr.2008.12.010.

Subramanian, M., Rao, S. R., Thacker, P., Chatterjee, S., & Karunagaran, D. (2014). MIR-29b downregulates canonical Wnt signaling by suppressing coactivators of β-catenin in human colorectal cancer cells. *Journal of Cellular Biochemistry*, *115*(11), 1974–1984. https://doi.org/10.1002/jcb.24869.

Sun, Z. F., Zhang, Z., Liu, Z., Qiu, B., Liu, K., & Dong, G. (2014). MicroRNA-335 inhibits invasion and metastasis of colorectal cancer by targeting ZEB2. *Medical Oncology*, *31*(6). https://doi.org/10.1007/s12032-014-0982-8.

Sun, G., Cheng, Y. W., Lai, L., Huang, T. C., Wang, J., Wu, X., … Yen, Y. (2016). Signature miRNAs in colorectal cancers were revealed using a bias reduction small RNA deep sequencing protocol. *Oncotarget*, *7*(4), 3857–3872. https://doi.org/10.18632/oncotarget.6460.

Sun, Y., Yang, B., Lin, M., Yu, H., Chen, H., & Zhang, Z. (2019). Identification of serum miR-30a-5p as a diagnostic and prognostic biomarker in colorectal cancer. *Cancer Biomarkers*, *24*(3), 299–305. https://doi.org/10.3233/CBM-182129.

Sun, S., Hang, T., Zhang, B., Zhu, L., Wu, Y., Lv, X., … Yao, H. (2019). MiRNA-708 functions as a tumor suppressor in colorectal cancer by targeting ZEB1 through Akt/mTOR signaling pathway. *American Journal of Translational Research*, *11*(9), 5338–5356. http://www.ajtr.org/files/ajtr0100808.pdf.

Sundaram, P., Hultine, S., Smith, L. M., Dews, M., Fox, J. L., Biyashev, D., … Thomas-Tikhonenko, A. (2011). p53-responsive miR-194 inhibits thrombospondin-1 and promotes angiogenesis in colon cancers. *Cancer Research*, *71*(24), 7490–7501. https://doi.org/10.1158/0008-5472.CAN-11-1124.

Suto, T., Yokobori, T., Yajima, R., Morita, H., Fujii, T., Yamaguchi, S., … Kuwano, H. (2014). MicroRNA-7 expression in colorectal cancer is associated with poor prognosis and regulates cetuximab sensitivity via EGFR regulation. *Carcinogenesis*, *36*(3), 338–345. https://doi.org/10.1093/carcin/bgu242.

Takahashi, M., Cuatrecasas, M., Balaguer, F., Hur, K., Toiyama, Y., Castells, A., … Goel, A. (2012). The clinical significance of MiR-148a as a predictive biomarker in patients with advanced colorectal cancer. *PLoS One*, *7*(10), e46684. https://doi.org/10.1371/journal.pone.0046684.

Takano, Y., Masuda, T., Iinuma, H., Yamaguchi, R., Sato, K., Tobo, T., … Mimori, K. (2017). Circulating exosomal microRNA-203 is associated with metastasis possibly via inducing tumor-associated macrophages in colorectal cancer. *Oncotarget*, *8*(45), 78598–78613. https://doi.org/10.18632/oncotarget.20009.

Tan, H. Y., Zheng, Y. B., & Liu, J. (2018). Serum miR-199a as a potential diagnostic biomarker for detection of colorectal cancer. *European Review for Medical and Pharmacological Sciences*, *22*(24), 8657–8663. https://www.europeanreview.org/wp/wp-content/uploads/8657-8663-.pdf.

Tan, Y., Lin, J. J., Yang, X. F., Gou, D. M., Fu, L. W., Li, F. R., & Yu, X. F. (2019). A panel of three plasma microRNAs for colorectal cancer diagnosis. *Cancer Epidemiology*, *60*, 67–76. https://doi.org/10.1016/j.canep.2019.01.015.

Tang, W., Zhu, Y., Gao, J., Fu, J., Liu, C., Liu, Y., … Cui, L. (2014). MicroRNA-29a promotes colorectal cancer metastasis by regulating matrix metalloproteinase 2 and E-cadherin via KLF4. *British Journal of Cancer*, *110*(2), 450–458. https://doi.org/10.1038/bjc.2013.724.

Tang, Q., Zou, Z., Zou, C., Zhang, Q., Huang, R., Guan, X., … Wang, X. (2015). MicroRNA-93 suppress colorectal cancer development via Wnt/β-catenin pathway downregulating. *Tumor Biology*, *36*(3), 1701–1710. https://doi.org/10.1007/s13277-014-2771-6.

Tang, S., Fu, H., Xu, Q., & Zhou, Y. (2019). MiR-20a regulates sensitivity of colorectal cancer cells to NK cells by targeting MICA. *Bioscience Reports*, *39*(7). https://doi.org/10.1042/BSR20180695, BSR20180695.

To, K. K. W., Leung, W. W., & Ng, S. S. M. (2017). A novel miR-203-DNMT3b-ABCG2 regulatory pathway predisposing colorectal cancer development. *Molecular Carcinogenesis*, *56*(2), 464–477. https://doi.org/10.1002/mc.22508.

To, K. K. W., Tong, C. W. S., Mingxia, W., & Cho, W. C. S. (2018). MicroRNAs in the prognosis and therapy of colorectal cancer: From bench to bedside. *World Journal of Gastroenterology*, *24*(27), 2949–2973. https://doi.org/10.3748/wjg.v24.i27.2949.

Toiyama, Y., Takahashi, M., Hur, K., Nagasaka, T., Tanaka, K., Inoue, Y., … Goel, A. (2013). Serum miR-21 as a diagnostic and prognostic biomarker in colorectal cancer. *Journal of the National Cancer Institute*, *105*(12), 849–859. https://doi.org/10.1093/jnci/djt101.

Toiyama, Y., Hur, K., Tanaka, K., Inoue, Y., Kusunoki, M., Boland, C. R., & Goel, A. (2014). Serum miR-200c is a novel prognostic and metastasis-predictive biomarker in patients with colorectal cancer. *Annals of Surgery*, *259*(4), 735–743. https://doi.org/10.1097/SLA.0b013e3182a6909d.

Toyota, M., Suzuki, H., Sasaki, Y., Maruyama, R., Imai, K., Shinomura, Y., & Tokino, T. (2008). Epigenetic silencing of microRNA-34b/c and B-cell translocation gene 4 is associated with CpG island methylation in colorectal cancer. *Cancer Research*, *68*(11), 4123–4132. https://doi.org/10.1158/0008-5472.CAN-08-0325.

Tsang, W. P., & Kwok, T. T. (2009). The miR-18a* microRNA functions as a potential tumor suppressor by targeting on K-Ras. *Carcinogenesis*, *30*(6), 953–959. https://doi.org/10.1093/carcin/bgp094.

Tsang, W. P., Ng, E. K. O., Ng, S. S. M., Jin, H., Yu, J., Sung, J. J. Y., & Kwok, T. T. (2010). Oncofetal H19-derived miR-675 regulates tumor suppressor RB in human colorectal cancer. *Carcinogenesis*, *31*(3), 350–358. https://doi.org/10.1093/carcin/bgp181.

Tsuchida, A., Ohno, S., Wu, W., Borjigin, N., Fujita, K., Aoki, T., … Kuroda, M. (2011). miR-92 is a key oncogenic component of the miR-17-92 cluster in colon cancer. *Cancer Science*, *102*(12), 2264–2271. https://doi.org/10.1111/j.1349-7006.2011.02081.x.

Tsunoda, T., Takashima, Y., Yoshida, Y., Doi, K., Tanaka, Y., Fujimoto, T., … Shirasawa, S. (2011). Oncogenic KRAS regulates miR-200c and miR-221/222 in a 3D-specific manner in colorectal cancer cells. *Anticancer Research*, *31*(7), 2453–2459.

Tyagi, A., Sharma, A. K., & Damodaran, C. (2020). A review on notch signaling and colorectal cancer. *Cell*, *9*(6). https://doi.org/10.3390/cells9061549.

Vicinus, B., Rubie, C., Stegmaier, N., Frick, V. O., Kölsch, K., Kauffels, A., … Glanemann, M. (2013). MiR-21 and its target gene CCL20 are both highly overexpressed in the microenvironment of colorectal tumors: Significance of their regulation. *Oncology Reports*, *30*(3), 1285–1292. https://doi.org/10.3892/or.2013.2580.

Vogt, M., Munding, J., Grüner, M., Liffers, S. T., Verdoodt, B., Hauk, J., … Hermeking, H. (2011). Frequent concomitant inactivation of miR-34a and miR-34b/c by CpG methylation in colorectal, pancreatic, mammary, ovarian, urothelial, and renal cell carcinomas and soft tissue sarcomas. *Virchows Archiv*, *458*(3), 313–322. https://doi.org/10.1007/s00428-010-1030-5.

Volinia, S., Calin, G. A., Liu, C. G., Ambs, S., Cimmino, A., Petrocca, F., … Croce, C. M. (2006). A microRNA expression signature of human solid tumors defines cancer gene targets. *Proceedings of the National Academy of Sciences of the United States of America*, *103*(7), 2257–2261. https://doi.org/10.1073/pnas.0510565103.

Vychytilova-Faltejskova, P., Pesta, M., Radova, L., Liska, V., Daum, O., Kala, Z., … Slaby, O. (2016). Genome-wide microRNA expression profiling in primary tumors and matched liver metastasis of patients with colorectal cancer. *Cancer Genomics and Proteomics*, *13*(4), 311–316. http://cgp.iiarjournals.org/content/13/4/311.full.pdf.

Vychytilova-Faltejskova, P., Radova, L., Sachlova, M., Kosarova, Z., Slaba, K., Fabian, P., … Slaby, O. (2016). Serum-based microRNA signatures in early diagnosis and prognosis prediction of colon cancer. *Carcinogenesis*, *37*(10), 941–950. https://doi.org/10.1093/carcin/bgw078.

Vychytilova-Faltejskova, P., Merhautova, J., Machackova, T., Gutierrez-Garcia, I., Garcia-Solano, J., Radova, L., … Slaby, O. (2017). MiR-215-5p is a tumor suppressor in colorectal cancer targeting EGFR ligand epiregulin and its transcriptional inducer HOXB9. *Oncogenesis*, *6*(11). https://doi.org/10.1038/s41389-017-0006-6, 399.

Wan, T. M. H., Lam, C. S. C., Ng, L., Chow, A. K. M., Wong, S. K. M., Li, H. S., … Pang, R. W. C. (2014). The clinicopathological significance of miR-133a in colorectal cancer. *Disease Markers*, *2014*. https://doi.org/10.1155/2014/919283, 919283.

Wang, H. (2020). MicroRNAs and apoptosis in colorectal cancer. *International Journal of Molecular Sciences*, *21*(15), 1–15. https://doi.org/10.3390/ijms21155353.

Wang, N., & Liu, W. (2018). Increased expression of miR-552 acts as a potential predictor biomarker for poor prognosis of colorectal cancer. *European Review for Medical and Pharmacological Sciences*, *22*(2), 412–416. http://www.europeanreview.org/.

Wang, L., & Yu, P. (2016). MiR-300 promotes proliferation and EMT-mediated colorectal cancer migration and invasion by targeting p53. *Oncology Reports*, *36*(6), 3225–3232. https://doi.org/10.3892/or.2016.5193.

Wang, X., Lam, E. K. Y., Zhang, J., Jin, H., & Sung, J. J. Y. (2009). MicroRNA-122a functions as a novel tumor suppressor downstream of adenomatous polyposis coli in gastrointestinal cancers. *Biochemical and Biophysical Research Communications*, *387*(2), 376–380. https://doi.org/10.1016/j.bbrc.2009.07.034.

Wang, Z., Li, Y., Kong, D., Ahmad, A., Banerjee, S., & Sarkar, F. H. (2010). Cross-talk between miRNA and Notch signaling pathways in tumor development and progression. *Cancer Letters*, *292*(2), 141–148. https://doi.org/10.1016/j.canlet.2009.11.012.

Wang, Q., Huang, Z., Ni, S., Xiao, X., Xu, Q., Wang, L., … Du, X. (2012). Plasma miR-601 and miR-760 are novel biomarkers for the early detection of colorectal cancer. *PLoS One*, *7*(9). https://doi.org/10.1371/journal.pone.0044398.

Wang, F., Ma, Y. L., Zhang, P., Shen, T. Y., Shi, C. Z., Yang, Y. Z., … Qin, H. L. (2013). SP1 mediates the link between methylation of the tumour suppressor miR-149 and outcome in colorectal cancer. *Journal of Pathology*, *229*(1), 12–24. https://doi.org/10.1002/path.4078.

Wang, J., Huang, S. K., Zhao, M., Yang, M., Zhong, J. L., Gu, Y. Y., … Huang, C. Z. (2014). Identification of a circulating microRNA signature for colorectal cancer detection. *PLoS One*, *9*(4). https://doi.org/10.1371/journal.pone.0087451.

Wang, S., Yang, M. H., Wang, X. Y., Lin, J., & Ding, Y. Q. (2014). Increased expression of miRNA-182 in colorectal carcinoma: An independent and tissue-speciic prognostic factor. *International Journal of Clinical and Experimental Pathology*, *7*(6), 3498–3503. http://www.ijcep.com/files/ij-cep0000358.pdf.

Wang, Y., Tang, Q., Li, M., Jiang, S., & Wang, X. (2014). MicroRNA-375 inhibits colorectal cancer growth by targeting PIK3CA. *Biochemical and Biophysical Research Communications*, *444*(2), 199–204. https://doi.org/10.1016/j.bbrc.2014.01.028.

Wang, L. L., Du, L. T., Li, J., Liu, Y. M., Qu, A. L., Yang, Y. M., … Wang, C. X. (2014). Decreased expression of miR-133a correlates with poor prognosis in colorectal cancer patients. *World Journal of Gastroenterology*, *20*(32), 11340–11346. https://doi.org/10.3748/wjg.v20.i32.11340.

Wang, J., Huang, S.-k., Zhao, M., Yang, M., Zhong, J.-l., Gu, Y.-y., … Huang, C.-z. (2014). Identification of a circulating microRNA signature for colorectal cancer detection. *PLoS One*, *9*(4), e87451. https://doi.org/10.1371/journal.pone.0087451.

Wang, J., Wang, H., Liu, A., Fang, C., Hao, J., & Wang, Z. (2015). Lactate dehydrogenase A negatively regulated by miRNAs promotes aerobic glycolysis and is increased in colorectal cancer. *Oncotarget*, *6*(23), 19456–19468. https://doi.org/10.18632/oncotarget.3318.

Wang, X. W., Xi, X. Q., Wu, J., Wan, Y. Y., Hui, H. X., & Cao, X. F. (2015). MicroRNA-206 attenuates tumor proliferation and migration involving the downregulation of NOTCH3 in colorectal cancer. *Oncology Reports*, *33*(3), 1402–1410. https://doi.org/10.3892/or.2015.3731.

Wang, H., Cao, F., Li, X., Miao, H., Jifu, E., Xing, J., & Fu, C. G. (2015). miR-320b suppresses cell proliferation by targeting c-Myc in human colorectal cancer cells. *BMC Cancer*, *15*(1). https://doi.org/10.1186/s12885-015-1728-5.

Wang, Z., Ma, B., Ji, X., Deng, Y., Zhang, T., Zhang, X., … Zhao, R. (2015). MicroRNA-378-5p suppresses cell proliferation and induces apoptosis in colorectal cancer cells by targeting BRAF. *Cancer Cell International*, *15*(1), 40. https://doi.org/10.1186/s12935-015-0192-2.

Wang, S., Xiang, J., Li, Z., Lu, S., Hu, J., Gao, X., … Zhu, H. (2015). A plasma microRNA panel for early detection of colorectal cancer. *International Journal of Cancer*, *136*(1), 152–161. https://doi.org/10.1002/ijc.28136.

Wang, J., Xu, J., Fu, J., Yuan, D., Guo, F., Zhou, C., & Shao, C. (2016). MIR-29a regulates radiosensitivity in human intestinal cells by targeting PTEN gene. *Radiation Research*, *186*(3), 292–301. https://doi.org/10.1667/RR14428.1.

Wang, Z. S., Zhong, M., Bian, Y. H., Mu, Y. F., Qin, S. L., Yu, M. H., & Qin, J. (2016). MicroRNA-187 inhibits tumor growth and invasion by directly targeting CD276 in colorectal cancer. *Oncotarget*, *7*(28), 44266–44276. https://doi.org/10.18632/oncotarget.10023.

Wang, S., Zeng, Y., Zhou, J. M., Nie, S. L., Peng, Q., Gong, J., & Huo, J. R. (2016). MicroRNA-1246 promotes growth and metastasis of colorectal cancer cells involving CCNG2 reduction. *Molecular Medicine Reports*, *13*(1), 273–280. https://doi.org/10.3892/mmr.2015.4557.

Wang, Y. X., Chen, Y. R., Liu, S. S., Ye, Y. P., Jiao, H. L., Wang, S. Y., … Ding, Y. Q. (2016). MiR-384 inhibits human colorectal cancer metastasis by targeting KRAS and CDC42. *Oncotarget*, *7*(51), 84826–84838. https://doi.org/10.18632/oncotarget.12704.

Wang, Y., Xue, J., Kuang, H., Zhou, X., Liao, L., & Yin, F. (2017). MicroRNA-1297 inhibits the growth and metastasis of colorectal cancer by suppressing cyclin D2 expression. *DNA and Cell Biology*, *36*(11), 991–999. https://doi.org/10.1089/dna.2017.3829.

Wang, S. J., Cao, Y. F., Yang, Z. Q., Jiang, Z. Y., Cai, B., Guo, J., … Gao, F. (2017). MicroRNA-93-5p increases multidrug resistance in human colorectal carcinoma cells by downregulating cyclin dependent kinase inhibitor 1A gene expression. *Oncology Letters*, *13*(2), 722–730. https://doi.org/10.3892/ol.2016.5463.

Wang, W., Qu, A., Liu, W., Liu, Y., Zheng, G., Du, L., … Chen, X. (2017). Circulating miR-210 as a diagnostic and prognostic biomarker for colorectal cancer. *European Journal of Cancer Care*, *26*, e12448. https://doi.org/10.1111/ecc.12448.

Wang, D., Li, Y., Zhang, C., Li, X., & Yu, J. (2018). MiR-216a-3p inhibits colorectal cancer cell proliferation through direct targeting COX-2 and ALOX5. *Journal of Cellular Biochemistry*, *119*(2), 1755–1766. https://doi.org/10.1002/jcb.26336.

Wang, Y. G., He, Q., Guo, S. Q., & Shi, Z. Z. (2019). Reduced serum MIR-98 predicts unfavorable clinical outcome of colorectal cancer. *European Review for Medical and Pharmacological Sciences*, *23*(19), 8345–8353. https://doi.org/10.26355/eurrev_201910_19145.

Wang, L., Zhao, Y., Xu, M., Zhou, F., & Yan, J. (2019). Serum miR-1301-3p, miR-335-5p, miR-28-5p, and their target B7-H3 may serve as novel biomarkers for colorectal cancer. *Journal of B.U.ON.*, *24*(3), 1120–1127. https://jbuon.com/archive/24-3-1120.pdf.

Wang, T., Ma, L., Li, W., Ding, L., & Gao, H. (2020). MicroRNA-498 reduces the proliferation and invasion of colorectal cancer cells via targeting Bcl-2. *FEBS Open Bio*, *10*(1), 168–175. https://doi.org/10.1002/2211-5463.12767.

Wang, D., Wang, X., Si, M., Yang, J., Sun, S., Wu, H., … Yu, X. (2020). Exosome-encapsulated miRNAs contribute to CXCL12/CXCR4-induced liver metastasis of colorectal cancer by enhancing M2 polarization of macrophages. *Cancer Letters*, *474*, 36–52. https://doi.org/10.1016/j.canlet.2020.01.005.

Wei, Z., Cui, L., Mei, Z., Liu, M., & Zhang, D. (2014). MiR-181a mediates metabolic shift in colon cancer cells via the PTEN/AKT pathway. *FEBS Letters*, *588*(9), 1773–1779. https://doi.org/10.1016/j.febslet.2014.03.037.

Wei, W. T., Nian, X. X., Wang, S. Y., Jiao, H. L., Wang, Y. X., Xiao, Z. Y., … Liao, W. T. (2017). miR-422a inhibits cell proliferation in colorectal cancer by targeting AKT1 and MAPK1. *Cancer Cell International*, *17*(1). https://doi.org/10.1186/s12935-017-0461-3, 91.

Wei, L., Chen, Z., Cheng, N., Li, X., Chen, J., Wu, D., … Wu, X. (2020). MicroRNA-126 inhibit viability of colorectal cancer cell by repressing mTOR induced apoptosis and autophagy. *Oncotargets and Therapy*, *13*, 2459–2468. https://doi.org/10.2147/OTT.S238348.

Weissmann-Brenner, A., Kushnir, M., Yanai, G. L., Aharonov, R., Gibori, H., Purim, O., … Brenner, B. (2012). Tumor microRNA-29a expression and the risk of recurrence in stage II colon cancer. *International Journal of Oncology*, *40*(6), 2097–2103. https://doi.org/10.3892/ijo.2012.1403.

Wightman, B., Ha, I., & Ruvkun, G. (1993). Posttranscriptional regulation of the heterochronic gene lin-14 by lin-4 mediates temporal pattern formation in C. elegans. *Cell*, *75*(5), 855–862. https://doi.org/10.1016/0092-8674(93)90530-4.

Wu, J., Qian, J., Li, C., Kwok, L., Cheng, F., Liu, P., … Lü, J. (2010). miR-129 regulates cell proliferation by downregulating Cdk6 expression. *Cell Cycle*, *9*(9), 1809–1818. https://doi.org/10.4161/cc.9.9.11535.

Wu, W., Yang, J., Feng, X., Wang, H., Ye, S., Yang, P., … Zhou, Y. (2013). MicroRNA-32 (miR-32) regulates phosphatase and tensin homologue (PTEN) expression and promotes growth, migration, and invasion in colorectal carcinoma cells. *Molecular Cancer*, *12*(1), 30. https://doi.org/10.1186/1476-4598-12-30.

Wu, C. W., Ng, S. C., Dong, Y., Tian, L., Ng, S. S. M., Leung, W. W., … Yu, J. (2014). Identification of microrna-135b in stool as a potential noninvasive biomarker for colorectal cancer and adenoma. *Clinical Cancer Research*, *20*(11), 2994–3002. https://doi.org/10.1158/1078-0432.CCR-13-1750.

Wu, Y., Song, Y., Xiong, Y., Wang, X., Xu, K., Han, B., … Zhou, L. (2017). MicroRNA-21 (Mir-21) promotes cell growth and invasion by repressing tumor suppressor PTEN in colorectal cancer. *Cellular Physiology and Biochemistry*, *43*(3), 945–958. https://doi.org/10.1159/000481648.

Wu, Q.-B., Sheng, X., Zhang, N., Yang, M.-W., & Wang, F. (2018). Role of microRNAs in the resistance of colorectal cancer to chemoradiotherapy. *Molecular and Clinical Oncology*, *8*(4), 523–527. https://doi.org/10.3892/mco.2018.1578.

Wu, Q. B., Chen, J., Zhu, J. W., Yin, X., You, H. Y., Lin, Y. R., & Zhu, H. Q. (2018). MicroRNA-125 inhibits RKO colorectal cancer cell growth by targeting VEGF. *International Journal of Molecular Medicine*, *42*(1), 665–673. https://doi.org/10.3892/ijmm.2018.3632.

Wu, F., Xing, T., Gao, X., & Liu, F. (2019). miR-501-3p promotes colorectal cancer progression via activation of Wnt/ß-catenin signaling. *International Journal of Oncology*, *55*(3), 671–683. https://doi.org/10.3892/ijo.2019.4852.

Wu, T., Mo, Y., & Wu, C. (2020). Prognostic values of CEA, CA19-9, and CA72-4 in patients with stages I-III colorectal cancer. *International Journal of Clinical and Experimental Pathology*, *13*, 1608–1614.

Xi, Y., Nakajima, G., Gavin, E., Morris, C. G., Kudo, K., Hayashi, K., & Ju, J. (2007). Systematic analysis of microRNA expression of RNA extracted from fresh frozen and formalin-fixed paraffin-embedded samples. *RNA*, *13*(10), 1668–1674. https://doi.org/10.1261/rna.642907.

Xiao, G., Tang, H., Wei, W., Li, J., Ji, L., & Ge, J. (2014). Aberrant expression of microRNA-15a and microRNA-16 synergistically associates with tumor progression and prognosis in patients with colorectal cancer. *Gastroenterology Research and Practice*, *2014*. https://doi.org/10.1155/2014/364549, 364549.

Xin, C., Zhang, H., & Liu, Z. (2014). MiR-154 suppresses colorectal cancer cell growth and motility by targeting TLR2. *Molecular and Cellular Biochemistry*, *387*(1–2), 271–277. https://doi.org/10.1007/s11010-013-1892-3.

Xing, Y., Jing, H., Zhang, Y., Suo, J., & Qian, M. (2020). MicroRNA-141-3p affected proliferation, chemosensitivity, migration and invasion of colorectal cancer cells by targeting EGFR. *International Journal of Biochemistry and Cell Biology*, *118*. https://doi.org/10.1016/j.biocel.2019.105643, 105643.

Xiong, Y., Zhang, Y. Y., Wu, Y. Y., Wang, X. D., Wan, L. H., Li, L., & Zhou, L. M. (2014). Correlation of over-expressions of miR-21 and Notch-1 in human colorectal cancer with clinical stages. *Life Sciences*, *106*(1–2), 19–24. https://doi.org/10.1016/j.lfs.2014.04.017.

Xiong, W., Yang, S., Zhang, W., Chen, Y., & Wang, F. (2019). MiR-761 inhibits colorectal cancer cell proliferation and invasion through targeting HDAC1. *Pharmazie*, *74*(2), 111–114. https://doi.org/10.1691/ph.2019.8756.

Xu, X. M., Qian, J. C., Deng, Z. L., Cai, Z., Tang, T., Wang, P., … Cai, J. P. (2012). Expression of miR-21, miR-31, miR-96 and miR-135b is correlated with the clinical parameters of colorectal cancer. *Oncology Letters*, *4*(2), 339–345. https://doi.org/10.3892/ol.2012.714.

Xu, X., Chen, R., Li, Z., Huang, N., Wu, X., Li, S., … Wu, S. (2015). MicroRNA-490-3p inhibits colorectal cancer metastasis by targeting TGFβR1. *BMC Cancer*, *15*(1), 1023. https://doi.org/10.1186/s12885-015-2032-0.

Xu, X. W., Zheng, B. A., Hu, Z. M., Qian, Z. Y., Huang, C. J., Liu, X. Q., & Wu, W. D. (2017). Circular RNA hsa_circ_000984 promotes colon cancer growth and metastasis by sponging miR-106b. *Oncotarget*, *8*(53), 91674–91683. https://doi.org/10.18632/oncotarget.21748.

Xu, M., Chen, X., Lin, K., Zeng, K., Liu, X., Xu, X., … Wang, S. (2019). LncRNA SNHG6 regulates EZH2 expression by sponging miR-26a/b and miR-214 in colorectal cancer. *Journal of Hematology & Oncology*, *12*(1), 3. https://doi.org/10.1186/s13045-018-0690-5.

Xu, J., Wu, G., Zhao, Y., Han, Y., Zhang, S., Li, C., & Zhang, J. (2020). Long noncoding RNA DSCAM-AS1 facilitates colorectal cancer cell proliferation and migration via miR-137/notch1 axis. *Journal of Cancer*, *11*(22), 6623–6632. https://doi.org/10.7150/jca.46562.

Xue, Q., Sun, K., Deng, H. J., Lei, S. T., Dong, J. Q., & Li, G. X. (2013). Anti-miRNA-221 sensitizes human colorectal carcinoma cells to radiation by upregulating PTEN. *World Journal of Gastroenterology*, *19*(48), 9307–9317. https://doi.org/10.3748/wjg.v19.i48.9307.

Yan, H. L., Xue, G., Mei, Q., Wang, Y. Z., Ding, F. X., Liu, M. F., … Sun, S. H. (2009). Repression of the miR-17-92 cluster by p53 has an important function in hypoxia-induced apoptosis. *EMBO Journal*, *28*(18), 2719–2732. https://doi.org/10.1038/emboj.2009.214.

Yan, S., Dang, G., Zhang, X., Jin, C., Qin, L., Wang, Y., … Duan, Q. (2017). Downregulation of circulating exosomal miR-638 predicts poor prognosis in colon cancer patients. *Oncotarget*, *8*(42), 72220–72226. https://doi.org/10.18632/oncotarget.19689.

Yan, S., Liu, G., Jin, C., Wang, Z., Duan, Q., Xu, J., & Xu, D. (2018). MicroRNA-6869-5p acts as a tumor suppressor via targeting TLR4/NF-κB signaling pathway in colorectal cancer. *Journal of Cellular Physiology*, *233*(9), 6660–6668. https://doi.org/10.1002/jcp.26316.

Yan, S., Jiang, Y., Liang, C., Cheng, M., Jin, C., Duan, Q., … Jin, P. (2018). Exosomal miR-6803-5p as potential diagnostic and prognostic marker in colorectal cancer. *Journal of Cellular Biochemistry*, *119*(5), 4113–4119. https://doi.org/10.1002/jcb.26609.

Yan, F., Ma, Y., Liu, L., Li, L., Deng, J., & Sun, J. H. (2020). Long noncoding RNA HOXD-AS1 promotes the proliferation, migration, and invasion of colorectal cancer via the miR-526b-3p/CCND1 axis. *Journal of Surgical Research*, *255*, 525–535. https://doi.org/10.1016/j.jss.2020.05.078.

Yan, Z., Bi, M., Zhang, Q., Song, Y., & Hong, S. (2020). LncRNA TUG1 promotes the progression of colorectal cancer via the miR-138-5p/ZEB2 axis. *Bioscience Reports*, *40*(6). https://doi.org/10.1042/BSR20201025, BSR20201025.

Yang, I. P., Tsai, H. L., Hou, M. F., Chen, K. C., Tsai, P. C., Huang, S. W., … Juo, S. H. H. (2012). MicroRNA-93 inhibits tumor growth and early relapse of human colorectal cancer by affecting genes involved in the cell cycle. *Carcinogenesis*, *33*(8), 1522–1530. https://doi.org/10.1093/carcin/bgs166.

Yang, S., Wu, B., Sun, H., Ji, F., Sun, T., Zhao, Y., & Zhou, D. (2016). Interrupted E2F1-miR-34c-SCF negative feedback loop by hyper-methylation promotes colorectal cancer cell proliferation. *Bioscience Reports*, *36*(1), e00293. https://doi.org/10.1042/BSR20150290.

Yang, I.-P., Tsai, H.-L., Miao, Z.-F., Huang, C.-W., Kuo, C.-H., Wu, J.-Y., … Wang, J.-Y. (2016). Development of a deregulating microRNA panel for the detection of early relapse in postoperative colorectal cancer patients. *Journal of Translational Medicine*, *14*(1), 108. https://doi.org/10.1186/s12967-016-0856-2.

Yang, I. P., Tsai, H. L., Miao, Z. F., Huang, C. W., Kuo, C. H., Wu, J. Y., … Wang, J. Y. (2016). Development of a deregulating microRNA panel for the detection of early relapse in postoperative colorectal cancer patients. *Journal of Translational Medicine*, *14*(1), 108. https://doi.org/10.1186/s12967-016-0856-2.

Yang, I. P., Tsai, H. L., Huang, C. W., Lu, C. Y., Miao, Z. F., Chang, S. F., … Wang, J. Y. (2016). High blood sugar levels significantly impact the prognosis of colorectal cancer patients through down-regulation of microRNA-16 by targeting Myb and VEGFR2. *Oncotarget*, *7*(14), 18837–18850. https://doi.org/10.18632/oncotarget.7719.

Yang, Y., Zhang, J., Chen, X., Xu, X., Cao, G., Li, H., & Wu, T. (2018). LncRNA FTX sponges miR-215 and inhibits phosphorylation of vimentin for promoting colorectal cancer progression. *Gene Therapy*, *25*(5), 321–330. https://doi.org/10.1038/s41434-018-0026-7.

Yang, L. P., Lin, Q., & Mu, X. L. (2019). MicroRNA-155 and FOXP3 jointly inhibit the migration and invasion of colorectal cancer cells by regulating ZEB2 expression. *European Review for Medical and Pharmacological Sciences*, *23*(14), 6131–6138. https://doi.org/10.26355/eurrev_201907_18426.

Yanlei, M., Zhang, P., Wang, F., Zhang, H., Yang, J., Peng, J., … Qin, H. (2012). miR-150 as a potential biomarker associated with prognosis and therapeutic outcome in colorectal cancer. *Gut*, *61*(10), 1447–1453. https://doi.org/10.1136/gutjnl-2011-301122.

Yao, Z., Xu, D., Li, Z., Yao, J., Pan, Z., & Chen, J. (2020). Long intergenic non-coding RNA LINC00662 contributes to malignant growth of colorectal cancer cells by upregulating c-myc via sponging microRNA-145. *Bioscience Reports*. https://doi.org/10.1042/BSR20200011, BSR20200011.

Yau, T. O., Wu, C. W., Dong, Y., Tang, C. M., Ng, S. S. M., Chan, F. K. L., … Yu, J. (2014). microRNA-221 and microRNA-18a identification in stool as potential biomarkers for the non-invasive diagnosis of colorectal carcinoma. *British Journal of Cancer*, *111*(9), 1765–1771. https://doi.org/10.1038/bjc.2014.484.

Yau, T. O., Wu, C. W., Tang, C. M., Chen, Y., Fang, J., Dong, Y., … Yu, J. (2016). MicroRNA-20a in human faeces as a non-invasive biomarker for colorectal cancer. *Oncotarget*, *7*(2), 1559–1568. https://doi.org/10.18632/oncotarget.6403.

Ye, Y. P., Wu, P., Gu, C.c., Deng, D.l., Jiao, H. L., Li, T. T., … Ding, Y. Q. (2016). MiR-450b-5p induced by oncogenic KRAS is required for colorectal cancer progression. *Oncotarget*, *7*(38), 61312–61324. https://doi.org/10.18632/oncotarget.11016.

Ye, J., Wang, Z., Zhao, J., Chen, W., Wu, D., Wu, P., & Huang, J. (2017). MicroRNA-141 inhibits tumor growth and minimizes therapy resistance in colorectal cancer. *Molecular Medicine Reports*, *15*(3), 1037–1042. https://doi.org/10.3892/mmr.2017.6135.

Yi, R., Li, Y., Wang, F., Gu, J., Isaji, T., Li, J., … Zhao, Y. (2016). Transforming growth factor (TGF) β1 acted through miR-130b to increase integrin α5 to promote migration of colorectal cancer cells. *Tumor Biology*, *37*(8), 10763–10773. https://doi.org/10.1007/s13277-016-4965-6.

Yi, M., Xu, L., Jiao, Y., Luo, S., Li, A., & Wu, K. (2020). The role of cancer-derived microRNAs in cancer immune escape. *Journal of Hematology & Oncology*, *13*(1), 25. https://doi.org/10.1186/s13045-020-00848-8.

Yin, Y., Song, M., Gu, B., Qi, X., Hu, Y., Feng, Y., … Huang, Z. (2016). Systematic analysis of key miRNAs and related signaling pathways in colorectal tumorigenesis. *Gene*, *578*(2), 177–184. https://doi.org/10.1016/j.gene.2015.12.015.

Yin, J., Shen, X., Li, M., Ni, F., Xu, L., & Lu, H. (2018). MiR-329 regulates the sensitivity of 5-FU in chemotherapyof colorectal cancer by targeting E2F1. *Oncology Letters*, *16*(3), 3587–3592. https://doi.org/10.3892/ol.2018.9121.

Yong, W., Zhuoqi, X., Baocheng, W., Dongsheng, Z., Chuan, Z., & Yueming, S. (2018). Hsa_circ_0071589 promotes carcinogenesis via the miR-600/EZH2 axis in colorectal cancer. *Biomedicine and Pharmacotherapy*, *102*, 1188–1194. https://doi.org/10.1016/j.biopha.2018.03.085.

Yu, H., Gao, G., Jiang, L., Guo, L., Lin, M., Jiao, X., … Huang, J. (2013). Decreased expression of miR-218 is associated with poor prognosis in patients with colorectal cancer. *International Journal of Clinical and Experimental Pathology*, *6*(12), 2904–2911. http://www.ijcep.com/files/ijcep1310038.pdf.

Yu, F. Y., Tu, Y., Deng, Y., Guo, C., Ning, J., Zhu, Y., … Ye, H. (2016). MiR-4500 is epigenetically downregulated in colorectal cancer and functions as a novel tumor suppressor by regulating HMGA2. *Cancer Biology and Therapy*, *17*(11), 1149–1157. https://doi.org/10.1080/15384047.2016.1235661.

Yuan, D., Li, K., Zhu, K., Yan, R., & Dang, C. (2015). Plasma miR-183 predicts recurrence and prognosis in patients with colorectal cancer. *Cancer Biology and Therapy*, *16*(2), 268–275. https://doi.org/10.1080/15384047.2014.1002327.

Yuan, Z., Baker, K., Redman, M. W., Wang, L., Adams, S. V., Yu, M., … Grady, W. M. (2017). Dynamic plasma microRNAs are biomarkers for prognosis and early detection of recurrence in colorectal cancer. *British Journal of Cancer*, *117*(8), 1202–1210. https://doi.org/10.1038/bjc.2017.266.

Zaharie, F., Muresan, M. S., Petrushev, B., Berce, C., Gafencu, G. A., Selicean, S., … Irimie, A. (2015). Exosome-carried microRNA-375 inhibits cell progression and dissemination via Bcl-2 blocking in colon cancer. *Journal of Gastrointestinal and Liver Diseases*, *24*(4), 435–443. https://doi.org/10.15403/jgld.2014.1121.244.375.

Zekri, A. R. N., Youssef, A. S. E. D., Lotfy, M. M., Gabr, R., Ahmed, O. S., Nassar, A., … Bahnassy, A. A. (2016). Circulating serum miRNAs as diagnostic markers for colorectal cancer. *PLoS One*, *11*(5), e0154130. https://doi.org/10.1371/journal.pone.0154130.

Zhai, H., Fesler, A., Ba, Y., Wu, S., & Ju, J. (2015). Inhibition of colorectal cancer stem cell survival and invasive potential by hsa-miR-140-5p mediated suppression of Smad2 and autophagy. *Oncotarget*, *6*(23), 19735–19746. https://doi.org/10.18632/oncotarget.3771.

Zhang, D., & Yang, N. (2019). MiR-335-5p inhibits cell proliferation, migration and invasion in colorectal cancer through downregulating LDHB. *Journal of B.U.ON.*, *24*(3), 1128–1136. https://jbuon.com/archive/24-3-1128.pdf.

Zhang, H., Li, Y., Huang, Q., Ren, X., Hu, H., Sheng, H., & Lai, M. (2011). MiR-148a promotes apoptosis by targeting Bcl-2 in colorectal cancer. *Cell Death and Differentiation*, *18*(11), 1702–1710. https://doi.org/10.1038/cdd.2011.28.

Zhang, J., Guo, H., Zhang, H., Wang, H., Qian, G., Fan, X., … Ge, S. (2011). Putative tumor suppressor miR-145 inhibits colon cancer cell growth by targeting oncogene friend leukemia virus integration 1 gene. *Cancer*, *117*(1), 86–95. https://doi.org/10.1002/cncr.25522.

Zhang, G. J., Xiao, H. X., Tian, H. P., Liu, Z. L., Xia, S. S., & Zhou, T. (2013). Upregulation of microRNA-155 promotes the migration and invasion of colorectal cancer cells through the regulation of claudin-1 expression. *International Journal of Molecular Medicine*, *31*(6), 1375–1380. https://doi.org/10.3892/ijmm.2013.1348.

Zhang, Y., Wang, X., Xu, B., Wang, B., Wang, Z., Liang, Y., … Jiang, B. (2013). Epigenetic silencing of miR-126 contributes to tumor invasion and angiogenesis in colorectal cancer. *Oncology Reports*, *30*(4), 1976–1984. https://doi.org/10.3892/or.2013.2633.

Zhang, G., Zhou, H., Xiao, H., Liu, Z., Tian, H., & Zhou, T. (2014). MicroRNA-92a functions as an oncogene in colorectal cancer by targeting PTEN. *Digestive Diseases and Sciences*, *59*(1), 98–107. https://doi.org/10.1007/s10620-013-2858-8.

Zhang, W., Zhang, T., Jin, R., Zhao, H., Hu, J., Feng, B., … Wang, M. (2014). MicroRNA-301a promotes migration and invasion by targeting TGFBR2 in human colorectal cancer. *Journal of Experimental & Clinical Cancer Research*, *33*(1). https://doi.org/10.1186/s13046-014-0113-6.

Zhang, J., Zhang, K., Bi, M., Jiao, X., Zhang, D., & Dong, Q. (2014). Circulating microRNA expressions in colorectal cancer as predictors of response to chemotherapy. *Anti-Cancer Drugs*, *25*(3), 346–352. https://doi.org/10.1097/CAD.0000000000000049.

Zhang, D., Zhou, J., & Dong, M. (2014). Dysregulation of microRNA-34a expression in colorectal cancer inhibits the phosphorylation of FAK via VEGF. *Digestive Diseases and Sciences*, *59*(5), 958–967. https://doi.org/10.1007/s10620-013-2983-4.

Zhang, X., Zhang, X., Liu, C., Jia, N., Li, X., & Xiao, J. (2014). MiR-224 promotes colorectal cancer cells proliferation via downregulation of P21WAF1/CIP1. *Molecular Medicine Reports*, *9*(3), 941–946. https://doi.org/10.3892/mmr.2014.1900.

Zhang, P., Ji, D. B., Han, H. B., Shi, Y. F., Du, C. Z., & Gu, J. (2014). Downregulation of miR-193a-5p correlates with lymph node metastasis and poor prognosis in colorectal cancer. *World Journal of Gastroenterology*, *20*(34), 12241–12248. https://doi.org/10.3748/wjg.v20.i34.12241.

Zhang, W., Zou, C., Pan, L., Xu, Y., Qi, W., Ma, G., … Jiang, P. (2015). MicroRNA-140-5p inhibits the progression of colorectal cancer by targeting VEGFA. *Cellular Physiology and Biochemistry*, *37*(3), 1123–1133. https://doi.org/10.1159/000430237.

Zhang, T., Cai, X., Li, Q., Xue, P., Chen, Z., Dong, X., & Xue, Y. (2016). Hsa-miR-875-5p exerts tumor suppressor function through down-regulation of EGFR in colorectal carcinoma (CRC). *Oncotarget*, *7*(27), 42225–42240. https://doi.org/10.18632/oncotarget.9944.

Zhang, P., Zuo, Z., Wu, A., Shang, W., Bi, R., Jin, Q., … Jiang, L. (2017). miR-600 inhibits cell proliferation, migration and invasion by targeting p53 in mutant p53-expressing human colorectal cancer cell lines. *Oncology Letters*, *13*(3), 1789–1796. https://doi.org/10.3892/ol.2017.5654.

Zhang, X. L., Xu, L. L., & Wang, F. (2017). Hsa_circ_0020397 regulates colorectal cancer cell viability, apoptosis and invasion by promoting the expression of the miR-138 targets TERT and PD-L1. *Cell Biology International*, *41*(9), 1056–1064. https://doi.org/10.1002/cbin.10826.

Zhang, W., Sun, Z., Su, L., Wang, F., Jiang, Y., Yu, D., … Liang, W. (2018). miRNA-185 serves as a prognostic factor and suppresses migration and invasion through Wnt1 in colon cancer. *European Journal of Pharmacology*, *825*, 75–84. https://doi.org/10.1016/j.ejphar.2018.02.019.

Zhang, Y., Guo, L., Li, Y., Feng, G. H., Teng, F., Li, W., & Zhou, Q. (2018). MicroRNA-494 promotes cancer progression and targets adenomatous polyposis coli in colorectal cancer. *Molecular Cancer*, *17*(1), 1. https://doi.org/10.1186/s12943-017-0753-1.

Zhang, H., Zhu, M., Shan, X., Zhou, X., Wang, T., Zhang, J., … Zhu, W. (2019). A panel of seven-miRNA signature in plasma as potential biomarker for colorectal cancer diagnosis. *Gene*, *687*, 246–254. https://doi.org/10.1016/j.gene.2018.11.055.

Zhang, Z. J., Zhang, Y. H., Qin, X. J., Wang, Y. X., & Fu, J. (2020). Circular RNA circDENND4C facilitates proliferation, migration and glycolysis of colorectal cancer cells through miR-760/GLUT1 axis. *European Review for Medical and Pharmacological Sciences*, *24*(5), 2387–2400. https://doi.org/10.26355/eurrev_202003_20506.

Zhang, Y., Zhou, Y. M., Zhang, Z. J., & Li, X. (2020). Mir-210 is a serological biomarker for predicting recurrence and prognosis of colon carcinoma patients with liver metastases after radiofrequency ablation treatment. *Cancer Management and Research*, *12*, 9077–9085. https://doi.org/10.2147/CMAR.S267731.

Zhang, N., Zhang, P. P., Huang, J. J., Wang, Z. Y., Zhang, Z. H., Yuan, J. Z., … Bai, J. (2020). Reduced serum exosomal miR-874 expression predicts poor prognosis in colorectal cancer. *European Review for Medical and Pharmacological Sciences*, *24*(2), 664–672. https://doi.org/10.26355/eurrev_202001_20043.

Zhao, J. P., & Chen, L. L. (2020). Circular rna mat2b induces colorectal cancer proliferation via sponging mir-610, resulting in an increased e2f1 expression. *Cancer Management and Research*, *12*, 7107–7116. https://doi.org/10.2147/CMAR.S251180.

Zhao, L., Yu, H., Yi, S., Peng, X., Su, P., Xiao, Z., … Shen, S. (2016). The tumor suppressor miR-138-5p targets PD-L1 in colorectal cancer. *Oncotarget*, *7*(29), 45370–45384. https://doi.org/10.18632/oncotarget.9659.

Zhao, L. D., Zheng, W. W., Wang, G. X., Kang, X. C., Qin, L., Ji, J. J., & Hao, S. (2016). Epigenetic silencing of miR-181b contributes to tumorigenicity in colorectal cancer by targeting RASSF1A. *International Journal of Oncology*, *48*(5), 1977–1984. https://doi.org/10.3892/ijo.2016.3414.

Zhao, S., Sun, H., Jiang, W., Mi, Y., Zhang, D., Wen, Y., … Yan, D. (2017). miR-4775 promotes colorectal cancer invasion and metastasis via the Smad7/TGFβ-mediated epithelial to mesenchymal transition. *Molecular Cancer*, *16*(1). https://doi.org/10.1186/s12943-017-0585-z.

Zhao, Y. J., Song, X., Niu, L., Tang, Y., Song, X., & Xie, L. (2019). Circulating exosomal mir-150-5p and mir-99b-5p as diagnostic biomarkers for colorectal cancer. *Frontiers in Oncology*, *9*. https://doi.org/10.3389/fonc.2019.01129.

Zhao, S., Mi, Y., Guan, B., Zheng, B., Wei, P., Gu, Y., … Li, D. (2020). Tumor-derived exosomal miR-934 induces macrophage M2 polarization to promote liver metastasis of colorectal cancer. *Journal of Hematology & Oncology*, *13*(1). https://doi.org/10.1186/s13045-020-00991-2.

Zheng, Y. B., Luo, H. P., Shi, Q., Hao, Z. N., Ding, Y., Wang, Q. S., … Tong, S. L. (2014). miR-132 inhibits colorectal cancer invasion and metastasis via directly targeting ZEB2. *World Journal of Gastroenterology*, *20*(21), 6515–6522. https://doi.org/10.3748/wjg.v20.i21.6515.

Zheng, L., Zhang, Y., Liu, Y., Zhou, M., Lu, Y., Yuan, L., … Li, X. (2015). MiR-106b induces cell radioresistance via the PTEN/PI3K/AKT pathways and p21 in colorectal cancer. *Journal of Translational Medicine*, *13*(1). https://doi.org/10.1186/s12967-015-0592-z.

Zheng, K., Zhou, X., Yu, J., Li, Q., Wang, H., Li, M., … Zhao, L. (2016). Epigenetic silencing of miR-490-3p promotes development of an aggressive colorectal cancer phenotype through activation of the Wnt/β-catenin signaling pathway. *Cancer Letters*, *376*(1), 178–187. https://doi.org/10.1016/j.canlet.2016.03.024.

Zheng, G. X., Qu, A. L., Yang, Y. M., Zhang, X., Zhang, S. C., & Wang, C. X. (2016). MiR-422a is an independent prognostic factor and functions as a potential tumor suppressor in colorectal cancer. *World Journal of Gastroenterology*, *22*(24), 5589–5597. https://doi.org/10.3748/wjg.v22.i24.5589.

Zheng, X., Ma, Y. F., Zhang, X. R., Li, Y., Zhao, H. H., & Han, S. G. (2020). Circ-0056618 promoted cell proliferation, migration and angiogenesis through sponging with miR-206 and upregulating CXCR4 and VEGF-A in colorectal cancer. *European Review for Medical and Pharmacological Sciences*, *24*(8), 4190–4202. https://doi.org/10.26355/eurrev_202004_20999.

Zhong, M., Bian, Z., & Wu, Z. (2013). MiR-30a suppresses cell migration and invasion through downregulation of PIK3CD in colorectal carcinoma. *Cellular Physiology and Biochemistry*, *31*(2–3), 209–218. https://doi.org/10.1159/000343362.

Zhou, Q., Zhu, Y., Wei, X., Zhou, J., Chang, L., Sui, H., … Bai, Y. (2016). MiR-590-5p inhibits colorectal cancer angiogenesis and metastasis by regulating nuclear factor 90/vascular endothelial growth factor A axis. *Cell Death & Disease*, *7*(10), e2413. https://doi.org/10.1038/cddis.2016.306.

Zhou, Y., Cheng, X., Wan, Y., Chen, T., Zhou, Q., Wang, Z., & Zhu, H. (2020). Microrna-421 inhibits apoptosis by downregulating caspase-3 in human colorectal cancer. *Cancer Management and Research*, *12*, 7579–7587. https://doi.org/10.2147/CMAR.S255787.

Zhou, C., Liu, H.s., Wang, F.w., Hu, T., Liang, Z.x., Lan, N., … Lan, P. (2020). circCAMSAP1 promotes tumor growth in colorectal cancer via the miR-328-5p/E2F1 axis. *Molecular Therapy*, *28*(3), 914–928. https://doi.org/10.1016/j.ymthe.2019.12.008.

Zhu, H., Dougherty, U., Robinson, V., Mustafi, R., Pekow, J., Kupfer, S., … Bissonnette, M. (2011). EGFR signals downregulate tumor suppressors miR-143 and miR-145 in western diet-promoted murine colon cancer: Role of G 1 regulators. *Molecular Cancer Research*, *9*(7), 960–975. https://doi.org/10.1158/1541-7786.MCR-10-0531.

Zhu, Y., Wu, G., Yan, W., Zhan, H., & Sun, P. (2017). miR-146b-5p regulates cell growth, invasion, and metabolism by targeting PDHB in colorectal cancer. *American Journal of Cancer Research*, *7*(5), 1136–1150. http://www.ajcr.us/files/ajcr0048026.pdf.

Zhu, M., Huang, Z., Zhu, D., Zhou, X., Shan, X., Qi, L. W., … Liu, P. (2017). A panel of microRNA signature in serum for colorectal cancer diagnosis. *Oncotarget*, *8*(10), 17081–17091. https://doi.org/10.18632/oncotarget.15059.

Zhu, D., Sun, Y., Zhang, D., Dong, M., Jiang, G., Zhang, X., & Zhou, J. (2018). miR-1 inhibits the progression of colon cancer by regulating the expression of vascular endothelial growth factor. *Oncology Reports*, *40*(2), 589–598. https://doi.org/10.3892/or.2018.6463.

Zhu, X., Ma, S. P., Yang, D., Liu, Y., Wang, Y. P., Lin, T., … Wang, X. L. (2018). MiR-142-3p suppresses cell growth by targeting CDK4 in colorectal cancer. *Cellular Physiology and Biochemistry*, *51*(4), 1969–1981. https://doi.org/10.1159/000495721.

Zhu, H., He, G., Wang, Y., Hu, Y., Zhang, Z., Qian, X., & Wang, Y. (2019). Long intergenic noncoding RNA 00707 promotes colorectal cancer cell proliferation and metastasis by sponging miR-206. *Oncotargets and Therapy*, *12*, 4331–4340. https://doi.org/10.2147/OTT.S198140.

Zhuoran, H., Xiaomin, W., & Jun, L. (2020). miR-101 suppresses colon cancer cell migration through regulation of EZH2. *Revista Española de Enfermedades Digestivas: Organo Oficial de la Sociedad Española de Patología Digestiva*, *113*(4), 255–260. https://doi.org/10.17235/reed.2020.6800/2019.

Zou, G., Wang, R., & Wang, M. (2019). Clinical response and prognostic significance of serum miR-497 expression in colorectal cancer. *Cancer Biomarkers*, *25*(1), 11–18. https://doi.org/10.3233/CBM-181902.

Zou, S. L., Chen, Y. L., Ge, Z. Z., Qu, Y. Y., Cao, Y., & Kang, Z. X. (2019). Downregulation of serum exosomal miR-150-5p is associated with poor prognosis in patients with colorectal cancer. *Cancer Biomarkers*, *26*(1), 69–77. https://doi.org/10.3233/CBM-190156.

Chapter 18

Pathophysiology roles and translational opportunities of miRNAs in lung cancer

Mattia Boeri, Gabriella Sozzi, and Orazio Fortunato
Tumor Genomics Unit, Department of Research, Fondazione IRCCS Istituto Nazionale dei Tumori, Milan, Italy

Pathophysiology

Cancer is a complex disease derived from the interaction between genetic and microenvironmental factors. MicroRNAs (miRNAs) are deregulated in lung cancer cells, causing the modulation of cellular ability such as proliferation, invasion, and aggressiveness. They are also involved in the modulation of the stromal cell phenotype. In this chapter, the functional role of miRNAs in the regulation of pathways and mechanisms of lung tumorigenesis will be discussed, focusing on miRNAs expressed by tumor cells or by the cells of the lung microenvironment.

MicroRNA expressed in lung tumor cells

Tumor-specific miRNAs and their involvement in lung tumorigenesis have been described, with particular emphasis on those miRNAs that have a critical role in the modulation of cancer hallmarks such as genetic instability, proliferation, invasion, epithelial to mesenchymal transition (EMT), tumor-derived angiogenesis, and apoptosis.

One of the most relevant factors involved in lung cancer is the epidermal growth factor receptor (EGFR), which is a transmembrane receptor tyrosine kinase activating epithelial cell proliferation. EGFR is overexpressed in 15%–30% of Non Small Cell Lung Cancer (NSCLC) cases and it is associated to a reduced patients survival, tumor progression and metastasis development (Li et al., 2015). An increasing number of miRNAs are found to be involved in EGFR signaling pathways and in drug resistance after anti-EGFR treatment. In NSCLC, miR-7 is considered an onco-miR and was found upregulated by the activation of EGFR in both in vitro and in vivo experiments. Conversely, miR-34 is described as a tumor-suppressor miRNA with an important role in blocking tumor growth and metastasis by directly targeting EGFR (Li et al., 2015).

Genetic alterations in the Kirsten rat sarcoma viral oncogene homolog (KRAS) are present in approximately 20–25% of lung adenocarcinoma and are associated with poor prognosis (Roncarati, Lupini, Shankaraiah, & Negrini, 2019).

This gene encodes for a small GTPase membrane-bound protein that regulates, through different effectors, several cellular mechanisms such as survival, proliferation, and differentiation. Several miRNAs such as the let-7 family and miR-193a-3p are described as repressors of KRAS that reduce lung cancer proliferation by inhibiting multiple cell cycle genes. Let-7 has an essential role in the Dicer-mediated miRNA maturation process (Roncarati et al., 2019).

The tumor-suppressor p53, frequently mutated in lung cancer cells, is an important regulator of different cell functions, including cell cycle, DNA repair, apoptosis, and senescence. Given its central role in many different pathways, p53 is finely regulated by several mechanisms. An essential regulatory network between miRNAs and p53 has been demonstrated (Feng, Zhang, Wu, & Hu, 2011).

In particular, mir-34 family members regulate cell cycle, apoptosis, and cellular senescence and their expression is finely controlled by p53. The described miRNA effects in lung cancer cells are mediated by the repression of the transcription of several targets such as Cyclin D1, BCL-2, and MYC and MET genes. Furthermore, miR-197 is found upregulated in lung cancer cells and it has demonstrated that modulated p53 pathways at different levels. Moreover, miR-197 controls the expression of two downstream targets, NOXA and BMF, and its inhibition induced apoptosis with a concomitant reduction of tumor growth (Feng et al., 2011).

Furthermore, the activity of p53 is negatively regulated by an ubiquitin-ligase, mouse double minute 2 (MDM2), through an autoregulatory feedback loop. Our previous work described two suppressor miRNAs, miR-660 and miR-486, downregulated in lung cancer tissues and involved in the p53 signaling pathway. In detail, miR-660 is able to reduce MDM2 levels

MicroRNA in Human Malignancies. https://doi.org/10.1016/B978-0-12-822287-4.00013-X

and stimulate miR-486 transcription by increasing p53 activity in lung cancer cells. The final effects of this combinatory miRNA modulation result in the reduction of tumor growth both in vitro and in vivo (Moro et al., 2019).

EMT is a critical event in tumor invasion and metastasis, characterized by phenotypic changes and cellular reprogramming during the transition from epithelial to mesenchymal cells. It is well known that miRNAs play an important role in the regulation of this molecular mechanism: the miR-200 family displays an important role in the modulation of EMT by targeting mainly zinc finger *E*-box binding homeobox (ZEB1) and ZEB2. Interestingly, miR200/ZEB modulates E-Cadherin transcription and controls the expression of multiple downstream genes involved in migration, invasion, and metastatic growth at distant sites. Finally, mir-221 and mir-222 are two homologous miRNAs that are overexpressed in lung cancer tissues; the tumor-suppressor genes PTEN and TIMP3 were directly regulated by these miRNAs, with a consequent improvement of cellular migration (Legras et al., 2017).

Unlike normal cells, tumor cells possess the ability to evade the cellular mechanism of apoptosis to promote uncontrolled cellular growth. In this regard, altered miRNA expression in lung cancer cells is linked to evasion from apoptosis and cell cycle arrest. Mir-21 is considered an oncogenic miRNA overexpressed in lung cancer, particularly in advanced disease. Mir-21 is an essential regulator of cellular mechanisms such as programmed cell death, cellular proliferation, and migration. This particular effect is mediated by the inhibition of negative regulators of ERK pathways with the activation of the apoptotic cascade. In lung cancer, the mir-17/92a cluster (mir-17, mir-18a, mir-19a, mir-20a, mir-19b, and mir-92) is frequently overexpressed and is regulated directly by c-MYC with involvement in the control of cellular proliferation. To this matter, mir-17 causes apoptosis through the direct targeting of E2F1. In addition, other targets for this miRNA's cluster are HIF-1a, PTEN, BCL2L11, CDKNA, and TSP-1, all fundamental genes for lung cancer development. In conclusion, the miR-15/16 cluster is frequently observed to be deleted in lung cancer and inversely correlated with Cyclin D1, D2, and E1 expression. This evidence suggests the pivotal role of these miRNAs in the control of the G0-G1 phase of the cell cycle. The functional effects of miR-15/16 to induce cell cycle arrest are also dependent on Rb, another important gene frequently mutated during lung tumorigenesis (Lima et al., 2011).

MicroRNA expressed in lung microenvironmental cells

The lung tumor microenvironment is composed of different cell types such as fibroblasts as well as endothelial and immune cells. The crosstalk between all these cells with cancer cells is responsible for tumor progression. For this reason, any altered miRNA expression in both stromal and tumor cells could drive tumorigenesis (Rupaimoole, Calin, Lopez-Berestein, & Sood, 2016).

One of the most important components of the lung microenvironment is the fibroblast that contributes to the malignant phenotype of lung cancer cells. Several miRNAs are described as critically involved in this crosstalk. MiR – 16 was found to be a regulator of the lung fibroblast secretome in an FGFR1-dependent manner (Andriani et al., 2018).

When miR-16 is downregulated, such as in lung cancer patients, the hepatocyte growth factor (HGF) is secreted by fibroblasts and promotes cancer cell migratory ability and reduced tumor growth in vivo by activating the cMet pathway. Also, miR-21 levels in the cancer-associated fibroblast (CAF) were observed to be inversely correlated with patient survival. Also, this miRNA was responsible for the modulation of the fibroblast's secretome. MiR-30 family members were described as downregulated in fibroblasts of premetastatic lungs with a role in the stabilization of pulmonary vessels. The replacement of miR-30s was proposed as an interesting therapeutic option for advanced lung cancer due to the disruption of the mechanism of premetastatic niche formation. Conversely, the reconstitution of miR-155 in normal fibroblasts resulted in the conversion of these cells to a CAF-like phenotype (Mitra et al., 2012).

Lung tumor growth is dependent on vascular supply and new vessel formation, a complex mechanism starting from the release of proangiogenic factors by tumor cells, and then endothelial activation and the formation of migration and capillary structures. Pro- and antiangiogenic factors, including miRNAs, are released by tumor cells to modulate the phenotype of endothelial cells, which exert an essential function in tumorigenesis by sustaining tumor growth and vessel formation (Salinas-Vera et al., 2019).

Mir-519c expression in lung cancer patients is associated with survival and was able to directly bind HIF-1α. The overexpression of this miRNA in lung cancer cells leads to a decrease in the tube structure formation ability of endothelial cells and a consequent reduction of tumor growth both in vitro and in vivo. Moreover, tumor cells under hypoxia conditions secreted miR-494, described as an important modulator of angiogenesis in lung cancer. In fact, the delivery of miR-494 through extracellular vesicles into endothelial cells enhanced migration and tube formation, promoting lung tumor growth (Salinas-Vera et al., 2019).

MiR-1 is downmodulated in lung endothelial cells. It regulated tumor progression, endothelial proliferation, and the secretome through the activation of ERK signaling. Dimitrova and colleagues showed that miR-143/145 expressed by lung microenvironmental cells promotes tumor growth through the stimulation of lung endothelial cell proliferation (Dimitrova et al., 2016).

Different types of immune cells are involved in the modulation of the lung tumor microenvironment. In particular, T lymphocytes, dendritic cells, and natural killers were considered key players in tumor suppression, whereas T regulatory cells and tumor-associated macrophages have immunosuppressive functions. Tumor cells secrete several factors to affect the functionality of immune cells. Among these factors, miR-23a is an important transcriptional repressor of BLIMP-1, fundamental for cytotoxicity and the differentiation of T cells. In lung cancer, TGF-β induces mir-23a upregulation and BLIMP-1 downmodulation in T cells, which promotes tumor growth. Dendritic cells (DCs) are key players responsible for the induction of antitumoral activity by the stimulation of cytotoxic T-cell responses. In the presence of lung tumors, miR-301 was upregulated in DCs, suppressing IL-12 secretion impairing antigen s-pecific Th1 polarization and tumor outcome (Lin et al., 2014).

NK cells can be considered one of the first immune cells that responds to tumor cell stimuli, promoting an antitumoral response through the secretion of IFN-γ, perforin, and granzyme B. Regarding NK activity, interestingly, TGF-β treatment is able to impair tumor cytolysis and abrogate perforin polarization. The authors revealed that TGF-β induced the upregulation of miR-183 with a concomitant reduction of DAP12 transcription. Furthermore, hypoxia increased the secretion by lung cancer cells of extracellular vesicles containing miR-23a. These hypoxic microvesicles exert immunosuppressive effects on NK cells by the transfer of TGF-β and miR-23 that was able to reduce CD107 expression in recipient cells (Berchem et al., 2016).

In response to several microenvironmental stimuli, macrophages could undergo into M1 or M2 differentiation with opposite immune responses. In our investigation, we observed that miR-320 is derived from activated neutrophils isolated from heavy smokers with a high risk of lung cancer and is able to promote the M2-like phenotype of lung macrophages. This miRNA increases intracellular levels of IL-10, VEGF, and CD163 and CD206 surface expression through downmodulation of the STAT4 protein (Fortunato et al., 2019).

MiR-103a contained in hypoxic extracellular vesicles mediated the switch of macrophages to an M2-like phenotype in patients with lung cancer. This effect was mediated by the reduction of PTEN levels and the activation of the AKT pathway and VEGF secretion (Hsu et al., 2018).

Tissutal miRNAs as diagnostic or prognostic biomarkers

MicroRNAs could be also used for the diagnosis of lung cancer (Table 18.1). The first study concerning the analysis of miRNA expression in lung cancer tissues with a diagnostic intent identified five miRNAs (*mir-155, mir−17-3p, let-7a-2, mir-145,* and *mir-21*) differently expressed. Among these miRNAs, the correlation of mir-155 and let-7a-2 with the overall survival of lung cancer patients identified a group of patients with poor prognosis (Yanaihara et al., 2006).

MiRNA expression levels could be considered an interesting marker also for the histology classification of lung cancers. MiR-205 was identified as upregulated in squamous cell carcinoma (SCC) cancer compared to adenocarcinoma (ADC), using a microarray approach analyzed in a cohort of 122 lung cancer patients. In that same analysis, the high expression of miR-21 and U6 was detected in lung cancers, and a threshold of 2.5 was appropriately defined, considering the normalized levels of these miRNAs able to distinguish SCC from non-SCC. This potentially diagnostic tool was further validated by qRT-PCR in FFPE specimens for the identification of SCC subtypes resulting in a sensitivity of 96% and a specificity of 90% in two independent sets (Aharonov et al., 2009).

TABLE 18.1 Tissutal miRNAs as diagnostic or prognostic markers in lung cancer.

microRNA	Lung cancer hystotype	Biomarker	Reference
miR-155 and let-7a-2	ADC and SCC	Diagnostic	Yanaihara et al. (2006)
MiR-205	SCC	Diagnostic	Aharonov et al. (2009)
miR-21 and miR-34a	Neuroendocrine	Diagnostic	Demes et al. (2016)
miR-31	ADC and SCC	Prognostic	Xiaogang et al. (2011)
mir-221,let-7a, miR-137, miR-372 and miR-182	ADC and SCC	Prognostic	Yu et al. (2008)
let-7e, miR-34a, miR-34c-5p, miR-25 and miR-191	SCC	Prognostic	Landi et al. (2010)

Another study reported miR-21 and miR-34a as having important roles in differentiating lung neuroendocrine tumors and could be possibly utilized as practical supportive markers (Demes, Aszyk, Bartsch, Schirren, & Fisseler-Eckhoff, 2016).

Similarly, a metaanalysis conducted among 14 independent studies on miRNA profiling reported that the upregulation of miR-210, miR-21, miR-31, and miR − 182 was able to identify SCC tissues, whereas miR-126 and miR-145 were the most downregulated miRNAs in ADC cancers (Guan, Yin, Li, Wu, & Zhou, 2012).

A comprehensive analysis of miRNA expression in Chinese lung cancer patients revealed the deregulation of 22 miRNAs compared to normal lung tissues. A classifier composed of five of these miRNAs (miR-210 and miR-182 upregulated; miR-486-5p, miR-30a, and miR-140-3p downregulated) was described to distinguish SCC from normal tissues in 34 lung cancer patients with an accuracy of 94.1%. This signature was also validated in another set of 26 patients with 96.2% accuracy. The prognostic values of this specific miRNA signature were also assessed by a total of 60 SCC lung cancer patients and confirmed in a validation set of 88 SCC subjects. Interestingly, miR-31 was demonstrated to be a strong prognostic biomarker, associated with overall survival (Xiaogang et al., 2011).

Much evidence supports the role of miRNAs as potential prognostic biomarkers in lung cancers. Interestingly, two groups reported two independent signatures associated with the clinical outcome of lung cancer patients, all composed of five microRNAs. In a work published in 2008, the authors identified two miRNAs (mir-221 and let-7a) directly associated and three miRNAs (miR-137, miR-372, and miR-182) inversely associated with patient survival in a training set composed of 56 lung cancer subjects. Based on these findings, a score that stratified lung cancer patients in high and low risk according to both relapse-free (14 months vs. not reached, $p = 0.003$) and overall survival (25 months vs. not reached, $p = 0.008$) was generated. Data were validated in a separate set of 62 lung cancer patients obtaining similar results (Yu et al., 2008).

A few years later, in a cohort of 290 lung cancer patients (165 ADC and 125 SCC) enrolled in the environment and fenetics in lung cancer etiology (EAGLE) trial, a five-miRNA signature (let-7e, miR-34a, miR-34c-5p, miR-25, and miR-191) was reported to predict survival only in squamous lung cancer patients, data also confirmed in an independent cohort of lung cancer specimens (Landi et al., 2010).

Interestingly, members of the let-7 family, described as fundamental miRNA in lung cancer development, were observed in both signatures. In this context, the expression levels of these particular miRNAs could be considered an interesting prognostic marker for resected lung cancer patients. Indeed, reduced let-7 levels were associated with short survival in a set of 143 lung cancer patients, with a hazard ratio (HR) of early deaths considering let-7 deregulation of 2.17 (95% confidence interval, 1.21–3.89) (Takamizawa et al., 2004).

Another group of miRNAs strongly associated with survival are those belonging to the 14q32 cluster. Among these, 11 miRNAs were found overexpressed in ADC patients with poor prognosis. Furthermore, miR-370, miR-411, and miR-376a were used to build a risk score discriminating 60 ADC patients according to both disease-free survival and overall survival with HRs of 3.32 and 3.37, respectively (Nadal et al., 2014).

Circulating miRNAs

As small molecules, miRNAs can be secreted by all cell types included in microvesicles or bound to protein complexes, such as Ago2. Once released into the extracellular space, these molecules can be taken up by other cells, where they can exploit their functions (Lv et al., 2014).

It is likely that circulating microRNAs are mainly released by hematopoietic or endothelial cells, but some of these molecules could also be derived from the tumor and its microenvironment (Rupaimoole et al., 2016).

Mir-320a is highly expressed in the plasma of lung cancer patients and, as described above, it is secreted into the bloodstream by neutrophils. It can be taken up by macrophages, where it is able to promote an M2-like protumorigenic phenotype through the downregulation of STAT4 (Fortunato et al., 2019).

Fabbri and coauthors reported that miR-21 and miR-29a secreted by tumor cells could bind the Toll-like receptor (TLR) family in immune cells, triggering a prometastatic inflammatory response enhancing tumor growth and metastasis formation (Fabbri et al., 2012).

Again, the downmodulation of miR-16, frequently observed in the plasma of lung cancer patients, could increase the protumorigenic potential of the lung fibroblast, as described above (Andriani et al., 2018).

Another microRNA frequently upmodulated in the plasma of lung cancer patients is the endothelial miR-126, which enhances angiogenesis by silencing Spred-1 in endothelial cells. MiRNA-126 may also have a role in cancer recipient cells where it is capable of promoting proliferation and migration while reducing apoptosis through the regulation of the signal transducers and activators of the transcription 3 (STAT3) signal pathway (Wang et al., 2008).

The muscle-enriched miR-486 is a mediator of protein degradation by inhibiting the Forkhead box O (FoxO) signaling pathway and preventing skeletal muscle atrophy. Thus, lower levels of miR-486 in the plasma of lung cancer patients are indicative of cachexia (Xu et al., 2012).

Circulating miRNAs for lung cancer diagnosis

Irrespective of the source of these molecules, an enormous number of publications have reported on the utility of circulating miRNAs as potential cancer biomarkers so far. A PubMed search using the text string "((plasma) OR (serum) OR (circulating)) AND (microRNA) AND (cancer) AND (biomarker)" revealed more than 4606 papers starting in 2008 (Fig. 18.1). Among these, 601 (13%) reported the utility of circulating miRNAs for the diagnosis and prognosis of lung cancer (Fig. 18.1). In this section, we will describe and compare four largely validated miRNA-based platforms for the diagnosis of lung cancer, considering studies with more than 1000 individuals.

The first two studies included patients and controls enrolled in low-dose computed tomography (LDCT) lung cancer screening trials. Given the high sensitivity of the technique, LDCT is considered the gold standard for lung cancer screening programs. In 2011, the results of the US National Lung Screening Trial (NLST) were published that compared LDCT vs. chest X-ray (CXR) on 53,454 high-risk subjects. The results showed a better performance of the LDCT in identifying the disease in its early stages with a 20% reduction of lung cancer specific mortality (National Lung Screening Trial Research Team, 2011). However, since the overdiagnosis was estimated around 18% and the high rate of false positive cases was around 96.4%, the implementation of screening programs with new diagnostic tools, such as circulating microRNAs is crucial. (Patz et al., 2014).

In 2011, two independent Italian studies for the first time described plasma and serum miRNA signatures able to discriminate patients and controls among volunteers enrolled in the INT-IEO and COSMOS screening trials (Bianchi et al., 2011; Boeri et al., 2011).

The two tested, named the microRNA signature classifier (MSC) and the miR-test, were composed of reciprocal ratios among 24 plasma miRNAs and 13 serum miRNAs normalized on six stable miRNAs, respectively. A few years later, both tests were validated in larger retrospective cohorts. The MSC test applied in 1085 volunteers enrolled in the MILD screening had 87% sensitivity and 81% specificity. In addition, the MSC also had a prognostic value and reduced the LDCT false-positive rate from 19.4% to 3.7% (Sozzi et al., 2014).

On the other hand, the miR-test in a cohort of 1008 lung cancer screening participants resulted in a 77.8% sensitivity and a 74.8% specificity with an overall accuracy of 74.9% (Montani et al., 2015).

Three large retrospective case–control studies published between 2018 and 2020 reported the results of circulating microRNA profiling on thousands of samples collected from the Caucasian and Asian populations. Lu et al. reported on a logistic regression model composed of six miRNAs (miR-17, miR-190b, miR-19a, miR-19b, miR-26b, and miR-375) based on a training set of 565 subjects and validated on an independent cohort of 461 between LC cases (n=258) and healthy controls (n=203). The six-miRNA panel resulted in a 0.868 AUC value, with 80% sensitivity and 80% specificity in the validation set (Lu et al., 2018).

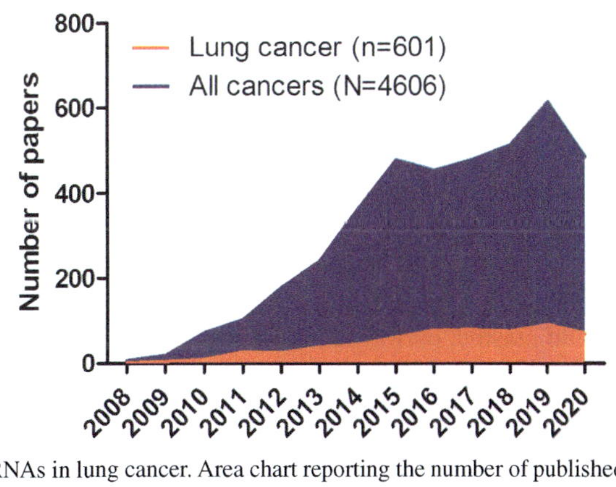

FIG. 18.1 Published paper regarding miRNAs in lung cancer. Area chart reporting the number of published papers each year, according to the PubMed search dated December 18, 2020, using the text strings "((plasma) OR (serum) OR (circulating)) AND (microRNA) AND (cancer) AND (biomarker)" and "((plasma) OR (serum) OR (circulating)) AND (microRNA) AND (lung cancer) AND (biomarker)."

A second large European study analyzed whole blood samples from 3046 patients including 606 patients with LC, 593 patients with other lung diseases, 883 patients with diseases not affecting the lung, and 964 controls. By dividing patients in the training (50%) and validation sets (50%), three signatures composed of a total of 27 miRNAs were established for three different scenarios: (1) patients with lung cancer vs. all other participants, (2) lung cancer patients vs. patients with nontumor lung diseases, and (3) patients with low stage tumors vs. all other participants. In the validation set of all three scenarios, the respective miRNA signature resulted in an overall accuracy higher than 93% (Fehlmann et al., 2020).

Similarly, the study from the Cancer Research Institute of Zhejiang, China, analyzed serum samples collected from 744 NSCLC cases (711 stage I and II) and 944 matched controls. Samples were divided into training, testing, and validation sets. Twenty-two candidate miRNAs were found differentially expressed in the training and testing sets. A five-miRNA model was developed that showed an 81.3% sensitivity for all cancer stages and an 82.9% sensitivity when only stages I and II with a 90.7% specificity were considered (Ying et al., 2020).

Circulating miRNAs as predictive and prognostic biomarkers in lung cancer patients

Given the origin and the role of circulating miRNAs, mainly released by hematopoietic cells and regulating the immune cell phenotype as described above, these molecules could be considered promising prognostic and predictive biomarkers in immunotherapy settings. In 2018, Constantini and coauthors analyzed several plasmatic biomarkers, including soluble PD-L1 and PD-L2, Interleukine-2, Interferon-gamma, Granzyme B, and microRNA profiling, in a cohort of 43 LC patients treated with Nivolumab. Concerning the miRNA analysis, they found that miR-320b and miR-375 were upmodulated in the plasma of lung cancer patients with early progression at 2 months (Costantini et al., 2018).

Soon after, a paper from a group in Norway reported that a seven-miRNA signature was associated with overall survival (OS) at 6 months in 20 + 31 LC patients treated with Nivolumab (Halvorsen et al., 2018).

In 2019, a prospective study performed at the Fondazione IRCCS Istituto Nazionale dei Tumori of Milan reported the results of the application of the MSC test previously described in the context of treatment of advanced LC patients with immune checkpoint inhibitors (ICIs). Indeed, the MSC test maintained its prognostic value in a cohort of 140 LC patients (multivariate HR, 0.33; 95%CI, 0.18–0.59; $P = 0.0002$) and was independent from PD-L1 tumor expression (Boeri et al., 2019).

A further study from the same group on 200 lung cancer patients described the possibility of integrating circulating microRNAs with other clinico and biochemical markers (ECOG PS, blood cell count, age, smoking history, and line of therapy) to create a prognostic score system, called DEMo, able to stratify advanced NSCLC patients treated with ICIs better than each single marker alone (Prelaj et al., 2020).

The median OS across the seven-risk groups identified by DEMo ranged from 29.7 to 1.5 months ($P < 0.0001$). Recently, the Yang team identified 27 sera miRNAs differentially expressed in 19 responders compared to 27 nonresponders. Ten highly expressed miRNAs were also able to stratify 80 LC patients according to both PFS (HR, 0.45; 95%CI, 0.25–0.76; $p < 0.001$) and OS (HR, 0.39; 95%CI, 0.15–0.68; $p < 0.001$) (Fan et al., 2020).

Similarly, Shukuya and coauthors identified in a cohort of 29 LC patients a set of miRNAs differentially expressed between responders and nonresponders. In a second cohort of 21 LC patients, the combination of three miRNAs (miR-199a-3p, miR-21-5p, and miR-28-5p) separates the two groups of patients with an area under the ROC curve of 0.925 (Shukuya et al., 2020).

Analysis of circulating miRNAs: Issues and solutions

The fast rate at which both novel biomarkers are identified and new technologies are implemented for circulating microRNAs makes it unavoidable that the standardization of protocols and the alignment of the general procedure are affected. In this regard, several groups have investigated different methodologies to better identify the more efficient, reproducible, accurate, and precise procedure. In 2014, the group headed by Vandesompele performed a microRNA quality control (miRQC) study comparing the three main technologies for the analysis of microRNAs: real-time reverse transcription quantitative PCR (RT-qPCR), microarray hybridization, and next-generation sequencing (NGS). The performances of 12 commercially available platforms were validated on both tissue and circulating miRNAs. Observations on tissue samples revealed a better reproducibility and specificity with NGS and microarray technologies, whereas the analysis of the low input of RNA samples, such as serum and plasma, performed more accurately, sensitively, and with a higher detection rate with RT-qPCR-based platforms (Mestdagh et al., 2014).

However, the most relevant criticism that emerged from the miRQC study was related to the reproducibility among different platforms: only 2 (3%) of 66 miRNAs were concordantly differentially expressed between the same two groups of samples in all platforms.

This particular concern, together with the heterogeneity of the collected samples (plasma, serum, whole blood, etc.), could explain the small overlapping microRNAs identified by different groups. As illustrated in Fig. 18.2, out of the 59

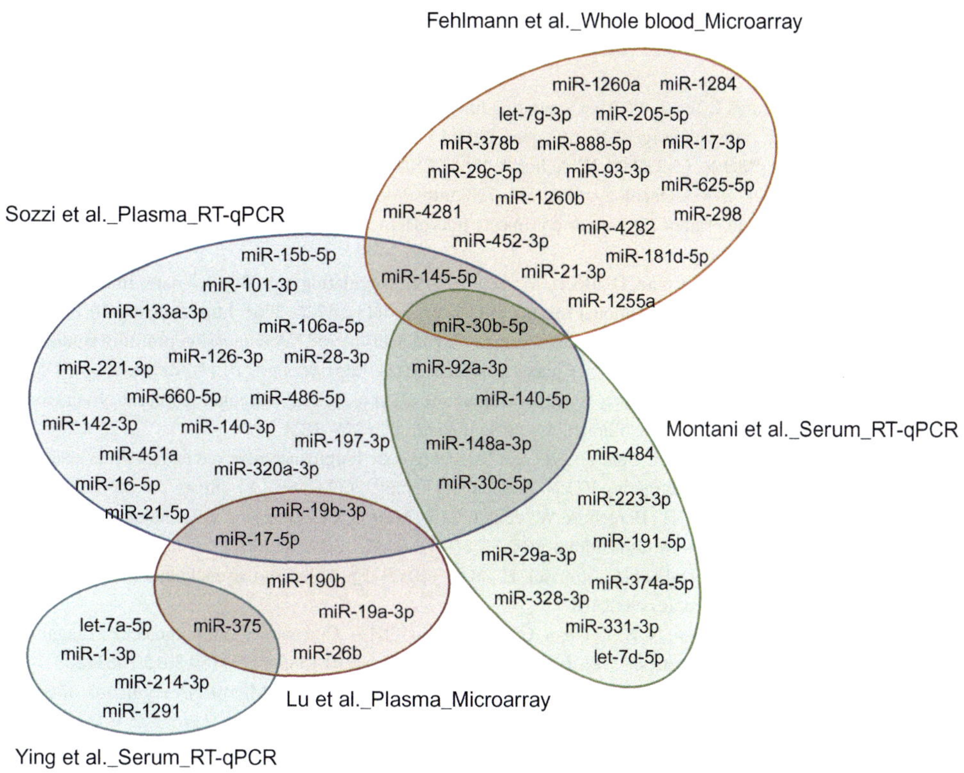

FIG. 18.2 Mirna as circulating biomarkers in lung cancer. Lists of microRNAs identified as diagnostic biomarkers by five large studies including more than 1000 individuals (Fehlmann et al., 2020; Lu et al., 2018; Montani et al., 2015; Sozzi et al., 2014; Ying et al., 2020). Each circle includes the microRNAs from each specific study identified as: "First author_type of sample analyzed_Technique adopted."

miRNAs overall selected by the five largest studies on circulating miRNAs as lung cancer diagnostic biomarkers (Fehlmann et al., 2020; Lu et al., 2018; Montani et al., 2015; Sozzi et al., 2014; Ying et al., 2020), only eight miRNAs (miR-140-5p, miR-145-5p, miR-148a-3p, miR-17-5p, miR-19b-3p, miR-30c-5p, miR-375, and miR-92a-3p) were shared by two studies while only miR-30b-5p was common among three studies.

References

Aharonov, R., Lebanony, D., Benjamin, H., Gilad, S., Ezagouri, M., Dov, A., et al. (2009). Diagnostic assay based on hsa-miR−205 expression distinguishes squamous from nonsquamous non-small-cell lung carcinoma. Journal of Clinical Oncology, 27(12), 2030–2037. https://doi.org/10.1200/JCO.2008.19.4134.

Andriani, F., Majorini, M. T., Mano, M., Landoni, E., Miceli, R., Facchinetti, F., et al. (2018). MiR-16 regulates the pro-tumorigenic potential of lung fibroblasts through the inhibition of HGF production in an FGFR-1- and MEK1-dependent manner. Journal of Hematology & Oncology, 11(1). https://doi.org/10.1186/s13045-018-0594-4.

Berchem, G., Noman, M. Z., Bosseler, M., Paggetti, J., Baconnais, S., Le Cam, E., et al. (2016). Hypoxic tumor-derived microvesicles negatively regulate NK cell function by a mechanism involving TGF-β and miR23a transfer. Oncoimmunology, 5(4). https://doi.org/10.1080/2162402X.2015.1062968.

Bianchi, F., Nicassio, F., Marzi, M., Belloni, E., Dall'Olio, V., Bernard, L., et al. (2011). A serum circulating miRNA diagnostic test to identify asymptomatic high-risk individuals with early stage lung cancer. EMBO Molecular Medicine, 3(8), 495–503. https://doi.org/10.1002/emmm.201100154.

Boeri, M., Milione, M., Proto, C., Signorelli, D., Russo, G. L., Galeone, C., et al. (2019). Circulating miRNAs and PD-L1 tumor expression are associated with survival in advanced NSCLC patients treated with immunotherapy: A prospective study. Clinical Cancer Research, 25(7), 2166–2173. https://doi.org/10.1158/1078-0432.CCR-18-1981.

Boeri, M, Verri, C, Conte, D, Roz, L, Modena, P, Facchinetti, F, et al. (2011). MicroRNA signatures in tissues and plasma predict development and prognosis of computed tomography detected lung cancer. *Proceedings of the National Academy of Sciences of the United States of America, 108*(9), 3713–3718. https://doi.org/10.1073/pnas.1100048108.

Costantini, A., Julie, C., Dumenil, C., Hélias-Rodzewicz, Z., Tisserand, J., Dumoulin, J., et al. (2018). Predictive role of plasmatic biomarkers in advanced non-small cell lung cancer treated by nivolumab. Oncoimmunology, 7(8). https://doi.org/10.1080/2162402X.2018.1452581.

Demes, M., Aszyk, C., Bartsch, H., Schirren, J., & Fisseler-Eckhoff, A. (2016). Differential miRNA-expression as an adjunctive diagnostic tool in neuroendocrine tumors of the lung. Cancers, 8(4). https://doi.org/10.3390/cancers8040038.

Dimitrova, N., Gocheva, V., Bhutkar, A., Resnick, R., Jong, R. M., Miller, K. M., et al. (2016). Stromal expression of miR-143/145 promotes neoangiogenesis in lung cancer development. Cancer Discovery, 188–201. https://doi.org/10.1158/2159-8290.CD-15-0854.

Fabbri, M., Paone, A., Calore, F., Galli, R., Gaudio, E., Santhanam, R., et al. (2012). MicroRNAs bind to Toll-like receptors to induce prometastatic inflammatory response. Proceedings of the National Academy of Sciences of the United States of America, 109(31), E2110–E2116. https://doi.org/10.1073/pnas.1209414109.

Fan, J., Yin, Z., Xu, J., Wu, F., Huang, Q., Yang, L., et al. (2020). Circulating microRNAs predict the response to anti-PD-1 therapy in non-small cell lung cancer. Genomics, 112(2), 2063–2071. https://doi.org/10.1016/j.ygeno.2019.11.019.

Fehlmann, T., Kahraman, M., Ludwig, N., Backes, C., Galata, V., Keller, V., et al. (2020). Evaluating the use of circulating MicroRNA profiles for lung cancer detection in symptomatic patients. JAMA Oncology, 6(5), 714–723. https://doi.org/10.1001/jamaoncol.2020.0001.

Feng, Z., Zhang, C., Wu, R., & Hu, W. (2011). Tumor suppressor p53 meets microRNAs. Journal of Molecular Cell Biology, 3(1), 44–50. https://doi.org/10.1093/jmcb/mjq040.

Fortunato, O., Borzi, C., Milione, M., Centonze, G., Conte, D., Boeri, M., et al. (2019). Circulating mir-320a promotes immunosuppressive macrophages M2 phenotype associated with lung cancer risk. International Journal of Cancer, 144(11), 2746–2761. https://doi.org/10.1002/ijc.31988.

Guan, P., Yin, Z., Li, X., Wu, W., & Zhou, B. (2012). Meta-analysis of human lung cancer microRNA expression profiling studies comparing cancer tissues with normal tissues. Journal of Experimental & Clinical Cancer Research, 31(1). https://doi.org/10.1186/1756-9966-31-54.

Halvorsen, A. R., Sandhu, V., Sprauten, M., Flote, V. G., Kure, E. H., Brustugun, O. T., et al. (2018). Circulating microRNAs associated with prolonged overall survival in lung cancer patients treated with nivolumab. *Acta Oncologica*, *57*, 1225–1231.

Hsu, Y. L., Hung, J. Y., Chang, W. A., Jian, S. F., Lin, Y. S., Pan, Y. C., et al. (2018). Hypoxic lung-cancer-derived extracellular vesicle MicroRNA-103a increases the oncogenic effects of macrophages by targeting PTEN. Molecular Therapy, 26(2), 568–581. https://doi.org/10.1016/j.ymthe.2017.11.016.

Landi, M. T., Zhao, Y., Rotunno, M., Koshiol, J., Liu, H., Bergen, A. W., et al. (2010). MicroRNA expression differentiates histology and predicts survival of lung cancer. Clinical Cancer Research, 430–441. https://doi.org/10.1158/1078-0432.CCR-09-1736.

Legras, A., Pécuchet, N., Imbeaud, S., Pallier, K., Didelot, A., Roussel, H., et al. (2017). Epithelial-to-mesenchymal transition and microRNAs in lung cancer. Cancers, 9(8). https://doi.org/10.3390/cancers9080101.

Li, C. M., Chu, W. Y., Wong, D. L., Tsang, H. F., Tsui, N. B. Y., Chan, C. M. L., et al. (2015). Current and future molecular diagnostics in non-small-cell lung cancer. Expert Review of Molecular Diagnostics, 15(8), 1061–1074. https://doi.org/10.1586/14737159.2015.1063420.

Lima, R. T., Busacca, S., Almeida, G. M., Gaudino, G., Fennell, D. A., & Vasconcelos, M. H. (2011). MicroRNA regulation of core apoptosis pathways in cancer. European Journal of Cancer, 47(2), 163–174. https://doi.org/10.1016/j.ejca.2010.11.005.

Lin, R., Chen, L., Chen, G., Hu, C., Jiang, S., Sevilla, J., et al. (2014). Targeting miR-23a in CD8+ cytotoxic T lymphocytes prevents tumor-dependent immunosuppression. Journal of Clinical Investigation, 124(12), 5352–5367. https://doi.org/10.1172/JCI76561.

Lu, S., Kong, H., Hou, Y., Ge, D., Huang, W., Ou, J., et al. (2018). Two plasma microRNA panels for diagnosis and subtype discrimination of lung cancer. Lung Cancer, 123, 44–51. https://doi.org/10.1016/j.lungcan.2018.06.027.

Lv, Z., Wei, Y., Wang, D., Zhang, C. Y., Zen, K., & Li, M. (2014). Argonaute 2 in cell-secreted microvesicles guides the function of secreted miRNAs in recipient cells. PLoS One, 9(7). https://doi.org/10.1371/journal.pone.0103599.

Mestdagh, P., Hartmann, N., Baeriswyl, L., Andreasen, D., Bernard, N., Chen, C., et al. (2014). Evaluation of quantitative mirnA expression platforms in the micrornA quality control (mirQC) study. Nature Methods, 11(8), 809–815. https://doi.org/10.1038/nMeth.3014.

Mitra, A. K., Zillhardt, M., Hua, Y., Tiwari, P., Murmann, A. E., Peter, M. E., et al. (2012). MicroRNAs reprogram normal fibroblasts into cancer-associated fibroblasts in ovarian cancer. *Cancer Discovery*, 1100–1108. https://doi.org/10.1158/2159-8290.CD-12-0206.

Montani, F., Marzi, M. J., Dezi, F., Dama, E., Carletti, R. M., Bonizzi, G., et al. (2015). MiR-test: A blood test for lung cancer early detection. Journal of the National Cancer Institute, 107(6). https://doi.org/10.1093/jnci/djv063.

Moro, M., Di Paolo, D., Milione, M., Centonze, G., Bornaghi, V., Borzi, C., et al. (2019). Coated cationic lipid-nanoparticles entrapping miR-660 inhibit tumor growth in patient-derived xenografts lung cancer models. Journal of Controlled Release, 308, 44–56. https://doi.org/10.1016/j.jconrel.2019.07.006.

Nadal, E., Zhong, J., Lin, J., Reddy, R. M., Ramnath, N., Orringer, M. B., et al. (2014). A microRNA cluster at 14q32 drives aggressive lung adenocarcinoma. Clinical Cancer Research, 20(12), 3107–3117. https://doi.org/10.1158/1078-0432.CCR-13-3348.

National Lung Screening Trial Research Team. (2011). Reduced lung-cancer mortality with low-dose computed tomographic screening. The New England Journal of Medicine, *365*, 395–409.

Patz, E. F., Pinsky, P., Gatsonis, C., Sicks, J. R. D., Kramer, B. S., Tammemägi, M. C., et al. (2014). Overdiagnosis in low-dose computed tomography screening for lung cancer. JAMA Internal Medicine, 174(2), 269–274. https://doi.org/10.1001/jamainternmed.2013.12738.

Prelaj, A., Proto, C., Russo, G. L., Signorelli, D., Ferrara, R., Mensah, M., et al. (2020). Integrating clinical and biological prognostic biomarkers in patients with advanced NSCLC treated with immunotherapy: The DEMo score system. Translational Lung Cancer Research, 9(3), 617–628. https://doi.org/10.21037/tlcr-20-231.

Roncarati, R., Lupini, R., Shankaraiah, R., & Negrini, M. (2019). The importance of microRNAs in RAS oncogenic activation in human cancer. Internet *Frontiers in Oncology*, *9*, 988.

Rupaimoole, R., Calin, G. A., Lopez-Berestein, G., & Sood, A. K. (2016). MiRNA deregulation in cancer cells and the tumor microenvironment. Cancer Discovery, 6(3), 235–246. https://doi.org/10.1158/2159-8290.CD-15-0893.

Salinas-Vera, Y. M., Marchat, L. A., Gallardo-Rincón, D., Ruiz-García, E., Astudillo-De la Vega, H., Echavarría-Zepeda, R., et al. (2019). AngiomiRs: MicroRNAs driving angiogenesis in cancer (review). International Journal of Molecular Medicine, 43(2), 657–670. https://doi.org/10.3892/ijmm.2018.4003.

Shukuya, T., Ghai, V., Amann, J. M., Okimoto, T., Shilo, K., Kim, T. K., et al. (2020). Circulating MicroRNAs and extracellular vesicle–containing MicroRNAs as response biomarkers of anti–programmed cell death protein 1 or programmed death-ligand 1 therapy in NSCLC. Journal of Thoracic Oncology, 15(11), 1773–1781. https://doi.org/10.1016/j.jtho.2020.05.022.

Sozzi, G., Boeri, M., Rossi, M., Verri, C., Suatoni, P., Bravi, F., et al. (2014). Clinical utility of a plasma-based miRNA signature classifier within computed tomography lung cancer screening: A correlative MILD trial study. Journal of Clinical Oncology, 32(8), 768–773. https://doi.org/10.1200/JCO.2013.50.4357.

Takamizawa, J., Konishi, H., Yanagisawa, K., Tomida, S., Osada, H., Endoh, H., et al. (2004). Reduced expression of the let-7 microRNAs in human lung cancers in association with shortened postoperative survival. Cancer Research, 64(11), 3753–3756. https://doi.org/10.1158/0008-5472.CAN-04-0637.

Wang, S., Aurora, A. B., Johnson, B. A., Qi, X., McAnally, J., Hill, J. A., et al. (2008). The endothelial-specific MicroRNA miR-126 governs vascular integrity and angiogenesis. Developmental Cell, 15(2), 261–271. https://doi.org/10.1016/j.devcel.2008.07.002.

Xiaogang, T., Wenyan, Q., Liang, Z., Jie, H., Baozhong, L., Cuiyan, Z., et al. (2011). A 5-MicroRNA signature for lung squamous cell carcinoma diagnosis and hsa-miR-31 for prognosis. Clinical Cancer Research, 17(21), 6802–6811. https://doi.org/10.1158/1078-0432.ccr-11-0419.

Xu, J., Li, R., Workeneh, B., Dong, Y., Wang, X., & Hu, Z. (2012). Transcription factor FoxO1, the dominant mediator of muscle wasting in chronic kidney disease, is inhibited by microRNA-486. Kidney International, 82(4), 401–411. https://doi.org/10.1038/ki.2012.84.

Yanaihara, N., Caplen, N., Bowman, E., Seike, M., Kumamoto, K., Yi, M., et al. (2006). Unique microRNA molecular profiles in lung cancer diagnosis and prognosis. Cancer Cell, 9(3), 189–198. https://doi.org/10.1016/j.ccr.2006.01.025.

Ying, L., Du, L., Zou, R., Shi, L., Zhang, N., Jin, J., et al. (2020). Development of a serum miRNA panel for detection of early stage non-small cell lung cancer. Proceedings of the National Academy of Sciences of the United States of America, 117(40), 25036–25042. https://doi.org/10.1073/pnas.2006212117.

Yu, S. L., Chen, H. Y., Chang, G. C., Chen, C. Y., Chen, H. W., Singh, S., et al. (2008). MicroRNA signature predicts survival and relapse in lung Cancer. Cancer Cell, 13(1), 48–57. https://doi.org/10.1016/j.ccr.2007.12.008.

Chapter 19

Pathophysiology roles and translational opportunities of miRNAs in mesothelioma

Elisabetta Casalone[a], Alessandra Allione[a], and Giuseppe Matullo[a,b]
[a]*Department of Medical Sciences, University of Turin, Turin, Italy,* [b]*Medical Genetics Unit, AOU Città della Salute e della Scienza, Turin, Italy*

Introduction

Malignant mesothelioma (MM) is an asbestos-related malignancy arising from mesothelial cells lining the serosal surfaces of pleural (80%–90%), peritoneum (10%–15%), pericardium, and tunica vaginalis (Pistolesi & Rusthoven, 2004).

Malignant pleural mesothelioma (MPM) is the most common form of MM and is strongly associated with asbestos exposure. In the past century, asbestos was widely used in several industrial applications for its chemical and physical properties. Despite the ban of asbestos in most industrial countries, MPM incidence is expected to increase over the next decades due to its long latency time after exposure and the continuing use of asbestos, mostly in emerging countries (Carbone et al., 2012).

To date, there is no screening program for the asbestos-exposed population and the lack of effective early diagnostic biomarkers along with the nonspecific symptoms, lead to a late diagnosis of MPM. The prognosis is poor with a median survival of less than 12 months, in consequence of poor responses to platinum-based therapy or the acquisition of a platinum-resistant phenotype (Sun, Vaynblat, & Pass, 2017). Survival is also influenced by histological subtypes. Mainly, MPM can be classified into three types: epithelioid, sarcomatoid, and biphasic mesothelioma (Pistolesi & Rusthoven, 2004). Epithelioid is the most common MPM type, and carries the most favorable prognosis.

The diagnosis of MPM relies on invasive and often difficult procedures involving the characterization of several factors, including Calretinin, Wilms tumor-1, and Cytokeratin 5/6 (Guidelines for the Diagnosis and Treatment of Malignant Pleural Mesothelioma, 2013). Because it usually occurs at clinically advanced stages when the tumor is less responsive to therapies, early and accurate MPM detection could allow new therapeutic approaches at a potentially treatable stage.

A growing amount of evidence has shown that deregulated microRNAs (miRNAs) can drive mesothelial transformation and the acquisition of malignant phenotypes.

In light of this scenario, miRNAs could be potential candidates as noninvasive diagnostic and prognostic biomarkers.

This chapter provides an overview of the available evidence on the involvement of miRNAs in MPM tumorigenesis such as diagnostic/prognostic biomarkers and the potential to exploit them for therapeutic purposes.

Role of miRNAs in the pathogenesis of malignant pleural mesothelioma.

Over the past 10 years, researchers have identified altered miRNA expression profiles in tumor tissue and cell lines, indicating a role of miRNAs in MPM initiation and oncogenic transformation; see Table 19.1.

Mounting evidence supports that miRNA changes in MPM are mostly characterized by general downregulation, suggesting a tumor-suppressor activity (Reid, Johnson, & van Zandwijk, 2020).

The levels of the miR-34 family were shown to be significantly downregulated in several tumors; it also was the first miRNA family that was demonstrated to be directly regulated by the tumor-suppressor p53 (Kubo et al., 2011; Welponer et al., 2020; Ziga & Tanja, 2014). MiR-34s comprise three processed miRNAs: miR-34a, miR-34,b and miR-34c; their decreased expression influences various processes such as angiogenesis, cell cycle progression, and apoptosis (Feng, Ge, Du, Zhang, & Liu, 2019; Hermeking, 2010). Indeed, miR-34s are transcriptionally activated by p53 and mediate downstream p53-functions, suggesting tumor-suppressing properties for this miR family.

Kubo and collaborators first reported that miR-34b/c downregulation is due to the epigenetic silencing of the gene promoter, observed in 85% of MM tumor samples as well as in MPM cell lines (Kubo et al., 2011). Overexpression of miR-34b/c in transfected MM cell lines triggered cell cycle arrest, inhibition of migration, and p53-mediated apoptosis (Kubo et al., 2011; Ueno et al., 2014).

MicroRNA in Human Malignancies. https://doi.org/10.1016/B978-0-12-822287-4.00024-4

TABLE 19.1 MiRNAs with biological activity in MPM.

microRNA	Target gene	Cellular function	References
miR-1	*HIF-1α, PIM1*	Inhibits cell proliferation, migration, proliferation	Amatya, Mawas, Kushitani, El-Din, and Takeshima (2016), Xu et al. (2013)
Let-7a	*RAS*	Inhibits MPM cellular proliferation, migration, and tumor growth	Lee et al. (2013), Khodayari, Mohammed, Goldberg, and Nasreen (2011)
miR-15a/b - miR-16	*BCL-2, CCDN1, PD-L1*	Inhibits growth of MPM cells, apoptosis	Reid et al. (2013), Kao et al. (2017)
miR-17-5p	*KCNMA1*	Inhibits cell migration	Cheng et al. (2016)
miR-18a-5p	*PIAS3*	Modulates growth inhibition	He et al. (2018)
miR-21-5p	*PARP1, mesothelin*	Modulates growth inhibition	Ueno et al. (2014), De Santi et al. (2017)
miR-24-3p	*CGN*	Regulates migration and tumor growth in mice	Oliveto et al. (2018)
miR-29c-5p	*DNMT1/3A*	Inhibits cell migration, proliferation	Pass et al. (2010), Walter et al. (2016)
miR-31-5p	*PPP6C*	Inhibits MPM cell growth, migration	Ivanov et al. (2010)
miR-34a/b/c	*c-MET, BCL-2*	Inhibits cell growth	Kubo et al. (2011), Tanaka et al. (2013), Maki et al. (2012), Menges et al. (2014)
miR-126	*IRS1, PDK, ACL*	Regulates cell metabolism, triggers TCA cycle, enhances gluconeogenesis	Tomasetti et al. (2014), Tomasetti et al. (2016), Andersen et al. (2015)
miR-137-3p	*YB-1*	Inhibits growth, migration, invasion	Johnson et al. (2018)
miR-145-5p	*OCT4*	Inhibits clonogenicity and migration	Cioce et al. (2014)
miR-182-5p; miR-183-5p	*FOXO1*	Inhibits MPM cell growth	Suzuki et al. (2018)
miR-193a-3p	*MCL-1; PD-L1*	Tumor suppressor	Williams et al. (2015)
miR-205-5p	*ZEB1, ZEB2*	Involved in EMT, affects cellular migration	Fassina et al. (2012)
miR-206-3p	*KRAS, CDK4, CCDN1*	Inhibits growth, migration, invasion	Singh et al. (2021)
miR-223-3p	*STMN1*	Inhibits migration; tubulin acethylation	Birnie et al. (2015)
miR-215-5p	*MDM2*	Inhibits growth	Singh et al. (2019)
miR-302b-3p	*MCL1*	Inhibits proliferation	Khodayari et al. (2016)
miR-379; miR-411	*IL-18*	Inhibits proliferation, invasion	Yamamoto et al. (2014)

An in vitro experiment showed that the inhibition of miR-34s increases cell viability and proliferation in normal mesothelial cells, inducing an oncogenic phenotype through target genes *c-MET* as well as *BCL-2*, frequently upregulated in MPM (Tanaka et al., 2013). Moreover, an in vivo study highlighted that the inactivation of *NF2* and *CDKN2A* in asbestos-exposed mice drove the development of highly aggressive MM, and *c-MET* activation occurs through a p53-miR34a-dependent mechanism, thereby enhancing the tumor spreading capability (Menges et al., 2014).

Although genetic alterations of *p53* are rare in MPM, the p53 response is impaired by the loss of upstream or downstream regulators, resulting in antiapoptotic effects. Like the miR-34 family, miR-215-5p is a p53-responsive miRNA, lost in MPM tumors and associated with a poor prognosis (Singh et al., 2019). Ectopic expression of miR-215-5p in MPM cells inhibited the malignant phenotype by decreasing MDM2 protein levels, and consequently inducing p53 activation by a positive feedback loop mechanism. The p53-dependent proapoptotic genes such as *CDKN1A*, *BAX*, and *PUMA* were all upregulated upon miR-215-5p transfection. The inhibitory effect was also demonstrated in vivo.

Cioce et al. reported that the hypermethylation of the miR-145 regulatory region is responsible for the low levels of this microRNA in both malignant mesothelial tissues and mesothelioma cell lines (Cioce et al., 2014). Thereby, the downregulation of miR-145 endowed the mesothelial cells with protumorigenic properties. miR-145 negatively regulates pluripotency transcription factors such as OCT4, SOX2, KLF4, and MYC (Xu, Papagiannakopoulos, Pan, Thomson, & Kosik, 2009). The downregulation of miR-145 increased the levels of OCT4, which regulated the expression of the epithelial to mesenchymal transition (EMT)-related gene *ZEB1*. Both *OCT4* and *ZEB1* were responsible for the maintenance of stemness, induction of EMT, and cancer chemoresistance (Cioce et al., 2014). MPM cells transfected with a miR-145 mimic triggered *OCT4* downregulation, reduced the tumorigenicity of cancer cells, and increased the sensitivity to chemotherapeutic drugs.

A large body of evidence underpinned the proposal that the chromosomal location of miRNAs is relevant to tumorigenesis (Calin et al., 2004). Indeed, Calin et al. reported that a significant percentage of miRNAs are frequently located in genomic fragile sites, known to be altered in cancer. In MPM, the homozygous deletion of the 9p21.3 chromosome region is known to be a hallmark of disease, and in this locus reside *CDKN2A* and MIR-31 genes (Ivanov et al., 2010). Ivanov et al. demonstrated that the lack of miR-31 expression in MPM cell lines was linked with the codeletion of the *CDKN2A* locus. A miR-31 target gene is *PPP6C*, a phosphatase associated with chemotherapy resistance and maintenance of chromosomal stability. The expression of miR-31 led to a reduction of *PPP6C* with the consequent inhibition of migration, invasion, and clonogenicity attenuation of the MM cells. These findings suggested that miR-31 held a tumor-suppressive property, implying the possibility of developing novel therapeutic agents against MPM and other cancers that exhibit 9p21.3 chromosome loss. Guled et al. first reported that deregulated miRNAs map to chromosomal regions frequently imbalanced in MPM (Guled et al., 2009). In addition to miR-31, the authors reported several miRNAs in frequently deleted loci associated with MPM: miR-34s and miR-429 located at 1p36 and miR-203 located at 14q32. Conversely, miR-30b* may act as an onco-miRNA mapping to the gained region at 8q24.

Changes in levels of the miR-15/16 family in MPM seem to be more frequent than those in other miRNAs, and these modifications are more common than single genetic events, such as the deletion of the *CDKN2A* locus. Reid et al. reported a consistent downregulation of the miR-15 family in MPM tumor specimens and cell lines that mirrors the changes seen in colon, ovary, lung, and prostate cancer (Bandi et al., 2009; Bhattacharya et al., 2009; Reid et al., 2013). Following miR-16 re-expression, MPM cells showed inhibition of cell growth and sensitization to gemcitabine and pemetrexed. Moreover, miR-16 mimics delivered to the MPM xenograft using EGFR-targeted minicells led to tumor regression and cell cycle arrest by regulating the expression of target genes *CCND1* and *BCL-2*. It is noteworthy that the expression levels of miR-16, along with miR-15 and miR193a-3p, were lower in nonepithelioid MPM with high PD-L1 expression, and in vitro functional studies showed that *PD-L1* could be a direct target of miR-15/16 and miR193a-3p (Kao et al., 2017). These data provide the rationale for potential future applications in the clinic by combining chemotherapy and immune checkpoint inhibitors with the restoration of miR-16. In agreement with this study, Cortez et al. reported that p53 regulated *PD-L1* via miR-34 in nonsmall cell lung cancer (NSCLC). Restoring miR-34 expression reduced PD-L1 expression in the tumor and antagonized T-cell exhaustion (Cortez et al., 2016).

Williams et al. analyzed MPM cells and tumor samples, suggesting that the loss of miR-15/16 in tumor tissue was due to transcriptional repression by c-MYC (Williams et al., 2019). The same group identified a significant downregulation of miR-193a-3p and miR-192 in MPM tumor specimens compared with normal pleura. However, only miR-193a-3p showed a role in MPM tumorigenesis both in vitro and in vivo. Restoring its expression caused growth inhibition and induction of apoptosis through the suppression, at least in part, of the apoptotic gene *MCL-1* (Williams et al., 2015). Despite the promoter methylation being the main cause of miR-193a silencing in NSCLC (Heller et al., 2012) and in acute myeloid leukemia (Gao et al., 2011), the authors ascribed the predominant cause of miR-193a-3p suppression in MPM cells to the genomic deletion of its locus (Williams et al., 2019). As a result of that, miR-193a-3p appeared to be involved in the pathogenesis of MPM and may represent a target for therapeutic intervention.

The Let-7 family has been deeply studied in cancer. A large body of evidence ascribed to this miRNA family a role in MPM tumorigenesis. As a matter of fact, in MPM and NSCLC cell lines, the upregulation of let-7a following activation of the Ephrin type A receptor inhibited the expression of RAS family proto-oncogenes, resulting in tumor growth inhibition, cell proliferation, and migration arrest (Khodayari et al., 2011).

The role of miR-223 was investigated by Birnie et al. in MPM cell lines as well as MPM tissues and cells from pleural effusions (Birnie et al., 2015). MiR-223 levels were significantly reduced in both the cellular component of the pleural effusion of MPM patients and in MPM tissue samples compared with noncancer specimens. MiR-223 targets *STMN1*, a microtubule regulator found overexpressed in several tumors, including MPM. Transfecting miR-223 into MPM cells reduced the STMN1 protein levels and cell motility while inducing tubulin acetylation. STMN1 is also regulated by JNK, whose re-expression of *JNK* in knockout murine fibroblasts decreased Stmn1 and miR-223 levels. The authors suggested a correlation between JNK-mediated miR-223 upregulation and *STMN1* downregulation, ascribing a potential tumor-suppressor role to miR-223.

MDM2 and *PARP1* were also reported to be miR-223 targets (Walter et al., 2016). Besides its role in DNA repair, *PARP1* has been found to induce apoptosis and cell death following asbestos exposure (Sydney & Gilbert, 2000). In this in vitro study, the authors reported a negative correlation between *PARP1* and miR-223 expression and a downregulation of miR-223 in MPM expressing *MDM2* (Walter et al., 2016).

Pass et al. studied a large series of surgical MPM specimens and identified significant downregulation for miR-29c (Pass et al., 2010). Transfecting MPM cell lines with the miR-29c mimic inhibited malignant cell proliferation, invasion, and colony formation. Moreover, its upregulation mediated epigenetic mechanisms through the downregulation of *DNMT1* and *DNMT3A* and the upregulation of demethylating genes. Higher levels of miR-29c were identified in epithelial subtypes and were associated with a better prognosis. No association was observed with asbestos exposure.

In addition to several novel MPM-associated miRNAs, a tumor-suppressor activity was also described for miR-1. Xu et al. observed a lower expression level of miR-1 in MPM tumors as compared with parietal normal pleural specimens (Xu et al., 2009). Re-expression of miR-1 in MPM cell lines resulted in the inhibition of the cellular proliferation rate and the induction of apoptosis (Amatya et al., 2016) by targeting *BCL-2* or an indirect effect on the apoptotic pathway (Cheng et al., 2018). Furthermore, the authors reported that in colorectal cancer, miR-1 suppressed cell proliferation by the inhibition of *SMAD3*. MiR-1 inhibited directly HIF-1α, preventing SMAD3/HIF-1α interaction and the expression of metabolic enzymes. Therefore, the miR-1/SMAD3 axis attenuates tumor glycolysis and inhibits proliferation (Xu et al., 2017).

Cancer cells need to rewire the metabolic programs fulfilling the bioenergetic demand of continuous cell growth. This mechanism represents an adaptive advantage of cancer cells, promoting uncontrolled cell division, cell growth, and survival. During metabolic reprogramming, miRNAs can contribute to keeping high levels of glycolysis and regulate directly or indirectly the expression of genes involved in metabolic machinery. De Santi and colleagues, analyzing tumor tissue compared to controls, identified a profile of miRNAs, including miR-126, miR-15b, miR-145, miR-185, miR-197, and miR-299, involved in the regulation of cell metabolism, MAPK, and p53 and Wnt signaling while miR-151a-5p regulated energy metabolism, reducing ATP production via mitochondrial dysfunction (De Santi et al., 2017).

In MPM cells, miR-126 induced complex metabolic reprogramming through the activation of the autophagic pathway (Tomasetti et al., 2014; Tomasetti et al., 2016). Ectopic expression of miR-126 suppressed IRS signaling, resulting in lower glucose uptake and the activation of the AMPK pathway as well as lysosomal autophagy. Additionally, miR-126 downregulated PDK and ACL activity, inducing citrate accumulation in the cytoplasm and stabilization of HIF1α. Overall, this metabolic reprogramming led to the loss of malignancy in MM.

Andersen et al. showed that DNA hypermethylation of the *EGFL7* promoter correlated with the downregulation of its mRNA and miR-126, leading to a reduction in MPM patient survival (Andersen et al., 2014). On the other hand, Gaetani et al. suggested a PARP1-mediated chromatin remodeling mechanism in miR-126 regulation (Gaetani et al., 2020).

It is noteworthy that the same group observed that asbestos induced miR-126 and miR-222 expression in a time-dependent manner, highlighting their involvement in MPM carcinogenesis (Gaetani et al., 2020).

Besides alterations of cell metabolism, MPM cells gained migratory and invasive properties associated with EMT (Fassina et al., 2012). In particular, the biphasic and sarcomatoid histotype are characterized by a mesenchymal phenotype with a consistent expression of N-cadherin and vimentin, αSMA on the cell surface, and S100A4 on the cytoskeleton (Acloque, Adams, Fishwick, Bronner-Fraser, & Nieto, 2009; Schramm et al., 2010).

The MiR-200 family has emerged as a key mediator in maintaining the epithelial phenotype. Loss of miR-205 expression was observed in nonepithelioid MPM by Fassina et al. (Fassina et al., 2012). Ectopic expression of miR-205 led to a reduction of ZEB1 and ZEB2, thus enhancing E-cadherin expression and inhibiting cell migration and proliferation.

Overall, most of the downregulated miRNAs seemed to have a tumor-suppressor activity regulating the key cellular response, including apoptosis, metabolism, and migration (Reid et al., 2020). However, a few miRNAs have been functionally characterized in vivo.

Despite most miRNAs with biological activity in MPM being described as downregulated, a few studies have also implied upregulated miRNAs in tumorigenesis. A significant upregulation of miR-182, miR-183, and miR-18a was found in mesothelioma cells as compared with normal mesothelial cells (Xu et al., 2013). MiR-182/183 may act as onco-miRNA downregulating *FOXO1*, which in turn targets p27 (Suzuki et al., 2018). FOXO1 is a transcription factor that plays an important role in cell cycle arrest and DNA repair. Furthermore, FOXO1 was reported to regulate apoptosis induced by asbestos in MT-2 cells and its downregulation was implicated in the acquisition of resistance to asbestos after continuous exposure (Matsuzaki et al., 2016). The same group provided evidence that the inhibition of miR-18a increased the sensitivity to cisplatin, inhibited cell migration, and promoted apoptosis (Suzuki et al., 2020).

miRNAs as biomarkers for MPM diagnosis

The major challenge in the treatment of MPM is the late diagnosis and the lack of effective therapies.

The potential role of miRNAs as diagnostic biomarker of MPM has been addressed by several studies, all of them confirming the widespread deregulation of miRNAs in tumor tissue (Russo et al., 2018) (Tables 19.2, 19.3).

Xu and colleagues compared MPM tissues and normal parietal pleural tissues by microarray (Xu et al., 2013). They identified 49 differentially expressed miRNAs; however they validated only miR-206, miR-1, mir-483-5p (downregulated), and miR-155 (upregulated).

Ak et al. showed the significant upregulation of 11 miRNAs (miR-484, miR-320, miR-744, let-7a, miR-20a, miR-193b, let-7d, miR-125a-5p, miR-92 s, miR-155, miR-152) in 18 MPM tissues, of which 11 were previously treated with chemotherapy, compared with six benign asbestos-related pleural effusion (BAPE) patients (Ak et al., 2015). ROC analysis showed that four miRNAs (miR-484, miR-320, let-7a, and miR-125a-5p) had AUC values ≥0.90.

TABLE 19.2 MicroRNAs differentially expressed in MPM compared with healthy or asbestos-exposed controls.

MicroRNA	Regulation in MPM vs nonneoplastic/healthy controls (Up/down)	Sample type	Sample size	References
let-7a	Up	Tumor tissue	$n=18$ MPM cases	Ak et al. (2015)
			$n=6$ benign asbestos-related pleural effusion	
miR-1	Down	Tumor tissue	$n=25$ MPM cases	Xu et al. (2013)
			$n=6$ controls	
miR-15b-5p	Down	Tumor tissue	$n=96$ MPM cases	De Santi et al. (2017)
			$n=10$ normal pleura	
miR-16	Down	Plasma	$n=32$ MPM	Mozzoni et al. (2017)
		Tumor tissue	$n=14$ asbestosis	
			$n=15$ other pulmonary noncancerous diseases	
miR-17	Down	Plasma	$n=32$ MPM	Mozzoni et al. (2017)
		Tumor tissue	$n=14$ asbestosis	
			$n=15$ other pulmonary noncancerous diseases	
miR-19a	Up	Tumor tissue	$n=51$ MM cases	Cappellesso et al. (2016)
		Pleural effusions	$n=40$ nonneoplastic controls	
			$n=29$ MM cases (PE cytology)	
			$n=24$ RMC controls (PE cytology)	
miR-19b	Up	Tumor tissue	$n=51$ MM cases	Cappellesso et al. (2016)
		Pleural effusions	$n=40$ nonneoplastic controls	
			$n=29$ MM cases (PE cytology)	
			$n=24$ RMC controls (PE cytology)	
miR-21	Up	Tumor tissue	$n=51$ MM cases	Cappellesso et al. (2016)
		Pleural effusions	$n=40$ nonneoplastic controls	
			$n=29$ MM cases (PE cytology)	
			$n=24$ RMC controls (PE cytology)	

Continued

TABLE 19.2 MicroRNAs differentially expressed in MPM compared with healthy or asbestos-exposed controls—cont'd

MicroRNA	Regulation in MPM vs nonneoplastic/healthy controls (Up/down)	Sample type	Sample size	References
miR-25	Up	Tumor tissue	n=51 MM cases	Cappellesso et al. (2016)
		Pleural effusions	n=40 nonneoplastic controls	
			n=29 MM cases (PE cytology)	
			n=24 RMC controls (PE cytology)	
miR-30e-3p	Down	Plasma (EVs)	n=23 MPM	Cavalleri et al. (2017)
			n=19 asbestos-exposed controls	
miR-32	Down	Tumor tissue	n=27 MPM cases	Santarelli et al. (2011)
			n=27 normal adjacent control tissue	
miR-32-3p	Up	Serum	n=10 MPM cases	Bononi et al. (2016)
			n=10 asbestos-exposed controls	
			n=10 healthy controls	
miR-103	Down	WBC	n=23 MPM	Weber, Johnen, Bryk, Jöckel, and Brüning (2012)
			n=17 asbestos-exposed controls	
			n=25 healthy controls	
miR-103a-3p	Down	Plasma (EVs)	n=23 MPM	Cavalleri et al. (2017)
			n=19 asbestos-exposed controls	
miR-103a-3p	Down	WBC	n=43 MPM	Weber et al. (2014)
			n=52 asbestos-exposed controls	
miR-125a-5p	Up	Tumor tissue	n=18 MPM cases	Ak et al. (2015)
			n=6 benign asbestos-related pleural effusion	
miR-126	Down	Plasma	n=32 MPM	Mozzoni et al. (2017)
		Tumor tissue	n=14 asbestosis	
			n=15 other pulmonary noncancerous diseases	
miR-126	Down	Serum	n=45 MPM cases	Tomasetti et al. (2012)
			n=20 NSCLC cases	
			n=56 controls	
miR-126	Down	Tumor tissue	n=27 MPM cases (cancer tissue and normal adjacent control)	Santarelli et al. (2011)
		Serum	n=44 MPM cases (serum)	
			n=196 asbestos-exposed (serum)	
			n=50 controls (serum)	
miR-126	Down	Tumor tissue	n=40 MPM cases	Andersen et al. (2014)
			n=14 control tissues	
miR-126	Down	Tumor tissue	n=51 MM cases	Cappellesso et al. (2016)
		Pleural effusions	n=40 nonneoplastic controls	
			n=29 MM cases (PE cytology)	
			n=24 RMC controls (PE cytology)	

TABLE 19.2 MicroRNAs differentially expressed in MPM compared with healthy or asbestos-exposed controls—cont'd

MicroRNA	Regulation in MPM vs nonneoplastic/healthy controls (Up/down)	Sample type	Sample size	References
miR-126-3p	Down	Tumor tissue	$n=96$ MPM cases $n=10$ normal pleura	De Santi et al. (2017)
miR-137	Up	Tumor tissue	$n=115$ MPM cases $n=23$ normal mesothelial tissue	Johnson et al. (2018)
miR-139-5p	Up	Pleural effusion	$n=26$ MPM $n=11$ lung adenocarcinoma $n=10$ benign samples	Birnie et al. (2019)
miR-143	Down	Tumor tissue	$n=40$ MPM cases $n=14$ control tissues	Andersen et al. (2014)
miR-143	Down	Pleural effusion	$n=26$ MPM $n=11$ lung adenocarcinoma $n=10$ benign samples	Birnie et al. (2019)
miR-145	Down	Tumor tissue	$n=40$ MPM cases $n=14$ control tissues	Andersen et al. (2014)
miR-145-5p	Down	Tumor tissue	$n=96$ MPM cases $n=10$ normal pleura	De Santi et al. (2017)
miR-155	Up	Tumor tissue	$n=25$ cases $n=6$ controls	Xu et al. (2013)
miR-185-5p	Down	Tumor tissue	$n=96$ MPM cases $n=10$ normal pleura	De Santi et al. (2017)
miR-192	Down	Tumor tissue	$n=120$ MPM cases $n=23$ normal pleural tissue	Williams et al. (2015)
miR-193a-3p	Down	Tumor tissue	$n=120$ MPM cases $n=23$ normal pleural tissue	Williams et al. (2015)
miR-197-3p	Down	Tumor tissue	$n=96$ MPM cases $n=10$ normal pleura	De Santi et al. (2017)
miR-197-3p	Up	Serum	$n=10$ MPM cases $n=10$ asbestos-exposed controls $n=10$ healthy controls	Bononi et al. (2016)
miR-200b	Down	Tumor tissue	$n=120$ MPM cases $n=23$ normal pleural tissue	Williams et al. (2015)
miR-200c	Down	Pleural effusion	$n=26$ MPM $n=11$ lung adenocarcinoma $n=10$ benign samples	Birnie et al. (2019)

Continued

TABLE 19.2 MicroRNAs differentially expressed in MPM compared with healthy or asbestos-exposed controls—cont'd

MicroRNA	Regulation in MPM vs nonneoplastic/healthy controls (Up/down)	Sample type	Sample size	References
miR-206	Down	Tumor tissue	n=25 cases	Xu et al. (2013)
			n=6 controls	
miR-210	Up	Pleural effusion	n=26 MPM	Birnie et al. (2019)
			n=11 lung adenocarcinoma	
			n=10 benign samples	
miR-299-5p	Down	Tumor tissue	n=96 MPM cases	De Santi et al. (2017)
			n=10 normal pleura	
miR-320	Up	Tumor tissue	n=18 MPM cases	Ak et al. (2015)
			n=6 benign asbestos-related pleural effusion	
miR-483-5p	Down	Tumor tissue	n=25 cases	Xu et al. (2013)
			n=6 controls	
miR-484	Up	Tumor tissue	n=18 MPM cases	Ak et al. (2015)
			n=6 benign asbestos-related pleural effusion	
miR-486	Down	Plasma	n=32 MPM	Mozzoni et al. (2017)
		Tumor tissue	n=14 asbestosis	
			n=15 other pulmonary noncancerous diseases	
miR-625-3p	Up	Plasma	n=15 MPM (plasma)	Kirschner et al. (2012)
		Serum	n=14 healthy controls (plasma)	
		Tumor tissue	n=10 MPM (serum)	
miR-652	Down	Tumor tissue	n=40 MPM cases	Andersen et al. (2014)
			n=14 control tissues	
miR-1281	Up	Serum	n=10 MPM cases	Bononi et al. (2016)
			n=10 asbestos-exposed controls	
			n=10 healthy controls	

Also, Andersen et al. profiled miRNAs levels in five patient-matched FFPE epithelioid tissues previously treated with chemotherapy, preoperative diagnostic biopsies of MPM, and nonneoplastic pleural tissues (Andersen et al., 2014). The authors reported a significant downregulation of miR-126, miR-652, miR-145, and miR-143 in both treated and preoperative MPM compared to nontreated tissues. Moreover, in the validation set, the ROC curve showed a sensitivity and specificity >80% for miR-145 and miR-652 and less than 80% for miR-143 and miR-126.

Even though miR-126, miR-145, and miR-143 have been reported in other studies, it is important to note that chemotherapy affects miRNA profiling, thereby this miRNA set cannot be used as a feasible diagnostic biomarker.

Santarelli and coauthors profiled the expression levels of 88 miRNAs involved in MPM cancerogenesis from 10 freshly collected biopsies of MPM and five healthy subjects (Santarelli et al., 2011). Among the miRNAs tested, only miR-126 significantly remained downregulated in the malignant tissues.

It should be noted that this study was limited to the analysis of a subset of miRNAs and their profiling in frozen samples may vary from those detected in FFPE samples. Furthermore, the authors tested the applicability of miR-126-3p as a diagnostic circulating biomarker for the early detection of MPM and risk-disease prediction in serum samples from 44

TABLE 19.3 MicroRNAs as biomarkers of differential diagnosis between MPM and other diseases.

MicroRNA signature	Sample type (for human studies)	Sample size	References
miR-200c, miR-141, miR-200b, miR-200a* miR-429, miR-203 and miR-205	Tumor tissue	n=100 MPM cases	Gee et al. (2010)
		n=32 Lung cancer	
miR-200c, miR-192, and miR193a-3p	Tumor tissue	n=33 MPM	Benjamin et al. (2010)
		n=210 other cancers	
miR-126	Serum	n=45 MPM cases	Tomasetti et al. (2012)
		n=20 NSCLC cases	
miR-136, miR-113, miR-72, miR-107, miR-79, miR-58, miR-18, miR-65, miR-12, miR-44	Tumor tissue	n=87 MPM cases (TCGA)	Martinez et al. (2019)
		n=497 lung adenocarcinoma (TCGA)	
		n=94 lung adenocarcinoma (BCCA)	
		n=1047 breast cancer (TCGA)	
miR-143, miR210, miR-200c	Pleural effusions	n=26 MPM	Birnie et al. (2019)
		n=11+40 lung adenocarcinoma	
		n=10 benign samples	
miR-130a	Tumor tissue	n=41 MM cases (tissue)	Cappellesso et al. (2012)
	Pleural effusions	n=40 lung AD (tissue)	
		n=26 MM cases (PE)	
		n=27 lung AD (PE)	

MPM cases, 196 asbestos-exposed defined as high-risk subjects, and 50 healthy controls. They observed that miR-126-3p could significantly differentiate the high-risk individuals from both healthy controls and the cancer group. It is noteworthy that the miRNA level in association with soluble mesothelin-related proteins (SMRPs) was able to identify subjects with high risk to develop the disease. The identification of tumor biomarkers used alone or in combination could greatly facilitate the surveillance procedure for cohorts of subjects exposed to asbestos.

Cappellesso et al. tested a panel of 15 miRNAs previously reported as candidate biomarkers in MPM (H2052 and H28) and normal mesothelial (MET-5A) cell lines (Cappellesso et al., 2016). Significant miRNAs were validated in 51 MPM and 40 benign pleural samples and then were tested in 29 MPM and 24 reactive mesothelial cell cytologic (RMC) specimens. The significant upregulation of miR-19a, miR-19b, and miR-21 and downregulation of miR-126 were identified in MPM samples. The best classifier for MPM diagnosis resulted from the combination of miR-21 and miR-126, achieving 86% and 87% of sensitivity and specificity, respectively.

To identify a pattern of miRNAs for differential diagnosis, Gee et al. investigated the miRNA expression profile in 15 MPM tissues and 10 lung adenocarcinomas, using a microarray platform (Gee et al., 2010). The authors validated the data by qPCR in 100 MPM tissues and 32 samples of lung tumors and reported a significantly lower expression of the miR-200 family (miR-200b, miR-200c, miR-203, miR-205), miR-141, and miR-429. This pattern of miRNAs was able to distinguish the different tumor samples with an accuracy greater than 0.8.

Cappellesso et al. identified three upregulated miRNAs (miR-130a, miR-193a, and miR-675) and three downregulated miRNAs (miR-141, miR-205, and miR-375) comparing MPM and lung adenocarcinoma tissues (Cappellesso et al., 2017). A validation analysis in 41 epithelioid MPM and 40 CE FFPE samples as well as 26 MPM and 27 CE cytological specimens revealed that only miR-103a was significantly overexpressed in both histological samples.

Despite the promising results, the quantification of miRNAs in tissue remains a challenge given the invasive nature of a biopsy.

Extracellular miRNAs have generated considerable interest for their clinical applications. One of the main features of miRNAs is their remarkable stability in biofluids; therefore, they represent a powerful tool for noninvasive diagnosis, prognosis, and risk stratification.

MiR-126-3p appears to be consistently downregulated in both the tissue and serum of MPM patients. It could be a valuable biomarker for risk prediction as well as to differentiate MPM patients from healthy controls and NSCLC (Tomasetti et al., 2012). Santarelli et al. showed that the combination of miR-126, methylated thrombomodulin promoter, and SMRPs significantly improved the diagnostic performance, distinguishing MPM patients from asbestos-exposed subjects and healthy controls, with an AUC of 0.857 (Santarelli et al., 2015). The accuracy and cancer specificity were confirmed in a validation cohort and in lung cancer patients.

An accurate diagnosis to distinguish MM from either another neoplasm or benign pleural diseases is clinically important to address therapy. Mozzoni et al. profiled miRNA expression in plasma and FFPE samples of MPM patients, individuals with asbestosis and healthy subjects (Mozzoni et al., 2017). miR-126, miR-16, miR-17, and miR-486 levels were downregulated in plasma samples both in patients with MPM and in asbestosis compared to controls.

Kirschner and collaborators identified an upregulation for circulating plasma miR-625-3p in MM cases compared to controls (Kirschner et al., 2012). ROC analysis showed that miR-625-3p levels were able to discriminate MPM patients and controls with an accuracy of 82.4%. MiR-625-3p was also found upregulated in serum and in tumor specimens of MM cases compared with asbestosis patients.

In a study by Bononi et al., a signature of three serum-circulating miRNAs was proposed as a potential new biomarker for MPM (Bononi et al., 2016). The authors reported a strong upregulation of miR-197-3p, miR-1281, miR-32-3p in MPM patients compared to healthy controls and an up-regulation of miR-1281 was also observed in asbestos-exposed workers compared with healthy controls.

Higher serum miR-548a-3p and miR-20a expression levels were able to distinguish MPM from asbestos-exposed individuals and healthy control groups, with 100% sensitivity and 87.5% specificity (Matboli et al., 2018).

MiRNAs can also be found circulating in serum/plasma packaged in extracellular vesicles (EVs) acting as signaling molecules even at long distances. MiRNAs detected in EVs are more stable and resistant to degradation in body fluids, during prolonged storage, and freeze–thaw cycles.

Cavalleri et al. profiled miRNAs in plasma EVs from MPM cases and individuals with a past exposure to asbestos using an OpenArray method (Cavalleri et al., 2017). ROC curve analysis showed a two-miRNA signature given by miR-103a-3p and miR-30e-3p, able to discriminate MPM from asbestos-exposed subjects (AUC 0.942).

This study was consistent with previous results reported by Weber and collaborators (Weber et al., 2012). In the first study, the authors identified a downregulation of miR-103 in the cellular fraction of human peripheral blood of 23 MPM cases and 17 asbestos-exposed ones, analyzing a panel of 328 miRNAs. Validation results showed a specificity of 76% and a sensitivity of 78% to discriminate MPM cases from controls while discrimination from asbestos-exposed subjects had a sensitivity of 83% and a specificity of 71%.

In a subsequent study, the authors assessed the combination of miR-103a-3p in the cellular blood fraction along with circulating mesothelin in plasma to enhance the biomarker performance to MPM diagnosis (Weber et al., 2014). The discrimination performance improved when both markers were taken together.

As described for miR-126, these results support the concept that combinations of biomarkers are needed for a reliable diagnosis of MPM. Weber et al. also measured miRNA levels in the plasma of 21 MPM patients and 21 asbestos-exposed controls using the TaqMan low density array, verifying the results by qRT-PCR (Weber et al., 2014). miR-132-3p was significantly downregulated in MPM patients showing 86% sensitivity and 61% specificity. The expression levels of miR-625-3p and miR-126-3p were additionally analyzed in the validation cohort: a significant downregulation in MPM cases compared with asbestos-exposed controls was observed only for miR-126, showing a sensitivity of 86% and a specificity of 59%, in contrast to Santarelli et al. (Santarelli et al., 2011).

The most discriminating miRNA combination was given by miR-126-3p and miR-132-3p, which generated an AUC of 0.88 (95% CI = 0.80–0.95) with a sensitivity of 77% and a specificity of 86%.

However, these findings were not confirmed as early diagnostic biomarkers for MPM even in asbestos-exposed individuals (Weber et al., 2019). The plasma miRNA profiles for miR-132-3p, miR-126-3p, and miR-103a-3p were explored in 17 prospectively collected MPM cases and 34 controls. All subjects were workers formerly exposed to asbestos and the group of cases developed MPM within about 8.9 months. The set of investigate miRNAs revealed 0% sensitivity and 98% specificity. Nevertheless, the authors stated that these profiles might still serve as possible markers for prognosis and response to therapy.

Cell-free miRNAs in pleural effusion can be exploited as potential biomarkers in particular to differentiate MPM from other lung malignancies.

To address this hypothesis, Birnie et al. analyzed miRNA in pleural effusion (PE) cells and supernatants by TaqMan OpenArray (Birnie et al., 2019). Only in PE cells, a signature of three miRNAs (miR-143, miR-210, and miR-200c) showed the potential to differentiate MPM from other PE-causing diseases with an AUC of 0.92.

One of the most ambitious aims for researchers is the identification of biomarkers that allow early diagnosis in the asbestos-exposed population. Ongoing and future studies using NGS approaches will clarify the effectiveness of microRNAs as novel biomarkers and their predictive value.

miRNAs as biomarkers for malignant pleural mesothelioma prognosis

The accurate prediction and monitoring of disease progression is of paramount importance for the appropriate clinical management of disease. MiRNAs can also be utilized as potential prognostic biomarkers (Table 19.4).

TABLE 19.4 MicroRNAs with prognostic value in MPM patients.

Aim of the study	miRNA signature	Sample type	Sample size	References
Overall Survival	Let-7c-5p+miR-151a-5p	Tumor tissue	*n*=52 MPM patients	De Santi et al. (2017)
			n=83 MPM TCGA (validation)	
			n=16 MPM patients (validation)	
Histological subtype	miR-205	Tumor tissue	*n*=58 MPM epithelioid	Fassina et al. (2012)
			n=26 MPM sarcomatoid	
			n=25 MPM biphasic	
Overall Survival	miR-126	Serum	*n*=45 MPM cases	Tomasetti et al. (2012)
Tumor staging	miR-126	Tumor tissue	*n*=27 MPM cases	Santarelli et al. (2011)
Tumor staging	miR-32	Tumor tissue	*n*=27 MPM cases	Santarelli et al. (2011)
Tumor staging	miR-335	Tumor tissue	*n*=27 MPM cases	Santarelli et al. (2011)
Overall survival	miR-21-5p+miR-23a-3p+miR-30e-5p+miR-221-3p+miR-222-3p+miR-31-5p	Tumor tissue	*n*=8 MPM long survivors	Kirschner et al. (2015)
			n=8 MPM short survivors	
Overall survival	mir-99a+let-7c+miR-125b	Tumor tissue	*n*=12 MPM long survivors	Truini et al. (2017)
			n=15 MPM short survivors	
			n=72 MPM cases (replication)	
			n=30 MPM cases (validation)	
Histological subtype	miR-17-5p, miR-21, miR-29a, miR-30c, miR-30e-5p, miR-106a, and miR-143	Tumor tissue	*n*=24 MPM cases	Busacca et al. (2010)
Overall survival	miR-17-5p and miR-30c	Tumor tissue	*n*=24 MPM cases	Busacca et al. (2010)
Time to progression	miR-29c*	Tumor tissue	*n*=44 MPM cases (training set)	Pass et al. (2010)
Overall survival			*n*=98 MPM cases (test set)	
Overall survival	miR-16	Plasma	*n*=32 MPM	Mozzoni et al. (2017)
		Tumor tissue	*n*=14 asbestosis	
			n=15 other lung diseases	
Overall survival	miR-137	Tumor tissue	*n*=115 MPM cases	Johnson et al. (2018)
Histological subtype	miR-31	Tumor tissue	*n*=25 MPM cases	Matsumoto, Nabeshima, Hamasaki, Shibuta, and Umemura (2014)
Overall survival			*n*=20 reactive mesothelial proliferation	

In a first study, Kirschner and colleagues identified a signature of six miRNAs (miR-21-5p, miR-23a-3p, miR-30e-5p, miR-221-3p, miR-222-3p, and miR-31-5p) predictive of higher overall survival with an accuracy of 92.3% in patients who underwent extrapleural pneumonectonomy (Kirschner et al., 2015). Subsequently, in a new cohort of MPM patients who received chemotherapy followed by extrapleural pneumonectomy and nontreated patients, the authors refined the miR-Score that includes miR-30e-5p and miR-221-3p less affected by chemotherapy (Kirschner, 2017). ROC curve analysis achieved an AUC of 0.77 (95% CI: 0.61–0.94) for nontreated patients and an AUC of 0.80 (95% CI: 0.64–0.96) for chemotherapy-treated patients.

A signature based on three miRNAs with lower expression levels (miR-99a, let-7c, and miR-125b) was correlated with poor overall survival in unresected MPM patients (Truini et al., 2017).

In a pilot study of 32 MPM, 14 subjects with asbestosis, and 15 with noncancerous pulmonary diseases, Mozzoni et al. observed a correlation of miR-16 and miR-486 in plasma and tissue. Moreover, the authors highlighted that MPM patients with a higher expression level of miR-16 both in plasma and tissue and miR-486 (tissue only) showed a higher overall survival (Mozzoni et al., 2017).

The analysis of 24 MPM specimens for the three histotypes yielded to the identification of miR-30c and miR-17-5p significantly downregulated in the sarcomatoid subtype and associated with better survival (Busacca et al., 2010). In the analysis of 38 epithelioid MPM, Mairinger and collaborators reported miR-335-5p and miR-566 associated with a worse prognosis and shortened progression-free survival (Truini et al., 2017). In contrast, the high expression of miR-1246 correlated with a better response and prolonged PFS as well as high levels of miR-146a-5p, miR-378a-3p, and miR-451a.

In a large study of MPM tumor samples, increased levels of miR-29c-5p have been identified in epithelioid subtypes and predicted a more favorable prognosis (Pass et al., 2010).

Contribution of miRNAs in drug MPM resistance

MPM showed remarkable resistance to current therapeutic treatments and new approaches are needed. Deregulated miRNAs may contribute to drug sensitivities of MPM cells.

Cioce et al. observed that the shift to the chemo-resistant phenotype within pemetrexed and cisplatin-treated MPM cell lines was linked to the upregulation of EMT and stemness-related genes, including *OCT4*. Indeed, miR-145-mimic expression antagonized the increase of *OCT4* induced by pemetrexed treatment (Cioce et al., 2014).

Suzuky et al. demonstrated that transfection with the miR-18a inhibitor reduced the proliferation and sensitized MPM cell lines to cisplatin (Suzuki et al., 2020).

Although the lack of miR-31 expression is associated with a poor prognosis and its re-expression resulted in the attenuation of the malignant phenotype, miR-31 expression enhanced resistance to standard chemotherapy (Moody, Lind, & Maher, 2017). Conversely, Moody et al. suggested that patients with miR-31 loss triggered chemosensitivity and may benefit from a platinum-based therapy.

Williams et al. reported that acquired drug resistance in MPM is associated with the suppression of miR-15/16 and miR-34a. In particular, miR-15a and miR-16 played a role in the development of acquired and intrinsic resistance in MPM cancer cells by modulating drug-induced apoptosis via the upregulation of *BCL-2* (Williams et al., 2019).

Maki et al. delved into the role of miR-34b/c in the cellular radiosensitivity of MPM cells by in vitro analyses (Maki et al., 2012). Transfection of miR-34b/c increased the radiosensitivity of MPM cell lines by suppressing *CDK4/6* and impaired DNA double-strand break repair by suppressing *CCND1*.

Therapeutic role of miRNAs

Based on the described data, MPM is characterized by widespread miRNA downregulation, which generally resulted in the overexpression of target genes. Due to their regulatory capabilities, miRNAs have been considered as one of the most interesting therapeutic tool for cancer treatment. Manipulation of miRNAs via their re-expression in the tissue may generate a pleiotropic effect, and multiple deregulated pathways can be potentially restored.

Despite several in vitro studies demonstrating the therapeutic benefit of restoring miRNA expression, only a few have proven to be effective in vivo (Reid et al., 2020).

A miRNA-based approach to activate p53 signaling could be a potential therapeutic strategy. Singh et al. investigated the mechanisms underlying miR-215-5p downregulation in MPM (Singh et al., 2019). The delivery of miR-215-5p to tumor xenografts complexed in an atelocollagen vehicle induced cell apoptosis via a mechanism of the MDM2-p53 positive feedback loop. Furthermore, the intrapleural administration of miR-215-5p in an orthotopic xenograft mouse model suppressed tumor growth and improved survival. It is noteworthy that miR-215-5p re-expression induced the activation of miR-145-5p in a p53-dipendent manner.

Cioce et al. directly assessed the anticancer effect of miR-145-5p in nude mice transplanted with cancer mesothelial cells (Cioce et al., 2014). Tumor formation occurred in all control mice but only in 2 of 8 micetreated with the miR-145-mimics.

Williams et al. demonstrated that the delivery of miR-193-3p mimics using EDVTM minicells inhibited tumor formation in a xenograft mouse model and increased apoptosis (Williams et al., 2015).

Despite several in vitro and in vivo studies aimed at the development of miRNA-based anticancer strategies, only miR-16 was tested in a phase 1 clinical trial. Based on the evidence that targeted delivery of miR-16 inhibited tumor growth in xenograft nude mice (Reid et al., 2013), the treatment aimed to restore the activity of miR-15/16 family.

This first-in-human, open-label, dose-escalation study assessed the efficacy and safety of miR-16 mimic packaged in TargomiR conjugated with EGFR-specific antibodies. The study enrolled 27 patients, 26 of whom received at least one TargomiR dose. In spite of the low toxicity, only a small proportion of patients achieved an objective response ($n=5\%$) (van Zandwijk et al., 2017).

Overall, these findings provided the rationale and evidenced the challenges for the development of miRNA-based anticancer strategies for improving the therapeutic efficacy in MPM care.

Conclusion and future perspectives

Despite the advancements in cancer research, MPM clinical progress is still insufficient.

Individuals with an environmental or past occupational exposure to asbestos may potentially benefit from sensitive and noninvasive biomarkers, allowing for both early detection and potential better outcomes. All presented studies highlighted the importance of miRNAs as potential diagnostic and prognostic biomarkers. However, from the technical and translational point of view, there are still several hurdles to overcome. The discordant results across the studies are most likely caused by the lack of standardized approaches for sample collection, miRNA quantification, choice of internal control, and analytical approach. Another limitation is the lacking of studies with large sample size and validation of results in different independent cohorts.

Furthermore, most of the studies profiled cell-free miRNAs levels in plasma/serum, making it challenging to specifically pinpoint their cell origin. The abundance of a tumor-derived miRNA can be diluted in blood by other cells secreting miRNAs. On the other hand, EV-derived miRNAs can be more specific, reflecting the status of parental cells. Thereby, the use of EVs might provide a reliable source for biomarker development and might represent an advantage for clinical use.

Overall, the presented findings implicate the p53 signaling as a key effector affecting the global miRNA downregulation (Reid et al., 2020). Indeed, p53 is involved in the maturation of primary miRNA transcripts, interacting with the Drosha complex. Second, the impairment of p53 affects target miRNAs, which in turn results in the alteration of cell proliferation. This evidence clearly supports the development of miRNA-based anticancer therapies in MPM by targeting multiple oncogenic pathways and increasing the efficacy of standard chemotherapy and immunotherapy. However, despite the promising results, many hurdles have yet to be overcome before the implementation of miRNA into clinics. One of the main issues is the use of an appropriate delivery system to protect miRNAs from nuclease degradation and reach target cells. In an attempt to address this challenge, EVs isolated from primary MPM cells and modified with a specific cargo may represent a potential vehicle for therapeutic miRNA delivery (Burgio et al., 2020).

In conclusion, the identification of novel differentially expressed miRNAs as well as the definition of their pathobiological functions are paving the way for new opportunities in the field of personalized medicine.

References

Acloque, H., Adams, M. S., Fishwick, K., Bronner-Fraser, M., & Nieto, M. A. (2009). Epithelial-mesenchymal transitions: The importance of changing cell state in development and disease. *Journal of Clinical Investigation*, *119*(6), 1438–1449. https://doi.org/10.1172/JCI38019.

Ak, G., Tomaszek, S. C., Kosari, F., Metintas, M., Jett, J. R., Metintas, S., et al. (2015). MicroRNA and mRNA features of malignant pleural mesothelioma and benign asbestos-related pleural effusion. *BioMed Research International*, *2015*. https://doi.org/10.1155/2015/635748.

Amatya, V. J., Mawas, A. S., Kushitani, K., El-Din, M. M. M., & Takeshima, Y. (2016). Differential microRNA expression profiling of mesothelioma and expression analysis of miR-1 and miR-214 in mesothelioma. *International Journal of Oncology*, *48*(4), 1599–1607. https://doi.org/10.3892/ijo.2016.3358.

Andersen, M., Grauslund, M., Ravn, J., Sørensen, J. B., Andersen, C. B., & Santoni-Rugiu, E. (2014). Diagnostic potential of miR-126, miR-143, miR-145, and miR-652 in malignant pleural mesothelioma. *Journal of Molecular Diagnostics*, *16*(4), 418–430. https://doi.org/10.1016/j.jmoldx.2014.03.002.

Andersen, M., Trapani, D., Ravn, J., Sorensen, J. B., Andersen, C. B., Grauslund, M., et al. (2015). Methylation-associated silencing of microRNA-126 and its host gene EGFL7 in malignant pleural mesothelioma. *Anticancer Research*, *35*(11), 6223–6229.

Bandi, N., Zbinden, S., Gugger, M., Arnold, M., Kocher, V., Hasan, L., et al. (2009). miR-15a and miR-16 are implicated in cell cycle regulation in a Rb-dependent manner and are frequently deleted or downregulated in non-small cell lung cancer. *Cancer Research*, *69*(13), 5553–5559. https://doi.org/10.1158/0008-5472.CAN-08-4277.

Benjamin, H., Lebanony, D., Rosenwald, S., Cohen, L., Gibori, H., Barabash, N., et al. (2010). A diagnostic assay based on microRNA expression accurately identifies malignant pleural mesothelioma. *The Journal of Molecular Diagnostics*, *12*(6), 771–779.

Bhattacharya, R., Nicoloso, M., Arvizo, R., Wang, E., Cortez, A., Rossi, S., et al. (2009). MiR-15a and MiR-16 control Bmi-1 expression in ovarian cancer. *Cancer Research*, *69*(23), 9090–9095. https://doi.org/10.1158/0008-5472.CAN-09-2552.

Birnie, K. A., Prêle, C. M., Musk, A. W. B., de Klerk, N., Lee, Y. C. G., Fitzgerald, D., et al. (2019). MicroRNA signatures in malignant pleural mesothelioma effusions. *Disease Markers*, *2019*, 8628612. https://doi.org/10.1155/2019/8628612.

Birnie, K. A., Yip, Y. Y., Ng, D. C. H., Kirschner, M. B., Reid, G., Prêle, C. M., et al. (2015). Loss of miR-223 and JNK signaling contribute to elevated stathmin in malignant pleural mesothelioma. *Molecular Cancer Research*, *13*(7), 1106–1118. https://doi.org/10.1158/1541-7786.MCR-14-0442.

Bononi, I., Comar, M., Puozzo, A., Stendardo, M., Boschetto, P., Orecchia, S., et al. (2016). Circulating microRNAs found dysregulated in ex-exposed asbestos workers and pleural mesothelioma patients as potential new biomarkers. *Oncotarget*, *7*(50), 82700–82711. https://doi.org/10.18632/oncotarget.12408.

Burgio, S., Noori, L., Gammazza, A. M., Campanella, C., Logozzi, M., Fais, S., et al. (2020). Extracellular vesicles-based drug delivery systems: A new challenge and the exemplum of malignant pleural mesothelioma. *International Journal of Molecular Sciences*, *21*(15), 1–29. https://doi.org/10.3390/ijms21155432.

Busacca, S., Germano, S., De Cecco, L., Rinaldi, M., Comoglio, F., Favero, F., et al. (2010). MicroRNA signature of malignant mesothelioma with potential diagnostic and prognostic implications. *American Journal of Respiratory Cell and Molecular Biology*, *42*(3), 312–319. https://doi.org/10.1165/rcmb.2009-0060OC.

Calin, G. A., Sevignani, C., Dumitru, C. D., Hyslop, T., Noch, E., Yendamuri, S., et al. (2004). Human microRNA genes are frequently located at fragile sites and genomic regions involved in cancers. *Proceedings of the National Academy of Sciences of the United States of America*, *101*(9), 2999–3004. https://doi.org/10.1073/pnas.0307323101.

Cappellesso, R., Galasso, M., Nicolè, L., Dabrilli, P., Volinia, S., & Fassina, A. (2017). miR-130A as a diagnostic marker to differentiate malignant mesothelioma from lung adenocarcinoma in pleural effusion cytology. *Cancer Cytopathology*, *125*(8), 635–643. https://doi.org/10.1002/cncy.21869.

Cappellesso, R., Nicolé, L., Caroccia, B., Guzzardo, V., Ventura, L., Fassan, M., et al. (2016). Young investigator challenge: MicroRNA-21/MicroRNA-126 profiling as a novel tool for the diagnosis of malignant mesothelioma in pleural effusion cytology. *Cancer Cytopathology*, *124*(1), 28–36. https://doi.org/10.1002/cncy.21646.

Carbone, M., Ly, B. H., Dodson, R. F., Pagano, I., Morris, P. T., Dogan, U. A., et al. (2012). Malignant mesothelioma: Facts, myths, and hypotheses. *Journal of Cellular Physiology*, *227*(1), 44–58. https://doi.org/10.1002/jcp.22724.

Cavalleri, T., Angelici, L., Favero, C., DIoni, L., Mensi, C., Bareggi, C., et al. (2017). Plasmatic extracellular vesicle microRNAs in malignant pleural mesothelioma and Asbestos-exposed subjects suggest a 2-miRNA signature as potential biomarker of disease. *PLoS One*, *12*(5). https://doi.org/10.1371/journal.pone.0176680.

Cheng, Y., Zhao, W., Zhang, X., Sun, L., Yang, H., Wang, Y., et al. (2018). Downregulation of microRNA-1 attenuates glucose-induced apoptosis by regulating the liver X receptor α in cardiomyocytes. *Experimental and Therapeutic Medicine*, *16*(3), 1814–1824. https://doi.org/10.3892/etm.2018.6388.

Cheng, Y. Y., Wright, C. M., Kirschner, M. B., Williams, M., Sarun, K. H., Sytnyk, V., et al. (2016). KCa1.1, a calcium-activated potassium channel subunit alpha 1, is targeted by miR-17-5p and modulates cell migration in malignant pleural mesothelioma. *Molecular Cancer*, *15*(1), 44.

Cioce, M., Ganci, F., Canu, V., Sacconi, A., Mori, F., Canino, C., et al. (2014). Protumorigenic effects of mir-145 loss in malignant pleural mesothelioma. *Oncogene*, *33*(46), 5319–5331. https://doi.org/10.1038/onc.2013.476.

Cortez, M. A., Ivan, C., Valdecanas, D., Wang, X., Peltier, H. J., Ye, Y., et al. (2016). PDL1 regulation by p53 via miR-34. *Journal of the National Cancer Institute*, *108*(1). https://doi.org/10.1093/jnci/djv303.

De Santi, C., Melaiu, O., Bonotti, A., Cascione, L., Di Leva, G., Foddis, R., et al. (2017). Deregulation of miRNAs in malignant pleural mesothelioma is associated with prognosis and suggests an alteration of cell metabolism. *Scientific Reports*, *7*, 3140.

Fassina, A., Cappellesso, R., Guzzardo, V., Dalla Via, L., Piccolo, S., Ventura, L., et al. (2012). Epithelial-mesenchymal transition in malignant mesothelioma. *Modern Pathology*, *25*(1), 86–99. https://doi.org/10.1038/modpathol.2011.144.

Feng, H., Ge, F., Du, L., Zhang, Z., & Liu, D. (2019). MiR-34b-3p represses cell proliferation, cell cycle progression and cell apoptosis in non-small-cell lung cancer (NSCLC) by targeting CDK4. *Journal of Cellular and Molecular Medicine*, *23*(8), 5282–5291. https://doi.org/10.1111/jcmm.14404.

Gaetani, S., Monaco, F., Alessandrini, F., Tagliabracci, A., Sabbatini, A., Bracci, M., et al. (2020). Mechanism of miR-222 and miR-126 regulation and its role in asbestos-induced malignancy. *International Journal of Biochemistry and Cell Biology*, *121*. https://doi.org/10.1016/j.biocel.2020.105700.

Gao, X. N., Lin, J., Li, Y. H., Gao, L., Wang, X. R., Wang, W., et al. (2011). MicroRNA-193a represses c-kit expression and functions as a methylation-silenced tumor suppressor in acute myeloid leukemia. *Oncogene*, *30*(31), 3416–3428. https://doi.org/10.1038/onc.2011.62.

Gee, G. V., Koestler, D. C., Christensen, B. C., Sugarbaker, D. J., Ugolini, D., Ivaldi, G. P., et al. (2010). Downregulated microRNAs in the differential diagnosis of malignant pleural mesothelioma. *International Journal of Cancer*, *127*(12), 2859–2869. https://doi.org/10.1002/ijc.25285.

Guidelines for the diagnosis and treatment of malignant pleural mesothelioma. (2013). https://doi.org/10.3978/j.issn.2072-1439.2013.11.28.

Guled, M., Lahti, L., Lindholm, P. M., Salmenkivi, K., Bagwan, I., Nicholson, A. G., et al. (2009). CDKN2A, NF2, and JUN are dysregulated among other genes by miRNAs in malignant mesothelioma - a mirna microarray analysis. *Genes Chromosomes and Cancer*, *48*(7), 615–623. https://doi.org/10.1002/gcc.20669.

He, T., McColl, K., Sakre, N., Chen, Y., Wildey, G., Dowlati, A., et al. (2018). Post-transcriptional regulation of PIAS3 expression by miR-18a in malignant mesothelioma. *Molecular Oncology*, *12*(12), 2124–2135.

Heller, G., Weinzierl, M., Noll, C., Babinsky, V., Ziegler, B., Altenberger, C., et al. (2012). Genome-wide miRNA expression profiling identifies miR-9-3 and miR-193a as targets for DNA methylation in non-small cell lung cancers. *Clinical Cancer Research*, *18*(6), 1619–1629. https://doi.org/10.1158/1078-0432.CCR-11-2450.

Hermeking, H. (2010). The miR-34 family in cancer and apoptosis. *Cell Death and Differentiation*, *17*(2), 193–199. https://doi.org/10.1038/cdd.2009.56.

Ivanov, S. V., Goparaju, C. M. V., Lopez, P., Zavadil, J., Toren-Haritan, G., Rosenwald, S., et al. (2010). Pro-tumorigenic effects of miR-31 loss in mesothelioma. *Journal of Biological Chemistry*, *285*(30), 22809–22817. https://doi.org/10.1074/jbc.M110.100354.

Johnson, T. G., Schelch, K., Cheng, Y. Y., Williams, M., Sarun, K. H., Kirschner, M. B., et al. (2018). Dysregulated expression of the microRNA miR-137 and its target YBX1 contribute to the invasive characteristics of malignant pleural mesothelioma. *Journal of Thoracic Oncology*, *13*(2), 258–272.

Kao, S. C., Cheng, Y. Y., Williams, M., Kirschner, M. B., Madore, J., Lum, T., et al. (2017). Tumor suppressor microRNAs contribute to the regulation of PD-L1 expression in malignant pleuralmesothelioma. *Journal of Thoracic Oncology*, *12*(9), 1421–1433. https://doi.org/10.1016/j.jtho.2017.05.024.

Khodayari, N., Mohammed, K. A., Goldberg, E. P., & Nasreen, N. (2011). EphrinA1 inhibits malignant mesothelioma tumor growth via let-7 microRNA-mediated repression of the RAS oncogene. *Cancer Gene Therapy*, *18*(11), 806–816. https://doi.org/10.1038/cgt.2011.50.

Khodayari, N., Mohammed, K. A., Lee, H., Kaye, F., Nasreen, N., et al. (2016). MicroRNA-302b targets Mcl-1 and inhibits cell proliferation and induces apoptosis in malignant pleural mesothelioma cells. *American Journal of Cancer Research*, *6*(9), 1996–2009.

Kirschner, M. (2017). Refinement of the prognostic miR-score for use in diagnostic specimens from chemo-naive malignant pleural mesothelioma patients. *Journal of Thoracic Oncology*, *12*(1).

Kirschner, M. B., Cheng, Y. Y., Armstrong, N. J., Lin, R. C. Y., Kao, S. C., Linton, A., et al. (2015). MiR-score: A novel 6-microRNA signature that predicts survival outcomes in patients with malignant pleural mesothelioma. *Molecular Oncology*, *9*(3), 715–726. https://doi.org/10.1016/j.molonc.2014.11.007.

Kirschner, M. B., Cheng, Y. Y., Badrian, B., Kao, S. C., Creaney, J., Edelman, J. J. B., et al. (2012). Increased circulating miR-625-3p: A potential biomarker for patients with malignant pleural mesothelioma. *Journal of Thoracic Oncology*, *7*(7), 1184–1191. https://doi.org/10.1097/JTO.0b013e3182572e83.

Kubo, T., Toyooka, S., Tsukuda, K., Sakaguchi, M., Fukazawa, T., Soh, J., et al. (2011). Epigenetic silencing of microRNA-34b/c plays an important role in the pathogenesis of malignant pleural mesothelioma. *Clinical Cancer Research*, *17*(15), 4965–4974. https://doi.org/10.1158/1078-0432.CCR-10-3040.

Lee, H. Y., Mohammed, K. A., Kaye, F., Sharma, P., Moudgil, B. M., Clapp, W. L., et al. (2013). Targeted delivery of let-7a microRNA encapsulated ephrin-A1 conjugated liposomal nanoparticles inhibit tumor growth in lung cancer. *International Journal of Nanomedicine*, *8*, 4481–4494.

Maki, Y., Asano, H., Toyooka, S., Soh, J., Kubo, T., Katsui, K., et al. (2012). MicroRNA miR-34b/c enhances cellular radiosensitivity of malignant pleural mesothelioma cells. *Anticancer Research*, *32*(11), 4871–4876. http://ar.iiarjournals.org/content/32/11/4871.full.pdf+html.

Martinez, V. D., Marshall, E. A., Anderson, C., Ng, K. W., Minatel, B. C., Sage, A. P., et al. (2019). Discovery of previously undetected microRNAs in mesothelioma and their use as tissue-of-origin markers. *American Journal of Respiratory Cell and Molecular Biology*, *61*(2), 266–268.

Matboli, M., Shafei, A. E., Azazy, A. E. M., Reda, M., El-Khazragy, N., Nagy, A. A., et al. (2018). Clinical evaluation of circulating miR-548a-3p and -20a expression in malignant pleural mesothelioma patients. *Biomarkers in Medicine*, *12*(2), 129–139. https://doi.org/10.2217/bmm-2017-0224.

Matsumoto, S., Nabeshima, K., Hamasaki, M., Shibuta, T., Umemura, T., et al. (2014). Upregulation of microRNA-31 associates with a poor prognosis of malignant pleural mesothelioma with sarcomatoid component. *Medical Oncology*, *31*(12), 303.

Matsuzaki, H., Lee, S., Maeda, M., Kumagai-Takei, N., Nishimura, Y., & Otsuki, T. (2016). FoxO1 regulates apoptosis induced by asbestos in the MT-2 human T-cell line. *Journal of Immunotoxicology*, *13*(5), 620–627. https://doi.org/10.3109/1547691X.2016.1143539.

Menges, C. W., Kadariya, Y., Altomare, D., Talarchek, J., Neumann-Domer, E., Wu, Y., et al. (2014). Tumor suppressor alterations cooperate to drive aggressive mesotheliomas with enriched cancer stem cells via a p53-miR-34a-c-met axis. *Cancer Research*, *74*(4), 1261–1271. https://doi.org/10.1158/0008-5472.CAN-13-2062.

Moody, H. L., Lind, M. J., & Maher, S. G. (2017). MicroRNA-31 regulates chemosensitivity in malignant pleural mesothelioma. *Molecular Therapy—Nucleic Acids*, *8*, 317–329. https://doi.org/10.1016/j.omtn.2017.07.001.

Mozzoni, P., Ampollini, L., Goldoni, M., Alinovi, R., Tiseo, M., Gnetti, L., et al. (2017). MicroRNA expression in malignant pleural mesothelioma and asbestosis: A pilot study. *Disease Markers*, *2017*. https://doi.org/10.1155/2017/9645940.

Oliveto, S., Alfieri, R., Miluzio, A., Scagliola, A., Secli, R. S., Gasparini, P., et al. (2018). A polysome-based microRNA screen identifies miR-24-3p as a novel promigratory miRNA in mesothelioma. *Cancer Research*, *78*(20), 5741–5753.

Pass, H. I., Goparaju, C., Ivanov, S., Donington, J., Carbone, M., Hoshen, M., et al. (2010). Hsa-miR-29c* is linked to the prognosis of malignant pleural mesothelioma. *Cancer Research*, *70*(5), 1916–1924. https://doi.org/10.1158/0008-5472.CAN-09-3993.

Pistolesi, M., & Rusthoven, J. (2004). Malignant pleural mesothelioma: Update, current management, and newer therapeutic strategies. *Chest*, *126*(4), 1318–1329. https://doi.org/10.1378/chest.126.4.1318.

Reid, G., Johnson, T. G., & van Zandwijk, N. (2020). Manipulating microRNAs for the treatment of malignant pleural mesothelioma: Past present and future. *Frontiers in Oncology*, *10*. https://doi.org/10.3389/fonc.2020.00105.

Reid, G., Pel, M. E., Kirschner, M. B., Cheng, Y. Y., Mugridge, N., Weiss, J., et al. (2013). Restoring expression of miR-16: A novel approach to therapy for malignant pleural mesothelioma. *Annals of Oncology*, *24*(12), 3128–3135. https://doi.org/10.1093/annonc/mdt412.

Russo, G. L., Tessari, A., Capece, M., Galli, G., De Braud, F., Garassino, M. C., et al. (2018). MicroRNAs for the diagnosis and management of malignant pleural mesothelioma: A literature review. *Frontiers in Oncology*, *8*. https://doi.org/10.3389/fonc.2018.00650.

Santarelli, L., Staffolani, S., Strafella, E., Nocchi, L., Manzella, N., Grossi, P., et al. (2015). Combined circulating epigenetic markers to improve mesothelin performance in the diagnosis of malignant mesothelioma. *Lung Cancer*, *90*(3), 457–464. https://doi.org/10.1016/j.lungcan.2015.09.021.

Santarelli, L., Strafella, E., Staffolani, S., Amati, M., Emanuelli, M., Sartini, D., et al. (2011). Association of MiR-126 with soluble mesothelin-related peptides, a marker for malignant mesothelioma. *PLoS One*, *6*(4). https://doi.org/10.1371/journal.pone.0018232.

Schramm, A., Opitz, I., Thies, S., Seifert, B., Moch, H., Weder, W., et al. (2010). Prognostic significance of epithelial-mesenchymal transition in malignant pleural mesothelioma. *European Journal of Cardio-Thoracic Surgery*, *37*(3), 566–572. https://doi.org/10.1016/j.ejcts.2009.08.027.

Singh, A., Bhattacharyya, N., Srivastava, A., Pruett, N., Ripley, R. T., Schrump, D. S., et al. (2019). MicroRNA-215-5p treatment suppresses mesothelioma progression via the MDM2-p53-signaling Axis. *Molecular Therapy*, *27*(9), 1665–1680. https://doi.org/10.1016/j.ymthe.2019.05.020.

Singh, A., Pruett, N., Pahwa, R., Mahajan, A. P., Schrump, D. S., & Hoang, C. D. (2021). MicroRNA-206 suppresses mesothelioma progression via the Ras signaling axis. *Molecular Therapy - Nucleic Acids*, *24*, 669–681.

Sun, H. H., Vaynblat, A., & Pass, H. I. (2017). Diagnosis and prognosis-review of biomarkers for mesothelioma. *Annals of Translational Medicine*, *5*(11). https://doi.org/10.21037/atm.2017.06.60.

Suzuki, R., Amatya, V. J., Kushitani, K., Kai, Y., Kambara, T., Fujii, Y., et al. (2020). Inhibition of miR-18a-3p reduces proliferation of mesothelioma cells and sensitizes them to cisplatin. *Oncology Letters*, *19*(6), 4161–4168. https://doi.org/10.3892/ol.2020.11504.

Suzuki, R., Amatya, V. J., Kushitani, K., Kai, Y., Kambara, T., & Takeshima, Y. (2018). miR-182 and miR-183 promote cell proliferation and invasion by targeting FOXO1 in mesothelioma. *Frontiers in Oncology*, *8*. https://doi.org/10.3389/fonc.2018.00446.

Sydney, S., & Gilbert, de M. (2000). Poly(ADP-ribose) polymerase-1: What have we learned from the deficient mouse model? Mutation Research/DNA Repair, 1–15. https://doi.org/10.1016/s0921-8777(00)00016-1.

Tanaka, N., Toyooka, S., Soh, J., Tsukuda, K., Shien, K., Furukawa, M., et al. (2013). Downregulation of microRNA-34 induces cell proliferation and invasion of human mesothelial cells. *Oncology Reports*, *29*(6), 2169–2174. https://doi.org/10.3892/or.2013.2351.

Tomasetti, M., Monaco, F., Manzella, N., Rohlena, J., Rohlenova, K., Staffolani, S., et al. (2016). MicroRNA-126 induces autophagy by altering cell metabolism in malignant mesothelioma. *Oncotarget*, *7*(24), 36338–36352. https://doi.org/10.18632/oncotarget.8916.

Tomasetti, M., Nocchi, L., Staffolani, S., Manzella, N., Amati, M., Goodwin, J., et al. (2014). MicroRNA-126 suppresses mesothelioma malignancy by targeting IRS1 and interfering with the mitochondrial function. *Antioxidants and Redox Signaling*, *21*(15), 2109–2125. https://doi.org/10.1089/ars.2013.5215.

Tomasetti, M., Staffolani, S., Nocchi, L., Neuzil, J., Strafella, E., Manzella, N., et al. (2012). Clinical significance of circulating miR-126 quantification in malignant mesothelioma patients. *Clinical Biochemistry*, *45*(7–8), 575–581. https://doi.org/10.1016/j.clinbiochem.2012.02.009.

Truini, A., Coco, S., Nadal, E., Genova, C., Mora, M., Dal Bello, M. G., et al. (2017). Downregulation of miR-99a/let-7c/miR-125b miRNA cluster predicts clinical outcome in patients with unresected malignant pleural mesothelioma. *Oncotarget*, *8*(40), 68627–68640. https://doi.org/10.18632/oncotarget.19800.

Ueno, T., Toyooka, S., Fukazawa, T., Kubo, T., Soh, J., Asano, H., et al. (2014). Preclinical evaluation of MicroRNA-34b/c delivery for malignant pleural mesothelioma. *Acta Medica Okayama*, *68*(1), 23–26. http://ousar.lib.okayama-u.ac.jp/file/52140/68_1_23.pdf.

van Zandwijk, N., Pavlakis, N., Kao, S. C., Linton, A., Boyer, M. J., Clarke, S., et al. (2017). Safety and activity of microRNA-loaded minicells in patients with recurrent malignant pleural mesothelioma: A first-in-man, phase 1, open-label, dose-escalation study. *The Lancet Oncology*, *18*(10), 1386–1396. https://doi.org/10.1016/S1470-2045(17)30621-6.

Walter, R. F. H., Vollbrecht, C., Werner, R., Wohlschlaeger, J., Christoph, D. C., Schmid, K. W., et al. (2016). microRNAs are differentially regulated between MDM2-positive and negative malignant pleural mesothelioma. *Oncotarget*, *7*(14), 18713–18721. https://doi.org/10.18632/oncotarget.7666.

Weber, D. G., Brik, A., Casjens, S., Burek, K., Lehnert, M., Pesch, B., et al. (2019). Are circulating microRNAs suitable for the early detection of malignant mesothelioma? Results from a nested case-control study. *BMC Research Notes*, *12*(1). https://doi.org/10.1186/s13104-019-4113-7.

Weber, D. G., Casjens, S., Johnen, G., Bryk, O., Raiko, I., Pesch, B., et al. (2014). Combination of MIR-103a-3p and mesothelin improves the biomarker performance of malignant mesothelioma diagnosis. *PLoS One*, *9*(12). https://doi.org/10.1371/journal.pone.0114483.

Weber, D. G., Johnen, G., Bryk, O., Jöckel, K. H., & Brüning, T. (2012). Identification of miRNA-103 in the cellular fraction of human peripheral blood as a potential biomarker for malignant mesothelioma - a pilot study. *PLoS One*, *7*(1). https://doi.org/10.1371/journal.pone.0030221.

Welponer, H., Tsibulak, I., Wieser, V., Degasper, C., Shivalingaiah, G., Wenzel, S., et al. (2020). The miR-34 family and its clinical significance in ovarian cancer. *Journal of Cancer*, *11*(6), 1446–1456. https://doi.org/10.7150/jca.33831.

Williams, M., Cheng, Y. Y., Kirschner, M. B., Sarun, K. H., Schelch, K., Winata, P., et al. (2019). Transcriptional suppression of the miR-15/16 family by c-Myc in malignant pleural mesothelioma. *Oncotarget*, *10*(41), 4125–4138. http://www.impactjournals.com/oncotarget/index.php?journal=oncotarget.

Williams, M., Kirschner, M. B., Cheng, Y. Y., Hanh, J., Weiss, J., Mugridge, N., et al. (2015). miR-193a-3p is a potential tumor suppressor in malignant pleural mesothelioma. *Oncotarget*, *6*(27), 23480–23495. https://doi.org/10.18632/oncotarget.4346.

Xu, N., Papagiannakopoulos, T., Pan, G., Thomson, J. A., & Kosik, K. S. (2009). MicroRNA-145 regulates OCT4, SOX2, and KLF4 and represses pluripotency in human embryonic stem cells. *Cell*, *137*(4), 647–658. https://doi.org/10.1016/j.cell.2009.02.038.

Xu, W., Zhang, Z., Zou, K., Cheng, Y., Yang, M., Chen, H., et al. (2017). MiR-1 suppresses tumor cell proliferation in colorectal cancer by inhibition of Smad3-mediated tumor glycolysis. *Cell Death and Disease*, *8*(5). https://doi.org/10.1038/cddis.2017.60;.

Xu, Y., Zheng, M., Merritt, R. E., Shrager, J. B., Wakelee, H. A., Kratzke, R. A., et al. (2013). MiR-1 induces growth arrest and apoptosis in malignant mesothelioma. *Chest*, *144*(5), 1632–1643. https://doi.org/10.1378/chest.12-2770.

Yamamoto, K., Seike, M., Takeuchi, S., Soeno, C., Miyanaga, A., Noro, R., et al. (2014). MiR-379/411 cluster regulates IL-18 and contributes to drug resistance in malignant pleural mesothelioma. *Oncology Reports*, *32*(6), 2365–2372.

Ziga, S., & Tanja, K. (2014). Data integration of 104 studies related with microRNA epigenetics revealed that miR-34 gene family is silenced by DNA methylation in the highest number of cancer types. *Discoveries*, e18. https://doi.org/10.15190/d.2014.10.

Chapter 20

Pathophysiology roles and translational opportunities of miRNAs in renal cancer

Dominik A. Barth and Martin Pichler

Research Unit of Non-Coding RNAs and Genome Editing in Cancer, Division of Clinical Oncology; Department of Internal Medicine; Comprehensive Cancer Center Graz; Medical University of Graz, Graz, Austria

Introduction

Kidney cancer accounts for approximately 295,000 newly diagnosed cancer cases each year (Hsieh et al., 2017), more than 90% of which are renal cell carcinomas (RCC). RCC is the 6th and 10th most frequently diagnosed cancer in men (5%) and women (3%), respectively (Capitanio et al., 2019; Hsieh et al., 2017). Only accounting for cancers of the urogenital system, RCC is the third and second most common malignancy in females and males, respectively (Siegel, Miller, & Jemal, 2019). Overall incident rates are increasing, which is attributed to incidental diagnosis in the course of abdominal imaging due to other reasons, among other factors (Capitanio et al., 2019). Generally, age-standardized incidence is higher in developed countries and varies strongly around the globe. Indeed, incidence rates are highest in European countries (9.8/100,000) and North America (12/100,000) (Capitanio et al., 2019). However, corresponding mortality rates have been constantly declining, which may also be connected to increasing incidental findings (Hsieh et al., 2017).

RCC is a heterogenous disease and is classified into several histological subtypes. Clear cell renal cell carcinoma (ccRCC) is the most prevalent subtype, accounting for up to 75% of RCCs, whereas papillary (pRCC) and chromophobe renal cell carcinoma (chRCC) are less frequent. pRCC accounts for roughly 15% of RCC cases and is further subdivided into pRCC type 1 and type 2, based on basophilic or eosinophilic staining in the histological analysis. About 5% of RCCs are chRCCs, yet there are several other less prevalent subtypes, including medullary, clear cell papillary, tubulocystic, and several hereditary and unclassified RCCs (Hsieh et al., 2017).

microRNAs (miRNAs) were shown to play a role in RCC carcinogenesis and related pathways and may represent future valuable clinical biomarkers in different stages of treatment and disease (Barth et al., 2019; Barth, Drula, et al., 2020; Braga, Fridman, Loginov, Dmitriev, & Morozov, 2019). This chapter aims to provide an overview of the field, give relevant examples of miRNAs in RCC pathogenesis, and highlight important translational aspects of miRNAs in RCC.

Pathophysiology

In the last decade, the major involvement of miRNAs in RCC pathogenesis and pathophysiology has been gradually revealed and the dysregulation of numerous different miRNAs has been shown (Braga et al., 2019). miRNAs may act as tumor-suppressive factors in RCC or enhance RCC development and progression as so-called oncomirs. They are involved in several important cancer-associated pathways and cellular processes and influence the regulation of angiogenesis, proliferation, apoptosis, immune evasion, migration, invasion, and metastasis (Braga et al., 2019; He, Chen, & Shi, 2018) (Fig. 20.1).

Angiogenesis

In ccRCC, the von Hippel–Lindau (VHL)/hypoxia-inducible factor (HIF) pathway is the main driver for enhanced angiogenesis in RCC pathogenesis (Hsieh et al., 2017). The tumor-suppressor gene VHL is located on the short arm of chromosome 3, a genomic region that is lost in up to 90% of ccRCC. A loss of the second copy by mechanisms including deletion and promoter hypermethylation results in reduced VHL-mediated ubiquitination of HIFs and subsequently reduced degradation of HIFs, leading to an accumulation of HIFs under normoxic conditions in cancer cells. This induces a cellular pseudohypoxic state favoring angiogenesis, the enhancement of hypoxia-related genes, and a shift toward

MicroRNA in Human Malignancies. https://doi.org/10.1016/B978-0-12-822287-4.00018-9

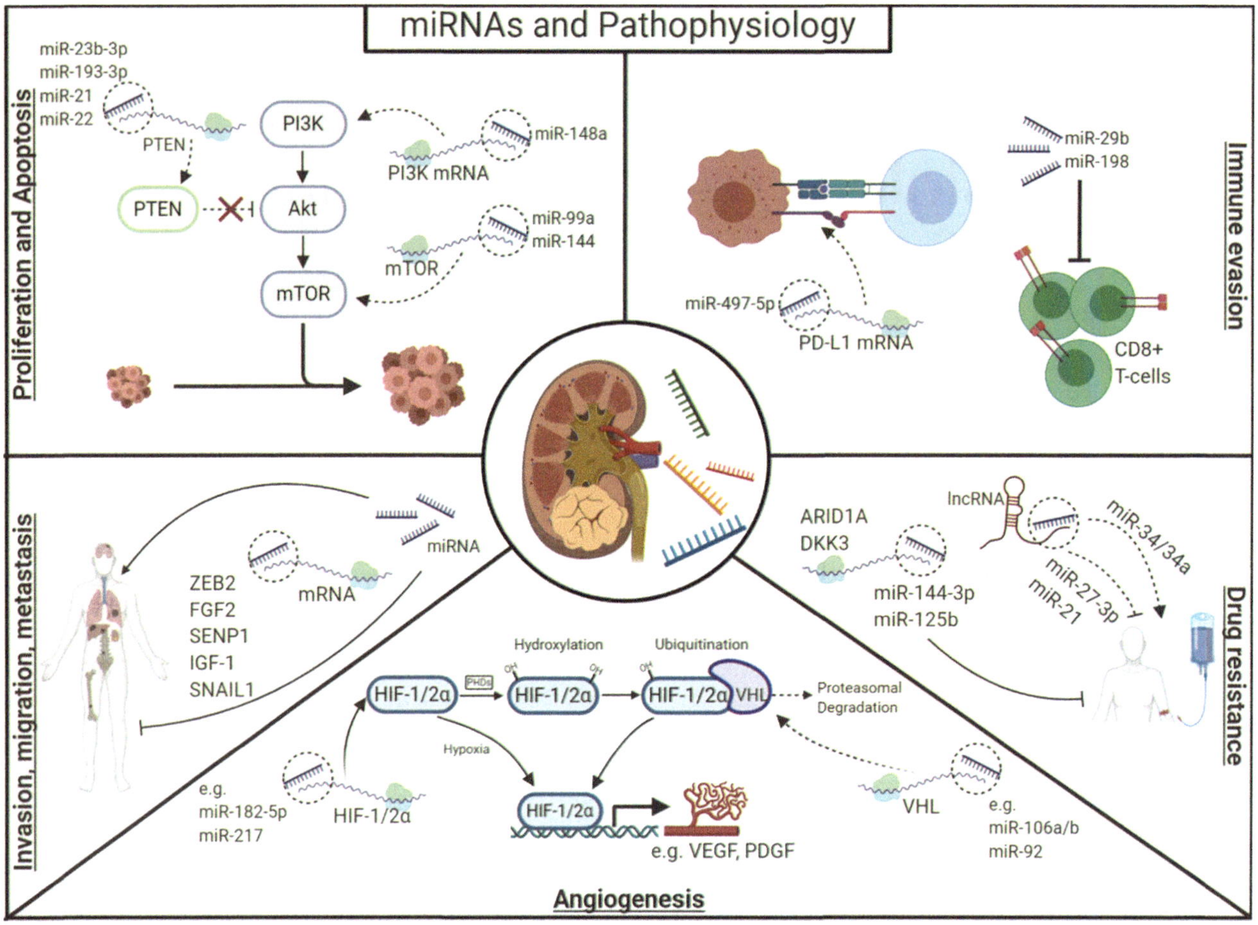

FIG. 20.1 miRNAs in RCC pathophysiology. PTEN–phosphatase and tensin homolog; *AKT,* protein kinase B; *ARID1A,* AT-rich interactive domain-containing protein 1A; *DKK3,* Dickkopf 3; *FGF2,* fibroblast growth factor 2; *HIF,* hypoxia inducible factor; *IGF-1,* insulin like growth factor; *lncRNA,* long noncoding RNA; *miRNA,* micro RNA; *mRNA,* messenger RNA; *mTOR,* mechanistic target of rapamycin; *PDGF,* platelet derived growth factor; *PD-L1,* programmed death ligand 1; *PI3K,* phosphoinositide 3-kinase; *SENP1,* sentrin-specific protease 1; *SNAIL,* Snail Family Zinc Finger 1; *VEGF,* vascular endothelial growth factor; *VHL,* von Hippel Lindau; *ZEB2,* zinc finger E box-binding homeobox 2. Created with BioRender.com.

anaerobic glycolysis (Warburg effect) (Hsieh et al., 2017; Linehan & Ricketts, 2019). Activation of the transcription factors HIF-1α and HIF-2α, among others, leads to an increased level of vascular endothelial growth factor (VEGF) and its associated receptor (VEGFR). Thus, inhibition of angiogenesis via the VEGF pathway is an important strategy in the treatment of RCC (Hsieh et al., 2017). To date, a large body of evidence supports the influence of miRNAs on angiogenesis regulation and various miRNAs may act as either enhancers (e.g., miR-210, which is characteristically upregulated in ccRCC, let-7, miR-126) or inhibitors (e.g., miR-221/miR-222 complex, miR-16, miR-195) of angiogenesis. Mechanisms include the direct or indirect regulation of prominent proangiogenic factors such as VEGF (Goradel et al., 2019; Tiwari, Mukherjee, & Dixit, 2018). Furthermore, miRNAs are regulated by the VHL/HIF axis as well, resulting in several positive and negative feedback loops (Braga et al., 2019).

In ccRCC, an inverse correlation between miR-221 and its target VEGFR2 was reported and its antiangiogenic effect was further confirmed in human umbilical vein endothelial cells (HUVEC). However, high miR-221 was associated with poor PFS and promotes proliferation in RCC cell lines (Braga et al., 2019; Khella et al., 2015). Further examples of confirmed miRNAs targeting VEGF in RCC are miR-125a-3p, miR-206, and miR-299-3p (Hou et al., 2019; Li et al., 2019; Shi, Zhang, Zhong, & Zhang, 2019). Another mechanism promoting angiogenesis in ccRCC is linked to the activation of the androgen receptor (AR), which upregulates miR-185-5p expression and thus induces two different effects on angiogenetic signaling. On the one hand, miR-185-5p directly targets VEGF-C. On the other hand, it transcriptionally increases HIF-2α expression by binding to its promoter region, thus resulting in higher VEGFA levels (Braga et al., 2019). miR-185 is especially downregulated in VHL-deficient ccRCC tissues and miR-185 overexpression was shown to enhance apoptosis in VHL-deficient cell lines. Furthermore, it is associated with increased levels of VEGFR-2 and promoted the receptor's downstream signaling (He et al., 2018).

miRNAs may also exert a regulatory function on angiogenesis by targeting VHL or HIFs directly, and subsequently either act as an oncogenic or tumor-suppressive factor. As an example, miR-182-5p is upregulated in the absence of functional VHL by the enzyme Dicer and could subsequently inhibit its target HIF-2α and angiogenesis because HIF-2α is an important proangiogenic regulator in VHL-deficient RCC (Braga et al., 2019; Schanza et al., 2017). miR-106a and miR-106b both target VHL in ccRCC. However, the role of miR-106a as an oncomir or tumor suppressor is not as clearly defined as it also targets multiple other pathways in ccRCC (Braga et al., 2019). The biological roles of miRNAs are also closely related to long noncoding RNAs (lncRNAs) as they could act as sponges for miRNAs in competitive endogenous RNA (ceRNA) interactions. This also is the case in the regulation of HIFs, as lncRNA HOX transcript antisense RNA (HOTAIR) suppresses miR-217, which would otherwise target and inhibit HIF-1α in RCC leading to enhanced proliferation and epithelial mesenchymal transition (EMT) (Barth, Prinz, et al., 2020; Braga et al., 2019). miRNA/lncRNA interactions are important mechanisms in the regulation of HIFs and have been observed across many cancer entities (Barth, Prinz, et al., 2020). Moreover, miRNAs encapsulated and transported in extracellular vesicles such as exosomes may also play a role in regulating angiogenesis in RCC. For instance, miR-92 may target VHL and therefore promote angiogenesis via an increased activation of HIFs (Grange, Brossa, & Bussolati, 2019).

Proliferation and apoptosis

The phosphoinositide 3-kinase/protein kinase B/mechanistic target of the rapamycin (PI3K/AKT/mTOR) pathway is an important signaling cascade in cancer development and progression, and is strongly associated with tumor cell proliferation, cell growth, and apoptosis evasion. It is especially important in RCC pathogenesis through its link with the HIF pathway through mTOR-mediated HIF modulation and is frequently genetically altered across several RCC subtypes, including ccRCC, pRCC, and chRCC (Braga et al., 2019; Guo et al., 2015). The activation of the PI3K/AKT/mTOR pathway through mutations is also a feasible biological explanation for the utility of mTOR inhibitors such as everolimus as a therapeutic option in RCC (Hsieh et al., 2017). In addition, several of the pathway's components are targeted and regulated by miRNAs in RCC. On the one hand, several miRNAs negatively regulate AKT2, whereas on the other hand, others promote AKT2 activation and further downstream signaling, thereby promoting proliferation and cell cycle activation. As for AKT2 inhibition, AKT2 is a confirmed target of miR-148a (Braga et al., 2019; Nogueira, Dias, Teixeira, & Medeiros, 2018). miR-182-5p indirectly inhibits AKT2 activation and the consecutive activation of the transcription factor forkhead box O3a (FOXO3a) by targeting Flotillin 1, thereby leading to G1 cell arrest and thus inhibited proliferation (FLOT1). Both, miR-148a and miR-182-5p are frequently downregulated in RCC (Braga et al., 2019) and miR-182-5p is additionally involved in negatively regulating HIF-2α, which underlines its role as a tumor suppressor in RCC carcinogenesis (Nogueira et al., 2018). As opposed to this, miR-122 enhances AKT2 and mTOR activation, thus acting as an oncomir in RCC (Braga et al., 2019).

miR-99a and miR-144 both target mTOR and are therefore suggested as tumor-suppressive miRNAs in RCC; however, results are partially conflicting despite mTOR being confirmed as a direct target in both cases. Yet both miRNAs were shown to inhibit tumor cell proliferation and arrest cells in the G1 phase, thus representing tumor-suppressive factors in RCC (Braga et al., 2019).

Apart from directly impacting the PI3K/AKT/mTOR pathway, miRNAs are also involved in the regulation of phosphatase and tensin homolog (PTEN). PTEN itself is an inhibitor of PI3K-induced AKT activation, thereby acting as a tumor suppressor. By targeting PTEN, miR-23b-3p, miR-193-3p, miR-21, and miR-22 exert an oncogenic function in regulating RCC cell proliferation (Braga et al., 2019; He et al., 2018; Nogueira et al., 2018). Increased miR-21 levels are further associated with higher stages and tumor grades in RCC (Braga et al., 2019).

Besides the PI3K/AKT/mTOR pathway, miRNAs may also influence other proliferation-regulating mechanisms. For instance, miR-106a-5p targets the histone lysine methyltransferase SET Domain Containing 2 (SETD2), which could stop cell cycle transition from the G0 to the G1 phase. As such, miR-106a-5p acts as an oncogenic factor in this case; however, as mentioned earlier, it has also shown tumor-suppressive abilities in other pathways in RCC (Braga et al., 2019).

Besides tumor cell growth, proliferation, and cell cycle activation, several miRNAs have also been linked to apoptosis. For instance, miR-486, which is also carried in extracellular vesicles in RCC, was observed to inhibit apoptosis by targeting apoptosis-related genes, including the PTEN/PI3K/AKT pathway, FOXO, and transforming growth factor beta (TGF-β)-activated kinase 1 (TAK1) (Grange et al., 2019).

Migration, invasion, and metastasis

Up to 30% of RCC patients present with metastasis at some point in the course of the disease (Braga et al., 2019), which consequently leads to a change in therapy toward palliative treatment options. Therefore, gaining a deep and detailed insight into the development of metastasis in RCC is crucial. EMT is among others characterized by tumor cells loosening intercellular bonds thereby eventually enabling tissue invasion and the development of metastasis. By targeting certain genes involved in EMT, miRNAs take a regulatory role and may exert either oncogenic or tumor-suppressive functions (Braga et al., 2019). For instance, zinc finger E box-binding homeobox 2 (ZEB2) leads to a loss of E-cadherin through transcriptional suppression and is targeted itself by miR-141 miR-30a, miR-192, miR-194, and miR-215, whereas the latter three additionally target the oncogenes mouse double minute 2 homolog (MDM2) and thymidylate synthetase (TYMS) in RCC (Braga et al., 2019; He et al., 2018). Interestingly, these tumor-suppressive miRNAs are frequently downregulated in RCC. Furthermore, miR-141 is also competitively sponged by lncRNA HOTAIR, which is a known driver of EMT in RCC and was already previously mentioned as an miRNA sponge for miR-217 and the subsequent activation of EMT (Braga et al., 2019). HOTAIR also sponges miR-203, which is characteristically downregulated in metastatic RCC by hypermethylation, thereby preventing it from targeting and subsequently suppressing fibroblast growth factor 2 (FGF2) and inhibiting EMT. Increasingly active miR-203 leads to a notable increase of tumor-suppressive and EMT preventive genes such as PTEN, E-cadherin, or p21 (Braga et al., 2019).

Importantly, miR-186 was associated with invasion and metastasis and moreover impacts tumor cell proliferation in RCC cells by inducing G0/G1 arrest. It is downregulated in RCC and its overexpression could inhibit tumor cell invasiveness by targeting sentrin-specific protease 1 (SENP1), thereby ultimately influencing the NF-kB pathway (Braga et al., 2019).

miR-182 was reported to inhibit its target insulin-like growth factor 1 (IGF-1) receptor in RCC cell lines, thus decreasing tumor cell viability, migration, and invasion. miR-182 was additionally downregulated in ccRCC tissue samples and patient blood samples (He et al., 2018).

Other miRNAs promoting RCC invasion and metastasis are miR-155, miR-720, miR-149-5p, and miR-149-3p, whereas miR-30e-3p was observed to act as an inhibitor of EMT by targeting Snail Family Zinc Finger 1 (SNAIL1) in ccRCC and miR-34a inhibited metastasis formation in vivo by targeting CD44 (Braga et al., 2019; He et al., 2018).

Immune evasion

Immune evasion is an important hallmark of cancer that has especially gained importance because of the success and introduction of immune checkpoint (such as CTLA4, PD-1 and PD-L1) inhibitor therapy in the treatment of melanoma and its subsequent and gradual implementation in the treatment of other cancers, including RCC (Kobayashi, Lim, & Yamaguchi, 2020). Several miRNAs were shown to play a role in the regulation of immune checkpoints and could therefore qualify as future predictive biomarkers or even therapeutic targets to achieve activation of the immune system against the tumor (Yang, Cao, Wang, Zhang, & Liu, 2018). In RCC, evidence of miRNAs regulating immune checkpoints is scarce. miR-497-5p was found to directly target PD-L1, thereby regulating its expression in RCC cell lines. Consequently, a decrease in miR-497-5p levels was associated with upregulated PD-L1 and moreover adverse OS in the Cancer Genome Atlas (TCGA) cohorts (Qu et al., 2019).

miR-187 is involved in the regulation of the B7-H3 immune checkpoint receptor in RCC as it directly targets its 3' UTR. B7-H3 can act as both a costimulating and a coinhibitory factor in cancer immunity (Omar et al., 2019; Yang et al., 2018). Moreover, its expression levels are decreased in ccRCC tissue and patient plasma, and low expression is associated with poor overall survival (OS). Consequently, it has potential as a future biomarker in RCC (Braga et al., 2019).

miRNAs may also directly influence cellular antitumor activity in RCC. Gigante et al. (2016) described an interesting role of miR-29b and miR-198 in in CD8 + T cells of RCC patients. By directly targeting Janus kinas 3 (JAK3) and MCL-1, increased levels of miR-29b and miR-198 result in a decreased antiapoptotic state and reduced proliferation in CD + T cells of RCC patients, consequently leading to their potential dysfunction and thereby facilitating immune evasion of tumor cells (Gigante et al., 2016). On the other hand, miR-133a and miR-148a enhance natural killer cell activation and cytotoxicity in RCC, resulting in increased tumor cell death (Omar et al., 2019). Taken together, although miRNAs play a role in tumor immunity, more research is needed to uncover further detailed mechanisms and to provide deeper insight into miRNA's involvement in tumor immunity and immune evasion of RCC, which could ultimately result in therapeutic advances.

Drug resistance

Intrinsic and acquired therapy resistance is a major obstacle in cancer care, resulting in tumor progression and shorter survival of cancer patients. In RCC, miRNAs were shown to participate in resistance mechanisms and could potentially help to overcome treatment resistance. For instance, miR-144-3p was reported to enhance resistance to sunitinib in RCC by targeting AT-rich interactive domain-containing protein 1A (ARID1A), which is a part of a chromatin-modeling complex and is associated with drug resistance, proliferation, and cell growth in several cancers (Xiao et al., 2017). Another interesting pathway that could explain the strong intrinsic chemoresistance of RCC involving miR-125b was discovered by Yan et al. (2019). The histone methyltransferase SET and MYND domain-containing protein 2 (SMYD2), which is overexpressed in RCC and associated with adverse prognosis, transcriptionally upregulates miR-125b, resulting in the inhibition of its associated target, Dickkopf 3 (DKK3). Eventually, the inhibition of both SMDY and miR-125b resulted in increased chemosensitivity though P-glycoprotein inhibition, suggesting a novel strategy to overcome chemoresistance in RCC (Yan et al., 2019).

Several miRNAs are parts of larger resistance mechanisms as they are part of lncRNA/miRNA interactions and are sponged by lncRNAs, preventing them from exerting their regulatory function in the development of treatment resistance. Important examples are miR-34 (sponged by lncRNA ARSR), miR-27-3p (lncRNA ADAMTS9-AS2), miR-34a (lncRNA NEAT1), and miR-21 (lncRNA GAS5), which are kept from binding their associated targets c-MET (miR-34/34a), AXL (miR-34), FOXO1 (miR-27-3p), and SOX5 (miR-21), respectively. If not sponged and suppressed by their associated lncRNA, miR-34 and miR-34a would act as suppressors of sunitinib or sorafenib resistance, respectively, whereas miR-27-3p and miR-21 promote chemo- or sorafenib resistance, respectively (Barth, Juracek, et al., 2020). miR-30a, which is only expressed at low levels in RCC, was additionally confirmed to promote sorafenib-induced apoptosis and autophagy by targeting Beclin-1 (Zheng et al., 2015).

Circulating miRNAs such as miR-31-5p carried in extracellular vesicles may also contribute to the development of drug resistance (Braga et al., 2019; He et al., 2018) and are discussed in a subsequent section.

Diagnostic, prognostic, and predictive value

A variety of miRNAs may offer additional value in RCC diagnosis, prognosis prediction, or even in predicting treatment success. Patients could potentially benefit from bringing miRNA-based biomarkers into the clinical routine. Depending on treatment settings and the underlying histological features, different individual miRNAs or miRNA panels measured from patient's tumor tissue or, as a recent development, liquid biopsies could potentially influence clinical decision making or answer certain important clinical questions. This section puts the focus on the significance and evidence of miRNA's diagnostic, prognostic, and predictive value in various treatment settings and histological subtypes in RCC. As a recently emerging field, the relevance and challenges of miRNA-based liquid biopsies in RCC are discussed in more detail in a subsequent chapter (Fig. 20.2).

miRNAs in RCC diagnosis

Because there are some significant differences in miRNA expression levels between histological subtypes in RCC, miRNAs may offer additional discriminatory value to histologically classify RCC subtypes. Even for experienced pathologists, it is not always easy to microscopically differentiate histological subtypes in RCC, as histological features may not always be clearly assigned to one particular RCC subtype and mixed histologies such as clear cell papillary RCC make the correct differentiation even more complex (Hsieh et al., 2017). Yet, the correct histological differentiation is crucial for both prognosis prediction and choosing the appropriate treatment for the patients. In these rather difficult and ambiguous cases, differentially expressed miRNA levels and diagnostic miRNA panels may be useful to assist pathologists in renal tumor diagnosis and subclassification (Barth et al., 2019). Most proposed miRNA panels aim to differentiate RCC subtypes by multistep approaches using several differential expressed miRNAs. For instance, to distinguish RCC patient samples into ccRCC, pRCC, chRCC, or the benign oncocytoma, Di Meo et al. (2018) established a two-step model using the expression levels of miR-221, miR-222, and miR-126 assessed by qRT-PCR and in situ hybridization. In their study, the relative downregulation of miR-221 and miR-222 in ccRCC and pRCC as compared to chRCC and oncocytoma is able to distinguish the first from the latter two RCC subtypes. In a second step, miR-126 expression successfully differentiated ccRCC from pRCC, whereas miRNAs miR-200b, miR-200c, and miR-222 could distinguish chRCC and oncocytoma with a sensitivity and specificity of 89.5% and 90.9% for miR-200b, 84% and 82% for miR-200c and 79% and 91% for miR-222, respectively. Interestingly, combinations of the three miRNAs did not result in a higher discriminatory capability (Di Meo et al., 2018). As a second example, Wach et al. (Wach et al., 2013) could even distinguish between pRCC type 1 and type 2 in their three-step approach using miR-145, miR-200c, miR-210, and miR-502-3p (healthy vs. malignant tissue); miR-145 and miR-503-3p (ccRCC vs. pRCC); and miR-210 and miRNA let-7c (pRCC type 1 vs. pRCC type 2). Without claim of completeness, further differentially expressed miRNAs that could be potentially used for RCC subtype discrimination are miR-126, miR-143, miR-155, miR-221, miR-203, miR-429, miR-629*, and miR-124. Although the two presented examples as well as other proposed approaches and diagnostic models show some similarities in the miRNAs used for the classification of RCC

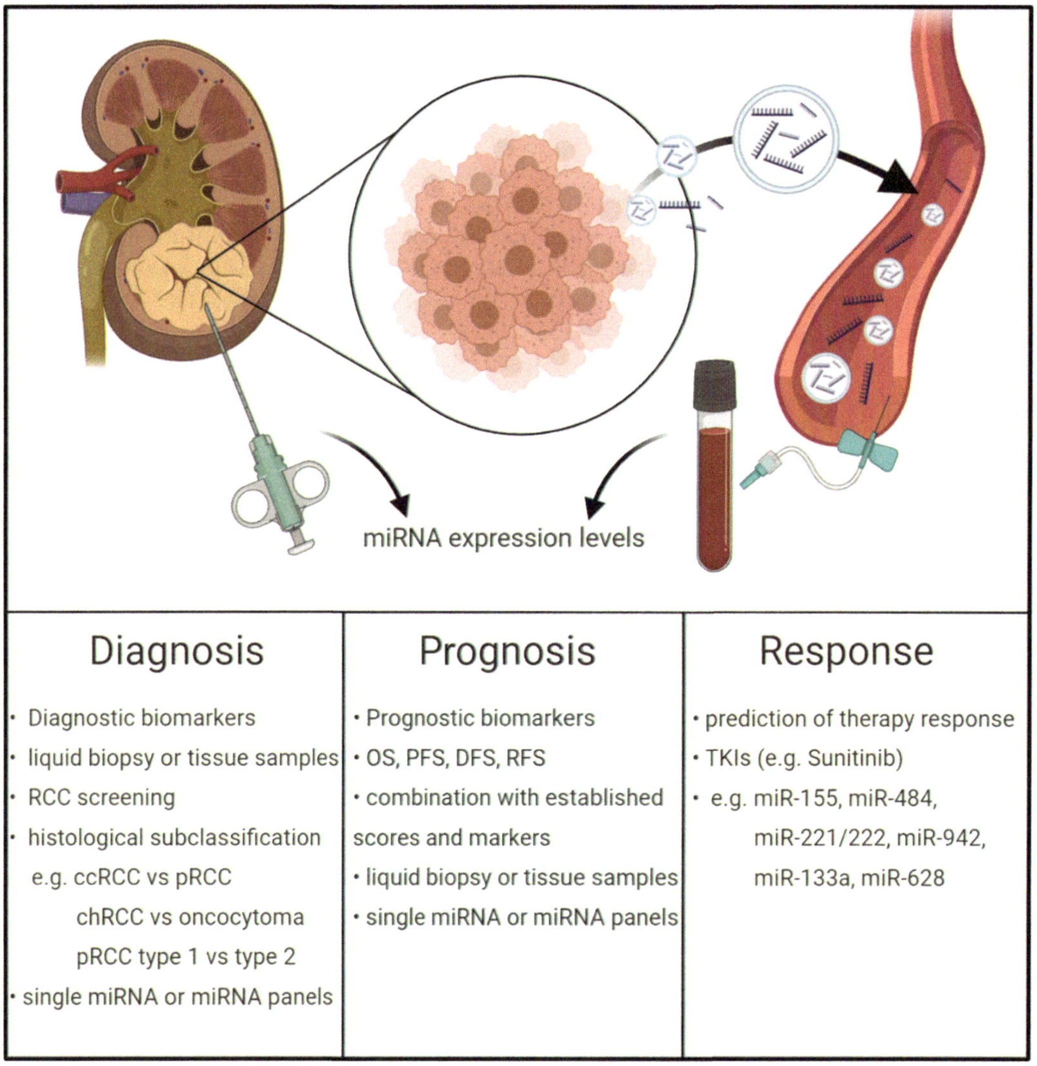

FIG. 20.2 Clinical potential of miRNAs in RCC miRNA expression levels may be obtained from tissue, blood, or urine (not depicted) samples. miRNAs may be actively secreted into the bloodstream, such as in extracellular vesicles or exosomes, or are passively released by cell death. miRNA expression levels may bring clinical value in diagnosis, prognosis, or response prediction. Abbreviations: *ccRCC*–clear cell renal cell carcinoma; *chRCC*–chromophobe renal cell carcinoma; *DFS*–disease free survival; *miRNA*–microRNA; *OS*–overall survival; *PFS*–progression free survival; *pRCC*–papillary renal cell carcinoma; *RFS*–recurrence free survival; *TKIs*–tyrosine kinase inhibitors. Created with BioRender.com.

subtypes (e.g., the miR-200 family and other miRNAs involved in RCC pathogenesis discussed in a previous section of the chapter), in order to achieve future clinical implementation of miRNA-assisted RCC diagnosis, further investigations and prospective trials are necessary as current evidence and proposed miRNA panels are widely lacking verification and analysis in larger or even prospective study cohorts.

Another important field of application for miRNAs in RCC diagnosis is their potential value as diagnostic biomarkers and a possible implementation in screening procedures. To date, there are no cost-effective and reliable markers or screenings for the early diagnosis of RCC available in the clinical routine, which would enable RCC diagnosis in potentially curative treatment settings more frequently. This is a gap that could be filled by miRNA expression levels, most conveniently by the recently emerging field of liquid biopsies and miRNA measurements out of easily accessible body fluids such as peripheral blood and urine (Barth, Drula, et al., 2020; Mytsyk et al., 2018; Oto et al., 2020).

For example, miR-210 has been frequently observed to be upregulated in ccRCC tissue as compared to healthy renal parenchyma and may thus be used to assist in the diagnostic work-up of RCC samples (Barth, Drula, et al., 2020). Moreover, a growing body of evidence also found miR-210 upregulated in the blood serum and plasma as well as in the urine of RCC patients with a significant recline in measured miR-210 levels after successful surgery, thereby supporting its role as a diagnostic biomarker (Barth, Drula, et al., 2020; Oto et al., 2020). Despite the relatively large number of studies on miR-210 in

RCC and its strong connection to the HIF pathway, at least for serum miR-210 the reported diagnostic abilities are diverging with an observed sensitivity and specificity ranging from 81% to 65% and 83%–70%, thus not concludingly confirming its utility in clinical practice (Barth, Drula, et al., 2020). Altered expression levels of numerous other miRNAs were found in the serum or plasma of RCC patients as compared to healthy controls and may therefore play a role in future RCC diagnostics. In general, the combined diagnostic power of several miRNAs in diagnostic panels appears to bring additional value to the use of only one single miRNA in RCC diagnosis (Barth, Drula, et al., 2020). For instance, the miR-224/miR-141 ratio could successfully find RCC patients with a sensitivity of more than 97% and a specificity of more than 98% (area under the curve; AUC=0.9898) (Chen et al., 2018). However, because verification studies are missing and study designs as well as technical procedures (e.g., normalization of gene expression) are not uniform, standardized further research is needed (Barth, Drula, et al., 2020).

As for urinary miRNAs in RCC diagnostics, increased urine levels of miRNAs miR-15a as well as all members of the let-7 family and decreased urine levels of miR-200a, miR-30c-5p, and miR-497 could significantly distinguish RCC patients from healthy controls (Oto et al., 2020). In addition, panels combining the diagnostic power of several miRNAs, such as miR-200a-3p, miR-34a-5p, and miR-365a-3p, were introduced (Oto et al., 2020).

miRNAs in RCC prognosis

Although many well-established and important clinical parameters and biomarkers–tumor stage and grade as well as clinical scores such as the Memorial Sloan Kettering Cancer Center (MSKCC) and International Metastatic RCC Database Consortium (IMDC) risk scores are undoubtedly important prognostic tools and thus play a role in the right treatment selection for RCC patients, additionally considering miRNA expression levels may bring important information to clinical practice. Several miRNAs are associated with these already established clinical markers. For example, the overexpression of miR-21 is associated with higher stages and tumor grades, whereas it is also connected to shorter disease-free (DFS) and overall survival (OS) while also exerting an oncogenic regulatory function on cell cycle, growth, and motility (Mytsyk et al., 2018; Tsiakanikas, Giaginis, Kontos, & Scorilas, 2018). Being part of several prognostic miRNA signatures for RCC, miR-21 is seen as a promising candidate marker in RCC prognosis prediction (Tsiakanikas et al., 2018). A large body of evidence also supports the prognostic utility of miR-210 and the adverse relationship of high expression levels with RFS, DFS, and OS, although there are also some conflicting results that could be attributed to technical differences and varying study cohorts (Barth, Drula, et al., 2020; Tsiakanikas et al., 2018). Nonetheless, adding miRNAs to already established biomarkers could further increase their prognostic capability as, for example, shown by a miRNA signature including miR-10b, miR-21, miR-143, miR-183, and miR-192, which were combined with the TNM stage and age of RCC patients to predict for OS. Tumor tissue expression levels of multiple other miRNAs have been successfully associated with the survival of RCC patients; some of the most promising candidates (e.g., miR-126, miR-155, miR-192, miR-194, miR-200c, miR-221) have been repeatedly confirmed in studies while other miRNAs still lack verification (Mytsyk et al., 2018; Tsiakanikas et al., 2018).

While miR-221 expression levels measured from tumor samples have already been associated with poor OS, miR-221 levels in RCC patient's blood plasma also proved to convey prognostic information (Barth, Drula, et al., 2020; Mytsyk et al., 2018). Moreover, when combined with tumor stage, age, and Fuhrman grade, miR-221 could better predict for OS than these clinico-pathological parameters alone, thereby improving their prognostic utility (Barth, Drula, et al., 2020).

In fact, liquid biopsies and miRNA expression levels in body fluids may also play an important role for prognosis prediction. Although to date no urinary miRNAs could be linked to RCC survival (Oto et al., 2020), a growing body of evidence points at blood-based miRNA biomarkers to predict for patient disease outcome. Besides miR-221, several other miRNAs, including the aforementioned miR-210, were reported to qualify for plasma or serum-based prognostic biomarkers in RCC (Barth, Drula, et al., 2020). As an example, Du et al. (Du et al., 2017) found high levels of plasma exosomal miRNA let-7i-5p to be independently associated with longer OS (HR=0.49, 95%CI 0.21–0.84, P=0.018) and its inclusion to the MSKCC risk score could significantly improve the score's ability. Further, circulating miRNAs with a proposed prognostic value include miR-224, miR-26a-1-3p, miR-615-3p, and miR-150 (Barth, Drula, et al., 2020).

Although up to 25% of kidney cancers are nonclear cell histological subtypes such as pRCC, chRCC or other rare or mixed histological subtypes (Hsieh et al., 2017), most available studies only focus on miRNAs as prognostic biomarkers in ccRCC. Thus, evidence on the prognostic utility of miRNA expression levels in nonclear cell RCCs is sparse and most studies fail to provide multivariate statistical analysis (Barth et al., 2019). In pRCC, decreased expression levels of miR-220c and miR-127 are significantly and independently associated with longer OS, although this significance was lost in the multivariate analysis of a validation cohort. Moreover, higher levels of miR-34a are an independent predictor for prolonged OS (HR=0.046, 95%CI 0.004–0.485, P=0.010) (Barth et al., 2019; Ge et al., 2015). Increased expressions of several miRNAs (miR-133a, miR-133b, miR-216a, miR-217, miR-1297, miR-211) showed significant associations with adverse OS outcome. However, the lack of multivariate analysis and validation in most cases does not currently support their use as clinically valuable prognostic biomarkers in pRCC (Barth et al., 2019). Upregulated miRNAs miR-1293 and miR-3199-2 are both related to improved progression-free survival (PFS); however, the same limitations for their clinical utility apply (Barth et al., 2019).

As for chRCC, miR-210, which was also previously mentioned in RCC diagnosis and HIF regulation, was demonstrated as an independent prognosticator for recurrence-free survival (RFS), as high expression levels were associated with reduced RFS. miR-210 is directly linked to the HIF pathway and is subsequently involved in angiogenesis, highlighting its role in RCC (Barth et al., 2019; Goradel et al., 2019; Tiwari et al., 2018).

Predictive value of miRNAs in RCC

For advanced and metastatic RCC, systemic therapies remain the treatment strategies of choice in these settings. As chemotherapy has no effect in RCC, targeted therapies with multikinase inhibitors such as sunitinib or sorafenib, which target vascular endothelial growth factor receptor (VEGFR), platelet-derived growth factor receptor (PDGFR), and Raf-kinase, among others, have been valuable therapeutics in palliative treatment settings (Hsieh et al., 2017). However, in recent years, treatment options for advanced and metastatic RCC have improved drastically with the introduction of immune checkpoint inhibitors and combinations with antiangiogenic targeted therapy. Although patient outcomes have improved, some patients do not show satisfying treatment response and most patients eventually experience disease progression (Hsieh et al., 2017). In order to select the right treatment with the considerable and predictable best response, miRNAs could potentially serve as important predictive biomarkers.

miRNAs were reported to have some value in predicting therapy response in patients treated with tyrosine kinase inhibitors (TKIs); however, current evidence does not support imminent use in clinical practice. Available studies mainly suffer from small sample sizes and in some cases show conflicting results between training and validation cohorts. Furthermore, methodological differences, such as the definition of poor and good responders or qRT-PCR vs. microarray screenings, add to the difficulties in comparability. Yet some overlap in the results suggests that miR-155, miR-484, miR-221/222, miR-942, miR-133a, and miR-628 have some value in predicting response to TKIs, including sunitinib and pazopanib (Kovacova, Poprach, Buchler, Cho, & Slaby, 2018).

In general, combinations of miRNAs to specific miRNA signatures and panels show better predictive ability than single miRNAs (Kovacova et al., 2018; Mytsyk et al., 2018).

To date, data on miRNAs for response prediction in patients treated with immune checkpoint mono- or combination therapy are missing and further research is warranted to keep miRNAs as predictive biomarkers in RCC up to the latest innovations in systemic treatment.

Circulating miRNAs

As of now, studies on circulating miRNAs in RCC are mainly limited to biomarker research for diagnosis, prognosis, and the prediction of treatment response, which have already been discussed in the previous section. Yet, the pathophysiological roles and the impact of altered circulating miRNA levels on cancer progression and cancer growth are barely elucidated. Further research in this direction would strengthen the biological background and subsequently add more importance to bringing circulating miRNAs in RCC to the clinics (Barth, Drula, et al., 2020).

Circulating miR-31-5p in extracellular vesicles was investigated by He et al. (2020), who reported significantly increased levels upon disease progression in RCC patients undergoing treatment with sorafenib as compared to the pretreatment expression levels. Both in vitro and in vivo experiments using extracellular vesicles containing elevated miR-31-5p levels from sorafenib-resistant RCC cells demonstrated the transmission of sorafenib resistance to former treatment-sensitive cells. miR-31-5p targets MutL homolog 1 (MLH1), resulting in its downregulation and increased sorafenib resistance by potentially increased genomic instability and the consecutive increased likelihood of treatment resistance-promoting mutations (He et al., 2020). As such, further investigation of circulating miRNAs could eventually help us to better understand and finally overcome treatment resistance in RCC (Barth, Drula, et al., 2020; He et al., 2020). Some other circulating miRNAs where the pathophysiological roles have been partly enlightened are miR-19b-3p and miR-224.

Exosomal miR-19b-3p may promote ccRCC progression and metastasis. Being secreted by ccRCC cancer metastatic stem cells, it could promote EMT as it is incorporated by cancer cells and targets the tumor-suppressor PTEN (Wang et al., 2019). Exosomal miR-224 was shown to increase RCC cell line proliferation and invasion in cell culture models and high levels of circulating exosomal miR-224 were moreover associated with significantly decreased cancer-specific survival (CSS), PFS, and OS. Thus, it could serve as a future prognostic biomarker in ccRCC (Barth, Drula, et al., 2020).

Although there are numerous studies investigating circulating miRNAs in RCC in diagnosis, prognosis, and response prediction and there are some promising miRNA candidates or miRNA panels, currently, circulating miRNAs are far from entering clinical practice and being considered in clinical decision making. There are a number of reasons for this. First, differences and missing standards in the normalization of miRNA measurements sometimes make it difficult to compare available studies even when investigating the very same miRNA. Differences in the study populations as well as limited sample sizes further aggravate this problem. Second, there is an ongoing debate as to whether serum or plasma should be used when establishing circulating miRNA-based diagnostics and the study landscape is heterogenous in this point, again impeding comparability between studies. RNA extraction yields a higher total RNA concentration in serum as compared to plasma. Third, sample processing has not been standardized and sample cooling and processing times could thus heavily impact experimental outcomes. Furthermore, differences in the experimental designs and methodological differences, such as qRT-PCR vs. microarrays, may influence the results and should be accounted for when comparing studies (Barth, Drula, et al., 2020).

Circulating miRNAs play a role in RCC pathogenesis and may be promising tools in RCC treatment, yet further research and effort in this field are needed.

Conclusions

miRNAs play an important role in RCC pathogenesis and progression, evidenced by an increasing number of studies in recent years. They also offer some potential as RCC therapeutics or in overcoming drug resistance. In addition, they are associated with therapy resistance and have great potential as future clinical biomarkers in diagnosis, prognosis, and treatment response prediction.

References

Barth, D. A., Drula, R., Ott, L., Fabris, L., Slaby, O., Calin, G. A., et al. (2020). Circulating non-coding RNAs in renal cell carcinoma-pathogenesis and potential implications as clinical biomarkers. *Frontiers in Cell and Developmental Biology*, *8*, 828. https://doi.org/10.3389/fcell.2020.00828.

Barth, D. A., Juracek, J., Slaby, O., Pichler, M., & Calin, G. A. (2020). lncRNA and mechanisms of drug resistance in cancers of the genitourinary system. *Cancers*, *12*(8). https://doi.org/10.3390/cancers12082148.

Barth, D. A., Prinz, F., Teppan, J., Jonas, K., Klec, C., & Pichler, M. (2020). Long-noncoding RNA (lncRNA) in the regulation of hypoxia-inducible factor (HIF) in Cancer. *Non-Coding RNA*, *6*(3). https://doi.org/10.3390/ncrna6030027.

Barth, D. A., Slaby, O., Klec, C., Juracek, J., Drula, R., Calin, G. A., et al. (2019). Current concepts of non-coding RNAs in the pathogenesis of non-clear cell renal cell carcinoma. *Cancers*, *11*(10). https://doi.org/10.3390/cancers11101580.

Braga, E. A., Fridman, M. V., Loginov, V. I., Dmitriev, A. A., & Morozov, S. G. (2019). Molecular mechanisms in clear cell renal cell carcinoma: Role of miRNAs and Hypermethylated miRNA genes in crucial oncogenic pathways and processes. *Frontiers in Genetics*, *10*, 320. https://doi.org/10.3389/fgene.2019.00320.

Capitanio, U., Bensalah, K., Bex, A., Boorjian, S. A., Bray, F., Coleman, J., et al. (2019). Epidemiology of renal cell carcinoma. *European Urology*, *75*(1), 74–84. https://doi.org/10.1016/j.eururo.2018.08.036.

Chen, X., Lou, N., Ruan, A., Qiu, B., Yan, Y., Wang, X., et al. (2018). miR-224/miR-141 ratio as a novel diagnostic biomarker in renal cell carcinoma. *Oncology Letters*, *16*(2), 1666–1674. https://doi.org/10.3892/ol.2018.8874.

Di Meo, A., Saleeb, R., Wala, S. J., Khella, H. W., Ding, Q., Zhai, H., et al. (2018). A miRNA-based classification of renal cell carcinoma subtypes by PCR and in situ hybridization. *Oncotarget*, *9*(2), 2092–2104. https://doi.org/10.18632/oncotarget.23162.

Du, M., Giridhar, K. V., Tian, Y., Tschannen, M. R., Zhu, J., Huang, C.-C., et al. (2017). Plasma exosomal miRNAs-based prognosis in metastatic kidney cancer. *Oncotarget*, *8*(38), 63703–63714. https://doi.org/10.18632/oncotarget.19476.

Ge, Y.-Z., Xu, L.-W., Xu, Z., Wu, R., Xin, H., Zhu, M., et al. (2015). Expression profiles and clinical significance of MicroRNAs in papillary renal cell carcinoma: A STROBE-compliant observational study. *Medicine*, *94*(16), e767. https://doi.org/10.1097/MD.0000000000000767.

Gigante, M., Pontrelli, P., Herr, W., Gigante, M., D'Avenia, M., Zaza, G., et al. (2016). miR-29b and miR-198 overexpression in CD8+ T cells of renal cell carcinoma patients down-modulates JAK3 and MCL-1 leading to immune dysfunction. *Journal of Translational Medicine*, *14*, 84. https://doi.org/10.1186/s12967-016-0841-9.

Goradel, N. H., Mohammadi, N., Haghi-Aminjan, H., Farhood, B., Negahdari, B., & Sahebkar, A. (2019). Regulation of tumor angiogenesis by microRNAs: State of the art. *Journal of Cellular Physiology*, *234*(2), 1099–1110. https://doi.org/10.1002/jcp.27051.

Grange, C., Brossa, A., & Bussolati, B. (2019). Extracellular vesicles and carried miRNAs in the progression of renal cell carcinoma. *International Journal of Molecular Sciences*, *20*(8). https://doi.org/10.3390/ijms20081832.

Guo, H., German, P., Bai, S., Barnes, S., Guo, W., Qi, X., et al. (2015). The PI3K/AKT pathway and renal cell carcinoma. *Journal of Genetics and Genomics = Yi Chuan Xue Bao*, *42*(7), 343–353. https://doi.org/10.1016/j.jgg.2015.03.003.

He, Y.-H., Chen, C., & Shi, Z. (2018). The biological roles and clinical implications of microRNAs in clear cell renal cell carcinoma. *Journal of Cellular Physiology*, *233*(6), 4458–4465. https://doi.org/10.1002/jcp.26347.

He, J., He, J., Min, L., He, Y., Guan, H., Wang, J., et al. (2020). Extracellular vesicles transmitted miR-31-5p promotes sorafenib resistance by targeting MLH1 in renal cell carcinoma. *International Journal of Cancer*, *146*(4), 1052–1063. https://doi.org/10.1002/ijc.32543.

Hou, P., Li, H., Yong, H., Chen, F., Chu, S., Zheng, J., et al. (2019). PinX1 represses renal cancer angiogenesis via the mir-125a-3p/VEGF signaling pathway. *Angiogenesis*, *22*(4), 507–519. https://doi.org/10.1007/s10456-019-09675-z.

Hsieh, J. J., Purdue, M. P., Signoretti, S., Swanton, C., Albiges, L., Schmidinger, M., et al. (2017). Renal cell carcinoma. *Nature Reviews. Disease Primers*, *3*, 17009. https://doi.org/10.1038/nrdp.2017.9.

Khella, H. W. Z., Butz, H., Ding, Q., Rotondo, F., Evans, K. R., Kupchak, P., et al. (2015). miR-221/222 are involved in response to Sunitinib treatment in metastatic renal cell carcinoma. *Molecular Therapy: The Journal of the American Society of Gene Therapy*, *23*(11), 1748–1758. https://doi.org/10.1038/mt.2015.129.

Kobayashi, Y., Lim, S.-O., & Yamaguchi, H. (2020). Oncogenic signaling pathways associated with immune evasion and resistance to immune checkpoint inhibitors in cancer. *Seminars in Cancer Biology*, *65*, 51–64. https://doi.org/10.1016/j.semcancer.2019.11.011.

Kovacova, J., Poprach, A., Buchler, T., Cho, W. C., & Slaby, O. (2018). MicroRNAs as predictive biomarkers of response to tyrosine kinase inhibitor therapy in metastatic renal cell carcinoma. *Clinical Chemistry and Laboratory Medicine*, *56*(9), 1426–1431. https://doi.org/10.1515/cclm-2017-0861.

Li, Y., Zheng, D., Pan, L., Dai, Y., Cai, S., Zhao, L., et al. (2019). Knockdown of TUG1 by shRNA inhibited renal cell carcinoma formation by miR-299-3p/VEGF axis in vitro and in vivo. *European Journal of Pharmacology*, *860*, 172536. https://doi.org/10.1016/j.ejphar.2019.172536.

Linehan, W. M., & Ricketts, C. J. (2019). The Cancer genome atlas of renal cell carcinoma: Findings and clinical implications. *Nature reviews. Urology*, *16*(9), 539–552. https://doi.org/10.1038/s41585-019-0211-5.

Mytsyk, Y., Dosenko, V., Skrzypczyk, M. A., Borys, Y., Diychuk, Y., Kucher, A., et al. (2018). Potential clinical applications of microRNAs as biomarkers for renal cell carcinoma. *Central European Journal of Urology*, *71*(3), 295–303. https://doi.org/10.5173/ceju.2018.1618.

Nogueira, I., Dias, F., Teixeira, A. L., & Medeiros, R. (2018). miRNAs as potential regulators of mTOR pathway in renal cell carcinoma. *Pharmacogenomics*, *19*(3), 249–261. https://doi.org/10.2217/pgs-2017-0160.

Omar, H. A., El-Serafi, A. T., Hersi, F., Arafa, E.-S. A., Zaher, D. M., Madkour, M., et al. (2019). Immunomodulatory MicroRNAs in cancer: Targeting immune checkpoints and the tumor microenvironment. *The FEBS Journal*, *286*(18), 3540–3557. https://doi.org/10.1111/febs.15000.

Oto, J., Plana, E., Sánchez-González, J. V., García-Olaverri, J., Fernández-Pardo, Á., España, F., et al. (2020). Urinary microRNAs: Looking for a new tool in diagnosis, prognosis, and monitoring of renal Cancer. *Current Urology Reports*, *21*(2), 11. https://doi.org/10.1007/s11934-020-0962-9.

Qu, F., Ye, J., Pan, X., Wang, J., Gan, S., Chu, C., et al. (2019). MicroRNA-497-5p down-regulation increases PD-L1 expression in clear cell renal cell carcinoma. *Journal of Drug Targeting*, *27*(1), 67–74. https://doi.org/10.1080/1061186X.2018.1479755.

Schanza, L.-M., Seles, M., Stotz, M., Fosselteder, J., Hutterer, G. C., Pichler, M., et al. (2017). MicroRNAs associated with Von Hippel-Lindau pathway in renal cell carcinoma: A comprehensive review. *International Journal of Molecular Sciences*, *18*(11). https://doi.org/10.3390/ijms18112495.

Shi, J., Zhang, D., Zhong, Z., & Zhang, W. (2019). lncRNA ROR promotes the progression of renal cell carcinoma through the miR-206/VEGF axis. *Molecular Medicine Reports*, *20*(4), 3782–3792. https://doi.org/10.3892/mmr.2019.10636.

Siegel, R. L., Miller, K. D., & Jemal, A. (2019). Cancer statistics, 2019. *CA: a Cancer Journal for Clinicians*, *69*(1), 7–34. https://doi.org/10.3322/caac.21551.

Tiwari, A., Mukherjee, B., & Dixit, M. (2018). MicroRNA key to angiogenesis regulation: MiRNA biology and therapy. *Current Cancer Drug Targets*, *18*(3), 266–277. https://doi.org/10.2174/1568009617666170630142725.

Tsiakanikas, P., Giaginis, C., Kontos, C. K., & Scorilas, A. (2018). Clinical utility of microRNAs in renal cell carcinoma: Current evidence and future perspectives. *Expert Review of Molecular Diagnostics*, *18*(11), 981–991. https://doi.org/10.1080/14737159.2018.1539668.

Wach, S., Nolte, E., Theil, A., Stöhr, C. T., Rau, T., Hartmann, A., et al. (2013). MicroRNA profiles classify papillary renal cell carcinoma subtypes. *British Journal of Cancer*, *109*(3), 714–722. https://doi.org/10.1038/bjc.2013.313.

Wang, L., Yang, G., Zhao, D., Wang, J., Bai, Y., Peng, Q., et al. (2019). CD103-positive CSC exosome promotes EMT of clear cell renal cell carcinoma: Role of remote MiR-19b-3p. *Molecular Cancer*, *18*(1), 86. https://doi.org/10.1186/s12943-019-0997-z.

Xiao, W., Lou, N., Ruan, H., Bao, L., Xiong, Z., Yuan, C., et al. (2017). Mir-144-3p promotes cell proliferation, metastasis, Sunitinib resistance in clear cell renal cell carcinoma by downregulating ARID1A. *Cellular Physiology and Biochemistry: International Journal of Experimental Cellular Physiology, Biochemistry, and Pharmacology*, *43*(6), 2420–2433. https://doi.org/10.1159/000484395.

Yan, L., Ding, B., Liu, H., Zhang, Y., Zeng, J., Hu, J., et al. (2019). Inhibition of SMYD2 suppresses tumor progression by down-regulating microRNA-125b and attenuates multi-drug resistance in renal cell carcinoma. *Theranostics*, *9*(26), 8377–8391. https://doi.org/10.7150/thno.37628.

Yang, Q., Cao, W., Wang, Z., Zhang, B., & Liu, J. (2018). Regulation of cancer immune escape: The roles of miRNAs in immune checkpoint proteins. *Cancer Letters*, *431*, 73–84. https://doi.org/10.1016/j.canlet.2018.05.015.

Zheng, B., Zhu, H., Gu, D., Pan, X., Qian, L., Xue, B., et al. (2015). MiRNA-30a-mediated autophagy inhibition sensitizes renal cell carcinoma cells to sorafenib. *Biochemical and Biophysical Research Communications*, *459*(2), 234–239. https://doi.org/10.1016/j.bbrc.2015.02.084.

Chapter 21

Pathophysiology roles and translational opportunities of miRNAs in pancreatic cancer

Sayra Dilmac[a] and Bulent Ozpolat[b]

[a]*Department of Histology and Embryology, School of Medicine, Akdeniz University, Antalya, Turkey,* [b]*Department of Experimental Therapeutics, The University of Texas MD Anderson Cancer Center, Houston, TX, United States*

Introduction

Pancreatic cancer (PC) has the highest mortality rate of all major cancers and is the fourth-leading cause of cancer mortality in the world (Chu, Goggins, & Fishman, 2017). Pancreatic cancer has two main tumor types: exocrine and pancreatic endocrine tumors (Ilic & Ilic, 2016). The average annual incidence rate of pancreatic cancer in the United States is 12.5 per 100,000. However, the mortality rate is higher with an average annual rate of 10.9 per 100,000. Pancreatic cancer is not easy to diagnose at an early stage and is generally found to be metastatic at initial diagnosis (Zhang, Sanagapalli, & Stoita, 2018). For this reason, survival rates are very poor.

Pathology of pancreatic cancer

Pancreatic cancer types can be divided into two larger categories: nonendocrine (exocrine) pancreatic cancer (95%), which includes adenocarcinoma, squamous, and colloid carcinomas, and neuroendocrine pancreatic cancer (5%), which develops from cells in the endocrine gland of the pancreas and secretes the hormones insulin and glucagon.

Pancreatic ductal adenocarcinoma (PDAC) accounts for about 90% of all pancreatic carcinomas. Therefore, the pathophysiology of pancreatic adenocarcinoma will be briefly summarized in this section. Nearly 60%–70% of PDACs are detected in the head of the pancreas while the rest are detected in the body (~ 15%) and tail (~ 15%).

The carcinogenesis of PDAC is a multistage process that involves alterations in well-characterized genes and a series of specific precursor lesions, ultimately leading to invasive carcinoma. PDAC in most cases evolves from preinvasive lesions named microscopic pancreatic intraepithelial neoplasia (PanIN). The other well-known precursors of pancreatic cancer are intraductal papillary mucinous neoplasms (IPMN) and mucinous cystic neoplasms (MCN) (McGuigan et al., 2018).

Pancreatic intraepithelial neoplasia (PanIN)

PanIN is a microscopic, noninvasive, flat or papillary epithelial small neoplasm within the pancreatic ducts. Because the PanIN size is smaller than 5 mm, it is difficult to identify by radiological methods. Low-grade PanINs (PanIN-1 or PanIN-2) are flat or papillary lesions. These are common and are particularly observed in populations over 50 years old (Fig. 21.1). High grade PanINs (PanIN 3), which are also called "*carcinoma in situ*," are usually papillary lesions characterized by loss of polarity, irregular stratification, and intraluminal necrosis (Fig. 21.1). It seems that high-grade PanINs are often associated with PDAC (Fig. 21.2) (Kim & Hong, 2018).

Intraductal papillary mucinous neoplasms (IPMN)

An IPMN is an epithelial neoplasm located in the main pancreatic duct and/or its branches that can produce mucin. IPMN is larger than PanIN (≥ 1 cm) and can be detected by imaging methods. IPMNs are most commonly located in the head of the pancreas, but they can occur anywhere in the pancreas (Fig. 21.1) (D'Angelica, Brennan, Suriawinata, Klimstra, & Conlon, 2004).

MicroRNA in Human Malignancies. https://doi.org/10.1016/B978-0-12-822287-4.00003-7

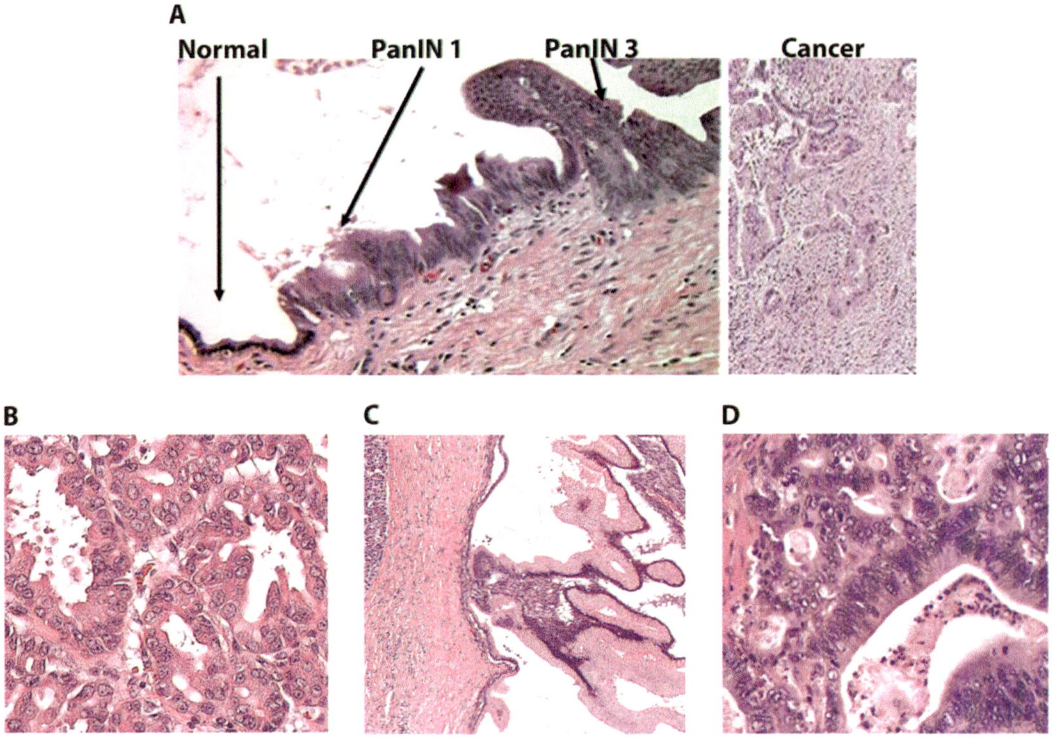

FIG. 21.1 Precursors of pancreatic cancer. A) Normal Pancreatic Duct, PanIN grade 1, PanIN grade 3 and Cancer, B) IPMN, C) Benign MCN, D) PDAC (*Wikipedia/Pancreatic Cancer; Wikipedia/Benign pancreatic mucinous cystic neoplasm*).

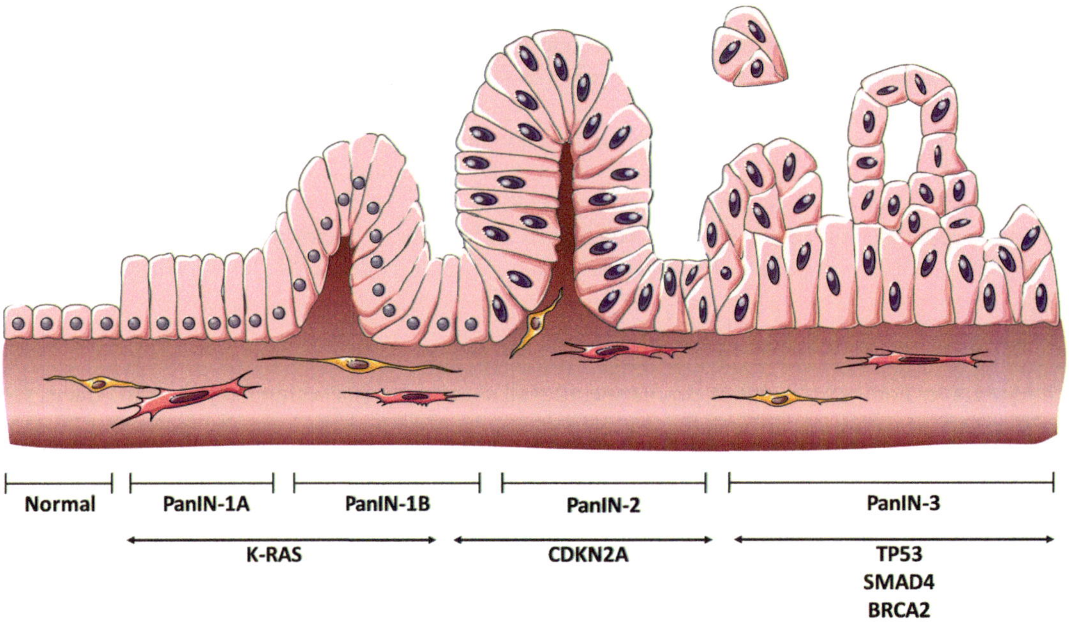

FIG. 21.2 Overview of PDAC development and progression. The precursor lesions (PanINs) go through a multistage process involving consecutive genetic mutations including KRAS oncogenic activation and multiple tumor suppressor silencing *(From Daoud, A. Z., Mulholland, E. J., Cole, G., & McCarthy, H. O. (2019). MicroRNAs in Pancreatic Cancer: Biomarkers, prognostic, and therapeutic modulators. BMC Cancer, 19(1), 1130. https://doi.org/10.1186/s12885-019-6284-y.)*

Mucinous cystic neoplasms (MCN)

MCNs are cystic neoplasms surrounded by mucin-producing epithelial cells. Unlike IPMNs, MCNs are not seen in the pancreatic duct system. MCNs usually occur in women over the age of 40, located on the body or tail of the pancreas. MCNs histologically consist of an epithelial lining and ovarian-type stroma components (Fig. 21.1). MCNs are classified as low or high grade according to the degree of cytological and structural atypia. Cytologically, MCNs and IPMNs are very similar. Therefore, only cytological properties are not sufficient to distinguish MCNs from IPMNs. However, the presence of subepithelial ovarian-type stroma is an important difference that distinguishes MCNs from IPMNs and is a distinctive feature of MCN (Kim & Hong, 2018).

Molecular basis of pancreatic cancer

Several well-defined driver genes are important in the tumorigenesis and progression of PDAC. Molecular changes consist of genetic and epigenetic rearrangements, including amplifications, deletions, translocations, inversions, frameshifts, and substitutions. Genetic changes include mutations in important genes, the activation of oncogenes, and the inactivation of tumor suppressors. Also, the epigenetic regulation of genes and alterations in the expression of noncoding RNAs (ncRNAs, such as microRNA, LncRNAs) play important roles in PDAC progression. In this section, the main molecular changes in pancreatic cancer will be discussed.

Roles of oncogenes and tumor-suppressor genes

The most common mutated gene in pancreatic cancer is *KRAS* (Kirsten rat sarcoma), which is detected in approximately 90%–95% of all pancreatic cancer cases. KRAS is a member of the RAS superfamily and regulates different cellular functions, mainly proliferation, differentiation, and survival (Malumbres & Barbacid, 2003). Oncogenic KRAS mutations are observed in ~30% of early neoplasms, and are present in about 95% of advanced PDAC. The *KRAS* mutation is detected in about 87% of PanIN2 and PanIN3 (Löhr, Klöppel, Maisonneuve, Lowenfels, & Lüttges, 2005). The most frequently observed genetic abnormalities in PanINs and invasive PDAC include the inactivation of tumor-suppressor genes (i.e., *TP53*, *SMAD*, and *CDKN2A*) as well as epigenetic alternations and chromosomal changes (rearrangements, losses, and amplifications).

The *TP53* tumor suppressor gene is mutated in ~50%–75% of PDAC patients. The inactivation of p53, which is a major mediator of cell-cycle arrest, leads to cell-cycle disruption and inhibition of apoptosis. The p53 mutation is not observed in low-grade PanIN but is seen in high-grade PanIN patients (Furukawa, 2009). p53 mutations promote not only tumor progression but also the prometastatic phenotype and metastasis in PDAC (Khan et al., 2017).

c-Myc is a proto-oncogene that affects cell growth, cell cycle progression, and cellular metabolism. Although the mutations of c-Myc alone do not affect the development of pancreatic tumors, the simultaneous presence of KRAS and c-Myc mutations is sufficient to trigger pancreatic tumorigenesis (Khan et al., 2017). In addition, achieving c-Myc protein stability by NAD-dependent deacetylase sirtuin-2 (SIRT2) or nuclear factor kappa B kinase subunit epsilon (IKKε) inhibitor has been shown to increase the occurrence of pancreatic cancer (Liu et al., 2013; Zubair et al., 2016).

CDKN2A (cyclin-dependent kinase inhibitor 2A) (mutated in 50%–90% of PDAC), also known as P16/INK4A, is an inhibitor of cyclin-dependent kinase 4 (CDK4). CDK4 interacts with cyclin D1 and supports cell proliferation. The progression of PanIN has been shown to be associated with loss of function of the tumor-suppressor molecule CDKN2A. Loss of CDKN2A expression causes low-grade PanIN (Khan et al., 2017).

SMAD4 is another tumor-suppressor protein associated with transforming growth factor-beta (TGFβ) and mutated in 20%–50% of PDAC. SMAD4 mutations promote cell proliferation by keeping the cell cycle at the G1 phase. The SMAD4 mutations have been associated with metastasis rather than tumor progression in PDAC. Also, the mutation of SMAD4 supports the sensitivity of PDAC cells to chemotherapy (Furukawa, 2009; Khan et al., 2017). Therefore, the loss of SMAD4 is considered a negative prognostic factor associated with epithelial-mesenchymal transition (EMT) and treatment failures.

The phosphatase and tensin homolog (PTEN) is another tumor suppressor that negatively regulates the mammalian target of rapamycin (mTOR) and PI3K signaling pathways, and its deletion or loss of expression is seen in pancreatic cancer. Loss of PTEN with KRAS mutation induces NF-κB activation, immune cell infiltration, and strong stromal activation, consequently supporting metastasis (Khan et al., 2017). Wartenberg et al. showed that the loss of PTEN was associated with metastasis and a decrease in the overall survival of PC patients. This study also suggested that the loss of PTEN leads to the activation of tumor-associated stromal cells, promoting pancreatic tumor cells and increasing the aggressiveness of PDAC (Wartenberg et al., 2016). Therefore, it suggests that a loss of PTEN activity is responsible not only for tumor progression but also for promoting the aggressive feature of tumor cells and metastasis.

Epigenetic regulation of pancreatic cancer

Previously, the pathogenesis of pancreatic cancer was attributed to genetic mutations; however, recent studies have demonstrated that the aberrant activation of epigenetic pathways contributes to the pathogenesis of PDAC. Epigenetic reorganization can enable the activation of oncogenes or the inhibition of tumor-suppressor genes. Thus, epigenetic changes in pancreatic cancer are considered possible biomarkers for early diagnosis and treatment (Khan et al., 2017). Epigenetic regulations include regulation through DNA methylation and acetylation. Altered endogenous miRNA expression has recently been linked to aberrations in gene expression, leading to promoting pancreatic cancer cell proliferation, survival, migration, metastasis, and drug resistance, and thus driving disease progression. In this section, special emphasis will be placed on noncoding RNAs, including microRNAs (miRNA).

miRNAs in pancreatic cancer

miRNAs (miRs) are small ncRNA molecules (ranging in size from 18 to 24 nucleotides) that regulate the transcription of targeted messenger RNA. More than 1900 precursor miRNAs and more than 2500 mature miRNAs have been identified in the human genome, and new ones continue to be identified. miRNAs play a role in many cellular processes, including the regulation of cellular differentiation, proliferation, metabolism, and apoptosis. A single miRNA may control the expression of multiple targets by specifically binding to the 3′ untranslated region (3′ UTR).

miRNAs are generated in the nucleus by the cleavage of long hairpin shaped pre-miRNAs by RNase II. Pre-miRNAs, which are capped, polyadenylated, and have a stem-loop structure, are transported to the cytoplasm by Exportin-5. In the cytoplasm, double-stranded pre-miRNA is cleaved by RNAse DICER into two single strands and converted into mature miRNA segments. miRNAs and Argonaute family proteins form the RNA-induced silencing complex (RISC), which leads to the suppression of the translation of the target mRNAs and their degradation. miRNAs are secreted to the systemic circulation in exosomes and transported to neighboring cells and other tissues and organs, providing an intercellular communication network.

miRNA expression is significantly dysregulated in various cancers, including PDAC, and has a broad impact on cellular events contributing to cancer cell proliferation, invasion, metastasis, drug resistance, angiogenesis, and hallmarks of cancer. miRNAs can be classified as oncogenic miRNAs and tumor-suppressor miRNAs with regard to their functions in carcinogenesis (Alles et al., 2019). Expression levels of many miRNAs are dysregulated in PDAC, promoting tumor growth and progression. Expression of oncogenic miRNAs is upregulated while tumor-suppressor miRNAs are often downregulated. Distinct miRNA expression profiles correlate with stages of malignant pancreatic disease; thus, some miRNAs may serve as potential biomarkers and therapeutic targets (Table 21.1).

KRAS mutations in the early stages in PanIN lesions affect the levels of specific miRNAs. The upregulation of miR-205, miR-200, and miR-21 is detected in early adenocarcinoma lesions through the use of a KRAS (G12D) mouse model. Furthermore, interaction with miRNAs is another mechanism by which mutant p53 exerts its effects. miR-155 and miR-27a, which have been shown to repress EGFR, are both suppressed by mutant p53.

microRNAs as prognostic markers

Pancreatic cancer is asymptomatic at the early stages and often detected at a late stage, resulting in poor prognosis and dismal patient survival. Therefore, the identification of pancreatic cancer-specific biomarkers is critical for early diagnosis and evaluation of the therapeutic response of the disease. Recent studies indicate that miRNAs have potential as diagnostic biomarkers and prognostic markers in PDAC due to the distinct miRNA expression profiles. miRNA expression patterns can be used to predict treatment failures and the response to therapy.

The miR-196b gene is located in an evolutionarily conserved region on chromosome 7 in the human genome between the HOXA9 and HOXA10 genes. Studies show that miR-196b targets the HOX family of molecules, the PI3K/AKT/mTOR signaling pathway, Bcr-Abl, and c-Myc. In addition, miR-196b has been reported to be a diagnostic and prognostic factor in digestive system cancers. Kanno et al. found that miR-196b expression is upregulated in pancreatic cancer tissues by miRNA profiling (Kanno et al., 2017). The expression of miR-196b is also found to be higher in cancer cells than in cancer stroma. The Kaplan-Meier analysis showed that high levels of miR-196b expression in stage II pancreatic cancer patients are inversely correlated with shorter patient survival and poor prognosis in PDAC patients (Kanno et al., 2017).

miR-203, which is located on chromosome 14, is known as a skin-specific miRNA. It plays a significant role in the regulation of embryonic epidermal differentiation and the formation of the protective layer of the skin. miR-203 has been shown to act as a tumor suppressor and regulate cell proliferation, invasion, angiogenesis, and tumor progression (Shao et al., 2017). A metaanalysis of the studies in patient tissues showed that the prognostic value of miR-203 expression varies

TABLE 21.1 MicroRNAs with altered expressions in pancreatic cancer.

miRNA	Normal pancreatic tissue levels	microRNA role in pancreatic cancer
miR-196	Low	Prognostic Marker
miR-203	Low	Prognostic Marker
miR-100	Low	Prognostic Marker
miR-182	Low	Prognostic Marker
miR-345-5p	High	Tumor Suppressor
miR-449a	High	Tumor Suppressor
miR-634	High	Tumor Suppressor
miR-873	High	Tumor Suppressor
miR-608	High	Tumor Suppressor
miR-296	High	Tumor Suppressor
miR-1290	Low	Oncogenic
miR-331-3p	Low	Oncogenic
miR-106a	Low	Oncogenic
miR-92b-3p	Low	Oncogenic
miR-301a	Low	Oncogenic
miR-221-3p	Low	Oncogenic
miR-122-5p	Low	Oncogenic

with different tumor types. A high miR-203 expression level in tumor tissues is associated with poor overall survival in colorectal, pancreatic, and ovarian cancers (Shao et al., 2017). Studies and analysis of TCGA data suggest that miR-203 expression may be an important prognostic marker in pancreatic cancer (Frampton et al., 2015; Shao et al., 2017).

miR-100, a member of the miR-99 family, is known to play an oncogenic or tumor-suppressor role in various human carcinomas. It has been suggested that miR-100 is involved in processes such as proliferation and epithelial-mesenchymal transition in pancreatic cancer, and may be a target for increasing the sensitivity of PDAC cells to chemotherapy and reducing metastasis in tumor models. Overexpression of miR-100 in PDAC tissue is associated with poor prognosis and chemotherapy resistance. Moreover, survival analysis has identified miR-100 as a negative prognostic factor in PDAC stage II patients. In addition, high levels of circulating miR-100 expression in patients with metastatic disease (stage IV) pancreatic cancer support the contention that it can be used as a noninvasive serum marker diagnostic (Stroese et al., 2018).

miR-182 is an oncogene microRNA with a dysregulated expression level in many human cancers. miR-182 overexpression induces tumor growth, invasion, and chemotherapeutic sensitivity of tumors. miR-182 has an abnormal expression in PanIN lesions and has been shown to play a role in the development of PDAC. Chen et al. found that circulating miR-182 is significantly increased in patients with PDAC. Also, it has a higher sensitivity and specificity in differentiating PDAC patients from healthy controls than traditional tumor markers such as CA19-9 (carbohydrate antigen 19-9). In addition, circulating miR-182 has proven to be an independent prognostic factor for PDAC patients. The results of the studies suggest that miR-182 may be a more reliable tumor marker for the diagnosis and prognosis of PDAC (Chen, Yang, Xiao, Zhu, & Li, 2014).

Tumor-suppressor microRNAs in pancreatic cancer

Dysregulated miRNAs in cancer cells can act as tumor-suppressor miRNAs, based on their capability to repress the expression of tumor-suppressor genes. Tumor-suppressor miRNAs modulate cancer cell proliferation, cell survival, metastasis, angiogenesis, tumor growth, and progression. Tumor-suppressor miRNAs are often downregulated in cancer cells, leading to increased activity in corresponding target oncogenes. For instance, miR-345-5p expression is downregulated in PDAC patient tumor tissues and cancer cells and restoration of its expression suppresses the proliferation, invasion, and progression of pancreatic cancer. Studies in in vivo pancreatic cancer models showed that the overexpression of miR-345-5p inhibits its tumor growth and reduces metastasis by suppressing CCL8 (Mou et al., 2019).

miR-449a expression is found to be decreased in pancreatic cancer tissues and cell lines (PANC-1, BxPC3, AsPC-1, and SW1990) compared to normal tissues and pancreas cells (HPNE), respectively. miR-449a inhibits proliferation and invasion in pancreatic cancer cells. It was shown that miR-449a binds directly to the Ataxia-telangiectasia group D complementing gene (ATDC) (Li, Liang, & Bai, 2018). Based on the results, miR-449a seems to act as a tumor suppressor by reducing pancreatic cancer proliferation and invasion.

miR-634 has been shown to act as a tumor-suppressor microRNA in glioma, gastric carcinoma, hepatocellular carcinoma, nasopharyngeal carcinoma, ovarian cancer, and cervical cancer. Low miR-634 expression in pancreatic cancer suggests its role as a tumor suppressor. Transfection of PANC1 with miR-634 mimics reduced cell viability, proliferation, invasion, and migration. On the other hand, the invasion and migration of pancreatic cancer cells increase after transfection with the miR-634 inhibitor. In addition, the miR-634 mimic was shown to reduce the expression of HSPA2, a protein coding gene important for cancer cell growth and metastasis, whereas the miR-634 inhibitor increased the protein expression of HSPA2, suggesting that miR-634 could reduce tumor growth by inhibiting HSPA2 (Chen, Wu, Zhao, & Zhao, 2019).

miR-873 is known to be an effective suppressor in many tumor types such as breast and ovarian cancers and glioma. miR-873 has been shown to play a role in cancer development by regulating tumor cell invasion, migration, proliferation, apoptosis, and sensitivity to chemotherapeutic drugs (Jin et al., 2018). We have demonstrated that miR-873 is associated with longer patient survival by the Kaplan-Meier survival analysis. We also found that miR-873 expression is markedly decreased or lost in pancreatic ductal adenocarcinoma cells and has a tumor-suppressor function by directly binding and inhibiting KRAS expression, leading to reduced cell invasion and migration in PANC1 and MiaPaCa-2 cells. An in vivo xenograft mouse model of PANC1 and MiaPaCa-2 demonstrated that therapeutically administered miR-873 encapsulated in lipid nanoparticles reduced KRAS expression in in vivo tumors and tumor growth, which is associated with reduced Ki-67 expression and increased apoptosis. Overall, this study provided evidence that miR-873 is an important tumor-suppressor miRNA in PDAC and can be used as a prognostic marker and therapeutic tool to target KRAS in PDAC (Mokhlis et al., 2019).

miR-608 upregulation is associated with longer overall patient survival, recurrence-free survival, and favorable prognosis. miR-608 targets and inhibits BRD4, known to be associated with c-Myc and Bcl-2, and induces apoptosis in the PANC1 and Bxpc-3 cell lines. Overall, studies showed that miRNA-608 plays a tumor-suppressor role in PDAC (Li et al., 2018; Li, Li, Ma, Wang, & Zhao, 2020).

Recent studies suggest that miR-296 acts as a tumor suppressor in nonsmall cell lung cancer, prostate cancer, and breast cancer. Li H. et al. found that miR-296 expression was reduced in pancreatic cancer cell lines compared to normal pancreatic cells. Ectopic expression of miR-296 reduced cell proliferation, cell migration, and invasion of PANC1 and Capan-2 cells by directly targeting AKT2 mRNA. These findings suggested that miR-296 has a tumor-suppressor role in pancreatic cancer pathogenesis by targeting AKT2 and the miR-296/AKT2 axis could be a novel therapeutic target in pancreatic cancer patients (Li, Li, Shi, & Chen, 2017).

Oncogenic microRNAs in pancreatic cancer

Oncogenic microRNAs promote cell proliferation, invasion, and tumor growth by inhibiting tumor-suppressor genes in human malignancies. For instance, miR-1290 expression is increased in patients with PanIN and PDAC. It has been observed that the miR-1290 expression level is increased in the blood/serum of pancreatic cancer patients. Increased miR-1290 expression is associated with histological differentiation and lymph node metastasis as well as increased cell proliferation and invasion by in vivo and in vitro studies. It was shown that miR-1290 binds directly to the 3′ UTR of IKappaB Kinase (IKK1) mRNA and suppresses its expression. According to these results, miR-1290 interacts with IKK1 and increases PDAC aggressiveness. These results suggest that miR-1290 could act as an oncogene and be a potential target for the diagnosis and treatment of PDAC (Ta et al., 2018).

miR-331-3p has been shown to increase proliferation, promote epithelial-mesenchymal transition, and facilitate metastasis in hepatocellular carcinoma (HCC). Chen X. et al. showed that miR-331-3p expression levels are increased in pancreatic cancer patient tumors and cell lines. Ectopic expression of miR-331-3p has been shown to promote cell proliferation and metastasis in pancreatic cancer cells. miR-331-3p binds directly to the suppression of tumorigenicity 7 like (ST7L), which is known to be associated with WNT2, and suppresses ST7L expression. In vivo inhibition of miR-331-3p expression in the mouse pancreatic cancer model suppresses the progression of pancreatic cancer, suggesting that miR-331-3p acts as an oncogene in PDAC and can be used as a potential diagnostic biomarker and drug target (Chen et al., 2018).

Increased expression of miR-106a has been shown to function as an oncogene in gastric cancer and colorectal cancer, and low levels of miR-106a expression correlate with the malignant progression of gliomas. In pancreatic cancer, miR-106a expression levels were found to be significantly higher in tumor samples than in normal adjacent pancreatic tissue. Increased miR-106a expression induces cell proliferation and invasion in pancreatic cancer cells (Li et al., 2014). It has also been observed that the expression of miR-106a leads to the induction of matrix metalloproteinases (MMP) and invasion in PDAC cells, suggesting that miR-106a promotes metastasis. miR-106a binds directly to tissue inhibitors of metalloproteinase 2 (TIMP2) and regulates the levels of MMP2 and MMP9 by targeting TIMP2. Li et al. showed that miR-106a induces pancreatic tumor formation by promoting pancreatic cancer proliferation, indicating that miR-106a may represent a potential molecular therapeutic target in pancreatic cancer (Li et al., 2014).

miR-92b-3p has been identified as an oncogenic ncRNA in glioma, glioblastoma, and nonsmall cell lung cancer. miR-92b-3p expression was found to be decreased in pancreatic cancer tissues when compared to noncancerous pancreatic tissues. Reduced miR-92b-3p expression correlates with reduced patient survival and increased lymph node metastasis. miR-92b-3p expression suppresses the proliferation and invasion of pancreatic cancer cells (AsPC-1, SW1990) through the inhibition of gamma-aminobutyric acid type A receptor subunit alpha3 (GABRA3). It was shown that the miR-92b-3p/GABRA3 axis regulates the AKT/mTOR and JNK pathways. Overexpressing miR-92b-3p suppresses AsPC-1 tumor growth in mouse models, suggesting that miR-92b-3p acts as a tumor suppressor and may be used as a diagnostic and prognostic biomarker in PDAC patients (Long et al., 2017).

miR-301a acts as an NF-κB activator and creates a proinflammatory microenvironment that facilitates tumor formation by activating the STAT3 signaling pathway. miR-301a is found to be increased in pancreatic cancer tissues compared to noncancerous tissues. miR-301a levels are also detected in the serum of pancreatic cancer patients. miR-301a increases the migration and invasion capability of PDAC cells (PANC1 and BXPC3) and induces angiogenic tube formation. miR-301a directly binds to the 3′ UTR of suppressor of cytokine signaling 5 (SOCS5) and suppresses the mRNA and protein levels, leading to the activation of JAK/STAT3 signaling (Hu et al., 2020). In the in vivo mouse pancreatic cancer model, the inhibition of miR-301a increases the level of SCOS5 in the tumor xenographs and decreases tumor size and metastasis, suggesting that miR-301a acts as an oncomiR and that the inhibition of miR-301a could be a potential therapeutic target to prevent pancreatic cancer progression (Hu et al., 2020).

Circulating microRNAs in pancreatic cancer

Circulating or extracellular miRNAs are endogenous ncRNAs that are secreted into the blood circulation. They are highly stable and participate in numerous cellular processes and are aberrantly expressed in various pathologies, including cancer. They are easily detected in peripheral blood, in apoptotic bodies, microvesicles, or exosomes and are thought to be involved in the tumor-promoting microenvironment, progression, and distant metastasis. Therefore, circulating miRNAs are considered useful markers in the early detection and prognosis of PDAC. Currently, considerable effort has been devoted to identify suitable circulating miRNAs as biomarkers in cancer.

miR-221-3p expression is highly upregulated in PDAC cell lines (SW190, CFPAC-1, PANC1, MiaPaCa-2) and induces cell proliferation. miR-221-3p is overexpressed in PDAC compared to nontumor tissues. An analysis of miR-221-3p expression levels in plasma found that miR-221-3p expression level in plasma released by PDAC tissues is associated with distant metastasis and progression of pancreatic cancer (Li et al., 2018).

miRNA profiling of serum from PDAC and healthy people identified several highly upregulated miRs, including miR-1273g-3p, miR-6126, and miR-122-5p, by ddPCR (droplet digital PCR). In addition, in the serum of patients with IPMN and chronic pancreatitis, there was no statistically significant difference in the expression of these miRs. However, the expression of miR-122-5p and miR-1273g-3p was significantly increased in patients with IPMN and chronic pancreatitis compared to healthy subjects (Mazza et al., 2020). Furthermore, the serum levels of miR-122-5p and miR-6126 were higher in patients with metastatic PDAC than patients without metastatic cancer, and the plasma levels of miR-122-5p were found to be associated with the aggressiveness of the tumor, poor overall survival, and poor prognosis. These studies suggested that the serum miR-122-5p level would provide an ideal marker for evaluating the course of the disease as well as new potential molecular targets for the development of targeted treatment strategies for PDAC (Mazza et al., 2020).

Conclusion

The major challenges in PDAC include the lack of effective biomarkers for early detection, and the absence of successful targeted therapeutic strategies, which contribute to dismal patient survival and prognosis. Recently, miRNAs have emerged as potential biomarkers for early diagnosis, predicting disease prognosis and molecular targets for precision medicine. Due to the ability of miRNAs to regulate the expression and signaling pathways of molecules involved in cancer cell proliferation, cell death, invasion and metastasis, angiogenesis, drug resistance, tumorigenesis, and progression, they may act as tumor suppressors or oncogenes in pancreatic cancer. Furthermore, endogenous miRNAs are secreted into circulation from cancer cells. Therefore, it is expected that in the near future, miRNAs will be essential players as biomarkers for early detection and provide a solid foundation for the development of targeted therapeutics for cancer treatment.

References

Alles, J., Fehlmann, T., Fischer, U., Backes, C., Galata, V., Minet, M., ... Meese, E. (2019). An estimate of the total number of true human miRNAs. *Nucleic Acids Research*, *47*(7), 3353–3364. https://doi.org/10.1093/nar/gkz097.

Chen, D., Wu, X., Zhao, J., & Zhao, X. (2019). MicroRNA-634 functions as a tumor suppressor in pancreatic cancer via directly targeting heat shock-related 70-kDa protein 2. *Experimental and Therapeutic Medicine*, *17*(5), 3949–3956. https://doi.org/10.3892/etm.2019.7433.

Chen, Q., Yang, L., Xiao, Y., Zhu, J., & Li, Z. (2014). Circulating microRNA-182 in plasma and its potential diagnostic and prognostic value for pancreatic cancer. *Medical Oncology (Northwood, London, England)*, *31*(11), 225. https://doi.org/10.1007/s12032-014-0225-z.

Chen, X., Luo, H., Li, X., Tian, X., Peng, B., Liu, S., ... Huang, X. (2018). miR-331-3p functions as an oncogene by targeting ST7L in pancreatic cancer. *Carcinogenesis*, *39*(8), 1006–1015. https://doi.org/10.1093/carcin/bgy074.

Chu, L. C., Goggins, M. G., & Fishman, E. K. (2017). Diagnosis and detection of pancreatic cancer. *Cancer Journal (Sudbury, Mass)*, *23*(6), 333–342. https://doi.org/10.1097/PPO.0000000000000290.

D'Angelica, M., Brennan, M. F., Suriawinata, A. A., Klimstra, D., & Conlon, K. C. (2004). Intraductal papillary mucinous neoplasms of the pancreas: An analysis of clinicopathologic features and outcome. *Annals of Surgery*, *239*(3), 400–408.

Frampton, A. E., Krell, J., Jamieson, N. B., Gall, T. M. H., Giovannetti, E., Funel, N., ... Stebbing, J. (2015). microRNAs with prognostic significance in pancreatic ductal adenocarcinoma: A meta-analysis. *European Journal of Cancer (Oxford, England: 1990)*, *51*(11), 1389–1404. https://doi.org/10.1016/j.ejca.2015.04.006.

Furukawa, T. (2009). Molecular pathology of pancreatic cancer: Implications for molecular targeting therapy. *Clinical Gastroenterology and Hepatology: The Official Clinical Practice Journal of the American Gastroenterological Association*, *7*(11 Suppl), S35–S39. https://doi.org/10.1016/j.cgh.2009.07.035.

Hu, H., Zhang, Q., Chen, W., Wu, T., Liu, S., Li, X., ... Lu, Z. (2020). MicroRNA-301a promotes pancreatic cancer invasion and metastasis through the JAK/STAT3 signaling pathway by targeting SOCS5. *Carcinogenesis*, *41*(4), 502–514. https://doi.org/10.1093/carcin/bgz121.

Ilic, M., & Ilic, I. (2016). Epidemiology of pancreatic cancer. *World Journal of Gastroenterology*, *22*(44), 9694–9705.

Jin, S., He, J., Li, J., Guo, R., Shu, Y., & Liu, P. (2018). MiR-873 inhibition enhances gefitinib resistance in non-small cell lung cancer cells by targeting glioma-associated oncogene homolog 1. *Thoracic Cancer*, *9*(10), 1262–1270. https://doi.org/10.1111/1759-7714.12830.

Kanno, S., Nosho, K., Ishigami, K., Yamamoto, I., Koide, H., Kurihara, H., ... Nakase, H. (2017). MicroRNA-196b is an independent prognostic biomarker in patients with pancreatic cancer. *Carcinogenesis*, *38*(4), 425–431. https://doi.org/10.1093/carcin/bgx013.

Khan, M. A. A., Azim, S., Zubair, H., Bhardwaj, A., Patel, G. K., Khushman, M., ... Singh, A. P. (2017). Molecular drivers of pancreatic cancer pathogenesis: Looking inward to move forward. *International Journal of Molecular Sciences*, *18*(4), 779. https://doi.org/10.3390/ijms18040779.

Kim, J. Y., & Hong, S.-M. (2018). Precursor lesions of pancreatic cancer. *Oncology Research and Treatment*, *41*(10), 603–610. https://doi.org/10.1159/000493554.

Li, F., Liang, J., & Bai, L. (2018). MicroRNA-449a functions as a tumor suppressor in pancreatic cancer by the epigenetic regulation of ATDC expression. *Biomedicine & Pharmacotherapy*, *103*, 782–789. https://doi.org/10.1016/j.biopha.2018.04.101.

Li, F., Xu, J.-W., Wang, L., Liu, H., Yan, Y., & Hu, S.-Y. (2018). MicroRNA-221-3p is up-regulated and serves as a potential biomarker in pancreatic cancer. *Artificial Cells, Nanomedicine, and Biotechnology*, *46*(3), 482–487. https://doi.org/10.1080/21691401.2017.1315429.

Li, H., Li, J., Shi, B., & Chen, F. (2017). MicroRNA-296 targets AKT2 in pancreatic cancer and functions as a potential tumor suppressor. *Molecular Medicine Reports*, *16*(1), 466–472. https://doi.org/10.3892/mmr.2017.6602.

Li, M., Li, T., Ma, W., Wang, X., & Zhao, G. (2020). MicroRNA-608 promotes apoptosis via BRD4 downregulation in pancreatic ductal adenocarcinoma. *Oncology Letters*, *19*(2), 1418–1426. https://doi.org/10.3892/ol.2019.11246.

Li, P., Xu, Q., Zhang, D., Li, X., Han, L., Lei, J., ... Wang, Z. (2014). Upregulated miR-106a plays an oncogenic role in pancreatic cancer. *FEBS Letters*, *588*(5), 705–712. https://doi.org/10.1016/j.febslet.2014.01.007.

Liu, P. Y., Xu, N., Malyukova, A., Scarlett, C. J., Sun, Y. T., Zhang, X. D., ... Liu, T. (2013). The histone deacetylase SIRT2 stabilizes Myc oncoproteins. *Cell Death and Differentiation*, *20*(3), 503–514. https://doi.org/10.1038/cdd.2012.147.

Löhr, M., Klöppel, G., Maisonneuve, P., Lowenfels, A. B., & Lüttges, J. (2005). Frequency of K-ras mutations in pancreatic intraductal neoplasias associated with pancreatic ductal adenocarcinoma and chronic pancreatitis: A meta-analysis. *Neoplasia (New York, N.Y.)*, *7*(1), 17–23.

Long, M., Zhan, M., Xu, S., Yang, R., Chen, W., Zhang, S., ... Wang, J. (2017). miR-92b-3p acts as a tumor suppressor by targeting Gabra3 in pancreatic cancer. *Molecular Cancer*, *16*(1), 167. https://doi.org/10.1186/s12943-017-0723-7.

Malumbres, M., & Barbacid, M. (2003). RAS oncogenes: The first 30 years. *Nature Reviews Cancer*, *3*(6), 459–465. https://doi.org/10.1038/nrc1097.

Mazza, T., Gioffreda, D., Fontana, A., Biagini, T., Carella, M., Palumbo, O., ... Tavano, F. (2020). Clinical significance of circulating miR-1273g-3p and miR-122-5p in pancreatic cancer. *Frontiers in Oncology*, *10*, 44. https://doi.org/10.3389/fonc.2020.00044.

McGuigan, A., Kelly, P., Turkington, R. C., Jones, C., Coleman, H. G., & McCain, R. S. (2018). Pancreatic cancer: A review of clinical diagnosis, epidemiology, treatment and outcomes. *World Journal of Gastroenterology*, *24*(43), 4846–4861. https://doi.org/10.3748/wjg.v24.i43.4846.

Mokhlis, H. A., Bayraktar, R., Kabil, N. N., Caner, A., Kahraman, N., Rodriguez-Aguayo, C., … Ozpolat, B. (2019). The modulatory role of MicroRNA-873 in the progression of KRAS-driven cancers. *Molecular Therapy - Nucleic Acids*, *14*, 301–317. https://doi.org/10.1016/j.omtn.2018.11.019.

Mou, T., Xie, F., Zhong, P., Hua, H., Lai, L., Yang, Q., & Wang, J. (2019). MiR-345-5p functions as a tumor suppressor in pancreatic cancer by directly targeting CCL8. *Biomedicine & Pharmacotherapy*, *111*, 891–900. https://doi.org/10.1016/j.biopha.2018.12.121.

Shao, Y., Gu, W., Ning, Z., Song, X., Pei, H., & Jiang, J. (2017). Evaluating the prognostic value of microRNA-203 in solid tumors based on a meta-analysis and the cancer genome atlas (TCGA) datasets. *Cellular Physiology and Biochemistry: International Journal of Experimental Cellular Physiology, Biochemistry, and Pharmacology*, *41*(4), 1468–1480. https://doi.org/10.1159/000470649.

Stroese, A. J., Ullerich, H., Koehler, G., Raetzel, V., Senninger, N., & Dhayat, S. A. (2018). Circulating microRNA-99 family as liquid biopsy marker in pancreatic adenocarcinoma. *Journal of Cancer Research and Clinical Oncology*, *144*(12), 2377–2390. https://doi.org/10.1007/s00432-018-2749-7.

Ta, N., Huang, X., Zheng, K., Zhang, Y., Gao, Y., Deng, L., … Zheng, J. (2018). miRNA-1290 promotes aggressiveness in pancreatic ductal adenocarcinoma by targeting IKK1. *Cellular Physiology and Biochemistry: International Journal of Experimental Cellular Physiology, Biochemistry, and Pharmacology*, *51*(2), 711–728. https://doi.org/10.1159/000495328.

Wartenberg, M., Centeno, I., Haemmig, S., Vassella, E., Zlobec, I., Galván, J. A., … Karamitopoulou, E. (2016). PTEN alterations of the stromal cells characterise an aggressive subpopulation of pancreatic cancer with enhanced metastatic potential. *European Journal of Cancer (Oxford, England: 1990)*, *65*, 80–90. https://doi.org/10.1016/j.ejca.2016.06.013.

Zhang, L., Sanagapalli, S., & Stoita, A. (2018). Challenges in diagnosis of pancreatic cancer. *World Journal of Gastroenterology*, *24*(19), 2047–2060. https://doi.org/10.3748/wjg.v24.i19.2047.

Zubair, H., Azim, S., Srivastava, S. K., Ahmad, A., Bhardwaj, A., Khan, M. A., … Singh, A. P. (2016). Glucose metabolism reprogrammed by overexpression of IKKε promotes pancreatic tumor growth. *Cancer Research*, *76*(24), 7254–7264.

Chapter 22

Extracellular vesicle-contained microRNAs in prostate cancer: From pathophysiology to emerging translational opportunities

Vera Constâncio[a,b], Rui Henrique[a,c,d], and Carmen Jerónimo[a,d]

[a]Cancer Biology and Epigenetics Group, IPO Porto Research Center, Portuguese Oncology Institute of Porto & Porto Comprehensive Cancer Center, Porto, Portugal, [b]Doctoral Programme in Biomedical Sciences, School of Medicine & Biomedical Sciences, University of Porto, Porto, Portugal, [c]Department of Pathology, Portuguese Oncology Institute of Porto, Porto, Portugal, [d]Department of Pathology and Molecular Immunology, Institute of Biomedical Sciences Abel Salazar, University of Porto, Porto, Portugal

Abbreviations

AUC	Area Under the Curve
BPH	Benign Prostatic Hyperplasia
DRE	Digital Rectal Examination
GS	Gleason Score
LN	Lymph Node
Sn%	Sensitivity
Sp%	Specificity
PCa	Prostate Cancer

Overview

Since the first report on microRNAs (miRNAs) in the early 2000s, a lot of ink has been spilled to shed light on the contribution of these small molecules for cancer pathogenesis. These small noncoding 21–24-base RNAs, which for a long time were considered junk RNA, are now widely recognized as key elements in posttranscriptional gene expression regulation, virtually linked to all biochemical processes, with tissue and developmental stage-specific expression (Iorio & Croce, 2009). Hence, it is understandable why miRNA deregulation is involved in several human diseases, including cancer.

Despite the complex behavior of miRNAs [one miRNA can potentially target several mRNAs, and one mRNA can be targeted by many miRNAs; and depending on the cellular context and the tumor type, the same miRNA may act as a tumor-suppressor miRNA (TSmiR) or as an oncogenic microRNA (oncomiR)], they have emerged as attractive biomarkers to be used in clinics as they fulfill most requirements of an optimal biomarker: they are stable, amenable to be detected in liquid biopsies, sensitive, specific, dynamic, easy, and relatively unexpensive to detect using conventional lab equipment (Sawyers, 2008).

MiRNAs can be secreted from cells by still unclear processes or cell death. Yet, they can circulate with very high stability enclosed into extracellular vesicles (EVs) or in a vesicle-free form incorporated in ribonucleoprotein complexes that protect them from RNase degradation (Boerrigter, Groen, Van Erp, Verhaegh, & Schalken, 2020). EVs are small particles (100–1000 nm in diameter) naturally released by cells and are key functional players in tumor progression, metastatic spread, and drug resistance (Linxweiler & Junker, 2020). EVs bear surface molecules that can target a recipient cell. Once attached, they may induce signaling interactions, be internalized or fused with a target cell's membrane to deliver its content, and modify the physiological state of the recipient cell (Tkach & Théry, 2016). Hence, the discovery that EV-contained miRNA released from one cell can shuttle information to other cells (adjacent or distant) represents a hallmark in miRNA studies. Moreover, because EVs are released into the circulation, they represent an ideal source for minimally invasive biomarkers carrying clinically useful information.

MicroRNA in Human Malignancies. https://doi.org/10.1016/B978-0-12-822287-4.00028-1

Notwithstanding the widespread use since the late 1980s of serum prostate-specific antigen (PSA) and, more recently, of nuclear magnetic resonance imaging for the tentative early detection of prostate cancer (PCa), it remains the second most common and the fifth deadliest cancer among men worldwide (Bray et al., 2018). This is due in part to disease heterogeneity, ranging from clinically indolent to extremely aggressive tumors, even in patients with apparently similar clinicopathological characteristics. Hence, although it is relatively easy to detect PCa, it is imperative to move toward personalized approaches capable of assessing the real threat that PCa poses to the patient's life, minimizing the risk of overdiagnosis and consequent overtreatment. In this context, EV-contained miRNA represents an opportunity to fill this significant clinical gap.

Herein, we will focus on the current state of the art about the known pathophysiological roles and emerging translational opportunities of EV miRNAs in PCa.

EV-contained miRNAs in PCa pathophysiology—What do we know so far?

Although the EV-contained miRNA profile is thought to resemble that of the corresponding parental cells (Linxweiler & Junker, 2020), differentially expressed miRNAs between EVs and parental PCa cell lines have been reported (Corcoran, Rani, & O'Driscoll, 2014; Hessvik, Phuyal, Brech, Sandvig, & Llorente, 2012; Worst et al., 2019). Moreover, not only do cancer cells release a much larger amount of EVs than normal cells (Hessvik et al., 2012), but distinct classes of miRNAs are also differently expressed in EVs derived from tumorigenic cells in comparison to their isogenic nontumorigenic counterpart. However, a strong correlation exists between intracellular miRNA expression levels of isogenic cell lines (Morello et al., 2013), suggesting differential miRNA packaging into EVs between benign and tumoral cell lines. Indeed, miRNAs with known oncogenic targets were present at lower levels in the tumorigenic cell line-derived EVs (Morello et al., 2013). Moreover, a distinct ncRNA composition was observed in a PCa cell line and its derived EVs. The predominant class of ncRNAs in EVs was mature miRNAs (66.25%), whereas snoRNA predominated in cells (60.50%). However, a significant amount of miRNA (32.38%) was also present, indicating that EVs were enriched in miRNAs (Worst et al., 2019). Consequently, this is an ideal marker to be studied in PCa-derived EVs.

Interestingly, Bhagirath et al. observed that the tumor-suppressor miR-1246 (inhibits EMT and cellular proliferation and promotes apoptosis) was downregulated in PCa tissues and cell lines, but overexpressed in serum EVs of PCa patients, with its expression also correlating with advanced disease (Bhagirath et al., 2018). Interestingly, this EV-contained miRNA expression decreased upon PCa cell treatment with an exosome inhibitor, and an inverse correlation was also observed between tissues and serum EV miRNA-1246 from xenograft mouse models. Thus, it was suggested that cells employ exosomes as vehicles to discard tumor-suppressor miRNAs, causing their upregulation in body fluids (Bhagirath et al., 2018).

Furthermore, Urabe et al. showed that miR-26a was involved in EV secretion from PCa cells (PC3/PC3M) by targeting SHC4, PFDN4, and CHORDC1. Suppression of these genes resulted in the inhibition of PCa progression in vivo, possibly by inhibiting EV communication. Indeed, when mice were injected with PCa cell line-derived EVs, the tumor size and weight of the gene-depleted mice were partially rescued (Urabe, Kosaka, Sawa, Yamamoto, et al., 2020). However, because the number and size of secreted EVs from different cell lines (PC3/PC3M and 22Rv1—all CRPC cell lines) were divergent, it was suggested that the EV secretion pathway may be dependent on specific cellular characteristics. Indeed, spermidine synthase (SRM) regulation by miR-1908 was identified as a potential EV secretion regulator in 22Rv1 PCa cells (Urabe, Kosaka, Sawa, Ito, et al., 2020).

Overall, considering all the above-mentioned studies indicating a selective release of EV-contained miRNAs by PCa cells, it is recognizable that these might indeed influence PCa progression by interacting with other local and distant cells. Hence, a summary of the currently described EV-contained miRNA pathophysiological roles in the PCa tumor microenvironment, bone metastasis development, and treatment resistance is depicted in Fig. 22.1 and Table 22.1.

Effects on tumor microenvironment

EVs have been implicated in tumor microenvironment interactions, and miRNAs may be in part responsible for this interplay. Indeed, large oncosomes were shown to induce cancer-associated fibroblast (CAF) migration, facilitating PCa progression by favoring the establishment of a more permissive tumor microenvironment, an effect that was enhanced by miR-1227 (Morello et al., 2013). Remarkably, increased EV shedding was observed in DIAPH3-silencing derived amoeboid PCa cells as well as from stimulation with prostate stroma-derived growth factor heparin-binding EGF-like growth factor (HB-EGF) combined with p38MAPK inhibition, a process mediated by ERK1/2 hyperactivation. EVs shed by DIAPH3-silenced PCa cells activated AKT1 and androgen signaling, increased the proliferation of recipient (autologous

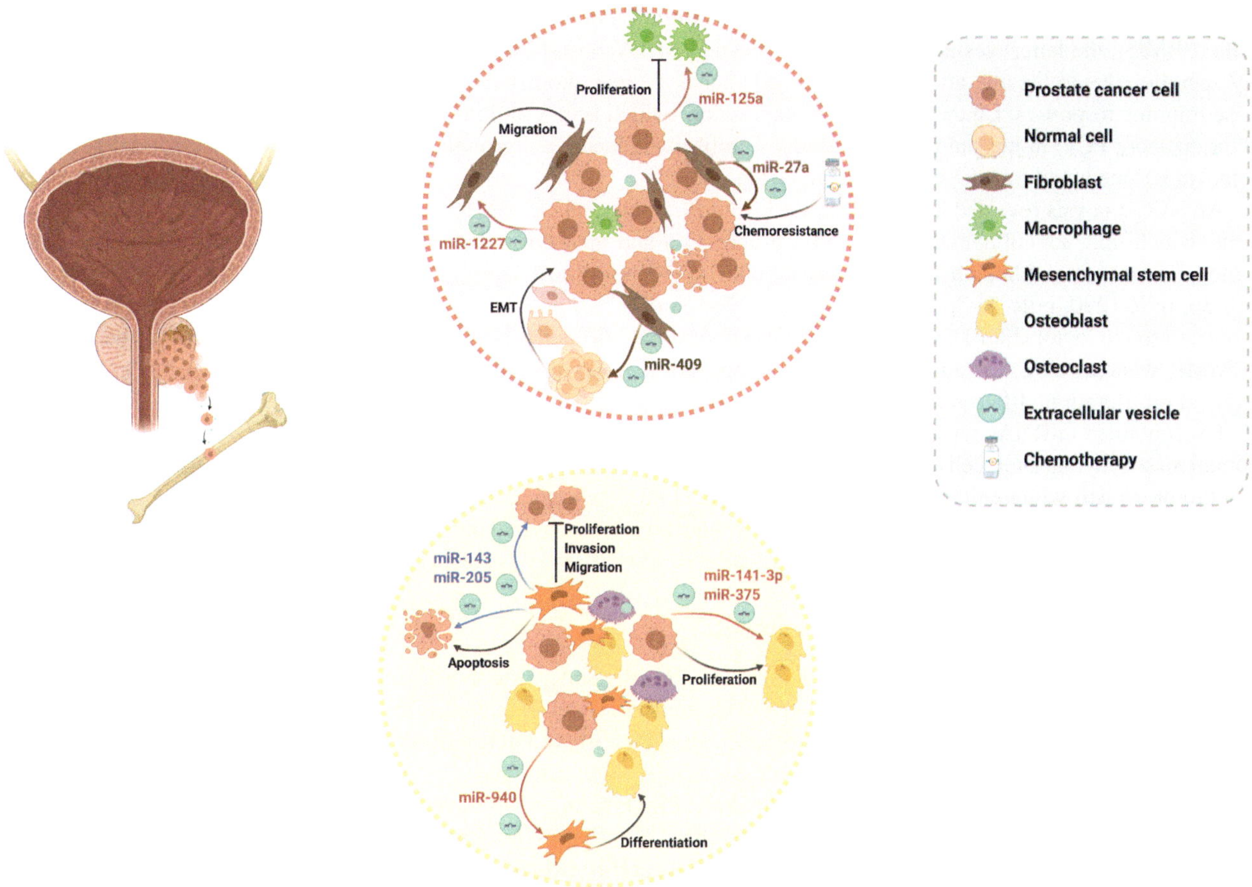

FIG. 22.1 Schematic overview of published extracellular vesicle-contained miRNA pathophysiological roles in PCa pathophysiology. Created with BioRender.com.

TABLE 22.1 Extracellular vesicle-contained microRNAs roles in PCa pathophysiology.

EV miRNA	Donor cell	Pathophysiologic role	Reference
miR-1227	PCa	↑ CAF migration	Morello et al. (2013)
miR-125a	PCa	↓ Macrophage proliferation	Kim et al. (2014)
miR-409	Stromal Fibroblasts	↑ PCa tumorigenesis, EMT, and stemness of epithelial cancer cells	Josson et al. (2015)
miR-141-3p	PCa	↑ Osteoblast activity	Ye et al. (2017)
miR-940	PCa	↑ Osteogenic differentiation of mesenchymal cells	Hashimoto et al. (2018)
miR-375	PCa	↑ Osteoblast activity	Li et al. (2019)
miR-27a	Fibroblasts	↑ PCa cell chemoresistance to cisplatin, doxorubicin, and docetaxel	Cao et al. (2019)
miR-143	Mesenchymal stem cells	↓ PCa proliferation, invasion, migration, and tumor growth ↑ Apoptosis	Che et al. (2019)
miR-205	Bone marrow mesenchymal stem cells	↓ PCa proliferation, invasion, migration, and tumor growth ↑ Apoptosis	Jiang et al. (2019)

Abbreviations: ↑—Promote; ↓—Inhibit; CAFs—Cancer-Associated Fibroblasts; EMT—Epithelial-Mesenchymal Transition; PCa—Prostate Cancer.

and heterologous) tumor cells, and suppressed the proliferation of human macrophages and peripheral blood mononuclear cells (PBMC); the latter was at least partially mediated by EV miR-125a (Kim et al., 2014). This suggests that enhanced EV secretion due to the alteration of the PCa cell phenotype may modify the tumor microenvironment and, thus, compromise immune responses. Likewise, miR-409 expression (which is elevated in human prostatic stromal tissues from higher Gleason score PCa) in normal prostate stromal fibroblasts induced cell growth, EMT, and the stemness of epithelial PCa cells via EV-mediated transfer (Josson et al., 2015).

An NGS analysis revealed that EVs released from PCa bulk and cancer stem cells (CSC) exhibit different miRNA patterns, which may act collaboratively for PCa progression and metastasis. Bioinformatic analysis indicated that the main biological process regulated by overexpressed bulk EV miRNAs (miR-7-5p, miR-20a-5p, miR-93-5p, miR-503-5p, miR-218-5p, miR-1290, miR-25-3p, miR-17-5p, miR-378d, miR-378c, miR-21-5p, miR-30c-5p, miR-125b-1-3p) were the fibroblast growth factor (FGF) pathway, epithelial proliferation, differentiation, migration, and EMT including the activation of MMP, whereas CSC EV miRNAs (miR-1307-5p, miR-139-5p, miR-148a-3p, miR-7641, miR-1307-3p, miR-183-5p) induced proliferation, differentiation to the epithelium, and angiogenesis (Sánchez et al., 2016).

EV-contained miRNAs released from stromal cells may also affect cancer progression. Indeed, miR-143 transfer from human mesenchymal stem cells (hMSC) to PCa cells through EVs inhibited PCa cell line proliferation, invasion, migration, and tumor growth while enhancing apoptosis through TFF3 downregulation (Che et al., 2019). Moreover, similar effects were obtained with EVs derived from human bone marrow mesenchymal stem cells (hBMSC) carrying miR-205 by the inhibition of RHPN2 (Jiang et al., 2019).

Effects on PCa bone metastasis development

Bone is the most frequent metastatic site of PCa. In a healthy state, osteoblasts and osteoclasts remain in homeostasis. However, increased osteoblast activity induced by PCa cells may lead to osteoblastic bone metastasis. Yet, the exact mechanisms involved in this phenomenon remain unknown. Ye et al. observed that miR-141-3p was increased in EVs from a PCa bone metastasis-derived cell line and that it could be transferred to osteoblasts through EVs. MiR-141-3p reduced the protein levels of its target gene DLC1 and activated p38MAPK signaling, consequently promoting bone formation (regulation of OPG/RANKL). Further in vivo studies indicated that EV miR-141-3p disclosed bone-target specificity and promoted osteoblast activity, suggesting that it could mediate cancer cell-to-osteoblast communication and playing a critical role in bone metastases formation (Ye et al., 2017). Additionally, miR-375 was found to be transferred from a PCa cell line to osteoblasts through EV, enhancing its proliferation. Moreover, miR-375 mimic-transfected osteoblasts overexpressed osteoprotegerin (OPG), osteopontin (OPN), run-related transcription factor 2 (RUNX2), and bone sialoprotein (BSP), which are all genes associated with osteoblast activity and differentiation (Li et al., 2019). Conversely, PCa-derived EVs were shown to inhibit osteoclast differentiation by repressing the NF-kB signaling pathway through miR-214 downregulation (a known regulator of bone homeostasis) (Duan et al., 2019).

Eight miRNAs (miR-99a-5p, miR-125b-2-3p, miR-141-3p, miR-200a-3p, miR-200c-3p, miR-940, and miR-1260a) were found to be highly expressed in EVs from osteoblastic phenotype-inducing PCa cell lines. Interestingly, miR-940 promoted the osteogenic differentiation of mesenchymal stem cells by targeting *ARHGAP1* and *FAM134A*. Remarkably, the implantation of an osteolytic phenotype-inducing breast cancer cell line overexpressing miR-940 induced osteoblastic lesions by facilitating the osteogenic differentiation of host mesenchymal cells, induced by the transfer of EVs from cancer cells to host stromal cells (Hashimoto et al., 2018). Overall, these findings suggest that the bone metastasis phenotype may be induced by EV-contained miRNAs shed by cancer cells.

Treatment resistance

At present, only a few reports are available regarding the effect of EV-encapsulated miRNAs in PCa resistance to therapy. Nonetheless, an array analysis revealed that several miRNAs were deregulated in EVs from chemo-resistant cell lines compared to those of their parental cell lines (Corcoran et al., 2014; Li et al., 2016). Moreover, EV miR-27a produced by primary prostate fibroblasts enhanced PCa cell chemoresistance to cisplatin, doxorubicin, and docetaxel by restraining the expression of the *P53* gene (Cao, Xu, & Zhao, 2019).

Translational opportunities of EV-contained miRNAs for prostate cancer management

Is there an optimal source of PCa EV-contained miRNA biomarkers?

Plasma and serum samples are undoubtedly the most studied source of EVs in PCa. Yet, due to the prostate location in the urological system, urine also has potential for EV biomarker discovery. Nonetheless, only a few studies have been conducted in PCa aiming to elucidate which source of miRNA has the best translational potential (diagnostic, prognostic, or predictive).

An analysis of urine EVs from four different fractions obtained by several centrifugation and filtration steps indicated that miRNAs are differentially distributed across the distinct fractions (Bryzgunova et al., 2016). From the tested miRNAs in all fractions, only miR-19b was diagnostically significant, discriminating PCa patients from healthy controls (HC) with 93% sensitivity and 100% specificity in total EV fraction, and with 79% sensitivity and 95% specificity in exosome-enriched fraction (Bryzgunova et al., 2016). Interestingly, an analysis of four miRNAs in paired urine and serum EVs displayed no correlation of miRNA levels between the two sources (Xu et al., 2017). Moreover, a study comparing a set of 12 miRNAs in plasma cell-free miRNAs, urine supernatant, and urine EV from patients with PCa, benign prostate hyperplasia (BPH), and HC demonstrated that urine EVs were the source of most of the differentially expressed miRNA ratios (Konoshenko, Lekchnov, et al., 2020). Remarkably, according to their algorithm, eight miRNAs detected in urine EV combined into six ratios (miR-125b/miR-30e, miR-200/miR-30e, miR-205/miR-30e, miR-31/miR-30e, miR-660/miR-30e, and miR-19b/miR-92a) correctly classified patients with PCa, BPH, and HC with 100% sensitivity and specificity (Konoshenko, Lekchnov, et al., 2020). The same miRNAs evaluated again in samples from PCa patients pre- and post-radical prostatectomy (RP) indicated that several miRNA ratios changed after RP in two or three biofluid fractions simultaneously. However, whereas changes in urine EV and urine supernatant were unidirectional, in plasma some miRNAs changed their expression in an opposite direction compared to urine EV and urine supernatant (Konoshenko, Bryzgunova, et al., 2020). Of note, these two studies compared urine EVs with plasma cell-free miRNA (without an EV isolation step), which raises the question of whether miRNAs derived from plasma EVs would achieve better performance in these patients. Indeed, data suggest that EV-incorporated and whole plasma cell-free miRNA profiles may differ and that the best miRNA biomarker source may be dependent on the miRNA analyzed. For instance, miR-375 could differentiate between PCa and BPH in whole plasma, whereas miR-200c-3p and miR-21-5p disclosed better performance in plasma EV, and EV-incorporated let-7a-5p levels were decreased in PCa patients with a Gleason score (GS) ≥ 8 vs. ≤ 6 (Endzeliņš et al., 2017). Comparisons between EV fraction and whole serum indicate that miR-141 expression in EVs outperforms whole serum when comparing its levels between PCa patients with those of BPH patients and HC. Higher EV-derived miR-141 expression was associated with elevated PSA, GS, advanced T-stage, and metastasis, discriminating systemic from localized disease with 80% sensitivity and 87% specificity, and outperforming serum PSA (Li et al., 2016). MiRNA expression differences observed across different biologic fluids may reflect the different pathways of miRNA cell release, eventually determining their lifetime in circulation as well as the different pathological processes taking place.

While the best EV source remains to be fully elucidated, and apart from the above-mentioned studies, several other reports on the diagnostic (Table 22.2) and prognostic (Table 22.3) potential of EV-contained miRNAs isolated from urine and plasma have been published.

Urine

MiR-574-3p, miR-141-5p, and miR-21-5p were shown to be overexpressed in urine-isolated EVs of PCa patients compared to HC, detecting PCa with 71%, 66%, and 46% sensitivity, respectively (Samsonov et al., 2016). Moreover, miR-21, miR-375, and let-7c were upregulated in urinary EVs of PCa patients compared to HC and between HC and low-risk PCa vs. intermediate and high-risk PCa (Foj et al., 2017). Also, members of the miR-188 cluster were elevated in post-DRE urine-derived EVs from high-grade compared to low-grade PCa and noncancer patients (Hasegawa et al., 2018).

Interestingly, a panel of three isomiRs (miRNAs with terminal end variations) with 3′ end modifications in urinary EVs was more PCa discriminatory than the mature form of the same miRNAs. Indeed, they identified PCa with 73% sensitivity and 88% specificity compared to 71% sensitivity and 72% specificity of the mature form (Koppers-Lalic et al., 2016), suggesting that isomiR evaluation might also be considered to improve biomarker performance.

EV discriminatory power may be increased by combining different classes of biomarkers as well as the standard ones. Hence, combined urine-derived EV miRNA-145 and serum PSA levels discriminated PCa from BPH more robustly than PSA alone, with this miRNA also increased in high Gleason score patients (Xu et al., 2017). Moreover, a panel comprising seven mRNAs (ANXA3, CD24, TMPRSS2-ERG, SLC45A3, FOLH1, HPN, ITSN1), two miRNAs (miR375-3p, miR574-3p), and the PCA3 value of post-DRE urine-isolated EVs plus patient age, serum PSA, prostate volume, and DRE results achieved 86% sensitivity and 93% specificity to discriminate PCa from benign conditions (Davey et al., 2020).

Plasma and serum

Of 12 miRNAs differentially expressed in plasma-derived EVs from PCa patients versus those of HC, miR-107 detected PCa with 67% sensitivity and 46% specificity, whereas miR-141 and miR-375 were associated with metastatic PCa in

TABLE 22.2 Extracellular vesicle-contained microRNAs with PCa detection potential.

EV miRNA	Biomarker performance			Number of cases/controls	EV source	Reference
	Sn%	Sp%	AUC			
miR-107	67%	46%	0.62	78 PCa/28 healthy controls	Plasma	(Bryant et al., 2012)
miR-574-3p	71%	–	0.85	10 PCa/10 healthy controls	Urine	(Samsonov et al., 2016)
miR-141-5p	66%	–	0.86			
miR-21-5p	46%	–	0.71			
miR-141	80%	87%	0.87	20 metastatic PCa/31 localized PCa	Serum	(Li, Ma, et al., 2016)
miR-21/miR-204/miR-375	73%	88%	0.87	48 PCa/26 healthy controls	Urine	(Koppers-Lalic et al., 2016)
miR-21/miR-204/miR-375 isomiRs	71%	72%	0.77			
miR-19b (total EV)	93%	100%	–	14 PCa/20 healthy controls	Urine	(Bryzgunova et al., 2016)
miR-19b (exosome-enriched EV fraction)	79%	95%	–			
miR-21	–	–	0.71	52 PCa/10 healthy controls	Urine	(Foj et al., 2017)
miR-375	–	–	0.79			
Let-7c	–	–	0.68			
miR-145/serum PSA	–	–	0.86	60 PCa/38 BPH	Urine	(Xu et al., 2017)
miR-200c-3p	–	–	0.68	50 PCa/22 BPH	Plasma	(Endzeliņš et al., 2017)
miR-21-5p	–	–	0.67			
Let-7a-5p	–	–	0.68	26 GS<=6/ 24 GS>8		
miR-1246	75%	100%	0.93	44 PCa (stage IV)/8 healthy controls	Serum	(Bhagirath et al., 2018)
	81%	59%	0.69	26 LN metastatic PCa/43 nonmetastatic PCa		
miR-125b/miR-30e; miR-200/miR-30e; miR-205/miR-30e; miR-660/miR-30e; miR-19b/miR92a	100%	100%	–	10 PCa/8 BPH/11 Healthy controls	Urine	(Konoshenko, Lekchnov, et al., 2020)
mRNA (ANXA3, CD24, TMPRSS2-ERG, SLC45A3, FOLH1, HPN, ITSN1)/miRNA (miR375-3p, miR574-3p)/ PCA3 value/age/serum PSA/ prostate volume/DRE results	86%	93%	0.96	28 PCa/28 Benign condition controls	Post-DRE Urine	(Davey et al., 2020)

TABLE 22.3 Extracellular vesicle-contained microRNAs associated with PCa prognosis.

EV miRNA	Prognostic association	Number of cases	Sample type	Reference
miR-375 miR-141	↑ metastatic PCa	47 Metastatic PCa/72 nonrecurrent PCa	Serum	Bryant et al. (2012)
miR-1290/miR-375/ ADT failure time/PSA levels at the time of CRPC	Associated with lower overall survival	100 CRPC	Plasma	Huang et al. (2015)
miR-141	↑ Higher PSA, Gleason score, advanced T-stage and metastasis	51 PCa	Serum	Li, Ma, et al. (2016)
miR-21 miR-375 let-7c	↑ Intermediate and high-risk PCa	13 Healthy controls and low-risk PCa/49 PCa	Urine	Foj et al. (2017)
miR-145	↑ High GS	34 GS6–7/26 GS>=8	Urine	Xu et al. (2017)
let-7a-5p	↓ High GS	26 GS<=6/24 GS>8	Plasma	Endzeliņš et al. (2017)
miR-188 cluster (miR-892c, miR-890, miR-888, miR-892a, miR-892b, miR-891b, miR-891a)	↑ High GS	9 GS8–9/16 GS6/15–healthy controls	Post-DRE urine	Hasegawa et al. (2018)
miR-1246	↑ Advanced pathological stage, positive metastasis and poor prognosis	67 PCa	Serum	Bhagirath et al. (2018)

Abbreviations: ↑—Upregulated in; ↓—Downregulated in; ADT—Androgen Deprivation Therapy; CRPC—Castration-Resistant Prostate Cancer; DRE—Digital Rectal Examination; GS—Gleason Score.

serum-derived EVs in PCa patients following RP (Bryant et al., 2012). Interestingly, although downregulated in PCa clinical tissues and cell lines, miR-1246 was found to be overexpressed in EVs of PCa patients, and its expression correlated with increased pathological grade, metastases, and poor prognosis (Bhagirath et al., 2018). The authors suggested that this miRNA may be a tumor suppressor in PCa that is released via EVs from tumor cells, causing its upregulation in bodily fluids (Bhagirath et al., 2018), thus allowing its detection in liquid biopsies. The fact that some miRNAs may be found differently expressed in EVs of metastatic patients supports the hypothesis of their usefulness in micrometastasis detection. Remarkably, the high-risk score of castration-resistant prostate cancer (CRPC) patients based on a multivariate model including plasma EV-derived miR-1290 and miR-375, androgen-deprivation therapy (ADT) failure time, and PSA levels at the time of CRPC presented a 2.58-fold higher risk of death than patients with a low-risk score (Huang et al., 2015).

Only a few studies, and with limited samples and follow-up, were conducted to explore the potential of EV-contained miRNA to predict therapy efficacy. Thus, one can only speculate about their long-term predictive value. Nevertheless, serum EV concentration was higher in PCa patients compared to HC and it further increased in postradiation (RT) samples. Both let-7a-5p and miR-215p expression increased after RT in high-risk PCa patients treated with curative-intent radiotherapy (RT) (Malla, Aebersold, & Pra, 2018). Moreover, high precarbon ion radiotherapy (CIRT) expression levels of miR-493-5p, miR-323a-3p, miR-411-5p, miR-494-3p, miR379-5p, miR-654-3p, miR-409-3p, miR-543, and miR-200c-3p predicted therapeutic benefits, whereas post-CIRT expression of miR-654-3p and miR-379-5p was also associated with CIRT efficacy (Yu et al., 2018).

EV miRNAs in clinical practice—What do we still need to accomplish?

Apart from some exceptions (e.g., miR-375, miR-21, or miR-141), the above-mentioned studies almost show no overlap of significantly deregulated miRNAs, which hinders the interpretation of the potential translational opportunities derived from EV-contained miRNAs in PCa. This can have various interlinked explanations. First, different EV isolation techniques were and are being used, with no consensus yet about which is the most effective one. Evidence suggests that the miRNA expression profile might differ between different EV isolation methods, despite similar efficiency (Mercadal et al., 2020). Although the most widely used methodology is ultracentrifugation, this is highly time consuming and requires specialized equipment, so it is not the most obvious choice for implementation in clinics. Hence, several studies have been conducted to compare different methods that are deemed to be faster and easier to perform (such as commercially available kits) to enable EV clinical implementation (Lee, Kwon, Kim, & Rhee, 2018; Li et al., 2015; Royo et al., 2016; Smith et al., 2018; Xu et al., 2017). Nonetheless, the best method remains to be identified. Furthermore, different miRNA expression analysis and miRNA preselection methods may also lead to discrepancies in the results. In this regard, appropriate EV-derived miRNA endogenous controls and standard normalization methods are notably lacking, hampering assay reproducibility. Remarkably, the development of more sensitive and specific technologies such as droplet digital PCR (ddPCR) may solve this issue by obviating the need for normalization. Lastly, only small-scale PCa cohorts have been used so far. Hence, large-scale independent and prospective validation studies are required to inquire about the real clinical impact of EV-encapsulated miRNAs.

Although a long way is foreseen to unravel all those issues, the already available results encourage the study of EV-contained miRNAs in PCa. Further discovery and integration of the pathophysiological roles with the translational opportunities may enable the personalized treatment of PCa patients and solve some of the current clinical gaps. For instance, because EVs are released by cells into the bloodstream to reach metastatic loci before cancer cells, EV-contained miRNAs might represent an opportunity to detect metastatic disease at an early stage (at the stages of premetastatic niches and micrometastasis), allowing for timely adjuvant treatment and reducing PCa lethality. Finally, the possibility to discriminate between aggressive and indolent tumors through the characterization of miRNA cargo associated with tumor progression might also contribute to improved patient stratification and therapy decisions.

Funding

VC is the recipient of a fellowship from "la Caixa" Foundation (ID 100010434). The fellowship code is LCF/BQ/DR20/11790013.

References

Bhagirath, D., Yang, T. L., Bucay, N., Sekhon, K., Majid, S., Shahryari, V., et al. (2018). MicroRNA-1246 is an exosomal biomarker for aggressive prostate cancer. *Cancer Research*, *78*(7), 1833–1844.

Boerrigter, E., Groen, L. N., Van Erp, N. P., Verhaegh, G. W., & Schalken, J. A. (2020). Clinical utility of emerging biomarkers in prostate cancer liquid biopsies. *Expert Review of Molecular Diagnostics*, *20*(2), 219–230.

Bray, F., Ferlay, J., Soerjomataram, I., Siegel, R. L., Torre, L. A., & Jemal, A. (2018). Global cancer statistics 2018: GLOBOCAN estimates of incidence and mortality worldwide for 36 cancers in 185 countries. *CA: A Cancer Journal for Clinicians*, *68*(6), 394–424.

Bryant, R. J., Pawlowski, T., Catto, J. W. F., Marsden, G., Vessella, R. L., Rhees, B., et al. (2012). Changes in circulating MicroRNA levels associated with prostate cancer. *British Journal of Cancer*, *106*(4), 768–774.

Bryzgunova, O. E., Zaripov, M. M., Skvortsova, T. E., Lekchnov, E. A., Grigor'eva, A. E., Zaporozhchenko, I. A., et al. (2016). Comparative study of extracellular vesicles from the urine of healthy individuals and prostate cancer patients. *PLoS One*, *11*(6), e0157566.

Cao, Z., Xu, L., & Zhao, S. (2019). Exosome-derived MiR-27a produced by PSC-27 cells contributes to prostate cancer chemoresistance through P53. *Biochemical and Biophysical Research Communications*, *515*(2), 345–351.

Che, Y., Xu, S., Shi, Y., Jiang, X., Ai, Q., Shi, Y., et al. (2019). Exosomes derived from MiR-143-overexpressing MSCs inhibit cell migration and invasion in human prostate cancer by downregulating TFF3. *Molecular Therapy-Nucleic Acids*, *18*(December), 232–244.

Corcoran, C., Rani, S., & O'Driscoll, L. (2014). MiR-34a is an intracellular and Exosomal predictive biomarker for response to docetaxel with clinical relevance to prostate cancer progression. *The Prostate*, *74*(13), 1320–1334.

Davey, M., Benzina, S., Savoie, M., Breault, G., Ghosh, A., & Ouellette, R. J. (2020). Affinity captured urinary extracellular vesicles provide MRNA and MiRNA biomarkers for improved accuracy of prostate cancer detection: A pilot study. *International Journal of Molecular Sciences*, *21*(21). https://doi.org/10.3390/ijms21218330.

Duan, Y., Tan, Z., Yang, M., Li, J., Liu, C., Wang, C., et al. (2019). PC-3-derived exosomes inhibit osteoclast differentiation by downregulating MiR-214 and blocking NF-KB signaling pathway. *BioMed Research International*, *2019*(April), 8650846.

Endzeliņš, E., Berger, A., Melne, V., Bajo-Santos, C., Soboļevska, K., Ābols, A., et al. (2017). Detection of circulating MiRNAs: Comparative analysis of extracellular vesicle-incorporated MiRNAs and cell-free MiRNAs in whole plasma of prostate cancer patients. *BMC Cancer*, *17*(1), 730.

Foj, L., Ferrer, F., Serra, M., Arévalo, A., Gavagnach, M., Giménez, N., et al. (2017). Exosomal and non-exosomal urinary MiRNAs in prostate cancer detection and prognosis. *The Prostate*, *77*(6), 573–583.

Hasegawa, T., Glavich, G. J., Pahuski, M., Aleena, S., John Semmes, O., Yang, L., et al. (2018). Characterization and evidence of the MiR-888 cluster as a novel cancer network in prostate. *Molecular Cancer Research*, *16*(4), 669–681.

Hashimoto, K., Ochi, H., Sunamura, S., Kosaka, N., Mabuchi, Y., Fukuda, T., et al. (2018). Cancer-secreted Hsa-MiR-940 induces an osteoblastic phenotype in the bone metastatic microenvironment via targeting ARHGAP1 and FAM134A. *Proceedings of the National Academy of Sciences of the United States of America*, *115*(9), 2204–2209.

Hessvik, N. P., Phuyal, S., Brech, A., Sandvig, K., & Llorente, A. (2012). Profiling of MicroRNAs in exosomes released from PC-3 prostate cancer cells. *Biochimica et Biophysica Acta*, *1819*(11–12), 1154–1163.

Huang, X., Yuan, T., Liang, M., Meijun, D., Xia, S., Dittmar, R., et al. (2015). Exosomal MiR-1290 and MiR 375 as prognostic markers in castration-resistant prostate cancer. *European Urology*, *67*(1), 33–41.

Iorio, M. V., & Croce, C. M. (2009). MicroRNAs in cancer: Small molecules with a huge impact. *Journal of Clinical Oncology: Official Journal of the American Society of Clinical Oncology*, *27*(34), 5848–5856.

Jiang, S., Mo, C., Guo, S., Zhuang, J., Huang, B., & Mao, X. (2019). Human bone marrow mesenchymal stem cells-derived MicroRNA-205-containing exosomes impede the progression of prostate cancer through suppression of RHPN2. *Journal of Experimental & Clinical Cancer Research*, *38*(1), 495.

Josson, S., Gururajan, M., Sung, S. Y., Hu, P., Shao, C., Zhau, H. E., et al. (2015). Stromal fibroblast-derived MiR-409 promotes epithelial-to-mesenchymal transition and prostate tumorigenesis. *Oncogene*, *34*(21), 2690–2699.

Kim, J., Morley, S., Le, M., Bedoret, D., Umetsu, D. T., Di Vizio, D., et al. (2014). Enhanced shedding of extracellular vesicles from amoeboid prostate cancer cells: Potential effects on the tumor microenvironment. *Cancer Biology & Therapy*, *15*(4), 409–418.

Konoshenko, M. Y., Bryzgunova, O. E., Lekchnov, E. A., Amelina, E. V., Yarmoschuk, S. V., Pak, S. V., et al. (2020). The influence of radical prostatectomy on the expression of cell-free MiRNA. *Diagnostics (Basel, Switzerland)*, *10*(8). https://doi.org/10.3390/diagnostics10080600.

Konoshenko, M. Y., Lekchnov, E. A., Bryzgunova, O. E., Zaporozhchenko, I. A., Yarmoschuk, S. V., Pashkovskaya, O. A., et al. (2020). The panel of 12 cell-free MicroRNAs as potential biomarkers in prostate neoplasms. *Diagnostics (Basel, Switzerland)*, *10*(1). https://doi.org/10.3390/diagnostics10010038.

Koppers-Lalic, D., Hackenberg, M., de Menezes, R., Misovic, B., Wachalska, M., Geldof, A., et al. (2016). Non-invasive prostate cancer detection by measuring MiRNA variants (IsomiRs) in urine extracellular vesicles. *Oncotarget*, *7*(16), 22566–22578.

Lee, J., Kwon, M. H., Kim, J. A., & Rhee, W. J. (2018). Detection of exosome MiRNAs using molecular beacons for diagnosing prostate cancer. *Artificial Cells, Nanomedicine, and Biotechnology*, *46*(sup3), S52–S63.

Li, S.-L., An, N., Liu, B., Wang, S.-Y., Wang, J.-J., & Ye, Y. (2019). Exosomes from LNCaP cells promote osteoblast activity through MiR-375 transfer. *Oncology Letters*, *17*(5), 4463–4473.

Li, Z., Ma, Y.-Y., Wang, J., Zeng, X.-F., Li, R., Kang, W., et al. (2016). Exosomal MicroRNA-141 is upregulated in the serum of prostate cancer patients. *Oncotargets and Therapy*, *9*, 139–148.

Li, M., Rai, A. J., Joel DeCastro, G., Zeringer, E., Barta, T., Magdaleno, S., et al. (2015). An optimized procedure for exosome isolation and analysis using serum samples: Application to cancer biomarker discovery. *Methods*, *87*(October), 26–30.

Li, J., Yang, X., Guan, H., Mizokami, A., Keller, E. T., Xu, X., et al. (2016). Exosome-derived MicroRNAs contribute to prostate cancer chemoresistance. *International Journal of Oncology*, *49*(2), 838–846.

Linxweiler, J., & Junker, K. (2020). Extracellular vesicles in urological malignancies: An update. *Nature Reviews Urology*, *17*(1), 11–27.

Malla, B., Aebersold, D. M., & Pra, A. D. (2018). Protocol for serum exosomal MiRNAs analysis in prostate cancer patients treated with radiotherapy. *Journal of Translational Medicine*, *16*(1), 223.

Mercadal, M., Herrero, C., López-Rodrigo, O., Castells, M., de la Fuente, A., Vigués, F., et al. (2020). Impact of extracellular vesicle isolation methods on downstream Mirna analysis in semen: A comparative study. *International Journal of Molecular Sciences*, *21*(17). https://doi.org/10.3390/ijms21175949.

Morello, M., Minciacchi, V. R., de Candia, P., Yang, J., Posadas, E., Kim, H., et al. (2013). Large oncosomes mediate intercellular transfer of functional MicroRNA. *Cell Cycle*, *12*(22), 3526–3536.

Royo, F., Diwan, I., Tackett, M. R., Zuñiga, P., Sanchez-Mosquera, P., Loizaga-Iriarte, A., et al. (2016). Comparative MiRNA analysis of urine extracellular vesicles isolated through five different methods. *Cancers*, *8*(12). https://doi.org/10.3390/cancers8120112.

Samsonov, R., Shtam, T., Burdakov, V., Glotov, A., Tsyrlina, E., Berstein, L., et al. (2016). Lectin-induced agglutination method of urinary exosomes isolation followed by mi-RNA analysis: Application for prostate cancer diagnostic. *The Prostate*, *76*(1), 68–79.

Sánchez, C. A., Andahur, E. I., Valenzuela, R., Castellón, E. A., Fullá, J. A., Ramos, C. G., et al. (2016). Exosomes from bulk and stem cells from human prostate cancer have a differential MicroRNA content that contributes cooperatively over local and pre-metastatic niche. *Oncotarget*, *7*(4), 3993–4008.

Sawyers, C. L. (2008). The cancer biomarker problem. *Nature*, *452*(7187), 548–552.

Smith, J. T., Wunsch, B. H., Dogra, N., Ahsen, M. E., Lee, K., Yadav, K. K., et al. (2018). Integrated nanoscale deterministic lateral displacement arrays for separation of extracellular vesicles from clinically-relevant volumes of biological samples. *Lab on a Chip*, *18*(24), 3913–3925.

Tkach, M., & Théry, C. (2016). Communication by extracellular vesicles: where we are and where we need to go. *Cell*, *164*(6), 1226–1232.

Urabe, F., Kosaka, N., Sawa, Y., Ito, K., Kimura, T., Egawa, S., et al. (2020). The MiR-1908/SRM regulatory Axis contributes to extracellular vesicle secretion in prostate cancer. *Cancer Science*, *111*(9), 3258–3267.

Urabe, F., Kosaka, N., Sawa, Y., Yamamoto, Y., Ito, K., Yamamoto, T., et al. (2020). MiR-26a regulates extracellular vesicle secretion from prostate cancer cells via targeting SHC4, PFDN4, and CHORDC1. *Science Advances*, *6*(18), eaay3051.

Worst, T. S., Previti, C., Nitschke, K., Diessl, N., Gross, J. C., Hoffmann, L., et al. (2019). MiR-10a-5p and MiR-29b-3p as extracellular vesicle-associated prostate cancer detection markers. *Cancers*, *12*(1). https://doi.org/10.3390/cancers12010043.

Xu, Y., Qin, S., An, T., Tang, Y., Huang, Y., & Zheng, L. (2017). MiR-145 detection in urinary extracellular vesicles increase diagnostic efficiency of prostate cancer based on hydrostatic filtration dialysis method. *The Prostate*, *77*(10), 1167–1175.

Ye, Y., Li, S.-L., Ma, Y.-Y., Diao, Y.-J., Yang, L., Ming-Quan, S., et al. (2017). Exosomal MiR-141-3p regulates osteoblast activity to promote the osteoblastic metastasis of prostate cancer. *Oncotarget*, *8*(55), 94834–94849.

Yu, Q., Li, P., Weng, M., Wu, S., Zhang, Y., Xue, C., et al. (2018). Nano-vesicles are a potential tool to monitor therapeutic efficacy of carbon ion radiotherapy in prostate cancer. *Journal of Biomedical Nanotechnology*, *14*(1), 168–178.

Chapter 23

Pathophysiology roles and translational opportunities of miRNAs in hepatocellular carcinoma

Francesca Fornari[a,b] and Laura Gramantieri[b,c]

[a]Department for Life Quality Studies, University of Bologna, Rimini, Italy, [b]Centre for Applied Biomedical Research - CRBA, University of Bologna, St. Orsola Hospital, Bologna, Italy, [c]Division of Internal Medicine, Hepatobiliary and Immunoallergic Diseases, IRCCS Azienda Ospedaliero-Universitaria di Bologna, Bologna, Italy

Introduction

Hepatocellular carcinoma (HCC) is the fifth most common cancer and it represents the second leading cause of cancer-related death worldwide (Torre et al., 2015). The main risk factors are the hepatitis B and C virus (HBV and HCV) infection, alcohol intake, and aflatoxin exposure. Beside these, metabolic syndrome and NAFLD are gaining more and more relevance, particularly in western countries (Petrick & McGlynn, 2019).

When diagnosed at early stages, HCC can be curatively treated either by surgery or locoregional approaches; however, the five-year recurrence rate approaches 70%, except for liver transplantation. At advanced stages, the expected 1-year survival rate is 25% (European Association for the Study of the Liver, 2018). The multi tyrosine-kinase inhibitor sorafenib has been the only FDA-approved drug for advanced HCC for 10 years. In 2017, novel first-line and second-line approved drugs such as oral tyrosine kinase and immune checkpoint inhibitors were approved (Li et al., 2019). Despite relevant efforts that have provided elements for a molecular classification of HCC, this tool never entered the clinical practice. This might also be ascribed to the fact that HCC diagnosis is performed by imaging procedures in the vast majority of cases, hampering biomarker identification. MicroRNAs (miRNAs) participate in liver carcinogenesis and they were shown to provide additional elements toward a molecularly based classification of HCC. Indeed, miRNA-based signatures accurately define prognostic subgroups of HCC with potential implication as prognostic and predictive biomarkers.

Herein, we will briefly describe the pathophysiologic roles of miRNAs, the aberrant expression of relevant miRNAs involved in liver carcinogenesis, and their diagnostic and prognostic values in HCC.

Pathophysiology

The role of miRNAs in liver physiology and organogenesis has been scarcely investigated, and most of what we know has been gathered from pathological conditions. MiR-122 represents the most abundant miRNA in hepatocytes, accounting for more than 70% of the total liver miRNA population. In physiologic conditions, it participates in lipid metabolism and its downregulation impairs cholesterol biosynthesis. A global reduction of miRNAs expression obtained by Dicer1 knockout (KO) in mice resulted in lipid and glucose metabolism impairment, steatosis, and a higher rate of HCC development. These events are at least in part due to miR-122 reduced expression. MiR-122 targets not only factors involved in lipid metabolism, but also factors involved in tumorigenesis (Esau et al., 2006; Gramantieri et al., 2007; Hsu et al., 2012). However, the comparison of *Dicer1* KO with miR-122 silenced livers showed a stronger tumor-suppressive activity of *Dicer1*, associated with even more impressive histopathologic changes, pointing to the role of other miRNAs in liver organogenesis and carcinogenesis. The consequences of Dicer1 disruption in the liver unravel a relevant role of Dicer1 and miRNAs in hepatocyte survival, metabolism, and transcription of cell cycle promoting and fetal liver genes, accounting for tumor development (Sekine et al., 2009). Sekine and coworkers reported the altered expression of genes responsible for liver zonation in *Dicer1* KO mice. Because *Dicer1* is essential for miRNA biogenesis by cleaving pre-miRNAs into mature miRNAs, these authors suggested the essential role of *Dicer1* itself in the suppression of periportal proteins, and the likely involvement of miRNAs

MicroRNA in Human Malignancies. https://doi.org/10.1016/B978-0-12-822287-4.00025-6

in this process (Sekine, Ogawa, McManus, Kanai, & Hebrok, 2009). Remarkably, miRNAs participate not only in hepatocyte development and the regulation of specific functions, but also they play a role in biliary lineage development, as shown for the miR-30 family (Le Guen, Friedman, & Hand, 2013).

Several feedback and feedforward loops involving miRNAs have been identified that influence hepatocyte differentiation, pathophysiologic responses to genotoxic hints and stressful conditions, and finally hepatocarcinogenesis. Some of these miRNA-based networks will be detailed below.

MicroRNA deregulation in HCC

HCC arises in a background of liver cirrhosis in 80%–90% of cases. Cirrhosis represents a risk factor for the development of premalignant lesions due to hepatocyte regeneration in an inflammatory context. These premalignant lesions may progress through a multistep process, leading to overt HCC. Interestingly, miRNAs are deregulated not only in cancer tissue but also in the surrounding liver, where cirrhosis and viral hepatitis greatly contribute to the transcriptional activation of precursor miRNAs. The high molecular heterogeneity of HCC and the use of different experimental approaches and control groups in profiling studies have limited the identification of univocal miRNA signatures to be used as diagnostic and prognostic tools (Fittipaldi et al., 2017). Despite the discrepancies in control groups, a global downregulation of miRNAs characterizes tumor tissue compared to the surrounding liver. Several miRNAs are consistently downregulated in different studies, such as let-7 and miR-30 family members, miR-26, miR-29a/b/c, miR-200a/b/c, and miR-122, highlighting their fundamental contribution to hepatocyte dedifferentiation and tumor progression.

MiRNA signatures were reported to discriminate HCC tissues from matched cirrhotic tissues (Gramantieri et al., 2007) or nontumor livers (Murakami et al., 2006). Moreover, a miRNA panel distinguished HCV- and HBV-related HCCs, suggesting HBV and HCV as triggers of their deregulated expression. A pathway analysis of targeted genes showed the prominent activation of cell death, DNA damage, and signal transduction pathways in HBV and the immune response, proteasome, cell cycle, and lipid metabolism in HCV-infected livers (Ura et al., 2009). However, a common panel of deregulated miRNAs was found in HCC independently from the concomitant viral infection, suggesting its participation in carcinogenesis irrespective of risk factors. Among these miRNAs, miR-221, miR-222, miR-21 are in common with a 12-miRNA signature that emerged by pairwise comparisons among normal livers, cirrhotic tissues, and tumor samples, likely contributing to liver tumorigenesis. This last study (Pineau et al., 2010) showed miR-106b, miR-21, miR-210, miR-221, miR-222, miR-224, miR-34a, miR-425, miR-519a, miR-93, and miR-96 upregulation as well as let-7c downregulation in both cirrhotic and tumor samples with respect to normal livers, representing early events in the malignant transformation of hepatocytes and stem or progenitor cells (Fig. 23.1).

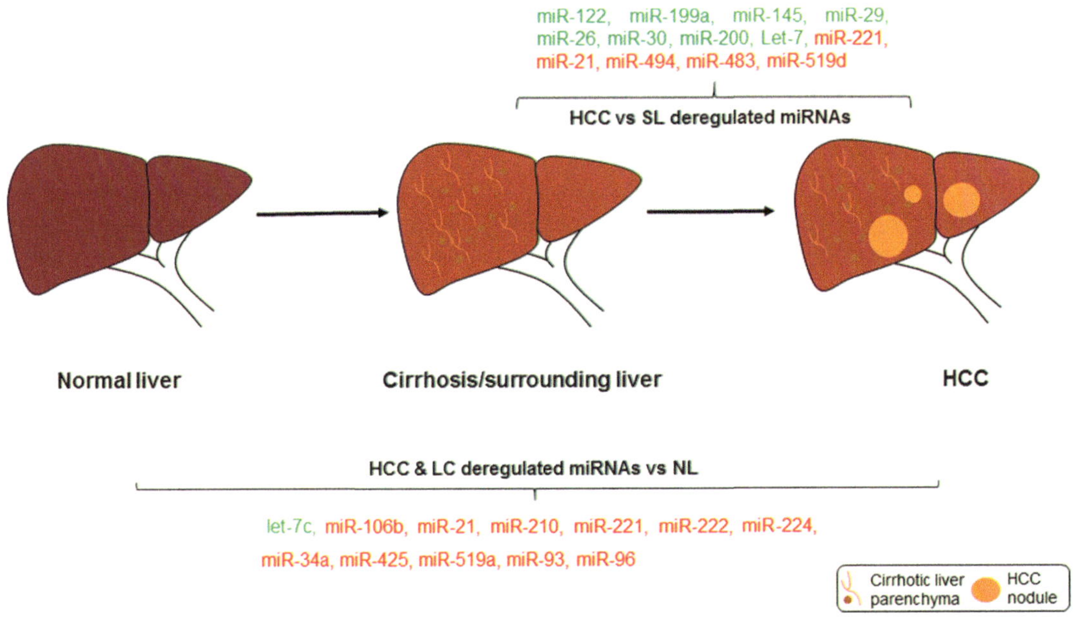

FIG. 23.1 Deregulated miRNAs in hepatocarcinogenesis. Overexpressed miRNAs in HCC are in red whereas dowregulated miRNAs are in green.

Herein, we will focus on miRNAs whose deregulated expression was confirmed in the vast majority of profiling studies, considering also their relative abundance in the liver. Indeed, the absolute expression of miRNAs represents a relevant point to be evaluated when figuring out the biologic relevance of miRNA expression changes. Among abundantly expressed miRNAs in liver tissue, miR-199a, miR-145, and miR-122 downregulation turned out to be a common event in HCC-specific signatures while miR-221 was upregulated (Gramantieri et al., 2007; Ura et al., 2009). Similarly, miR-494 was reported as upregulated in nearly 25% of HCCs and was associated with specific tumor characteristics (Luk et al., 2011).

A common feature of miRNAs is the concomitant targeting of multiple RNA molecules, thus modulating the cell phenotype by affecting several pathways, sometimes with opposite functions. These events are commonly endowed in feedback and feedforward loops aimed at controlling physiological challenges. The impairment of such mechanisms may contribute to cancer development. Remarkably, oncogenic drivers are linked not only to aberrant miRNA expression, but also to their targets expression, which differs in each context.

Hereafter, we will briefly resume the pathophysiology of the most studied HCC-specific miRNAs to depict some representative examples of their pleomorphic functions.

MiR-122

MiR-122 is a hepato-specific miRNA and its implication in liver physiopathology and carcinogenesis appeared evident in both mouse (Kutay et al., 2006) and human HCCs (Gramantieri et al., 2007; Ura et al., 2009). It is involved in lipid metabolism and HCV replication (Jopling, Yi, Lancaster, Lemon, & Sarnow, 2005). It was the first miRNA to be silenced in vivo, demonstrating the feasibility and safety of antagomiR-based treatments in mice and nonhuman primates (Elmen et al., 2008; Krutzfeldt et al., 2005) as well as HCV-infected patients (Janssen et al., 2013), confirming the liver as an ideal organ for oligonucleotide-based systemic delivery. Interestingly, miR-122 was the most downregulated miRNA due to DNA hypermethylation in the IDH1/2-mutated and IDH-like patient subgroups, characterized by a poorly differentiated and hepatic stem cell-like phenotype and poor prognosis (Cancer Genome Atlas Research Network, 2017). Indeed, IDH1/2 mutations often cause a gain-of-function activity, leading to the production of an oncometabolite that influences the epigenetic program of cancer cells, linking altered cellular metabolism to the aberrant epigenetic regulation of tumor-suppressor genes. On the other side, miR-122 itself modulates genes involved in the regulation of glucose metabolism such as *Pyruvate Kinase M2* (*PKM2*) influencing doxorubicin resistance in HCC cells (Pan et al., 2016).

Target gene identification is a fundamental step in the knowledge of miRNA functionality and contribution to cancer development and progression. Cyclin G1, a p53 transcriptional target regulating the G2/M phase transition, was first validated as a miR-122 target gene (Gramantieri et al., 2007). It influences cell cycle progression, invasion capabilities, doxorubicin sensitizationn and HBV replication (Fornari et al., 2009; Wang et al., 2012) by establishing a miR-122/CCNG1/p53 positive feedback loop. MiR-122 also plays a fundamental role in HCV replication, justifying its therapeutic exploitation (Janssen et al., 2013). Interestingly, the miR-122* passenger strand was downregulated in HCC and contributed to p53 activation through *MDM2* targeting, exerting a tumor-suppressive activity as well. Indeed, though less expressed than its counterpart, miR-122* still exceeds the level of several other miRNAs in the liver, and it was proposed as a therapeutic target in several cancer types (Simerzin et al., 2016). MiRNAs often establish complex autoregulatory networks with their own targets, such as the one described by Zeng and coworkers that revealed CCAAT/enhancer-binding protein alpha (C/EBPα) as a transcriptional activator of miR-122 in HCC. GSK-3β–C/EBPα–miR-122–IGF-1R regulatory circuitry was demonstrated to maintain GSK-3β switched off, leading to miR-122 downregulation and IGF-1R pathway hyperactivation that contributes to anchorage-independent growth and in vivo tumorigenesis (Zeng et al., 2010). An entangled miR-122/CUX1/miR-214/ZBT20/AFP regulatory cascade interconnected miR-122 downregulation to AFP elevation and to an aggressive phenotype, suggesting AFP as a surrogate noninvasive marker of miR-122 deregulation in HCC to be possibly employed in a miRNA-based therapeutic perspective (Kojima et al., 2011). However, not all the cases with miR-122 downregulation showed elevated AFP levels, and vice versa. Therefore, despite the correlation of AFP with molecular subclasses (Boyault et al., 2007; Hoshida et al., 2009) and the importance of biomarker discovery linking molecular drivers to therapeutic options, caution should be used for patient classification, which remains an unsolved issue in HCC.

MiR-199

The miR-199a hairpin precursor maps to both chromosome 19 (miR-199a-1) and chromosome 1 (miR-199a-2) as a consequence of gene duplication, and gives rise to two mature miRNA sequences, miR-199a-5p and miR-199a-3p deriving from the 5′ and 3′ arms of the precursor sequence, respectively. Profiling studies revealed the downregulation of both mature and precursor isoforms in several cancer types, including HCC (Gramantieri et al., 2007; Ichimi et al., 2009; Iorio et al., 2007). Cancer-derived cell lines express these miRNAs at very low, near undetectable, levels. A functional analysis displayed that both miRNAs, and especially miR-199a-3p, were able to induce apoptotic cell death once overexpressed in cancer cell lines. *MET* proto-oncogene was identified as the first target gene of miR-199a-3p, also affecting the expression and phosphorylation pattern of the *ERK2* downstream effector (Kim et al., 2008). *MET* targeting was confirmed in HCC cells and was primarily responsible for decreased invasion capabilities following miR-199a-3p restoration. In addition, we identified mTOR as a miR-199a-3p target in HCC cells, where it mediated the inhibition of cell proliferation and the increased apoptotic rate in doxorubicin-treated cells and hypoxic conditions. An inverse correlation between miR-199a-3p and mTOR levels was displayed in HCC tissues, confirming the influence of miR-199a-3p on mTOR overexpression sustaining cell survival and malignant clone expansion in low nutrient and hypoxic microenvironments, occurring in large tumor nodules (Fornari et al., 2010). A good correlation exists between the two mature isoforms. Despite their completely different sequences and regulation of different targets, they both influence the EMT phenotype of HCC cells by directly (miR-199a-5p) and indirectly (miR-199a-3p) targeting E-cadherin.

Metabolic reprogramming is emerging as an important hallmark of cancer aiding neoplastic cells to shift from oxidative phosphorylation to glycolysis, conferring them an advantage to proliferate in a hypoxic context. Zhang and coworkers reported the induction of the RNA-binding protein HuR in a hypoxic environment and its ability to interfere with miR-199a primary transcript maturation, blocking its processing to mature miRNAs. In turn, miR-199a downregulation is a crucial event in HCC cells, accelerating the glycolysis rate due to the regulation of key players of the Warburg effect, hesokinase 2 (HK2) and PKM2. Systemic delivery of miR-199a mimics reduces glucose uptake and tumor growth in a xenograft model, demonstrating the interference of cancer cell metabolism as a promising miRNA-mediated anticancer approach in HCC (Zhang et al., 2015). Strikingly, the mutual interplay between miR-199a-5p and HIF-1α, the master regulator of cell survival during hypoxia, establishes a feedforward loop that contributes to a metabolic shift, sustaining cancer cell proliferation and favoring the selection of aggressive clones responsible for poor prognosis.

MiR-145

Profiling studies revealed miR-145 downregulation in many cancer types, including HCC (Gramantieri et al., 2007; Iorio et al., 2005, 2007). Interestingly, miR-145 is transcriptionally regulated by p53 and takes part in its regulatory network by directly repressing the c-Myc proto-oncogene, functioning as a *trait d'union* between the guardian of the genome and a frequently activated oncogene in human tumors (Sachdeva et al., 2009). Interestingly, miR-145 and miR-199b deregulation is an early event in HBV-associated HCCs that is maintained along multiple steps of carcinogenesis, possibly contributing to disease progression in infected livers. MiR-145 restoration in HCC cell lines determines the arrest of cell proliferation and a reduction of cell invasion and migration, irrespective of their *TP53* status (Gao et al., 2011). Its high expression in nontumor livers makes miR-145 an ideal candidate for mimic replacement therapies in HCC patients. Despite miR-145 downregulation in many HCCs, some tumors display a physiologic expression and, surprisingly, miR-145-5p positively correlates with onco-miR-483-3p in liver cancers but not in surrounding livers. In agreement with this "Dr. Jekyll and Mr. Hyde" behavior, miR-145 activation in p53 WT cells is sufficient for the selection of apoptosis-resistant cell clones that selectively overexpress miR-483-3p in high glucose conditions only, suggesting that in permissive growth conditions miR-145 might acquire an oncogenic function that favors the survival and expansion of malignant hepatocytes. MiR-483-3p suppression in these cell clones is sufficient for triggering their apoptotic cell death, clearing the way for a combined miRNA-based therapeutic approach in p53 WT HCCs (Lupini et al., 2016). Similar to protein-coding genes, several mechanisms contribute to miRNA regulation such as transcription factors, epigenetic events, chromosomal rearrangements, antisense RNAs, and molecules controlling their maturation process. Long noncoding RNAs (lncRNAs) are a class of noncoding RNAs more than 200 nucleotides long that plays a central role in cell physiology and whose deregulation takes part in neoplastic transformation. Through complementary binding sites, lncRNAs might act as competitive endogenous RNAs (ceRNAs) sequestering miRNA molecules and, consequently, increasing their target gene availability. Colon cancer-associated transcript 2 (CCAT2) was found upregulated in the nucleus of HCC cell lines and tissues. Rather than acting as a ceRNA, it does exert its oncogenic function interfering with miR-145 biogenesis, finally leading to a decrease of the mature miR-145 isoform in the cytoplasm of HCC cells. Niu et al. demonstrated that miR-145 acts as a critical mediator of CCAT2, decreasing cancer cell proliferation and metastasis through *MDM2* targeting and consequent p53/p21 axis activation. Therefore, miR-145 establishes a positive feedback loop with p53 sustaining its activation by inhibiting its principal negative regulator, *MDM2*.

In physiologic conditions, CCAT2 is probably responsible for miR-145 fine regulation counteracting p53 transcriptional modulation, whereas in cancer cells both p53 mutations and CCAT2 overexpression contribute to miR-145 downregulation and tumor aggressiveness (Niu et al., 2020).

MiR-221

MiR-221 is an oncogenic miRNA, upregulated in many solid tumors including HCC. The association of miR-221 levels with HCC multifocality and a reduced time to recurrence (TTR) in HCC patients highlights its contribution to tumor progression and metastasis and reinforces its possible role as a therapeutic target (Callegari et al., 2012). Multiple miR-221 targets have been identified, demonstrating its influence on malignant transformation in HCC. Indeed, miR-221 promotes cell proliferation by targeting *DDIT4* (Pineau et al., 2010) and invasion by targeting *PTEN* and *TIMP3* (Garofalo et al., 2009). It also triggers cell cycle progression by modulating *CDKN1B*/p27 and *CDKN1C*/p57 (Fornari et al., 2008) while it hampers apoptosis and anoikis by targeting *CASP3* and *BMF,* respectively (Fornari et al., 2017; Gramantieri et al., 2009). Remarkably, miR-221 functions in nonneoplastic conditions may also be envisaged. By targeting *MDM2*, miR-221 is responsible for p53 upregulation, which guarantees a physiologic response to stressful and genotoxic hints by activating tumor suppression (Fornari et al., 2014). In the neoplastic counterpart, when p53 is nonfunctional, *MDM2* inhibition is ineffective on p53, thus antiapoptotic, proproliferating, and promigrating activities prevail. As confirmation, miR-221 overexpression does not accelerate cell proliferation of all HCC cell lines but, on the contrary, it decreases cell cycle progression and doxorubicin resistance of HepG2 cells bearing a p53 WT isoform through *MDM2* targeting and p53/p21 axis activation (Fornari et al., 2014). As a proof of principle, miR-221 overexpression in p53-silenced HepG2 cells or p53-null Hep3B cells led to an increase of the S-phase cell population with a consistent decrement of p27 expression, letting miR-221 oncogenic behavior prevail in p53 mutated cell contexts. Therefore, we have suggested that miR-221 might behave as a tumor suppressor (TS) gene or as an oncogene, depending on p53 status. This dual role is particularly relevant when a miRNA-silencing approach is to be considered, emphasizing the importance of a molecularly driven HCC classification (Fig. 23.2).

The concomitant targeting of several pathways or multiple steps of the same pathway needs to be considered when trying to figure out the total effects of miRNA modulation strategies. Indeed, the basal expression of each target and its constitutive relevance in specific contexts play pivotal roles not only in the case of miR-221, but also for other deregulated miRNAs.

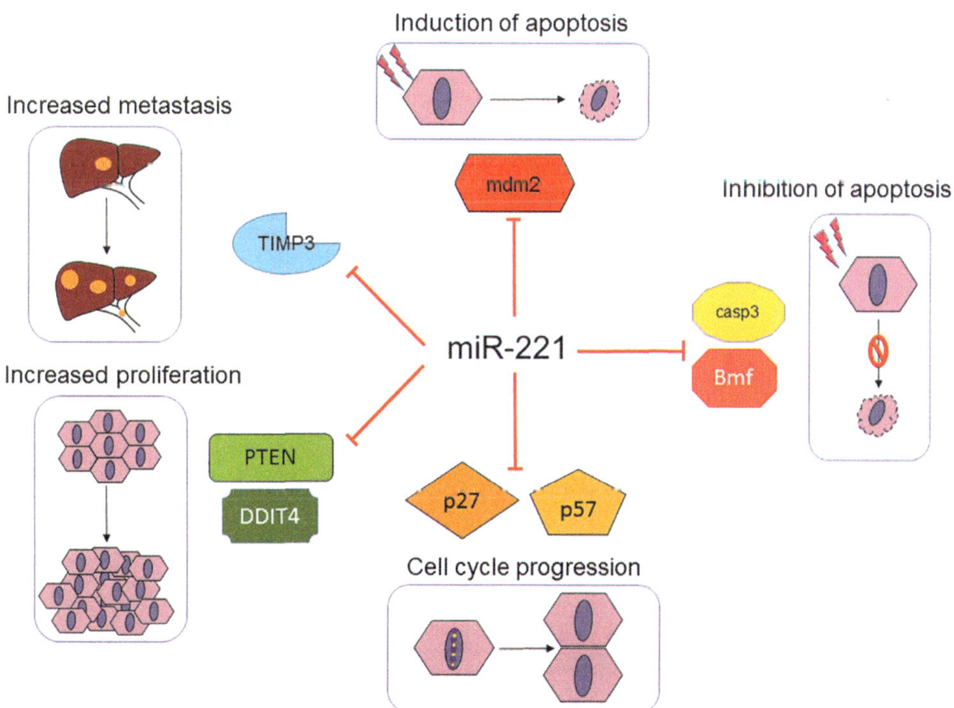

FIG. 23.2 Principal target genes of miR-221 and its biological functions.

MiR-30e and miR-519d

When dealing with oncogenic miRNA functions, we need to note that the cause-effect relationships between miRNAs and target genes and the possible onset of autoregulatory feedback loops are complex events. The multiplicity of targets, sometimes with opposite functions, clearly tells us that the biologic effect of miRNAs relies upon the fine balance between the modulation of multiple molecules as well as their relevance in driving and sustaining malignant transformation.

The same is also true for those miRNAs that were found downregulated in different tumor types showing an apparent TS activity, underscoring the "duality" of the miRNA role in different genetic backgrounds. MiR-30e-3p, a member of the miR-30 family containing six miRNAs located in three bicistronic regions, is downregulated in most HCCs and is associated with p53 mutation. Strikingly, it plays a dual role in the same cancer type showing an opposite behavior in HCC, primarily depending on p53 status. In p53 WT contexts, miR-30e-3p behaves as a TS miRNA through *MDM2* regulation and activation of the p53/p21 pathway, causing cell cycle arrest and drug sensitization; on the contrary, it acts as an oncogene in p53-mutated cells where PTEN and p27 inhibition prevails (Gramantieri et al., 2020).

Another example explaining the relevance of the molecular background on miRNA function is represented by miR-519d. This miRNA belongs to the chromosome 19 miRNA cluster (C19MC), a primate-specific miRNA cluster associated with a peculiar placental expression. DNA hypomethylation and p53 transcriptional regulation are among drivers of miR-519d aberrant expression with a 10-fold increase in half of HCCs. MiR-519d promotes cell proliferation, invasion, and drug sensitization in HCC cell lines through *CDKN1A*/p21, *AKT3,* and *PTEN* targeting (Fornari et al., 2012). Notably, miR-519d contemporaneously targets an oncogene (AKT3) and a TS gene (PTEN) belonging to the same signaling cascade (Fig. 23.3). Thus, the constitutive expression of each target dictates the biological effects of miR-519d in each context. This may explain its dual role in other cancer types (Flor & Bullerdiek, 2012) and highlights the importance of an accurate assessment of the cellular background before therapeutic exploitations.

These findings outline how miRNA functions cannot be limited to simple relationships with their direct targets and pinpoint that their final effects are the result of complex networks where the molecular background plays a fundamental role.

MiR-494 and miR-200a/b

The 14q32 miRNA cluster is located in the DLK1-DIO3 locus, which is conserved among placental mammals and contains imprinted genes, noncoding RNAs, and multiple snoRNAs and miRNAs. Site-specific gene targeting demonstrated the activation of this region as a driver event leading to HCC in 100% of mice (Wang et al., 2012). Similarly, this miRNA cluster is upregulated in both MYC and RAS oncogene-driven mouse models as well as in a subset (25%–30%) of human HCCs with

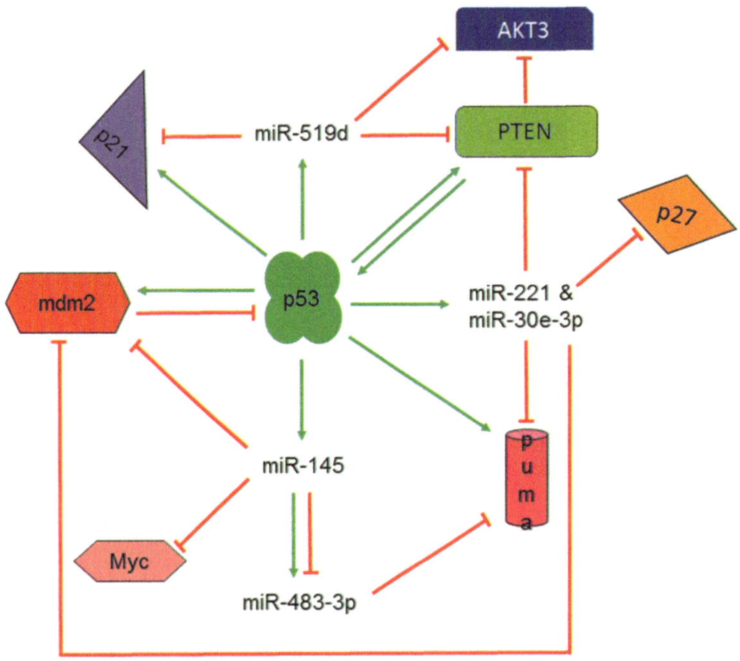

FIG. 23.3 MicroRNA/p53/MDM2 feedback loops in HCC.

elevated AFP levels, a stem cell-like phenotype, and poor survival (Luk et al., 2011). MiR-494 is a potent inducer of cell cycle progression through direct targeting of MCC (mutated in colorectal cancer) and *CDKN1B*/p27. It regulates invasion and apoptosis in stressful conditions by inhibiting *PTEN* and *BBC3*/PUMA. In particular, PTEN silencing mediates AKT/mTOR pathway activation and is responsible for sorafenib resistance. MiR-494 antagonists reduce the tumor burden in vivo and increase sorafenib activity in a chemically induced HCC (DEN-HCC) rat model (Lim et al., 2014; Pollutri et al., 2018). The upregulation of miRNAs belonging to *Igf2* and *Dlk1-Dio3* imprinted loci was reported in miR-122 KO mice and might represent a precancerous event linking HCC development to miR-122 deregulated expression (Valdmanis et al., 2018).

Notably, miRNAs located in imprinted loci are subjected to strong epigenetic regulation and several mechanisms often contribute to their deregulation in human diseases, likely explaining the association between their aberrant expression and well-defined cancer subgroups as evidenced by molecular classification studies. Complex epigenetic mechanisms and regulatory loops also contribute to miR-200b/a/429 expression by chemical modifications of DNA and histone residues close to their promoter region. These miRNA family members are downregulated in human cancers where their aberrant expression favors mesenchymal transition and metastasis. A global downregulation of H3 acetylation was found in HCC modulating the transcription of many genes involved in carcinogenesis. Reduced H3 acetylation levels were found at the promoter of miR-200a, causing its decreased expression by blocking Sp1-mediated transcription. Intriguingly, miR-200a downregulation is responsible itself for elevated HDAC4 levels through the establishment of a feedforward loop that sustains global H3 deacetylation leading to the repression of important cell cycle controller genes, such as p21, triggering cell proliferation and invasion (Yuan et al., 2011). Again, the ceRNA activity of oncogenic lncRNAs may contribute to miRNA deregulation by sequestering them through complementary sequences, increasing the cellular concentration of their own targets. Yuan et al. (Yuan et al., 2014) reported the role of a TGF-β-activated lncRNA (lncRNA-ATB) in HCC metastasis by identifying its ceRNA role against miR-200 family members leading to the upregulation of EMT key mediator factors, ZEB1 and ZEB2, where increased expression is associated with HCC poor prognosis and higher metastasis propensity.

Diagnostic, prognostic, and predictive values of miRNAs

When compared to mRNA profiles, miRNAs display a high predictive potential as well as a diagnostic and prognostic impact in human cancer (Lu et al., 2005). Indeed, one of the advantage of miRNA-based profiling with respect to transcriptomic analysis resides in the limited number of genes able to discriminate different tumor phenotypes, aiding a more reliable translation into the clinics. In addition, the small size of miRNAs makes them ideal candidates to be investigated in biologic samples, including formalin-fixed paraffin-embedded (FFPE) tissues and circulating fluids. MiRNAs are far more stable than messenger RNAs, decreasing the risk of loss of information or unreliability/irreproducibility of profiling data related to poor RNA quality. These are among the main reasons moving the scientific community to focus on miRNA investigations when searching for novel biomarkers. While diagnostic purposes have mainly been addressed by testing circulating miRNAs, the molecular classification and prognostication of HCC have mainly been studied at the tissue level.

Intrahepatic metastasis and tumor recurrence are the main causes of dismal prognosis in curatively treated HCCs. The identification of molecular profiles able to predict the propensity toward recurrence (either metastasis or early recurrence) would be of great utility in terms of treatment and follow-up personalization. A 20-miRNA signature from tumor tissue predicts venous metastasis, overall survival and recurrence, and represents an independent prognostic predictor for HCC patients. Conversely, miRNA signatures from the surrounding tissue are not informative, demonstrating tumor samples as a better source for miRNA profiling studies (Budhu et al., 2008). Interestingly, three members of the miR-30 family (miR-30c-1, miR-30a, miR-30e) represent the most prominent miRNA hubs in solid tumors and are among the most downregulated miRNAs in metastatic tissues, which is in agreement with their regulation of cell invasion and EMT-associated molecules (Deng et al., 2017; Liu, Tu, & Liu, 2014). MiR-122 is lost in metastatic HCCs and its low levels are associated with poor prognosis and decreased overall survival, demonstrating its diagnostic and prognostic potential for HCC progression (Coulouarn, Factor, Andersen, Durkin, & Thorgeirsson, 2009). Taking advantage of reliable miRNA quantification in biologic samples, miRNA expression can be easily assessed in different areas of FFPE specimens following laser microdissection. As an example, miR-139 is downregulated not only in liver tumors but also in venous metastases and its low expression is associated with advanced pathologic features, shorter disease-free survival, and overall survival. MiR-139 regulates cell invasion and metastasis through the inhibition of Rho-associated coiled-coil containing protein kinase 2 (ROCK2) but does not influence the cell growth of HCC-derived cell lines (Wong et al., 2011). MiRNAs exert their function by the fine modulation of hundreds, or even thousands, of downstream targets; on the other hand, a single gene is contemporaneously regulated by several miRNAs. Because of its importance in cell motility, the Rho/ROCK pathway is tightly regulated and is targeted by several miRNAs. Because many cancer-associated miRNAs reside in fragile sites of the human genome, the simultaneous deregulation of miRNAs is a common event in neoplastic tissues. MiR-124 is downregulated in two thirds of

HCCs and influences intra- and extrahepatic metastasis in preclinical models by targeting ROCK2. Low miR-124 expression is associated with increased tumor size and multiplicity, higher tumor stage, distant metastasis, and short disease-free survival; it is also an independent prognostic factor for poor survival (Zheng et al., 2012). Some miRNAs regulate multiple cellular processes such as proliferation, invasion, and apoptosis, whereas others are mainly implicated in the fine-tuning of well-defined biologic functions. This is the case of miR-29b, whose decreased expression in HCC is associated with shorter disease-free survival and diminished apoptotic rate due to the enhanced expression of the antiapoptotic members of mitochondrion-dependent cell death, Bcl-2 and Mcl-1. Because of chemotherapy sensitization, miR-29b represents both an interesting prognostic factor and a therapeutic target triggering tumor shrinkage and drug response (Xiong et al., 2010).

The polycistronic miR-17-92 cluster was initially found overexpressed in B-cell lymphomas and the proto-oncogene c-Myc was demonstrated to be its master regulator, suggesting the oncogenic role of this miRNA cluster in human malignancies (O'Donnell, Wentzel, Zeller, Dang, & Mendell, 2005). Regarding HCC, several miR-17-92 members are upregulated in tumor tissue with respect to the surrounding liver. Conversely, miR-19a downregulation was described in venous metastasis and miR-19b is reduced in patients experiencing a tumor recurrence within two years after surgery, showing a prognostic relevance in HCC patients with vascular invasion or multifocal diseases (Hung, Yen, Tsai, Su, & Yen, 2015).

MiRNAs upregulated in tumor tissues are considered oncomiRNAs. Notwithstanding, in some cases, they can increase drug sensitization as in the case of miR-519d that favors an anticancer response to doxorubicin and rapamycin (Fornari et al., 2012). The 46-miRNA members of the C19MC are consistently upregulated in HCCs from The Cancer Genome Atlas (TCGA) database and their increased expression is associated with DNA hypomethylation. This is in line with the contribution of epigenetic mechanisms to the transcriptional activation of miRNA clusters located in imprinted loci. Three miRNAs of this cluster (mir-512-1, miR-516a-1, and miR-519a-2) positively correlate with the T-stage and are negatively associated with overall survival, contributing to increased tumor burden through the regulation of multiple molecules belonging to the MAPK pathway (Rui et al., 2020). In addition, C19MC miRNA members distinguish early HCCs from dysplastic nodules, representing an interesting tool aiding differential diagnosis at very early stages (Augello et al., 2012).

MiR-487 overexpression defines a group of nodular HCCs (NHCCs) showing high metastatic potential. This miRNA, belonging to the 14q32 miRNA cluster, has indeed a key role in tumor progression accelerating both cell proliferation and invasion through *PIK3R1* and *SPRED2* targeting, activating Akt and MAPK signaling, respectively. In human cohorts, miR-487 is associated with multinodularity, microvascular invasion, and capsular formation, and its upregulation represents an independent prognostic marker for reduced survival and early recurrence, predicting poor postoperative outcome (Chang et al., 2017). Therefore, members of the DLK1-DIO3 miRNA cluster are promising candidates for the molecular classification of HCC subgroups that might benefit from anti-miRNA-based therapies.

The presence of cancer stem cells (CSCs) represents a hallmark of poor prognosis linked to tumor relapse, metastasis, and the drug-resistant phenotype. Stem cell features such as EpCAM and CD133 expression characterize more than one-third of HCCs as well as the activation of signaling cascades triggered by octamer 4 (Oct4), Wnt/β-catenin, and Notch pathways. In particular, the Wnt/β-catenin axis is frequently activated in non-HBV well-differentiated HCCs due to the presence of *CTNNB1* activating mutations (11%–37% of cases) or the deletion/inactivation of negative control genes, *AXIN1* (10% of cases) and *APC* (1%–2% of cases). MiR-1246 is among the most upregulated miRNAs in CD133-positive HCC cells and triggers Wnt/β-catenin pathway activation through AXIN2 and GSK3β direct inhibition. This leads to β-catenin nuclear translocation and transcriptional activation of target genes, enhancing tumor sphere formation, invasion, and in vivo tumorigenesis. The stem-associated gene Oct4 binds to the miR-1246 promoter and transactivates its expression, triggering the epigenetic activation of the Wnt/β-catenin axis. Intriguingly, miR-1246 upregulation and constitutive *CTNNB1* mutations are mutually exclusive events in HCC, accounting for Wnt/β-catenin activation in nonmutated tumors. MiR-1246 is associated with high AFP expression and is an independent prognostic marker for both overall and disease-free survival. In addition, because of its interaction with several actors of the Wnt pathway, miR-1246 represents an interesting therapeutic target for stem-related HCCs (Chai et al., 2016). Gene classifiers obtained by miRNA and/or gene combinations are often better predictive tools than single miRNAs and identify specific tumor subtypes. Interestingly, bioinformatics models predict gene regulatory networks (GRNs) that link different genes/miRNAs to each other. Moreover, mathematical models might uncover the impact of therapeutic agents with respect to all the members of the GRN itself, which is of utmost importance to predict possible events occurring during drug resistance. In particular, let-7b and its target gene *LIN28B*, whose expression is often deregulated in cancers, establish a GRN with *CTNNB1*, *SMARCA4*, *SOX9*, *TP53,* and *MYC*. High levels of this GRN are associated with hyperproliferation, dedifferentiation, and HGF/MET pathway activation while they negatively correlate with survival, suggesting the involvement of this GRN in the proliferative HCC class. Strikingly, this GRN links important driver genes whose mutational index is among the highest in HCC, such as *CTNNB1* and *TP53*, or whose prooncogenic role is well known in both preclinical models and HCC subtypes, such as c-Myc, or with a key role in the stem cell phenotype such as SOX9.

In conclusion, miRNA signatures provide diagnostic and prognostic information overcoming mRNA-based signatures. MiRNA aberrant expression is peculiar of well-defined HCC subclasses and this aspect not only has important prognostic implications but also therapeutic value, offering possible combined treatment approaches in well-defined subgroups.

Circulating miRNAs

Circulating miRNAs have attracted much attention as possible noninvasive biomarkers in HCC patients. Indeed, according to current guidelines (European Association for the Study of the Liver, 2018), the vast majority of HCCs are diagnosed by imaging techniques, making biopsy unnecessary. This has hampered the development of morphological and molecular correlations with clinical course, prognostication, and response to treatments. Given the risk of adverse events following a liver biopsy in patients with advanced HCC, a biopsy might even be considered unethical in certain patients. Thus, research has focused on noninvasive biomarkers providing additional information over imaging. Notably, the use of AFP in HCC surveillance programs is debated and the European Association for the Study of the Liver (EALS) no longer recommends it (Foerster & Galle, 2019), leaving HCC without serum diagnostic biomarkers. Indeed, AFP misses one-third of patients with small HCC (<3cm). Similarly, increased des-γ-carboxy prothrombin (DCP) can be found in only 44%–47% of patients with HCCs less than 3cm (Abdalla & Haj-Ahmad, 2012). Given their abundance, stability, ease, and economic detection in circulating fluids, miRNAs have been extensively studied. As remarked for tissue studies, circulating miRNA profiles in HCC patients turned out to be very heterogeneous. The inconsistency across studies may be ascribed to differences in analytical procedures, study design, and patient populations (different etiologies, HCC stage, presence of cirrhosis, previous, and ongoing treatments, etc.). Notwithstanding this, many miRNAs were reported to improve the diagnostic accuracy in HCC, not only as single tests but also when combined with other miRNAs or with other circulating biomarkers such as AFP. The identification of consensus reference genes and analytical protocols represents the first technical matter, whereas the employment of different control groups represents the second issue. The presence of multiple risk factors further complicates the establishment of univocal controls. The same goes for the presence and stage of the underlying chronic liver disease that influence miRNA expression profiles. As an example, circulating miR-122 and miR-21 levels are higher in HCC patients with respect to healthy volunteers, but their serum levels are even increased in patients with HBV-related hepatitis; therefore, depending on the reference group, these two miRNAs display opposite changes in the HCC subgroup (Xu et al., 2011). Circulating miRNA signatures in HBV- and HCV-infected patients included different miRNAs. Sidhu et al. reported that eight circulating miRNAs (miR-20a-5p, miR-25-3p, miR-30a-5p, miR-92a-3p, miR-132-3p, miR-185-5p, miR-320a, and miR-324-3p) are overexpressed in HBV-positive HCC patients compared with HBV-positive cancer-free controls (Sidhu, Kapoor, Pandey, & Kumar, 2015). Huang et al. performed a metaanalysis (Huang et al., 2016) to evaluate the possible contribution of circulating miRNAs in the early diagnosis of HCC. They identified miR-21 and miR-122 as the most promising miRNAs to distinguish patients with HCC from healthy controls, irrespective of risk factors.

The nature of biologic samples also has to be taken into account when circulating biomarkers are considered; indeed, variations in terms of the miRNA spectrum and abundance are well known between serum and plasma. We identified several promising circulating miRNAs to be further tested as putative diagnostic biomarkers. In plasma samples, miR-101-3p, miR-106b-3p, miR-1246, and miR-411-5p were aberrantly expressed in HCC patients. In particular, a classifier consisting of plasmatic levels of miR-101-3p, miR-1246, and miR-106b-3p discriminated HCC from cirrhotic patients (AUC=0.99) as well as from healthy controls (AUC=1.00). When we analyzed the diagnostic accuracy of the same miRNA panel in serum instead of plasma, the combination of miR-101-3p and miR-106b-3p exhibited the best diagnostic accuracy (AUC=0.96). Remarkably, serum levels show the same trend of plasma levels only for miR-1246 whose increased expression is retained in HCC versus cirrhotic patients while miR-101-3p displayed an opposite modulation that was upregulated in plasma and downregulated in the sera of HCC patients (Moshiri et al., 2018). These aspects should be carefully considered when evaluating the best option for each clinical need. When we analyzed the serum of cirrhotic patients, most of which had HCV infection, the best performance in the diagnostic setting was displayed by miR-939, miR-595, and miR-519d, which differentiated cirrhotic patients with and without HCC (Fornari et al., 2015). An exosomal secretion was observed in the case of miR-519d, miR-21, miR-221, and miR-1228. Exosomes are gaining much relevance both in tumor biology as well as in the identification of noninvasive biomarkers. Exosomes are extracellular vesicles of 30–120nm size. They are actively secreted by donor cells and are internalized by recipient cells, thus acting as mediators of cell-cell communication even between distant cells or tissues. Concerning cancer, microvescicles and exosomes play key roles in cell-to-cell communication regulating tumor cell proliferation, mobility, and colonization of distant sites as well as interaction with inflammatory and stromal cells influencing the tumor microenvironment and the metastatic niche. Exosomes released from HCC cells display a peculiar content in terms of proteins, mRNAs, and small RNAs. In particular, an electrophoresis band smaller than 200 base pairs appears evident in the RNA from the exosomal fraction, outlining the enrichment of small RNAs in secreted vesicles. MiRNA profiling from exosomes and donor cells highlights, on one hand, a different pattern of miRNA expression in intra- and extracellular compartments. On the other hand, the similar miRNA content in exosomes from different HCC cells suggests common selective mechanisms for miRNA enrichment into these microparticles. The biogenesis and maturation of exosomes are central steps mediating intercellular relationships between multiple cell types, due to the peculiar incorporation of intracellular cargo molecules that are tissue/organ-specific and might reflect the pathophysiologic state of the entire organism. Interestingly, miRNAs embedded in exosomes secreted from donor cells can be internalized by recipient cells, where they can exert an epigenetic function regulating the expression of target genes (Kogure, Lin, Yan, Braconi, & Patel, 2011). Highly metastatic HCC cells specifically incorporate miR-1247-3p into exosomes during their maturation program inside multivesicular bodies that, once secreted, can be internalized by fibroblasts provoking their transformation to cancer-associated-fibroblasts (CAFs) through Beta-1,4-Galactosyltransferase 3 (B4GALT3) suppression that promotes β1-integrin-NF-κB signaling activation. Activated CAFs secrete proinflammatory cytokines that, in turn, modify the tumor cell phenotype, resulting in enhanced expression of stemness and EMT genes, increased spheroid-growth and migration capabilities as well as sorafenib resistance. In line, HCC patients display higher exosomal miR-1247-3p levels that are even higher in patients with lung metastasis (Fang et al., 2018).

The integrity of endothelial cells is critical to maintain vascular permeability and its dysfunction contributes to tumor spreading. Conditioned media from highly invasive HCC cells is enriched in miR-103 and it dysregulates HUVEC monolayer permeability by inhibiting molecules involved in tight junction and cell-cell contact such as VE-cadherin, p120-catenin, and zonula occludens 1. In line, miR-103 is the most abundant miRNA in sera from poor prognosis HCCs and represents an independent predictor for shorter recurrence-free survival. Orthotopic animal models confirmed an increase of vascular permeability leading to enhanced circulating tumor cells as well as intra- and extrahepatic metastatic tumor foci, demonstrating the key role of exosomes in cellular crosstalk and tumor aggressiveness (Fang et al., 2018).

In conclusion, despite the inconsistency across studies, which seems to be ascribed to the heterogeneity of the disease and to the different analytical approaches, circulating miRNAs appear as promising noninvasive biomarkers. The studies performed so far let us hypothesize that specific miRNAs might be more informative in defined clinical settings, identified on the basis of risk factors, stage of the disease, and previous treatments. The concomitant evaluation of miRNA panels or other traditional biomarkers might enhance their diagnostic performance. Beside the role of biomarkers, the exosomal secretion of miRNAs also provides biological data to deepen our knowledge of tumor-associated mechanisms.

MicroRNAs as biomarkers of response to treatments

Predictive tools able to point patients toward the optimal therapeutic strategy are an unmet clinical need in advanced HCC. Several studies have reported the encouraging performance of single miRNAs as well as miRNA combinations.

Basal miR-181a-5p serum levels are informative of disease control in patients taking sorafenib and lower circulating levels are associated with prolonged overall survival, indicating miR-181a-5p as a good candidate to improve patient selection (Nishida et al., 2017). In line with its role as a tumor-suppressor miRNA, miR-199a/b-3p serum levels are decreased in HCC patients with respect to healthy controls and a negative correlation exists between circulating miRNA levels and tumor size. MiR-199a/b-3p is also a good predictor of TACE response, discriminating patients at risk for locoregional treatment failure that might benefit from other therapies. Interestingly, miR-199a/b-3p circulating levels decrease in responder patients over time, suggesting a possible hypoxia-related mechanism in the regulation of miRNA secretion or expression (Luo et al., 2016).

Tumor recurrence after liver transplantation is a major challenge affecting the performance of this curative intervention in HCC patients. The identification of early biomarkers predicting at-risk patients might help treatment and follow-up personalization. The elevation of circulating miR-148a, miR-1246, and miR-1290 at 2 h after portal vein reperfusion is associated with HCC recurrence. In particular, miR-1246 is an independent predictor of both overall and disease free-survival and its serum levels are correlated with liver injury, as indicated by increased AST and ALT levels. Strikingly, injury-associated macrophage activation led to an increase of miR-1246 secretion in cell culture supernatant, outlining the possibility that tumor-associated cells might be an alternative source of circulating miRNAs in pathologic conditions contributing to tumor development and evolution (Ng et al., 2016).

MiR-30e-3p serum levels discriminate between sorafenib responder and nonresponder patients at the 2-month follow-up. Higher miR-30e-3p levels associated with sorafenib resistance in both HCC patients and animal models. Notably, *TP53* is activated by miR-30e-3p through *MDM2* targeting and, in turn, it establishes a positive feedback loop by regulating both miR-30e transcription and maturation. This highlights the presence of entangled circuits that mirror the intrinsic nature of miRNAs as fine regulators of gene expression. Remarkably, p53 participates in miR-30e-3p exosomal excretion, thus modulating the response to treatments at multiple levels by interfering with the miR-30e-3p/*MDM2*/*TP53* axis and intracellular adaptation to stressful conditions as well as by triggering exosomal secretion (Gramantieri et al., 2020).

Interestingly, the baseline levels of nine miRNAs with a known role in hepatocarcinogenesis (MIR30A, MIR122, MIR125B, MIR200A, MIR374B, MIR15B, MIR107, MIR320, MIR645) predict the overall survival of HCC patients treated with regorafenib, demonstrating the high versatility of this class of small RNAs for patient stratification to second- and third-line anticancer agents (Teufel et al., 2019). Moreover, top gene sets associated with these miRNAs whose expression identifies an improvement in regorafenib treatment are in common with the Boyault (Boyault et al., 2007) and Chiang (Chiang, 2008 #98) HCC subgroups coinciding with the S3 subtype from the Hoshida classification (Hoshida et al., 2009), characterized by well-differentiated, small-sized tumors with a hepatocyte-like phenotype.

As observed above, exosomal miRNA trafficking has been identified as a possible mechanism driving resistance to anticancer treatment. Several mechanisms are involved in the regulation of exosome content and production during stressful conditions, including *TP53* and *AKT* pathways, modulating intracellular RNA and protein abundance as well as cell communication. This aspect is particularly interesting when anticancer treatments are concerned for at least two main reasons. The first one refers to the possible discovery of mechanisms involved in drug resistance while the second one concerns the identification of possible biomarkers of early response/tumor escape.

Strikingly, oncomiR-221 secretion is mediated by exosomes in HCC patients (Fornari et al., 2015) and sorafenib treatment modulates its extrusion in preclinical models interfering with AKT phosphorylation. In HCC-derived cells sensitive to sorafenib, a strong increase of the miR-221 exosomal fraction occurs, and it is associated with AKT pathway inactivation by sorafenib. On the contrary, in highly resistant cells, no change in miR-221 secretion is registered following sorafenib administration and in these settings, the AKT axis inactivation does not occur. PI3K/AKT pathway inactivation leads to an miR-221 increase in cell culture supernatant, emphasizing the relevance of molecular mechanisms regulating miRNA extrusion and, consequently, drug response. In sorafenib-treated HCC patients, miR-221 serum levels increase in responder patients only. In addition, basal miR-221 serum levels predict sorafenib response, possibly aiding patient stratification to first-line agents in HCC (Fornari et al., 2017).

The early identification of acquired drug resistance is still an unsolved clinical issue both for first- and second-line agents. Liquid biopsies are easy to collect at different time points from the beginning of treatment, therefore representing ideal biologic specimens for biomarker discovery to help predict tumor escape in the early phases.

Therapeutic use of microRNAs

In the last two decades, great efforts have been made for the molecular classification of this heterogeneous tumor. The clinical challenge is to address HCC subclasses to the best therapeutic intervention based on genetic background and pathway activation. Genomic alterations and transcriptomic analysis classified HCCs into several subgroups based on common clinical and genetic characteristics (Boyault et al., 2007; Hoshida et al., 2009). Due to the high specificity of miRNA profiling in tumors, miRNAs were included in HCC classifications. Intriguingly, unsupervised cluster analysis of miRNA expression grouped samples based on histological features, risk factors, and genetic mutations, confirming the miRNA role in liver tumorigenesis and suggesting their use as diagnostic markers. Toffanin et al. identified three main HCC subgroups based on miRNA profiling. Cluster A was associated with β-catenin mutations, low AFP levels, and a good prognosis, whereas cluster B was mainly associated with INF-related genes. Cluster C was characterized by IGF and AKT/mTOR pathway activation and, in particular, cluster C2 showed the upregulation of several members of C19MC while subclass C3 was characterized by the overexpression of the 14q32 miRNA cluster, EpCAM positivity, and high AFP levels (Toffanin et al., 2011).

Because the liver represents an ideal organ for miRNA-based therapeutic approaches due to the high uptake of peripherally administered miRNAs as well as their low toxicity, both in animals and in humans, several preclinical studies were aimed at confirming the effectiveness of such treatments. MiR-122 was the first miRNA to be studied in preclinical models and to be exploited in humans. Anti-miR-221 therapeutic exploitation represents also a promising strategy for HCC treatment and prevention (Park et al., 2011) and the development of suitable animal models is of utmost importance, aiding translation into the clinics. The establishment of a liver-specific miR-221 transgenic mouse demonstrated for the first time that miR-221 overexpression is a sufficient condition to induce liver tumorigenesis in a cirrhotic background and revealed its usefulness as a preclinical tool to assay miRNA-based therapeutic and preventive approaches (Callegari et al., 2012, 2019). A further aspect to be considered for the implementation of therapeutic options is miRNA relative abundance in the organ of interest, especially when an miRNA-replacement strategy is taken into account. In the liver, nine miRNAs represent 88.2% of the total miRNA population and miR-199a/b-3p is the third most highly expressed, accounting for 4.9% of the entire miRNome. MiR-199a/b-3p is downregulated in almost all HCC specimens due to altered histone modifications and its low expression correlates with decreased tumor-free and overall survival in HCC patients (Hou et al., 2011). c-Met, mTOR, CD44, and PAK4 have been identified among miR-199a-3p target genes, leading to the activation of signaling cascades with a key role in tumor proliferation, invasion, and drug resistance (Fornari et al., 2010; Henry et al., 2010; Hou et al., 2011; Kim et al., 2008). Different replacement strategies (e.g., cholesterol-conjugated miRNA mimics, AAV-mediated gene delivery, and nanoparticle formulations) and animal models (PDX, xenograft, orthotopic, and TG221 mice) have been employed to investigate miR-199a-3p restoration in HCC, showing a safety profile and consistent anticancer activity (Callegari et al., 2013, 2018; Hou et al., 2011; Varshney et al., 2018). Similarly, miR-26a is downregulated in HCC specimens and animal models (Fornari et al., 2019), and highly expressed in normal livers, making it an ideal candidate for proof-of-concept studies underscoring the therapeutic benefit of the miRNA-replacement option. Due to the discontinuous architecture of its sinusoidal capillary network, the liver is a well-suited organ for testing alternative therapeutic approaches and delivery systems. In addition, continuous improvement of viral vectors, oligonucleotide stability, and carrier molecules facilitated the targeted delivery and long-term expression of miRNA mimics and antgomiRs without systemic toxicity. Adeno-associated vectors (Lu et al., 2005) showed a high transduction efficiency of hepatocytes and turned out to be a powerful strategy for miRNA replacement in liver-specific MYC transgenic mice, inducing tumor regression following miR-26a restoration in cancer cells (Kota et al., 2009). The reexpression of a single miRNA could provide a consistent therapeutic benefit, abolishing tumor cell proliferation and inducing apoptosis of transformed cells but not of normal hepatocytes. This holds promise as a safe and efficient option, especially in advanced patients whose liver functionality is highly compromised.

MiRNA-based therapeutic approaches are feasible and safe in humans as proved by a phase II clinical trial on HCV-infected patients treated with an anti-miR-122-based formulation (Miravirsen) that showed a good efficacy in terms of HCV replication impairment (Janssen et al., 2013). Nevertheless, these treatments must take into account the tumor genetic background and patient comorbidities. Encouraging results were obtained in animals and were corroborated by molecular data, showing the efficacy of miR-34 restoration with an miR-34 mimic in a subgroup of patients. This approach also increased sorafenib efficacy (Beg et al., 2017). Despite the optimistic data, the first in-human phase I trial in drug refractory solid tumors investigating an miR-34a mimic-liposomal formulation (MRX34) displayed severe adverse events (AEs), requiring its suspension. These adverse events were possibly immune-mediated, or triggered by excessive RNA concentration and/or delivery systems (Hong et al., 2020). These findings emphasize the need to refine delivery systems with a careful incorporation of genetic background into patient selection.

References

Abdalla, M. A., & Haj-Ahmad, Y. (2012). Promising candidate urinary MicroRNA biomarkers for the early detection of hepatocellular carcinoma among high-risk hepatitis C virus Egyptian patients. *Journal of Cancer*, *3*, 19–31.

Augello, C., Vaira, V., Caruso, L., Destro, A., Maggioni, M., Park, Y. N., et al. (2012). MicroRNA profiling of hepatocarcinogenesis identifies C19MC cluster as a novel prognostic biomarker in hepatocellular carcinoma. *Liver International*, *32*, 772–782.

Beg, M. S., Brenner, A. J., Sachdev, J., Borad, M., Kang, Y. K., Stoudemire, J., et al. (2017). Phase I study of MRX34, a liposomal miR-34a mimic, administered twice weekly in patients with advanced solid tumors. *Investigational New Drugs*, *35*, 180–188.

Boyault, S., Rickman, D. S., de Reynies, A., Balabaud, C., Rebouissou, S., Jeannot, E., et al. (2007). Transcriptome classification of HCC is related to gene alterations and to new therapeutic targets. *Hepatology*, *45*, 42–52.

Budhu, A., Jia, H. L., Forgues, M., Liu, C. G., Goldstein, D., Lam, A., et al. (2008). Identification of metastasis-related microRNAs in hepatocellular carcinoma. *Hepatology*, *47*, 897–907.

Callegari, E., D'Abundo, L., Guerriero, P., Simioni, C., Elamin, B. K., Russo, M., et al. (2018). miR-199a-3p modulates MTOR and PAK4 pathways and inhibits tumor growth in a hepatocellular carcinoma transgenic mouse model. *Molecular Therapy- -Nucleic Acids*, *11*, 485–493.

Callegari, E., Domenicali, M., Shankaraiah, R. C., D'Abundo, L., Guerriero, P., Giannone, F., et al. (2019). MicroRNA-based prophylaxis in a mouse model of cirrhosis and liver cancer. *Molecular Therapy- -Nucleic Acids*, *14*, 239–250.

Callegari, E., Elamin, B. K., D'Abundo, L., Falzoni, S., Donvito, G., Moshiri, F., et al. (2013). Anti-tumor activity of a miR-199-dependent oncolytic adenovirus. *PLoS One*, *8*, e73964.

Callegari, E., Elamin, B. K., Giannone, F., Milazzo, M., Altavilla, G., Fornari, F., et al. (2012). Liver tumorigenicity promoted by microRNA-221 in a mouse transgenic model. *Hepatology*, *56*, 1025–1033.

Cancer Genome Atlas Research Network. (2017). Comprehensive and integrative genomic characterization of hepatocellular carcinoma. *Cell*, *169*, 1327–1341 (e23).

Chai, S., Ng, K. Y., Tong, M., Lau, E. Y., Lee, T. K., Chan, K. W., et al. (2016). Octamer 4/microRNA-1246 signaling axis drives Wnt/beta-catenin activation in liver cancer stem cells. *Hepatology*, *64*, 2062–2076.

Chang, R. M., Xiao, S., Lei, X., Yang, H., Fang, F., & Yang, L. Y. (2017). miRNA-487a promotes proliferation and metastasis in hepatocellular carcinoma. *Clinical Cancer Research*, *23*, 2593–2604.

Coulouarn, C., Factor, V. M., Andersen, J. B., Durkin, M. E., & Thorgeirsson, S. S. (2009). Loss of miR-122 expression in liver cancer correlates with suppression of the hepatic phenotype and gain of metastatic properties. *Oncogene*, *28*, 3526–3536.

Deng, L., Tang, J., Yang, H., Cheng, C., Lu, S., Jiang, R., et al. (2017). MTA1 modulated by miR-30e contributes to epithelial-to-mesenchymal transition in hepatocellular carcinoma through an ErbB2-dependent pathway. *Oncogene*, *36*, 3976–3985.

Elmen, J., Lindow, M., Schutz, S., Lawrence, M., Petri, A., Obad, S., et al. (2008). LNA-mediated microRNA silencing in non-human primates. *Nature*, *452*, 896–899.

Esau, C., Davis, S., Murray, S. F., Yu, X. X., Pandey, S. K., Pear, M., et al. (2006). miR-122 regulation of lipid metabolism revealed by in vivo antisense targeting. *Cell Metabolism*, *3*, 87–98.

European Association for the Study of the Liver. (2018). EASL clinical practice guidelines: Management of hepatocellular carcinoma. *Journal of Hepatology*, *69*, 182–236.

Fang, T., Lv, H., Lv, G., Li, T., Wang, C., Han, Q., et al. (2018). Tumor-derived exosomal miR-1247-3p induces cancer-associated fibroblast activation to foster lung metastasis of liver cancer. *Nature Communications*, *9*, 191.

Fang, J. H., Zhang, Z. J., Shang, L. R., Luo, Y. W., Lin, Y. F., Yuan, Y., et al. (2018). Hepatoma cell-secreted exosomal microRNA-103 increases vascular permeability and promotes metastasis by targeting junction proteins. *Hepatology*, *68*, 1459–1475.

Fittipaldi, S., Vasuri, F., Bonora, S., Degiovanni, A., Santandrea, G., Cucchetti, A., et al. (2017). miRNA signature of hepatocellular carcinoma vascularization: How the controls can influence the signature. *Digestive Diseases and Sciences*, *62*, 2397–2407.

Flor, I., & Bullerdiek, J. (2012). The dark side of a success story: microRNAs of the C19MC cluster in human tumours. *The Journal of Pathology*, *227*, 270–274.

Foerster, F., & Galle, P. R. (2019). Comparison of the current international guidelines on the management of HCC. *JHEP Reports*, *1*, 114–119.

Fornari, F., Ferracin, M., Trere, D., Milazzo, M., Marinelli, S., Galassi, M., et al. (2015). Circulating microRNAs, miR-939, miR-595, miR-519d and miR-494, Identify Cirrhotic Patients with HCC. *PLoS One*, *10*, e0141448.

Fornari, F., Gramantieri, L., Callegari, E., Shankaraiah, R. C., Piscaglia, F., Negrini, M., et al. (2019). MicroRNAs in Animal Models of HCC. *Cancers*, 11(12).

Fornari, F., Gramantieri, L., Ferracin, M., Veronese, A., Sabbioni, S., Calin, G. A., et al. (2008). MiR-221 controls CDKN1C/p57 and CDKN1B/p27 expression in human hepatocellular carcinoma. *Oncogene*, *27*, 5651–5661.

Fornari, F., Gramantieri, L., Giovannini, C., Veronese, A., Ferracin, M., Sabbioni, S., et al. (2009). MiR-122/cyclin G1 interaction modulates p53 activity and affects doxorubicin sensitivity of human hepatocarcinoma cells. *Cancer Research*, *69*, 5761–5767.

Fornari, F., Milazzo, M., Chieco, P., Negrini, M., Calin, G. A., Grazi, G. L., et al. (2010). MiR-199a-3p regulates mTOR and c-Met to influence the doxorubicin sensitivity of human hepatocarcinoma cells. *Cancer Research*, *70*, 5184–5193.

Fornari, F., Milazzo, M., Chieco, P., Negrini, M., Marasco, E., Capranico, G., et al. (2012). In hepatocellular carcinoma miR-519d is up-regulated by p53 and DNA hypomethylation and targets CDKN1A/p21, PTEN, AKT3 and TIMP2. *The Journal of Pathology*, *227*, 275–285.

Fornari, F., Milazzo, M., Galassi, M., Callegari, E., Veronese, A., Miyaaki, H., et al. (2014). p53/mdm2 feedback loop sustains miR-221 expression and dictates the response to anticancer treatments in hepatocellular carcinoma. *Molecular Cancer Research*, *1*, 203–216.

Fornari, F., Pollutri, D., Patrizi, C., La Bella, T., Marinelli, S., Casadei Gardini, A., et al. (2017). In hepatocellular carcinoma miR-221 modulates sorafenib resistance through inhibition of caspase-3-mediated apoptosis. *Clinical Cancer Research*, *23*, 3953–3965.

Gao, P., Wong, C. C., Tung, E. K., Lee, J. M., Wong, C. M., & Ng, I. O. (2011). Deregulation of microRNA expression occurs early and accumulates in early stages of HBV-associated multistep hepatocarcinogenesis. *Journal of Hepatology*, *54*, 1177–1184.

Garofalo, M., Di Leva, G., Romano, G., Nuovo, G., Suh, S. S., Ngankeu, A., et al. (2009). miR-221&222 regulate TRAIL resistance and enhance tumorigenicity through PTEN and TIMP3 downregulation. *Cancer Cell*, *16*, 498–509.

Gramantieri, L., Ferracin, M., Fornari, F., Veronese, A., Sabbioni, S., Liu, C. G., et al. (2007). Cyclin G1 is a target of miR-122a, a microRNA frequently down-regulated in human hepatocellular carcinoma. *Cancer Research*, *67*, 6092–6099.

Gramantieri, L., Fornari, F., Ferracin, M., Veronese, A., Sabbioni, S., Calin, G. A., et al. (2009). MicroRNA-221 targets Bmf in hepatocellular carcinoma and correlates with tumor multifocality. *Clinical Cancer Research*, *15*, 5073–5081.

Gramantieri, L., Pollutri, D., Gagliardi, M., Giovannini, C., Quarta, S., Ferracin, M., et al. (2020). MiR-30e-3p influences tumor phenotype through MDM2/TP53 axis and predicts sorafenib resistance in hepatocellular carcinoma. *Cancer Research*, *80*, 1720–1734.

Henry, J. C., Park, J. K., Jiang, J., Kim, J. H., Nagorney, D. M., Roberts, L. R., et al. (2010). miR-199a-3p targets CD44 and reduces proliferation of CD44 positive hepatocellular carcinoma cell lines. *Biochemical and Biophysical Research Communications*, *403*, 120–125.

Hong, D. S., Kang, Y. K., Borad, M., Sachdev, J., Ejadi, S., Lim, H. Y., et al. (2020). Phase 1 study of MRX34, a liposomal miR-34a mimic, in patients with advanced solid tumours. *British Journal of Cancer*, *122*, 1630–1637.

Hoshida, Y., Nijman, S. M., Kobayashi, M., Chan, J. A., Brunet, J. P., Chiang, D. Y., et al. (2009). Integrative transcriptome analysis reveals common molecular subclasses of human hepatocellular carcinoma. *Cancer Research*, *69*, 7385–7392.

Hou, J., Lin, L., Zhou, W., Wang, Z., Ding, G., Dong, Q., et al. (2011). Identification of miRNomes in human liver and hepatocellular carcinoma reveals miR-199a/b-3p as therapeutic target for hepatocellular carcinoma. *Cancer Cell*, *19*, 232–243.

Hsu, S. H., Wang, B., Kota, J., Yu, J., Costinean, S., Kutay, H., et al. (2012). Essential metabolic, anti-inflammatory, and anti tumorigenic functions of miR-122 in liver. *The Journal of Clinical Investigation*, *122*, 2871–2883.

Huang, J. T., Liu, S. M., Ma, H., Yang, Y., Zhang, X., Sun, H., et al. (2016). Systematic review and meta-analysis: Circulating miRNAs for diagnosis of hepatocellular carcinoma. *Journal of Cellular Physiology*, *231*, 328–335.

Hung, C. L., Yen, C. S., Tsai, H. W., Su, Y. C., & Yen, C. J. (2015). Upregulation of MicroRNA-19b predicts good prognosis in patients with hepatocellular carcinoma presenting with vascular invasion or multifocal disease. *BMC Cancer*, *15*, 665.

Ichimi, T., Enokida, H., Okuno, Y., Kunimoto, R., Chiyomaru, T., Kawamoto, K., et al. (2009). Identification of novel microRNA targets based on microRNA signatures in bladder cancer. *International Journal of Cancer*, *125*, 345–352.

Iorio, M. V., Ferracin, M., Liu, C. G., Veronese, A., Spizzo, R., Sabbioni, S., et al. (2005). MicroRNA gene expression deregulation in human breast cancer. *Cancer Research*, *65*, 7065–7070.

Iorio, M. V., Visone, R., Di Leva, G., Donati, V., Petrocca, F., Casalini, P., et al. (2007). MicroRNA signatures in human ovarian cancer. *Cancer Research*, *67*, 8699–8707.

Janssen, H. L., Reesink, H. W., Lawitz, E. J., Zeuzem, S., Rodriguez-Torres, M., Patel, K., et al. (2013). Treatment of HCV infection by targeting microRNA. *The New England Journal of Medicine*, *368*, 1685–1694.

Jopling, C. L., Yi, M., Lancaster, A. M., Lemon, S. M., & Sarnow, P. (2005). Modulation of hepatitis C virus RNA abundance by a liver-specific MicroRNA. *Science*, *309*, 1577–1581.

Kim, S., Lee, U. J., Kim, M. N., Lee, E. J., Kim, J. Y., Lee, M. Y., et al. (2008). MicroRNA miR-199a* regulates the MET proto-oncogene and the downstream extracellular signal-regulated kinase 2 (ERK2). *The Journal of Biological Chemistry*, *283*, 18158–18166.

Kogure, T., Lin, W. L., Yan, I. K., Braconi, C., & Patel, T. (2011). Intercellular nanovesicle-mediated microRNA transfer: A mechanism of environmental modulation of hepatocellular cancer cell growth. *Hepatology*, *54*, 1237–1248.

Kojima, K., Takata, A., Vadnais, C., Otsuka, M., Yoshikawa, T., Akanuma, M., et al. (2011). MicroRNA122 is a key regulator of alpha-fetoprotein expression and influences the aggressiveness of hepatocellular carcinoma. *Nature Communications*, *2*, 338.

Kota, J., Chivukula, R. R., O'Donnell, K. A., Wentzel, E. A., Montgomery, C. L., Hwang, H. W., et al. (2009). Therapeutic microRNA delivery suppresses tumorigenesis in a murine liver cancer model. *Cell*, *137*, 1005–1017.

Krutzfeldt, J., Rajewsky, N., Braich, R., Rajeev, K. G., Tuschl, T., Manoharan, M., et al. (2005). Silencing of microRNAs in vivo with 'antagomirs'. *Nature*, *438*, 685–689.

Kutay, H., Bai, S., Datta, J., Motiwala, T., Pogribny, I., Frankel, W., et al. (2006). Downregulation of miR-122 in the rodent and human hepatocellular carcinomas. *Journal of Cellular Biochemistry*, *99*, 671–678.

Le Guen, C. L., Friedman, J. R., & Hand, N. J. (2013). Novel targets of miR-30, a microRNA required for biliary development. *F1000Res*, *2*, 197.

Li, D., Sedano, S., Allen, R., Gong, J., Cho, M., & Sharma, S. (2019). Current treatment landscape for advanced hepatocellular carcinoma: Patient outcomes and the impact on quality of life. *Cancers*, *11*(6).

Lim, L., Balakrishnan, A., Huskey, N., Jones, K. D., Jodari, M., Ng, R., et al. (2014). MicroRNA-494 within an oncogenic microRNA megacluster regulates G1/S transition in liver tumorigenesis through suppression of mutated in colorectal cancer. *Hepatology*, *59*, 202–215.

Liu, Z., Tu, K., & Liu, Q. (2014). Effects of microRNA-30a on migration, invasion and prognosis of hepatocellular carcinoma. *FEBS Letters*, *588*, 3089–3097.

Lu, J., Getz, G., Miska, E. A., Alvarez-Saavedra, E., Lamb, J., Peck, D., et al. (2005). MicroRNA expression profiles classify human cancers. *Nature*, *435*, 834–838.

Luk, J. M., Burchard, J., Zhang, C., Liu, A. M., Wong, K. F., Shek, F. H., et al. (2011). DLK1-DIO3 genomic imprinted microRNA cluster at 14q32.2 defines a stemlike subtype of hepatocellular carcinoma associated with poor survival. *The Journal of Biological Chemistry*, *286*, 30706–30713.

Luo, Z., Feng, C., Hu, P., Chen, Y., He, X. F., Li, Y., et al. (2016). Serum microRNA-199a/b-3p as a predictive biomarker for treatment response in patients with hepatocellular carcinoma undergoing transarterial chemoembolization. *Oncotargets and Therapy*, *9*, 2667–2674.

Lupini, L., Pepe, F., Ferracin, M., Braconi, C., Callegari, E., Pagotto, S., et al. (2016). Over-expression of the miR-483-3p overcomes the miR-145/TP53 pro-apoptotic loop in hepatocellular carcinoma. *Oncotarget*, *7*, 31361–31371.

Moshiri, F., Salvi, A., Gramantieri, L., Sangiovanni, A., Guerriero, P., De Petro, G., et al. (2018). Circulating miR-106b-3p, miR-101-3p and miR-1246 as diagnostic biomarkers of hepatocellular carcinoma. *Oncotarget*, *9*, 15350–15364.

Murakami, Y., Yasuda, T., Saigo, K., Urashima, T., Toyoda, H., Okanoue, T., et al. (2006). Comprehensive analysis of microRNA expression patterns in hepatocellular carcinoma and non-tumorous tissues. *Oncogene*, *25*, 2537–2545.

Ng, K. T., Lo, C. M., Wong, N., Li, C. X., Qi, X., Liu, X. B., et al. (2016). Early-phase circulating miRNAs predict tumor recurrence and survival of hepatocellular carcinoma patients after liver transplantation. *Oncotarget*, *7*, 19824–19839.

Nishida, N., Arizumi, T., Hagiwara, S., Ida, H., Sakurai, T., & Kudo, M. (2017). MicroRNAs for the prediction of early response to sorafenib treatment in human hepatocellular carcinoma. *Liver Cancer*, *6*, 113–125.

Niu, C., Wang, L., Ye, W., Guo, S., Bao, X., Wang, Y., et al. (2020). CCAT2 contributes to hepatocellular carcinoma progression via inhibiting miR-145 maturation to induce MDM2 expression. *Journal of Cellular Physiology*, *235*, 6307–6320.

O'Donnell, K. A., Wentzel, E. A., Zeller, K. I., Dang, C. V., & Mendell, J. T. (2005). c-Myc-regulated microRNAs modulate E2F1 expression. *Nature*, *435*, 839–843.

Pan, C., Wang, X., Shi, K., Zheng, Y., Li, J., Chen, Y., et al. (2016). MiR-122 reverses the doxorubicin-resistance in hepatocellular carcinoma cells through regulating the tumor metabolism. *PLoS One*, *11*, e0152090.

Park, J. K., Kogure, T., Nuovo, G. J., Jiang, J., He, L., Kim, J. H., et al. (2011). miR-221 silencing blocks hepatocellular carcinoma and promotes survival. *Cancer Research*, *71*, 7608–7616.

Petrick, J. L., & McGlynn, K. A. (2019). The changing epidemiology of primary liver cancer. *Current Epidemiology Reports*, *6*, 104–111.

Pineau, P., Volinia, S., McJunkin, K., Marchio, A., Battiston, C., Terris, B., et al. (2010). miR-221 overexpression contributes to liver tumorigenesis. *PNAS*, *107*, 264–269.

Pollutri, D., Patrizi, C., Marinelli, S., Giovannini, C., Trombetta, E., Giannone, F. A., et al. (2018). The epigenetically regulated miR-494 is associated with stem-cell phenotype and induces sorafenib resistance in hepatocellular carcinoma. *Cell Death & Disease*, *9*, 4.

Rui, T., Xu, S., Zhang, X., Huang, H., Feng, S., Zhan, S., et al. (2020). The chromosome 19 microRNA cluster, regulated by promoter hypomethylation, is associated with tumour burden and poor prognosis in patients with hepatocellular carcinoma. *Journal of Cellular Physiology*, *235*, 6103–6112.

Sachdeva, M., Zhu, S., Wu, F., Wu, H., Walia, V., Kumar, S., et al. (2009). p53 represses c-Myc through induction of the tumor suppressor miR-145. *PNAS*, *106*, 3207–3212.

Sekine, S., Ogawa, R., Ito, R., Hiraoka, N., McManus, M. T., Kanai, Y., et al. (2009). Disruption of Dicer1 induces dysregulated fetal gene expression and promotes hepatocarcinogenesis. *Gastroenterology*, *136*, 2304–2315 (e1–4).

Sekine, S., Ogawa, R., McManus, M. T., Kanai, Y., & Hebrok, M. (2009). Dicer is required for proper liver zonation. *The Journal of Pathology*, *219*, 365–372.

Sidhu, K., Kapoor, N. R., Pandey, V., & Kumar, V. (2015). The "macro" world of microRNAs in hepatocellular carcinoma. *Frontiers in Oncology*, *5*, 68.

Simerzin, A., Zorde-Khvalevsky, E., Rivkin, M., Adar, R., Zucman-Rossi, J., Couchy, G., et al. (2016). The liver-specific microRNA-122*, the complementary strand of microRNA-122, acts as a tumor suppressor by modulating the p53/mouse double minute 2 homolog circuitry. *Hepatology*, *64*, 1623–1636.

Teufel, M., Seidel, H., Kochert, K., Meinhardt, G., Finn, R. S., Llovet, J. M., et al. (2019). Biomarkers associated with response to regorafenib in patients with hepatocellular carcinoma. *Gastroenterology*, *156*, 1731–1741.

Toffanin, S., Hoshida, Y., Lachenmayer, A., Villanueva, A., Cabellos, L., Minguez, B., et al. (2011). MicroRNA-based classification of hepatocellular carcinoma and oncogenic role of miR-517a. *Gastroenterology*, *140*, 1618–1628 (e16).

Torre, L. A., Bray, F., Siegel, R. L., Ferlay, J., Lortet-Tieulent, J., & Jemal, A. (2015). Global cancer statistics, 2012. *CA: A Cancer Journal for Clinicians*, *65*, 87–108.

Ura, S., Honda, M., Yamashita, T., Ueda, T., Takatori, H., Nishino, R., et al. (2009). Differential microRNA expression between hepatitis B and hepatitis C leading disease progression to hepatocellular carcinoma. *Hepatology*, *49*, 1098–1112.

Valdmanis, P. N., Kim, H. K., Chu, K., Zhang, F., Xu, J., Munding, E. M., et al. (2018). miR-122 removal in the liver activates imprinted microRNAs and enables more effective microRNA-mediated gene repression. *Nature Communications*, *9*, 5321.

Varshney, A., Panda, J. J., Singh, A. K., Yadav, N., Bihari, C., Biswas, S., et al. (2018). Targeted delivery of microRNA-199a-3p using self-assembled dipeptide nanoparticles efficiently reduces hepatocellular carcinoma in mice. *Hepatology*, *67*, 1392–1407.

Wang, S., Qiu, L., Yan, X., Jin, W., Wang, Y., Chen, L., et al. (2012). Loss of microRNA 122 expression in patients with hepatitis B enhances hepatitis B virus replication through cyclin G(1) -modulated P53 activity. *Hepatology*, *55*, 730–741.

Wang, P. R., Xu, M., Toffanin, S., Li, Y., Llovet, J. M., & Russell, D. W. (2012). Induction of hepatocellular carcinoma by in vivo gene targeting. *PNAS*, *109*, 11264–11269.

Wong, C. C., Wong, C. M., Tung, E. K., Au, S. L., Lee, J. M., Poon, R. T., et al. (2011). The microRNA miR-139 suppresses metastasis and progression of hepatocellular carcinoma by down-regulating rho-kinase 2. *Gastroenterology*, *140*, 322–331.

Xiong, Y., Fang, J. H., Yun, J. P., Yang, J., Zhang, Y., Jia, W. H., et al. (2010). Effects of microRNA-29 on apoptosis, tumorigenicity, and prognosis of hepatocellular carcinoma. *Hepatology*, *51*, 836–845.

Xu, J., Wu, C., Che, X., Wang, L., Yu, D., Zhang, T., et al. (2011). Circulating microRNAs, miR-21, miR-122, and miR-223, in patients with hepatocellular carcinoma or chronic hepatitis. *Molecular Carcinogenesis*, *50*, 136–142.

Yuan, J. H., Yang, F., Chen, B. F., Lu, Z., Huo, X. S., Zhou, W. P., et al. (2011). The histone deacetylase 4/SP1/microrna-200a regulatory network contributes to aberrant histone acetylation in hepatocellular carcinoma. *Hepatology*, *54*, 2025–2035.

Yuan, J. H., Yang, F., Wang, F., Ma, J. Z., Guo, Y. J., Tao, Q. F., et al. (2014). A long noncoding RNA activated by TGF-beta promotes the invasion-metastasis cascade in hepatocellular carcinoma. *Cancer Cell*, *25*, 666–681.

Zeng, C., Wang, R., Li, D., Lin, X. J., Wei, Q. K., Yuan, Y., et al. (2010). A novel GSK-3 beta-C/EBP alpha-miR-122-insulin-like growth factor 1 receptor regulatory circuitry in human hepatocellular carcinoma. *Hepatology*, *52*, 1702–1712.

Zhang, L. F., Lou, J. T., Lu, M. H., Gao, C., Zhao, S., Li, B., et al. (2015). Suppression of miR-199a maturation by HuR is crucial for hypoxia-induced glycolytic switch in hepatocellular carcinoma. *The EMBO Journal*, *34*(21), 2671–2685.

Zheng, F., Liao, Y. J., Cai, M. Y., Liu, Y. H., Liu, T. H., Chen, S. P., et al. (2012). The putative tumour suppressor microRNA-124 modulates hepatocellular carcinoma cell aggressiveness by repressing ROCK2 and EZH2. *Gut*, *61*, 278–289.

Chapter 24

Pathophysiology roles and translational opportunities of miRNAs in ovarian cancer

Ben Davidson[a,b] and Reuven Reich[c,*]

[a]*Department of Pathology, Oslo University Hospital, Norwegian Radium Hospital, Oslo, Norway,* [b]*University of Oslo, Faculty of Medicine, Institute of Clinical Medicine, Oslo, Norway,* [c]*Institute of Drug Research, School of Pharmacy, Faculty of Medicine, The Hebrew University of Jerusalem, Jerusalem, Israel*

Introduction

Ovarian cancer (OC), the most aggressive gynecologic cancer, was predicted to be diagnosed in 21,750 women and lead to 13,940 fatalities in the United States in 2020 (Siegel, Miller, & Jemal, 2020). Globally, this cancer was diagnosed in 295,414 women and resulted in 184,799 deaths in 2018, making it the eighth most common and eighth most lethal cancer in women (Bray et al., 2018). Novel targeted therapies, such as PARP inhibitors, have in recent years extended survival, particularly of patients carrying mutations in *BRCA* and related DNA repair genes. However, prognosis remains dismal for the majority of patients, particularly when the disease is diagnosed at an advanced stage (Lheureux, Braunstein, & Oza, 2019).

The majority of OC are carcinomas. These constitute a heterogeneous group of tumors that differ in their morphology, immunohistochemistry (IHC) profile, and genetic make-up. The main histotypes are high-grade serous carcinoma (HGSC), low-grade serous carcinoma (LGSC), clear cell carcinoma (CCC), endometrioid carcinoma (EC), and mucinous carcinoma (MC). Despite being labeled "ovarian carcinoma," the majority of HGSC originate from the distal (fimbrial) end of the fallopian tube, whereas CCC and EC are endometriosis-related tumors that do not have a true precursor in the ovary (reviewed in Prat, D'Angelo, & Espinosa, 2018).

In view of the aggressive clinical course of advanced-stage OC, early diagnosis is critical for curing this disease. However, studies in which the diagnostic role of miRNA signatures in tissue specimens or in body fluids (blood, urine, serous effusions) has been assessed have not yielded a reproducible panel for the early diagnosis of this cancer. It is very likely that the main reason for this disappointing outcome is the fact that a considerable number of these studies have included ovarian carcinomas of different histotypes, in some studies without a pathology review, resulting in little consensus. The same has been true for studies focusing on the biological and clinical role of microRNAs (miRNAs) in OC (reviewed in Katz, Tropé, Reich, & Davidson, 2015; Mandilaras, Vernon, Meryet-Figuière, et al., 2017).

In recent years, an increasing number of studies focusing on the role of miRNAs in OC has analyzed only one histotype, often HGSC, which constitute 70% of ovarian carcinomas, thereby generating data that are more likely to be reproducible. This chapter focuses on publications from the years 2017–2020 that present novel findings related to the pathophysiology and clinical relevance of miRNAs in OC. Studies in which clinical specimens have been studied without reclassification based on the 2014 World Health Organization (WHO) book guidelines and studies based on public datasets with no such review are not cited or discussed in this chapter.

Pathophysiology

Several recent studies have reported novel data with respect to the role of miRNAs in the pathophysiology of OC, some of which additionally assessed the clinical relevance of the findings in clinical specimens.

The majority, although not all HGSCs, have been shown to develop from precursor lesions termed "serous tubal intraepithelial carcinoma" (STIC) in the fallopian tube (reviewed in Prat et al., 2018). Ducie et al. compared the molecular features of HGSC associated with STIC to tumors in which no STIC was found. miRNA profiling data were available for

*R.R. is affiliated with the David R. Bloom Center for Pharmacy and the Adolf and Klara Brettler Center for Research in Molecular Pharmacology and Therapeutics at The Hebrew University of Jerusalem, Israel.

MicroRNA in Human Malignancies. https://doi.org/10.1016/B978-0-12-822287-4.00004-9

95 of the 96 cases studied. Twenty-four miRNAs were differentially expressed in these two groups. However, only three of the latter (miR-21, miR-22, and miR-25) were reported to be highly expressed in HGSC, suggesting that differences were generally of limited significance (Ducie, Dao, Considine, et al., 2017).

To et al. compared highly metastatic (HM) and nonmetastatic (NM) clones of HEY-A8 ovarian cancer cells. HM cells had a higher expression of cancer stem cell (CSC) markers and formed spheroids more efficiently. Proteome analysis identified β-catenin as an overexpressed protein in the HM group, with activation of its network in pathway analysis. β-Catenin silencing by siRNA in HM clones resulted in increased mRNA and protein levels of Dicer, involved in miRNA processing. This resulted in reduced migration as well as upregulation of 142 miRNAs, including the miR-29 family members miR-29a-5p, 29b-3p, and 29c-3p. Treatment of HM cells by miR-29 mimics suppressed migration and reduced the levels of the epithelial-to-mesenchymal transition (EMT)-related molecules Snail and N-cadherin. β-catenin silencing by shRNA suppressed metastasis in vivo. The β-catenin signature was associated with the mesenchymal molecular type of HGSC in an analysis of the TCGA database (To et al., 2017).

Givel et al. studied the role of cancer-associated fibroblasts (CAF) in OC, focusing on the molecularly defined mesenchymal subtype reported to be clinically aggressive by Tothill, Tinker, George, et al. (2008). The authors performed a flow cytometry analysis of fresh specimens, applying the fibroblast markers FAP (fibroblast activation protein), CD29 (β1 integrin), SMA (α-smooth muscle actin), FSP1 (fibroblast-specific protein 1), PDGFRβ (platelet-derived growth factor receptor-β), and caveolin as well as EpCAM, CD45, and CD31 as markers of epithelial cells, leukocytes, and endothelial cells, respectively. Four classes of CAF, termed CAF-S1 to CAF-S4, were defined based on CD29, FAP, FSP1, and SMA expression levels. The CAF-S1 and CAF-S4 groups were regarded as activated, CAF-S2 and CAF-S3 as nonactivated, and the former were often detected in HGSC specimens in a validation series using IHC and immunofluorescence, with CAF-S1 predominating in the mesenchymal subtype. HGSC enriched with CAF-S1 had higher counts of immune-suppressing FOXP3+ T lymphocytes. RNA sequencing comparing the CAF-S1 and CAF-S4 groups showed the overexpression of genes encoding extracellular matrix (ECM) components and proteins involved in immune regulation, including the chemokine CXCL12β, and the latter was shown to regulate T-cell infiltration in the CAF-S1 group by attracting CD4+CD25+T lymphocytes. In the CAF-S4 fibroblasts, the downregulation of *CXCL12β* mRNA was shown to be mediated by miR-141 and miR-200a, which thereby appear to affect the fibroblast class and immune system activation in HGSC (Givel, Kieffer, Scholer-Dahirel, et al., 2018).

Wang et al. investigated the role of miRNAs in regulating the mTOR signaling pathway in HGSC. The tumor-suppressor tuberous sclerosis complex 1 (TSC1) was underexpressed in HGSC compared to the fallopian tube epithelium and lower TSC1 levels were associated with a more advanced stage. miR-130a suppressed TSC1 expression in vitro and its levels, as opposed to TSC1, were higher in HGSC compared to the fallopian tube epithelium. Overexpression of miR-130a in A2780 cells increased colony formation, anchorage-independent growth, invasion, proliferation, and EMT, whereas the reexpression of TSC1 suppressed colony formation and proliferation. The overexpression of miR-130a additionally suppressed autophagy due to rapamycin treatment or starvation. miR-130a was positively regulated by NF-κB. Finally, the overexpression of miR-130a promoted tumor growth in vivo in nude mice (Wang, Zhang, Tang, et al., 2017).

Gharpure and coworkers analyzed the role of fatty acid binding protein 4 (FABP4) in OC. FABP4 silencing by siRNA in HeyA8 MDR cells suppressed migration and invasion, with an opposite effect observed upon FABP4 reexpression. Comparable results were seen in OVCAR-5 cells. A higher tumor burden, a higher number of metastases, and more invasive growth were seen in A2780-ip1 cells treated with the FABP4-expressing vector compared to those with the control vector. An analysis of the TCGA dataset identified miR-409-3p as the negative regulator of FABP4, and levels of this miRNA were lower in OC cell lines compared to nontransformed ovarian epithelial cells (HIO180). An inverse association was observed between miR-409-3p and FABP4 levels, and shown mechanistically to result from suppression of the latter by the former. miR-409-3p overexpression in OC cells suppressed migration and invasion in vitro and tumor growth in vivo. Hypoxia and treatment by bevacizumab, the latter also inducing hypoxia, resulted in reduced miR-409-3p and increased FABP4 levels. FABP4 further affected lipid metabolism and higher FABP4 expression was significantly associated with poor survival. Finally, tamoxifen was shown to negatively regulate FABP4 and migration (Gharpure, Pradeep, Sans, et al., 2018).

Chen et al. studied the role of miRNAs in regulating angiogenesis. An analysis of an HGSC dataset identified six miRNAs (miR-125b-5p, miR-199a-5p, miR-218-5p, miR-199b-5p, miR-204-5p, miR-497-5p) that were overexpressed in tumors with high microvessel counts compared with those with low counts. In an analysis of the TCGA dataset, higher miR-218-5p levels were associated with poor disease-free survival (DFS), whereas high miR-204-5p levels were significantly related to both poor DFS and overall survival (OS). miR-204-5p silencing in endothelial cells did not affect tube formation. However, significantly reduced tube formation and migration were seen in endothelial cells cultured with a medium from HeyA8-MDR cells in which miR-204-5p was silenced compared to controls. The opposite effect was observed when endothelial cells were exposed to a medium from SKOV3ip1 cells in which miR-204-5p was overexpressed. The effect of miR-204-5p was studied in vivo using the rHDL nanovector as a delivery tool. In this experiment, miR-204-5p inhibition, alone or in combination with paclitaxel, suppressed tumor growth. A gene expression microarray of HeyA8-MDR cells that underwent miR-204-5p inhibition identified three genes—thioredoxin interacting protein (TXNIP), pentraxin 3 (PTX3), and thrombospondin 1 (THBS1)—with antiangiogenic effects that were overexpressed following miR-204-5p inhibition, a finding confirmed by qRT-PCR in OC cell lines as well as at the protein level for THBS1. THBS1 silencing in HeyA8-MDR cells increased tube formation in endothelial cells, confirming its antiangiogenesis role (Chen, Mangala, Mooberry, et al., 2019).

Muys and coworkers performed RNA sequencing comparing A2780 OC cells with the stable expression of miR-450a via lentiviral transduction to control cells. Among the 100 most downregulated genes were several genes associated with EMT, including *VIM*, *TWIST1,* and *SERPINE1*, whereas epithelial genes such as *CDH1* were overexpressed, data validated by qPCR. Overexpression of miR-450a suppressed migration and invasion and increased anoikis in A2780 and SKOV3 OC cells, reduced colony formation in A2780 cells, and inhibited tumor growth in vivo in an NOD/SCID mouse model. Similar effects were found for miR-450b. Among miR-450a targets were several mRNAs encoding mitochondrial genes, including *ACO2*, *TIMMDC1*, *ATP5B,* and *MT-ND2*. *ACO2* and *ATP5B* levels were inversely related to the miR-450a level in an analysis of the TCGA dataset, and higher *TIMMDC1* levels were associated with shorter DFS. In mitochondria, miR-450a modulated glutaminolysis and glucose metabolism, with an overall role as a tumor suppressor (Muys, Sousa, Plaça, et al., 2019).

The study by Srivastava and coworkers focused on the role of miR-328-3p in OC. A comparison of the Kuramochi and SKOV3 cell lines, cultured as spheres to enhance CSC characteristics, with their corresponding bulk cells identified 12 upregulated and five downregulated miRNAs in cancer stem-like cells (CSLC), of which miR-328-3p was the most increased. High expression of this miRNA was also observed in CSLC from clinical HGSC specimens and cell lines of other cancers. miR-328-3p inhibition in OV2008 and OVCAR4 OC cells reduced the $ALDH^+$ cell subpopulation as well as the sphere formation. In OV2008 and OVCAR4 cells cultured as spheroids, miR-328-3p inhibition suppressed the mRNA expression of the CSC markers *SOX2*, *OCT4,* and *NANOG*. Overexpression of miR-328-3p in OV2008 cells led to the downregulation of 51 genes, among them *DDB2*, encoding DNA damage binding protein 2, previously shown by the authors to reduce CSC abundance. DDB2 downregulation restored spheroid formation following suppression by miR-328-3p inhibition. Reduced ERK1/2 phosphorylation in CSLC was associated with an increased miR-328-3p level, and the functional link between these observations was shown using the ERK inhibitor U0126 and the ERK activator phorbol 12-myristate 13-acetate (PMA). ERK and miR-328-3p levels regulated CSLC characteristics in OV2008 and OVCAR4 cells. Depletion of reactive oxygen species (ROS) resulted in reduced ERK phosphorylation and an increase in miR-328-3p levels, with the opposite effect following treatment with H_2O_2 and the latter effect abrogated by treatment with U0126. miR-328-3p inhibition further suppressed tumor growth in vivo in nude mice (Srivastava, Banerjee, Cui, et al., 2019).

The majority of HGSCs have their origin in the fimbrial region of the fallopian tube, and a preinvasive lesion termed serous tubal intraepithelial carcinoma (STIC) can be identified in many of these cases. Knarr et al. studied the role of miR-181a in transforming the fallopian tube epithelium. The stable overexpression of miR-181a (pmiR-181a) in three immortalized cell lines derived from fallopian tube secretory epithelial cells (FTSECs) resulted in the loss of 2D contact inhibition, increased viability, clonogenicity, and the formation of anchorage-independent colonies as well as an increase in the G2/M subpopulation and in the fraction of cells with > 4 N content, reflecting aneuploidy and suggesting chromosomal instability. pmiR-181a cells additionally had an increase in structural nuclear defects, in the frequency of mitotic and cytokinetic (MITOC) defects, in gene signatures associated with DNA damage and genomic instability, and in DNA damage as measured by γH2AX foci. An analysis of copy number variation (CNV) showed increases in the loss of heterozygosity (LOH) and duplication events in pmiR-181a cells, providing further evidence of DNA damage. FTSECs with miR-181a overexpression formed tumors in vivo in 9/10 subcutaneous sites, compared to none for cells with miR-181a scramble control (pscram-miR) and one tumor for cells overexpressing both miR-181a and an miR-181a-specific antagomiR (pmiR-181a + antimiR), with a similar difference observed for intraperitoneal tumor nodule and ascite formation. Global mRNA transcriptome analysis of pmiR-181a, pscram-miR, and pmiR-181a + antimiR cells identified differentially expressed genes, including 25 that were functionally related to the cellular changes observed in miR-181a cells. The majority of these changes were experimentally shown to be related to silencing of *RB1*, encoding for the retinoblastoma protein. However, the downregulation of STING (stimulator-of-interferon-genes) was additionally necessary for the transformation of miR-181a cells. STING knockdown in miR-181a cells induced proliferation, clonogenicity, anchorage-independent growth, G2/M increase, aneuploidy, nuclear membrane ruptures, and γH2AX foci. Reversal of these changes as well as senescence and cell death were related to STING overexpression. Finally, in the TCGA dataset, miR-181a and STING levels were inversely related, and high miR-181a levels were associated with reduced infiltration of lymphocytes and M1 macrophages (Knarr, Avelar, Sekhar, et al., 2020).

Dasari et al. compared the miRNA profiles of primary and metastatic OC (n=42; 40 HGSC and 2 CCC) using next-generation mRNA seq. The authors additionally used an organotypic 3D culture of the omentum with the Kuramochi, OVCAR4, and OVCAR8 cell lines. Among 26 miRNAs downregulated in clinical metastases, only miR-4454 was additionally reduced in all three cell lines in the omentum model. Overexpression of miR-4454 reduced migration, proliferation, and colony formation in Kuramochi, OVCAR3, OVCAR4, and OVCAR8 cells compared to controls transfected with scrambled oligos, with an opposite effect following miR-4454 inhibition. OVCAR8 colony formation in the omental model was similarly reduced by miR-4454, as were tumor weight and metastasis formation in vivo. Coculturing of OC cells with normal omental fibroblasts or cancer-associated fibroblasts reduced miR-4454 levels. SPARC, BAG5, and SOX4 were identified as miR-4454 targets, and SPARC or BAG5 silencing in Kuramochi, OVCAR4, and OVCAR8 cells reduced migration, proliferation, and colony formation. Specifically, reduced BAG5 levels resulted in increased apoptosis and reduced SPARC levels suppressed FAK phosphorylation (Dasari, Pandhiri, Grassi, et al., 2020).

Extracellular miRNAs and prediction of treatment response

Several studies focusing on extracellular miRNAs and/or treatment response are discussed below. Some of these studies have also investigated pathophysiology, though the emphasis on clinical end points is more central.

Kanlikilicer et al. studied the role of exosomal miR-1246 in OC, based on a previous study that showed this miRNA to be the most upregulated one in OC exosomes compared to the cells of origin. miR-1246 was further overexpressed in OC cells compared to normal ovarian surface epithelium (OSE) as well as OC exosomes compared to exosomes from OSE. An analysis of 138 tumors from the TCGA OC cohort showed a significant association between high miR1246 levels and shorter OS. Transfection of SKOV3ip1 OC cells with an miR-1246 mimic affected signaling pathways, including PDGFR signaling, with a decrease in caveolin-1, a protein that inhibits PDGFRβ kinase activity. Caveolin-1 expression was associated with longer OS in the TCGA dataset. miR-1246 directly suppressed caveloin-1 through binding to its 3′ UTR. Caveolin-1 levels were lower in multidrug- and paclitaxel-resistant HeyA8-MDR and SKOV3-TR OC cells compared to the chemo-sensitive cell lines HeyA8 and SKOV3-ip1, and inversely related to miR-1246 levels in all four cell lines. Caveolin-1 silencing increased PDGFβ levels, thereby demonstrating that the effect of miR-1246 on PDGFβ was mediated via caveolin-1. A combination of paclitaxel and miR-1246 inhibition resulted in more effective suppression of tumor growth compared to paclitaxel alone in vivo, and miR-1246 inhibition was shown to reduce proliferation, increase caveolin-1 expression, and suppress PDGFRβ expression. miR-1246-induced resistance to paclitaxel was via reduced caveolin-1 and increased p-glycoprotein expression. In tumors that underwent treatment with a combination of paclitaxel and miR-1246 inhibition in the in vivo model, the number of M2 macrophages was lower, and these macrophages had lower levels of caveolin-1 and higher levels of p-glycoprotein compared to other macrophage classes, suggesting a role in paclitaxel resistance. miR-1246 levels were significantly higher when M2 macrophages were cocultured with SKOV3-ip1 OC cells compared to M2 macrophages cultured alone. M2-type macrophages treated with exosomes from SKOV3-ip1 cells had significantly lower caveolin-1 mRNA levels compared to controls, a difference that was not found when SKOV3-ip1 cells were treated with the exosome inhibitor GW4869 (Kanlikilicer, Bayraktar, Denizli, et al., 2018).

Alharbi and coworkers analyzed the miRNA profiles of extracellular vesicles (EV) and their parent cells as well as their relationship to carboplatin treatment. A comparison between chemo-sensitive OC cell lines with low metastatic capacity (termed low) and chemo-resistant and highly metastatic cell lines (termed high) identified 12 differentially expressed miRNAs (miR-142-5p, miR-891a-5p, miR-142-3p, miR-4284, miR-3919, miR-3152-5p, miR-3681-5p, let-7a-2-3p, miR-582-3p, miR-582-5p, miR-21-3p, and miR-21-5p) that were upregulated in the latter group. Carboplatin treatment of the high group cells increased the expression of miR-21-3p, miR-21-5p, miR-891a-5p, miR-142-5p, and miR-142-3p compared to non-treated controls, and increased miR-21-5p and miR-142-5p expression in the low group. A comparative analysis of miR-21-3p, miR-21-5p, miR-891a-5p, miR-142-5p, and miR-142-3p in EV vs. the parent OC cells identified miR-21-3p, miR-21-5p, and miR-891a-5p as overexpressed in the former in the high group, whereas miR-891a-5p, miR-142-5p, and miR-142-3p were overexpressed in EV compared to the parent cells in the low group. Overexpression of miR-21-5p, miR-21-3p, and miR-891-5p by mimics significantly increased cell viability in response to carboplatin in CAOV-3 cells. Higher miR-891-5p levels were detected in EVs isolated from the serum of OC patients who experienced disease recurrence compared to those with no recurrence in a small series totaling 17 patients (Alharbi, Sharma, Guanzon, et al., 2020).

The study by Rodriguez-Aguayo and coworkers focused on the interaction between the RNA-binding protein hnRNPA1 and the miRNAs miR-15a-5p and miR-25-3p. Reduced hnRNPA1 expression was found in SKOV3-TR and HeyA8-MDR chemo-resistant OC cells compared to their parental cells, though not in A2780-CP20 cells compared to the parental line. Target prediction models identified 42 targets of hnRNPA1, of which two (miR-25-3p and miR-15a-5p) were overexpressed in at least two chemo-resistant OC cell lines compared to the parental cells. A luciferase assay showed that hnRNPA1 is regulated by miR-25-3p and miR-15a-5p. Transfection of HeyA8 cells with miR-15a-5p, miR-25-3p, or both led to reduced hnRNPA1 mRNA and protein levels compared to cells transfected with a control mimic. siRNA silencing of hnRNPA1 increased KRAS expression, and the negative regulatory effect of hnRNPA1 on KRAS was mediated via miR-18a-3p. miR-25-3p, and miR-15a-5p inhibition in SKOV3-TR. HeyA8-MDR cells reduced proliferation, colony formation, migration, and invasion, whereas hnRNPA1 silencing increased the expression of MMPs (MMP-1 and MMP-9) and angiogenic factors (FGF2, VEGF C, and VEGF-D). A combination of docetaxel with anti-miR-15a-5p or anti-miR-25-3p suppressed tumor growth in vivo, an effect that was mediated by increased apoptosis, reduced proliferation, and reduced angiogenesis. In an analysis of the TCGA database, patients with tumors with high miR-18a-3p and low KRAS levels had significantly longer OS compared to those with the inverse association. Similarly, low levels of KRAS and either miR-25-3p or miR-15a-5p were significantly associated with longer OS, as was the combination of high miR-18a-3p, low KRAS levels, and low miR-25-3p levels compared to the inverse relationship (Rodriguez-Aguayo, Monroig, Redis, et al., 2017).

Hisamatsu et al. studied the RMUG-L-ip1 MC cell line using a kinome-based siRNA library screen with the objective of identifying candidate molecules involved in oxaliplatin resistance. The authors identified PRKRA, an activator of PKR kinase, as one of the top five genes whose silencing sensitized MC cells to oxaliplatin, and the combination of *PRKRA* siRNA and treatment with oxaliplatin reduced viability significantly more than control siRNA + oxaliplatin. The latter effect was mediated through reduced colony formation and increased apoptosis, as measured by caspase-3 cleavage, in RMUG-L-ip1 and RMUG-S-ip1 cells in vitro. The role of PRKRA in response to oxaliplatin was similarly demonstrated in the non-MC cell line A2780 by *PRKRA* silencing and by overexpression of its protein product PACT. Similarly, combination therapy with *PRKRA* siRNA and oxaliplatin resulted in lower tumor weight and fewer tumor nodules compared to control siRNA plus oxaliplatin in vivo through reduced proliferation and increased apoptosis. PACT was found to negatively regulate miR515-3p and miR-519-3p biogenesis through binding of Dicer in MC cells. Exposure to an miR-515 inhibitor increased oxaliplatin resistance in MC cells. Interrogation of the lists of downregulated genes in miR-5153p-overexpressing MC cells and *PRKRA*-silenced MC cells identified AXL as a potential molecule mediating chemosensitivity. Transfection with an miR-515 mimic reduced the AXL level, with a similar effect following *PRKRA* siRNA treatment, and a luciferase assay identified AXL as a target of miR-515-3p. AXL silencing reduced viability in MC cells (Hisamatsu, McGuire, Wu, et al., 2019).

Gunaratne et al. investigated mechanisms by which the effect of p53 loss through *TP53* mutation could be circumvented via activation of p63, another tumor suppressor. Overexpression of miR-130b, a downstream target of p63, resulted in reduced migration and invasion in some of the OC cell lines studied as well as impaired spheroid formation in a 3D model. Transient transfection of OC cells with miR-130b increased p53, p21, mouse double minute 2 homolog (MDM2), p63, phosphatase, and tensin homolog (PTEN), and RB1 transcript levels in HEYA8 cells, whereas only p63, MDM2, and PTEN levels were increased in OVCAR8 cells. The protein level of TAp63 (transactivation domain of tumor protein p63) was increased in both cell lines, as was the BCL2 inhibitor BIM. In HEYA8 cells, miR-130b further upregulated caspase-3, p21, p27, and cyclin-dependent kinase 6 (CDK6) and downregulated CDK4 and the p53-upregulated modulator of apoptosis (PUMA), whereas mutant p53 and CDK6 were upregulated and CDK4 and p21 downregulated in OVCAR8 cells. A luciferase assay showed p63 to be a target of miR-130b. Transfection with TAp63 reduced cell viability. In an analysis of the TCGA database, patients who had tumors with high miR-130b levels had significantly longer overall survival compared to those with lower miR-130b levels. Mice injected with OVCAR8 cells that were treated with 1,2-dioleoyl-sn-glycero-3-phosphatidylcholine (DOPC)-miR-130b showed tumor regression and longer survival compared to controls, with the most pronounced effect observed in a combination of the two treatments. A comparable effect was seen in vivo in a HEYA8 model using the scL nanocomplex as the carrier for miR-130b (Gunaratne, Pan, Rao, et al., 2019).

Zhao and coworkers investigated the role of losartan, an angiotensin system inhibitor that blocks angiotensin II receptor type 1 (AT1), in treating OC. Losartan treatment significantly reduced collagen and hyaluronan levels as well as the number of intratumoral αSMA-positive stromal cells, and the expression of collagens I III, αSMA, and integrin chains β3 and β6 in tumors formed by SKOV3ip1 and Hey-A8 cells in a mouse orthotopic model. Losartan treatment additionally reduced hypoxia and improved the delivery of doxorubicin to the tumor by improving perfusion. Combining losartan with paclitaxel in the above mouse model increased the response to paclitaxel, with reduced ascite volume, reduced tumor cell proliferation, and increased apoptosis. Losartan treatment preserved the normal form of the lymph vessels and improved lymph flow, as opposed to their collapse in control mice, which induces ascites accumulation. An miRNA array showed significant changes in the levels of several miRNAs following treatment with losartan, including miR-133a-3p (miR-133), which potentially targets the collagen genes *COL1A1*, *COL5A3,* and *COL6A3*. Overexpression of miR-133 did not change *COL1A1* mRNA levels, but it significantly reduced collagen I protein levels. The binding of miR-133 to *COL1A1* was confirmed by a luciferase assay. However, miR-133 overexpression in SKOV3ip1 cells in a peritoneal mouse model did not significantly affect tumor growth, ascites formation, or collagen I expression (Zhao, Cao, Melamed, et al., 2019).

Concluding remarks

RNA, including noncoding RNA, is widely recognized to have a central role in regulating physiological and pathological roles, and miRNAs are major effectors of these regulatory networks (Goodall & Wickramasinghe, 2021). OC research has in recent years continued to improve our understanding of the pathophysiological effects of miRNAs on tumor progression and chemotherapy response, thereby identifying potential therapeutic targets for intervention. It should nevertheless be noted that no miRNA is at present widely accepted as a diagnostic, predictive, or prognostic marker that can be used in the clinical management of OC, and that targeting of miRNAs has yet to be implemented as a treatment modality in clinical practice, despite considerable improvement in delivery models. This transition into a clinically relevant application requires analyses of well-characterized patient material of uniform histology, preferably in multiinstitutional studies, using tools such as patient-derived xenografts, which are more true to the biology of the clinical tumor. It is hoped that this may be achieved in the near future.

References

Alharbi, M., Sharma, S., Guanzon, D., et al. (2020). miRNa signature in small extracellular vesicles and their association with platinum resistance and cancer recurrence in ovarian cancer. *Nanomedicine*, *28*, 102207.

Bray, F., Ferlay, J., Soerjomataram, I., Siegel, R. L., Torre, L. A., Jemal, A., et al. (2018). GLOBOCAN estimates of incidence and mortality worldwide for 36 cancers in 185 countries. *CA: A Cancer Journal for Clinicians*, *68*(2018), 394–424.

Chen, X., Mangala, L. S., Mooberry, L., et al. (2019). Identifying and targeting angiogenesis-related microRNAs in ovarian cancer. *Oncogene*, *38*(33), 6095–6108.

Dasari, S., Pandhiri, T., Grassi, T., et al. (2020). Signals from the metastatic niche regulate early and advanced ovarian cancer metastasis through miR-4454 downregulation. *Molecular Cancer Research*, *18*(8), 1202–1217.

Ducie, J., Dao, F., Considine, M., et al. (2017). Molecular analysis of high-grade serous ovarian carcinoma with and without associated serous tubal intra-epithelial carcinoma. *Nature Communications*, *8*(1), 990.

Gharpure, K. M., Pradeep, S., Sans, M., et al. (2018). FABP4 as a key determinant of metastatic potential of ovarian cancer. *Nature Communications*, *9*(1), 2923.

Givel, A. M., Kieffer, Y., Scholer-Dahirel, A., et al. (2018). miR200-regulated CXCL12β promotes fibroblast heterogeneity and immunosuppression in ovarian cancers. *Nature Communications*, *9*(1), 1056.

Goodall, G. J., & Wickramasinghe, V. O. (2021). RNA in cancer. *Nature Reviews Cancer*, *21*(1), 22–36. https://doi.org/10.1038/s41568-020-00306-0.

Gunaratne, P. H., Pan, Y., Rao, A. K., et al. (2019). Activating p53 family member TAp63: A novel therapeutic strategy for targeting p53-altered tumors. *Cancer*, *125*(14), 2409–2422.

Hisamatsu, T., McGuire, M., Wu, S. Y., et al. (2019). *PRKRA*/PACT expression promotes chemoresistance of mucinous ovarian cancer. *Molecular Cancer Therapeutics*, *18*(1), 162–172.

Kanlikilicer, P., Bayraktar, R., Denizli, M., et al. (2018). Exosomal miRNA confers chemo resistance via targeting Cav1/p-gp/M2-type macrophage axis in ovarian cancer [published correction appears in EBioMedicine. 2020 Feb;52:102630]. *eBioMedicine*, *38*, 100–112.

Katz, B., Tropé, C. G., Reich, R., & Davidson, B. (2015). MicroRNAs in ovarian cancer. *Human Pathology*, *46*(9), 1245–1256.

Knarr, M., Avelar, R. A., Sekhar, S. C., et al. (2020). miR-181a initiates and perpetuates oncogenic transformation through the regulation of innate immune signaling. *Nature Communications*, *11*(1), 3231.

Lheureux, S., Braunstein, M., & Oza, A. M. (2019). Epithelial ovarian cancer: Evolution of management in the era of precision medicine. *CA: A Cancer Journal for Clinicians*, *69*, 280–304.

Mandilaras, V., Vernon, M., Meryet-Figuière, M., et al. (2017). Updates and current challenges in microRNA research for personalized medicine in ovarian cancer. *Expert Opinion on Biological Therapy*, *17*(8), 927–943. https://doi.org/10.1080/14712598.2017.1340935.

Muys, B. R., Sousa, J. F., Plaça, J. R., et al. (2019). miR-450a acts as a tumor suppressor in ovarian cancer by regulating energy metabolism. *Cancer Research*, *79*(13), 3294–3305.

Prat, J., D'Angelo, E., & Espinosa, I. (2018). Ovarian carcinomas: At least five different diseases with distinct histological features and molecular genetics. *Human Pathology*, *80*, 11–27.

Rodriguez-Aguayo, C., Monroig, P. D. C., Redis, R. S., et al. (2017). Regulation of hnRNPA1 by microRNAs controls the miR-18a-*K-RAS* axis in chemotherapy-resistant ovarian cancer. *Cell Discovery*, *3*, 17029 (Published 2017 Sep 12) https://doi.org/10.1038/celldisc.2017.29.

Siegel, R. L., Miller, K. D., & Jemal, A. (2020). Cancer statistics, 2020. *CA: A Cancer Journal for Clinicians*, *70*(1), 7–30.

Srivastava, A. K., Banerjee, A., Cui, T., et al. (2019). Inhibition of miR-328-3p impairs cancer stem cell function and prevents metastasis in ovarian cancer. *Cancer Research*, *79*(9), 2314–2326.

SKY, T., Mak, A. S. C., Eva Fung, Y. M., et al. (2017). β-Catenin downregulates dicer to promote ovarian cancer metastasis. *Oncogene*, *36*(43), 5927–5938. https://doi.org/10.1038/onc.2017.185.

Tothill, R. W., Tinker, A. V., George, J., et al. (2008). Novel molecular subtypes of serous and endometrioid ovarian cancer linked to clinical outcome. *Clinical Cancer Research*, *14*(16), 5198–5208.

Wang, Y., Zhang, X., Tang, W., et al. (2017). miR-130a upregulates mTOR pathway by targeting TSC1 and is transactivated by NF-κB in high-grade serous ovarian carcinoma. *Cell Death and Differentiation*, *24*(12), 2089–2100.

Zhao, Y., Cao, J., Melamed, A., et al. (2019). Losartan treatment enhances chemotherapy efficacy and reduces ascites in ovarian cancer models by normalizing the tumor stroma. *Proceedings of the National Academy of Sciences of the United States of America*, *116*(6), 2210–2219.

Chapter 25

Pathophysiology roles and translational opportunities of miRNAs in bladder cancer

Barbara Pardini[a,b] and Giuseppe Matullo[c,d]

[a]Italian Institute for Genomic Medicine, Candiolo, Italy, [b]Candiolo Cancer Institute, FPO-IRCCS, Candiolo, Italy, [c]Department of Medical Sciences, University of Turin, Turin, Italy, [d]Medical Genetics Unit, AOU Città della Salute e della Scienza, Turin, Italy

Bladder cancer epidemiology

Bladder cancer (BC) is the most frequent malignancy of the urinary tract, excluding prostate cancer, with 81,400 new cases estimated in 2020 in the United States (Siegel, Miller, & Jemal, 2020). Overall, BC constitutes the fifth-leading cancer in terms of mortality and morbidity (Enokida, Yoshino, Matsushita, & Nakagawa, 2016).

BC is a multifactorial disease in which both exogenous and endogenous risk factors contribute to development. The main risk factor for this type of cancer is smoking: smokers have 2.5 times increased risk of developing BC in comparison with nonsmokers (Cumberbatch, Rota, Catto, & La Vecchia, 2016). BC risk rises also with age, with a peak between 50 and 70, and is three times more common in men than in women. In males, BC represents the fourth-most common tumor with 62,100 new cases expected in 2020 in the United States (Siegel et al., 2020).

BC risk factors include genetic and molecular abnormalities, chemical or environmental exposures and chronic irritation, occupational exposures to aromatic amines and polycyclic aromatic hydrocarbons, *Schistosoma haematobium* infections, and exposure to ionizing radiation (Comperat et al., 2019).

Bladder cancer classification and diagnosis

Based on tissue infiltration, BC is classified into two macrocategories: nonmuscle-invasive BC (NMIBC; stages Ta and T1), which constitutes 75% of total cases, and muscle-invasive BC (MIBC; stages T2-T4) (Anastasiadis & de Reijke, 2012; Babjuk et al., 2019; Comperat et al., 2019; Qie et al., 2016; Witjes et al., 2020). BC has an unfavorable prognosis as about 70% of cases recur regardless of the treatment performed (surgery, chemotherapy, or a combination of the two) (Enokida et al., 2016). Overall, the 5-year relative survival rate for BC is 77% (Siegel et al., 2020).

The initial symptoms for suspected BC are either macroscopic hematuria or urinary tract symptoms together with microscopic hematuria. The current gold standard for diagnosis is cystoscopy and tissue biopsy, an invasive and costly measure that also includes patient discomfort and potential adverse effects (Stav et al., 2004). Cystoscopy is effective in detecting BC: the sensitivity and specificity range from 62% to 84% and 43% to 98%, depending on the type, stage, and tumor grading. The low sensitivity is mainly for carcinomas in situ (Tis) and is operator-dependent, especially for the detection of recurrences.

Cystoscopy is an invasive procedure and patients often experience pain during urination, hematuria, and infections (Oeyen et al., 2019). The same approach is also applied for monitoring patients after treatment to evaluate the BC recurrence or progression of the disease. BC monitoring is a life-long invasive surveillance with high costs for the National health systems that make BC as one of the most expensive tumors (Matullo, Naccarati, & Pardini, 2016).

Urine cytology is the current noninvasive diagnostic tool used as an adjuvant in the diagnosis of new cases. Voided urine is analyzed for exfoliated cancer cells; therefore, there is no risk for patients and the exam can be repeated several times. However, it has limited utility because of the relatively low specificity, especially in low-grade tumors (16%) (Babjuk et al., 2017). Replacing the invasive and costly cystoscopy procedure is one of the main challenges in biomarker development for BC (Oeyen et al., 2019).

MicroRNA in Human Malignancies. https://doi.org/10.1016/B978-0-12-822287-4.00002-5

Over the past three decades, very few advances in BC clinical management have been added; however, in the last years, we have inverted this trend thanks to the improvement of molecular biology techniques. Large-scale gene expression and sequencing efforts have brought clinicians and researchers toward better targeted treatments and effective immunotherapies. Sequencing studies have also added new insights into the classification of several distinct molecular signatures and the determination of BC subtypes, providing a more accurate prediction of disease progression and therapeutic response (reviewed in detail in Tran, Xiao, Agarwal, Duex, & Theodorescu, 2020; Fig. 25.1).

Among NMIBCs, the majority of patients are treated with a transurethral removal of the bladder tumor (TURBT), which may be followed by intravesical chemotherapy or biological therapy with bacillus Calmette-Guerin (BCG) (Qie et al., 2016). The vast majority of patients with stage I BC (91%) and two-thirds of those with stage II are diagnosed and treated with TURBT with or without chemotherapy and/or radiation. In contrast, 68% of patients with stage III BC receive a cystectomy with or without chemotherapy and/or radiation. Chemotherapy is usually the first treatment for cancers that have metastasized but other treatments might be used as well. For patients with advanced BC or those experiencing a failure of first-line chemotherapy, there is no clear definition of a second-line treatment approach and in the last 20 years, no significant progress has been made regarding new therapies or early diagnosis methods for BC (Comperat et al., 2019; Oeyen et al., 2019; Stav et al., 2004).

BC is one of the most frequently mutated cancers in humans after lung and skin cancers. The most frequent alterations observed in BC are mutations in the gene encoding for telomerase reverse transcriptase (*TERT*), which is mutated in 70%–80% of patients with BC (Tran et al., 2020) (Fig. 25.1). Several oncogenes (*TP63*, *EGFR*, and *PIK3CA*) and tumor-suppressor genes (*TP53*, *RB1*, and *PTEN*) also play roles in the genetic pathogenesis of BC and are frequently mutated in this type of tumor (Matullo et al., 2016).

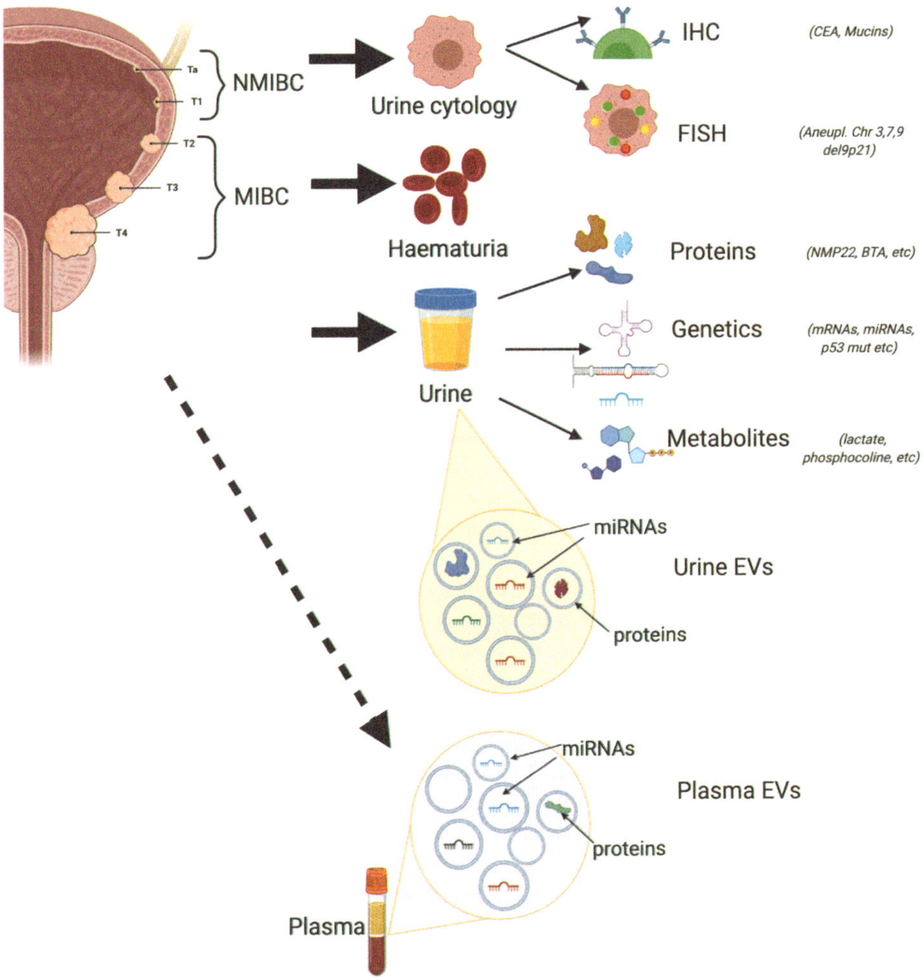

FIG. 25.1 Overview of the applied methodologies for bladder cancer and diagnosis as well as recognized biomarkers. Created with BioRender.com. No permission required.

Increasing evidence indicates that microRNAs (miRNAs) are also altered in BC tumor initiation, progression, and metastasis and could be used as potential biomarkers for results altered in 70%–80% early diagnosis of BC, prognostic markers for its recurrence, and predictive markers for response/overall survival (Sabo et al., 2020) (Fig. 25.1).

In this chapter, we will evaluate the role of miRNAs in the pathophysiology of BC but also as diagnostic, prognostic, and predictive tools. In the last part, we will highlight the role of circulating miRNAs in BC.

MicroRNAs altered in BC and relevant for BC pathophysiology

Several miRNAs have been reported to be altered and associated with BC. Because miRNAs have the capability to regulate the expression of a large portion of target genes, knowing those miRNAs involved in bladder tumorigenesis could help to understand which genes and relative pathways are involved in the process. Table 25.1 summarizes those miRNAs involved in BC development with their relative target genes. Knowing the miRNAs altered in BC could allow using them as biomarkers for diagnosis but also could provide a strategy to develop new therapeutics.

The altered expression levels of miRNAs observed in BC can have important effects on the modification of the protein expression relevant for the development of this cancer. It is commonly accepted that specific miRNAs targeting oncogenes work as tumor-suppressor (TS) miRNAs. Also, the term oncomiRNAs (or oncomiRs) describes those species that target and regulate the expression of tumor-suppressor genes. To date, the majority of altered miRNAs are tumor suppressors because of their general downregulation in BC and the fact that they target important oncogenes (Table 25.1). The most relevant miRNAs of this type include miR-1, miR-16, miR-24, miR-101, miR-125b, miR-129, and miR-133a (Enokida et al., 2016).

The role of tumor-suppressor miR-331-3p has been demonstrated recently (Morita et al., 2018). This miRNA is downregulated in BC tissues, especially in nonmuscle-invasive tumors in which the upregulation of *NACC1*, a gene regulated by miR-331-3p and involved in cell proliferation and tumor invasion, is often observed. It seems that miR-331-3p and its target *NACC1* are fundamental in the phases of passage from noninvasive to invasive BC.

Interestingly, the number of upregulated miRNAs in BC is low. Among these oncomiRs of note are miR-9, miR-10b, miR-19a, and miR-96, which have among their targets important genes such as *PTEN* and *CDK1A* (Enokida et al., 2016).

miR-188-5p and miR-141-3p are often downregulated in some cancers (such as gastric, liver, or malignant gliomas) but could also be upregulated in certain types of solid tumors (such as miR-141-3p for cervical and prostatic cancers). In BC, it has been demonstrated that these two miRNAs work in a synergistic way as oncomiRs because they are upregulated in BC tissues and cells with a consequent activation of proliferation, migration, and invasion pathways (Yang & Wang, 2019).

As with many other cancers, in BC tissues a general dysregulation of let-7 family members and an upregulation of the well-known oncomiR miR-21 have been observed (Boyerinas, Park, Hau, Murmann, & Peter, 2010; Calderaro et al., 2014; Gottardo et al., 2007; Jerome et al., 2007; Kozinn et al., 2013; Lei et al., 2015; Li et al., 2013). miR-21 is fundamental in the regulation of cell proliferation and migration via *PTEN* and *P53,* which constitute the most important pathway implicated in BC (Lei et al., 2015; Witjes et al., 2020).

miRNAs are known to be spread in the human genome in clusters. The simultaneous expression of miRNAs of the same cluster and targeting common genes could help in the explanation of their role in bladder carcinogenesis. For example, miR-145 and miR-143, frequently downregulated in BC (Braicu et al., 2019), form a cluster on chromosome 5q and are coexpressed because they share the same promoter (Iio, Nakagawa, Hirata, Naoe, & Akao, 2010). Both target *SERPINE1,* which is always upregulated and promotes cell invasiveness in BC (Li et al., 2019; Villadsen et al., 2012). In addition, miR-145 represses *MYC* and is implicated in the *P53* regulatory network because it could be transcriptionally induced directly from p53 (Sachdeva et al., 2009).On the other hand, miR-143 also interacts with RAS, ERK, p53, Rb, and E2F (Braicu et al., 2019).

Other interesting clusters are miR-23b/27b, implicated in the modulation of *EFGR/MET* (Chiyomaru et al., 2015), and miR-195/497, regulating the survivin encoding gene, *BIRC5* (Itesako et al., 2014).

The involvement of miRNAs in BC is tightly related to the gene(s) they target. A large proportion of the genes that were validated as targets of dysregulated miRNAs in BC are key regulators of cell cycle and apoptosis. For example, *BCL2L* and *BIRC5*, key inhibitors of apoptosis, are targeted by miR-133b, miR-203, miR-195, and miR-497, frequently downregulated in BC. On the other side, all miRNAs targeting caspases (such as *CASP2* in the case of miR-708) could work as oncomiRs because their upregulation blocks the important pathway of activation of caspases that has a central role in the apoptosis pathway (Monteleone & Lutz, 2017). Other genes targeted by miRNAs altered in BC are those involved in cell cycle control, including *CDC25A*, *CDK4*, *CDK6,* and *CCND1*.

Another relevant category of genes involved in the pathogenesis of BC is represented by those involved in the epithelial-mesenchimal transition (EMT) and their relative targeting miRNAs altered in BC. Important progressors of EMT such as *ZEB1*, *ZEB2,* and *ERFFI1* are upregulated in BC as a consequence of the downregulation of miR-23b, miR-138,

TABLE 25.1 microRNAs involved in bladder cancer development and their validated target genes as extracted from bladder cancer studies (*Onco*, oncomiRNA; *TS*, tumor suppressor).

miRNAs	Locus	Clustered miRNAs	Type	Validated target(s)	References
miR-1	20q13.33 18q11.2	– miR-133a-1	TS	*LASP1 PNP* *PTMA* *SRSF9* *TAGLN2*	Enokida et al. (2016)
miR-16	13q14.2 3q25.33	miR-15a miR-15b	TS	*CCND1*	Enokida et al. (2016)
miR-23b	9q22.32	miR-27b, miR-3074, miR-24-1	TS	*EGFR MET* *ZEB1*	Enokida et al. (2016)
miR-24	9q22.32 19p13.12	miR-23b, miR-27b, miR-3074 miR-23a, miR-27a	TS	*CARD10* *FOXM1*	Enokida et al. (2016)
miR-27a	19p13.12	miR-23a, miR-24-2	TS	*RUNX1* *SLC7A11*	Enokida et al. (2016)
miR-27b	9q22.32	miR-23b, miR-24-1, miR-3074	TS	*DROSHA* *EGFR MET*	Enokida et al. (2016)
miR-29c	1q32.2	miR-29b-2	TS	*CDK6*	Enokida et al. (2016)
miR-30a	6q13	–	TS	*NOTCH1*	Enokida et al. (2016)
miR-34a	1p36.22	–	TS	*HNF4G* *CD44* *NOTCH1*	Enokida et al. (2016)
miR-99a	21q21.1	Let-7c	TS	*FGFR3*	Enokida et al. (2016)
miR-100	11q24.1	Let-7a-2 miR-10526	TS	*MTOR*	Enokida et al. (2016)
miR-101	1p31.3 9p24.1	miR-3671 –	TS	*COX2 MET* *VEGFC*	Enokida et al. (2016)
miR-124-3p	20q13.33 8p23.1 8q12.3	–	TS	*ROCK1* *CDK4*	Enokida et al. (2016)
miR-125b	11q24.1 21q21.1	– –	TS	*E2F3* *MMP13* *SPHK1*	Enokida et al. (2016)
miR-128	2q21.3 3p22.3	– –	TS	*VEGFC*	Enokida et al. (2016)
miR-129	7q32.1 11p11.2	– –	TS	*GALNT1* *SOX4*	Enokida et al. (2016)
miR-133a	18q11.2 20q13.33	miR-1-2 –	TS	*EGFR* *FSCN1* *GSTP1* *LASP1 PNP* *PTMA TAGLN2*	Enokida et al. (2016)
miR-133b	6p12.2	miR-206	TS	*AKT1* *BCL2L2* *EGFR*	Enokida et al. (2016) and Braicu et al. (2019)
miR-135a	3p21.2 12q23.1	– –	TS	*FOXO1*	Enokida et al. (2016)
miR-138	3p21.32 16q13	– –	TS	*ZEB2*	Enokida et al. (2016)

TABLE 25.1 microRNAs involved in bladder cancer development and their validated target genes as extracted from bladder cancer studies (Onco, oncomiRNA; TS, tumor suppressor)—cont'd

miRNAs	Locus	Clustered miRNAs	Type	Validated target(s)	References
miR-139-5p	11q13.4	–	TS	*CX43*	Braicu et al. (2019) and Chi et al. (2019)
miR-143	5q32	miR-145	TS	*SERPINE1* *AKT*	Enokida et al. (2016) and Braicu et al. (2019)
miR-144-5p/3p	17q11.2	miR-451a, miR-451b, miR-4732	TS	*CCNE1* *CCNE2* *CDC25A* *PKMYT1*	Enokida et al. (2016)
miR-145	5q32	miR-143	TS	*CBFB* *CLINT1 FSCN1 ILK* *PAK1* *PPP3CA* *SERPINE1* *SOCS7* *IGF1R*	Enokida et al. (2016) and Braicu et al. (2019)
miR-186	1p31.1	–	TS	*HMGN5*	Enokida et al. (2016)
miR-193a-3p	17q11.2	–	TS	*LOXL4* *PSEN1* *HOXC9*	Enokida et al. (2016)
miR-195	17p13.1	miR-497	TS	*BIRC5* *CDC42* *CDK4* *GLUT3* *WNT7A*	Enokida et al. (2016)
miR-200b	1p36.33	miR-200a, miR-429	TS Onco	*MMP16*	Enokida et al. (2016) and Braicu et al. (2019)
miR-200c	12p13.31	miR-141	TS Onco	*BMI1 E2F3*	Enokida et al. (2016) and Braicu et al. (2019)
miR-203a	14q32.33	miR-203b	TS	*BCL2L2* *BIRC5*	Enokida et al. (2016)
miR-214	1q24.3	miR-199a, miR-3120	TS	*PDRG1*	Enokida et al. (2016)
miR-218	4p15.31 5q34	– –	TS	*BMI1* *LASP1*	Enokida et al. (2016)
miR-221	Xp11.3	miR-222	TS	*STMN1*	Enokida et al. (2016)
miR-320a	8p21.3	–	TS	*ITGB3*	Enokida et al. (2016)
miR-320c	18q11.2	–	TS	*CDK6*	Enokida et al. (2016)
miR-331-3p	12q22	miR-3685	TS	*NACC1*	Morita et al. (2018)
miR-449a	5q11.2	miR-449b, miR-449c	TS	*CDC25A*	Enokida et al. (2016)
miR-485-5p	14q32.31	miR-381, miR-487b, miR-539, miR-889, miR-544a, miR-655, miR-487a, miR-382, miR-134, miR-668, miR-323b, miR-154, miR-496, miR-377, miR-541, miR-409	TS	*HMGA2*	Enokida et al. (2016)
miR-490-5p	7q33	–	TS	*FOS*	Enokida et al. (2016)

Continued

TABLE 25.1 microRNAs involved in bladder cancer development and their validated target genes as extracted from bladder cancer studies (Onco, oncomiRNA; TS, tumor suppressor)—cont'd

miRNAs	Locus	Clustered miRNAs	Type	Validated target(s)	References
miR-493	14q32.2	miR-337, miR-665	TS	*FZD4* *RHOC*	Enokida et al. (2016)
miR-497	17p13.1	miR-195	TS	*BIRC5* *WNT7A*	Enokida et al. (2016)
miR-574-3p	4p14	–	TS	*MESDC1*	Enokida et al. (2016)
miR-576-3p	4q25	–	TS	*CCND1*	Enokida et al. (2016)
miR-590-3p	7q11.23	–	TS	*TFAM*	Enokida et al. (2016)
miR-1182	1q42.2	–	TS	*TERT*	Enokida et al. (2016)
miR-9	1q22 5q14.3 15q26.1	– – –	Onco	*CBX7* *CERS2*	Enokida et al. (2016)
miR-10b	2q31.1	–	Onco	*HOXD10* *KLF4*	Enokida et al. (2016)
miR-19a	13q31.3	miR-17, miR-18a, miR-20a, miR-19b-1, miR-92a-1	Onco	*PTEN*	Enokida et al. (2016)
miR-19b	13q31.3 Xq26.3	miR-17, miR-18a, miR-19a, miR-20a, miR-92a-1, miR-106a, miR-18b, miR-20b, miR-363, miR-92a-2	Onco	*Not defined*	Braicu et al. (2019)
miR-21	17p23.1	–	Onco	*Not defined*	Braicu et al. (2019), Calderaro et al. (2014), Jerome, Laurie, Louis, and Pierre (2007), and Lei et al. (2015)
miR-23a	19p13.12	miR-27a, miR-24-2	Onco	*Not defined*	Braicu et al. (2019)
miR-96	7q32.2	miR-182, miR-183	Onco	*CDK1A*	Enokida et al. (2016)
miR-141-3p	12p13.31	miR-200c	Onco	*Not defined*	Yang and Wang (2019)
miR-150	19q13.33	–	Onco	*PDCD4*	Enokida et al. (2016)
miR-155	21q21.3	–	Onco	*DMTF1*	Enokida et al. (2016)
miR-181b	1q32.1 9q33.3	miR-181a-1 miR-181a-2	Onco	*Not defined*	Braicu et al. (2019)
miR-181d	19p13.12	miR-181c	Onco	*Not defined*	Braicu et al. (2019)
miR-182-5p	7q32.2	miR-183 miR-96	Onco	*RECK* *SMAD4*	Enokida et al. (2016)
miR-188-5p	Xp11.23	miR-532, miR-500a, miR-362, miR-501, miR-500b, miR-660	Onco	*Not defined*	Yang and Wang (2019)
miR-191	3p21.31	miR-425	Onco	*Not defined*	Braicu et al. (2019)
miR-200a	1p36.33	miR-200b, miR-429	Onco	*Not defined*	Braicu et al. (2019)
miR-205	1q32.2	–	Onco	*Not defined*	Braicu et al. (2019)
miR-429	1p36.33	miR-200a, miR-200b	Onco	*Not defined*	Braicu et al. (2019)
miR-708	11q14.1	–	Onco	*CASP2*	Enokida et al. (2016)
miR-940	16p13.3	miR-3677 miR-4717	Onco	*INPP4A* *GSK3β*	Wang, Wu, Huang, and Chen (2018)
Let7g	3p21.2	–	Onco	*Not defined*	Braicu et al. (2019)

and miR-200c. The miR-200 family, in general, is known to be bound to EMT (Abba, Patil, Leupold, & Allgayer, 2016). Regarding the miR-200 family, it has been shown that the expression levels of these members (mainly miR-200c and miR-141) showed an inverse relation with the expression levels of *E-cadherin* and *ZEB1*, providing some prospective therapeutic targets (Tamagawa et al., 2014) and markers of prognosis (Amuran, Eyuboglu, Tinay, & Akkiprik, 2018). Interestingly, the miR-200 family is often upregulated in BC while for other cancers, it is more frequently downregulated (such as in renal cancer Kurozumi, Goto, Okato, Ichikawa, & Seki, 2017).

Of note, there are some inconsistencies regarding the direction of alteration of some miRNAs in BC. For example, there are discordant data for miR-205 and miR-200c because in some studies, researchers observed a downregulation of these miRNAs in BC while in others, they were upregulated (Catto et al., 2011; Gottardo et al., 2007; Mahdavinezhad et al., 2015; Qie et al., 2016; Sun et al., 2015; Tran et al., 2013). It is not uncommon that the same miRNAs could work as tumor suppressors or oncogenes in different types of solid tumors. On the other side, it is not common to have a discordant role for the same type of tumor. Among the possible reasons for these observed discrepancies could be a different study population (including, for example, a mixed set of NMIBCs and MIBCs in different proportions), an insufficient sample size of the population of study, or the use of different reference genes for the evaluation of expression levels.

One of the most investigated BC markers is the family of matrix metalloproteinases (MMPs)(Gobin et al., 2019). MMPs in cancer facilitate tumor cell invasion and metastasis because they are responsible for the degradation of the extracellular matrix. MMP2, MMP9, and MMP16 are specific MMPs whose expression is modulated by miR-200b and miR-125b (Abba, Patil, & Allgayer, 2014). In addition, miR-139, frequently downregulated in BC, also targets MMPs (Jia et al., 2013; Yonemori et al., 2016). MMPs are ubiquitously upregulated across different cancer types with some of them closely bound to specific ones. Interestingly, MMPs are highly investigated for their potential utility as pan-cancer biomarkers (mainly MMP11 and MMP13, which are altered in several cancers), especially considering that they could be detected noninvasively in body fluids such as urine (see, for example, MMP23) (Allione et al., 2018).

miRNAs have also been involved in the modulation of response of cancer cells to chemotherapeutic agents. miR-193a-3p seems to enhance the chemoresistance of BC cells modulating the oxidative stress pathway (Deng et al., 2014). Other miRNAs involved in multidrug resistance are miR-22-3p, miR-196a-5p, and miR-294. Interestingly, some miRNAs are more specific of the therapy response to a single agent (such as miR-193b, miR-203, and miR-218 for cisplatin or miR-143 and miR-145 for gemcitabine) (Taheri, Shirvani-Farsani, Ghafouri-Fard, & Omrani, 2020).

MicroRNAs as diagnostic, prognostic, and predictive biomarkers

In the last 15 years, research has advanced in the discovery of new potential biomolecules involved in the development and diagnosis of BC to be used as biomarkers for the diagnosis and monitoring of this cancer. Despite all these efforts, the histological examination of biopsy tissues has remained the gold standard. Despite their high reliability, biopsies are unpleasant and risky for patients; they are also expensive, time-consuming, and require a certain expertise from the pathologist (Sole, Arnaiz, Manterola, Otaegui, & Lawrie, 2019). Consequently, researchers have recently investigated liquid biopsies to develop methodologies for new minimally invasive or noninvasive biomarkers for a more specific diagnosis, precise prognosis, and prediction of response to therapy (Linxweiler & Junker, 2020).

Cytology is currently the most frequently applied noninvasive procedure. It is a highly specific method (specificity range 78%–100%); however, it is observer-dependent and displays a variable and low sensitivity (34%–84% for high-grade tumors and 12%–26% for low-grade BCs). There are quite a few alternative biomarker-based tests for the noninvasive diagnosis of BC (via blood-derived fluids or urine) but none of them have reached routine clinical practice because of their inadequate performances. For most of them, the sensitivity is higher than the cytology but they do not perform as good as cytology in terms of specificity (Chou et al., 2015; Erdmann et al., 2020). The main challenge in the next years will be to find biomarkers that are highly specific, sensitive, and cost-efficient for the noninvasive diagnosis of BC as well as its screening and monitoring. Similar biomarkers could potentially replace the excessive number of invasive and costly cystoscopies.

The main advantage of employing miRNAs as BC biomarkers is the high stability of these transcripts, which remain unaltered during TURBT. Moreover, for miRNA profiling and validation, there are a full range of tools and methods (microarray, next-generation sequencing, nanostring, qRT-PCR, or in situ hybridization) that makes the analyses feasible on a large scale. In this sense, searching for circulating nucleic acids like miRNAs in noninvasive biofluids represents a better option for patients because miRNAs are stable and measurable in specimens such as plasma and urine. In the specific case of BC, the tumor is in close contact with urine; thus, urinary biomarkers are considered the most promising noninvasive tools for improving the diagnosis and monitoring of this type of cancer.

Analyses on circulating miRNAs have some advantages such as easy sampling (even multiple times), reduction of pain and stress, and reduction of cost (Pardini, Sabo, Birolo, & Calin, 2019). Especially in the context of heterogeneous tumors such as BC, the analysis of circulating miRNAs may better characterize the tumor as a whole while classic biopsies, analyzing only one specific fragment, might miss some characteristics (McGranahan & Swanton, 2015; Pardini et al., 2019). In addition, the option of repeated samplings permits better monitoring of the disease progression or treatment efficacy for a better intervention and dynamic treatment of the patient, which is the fundamental basis of personalized medicine (Chang et al., 2017). Moreover, extracellular miRNAs can be delivered to target cells as free-circulating molecules (often associated with proteins, especially Argonaute 2) or loaded into extracellular vesicles (EVs) such as exosomes, microvesicles, and apoptotic bodies. They can also behave with a hormone-like mechanism acting as autocrine, paracrine, and/or endocrine regulators to modulate cellular activities (Anfossi, Fu, Nagvekar, & Calin, 2018; Pardini & Calin, 2019).

Freely circulating miRNAs in urine as biomarkers for BC diagnosis

There is much interest in developing noninvasive means for profiling miRNAs as biomarkers in biofluids such as blood and urine. The use of whole urine as a substrate for miRNA profiling has been reported with several other recent investigations that have included the use of miRNA profiles from urine sediment (Mengual et al., 2013; Miah et al., 2012; Shimizu et al., 2013; Wang et al., 2012). Also, specific miRNAs have been examined in cell-free urine (Kim et al., 2013; Oliveira et al., 2020), demonstrating the feasibility of a clinical role for miRNAs in patient stratification. A summary of circulating miRNA as BC biomarkers is listed in Table 25.2.

Andreu and colleagues analyzed miRNAs freely circulating in 43 urine samples from 34 BC and nine nonsmoker healthy volunteers. Twenty-six miRNAs had altered expression levels, with most of them downregulated in high-grade BC. Among the miRNAs selected for further validation, the authors found that miR-375 was significantly downregulated in high-grade BC patients and miR-146a was significantly upregulated in low-grade BC (Andreu et al., 2017). The same authors analyzed miRNAs in urine EVs for BC recurrence and reported that miR-146a, miR-194, and let-7c levels were lower in recurrent low-grade cancer patients while miR-30c-2 was upregulated in low-grade relapsing cases (Andreu et al., 2017). Although promising, these results were not statistically significant, so larger studies are recommended.

Interestingly, miR-146a-5p levels might be a candidate for a noninvasive diagnostic marker for BC. Sasaki and colleagues found increased levels of this miRNA in urine samples of BC patients compared to controls, with its expression levels reduced in the urine of patients with BC after TURBT (Sasaki et al., 2016).

A panel of six miRNAs (miR-16, miR-21, miR-34a, miR-200c, miR-205, and miR-221) was described by Sapre et al. with an impressive ability to detect active BC (sensitivity of the recurrence was estimated at 88%, specificity at 48%) (Sapre et al., 2016). Other urinary miRNAs identified as biomarkers of BC were miR-125b, miR-30b, miR-204, miR-99a, and miR-532-3p, all significantly downregulated in urine supernatants of 46 BC and 13 healthy controls (Pospisilova et al., 2016).

The urinary expression levels of miR-141 and miR-200b could distinguish patients with lymph node metastasis from those who were lymph node negative (Liu et al., 2015), confirming the relevant role of these miRNAs. The loss of miR-141 and miR-200b increases the invasion and migration capacities of BC cell lines and upregulates MMP2, MMP9, vimentin, and N-cadherin while downregulating E-cadherin expression.

Decreased expression of extracellular miR-214 in urine was associated with a high tumor stage, a high lymph node status, and a higher grade, age, and history of NMIBC. The urinary cell-free miR-214 not only could work as a diagnostic biomarker, but could also serve as an independent prognostic predictor of recurrence-free survival (RFS) and overall survival (OS) in patients with MIBC (Wang, Zhang, et al., 2015).

The number of studies on urinary miRNAs as BC biomarkers is quite large and the majority of them tested specific miRNAs by quantitative PCR (qPCR) with a candidate miRNA approach. The specimens that were validated in at least two independent cohorts were miR-16, miR-21, and miR-200c (identified as a diagnostic panel by Sapre et al., 2016), but also miR-140-5p, miR-199a-3p, miR-93, miR-652, miR-1305, miR-224, miR-96, and miR-766 (inserted in a model by Urquidi et al., 2016).

Increased expression levels of miR-31-5p, miR-93-5p, and miR-191-5p were identified also in the urine of BC patients with miR-93-5p and miR-31-5p being sensitive enough to detect BC and some histopathological subgroups (Juracek et al., 2018). This study was particularly interesting because it introduced a genome-wide approach based on next-generation sequencing in the study of miRNAs. We also applied this type of method. and identified three miRNAs (miR-30a-5p, let-7c-5p, and miR-486-5p) altered in the urine of BC cases showing high accuracy in the discrimination of cases from controls. Moreover, specific signatures of urinary miRNAs could clearly distinguish not only NMIBC from MIBC but also specific subtypes such as grade I-II and grade III (Pardini et al., 2018).

TABLE 25.2 Circulating microRNAs as biomarkers for bladder cancer diagnosis (*EVs*, extracellular vesicles).

miRNA(s)	Direction of expression	Biofluid(s)/matrix	References
miR-375 miR-146a miR-146a, miR-194, and let-7c miR-30c-2	Downregulated in high-grade BC patients Upregulated in low-grade BC Downregulated in low-grade recurrent BC patients Upregulated in low grade relapsing BC	Urine EVs (exosomes)	Andreu et al. (2017)
miR-21-5p	Upregulated in BC patients	Urine EVs (exosomes)	Matsuzaki et al. (2017)
miR-146a-5p	Upregulated in BC patients Downregulated in BC patients after resection	Urine	Sasaki et al. (2016)
miR-16, miR-21, miR-34a, miR-200c, miR-205, and miR-221	Upregulated in BC patients	Urine	Sapre et al. (2016)
miR-146b-5p, miR-155-5p	Upregulated in MIBC versus NMIBC	Urine EVs	Baumgart et al. (2019)
miR-125b, miR-30b, miR-204, miR-99a, and miR-532-3p	Downregulated in BC patients	Cell-free urine	Pospisilova et al. (2016)
miR-141, miR-200b	Differentially expressed in patients with lymph node metastasis or with lymph node negative	Urine	Liu et al. (2015)
miR-214	Downregulated in NMIBC patients with high tumor stage, high lymph node status, high grade	Cell-free urine	Wang et al. (2015)
miR-140-5p, miR-199a-3p, miR-93, miR-652, miR-1305, miR-224, miR-96, miR-766	A 25-miRNA model showed the best performance of discrimination but the reduced version of the model including 10 miRNAs was efficient, although resulted in some loss of performance	Urine sediments	Urquidi et al. (2016)
miR-31-5p, miR-93-5p and miR-191-5p	Upregulated in BC patients	Urine	Juracek et al. (2018)
miR-30a-5p let-7c-5p, miR-486-5p	Downregulated in BC patients Upregulated in BC patients	Cell-free urine	Pardini et al. (2018)
miR-7-5p, miR-22-3p, miR-29a-3p, miR-126-5p, and miR-375 miR-200a-3p, miR-423-5p	Upregulated in BC patients Downregulated in BC patients	Cell-free urine	Du et al. (2017)
miR-26a, miR-93, miR-191, and miR-940	Upregulated in BC patients	Cell-free urine, urine sediments, and urinary EVs	Long et al. (2015)
miR-125b, miR-145, and miR-221 miR-96, miR-126, and miR-183	Downregulated in BC patients Upregulated in BC patients	Urine sediments	Erdmann et al. (2020)
miR-125b-5p, miR-142-3p, miR-140-5p, miR-204-5p miR-18a-3p, miR-25-3p, miR-92a-3p, miR-187-3p	Downregulated in BC patients Upregulated in BC patients	Urine sediments	Mengual et al. (2013)
miR-126-3p	Upregulated in BC patients	Plasma EVs	Sabo et al. (2020)
miR148b-3p	Upregulated in BC patients	Serum	Jiang et al. (2015) and Wang et al. (2015)
miR-141	Upregulated in BC patients	Serum	Kriebel et al. (2015) and Scheffer et al. (2014)
miR-27a-3p	Downregulated in BC patients	Serum	Jiang et al. (2015)
miR-92a-3p	Downregulated in BC patients	Plasma	Adam et al. (2013) and Motawi, Rizk, Ibrahim, and Ibrahim (2016)
miR-99a	Downregulated in BC patients	Plasma	Feng et al. (2014)
miR-100	Downregulated in BC patients	Plasma	Motawi et al. (2016)

Another group of researchers sequenced urinary cell-free miRNAs of BC samples and healthy controls to identify diagnostic and prognostic candidate biomarkers. Among them, five miRNAs (miR-7-5p, miR-22-3p, miR-29a-3p, miR-126-5p, and miR-375) were significantly upregulated and two (miR-200a-3p, miR-423-5p) were downregulated in BC samples compared with controls, with miR-22-3p and miR-200a-3p identified as independent factors for tumor recurrence in NMIBC (Du et al., 2017).

Recently, Long et al. (2015) profiled miRNA expression levels within cell-free urine supernatant and found that some miRNAs (single or in groups) were able to detect BC. Interestingly, a classifier based on miR-26a, miR-93, miR-191, and miR-940 had high sensibility and specificity in recognizing cancer patients. The authors also tested the presence of the same four miRNAs on urinary EV pellets. The expression patterns differentiating BC cases from controls varied between urine supernatants and EVs, with all of them expressed at higher levels in the cell-free portion (Long et al., 2015).

Finally, an alternative application to search for biomarkers in urine is to measure miRNAs present in the cells released during bladder epithelium exfoliation. In this sense, Ingelmo-Torres et al. identified eight miRNAs (miR-25-3p, miR-18a-3p, miR-92a-3p, miR-140-5p, miR-125b-5p, miR-142-3p, miR-204-5p, and miR-187-3p) in urine pellets that could work as BC diagnostic biomarkers (Mengual et al., 2013) but also as prognostic ones (Ingelmo-Torres et al., 2017). A recent paper from Erdmann and colleagues reported the potential role of miR-96, miR-125b, miR-126, miR-145, miR-183 and miR-2214 as diagnostic biomarkers in the urine sediments of BC patients. Interestingly, the authors demonstrated the particularly high potentiality of miR-125b, miR-145, miR-183, and miR-2214 in combination with cytology on voided urine for a better identification of BC cases in a noninvasive way (Erdmann et al., 2020).

From this long list of studies, the importance of the selection of the matrix for biomarker analyses appears quite evident. A validated strategy or a preferred matrix (whole urine, cell-free urine, or urine-derived sediments) has not been elected and all studies are equally valid. Certainly, the easiest analyses approach with less manipulation could help in reducing bias and, more importantly, could be of easiest application in clinics. Another important aspect emerging is that it is quite improbable that an individual miRNA could be ascribed as a single and strong biomarker for BC diagnosis, prognosis, and/or monitoring. The best prediction model or the most promising biomarker is a combination of miRNA signatures to be detected in a noninvasive way. Notably, a whole genome, hypothesis-free approach, such as those based on microarrays or next generation sequencing, is very important to have the wider view of investigated biomarkers and the possibility to implement the analyses with mathematical models.

Circulating miRNAs as biomarkers for the prognosis and therapy of BC

While there is an extensive number of studies attempting to identify miRNAs as biomarkers of BC diagnosis, there is a lack of information on miRNAs as biomarkers of prognosis and therapy for this type of cancer (Fig. 25.2). In an attempt to find prognostic markers for BC, Yin et al. identified in the TCGA-BCLA dataset a 21-miRNA signature that was significantly correlated with the OS of BC patients. Interestingly, this panel was also associated with the BC type and stage (Yin et al., 2019). This study and others were performed on BC tissue pairs. In general, the most promising miRNAs associated with recurrence and survival were miR-9, miR-141, miR-143, miR-154, miR-200a, miR-200b, and miR-182 (summarized in Taheri et al., 2020).

Other researchers focused more on urinary miRNAs. For instance, the combination of high miR-152/low miR-3187-3p expression levels was associated with poor recurrence-free survival of NMIBC patients. In particular, miR-152 levels seem to be associated with tumor recurrence (Jiang et al., 2015). Yun et al. identified miR-200a as a potential biomarker of recurrence in urine, but a validation in an independent cohort is still missing (Yun et al., 2012). Other studies reporting the prognostic role of urinary miRNAs in BC identified miR-214, miR-26b-5p, and miR-20b as unfavorable prognostic factors if upregulated (all the studies available are summarized in Taheri et al., 2020). The great improvement in the field of BC research in recent years will find utility in assessing a better therapeutical strategy to cure the cancer and decrease adverse events.

miRNAs as biomarkers of BC in other biofluids than urine

Besides the evident advantage of analyzing circulating miRNAs in urine as BC biomarkers, there are some preanalytical/analytical issues with urine samples that include hematuria contamination and the nonstandardized selection of urine fractions (including noncentrifuged urine, urine sediments, supernatant following centrifugation, and exosome preparation). In this sense, plasma/serum samples, at least theoretically, could have similar issues but be easier to control.

A recent review by Tolle et al. summarized the available studies performed on blood derivates for BC biomarker discovery (Tolle, Blobel, & Jung, 2018). About 20–22 miRNAs in plasma/serum have been identified as potential significant diagnostic markers, but only miR-92a-3p, miR-99a, and miR-100 in plasma have been reported in at least two independent

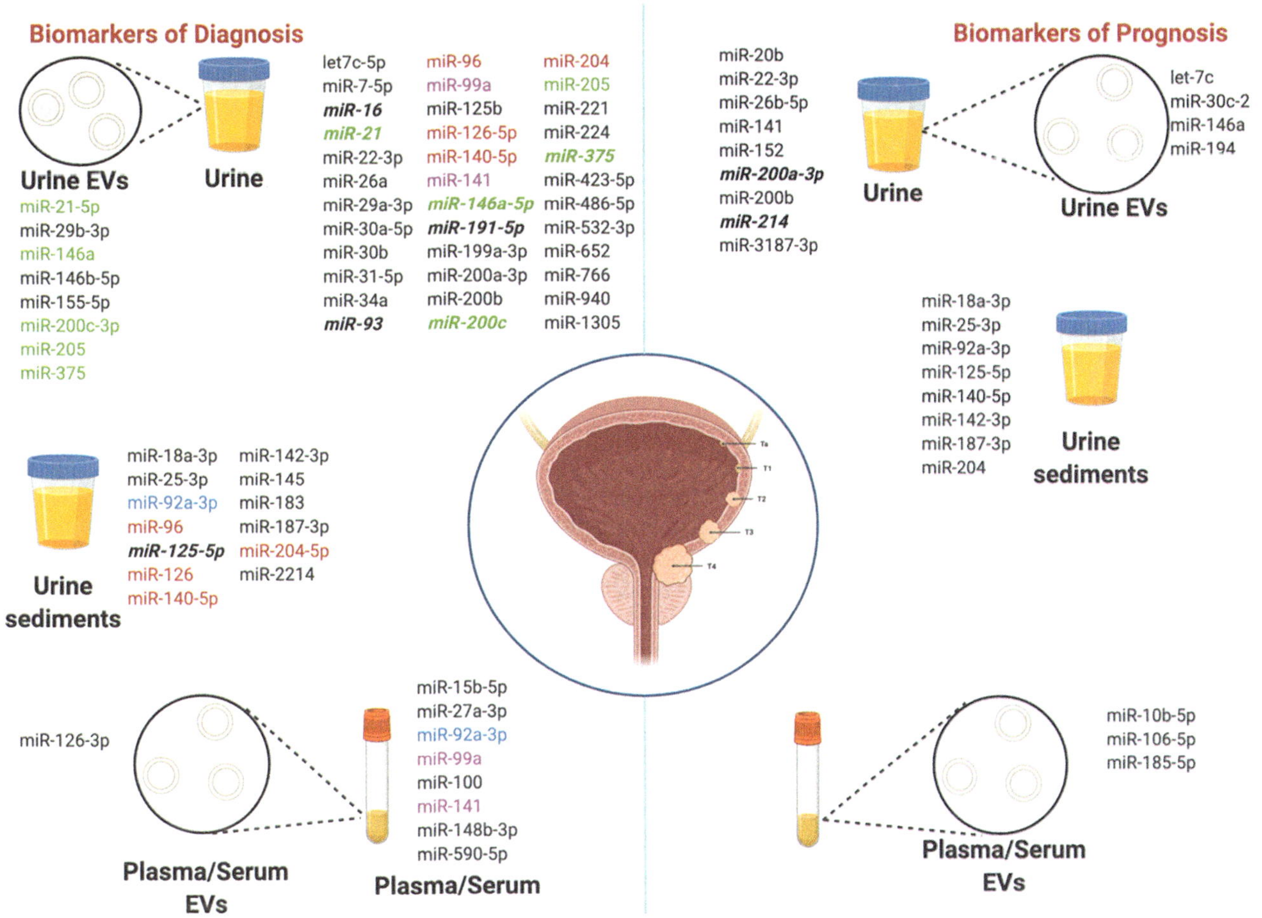

FIG. 25.2 miRNAs from different noninvasive biofluids associated with BC diagnosis or prognosis. *Bold* and *italics* show miRNAs detected in multiple studies; in *red*, miRNAs in common between native urine and sediments; in *green*, miRNAs in common between native urine and urinary EVs; in *blue*, miRNAs in common between urine sediments and plasma/serum; and in *purple*, miRNAs in common between native urine and plasma/serum. Created with BioRender.com. No permission required.

studies (Adam et al., 2013; Feng et al., 2014; Motawi et al., 2016). In serum, the most prominent markers are miR-148b-3p and miR-141 (upregulated in BC patients) and miR-27a-3p (downregulated) (Jiang et al., 2015; Kriebel et al., 2015; Scheffer et al., 2014; Wang, Liu, et al., 2015).

Importantly, four miRNAs (miR-25-3p, miR-33b-5p, miR-92a-3p, and miR-194) may be influenced by hemolysis or could be detectable in blood cells; therefore, the hemolysis factor should be carefully considered when analyzing blood-derived biofluids. Due to a lack of consistent results, none of these miRNA panels has been applied in practice (Tolle et al., 2018). A recent work found two miRNAs (miR-15b-5p, miR-590-5p) upregulated in plasma samples from BC patients that could be potential diagnostic candidates (Tolle, Buckendahl, & Jung, 2019). These two miRNAs were also upregulated in BC in previous studies (Ratert et al., 2013; Song et al., 2010). The authors controlled carefully for all possible hemolysis contaminations and performed an independent validation. Interestingly, other researchers reported miR-15b-5p and miR-590-5p as being among the top 25 miRNAs with highest concentrations in BC tissue, white-blood cells, urine, and plasma only in patients with BC but not in controls.

Although results are quite promising, improvements are required mostly on expanding the reproducibility of the biomarkers detected and the sample size.

Circulating miRNAs in extracellular vesicles

EVs are a heterogeneous class of lipid bilayered particles released from all types of cells into the extracellular space; they serve a variety of functions, including intercellular communication. EVs are classified according to their size and their mechanisms of biogenesis and secretion in exosomes, microvesicles, and apoptotic bodies. Exosomes, the smallest vesicles

(30–150 nm), originate by the inward budding of intracellular endosomes, which are then converted into multivesicular bodies that fuse with the cellular membrane and release their cargo into the extracellular space. On the other hand, microvesicles (typically larger than exosomes, 100–1000 nm) are shed directly from the bleebing of the outward cellular membrane. Finally, apoptotic bodies (1000–5000 nm) are derived from cells undergoing programmed cell death (Pardini & Calin, 2019).

When EVs reach target cells, they bind to them or are incorporated to release their content (miRNAs, proteins, lipids, etc.). Interestingly, the specific role of EVs in physiological or pathological processes is determined by specific markers on the EV membrane and by their content. EVs contain a variety of biomolecules including miRNAs, and their utility as reservoirs for biomarkers of oncogenesis has recently been discussed (Pardini & Calin, 2019).

It is now accepted that extracellular/circulating miRNAs, transported into EVs, are optimal biomarkers for diseases and play a role in intercellular communication (Linxweiler & Junker, 2020; Pardini et al., 2019; Pardini & Calin, 2019). Precursors and mature miRNAs, in fact, can be secreted in EVs and transferred to recipient cells where they are functionally active (Mittelbrunn et al., 2011). EVs can be isolated in all biofluids, including plasma/serum, saliva, and urine. The presence of EVs in urine was first reported in 2004 by Pisitkun, Shen, and Knepper (2004). Since then, the importance of urinary EVs in urologic malignancies has been highlighted, as they harbor markers of BC (Chen et al., 2012) but also prostate cancer and renal disease (Pardini et al., 2019).

Investigations on urine EVs are predominant in BC research, as urine is in permanent contact with the tumor (Table 25.2, Fig. 25.2). For instance, miR-146b-5p and miR-155-5p transported in urine EVs have been described as good biomarkers to distinguish NMIBC and MIBC (Baumgart et al., 2019). Interestingly, it has been shown that specific miRNAs (miR-205, miR-200c-3p, and miR-29b-3p) measured in urinary EVs reflected the expression levels measured in bladder tumor tissues (Armstrong, Green, Seigne, Schned, & Marsit, 2015). Another research group has identified miR-375 and miR-146a as diagnostic markers for high- and low-grade BC, respectively (Andreu et al., 2017).

Matsuzaki and colleagues identified miR-21-5p as being upregulated in BC and as the most potent biomarker in urinary EVs among six other miRNAs altered in the same matrix. In particular, miR-21-5p expression levels were also particularly high in EVs from BC patients with negative urine cytology (Matsuzaki et al., 2017). Serum and plasma could also be a good source of EVs. In the case of BC research, it is possible that BC cells, especially in the most aggressive form, can release EVs into the circulatory system.

To the best of our knowledge, the only available study performed so far on miRNAs released in plasma EVs as BC biomarkers is the one from our group (Sabo et al., 2020). In general, miRNA expression levels and variability are lower in EVs compared to freely circulating molecules. However, it was demonstrated that miR-126-3p was upregulated both in plasma EVs and the urine of BC patients when compared to controls. In plasma EVs, the progressive decrease of miR-4508 levels was also observed going from control to high-risk BC. This implicates the important role of miRNAs contained in plasma EVs in the determination of the spread of the tumorigenesis signal, especially in advanced BC. Some of the miRNAs contained in EVs also seem to be potential biomarkers for prognosis and survival. BC cases showing low expression levels of miR-185-5p and miR-106a-5p or high expression of miR-10b-5p in their plasma EVs were associated with shorter survival (Sabo et al., 2020). This is just the beginning of a very promising field that will certainly bring new insights in the near future.

In the last year, we have seen an explosion of research based on EVs and their content as cancer biomarkers. Nevertheless, further investigations are still needed from bench to bedside, especially for the demonstration of clinical effectiveness. For the moment, in the available studies there are a number of limitations. First, there are technical problems related to EV separation and characterization. Moreover, a consensus in EV nomenclature is still missing as well as fast, reproducible, and effective separation methods to improve the standardization and reproducibility between studies. The actual most approved method for EV separation (based on ultracentrifugation procedures) is time-consuming, laborious, and needs large amounts of sample material and specific equipment not retrievable in clinical laboratories. Other procedures that combine immuno-affinity, sieving, and trapping to concentrate EVs have been proposed but they damage their structure and have low recovery rates.

Another source of criticism that hampers EV application as a source of biomarkers in BC relies on the fact that most studies typically have a small and often heterogeneous cohort of BC patients. After the identification of robust candidate biomarkers in EVs, additional external validation in large independent multiinstitutional studies are required to establish their value as useful BC biomarkers.

Conclusions and perspectives

In the last 20 years of research, miRNAs have been demonstrated to have an important role in the carcinogenesis of urothelial cancer (Fig. 25.2). In general, a downregulation of several miRNAs with a consequent upregulation of genes targeted by those miRNAs has been observed. The large number of altered miRNAs identified in BC has opened research on these molecules as biomarkers.

BC is the most expensive tumor to treat per patient, per year, as well as per lifetime because of the high costs associated with surveillance for recurrence of the disease. A cost-effective, highly sensitive method for BC detection is one of the main challenges in the prevention of this cancer characterized by high morbidity and recurrence. The possibility of developing specific and sensitive biomarkers of disease at a low cost could enable the detection of a larger proportion of tumors, especially in the at-risk population and when those tumors are at their most treatable. BC tumors usually manifest abnormalities detectable in urine. Urine cytology is the most specific test for disease recurrence, but it is poorly sensitive, particularly for low-grade disease. Detection strategies have been developed to improve sensitivity, but unfortunately these assays are still skewed toward the detection of high-grade or specific subsets of the disease. They are also expensive and do not have sufficient sensitivity or specificity to replace cystoscopy as a standard surveillance method.

miRNAs possess peculiarity of stability and reproducibility as well as the quality to be measurable not only in cells and tissues but also in biofluids such as urine, which have the best matrix collectable noninvasively for BC diagnosis. All these listed are good premises for robust biomarkers. However, the heterogeneity of studies available and the clinicopathological variability of BC do not currently allow recommending a useful miRNA marker as a diagnostic or prognostic tool in BC. So far, the studies performed did not carefully consider diagnostic and prognostic biomarker guidelines and missed internal and external validation of data. Future large multicenter investigations are necessary to overcome the deficiencies in these studies to prove the usefulness of circulating miRNAs as robust biomarkers for BC.

Finally, the long list of miRNAs involved in BC constitutes just the beginning of a new era. Several miRNA-targeted therapeutics have reached clinical development for treating several types of cancer. miRNAs that have revealed an important role in BC such as miR-34a, miR-143, and miR-145 have been already applied successfully in preclinical models for the treatment of other solid tumors (Rupaimoole & Slack, 2017). Therefore, we are not too far away from applications in bladder cancer research.

References

Abba, M., Patil, N., & Allgayer, H. (2014). MicroRNAs in the regulation of MMPs and metastasis. *Cancers (Basel)*, *6*, 625–645.

Abba, M. L., Patil, N., Leupold, J. H., & Allgayer, H. (2016). MicroRNA regulation of epithelial to mesenchymal transition. *Journal of Clinical Medicine*, *5*, 8.

Adam, L., Wszolek, M. F., Liu, C. G., Jing, W., Diao, L., Zien, A., … Dinney, C. P. (2013). Plasma microRNA profiles for bladder cancer detection. *Urologic Oncology*, *31*, 1701–1708.

Allione, A., Pardini, B., Viberti, C., Giribaldi, G., Turini, S., Di Gaetano, C., … et al. (2018). MMP23B expression and protein levels in blood and urine are associated with bladder cancer. *Carcinogenesis*, *39*, 1254–1263.

Amuran, G. G., Eyuboglu, I. P., Tinay, I., & Akkiprik, M. (2018). New insights in bladder cancer diagnosis: Urinary miRNAs and proteins. *Medical Sciences*, *6*, 113.

Anastasiadis, A., & de Reijke, T. M. (2012). Best practice in the treatment of nonmuscle invasive bladder cancer. *Therapeutic Advances in Urology*, *4*, 13–32.

Andreu, Z., Otta Oshiro, R., Redruello, A., Lopez-Martin, S., Gutierrez-Vazquez, C., Morato, E., … Yanez-Mo, M. (2017). Extracellular vesicles as a source for non-invasive biomarkers in bladder cancer progression. *European Journal of Pharmaceutical Sciences*, *98*, 70–79.

Anfossi, S., Fu, X., Nagvekar, R., & Calin, G. A. (2018). MicroRNAs, regulatory messengers inside and outside cancer cells. *Advances in Experimental Medicine and Biology*, *1056*, 87–108.

Armstrong, D. A., Green, B. B., Seigne, J. D., Schned, A. R., & Marsit, C. J. (2015). MicroRNA molecular profiling from matched tumor and bio-fluids in bladder cancer. *Molecular Cancer*, *14*, 194.

Babjuk, M., Bohle, A., Burger, M., Capoun, O., Cohen, D., Comperat, E. M., … et al. (2017). EAU guidelines on non-muscle-invasive urothelial carcinoma of the bladder: Update 2016. *European Urology*, *71*, 447–461.

Babjuk, M., Burger, M., Comperat, E. M., Gontero, P., Mostafid, A. H., Palou, J., … et al. (2019). European association of urology guidelines on non-muscle-invasive bladder Cancer (TaT1 and carcinoma in situ)—2019 update. *European Urology*, *76*, 639–657.

Baumgart, S., Meschkat, P., Edelmann, P., Heinzelmann, J., Pryalukhin, A., Bohle, R., … Junker, K. (2019). MicroRNAs in tumor samples and urinary extracellular vesicles as a putative diagnostic tool for muscle-invasive bladder cancer. *Journal of Cancer Research and Clinical Oncology*, *145*, 2725–2736.

Boyerinas, B., Park, S. M., Hau, A., Murmann, A. E., & Peter, M. E. (2010). The role of let-7 in cell differentiation and cancer. *Endocrine-Related Cancer*, *17*, F19–F36.

Braicu, C., Buiga, R., Cojocneanu, R., Buse, M., Raduly, L., Pop, L. A., … et al. (2019). Connecting the dots between different networks: miRNAs associated with bladder cancer risk and progression. *Journal of Experimental & Clinical Cancer Research*, *38*, 433.

Calderaro, J., Rebouissou, S., de Koning, L., Masmoudi, A., Herault, A., Dubois, T., … et al. (2014). PI3K/AKT pathway activation in bladder carcinogenesis. *International Journal of Cancer*, *134*, 1776–1784.

Catto, J. W., Alcaraz, A., Bjartell, A. S., De Vere White, R., Evans, C. P., Fussel, S., … et al. (2011). MicroRNA in prostate, bladder, and kidney cancer: A systematic review. *European Urology*, *59*, 671–681.

Chang, Y., Tolani, B., Nie, X., Zhi, X., Hu, M., & He, B. (2017). Review of the clinical applications and technological advances of circulating tumor DNA in cancer monitoring. *Therapeutics and Clinical Risk Management*, *13*, 1363–1374.

Chen, C. L., Lai, Y. F., Tang, P., Chien, K. Y., Yu, J. S., Tsai, C. H., … et al. (2012). Comparative and targeted proteomic analyses of urinary microparticles from bladder cancer and hernia patients. *Journal of Proteome Research*, *11*, 5611–5629.

Chi, Q., Wang, Z. Y., Li, H. Y., Song, D. B., Xu, H., Ma, G., … Li, X. M. (2019). Tumor-suppressor microRNA-139-5p restrains bladder cancer cell line ECV-304 properties via targeting Connexin 43. *Chinese Medical Journal*, *132*, 2354–2361.

Chiyomaru, T., Seki, N., Inoguchi, S., Ishihara, T., Mataki, H., Matsushita, R., … et al. (2015). Dual regulation of receptor tyrosine kinase genes EGFR and c-Met by the tumor-suppressive microRNA-23b/27b cluster in bladder cancer. *International Journal of Oncology*, *46*, 487–496.

Chou, R., Gore, J. L., Buckley, D., Fu, R., Gustafson, K., Griffin, J. C., … Selph, S. (2015). Urinary biomarkers for diagnosis of bladder cancer: A systematic review and meta-analysis. *Annals of Internal Medicine*, *163*, 922–931.

Comperat, E. M., Burger, M., Gontero, P., Mostafid, A. H., Palou, J., Rouppret, M., … et al. (2019). Grading of urothelial carcinoma and the new "World Health Organisation classification of tumours of the urinary system and male genital organs 2016". *European Urology Focus*, *5*, 457–466.

Cumberbatch, M. G., Rota, M., Catto, J. W., & La Vecchia, C. (2016). The role of tobacco smoke in bladder and kidney carcinogenesis: A comparison of exposures and meta-analysis of incidence and mortality risks. *European Urology*, *70*, 458–466.

Deng, H., Lv, L., Li, Y., Zhang, C., Meng, F., Pu, Y., … et al. (2014). miR-193a-3p regulates the multi-drug resistance of bladder cancer by targeting the LOXL4 gene and the oxidative stress pathway. *Molecular Cancer*, *13*, 234.

Du, L., Jiang, X., Duan, W., Wang, R., Wang, L., Zheng, G., … et al. (2017). Cell-free microRNA expression signatures in urine serve as novel noninvasive biomarkers for diagnosis and recurrence prediction of bladder cancer. *Oncotarget*, *8*, 40832–40842.

Enokida, H., Yoshino, H., Matsushita, R., & Nakagawa, M. (2016). The role of microRNAs in bladder cancer. *Investigative and Clinical Urology*, *57*(Suppl. 1), S60–S76.

Erdmann, K., Salomo, K., Klimova, A., Heberling, U., Lohse-Fischer, A., Fuehrer, R., … et al. (2020). Urinary microRNAs as potential markers for non-invasive diagnosis of bladder cancer. *International Journal of Molecular Sciences*, *21*, 3814.

Feng, Y., Kang, Y., He, Y., Liu, J., Liang, B., Yang, P., & Yu, Z. (2014). microRNA-99a acts as a tumor suppressor and is down-regulated in bladder cancer. *BMC Urology*, *14*, 50.

Gobin, E., Bagwell, K., Wagner, J., Mysona, D., Sandirasegarane, S., Smith, N., … She, J. X. (2019). A pan-cancer perspective of matrix metalloproteases (MMP) gene expression profile and their diagnostic/prognostic potential. *BMC Cancer*, *19*, 581.

Gottardo, F., Liu, C. G., Ferracin, M., Calin, G. A., Fassan, M., Bassi, P., … et al. (2007). Micro-RNA profiling in kidney and bladder cancers. *Urologic Oncology*, *25*, 387–392.

Iio, A., Nakagawa, Y., Hirata, I., Naoe, T., & Akao, Y. (2010). Identification of non-coding RNAs embracing microRNA-143/145 cluster. *Molecular Cancer*, *9*, 136.

Ingelmo-Torres, M., Lozano, J. J., Izquierdo, L., Carrion, A., Costa, M., Gomez, L., … Mengual, L. (2017). Urinary cell microRNA-based prognostic classifier for non-muscle invasive bladder cancer. *Oncotarget*, *8*, 18238–18247.

Itesako, T., Seki, N., Yoshino, H., Chiyomaru, T., Yamasaki, T., Hidaka, H., … et al. (2014). The microRNA expression signature of bladder cancer by deep sequencing: The functional significance of the miR-195/497 cluster. *PLoS One*, *9*, e84311.

Jerome, T., Laurie, P., Louis, B., & Pierre, C. (2007). Enjoy the silence: The story of let-7 microRNA and cancer. *Current Genomics*, *8*, 229–233.

Jia, A. Y., Castillo-Martin, M., Domingo-Domenech, J., Bonal, D. M., Sanchez-Carbayo, M., Silva, J. M., & Cordon-Cardo, C. (2013). A common MicroRNA signature consisting of miR-133a, miR-139-3p, and miR-142-3p clusters bladder carcinoma in situ with normal umbrella cells. *The American Journal of Pathology*, *182*, 1171–1179.

Jiang, X., Du, L., Wang, L., Li, J., Liu, Y., Zheng, G., … et al. (2015). Serum microRNA expression signatures identified from genome-wide microRNA profiling serve as novel noninvasive biomarkers for diagnosis and recurrence of bladder cancer. *International Journal of Cancer*, *136*, 854–862.

Juracek, J., Peltanova, B., Dolezel, J., Fedorko, M., Pacik, D., Radova, L., … Stanik, M. (2018). Genome-wide identification of urinary cell-free microRNAs for non-invasive detection of bladder cancer. *Journal of Cellular and Molecular Medicine*, *22*, 2033–2038.

Kim, S. M., Kang, H. W., Kim, W. T., Kim, Y. J., Yun, S. J., Lee, S. C., & Kim, W. J. (2013). Cell-free microRNA-214 from urine as a biomarker for non-muscle-invasive bladder cancer. *Korean Journal of Urology*, *54*, 791–796.

Kozinn, S. I., Harty, N. J., Delong, J. M., Deliyiannis, C., Logvinenko, T., Summerhayes, I. C., … Rieger-Christ, K. M. (2013). MicroRNA profile to predict gemcitabine resistance in bladder carcinoma cell lines. *Genes & Cancer*, *4*, 61–69.

Kriebel, S., Schmidt, D., Holdenrieder, S., Goltz, D., Kristiansen, G., Moritz, R., … Ellinger, J. (2015). Analysis of tissue and serum microRNA expression in patients with upper urinary tract urothelial cancer. *PLoS One*, *10*, e0117284.

Kurozumi, A., Goto, Y., Okato, A., Ichikawa, T., & Seki, N. (2017). Aberrantly expressed microRNAs in bladder cancer and renal cell carcinoma. *Journal of Human Genetics*, *62*, 49–56.

Lei, M., Xie, W., Sun, E., Sun, Y., Tian, D., Liu, C., … Liu, L. (2015). microRNA-21 regulates cell proliferation and migration and cross talk with PTEN and p53 in bladder cancer. *DNA and Cell Biology*, *34*, 626–632.

Li, X., Dong, P., Wei, W., Jiang, L., Guo, S., Huang, C., … Xie, D. (2019). Overexpression of CEP72 promotes bladder urothelial carcinoma cell aggressiveness via epigenetic CREB-mediated induction of SERPINE1. *The American Journal of Pathology*, *189*, 1284–1297.

Li, Y., Liu, H., Lai, C., Du, X., Su, Z., & Gao, S. (2013). The Lin28/let-7a/c-Myc pathway plays a role in non-muscle invasive bladder cancer. *Cell and Tissue Research*, *354*, 533–541.

Linxweiler, J., & Junker, K. (2020). Extracellular vesicles in urological malignancies: An update. *Nature Reviews Urology*, *17*, 11–27.

Liu, W., Qi, L., Lv, H., Zu, X., Chen, M., Wang, J., … Li, Y. (2015). MiRNA-141 and miRNA-200b are closely related to invasive ability and considered as decision-making biomarkers for the extent of PLND during cystectomy. *BMC Cancer*, *15*, 92.

Long, J. D., Sullivan, T. B., Humphrey, J., Logvinenko, T., Summerhayes, K. A., Kozinn, S., … et al. (2015). A non-invasive miRNA based assay to detect bladder cancer in cell-free urine. *American Journal of Translational Research*, *7*, 2500–2509.

Mahdavinezhad, A., Mousavibahar, S. H., Poorolajal, J., Yadegarazari, R., Jafari, M., Shabab, N., & Saidijam, M. (2015). Association between tissue miR-141, miR-200c and miR-30b and bladder cancer: A matched case-control study. *Urology Journal*, *12*, 2010–2013.

Matsuzaki, K., Fujita, K., Jingushi, K., Kawashima, A., Ujike, T., Nagahara, A., … et al. (2017). MiR-21-5p in urinary extracellular vesicles is a novel biomarker of urothelial carcinoma. *Oncotarget*, *8*, 24668–24678.

Matullo, G., Naccarati, A., & Pardini, B. (2016). MicroRNA expression profiling in bladder cancer: The challenge of next-generation sequencing in tissues and biofluids. *International Journal of Cancer*, *138*, 2334–2345.

McGranahan, N., & Swanton, C. (2015). Biological and therapeutic impact of intratumor heterogeneity in cancer evolution. *Cancer Cell*, *27*, 15–26.

Mengual, L., Lozano, J. J., Ingelmo-Torres, M., Gazquez, C., Ribal, M. J., & Alcaraz, A. (2013). Using microRNA profiling in urine samples to develop a non-invasive test for bladder cancer. *International Journal of Cancer*, *133*, 2631–2641.

Miah, S., Dudziec, E., Drayton, R. M., Zlotta, A. R., Morgan, S. L., Rosario, D. J., … Catto, J. W. (2012). An evaluation of urinary microRNA reveals a high sensitivity for bladder cancer. *British Journal of Cancer*, *107*, 123–128.

Mittelbrunn, M., Gutierrez-Vazquez, C., Villarroya-Beltri, C., Gonzalez, S., Sanchez-Cabo, F., Gonzalez, M. A., … Sanchez-Madrid, F. (2011). Unidirectional transfer of microRNA-loaded exosomes from T cells to antigen-presenting cells. *Nature Communications*, *2*, 282.

Monteleone, N. J., & Lutz, C. S. (2017). miR-708-5p: A microRNA with emerging roles in cancer. *Oncotarget*, *8*, 71292–71316.

Morita, K., Fujii, T., Itami, H., Uchiyama, T., Nakai, T., Hatakeyama, K., … et al. (2018). NACC1, as a target of microRNA-331-3p, regulates cell proliferation in urothelial carcinoma cells. *Cancers (Basel)*, *10*, 347.

Motawi, T. K., Rizk, S. M., Ibrahim, T. M., & Ibrahim, I. A. (2016). Circulating microRNAs, miR-92a, miR-100 and miR-143, as non-invasive biomarkers for bladder cancer diagnosis. *Cell Biochemistry and Function*, *34*, 142–148.

Oeyen, E., Hoekx, L., De Wachter, S., Baldewijns, M., Ameye, F., & Mertens, I. (2019). Bladder cancer diagnosis and follow-up: The current status and possible role of extracellular vesicles. *International Journal of Molecular Sciences*, *20*, 821.

Oliveira, M. C., Caires, H. R., Oliveira, M. J., Fraga, A., Vasconcelos, M. H., & Ribeiro, R. (2020). Urinary biomarkers in bladder cancer: Where do we stand and potential role of extracellular vesicles. *Cancers (Basel)*, *12*, 1400.

Pardini, B., & Calin, G. A. (2019). MicroRNAs and long non-coding RNAs and their hormone-like activities in cancer. *Cancers (Basel)*, *11*, 378.

Pardini, B., Cordero, F., Naccarati, A., Viberti, C., Birolo, G., Oderda, M., … et al. (2018). microRNA profiles in urine by next-generation sequencing can stratify bladder cancer subtypes. *Oncotarget*, *9*, 20658–20669.

Pardini, B., Sabo, A. A., Birolo, G., & Calin, G. A. (2019). Noncoding RNAs in extracellular fluids as cancer biomarkers: The new frontier of liquid biopsies. *Cancers (Basel)*, *11*, 1170.

Pisitkun, T., Shen, R. F., & Knepper, M. A. (2004). Identification and proteomic profiling of exosomes in human urine. *Proceedings of the National Academy of Sciences of the United States of America*, *101*, 13368–13373.

Pospisilova, S., Pazourkova, E., Horinek, A., Brisuda, A., Svobodova, I., Soukup, V., … et al. (2016). MicroRNAs in urine supernatant as potential non-invasive markers for bladder cancer detection. *Neoplasma*, *63*, 799–808.

Qie, Y., Hu, H., Tian, D., Zhang, Y., Xie, L., Xu, Y., & Wu, C. (2016). The value of extensive transurethral resection in the diagnosis and treatment of nonmuscle invasive bladder cancer with respect to recurrence at the first follow-up cystoscopy. *Oncotargets and Therapy*, *9*, 2019–2025.

Ratert, N., Meyer, H. A., Jung, M., Lioudmer, P., Mollenkopf, H. J., Wagner, I., … et al. (2013). miRNA profiling identifies candidate mirnas for bladder cancer diagnosis and clinical outcome. *The Journal of Molecular Diagnostics*, *15*, 695–705.

Rupaimoole, R., & Slack, F. J. (2017). MicroRNA therapeutics: Towards a new era for the management of cancer and other diseases. *Nature Reviews. Drug Discovery*, *16*, 203–222.

Sabo, A. A., Birolo, G., Naccarati, A., Dragomir, M. P., Aneli, S., Allione, A., … et al. (2020). Small non-coding RNA profiling in plasma extracellular vesicles of bladder cancer patients by next-generation sequencing: Expression levels of miR-126-3p and piR-5936 increase with higher histologic grades. *Cancers (Basel)*, *12*, 1507.

Sachdeva, M., Zhu, S., Wu, F., Wu, H., Walia, V., Kumar, S., … Mo, Y. Y. (2009). p53 represses c-Myc through induction of the tumor suppressor miR-145. *Proceedings of the National Academy of Sciences of the United States of America*, *106*, 3207–3212.

Sapre, N., Macintyre, G., Clarkson, M., Naeem, H., Cmero, M., Kowalczyk, A., … Hovens, C. M. (2016). A urinary microRNA signature can predict the presence of bladder urothelial carcinoma in patients undergoing surveillance. *British Journal of Cancer*, *114*, 454–462.

Sasaki, H., Yoshiike, M., Nozawa, S., Usuba, W., Katsuoka, Y., Aida, K., … et al. (2016). Expression level of urinary MicroRNA-146a-5p is increased in patients with bladder cancer and decreased in those after transurethral resection. *Clinical Genitourinary Cancer*, *14*, e493–e499.

Scheffer, A. R., Holdenrieder, S., Kristiansen, G., von Ruecker, A., Muller, S. C., & Ellinger, J. (2014). Circulating microRNAs in serum: Novel biomarkers for patients with bladder cancer? *World Journal of Urology*, *32*, 353–358.

Shimizu, T., Suzuki, H., Nojima, M., Kitamura, H., Yamamoto, E., Maruyama, R., … et al. (2013). Methylation of a panel of microRNA genes is a novel biomarker for detection of bladder cancer. *European Urology*, *63*, 1091–1100.

Siegel, R. L., Miller, K. D., & Jemal, A. (2020). Cancer statistics, 2020. *CA: A Cancer Journal for Clinicians*, *70*, 7–30.

Sole, C., Arnaiz, E., Manterola, L., Otaegui, D., & Lawrie, C. H. (2019). The circulating transcriptome as a source of cancer liquid biopsy biomarkers. *Seminars in Cancer Biology*, *58*, 100–108.

Song, T., Xia, W., Shao, N., Zhang, X., Wang, C., Wu, Y., … Li, H. (2010). Differential miRNA expression profiles in bladder urothelial carcinomas. *Asian Pacific Journal of Cancer Prevention*, *11*, 905–911.

Stav, K., Leibovici, D., Goren, E., Livshitz, A., Siegel, Y. I., Lindner, A., & Zisman, A. (2004). Adverse effects of cystoscopy and its impact on patients' quality of life and sexual performance. *The Israel Medical Association Journal*, *6*, 474–478.

Sun, X., Du, P., Yuan, W., Du, Z., Yu, M., Yu, X., & Hu, T. (2015). Long non-coding RNA HOTAIR regulates cyclin J via inhibition of microRNA-205 expression in bladder cancer. *Cell Death & Disease*, *6*, e1907.

Taheri, M., Shirvani-Farsani, Z., Ghafouri-Fard, S., & Omrani, M. D. (2020). Expression profile of microRNAs in bladder cancer and their application as biomarkers. *Biomedicine & Pharmacotherapy*, *131*, 110703.

Tamagawa, S., Beder, L. B., Hotomi, M., Gunduz, M., Yata, K., Grenman, R., & Yamanaka, N. (2014). Role of miR-200c/miR-141 in the regulation of epithelial-mesenchymal transition and migration in head and neck squamous cell carcinoma. *International Journal of Molecular Medicine*, *33*, 879–886.

Tolle, A., Blobel, C. C., & Jung, K. (2018). Circulating miRNAs in blood and urine as diagnostic and prognostic biomarkers for bladder cancer: An update in 2017. *Biomarkers in Medicine*, *12*, 667–676.

Tolle, A., Buckendahl, L., & Jung, K. (2019). Plasma miR15b5p and miR5905p for distinguishing patients with bladder cancer from healthy individuals. *Oncology Reports*, *42*, 1609–1620.

Tran, M. N., Choi, W., Wszolek, M. F., Navai, N., Lee, I. L., Nitti, G., … et al. (2013). The p63 protein isoform DeltaNp63alpha inhibits epithelial-mesenchymal transition in human bladder cancer cells: Role of MIR-205. *The Journal of Biological Chemistry*, *288*, 3275–3288.

Tran, L., Xiao, J. F., Agarwal, N., Duex, J. E., & Theodorescu, D. (2020). Advances in bladder cancer biology and therapy. *Nature Reviews. Cancer*, *21*, 104–121.

Urquidi, V., Netherton, M., Gomes-Giacoia, E., Serie, D. J., Eckel-Passow, J., Rosser, C. J., & Goodison, S. (2016). A microRNA biomarker panel for the non-invasive detection of bladder cancer. *Oncotarget*, *7*, 86290–86299.

Villadsen, S. B., Bramsen, J. B., Ostenfeld, M. S., Wiklund, E. D., Fristrup, N., Gao, S., … et al. (2012). The miR-143/−145 cluster regulates plasminogen activator inhibitor-1 in bladder cancer. *British Journal of Cancer*, *106*, 366–374.

Wang, G., Chan, E. S., Kwan, B. C., Li, P. K., Yip, S. K., Szeto, C. C., & Ng, C. F. (2012). Expression of microRNAs in the urine of patients with bladder cancer. *Clinical Genitourinary Cancer*, *10*, 106–113.

Wang, L., Liu, Y., Du, L., Li, J., Jiang, X., Zheng, G., … et al. (2015). Identification and validation of reference genes for the detection of serum microRNAs by reverse transcription-quantitative polymerase chain reaction in patients with bladder cancer. *Molecular Medicine Reports*, *12*, 615–622.

Wang, R., Wu, Y., Huang, W., & Chen, W. (2018). MicroRNA-940 targets INPP4A or GSK3beta and activates the Wnt/beta-catenin pathway to regulate the malignant behavior of bladder cancer cells. *Oncology Research*, *26*, 145–155.

Wang, J., Zhang, X., Wang, L., Dong, Z., Du, L., Yang, Y., … Wang, C. (2015). Downregulation of urinary cell-free microRNA-214 as a diagnostic and prognostic biomarker in bladder cancer. *Journal of Surgical Oncology*, *111*, 992–999.

Witjes, J. A., Bruins, H. M., Cathomas, R., Comperat, E. M., Cowan, N. C., Gakis, G., … et al. (2020). European association of urology guidelines on muscle-invasive and metastatic bladder cancer: Summary of the 2020 guidelines. *European Urology*, *79*, 82–104.

Yang, X., & Wang, P. (2019). MiR-188-5p and MiR-141-3p influence prognosis of bladder cancer and promote bladder cancer synergistically. *Pathology, Research and Practice*, *215*, 152598.

Yin, X. H., Jin, Y. H., Cao, Y., Wong, Y., Weng, H., Sun, C., … Zeng, X. T. (2019). Development of a 21-miRNA signature associated with the prognosis of patients with bladder cancer. *Frontiers in Oncology*, *9*, 729.

Yonemori, M., Seki, N., Yoshino, H., Matsushita, R., Miyamoto, K., Nakagawa, M., & Enokida, H. (2016). Dual tumor-suppressors miR-139-5p and miR-139-3p targeting matrix metalloprotease 11 in bladder cancer. *Cancer Science*, *107*, 1233–1242.

Yun, S. J., Jeong, P., Kim, W. T., Kim, T. H., Lee, Y. S., Song, P. H., … Kim, W. J. (2012). Cell-free microRNAs in urine as diagnostic and prognostic biomarkers of bladder cancer. *International Journal of Oncology*, *41*, 1871–1878.

Chapter 26

Pathophysiology roles and translational opportunities of miRNAs in cutaneous melanoma

Manuela Ferracin, Elisabetta Broseghini, and Emi Dika
Department of Experimental, Diagnostic, and Specialty Medicine, University of Bologna, Bologna, Italy

Cutaneous melanoma

Incidence and epidemiology

Melanoma is a malignancy of melanocytes, pigment-producing cells that are predominantly located in the skin. Melanoma accounts for 3%–5% of all cutaneous cancers but determines approximately 75% of all skin cancer deaths (Shenenberger, 2012). Cutaneous melanoma (CM) affects principally fair-skinned populations, with variable incidence attributable to latitude and differences in sun exposure. The geriatric population is principally affected by this tumor and the median age at diagnosis is 65 years, but this neoplasm is also frequently diagnosed in adolescent and young adult populations (Erdmann et al., 2013). Melanoma incidence increases linearly starting from the age of 25 years until 50 (Matthews, Li, Qureshi, Weinstock, & Cho, 2017; NIH & SEER, 2019). Melanoma affects women and men indifferently, but men are generally more affected. The incidence rates in the world are 3.5% for men and 2.9% for women (WHO, 2018). When age is considered, adolescent and young adult women are more susceptible to melanoma than men. However, after the age of 40, the rates reverse, and melanoma incidence among men is greater than in women (Weir et al., 2011).

In the past decades, the worldwide incidence of cutaneous melanoma has been increasing more rapidly than any other cancer type. Worldwide in 2018, a total of 287,723 new cases (comprehensive of all ages and both sexes) was reported (WHO, 2018). The incidences are higher in the Caucasian population (Guy et al., 2015; Viros et al., 2008) and about 85% of the CM cases that annually occur in the world affect the populations of North America, Europe, and Oceania. Moreover, the world's highest incidence rates are registered in Australia (30–50/100.000 people per year) while in Europe, the incidence rate is 11.2/100.000. Very low incidence rates (below 0.5/100.000 new cases per year) are estimated in Southeastern Asia and South-Central Asia (Arnold et al., 2014; Ferlay et al., 2015).

The 5-year survival is about 95%–98%, but when regional or distant metastases are present, this rate decreases dramatically to 63% and 20%, respectively (Matthews, Li, Qureshi, Weinstock, & Cho, 2017; NIH & SEER, 2019; Siegel et al., 2018).

Melanoma etiology

The pathogenesis of cutaneous melanoma is complex. CM is considered a multifactorial disease, which results from a combination of environmental, phenotypic, and genetic factors such as ultraviolet (UV) exposure, fair phototypes, multiple dysplastic nevi, and a positive family/personal history of cutaneous melanoma.

Environmental and phenotypic risk factors

Among the environmental factors, the most common risk factor is UV radiation from excessive sun exposure or the use of tanning beds. Several studies have underlined the direct association between sun exposure and increasing risk of melanoma, as the sun is a major source of UV radiation (Elwood & Koh, 1994; Newton-Bishop et al., 2011). Melanocytes are resistant to UV-induced apoptosis; however, the growth and accumulation of genetic mutations lead to their transformation in melanoma (Gilchrest et al., 1999). Skin exposure to UV radiation, especially UV-B, leads to severe damage in DNA and melanocyte functions, inducing direct alterations (thymine dimers, pyrimidine dimers, cyclobutane pyrimidine dimers, and

MicroRNA in Human Malignancies. https://doi.org/10.1016/B978-0-12-822287-4.00022-0

UV-endonuclease-sensitive sites) in double-stranded DNA and indirect DNA alterations (consequence of reactive oxygen species generation) (Bais et al., 2018).

Many other factors, such as the level and type of sun protection, chronic or intermittent UV exposure, and a history of sunburn in childhood, are all related to the amount of UV radiation necessary to increase the melanoma risk. Intense and intermittent sun exposure appears to be especially related to developing cutaneous melanoma while a chronical pattern of exposure is associated with nonmelanoma skin cancer or actinic keratosis disease (Gandini et al., 2005b; Wojcik et al., 2019).

The rise in melanoma incidence is driven also by the increasing use of tanning sunbeds, an artificial source of UV radiation. Indeed, people who frequently use tanning sunbeds expose themselves to a higher amount of UVA in comparison to sunbathing or outdoor activity exposure (Arnold et al., 2018). Furthermore, a positive association was found with episodes of sunburn and melanoma. Sunburn is considered an acute sun exposure. Recurring episodes of sunburn in childhood (under 15 years) correlate with a highest risk of melanoma (Naldi et al., 2000; Newton-Bishop et al., 2011; Wojcik et al., 2019).

Other environmental risk factors for melanoma have been demonstrated to be the skin pigmentation phenotype (fair or light skin) and the presence of atypical and multiple nevi.

Reduced pigmentation is a risk factor for melanoma. Fair-skinned individuals are more likely to develop melanoma than more darkly pigmented individuals. A light skin color confers approximately a two-fold increase in risk, and red hair confers a two-fold increase in risk compared to black or dark brown hair. Also, extensive freckling is associated with a two-fold risk, which increases to three-fold in people under 40 (Bliss et al., 1995). Finally, the risk increases also in people with a specific phototype that includes red hair, freckles, fair skin, light eyes, and sun sensitivity (Gandini et al., 2005c; Titus-Ernstoff et al., 2005).

The number of melanocytic nevi is an important risk factor. Nevi may be congenital or acquired. Sometimes these moles (nevi) have a suspicious dysplastic morphology and they are clinically called "atypical" or "dysplastic." Atypical nevi are characterized by disproportional dimensions (large nevi with diameter >5 mm or giant nevi with diameter >20 cm), variable pigmentation, and irregular and indistinct borders. Indeed, melanoma risk varies according to the number and size of nevi. Approximatively 36% of melanoma cases develops in patients with a preexisting dysplastic nevus (Purdue et al., 2005) (Gandini, Sera, Cattaruzza, Pasquini, Abeni, et al., 2005a; Naldi et al., 2000). People with >100 nevi have a significantly higher risk of melanoma and those who develop nevus-associated melanomas show a distinct risk profile that differs from that of patients with de novo melanomas (Haenssle et al., 2016). In these nevus-associated melanoma patients, the tumors are generally located on the trunk and belong to the superficial spreading subtype (Purdue et al., 2005).

Hereditary genetic alterations in melanoma

In addition to environmental and phenotypic risk factors, germline genetic alterations contribute to the increase of melanoma risk (Shain et al., 2015). Indeed, a familial history of melanoma is associated with an increased risk of tumor development. In fact, about 10% of melanomas develop in patients with a familial history of melanoma: the presence of a germline mutation constitutes a risk factor for the hereditary form of the disease (Colombino et al., 2014) (Gandini, Sera, Cattaruzza, Pasquini, Zanetti, et al., 2005c; Pasquali et al., 2015). Several studies demonstrated the presence of a clear pattern of autosomal dominant inheritance with multiple family members affected in various generations. One of the principal germline mutations was found in CDKN2A (Helgadottir et al., 2017) (Tsao & Niendorf, 2004). Less common melanoma susceptibly genes include CDK4, TERT, MITF, BAP1, and POT1 (Read, Wadt, & Hayward, 2016).

Cyclin dependent kinase inhibitor 2A (*CDKN2A*) is located on chromosome 9p21.3 and can be identified as a high penetrance gene that predisposes to melanoma. This gene is commonly altered in patients with a familial history of melanoma (Goldstein et al., 2007). *CDKN2A* gene encodes two unrelated proteins in different reading frames arising from alternatively spliced transcripts: the p16 inhibitor of cyclin-dependent kinase 4 ($p16^{INK4a}$) and the p14 alternate reading frame ($p14^{ARF}$). $p16^{INK4a}$ inhibits cyclin-dependent kinase 4 (CDK4) and prevents that cell from entering cycle S-phase while $p14^{ARF}$ is a positive regulator of P53 and therefore a loss of $p14^{ARF}$ correlates with accumulation of DNA damage. Mutations in *CDKN2A* can affect either only one protein ($p16^{INK4a}$ or $p14^{ARF}$) or both, depending on which exon is interested by the mutation (Goldstein et al., 2007). An autosomal dominant inheritance of germline *CDKN2A* mutations is found in approximately 20%–40% of familial melanomas. This germline *CDKN2A* mutation is the strongest genetic risk factor for cutaneous melanoma, and some mutations are found also in 5%–15% of multiple primary melanoma (MPM) cases (Helgadottir et al., 2017).

Cyclin dependent kinase 4 *(CDK4)* is located on chromosome 12q14.1 and encodes for a member of the Ser/Thr kinase family. The CDK4 protein is active during cell cycle G1 phase and physiologically its product is negatively regulated p16. Mutations on the *CDK4* gene impact the same pathway as *CDKN2A* mutations. Germline mutations in *CDK4* genes have been identified on codon 24 in exon 2. These mutations cause the substitution of arginine with histidine (R24H) or cysteine (R24C), resulting in a CDK4 oncoprotein insensitive to p16 inhibition. No particular differences in phenotypic characteristics, such as age at diagnosis and number of melanomas, are found in families carrying *CDKN2A* or *CDK4* mutations (Nikolaou et al., 2011).

Protection of telomeres 1 (*POT1*) is located on chromosome 7q31.33 and belongs to the complex of Shelterin (also known as Telosome) that controls telomere length and protects chromosome extremities from illegitimate recombination and abnormal chromosome segregation. Most germline mutations found in this gene affect oligonucleotide/oligosaccharide binding (OB) fold domains, which are crucial for the binding of POT1 to telomeric single-stranded DNA (Flynn & Zou, 2010). Melanoma-associated *POT1* mutations include a Tyr89Cys variant of the N-terminal OB domain. Familial melanoma pedigrees negative for *CDKN2A* and *CDK4* mutations were found to have a variant in *POT1* in 4% of the melanoma families (Robles-Espinoza et al., 2014).

Telomerase reverse transcriptase (*TERT*) is located on chromosome 5p15.33 and encodes for an enzyme with reverse transcriptase activity that maintains telomere ends by the addition of the telomere repeat TTAGGG, playing an important role in cellular senescence. The main *TERT* mutation is transversion T->G in the gene promoter. This mutation was first identified in familial melanomas using high-throughput sequencing of four affected and four nonaffected individuals in a 14-case German family (Horn et al., 2013). This TERT mutation appears to be linked with a high risk for early onset melanoma and other cancers.

BRCA1-associated protein-1 (*BAP1*) is located on chromosome 3p21 and is a tumor-suppressor gene with a physiological deubiquitinase function, which was recently found mutated in cutaneous melanoma. Several studies demonstrated that *BAP1* germline mutations may be associated with tumor predisposition syndrome (TPDS) and that it is linked with various cancers (Wadt et al., 2015) (Wiesner et al., 2011). *BAP1* germline mutations in melanoma seem to be related to the presence of little pink or tan papules and nodules termed "melanocytic BAP1-mutated atypical intradermal tumors" (MBAITs) (Wiesner et al., 2012, 2011). *BAP1* functional inactivation seems to contribute to a small proportion of sporadic cutaneous melanoma, with an absence of *BAP1* expression described in approximately 5% of tumors (Murali et al., 2013).

Microphthalmia-associated transcription factor (*MITF*) is located on chromosome 3p13 and has recently been identified as a medium penetrance gene predisposing to melanoma. *MITF* is a gene that encodes a transcription factor of the basic-helix–loop–helix-leucine zipper type. MITF is a major regulator of melanocyte development and has a crucial role in melanoma oncogenesis (Möller et al., 2019). The substitution of glutamic acid at position 318 with lysine, called *MITF* E318K, is identified as a recurrent germline mutation in melanoma. This mutation acts as a gain-of-function mutation, decreasing the SUMOylation effect with a consequent increased transcriptional function of the MITF protein (Bertolotto et al., 2016). Frequently, the *MITF* E318K mutation was correlated with a particular phenotype, comprising a nonblue eye color, an increased number of nevi, and multiple primary melanomas (Bertolotto et al., 2016; Sturm et al., 2014).

Somatic genetic alterations in melanoma

Most mutations in melanoma are attributable to random acquired mutations, which can contribute to cancer development and metastasis progression. Somatic mutations in melanoma occur with a certain frequency in the genes that are correlated with mitogen-activated protein kinase (MAPK) pathway activation. At least three MAPK families have been characterized, namely classical MAPK (originally called extracellular signal-regulated kinases, ERK), C-Jun N-terminal kinase/stress-activated protein kinase (JNK/SAPK), and p38 kinase. MAP kinases lie within protein kinase cascades. Each cascade consists of at least three enzymes that are activated in series: an MAPK kinase kinase kinase (MAPKKK), an MAPK kinase kinase (MAPKK), and an MAP kinase (MAPK) (Wei & Tu, 2002). The MAPK/ERK pathway includes a chain of protein in the cells that transmits a signal from a receptor protein located on the surface of the cells to the DNA in the nucleus. The signal starts when a signaling molecule, namely a ligand, binds the receptor on the surface and ends with the activation/repression of a specific set of genes with a central role in cell proliferation, differentiation, and survival (Raman et al., 2007).

The phosphoinositol-3-kinase/protein kinase B (PI3K/Akt) pathway also takes control of cellular homeostasis. Specifically, the PI3K Akt pathway is an intracellular signal transduction pathway that promotes metabolism, proliferation, cell survival, growth, and angiogenesis in response to extracellular signals. PI3K activation phosphorylates and further activates Akt, which localizes to the cellular membrane. Activated Akt activates cAMP-response element binding protein and phosphatidylinositol 3-phosphate (PI3KP), the mammalian target of rapamycin (mTOR); inhibits cyclin-dependent kinase inhibitor 1B (CDKN1B, also known as p27); and localizes Forkhead box protein O (FOXO) in the cytoplasm (Xie et al., 2019).

The MAPK and PI3K/AKT pathways are pivotal players in the biology of different cancer types, including melanoma (Carlino et al., 2015).

The most frequent genes of these pathways with somatic mutations in melanoma include B-Raf proto-oncogene (also known as v-Raf murine sarcoma viral oncogene homolog B, *BRAF*), neuroblastoma RAS viral oncogene homolog (*N-RAS*), neurofibromin 1 (*NF1*), proto-oncogene receptor tyrosine kinase (*KIT*), and the phosphatase and tensin homolog (*PTEN*). *CDKN2A* and *MITF* genes are found mutated also in melanoma with somatic alterations.

The *BRAF* gene is located on chromosome 7q34 and encodes for serine/threonine-protein kinase B-Raf (BRAF), which is composed of 766 amino acids and three domains: one catalytic domain involved in MEK (also known as mitogen-activated protein kinase kinase and MAP2K) phosphorylation and two regulatory ones. Physiologically, the catalytic domain keeps the protein in its inactivate conformation while the active conformation induces the activation of MEK and consequently of ERK, the final effector of the pathway (Davies et al., 2002). Up to 50% of somatic mutations derived from intermittent sun exposure damage are present on the *BRAF* gene (Ponti et al., 2017). BRAF mutation takes place in 71%–75% of melanomas (Ponti et al., 2017) and it occurs in the catalytic domain, where hydrophobic valine is replaced by hydrophilic glutamic acid. Also, the kinase becomes constitutively active and insensitive to negative feedback mechanisms, inducing angiogenesis, evading apoptosis, and promoting invasion and metastasis (Maurer et al., 2011; Pratilas et al., 2009).

Frequently, gain-of-function mutations are in melanoma *BRAF* mutations: the main one is a missense mutation that determinates amino acid substitution at position 600, named BRAF V600E (it covers up to 90% of all BRAF mutations) (Cheng et al., 2018). Some studies report the BRAF V600E mutation also in common and atypical nevi (Poynter et al., 2006; Shain & Bastian, 2016) and the rate of BRAF V600E mutation is even lower in melanoma in comparison with benign nevi (Seitz-Alghrouz et al., 2018). This evidence sheds light on the probable role of classic BRAF V600E as an early step in melanocytic transformation, but starting from a precursor benign lesion, this mutation alone is not relevant in promoting tumor progression (Uribe et al., 2003). Finally, beyond the classical BRAF V600E mutation, there are various other point mutations in the same position (e.g., V600K, V600D, V600R) with the same effect of increasing BRAF kinase activity. Less than 1% is accounted for in other mutational sites (Cheng et al., 2018). Patients with BRAF-mutated melanoma are frequently younger than other melanoma patients (both in terms of diagnosis of primary melanoma and metastasis) and they often develop tumors with superficial spreading or nodular histology (Long et al., 2011).

N-RAS is located on chromosome 1p13.2 and is mutated in about 15%–30% of melanomas. N-RAS belongs to the RAS family and is also involved in the MAPK pathway (Jakob et al., 2012). This gene encodes for a small GTP binging protein (189 amino acids) that becomes active when bound to GTP and becomes inactivated once it hydrolyzes GTP to GDP (Platz et al., 2008). *N-RAS* mutations are linked with chronic sun exposure (Van Dijk et al., 2005). *N-RAS* is frequently interested by missense mutations, especially in exon 2 at codon 61, named Q61- (that accounts for 80% of NRAS mutations in melanoma). These mutations involve the substitution of glutamine with lysine (Q61K), arginine (named Q61R), or leucine (Q61L), which impairs GTP hydrolysis (Burd et al., 2014; Omholt et al., 2002). Consequently, the protein becomes constitutively active and can drive both MAPK and PIK3/AKT pathways with a central role in cell-cycle dysregulation, prosurvival mechanisms, and cellular proliferation (Hodis et al., 2012). Generally, *N-RAS* and *BRAF* mutations are mutually exclusive in melanoma (Goel et al., 2006).

It was observed that patients with N-RAS Q61- mutations tend to be older (> 55 years) than patients with the BRAF V600E mutation. This mutation is associated with the presence of lesions located at the extremities, a lower rate of ulceration, thicker primary tumors, and a higher rate of mitosis than in the BRAF mutation (Muñoz-Couselo et al., 2017). *N-RAS* mutations are considered independent adverse prognostic factors. In some studies, their association was underlined with shorter overall survival (OS) in comparison to tumors with BRAF mutations when the diagnosis occurs in an advanced stage of disease (Jakob et al., 2012; Omholt et al., 2003). In contrast, in an early stage of the disease, *N-RAS* mutations have no impact on OS (Ellerhorst et al., 2011).

NF-1 is located on chromosome 17q11.2 and is a tumor-suppressor gene that encodes a protein of 2839 amino acids, which inactivates RAS downstream signaling by catalyzing the hydrolysis of RAS-GTP to RAS-GDP. *NF-1* mutations can cooccur with BRAF or N-RAS mutations in human melanoma. NF-1 loss of function leads to the hyperactivation of RAS proteins and stimulates both MAPK and PIK3/Akt pathway signaling (Maertens et al., 2013). NF-1 mutations are the third most common driver mutation in melanoma, reported in about 14% of cases (Manzano et al., 2016); NF-1 mutant melanomas show a higher tumor mutational burden that other subtypes and a robust UV-derived mutational signature (Cirenajwis et al., 2017).

KIT is located on chromosome 4q12 and encodes for a transmembrane receptor of 976 amino acids with tyrosine kinase activity that binds the stem cell factor and allows activation of the intracellular kinase domain (Shtivelman et al., 2014). *KIT* mutation occurs in 3% of all melanomas and is mutually exclusive with BRAF and N-RAS mutations (Carvajal et al., 2015). It is especially found in 15% of acral and mucosal subtypes of melanoma (Curtin et al., 2006; Vita et al., 2014). The main mutations are L576P (lysine to proline substitution on 576 codon) localized to exon 11 and K642E (methionine to glutamic acid substitution on 642 codon) localized to exon 13 (Vita et al., 2014). These somatic mutations in the *KIT* gene result in a gain of function, leading to a constitutive activation of kinase activity, inducing both MAPK and PI3K/AKT pathways (Shtivelman et al., 2014).

PTEN is located on chromosome 10q23.31 and is a tumor-suppressor gene that encodes for the phosphatidyl-inositol-3,4,5-triphosphate 3- phosphatase (PTEN) protein; it is a key regulator of the PI3K signaling pathway. The most frequent *PTEN* mutations are frameshift or chromosomal deletion and they occur in up to 30% of cutaneous melanomas (Wu et al., 2003). Studies showed that *PTEN* mutations are more prevalent in invasive and metastatic melanomas than in primary tumors (Tsao et al., 2004). *PTEN* mutation often coexists with the *BRAF* mutation and contributes to activating the PI3K/AKT pathway, but it is mutually exclusive with the *N-RAS* mutation, which independently activates the same pathway (Nogueira et al., 2010; Tsao et al., 2004). Epigenetic mechanisms are also implicated in the regulation of PTEN (Mirmohammadsadegh et al., 2006). Loss or partial loss of gene function in tumoral cells seems to contribute to melanoma development, increasing Akt activity, stimulating mitogen signaling, and decreasing apoptosis (Giles et al., 2019).

A genomic classification of melanoma has been established that classified melanomas into four subtypes based on the pattern of the most prevalent significantly mutated genes: mutant *BRAF*, mutant *RAS*, mutant *NF1*, and triple-WT (wild-type). The triple-WT subtype is enriched with *KIT* mutations, focal amplifications, and complex structural rearrangements. The first subgroup presents (mutant) *BRAF* and *MITF* amplification and patients are generally younger compared to the other subgroups. The mutant *RAS* subgroup is characterized by MAPK activation and RAC-gamma serine/threonine-protein kinase (Akt3) overexpression. In the mutant *NF1* subgroup, patients are older and there is a higher mutation burden, which is the approximate amount of gene mutation that occurs in the genome of a cancer cell. Finally, the triple wild-type subgroup lacks a UV signature and manifests a higher copy number and complex rearrangements (Akbani et al., 2015).

The *MITF* gene is also mutated in sporadic melanoma and can generate different isoforms as a consequence of alternative splicing and use of alternative promoters: *M-MITF* is the central isoform in melanocyte development and carcinogenesis (Wellbrock & Arozarena, 2015). This form is found in about 80% of melanomas and is present at different stages of tumor progression (Wellbrock & Arozarena, 2015). Metastatic melanoma acquires *MITF* mutation in 20% of cases (Shtivelman et al., 2014). Additionally, *MITF* is important for the progression of melanoma carrying the *BRAF* V600E mutation (Lister et al., 2014).

Pathological staging

According to the last World Health Organization (WHO) classification, (Elder, Massi, Scoyler, & Willemze, 2017), the American Joint Committee on Cancer (AJCC) developed the TNM classification system for cutaneous melanomas considering some relevant features related to the different clinical stagings of the tumor (e.g., Breslow thickness, ulceration, metastases) (Amin et al., 2017). The clinical stage of the tumor is determined after a biopsy and/or a clinical assessment of regional lymph nodes.

T category

The T category of melanoma is determined primarily by measuring the thickness of the melanoma (from T1 to T4), and secondly by defining the ulceration status (a or b). Tumor thickness is measured from the top of the granular layer to the deepest invasive cell across the broad base of the tumor (in the dermis or subcutis) (Amin et al., 2017). Generally, higher values of Breslow thickness correspond to poorer prognosis (Gimotty et al., 2007; Green et al., 2012). Ulceration is defined from the total absence of an intact epidermis, in particular of the stratum corneum and basement membrane of the dermo-epidermal junction, and from the presence of an associated host reaction (with fibrin deposition, neutrophils) above the primary melanoma (Amin et al., 2017; Spatz et al., 2003).

Specifically, T1 includes melanomas with a thickness ≤ 1 mm, T2 with a thickness higher than 1 mm and lower than 2 mm, T3 has a thickness between 2 and 4 mm, and finally T4 is higher than 4 mm. "TX" is used when the primary tumor thickness cannot be assessed while "T0" is used when there is no evidence of a primary tumor (unknown primary tumor or completely regressed melanoma). Melanoma in situ (Tis), is used when melanocytes proliferate remaining confined within the epidermal layer of the skin only and not invading the underlying dermis and deeper tissues (Gershenwald & Scolyer, 2018). T1 tumors are further classified into T1a and T1b groups. T1a has a Breslow <0.8 mm and no ulceration while T1b consists of melanomas with a thickness <0.8 mm with ulceration or melanomas with a thickness between 0.8 and 1 mm with or without ulceration. Patients with a thickness ≤ 0.8 mm have a better prognosis than patients with a melanoma ≥ 0.8 or 1.0 mm (Lo et al., 2018). For the other T classes (T2, T3, T4), each T number is followed by the letter "a" when the melanoma is without ulceration; otherwise, when there is ulceration, the T number is followed by the letter "b" (Gershenwald & Scolyer, 2018).

N category

The N category reflects the number of tumor-involved regional lymph nodes (N1 – 3). In addition, the presence or absence of in transit, satellite, and/or microsatellite metastases is used in subclassification (a-c). Specifically, to enhance clarity, the eighth edition of melanoma staging defines the regional disease as clinically occult (subgroup a), namely detected by a sentinel lymph node (SLN) biopsy; and clinical evident (subgroup b), when detected by clinical examination or radiographic imaging. The subgroup c indicates the presence of microsatellites, satellites, or in transit metastases.

NX is used when regional lymph nodes are not assessed, for example when an SLN biopsy is not performed or regional lymph nodes have been previously removed for other reasons. There is an exception: the pathological N category is not required for T1 melanomas. N0 is used when no regional metastases are detected (Gershenwald & Scolyer, 2018).

N1a is used when one clinically occult tumor-involved regional lymph node is detected (detected by SLN biopsy), N1b for one clinically detected tumor-involved regional lymph node, and N1c for cases where there is not a regional lymph node disease, but in transit, satellite, and/or microsatellite metastases are present. N2a tumors have two or three clinical occult tumor-involved regional lymph nodes; N2b two or three tumor-involved regional lymph nodes, at least one of which was clinically detected; and N2c has one tumor-involved regional lymph node, either clinically occult or detected, and the presence of in transit, satellite, and/or microsatellite metastases. N3a is used when four or more tumor-involved regional lymph nodes are detected but clinically occult; N3b when four or more positive regional lymph nodes are clinically occult but at least one is clinically detected, or there is any number of matted nodes; and N3c has two or more positive regional lymph nodes that are clinically occult or detected and/or the presence of any number of matted nodes, and there are in transit, satellite, and/or microsatellite metastases (Gershenwald & Scolyer, 2018).

M category

The M category defines the absence or presence of distant metastasis (0–1), and the site(s) of development of metastasis (a-d). The latest melanoma classification has also maintained the analysis of serum LDH levels as a classification feature, and it reported using the following suffix for the M category: (0) LDH not elevated, (1) LDH elevated (Gershenwald & Scolyer, 2018).

M0 melanomas show no evidence of metastasis while M1 is used when there is evidence of distant metastasis. The letter "a" is used when the localization of the metastases is skin, soft tissue including muscle, and/or nonregional lymph nodes; the letter "b" is used for distant metastasis localized in the lungs with or without an M1a site of disease; "c" is used for distant metastasis localized in non-aCNS visceral sites with or without M1a or M1b sites of disease; and finally "d" defines distant metastasis in CNS with or without M1a, M1b, or M1c sites of disease (Gershenwald & Scolyer, 2018).

Stage groups

Melanoma staging is based on the degree of disease spread: localized (stage I and II), regional (stage III), and distant (stage IV). Stage 0 correspond to melanoma in situ, namely Tis. Stage I and II are N0 and M0. Therefore, their stage changes depending on the T value. Specifically, stage IA includes T1a and T1b, stage IB includes T2a, stage IIA includes T2b and T3a, stage IIB includes T3b and T4a, and IIC corresponds to T4b. Stage III has no metastasis (M0) and is divided into four subgroups: IIIA (T1a/b-T2a, N1a or N3a), IIIB (T1a/b-T2a, N1b/c or N2b; T2b/T3a, N1a-N2b; T0, N1b/c), IIIC (T1a-T3a, N2c or N3a/b/c; T3b/T4a, any N N1; T4b, N1a-N2c; T0, N2b/c or N3b/c), and IIID (T4b, N3a/b/c). Finally, stage IV includes any T, any N, and distant metastasis (M1) (Gershenwald & Scolyer, 2018).

Other very important histologic features in tumor prognosis are regression and the presence of tumor-infiltrating lymphocytes (TILs). The histopathologic classification of TILs consists of: 1) absent TIL infiltrate: no lymphocytes are present or, if present, they do not interact with the tumor cells; 2) nonbrisk TIL infiltrate: focal areas of lymphocytic infiltration in the tumor (isolated, multifocal, or segmental); and3) brisk TIL infiltrate: TIL infiltration covers the entire base of the tumor (Amin et al., 2017).

Finally, regression is characterized by the loss of thickness of the lesions with scarring fibrosis and the presence of inflammation between an apparently separate nodule and the primary tumor (rather than normal stroma); if these findings are present, then the nodule is considered to be an extension of the primary tumor and not a micrometastasis. The relationship between regression and prognosis is controversial (Bhoyrul et al., 2019; Gershenwald et al., 2017).

Diagnosis and prognosis

The early detection of melanoma greatly increases the chances of survival. For this reason, early recognition, detection, and treatment remain crucial (Rigel et al., 2010). One of the major advantages of melanoma is the cutaneous location, which allows a noninvasive skin exam to contribute to an early diagnosis (Rigel et al., 2010).

Before 1980, the diagnosis was made by identifying clinically macroscopic features. After that, ABCDE criteria were adopted to give the physicians a simple, objective, and adequate tool for morphological diagnosis, evaluating and differentiating common moles from cutaneous lesions most suspicious for melanoma development. ABCDE is an acronym for "asymmetry, border irregularity, color variegation, diameter, evolving." These criteria increase the sensitivity of the skin examination from 57% to 90% (Robinson & Turrisi, 2006).

Recently, dermoscopes (or dermatoscopes, epiluminescent microscope) began being used. These are high-resolution devices that allow increasing accuracy in melanoma detection. It is a noninvasive technique that uses a handheld optical device that permits the observation of structures not visible to the naked eye (Neila & Soyer, 2011). Nowadays, to perform a correct screening, three-dimensional (3D) technologies can be used. It is possible to obtain total-body images through a system of many cameras that takes photos of the entire body; you can then retrieve images in a digital 3D model (Rayner et al., 2018). The model shows every nevus or spot on the body, and in this way, doctors can observe every minimal change in the early stage of melanoma. This can also permit just monitoring patients at high risk due to their personal or family history or those with multiple dysplastic nevi that can be difficult to monitor without technology (Rayner et al., 2018). The diagnostic accuracy and sensitivity for skin tumors have been improved with the development of many noninvasive in vivo techniques, including ultrasound, confocal scanning laser microscopy, magnetic resonance imaging, and optical coherence tomography (Branzan et al., 2007).

However, the histological examination remains the gold standard for diagnosis. A definitive diagnosis of melanoma requires a biopsy that is analyzed by an expert pathologist. Diagnostic criteria for melanoma include architectural and cytologic aspects of the tumor, which prove to be fundamental to define the diagnosis, staging, and prognosis (Smoller, 2006). As the molecular mechanisms driving melanoma progression have become better understood, molecular biomarkers are utilized to assist in the recognition of melanoma, and immunohistochemistry (IHC) has been increasingly utilized, especially for the interpretation of difficult cases (Kim & Meehan, 2017). IHC tests for biomarkers expressed in or on the surface of tumor cells.

For melanoma, the most used biomarkers for diagnostic purposes are melanocytic markers, which are used to determine if an ambiguous lesion has a melanocytic origin and if it expresses proteins involved in melanosome biogenesis, melanin synthesis, or melanocyte differentiation. Melan-A (also known as MART-1, melanoma antigen recognized by T-cells) is encoded by the *MLANA* gene and is a melanosomal differentiation antigen that is recognized by cytotoxic T-cells. Another used melanocytic antigen is gp100 that is recognized by the antibody HMB-45 (human melanoma black). Gp100 is a melanocyte lineage-specific antigen, which is recognized by melanoma-derived tumor-infiltrating lymphocytes. It plays a role in melanin polymerization, melanosome biogenesis, and melanogenesis (Bakker et al., 1994). Other common melanocytic markers used for melanoma diagnosis include the S100 protein, MITF, tyrosinase, and SRY-box transcription factor 10 (SOX10) (Davis et al., 2019). However, melanocytic markers stain all melanocytes and thus they do not discriminate between melanoma and benign melanocytes (Davis et al., 2019).

For this reason, proliferative markers are used to evaluate cell cycle activity in a lesion. Proliferative markers are used as diagnostic and prognostic biomarkers. In fact, proliferative markers allow determining the number of cells in the cell cycle: benign melanocytic lesions have a less proliferative rate while malignant melanomas have a high proliferative index. The most used proliferative biomarkers are Ki-67 and phosphohistone H3 (PHH3). Ki-67 correlates with cell proliferation. It is a protein involved in ribosomal transcription, and is highly expressed in the most aggressive melanomas (Henrique et al., 2000). PHH3 is involved in histone modification and detects cells actually undergoing mitosis; in fact it is associated with chromatin condensation in the late G_2/M phase of the cell cycle (Ladstein et al., 2012). The limitation of these biomarkers is their lack of specificity for melanocytes (Davis et al., 2019).

As prognostic biomarkers, there are also some biomarkers detected in patient blood, plasma, or serum. Lactate dehydrogenase (LDH) is used in the AJCC melanoma staging system. LDH catalyzes the conversion of pyruvate into lactate in hypoxic or anoxic conditions, namely when there is a partial lack or the complete depletion of oxygen, respectively. Oxygen deficiency is a frequent condition in the tumor microenvironment. LDH is an independent prognostic indicator for malignant melanoma; in fact, elevated LDH is recognized as an adverse prognostic indicator for patients with stage IV melanoma and poor survival (Palmer et al., 2011).

Breslow thickness, ulceration, the presence of metastases in regional lymph nodes, or distant site metastases correlate with the prognosis (McGovern et al., 1982; Sprakel et al., 2003; Vollmer & Seigler, 2001). Notably, it remains evident that the prognosis is closely correlated to the thickness, which usually increases with tumor progression (Breslow, 1975). On the other hand, the presence of TILs is an index of favorable prognosis (Fu et al., 2019).

A sentinel lymph node (SLN) biopsy is associated with tumor prognosis beyond the classic evaluation of the biopsy (Faries et al., 2017). Regional lymph nodes are the most common site of initial metastasis in CM patients. Mapping SLNs is a minimally invasive method to find metastases also in patients with the clinically occult disease. The presence of metastases or micrometastases in SLN provides prognostic information needed by the clinician to decide the treatment plan (Wong et al., 2018).

Treatment

Melanoma therapeutic approaches include surgical resection, chemotherapy, targeted therapy, and immunotherapy. Depending on the patient's health, stage, and location of the tumor, single agents or combined therapy can be used. The principal treatment for early stage melanoma, namely stage I and II, is surgery. Treating metastatic melanoma still remains an important challenge. The efficiency of the treatment can be impaired by the development of drug resistance (Domingues et al., 2018).

Surgery and radiotherapy

Surgery is used for early melanoma treatment. After the complete excision of a primary melanoma, radiation therapy (RT) can be used as adjuvant therapy. In fact, postoperative radiotherapy can reduce the risk of local recurrence in high-risk clinical situations. In metastatic melanoma, radiotherapy leads to better outcomes; however, it does not affect survival (Strojan, 2010). Stage III melanomas can be treated surgically in combination with adjuvant treatment, such as targeted therapy and immunotherapy. The adjuvant-targeted therapy with BRAF inhibitors, such as dabrafenib and trametinib, has improved survival outcomes in BRAF-mutated (V600E and V600K) melanomas. Adjuvant immunotherapy with a high dose of CTLA-4 inhibitor (ipilimumab) improves overall survival while therapy of PD-1 inhibitors (nivolumab and pembrolizumab) increases relapse-free survival (Franke & van Akkooi, 2019).

Metastatic melanoma is highly unlikely to be cured by surgery due to the high number of metastases or the low accessibility and difficulty of detecting small metastatic lesions (Davis et al., 2019).

Chemotherapy

For melanoma patients with progressive, refractory, or relapsed disease, surgical treatment alone is not curative and so drug therapies are used. Until recently, the only available treatment for patients with metastatic melanoma was chemotherapy. The most commonly used chemotherapeutic agents are temozolomide (TMZ) and dacarbazine (DTIC), alkylating agents that block or slow cell proliferation by inhibiting DNA synthesis. These agents have been used alone or in combination (Bhatia et al., 2009). More recent drug discoveries have greatly improved the prognosis for patients with metastatic melanoma. Nowadays, chemotherapy is only used after more effective treatments, such as targeted therapy and immunotherapy, fail (Varrone & Caputo, 2020).

Targeted therapy

The MAPK pathway is critically involved in melanoma pathogenesis, and its components are frequently mutated in melanoma cells. This provides the base for targeted therapies in melanoma. Among them, the most effective include the BRAF inhibitors, vemurafenib and dabrafenib, which were approved in 2011 and 2013, respectively, for the treatment of metastatic and unresectable BRAF-mutated melanomas (Davis et al., 2019). Another druggable component of MAPK signaling is mitogen-activated protein kinase kinase 7 (MAP2K7), also known as MEK. Selective MEK inhibitors inhibit growth and induce apoptosis in BRAF- and NRAS-mutant melanoma cell lines. MEK inhibitor therapy is used in combination with a BRAF inhibitor. The combination of BRAF and MEK inhibitors is more effective and less toxic than treatment with a BRAF inhibitor alone, and it is the standard of care for patients with BRAF-mutated melanoma. The MEK inhibitor trametinib is approved in combination with the BRAF inhibitor dabrafenib while the MEK inhibitor cobimetinib is approved in combination with the BRAF inhibitor vemurafenib (Grimaldi et al., 2017).

However, continuous treatment with BRAF and MEK inhibitors agents is associated with the onset of drug resistance (Kakadia et al., 2018). Patients with BRAF-mutant melanoma have a novel treatment option available, namely the combination of a new generation of BRAF (Encorafenib) and MEK (Binimetinib) inhibitors. Compared to the dabrafenib and trametinib combination, the new combined treatments have a similar response rate but a longer duration of response (Koelblinger et al., 2018).

Immunotherapy

In 1992, the first immune therapy for metastatic melanoma based on interleukein-2 (IL2) treatment was approved. IL-2 is a T-cell growth factor that promotes the expansion of antimelanoma CD8+ T-cells but is also highly toxic (Davis et al., 2019).

To date, the most effective immunotherapy in metastatic melanoma is based on immune checkpoint inhibitors, namely antibodies against programmed cell death protein 1 (PD1), programmed death-ligand 1/2 (PD-L1/2), and cytotoxic T-lymphocyte-associated protein 4 (CTLA-4). Among the immune checkpoint inhibitor drugs that have been approved in melanoma treatment, we mention ipilimumab (anti-CTLA-4 antibody), nivolumab and pembrolizumab (anti-PD-1 antibodies) (Topalian et al., 2015), and atezolizumab (anti-PD-L1 antibodies). The combination of ipilimumab and nivolumab is preferred to the mono-treatment in immunotherapy, but there is an increase of drug-related toxicity (Abdel-Wahab, Shah, &

Suarez-Almazor, 2016). The side effects are usually immune-related inflammatory conditions of the skin, gastrointestinal system, and endocrine organs, and these lead to the termination of the treatment (Brahmer et al., 2018). In addition, many patients develop drug resistance.

Immunotherapy with anti-PD-1 drugs leads to lower response rates but a longer response duration if compared to BRAF/MEK inhibitors. The combination of targeted therapy and immunotherapy is suggested for patients with stage IV melanoma. BRAF and MEK inhibitors combined with anti-PD-1 agents can improve the antitumor activity (Deken et al., 2016; Homet Moreno et al., 2016). The use of targeted and immunotherapy therapies has improved survival for most patients, and they are now the preferred approaches for patients with metastatic melanoma.

Melanoma onset

Melanoma originates from an altered proliferation of melanocytes. Melanocytes derive from neural crest cells and several transcription factors, including Forkhead box D3 (FOXD3), SRY-box transcription factor 10 (SOX10), paired box 3 (PAX3), and microphthalmia-associated transcription factor (MITF), cooperate in the differentiation to melanocytes. Melanocytes represent a skin minority cell population with a low proliferative rate. They produce melanin, the brown-black skin pigment, and provide this pigment to keratinocytes. Proliferation and melanin production are promoted by UV. Alterations in specific pathways are associated with the malignant transformation of melanocytes, namely melanoma (Sommer, 2011).

For melanoma progression, there are two principal models, one linear and one nonlinear. The first one was proposed by Clark who described the melanomagenesis as a linear process with six lesional steps. The first step is the acquisition of a melanocytic nevus; the second a melanocytic nevus with lentiginous melanocytic hyperplasia (aberrant differentiation); the third a melanocytic nevus with aberrant differentiation and melanocytic nuclear atypia (melanocytic dysplasia); the fourth the radial growth phase of primary melanoma; the fifth the vertical growth phase of primary melanoma; and finally metastatic melanoma. This is known as the Clark model for melanoma progression (Greene et al., 1984). This model has been significantly improved with more recent discoveries (Fig. 26.1).

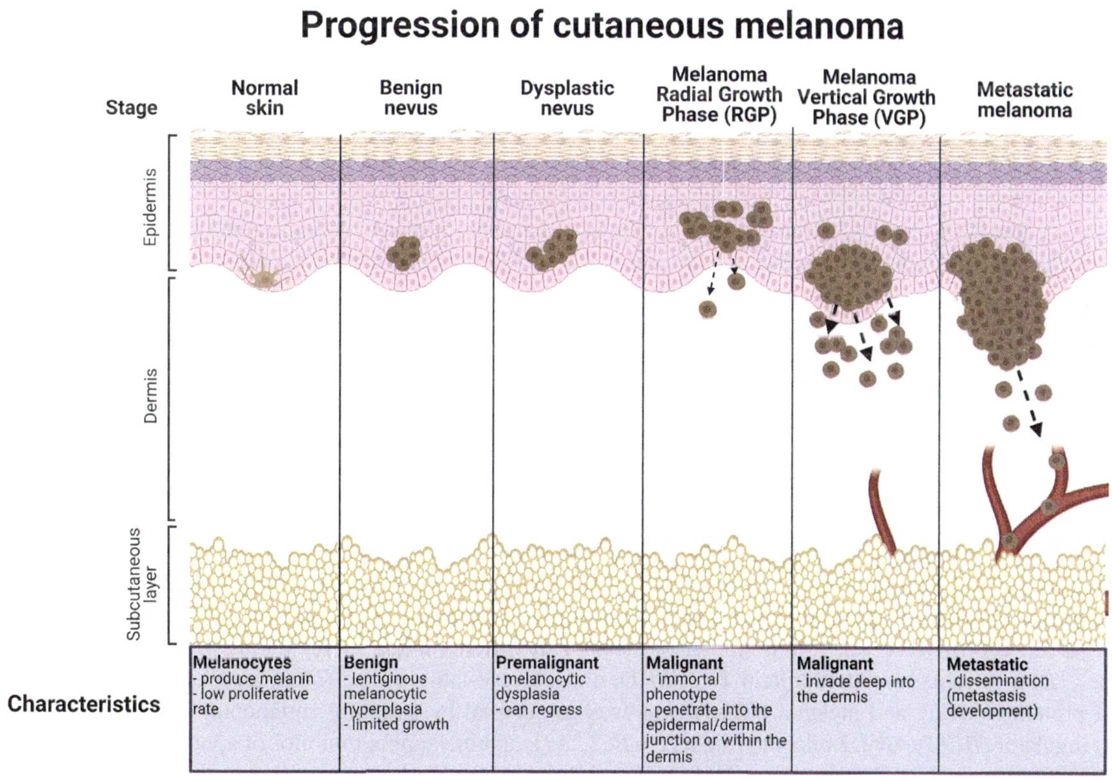

FIG. 26.1 Progression of cutaneous melanoma. According to the Clark model, melanoma progression starts from the transformation of normal melanocytes to benign nevi, which can proliferate to a premalignant lesion, namely dysplastic nevus. In the radial growth phase (RGP), melanoma cells proliferate toward the epidermis, followed by the invasion of the dermis in the vertical growth phase (VGP). The last stage consists of the formation of metastatic melanoma, which spreads in the bloodstream to invade other organs. Other models postulate that malignant cutaneous melanomas can arise directly as RGP/VGP tumors from a single transformed melanocyte. Figure created with BioRender.com.

In the linear model, the first stage is the formation of a nevus, which is due to benign proliferation and aggregation of melanocytes. Some of them are referred to as dysplastic nevi (atypical moles), when they present an altered growth pattern. The transition from normal melanocytes to dysplastic nevi is often due to the oncogenic mutation of v-Raf murine sarcoma viral oncogene homolog B (BRAF) or the activation of protein kinase B (PKB, also known as AKT) (Rastrelli et al., 2014). The uncontrolled and continuous proliferation allows the melanocyte to penetrate into the epidermal/dermal junction or within the dermis by a radial growth phase (RGP). In this step, melanocytes activate the human telomerase reverse transcriptase (hTERT) and acquire an immortal phenotype (Bennett, 2016). The intermediate neoplastic phase (melanoma in situ) is characterized by alteration of apoptosis and cellular survival, usually as a consequence of mutations or epigenetic silencing of cyclin dependent kinase inhibitor 2A (CDKN2A, also known as $p16^{INK4a}$), impair of $p16^{INK4a}$-retinoblastoma (Rb) pathway (Hussussian et al., 1994), amplification of cyclin D1 (CCND1), or inactivation of p53. In the final stage of progression, the tumor develop a vertical growth phase (VGP), which allows invasion deep into the dermis and metastasis development.

This phase requires alterations that could repress apoptosis, allowing survival far from the keratinocytes. Activating mutations in the neuroblastoma RAS viral (v-ras) oncogene homolog (NRAS) and t the yrosine-protein kinase Kit (c-KIT) contribute to neoplastic transformation. The loss of PTEN, gene silencing of p16 CDKN2A, the activation of the phosphoinositide 3-kinase (PI3K)-AKT pathway, and the amplification of the microphthalmia transcription factor (MITF) gene promote the acquisition of a malignant phenotype with melanoma progression. A combination of these processes represents a linear model of melanoma genesis (Rastrelli et al., 2014). In addition, other alterations have been involved in the progression from RGP to VGP, including the loss of epithelial (E)-cadherin and the aberrant expression of neural (N)-cadherin and αVβ3 integrin (Miller & Mihm, 2006).

The second model suggests that the malignant melanoma can originate from each of the above-mentioned steps, without necessarily passing through all of them. This nonlinear model has been supported by the demonstration that melanoma stem cells are involved in melanoma origin, progression, and metastasis (Schatton & Frank, 2008). Specifically, alterations in melanoma stems cells result in progression to melanoma cells in the superficial or vertical growth phase, or the direct onset of metastatic cells (Rastrelli et al., 2014).

However, our increasing knowledge of epigenetic, genetic, and transcriptomic mechanisms affecting cancer development suggests a more complex picture of melanoma onset, progression, and metastasis.

In this chapter, we will take into account the contribution of microRNAs to the biology and pathogenesis of melanoma, as well as their role as disease biomarkers.

MicroRNA

In melanoma, microRNAs (miRNAs) are involved in various tumoral processes: melanoma genesis, proliferation, apoptosis, invasion, metastasis, and immune response (Riefolo et al., 2019). All these mechanisms are linked with the imbalance of an extensive pool of up- and/or downregulated miRNAs (Negrini et al., 2007; Zhang et al., 2007).

For a better understanding of the functional role of miRNAs in melanoma, we organized the main experimental discoveries according to the cancer hallmark (Hanahan & Weinberg, 2011) that is most influenced by the miRNA, namely proliferation and cell cycle, apoptosis, invasion and metastasis, immune system, hypoxia and metabolism, and drug resistance. We also revised the main findings on miRNAs associated with response to treatment. Finally, we reported circulating miRNAs that can be used as melanoma biomarkers.

microRNAs, melanogenesis, and melanoma genesis

Melanogenesis is the process through which the pigment melanin is produced in melanosomes by melanocytes. Microphthalmia-associated transcription factor (MITF) is a master regulator of melanocyte development and melanogenesis (Levy et al., 2006). MITF acts as a promoter of melanocyte differentiation by regulating pigmentation genes such as tyrosinase (*TRY)*, tyrosinase related protein 1 (*TYRP1)*, dopachrome tautomerase (*DCT,* also known as *TYRP2)*, premelanosome protein (*PMEL)*, and melan-a (*MLANA)* as well as survival by targeting antiapoptotic genes, including BCL2 apoptosis regulator (*BCL2)*, BCL2 related protein A1 (*BCL2A1)*, and melanoma inhibitor of apoptosis (*ML-IAP*). In melanoma, MITF plays a dual role in the cell cycle. In fact, it induces the transcription of cell cycle promoting factors, such as cyclin dependent kinase 2 (*CDK2)* and T-box transcription factor 2 (*TBX2*), but also of genes involved in cell cycle arrest, such as cyclin-dependent kinase inhibitor 1A (*CDKN1A*, also known as $p21^{CIP1}$) and cyclin-dependent kinase inhibitor 2A (*CDKN2A*, also known as $p16^{INK4a}$), which is often mutated in melanoma. Moreover, MITF regulates motility and invasion by transcribing the proto-oncogene hepatocyte growth factor receptor (c-MET) and zinc finger protein SNAI2 (SNAI2), and the melanocyte metabolism by regulating peroxisome proliferator-activated receptor gamma coactivator 1-alpha (PGC-1-alpha) (Hsiao & Fisher, 2014; Kawakami & Fisher, 2017).

Amplification or specific mutations of MITF are oncogenic and detected both in familial and sporadic melanomas (Kawakami & Fisher, 2017). In melanoma, MITF acts in a cancer-promoting or cancer-suppressing manner, which depends on the fine balance between proapoptotic and proliferative effects (Uzdensky et al., 2013). MITF regulates the expression of some miRNAs and many miRNAs have been found to target MITF (Fig. 26.2A).

An important melanocytic lineage-specific miRNA regulated by MITF is miR-211-5p. In human melanocytes, MITF regulates the transcription of miR-211-5p and its host gene, the transient receptor potential cation channel subfamily M member 1 (*TRPM1*) gene. miR-211-5p has been found to be one of the most differentially expressed miRNAs between normal human melanocytes and melanoma cell lines (Levy et al., 2010; Mazar et al., 2010; Xu et al., 2012) or tumors. TRPM1 and miR-211-5p expression is frequently reduced or lost during the transition from nevi to primary melanoma and downregulated even further in malignant melanoma. In addition, a high TRPM1 level has been associated with longer disease-free survival in primary melanoma patients (Hammock et al., 2006).

The functional role of miR-211-5p in melanoma cell lines has been extensively studied. The miR-211-5p overexpression enhances the total number of melanosomes, resulting in increased pigmentation. miR-211 inhibits the MAPK/ERK pathway, which leads to the repression of MITF and promotes the switch from glycolysis to oxidative phosphorylation through PGC1α and mitochondrial biogenesis. miR-211-5p action induces a more differentiated phenotype (Vitiello et al., 2017).

Several miRNAs are involved in MITF regulation, including miR-137-3p, miR-148a-3p, miR-148b-3p, miR-101–3p, miR-26a-5p, miR-218-5p, miR-340-5p miR-182-5p, and miR-155-5p.

miR-137-3p is significantly downregulated in melanoma tissues compared with normal pigmented nevi (Qi, Wang, Chen, & Shang, 2018). miR-137-3p targets MITF and downregulates multiple oncogenic mRNAs, such as MET proto-oncogene receptor tyrosin kinase (c-MET), Y box-binding protein 1 (YB1), enhancer of zeste homolog 2 (EZH2), and phosphatidylinositol 3 kinase regulatory 3 (PIK3R3). Its protective role is demonstrated by the fact that a low expression of miR-137-3p correlates with the poor survival of stage IV melanoma patients (Bemis et al., 2008; Haflidadottir et al., 2010; Luo, Tetteh, et al., 2013). Together with miR-137-3p, both miR-148a-3p and miR-148b-3p directly target MITF through a binding site found in the MITF 3' UTR sequence (Haflidadottir et al., 2010). Interestingly, it was shown that the MITF protein binds to the promoter of miR-148b-3p (Ozsolak et al., 2008), suggesting a negative feedback regulation where a high level of MITF leads to a major activation of miR-148b-3p, which in turn decreases MITF at desirable levels. Similarly, the low expression of miR-101–3p correlates with poor survival in stage IV melanoma patients. miR-101–3p directly targets MITF and EZH2, resulting in inhibition of invasion and proliferation (Luo, Merz, et al., 2013).

In melanoma cells, miR-26a-5p binds the 3' UTR of MITF and downregulates the expression of MITF and of mitogen-activated protein kinase kinase kinase kinase 3 (MAP4K3), inducing apoptosis and reducing cell viability. Moreover, miR-26a-5p significantly delays melanoma growth in vivo (Qian et al., 2017).

MITF is also targeted by miR-218-5p and miR-340-5p, resulting in the suppression of MITF expression. miR-218-5p is downregulated in melanoma compared to benign nevi (Gencia et al., 2020). miR-218-5p strongly reduces MITF expression, leading to the inhibition of melanogenesis and tyrosinase activity and to the induction of depigmentation. In both melanoma cell lines and human melanocytes, miR-218-5p is negatively correlated with MITF (Guo et al., 2014). miR-340-5p binds two target sites on the 3' UTR of MITF, leading to the degradation of MITF. However, these two bindings can be prevented by coding region determinant-binding protein (CRD-BP), which is highly expressed in melanoma, resulting in the stabilization, elevated expression, and transcriptional activity of MITF (Goswami et al., 2015).

miR-182-5p is located in a frequently amplified chromosomal locus (7q31–34) in melanoma; its expression is commonly high in human melanoma and melanoma cell lines and increases in melanoma metastatic tumors (Segura et al., 2009). The silencing of this miRNAs decreases the invasive capacity and induces apoptosis in melanoma cells. On the other hand, its overexpression leads to an anchorage-independent growth and stimulates migration and cell survival. miR-182-5p negatively correlates with its direct targets, MITF and FOXO3, while the enhanced exprcssion of MITF and FOXO3 blocks the invasive effects of miR-182-5p. Interestingly, the overexpression of MITF suppresses the miR-182-dependent proinvasive effect (Segura et al., 2009).

Finally, miR-155-5p targets MITF in melanoma cells and is involved in the repression of MITF induced by interleukins 1 beta (IL-1ß) (Arts et al., 2015). In fact, the expressions of MITF and melanocyte differentiation antigens are strongly reduced by interleukins 1 alfa (IL-1α) and IL-1ß produced by stromal cells or by tumor cells themselves. This may represent a possible mechanism of melanoma immune escape (Kholmanskikh et al., 2010). In a mouse melanoma model, a strong negative correlation was observed between MITF and miR-155-5p levels, suggesting an oncogenic role of miR-155-5p (Arts et al., 2015).

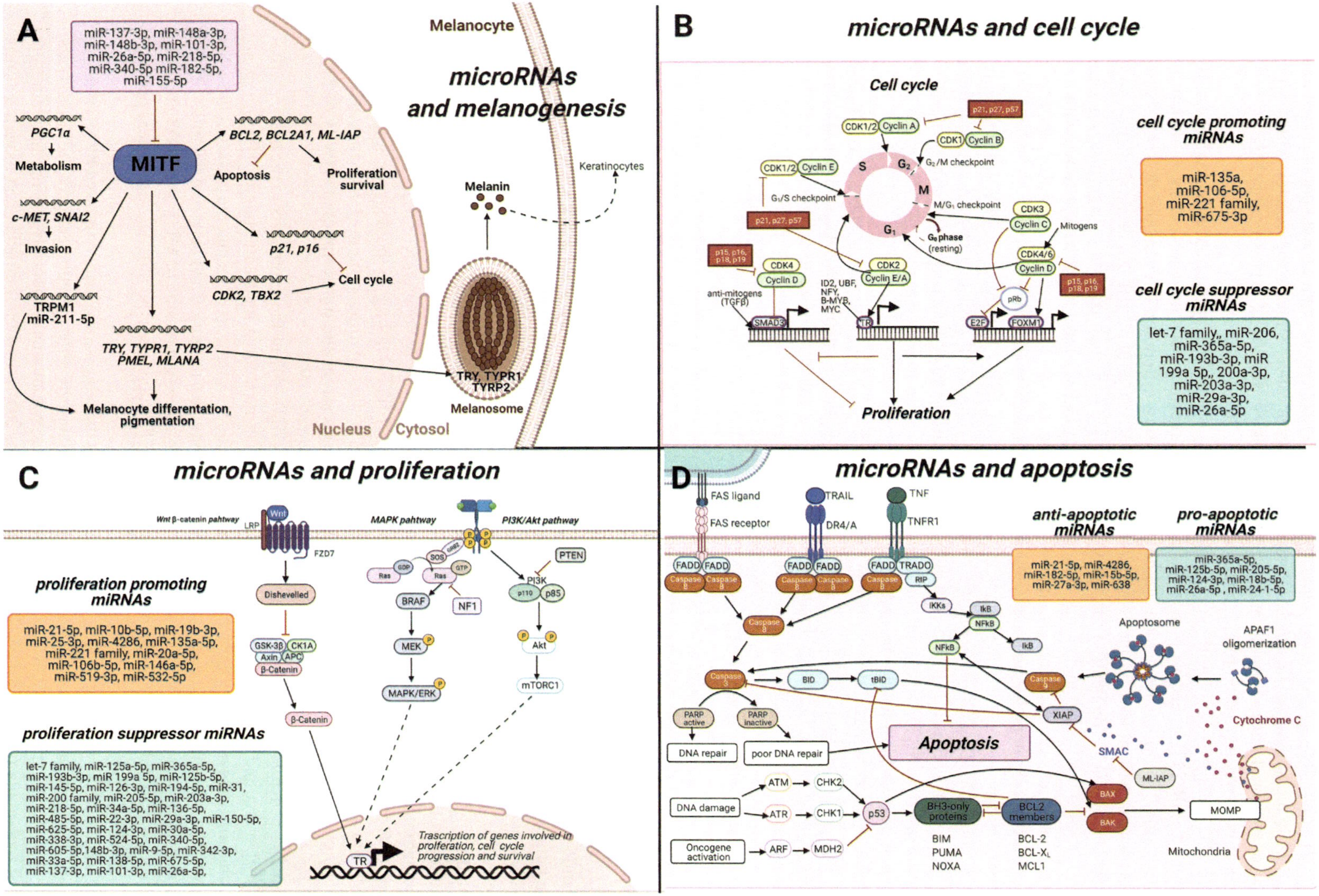

FIG. 26.2 microRNAs in melanogenesis, cell cycle, proliferation, and programmed cell death. (A) MITF is a master regulator of melanocyte development and melanogenesis, and regulates genes involved in melanocyte differentiation, pigmentation, cell cycle, apoptosis, proliferation and survival, metabolism, and invasion. The expression of MITF is regulated by several microRNAs (violet box). (B) Cell cycle phases: S-phase (S), mitosis (M); transition phases (G1 and G2), whereas G0 indicate quiescent cells. Several microRNAs regulate cell cycle progression by targeting proteins involved in the promotion of the cell cycle, such as cyclins and cyclin dependent kinases (CDKs), or involved in the arrest of the cell cycle, including p15, p16, p18, p18, p21, p27, and p57. (C) Proliferation is induced by several cellular pathways, including Wnt-β-catenin and the MAPK and PI3K/Akt pathways. microRNAs can promote or repress the expression of proteins involved in cell proliferation by acting as oncomiRs (orange box) or tumor-suppressor miRNAs (green box). (D) Apoptosis can be activated by extrinsic or intrinsic signals. The extrinsic pathway is induced when death ligands bind death receptors, which lead to the activation of initiator caspases such as caspase 8 and the consequent activation of effector caspases, such as caspases 3 and 9, thus promoting apoptotic cell death. The intrinsic pathway induces mitochondrial outer membrane permeabilization (MOMP) resulting in the release of cytochrome *c* from the mitochondria. The formation of the apoptosome promotes caspase 9 activation, which actives caspase 3 to induce cell death. microRNAs can act as oncomiRs, that is, protecting from apoptosis (orange box), or tumor suppressors, that is, stimulating apoptosis (green box). Figure created with BioRender.com.

microRNAs in cell cycle and proliferation

Cell cycle dysregulation can lead to uncontrolled proliferation and thus carcinogenesis (Golias, Charalabopoulos, & Charalabopoulos, 2004). Unrestrained cell proliferation is an indicator of tumor progression and it is related to the tumor growth rate. Melanomas with high growth rates are considered high-risk tumors (Pierard, 2012). miRNAs can induce melanoma cell proliferation by inducing proliferative signals or by repressing growth-suppressive targets (Fig. 26.2C). Proliferation is closely linked to the cell cycle, and several miRNAs modulate the cell cycle regulators (Fig. 26.2B). miRNAs can target both cyclin-dependent kinase (CDK) inhibitors or transcriptional repressors of the retinoblastoma family, such as Rb (retinoblastoma protein), thus facilitating cell cycle entry and progression (Bueno & Malumbres, 2011).

miRNAs that stimulate proliferation and/or cell cycle include miR-21-5p, miR-10b-5p, miR-19b-3p, miR-25-3p, miR-4286, miR-135a-5p, miR-221 family, miR-20a-5p, miR-106b-5p, miR-146a-5p, miR-519d-3p, miR-532-5p, and miR-675-3p.

miR-21-5p is an important cancer-related miRNA that is upregulated in primary lesions with histological atypia and high mitotic activity (Grignol et al., 2011), melanoma cell lines compared to melanocytes (Satzger et al., 2012), and primary melanoma compared to benign nevi (Grignol et al., 2011). Its expression correlates with Breslow thickness (Dika et al., 2020), advanced clinical stage, and poor survival (Jiang et al., 2012). miR-21-5p plays important roles in several cellular processes, including proliferation, migration, and apoptosis. In fact, the depletion of miR-21-5p in melanoma cells increases programmed cell death 4 (PDCD4), PTEN, and BTG family member 2 (BTG2) expression, resulting in the inhibition of proliferation and migration and in the promotion of apoptosis (Yang et al., 2011).

miR-10b-5p is significantly upregulated in melanoma tumors and cell lines, and especially in metastases compared to their matched primary. Its expression correlates with a poor overall survival of melanoma patients (Saldanha et al., 2016). Inhibition of miR-10b-5p prevents melanoma cell proliferation, migration, and invasion in vitro and suppresses tumor growth in vivo by targeting the itchy E3 ubiquitin protein ligase (ITCH) and by regulating the Wnt/β-catenin pathway (Wang et al., 2019).

miR-19b-3p expression is higher in most melanoma cell lines compared with normal melanocyte cells. miR-19b-3p functions as an oncomiR; in fact, it regulates human telomerase reverse transcriptase (hTERT) and targets paired-like homeodomain1 (PITX1), promoting cell proliferation (Ohira et al., 2015).

miR-25-3p is significantly upregulated in melanoma cell lines and clinical tissues. The downregulation of miR-25-3p strongly suppresses proliferation and invasion of melanoma cells in vitro while high expression enhances melanoma proliferation and migration by targeting Dickkopf-related protein 3 (DKK3) (Huo et al., 2016) and RNA-binding motif protein 47 (RBM47) (Jiang and Liu, 2018).

Melanoma exhibits upregulated expression of miR-4286 compared to benign melanocytes. miR-4286 promotes proliferation and protects from apoptosis. Suppression of miR-4286 expression leads to the alteration of its targets that are implicated in proliferation and apoptosis pathways, including folylpolyglutamate synthase (FPGS), RNA polymerase I-specific transcription initiation factor (RRN3), apelin (APLN), G-protein-coupled receptor 55 (GPR55), and high-mobility group A1 protein (HMGA1) (Komina et al., 2016).

Malignant melanoma tissue and cell lines have high levels of miR-135a-5p, which promotes cell proliferation and progression of the cell cycle. The ectopic expression of miR-135a-5p inhibits the FOXO1 protein, leading to the upregulation of CCND1 and the downregulation of CDKN1A/B through the AKT pathway (Ren et al., 2015).

The miR-221 family includes miR-221–3p and miR-222-3p and regulates the progression of melanoma through the downmodulation of CDKN1B (also known as p27) and c-KIT, resulting in enhanced proliferation and a differentiation blockade of the melanoma cells, respectively (Felicetti et al., 2008).

miR-17/92 cluster members are highly expressed in melanoma cell lines compared to normal human epidermal melanocytes, and the expression is even higher in metastatic cell lines (Mueller et al., 2009). miR-20a-5p belongs to the miR-17-92 family, and with miR-17-5p directly targets the adenosine deaminase RNA specific (*ADAR1*) gene, enhancing melanoma aggressiveness (Greenberg et al., 2014). ADAR1 performs adenosine-to-inosine (A-to I) RNA editing and is frequently reduced in metastatic melanomas. The depletion of ADAR1 affects cell morphology, promotes proliferation in vitro, and markedly enhances the tumorigenicity in vivo (Nemlich et al., 2013). However, a functional study has shown that only miR-20a-5p, but not miR-17-5p, overexpression inhibits cell proliferation in vitro (Greenberg et al., 2014).

miR-106b-5p has been found overexpressed in primary melanoma compared to dysplastic nevi, and in metastatic melanoma compared to primary melanoma. miR-106b-5p is implicated in the progression of cutaneous melanoma and high level of this miRNA correlated with poor prognosis (Lin et al., 2015). miR-106b-5p promotes cell cycle progression by targeting PTEN (Chen et al., 2018).

miR-146a-5p-5p expression correlates with Breslow thickness in melanoma (Dika et al., 2020). miR-146a-5p induces melanoma cell growth by its direct targets, the lunatic fringe (LFNG) and protein numb homolog NUMB, which operate on the NOTCH/PTEN/Akt pathway (Raimo et al., 2016).

Metastatic melanoma tissues upregulate miR-519d-3p, whose overexpression enhances cell proliferation, migration, invasion, and adhesion in vitro and lung metastatic ability in vivo by targeting the EPH receptor A4 (EphA4) (Hua et al., 2018). miR-532-5p has been shown to be significantly overexpressed in melanoma lines and metastatic melanoma tumors compared to normal melanocytes and primary melanomas, respectively. miR-532-5p promotes melanoma progression by targeting RUNX family transcription factor 3 (RUNX3) (Kitago et al., 2009).

miR-675-3p acts as an oncomiR. In fact, it overexpressed in metastatic melanoma compared to primary melanoma, and its expression correlates with poorer overall patient survival, tumor histologic grade, and Clark level. miR-675-3p promotes cell cycle progression resulting in the stimulation of cell proliferation (Zhao et al., 2021). Differently to miR-675-3p, miR-675-5p expression is reduced in melanoma cell lines and clinical specimens. The overexpression of miR-675-5p impairs cell proliferation and invasion in melanoma by targeting metadherin (MTDH) (Liu, Jin, et al., 2018).

There are several miRNAs that can inhibit proliferation and/or induce cell cycle arrest by downregulating multiple components of the cell cycle machinery (Bueno & Malumbres, 2011). These include miR-675-5p, let-7 family, miR-125a-5p, miR-206, miR-365a-5p, miR-193b-3p, miR-199a-5p, miR-125b-5p, miR-145-5p, miR-126-3p/miR-126-5p, miR-194-5p, miR-31, miR-200 family, miR-205-5p, miR-203a-3p, miR-218-5p, miR-34a-5p, miR-136-5p, miR-485-5p, miR-22-3p, miR-29a-3p, miR-150-5p, miR-625-5p, miR-124-3p, miR-30a-5p, miR-338-3p, miR-524-5p, miR-340-5p, miR-605-5p, miR-148b-3p, miR-9-5p, miR-342-3p, miR-33a-5p, miR-138-5p, miR-137-3p, miR-101–3p, miR-26a-5p, and miR-155-5p.2.

The let-7 family is an important miRNA family involved in the regulation of cell proliferation. The expression of let-7a-5p and let-7b is significantly decreased in primary melanoma compared to melanocytic nevi (Schultz et al., 2008). let-7a-5p inhibits proliferation and invasiveness and halts the cell cycle at the G_1/G_0 phase in malignant melanoma cells lines (Qian et al., 2017). The overexpression of let-7b-5p in melanoma cell lines decreased the expression of CCND1, cyclin D3 (CCND3), and cyclin A (CCNA), which are involved in the inhibition of the tumor-suppressor Rb and in the promotion of proliferation in melanoma (Schultz et al., 2008).

miR-125a-5p is less expressed in melanoma cells compared with melanocytes. A direct target of miR-125a-5p is the let-7 suppressor Lin-28 homolog B (Lin28B). Lin28B blocks the let-7 family miRNAs and activates the transforming growth factor (TGF)-β signaling pathway, thereby promoting melanoma tumor growth. This suggests a protective role of miR-125a-5p in blocking melanoma growth by targeting Lin28B (Negrini et al., 2007; Zhang et al., 2015).

Downregulated levels of miR-206 have been found in melanoma cells compared with normal melanocytes. miR-206 induces G_1 arrest in multiple melanoma cell lines by targeting cyclin-dependent kinase 4 (CDK4), CCND1, and cyclin C (CCNC) (Georgantas et al., 2014). CCND1 is also a direct target of miR-365a-5p and miR-193b-3p. miR-365a-5p expression is downregulated in melanoma cells and its overexpression strongly inhibits cell proliferation and induces cell cycle arrest (Zhu et al., 2018). miR-193b-3p expression is lower in metastatic melanoma than benign nevi. The overexpression of miR-193b-3p in melanoma cell lines inhibits proliferation (Chen et al., 2010). In addition, miR-193b-3p downregulates the myeloid cell leukemia protein 1 (MCL-1) in melanoma cells (J. Chen et al., 2011).

miR-199a-5p in tumor melanoma tissues is downregulated compared to normal controls, and the low expression has been correlated to a high histological grade and advanced tumor stage. The overexpression of miR-199a-5p inhibits cell proliferation and arrests the cell cycle in the G_1 phase by targeting HIF1α (Yang et al., 2016).

miR-125b-5p is downregulated in melanoma relative to normal skin (Holst et al., 2011), in melanoma cell lines compared to human epidermal melanocytes (Kappelmann et al., 2013; J. Zhang et al., 2014), in atypical nevi in comparison with common nevi (Holst et al., 2011), in metastatic melanoma vs. primary melanoma (Beatriz et al., 2018), and in melanoma with lymph node involvement (N+) compared with melanoma without lymph node involvement (N0) (Glud et al., 2010). Its downregulation is associated with a worse prognosis in the keratin expressor melanoma subtype while its overexpression in the MITF-low melanoma subgroup is associated with better survival (Beatriz et al., 2018). The overexpression of miR-125b-5p resulted in decreased proliferation and cell cycle arrest (Nyholm et al., 2014).

miR-145-5p is downregulated in melanoma cells compared to normal human primary melanocytes (Dynoodt et al., 2013), and its ectopic expression in melanoma cell lines results in the inhibition of cell growth by targeting the MYC proto-oncogene bHLH transcription factor (c-MYC) (Noguchi et al., 2012). In fact, the inhibition of c-MYC leads to a proliferative arrest of melanoma cells through cellular crisis manifested by growth in size, multiple nuclei, induction of senescence-associated beta-galactosidase, vacuolated cytoplasm, and apoptosis (Biroccio et al., 2003).

Both miR-126-3p and miR-126-5p are downregulated in melanoma, especially in metastatic melanoma. The restored expression of miR-126-3p/126-5p inhibits cell proliferation, invasion in vitro, and melanoma growth and dissemination in vivo. On the other hand, the opposite effect is observed when miR-126-3p/126-5p is silenced by two metalloproteases, namely A disintegrin and a metalloprotease 9 (ADAM9) and matrix metallopeptidase 7 (MMP7), which play important roles in melanoma progression (Felli et al., 2013).

miR-194-5p is downregulated in melanoma cells compared to normal human primary melanocytes (Mueller et al., 2009). The overexpression of miR-194-5p inhibits cell proliferation through the PI3K/AKT/FoxO3a signaling pathway (Bai, Zhang, Long, et al., 2017), and the negative regulation of the GEF-H1 Rho guanine nucleotide exchange factor 2/Ras homolog family member A (GEF-H1/RhoA) pathway. Moreover, miR-194-5p expression is negatively associated with tumor-node-metastasis (TNM) stages (Guo et al., 2016).

Similarly, miR-31-5p is involved in PI3K/AKT signaling pathway regulation. miR-31-5p modulates tumor cell growth and chemosensitivity by suppressing SRY-box transcription factor 10 (SOX10). miR-31-5p is downregulated in melanoma, and its expression is positively associated with a better overall survival. A high level of SOX10 promotes melanoma cell proliferation and chemotherapy resistance both in vitro and in vivo and it was associated with poor overall survival (Zheng et al., 2018).

miR-200 family members are usually downregulated in melanoma. miR-200a-3p, miR-200b-3p, miR-200c-3p, and miR-429 are decreased in metastatic melanoma compared to benign nevi. In addition, miR-200a-3p and miR-200b-3p are downregulated in primary melanoma compared to benign nevi (Xu et al., 2012) and miR-200c-3p expression decreases progressively in primary and metastatic melanomas. By targeting Golgi membrane protein 1 (GOLM1), miR-200a-3p inhibits the PI3K/Akt signaling pathway, resulting in the inhibition of cell proliferation, migration, and invasion (Chen et al., 2019). miR-200a-3p represses CDK6 expression leading to a reduction of the response of CDK4/6 inhibitors in highly proliferative metastatic melanoma. In fact, the overexpression of miR-200a-3p causes cell cycle arrest by targeting CDK6. The downregulation of miR-200a-3p leads to an increase of CDK6 and to a more significant response to CDK4/6 inhibitors. However, the downregulation of miR-200a-3p is also associated with disease progression and a higher number of lymph node metastases (Bustos et al., 2017). miR-200b-3p is less expressed in melanoma tissues and cell lines compared to benign nevi. SMAD family member 2 (SMAD2) is targeted by miR-200b-3p, but it can promote proliferation when miR-200b-3p is blocked by the long noncoding (lncRNA) nuclear enriched abundant transcript 1 (NEAT1) (Zhou et al., 2020). miR-200c-3p overexpression results in the inhibition of cell proliferation and migration by targeting the B lymphoma Mo-MLV insertion region 1 homolog (BMI1). By targeting BMI1, miR-200c inhibits proliferation and migration (Liu, Tetzlaff, Cui, & Xu, 2012). The downregulation of miR-200c-3p in melanoma predicts shorter survival (Beatriz et al., 2018). miR-429 acts as a tumor suppressor in melanoma cells; its overexpression reduced AKT serine/threonine kinase 1 (AKT1), resulting in the inhibition of cell proliferation and migration (Huang et al., 2019).

miR-205-5p and miR-203a-3p are skin-specific miRNAs. miR-205-5p is downregulated in metastatic and primary melanomas compared with benign nevi (Xu et al., 2012) and the downregulation is associated with worse clinical outcome (Hanna et al., 2012). The overexpression of miR-205-5p in cell lines inhibits anchorage-independent colony formation, reducing survival capability (Franken et al., 2006; Liu et al., 2012). It is also correlated with zinc-finger E-box binding homeobox 2 (ZEB2) downregulation, E-cadherin upregulation (Liu, Tetzlaff, Liu, et al., 2012), and suppression of cell proliferation by targeting EF2 transcription factor 1 (E2F1) and EF2 transcription factor 5 (E2F5) (Dar et al., 2011). miR-203a-3p has been found to be downregulated in melanoma, especially in metastatic melanoma (Lohcharoenkal et al., 2018). In particular, the downregulation of miR-203a-3p has been inversely correlated with the BMI1 level. miR-203a-3p inhibits proliferation by targeting BMI1, reducing invasion and tumor sphere formation (Chang et al., 2015). In addition, the ectopic expression of miR-203a-3p in melanoma cells reduced the expression of EF2 transcription factor 3a (E2F3a) and EF2 transcription factor 3b (E2F3b), inhibiting cell growth and promoting cell cycle arrest and senescence (Noguchi et al., 2012).

miR-218-5p is a tumor-suppressor miRNA in melanoma. It is downregulated in primary and metastatic melanoma compared to nevi. In fact, miR-218-5p overexpression leads to the inhibition of proliferation, migration, and invasion in melanoma cell lines by targeting the cellular inhibitor of PP2A (CIP2A) and BMI1 (Wei et al., 2014).

The miR-34 family is regulated by p53. In human melanoma cell lines, the overexpression of the miR-34 family inhibits growth and invasion in cells expressing the wild-type p53 gene. miR-34a-5p was downregulated in metastatic melanoma compared to melanoma in situ, nevi, and normal melanocytes. The high expression of miR-34a-5p blocks proliferation and metastasis development by targeting flotillin 2 (FLOT2) (Yamazaki et al., 2012).

Some proliferation suppressor miRNAs are associated with Wnt/β-catenin signaling. miR-136-5p expression levels are decreased in mouse melanoma cells and are linked to the progression of melanoma. The action of miR-136-5p results in the suppression of proliferation, migration, and invasion as well as the promotion of apoptosis by targeting PMEL and downregulating the Wnt signaling pathway (Wang et al., 2017). A significant downregulation of miR-485-5p has been described in melanoma tissue and cell lines compared to controls. The overexpression of miR-485-5p inhibits proliferation and invasion through the downregulation of Frizzled7 (FZD7), thus demonstrating a role of miR-485-5p in the regulation of the Wnt pathway (Wu et al., 2017). miR-22-3p is decreased in melanoma tissues and its downregulation correlates with adverse clinicopathologic features and poor prognosis. miR-22-3p inhibits proliferation by regulating the Wnt/β-catenin pathway and by targeting formin-like 2 (FMNL2) (Shi et al., 2019).

In melanoma cell lines, IFN-γ induces via STAT1 an upregulation of the miR-29 family. IFN-γ-promotes G_1-arrest by downregulating CDK6, which is a direct target of miR-29a-3p in melanoma cells. miR-29a-3p expression is negatively correlated with the proliferation rate (Schmitt et al., 2012). In addition, miR-29a-3p is downregulated in melanoma cells. The overexpression of miR-29a-3p effectively inhibits proliferation, migration, and invasion while promoting apoptosis by targeting BMI1. In addition, miR-29a-3p blocks the Wnt/β-catenin and NF-κB pathways (Xiong et al., 2018).

miR-150-5p directly targets the 3' UTR of sine oculis homeobox 1 (SIX1) in melanoma cells and this results in the suppression of melanoma cell proliferation, migration, and invasion (X. Yang et al., 2019). The expression of miR-150-5p is associated with a good prognosis (Tembe et al., 2015). miR-625-5p is downregulated in melanoma compared with normal tissue. Its overexpression inhibits proliferation and glycolysis by targeting pyruvate kinase (PKM2) (Zhang et al., 2019).

The expression of miR-124-3p in melanoma tissue is lower than that in normal skin tissue while the expression of its target RalA-binding protein 1 (RALBP1, also called RLIP76) is higher in melanoma than in normal skin tissue. The degradation of RALBP1 by miR-124-3p leads to the inhibition of proliferation, migration, and invasion of melanoma cells while promoting melanoma cell apoptosis (Zhang, Han, & Xu, 2016). miR-124-3p inhibits proliferation and invasion also by targeting versican (VCAN) (P. Yang et al., 2017) and receptor of activated protein C kinase 1 (RACK1) (Shen et al., 2019). The expression levels of miR-124-3p are negatively correlated with the advanced stage of melanoma (Shen et al., 2019). In addition, decreased expression is associated with increased tumor thickness, advanced clinical stage, and node metastasis (P. Yang et al., 2017).

miR-30a-5p is downregulated in melanoma tissues and cell lines. miR-30a-5p overexpression results in the significant inhibition of proliferation, migration, and invasion of melanoma cells by targeting sex determining region Ybox 4 (SOX4). Moreover, miR-30a-5p delays tumor growth in vivo (Liu, Sun, et al., 2018).

miR-338-3p expression levels have been quantified in melanomas and adjacent nontumor tissues and an aberrant downregulation in melanoma has been shown. A low expression of miR-338-3p is significantly correlated to advanced clinical stage and lymph node metastasis. The overexpression of miR-338-3p inhibits cell proliferation and migration by targeting metastasis-associated in colon cancer 1 (MACC1) (Chunhua et al., 2019). miR-524-5p is downregulated in melanoma cells with an active MAPK/ERK pathway. miR-524-5p directly binds two components of this pathway, BRAF and ERK2. The overexpression of miR-524-5p reduces proliferation (Liu et al., 2014). Although miR-340-5p is highly expressed in melanoma compared to melanocytes, it plays a tumor-suppressor role. In fact, miR-340-5p inhibits proliferation and invasion by targeting the MAPK pathway (Poenitzsch Strong et al., 2014).

Melanoma cell lines and tissues have low expression levels of miR-605-5p. miR-605-5p represses cell growth in vitro and in vivo by targeting inositol polyphosphate 4-phosphatase type II (INPP4B), which promotes cell proliferation via serine/threonine-protein kinase Sgk3 (SGK3) signaling in melanoma (L. Chen et al., 2017).

miR-148b-3p is downregulated in melanoma and its overexpression can reduce the proliferation, migration, and invasion of melanoma cells by targeting sirtuin 7 (SIRT7). miR-148b-3p could significantly inhibit tumor growth and metastasis development of melanoma in vivo (Sun et al., 2020, #750).

miR-9-5p is downregulated in malignant melanoma tissues when compared to adjacent nontumor tissues. The ectopic expression of miR-9-5p significantly inhibits proliferation migration by targeting sirtuin 1 (SIRT1) (Bu et al., 2017). miR-342-3p is significantly downregulated in melanoma tissues and cells. miR-342-3p overexpression inhibits proliferation and invasion of melanoma by targeting zinc-finger *E*-box-binding homeobox 1 (ZEB1) (Shi et al., 2018). miR-33a-5p is downregulated in melanoma cells, and its expression can inhibit proliferation by directly targeting cyclin-dependent kinase 16 (CDK16, also known as PCTAIRE1) (Tian et al., 2016) and hypoxia-inducible factor 1-alpha (HIF1α) (Zhou et al., 2015). miR-138-5p is significantly downregulated in melanoma cells compared to normal melanocytes. Its overexpression inhibits proliferation by targeting HIF-1α (Chen et al., 2016) and the PI3K/AKT signaling pathway (Fanjun et al., 2017).

Among the MITF-related miRNAs described above, some were reported to play a role in cell proliferation: miR-137-3p, miR-101–3p, miR-26a-5p, and miR-155-5p. miR-137-3p mediates the inhibition of cell proliferation by targeting c-Met, YB1, MITF, EZH2 (Luo, Tetteh, et al., 2013), glyoxalase system member glyoxalase 1 (GLO1) (Lv et al., 2018), and p21-activated kinase 2 (PAK2) (Hao et al., 2015). Similarly, miR-101–3p inhibits proliferation through the downregulation of MITF and EZH2. Despite this observation, miR-101–3p is not considered to be a classical tumor suppressor because of its low expression in human melanocytes (Luo, Tetteh, et al., 2013). In addition, a long noncoding RNA, namely small nucleolar RNA host gene 6 (SNHG6), directly binds miR-101–3p, resulting in the upregulation of miR-101–3p target Ras-related protein Rap-2b (RAP2B), and a consequent promotion of melanoma progression (Hong et al., 2020). miR-26a-5p represses proliferation and invasiveness and blocks the cell cycle at the G_1/G_0 phase in malignant melanoma cells lines. Reduction of cell viability is due to MITF targeting. In addition, using a murine model, it has been shown that miR-26a-5p significantly retarded the growth of melanoma tumors in vivo (Qian et al., 2017). miR-26a-5p also targets the short transient receptor potential channel 3 (TRPC3), resulting in the inhibition of proliferation and the promotion of apoptosis. However, miR-26a-5p can be sequestered by a long noncoding RNA, namely small nucleolar RNA host gene 5 (SNHG5), which leads to the upregulation of TRPC3 and proliferation and the suppression of apoptosis (Gao et al., 2019).

In vitro experiments demonstrated that the overexpression of miR-155-5p inhibits proliferation and promotes apoptosis (Levati et al., 2009) through targeting the oncogene v-ski avian sarcoma viral oncogene homolog (SKI) (Levati et al., 2011). However, it was observed that miR-155-5p is upregulated in primary and metastatic melanoma (Grignol et al., 2011; Philippidou et al., 2010; Segura et al., 2010), and melanoma with a positive sentinel lymph node biopsy compared with a negative sentinel biopsy (Grignol et al., 2011). In addition, in borderline melanocytic tumors, miR-155-5p-5p expression is associated with Breslow thickness (Grignol et al., 2011).

The role of other miRNAs in regulating melanoma proliferation is not completely clear, as different studies report opposite roles. These studies are not mentioned in this chapter.

microRNAs in apoptosis and autophagy

Uncontrolled cell growth is strongly connected with programmed cell death. MicroRNAs target antiapoptotic proteins, including BCL2 and BCL2-like 1, in melanoma. Furthermore, p53 activity is positively and negatively regulated by specific miRNAs (Feng et al., 2011; Liu et al., 2017). miRNAs that act as antiapoptotic factors include miR-21-5p, miR-4286, miR-182-5p, miR-15b-5p, miR-27a-3p and miR-638 while proapoptotic miRNAs are miR-365a-5p, miR-125b-5p, miR-205-5p, miR-124-3p, miR-26a-5p, miR-29a-3p, miR-137-3p, miR-485-5p, miR-155-5p, miR-18b-5p, and miR-24-1-5p (Fig. 26.2D).

miR-21-5p inhibits apoptosis by targeting PDCD4, PTEN, and BTG2 and affecting BCL-2 and pAKT (L. Jiang et al., 2012; C. H. Yang et al., 2011). miR-4286 suppresses apoptosis by targeting FPGS, RRN3, APLN, GPR55, and HMGA1 (Komina et al., 2016). miR-182-5p targets MITF and FOXO3; the silencing of miR-182-5p promotes apoptosis (Segura et al., 2009).

miR-15b-5p is significantly upregulated in melanoma compared to nevi, suggesting an oncogenic role. In fact, its downregulation is associated with an increase of caspase 3 and 7 as well as annexin V, but not Bcl-2 as described in other cancer types. Moreover, the high expression of this miRNA is associated with poor recurrence-free survival and overall survival (Satzger et al., 2010).

Melanoma tissue showed upregulated levels of miR-27a-3p compared to normal tissue. In melanoma cells, miR-27a-3p expression is increased whereas spleen tyrosine kinase (SYK) expression is downregulated, resulting in the inhibition of the mammalian target of the rapamycin (mTOR) signaling pathway. miR-27a-3p is positively correlated with tumor stage and lymph node metastasis of melanoma tissues. Mechanistically, miR-27a-3p targets SYK and the downregulation of miR-27a-3p induces autophagy and apoptosis of melanoma cells (Tang et al., 2019).

Metastatic lesions showed higher levels of miR-638 compared with primary melanoma. Its role is to protect melanoma cells from apoptosis and autophagy by downregulating p53-inducible nuclear protein 2 (TP53INP2). miR-638 expression positively correlates with primary tumor thickness (Bhattacharya et al., 2015).

Among the miRNAs with a proapoptotic action, some were already described as regulators of cell proliferation, including miR-365a-5p, miR-125b-5p, miR-205-5p, miR-124-3p, miR-26a-5p, miR-29a-3p, miR-137-3p, miR-485-5p, and miR-155-5p.

Melanoma cells downregulate miR-365a-5p, and the overexpression of miR-365a-5p promotes apoptosis by targeting BCL-2 (Zhu et al., 2018). miR-125b-5p affects senescence and apoptosis; in fact, miR-125b-5p increased levels of p27, p53; and p21, and consequently induced senescence (Nyholm et al., 2014). miR-205-5p is considered a tumor suppressor for promoting apoptosis. It was suggested that the downregulation of miR-205-5p in metastatic melanomas may lead to the activation of E2F1 and the inhibition of Rb (Dar et al., 2011). Apoptosis can be repressed also by miR-124-3p, which targets RALBP1 (Zhang, Han, & Xu, 2016). miR-26a-5p is downregulated in melanoma cells compared to normal melanocytes. In addition, the replacement of miR-26a-5p promoted cell death by targeting directly the antiapoptotic protein, silencer of death domains (SODD) (Reuland et al., 2013). Moreover, miR-26a-5p inhibits proliferation and invasiveness and halts the cell cycle at the G_1/G_0 phase as well as promotes apoptosis by targeting MITF in malignant melanoma cells lines (Qian et al., 2017). SNHG5 blocks miR-26a-5p, thus leading to the upregulation of miR-26a-5p target TRPC3 and resulting in inhibition of apoptosis (Gao et al., 2019). Apoptosis is promoted by the inhibition of BMI1 mediated by the overexpression of miR-29a-3p (Xiong et al., 2018). miR-137-3p can decrease BCL2, probably through the downregulation of MITF, resulting in the promotion of intrinsic apoptosis in melanoma cell lines (Luo, Tetteh, et al., 2013). miR-485-5p promotes apoptosis and inhibits epithelial-mesenchymal transition by targeting paired related homeobox 1 (PRRX1) (X. Wu & Bao, 2021). The overexpression of miR-155-5p inhibits proliferation and promotes apoptosis by targeting SKI (Levati et al., 2011).

Other miRNAs with a tumor-suppressor role associated with apoptosis are miR-18b-5p and miR-24-1-5p. miR-18b-5p has been found to be downregulated in melanoma compared with benign nevi and its downregulation is associated with decreased p53 activity. Specifically, miR-18b-5p targets the mouse double minute 2 homolog (MDM2), thus increasing p53 and promoting apoptosis (Dar et al., 2013). miR-24-1-5p is significantly downregulated in malignant melanoma tissues compared to adjacent normal tissues. The overexpression of miR-24-1-5p significantly enhances autophagy-associated proteins such as microtubule-associated protein 1A/1B-light chain 3 (LC3) -II/I ratio and beclin-1 expression and downregulates apoptosis-associated proteins, including Bcl-2 and Bcl-xL. In addition, miR-24-1-5p targets ubiquitin D (UBD), promoting autophagy and the apoptosis of malignant melanoma cells through activation of the JNK signaling pathway (Xiao et al., 2017).

Autophagy is a survival program that is activated by tumor cells in response to stressful conditions such as hypoxia or starvation for the maintenance of metabolic homeostasis and viability (Russell et al., 2014). Unbalanced autophagy can lead to programmed cell death. In cancer, autophagy can be pro- or antitumorigenic. Currently, it seems that autophagy has a tumor-suppressive role in the early stages of cancer while a tumor-promoting action in established tumors (Abibatou & Weeraratna, 2016). For this reason, the following microRNAs are divided into proautophagic and antiautophagic.

miR-216b-5p targets proteins involved in autophagy, such as beclin-1 (BECN1), autophagy related 5 (ATG5), and UV radiation resistance associated (UVRAG). In melanoma, the downregulation of miR-216b-5p is due to BRAF inhibition and results in the promotion of autophagy (Luo et al., 2018a). The silencing of miR-27a-3p promotes autophagy and apoptosis of melanoma cells by upregulating SYK expression and activating the mTOR signaling pathway (Tang et al., 2019). The miR-290-295 cluster is strongly upregulated in malignant melanoma cell lines. The overexpression of this cluster does not affect cell proliferation, migration, or anchorage-independent growth, but provides resistance to glucose starvation by targeting proteins involved in autophagy. Mechanistically, the miR-290-295 cluster downregulates autophagy related 7 (ATG7) and Unc-51 like autophagy activating kinase 1 (ULK1), leading to the inhibition of autophagy induced by glucose starvation (Chen, Liersch, & Detmar, 2012).

Autophagy can be promoted by the activation of the JNK signaling pathway mediated by the expression of miR-24-1-5p, which increases autophagy-related proteins, namely LC3II/I ratio and Beclin1, and downregulates Bcl2, BclxL, and UBD (Xiao et al., 2017).

The downregulation of miR-137-3p enhances erastin-induced ferroptosis, which is an iron-dependent cell death driven by small molecules or conditions that promote lipid-based ROS accumulation (Luo et al., 2018b). miR-9-5p affects ferroptosis by targeting glutamic-oxaloacetic transaminase (GOT1) in melanoma (K. Zhang et al., 2018).

microRNAs in invasion, angiogenesis, and metastasis

During melanoma progression, tumor cells acquire new abilities, including the capacity to migrate and go through a dedifferentiation program called epithelial-mesenchymal transition (EMT). EMT results from the loss of cell-to-cell junctions, induced by the loss of *E*-cadherin. Some transcription factors mediate this process, including zinc finger protein SNAI1 (SNAIL), snail family transcriptional repressor 2 (SNAIL2, also known as SLUG), zinc finger E-box binding homeobox 2 (ZEB2, also known as SPI1),and E2A (also known as TCF3, transcription factor 3). EMT is regulated by proteins such as transforming growth factor beta (TGFβ), epidermal growth factor (EGF), platelet-derived growth factor (PDGF), MAPK, PI3K/AKT, SMADs, Ras homolog gene family, member B (RHOB), β-catenin, lymphoid enhancer-binding factor (LEF), RAS, proto-oncogene FOS (c-FOS), integrin β4, and integrin α5 (Tse & Kalluri, 2007).

The EMT process is fundamental for the development of metastasis. Melanocytes derive from neural crest precursors, but EMT has been reported to occur in melanoma cells, where it promotes the metastatic phenotype of malignant melanocytes (Alonso et al., 2007). Indeed, a reorganization in the expression of embryonic EMT inducers drives the development of malignancy. SNAIL2 and ZEB2 are expressed in normal melanocytes and act as tumor-suppressor proteins by activating an MITF-dependent melanocyte differentiation program. However, in response to NRAS-BRAF activation, the EMT inducer network switches in favor of twist-related protein 1 (TWIST1) and ZEB1 (Caramel et al., 2013). Together with ZEB1, ZEB2 and MITF form a transcriptional network that regulates melanogenesis, melanoma progression, and migration (Denecker et al., 2014).

Several miRNAs are involved in the regulation of EMT with a promoting action, such as miR-576-5p, miR-182-5p, miR-21-5p, miR-10b-5p, miR-25-3p,miR-519d-3p, miR-221–3p/222-3p, miR-17-5p, miR-30b-5p/30d-5p, miR-224/miR-452 cluster, miR-373-3p, miR-638, miR-1246, miR-106b-5p, miR-1908-5p, miR-199a-5p, and miR-214-3p (Fig. 26.3A).

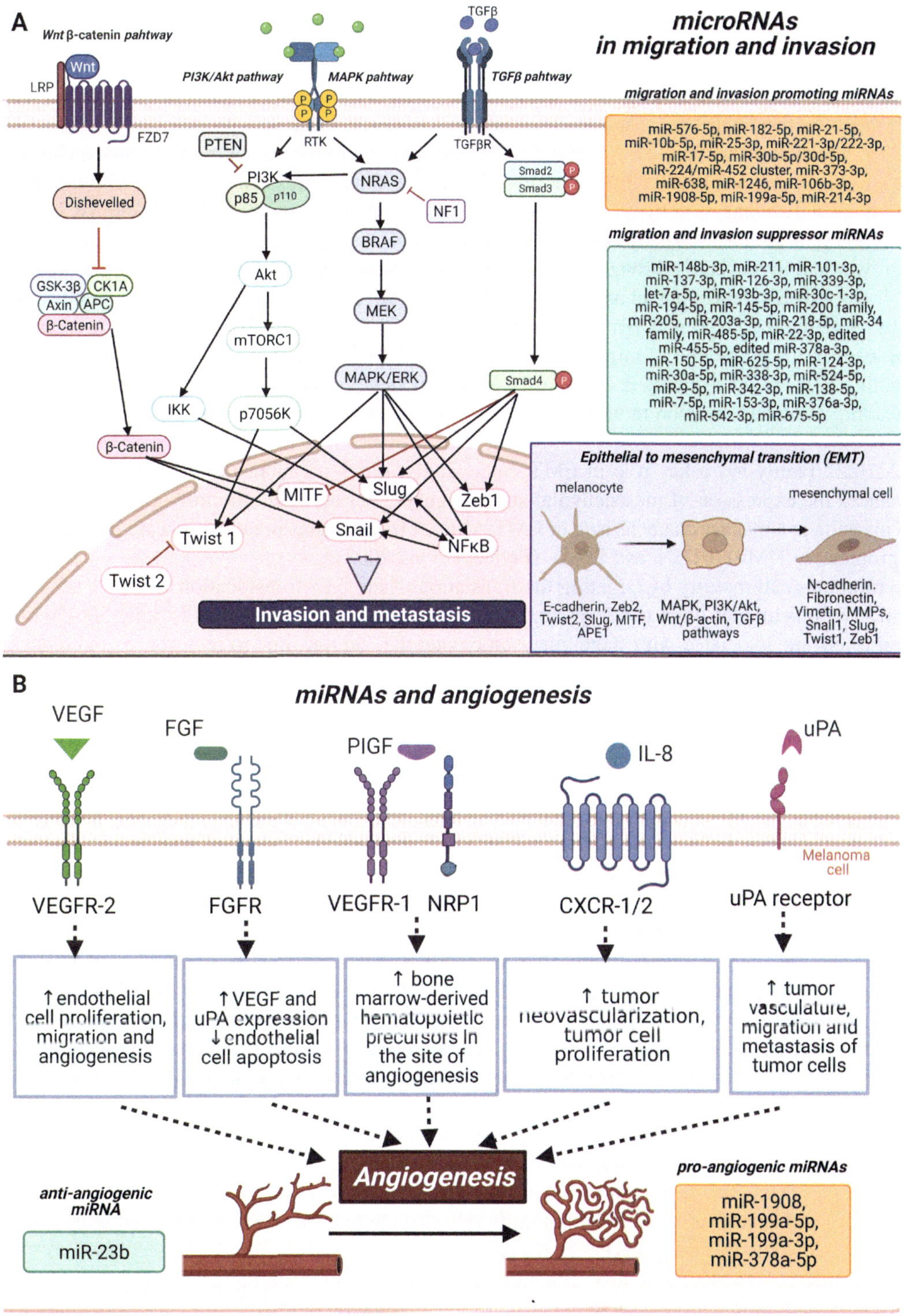

FIG. 26.3 microRNAs in melanoma migration, invasion, and angiogenesis. (A) Melanoma cells activate the epithelial to mesenchymal transition (EMT) pathway to promote migration, invasion, and metastasis. Several pathways, including Wnt β-catenin, PI3K/Akt, MAPK, and TGFβ, regulate EMT. By targeting proteins from these pathways, miRNAs promote (orange box) or suppress (green box) the migration and invasion of melanoma cells, acting as oncomiRs and tumor-suppressor miRNAs, respectively. (B) Angiogenesis is a fundamental hallmark for melanoma progression and dissemination. Malignant melanoma cells and endothelial cells present on their surface growth factor receptors that promote angiogenesis and are regulated by miRNAs. Some miRNAs are implicated in angiogenesis with a promoting (orange box) or suppressor role (green box). Figure created with BioRender.com.

An interesting study tested a large panel of miRNA mimics to assess their effect on melanoma cell invasion. They identified the most effective miRNAs in promoting cell invasion (miR-576-5p, miR-182-5p, miR-21-5p, and miR-214-3p) and those that were most effective in preventing cell invasion (miR-339-3p, miR-211-5p, miR-101–3p, miR-126-3p, miR-126-5p) (Weber et al., 2016). These findings were supported by functional validations in this and other studies.

Among the miRNAs that contribute to the acquisition of invasive and metastatic properties, including EMT, we recognize miR-576-5p, which is overexpressed in melanoma cells compared to normal human primary melanocytes. In melanoma cell lines, miR-576-5p acts as a tumor-promoting oncomiR and its expression correlates with the invasion score. In fact, miR-576-5p accelerates invasion by regulating B-cell CLL/lymphoma 9 protein (BCL9) and MCL-1 (Kordass et al., 2018). miR-182-5p promotes metastasis by targeting MITF and FOXO3 (Segura et al., 2009). miR-21-5p upregulation promotes tumor growth and invasiveness through the downregulation of the tissue inhibitor of metalloproteinase-3 (TIMP3), PTEN, PDCD4, and BTG2 (Del Campo et al., 2015; C. H. Yang et al., 2011). miR-10b-5p expression promotes migration in melanoma by targeting the ITCH and Wnt/β-catenin pathways (S. Wang et al., 2019). miR-25-3p expression is high in melanoma and leads to the promotion of migration by regulating DKK3 through the WNT/β-catenin pathway (Huo et al., 2016) and targeting RBM47 (Q. Q. Jiang & Liu, 2018). Metastatic lesions have higher expression levels of miR-519d-3p, which promotes migration and invasion in vitro and lung metastasis development in vivo by targeting EphA4 (Hua et al., 2018).

The miR-221/222 family has a key role in EMT. Specifically, it downregulates the expression of epithelial-specific genes and enhances the expression of mesenchymal-specific genes. Furthermore, the miR-221/222 family increases cell migration and invasion in a manner characteristic of EMT. MiR-221/222 targets in melanoma are p27, C-KIT, tissue inhibitor of metalloproteinase (TIMP), PTEN, and BRAF (Felicetti et al., 2008).

miR-17-5p promotes cell motility by targeting the translation of the ETS translocation variant 1 (ETV1) protein, suggesting that miR-17-5p is implicated in the development of metastasis (Cohen et al., 2015). miR-30b-5p and miR-30d-5p high expression levels are associated with stage, metastatic potential, shorter time to recurrence, and reduced overall survival in melanoma. The expression of miR-30b-5p/30d-5p promotes the metastatic behavior of melanoma cells by directly targeting polypeptide *N*-acetylgalactosaminyltransferase 7 (GALNT7). The inhibition of GALNT7 leads to enhanced invasion and immunosuppression. Specifically, there is the promotion of the synthesis of the immunosuppressive cytokine IL-10 and a reduction of immune cell activation and recruitment (Gaziel-Sovran et al., 2011).

The miR-224/miR-452 cluster (miR-224-5p and miR-452-5p) is significantly upregulated in metastatic melanoma and cells compared to primary melanoma and noninvasive melanoma cells, respectively. In advanced melanoma and invasive/metastatic cell lines, the upregulation of the miR-224/miR-452 cluster is associated with high levels of E2F1. E2F1 actives the miR-224/miR-452 cluster through transactivation of the gamma-aminobutyric acid receptor subunit epsilon (GABRE). The miR-224/miR-452 cluster promotes cytoskeletal rearrangements, migration, invasion, and EMT by targeting the thioredoxin interacting protein (TXNIP) (Knoll et al., 2014). miR-373-3p functions as an oncomiR and is upregulated in melanoma tissues and cell lines compared with nevi and normal melanocytes. The expression of miR-373-3p inhibits salt-inducible kinase 1 (SIK1), promoting melanoma cell migration (X. Bai et al., 2018). miR-638 is upregulated in metastatic melanoma compared with primary melanoma and it promotes metastasis and protects from apoptosis by regulating TP53INP2 (Bhattacharya et al., 2015). miR-1246 is higher expressed in melanoma tissues and cells compared to normal tissue. miR-1246 promotes cell viability and metastasis in melanoma by directly targeting Forkhead box protein A2 (FOXA2) (Yu et al., 2020).

Exosomal miR-106b-5p promotes EMT, migration, invasion, and adhesion of melanocytes in vitro and lung metastasis in vivo by targeting EphA4 to activate the ERK pathway (Luan et al., 2021). miR-1908-5p, miR-199a-5p, and miR-199a-3p function as endogenous promoters of metastatic invasion, angiogenesis, and colonization in melanoma by targeting apolipoprotein E (ApoE) and DnaJ heat shock protein family (Hsp40) member A4 (DNAJA4). Their expression is associated with a reduction of metastasis-free survival (Pencheva et al., 2012).

Suppression of miR-214-3p blocks the dissemination of melanoma, and this effect can be improved with the upregulation of miR-148b-3p. The combination of these two events, downregulation of miR-214-3p and upregulation of miR-148b-3p, leads to inhibition of the passage of malignant cells through the blood vessel endothelium by targeting cell adhesion molecules integrin alfa 5 (ITGA5) and the activated leukocyte cell adhesion molecule (ALCAM) (Orso et al., 2016). miR-214-3p is upregulated in melanoma (Orso et al., 2016) and contributes to melanoma progression and metastasis development by targeting transcription factor AP-2 gamma (TFAP2C) (Penna et al., 2011). miR-148b-3p inhibits migration and invasion also by targeting SIRT7, also reducing tumor growth and metastasis in vivo (Sun et al., 2020).

Several miRNAs that inhibit EMT, invasion, and metastasis have been identified over the years, including miR-148b-3p, miR-211-5p, miR-101–3p, miR-137-3p, miR-126-3p, miR-339-3p, let-7a-5p, miR-193b-3p, miR-30c-1-3p, miR-194-5p, miR-145-5p, miR-200 family, miR-205-5p, miR-203a-3p, miR-342-3p, miR-218-5p, miR-29a-3p, miR-34 family, miR-485-5p, miR-22-3p, miR-455-5p (upon A-I editing), miR-378a-3p (upon A-I editing), miR-150-5p, miR-625-5p, miR-124-3p, miR-30a-5p, miR-338-3p, miR-524-5p, miR-340-5p, miR-9-5p, miR-138-5p, miR-7-5p, miR-153-3p, miR-376a-3p, miR-542-3p, and miR-675-5p.

miR-211-5p is the melanocytic lineage-specific miRNA regulated by MITF that can also affect invasion. miR-211-5p regulates melanoma adhesion by targeting NUAK family SNF1-like kinase 1 (NUAK1), inhibiting melanoma invasion (Bell et al., 2014). A reduction of miR-211-5p expression enhanced melanoma invasiveness 10-fold while overexpression repressed the invasive potential of melanoma cells by downregulating transforming growth factor beta (TGFβ). However, miR-211-5p does not have any effect on the growth rate (Levy et al., 2010). In addition, miR-211-5p inhibits migration and invasion (Xu et al., 2012) by targeting the brain-specific homeobox/POU domain protein (BRN2) (Boyle et al., 2011) and potassium calcium-activated channel subfamily M alpha 1 (KCNMA1) (Mazar et al., 2010).

In addition to its roles in proliferation and melanogenesis, miR-101–3p represses the invasion of melanoma cells due to the inhibition of its target EZH2 and MITF genes (Luo, Merz, et al., 2013). The expression of miR-137-3p in melanoma results in inhibition of proliferation and invasion by the downregulation of MITF, c-Met, YB-1, and EZH2 (Luo, Tetteh, et al., 2013). In addition, miR-137-3p inhibited melanoma cell migration and invasion by targeting phosphoinositide-3-kinase regulatory subunit 3 (PIK3R3) (Qi et al., 2018).

miR-126-3p is upregulated during early progression but then downregulated in metastatic colonization (Mueller et al., 2009). The expression of miR-126-3p in advanced melanomas improves the antitumor activity of PIK-75, an inhibitor of the PI3K/AKT pathway, and vemurafenib (Pedini et al., 2019). Moreover, in dabrafenib-resistant melanoma cells, the restoration of miR-126-3p can inhibit proliferation, cell cycle progression, the phosphorylation of ERK1/2 and/or AKT, invasiveness, and VEGF-A (vascular endothelial growth factor) and ADAM9 expression as well as increase dabrafenib sensitivity (Caporali et al., 2019).

miR-339-3p is downregulated in melanoma compared to normal human epidermal melanocytes. The tumor-suppressive effects of miR-339-3p in melanoma have been tested in vitro and in vivo, where it has been found that miR-339-3p blocks cell invasion without affecting cell survival. miR-339-3p expression leads to the inhibition of cell invasion by targeting MCL-1, also confirmed by a reduction of tumor nodules in the lung in vivo (Weber et al., 2016). let-7a-5p is downregulated in melanoma cells compared with melanocytes. Let-7a-5p targets, such as NRAS and human integrin β3 (ITGB3), have a well-documented role in melanoma progression and invasion, indicating that this miRNA has a tumor-suppressive role in melanoma (Müller & Bosserhoff, 2008). miR-193b-3p and miR-30c-1-3p act as tumor-suppressive miRNAs in vitro. In fact, they both repress invasion by regulating BCL9 transcription coactivator (BCL9) and miR-193b-3p also by targeting MCL-1 (Kordass et al., 2018).

GEF–H1, a regulator of RhoA, promotes the growth and invasion of melanoma. The overexpression of miR-194-5p inhibits the GEF-H1/RhoA pathway, repressing cell proliferation and metastasis development (Guo, Hui, et al., 2016). miR-145-5p exhibits antitumorigenic activity in melanoma. Mechanistically, miR-145-5p the regulates NF-κB signaling pathway by targeting the Toll-like receptor 4 (TLR4) in malignant melanoma. That leads to the inhibition of tumor occurrence and metastasis (Jin et al., 2019).

miR-200 family members play important roles in the EMT process. Specifically, miR-200a-3p targets GOLM1, inhibiting the PI3K/Akt signaling pathway and EMT (Chen et al., 2019). In melanoma, it was shown that the NEAT1 directly sponges miR-200b-3p, resulting in the upregulation of SMAD2. Both NEAT1 and SMAD2 promote proliferation and invasion in vivo and in vitro (W. J. Zhou et al., 2020). The overexpression of miR-200c-3p leads to an inhibition of BMI1 and an increase of E-cadherin, resulting in the repression of cell proliferation and migration capacity (Liu, Tetzlaff, Cui, & Xu, 2012). In melanoma, miR-429 inhibits cancer cell proliferation and migration by targeting AKT1 (Huang et al., 2019).

The loss of miR-205-5p is associated with melanoma progression. Specifically, miR-205 overexpression leads to a decreased expression of ZEB2 mRNA and protein, resulting in the inhibition of cell migration and F-actin polymerization. The downregulation of miR-205-5p results also in a reduction of E-cadherin expression (Liu, Tetzlaff, Liu, et al., 2012). The overexpression of miR-203a-3p leads to the inhibition of cell motility, colony formation, sphere formation, and angiogenesis. In vivo, xenograft tumor growth is inhibited and lymph node and lung metastasis are reduced by the action of miR-203a-3p, which targets SNAIL2. Metastatic melanomas have low expression levels of miR-203a-3p, which are associated with poor overall survival (Lohcharoenkal et al., 2018). In addition to suppressing proliferation, the overexpression of miR-203a-3p leads to the inhibition of the invasiveness in melanoma by targeting BMI1 (Chang et al., 2015). miR-342-3p represses migration and invasion by directly targeting ZEB1 (Q. Shi et al., 2018).

miR-218-5p regulates proliferation, migration, and invasion of melanoma cell lines by targeting CIP2A and BMI1 (Y. Wei et al., 2014). miR-29a-3p overexpression leads to the suppression of migration and invasion by targeting BMI1 (Xiong et al., 2018).

miR-34a-5p acts as a tumor-suppressor miRNA that inhibits metastasis development by targeting FLOT2 (Liu, Xie, et al., 2015). miR-34c-5p is downregulated in melanoma compared to benign nevi (Sand et al., 2013). miR-34c-5p and miR-34b-3p have been identified as regulators of the tyrosine kinase receptor for hepatocyte growth factor, encoded by the MET oncogene, known to be one of the promoters of invasive growth by enhancing cell motility and invasion. These two miRNAs act as tumor-suppressor miRNAs through the regulation of MET-induced migration in melanoma-derived primary cells (Migliore et al., 2008). miR-485-5p inhibits migration and invasion by targeting FZD7 and PRRX1 (Wu et al., 2017; Wu & Bao, 2021).

Metastasis-associated lung adenocarcinoma transcript 1 (MALAT1) is overexpressed in melanoma, where it acts as an oncogene. MALAT1 promotes the growth and metastasis of melanoma cells though sponging miR-22-3p that leads to the release of specific targets of miR-22-3p, including matrix metallopeptidase 14 (MMP14) and SNAIL. That causes the activation of matrix metallopeptidase 2 (MMP2), the downregulation of E-cadherin, and the promotion of EMT (Luan et al., 2016). In addition, miR-22-3p inhibits EMT by targeting FMNL2 (L. Shi et al., 2019).

ADAR1 is downregulated in metastatic melanoma, and its silencing induces melanoma cell invasiveness by the upregulation of integrin subunit beta 3 (ITGB3). ADAR1 controls ITGB3 expression through miR-22-3p and paired box 6 (PAX6) (Nemlich et al., 2018). ADAR regulates miR-455-5p, which has two A-to-I RNA-editing sites. The function of edited miR-455-5p is opposite from that of the unedited form. In fact, wild-type miR-455-5p promotes melanoma growth and metastasis in vivo by targeting cytoplasmic polyadenylation element binding protein 1 (CPEB1). On the other hand, the edited form represses melanoma growth and metastasis in vivo (Shoshan et al., 2015). miR-378a-3p undergoes A-to-I editing only in the nonmetastatic melanoma cells, but not in metastatic melanoma cells. Only the edited form prevents progression to the malignant phenotype by binding the parvin alpha (PARVA) oncogene and thus inhibiting melanoma metastasis in vivo (Velazquez-Torres et al., 2018).

miR-150-5p has been found as low expressed in melanoma tissues and cell lines. miR-150-5p significantly inhibited the proliferation, migration, and invasion of melanoma cells by downregulating MYB (Sun et al., 2019) and inhibiting sine oculis homeobox 1 (SIX1) (X. Yang et al., 2019). MiR-625-5p is less expressed in malignant melanoma and its expression inhibits migration in malignant melanoma. In malignant melanoma, the expression level of miR-625-5p is inversely correlated with SRY-box 2 (SOX2) (Fang et al., 2017). miR-124-3p leads to the suppression of proliferation and invasion by inhibiting RALBP1(Zhang, Han, & Xu, 2016), RACK1 (Shen et al., 2019) and VCAN (P. Yang et al., 2017).

Melanoma tissues and cell lines show reduced expression levels of miR-30a-5p and high levels of SRY-box transcription factor 4 (SOX4). The increased expression of miR-30a-5p inhibits SOX4, resulting in the repression of proliferation, migration, and invasion in vitro (Liu, Sun, et al., 2018). In addition, in melanoma, the overexpression of miR-30a-5p inhibits metastasis in vivo by targeting ZEB2 and E-cadherin (Noori et al., 2018).

Migration and invasion in human malignant melanoma can be suppressed by the action of miR-338-3p, which targets the MET transcription regulator MACC1 (MACC1). In addition, decreased miR-338-3p expression is significantly correlated with lymph node metastasis and advanced clinical stage (Chunhua et al., 2019). miR-524-5p is implicated in the MAPK signaling pathway; in fact, it targets BRAF and ERK2, leading to the inhibition of proliferation and cell migration (Liu et al., 2014). Migration and invasion can be repressed by miR-340-5p, which targets the MAPK pathway (Poenitzsch Strong et al., 2014).

A low expression miR-9-5p has been observed in metastatic melanoma compared with primary melanoma. Its role consists of metastasis inhibition by targeting SNAIL and NF-κB1 and by the upregulation of E-cadherin (Liu et al., 2012). Moreover, miR-9-5p represses proliferation and migration of malignant melanoma cells via targeting SIRT1 (Bu et al., 2017).

Melanoma tissues present a low expression of miR-138-5p and a high expression of hypoxia-inducible factor 1α (HIF1α). It was shown that miR-138-5p^{low}/HIF1α^{high} signatures were predominant in stage III/IV of melanoma. The role of miR-138 is to inhibit migration and invasion in vitro, and metastasis in vivo by targeting HIF1α (Chen et al., 2016; Qiu et al., 2019).

The role of miR-7-5p has been studied in melanoma cell lines, as this miRNA is downregulated in metastatic *v.s* primary melanoma-derived cell lines. Furthermore, miR-7-5p inhibits cell migration and invasion by directly targeting insulin receptor substrate-2 (IRS-2), resulting in the reduction of AKT activity (Giles et al., 2013). miR-153-3p is downregulated in melanoma tissue and cell lines compared to para-tumor tissue and normal melanocytes. It was observed that miR-153-3p inhibits invasion by the repression of its direct target SNAIL (Zeng et al., 2017). miR-376a-3p belongs to an miRNA cluster on chromosome 14q32, which is silenced in melanoma. mir-376a-3p targets the insulin-like growth factor 1 receptor (IGF1R), leading to a reduction of migration (Zehavi et al., 2012). miR-542-3p is significantly downregulated in melanoma cell lines and tissues. miR-542-3p targets proto-oncogene serine/threonine protein kinase (PIM1), leading to the inhibition of melanoma cell migration, invasion, and EMT in vitro and lung metastasis in vivo (Rang et al., 2016). Metadherin

(MTDH) is a direct target of miR-675-5p and the inhibition of MTDH leads to the repression of cell proliferation and invasion in melanoma (Liu, Jin, et al., 2018).

The roles of miR-146a-5p, miR-125b-5p, and miR-199a-3p in melanoma migration and invasion are unclear and controversial. miR-146a-5p promotes tumor growth, but it blocks dissemination. The suppression of metastasis formation has been linked to the decreased expression of its two targets, integrin alpha-V precursor (ITGAV) and Rho-associated protein kinase 1 (ROCK1) (Raimo et al., 2016). However, it was observed that miR-146a-5p promotes cell migration and invasion in melanoma by directly targeting SMAD family member 4 (SMAD4) (Pu et al., 2018).

miR-125b-5p acts as a tumor suppressor and decreases cell migration, probably by the downregulation of its direct targets c-Jun, mitogen-activated protein kinase kinase kinase 11 (MAP3K11, also known as MLK3), and mitogen-activated protein kinase kinase 7 (MMK7) (Kappelmann et al., 2013). In addition, miR-125b-5p inhibits cell invasion and EMT by targeting integrin alpha-9 precursor ITGA9 (J. Zhang et al., 2016). However, it seems that the inhibition of miR-125b-5p reduces migratory and invasive capacities without affecting cell proliferation in vitro. It was observed that the neural precursor cell expressed developmentally downregulated protein 9 (NEDD9), which modulates melanoma cell migration and invasion, is a direct target of miR-125b-5p. In addition, the transcription of miR-125b-5p is induced by transcription factor 4 (TCF4), which is associated with EMT (Rambow et al., 2016).

Angiogenesis plays an important role in melanoma growth and metastasis dissemination. Malignant melanoma cells and endothelial cells present on their surface several growth factor receptors that lead to the promotion of angiogenesis. These growth factors include vascular endothelial growth factor (VEGF), fibroblast growth factor (FGF), placental growth factor (PIGF), interleukin 8 (IL-8), and urokinase-type plasminogen activator (uPA). In addition to growth factor receptors, integrins and matrix metalloproteinases are also important for melanoma growth and angiogenesis. For example, $\alpha v\beta 3$ integrin is important for the progression of cutaneous melanomas from the benign radial growth phase to the metastatic vertical growth phase. Matrix metallopeptidase 2 (MMP-2) and matrix metallopeptidase 9 (MMP-9) are extensively studied and have been correlated with the invasive and metastatic phenotypes of melanomas (Mahabeleshwar & Byzova, 2007).

miR-1908-5p, miR-199a-5p, and miR-199a-3p are implicated in the angiogenic process. In fact, miR-1908-5p, miR-199a-5p, and miR-199a-3p target ApoE and DNAJA4, leading to a promotion of metastatic invasion, angiogenesis, and colonization in melanoma (Pencheva et al., 2012). However, 199a-3p also acts as a tumor-suppressor miRNA through the regulation of MET-induced migration in melanoma-derived primary cells (Migliore et al., 2008). Moreover, in a metastatic melanoma model, miR-199a-3p was silenced by promoter methylation. The expression of miR-199a-3p was restored by promoter demethylation using 5-aza-2′-deoxycytidine, thus preventing in vitro 3D invasiveness of metastatic melanoma cells and reducing lung metastasis formation in vivo (Desjobert et al., 2019).

In addition, miR-378a-5p regulates the suppressor of fused homolog (SUFU), nuclear fusion protein FUS1 (FUS-1), Krueppel-like factor 9 (KLF9), STAM-binding protein (STAMBP), and homeobox D10 (HOXD10), facilitating the ability of melanoma cells to form de novo vasculogenic structures. miR-378a-5p is an oncomiR, which is overexpressed in metastatic melanoma compared to primary melanoma and melanoma in situ (Tupone et al., 2020).

An antiangiogenic miRNA is miR-23b-3p that is downregulated in melanoma compared to benign nevi (Kozubek et al., 2013) and inhibits angiogenesis by targeting nicotinamide phosphoribosyltransferase (NAMPT), thus impairing the NF-κB signaling pathway (Lv et al., 2020).

microRNAs in hypoxia and metabolism

In order to survive, tumor cells need to acquire new capacities, and one of them is the ability to adapt and survive in a hostile microenvironment. An issue that the tumor needs to address is the reduction of oxygen concentration, hypoxia, in the inner tumor area. miRNAs regulate several metabolic processes, specifically in response to hypoxic stimuli, and also mediate the metabolic switch from an oxidative to a glycolytic metabolism. Hypoxia-induced transcription factors, such as HIF1α and hypoxia-induced transcription factor 2α (HIF2α), are widely studied and are also direct targets of some miRNAs (Kulshreshtha et al., 2007). Among them is miR-210-3p, which inhibits the max-binding protein (MNT) that leads to the promotion of cell cycle and tumor growth in the absence of oxygen (Z. Zhang et al., 2009). Melanoma has an upregulated expression level of miR-210-3p compared to melanocytic nevi (Satzger et al., 2010). The expression of miR-210-3p can be induced by hypoxia in melanoma and promotes immune escape (Noman et al., 2012). On normoxic cells, miR-210-3p, miR-224-5p, miR-452-5p, and miR-218-5p are upregulated by HIF1α, resulting in the activation of an oxygen deprivation responsive target, such as BCL2/adenovirus E1B (BNIP3) and cyclic AMP-dependent transcription factor (ATF3) (Hwang et al., 2014).

Melanoma tissues present low expression of miR-138-5p and high expression of HIF1α. It has been observed that low levels of miR-138-5p and high levels of HIF1α are predominant in stage III/IV of melanoma. miR-138-5p inhibits migration and invasion in vitro as well as metastasis in vivo by targeting HIF1α (Chen et al., 2016; Qiu et al., 2019). Under hypoxic conditions, miR-340-5p is downregulated while its direct target, namely ATP-binding cassette subfamily B member 5 (ABCB5), is upregulated (Wozniak et al., 2015). Additionally, miR-33a-5p and miR-18b-5p directly target HIF1α; however, they act as tumor-suppressive miRNAs. In fact, their expression inhibits cell growth and is generally associated with a better prognosis. In addition, miR-18b-5p expression causes cell cycle arrest through glycolysis inhibition (Chen et al., 2016). In addition to the previously described functions, miR-211-5p targets pyruvate dehydrogenase lipoamide kinase isozyme 4 (PDK4) and increases oxygen consumption (Mazar et al., 2016).

Despite the presence of O2, tumors perform a metabolic switch in favor of a high rate of glycolysis followed by lactic acid fermentation in the cytosol, rather than glycolysis followed by oxidation of pyruvate in mitochondria, a process known as the Walburg effect (Heiden et al., 2009). miRNAs can affect this equilibrium, favoring the glycolytic metabolism and inhibiting oxidative phosphorylation, or vice versa.

In melanoma, miR-150-5p reduces glycolysis by inhibiting the glucose uptake, lactate production, ATP generation, and extracellular acidification rate (ECAR) while increasing the oxygen consumption rate (OCR) (X. Yang et al., 2019). miR-625-5p overexpression inhibits proliferation and glycolysis by targeting PKM2 (H. Zhang et al., 2019).

Let-7a-5p is an example of a tumor-suppressive miRNA in melanoma cells because it targets glucose-6-phosphate dehydrogenase (G6PD), inosine monophosphate dehydrogenase (IMPDH2), fatty acid synthase (FASN), stearoyl-CoA desaturase (SCD), and 4-phosphopantetheinyl transferase (AASDHPPT), which are anabolic enzymes. This favors oxidative phosphorylation and induces oxidative stress (Serguienko et al., 2015). miR-211-5p inhibits the ERK pathway, which results in the repression of MITF, and promotes the switch from glycolysis to oxidative phosphorylation through PGC1α and mitochondrial biogenesis (Vitiello et al., 2017). The overexpression of miR-138-5p reduces glucose consumption and lactic acid production, inhibiting glycolysis by targeting HIF1α (Chen, Cao, et al., 2016).

Targeting the glutamine metabolism sensitizes melanoma cells to TNF-related apoptosis-inducing ligand (TRAIL)-induced death (Qin et al., 2010). miR-605-5p inhibits melanoma progression and glutamine catabolism by targeting glutaminase (GLS) (Lu et al., 2020). Similarly, miR-137-3p acts as a tumor suppressor and inhibits proliferation and glutamine catabolism of malignant melanoma by targeting GLS (Luan et al., 2018). In addition, miR-137-3p targets the glutamine transporter solute carrier family 1 member 5 (SLC1A5), which is a glutamine transporter, leading to a reduction of glutamine uptake and malondialdehyde (MDA) accumulation (Luo, Wu, Zhang, Wang, Zhang, et al., 2018b).

microRNAs and melanoma microenvironment

The role of the immune system in melanoma has been intensely studied. Immune surveillance acts against melanoma, especially as a consequence of T-cell activity. Many factors can affect this system, allowing melanoma cells to escape from immune surveillance (Mahmoud et al., 2017), and these mechanisms were recently targeted by immune checkpoint inhibitors such as anti-CTLA4 and anti-PD1 (Postow et al., 2015). miRNAs can also regulate immune system members by affecting the activity of key elements of both innate and adaptative immunity (Fig. 26.4A). The innate components of the immune system, especially cells from the myeloid lineage, have an important function in melanoma (Ramirez-Montagut et al., 2003). miRNAs regulate immune responses mediated by natural killer (NK) cells, macrophages, and myeloid-derived suppressor cells (MDSC). Some miRNAs present an immune evasive or suppressive activity (Mahmoud et al., 2017).

The expression of miR-21-5p, miR-29a-3p, miR-142-3p, and miR-223-3p drives M2 macrophage reprogramming to tumor-infiltrating myeloid cells (TIMs), which acquire protumor and proangiogenic properties that lead to metastasis development. miR-21-5p, miR-29a-3p, miR-142-3p, and miR-223-3p are regulated by the colony stimulating factor 1/proto-oncogene 2 and the transcription factor (CSF1/ETS2) pathway, a mediator of M2 reprogramming. In addition, miR-21-5p and miR-29a-3p affect tumor growth and metastasis by targeting antiangiogenic factors involved in M1 polarization: programmed cell death 4 (PDCD4), protein sprouty homolog 1 (SPRY1), and TIMP3 are regulated by miR-21-5p while miR-29a-3p targets collagen alpha-2(IV) (COL4A2), osteonectin (SPARC),and TIMP3 (Mathsyaraja et al., 2015).

Natural killer group 2 member D (NKG2D) is one of the most potent activating receptors expressed on the surface of NK cells. NKG2D interacts with several human stress-inducible ligands: the transmembrane proteins MHC class I chain-related A and B (MICA and MICB), and 6 UL-16 binding proteins (ULBPs 1–6) (Raneros et al., 2014). miR-376a-3p and miR-433-3p are regulators of MICB. Analysis in melanoma cells has shown that levels of miR-376a-3p are decreased and those of miR-433-3p are increased compared to melanocytes, indicating that the balance of specific miRNAs determines the outcome on MICB expression (Nachmani et al., 2010). NKG2D ligands (NKG2DL) bind NKG2D receptors and constitute sensitizing factors to tumor cell killing by NK cells. However, tumor cells can release NKG2DL to evade NKG2D-mediated immune surveillance (Groh et al., 2002). UL-16 binding protein 2 (ULBP2) belongs to the NKG2DL family and is regulated by miR-34a-5p/34c-5p and miR-499a/c through binding to its 3′ UTR, suggesting a role of these miRNAs in

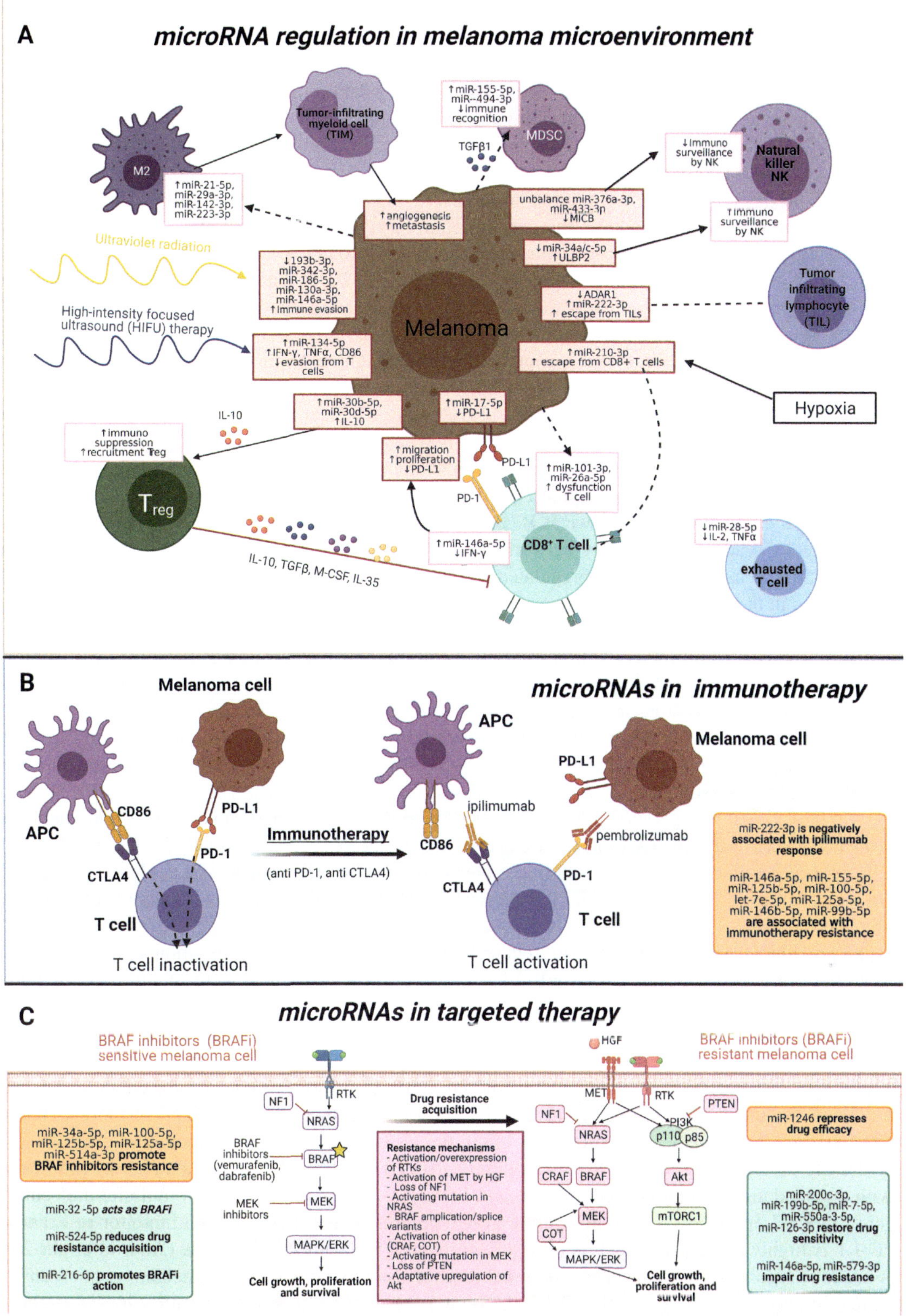

FIG. 26.4 microRNAs in melanoma microenvironment, immunotherapy, and targeted therapy. (A) To survive, melanoma cells need to escape from immune system surveillance. microRNAs regulate the crosstalk between melanoma and its microenvironment, including immune cells. Specifically, miRNAs induce immune evasion or improve immune surveillance by affecting both tumor and immune cell behavior. (B) Immunotherapy is based on the blockage of T cell receptors, such as PD1 or CTLA4, resulting in the activation of CD8+ T cells against the tumor. Treatment with immune checkpoint inhibitors is effective in melanoma but leads to the acquisition of immunotherapy resistance. The response to the treatment and acquisition of immunotherapy resistance in melanoma can be regulated by miRNAs. (C) Melanoma with mutated and activated BRAF can be treated with targeted therapy, including BRAF and/or MEK inhibitors. Some miRNAs affect the inhibitor activity by promoting their efficacy (green box, in the left) or impairing efficacy and inducing drug resistance (orange box). However, drug sensitivity in drug-resistant cells can be restored by the action of specific miRNAs (green box, in the right). Figure created with BioRender.com.

the process of melanoma cell recognition by NK cells. The downregulation of miR-34a-5p/34c-5p occurs frequently in cancer and leads to an upregulation of ULBP2, resulting in increased tumor immune surveillance by NK cells (Heinemann et al., 2012). ULBP2 is also a prognostic marker in human malignant melanoma: high levels of this gene in the sera of patients were correlated with poor prognosis (Paschen et al., 2009).

There are two miRNAs that induce an immune-suppressive phenotype in MDSCs: miR-155-5p and miR-494-3p. miR-155-5p upregulation in MDSCs induces the immune-suppressive phenotype by the inhibition of the suppressor of cytokine signaling 1 (SOCS1), a negative regulator of the Janus tyrosine kinase/signal transducer and activator of the transcription (JAK/STAT) pathway (Chen et al., 2015; Gulei et al., 2019). In addition, miR-155-5p upregulation is favored by IL-1β, and this mechanism can be a way to avoid immune recognition (Arts et al., 2015). miR-494 induces C-X-C chemokine receptor type 4 (CXCR4)-mediated chemotaxis and affects the survival of MDSCs through PTEN inhibition. The miR-494 expression in MDSCs is promoted by melanoma tumor cells through TGFβ1 secretion (Liu et al., 2012).

Ultraviolet radiation (UVR) is the main environmental risk factor for cutaneous melanoma and UVR-induced inflammation can promote immune evasion (Hodis et al., 2012). It was found that miR-193b-3p, miR-342-3p, miR-186-5p, miR-130a-3p, and miR-146a-5p are significantly downregulated after UV exposure, leading to an increase of immune evasive molecules such as C–C motif chemokine ligand 2 (CCL2), C–C motif chemokine ligand 8 (CCL8), PD1, and inducible T cell costimulator ligand (ICOSLG, also known as B7H2) (Sha et al., 2016).

miRNAs can affect the adaptive immunity network by regulating T cell activity, thus acting as immunosuppressive miRNAs. miR-30b-5p and miR-30d-5p are upregulated in melanoma, and their expression leads to the downregulation of GALNT7, resulting in the high expression immune suppressive IL-10 cytokine, which induces an immune-suppressive microenvironment. The upregulation of miR-30d-5p expression induces the immune-suppressive properties of lung metastases marked by enhanced infiltration with regulatory T cells (Treg) (Gaziel-Sovran et al., 2011).

Immune escape can be promoted also by a hypoxic microenvironment (Noman et al., 2011). In melanoma, miR-210-3p expression is induced by HIF, resulting in escape from cell lysis by antigen-specific cytotoxic T lymphocytes (CTL or CD8+ T cells) (Noman et al., 2012). Mechanistically, miR-210-3p acts as an immunosuppressive miRNA by inhibiting protein tyrosine phosphatase nonreceptor type 1 (PTPN1), homeobox protein Hox-A1 (HOXA1), and tumor protein p53-inducible protein 11 (TP53I1) (Noman et al., 2012).

Another target in the immune regulation process mediated by miRNAs in melanoma is ADAR1 (Nemlich et al., 2013), which performs adenosine-to-inosine (A-to-I) RNA editing. ADAR1 seems to be important for the recognition of tumor cells by T cells, in a cell-contact intercellular adhesion molecule 1 (ICAM1) dependent. The depletion of ADAR1 induces resistance to tumor-infiltrating lymphocytes (TILs). Specifically, ADAR1 regulates miR-222-3p that in turn inhibits ICAM1 (Galore-Haskel et al., 2015).

In melanoma preclinical models, high-intensity focused ultrasound (HIFU) therapy promotes immune response to the tumor that is miRNA-dependent. Mediated by miR-134-5p, HIFU leads to an increment of interferon gamma (IFN-γ) gamma, TNFα, and cluster of differentiation 86 (CD86, also known as B7–2) expression in tumor tissue (Yuan et al., 2015). CD86 is a direct target of miR-134-5p and when CD86 is lacking, there is an impairment in T lymphocyte activation (Zheng et al., 2004).

miRNAs regulate the expression of immune checkpoints. miR-28-5p expression is reduced in about one-third of exhausted T cells in melanoma. The restoration of miR-28-5p in these cells reverts the T cell phenotype, restoring IL-2 and TNF-α production. Mechanistically, miR-28-5p binds the 3′ UTR of T-cell immunoglobulin and mucin-domain containing-3 (TIM3), the B- and T-lymphocyte attenuator precursor (BTLA), and PD1 (Li et al., 2016). miR-17-5p has been associated with the regulation of programmed death-ligand 1 (PD-L1), which is a checkpoint inhibitor molecule. PD-L1 is a direct target of miR-17-5p. In BRAF- and mitogen-activated protein kinase kinase (MEK)-inhibitor-resistant melanoma cell lines, PD-L1 is highly expressed, which is inversely correlated with miR-17-5p levels (Audrito et al., 2017). miR-146a-5p is a negative regulator of immune activation and its expression is upregulated in melanoma microenvironmental tissue. miR-146a-5p plays a central role in the STAT1/IFN-γ axis-. Specifically, in T cells, miR-146a-5p targets Stat1 and Stat1-regulated cytokine, namely IFN- γ, affecting PD-L1 levels and migration, proliferation, and metabolism of melanoma cells (Mastroianni et al., 2019).

In the tumor microenvironment, melanoma can impair T cell functionality by targeting EZH2 via glycolysis restriction. In fact, melanoma stimulates the high expression of miR-101–3p and miR-26a-5p in effector T cells, resulting in the inhibition of EZH2, the limitation of aerobic glycolysis, and the dysfunction of the T cell. Specifically, EZH2 promotes the Notch pathway by suppressing NUMB and F-box/WD repeat-containing protein 7 (FBXW7), which are Notch repressors, as well as T cell polyfunctional cytokine expression and T cell survival through BCL-2 signaling (Zhao et al., 2016).

microRNAs as modulators of therapeutic response in melanoma

One of the main therapeutic opportunities for patients with BRAF or NRAS mutated melanomas is constituted by targeted therapies. However, continuous treatment with target-based drugs risks failure caused by the acquisition of drug resistance. The expression of specific miRNAs in melanoma cells could affect the response to target treatments, some of which are able to induce drug resistance and others to restore drug sensitivity (Fig. 26.1C).

BRAF inhibitors (BRAFi) have been developed to treat metastatic melanoma patients with BRAF mutations, but resistance to therapy eventually occurs and leads to disease progression. Similarly, treatment with MEK inhibitors (MEKi) can produce resistance to treatment after initial regression (Varrone & Caputo, 2020). A new generation of BRAFi and MEK1 as well as their combination is a most promising opportunity. Melanomas gain resistance to BRAFi by acquiring novel mutations and altering the pattern of cytokine production. Some miRNAs are associated with drug resistance or sensitivity in the context of BRAFi and BRAFi/MEKi.

Treatment with vemurafenib (BRAFi) leads to an increase of chemokine monocyte chemoattractant protein-1 (CCL2), which acts as an autocrine growth factor in melanoma and upregulates miR-34a-5p, miR-100-5p, and miR-125b-5p, which block apoptosis and promote drug resistance (Vergani et al., 2016). miR-100-5p and miR-125b-5p have been found upregulated in aggressive melanoma cells (Rambow et al., 2016). miR-125a-5p is upregulated by TGFβ signaling in melanoma cells and tissues upon BRAFi resistance acquisition. In fact, its overexpression suppresses proapoptotic components of the intrinsic apoptosis pathway, including Bcl-2 homologous antagonist killer (BAK1) and mixed lineage kinase 3 (MLK3, also known as MAP3K11), promoting MAPK signaling (Koetz-Ploch et al., 2017).

miR-514a-3p affects the sensitivity of BRAF inhibitors. In fact, miR-514a-3p regulates neurofibromatosis type 1 (NF1) and induces the activation of the MAPK pathway. The overexpression of miR-514a-3p confers resistance to BRAF inhibitors (Stark et al., 2015). miR-1246 significantly reduces the antiproliferative effects of the BRAFi PLX4720 in BRAFi-resistant melanoma cells by the downregulation of p-ERK (Kim et al., 2017).

Resistant melanoma tumors and cell lines downregulate miR-200c-3p, with a consequent upregulation of its targets, including BMI1, ZEB2, tubulin beta 3 class III (TUBB3), ATP-binding cassette subfamily G member 5 (ABCG5), and multidrug resistance protein 1 (MDR1). The restoration of miR-200c-3p confers the sensitivity of BRAF inhibitors by suppressing the PI3K/AKT and MAPK signaling pathways (Liu et al., 2015). BRAFi-resistant melanoma cells downregulate miR-199b-5p, leading to VEGF release and to the acquisition of a proangiogenic status. High levels of miR-199b-5p can reverse the proangiogenic properties of drug-resistant melanoma cells (Fattore et al., 2019).

miR-7-5p is downregulated in vemurafenib=resistant melanoma cells. The restoration of miR-7-5p expression can reverse the resistance of BRAFi in melanoma by regulating epidermal growth factor receptor (EGFR), insulin-like growth factor-1 receptor (IGF-1R), and c-RAF, inhibiting the MAPK and PI3K/AKT signaling pathways (Sun et al., 2016). Also, miR-550a-3-5p can reverse BRAFi resistance and function as a tumor suppressor. Mechanistically, miR-550a-3-5p targets yes-associated protein 1 (YAP) with the reduction of AKT activity and restores sensitivity to vemurafenib (Choe et al., 2018). miR-126 3p expression is significantly downregulated in cells resistant to dabrafenib, a BRAF inhibitor. The expression of miR-126-3p in the drug-resistant cells inhibits proliferation, cell cycle progression, and invasiveness, and increases dabrafenib sensitivity by targeting ADAM9 and VEGF-A (Caporali et al., 2019).

Malignant tumors frequently show low expression of miR-32-5p and an upregulation of its targets, including NRAS, PI3K, and MCL-1. miR-32-5p expression inhibits tumor growth. In addition, the combination of miR-32-5p and vemurafenib is more effective than either vemurafenib or miR-32-5p treatment alone. miR-32-5p has the same effect as sabutoclax; in fact, they both target MCL-1 (Mishra et al., 2016).

miR-524-5p reduces the progression of the BRAF inhibitor-resistant melanoma. Specifically, the overexpression of miR-524-5p inhibits proliferation, anchorage-independent growth, as well as the migratory and invasive capacities of BRAF inhibitor-resistant melanoma cells. In addition, the introduction of miR-524-5p decreases the development of BRAF inhibitor-resistant melanoma by downregulation of the MAPK/ERK signaling pathway (Nguyen et al., 2020).

miR-216b-5p regulates autophagy by targeting BECN1, ATG5, and UVRAG, and promotes the antitumor activity of vemurafenib in BRAF(V600E) melanoma cells (Luo, Wu, Zhang, Wang, Wu, et al., 2018a). miR-378a-5p promotes angiogenesis and its expression is increased in melanoma cells resistant to trametinib (MEK inhibitor), and decreased in response to drug treatment (Tupone et al., 2020). miR-146a-5p expression is reduced in BRAF/MEKi-resistant melanoma cells and tissues. The overexpression of miR-146a-5p impairs melanoma resistance by targeting prostaglandin-endoperoxide synthase 2 (COX2) and regulating NFkB-mediated inflammatory mediators (Vergani et al., 2020).

The roles of miR-204-5p and miR-211-5p are controversial. Similarly to miR-211-5p, miR-204 is normally considered a tumor-suppressor miRNA. miR-204-5p is upregulated in melanocytes compared to melanoma, and its expression is associated with a better prognosis (Galasso et al., 2018). On the one hand, it seems that vemurafenib induces the expression of miR-204-5p and miR-211-5p, which behave as tumor-suppressor miRNAs. STAT3 regulates the expression of miR-204-5p in amelanotic melanoma cells, where miR-204-5p functions as an effector of vemurafenib's antimotility effect by regulating AP-1 complex subunit sigma-2 (AP1S2). The expression of miR-211-5p is induced in melanotic melanoma cells, where it targets ER degradation-enhancing alpha-mannosidase-like protein 1 (EDEM1) and impairs the degradation of tyrosinase (TYR) via the ER-associated degradation (ERAD) pathway. miR-211-5p is an effector of vemurafenib's propigmentation activity (Vitiello et al., 2017). On the other hand, vemurafenib-resistant cells increase miR-204-5p and miR-211-5p expression. The expression of miR-204-5p and miR-211-5p confers vemurafenib resistance in drug-naïve human melanoma cells by inducing Ras and MAPK upregulation (Díaz-Martínez et al., 2018). In addition, the deficiency of miR-211-5p represses growth and invasion by inhibition of PI3K signaling and promotion of metabolic vulnerability. Vemurafenib-resistant melanoma cells upregulate miR-211-5p while the loss of miR-211-5p confers drug sensibility (Sahoo et al., 2019). miR-211-5p promotes melanoma growth driven by BRAFV600E and resistance to BRAK/MEK inhibitors by targeting dual-specificity phosphatase 6 (DUSP6) with a consequent upregulation of extracellular signal-regulated kinase (ERK) 5 signaling (Lee et al., 2021). Moreover, vemurafenib induces the upregulation of miR-211-5p in melanoma cells and in tumor-derived extracellular vesicles (EVs). The expression of miR-211-5p in melanoma cells reduced the sensitivity to vemurafenib treatment while the inhibition suppresses the proliferation in the vemurafenib-resistant cell line (Lunavat et al., 2017).

Low levels of miR-579-3p are associated with a poor prognosis; in fact, the expression decreases progressively from nevi to III/IV melanoma samples. In addition, BRAF/MEK inhibitor-resistant melanoma cell lines downregulate miR-579-3p. The overexpression of miR-579-3p impairs drug resistance in melanoma cells (Fattore et al., 2016). Specifically, STAT3 downregulated miR-579-3p, promoting resistance to vemurafenib (X. Wang et al., 2018).

Metastatic melanoma can be treated with immunotherapies. Some miRNAs regulate directly or indirectly the expression of immune checkpoints, acting on tumor cells or immune cells, respectively (Motti et al., 2020) (Fig. 26.4B). The expression of miR-222-3p in melanoma is significantly lower in patients responsive to ipilimumab than patients that have no clinical benefit from treatment. Ipilimumab is a monoclonal antibody that works to activate the immune system by targeting CTLA-4, a protein receptor that inhibits the immune response. miR-222-3p may be a biomarker for the prediction of response to ipilimumab (Galore-Haskel et al., 2015).

Anti-PD-1 treatment increases miR-155-5p expression in CD8+ T-cells. miR-155-5p overexpression enhances exhausted CD8+ T-cell persistence and the expression is positively correlated with CD8+ T-cell infiltration (Martinez-Usatorre et al., 2019).

Looking at exosomal miRNAs in sera from melanoma patients, higher levels of miR-532-5p and miR-106b-5p have been found in melanoma patients with stage III-IV disease compared to patients with stage I-II, and in untreated patients compared to patients treated with pembrolizumab (Tengda et al., 2018), which is a humanized antibody that targets the PD-1 receptor of lymphocytes.

In melanoma patients, a set of microRNAs (miR-146a-5p, miR-155-5p, miR-125b-5p, miR-100-5p, let-7e-5p, miR-125a-5p, miR-146b-5p, miR-99b-5p) is associated with myeloid-derived suppressor cells (MDSCs) and resistance to treatment with immune checkpoint inhibitors, namely ipilimumab and nivolumab. Specifically, melanoma EVs release these miRNAs that promote the conversion of monocytes into MDSCs. Higher circulating levels of these miRNAs cluster with shorter progression-free survival and overall survival in patients treated with ipilimumab and nivolumab, but not in those treated with BRAF/MEK inhibitors (Huber et al., 2018).

Metastatic melanoma patients have received combination targeted therapy with the mTOR inhibitor temsirolimus and the anti-VEGF antibody bevacizumab, and the miRNA expression has been analyzed pretreatment and posttreatment in human metastatic melanoma tissue samples. Fifteen miRNAs have been found upregulated in the posttreatment (miR-125b-5p, miR-320a-5p, miR-320b, miR-320c, miR-320d, miR-320e, let-7b-5p, let-7c-5p, miR-10b-5p, miR-29c-3p, miR-100-5p, miR-145-5p, miR-140-3p, miR-99a, miR-4328), and 12 of these 15 are known tumor-suppressor miRNAs (Wagenseller et al., 2013).

Metformin, an oral biguanide drug used for diabetes, significantly suppresses melanoma cancer cell motility and growth through inducing cell cycle arrest at the G_2/M phase and apoptosis. Mechanistically, metformin upregulates miR-192-5p and miR-584-3p that target epidermal growth factor (EGF) containing fibulin-like extracellular matrix protein 1 (EFEMP1) and an isoform of the secretory carrier membrane proteins (SCAMP3), respectively (Tseng et al., 2019).

In melanoma, miR-30a-5p enhances chemoresistance to cisplatin by targeting the insulin-like growth factor 1 receptor (IGF1R). The downregulation of IGF1R leads to cisplatin resistance and cell cycle arrest upon activation of P-AKT and p53. In fact, it seems that G_2/M phase cell arrest could be a protective mechanism to prevent replication of cisplatin-damaged DNA (Li et al., 2018). On the contrary, miR-31-5p functions as protective miRNA; in fact, miR-31-5p is downregulated in melanoma but its enforced expression increases chemosensitivity. Specifically, miR-31-5p targets and suppresses

SOX10. High expression of SOX10 is associated with melanoma cell proliferation and chemotherapy resistance via the PI3K/AKT signaling pathway (Zheng et al., 2018). miR-211-5p seems to be downregulated in melanoma due to DNA hypermethylation, and its overexpression can restore cisplatin susceptibility in cisplatin-resistant cells. It was also observed that low expression levels of miR-211-5p correlated with a poor prognosis in a 38-melanoma cohort (N. Li et al., 2019). The overexpression of miR-203a-3p sensitizes malignant melanoma cells to temozolomide chemotherapy by regulating glutaminase (GLS) (Chang et al., 2017).

In melanoma patients treated with chemotherapy, specifically with carboplatin/paclitaxel, miR-659-3p has been found to be positively associated with progression-free survival. Indeed, high levels of this miRNA can discriminate patients with responsive disease from patients with stable disease. Mechanistically, miR-659-3p regulates nuclear factor 1 X-type (NFIX), which interacts with c-Jun and AP-1 (Villaruz et al., 2015).

Circulating microRNAs as melanoma biomarkers

Cell-free microRNAs can be released by the tumor in the bloodstream, where they remain stable and are resistant to RNAse activity. A fraction of cell-free miRNAs is bound to macromolecular complexes, including cholesterol and lipoproteins, Argonaute 2 proteins and EVs, which can also carry miRNAs inside. EVs are one of the main sources of miRNAs in the blood (Cheng et al., 2014). miRNAs can be measured in a noninvasive way, in a so-called "liquid biopsy" approach. Significant alterations in miRNA levels have been detected in the serum and plasma of cancer patients compared to healthy controls, suggesting that circulating miRNAs could be used as diagnostic biomarkers (Mitchell et al., 2008) as well as markers of cancer diagnosis and prognosis (Ferracin & Negrini, 2015). The main technical issues associated with circulating biomarkers are related to the impact of sample collection and processing in establishing the tested miRNA fraction, and therefore their level. It is mandatory to compare miRNA levels across samples with similar processing.

The management of melanoma depends greatly on early diagnosis. In fact, melanoma is almost 100% curable by surgical resection when it is detected at early, nonmetastatic stages. Therefore, it is essential to identify diagnostic biomarkers. Another promising application of circulating miRNAs is the early detection of recurrence. We report here the main studies on diagnostic and prognostic cell-free miRNA biomarkers in melanoma in Table 26.1.

Among the plasmatic miRNAs proposed for early diagnosis, miR-320a-5p has been demonstrated as a plasma melanoma biomarker in comparison with other main types of solid tumors and healthy subjects (Ferracin et al., 2015). Plasma levels of miR-21-5p were found to be elevated in melanoma. In addition, miR-21-5p is associated with tumor burden (Ferracin et al., 2015; Saldanha et al., 2013).

A 16-miRNA diagnostic signature (miR-186, let-7d-3p, miR-18a-3p, miR-145-5p, miR-99a-5p, miR-664a-3p, miR-501-5p, miR-378a-5p, miR-29c-5p, miR-1280, miR-365a-3p, miR-1249-3p, miR-328-3p, miR-422a, miR-30d-5p, and miR-17-3p) has been found in blood cells and distinguishes metastatic melanoma patients (stages III and IV) from healthy control individuals (Leidinger et al., 2010). Of note, miR-1280 was later recognized as a tRNA fragment. Similarly, a 38-miRNA signature (MEL38) has been identified in plasma samples and is able to distinguish between melanoma patients (stages I-IV) and healthy controls. In this signature, 19 miRNAs are upregulated in melanoma: miR-301a-3p, miR-424-5p, miR-27a-3p, miR-34a-5p, miR-497-5p, miR-299-3p, miR-152-3p, miR-1910-5p, miR-181b-5p, miR-548a-5p, miR-454-3p, miR-4532, miR-1537-3p, miR-1258, miR-431-5p, miR-450a-5p, miR-2682-5p, miR-337-5p, and miR-154-5p. The other 19 miRNAs are downregulated in melanoma: miR-205-5p, miR-548L, miR-1269a, miR-624-3p, miR-138-5p, miR-1-5p, miR-3928-3p, miR-3131, miR-1973, miR-520d-3p, miR-548ad-3p, miR-553, miR-764, miR-1302, miR-522-3p, miR-1264, miR-1306-5p, miR-219a-2-3p, and miR-4787-3p (Van Laar, Lincoln, & Van Laar, 2018). Also in this signature, miR-4532 was later identified as a 28S rRNA fragment.

Comparing plasma samples of melanoma patients with those of healthy controls, miR-15b-5p, miR-149-3p, and miR-150-5p were found increased in melanoma compared with controls while miR-193a-3p and miR-524-5p are downregulated. The diagnostic value of microRNAs is improved when considering the combination of miR-149-3p, miR-150-5p, and miR-193a-3p (Fogli et al., 2017). The detection of miR-185-5p and miR-1246 in plasma discriminated patients with metastatic melanoma from healthy individuals (Armand-Labit et al., 2016).

In addition to plasma, circulating miRNA biomarkers can be identified in serum. miR-29c-5p and miR-324-3p expression levels are lower in the serum of stage IV metastatic patients relative to healthy controls (Greenberg et al., 2013). In serum samples of melanoma patients, the levels of miR-122-5p and miR-3201 are higher than in serum samples of healthy people (Margue et al., 2015).

A panel of 17 miRNAs (MELmiR-17) has been identified in melanoma tissues (stage III and stage IV) and serum samples (stages I/II, stage III, stage IV, healthy control). In melanoma tissues, the MELmiR-17 panel was able to predict the stage, recurrence, and survival. In addition, a seven-miRNA panel (MELmiR-7) detected the presence of melanoma by

TABLE 26.1 Circulating microRNAs as melanoma biomarkers.

microRNA/miRNA signature	Type of biomarker	Expression	Sample	Reference
miR-186, let-7d-3p, miR-18a-3p, miR-145-5p, miR-99a-5p, miR-664a-3p, miR-501-5p, miR-378a-5p, miR-29c-5p, miR-1280[a], miR-365a-3p, miR-1249-3p, miR-328-3p, miR-422a, miR-30d-5p, miR-17-3p	Diagnostic	Dysregulated in metastatic melanoma patients (stages III and IV) vs. healthy control individuals	Blood cells	(Leidinger et al., 2010)
miR-320a-5p	diagnostic	upregulated in melanoma vs. other tumors and healthy controls	plasma	(Ferracin et al., 2015)
miR-301a-3p, miR-424-5p, miR-27a-3p, miR-34a-5p, miR-497-5p, miR-299-3p, miR-152-3p, miR-1910-5p, miR-181b-5p, miR-548a-5p, miR-454-3p, miR-4532, miR-1537-3p, miR-1258,[b] miR-431-5p, miR-450a-5p, miR-2682-5p, miR-337-5p, miR-154-5p	Diagnostic	Upregulated in melanoma patients vs. healthy controls	Plasma	(Van Laar, Lincoln, & Van Laar, 2018)
miR-205-5p, miR-548l, miR-1269a, miR-624-3p, miR-138-5p, miR-1-5p, miR-3928-3p, miR-3131, miR-1973, miR-520d-3p, miR-548ad-3p, miR-553, miR-764, miR-1302, miR-522-3p, miR-1264, miR-1306-5p, miR-219a-2-3p, miR-4787-3p	Diagnostic	Downregulated in melanoma patients vs. healthy controls	Plasma	(Van Laar, Lincoln, & Van Laar, 2018)
miR-15b-5p, miR-149-3p, miR-150-5p	Diagnostic	Upregulated in melanoma patients vs. healthy controls	Plasma	(Fogli et al., 2017)
miR-193a-3p, miR-524-5p	Diagnostic	Downregulated in melanoma patients vs. healthy controls	Plasma	(Fogli et al., 2017)
miR-185-5p, miR-1246	Diagnostic	Dysregulated in metastatic patients vs. healthy controls	Plasma	(Armand-Labit et al., 2016)
miR-29c-5p, miR-324-3p	Diagnostic	Downregulated in metastatic melanoma patients (stage IV) vs. healthy controls	Serum	(Greenberg et al., 2013)
miR-122-5p, miR-3201	Diagnostic	Upregulated in melanoma patients vs. healthy controls	Serum	(Margue et al., 2015)
miR-16-5p, miR-211-5p	Diagnostic	Upregulated in melanoma patients (only stage IV) vs. healthy controls	Serum	(Stark et al., 2015)
miR-16-5p	Diagnostic	Downregulated in melanoma patients vs. healthy controls	Serum	(Guo et al., 2016)
miR-4487, miR-4706, miR-4731, miR-509-3p, miR-509-5p	Diagnostic	Downregulated in melanoma patients vs. healthy controls	Serum	(Stark et al., 2015)
miR-17-5p, miR-19a-3p, miR-21-5p, miR-126-3p, miR-149-5p	Diagnostic	Upregulated in metastatic sporadic melanoma patients vs. healthy controls	Plasma-derived exosomes	(Pfeffer et al., 2015)

miR-125b-5p	Diagnostic	Downregulated in advanced melanoma patients vs. healthy controls	Serum-derived exosomes	(Alegre et al., 2014)
miR-150-5p, miR-199a-5p, miR-424-5p	Prognostic	Upregulated in melanoma patients with high risk of recurrence	Serum	(Friedman et al., 2012)
miR-15b-5p, miR-33a-5p	Prognostic	Downregulated in melanoma patients with high risk of recurrence	Serum	(Friedman et al., 2012)
miR-150-5p, miR-30d-5p, miR-15b-5p, miR-425-5p	Prognostic	Associated with recurrence-free and overall survival	Serum	(Fleming et al., 2015)
miR-150-5p, miR-9-5p, miR-145-5p, miR-155-5p, miR-203a-3p, miR-205-5p	Prognostic	Dysregulated in metastatic melanoma patients vs. nonmetastatic melanoma patients	Serum	(Shiiyama et al., 2013)
miR-10b-5p	Prognostic	Associated with lymph node metastasis, advanced clinical stage, and shortened survival rate	Serum	Bai, Zhang, Si, et al. (2017)
miR-221-3p	Prognostic	Downregulated in metastatic melanoma	Serum	(Kanemaru et al., 2011)
miR-21-5p	Diagnostic/prognostic	Upregulated in melanoma patients vs. healthy controls; associated with tumor burden	Plasma	(Ferracin et al., 2015; Saldanha et al., 2013)
miR-210-3p	Diagnostic/prognostic	Upregulated in melanoma patients vs. healthy controls; high expression correlates with poor prognosis	Plasma	(Ono et al., 2015)
miR-206	Diagnostic/prognostic	Downregulated in melanoma patients vs. healthy controls; low expression correlates with poor prognosis	Serum	(Tian et al., 2015)

[a] *Later removed from miRBase.*

[b] *Later removed from miRBase.*

serum analysis: miR-4487, miR-4706, miR-4731, miR-509-3p and miR-509-5p were reduced while miR-16-5p and miR-211-5p (stage IV only) were increased in melanoma patients (Stark et al., 2015).

An exosome-associated miRNA pool has been investigated between melanoma patients and controls. miR-125b-5p levels are significantly lower in the serum exosomes of patients with advanced melanoma compared to serum exosomes of disease-free patients and healthy controls (Alegre et al., 2014). Plasma-derived exosomal miRNAs have been analyzed and miR-17-5p, miR-19a-3p, miR-21-5p, miR-126-3p, and miR-149-5p are expressed at higher levels in patients with metastatic sporadic melanoma as compared with familial melanoma patients or unaffected control subjects (Pfeffer et al., 2015).

Circulating miRNA can also be used to provide prognostic information. A good prognostic biomarker should be able to discriminate stage I/II and stage III/IV melanoma patients, and predict which patients are likely to develop metastasis. In addition, it should correlate with poor prognosis and classify patients into groups at high and low risk of recurrence after the surgical resection of a nonmetastatic melanoma.

A panel of five circulating miRNAs has identified melanoma patients with a high risk of recurrence. Specifically, the downregulation of miR-15b-5p and miR-33a-5p and the upregulation of miR-150-5p, miR-199a-5p, and miR-424-5p in melanoma serum samples are associated with a high risk of recurrence (Friedman et al., 2012). A predictive panel of four miRNAs (miR-150-5p, miR-30d-5p, miR-15b-5p, and miR-425-5p) has been identified in serum samples. This panel was superior to conventional staging criteria in predicting recurrence-free and overall survival. Moreover, miR-15b-5p levels were demonstrated to be useful for the early monitoring of recurrence in melanoma patients (Fleming et al., 2015).

To identify metastatic melanoma, a panel of six serum circulating miRNAs was proposed, namely miR-150-5p, miR-9-5p, miR-145-5p, miR-155-5p, miR-203a-3p, and miR-205-5p, which are differently expressed between metastatic and nonmetastatic melanoma patients (Shiiyama et al., 2013).

The upregulation of miR-10b-5p in serum has been positively associated with enhanced lymph node metastasis, advanced clinical stage, and shortened survival rate (Bai, Zhang, Long, et al., 2017).

The serum expression of miR-16-5p is lower in melanoma patients than in healthy controls, and the decrease of miR-16-5p has been correlated with an increase in tumor thickness, ulceration status, AJCC stage, and tissue MKI67 marker of proliferation Ki-67 (MKI67, also known as Ki-67) expression (Guo et al., 2016). miR-221-3p has been found to be increased in the serum of patients with metastatic melanoma, decreasing after excision and increasing with disease recurrence. miR-221–3p serum expression differentiates melanoma in situ from stage I-IV, and correlates with Breslow thickness (Kanemaru et al., 2011).

Similarly to miR-21-5p, there are miRNAs that can be both diagnostic and prognostic biomarkers, such as miR-210-3p and miR-206. For example, plasma levels of miR-210-3p are higher in individuals with melanoma and an increased level of miR-210-3p can predict disease recurrence. Furthermore, the upregulation of miR-210-3p in plasma was correlated with a poor prognosis (Ono et al., 2015). The serum levels of miR-206 are significantly lower in melanoma compared to healthy controls. Moreover, patients with low serum miR-206 levels present two or more sites of metastases and had a shorter 5-year overall and disease-free survival than melanoma patients with a high miR-206 level (Tian et al., 2015).

References

Abdel-Wahab, N., Shah, M., & Suarez-Almazor, M. E. (2016). Adverse events associated with immune checkpoint blockade in patients with cancer: A systematic review of case reports. *PLoS One*. https://doi.org/10.1371/journal.pone.0160221, e0160221.

Abibatou, N., & Weeraratna, A. T. (2016). Autophagy- An emerging target for melanoma therapy. *F1000Research*, 1888. https://doi.org/10.12688/f1000research.8347.1.

Akbani, R., Akdemir, K. C., Aksoy, B. A., Albert, M., Ally, A., Amin, S. B., et al. (2015). Genomic classification of cutaneous melanoma. *Cell*, *161*(7), 1681–1696. https://doi.org/10.1016/j.cell.2015.05.044.

Alegre, E., Sanmamed, M. F., Rodriguez, C., Carranza, O., Martín-Algarra, S., & González, A. (2014). Study of circulating MicroRNA-125b levels in serum exosomes in advanced melanoma. *Archives of Pathology and Laboratory Medicine*, *138*(6), 828–832. https://doi.org/10.5858/arpa.2013-0134-OA.

Alonso, S. R., Tracey, L., Ortiz, P., Pérez-Gómez, B., Palacios, J., Pollán, M., et al. (2007). A high-throughput study in melanoma identifies epithelial-mesenchymal transition as a major determinant of metastasis. *Cancer Research*, *67*(7), 3450–3460. https://doi.org/10.1158/0008-5472.CAN-06-3481.

Amin, M. B., Greene, F. L., Edge, S. B., Compton, C. C., Gershenwald, J. E., Brookland, R. K., et al. (2017). The eighth edition AJCC cancer staging manual: Continuing to build a bridge from a population-based to a more "personalized" approach to cancer staging. *CA: a Cancer Journal for Clinicians*, *67*(2), 93–99. https://doi.org/10.3322/caac.21388.

Armand-Labit, V., Meyer, N., Casanova, A., Bonnabau, H., Platzer, V., Tournier, E., et al. (2016). Identification of a circulating MicroRNA profile as a biomarker of metastatic cutaneous melanoma. *Acta Dermato-Venereologica*, 29–34. https://doi.org/10.2340/00015555-2156.

Arnold, M., Holterhues, C., Hollestein, L. M., Coebergh, J. W. W., Nijsten, T., Pukkala, E., et al. (2014). Trends in incidence and predictions of cutaneous melanoma across Europe up to 2015. *Journal of the European Academy of Dermatology and Venereology*, *28*(9), 1170–1178. https://doi.org/10.1111/jdv.12236.

Arnold, M., Kvaskoff, M., Thuret, A., Guénel, P., Bray, F., & Soerjomataram, I. (2018). Cutaneous melanoma in France in 2015 attributable to solar ultraviolet radiation and the use of sunbeds. *Journal of the European Academy of Dermatology and Venereology*, *32*(10), 1681–1686. https://doi.org/10.1111/jdv.15022.

Arts, N., Cane, S., Hennequart, M., Lamy, J., Bommer, G., Eynde, et al. (2015). microRNA-155, induced by interleukin-1ss, represses the expression of microphthalmia-associated transcription factor (MITF-M) in melanoma cells. *PLoS One*, *10*(4).

Audrito, V., Serra, S., Stingi, A., Orso, F., Gaudino, F., Bologna, C., et al. (2017). PD-L1 up-regulation in melanoma increases disease aggressiveness and is mediated through miR-17-5p. *Oncotarget*, *8*(9), 15894–15911. https://doi.org/10.18632/oncotarget.15213.

Bai, M., Zhang, H., Si, L., Yu, N., Zeng, A., & Zhao, R. (2017). Upregulation of serum miR-10b is associated with poor prognosis in patients with melanoma. *Journal of Cancer*, *8*(13), 2487–2491. https://doi.org/10.7150/jca.18824.

Bai, M., Zhang, M., Long, F., Yu, N., Zeng, A., & Zhao, R. (2017). Circulating microRNA-194 regulates human melanoma cells via PI3K/AKT/Fox03a and p53/p21 signaling pathway. *Oncology Reports*, *37*(5), 2702–2710. https://doi.org/10.3892/or.2017.5537.

Bai, X., Yang, M., & Xu, Y. (2018). MicroRNA-373 promotes cell migration via targeting salt-inducible kinase 1 expression in melanoma. *Experimental and Therapeutic Medicine*, *16*(6), 4759–4764. https://doi.org/10.3892/etm.2018.6784.

Bais, A. F., Lucas, R. M., Bornman, J. F., Williamson, C. E., Sulzberger, B., Austin, A. T., et al. (2018). Environmental effects of ozone depletion, UV radiation and interactions with climate change: UNEP environmental effects assessment panel, update 2017. *Photochemical and Photobiological Sciences*, *17*(2), 127–179. https://doi.org/10.1039/c7pp90043k.

Bakker, A. B. H., Schreurs, M. W. J., de Boer, A. J., Kawakami, Y., Rosenberg, S. A., Adema, G. J., et al. (1994). Melanocyte lineage-specific antigen gpl00 is recognized by melanoma-derived tumor-infiltrating lymphocytes. *Journal of Experimental Medicine*, *179*(3), 1005–1009. https://doi.org/10.1084/jem.179.3.1005.

Beatriz, S.-S., Carolina, M.-C., González-Muñoz, J. F., Amelia, M., Liria, T., & Carlos, M. (2018). Downregulation of intratumoral expression of miR-205, miR-200c and miR-125b in primary human cutaneous melanomas predicts shorter survival. *Scientific Reports*. https://doi.org/10.1038/s41598-018-35317-3.

Bell, R. E., Khaled, M., Netanely, D., Schubert, S., Golan, T., Buxbaum, A., et al. (2014). Transcription factor/microRNA axis blocks melanoma invasion program by miR-211 targeting NUAK1. *Journal of Investigative Dermatology*, *134*(2), 441–451. https://doi.org/10.1038/jid.2013.340.

Bemis, L. T., Chen, R., Amato, C. M., Classen, E. H., Robinson, S. E., Coffey, D. G., et al. (2008). MicroRNA-137 targets microphthalmia-associated transcription factor in melanoma cell lines. *Cancer Research*, *68*(5), 1362–1368. https://doi.org/10.1158/0008-5472.CAN-07-2912.

Bennett, D. C. (2016). Genetics of melanoma progression: The rise and fall of cell senescence. *Pigment Cell & Melanoma Research*, *29*(2), 122–140. https://doi.org/10.1111/pcmr.12422.

Bertolotto, C., Lesueur, F., Giuliano, S., Strub, T., Lichy, Bille, K., et al. (2016). A SUMOylation-defective MITF germline mutation predisposes to melanoma and renal carcinoma. *Nature*, *531*(7592).

Bhatia, S., Tykodi, S. S., & Thompson, J. A. (2009). Treatment of metastatic melanoma: An overview. *Oncology*, *23*(6), 488–496.

Bhattacharya, A., Schmitz, U., Raatz, Y., Schönherr, M., Kottek, T., Schauer, M., et al. (2015). miR-638 promotes melanoma metastasis and protects melanoma cells from apoptosis and autophagy. *Oncotarget*, *6*(5), 2966–2980. https://doi.org/10.18632/oncotarget.3070.

Bhoyrul, B., Brent, G., Elliott, F., McLorinan, J., Wilson, A., Peach, H., et al. (2019). Pathological review of primary cutaneous malignant melanoma by a specialist skin cancer multidisciplinary team improves patient care in the UK. *Journal of Clinical Pathology*, *72*(7), 482–486. https://doi.org/10.1136/jclinpath-2019-205767.

Biroccio, A., Amodei, S., Antonelli, A., Benassi, B., & Zupi, G. (2003). Inhibition of c-Myc oncoprotein limits the growth of human melanoma cells by inducing cellular crisis. *Journal of Biological Chemistry*, *278*(37), 35693–35701. https://doi.org/10.1074/jbc.M304597200.

Bliss, J. M., Ford, D., Swerdlow, A. J., Armstrong, B. K., Cristofolini, M., Elwood, J. M., et al. (1995). Risk of cutaneous melanoma associated with pigmentation characteristics and freckling: Systematic overview of 10 case-control studies. *International Journal of Cancer*, *62*(4), 367–376. https://doi.org/10.1002/ijc.2910620402.

Boyle, G. M., Woods, S. L., Bonazzi, V. F., Stark, M. S., Hacker, E., Aoude, L. G., et al. (2011). Melanoma cell invasiveness is regulated by miR-211 suppression of the BRN2 transcription factor. *Pigment Cell & Melanoma Research*, *24*(3), 525–537. https://doi.org/10.1111/j.1755-148X.2011.00849.x.

Brahmer, J. R., Lacchetti, C., Schneider, B. J., Atkins, M. B., Brassil, K. J., Caterino, J. M., et al. (2018). Management of immune-related adverse events in patients treated with immune checkpoint inhibitor therapy: American society of clinical oncology clinical practice guideline. *Journal of Clinical Oncology*, *36*(17), 1714–1768. https://doi.org/10.1200/JCO.2017.77.6385.

Branzan, A. L., Landthaler, M., & Szeimies, R. M. (2007). In vivo confocal scanning laser microscopy in dermatology. *Lasers in Medical Science*, *22*(2), 73–82. https://doi.org/10.1007/s10103-006-0416-8.

Breslow, A. (1975). Tumor thickness, level of invasion and node dissection in stage I cutaneous melanoma. *Annals of Surgery*, *182*(5), 572–575. https://doi.org/10.1097/00000658-197511000-00007.

Bu, P., Luo, C., He, Q., Yang, P., Li, X., & Xu, D. (2017). MicroRNA-9 inhibits the proliferation and migration of malignant melanoma cells via targeting sirituin 1. *Experimental and Therapeutic Medicine*, *14*(2), 931–938. https://doi.org/10.3892/etm.2017.4595.

Bueno, M. J., & Malumbres, M. (2011). MicroRNAs and the cell cycle. *Biochimica et Biophysica Acta - Molecular Basis of Disease*, *1812*(5), 592–601. https://doi.org/10.1016/j.bbadis.2011.02.002.

Burd, C. E., Liu, W., Huynh, M. V., Waqas, M. A., Gillahan, J. E., Clark, K. S., et al. (2014). Mutation-specific RAS oncogenicity explains NRAS codon 61 selection in melanoma. *Cancer Discovery*, *4*(12), 1418–1429. https://doi.org/10.1158/2159-8290.CD-14-0729.

Bustos, M. A., Ono, S., Marzese, D. M., Oyama, T., Iida, Y., Cheung, G., et al. (2017). MiR-200a regulates CDK4/6 inhibitor effect by targeting CDK6 in metastatic melanoma. *Journal of Investigative Dermatology*, *137*(9), 1955–1964. https://doi.org/10.1016/j.jid.2017.03.039.

Caporali, S., Amaro, A., Levati, L., Alvino, E., Lacal, P. M., Mastroeni, S., et al. (2019). MiR-126-3p down-regulation contributes to dabrafenib acquired resistance in melanoma by up-regulating ADAM9 and VEGF-A. *Journal of Experimental & Clinical Cancer Research*, *38*(1). https://doi.org/10.1186/s13046-019-1238-4.

Caramel, J., Papadogeorgakis, E., Hill, L., Browne, G. J., Richard, G., Wierinckx, A., et al. (2013). A switch in the expression of embryonic EMT-inducers drives the development of malignant melanoma. *Cancer Cell*, *24*(4), 466–480. https://doi.org/10.1016/j.ccr.2013.08.018.

Carlino, M. S., Long, G. V., Kefford, R. F., & Rizos, H. (2015). Targeting oncogenic BRAF and aberrant MAPK activation in the treatment of cutaneous melanoma. *Critical Reviews in Oncology/Hematology*, *96*(3), 385–398. https://doi.org/10.1016/j.critrevonc.2015.08.021.

Carvajal, R. D., Lawrence, D. P., Weber, J. S., Gajewski, T. F., Gonzalez, R., Lutzky, J., et al. (2015). Phase II study of nilotinib in melanoma harboring KIT alterations following progression to prior KIT inhibition. *Clinical Cancer Research*, *21*(10), 2289–2296. https://doi.org/10.1158/1078-0432.CCR-14-1630.

Chang, X., Sun, Y., Han, S., Zhu, W., Zhang, H., & Lian, S. (2015). MiR-203 inhibits melanoma invasive and proliferative abilities by targeting the polycomb group gene BMI1. *Biochemical and Biophysical Research Communications*, *456*(1), 361–366. https://doi.org/10.1016/j.bbrc.2014.11.087.

Chang, X., Zhu, W., Zhang, H., & Lian, S. (2017). Sensitization of melanoma cells to temozolomide by overexpression of microRNA 203 through direct targeting of glutaminase-mediated glutamine metabolism. *Clinical and Experimental Dermatology*, *42*(6), 614–621. https://doi.org/10.1111/ced.13119.

Chen, J., Feilotter, H. E., Paré, G. C., Zhang, X., Pemberton, J. G. W., Garady, C., et al. (2010). MicroRNA-193b represses cell proliferation and regulates cyclin D1 in melanoma. *American Journal of Pathology*, *176*(5), 2520–2529. https://doi.org/10.2353/ajpath.2010.091061.

Chen, J., Zhang, X., Lentz, C., Abi-Daoud, M., Paré, G. C., Yang, X., et al. (2011). MiR-193b regulates Mcl-1 in melanoma. *American Journal of Pathology*, *179*(5), 2162–2168. https://doi.org/10.1016/j.ajpath.2011.07.010.

Chen, Y., Liersch, R., & Detmar, M. (2012). The miR-290-295 cluster suppresses autophagic cell death of melanoma cells. *Scientific Reports*, *2*, 808. https://doi.org/10.1038/srep00808.

Chen, S., Wang, L., Fan, J., Ye, C., Dominguez, D., Zhang, Y., et al. (2015). Host miR155 promotes tumor growth through a myeloid-derived suppressor cell-dependent mechanism. *Cancer Research*, *75*(3), 519–531. https://doi.org/10.1158/0008-5472.CAN-14-2331.

Chen, Y., Cao, K., Wangi, S., Chen, J., He, B., He, G., et al. (2016). MicroRNA-138 suppresses proliferation, invasion and glycolysis in malignant melanoma cells by targeting HIF-1α. *Experimental and Therapeutic Medicine*, *11*(6), 2513–2518. https://doi.org/10.3892/etm.2016.3220.

Chen, Y., Zhang, Z., Luo, C., Chen, Z., & Zhou, J. (2016). MicroRNA-18b inhibits the growth of malignant melanoma via inhibition of HIF-1α-mediated glycolysis. *Oncology Reports*, *36*(1), 471–479. https://doi.org/10.3892/or.2016.4824.

Chen, L., Cao, Y., Rong, D., Wang, Y., & Cao, Y. (2017). MicroRNA-605 functions as a tumor suppressor by targeting INPP4B in melanoma. *Oncology Reports*, *38*(2), 1276–1286. https://doi.org/10.3892/or.2017.5740.

Chen, X. E., Chen, P., Chen, S. S., Ma, T., Shi, G., Zhou, Y., et al. (2018). MIR-106b-5p promotes cell cycle progression of malignant melanoma by targeting PTEN. *Oncology Reports*, *39*(1), 331–337. https://doi.org/10.3892/or.2017.6099.

Chen, W. Y., Xu, Y. Y., & Zhang, X. Y. (2019). Targeting GOLM1 by microRNA-200a in melanoma suppresses cell proliferation, invasion and migration via regulating PI3K/Akt signaling pathway and epithelial-mesenchymal transition. *European Review for Medical and Pharmacological Sciences*, *23*(16), 6997–7007. https://doi.org/10.26355/eurrev_201908_18740.

Cheng, L., Sharples, R. A., Scicluna, B. J., & Hill, A. F. (2014). Exosomes provide a protective and enriched source of miRNA for biomarker profiling compared to intracellular and cell-free blood. *Journal of Extracellular Vesicles*, *3*(1). https://doi.org/10.3402/jev.v3.23743.

Cheng, L., Lopez-Beltran, A., Massari, F., Maclennan, G. T., & Montironi, R. (2018). Molecular testing for BRAF mutations to inform melanoma treatment decisions: A move toward precision medicine. *Modern Pathology*, *31*(1), 24–38. https://doi.org/10.1038/modpathol.2017.104.

Choe, M. H., Yoon, Y., Kim, J., Hwang, S. G., Han, Y. H., & Kim, J. S. (2018). MiR-550a-3-5p acts as a tumor suppressor and reverses BRAF inhibitor resistance through the direct targeting of YAP article. *Cell Death & Disease*, *9*(6). https://doi.org/10.1038/s41419-018-0698-3.

Chunhua, Z., Hui, L., Junling, W., Jibei, Z., & Xiaoqian, H. (2019). MicroRNA-338-3p suppresses cell proliferation, migration and invasion in human malignant melanoma by targeting MACC1. *Experimental and Therapeutic Medicine*. https://doi.org/10.3892/etm.2019.7644.

Cirenajwis, H., Lauss, M., Ekedahl, H., Törngren, T., Kvist, A., Saal, L. H., et al. (2017). NF1-mutated melanoma tumors harbor distinct clinical and biological characteristics. *Molecular Oncology*, *11*(4), 438–451. https://doi.org/10.1002/1878-0261.12050.

Cohen, R., Greenberg, E., Nemlich, Y., Schachter, J., & Markel, G. (2015). miR-17 regulates melanoma cell motility by inhibiting the translation of ETV1. *Oncotarget*, *6*(22), 19006–19016. https://doi.org/10.18632/oncotarget.4147.

Colombino, M., Sini, M. C., Lissia, A., De Giorgi, V., Stanganelli, I., Ayala, F., et al. (2014). Discrepant alterations in main candidate genes among multiple primary melanomas. *Journal of Translational Medicine*, *12*(1). https://doi.org/10.1186/1479-5876-12-117.

Curtin, J. A., Busam, K., Pinkel, D., & Bastian, B. C. (2006). Somatic activation of KIT in distinct subtypes of melanoma. *Journal of Clinical Oncology*, *24*(26), 4340–4346. https://doi.org/10.1200/JCO.2006.06.2984.

Dar, A. A., Majid, S., De Semir, D., Nosrati, M., Bezrookove, V., & Kashani-Sabet, M. (2011). miRNA-205 suppresses melanoma cell proliferation and induces senescence via regulation of E2F1 protein. *Journal of Biological Chemistry*, *286*(19), 16606–16614. https://doi.org/10.1074/jbc.M111.227611.

Dar, A. A., Majid, S., Rittsteuer, C., De Semir, D., Bezrookove, V., Tong, S., et al. (2013). The role of miR-18b in MDM2-p53 pathway signaling and melanoma progression. *Journal of the National Cancer Institute*, *105*(6), 433–442. https://doi.org/10.1093/jnci/djt003.

Davies, H., Bignell, G. R., Cox, C., Stephens, P., Edkins, S., Clegg, S., et al. (2002). Mutations of the BRAF gene in human cancer. *Nature*, *417*(6892), 949–954. https://doi.org/10.1038/nature00766.

Davis, L. E., Shalin, S. C., & Tackett, A. J. (2019). Current state of melanoma diagnosis and treatment. *Cancer Biology and Therapy*, *20*(11), 1366–1379. https://doi.org/10.1080/15384047.2019.1640032.

Deken, M. A., Gadiot, J., Jordanova, E. S., Lacroix, R., van Gool, M., Kroon, P., et al. (2016). Targeting the MAPK and PI3K pathways in combination with PD1 blockade in melanoma. *OncoImmunology*, *5*(12). https://doi.org/10.1080/2162402X.2016.1238557.

Del Campo, S. E. M., Latchana, N., Levine, K. M., Grignol, V. P., Fairchild, E. T., Jaime-Ramirez, A. C., et al. (2015). MiR-21 enhances melanoma invasiveness via inhibition of tissue inhibitor of metalloproteinases 3 expression: In vivo effects of MiR-21 inhibitor. *PLoS One*, *10*(1). https://doi.org/10.1371/journal.pone.0115919.

Denecker, G., Vandamme, N., Akay, O., Koludrovic, D., Taminau, J., Lemeire, K., et al. (2014). Identification of a ZEB2-MITF-ZEB1 transcriptional network that controls melanogenesis and melanoma progression. *Cell Death and Differentiation*, *21*(8), 1250–1261. https://doi.org/10.1038/cdd.2014.44.

Desjobert, C., Carrier, A., Delmas, A., Marzese, D. M., Daunay, A., Busato, F., et al. (2019). Demethylation by low-dose 5-aza-2′-deoxycytidine impairs 3D melanoma invasion partially through miR-199a-3p expression revealing the role of this miR in melanoma. *Clinical Epigenetics*, *11*(1).

Díaz-Martínez, M., Benito-Jardon, L., Alonso, L., Koetz-Ploch, L., Hernando, E., & Teixido, J. (2018). miR-204-5p and miR-211-5p contribute to BRAF inhibitor resistance in melanoma. *Cancer Research*, *78*(4), 1017–1030. https://doi.org/10.1158/0008-5472.CAN-17-1318.

Dika, E., Riefolo, M., Porcellini, E., Broseghini, E., Ribero, S., Senetta, R., et al. (2020). Defining the prognostic role of MicroRNAs in cutaneous melanoma. *Journal of Investigative Dermatology*, *140*(11), 2260–2267. https://doi.org/10.1016/j.jid.2020.03.949.

Domingues, B., Lopes, J., Soares, P., & Pópulo, H. (2018). Melanoma treatment in review. *ImmunoTargets and Therapy*, *7*. https://doi.org/10.2147/ITT.S134842. 29922629.

Dynoodt, P., Speeckaert, R., De Wever, O., Chevolet, I., Brochez, L., Lambert, J., et al. (2013). MiR-145 overexpression suppresses the migration and invasion of metastatic melanoma cells. *International Journal of Oncology*, *42*(4), 1443–1451. https://doi.org/10.3892/ijo.2013.1823.

Elder, D., Massi, D., Scoyler, R. A., & Willemze, R. (2017). WHO classification of skin tumor. *Vol. 11. WHO classification of tumours* (4th). IARC.

Ellerhorst, J. A., Greene, V. R., Ekmekcioglu, S., Warneke, C. L., Johnson, M. M., Cooke, C. P., et al. (2011). Clinical correlates of NRAS and BRAF mutations in primary human melanoma. *Clinical Cancer Research*, *17*(2), 229–235. https://doi.org/10.1158/1078-0432.CCR-10-2276.

Elwood, J. M., & Koh, H. K. (1994). Etiology, epidemiology, risk factors, and public health issues of melanoma. *Current Opinion in Oncology*, *6*(2), 179–187. https://doi.org/10.1097/00001622-199403000-00011.

Erdmann, F., Lortet-Tieulent, J., Schüz, J., Zeeb, H., Greinert, R., Breitbart, E. W., et al. (2013). International trends in the incidence of malignant melanoma 1953-2008-are recent generations at higher or lower risk? *International Journal of Cancer*, *132*(2), 385–400. https://doi.org/10.1002/ijc.27616.

Fang, W., Fan, Y., Fa, Z., Xu, J., Yu, H., Li, P., et al. (2017). microRNA-625 inhibits tumorigenicity by suppressing proliferation, migration and invasion in malignant melanoma. *Oncotarget*, *8*(8), 13253–13263. https://doi.org/10.18632/oncotarget.14710.

Fanjun, M., Yuxia, Z., Li, X., Juan, W., & Zhiyu, W. (2017). Clinical significance of miR-138 in patients with malignant melanoma through targeting of PDK1 in the PI3K/AKT autophagy signaling pathway. *Oncology Reports*, *38*(3), 1655–1662. https://doi.org/10.3892/or.2017.5838.

Faries, B., Thompson, J. F., Cochran, A. J., Andtbacka, R. H., Mozzillo, N., Zager, J. S., et al. (2017). Completion dissection or observation for sentinel-node metastasis in melanoma. *New England Journal of Medicine*, *376*(23), 2211–2222. https://doi.org/10.1056/NEJMoa1613210.

Fattore, L., Mancini, R., Acunzo, M., Romano, G., Laganà, A., Pisanu, M. E., et al. (2016). miR-579-3p controls melanoma progression and resistance to target therapy. *Proceedings of the National Academy of Sciences of the United States of America*, *113*(34), E5005–E5013. https://doi.org/10.1073/pnas.1607753113.

Fattore, L., Ruggiero, C. F., Pisanu, M. E., Liguoro, D., Cerri, A., Costantini, S., et al. (2019). Reprogramming miRNAs global expression orchestrates development of drug resistance in BRAF mutated melanoma. *Cell Death and Differentiation*, *26*(7), 1267–1282. https://doi.org/10.1038/s41418-018-0205-5.

Felicetti, F., Errico, M. C., Bottero, L., Segnalini, P., Stoppacciaro, A., Biffoni, M., et al. (2008). The promyelocytic leukemia zinc finger microRNA-221/-222 pathway controls melanoma progression through multiple oncogenic mechanisms. *Cancer Research*, *68*(8), 2745–2754. https://doi.org/10.1158/0008-5472.CAN-07-2538.

Felli, N., Felicetti, F., Lustri, A. M., Errico, M. C., Bottero, L., Cannistraci, A., et al. (2013). miR-126&126* restored expressions play a tumor suppressor role by directly regulating ADAM9 and MMP7 in melanoma. *PLoS One*, *8*(2). https://doi.org/10.1371/journal.pone.0056824.

Feng, Z., Zhang, C., Wu, R., & Hu, W. (2011). Tumor suppressor p53 meets microRNAs. *Journal of Molecular Cell Biology*, *3*(1), 44–50. https://doi.org/10.1093/jmcb/mjq040.

Ferlay, J., Soerjomataram, I., Dikshit, R., Eser, S., Mathers, C., Rebelo, M., et al. (2015). Cancer incidence and mortality worldwide: Sources, methods and major patterns in GLOBOCAN 2012. *International Journal of Cancer*, *136*(5), E359–E386. https://doi.org/10.1002/ijc.29210.

Ferracin, M., & Negrini, M. (2015). Micromarkers 2.0: An update on the role of microRNAs in cancer diagnosis and prognosis. *Expert Review of Molecular Diagnostics*, *15*(10), 1369–1381. https://doi.org/10.1586/14737159.2015.1081058.

Ferracin, M., Lupini, L., Salamon, I., Saccenti, E., Zanzi, M. V., Rocchi, A., et al. (2015). Absolute quantification of cell-free microRNAs in cancer patients. *Oncotarget*, *6*(16), 14545–14555. https://doi.org/10.18632/oncotarget.3859.

Fleming, N. H., Zhong, J., Da Silva, I. P., De Miera, E. V. S., Brady, B., Han, S. W., et al. (2015). Serum-based miRNAs in the prediction and detection of recurrence in melanoma patients. *Cancer*, *121*(1), 51–59. https://doi.org/10.1002/cncr.28981.

Flynn, R. L., & Zou, L. (2010). Oligonucleotide/oligosaccharide-binding fold proteins: A growing family of genome guardians. *Critical Reviews in Biochemistry and Molecular Biology*, *45*(4), 266–275. https://doi.org/10.3109/10409238.2010.488216.

Fogli, S., Polini, B., Carpi, S., Pardini, B., Naccarati, A., Dubbini, N., et al. (2017). Identification of plasma microRNAs as new potential biomarkers with high diagnostic power in human cutaneous melanoma. *Tumor Biology*, *39*(5). https://doi.org/10.1177/1010428317701646.

Franke, V., & van Akkooi, A. C. J. (2019). The extent of surgery for stage III melanoma: How much is appropriate? *The Lancet Oncology*, *20*(3), e167–e174. https://doi.org/10.1016/S1470-2045(19)30099-3.

Franken, N. A. P., Rodermond, H. M., Stap, J., Haveman, J., & van Bree, C. (2006). Clonogenic assay of cells in vitro. *Nature Protocols*, *1*(5), 2315–2319. https://doi.org/10.1038/nprot.2006.339.

Friedman, E. B., Shang, S., de Miera, E. V., Fog, J. U., Teilum, M. W., Ma, M. W., et al. (2012). Serum microRNAs as biomarkers for recurrence in melanoma. *Journal of Translational Medicine*, *10*(1). https://doi.org/10.1186/1479-5876-10-155.

Fu, Q., Chen, N., Ge, C., Li, R., Li, Z., Zeng, B., et al. (2019). Prognostic value of tumor-infiltrating lymphocytes in melanoma: A systematic review and meta-analysis. *OncoImmunology*, *8*(7). https://doi.org/10.1080/2162402X.2019.1593806.

Galasso, M., Morrison, C., Minotti, L., Corrà, F., Zerbinati, C., Agnoletto, C., et al. (2018). Loss of miR-204 expression is a key event in melanoma. *Molecular Cancer*, *17*(1). https://doi.org/10.1186/s12943-018-0819-8.

Galore-Haskel, G., Nemlich, Y., Greenberg, E., Ashkenazi, S., Hakim, M., Itzhaki, O., et al. (2015). A novel immune resistance mechanism of melanoma cells controlled by the ADAR1 enzyme. *Oncotarget*, *6*(30), 28999–29015. https://doi.org/10.18632/oncotarget.4905.

Gandini, S., Sera, F., Cattaruzza, M. S., Pasquini, P., Abeni, D., Boyle, P., et al. (2005a). Meta-analysis of risk factors for cutaneous melanoma: I. common and atypical naevi. *European Journal of Cancer*, *41*(1), 28–44. https://doi.org/10.1016/j.ejca.2004.10.015.

Gandini, S., Sera, F., Cattaruzza, M. S., Pasquini, P., Picconi, O., Boyle, P., et al. (2005b). Meta-analysis of risk factors for cutaneous melanoma: II. Sun exposure. *European Journal of Cancer*, *41*(1), 45–60. https://doi.org/10.1016/j.ejca.2004.10.016.

Gandini, S., Sera, F., Cattaruzza, M. S., Pasquini, P., Zanetti, R., Masini, C., et al. (2005c). Meta-analysis of risk factors for cutaneous melanoma: III. Family history, actinic damage and phenotypic factors. *European Journal of Cancer*, *41*(14), 2040–2059. https://doi.org/10.1016/j.ejca.2005.03.034.

Gao, J., Zeng, K., Liu, Y., Gao, L., & Liu, L. (2019). Lncrna snhg5 promotes growth and invasion in melanoma by regulating the mir-26a-5p/trpc3 pathway. *Oncotargets and Therapy*, *12*, 169–179. https://doi.org/10.2147/OTT.S184078.

Gaziel-Sovran, A., Segura, M. F., Di Micco, R., Collins, M. K., Hanniford, D., Vega-Saenz de Miera, E., et al. (2011). MiR-30b/30d regulation of GalNAc transferases enhances invasion and immunosuppression during metastasis. *Cancer Cell*, *20*(1), 104–118. https://doi.org/10.1016/j.ccr.2011.05.027.

Gencia, I., Baderca, F., Avram, S., Gogulescu, A., Marcu, A., Seclaman, E., et al. (2020). A preliminary study of microRNA expression in different types of primary melanoma. *Bosnian Journal of Basic Medical Sciences*, *20*(2), 197–208. https://doi.org/10.17305/bjbms.2019.4271.

Georgantas, R. W., Streicher, K., Luo, X., Greenlees, L., Zhu, W., Liu, Z., et al. (2014). MicroRNA-206 induces G1 arrest in melanoma by inhibition of CDK4 and cyclin D. *Pigment Cell & Melanoma Research*, *27*(2), 275–286. https://doi.org/10.1111/pcmr.12200.

Gershenwald, J. E., & Scolyer, R. A. (2018). Melanoma staging: American joint committee on Cancer (AJCC) 8th edition and beyond. *Annals of Surgical Oncology*, *25*(8), 2105–2110. https://doi.org/10.1245/s10434-018-6513-7.

Gershenwald, J. E., Scolyer, R. A., Hess, K. R., Sondak, V. K., Long, G. V., Ross, M. I., et al. (2017). Melanoma staging: Evidence-based changes in the American joint committee on Cancer eighth edition cancer staging manual. *CA: a Cancer Journal for Clinicians*, *67*(6), 472–492. https://doi.org/10.3322/caac.21409.

Gilchrest, B. A., Eller, M. S., Geller, A. C., & Yaar, M. (1999). The pathogenesis of melanoma induced by ultraviolet radiation. *New England Journal of Medicine*, *340*(17), 1341–1348. https://doi.org/10.1056/NEJM199904293401707.

Giles, K. M., Brown, R. A. M., Epis, M. R., Kalinowski, F. C., & Leedman, P. J. (2013). MiRNA-7-5p inhibits melanoma cell migration and invasion. *Biochemical and Biophysical Research Communications*, *430*(2), 706–710. https://doi.org/10.1016/j.bbrc.2012.11.086.

Giles, K. M., Rosenbaum, B. E., Berger, M., Izsak, A., Li, Y., Illa Bochaca, I., et al. (2019). Revisiting the clinical and biologic relevance of partial PTEN loss in melanoma. *Journal of Investigative Dermatology*, *139*(2), 430–438. https://doi.org/10.1016/j.jid.2018.07.031.

Gimotty, P. A., Elder, D. E., Fraker, D. L., Botbyl, J., Sellers, K., Elenitsas, R., et al. (2007). Identification of high-risk patients among those diagnosed with thin cutaneous melanomas. *Journal of Clinical Oncology*, *25*(9), 1129–1134. https://doi.org/10.1200/JCO.2006.08.1463.

Glud, M., Rossing, M., Hother, C., Holst, L., Hastrup, N., Nielsen, F. C., et al. (2010). Downregulation of miR-125b in metastatic cutaneous malignant melanoma. *Melanoma Research*, *20*(6), 479–484. https://doi.org/10.1097/CMR.0b013e32833e32a1.

Goel, V. K., Lazar, A. J. F., Warneke, C. L., Redston, M. S., & Haluska, F. G. (2006). Examination of mutations in BRAF, NRAS, and PTEN in primary cutaneous melanoma. *Journal of Investigative Dermatology*, *126*(1), 154–160. https://doi.org/10.1038/sj.jid.5700026.

Goldstein, A. M., Chan, M., Harland, M., Hayward, N. K., Demenais, F., Bishop, D. T., … Melanoma Genetics Consortium (GenoMEL). (2007). Features associated with germline CDKN2A mutations: A GenoMEL study of melanoma-prone families from three continents. *Journal of Medical Genetics*, *44*(2), 99–106. https://doi.org/10.1136/jmg.2006.043802.

Golias, C. H., Charalabopoulos, A., & Charalabopoulos, K. (2004). Cell proliferation and cell cycle control: a mini review. *Int J Clin Pract*, *58*(12), 1134–1341. https://doi.org/10.1111/j.1742-1241.2004.00284.x.

Goswami, S., Tarapore, R. S., Strong, A. M. P., TeSlaa, J. J., Grinblat, Y., Setaluri, V., et al. (2015). MicroRNA-340-mediated degradation of microphthalmia-associated transcription factor (MITF) mRNA is inhibited by coding region determinant-binding protein (CRD-BP). *Journal of Biological Chemistry*, *290*(1), 384–395. https://doi.org/10.1074/jbc.M114.590158.

Green, A. C., Baade, P., Coory, M., Aitken, J. F., & Smithers, M. (2012). Population-based 20-year survival among people diagnosed with thin melanomas in Queensland, Australia. *Journal of Clinical Oncology*, *30*(13), 1462–1467. https://doi.org/10.1200/JCO.2011.38.8561.

Greenberg, E., Besser, M. J., Ben-Ami, E., Shapira-Frommer, R., Itzhaki, O., Zikich, D., et al. (2013). A comparative analysis of total serum miRNA profiles identifies novel signature that is highly indicative of metastatic melanoma: A pilot study. *Biomarkers*, *18*(6), 502–508. https://doi.org/10.3109/1354750X.2013.816777.

Greenberg, E., Hajdu, S., Nemlich, Y., Cohen, R., Itzhaki, O., Jacob-Hirsch, J., et al. (2014). Differential regulation of aggressive features in melanoma cells by members of the miR-17-92 complex. *Open Biology*, *4*(JUNE). https://doi.org/10.1098/rsob.140030.

Greene, M. H., Elder, D. E., Guerry, D., Epstein, M. N., Greene, M. H., & Van Horn, M. (1984). A study of tumor progression: The precursor lesions of superficial spreading and nodular melanoma. *Human Pathology*, *15*(12), 1147–1165. https://doi.org/10.1016/S0046-8177(84)80310-X.

Grignol, V., Fairchild, E. T., Zimmerer, J. M., Lesinski, G. B., Walker, M. J., Magro, C. M., et al. (2011). MiR-21 and miR-155 are associated with mitotic activity and lesion depth of borderline melanocytic lesions. *British Journal of Cancer*, *105*(7), 1023–1029. https://doi.org/10.1038/bjc.2011.288.

Grimaldi, A. M., Simeone, E., Festino, L., Vanella, V., Strudel, M., & Ascierto, P. A. (2017). MEK inhibitors in the treatment of metastatic melanoma and solid tumors. *American Journal of Clinical Dermatology*, *18*(6), 745–754. https://doi.org/10.1007/s40257-017-0292-y.

Groh, V., Wu, J., Yee, C., & Spies, T. (2002). Tumour-derived soluble MIC ligands impair expression of NKG2D and T-cell activation. *Nature*, *419*(6908), 734–738. https://doi.org/10.1038/nature01112.

Gulei, D., Raduly, L., Broseghini, E., Ferracin, M., & Berindan-Neagoe, I. (2019). Berindan-Neagoe I. The extensive role of miR-155 in malignant and non-malignant diseases. *Molecular Aspects of Medicine*, *70*. https://doi.org/10.1016/j.mam.2019.09.004.

Guo, J., Zhang, J. F., Wang, W. M., Cheung, F. W. K., Lu, Y. F., Ng, C. F., et al. (2014). MicroRNA-218 inhibits melanogenesis by directly suppressing microphthalmia-associated transcription factor expression. *RNA Biology*, *11*(6). https://doi.org/10.4161/rna.28865.

Guo, S., Guo, W., Li, S., Dai, W., Zhang, N., Zhao, T., et al. (2016). Serum miR-16: A potential biomarker for predicting melanoma prognosis. *Journal of Investigative Dermatology*, *136*(5), 985–993. https://doi.org/10.1016/j.jid.2015.12.041.

Guo, S., Guo, W., Li, S., Dai, W., Zhang, N., Zhao, T., … Li, C. (2016). Serum miR-16: A potential biomarker for predicting melanoma prognosis. *Journal of Investigative Dermatology*, *136*(5), 985–993. https://doi.org/10.1016/j.jid.2015.12.041.

Guo, B., Hui, Q., Zhang, Y., Chang, P., & Tao, K. (2016). MIR 194 is a negative regulator of GEF-H1 pathway in melanoma. *Oncology Reports*, *36*(4), 2412–2420. https://doi.org/10.3892/or.2016.5020.

Guy, G. P., Jr., Thomas, C. C., Thompson, T., Watson, M., Massetti, G. M., Richardson, L. C., et al. (2015). Vital signs: Melanoma incidence and mortality trends and projections - United States. *MMWR. Morbidity and Mortality Weekly Report*, *64*(21), 591–596.

Haenssle, H. A., Mograby, N., Ngassa, A., Buhl, T., Emmert, S., Schön, M. P., et al. (2016). Association of patient risk factors and frequency of nevus-associated cutaneous melanomas. *JAMA Dermatology*, *152*(3), 291–298. https://doi.org/10.1001/jamadermatol.2015.3775.

Haflidadottir, B., Bergsteinsdottir, K., Praetorius, C., & Steingrimsson, E. (2010). miR-148 regulates Mitf in melanoma cells. *PLoS One*, *5*(7).

Hammock, L., Cohen, C., Carlson, G., Murray, D., Ross, J. S., Sheehan, C., et al. (2006). Chromogenic in situ hybridization analysis of melastatin mRNA expression in melanomas from American joint committee on cancer stage I and II patients with recurrent melanoma. *Journal of Cutaneous Pathology*, *33*(9), 599–607. https://doi.org/10.1111/j.1600-0560.2006.00501.x.

Hanahan, D., & Weinberg, R. A. (2011). Hallmarks of cancer: The next generation. *Cell*, *144*(5), 646–674. https://doi.org/10.1016/j.cell.2011.02.013.

Hanna, J. A., Hahn, L., Agarwal, S., & Rimm, D. L. (2012). In situ measurement of miR-205 in malignant melanoma tissue supports its role as a tumor suppressor microRNA. *Laboratory Investigation*, *92*(10), 1390–1397. https://doi.org/10.1038/labinvest.2012.119.

Hao, S., Luo, C., Abukiwan, A., Wang, G., He, J., Huang, L., et al. (2015). MiR-137 inhibits proliferation of melanoma cells by targeting PAK2. *Experimental Dermatology*, *24*(12), 947–952. https://doi.org/10.1111/exd.12812.

Heiden, M. G. V., Cantley, L. C., & Thompson, C. B. (2009). Understanding the Warburg effect: The metabolic requirements of cell proliferation. *Science*, *324*(5930), 1029–1033. https://doi.org/10.1126/science.1160809.

Heinemann, A., Zhao, F., Pechlivanis, S., Eberle, J., Steinle, A., Diederichs, S., et al. (2012). Tumor suppressive microRNAs miR-34a/c control cancer cell expression of ULBP2, a stress-induced ligand of the natural killer cell receptor Nkg2*d*. *Cancer Research*, *72*(2), 460–471. https://doi.org/10.1158/0008-5472.CAN-11-1977.

Helgadottir, H., Tuominen, R., Olsson, H., Hansson, J., & Höiom, V. (2017). Cancer risks and survival in patients with multiple primary melanomas: Association with family history of melanoma and germline CDKN2A mutation status. *Journal of the American Academy of Dermatology*, *77*(5), 893–901. https://doi.org/10.1016/j.jaad.2017.05.050.

Henrique, R., Azevedo, R., Bento, M. J., Domingues, J. C., Silva, C., & Jerónimo, C. (2000). Prognostic value of Ki-67 expression in localized cutaneous malignant melanoma. *Journal of the American Academy of Dermatology*, *43*(6), 991–1000. https://doi.org/10.1067/mjd.2000.109282.

Hodis, E., Watson, I. R., Kryukov, G. V., Arold, S. T., Imielinski, M., Theurillat, J. P., et al. (2012). A landscape of driver mutations in melanoma. *Cell*, *150*(2), 251–263. https://doi.org/10.1016/j.cell.2012.06.024.

Holst, L. M. B., Kaczkowski, B., Glud, M., Futoma-Kazmierczak, E., Hansen, L. F., & Gniadecki, R. (2011). The microRNA molecular signature of atypic and common acquired melanocytic nevi: Differential expression of miR-125b and let-7c. *Experimental Dermatology*, *20*(3), 278–280. https://doi.org/10.1111/j.1600-0625.2010.01163.x.

Homet Moreno, B., Mok, S., Comin-Anduix, B., Hu-Lieskovan, S., & Ribas, A. (2016). Combined treatment with dabrafenib and trametinib with immune-stimulating antibodies for BRAF mutant melanoma. *OncoImmunology*, *5*(7). https://doi.org/10.1080/2162402X.2015.1052212.

Hong, Z., Lingqiao, L., Yingqian, W., & Dewei, W. (2020). Long non-coding RNA SNHG6 promotes tumorigenesis in melanoma cells via the microRNA-101-3p/RAP2B axis. *Oncology Letters*. https://doi.org/10.3892/ol.2020.12186.

Horn, S., Figl, A., Rachakonda, P. S., Fischer, C., Sucker, A., Gast, A., et al. (2013). TERT promoter mutations in familial and sporadic melanoma. *Science*, *339*(6122), 959–961. https://doi.org/10.1126/science.1230062.

Hsiao, J. J., & Fisher, D. E. (2014). The roles of microphthalmia-associated transcription factor and pigmentation in melanoma. *Archives of Biochemistry and Biophysics*, *563*, 28–34. https://doi.org/10.1016/j.abb.2014.07.019.

Hua, K. T., Hong, J. B., Sheen, Y. S., Huang, H. Y., Huang, Y. L., Chen, J. S., et al. (2018). miR-519d promotes melanoma progression by downregulating EphA4. *Cancer Research*, *78*(1), 216–229. https://doi.org/10.1158/0008-5472.CAN-17-1933.

Huang, D., Wang, F., Wu, W., Lian, C., & Liu, E. (2019). MicroRNA-429 inhibits cancer cell proliferation and migration by targeting the AKT1 in melanoma. *Cancer Biomarkers*, *26*(1), 63–68. https://doi.org/10.3233/CBM-190289.

Huber, V., Vallacchi, V., Fleming, V., Hu, X., Cova, A., Dugo, M., et al. (2018). Tumor-derived microRNAs induce myeloid suppressor cells and predict immunotherapy resistance in melanoma. *Journal of Clinical Investigation*, *128*(12), 5517–5530. https://doi.org/10.1172/JCI98060.

Huo, J., Zhang, Y., Li, R., Wang, Y., Wu, J., & Zhang, D. (2016). Upregulated microRNA-25 mediates the migration of melanoma cells by targeting DKK3 through the WNT/β-catenin pathway. *International Journal of Molecular Sciences*, *17*(11). https://doi.org/10.3390/ijms17111124.

Hussussian, C. J., Struewing, J. P., Goldstein, A. M., Higgins, P. A. T., Ally, D. S., Sheahan, M. D., et al. (1994). Germline p16 mutations in familial melanoma. *Nature Genetics*, *8*(1), 15–21. https://doi.org/10.1038/ng0994-15.

Hwang, H. W., Baxter, L. L., Loftus, S. K., Cronin, J. C., Trivedi, N. S., Borate, B., et al. (2014). Distinct microRNA expression signatures are associated with melanoma subtypes and are regulated by HIF1A. *Pigment Cell & Melanoma Research*, *27*(5), 777–787. https://doi.org/10.1111/pcmr.12255.

Jakob, J. A., Bassett, R. L., Ng, C. S., Curry, J. L., Joseph, R. W., Alvarado, G. C., et al. (2012). NRAS mutation status is an independent prognostic factor in metastatic melanoma. *Cancer*, *118*(16), 4014–4023. https://doi.org/10.1002/cncr.26724.

Jiang, Q. Q., & Liu, W. B. (2018). MiR-25 promotes melanoma progression by regulating RNA binding motif protein 47. *Médecine Sciences*, *34*, 59–65. https://doi.org/10.1051/medsci/201834f111.

Jiang, L., Lv, X., Li, J., Li, J., Li, X., Li, W., et al. (2012). The status of microRNA-21 expression and its clinical significance in human cutaneous malignant melanoma. *Acta Histochemica*, *114*(6), 582–588. https://doi.org/10.1016/j.acthis.2011.11.001.

Jin, C., Wang, A., Liu, L., Wang, G., Li, G., & Han, Z. (2019). miR-145-5p inhibits tumor occurrence and metastasis through the NF-κB signaling pathway by targeting TLR4 in malignant melanoma. *Journal of Cellular Biochemistry*, *120*(7), 11115–11126. https://doi.org/10.1002/jcb.28388.

Kakadia, S., Yarlagadda, N., Awad, R., Kundranda, M., Niu, J., Naraev, B., et al. (2018). Mechanisms of resistance to BRAF and MEK inhibitors and clinical update of us food and drug administration-approved targeted therapy in advanced melanoma. *Oncotargets and Therapy*, *11*, 7095–7107. https://doi.org/10.2147/OTT.S182721.

Kanemaru, H., Fukushima, S., Yamashita, J., Honda, N., Oyama, R., Kakimoto, A., et al. (2011). The circulating microRNA-221 level in patients with malignant melanoma as a new tumor marker. *Journal of Dermatological Science*, *61*(3), 187–193. https://doi.org/10.1016/j.jdermsci.2010.12.010.

Kappelmann, M., Kuphal, S., Meister, G., Vardimon, L., & Bosserhoff, A. K. (2013). MicroRNA miR-125b controls melanoma progression by direct regulation of c-Jun protein expression. *Oncogene*, *32*(24), 2984–2991. https://doi.org/10.1038/onc.2012.307.

Kawakami, A., & Fisher, D. E. (2017). The master role of microphthalmia-associated transcription factor in melanocyte and melanoma biology. *Laboratory Investigation*, *97*(6), 649–656. https://doi.org/10.1038/labinvest.2017.9.

Kholmanskikh, O., Van Baren, N., Brasseur, F., Ottaviani, S., Vanacker, J., Arts, N., et al. (2010). Interleukins 1α and 1β secreted by some melanoma cell lines strongly reduce expression of MITF-M and melanocyte differentiation antigens. *International Journal of Cancer*, *127*(7), 1625–1636. https://doi.org/10.1002/ijc.25182.

Kim, R. H., & Meehan, S. A. (2017). Immunostain use in the diagnosis of melanomas referred to a tertiary medical center: A 15-year retrospective review (2001–2015). *Journal of Cutaneous Pathology*, *44*(3), 221–227. https://doi.org/10.1111/cup.12867.

Kim, J. H., Ahn, J. H., & Lee, M. (2017). Upregulation of microRNA-1246 is associated with BRAF inhibitor resistance in melanoma cells with mutant BRAF. *Cancer Research and Treatment*, *49*(4), 947–959. https://doi.org/10.4143/crt.2016.280.

Kitago, M., Martinez, S. R., Nakamura, T., Sim, M. S., & Hoon, D. S. B. (2009). Regulation of RUNX3 tumor suppressor gene expressionin cutaneous melanoma. *Clinical Cancer Research*, *15*(9), 2988–2994. https://doi.org/10.1158/1078-0432.CCR-08-3172.

Knoll, S., Fürst, K., Kowtharapu, B., Schmitz, U., Marquardt, S., Wolkenhauer, O., et al. (2014). E2F1 induces miR-224/452 expression to drive EMT through TXNIP downregulation. *EMBO Reports*, *15*(12), 1315–1329. https://doi.org/10.15252/embr.201439392.

Koelblinger, P., Thuerigen, O., & Dummer, R. (2018). Development of encorafenib for BRAF-mutated advanced melanoma. *Current Opinion in Oncology*, *30*(2), 125–133. https://doi.org/10.1097/CCO.0000000000000426.

Koetz-Ploch, L., Hanniford, D., Dolgalev, I., Sokolova, E., Zhong, J., Díaz-Martínez, M., et al. (2017). MicroRNA-125a promotes resistance to BRAF inhibitors through suppression of the intrinsic apoptotic pathway. *Pigment Cell & Melanoma Research*, *30*(3), 328–338. https://doi.org/10.1111/pcmr.12578.

Komina, A., Palkina, N., Aksenenko, M., Tsyrenzhapova, S., & Ruksha, T. (2016). Antiproliferative and pro-apoptotic effects of MiR-4286 inhibition in melanoma cells. *PLoS One*, *11*(12). https://doi.org/10.1371/journal.pone.0168229.

Kordass, T., Weber, C. E. M., Eisel, D., Pane, A. A., Osen, W., & Eichmüller, S. B. (2018). miR-193b and miR-30c-1* inhibit, whereas miR-576-5p enhances melanoma cell invasion in vitro. *Oncotarget*, 32507–32522. https://doi.org/10.18632/oncotarget.25986.

Kozubek, J., Ma, Z., Fleming, E., Duggan, T., Wu, R., Shin, D. G., et al. (2013). In-depth characterization of microRNA transcriptome in melanoma. *PLoS One*, *8*(9). https://doi.org/10.1371/journal.pone.0072699.

Kulshreshtha, R., Ferracin, M., Wojcik, S. E., Garzon, R., Alder, H., Agosto-Perez, F. J., et al. (2007). A microRNA signature of hypoxia. *Molecular and Cellular Biology*, *27*(5), 1859–1867. https://doi.org/10.1128/MCB.01395-06.

Ladstein, R. G., Bachmann, I. M., Straume, O., & Akslen, L. A. (2012). Prognostic importance of the mitotic marker phosphohistone H3 in cutaneous nodular melanoma. *Journal of Investigative Dermatology*, *132*(4), 1247–1252. https://doi.org/10.1038/jid.2011.464.

Lee, B., Sahoo, A., Sawada, J., Marchica, J., Sahoo, S., Layng, F. I. A. L., et al. (2021). MicroRNA-211 modulates the DUSP6-ERK5 signaling Axis to promote BRAFV600E-driven melanoma growth in vivo and BRAF/MEK inhibitor resistance. *Journal of Investigative Dermatology*, *141*(2), 385–394. https://doi.org/10.1016/j.jid.2020.06.038.

Leidinger, P., Keller, A., Borries, A., Reichrath, J., Rass, K., Jager, S. U., et al. (2010). High-throughput miRNA profiling of human melanoma blood samples. *BMC Cancer*, *10*. https://doi.org/10.1186/1471-2407-10-262.

Levati, L., Alvino, E., Pagani, D., Arcelli, D., Caporaso, P., Bondanza, S., et al. (2009). Altered expression of selected microRNAs in melanoma: Antiproliferative and proapoptotic activity of miRNA-155. *International Journal of Oncology*, *35*(2), 393–400. https://doi.org/10.3892/ijo_00000352.

Levati, L., Pagani, E., Romani, S., Castiglia, D., Piccinni, E., Covaciu, C., et al. (2011). MicroRNA-155 targets the SKI gene in human melanoma cell lines. *Pigment Cell & Melanoma Research*, *24*(3), 538–550. https://doi.org/10.1111/j.1755-148X.2011.00857.x.

Levy, C., Khaled, M., & Fisher, D. E. (2006). MITF: Master regulator of melanocyte development and melanoma oncogene. *Trends in Molecular Medicine*, *12*(9), 406–414. https://doi.org/10.1016/j.molmed.2006.07.008.

Levy, C., Khaled, M., Iliopoulos, D., Janas, M. M., Schubert, S., Pinner, S., et al. (2010). Intronic miR-211 assumes the tumor suppressive function of its host gene in melanoma. *Molecular Cell*, *40*(5), 841–849. https://doi.org/10.1016/j.molcel.2010.11.020.

Li, Q., Johnston, N., Zheng, X., Wang, H., Zhang, X., Gao, D., et al. (2016). miR-28 modulates exhaustive differentiation of T cells through silencing programmed cell death-1 and regulating cytokine secretion. *Oncotarget*, *7*(33), 53735–53750. https://doi.org/10.18632/oncotarget.10731.

Li, Y., Zhang, J., Liu, Y., Zhang, B., Zhong, F., Wang, S., et al. (2018). MiR-30a-5p confers cisplatin resistance by regulating IGF1R expression in melanoma cells. *BMC Cancer*, *18*(1). https://doi.org/10.1186/s12885-018-4233-9.

Li, N., Liu, Y., Pang, H., Lee, D., Zhou, Y., & Xiao, Z. (2019). Methylation-mediated silencing of MicroRNA-211 decreases the sensitivity of melanoma cells to cisplatin. *Medical Science Monitor*, *25*, 1590–1599. https://doi.org/10.12659/MSM.911862.

Lin, N., Zhou, Y., Lian, X., & Tu, Y. (2015). Expression of microRNA-106b and its clinical significance in cutaneous melanoma. *Genetics and Molecular Research*, *14*(4), 16379–16385. https://doi.org/10.4238/2015.December.9.6.

Lister, J. A., Capper, A., Zeng, Z., Mathers, M. E., Richardson, J., Paranthaman, K., et al. (2014). A conditional zebrafish MITF mutation reveals MITF levels are critical for melanoma promotion vs. regression in vivo. *Journal of Investigative Dermatology*, *134*(1), 133–140. https://doi.org/10.1038/jid.2013.293.

Liu, S., Kumar, S. M., Lu, H., Liu, A., Yang, R., Pushparajan, A., et al. (2012). MicroRNA-9 up-regulates E-cadherin through inhibition of NF-κB1-Snail1 pathway in melanoma. *Journal of Pathology*, *226*(1), 61–72. https://doi.org/10.1002/path.2964.

Liu, Y., Lai, L., Chen, Q., Song, Y., Xu, S., Feng, M., et al. (2012). MicroRNA-494 is required for the accumulation and functions of tumor-expanded myeloid-derived suppressor cells via targeting of PTEN. *Journal of Immunology*, *188*(11), 5500–5510. https://doi.org/10.4049/jimmunol.1103505.

Liu, S., Tetzlaff, M. T., Liu, A., Liegl-Atzwanger, B., Guo, J., & Xu, X. (2012). Loss of microRNA-205 expression is associated with melanoma progression. *Laboratory Investigation*, *92*(7), 1084–1096. https://doi.org/10.1038/labinvest.2012.62.

Liu, S., Tetzlaff, M. T., Cui, R., & Xu, X. (2012). MiR-200c inhibits melanoma progression and drug resistance through down-regulation of Bmi-1. *American Journal of Pathology*, *181*(5), 1823–1835. https://doi.org/10.1016/j.ajpath.2012.07.009.

Liu, S. M., Lu, J., Lee, H. C., Chung, F. H., & Ma, N. (2014). miR-524-5p suppresses the growth of oncogenic BRAF melanoma by targeting BRAF and ERK2. *Oncotarget*, *5*(19), 9444–9459. https://doi.org/10.18632/oncotarget.2452.

Liu, S., Tetzlaff, M. T., Wang, T., Yang, R., Xie, L., Zhang, G., et al. (2015). miR-200c/Bmi1 axis and epithelial-mesenchymal transition contribute to acquired resistance to BRAF inhibitor treatment. *Pigment Cell & Melanoma Research*, *28*(4), 431–441. https://doi.org/10.1111/pcmr.12379.

Liu, R., Xie, H., Luo, C., Chen, Z., Zhou, X., Xia, K., et al. (2015). Identification of FLOT2 as a novel target for microRNA-34a in melanoma. *Journal of Cancer Research and Clinical Oncology*, *141*(6), 993–1006. https://doi.org/10.1007/s00432-014-1874-1.

Liu, J., Zhang, C., Zhao, Y., & Feng, Z. (2017). MicroRNA control of p53. *Journal of Cellular Biochemistry*, *118*(1), 7–14. https://doi.org/10.1002/jcb.25609.

Liu, K., Jin, J., Rong, K., Zhuo, L., & Li, P. (2018). MicroRNA-675 inhibits cell proliferation and invasion in melanoma by directly targeting metadherin. *Molecular Medicine Reports*, *17*(2), 3372–3379. https://doi.org/10.3892/mmr.2017.8264.

Liu, E., Sun, X., Li, J., & Zhang, C. (2018). miR-30a-5p inhibits the proliferation, migration and invasion of melanoma cells by targeting SOX4. *Molecular Medicine Reports*, *18*(2), 2492–2498. https://doi.org/10.3892/mmr.2018.9166.

Lo, S. N., Scolyer, R. A., & Thompson, J. F. (2018). Long-term survival of patients with thin (T1) cutaneous melanomas: A breslow thickness cut point of 0.8 mm separates higher-risk and lower-risk tumors. *Annals of Surgical Oncology*, *25*(4), 894–902. https://doi.org/10.1245/s10434-017-6325-1.

Lohcharoenkal, W., Das Mahapatra, K., Pasquali, L., Crudden, C., Kular, L., Akkaya Ulum, Y. Z., et al. (2018). Genome-wide screen for MicroRNAs reveals a role for miR-203 in melanoma metastasis. *Journal of Investigative Dermatology*, *138*(4), 882–892. https://doi.org/10.1016/j.jid.2017.09.049.

Long, G. V., Menzies, A. M., Nagrial, A. M., Haydu, L. E., Hamilton, A. L., Mann, G. J., et al. (2011). Prognostic and clinicopathologic associations of oncogenic BRAF in metastatic melanoma. *Journal of Clinical Oncology*, *29*(10), 1239–1246. https://doi.org/10.1200/JCO.2010.32.4327.

Lu, R., Zhang, X., Li, X., & Wan, X. (2020). Circ_0016418 promotes melanoma development and glutamine catabolism by regulating the miR-605-5p/GLS axis. *International Journal of Clinical and Experimental Pathology*, *13*(7).

Luan, W., Li, L., Shi, Y., Bu, X., Xia, Y., Wang, J., Djangmah, H. S., Liu, X., You, Y., & Xu, B. (2016). Long non-coding RNA MALAT1 acts as a competing endogenous RNA to promote malignant melanoma growth and metastasis by sponging miR-22. Oncotarget, 7(39), 63901–63912. doi:10.18632/oncotarget.11564.

Luan, W., Zhou, Z., Zhu, Y., Xia, Y., Wang, J., & Xu, B. (2018). miR-137 inhibits glutamine catabolism and growth of malignant melanoma by targeting glutaminase. *Biochemical and Biophysical Research Communications*, *495*(1), 46–52. https://doi.org/10.1016/j.bbrc.2017.10.152.

Luan, W., Ding, Y., Xi, H., Ruan, H., Lu, F., Ma, S., et al. (2021). Exosomal miR-106b-5p derived from melanoma cell promotes primary melanocytes epithelial-mesenchymal transition through targeting EphA4. *Journal of Experimental & Clinical Cancer Research*, *40*(1). https://doi.org/10.1186/s13046-021-01906-w.

Lunavat, T. R., Cheng, L., Einarsdottir, B., Bagge, O., Muralidharan, R., Veppil, et al. (2017). BRAF(V600) inhibition alters the microRNA cargo in the vesicular secretome of malignant melanoma cells. *Proceedings of the National Academy of Sciences*, *114*(29).

Luo, C., Merz, P. R., Chen, Y., Dickes, E., Pscherer, A., Schadendorf, D., et al. (2013). MiR-101 inhibits melanoma cell invasion and proliferation by targeting MITF and EZH2. *Cancer Letters*, *341*(2), 240–247. https://doi.org/10.1016/j.canlet.2013.08.021.

Luo, C., Tetteh, P. W., Merz, P. R., Dickes, E., Abukiwan, A., Hotz-Wagenblatt, A., et al. (2013). MiR-137 inhibits the invasion of melanoma cells through downregulation of multiple oncogenic target genes. *Journal of Investigative Dermatology*, *133*(3), 768–775. https://doi.org/10.1038/jid.2012.357.

Luo, M., Wu, L., Zhang, K., Wang, H., Wu, S., O'Connell, D., et al. (2018a). miR-216b enhances the efficacy of vemurafenib by targeting Beclin-1, UVRAG and ATG5 in melanoma. *Cellular Signalling*, *42*, 30–43. https://doi.org/10.1016/j.cellsig.2017.09.024.

Luo, M., Wu, L., Zhang, K., Wang, H., Zhang, T., Gutierrez, L., et al. (2018b). miR-137 regulates ferroptosis by targeting glutamine transporter SLC1A5 in melanoma. *Cell Death and Differentiation*, *25*(8), 1457–1472. https://doi.org/10.1038/s41418-017-0053-8.

Lv, N., Hao, S., Luo, C., Abukiwan, A., Hao, Y., Gai, F., et al. (2018). miR-137 inhibits melanoma cell proliferation through downregulation of GLO1. *Science China. Life Sciences*, *61*(5), 541–549. https://doi.org/10.1007/s11427-017-9138-9.

Lv, R., Yu, J., & Sun, Q. (2020). Anti-angiogenic role of microRNA-23b in melanoma by disturbing NF-κB signaling pathway targeted inhibition of NAMPT. *Future Oncology*, *16*(10), 541–558. https://doi.org/10.2217/fon-2019-0699.

Maertens, O., Johnson, B., Hollstein, P., Frederick, D. T., Cooper, Z. A., Messiaen, L., et al. (2013). Elucidating distinct roles for NF1 in melanomagenesis. *Cancer Discovery*, *3*(3), 339–349. https://doi.org/10.1158/2159-8290.CD-12-0313.

Mahabeleshwar, G. H., & Byzova, T. V. (2007). Angiogenesis in Melanoma. *Seminars in Oncology*, *34*(6), 555–565. https://doi.org/10.1053/j.seminoncol.2007.09.009.

Mahmoud, F., Shields, B., Makhoul, I., Avaritt, N., Wong, H. K., Hutchins, L. F., et al. (2017). Immune surveillance in melanoma: From immune attack to melanoma escape and even counterattack. *Cancer Biology and Therapy*, *18*(7), 451–469. https://doi.org/10.1080/15384047.2017.1323596.

Manzano, J. L., Layos, L., Buges, C., De Los Llanos Gil, M., Vila, L., Martinez-Balibrea, E., … et al. (2016). Resistant mechanisms to BRAF inhibitors in melanoma. *Annals of Translational Medicine*, *4*, 237. https://doi.org/10.21037/atm.2016.06.07.

Margue, C., Reinsbach, S., Philippidou, D., Beaume, N., Walters, C., Schneider, J. G., et al. (2015). Comparison of a healthy miRNome with melanoma patient miRNomes: Are microRNAs suitable serum biomarkers for cancer? *Oncotarget*, *6*(14), 12110–12127. https://doi.org/10.18632/oncotarget.3661.

Martinez-Usatorre, A., Sempere, L. F., Carmona, S. J., Carretero-Iglesia, L., Monnot, G., Speiser, D. E., et al. (2019). MicroRNA-155 expression is enhanced by T-cell receptor stimulation strength and correlates with improved tumor control in melanoma. *Cancer Immunology Research*, *7*(6), 1013–1024. https://doi.org/10.1158/2326-6066.CIR-18-0504.

Mastroianni, J., Stickel, N., Andrlova, H., Hanke, K., Melchinger, W., Duquesne, S., et al. (2019). MiR-146a controls immune response in the melanoma microenvironment. *Cancer Research*, *79*(1), 183–195. https://doi.org/10.1158/0008-5472.CAN-18-1397.

Mathsyaraja, H., Thies, K., Taffany, D. A., Deighan, C., Liu, T., Yu, L., et al. (2015). CSF1-ETS2-induced microRNA in myeloid cells promote metastatic tumor growth. *Oncogene*, *34*(28), 3651–3661. https://doi.org/10.1038/onc.2014.294.

Matthews, N. H., Li, W. Q., Qureshi, A. A., Weinstock, M. A., & Cho, E. (2017). Epidemiology of melanoma. *Cutaneous melanoma: Etiology and therapy [Internet]*. Brisbane: Codon Publications.

Maurer, G., Tarkowski, B., & Baccarini, M. (2011). Raf kinases in cancer-roles and therapeutic opportunities. *Oncogene*, *30*(32), 3477–3488. https://doi.org/10.1038/onc.2011.160.

Mazar, J., de Young, K., Khaitan, D., Meister, E., Almodovar, A., Goydos, J., et al. (2010). The regulation of miRNA-211 expression and its role in melanoma cell invasiveness. *PLoS One*, *5*(11). https://doi.org/10.1371/journal.pone.0013779.

Mazar, J., Qi, F., Lee, B., Marchica, J., Govindarajan, S., Shelley, J., et al. (2016). MicroRNA 211 functions as a metabolic switch in human melanoma cells. *Molecular and Cellular Biology*, *36*(7), 1090–1108. https://doi.org/10.1128/MCB.00762-15.

McGovern, V. J., Shaw, H. M., Milton, G. W., & McCarthy, W. H. (1982). Ulceration and prognosis in cutaneous malignant melanoma. *Histopathology*, *6*(4), 399–407. https://doi.org/10.1111/j.1365-2559.1982.tb02737.x.

Migliore, C., Petrelli, A., Ghiso, E., Corso, S., Capparuccia, L., Eramo, A., et al. (2008). MicroRNAs impair MET-mediated invasive growth. *Cancer Research*, *68*(24), 10128–10136. https://doi.org/10.1158/0008-5472.CAN-08-2148.

Miller, A. J., & Mihm, M. C. (2006). Melanoma. *New England Journal of Medicine*, *355*(1), 10–65. https://doi.org/10.1056/NEJMra052166.

Mirmohammadsadegh, A., Marini, A., Nambiar, S., Hassan, M., Tannapfel, A., Ruzicka, T., et al. (2006). Epigenetic silencing of the PTEN gene in melanoma. *Cancer Research*, *66*(13), 6546–6552. https://doi.org/10.1158/0008-5472.CAN-06-0384.

Mishra, P. J., Mishra, P. J., & Merlino, G. (2016). Integrated genomics identifies miR-32/MCL-1 pathway as a critical driver of melanomagenesis: Implications for mirreplacement and combination therapy. *PLoS One*, *11*(11). https://doi.org/10.1371/journal.pone.0165102.

Mitchell, P. S., Parkin, R. K., Kroh, E. M., Fritz, B. R., Wyman, S. K., Pogosova-Agadjanyan, E. L., et al. (2008). Circulating microRNAs as stable blood-based markers for cancer detection. *Proceedings of the National Academy of Sciences of the United States of America*, *105*(30), 10513–10518. https://doi.org/10.1073/pnas.0804549105.

Möller, K., Sigurbjornsdottir, S., Arnthorsson, A. O., Pogenberg, V., Dilshat, R., Fock, V., et al. (2019). MITF has a central role in regulating starvation-induced autophagy in melanoma. *Scientific Reports*, *9*(1). https://doi.org/10.1038/s41598-018-37522-6.

Motti, M. L., Minopoli, M., Di Carluccio, G., Ascierto, P. A., & Carriero, M. V. (2020). Micrornas as key players in melanoma cell resistance to mapk and immune checkpoint inhibitors. *International Journal of Molecular Sciences*, *21*(12), 1–16. https://doi.org/10.3390/ijms21124544.

Mueller, D. W., Rehli, M., & Bosserhoff, A. K. (2009). MiRNA expression profiling in melanocytes and melanoma cell lines reveals miRNAs associated with formation and progression of malignant melanoma. *Journal of Investigative Dermatology*, *129*(7), 1740–1751. https://doi.org/10.1038/jid.2008.452.

Müller, D. W., & Bosserhoff, A. K. (2008). Integrin β3 expression is regulated by let-7a miRNA in malignant melanoma. *Oncogene*, *27*(52), 6698–6706. https://doi.org/10.1038/onc.2008.282.

Muñoz-Couselo, E., Adelantado, E. Z., Ortiz, C., García, J. S., & Perez-Garcia, J. (2017). NRAS-mutant melanoma: Current challenges and future prospect. *Oncotargets and Therapy*, *10*, 3941–3947. https://doi.org/10.2147/OTT.S117121.

Murali, R., Wilmott, J. S., Jakrot, V., Al-Ahmadie, H. A., Wiesner, T., McCarthy, S. W., et al. (2013). BAP1 expression in cutaneous melanoma: A pilot study. *Pathology*, *45*(6), 606–609. https://doi.org/10.1097/PAT.0b013e3283653818.

Nachmani, D., Lankry, D., Wolf, D. G., & Mandelboim, O. (2010). The human cytomegalovirus microRNA miR-UL112 acts synergistically with a cellular microRNA to escape immune elimination. *Nature Immunology*, *11*(9), 806–813. https://doi.org/10.1038/ni.1916.

Naldi, L., Imberti, G. L., Parazzini, F., Gallus, S., La Vecchia, C., Cainelli, T., et al. (2000). Pigmentary traits, modalities of sun reaction, history of sunburns, and melanocytic nevi as risk factors for cutaneous malignant melanoma in the italian population: Results of a collaborative case-control study. *Cancer*, *88*(12), 2703–2710. https://doi.org/10.1002/1097-0142(20000615)88:12<2703::AID-CNCR8>3.0.CO;2-Q.

Negrini, M., Ferracin, M., Sabbioni, S., & Croce, C. M. (2007). MicroRNAs in human cancer: From research to therapy. *Journal of Cell Science*, *120*(11), 1833–1840. https://doi.org/10.1242/jcs.03450.

Neila, J., & Soyer, H. P. (2011). Key points in dermoscopy for diagnosis of melanomas, including difficult to diagnose melanomas, on the trunk and extremities. *Journal of Dermatology*, *38*(1), 3–9. https://doi.org/10.1111/j.1346-8138.2010.01131.x.

Nemlich, Y., Greenberg, E., Ortenberg, R., Besser, M. J., Barshack, I., Jacob-Hirsch, J., et al. (2013). MicroRNA-mediated loss of ADAR1 in metastatic melanoma promotes tumor growth. *Journal of Clinical Investigation*, *123*(6), 2703–2718. https://doi.org/10.1172/JCI62980.

Nemlich, Y., Baruch, E. N., Besser, M. J., Shoshan, E., Bar-Eli, M., Anafi, L., et al. (2018). ADAR1-mediated regulation of melanoma invasion. *Nature Communications*, *9*(1). https://doi.org/10.1038/s41467-018-04600-2.

Newton-Bishop, J. A., Chang, Y. M., Elliott, F., Chan, M., Leake, S., Karpavicius, B., et al. (2011). Relationship between sun exposure and melanoma risk for tumours in different body sites in a large case-control study in a temperate climate. *European Journal of Cancer*, *47*(5), 732–741. https://doi.org/10.1016/j.ejca.2010.10.008.

Nguyen, M. H. T., Lin, C. H., Liu, S. M., Miyashita, A., Ihn, H., Lin, H., et al. (2020). miR-524-5p reduces the progression of the BRAF inhibitor-resistant melanoma. *Neoplasia (United States)*, *22*(12), 789–799. https://doi.org/10.1016/j.neo.2020.10.009.

NIH, & SEER. (2019). *Cancer stat facts: Melanoma of the skin*.

Nikolaou, V., Kang, X., Stratigos, A., Gogas, H., Latorre, M. C., Gabree, M., et al. (2011). Comprehensive mutational analysis of CDKN2A and CDK4 in Greek patients with cutaneous melanoma. *British Journal of Dermatology*, *165*(6), 1219–1222. https://doi.org/10.1111/j.1365-2133.2011.10551.x.

Noguchi, S., Mori, T., Hoshino, Y., Yamada, N., Nakagawa, T., Sasaki, N., et al. (2012). Comparative study of anti-oncogenic MicroRNA-145 in canine and human malignant melanoma. *Journal of Veterinary Medical Science*, *74*(1), 1–8. https://doi.org/10.1292/jvms.11-0264.

Noguchi, S., Mori, T., Otsuka, Y., Yamada, N., Yasui, Y., Iwasaki, J., et al. (2012). Anti-oncogenic microRNA-203 induces senescence by targeting E2F3 protein in human melanoma cells. *Journal of Biological Chemistry*, *287*(15), 11769–11777. https://doi.org/10.1074/jbc.M111.325027.

Nogueira, C., Kim, K. H., Sung, H., Paraiso, K. H. T., Dannenberg, J. H., Bosenberg, M., et al. (2010). Cooperative interactions of PTEN deficiency and RAS activation in melanoma metastasis. *Oncogene*, *29*(47), 6222–6232. https://doi.org/10.1038/onc.2010.349.

Noman, M. Z., Messai, Y., Carré, T., Akalay, I., Méron, M., Janji, B., et al. (2011). Microenvironmental hypoxia orchestrating the cell stroma cross talk, tumor progression and antitumor response. *Critical Reviews in Immunology*, *31*(5), 357–377. https://doi.org/10.1615/critrevimmunol.v31.i5.10.

Noman, M. Z., Buart, S., Romero, P., Ketari, S., Janji, B., Mari, B., et al. (2012). Hypoxia-inducible miR-210 regulates the susceptibility of tumor cells to lysis by cytotoxic T cells. *Cancer Research*, *72*(18), 4629–4641. https://doi.org/10.1158/0008-5472.CAN-12-1383.

Noori, J., Sharifi, M., & Haghjooy Javanmard, S. (2018). miR-30a inhibits melanoma tumor metastasis by targeting the E-cadherin and zinc finger E-box binding homeobox 2. *Advanced Biomedical Research*, *7*.

Nyholm, A. M., Lerche, C. M., Manfé, V., Biskup, E., Johansen, P., Morling, N., et al. (2014). MiR-125b induces cellular senescence in malignant melanoma. *BMC Dermatology*, *14*. https://doi.org/10.1186/1471-5945-14-8.

Ohira, T., Naohiro, S., Nakayama, Y., Osaki, M., Okada, F., Oshimura, M., et al. (2015). MiR-19b regulates hTERT mRNA expression through targeting PITX1 mRNA in melanoma cells. *Scientific Reports*, *5*. https://doi.org/10.1038/srep08201.

Omholt, K., Karsberg, S., Platz, A., Kanter, L., Ringborg, U., & Hansson, J. (2002). Screening of N-ras codon 61 mutations in paired primary and metastatic cutaneous melanomas: Mutations occur early and persist throughout tumor progression. *Clinical Cancer Research*, *8*(11), 3468–3474.

Omholt, K., Platz, A., Kanter, L., Ringborg, U., & Hansson, J. (2003). NRAS and BRAF mutations arise early during melanoma pathogenesis and are preserved throughout tumor progression. *Clinical Cancer Research*, *9*(17), 6483–6488.

Ono, S., Oyama, T., Lam, S., Chong, K., Foshag, L. J., & Hoon, D. S. B. (2015). A direct plasma assay of circulating microRNA-210 of hypoxia can identify early systemic metastasis recurrence in melanoma patients. *Oncotarget*, *6*(9), 7053–7064. https://doi.org/10.18632/oncotarget.3142.

Orso, F., Quirico, L., Virga, F., Penna, E., Dettori, D., Cimino, D., et al. (2016). Mir-214 and mir-148b targeting inhibits dissemination of melanoma and breast cancer. *Cancer Research*, *76*(17), 5151–5162. https://doi.org/10.1158/0008-5472.CAN-15-1322.

Ozsolak, F., Poling, L. L., Wang, Z., Liu, H., Liu, X. S., Roeder, R. G., et al. (2008). Chromatin structure analyses identify miRNA promoters. *Genes and Development*, *22*(22), 3172–3183. https://doi.org/10.1101/gad.1706508.

Palmer, S. R., Erickson, L. A., Ichetovkin, I., Knauer, D. J., & Markovic, S. N. (2011). Circulating serologic and molecular biomarkers in malignant melanoma. *Mayo Clinic Proceedings*, *86*(10), 981–990. https://doi.org/10.4065/mcp.2011.0287.

Paschen, A., Sucker, A., Hill, B., Moll, I., Zapatka, M., Xuan, D. N., et al. (2009). Differential clinical significance of individual NKG2D ligands in melanoma: Soluble ULBP2 as an indicator of poor prognosis superior to S100B. *Clinical Cancer Research*, *15*(16), 5208–5215. https://doi.org/10.1158/1078-0432.CCR-09-0886.

Pasquali, E., García-Borrón, J. C., Fargnoli, M. C., Gandini, S., Maisonneuve, P., Bagnardi, V., et al. (2015). MC1R variants increased the risk of sporadic cutaneous melanoma in darker-pigmented Caucasians: A pooled-analysis from the M-SKIP project. *International Journal of Cancer*, *136*(3), 618–631. https://doi.org/10.1002/ijc.29018.

Pedini, F., De Luca, G., Felicetti, F., Puglisi, R., Boe, A., Arasi, M. B., et al. (2019). Joint action of miR-126 and MAPK/PI3K inhibitors against metastatic melanoma. *Molecular Oncology*, *13*(9), 1836–1854. https://doi.org/10.1002/1878-0261.12506.

Pencheva, N., Tran, H., Buss, C., Huh, D., Drobnjak, M., Busam, K., et al. (2012). Convergent multi-miRNA targeting of ApoE grives LRP1/LRP8-dependent melanoma metastasis and angiogenesis. *Cell*, *151*(5), 1068–1082. https://doi.org/10.1016/j.cell.2012.10.028.

Penna, E., Orso, F., Cimino, D., Tenaglia, E., Lembo, A., Quaglino, E., et al. (2011). MicroRNA-214 contributes to melanoma tumour progression through suppression of TFAP2C. *EMBO Journal*, *30*(10), 1990–2007. https://doi.org/10.1038/emboj.2011.102.

Pfeffer, S., Grossmann, K. F., Cassidy, P. B., Yang, C. H., Fan, M., Kopelovich, L., … Pfeffer, L. M. (2015). Detection of exosomal miRNAs in the plasma of melanoma patients. *Journal of Clinical Medicine*, 2012–2027. https://doi.org/10.3390/jcm4121957.

Philippidou, D., Schmitt, M., Moser, D., Margue, C., Nazarov, P. V., Muller, A., et al. (2010). Signatures of MicroRNAs and selected MicroRNA target genes in human melanoma. *Cancer Research*, *70*(10), 4163–4173. https://doi.org/10.1158/0008-5472.CAN-09-4512.

Pierard, G. E. (2012). Cell proliferation in cutaneous malignant melanoma: Relationship with neoplastic progression. *ISRN Dermatology*, *2012*. https://doi.org/10.5402/2012/828146, 828146.

Platz, A., Egyhazi, S., Ringborg, U., & Hansson, J. (2008). Human cutaneous melanoma; a review of NRAS and BRAF mutation frequencies in relation to histogenetic subclass and body site. *Molecular Oncology*, *1*(4), 395–405. https://doi.org/10.1016/j.molonc.2007.12.003.

Poenitzsch Strong, A. M., Setaluri, V., & Spiegelman, V. S. (2014). MicroRNA-340 as a modulator of RAS-RAF-MAPK signaling in melanoma. *Archives of Biochemistry and Biophysics*, *563*, 118–124. https://doi.org/10.1016/j.abb.2014.07.012.

Ponti, G., Manfredini, M., Greco, S., Pellacani, G., Depenni, R., Tomasi, A., et al. (2017). BRAF, NRAS and c-KIT advanced melanoma: Clinico-pathological features, targeted-therapy strategies and survival. *Anticancer Research*, *37*(12), 7043–7048. https://doi.org/10.21873/anticanres.12175.

Postow, M. A., Chesney, J., Pavlick, A. C., Robert, C., Grossmann, K., McDermott, D., et al. (2015). Nivolumab and ipilimumab versus ipilimumab in untreated melanoma. *New England Journal of Medicine*, *372*(21), 2006–2017. https://doi.org/10.1056/NEJMoa1414428.

Poynter, J. N., Elder, J. T., Fullen, D. R., Nair, R. P., Soengas, M. S., Johnson, T. M., et al. (2006). BRAF and NRAS mutations in melanoma and melanocytic nevi. *Melanoma Research*, *16*(4), 267–273. https://doi.org/10.1097/01.cmr.0000222600.73179.f3.

Pratilas, C. A., Taylor, B. S., Ye, Q., Viale, A., Sander, C., Solit, D. B., et al. (2009). V600EBRAF is associated with disabled feedback inhibition of RAF-MEK signaling and elevated transcriptional output of the pathway. *Proceedings of the National Academy of Sciences of the United States of America*, *106*(11), 4519–4524. https://doi.org/10.1073/pnas.0900780106.

Pu, W., Shang, Y., Shao, Q., & Yuan, X. (2018). miR-146a promotes cell migration and invasion in melanoma by directly targeting SMAD4. *Oncology Letters*, *15*(5), 7111–7117. https://doi.org/10.3892/ol.2018.8172.

Purdue, M. P., From, L., Armstrong, B. K., Kricker, A., Gallagher, R. P., McLaughlin, J. R., et al. (2005). Etiologic and other factors predicting nevus-associated cutaneous malignant melanoma. *Cancer Epidemiology, Biomarkers & Prevention*, *14*(8), 2015–2022. https://doi.org/10.1158/1055-9965.EPI-05-0097.

Qi, J., Wang, W.-W., Chen, W., & Shang, A. Q. (2018). Mechanism of miR-137 regulating migration and invasion of melanoma cells by targeting PIK3R3 gene. *Journal of Cellular Biochemistry*, *120*(5), 8393–8400. https://doi.org/10.1002/jcb.28124.

Qian, H., Yang, C., & Yang, Y. (2017). MicroRNA-26a inhibits the growth and invasiveness of malignant melanoma and directly targets on MITF gene. *Cell Death Discovery*, *3*(1). https://doi.org/10.1038/cddiscovery.2017.28.

Qin, J. Z., Xin, H., & Nickoloff, B. J. (2010). Targeting glutamine metabolism sensitizes melanoma cells to TRAIL-induced death. *Biochemical and Biophysical Research Communications*, *398*(1), 146–152. https://doi.org/10.1016/j.bbrc.2010.06.057.

Qiu, H., Chen, F., & Chen, M. (2019). MicroRNA-138 negatively regulates the hypoxia-inducible factor 1? To suppress melanoma growth and metastasis. *Biology Open*, *8*(8). https://doi.org/10.1242/bio.042937.

Raimo, M., Orso, F., Grassi, E., Cimino, D., Penna, E., De Pittà, C., et al. (2016). miR-146a exerts differential effects on melanoma growth and metastatization. *Molecular Cancer Research*, 548–562. https://doi.org/10.1158/1541-7786.MCR-15-0425-T.

Raman, M., Chen, W., & Cobb, M. H. (2007). Differential regulation and properties of MAPKs. *Oncogene*, *26*(22), 3100–3112. https://doi.org/10.1038/sj.onc.1210392.

Rambow, F., Bechadergue, A., Luciani, F., Gros, G., Domingues, M., Bonaventure, J., et al. (2016). Regulation of melanoma progression through the TCF4/miR-125b/NEDD9 Cascade. *Journal of Investigative Dermatology*, *136*(6), 1229–1237. https://doi.org/10.1016/j.jid.2016.02.803.

Ramirez-Montagut, T., Turk, M. J., Wolchok, J. D., Guevara-Patino, J. A., & Houghton, A. N. (2003). Immunity to melanoma: Unraveling the relation of tumor immunity and autoimmunity. *Oncogene*, *22*(20), 3180–3187. https://doi.org/10.1038/sj.onc.1206462.

Raneros, A. B., Suarez-Álvarez, B., & López-Larrea, C. (2014). Secretory pathways generating immunosuppressive NKG2D ligands: New targets for therapeutic intervention. *OncoImmunology*, *3*(4). https://doi.org/10.4161/onci.28497.

Rang, Z., Yang, G., Wang, Y. W., & Cui, F. (2016). MIR-542-3p suppresses invasion and metastasis by targeting the proto-oncogene serine/threonine protein kinase, PIM1, in melanoma. *Biochemical and Biophysical Research Communications*, *474*(2), 315–320. https://doi.org/10.1016/j.bbrc.2016.04.093.

Rastrelli, M., Tropea, S., Pigozzo, J., Bezzon, E., Campana, L. G., Stramare, R., et al. (2014). Melanoma m1: Diagnosis and therapy. *In Vivo*, *28*(3), 273–285. http://iv.iiarjournals.org/content/28/3/273.full.pdf+html.

Rayner, J. E., Laino, A. M., Nufer, K. L., Adams, L., Raphael, A. P., Menzies, S. W., et al. (2018). Clinical perspective of 3D total body photography for early detection and screening of melanoma. *Frontiers in Medicine*, *5*(MAY). https://doi.org/10.3389/fmed.2018.00152.

Read, J., Wadt, K. A. W., & Hayward, N. K. (2016). Melanoma genetics. *Journal of Medical Genetics*, *53*(1), 1–14. https://doi.org/10.1136/jmedgenet-2015-103150.

Ren, J. W., Li, Z. J., & Tu, C. (2015). MiR-135 post-transcriptionally regulates FOXO1 expression and promotes cell proliferation in human malignant melanoma cells. *International Journal of Clinical and Experimental Pathology*, *8*(6), 6356–6366. http://www.ijcep.com/files/ijcep0008832.pdf.

Reuland, S. N., Smith, S. M., Bemis, L. T., Goldstein, N. B., Almeida, A. R., Partyka, K. A., et al. (2013). MicroRNA-26a is strongly downregulated in melanoma and induces cell death through repression of silencer of death domains (SODD). *Journal of Investigative Dermatology*, *133*(5), 1286–1293. https://doi.org/10.1038/jid.2012.400.

Riefolo, M., Porcellini, E., Dika, E., Broseghini, E., & Ferracin, M. (2019). Interplay between small and long non-coding RNAs in cutaneous melanoma: A complex jigsaw puzzle with missing pieces. *Molecular Oncology*, *13*(1), 74–98. https://doi.org/10.1002/1878-0261.12412.

Rigel, D. S., Russak, J., & Friedman, R. (2010). The evolution of melanoma diagnosis: 25 years beyond the ABCDs. *CA: a Cancer Journal for Clinicians*, *60*(5), 301–316. https://doi.org/10.3322/caac.20074.

Robinson, J. K., & Turrisi, R. (2006). Skills training to learn discrimination of ABCDE criteria by those at risk of developing melanoma. *Archives of Dermatology*, *142*(4), 447–452. https://doi.org/10.1001/archderm.142.4.447.

Robles-Espinoza, C. D., Harland, M., Ramsay, A. J., Aoude, L. G., Quesada, V., Ding, Z., et al. (2014). POT1 loss-of-function variants predispose to familial melanoma. *Nature Genetics*, *46*(5), 478–481. https://doi.org/10.1038/ng.2947.

Russell, R. C., Yuan, H. X., & Guan, K. L. (2014). Autophagy regulation by nutrient signaling. *Cell Research*, *24*(1), 42–57. https://doi.org/10.1038/cr.2013.166.

Sahoo, A., Sahoo, S. K., Joshi, P., Lee, B., & Perera, R. J. (2019). MicroRNA-211 loss promotes metabolic vulnerability and BRAF inhibitor sensitivity in melanoma. *Journal of Investigative Dermatology*, *139*(1), 167–176. https://doi.org/10.1016/j.jid.2018.06.189.

Saldanha, G., Potter, L., Shendge, P., Osborne, J., Nicholson, S., Yii, N., et al. (2013). Plasma microRNA-21 is associated with tumor burden in cutaneous melanoma. *Journal of Investigative Dermatology*, *133*(5), 1381–1384. https://doi.org/10.1038/jid.2012.477.

Saldanha, G., Elshaw, S., Sachs, P., Alharbi, H., Shah, P., Jothi, A., et al. (2016). microRNA-10b is a prognostic biomarker for melanoma. *Modern Pathology*, *29*(2), 112–121. https://doi.org/10.1038/modpathol.2015.149.

Sand, M., Skrygan, M., Sand, D., Georgas, D., Gambichler, T., Hahn, S. A., et al. (2013). Comparative microarray analysis of microRNA expression profiles in primary cutaneous malignant melanoma, cutaneous malignant melanoma metastases, and benign melanocytic nevi. *Cell and Tissue Research*, *351*(1), 85–98. https://doi.org/10.1007/s00441-012-1514-5.

Satzger, I., Mattern, A., Kuettler, U., Weinspach, D., Voelker, B., Kapp, A., et al. (2010). MicroRNA-15b represents an independent prognostic parameter and is correlated with tumor cell proliferation and apoptosis in malignant melanoma. *International Journal of Cancer*, *126*(11), 2553–2562. https://doi.org/10.1002/ijc.24960.

Satzger, I., Mattern, A., Kuettler, U., Weinspach, D., Niebuhr, M., Kapp, A., et al. (2012). microRNA-21 is upregulated in malignant melanoma and influences apoptosis of melanocytic cells. *Experimental Dermatology*, *21*(7), 509–514. https://doi.org/10.1111/j.1600-0625.2012.01510.x.

Schatton, T., & Frank, M. H. (2008). Cancer stem cells and human malignant melanoma. *Pigment Cell & Melanoma Research*, *21*(1), 39–55. https://doi.org/10.1111/j.1755-148X.2007.00427.x.

Schmitt, M., Philippidou, D., Reinsbach, S., Margue, C., Wienecke-Baldacchino, A., Nashan, D., et al. (2012). Interferon-gamma-induced activation of signal transducer and activator of transcription 1 (STAT1) up-regulates the tumor suppressing microRNA-29 family in melanoma cells. *Cell Communication and Signaling: CCS*, *10*(1).

Schultz, J., Lorenz, P., Gross, G., Ibrahim, S., & Kunz, M. (2008). MicroRNA let-7b targets important cell cycle molecules in malignant melanoma cells and interferes with anchorage-independent growth. *Cell Research*, *18*(5), 549–557. https://doi.org/10.1038/cr.2008.45.

Segura, M. F., Hanniford, D., Menendez, S., Reavie, L., Zou, X., Alvarez-Diaz, S., et al. (2009). Aberrant miR-182 expression promotes melanoma metastasis by repressing FOXO3 and microphthalmia-associated transcription factor. *Proceedings of the National Academy of Sciences of the United States of America*, *106*(6), 1814–1819. https://doi.org/10.1073/pnas.0808263106.

Segura, M. F., Belitskaya-Lévy, I., Rose, A. E., Zakrzewski, J., Gaziel, A., Hanniford, D., et al. (2010). Melanoma MicroRNA signature predicts post-recurrence survival. *Clinical Cancer Research*, 1577–1586. https://doi.org/10.1158/1078-0432.CCR-09-2721.

Seitz-Alghrouz, R., Hidalgo, J. V., Kayser, C., Kreutz, C., Technau-Hafsi, K., Diaz, C., et al. (2018). BRAF V600E mutations in nevi and melanocytic tumors of uncertain malignant potential. *Journal of Investigative Dermatology*, *138*(11), 2489–2491. https://doi.org/10.1016/j.jid.2018.04.035.

Serguienko, A., Grad, I., Wennerstrøm, A. B., Meza-Zepeda, L. A., Thiede, B., Stratford, E. W., et al. (2015). Metabolic reprogramming of metastatic breast cancer and melanoma by let-7a microRNA. *Oncotarget*, *6*(4), 2451–2465. https://doi.org/10.18632/oncotarget.3235.

Sha, J., Gastman, B. R., Morris, N., Mesinkovska, N. A., Baron, E. D., Cooper, K. D., et al. (2016). The response of microRNAs to solar UVR in skin-resident melanocytes differs between melanoma patients and healthy persons. *PLoS One*, *11*(5). https://doi.org/10.1371/journal.pone.0154915.

Shain, A. H., & Bastian, B. C. (2016). From melanocytes to melanomas. *Nature Reviews Cancer*, *16*(6), 345–358. https://doi.org/10.1038/nrc.2016.37.

Shain, A. H., Yeh, I., Kovalyshyn, I., Sriharan, A., Talevich, E., Gagnon, A., et al. (2015). The genetic evolution of melanoma from precursor lesions. *New England Journal of Medicine*, *373*(20), 1926–1936. https://doi.org/10.1056/NEJMoa1502583.

Shen, C., Hua, H., Gu, L., Cao, S., Cai, H., Yao, X., et al. (2019). MiR-124 functions as a melanoma tumor suppressor by targeting RACKI. *Oncotargets and Therapy*, *12*, 9975–9986. https://doi.org/10.2147/OTT.S225120.

Shenenberger, D. W. (2012). Cutaneous malignant melanoma: A primary care perspective. *American Family Physician*, *85*(2), 161–168. http://www.aafp.org/afp/2012/0115/p161.pdf.

Shi, Q., He, Q., & Wei, J. (2018). MicroRNA-342 prohibits proliferation and invasion of melanoma cells by directly targeting zinc-finger E-box-binding homeobox 1. *Oncology Research*, *26*(9), 1447–1455. https://doi.org/10.3727/096504018X15193823766141.

Shi, L., Huo, J. W., Chen, S. S., Xue, J. X., Gao, W. Y., Li, X. Y., et al. (2019). MicroRNA-22 targets FMNL2 to inhibit melanoma progression via the regulation of the Wnt/β-catenin signaling pathway and epithelial-mesenchymal transition. *European Review for Medical and Pharmacological Sciences*, *23*(12), 5332–5342. https://doi.org/10.26355/eurrev_201906_18200.

Shiiyama, R., Fukushima, S., Jinnin, M., Yamashita, J., Miyashita, A., Nakahara, S., et al. (2013). Sensitive detection of melanoma metastasis using circulating microRNA expression profiles. *Melanoma Research*, *23*(5), 366–372. https://doi.org/10.1097/CMR.0b013e328363e485.

Shoshan, E., Mobley, A. K., Braeuer, R. R., Kamiya, T., Huang, L., Vasquez, M. E., et al. (2015). Reduced adenosine-to-inosine miR-455-5p editing promotes melanoma growth and metastasis. *Nature Cell Biology*, *17*(3), 311–321. https://doi.org/10.1038/ncb3110.

Shtivelman, E., Davies, M. A., Hwu, P., Yang, J., Lotem, M., Oren, M., et al. (2014). Pathways and therapeutic targets in melanoma. *Oncotarget*, *5*(7), 1701–1752. https://doi.org/10.18632/oncotarget.1892.

Siegel, R. L., Miller, K. D., & Jemal, A. (2018). Cancer statistics, 2018. *CA: a Cancer Journal for Clinicians*, *68*(1), 7–30. https://doi.org/10.3322/caac.21442.

Smoller, B. R. (2006). Histologic criteria for diagnosing primary cutaneous malignant melanoma. *Modern Pathology*, *19*, 34–40. https://doi.org/10.1038/modpathol.3800508.

Sommer, L. (2011). Generation of melanocytes from neural crest cells. *Pigment Cell & Melanoma Research*, *24*(3), 411–421. https://doi.org/10.1111/j.1755-148X.2011.00834.x.

Spatz, A., Cook, M. G., Elder, D. E., Piepkorn, M., Ruiter, D. J., & Barnhill, R. L. (2003). Interobserver reproducibility of ulceration assessment in primary cutaneous melanomas. *European Journal of Cancer*, *39*(13), 1861–1865. https://doi.org/10.1016/S0959-8049(03)00325-3.

Sprakel, B., Stenschke, F., Unnewehr, M., Ladas, A., & Senninger, N. (2003). Prognose beim malignen Melanom nach Dissektion regionärer Lymphknotenmetastasen. *Chirurg*, *74*(1), 55–60. https://doi.org/10.1007/s00104-002-0548-8.

Stark, M. S., Klein, K., Weide, B., Haydu, L. E., Pflugfelder, A., Tang, Y. H., et al. (2015). The prognostic and predictive value of melanoma-related MicroRNAs using tissue and serum: A MicroRNA expression analysis. *eBioMedicine*, *2*(7), 671–680. https://doi.org/10.1016/j.ebiom.2015.05.011.

Stark, M. S., Bonazzi, V. F., Boyle, G. M., Palmer, J. M., Symmons, J., Lanagan, C. M., et al. (2015). miR-514a regulates the tumour suppressor NF1 and modulates BRAFi sensitivity in melanoma. *Oncotarget*, *6*(19), 17753–17763. https://doi.org/10.18632/oncotarget.3924.

Strojan, P. (2010). Role of radiotherapy in melanoma management. *Radiology and Oncology*, *44*(1), 1–12. https://doi.org/10.2478/v10019-010-0008-x.

Sturm, R. A., Fox, C., McClenahan, P., Jagirdar, K., Ibarrola-Villava, M., Banan, P., et al. (2014). Phenotypic characterization of nevus and tumor patterns in MITF E318K mutation carrier melanoma patients. *Journal of Investigative Dermatology*, *134*(1), 141–149. https://doi.org/10.1038/jid.2013.272.

Sun, X., Li, J., Sun, Y., Zhang, Y., Dong, L., Shen, C., et al. (2016). miR-7 reverses the resistance to BRAFi in melanoma by targeting EGFR/IGF-1R/CRAF and inhibiting the MAPK and PI3K/AKT signaling pathways. *Oncotarget*, *7*(33), 53558–53570. https://doi.org/10.18632/oncotarget.10669.

Sun, X., Zhang, C., Cao, Y., & Liu, E. (2019). MiR-150 suppresses tumor growth in melanoma through downregulation of myb. *Oncology Research*, *27*(3), 317–323. https://doi.org/10.3727/096504018X15228863026239.

Sun, R., Guo, M., Fan, X., Meng, Q., Yuan, D., Yang, X., et al. (2020). MicroRNA-148b inhibits the malignant biological behavior of melanoma by reducing sirtuin 7 expression levels. *BioMed Research International*, *2020*. https://doi.org/10.1155/2020/9568976.

Tang, H., Xu, X., Xiao, W., Liao, Y., Xiao, X., Li, L., et al. (2019). Silencing of microRNA-27a facilitates autophagy and apoptosis of melanoma cells through the activation of the SYK-dependent mTOR signaling pathway. *Journal of Cellular Biochemistry*, *120*(8), 13262–13274. https://doi.org/10.1002/jcb.28600.

Tembe, V., Schramm, S. J., Stark, M. S., Patrick, E., Jayaswal, V., Tang, Y. H., et al. (2015). MicroRNA and mRNA expression profiling in metastatic melanoma reveal associations with BRAF mutation and patient prognosis. *Pigment Cell & Melanoma Research*, *28*(3), 254–266. https://doi.org/10.1111/pcmr.12343.

Tengda, L., Shuping, L., Mingli, G., Jie, G., Yun, L., Weiwei, Z., et al. (2018). Serum exosomal microRNAs as potent circulating biomarkers for melanoma. *Melanoma Research*, *28*(4), 295–303. https://doi.org/10.1097/CMR.0000000000000450.

Tian, R., Liu, T., Qiao, L., Gao, M., & Li, J. (2015). Decreased serum microRNA-206 level predicts unfavorable prognosis in patients with melanoma. *International Journal of Clinical and Experimental Pathology*, *8*(3), 3097–3103. http://www.ijcep.com/files/ijcep0005554.pdf.

Tian, F., Wei, H., Tian, H., Qiu, Y., & Xu, J. (2016). miR-33a is downregulated in melanoma cells and modulates cell proliferation by targeting PCTAIRE1. *Oncology Letters*, *11*(4), 2741–2746. https://doi.org/10.3892/ol.2016.4321.

Titus-Ernstoff, L., Perry, A. E., Spencer, S. K., Gibson, J. J., Cole, B. F., & Ernstoff, M. S. (2005). Pigmentary characteristics and moles in relation to melanoma risk. *International Journal of Cancer*, *116*(1), 144–149. https://doi.org/10.1002/ijc.21001.

Topalian, S. L., Drake, C. G., & Pardoll, D. M. (2015). Immune checkpoint blockade: A common denominator approach to cancer therapy. *Cancer Cell*, *27*(4), 450–461. https://doi.org/10.1016/j.ccell.2015.03.001.

Tsao, H., & Niendorf, K. (2004). Genetic testing in hereditary melanoma. *Journal of the American Academy of Dermatology*, *51*(5), 803–808. https://doi.org/10.1016/j.jaad.2004.04.045.

Tsao, H., Goel, V., Wu, H., Yang, G., & Haluska, F. G. (2004). Genetic interaction between NRAS and BRAF mutations and PTEN/MMAC1 inactivation in melanoma. *Journal of Investigative Dermatology*, *122*(2), 337–341. https://doi.org/10.1046/j.0022-202X.2004.22243.x.

Tse, J. C., & Kalluri, R. (2007). Mechanisms of metastasis: Epithelial-to-mesenchymal transition and contribution of tumor microenvironment. *Journal of Cellular Biochemistry*, *101*(4), 816–829. https://doi.org/10.1002/jcb.21215.

Tseng, H. W., Li, S. C., & Tsai, K. W. (2019). Metformin treatment suppresses melanoma cell growth and motility through modulation of microRNA expression. *Cancers*, *11*(2). https://doi.org/10.3390/cancers11020209.

Tupone, M. G., D'Aguanno, S., Di Martile, M., Valentini, E., Desideri, M., Trisciuoglio, D., et al. (2020). microRNA-378a-5p iS a novel positive regulator of melanoma progression. *Oncogenesis*, *9*(2). https://doi.org/10.1038/s41389-020-0203-6.

Uribe, P., Wistuba, I. I., & González, S. (2003). A frequent event in benign, atypical, and malignant melanocytic lesions of the skin. *American Journal of Dermatopathology*, *25*(5), 365–370. https://doi.org/10.1097/00000372-200310000-00001.

Uzdensky, A. B., Demyanenko, S. V., & Bibov, M. Y. (2013). Signal transduction in human cutaneous melanoma and target drugs. *Current Cancer Drug Targets*, *13*(8), 843–866. https://doi.org/10.2174/1568009611313080004.

Van Dijk, M. C. R. E., Bernsen, M. R., & Ruiter, D. J. (2005). Analysis of mutations in B-RAF, N-RAS, and H-RAS genes in the differential diagnosis of Spitz nevus and spitzoid melanoma. *American Journal of Surgical Pathology*, *29*(9), 1145–1151. https://doi.org/10.1097/01.pas.0000157749.18591.9e.

Van Laar, R., Lincoln, M., & Van Laar, B. (2018). Development and validation of a plasma-based melanoma biomarker suitable for clinical use. *British Journal of Cancer*, 857–866. https://doi.org/10.1038/bjc.2017.477.

Varrone, F., & Caputo, E. (2020). The miRNAs role in melanoma and in its resistance to therapy. *International Journal of Molecular Sciences*, *21*(3). https://doi.org/10.3390/ijms21030878.

Velazquez-Torres, G., Shoshan, E., Ivan, C., Huang, L., Fuentes-Mattei, E., Paret, H., et al. (2018). A-to-I miR-378a-3p editing can prevent melanoma progression via regulation of PARVA expression. *Nature Communications*, *9*(1). https://doi.org/10.1038/s41467-018-02851-7.

Vergani, E., Di Guardo, L., Dugo, M., Rigoletto, S., Tragni, G., Ruggeri, R., et al. (2016). Overcoming melanoma resistance to vemurafenib by targeting CCL2-induced miR-34a, miR-100 and miR-125b. *Oncotarget*, *7*(4), 4428–4441. https://doi.org/10.18632/oncotarget.6599.

Vergani, E., Dugo, M., Cossa, M., Frigerio, S., Di Guardo, L., Gallino, G., et al. (2020). MiR-146a-5p impairs melanoma resistance to kinase inhibitors by targeting COX2 and regulating NFkB-mediated inflammatory mediators. *Cell Communication and Signaling: CCS*, *18*(1). https://doi.org/10.1186/s12964-020-00601-1.

Villaruz, L. C., Huang, G., Romkes, M., Kirkwood, J. M., Buch, S. C., Nukui, T., et al. (2015). MicroRNA expression profiling predicts clinical outcome of carboplatin/paclitaxelbased therapy in metastatic melanoma treated on the ECOG-ACRIN trial E2603. *Clinical Epigenetics*, *7*(1). https://doi.org/10.1186/s13148-015-0092-2.

Viros, A., Fridlyand, J., Bauer, J., Lasithiotakis, K., Garbe, C., Pinkel, D., et al. (2008). Improving melanoma classification by integrating genetic and morphologic features. *PLoS Medicine*, *5*(6), 0941–0952. https://doi.org/10.1371/journal.pmed.0050120.

Vita, M., Tisserand, J. C., De Beauchêne, I. C., Panel, N., Tchertanov, L., Agopian, J., et al. (2014). Characterization of S628N A novel KIT mutation found in a metastatic melanoma. *JAMA Dermatology*, *150*(12), 1345–1349. https://doi.org/10.1001/jamadermatol.2014.1437.

Vitiello, M., Tuccoli, A., D'Aurizio, R., Sarti, S., Giannecchini, L., Lubrano, S., et al. (2017). Context-dependent miR-204 and miR-211 affect the biological properties of amelanotic and melanotic melanoma cells. *Oncotarget*, *8*(15), 25395–25417. https://doi.org/10.18632/oncotarget.15915.

Vollmer, R. T., & Seigler, H. F. (2001). Using a continuous transformation of the breslow thickness for prognosis in cutaneous melanoma. *American Journal of Clinical Pathology*, *115*(2), 205–212. https://doi.org/10.1309/WAVR-560R-NU5E-4Q96.

Wadt, K. A. W., Aoude, L. G., Johansson, P., Solinas, A., Pritchard, A., Crainic, O., et al. (2015). A recurrent germline BAP1 mutation and extension of the BAP1 tumor predisposition spectrum to include basal cell carcinoma. *Clinical Genetics*, *88*(3), 267–272. https://doi.org/10.1111/cge.12501.

Wagenseller, A. G., Shada, A., D'Auria, K. M., Murphy, C., Sun, D., Molhoek, K. R., et al. (2013). MicroRNAs induced in melanoma treated with combination targeted therapy of temsirolimus and bevacizumab. *Journal of Translational Medicine*, *11*(1). https://doi.org/10.1186/1479-5876-11-218.

Wang, J. J., Li, Z. F., Li, X. J., Han, Z., Zhang, L., & Liu, Z. J. (2017). Effects of microRNA-136 on melanoma cell proliferation, apoptosis, and epithelial–mesenchymal transition by targetting PMEL through the Wnt signaling pathway. *Bioscience Reports*, *37*(5). https://doi.org/10.1042/BSR20170743.

Wang, X., Qu, H., Dong, Y., Wang, G., Zhen, Y., & Zhang, L. (2018). Targeting signal-transducer-and-activator-of-transcription 3 sensitizes human cutaneous melanoma cells to BRAF inhibitor. *Cancer Biomarkers*, *23*(1), 67–77. https://doi.org/10.3233/CBM-181365.

Wang, S., Wu, Y., Xu, Y., & Tang, X. (2019). miR-10b promoted melanoma progression through Wnt/β-catenin pathway by repressing ITCH expression. *Gene*, *710*, 39–47. https://doi.org/10.1016/j.gene.2019.05.043.

Weber, C. E. M., Luo, C., Hotz-Wagenblatt, A., Gardyan, A., Kordaß, T., Holland-Letz, T., et al. (2016). miR-339-3p is a tumor suppressor in melanoma. *Cancer Research*, *76*(12), 3562–3571. https://doi.org/10.1158/0008-5472.CAN-15-2932.

Wei, Z., & Tu, L. H. (2002). MAPK signal pathways in the regulation of cell proliferation in mammalian cells. *Cell Research*, 9–18. https://doi.org/10.1038/sj.cr.7290105.

Wei, Y., Du, Y., Chen, X., Li, P., Wang, Y., Zang, W., et al. (2014). Expression patterns of microRNA-218 and its potential functions by targeting CIP2A and BMI1 genes in melanoma. *Tumor Biology*, *35*(8), 8007–8015. https://doi.org/10.1007/s13277-014-2079-6.

Weir, H., Marrett, L. D., Cokkinides, V., Barnholtz-Sloan, J., Patel, P., Tai, E., et al. (2011). Melanoma in adolescents and young adults (ages 15-39 years): United States. *Journal of the American Academy of Dermatology*, *65*(5 Suppl 1), 38–49.

Wellbrock, C., & Arozarena, I. (2015). Microphthalmia-associated transcription factor in melanoma development and MAP-kinase pathway targeted therapy. *Pigment Cell & Melanoma Research*, *28*(4), 390–406. https://doi.org/10.1111/pcmr.12370.

WHO. (2018). Cancer fact sheet: Melanoma of skin. *Cancer Today*.

Wiesner, T., Obenauf, A. C., Murali, R., Fried, I., Griewank, K. G., Ulz, P., et al. (2011). Germline mutations in BAP1 predispose to melanocytic tumors. *Nature Genetics*, *43*(10), 1018–1022. https://doi.org/10.1038/ng.910.

Wiesner, T., Fried, I., Ulz, P., Stacher, E., Popper, H., Murali, R., et al. (2012). Toward an improved definition of the tumor spectrum associated with BAP1 germline mutations. *Journal of Clinical Oncology*, *30*(32), e337–e340. https://doi.org/10.1200/JCO.2011.41.2965.

Wojcik, K. Y., Escobedo, L. A., Wysong, A., Heck, J. E., Ritz, B., Hamilton, A. S., et al. (2019). High birth weight, early UV exposure, and melanoma risk in children, adolescents, and young adults. *Epidemiology*, *30*(2), 278–284. https://doi.org/10.1097/EDE.0000000000000963.

Wong, S. L., Faries, M. B., Kennedy, E. B., Agarwala, S. S., Akhurst, T. J., Ariyan, C., et al. (2018). Sentinel lymph node biopsy and management of regional lymph nodes in melanoma: American society of clinical oncology and society of surgical oncology clinical practice guideline update. *Journal of Clinical Oncology*, *36*(4), 399–413. https://doi.org/10.1200/JCO.2017.75.7724.

Wozniak, M., Sztiller-Sikorska, M., & Czyz, M. (2015). Diminution of miR-340-5p levels is responsible for increased expression of ABCB5 in melanoma cells under oxygen-deprived conditions. *Experimental and Molecular Pathology*, *99*(3), 707–716. https://doi.org/10.1016/j.yexmp.2015.11.014.

Wu, X., & Bao, H. (2021). Tumor suppressive microRNA-485-5p targets PRRX1 in human skin melanoma cells, regulating epithelial-mesenchymal transition and apoptosis. *Cell Biology International*, *45*(7), 1404–1414. https://doi.org/10.1002/cbin.11575.

Wu, H., Goel, V., & Haluska, F. G. (2003). PTEN signaling pathways in melanoma. *Oncogene*, *22*(20), 3113–3122. https://doi.org/10.1038/sj.onc.1206451.

Wu, J., Li, J., Ren, J., & Zhang, D. (2017). MicroRNA-485-5p represses melanoma cell invasion and proliferation by suppressing Frizzled7. *Biomedicine and Pharmacotherapy*, *90*, 303–310. https://doi.org/10.1016/j.biopha.2017.03.064.

Xiao, Y., Diao, Q., Liang, Y., Peng, Y., & Zeng, K. (2017). MicroRNA-24-1-5p promotes malignant melanoma cell autophagy and apoptosis via regulating ubiquitin D. *Molecular Medicine Reports*, *16*(6), 8448–8454. https://doi.org/10.3892/mmr.2017.7614.

Xie, Y., Shi, X., Sheng, K., Han, G., Li, W., Zhao, Q., et al. (2019). PI3K/Akt signaling transduction pathway, erythropoiesis and glycolysis in hypoxia (review). *Molecular Medicine Reports*, *19*(2), 783–791. https://doi.org/10.3892/mmr.2018.9713.

Xiong, Y., Liu, L., Qiu, Y., & Liu, L. (2018). MicroRNA-29a inhibits growth, migration and invasion of melanoma A375 cells in vitro by directly targeting BMI1. *Cellular Physiology and Biochemistry*, *50*(1), 332–341. https://doi.org/10.1159/000494015.

Xu, Y., Brenn, T., Brown, E. R. S., Doherty, V., & Melton, D. W. (2012). Differential expression of microRNAs during melanoma progression: MiR-200c, miR-205 and miR-211 are downregulated in melanoma and act as tumour suppressors. *British Journal of Cancer*, *106*(3), 553–561. https://doi.org/10.1038/bjc.2011.568.

Yamazaki, H., Chijiwa, T., Inoue, Y., Abe, Y., Suemizu, H., Kawai, K., et al. (2012). Overexpression of the miR-34 family suppresses invasive growth of malignant melanoma with the wild-type p53 gene. *Experimental and Therapeutic Medicine*, *3*(5), 793–796. https://doi.org/10.3892/etm.2012.497.

Yang, C. H., Yue, J., Pfeffer, S. R., Handorf, C. R., & Pfeffer, L. M. (2011). MicroRNA miR-21 regulates the metastatic behavior of B16 melanoma cells. *Journal of Biological Chemistry*, *286*(45), 39172–39178. https://doi.org/10.1074/jbc.M111.285098.

Yang, X., Lei, S., Long, J., Liu, X., & Wu, Q. (2016). MicroRNA-199a-5p inhibits tumor proliferation in melanoma by mediating HIF-1α. *Molecular Medicine Reports*, *13*(6), 5241–5247. https://doi.org/10.3892/mmr.2016.5202.

Yang, P., Bu, P., & Li, C. (2017). miR-124 inhibits proliferation, migration and invasion of malignant melanoma cells via targeting versican. *Experimental and Therapeutic Medicine*, *14*(4), 3555–3562. https://doi.org/10.3892/etm.2017.4998.

Yang, X., Zhao, H., Yang, J., Ma, Y., Liu, Z., Li, C., et al. (2019). MiR-150-5p regulates melanoma proliferation, invasion and metastasis via SIX1-mediated Warburg effect. *Biochemical and Biophysical Research Communications*, *515*(1), 85–91. https://doi.org/10.1016/j.bbrc.2019.05.111.

Yu, Y., Yu, F., & Sun, P. (2020). MicroRNA-1246 promotes melanoma progression through targeting FOXA2. *Oncotargets and Therapy*, *13*, 1245–1253. https://doi.org/10.2147/OTT.S234276.

Yuan, S. M., Li, H., Yang, M., Zha, H., Sun, H., Li, X. R., et al. (2015). High intensity focused ultrasound enhances anti-tumor immunity by inhibiting the negative regulatory effect of miR-134 on CD86 in a murine melanoma model. *Oncotarget*, *6*(35), 37626–37637. https://doi.org/10.18632/oncotarget.5285.

Zehavi, L., Avraham, R., Barzilai, A., Bar-Ilan, D., Navon, R., Sidi, Y., et al. (2012). Silencing of a large microRNA cluster on human chromosome 14q32 in melanoma: Biological effects of mir-376a and mir-376c on insulin growth factor 1 receptor. *Molecular Cancer*, *11*. https://doi.org/10.1186/1476-4598-11-44.

Zeng, H. F., Yan, S., & Wu, S. F. (2017). MicroRNA-153-3p suppress cell proliferation and invasion by targeting SNAI1 in melanoma. *Biochemical and Biophysical Research Communications*, *487*(1), 140–145. https://doi.org/10.1016/j.bbrc.2017.04.032.

Zhang, B., Pan, X., Cobb, G. P., & Anderson, T. A. (2007). microRNAs as oncogenes and tumor suppressors. *Developmental Biology*, *302*(1), 1–12. https://doi.org/10.1016/j.ydbio.2006.08.028.

Zhang, Z., Sun, H., Dai, H., Walsh, R. M., Imakura, M., Schelter, J., et al. (2009). MicroRNA miR-210 modulates cellular response to hypoxia through the MYC antagonist MNT. *Cell Cycle*, *8*(17), 2756–2768. https://doi.org/10.4161/cc.8.17.9387.

Zhang, J., Lu, L., Xiong, Y., Qin, W., Zhang, Y., Qian, Y., et al. (2014). MLK3 promotes melanoma proliferation and invasion and is a target of microRNA-125b. *Clinical and Experimental Dermatology*, *39*(3), 376–384. https://doi.org/10.1111/ced.12286.

Zhang, Z., Zhang, S., Ma, P., Jing, Y., Peng, H., Gao, W. Q., et al. (2015). Lin28B promotes melanoma growth by mediating a microRNA regulatory circuit. *Carcinogenesis*, *36*(9), 937–945. https://doi.org/10.1093/carcin/bgv085.

Zhang, J., Na, S., Liu, C., Pan, S., Cai, J., & Qiu, J. (2016). MicroRNA-125b suppresses the epithelial–mesenchymal transition and cell invasion by targeting ITGA9 in melanoma. *Tumor Biology*, *37*(5), 5941–5949. https://doi.org/10.1007/s13277-015-4409-8.

Zhang, D., Han, Y., & Xu, L. (2016). Upregulation of miR-124 by physcion 8-O-β-glucopyranoside inhibits proliferation and invasion of malignant melanoma cells via repressing RLIP76. *Biomedicine & Pharmacotherapy*, *84*, 166–176. https://doi.org/10.1016/j.biopha.2016.09.022.

Zhang, K., Wu, L., Zhang, P., Luo, M., Du, J., Gao, T., et al. (2018). miR-9 regulates ferroptosis by targeting glutamic-oxaloacetic transaminase GOT1 in melanoma. *Molecular Carcinogenesis*, *57*(11), 1566–1576. https://doi.org/10.1002/mc.22878.

Zhang, H., Feng, C., Zhang, M., Zeng, A., Si, L., Yu, N., et al. (2019). miR-625-5p/PKM2 negatively regulates melanoma glycolysis state. *Journal of Cellular Biochemistry*, *120*(3), 2964–2972. https://doi.org/10.1002/jcb.26917.

Zhao, E., Maj, T., Kryczek, I., Li, W., Wu, K., Zhao, L., et al. (2016). Cancer mediates effector T cell dysfunction by targeting microRNAs and EZH2 via glycolysis restriction. *Nature Immunology*, *17*(1), 95–103. https://doi.org/10.1038/ni.3313.

Zhao, C. C., Guo, H., Wang, Y., & Li, J. H. (2021). Comprehensive upstream and downstream regulatory analyses identify miR-675-3p as a potential prognostic biomarker in melanoma. *Human Cell*, *34*(2), 654–666. https://doi.org/10.1007/s13577-020-00473-0.

Zheng, Y., Manzotti, C. N., Liu, M., Burke, F., Mead, K. I., & Sansom, D. M. (2004). CD86 and CD80 differentially modulate the suppressive function of human regulatory T cells. *Journal of Immunology*, *172*(5), 2778–2784. https://doi.org/10.4049/jimmunol.172.5.2778.

Zheng, Y., Sun, Y., Liu, Y., Zhang, X., Li, F., Li, L., & et al. (2018). The miR-31-SOX10 axis regulates tumor growth and chemotherapy resistance of melanoma via PI3K/AKT pathway. Biochemical and Biophysical Research Communications.

Zhou, J., Xu, D., Xie, H., Tang, J., Liu, R., Li, J., et al. (2015). miR-33a functions as a tumor suppressor in melanoma by targeting HIF-1α. *Cancer Biology and Therapy*, *16*(6), 846–855. https://doi.org/10.1080/15384047.2015.1030545.

Zhou, W. J., Wang, H. Y., Zhang, J., Dai, H. Y., Yao, Z. X., Zheng, Z., et al. (2020). NEAT1/miR-200b-3p/SMAD2 axis promotes progression of melanoma. *Aging*, *12*(22), 22759–22775. https://doi.org/10.18632/aging.103909.

Zhu, Y., Wen, X., & Zhao, P. (2018). Microrna-365 inhibits cell growth and promotes apoptosis in melanoma by targeting BCL2 and cyclin D1 (CCND1). *Medical Science Monitor*, *24*, 3679–3692. https://doi.org/10.12659/MSM.909633.

Chapter 27

Pathophysiology roles and translational opportunities of miRNAs in brain tumors

Vlad Moisoiu[a], George E.D. Petrescu[b], and Mihnea P. Dragomir[c]

[a]*Faculty of Physics, Babeş-Bolyai University, Cluj-Napoca, Romania,* [b]*Department of Neurosurgery, Bagdasar-Arseni Clinical Emergency Hospital, Bucharest, Romania,* [c]*Department of Pathology, Charité-Universitätsmedizin Berlin, Corporate Member of Freie Universität Berlin, Humboldt-Universität zu Berlin and Berlin Institute of Health, Berlin, Germany*

Introduction

Cancer is a genetic and epigenetic disorder resulting from a series of hijacked intracellular and intercellular molecular circuits that under normal conditions are responsible for cell survival, cell fate, and genome maintenance (Vogelstein et al., 2013).

The evolution from unicellular organisms to multicellular ones implies that the benefits of individual cells must be subjugated to the benefits of an organism as a whole. Cancer represents a reversion of this process, such that individual cells threaten the very existence of the organism to which they are a part. They do so by activating transcriptional programs that should be inaccessible under normal circumstances (Trigos, Pearson, Papenfuss, & Goode, 2018).

In addition to the reversal of a phylogenetic process, cancer cells also reverse ontogenetic processes by running a set of embryologic transcription programs that allows unrestrained growth and immortality. This is particularly true for a subset of cancer cells termed cancer stem cells that exhibit a stem-like phenotype. These cancer stem cells sustain malignant proliferation akin to the way normal stem cells sustain tissue architecture. MicroRNAs (miRNAs), which are short noncoding transcripts that regulate gene expression at the posttranscriptional level, were initially discovered as key regulators of embryogenesis. Hence, it is not surprising that they also play an important role in cancer and in the phenotypic delineation of cancer stem cells.

The brain represents the most complex differentiated tissue, being composed of hundreds of different cell types that synchronously cooperate. MiRNAs play an important role in achieving this histologic complexity by fine-tuning and reinforcing transcriptional programs that underlie the multitude of cell types. In the case of brain tumors, these miRNA-based circuits are perturbed and contribute to the malignant process either directly (i.e., at the genomic level) or indirectly (i.e., at the posttranscriptional level).

In addition to their cell intrinsic functions, miRNAs are important players in intercellular communication. Thus, miRNAs can also exert their effect on adjacent cells by being secreted in the extracellular space. In this regard, miRNAs act akin to hormones. Two populations of miRNAs exist in the extracellular space: a population of miRNAs associated with proteins (especially AGO2) and a population of miRNAs included in extracellular vesicles (especially exosomes). In addition, miRNAs can be readily exchanged between cells through intercellular junctions. In the case of cancer, the exchange of miRNAs concerns both homotypic interactions among cancer cells as well as heterotypic interactions between cancer cells and the multitude of populations that comprise the tumor microenvironment (TME), such as endothelial cells and immune cells.

This molecular "conversation" between cells through miRNAs can be detected in biofluids (e.g., cerebrospinal fluid) through a series of techniques, a possibility that has sparked immense excitement toward the development of miRNA-based liquid biopsy tools for diagnostic, prognostic, and predictive purposes as well as for the implementation of miRNA-based therapeutic strategies (Fig. 27.1). In fact, the implementation of miRNA-based liquid biopsy tools would be particularly useful in brain tumors by replacing the need for traditional biopsies.

In this chapter, we discuss the emerging landscape concerning the roles of miRNAs in brain tumors. Because of space restraints, the chapter focuses on high-grade gliomas and glioblastoma, which are the most prevalent types of brain malignancy in the adult population and the most well-studied types of brain tumors, but the field is advancing at a high pace

MicroRNA in Human Malignancies. https://doi.org/10.1016/B978-0-12-822287-4.00012-8

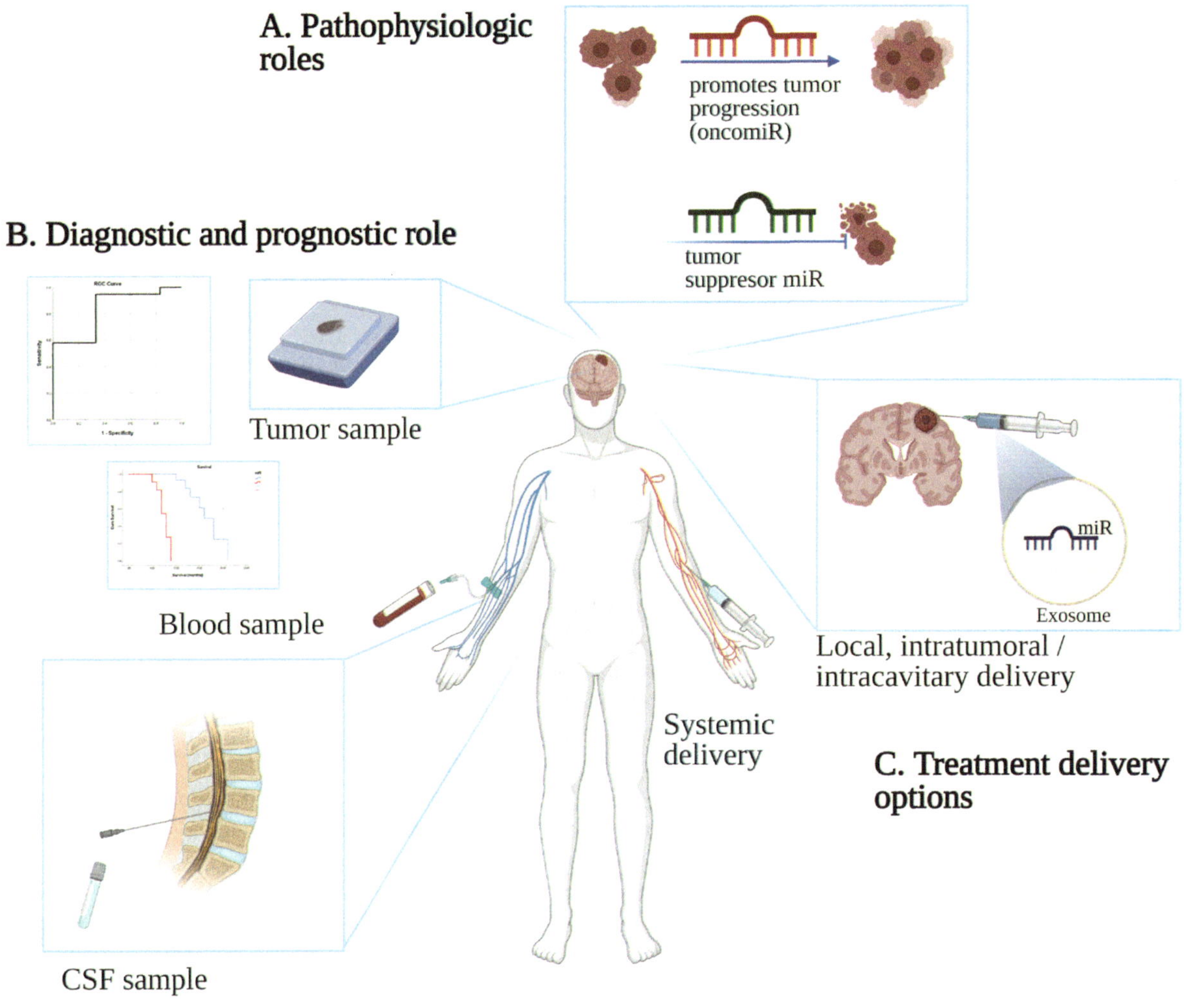

FIG. 27.1 The roles of miRNAs in brain tumors. The figure was created with biorender.com. (A) Pathophysiologic roles: miRNAs can act as oncomiRs, thus promoting tumorigenesis, or can be tumor-suppressor miRNAs and consequently prevent the oncogenic process. (B) Diagnostic and prognostic roles: miRNA expression in tumor samples obtained from surgical procedures or biopsies could be analyzed to diagnose a certain brain tumor or assess the prognoses. Moreover, liquid biopsies from blood or cerebrospinal fluid (CSF) are an even better, less-invasive, and quicker approach for diagnoses, prognoses, and recurrence assessment. (C) Treatment delivery options: miRNA therapies (inhibition of oncogenic miRNAs or restoration of tumor-suppressor miRNAs) could be achieved either by systemic delivery or through direct, local delivery, following tumor resection or through guided procedures.

across multiple tumor types. To accommodate the discussion concerning the roles of miRNAs in cancer onset and progression, we propose four layers of complexity that explain the remarkable resilience of cancer: (1) perturbed cellular circuitry; (2) tumor architecture; (3) homotypic interactions among cancer cells, and (4) heterotypic interactions between cancer cells and the surrounding TME. After discussing the roles of miRNAs in each of the four layers of complexity, we end the chapter by discussing the translational potential of miRNAs, including miRNA-based therapies and miRNA-based liquid biopsy tools for diagnostic, prognostic, and predictive purposes.

MiRNAs and perturbed cellular circuits in gliomas and glioblastomas

Perturbations in the tumor suppressors and proto-oncogenes of the cellular circuits are responsible for cancer onset and progression. Dissecting the molecular underpinning of these perturbations through international consortia such as The Cancer Genome Atlas (TCGA) and International Cancer Genome Consortium (ICGC) has been one of the most remarkable achievements of biomedical research (Campbell et al., 2020). Among the multitude of mechanisms coopted by cancer cells for securing a growth advantage, miRNAs represent one of the newest additions. Cancer cells can hijack the miRNA circuit both directly (i.e., at the genomic level) and indirectly at the posttranscriptional level.

Directly, it has been revealed that genetic events affecting miRNA loci can indeed lead to cancer. Actually, the first miRNAs ever associated with cancer are located in a frequently deleted region on chromosome 13q14, a genomic event that can by itself drive chronic lymphocytic leukemia (Calin et al., 2002). Other examples of direct involvement of miRNAs in cancer include mutations that alter the targeting capacity of miRNAs (seed region mutations) or amplifications of miRNA genes (Urbanek-Trzeciak et al., 2020). Indirectly, there is a significant body of evidence demonstrating that the network of miRNAs is profoundly perturbed in cancer cells through a multitude of mechanisms that we shall explore in this chapter, such as incorrect editing or miRNA-sponge inactivation (Van Roosbroeck et al., 2019).

The paradigm of oncogenes and tumor suppressors has been extended also to miRNAs. The number of validated miRNAs that are perturbed in brain malignancies is continuously growing, but the exact function of many of them is often unknown (Petrescu, Sabo, Torsin, Calin, & Dragomir, 2019). Several studies have shown that miRNAs can act as bona fide driver genes in the case of both gliomas and glioblastomas. Parker et al. and simultaneously Singh et al. reported that a small subset of patients with glioblastoma harbors an oncogenic recurrent tandem duplication on chromosome 4p16.3, leading to a fusion gene between fibroblast growth factor receptor 3 on the one hand and the transforming acidic coiled-coil containing protein 3 on the other hand (*FGFR3-TACC3*) (Parker et al., 2013; Singh et al., 2012). The resulting fusion transcript lacks the 3′ untranslated region (UTR) of the wild-type *FGFR3*. The 3′ UTR of FGFR3 mRNA harbors a target site for miR-99a, meaning that the fusion transcript is able to escape miR-99a downregulation. In vitro data revealed that cell proliferation and anchorage-independent cell growth are promoted by FGFR3-TACC3 through extracellular signal-regulated kinases (ERK) and signal transducer and activator of transcription 3 (STAT3) signaling while also impacting survival in vivo.

Huse et al. (2009) and Kim et al. (2010) determined that some high-grade gliomas exhibit an amplified miR-26a at the DNA level, leading to its upregulation. Downstream targets of miR-26a include known tumor suppressors such as the phosphatase and tensin homolog (PTEN) and retinoblastoma 1 (RB1). The TCGA data showed that miR-26a upregulation is mutually exclusive with the homozygous loss of PTEN, suggesting a redundancy in their function. However, miR-26a upregulation was often associated with the monoallelic loss of PTEN. This result suggests that the amplification of miR-26a silences residual PTEN expression in $PTEN^{+/}$ tumors, acting akin to a genomic loss of heterozygosity event. In support of this interpretation, the authors showed in mice harboring $PTEN^{+/}$ tumors that the forced expression of miR-26a prevented the cancer cells from losing the remaining PTEN allele while in the absence of miR-26a, glioblastoma cells lost heterozygosity for gaining an additional growth advantage. A similar role of miR-26a was also proved for RB1 by Kim et al. (2010).

While direct genomic events affecting miRNAs have been described in few patients, most of the time miRNAs contribute indirectly to the malignant process, being coopted by driving genetic events to propagate their effects along the entire cellular circuitry. Sumazin et al. performed an in-depth network analysis of RNA-RNA interactions in glioblastoma and uncovered a mechanism for achieving PTEN downregulation based on miRNA sponge deletion (Sumazin et al., 2011). The authors shed light on a series of 13 transcripts, including RB1 and hypoxia-inducible factor 1α (HIF1α), that buffer the miRNA-based downregulation of PTEN by exhibiting the same miRNA targeting sequences as PTEN, thus acting as a sort of endogenous miRNA sponge. Consequently, the deletion of these miRNA sponges increases the number of miRNAs available for targeting PTEN, leading to its downregulation, a mechanism that seems to operate in around 80% of glioblastoma cases that are PTEN proficient. In addition to this miRNA-sponge deletion mechanism, the authors also unveiled two nonsponge modulators of PTEN, namely the partner and localizer of BRCA2 (PALB2) and Wnt family member 7A (WNT7A), which drive PTEN downregulation by perturbing the miRNA-mediated silencing process (Sumazin et al., 2011).

Another miRNA-based mechanism of disrupting the intracellular circuitry in cancer is through perturbed posttranscriptional editing. Choudhury et al. explored a cancer-promoting mechanism in glioblastoma based on the attenuated adenosine-to-inosine editing of miR-376a-5p (previously called miR-376a*) (Choudhury et al., 2012). Adenosine deaminases acting on RNA (ADAR) are a class of enzymes that bind double-stranded RNA, converting adenosine to inosine by deamination. During translation, inosine is read as guanosine such that ADAR acts as a type of posttranscriptional editing. In the normal brain, there is a need for an almost 100% editing of a certain position in the AMPA-type glutamate receptor B subunit that regulates Ca^{2+} inflow. A series of studies demonstrated that ADAR downregulation has a prooncogenic role in glioblastoma, and that it is associated with decreased survival (Zhang et al., 2018). Choudhury et al. showed that miR-376a-5p, which is a target of ADAR1, is underedited in human glioblastoma samples. Through a series of elegant experiments, they went on to show that the downstream target of unedited miR-376a is Ras-related protein A (*RAP2A*), which is a member of the Rous avian sarcoma (RAS) oncogene family, while the downstream target of edited miR-376a-5p is the autocrine motility factor receptor (*AMFR*). The effect of attenuated adenosine-to-inosine editing in glioma is twofold: increased inhibition of the migration-restraining effect of RAP2A because of the build-up of unedited miR-376a-5p, paired with enhanced cell migration by failure in blocking AMFR because of the decrease of edited miR-376a-5p levels (Choudhury et al., 2012).

Another example of perturbed miRNA editing operating in glioblastoma is the underediting of miR-589-3p by ADAR2. This aberrant process leads to increased cell proliferation via protocadherin 2 (PCDH2) downregulation and increased invasion and migration via ADAM metallopeptidase domain 12 (ADAM12) upregulation. Moreover, glioblastoma is characterized also by the underediting of miR-221/222 and miR-21 by ADAR2, which increases cell proliferation by downregulating Cyclin-dependent kinase inhibitor 1B (p27Kip1) (see Torsin et al., 2021 for review).

Jiang et al. demonstrated how the oncomiRNA miR-30e-3p (previously called miR-30e*) promotes glioma cell invasiveness by disrupting the negative feedback loop between nuclear factor κB (NF-κB) and the inhibitor of NF-κB (IκBα) (Jiang et al., 2012). The overactivation of NF-κB is featured by a multitude of cancer types, including gliomas. Under normal circumstances, NF-κB signaling is regulated tightly through a negative feedback loop: NF-κB is restrained in the cytoplasm by IκBs, which prevents its translocation to the nucleus. Triggering NF-κB signaling requires IκB to be degraded via the ubiquitine/proteasome pathway, enabling the translocation of NF-κB in the nucleus where it acts as a transcription factor. Importantly, IκB is among the genes upregulated by NF-κB, which then exerts its inhibiting effect on NF-κB signaling, thus closing the negative feedback loop. Using in vitro data, the authors showed that miR-30e-3p suppresses the expression of IκBα by directly targeting the IκBα 3′-UTR. In primary human glioma cells, miR-30e-3p is upregulated and is associated with poor survival, providing evidence that miR-30e-3p acts as an oncomiRNA in glioma. In a subsequent study, the same group of authors showed that besides miR-30e-3p, miR-182 is another oncomiRNA upregulated in gliomas that can overactivate the NF-κB pathway (Song et al., 2012). Thus, the authors showed that miR-182 expression was increased in a subtype of gliomas with relatively lower miR-30e-3p expression, and that it drives NF-κB signaling by targeting and inhibiting the members of the deubiquitinase pathway responsible for inhibiting IκBα degradation.

These examples, which are far from being exhaustive, underscore the multitude of mechanisms by which miRNAs can contribute to the gliomagenesis process and the great unknown that still lies ahead.

miRNAs and the glioma architecture

The advent of single cell technologies has demonstrated beyond any doubt that cancer is comprised of a multitude of populations of cells with distinct roles, an architecture that often mimics the hierarchy of the tissue from which the cancer emerged; this phenomenon is termed developmental coherence (Suvà & Tirosh, 2020). Arguably, the most important cells in a tumor are cancer stem cells, which are defined based on their ability to recapitulate tumors in vivo and the expression of a series of stem cell transcription factors such as SRY-Box transcription factor 2 (SOX2), octamer-binding transcription factor 4 (OCT4), NANOG, oligodendrocyte transcription factor (OLIG2), MYC, BES1-interacting Myc-like protein 1 (BMI1), and NESTIN as well as cell surface markers such as CD24, CD44, CD133, and epidermal growth factor receptor (EGFR). The cancer stem cell phenotype is dependent on a series of cell intrinsic genetic, epigenetic, and metabolic factors as well as cell extrinsic factors (e.g., hypoxia).

In the case of isocitrate dehydrogenase (IDH)-mutant gliomas and histone 3 K27M (H3K27M)-mutated gliomas, the hierarchy of cells is reigned by a population of rapidly dividing cells with a stem-like phenotype that is radio- and chemotherapy-resistant and that sustains proliferation of the whole tumor by giving rise to two populations of cells with astrocyte and oligodendrocyte-like phenotypes (Suvà & Tirosh, 2020). In contrast, IDH-wild type glioblastoma is characterized by four populations of cells, namely neural progenitor-like cells, oligodendrocyte progenitor-like cells, astrocyte-like cells, and mesenchymal-like cells, of which all, except astrocyte-like cells, present stem features (Suvà & Tirosh, 2020). The existence of more than one population with stem-like features is termed stem cell multiplicity. Each population of cells is preferentially expressing a specific stem cell marker, with CD24 being highest in neural progenitor-like cells, CD133 in oligodendrocyte progenitor-like cells, EGFR in astrocyte-like cells, and CD44 in mesenchymal-like cells. NESTIN, which is a transcription marker known to orchestrate stemness in the brain, seems to be specifically expressed in astrocyte-like cells.

The execution of stem cell programs relies also on the coordinated expression of a series of miRNAs, a phenomenon that is also true in the case of glioma stem cells. There is a paradigm in glioma research that hypoxia is an important cell extrinsic element that reprograms cancer cells toward a stem cell phenotype, a process that is mediated by hypoxia-inducible factor (HIF), among other mechanisms (Kulshreshtha et al., 2007). Existing data imply that a network of perturbed miRNAs plays a role in the phenotypic delineation of cancer stem cells, a process that is enhanced by the hypoxic niche.

A study by Mathew et al. revealed how miR-218 is a tumor-suppressor miRNA linking hypoxia with stemness in TCGA mesenchymal-type glioblastoma (Mathew et al., 2014). Thus, the authors showed how downregulated miR-218 fails to restrain several receptor tyrosine kinases (RTKs), which in turn trigger the expression of HIF-related signaling. TCGA data showed that miR-218 is downregulated in mesenchymal-type glioblastoma. Forced expression of miR-218 leads to temozolomide sensitivity in several cell lines and the upregulation of several RTKs (EGFR, phospholipase C γ1—PLCγ1, phosphatidylinositol 3-kinase—PIK3C2A, serine/threonine-protein kinase A-ARAF). It has been previously shown that multiple components of RTK signaling promote HIF expression, which in turn influences tumor angiogenesis and response to therapy. To demonstrate the link between miR-218 and HIF signaling, the authors showed that stably expressed miR-218 in glioblastoma cell lines inhibited the transcription of HIF2α-regulated genes (e.g., transforming growth factor α (TGF-α) and vasculo-endothelial growth factor (VEGF)). In contrast, HIF1α-regulated genes were not inhibited by miR-218 treatment. By analyzing the promoter region of HIF2α, the authors validated a promoter for jun protooncogene (c-JUN), which is a transcription factor part of the RTK signaling. To validate their findings in human samples, the authors analyzed expression data from publicly available databases and showed that mesenchymal glioblastomas with low miR-218 levels are associated with an elevated expression of an HIF metagene, a relation that was not seen in TCGA proneural-type glioblastoma. Thus, these results suggest that the miR-218-RTK-HIF2α signaling pathway operates preferentially in mesenchymal glioblastoma.

In nontransformed cells, pluripotency is equivalent to the expression of four transcription factors, known as Yamanaka factors (SOX2, OCT4, KLF3, MYC), that are also expressed in cancer stem cells. The downstream signaling of two Yamanaka stem cell factors (SOX2 and OCT4) was described by Lopez-Bertoni et al. in the case of glioblastoma, showing that miRNAs are key components of this circuit. Coexpressing transgenic OCT4 and SOX2 in glioblastoma cell lines results in enhanced neurosphere formation, resistance to radiotherapy, and induction of stem cell markers BMI1 and OLIG2. Searching for miRNAs that might mediate this phenotype, the authors identified the downregulation of miR-148a and miR-296-5p and the upregulation of miR-486-5p as putative culprits. Focusing on miR-296-5p, they explored the mechanism by which SOX2 and OCT4 induced miR-296-5p downregulation, which orchestrates the stem cell phenotype (Lopez-Bertoni et al., 2016). The authors found that SOX2 drives miR-296-5p downregulation by DNMT-dependent promoter CpG methylation, with DNMTA1 exhibiting a SOX2 binding site in its promoter. Bisulfite sequencing revealed that the CpG methylation promoter of miR-296-5p increased from 50% to 83% after the forced expression of SOX2 and OCT4. The forced expression of miR-296-5p resulted in the downregulation of several stem cell markers, including SOX2, while the reverse was true when miR-296-5p expression was inhibited using an miR-296-5p sponge. The fact that the inhibition of the same miRNA also led to the upregulation of SOX2 was extremely interesting, as this hints toward a closed loop signaling driving the glioblastoma stemness. Searching for the mechanisms by which downregulated miR-296-5p upregulates SOX2, the authors showed that miR-296-5p targets the 3′ UTR of high mobility group A1 (HMGA1), which is a known modulator of SOX2 expression. Thus, HMGA1 modulates the chromatin architecture of the *SOX2* promoter by displacing histone H1, thus contributing to SOX2 expression (Shah et al., 2012). The SOX2 promoter contains two regions with putative HMGA1 binding sites. Using chromatin immunoprecipitation, the authors showed that the forced expression of *HMGA1* increased its binding to the SOX2 promoter at the expense of histone H1. In agreement with this result, the forced expression of miR-296-5p, which directly targets HMGA1, leads to a decrease in HMGA1 binding to the SOX2 promoter while exhibiting increased H1 binding to the SOX2 promoter. In conclusion, these results show that there is an extensive miR-296-5p-dependent circuit operating in glioblastoma by which SOX2 drives its own expression, thus sustaining and reinforcing stemness.

miRNAs and homotypic interactions between glioma cells

Besides the cellular complexity of cancer cells, there is an additional layer of complexity represented by a network of perturbed intercellular communication. This takes place through a multitude of mechanisms, including through soluble factors, intercellular junctions (e.g., gap junctions), and extracellular vesicles (e.g., exosomes), with miRNAs having the capacity to contribute to all these processes. Thus, channels of miRNA exchange include secretion of miRNAs followed by endocytosis in the recipient cells or the transfer of miRNAs via exosomes and other extracellular vesicles. In addition, miRNAs can be exchanged directly between adjacent cells via gap junctions.

The perturbed circuitry of cells regards different populations of cancer cells (the so-called homotypic interactions) as well as a complex series of heterotypic interactions between cancer cells and nonmalignant cells that comprises the TME, such as immune cells, neurons, and cells that are part of the tumor vasculature and lymphatics (Dragomir, Kopetz, Ajani, & Calin, 2020).

Underscoring the importance of homotypic intercellular communication, it has been revealed that astrocytic tumors (including glioblastoma) exhibit a network of ultralong membrane protrusions that allows the interconnection of cells over long distances via gap junctions (Osswald et al., 2015). Network-connected glioblastoma cells exhibit increased stem-like behavior and are resistant to radiotherapy (Xie et al., 2020). By exploring the permeability of gap junctions, Peng et al. showed that gap junctions formed by Connexin 43 (CX43) (but not CX32 or 37) are permeable to miRNAs, a finding that has important implications for therapy using miRNAs (Peng et al., 2019). Thus, the transfer of exogenous miRNAs tagged with a fluorescent dye was abrogated upon CX43 knockdown while the upregulation of CX43 had an opposing effect. To demonstrate the clinical implications of their findings, the authors employed miR-34a, which is a well-known tumor-suppressor gene, as a proof of concept for miRNA therapy. The results showed that the antitumorigenic effect of miR-34a on glioblastoma cells was dependent on CX43 while the overexpression of CX43 led to enhanced sensitivity to miR-34a, allowing the diffusion of the tumor-suppressor miRNA from one cell to another. The possibility of using gap junctions for expanding the therapeutic effects of miRNAs has also been explored by Zhang et al. in the case of miR-124-3p, which is another tumor-suppressor miRNA with therapeutic potential (Suzhi et al., 2015). The authors showed by using coculturing and patch clamp assays that there was specific transmission of miR-124-3p through gap junctions between adjacent glioblastoma cells. This finding was also verified in vivo and expanded the therapeutic potential of miR-124-3p.

In addition to miRNA exchange via gap junctions, communication between glioblastoma cells through exosomal miRNAs also occurs. Bronisz et al. demonstrated that glioblastoma-derived extracellular vesicles promote invasion and proliferation of adjacent cells (Bronisz et al., 2014). In parallel, the authors showed that miR-1 is downregulated in glioblastoma cell lines and in human glioblastoma samples, suggesting that this miRNA could work as a tumor suppressor. Overexpressing miR-1 abrogated the protumorigenic effects of extracellular vesicles, a finding that suggests that miR-1 might have therapeutic potential. In vitro data showed that the downstream target of miR-1 is Annexin 2 (ANXA2), which is a known oncogenic driver that induces proliferation, invasion, and angiogenesis. Concordantly with the in vitro data, TCGA data and western blot analyses of human glioblastoma samples showed that the high expression of *ANXA2* is significantly associated with better outcomes in glioblastoma and is inversely correlated with miR-1 levels. These results suggest that the miR-1/ANXA2 axis is an important antitumorigenic mechanism that is downregulated in glioblastoma and that the exchange of miRNA-containing extracellular vesicles might aid in the use of miR-1 as a therapeutic agent.

A study by Yin et al. implicated the exosomal transfer of miRNAs as a mechanism to disseminate resistance to temozolomide in glioblastoma (Yin et al., 2019). MiR-1238 levels were higher in secondary glioblastoma samples compared to primary glioblastoma, a finding that could be replicated in cell lines in which resistance to temozolomide was induced by gradual exposure. In addition, higher levels of miR-1238 were also found in the sera of patients with secondary glioblastoma compared to patients with primary glioblastoma, a finding that suggests that miR-1238 might have predictive value. Exosomes from temozolomide-resistant cells were enriched for miR-1238 and they induced resistance in temozolomide-sensitive recipient cells. To gain insights concerning the mechanism by which miR-1238 upregulation confers resistance to temozolomide, the authors performed an in silico screening for putative targets, which identified Caveolin 1 (*CAV1*) as the most probable downstream target of miR-1238. CAV1 is the most prominent member of the Caveolin family and was shown to play a protumorigenic role in glioblastoma by stabilizing EGFR in an inactive conformation that prevents autophosphorylation and subsequent activation of EGFR. The authors showed that by targeting CAV1 and blocking its inhibitory effect on EGFR, miR-1238 results in the activation of several pathways downstream of EGFR, including phosphoinositide 3-kinase (PI3K), protein kinase B (PKB), and mammalian target of rapamycin (mTOR). Finally, the authors were able to replicate miR-1238-induced resistance to temozolomide in nude mice by orthotopically injecting glioblastoma cells that were transfected with a lentivirus carrying the miR-1238 inhibitor or that were exposed to exosomes from temozolomide-resistant cells. These results demonstrate that homotypic interactions among cancer cells rely on the exchange of miRNAs via gap junctions and exosomes, information with important potential clinical implications.

miRNAs and heterotypic interactions of gliomas and glioblastoma

The evidence demonstrating miRNA exchange between cancer cells and the surrounding TME is considerable. In the case of gliomas and glioblastomas, the exchange of miRNAs directly through extracellular vesicles or via intercellular junctions has been described in the case of astrocytes, oligodendrocytes, microglial cells, neurons, endothelial cells, tumor-associated macrophages and monocytes, mesenchymal cells, and lymphocytes (reviewed in Buruiană et al., 2020). The mechanisms that govern the multitude of heterotypic interactions that sustain the malignant growth are poorly understood. Here, we discuss some studies concerning the interaction between glioma cells and macrophages, and between glioblastoma cells and endothelial cells. Although far from being exhaustive, these studies are instructive as examples of heterotypic interactions in gliomas and glioblastomas.

Gliomas and glioblastomas are paradigmatic for lymphocyte depletion and the enrichment of immunosuppressive, mostly bone-marrow-derived macrophages. Qian et al. demonstrated that hypoxic glioma-derived exosomes (H-GDEs) promote the M2 (tumorigenic) polarization of macrophages, compared to normoxic glioma-derived exosomes (GDEs) (Qian et al., 2020). Moreover, macrophages exposed to H-GDEs increased the proliferation, migration, and invasion of glioma cells in vitro. The authors also demonstrated that by implanting glioma cells and macrophages with H-GDEs in mice, these macrophages had higher CD163 levels (M2 marker) and the gliomas were more aggressive, compared to macrophages treated with GDE. Through miRNA sequencing, Qian et al. found that miR-1246 expression levels were increased in H-GDEs and that macrophages exposed to H-GDEs also had higher levels of miR-1246 compared to those exposed to GDE or PBS, suggesting that GDEs could pass on miR-1246 to macrophages. Next, by investigating the level of miR-1246 in cerebrospinal fluid from glioma patients, they observed that in the preoperative setting, GBM patients exhibit higher levels compared to low-grade gliomas, and that the level decreased following tumor resection, proving that miR-1246 might be a relevant marker for diagnosis and prognosis. By using miR-1246 mimics, they increased the levels of M2 markers in macrophages while miR-1246 inhibitors had an antagonistic consequence. Furthermore, the authors demonstrated that macrophages with an increased miR-1246 expression had greater proliferation, migratory, and invasion capacities. The authors determined that telomeric repeat-binding factor 2-interacting protein 1 (*TERF2IP*) was downregulated by high levels of miR-1246 and upregulated by low levels, validating *TERF2IP* as a target of miR-1246. Finally, the authors determined that H-GDEs stimulated the expression of STAT3 and decreased that of NF-κB. They also determined that increasing the miR-1246 levels or inhibiting *TERF2IP* had the same results, implying that H-GDEs with miR-1246 promote the M2 phenotype of macrophages by inhibiting TERF2IP and subsequently activating the STAT3 and suppressing the NF-κB pathways.

The exchange of miRNAs between glioblastoma and endothelial cells occurs through both gap junctions and exosomes. Thuringer et al. showed that these two types of cells form functional gap junctions that allow the transfer of antitumorigenic miR-145 from endothelial cells but also the transfer of proangiogenic miR-5096 from glioblastoma cells to the endothelial cells (Thuringer et al., 2016). The authors first showed that, when cultured separately, miR-145 was mostly expressed in endothelial cells while miR-5096 was mostly expressed in a glioblastoma cell line. However, after coculturing glioblastoma and endothelial cells, significant levels of miR-145 and miR-5096 could be detected in both cell types, suggesting that a reciprocal transfer of these two miRNAs occurred. However, this transfer of miRNAs did not occur in transwell conditions, a finding that pointed toward direct transfer via gap junctions as the most likely transfer mechanism. The authors confirmed this was the case because no transfer of miRNAs occurred when glioblastoma and endothelial cells were cocultured in the presence of carbenoxolone, an inhibitor of gap junction intercellular communication. The same results were observed when CX43 was knocked down in glioblastoma cells. Using an in vitro tube formation assay, the authors showed that upregulating miR-5096 in glioblastoma cells enhanced angiogenesis while the opposite was true when miR-5096 was inhibited in glioblastoma cells. However, this assay also showed that expressing miR-145 in endothelial cells resulted in the downregulation of angiogenesis. Taken together, these results suggest that the balance between miR-5096 and miR-145 regulates angiogenesis in endothelial cells and that glioblastoma cells seek to tilt this balance toward enhanced angiogenesis by transferring miR-5096 to endothelial cells via gap junctions. Interestingly, the VEGF expression was not influenced by miR-5096, showing that the proangiogenic effect of miR-5096 occurs through another, yet unknown mechanism.

The exosomal transfer of miRNAs between glioblastoma stem cells and endothelial cells was demonstrated by Sun et al. (2017). The authors sorted glioblastoma stem cells using CD133 as a marker of stemness. In a series of experiments consisting of the incubation of human brain endothelial cells with exosomes from glioblastoma stem cells transfected with miR-21 mimics and scramble miRNA, the authors showed that exosomes were enriched in miR-21 and able to upregulate the VEGF/VEGFR axis in endothelial cells, leading to enhanced migration and tube-forming capabilities.

Translating brain miRNAs in the clinical setting

Given the ease with which miRNAs can be assayed, there has been a surge of interest toward developing miRNA-based diagnostic, predictive, and prognostic markers. As of March 2021, there are seven trials registered on clinicaltrials.gov exploring miRNA-based liquid biopsy strategies under various phases of development. A compendium of potential miRNAs for application in glioma and glioblastoma invasive/noninvasive diagnoses and prognoses is illustrated in Table 27.1 (Akers et al., 2013; Cui et al., 2019; Ohno et al., 2019; Srinivasan, Patric, & Somasundaram, 2011; Wang et al., 2019).

The findings presented in this chapter clearly demonstrate that miRNAs have the potential of becoming therapeutic adjuvants for glioblastomas. Most probably, miRNA therapy will not be used solely but in combination with already approved adjuvant therapies. Mouse model studies have shown that miRNA inhibitors (for overexpressed oncomiRNAs) and miRNA mimetics (for downregulated tumor-suppressor miRNAs) can inhibit gliomagenesis. MiRNA mimetics are double-stranded RNA duplexes that after delivery are incorporated in the RISC and guide it toward the target mRNA, the template of an oncogenic protein. MiRNA inhibitors are more diverse, including anti-miRNA oligonucleotides, locked nucleic acid, antagomirs, miRNA sponges (including artificial circRNAs), miRNA masks, and small molecule inhibitors of miRNAs (Dragomir, Moisoiu et al., 2020). In vivo studies proved that these molecules are effective but great challenges remain regarding the mechanisms to deliver them across the blood-brain barrier and the collateral effect on adjacent healthy cells. Several new delivery methods were recently developed that could overcome these limitations; for more detail please refer to

TABLE 27.1 Representative examples of diagnostic and prognostic miRNAs.

Cancer stage	miRNA	Source	Number of patients	Method of analysis	Time point of analysis	Diagnostic	Prognostic	Reference
GBM	miR-4516	FFPE	268	nCounter human miRNA v3a	postop.	No	Increased levels are associated with poor prognosis; HR (95% CI) = 1.49 (1.12–1.99)	Cui et al. (2019)
GBM	10 miRNAs signature (miR-20a, miR-106a, miR-17-5p, miR-31, miR-222, miR-148a, miR-221, miR-146b, miR-200b, miR-193a)	FF tissue	222	Agilent 8x15K human miRNA microarray	postop.	No	Modified miRNA signature is associated with poor prognosis; HR (95% CI) = 2.0 (1.4–2.8)	Srinivasan et al. (2011)
Diffuse glioma	Model of three miRNAs (miR-4763-3p, miR-1915-3p, miR-3679-5p)	Serum	157	Microarray	postop.	Sn (95% CI) = 96% (92%–100%); Sp (95% CI) = 97% (95%–97%) AUC (95% CI) = 0.99 (0.99–1.00)	No	Ohno et al. (2019)
GBM	miR-21	CSF	13 GBM vs. 14 noncancer controls followed by 14 GBM vs. 15 noncancer controls	qRT-PCR	periop.	Sn = 86% Sp = 93% AUC = 0.91	No	Akers et al. (2013)
Glioma	miR-214	Serum	100 (38 LGG, 62 HGG)	qRT-PCR	preop.	Sn = 90%, Sp = 71% AUC (95% CI) = 0.885 (0.833–0.926)	Poor prognosis for increased levels of miRNA-214 (log-rank test = 25.699, $p < 0.001$)	Wang et al. (2019)

AUC = area under the curve; CSF = cerebrospinal fluid; FF = fresh-frozen; FFPE = formalin-fixed paraffin-embeded; GBM = glioblastoma; Sn = sensitivity; Sp = specificity.

the review of Petrescu et al. regarding this topic (Petrescu et al., 2019). Additionally, it is important to mention that for multiple instances, it is necessary to target several miRNAs to restore the hijacked miRNA regulatory network. This element needs to be taken into consideration when developing miRNA therapies for gliomas. Of note, as a much less immunogenic environment, the brain is very attractive for testing these new therapies, and this can cause extreme immunologic reactions.

Regarding the diagnostic and prognostic implementation of miRNAs, several points need to be raised. First, there is an increased need to standardize the technologies used for the extraction and detection of miRNAs from tissues and/or bodily fluids. Kopkova et al. showed that miRNA levels from CSF samples from GBM or control patients differ based on the RNA extraction kit used (Kopkova et al., 2018). Moreover, the authors suggest that RNA-Seq and digital PCR are the most appropriate methods for evaluating miRNA expression from cerebrospinal fluid samples. Second, in the case of the development of diagnostic and prognostic tools, probably the most reliable ones will include multiple miRNA molecules, which will reflect the deregulated miRNA network. Third, it would be very interesting to analyze those miRNAs that reflect the local communication between tumor cells and the TME, which would add important details regarding the tumor evolution. Most probably, this will be possible using imaging techniques, currently under development, that can quantify miRNAs in their spatial context. Such technologies will add a new layer of complexity that will provide a better understanding of the mechanism through which cancer cells use miRNAs to hijack the extracellular regulatory mechanism.

In conclusion, we believe that the last 20 years of miRNA research in brain tumors has offered the foundation for developing new diagnostic and therapeutic procedures that, in the near future, can be translated into clinical practice.

References

Akers, J. C., Ramakrishnan, V., Kim, R., Skog, J., Nakano, I., Pingle, S., et al. (2013). miR-21 in the extracellular vesicles (EVs) of cerebrospinal fluid (CSF): A platform for glioblastoma biomarker development. *PLoS One*, *8*(10). https://doi.org/10.1371/journal.pone.0078115, e78115.

Bronisz, A., Wang, Y., Nowicki, M. O., Peruzzi, P., Ansari, K. I., Ogawa, D., et al. (2014). Extracellular vesicles modulate the glioblastoma microenvironment via a tumor suppression signaling network directed by miR-1. *Cancer Research*, *74*(3), 738. https://doi.org/10.1158/0008-5472.CAN-13-2650.

Buruiană, A., Florian, Ș. I., Florian, A. I., Timiș, T.-L., Mihu, C. M., Miclăuș, M., et al. (2020). The roles of miRNA in glioblastoma tumor cell communication: Diplomatic and aggressive negotiations. *International Journal of Molecular Sciences*, *21*(6). https://doi.org/10.3390/ijms21061950.

Calin, G. A., Dumitru, C. D., Shimizu, M., Bichi, R., Zupo, S., Noch, E., et al. (2002). Frequent deletions and down-regulation of micro- RNA genes miR15 and miR16 at 13q14 in chronic lymphocytic leukemia. *Proceedings of the National Academy of Sciences of the United States of America*, *99*(24), 15524. https://doi.org/10.1073/pnas.242606799.

Campbell, P. J., Getz, G., Korbel, J. O., Stuart, J. M., Jennings, J. L., Stein, L. D., et al. (2020). Pan-cancer analysis of whole genomes. *Nature*, *578*(7793), 82–93. https://doi.org/10.1038/s41586-020-1969-6.

Choudhury, Y., Tay, F. C., Lam, D. H., Sandanaraj, E., Tang, C., Ang, B.-T., et al. (2012). Attenuated adenosine-to-inosine editing of microRNA-376a* promotes invasiveness of glioblastoma cells. *The Journal of Clinical Investigation*, *122*(11), 4059–4076. https://doi.org/10.1172/JCI62925.

Cui, T., Bell, E. H., McElroy, J., Becker, A. P., Gulati, P. M., Geurts, M., et al. (2019). miR-4516 predicts poor prognosis and functions as a novel oncogene via targeting PTPN14 in human glioblastoma. *Oncogene*, *38*(16), 2923–2936. https://doi.org/10.1038/s41388-018-0601-9.

Dragomir, M. P., Kopetz, S., Ajani, J. A., & Calin, G. A. (2020). Non-coding RNAs in GI cancers: From cancer hallmarks to clinical utility. *Gut*, *69*(4), 748. https://doi.org/10.1136/gutjnl-2019-318279.

Dragomir, M. P., Moisoiu, V., Manaila, R., Pardini, B., Knutsen, E., Anfossi, S., et al. (2020). A holistic perspective: Exosomes shuttle between nerves and immune cells in the tumor microenvironment. *Journal of Clinical Medicine*, *9*(11). https://doi.org/10.3390/jcm9113529.

Huse, J. T., Brennan, C., Hambardzumyan, D., Wee, B., Pena, J., Rouhanifard, S. H., et al. (2009). The PTEN-regulating microRNA miR-26a is amplified in high-grade glioma and facilitates gliomagenesis in vivo. *Genes & Development*, *23*(11), 1327–1337. http://genesdev.cshlp.org/content/23/11/1327.abstract.

Jiang, L., Lin, C., Song, L., Wu, J., Chen, B., Ying, Z., et al. (2012). MicroRNA-30e* promotes human glioma cell invasiveness in an orthotopic xenotransplantation model by disrupting the NF-κB/IκBα negative feedback loop. *The Journal of Clinical Investigation*, *122*(1), 33–47. https://doi.org/10.1172/JCI58849.

Kim, H., Huang, W., Jiang, X., Pennicooke, B., Park, P. J., & Johnson, M. D. (2010). Integrative genome analysis reveals an oncomir/oncogene cluster regulating glioblastoma survivorship. *Proceedings of the National Academy of Sciences of the United States of America*, *107*(5), 2183. https://doi.org/10.1073/pnas.0909896107.

Kopkova, A., Sana, J., Fadrus, P., Machackova, T., Vecera, M., Vybihal, V., et al. (2018). MicroRNA isolation and quantification in cerebrospinal fluid: A comparative methodical study. *PLoS One*, *13*(12). https://doi.org/10.1371/journal.pone.0208580, e0208580.

Kulshreshtha, R., Ferracin, M., Wojcik, S. E., Garzon, R., Alder, H., Agosto-Perez, F. J., et al. (2007). A MicroRNA signature of hypoxia. *Molecular and Cellular Biology*, *27*(5), 1859. https://doi.org/10.1128/MCB.01395-06.

Lopez-Bertoni, H., Lal, B., Michelson, N., Guerrero-Cázares, H., Quiñones-Hinojosa, A., Li, Y., et al. (2016). Epigenetic modulation of a miR-296-5p:HMGA1 axis regulates Sox2 expression and glioblastoma stem cells. *Oncogene*, *35*(37), 4903–4913. https://doi.org/10.1038/onc.2016.22.

Mathew, L. K., Skuli, N., Mucaj, V., Lee, S. S., Zinn, P. O., Sathyan, P., et al. (2014). miR-218 opposes a critical RTK-HIF pathway in mesenchymal glioblastoma. *Proceedings of the National Academy of Sciences of the United States of America*, *111*(1), 291. https://doi.org/10.1073/pnas.1314341111.

Ohno, M., Matsuzaki, J., Kawauchi, J., Aoki, Y., Miura, J., Takizawa, S., et al. (2019). Assessment of the diagnostic utility of serum MicroRNA classification in patients with diffuse glioma. *JAMA Network Open*, *2*(12), e1916953. https://doi.org/10.1001/jamanetworkopen.2019.16953.

Osswald, M., Jung, E., Sahm, F., Solecki, G., Venkataramani, V., Blaes, J., et al. (2015). Brain tumour cells interconnect to a functional and resistant network. *Nature*, *528*(7580), 93–98. https://doi.org/10.1038/nature16071.

Parker, B. C., Annala, M. J., Cogdell, D. E., Granberg, K. J., Sun, Y., Ji, P., et al. (2013). The tumorigenic FGFR3-TACC3 gene fusion escapes miR-99a regulation in glioblastoma. *The Journal of Clinical Investigation*, *123*(2), 855–865. https://doi.org/10.1172/JCI67144.

Peng, Y., Wang, X., Guo, Y., Peng, F., Zheng, N., He, B., et al. (2019). Pattern of cell-to-cell transfer of microRNA by gap junction and its effect on the proliferation of glioma cells. *Cancer Science*, *110*(6), 1947–1958. https://doi.org/10.1111/cas.14029.

Petrescu, G. E. D., Sabo, A. A., Torsin, L. I., Calin, G. A., & Dragomir, M. P. (2019). MicroRNA based theranostics for brain cancer: Basic principles. *Journal of Experimental & Clinical Cancer Research*, *38*(1), 231. https://doi.org/10.1186/s13046-019-1180-5.

Qian, M., Wang, S., Guo, X., Wang, J., Zhang, Z., Qiu, W., et al. (2020). Hypoxic glioma-derived exosomes deliver microRNA-1246 to induce M2 macrophage polarization by targeting TERF2IP via the STAT3 and NF-κB pathways. *Oncogene*, *39*(2), 428–442. https://doi.org/10.1038/s41388-019-0996-y.

Shah, S. N., Kerr, C., Cope, L., Zambidis, E., Liu, C., Hillion, J., et al. (2012). HMGA1 reprograms somatic cells into pluripotent stem cells by inducing stem cell transcriptional networks. *PLoS One*, *7*(11). https://doi.org/10.1371/journal.pone.0048533, e48533.

Singh, D., Chan, J. M., Zoppoli, P., Niola, F., Sullivan, R., Castano, A., et al. (2012). Transforming fusions of FGFR and TACC genes in human glioblastoma. *Science*, *337*(6099), 1231. https://doi.org/10.1126/science.1220834.

Song, L., Liu, L., Wu, Z., Li, Y., Ying, Z., Lin, C., et al. (2012). TGF-β induces miR-182 to sustain NF-κB activation in glioma subsets. *The Journal of Clinical Investigation*, *122*(10), 3563–3578. https://doi.org/10.1172/JCI62339.

Srinivasan, S., Patric, I. R. P., & Somasundaram, K. (2011). A ten-microRNA expression signature predicts survival in glioblastoma. *PLoS One*, *6*(3). https://doi.org/10.1371/journal.pone.0017438, e17438.

Sumazin, P., Yang, X., Chiu, H.-S., Chung, W.-J., Iyer, A., Llobet-Navas, D., et al. (2011). An extensive MicroRNA-mediated network of RNA-RNA interactions regulates established oncogenic pathways in glioblastoma. *Cell*, *147*(2), 370–381. https://doi.org/10.1016/j.cell.2011.09.041.

Sun, X., Ma, X., Wang, J., Zhao, Y., Wang, Y., Bihl, J. C., et al. (2017). Glioma stem cells-derived exosomes promote the angiogenic ability of endothelial cells through miR-21/VEGF signal. *Oncotarget*, *8*(22). https://www.oncotarget.com/article/16661/text/.

Suvà, M. L., & Tirosh, I. (2020). The glioma stem cell model in the era of single-cell genomics. *Cancer Cell*, *37*(5), 630–636. https://doi.org/10.1016/j.ccell.2020.04.001.

Suzhi, Z., Liang, T., Yuexia, P., Lucy, L., Xiaoting, H., Yuan, Z., et al. (2015). Gap junctions enhance the Antiproliferative effect of MicroRNA-124-3p in glioblastoma cells. *Journal of Cellular Physiology*, *230*(10), 2476–2488. https://doi.org/10.1002/jcp.24982.

Thuringer, D., Boucher, J., Jego, G., Pernet, N., Cronier, L., Hammann, A., et al. (2016). Transfer of functional microRNAs between glioblastoma and microvascular endothelial cells through gap junctions. *Oncotarget*, *7*(45). https://www.oncotarget.com/article/12136/text/.

Torsin, L. I., Petrescu, G. E. D., Sabo, A. A., Chen, B., Brehar, F. M., Dragomir, M. P., et al. (2021). Editing and chemical modifications on non-coding RNAs in cancer: A new tale with clinical significance. *International Journal of Molecular Sciences*, *22*(2). https://doi.org/10.3390/ijms22020581.

Trigos, A. S., Pearson, R. B., Papenfuss, A. T., & Goode, D. L. (2018). How the evolution of multicellularity set the stage for cancer. *British Journal of Cancer*, *118*(2), 145–152. https://doi.org/10.1038/bjc.2017.398.

Urbanek-Trzeciak, M. O., Galka-Marciniak, P., Nawrocka, P. M., Kowal, E., Szwec, S., Giefing, M., et al. (2020). Pan-cancer analysis of somatic mutations in miRNA genes. *eBioMedicine*, *61*. https://doi.org/10.1016/j.ebiom.2020.103051.

Van Roosbroeck, K., Bayraktar, R., Calin, S., Bloehdorn, J., Dragomir, M. P., Okubo, K., et al. (2019). The involvement of microRNA in the pathogenesis of Richter syndrome. *Haematologica*, *104*(5), 1004–1015. https://doi.org/10.3324/haematol.2018.203828.

Vogelstein, B., Papadopoulos, N., Velculescu, V. E., Zhou, S., Diaz, L. A., & Kinzler, K. W. (2013). Cancer genome landscapes. *Science*, *339*(6127), 1546. https://doi.org/10.1126/science.1235122.

Wang, J., Che, F., Zhang, J., Zhang, M., Xiao, S., Liu, Y., et al. (2019). Diagnostic and prognostic potential of serum cell-free microRNA-214 in glioma. *World Neurosurgery*, *125*, e1217–e1225. https://doi.org/10.1016/j.wneu.2019.02.009.

Xie, R., Kessler, T., Grosch, J., Hai, L., Venkataramani, V., Huang, L., et al. (2020). Tumor cell network integration in glioma represents a stemness feature. *Neuro-Oncology*. https://doi.org/10.1093/neuonc/noaa275.

Yin, J., Zeng, A., Zhang, Z., Shi, Z., Yan, W., & You, Y. (2019). Exosomal transfer of miR-1238 contributes to temozolomide-resistance in glioblastoma. *eBioMedicine*, *42*, 238–251. https://doi.org/10.1016/j.ebiom.2019.03.016.

Zhang, Y., Wang, K., Zhao, Z., Sun, S., Zhang, K., Huang, R., et al. (2018). ADAR3 expression is an independent prognostic factor in lower-grade diffuse gliomas and positively correlated with the editing level of GRIA2Q607. *Cancer Cell International*, *18*(1), 196. https://doi.org/10.1186/s12935-018-0695-8.

Chapter 28

Pathophysiology roles and translational opportunities of miRNAs in sarcomas

Laura Pazzaglia and Katia Scotlandi
Experimental Oncology Laboratory, IRCCS Istituto Ortopedico Rizzoli, Bologna, Italy

Introduction

Sarcomas are a heterogeneous group of malignant neoplasms that originate in bone or soft tissues, such as muscle, cartilage, and connective tissues (Picci, Manfrini, Fabbri, Gambarotti, & Vanel, 2016). As a whole, these tumors match the formal definition of rare tumors, as they characterized by an incidence of approximately five cases out of 100,000 (Gatta et al., 2011). The umbrella of sarcomas includes approximately 70 subtypes, each defined by distinct morphology and genetics that often translate into a specific clinical behavior as well as into specific therapeutic approaches. The recent publication of the new WHO classification of soft tissue tumors and bone provides a refinement of the diagnostic criteria and represents a major step toward improved standardization of diagnosis (WHO Classification of Tumours Editorial Board, 2020).

Traditionally, sarcomas are classified into bone or soft tissue sarcomas according to their site of origin. Genetically, they are classified into sarcomas characterized by the presence of specific, reciprocal chromosomal translocations that give rise to fusion transcripts or into sarcomas with a very rearranged and unbalanced karyotype, characterized by multiple and complex genetic alterations (Taylor et al., 2011). These categories, however, do not fully reflect the genetic diversity of the different tumors. Widespread genome- and epigenome-wide profiling has just started to reveal how heterogeneous sarcomas are at the molecular/genetic level, and to appreciate the impact of epigenetic alterations that co-occurring with genomic alterations and mutations lead to crucial modification of lineage-specific signaling and differentiation pathway. Rarity, heterogeneity, genetic diversity, and complexity are all factors that have severely limited significant therapeutic advancements for these tumors. Overall, the management of malignant sarcomas is still clinically challenging and awaits the identification of regulators that can be therapeutically used as diagnostic and/or prognostic biomarkers or as novel drug targets to improve patient outcomes.

In this scenario, a growing amount of evidence has indicated a role for noncoding RNAs in the pathogenesis and progression of bone and soft tissue sarcomas. This chapter provides an overview of the miRNAs that have a role as diagnostic and/or prognostic/predictive factors or that may act as potential novel guidance for medical treatments.

Role of miRNAs in the pathogenesis and progression of bone sarcomas

Osteosarcoma (OS) and Ewing sarcoma (EWS) are the two most frequent primary bone sarcomas in children and young adults while chondrosarcoma (CHS) is typical of adults or elderly patients. The involvement of miRNA in the three most common primary tumors is summarized here.

Osteosarcoma

Osteosarcoma (OS) is a malignant tumor composed of mesenchymal cells producing osteoid and immature bone. The conventional form is high-grade OS, which represents around 90% of all OSs (Picci, 2020). A recent comprehensive analysis of high-grade OS genomic alterations through the application of genome-wide sequencing techniques has pointed out how in 77% of OS, genomic complexity follows chromothripsis (Cortés-Cirano et al., 2020), a catastrophic mutational process during which thousands of clustered chromosomal rearrangements happen in a single event (Kansara, Teng, Smyth, & Thomas, 2014). Chromothripsis is thought to be largely responsible for the extensive chromosomal rearrangements and the widespread mutational load that characterize OS. Unlike other tumors in which specific oncogene addictions and actionable mutations have been identified, the genetic heterogeneity of OS has hampered the identification of specific, common

MicroRNA in Human Malignancies. https://doi.org/10.1016/B978-0-12-822287-4.00017-7

oncogenic drivers that could be exploited for therapy. OS lacks pharmacologically tractable DNA alterations and its therapy is still entrapped with the use of conventional treatments based on local surgical control of the disease combined with high-dose chemotherapy. For details, please refer to a recent review (Scotlandi, Hattinger, Pellegrini, Gambarotti, & Serra, 2020).

The identification of patients who would respond to systemic treatments is, therefore, a first requirement to provide evidence-based, more personalized medicine for children, adolescents, and young adults affected by this disease. Numerous studies have supported the involvement of miRNAs in OS pathogenesis and progression through the regulation of diverse mechanisms They have also indicated miRNA signatures that may be invaluable in the early diagnosis of OS or to provide a more accurate prognosis. The literature is ample and the results are very different among the different studies. The lack of clinical validation for experimental evidence or the absence of clinical studies in a large series of patients prevents a clear picture. Here, we summarize the most consistent findings that connect miRNA expression to different biological functions, offering some examples in the intricate scenario of the miRNA landscape in OS.

Single nucleotide polymorphisms (SNPs) in genes that codify for proteins involved in the miRNA processing pathway may affect miRNAs and their target genes and, therefore, may be associated with cancer development and progression. Three SNPs in RISC complex genes were found to be involved in OS susceptibility (Bilbao-Aldaturriga et al., 2015) while a more recent study identified four polymorphisms in pre-miRNAs at the 14q32 miRNA cluster that were associated with OS risk in the Spanish population but not in the Slovenian population (Martin-guerrero et al., 2018). The 14q32 chromosomal band uniquely contains a very large cluster (> 100) of noncoding RNAs, including snoRNAs, microRNAs, and long noncoding RNAs; it is the largest known miRNA cluster in the genome (54 miRNAs). Genetic defects at 14q32 have been associated with severe developmental abnormalities, suggesting a very tight regulatory role in early tissue growth and differentiation. Downregulation of 14q32 miRNA expression contributes to the biological behavior of OS, and the quantification of representative transcripts from this family, such as miR-382, miR-134, and miR-544, were indeed found to act as prognostic and predictive biomarkers (Kelly et al., 2013; Sarver et al., 2013; Thayanithy et al., 2012). Of note, the 14q32 miRNA confers prognostic value independent of chemotherapy response (Kelly et al., 2013), and its evaluation may be combined with pathologic necrosis to allow for better treatment stratification. More recently, miRNA sets on 14q32 showed strong discriminatory power for recurrence and survival in three clinical independent datasets (Hill et al., 2017). This replication is quite significant when one considers the very different characteristics of the studies (different array platforms, frozen versus paraffin tissue material, somewhat different clinical characteristics). New studies will be required, in larger multiinstitutional cohorts, to select the optimal subset of miRNAs among those present at 14q32 for diagnostic/prognostic purposes. The value of the 14q32 miRNAs as a function of OS aggressiveness was also confirmed in experimental models, indicating a group of miRNAs that deserves attention. 14q32 miRNAs should probably be considered functionally interrelated, and analyzed via clustering. Indeed, a networking analysis of these miRNAs suggested possible therapeutic intervention, including the use of histone deacetylase (HDAC) inhibitors, proteasome inhibitors, or mTOR inhibitors, some of which are already in clinical or preclinical development in OS (Perry et al., 2014).

However, the biological functions of most of the differentially expressed miRNAs in OS are still unclear. Fig. 28.1 summarizes miRNAs that are involved in the regulation of the major biological processes in OS.

Zhu et al. reported that miR-29b acted as a tumor suppressor by targeting CDK6 in OS cells (Zhu et al., 2016). In addition, miR-29b inhibits resistance to methotrexate and promotes apoptosis by targeting COL3A1 and MCL1 (Xu, Li, Zhu, Xu, & Xu, 2018) while it increases cell sensitivity to doxorubicin by targeting matrix metalloproteinase 9 (MMP-9) (Luo et al., 2019). miR-29b was found to be part of an miRNA signature associated with the pathogenesis of OS as well as critical pretreatment biomarkers of metastasis and responsiveness to therapy (Jones et al., 2012).

Numerous miRNAs are reported to be involved in OS chemo-resistance. The miR-34 family, which includes miR-34a, miR-34b, and miR-34c, is known to inhibit tumorigenesis. In OS, the decreased expression of miR-34a supports cell proliferation by targeting genes such as SIRT1, BCL2, c-MET, and CDK6 (Zhao et al., 2016) and correlates with a worse prognosis in patients (Wang et al., 2015). Upregulation of miR-34a expression enhances the chemo-sensitivity of OS cells to doxorubicin, methotrexate, gemcitabine (Zhou et al., 2016), and etoposide (Novello et al., 2014), thus offering therapeutic perspectives. miR-221, miR-101, miR-22, miR-155 (Wang et al., 2017), miR-124 (Chen et al., 2015), miR-200c (Liu et al., 2017) and miR-15b (Duan et al., 2017) have been proved to participate in cisplatin and doxorubicin chemo-resistance. miR-124 and miR-15b both act through the regulation of genes involved in DNA damage responses (ATMIN, PARP1, WEE1). For details please refer to the review (Chen et al., 2015; Chen, Wang, Zheng, Hua, & Cai, 2019). Other studies showed that the overexpression of miR-215 causes methotrexate resistance by inhibiting denticle-less protein homologue (DTL) (Song et al., 2010). miR-33a expression is high in chemo-resistant OS tissues, and it induces OS chemo-resistance by targeting TWIST (Zhou et al., 2014) while the suppression of the miR-645/IFIT2 axis by lncRNA LINC00161 sensitizes OS cells to cisplatin (Wang et al., 2016). Interestingly, the upregulation of miR-214, a frequent event in OS, was also found to

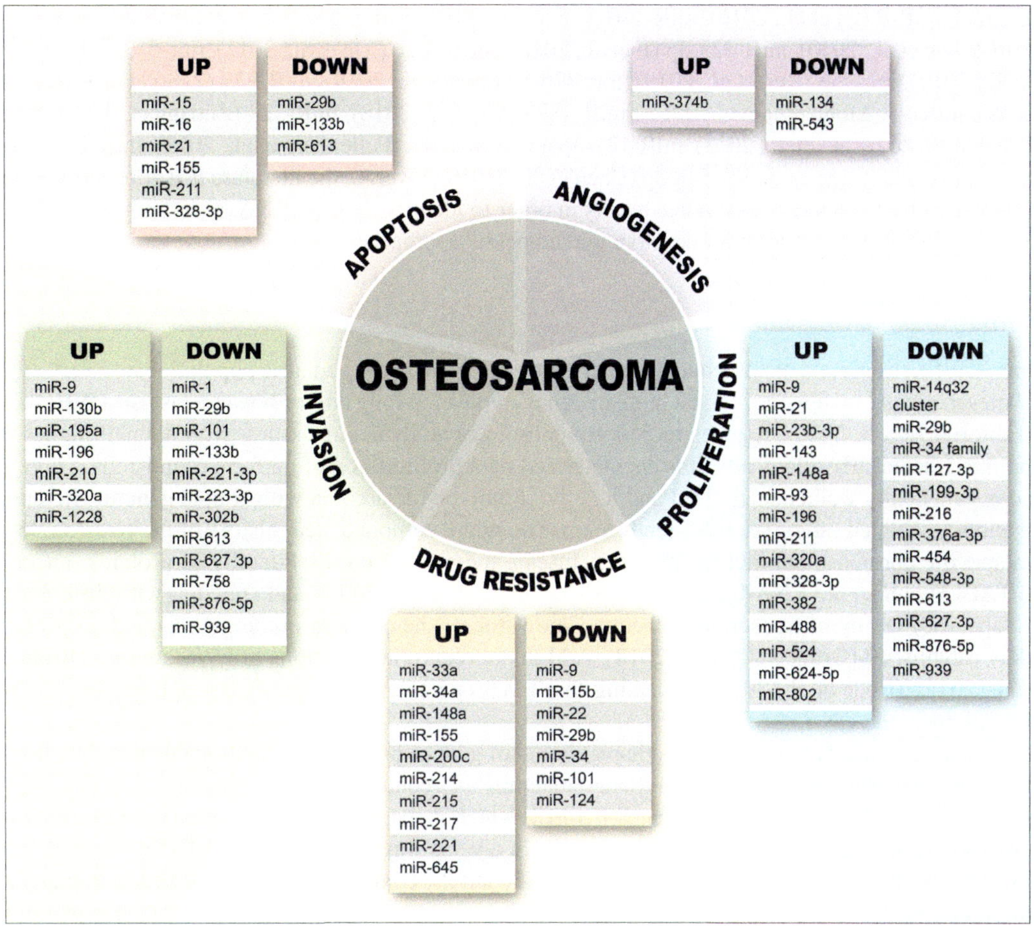

FIG. 28.1 Main miRNAs involved in many biological functions in osteosarcoma. *DOWN*, downexpressed miRNA; *UP*, overexpressed miRNA.

contribute to radio-resistance by regulating the PHLDA2/Akt pathway (Li et al., 2019). Thus, the miR-214/PHLDA2/Akt axis may be a potential target for intervention.

A variety of miRNAs have been shown to participate in the regulation of cell proliferation/survival/invasion/angiogenesis. miR-199a-3p was reported to play an important role in OS cell proliferation and migration through the downregulation of mTOR, STAT3, and CD44 (Duan et al., 2011; Gao et al., 2015). Additionally, a comprehensive metaanalysis of miR-199a-3p targeted genes showed that 391 genes were considered targets of miR-199a-3p in OS, and the enrichment analysis indicated that these targets were mainly enriched in proteoglycans in cancer and in spliceosomes (Huang et al., 2019). miR199a-3p, together with miR-21, -34a, -143, -148a, -195a, -and -382 also regulates the MAPK and PI3K/Akt-signaling pathways in OS (Namløs et al., 2012).

Many other studies have reported how miRNAs play multiple roles in the biological behaviors of OS, including their involvement in mediating the communication between OS cells and the surrounding tumor microenvironment. mRNAs involved in the regulation of developmental pathways, such as Notch, Wnt, and TGFBeta, have been shown to impact cell initiation and development at different levels. For details please refer to the recent review (Lei, Junxin, Yongcan, Xiaoguang, & Binsheng, 2020). It is known that mesenchymal stem cells, immune cells, and cancer-associated fibroblasts (CAF) may favor the growth of tumors. Extracellular vesicles (EVs) from mesenchymal stem cells were found to contain miR-195, miR-124, and miR-148a and affect cell migration and the response to doxorubicin (Vallabhaneni et al., 2016). CAFs produce EVs with miR-1228 to promote the migration and invasion of OS cells by targeting the protein suppressor of cancer cell invasion (SCAI) (Wang, Wu, Gu, & Jiang, 2019). This area is still under active investigation and new data will emerge in the near future with great potential for the definition of novel biomarkers of progression.

Additional miRNAs have been sporadically described in the regulation of these crucial biological processes and are schematized in Fig. 28.1 (For details please refer (Cai et al., 2012), mir-211 (Pei et al., 2019), miR-328-3p (Shi et al., 2019), miR-133b (Gao et al., 2018), mir-613 (Zhu et al., 2018), miR-374b (Liao et al., 2016), miR-134 (Zhang et al., 2018), miR-543

(Zhang et al., 2017), miR-9 (Qi et al., 2016), miR-196 (Cao et al., 2018), miR-320a (Wang et al., 2020), miR-210 (Liu et al., 2020), mir-101 (Zhang et al., 2020), miR-223-3p (Ji et al., 2018), miR-302b (Xie et al., 2017), mir-613 (Zhu et al., 2018), mir-627-3p (He et al., 2019), mir-758 (Ren et al., 2019), mir-876-5p (Xie et al., 2019), miR-939 (Zhao et al., 2019), miR-23b-3p (Zhu et al., 2019), miR-93 (Montanini et al., 2012), miR-488 (Qiu et al., 2018), miR-524 (Zhuang et al., 2018), miR-624-5p (Luo et al., 2019), miR-802 (Cao et al., 2013), miR-127-3p and miR-376a (Fellenberg et al., 2019), miR-216 (Quambo et al., 2017), miR-454 (Niu et al., 2015), and miR-548-3p (Chen et al., 2019)). Overall, the literature that connects OS with miRNAs is very heterogenous. The challenge is to validate results in order to identify those miRNAs that deserve to be selected as the best candidates for clinical studies and/or for developing therapeutic approaches.

Ewing sarcoma

Ewing sarcoma (EWS), the second most frequent primary tumor of bone in the pediatric population, is a very aggressive, undifferentiated mesenchymal malignancy with a high tendency to develop lung and/or bone metastasis (WHO Classification of Tumours Editorial Board, 2020). It arises more frequently in bone than soft tissues, with a small higher incidence in males compared to females. Histologically, it is characterized by a proliferation of uniform round cells.

The tumor is genetically well characterized and it is the prototype of sarcoma with simple genetic alterations. As with other translocation-associated sarcomas, EWS tends to arise de novo, without a clear history of progression. Its main driver alteration is a specific chromosomal translocation that fuses a member of the FET family of proteins (encoded by FUS, EWSR1, and TAF15), which are RNA-binding proteins involved in transcription and splicing, with different members of the ETS (E26-specific) family of transcription factors. The value of the chimera in the genesis and progression of EWS has been widely established (Grünewald et al., 2018). Additional recurrent mutations in EWS are rare (Brohl et al., 2014; Crompton et al., 2014; Tirode et al., 2014), indicating that EWS occurs and progresses through epigenomic mechanisms rather than alterations in the DNA.

Considering the role of miRNA in the epigenetic regulation of gene expression, it is reasonable that the oncogenic fusion transcripts affect the expression of miRNAs in EWS, and that miRNAs contribute to the chimera-driven oncogenic reshaping of the genetic program in EWS cells. In particular, EWS-FLI, which is the most common chimera (85% of cases) (Delattre et al., 1992), was shown to affect noncoding RNA expression in diverse ways. Riggi et al. (2010) first demonstrated the existence of a regulatory feedback loop among EWS-FLI, miR-145, and the common target gene SOX2 to initiate mesenchymal stem cell reprogramming toward EWS cancer stem cells. Ban et al. (2011) performed a genome-wide analysis of microRNAs affected by RNAi-mediated silencing of EWS-FLI in EWS cell lines, and differentially expressed between primary EWS and mesenchymal progenitor cells. This approach identified a signature of 29 differentially expressed miRNAs across all comparison groups. MiR-145 was the miRNA most consistently changed, namely repressed by EWS/FLI, and underexpressed in EWS compared to normal mesenchymal stem cells, the presumed precursor of EWS (Tirode et al., 2007). RNA-145 is known to inhibit stem cell transcription factors, such as Oct4, Sox2, Klf4, and Myc (Sachdeva et al., 2009; Xu, Papagiannakopoulos, Pan, Thomson, & Kosik, 2009), and has a role in the maturation and differentiation of stem cells (Xu et al., 2009). miR-145 was also shown to directly repress EWS-FLI (Ban et al., 2011; Riggi et al., 2010), thus suggesting the existence in EWS cells of a feedback loop that may on one site increase the levels of EWS-FLI protein in the cells, and on the other site may inhibit the terminal differentiation of mesenchymal stem cells. Several other attempts were made to identify EWS-associated miRNA alterations through the modulation of its oncogenic driver EWS-FLI (De Vito et al., 2011; Franzetti et al., 2013; McKinsey et al., 2011). While some miRNAs were identified in multiple studies, many miRNA changes were specific to a given study. Differences in profiling platforms, different controls, shRNAs, and culture conditions as well as the depth of the EWS/FLI knockdown may represent critical variables and should impose caution in the interpretation of data that are not cross-validated. For a detailed list of EWS-FLI-associated miRNA alterations found in the different studies, please refer to Dylla, Moore, and Jedlicka (2013). In general, a comparison of miRNA alterations includes the fairly consistent upregulation of members of the paralogous 17~92a, 106b~25, and 106a~363 "oncomiR" clusters, and the downregulation of miR-145. In addition, EWS-FLI was found to repress the tumor-suppressor let-7 family (De Vito et al., 2011); the miR-30a-5p (Franzetti et al., 2013), which contributes to the control of the EWS hallmark CD99 (Pasello, Manara, & Scotlandi, 2018); and a group of miRs (100, 125b, 22, 221/222, 27a, and 29a) (McKinsey et al., 2011) that negatively regulates the expression of several prooncogenic components of the IGF pathway. Both direct and indirect mechanisms, including alterations in miRNA processing and maturation (De Vito et al., 2011; Morlando et al., 2012; Sohn, Park, Kang, & Wu, 2012), appear to contribute to transcriptional miRNA regulation induced by the chimera; further studies are necessary to better clarify this relationship.

Overall, these studies have revealed the importance of miRNAs in the regulation of crucial players of EWS malignancy, such as the oncogenic drivers EWS-FLI, CD99, and the IGF pathway (for details please refer to (Manara, Pasello, & Scotlandi, 2018; Mancarella & Scotlandi, 2018), and have suggested the potential value of miRNA-based therapeutics.

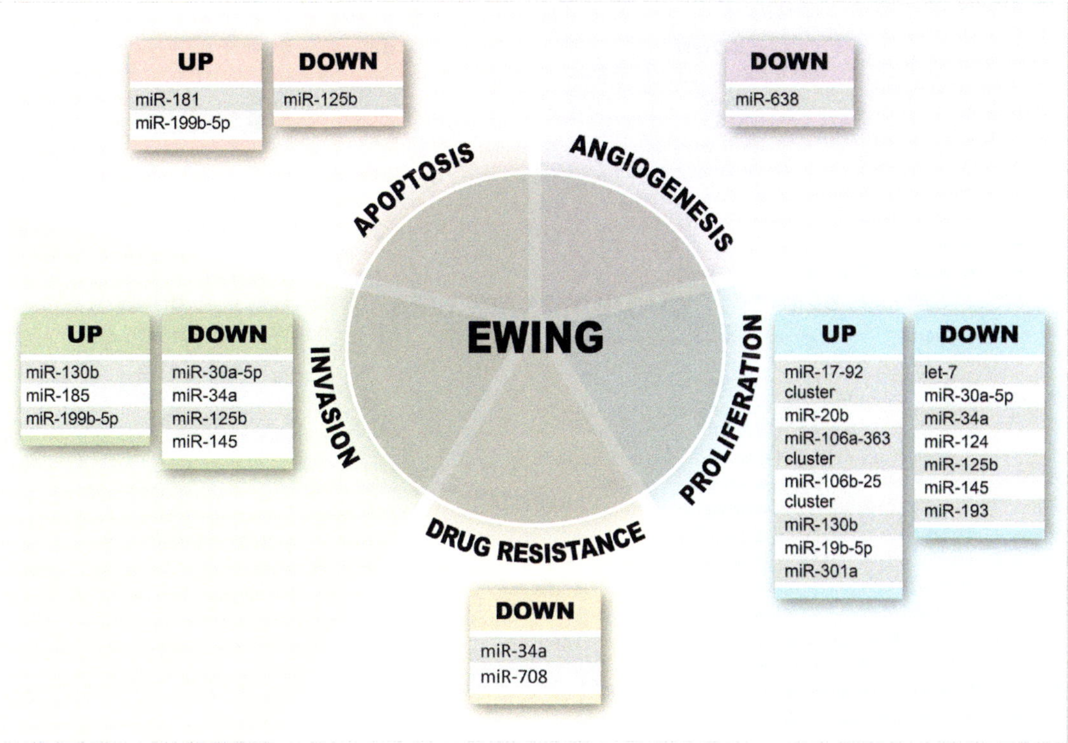

FIG. 28.2 Main miRNAs involved in many biological functions in Ewing sarcoma. *DOWN*, downexpressed miRNA; *UP*, overexpressed miRNA.

Another EWS/FLI-downregulated miRNA identified in the studies of Ban et al. (2011) was miR-708. This miRNA was shown to modulate the chemotherapy responsiveness of EWS cells in vitro (Robin et al., 2012), but does not represent a prognostic predictor in EWS clinical samples (Roberto et al., 2019). In contrast, low levels of miR-34a were identified as a mediator of EWS cell chemo-sensitivity in experimental models and a robust predictor of tumor aggressiveness and early relapse in two independent clinical series of patients (Nakatani et al., 2012). High expression of miR-34a in localized tumors was significantly related to better event-free and overall survival and the expression of miR-34a was lower in metastases than in primary tumors (Marino et al., 2014). In addition, miR-34a expression was higher in the blood of localized EWS patients than in the blood of metastatic EWS patients (Sciandra et al., 2020). Functionally, miR-34a can antagonize many different oncogenic processes by regulating genes that affect various cellular pathways (Zhang, Liao, & Tang, 2019). It is transcriptionally induced by p53 and is an important effector in the execution of p53 signaling because it represses Bcl-2, SIRT1, and YY1. In EWS, Nakatani et al. (2012) showed that miR-34a inhibits cell growth and sensitizes cells to vincristine and doxorubicin. Consistent with the established role of p53 as a regulator of miR-34a expression, miR-34a levels were found to be low in EWS cell lines with p53 inactivating mutations. Thus, miR-34a represents a validated biomarker with potential real clinical utility. Assaying miR-34a levels in tumor specimens and in the blood is feasible (Marino et al., 2014; Sciandra et al., 2020) and may allow early risk stratification of EWS patients. Restoration of miR-34a activity may be also potentially useful to decrease malignancy and increase tumor sensitivity to drugs currently used in the treatment of these patients. Other remarkable miRNAs described to be involved in the regulation of Ewing sarcoma progression are: miR-181 (Kawano et al., 2018), miR-199b-5p (Li et al., 2015), miR-638 (Zhou et al., 2018), miR-130b (Satterfield et al., 2017), miR-185 (Zhang et al., 2018), miR-20b (Kawano et al., 2017), miR-301a (Kawano et al., 2016), miR-124 (Li et al., 2017), and miR-193b (Moore et al., 2017)

Fig. 28.2 summarizes miRNAs that are involved in the regulation of the major biological processes in EWS.

Chondrosarcoma

Chondrosarcoma (CHS) is the second most common primary malignant tumor of the bone. It is a malignancy of adulthood and older age and occurs at any bone formed by endochondral ossification. This category comprises different entities, ranging from locally aggressive to high-grade malignant tumors (WHO Classification of Tumours Editorial Board, 2020). Around 90% are central and secondary peripheral CHS, either presenting as de novo in the medulla of bone or arising as a

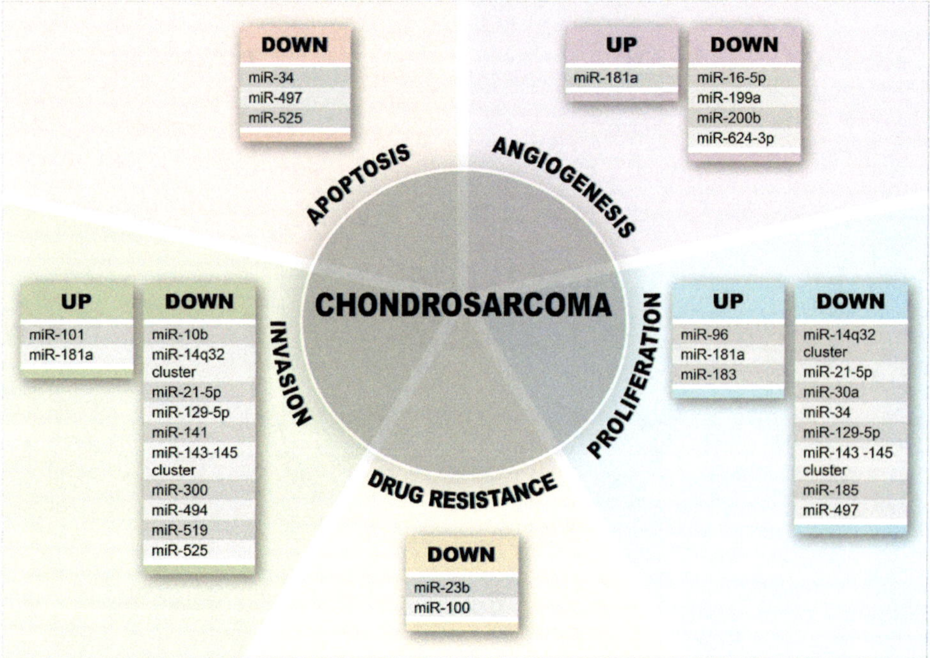

FIG. 28.3 Main miRNAs involved in many biological functions in Chondrosarcoma. *DOWN*, downexpressed miRNA; *UP*, overexpressed miRNA.

secondary tumor from preexisting benign lesions. The remaining 10% are rare variants of CHS, which include periosteal, clear cell, mesenchymal, and synovial CHS.

Differential expression and regulation of miRNA in CHS is an area of recent investigation (Nicolle et al., 2019; Parafioriti et al., 2020). Fig. 28.3 summarizes miRNAs that are involved in the regulation of the major biological processes in CHS.

The microRNA classification of 102 cartilage tumors revealed the importance of the loss of expression of the 14q32 locus in defining the level of malignancy. Inhibition of the microRNAs located in the 14q32 locus was previously found to regulate proliferation in OS (Kelly et al., 2013; Sarver et al., 2013; Thayanithy et al., 2012). Annotation analysis of the genes correlating to the expression of the 14q32 microRNA cluster revealed that 14q32 loss of expression was associated with the downregulation of pathways involved in extracellular matrix interactions and remodeling of the tumor cell environment. In addition, patients characterized by the loss of expression of the 14q32 locus were associated with the poorest prognosis.

Another cluster that inversely correlates with tumor grade was the miR-143/145 cluster levels (Urdinez et al., 2020) (Urdinez et al., 2020). These miRNAs act as oncosuppressors through depletion of the direct target fascin1, a malignancy-promoting factor associated with aggressive CHS progression. Several other studies reported the involvement of mir-21-5p, which inhibits the progression of human CHS by regulating the CCR7/STAT3/NF-κB pathway (Li, Song, et al., 2019; Li, Yang, Xu, He, & Zhang, 2021); miRNA-525, which enhances CHS malignancy by targeting F-spondin 1 and activating FAK/Src/PI3K/Akt signaling (Liu et al., 2019); and miR-129b-5p, which acts as a tumor suppressor by inhibiting the proliferation and invasion of CHS cells through regulation of the SOX4/Wnt/β-Catenin signaling pathway (Zhang et al., 2019; Zhang, Li, Song, & Wang, 2017) and by favoring cell chemo-sensitivity to doxorubicin through direct targeting of the ErbB2-regulated glucose metabolism (Tang et al., 2016).

CHS is known to be resistant to conventional chemo- and radio-therapy and very limited information is available regarding the mechanisms of metastasis formation. Overexpression of miR-23b was found to suppress the Src-Akt pathway, leading to sensitization to cisplatin (Huang et al., 2019; Huang, Chen, Yang, Tang, & Pan, 2017). Brain-derived neurotrophic factor (BDNF) is known to promote metastasis in human CHS cells, acting at least partly through the suppression of miR-624-3p (Lin et al., 2017). Other sporadic evidence indicated a role for miR-497 in apoptosis (Lu et al., 2016); miR-181a, miR-16-5p, miR-199a and 200b in angiogenesis (Chen et al., 2019; Liu et al., 2014; Liu et al., 2015; Sun et al., 2019); miR-100 in drug resistance (Zhu et al., 2014) while miR-101, miR-141 , miR-300, miR-494, miR-30a, miR-519, miR-185, miR-96 and miR-183 in proliferation and/or invasion mechanisms (Horng et al., 2014; Li et al., 2015; Palmini et al., 2017; Tsai et al., 2017; Wu et al., 2016; Yoshitaka et al., 2013).

Although all these findings need validation, any additional information may help in the development of novel strategies for the treatment of a tumor that is still highly refractory to current therapy.

Role of miRNAs in the pathogenesis and progression of soft tissue sarcomas

Soft-tissue sarcomas (STS) are a group of rare mesenchymal malignancies, genetically heterogeneous, that account for about 1% of adult malignancies. Liposarcoma, synovial sarcoma, leiomyosarcoma, and undifferentiated pleomorphic sarcoma are the most common histotypes (WHO Classification of Tumours Editorial Board, 2020).

The genetic abnormalities in STS are very heterogenous and tumors are generally only moderately sensitive to chemotherapy. In fact, despite some advances in diagnostic and treatment methods that have improved the survival of patients affected by these diseases, the clinical outcome of patients with high-risk/metastatic tumors remains strongly unfavorable (survival less than 50%) and in desperate need of better understanding of the molecular features that characterize the various STS.

Here, we emphasize the most significant evidence regarding the alteration of miRNAs in the four principal subtypes of STS (Fig. 28.4). For further details, please refer to the review (Smolle et al., 2017).

Liposarcoma (LPS) represents approximately 20% of all STS and is subclassified into four distinct subsets: well-differentiated LPS and dedifferentiated LPS, both characterized by chromosome 12q13-15 amplification in ~90% of cases; myxoid/round cell LPS; and pleomorphic LPS (WHO Classification of Tumours Editorial Board, 2020).

miR-155 is significantly overexpressed in myxoid/round cell, dedifferentiated, and pleomorphic LPS as compared with normal adipose tissue. Originally found to direct target casein kinase 1α (CK1α), which enhances β-catenin signaling and cyclin D1 expression (Zhang et al., 2012; Zhang, Li, et al., 2017), miR-155 represents a novel independent indicator of unfavorable prognosis in LPS (Kapodistrias et al., 2017). Of note, plasma levels of miR-155 are increased in patients with dedifferentiated LPS compared to healthy control patients and the evaluation of this miRNA may help the diagnosis of this LPS subtype (Boro, Bauer, Born, & Fuchs, 2016). Other upregulated miRNAs include: i. miR-26a-2, which is located in the 12q13-q22 gene region and is overexpressed in well-differentiated and dedifferentiated LPS (Lee et al., 2013) where it enhances cellular proliferation, survival, and invasion and correlates with poor patient survival; and ii. miR-135b, which is highly expressed in the myxoid/round cell LPS and promotes invasion in vitro and metastasis in vivo through the downregulation of thrombospondin 2 and the enhanced degradation of the extracellular matrix by the increased matrix metalloproteinase 2 (MMP2) activity (Nezu et al., 2016). Among the underexpressed miRNAs, a central role was shown for miR-143 in well-differentiated LPS and dedifferentiated LPS. miR-143 is strongly downregulated in these tumors compared to

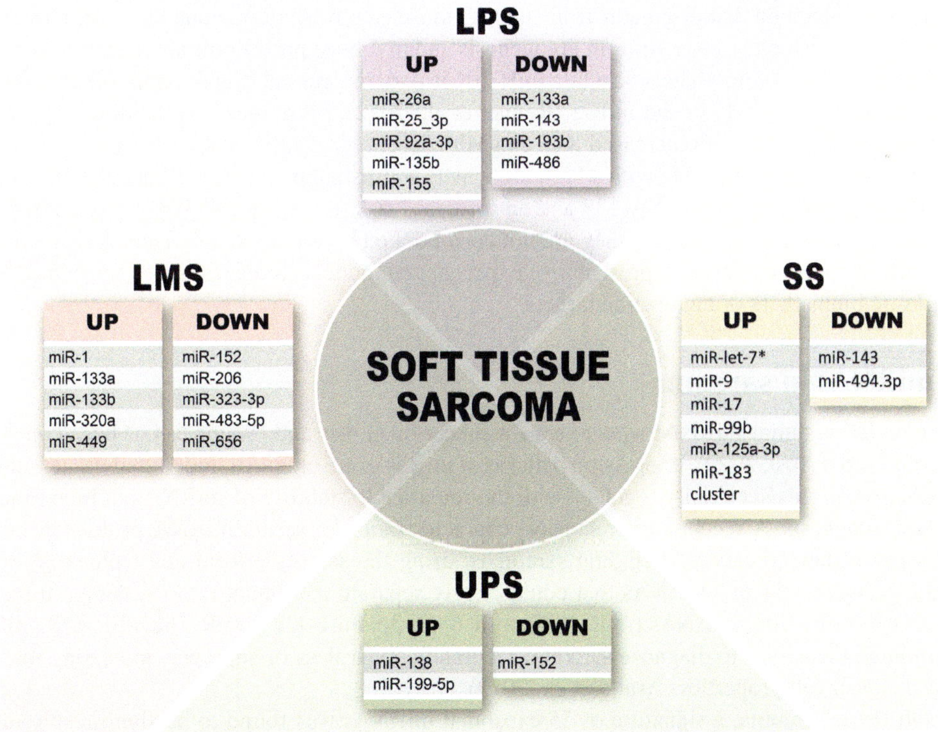

FIG. 28.4 Main miRNAs involved in the principal subtypes of soft tissue sarcoma: liposarcoma (LPS); synovial sarcoma (SS); leiomyosarcoma (LMS); and undifferentiated pleomorphic sarcoma (UPS). *DOWN*, downexpressed miRNA; *UP*, overexpressed miRNA.

normal fat (Ugras et al., 2011) and its reexpression inhibits proliferation and induces apoptosis by targeting several different genes involved in the regulation of these processes, such as BCL2, PRC1, TOP2A, and PLK1 (Ugras et al., 2011). In addition, mir-133a and miR-193b was found to be significantly downregulated in dedifferentiated LPS (Mazzu et al., 2017; Yu et al., 2018). In particular miR-193b was found to function as a tumor suppressor by targeting multiple key oncogenes, including PDGFRβ, SMAD4, and YAP1 (Mazzu et al., 2019; Mazzu, Hu, Shen, Tuschl, & Singer, 2019). In myxoid LPS, which is characterized by t(12;16) (q13;p11) translocation and the expression of TLS-CHOP chimeric oncoprotein, miR-486 was found to be repressed by the chimera, in line with its oncosuppressor role (Borjigin et al., 2012). More recently, dedifferentiated LPS cells were found to release microvesicles enriched in two miRNAs, miR-25–3p and miR-92a-3p, which led to the activation of the NF-κB pathway in recipient macrophages (Casadei & Pollock, 2020). Exosome-derived miR-25-3p and miR-92a-3p were demonstrated to stimulate LPS progression (Casadei et al., 2017) by affecting the communication between tumor cells and their microenvironment. Information in this field is rapidly expanding and other studies will help our understanding of the interplay between immune and LPS cells.

Synovial sarcoma (SS) accounts for approximately 5%–10% of STS and is genetically characterized by the chromosomal translocation (X;18) (p11;q11). The fusion of *SYT* on chromosome 18 with either *SSX1* or *SSX2* leads to a fusion gene that is known to be the major oncogenic driver of this tumor (Sandberg & Bridge, 2002).

Differential miRNA profiling in SS has identified recurrent alterations. Subramanian and colleagues found a significant downregulation of miR-143 (Subramanian et al., 2008) while a comprehensive miRNA expression analysis in SS indicates a unique miRNA expression pattern compared to other STS that includes overexpression of the miR-200 family (Renner et al., 2012) let-7e, miR-99b, and miR-125a-3p (Hisaoka et al., 2011). HMGA2 and SMARCA5 were identified as targets of let-7e and miR-99b, respectively. HMGA2 is involved in chromatin remodeling and has been implicated in various mesenchymal tumors, including LPS and leiomyosarcoma (Hisaoka et al., 2011). SMARCA5 encodes a protein involved in the SWI/SNF complex that has also been shown to be implicated in the malignant transformation of SS (Kato et al., 2002). Downregulation of SMARCA5 by miR-99b may result in deregulated functioning of the SWI/SNF complex, contributing to synovial sarcomagenesis. In addition, the miR-183 cluster, which is composed of miR-183, miR-96, and miR-182, was also found to be increased in SS (Sarver, Li, & Subramanian, 2010) while other studies identified a role for miR-9 (Xu et al., 2019), miR-17 (Minami et al., 2014a), and miR-494.3p (Pazzaglia et al., 2019).

Few data are available for leiomyosarcoma (LMS) and *undifferentiated pleomorphic sarcoma (UPS)*. Guled M identified a signature of 38 and 46 miRNAs that classified UPS and LMS samples, respectively, into separate groups. In particular, high levels of miR-199-5p characterized UPS and miR-320a LMS (Guled et al., 2014).

LMS is an aggressive sarcoma that originates from smooth muscle cells. Subramanian et al. demonstrated that miR-1, mir-133a, and mir-133b, which play major roles in myogenesis and myoblast proliferation, are significantly overexpressed in LMS in comparison to normal smooth muscle while miR-206 is underexpressed (Subramanian et al., 2008). Renner et al. confirmed the upregulation of miR-133a and miR-1 (Renner et al., 2012). More recently, the downregulation of miR-152 in LMS was found to be associated with increased KIT and MET levels (Pazzaglia et al., 2017).

UPS is among the most aggressive STS, with a high propensity for distant metastasis. Wong et al. identified an association between the higher expression of miRNA-138 and an increased risk of metastasis (Wong et al., 2015). In general, all the reported data are sporadic and, frequently, they are not corroborated by other studies. Thus, larger studies in a clinical series of patients or functional analysis in representative experimental models are necessary to select miRNAs with potential value as diagnostic biomarkers or therapeutic targets.

Circulating miRNAs in sarcoma

The relevance of circulating tumor miRNAs as markers for cancer diagnosis and prognosis has been considered in several types of cancer (Andersen & Tost, 2020). Expression patterns of miRNAs are unique to individual tissues and differ between cancer and normal tissues. This specificity together with the remarkable stability of miRNAs in blood make them useful biomarkers in cancer diagnosis. A recent comprehensive characterization of serum miRNA profiles in more than 10,000 patients with 13 types of human cancer, including sarcomas, using the same platform and technology to evaluate each sample identified a group of selected miRNAs that could clearly separate malignant tumors, benign tumors, and healthy controls. The levels of seven serum miRNAs (miR-4736, miR-6836-3p, miR-4281, miR-762, miR-658, miR-4649-5p, and miR-4665-3p) exhibited the ability to diagnose sarcoma accurately, regardless of subtypes, indicating that serum miRNA profiles may reflect cancer cell properties (Asano et al., 2019).

Considering soft tissue sarcoma, a signature of 55 exosomal miRNAs was found to be significantly modulated in the desmoplastic small round cell tumors compared with healthy controls, and they characterize disease status (Colletti et al., 2019). Fricke and colleagues identified a blood-borne miRNA signature in SS (Fricke et al., 2015) and, more recently, serum miR-92b-3p was found to discriminate patients with SS from other STS patients (Uotani et al., 2017). In LPS, circulating vesicular miR-25-3p and miR-92a-3p can reliably distinguish between LPS patients and healthy controls. Functionally, exosome-secreted miR-25-3p and miR-92a-3p were found to be involved in the communication between LPS cells and the surrounding microenvironment, stimulating the secretion of IL6 from macrophages and participating to the protumoral inflammatory process (Casadei et al., 2017). The diagnostic and prognostic potential of tumor-suppressor miRNA in the blood of patients has also been demonstrated. As an example, in rhabdomyosarcoma, reduced levels of circulating miR-26a and miR-30b/c were observed in the blood of children affected by this malignancy. miR-26a absolute plasma levels were associated with fusion status and adverse outcome (Tombolan et al., 2020).

Considering bone sarcomas, circulating levels of the tumor-suppressor miR34a inversely correlated with tumor volume in EWS. In addition, miR-34a expression was lower in the blood of metastatic EWS patients than in the blood of localized EWS patients (Sciandra et al., 2020). In another study on Chinese children affected by EWS, the expression levels of serum miR-125b were found to be decreased in EWS patients when compared with healthy controls and patients who had a poor response to chemotherapy presented a significant downregulation of miR-125b (Nie, Ren, Ma, Xi, & Han, 2015). Most of the available data are, however, on OS. Many studies have tested the relevance of circulating miRNAs in the peripheral blood of OS patients and have shown that circulating miRNAs can be specifically associated with OS and utilized in early diagnosis, prediction of chemosensitivity, or the detection of metastasis. The results are promising but do not provide evidence of general value because most of the data are sporadic and sometimes controversial. For example, Wang, Wang, Tan, Wang, and Yuan (2015) reported the downregulation of miR-152 as a biomarker of diagnosis and prognosis, Wang, Ji, Dai, Xie, and Yuan (2015) indicated the overexpression of miR-191 as a potential serum marker, and Monterde-Cruz et al. (2018) reported the diagnostic and prognostic accuracy of miRNA-215-5p and 642a-5p. Other studies suggest that this tumor is associated with increased levels of circulating miRNAs (including 14q32 miRNAs, miR-95-3p, miR-300, miR-101, miR-542-3p, miR-196a, miR-196b, and miR-491) and with decreased levels of miR-17, miR-497, miR-106a-5p, miR-16-5p, miR-20a-5p, miR-425-5p, miR-451a, miR-25-3p, and miR-139-5p. Of these, serum miR-199a-5p concentrations are significantly higher in OS patients than in controls and are significantly lower in postoperative samples than in preoperative samples (Zhou et al., 2015; Zhou et al., 2016). A genome-wide miRNA profile on a cohort of OS obtained from the Children's Oncology Group and control plasma samples identified a total of 56 upregulated miRNAs and 164 downregulated miRNAs in OS blood samples when compared to controls. Lower expressions of miR-21 and miR-221 significantly correlated with a poor outcome (Nakka et al., 2017). The comparison of patients with metastatic OS with patients without metastasis indicated how miR-27a, -195-5p, -199a-3p, and -221 are increased, while miR-34b, -95-3p, -133b, -195-5p, -206, -223-5p, -326, -375, and -491 are decreased in the plasma and serum of metastatic OS patients compared to patients without metastasis. For details, please refer to Sasaki, Osaki, and Okada (2019) and Gao, Wang, Zhang, and Zhang (2020) (Duan et al., 2011). This heterogenous scenario may be caused by the small clinical samples that were considered in the different studies, the lack of validation or cross-validation in multicenter studies, and the absence of standardized procedures for miRNA detection and data analysis. More recently, a metaanalysis that included 31 articles about 2634 OS patients and 1715 healthy controls showed that: (i) the circulating miRNAs have a high accuracy in diagnosing OS; (ii) miRNA clusters have higher diagnostic accuracy than single miRNA; and (iii) miRNAs in plasma were more reliable than those in serum (Gao et al., 2020).

In general, while studies have identified the differential expression of many circulating miRNAs in patients with sarcomas compared to healthy controls and have supported their potential value as noninvasive biomarkers to assist in the early diagnosis of sarcomas, a consensus set of prognostic miRNAs has yet to be validated. Therefore, prospective, multicenter studies harmonized for technical procedures and data analysis should be encouraged to provide high-quality evidence.

MiRNA and therapeutic approaches in sarcoma

The ability of miRNAs to control cellular processes by regulating various targets underscores their potential as a therapeutic tool for cancer treatment. OncomiRs are frequently overexpressed in human cancers and several types of inhibitors, including anti-miR oligonucleotide, locked nucleic acid, miRNA sponges, and small molecules, have been tested (Ebert, Neilson, & Sharp, 2007; Lima, Cerqueira, Figueiredo, Oliveira, & Azevedo, 2018). In contrast, the miRNA mimetic agent therapy aims to restore the expression and functions of tumor-suppressive miRNAs. Accumulating evidence shows the efficiency of miRNA mimic therapy in in vitro and in vivo models. For instance, polymeric nanoparticle-based delivery of microRNA-199a-3p was found to inhibit the proliferation and growth of OS cells (Zhang et al., 2015). Many other preclinical studies have indicated the powerful therapeutic potential of restoring the functionality of the tumor-suppressor miR-34a, strongly supporting the use of miR-34a mimics as therapeutics (Li et al., 2020; Nakatani et al., 2012; Vares et al., 2020). However, when MRX34, a special amphoteric lipid nanoparticle filled with miR-34 mimics, was tested in a clinical trial (NCT02862145), including sarcomas (Beg et al., 2017), the agent was withdrawn soon because of the occurrence of immune-related adverse events (Hong et al., 2020). It remains unclear whether the clinical effects of MRX34 are related to the specific gene-suppressing activity of miR-34a, to a nonspecific inflammatory effect of MRX34 formulation, or to some other mechanism. Effective delivery of these RNA constructs also remains an unresolved challenge and demands technical innovation. Although further clinical studies of MRX34 are not planned, these studies demonstrate a proof of concept for miRNA-based therapeutics in cancer. Other strategies include the use of agents that impact tumor cell behavior through the modulation of miRNAs. As an example, magnoflorine, an essential type of alkaloid that possesses antitumor activity in multiple cancers, was reported to inhibit the malignant phenotypes and increased cisplatin sensitivity of OS cells via modulating the miR-410-3p/HMGB1/NF-κB pathway (Wang et al., 2020). A similar effect seems to occur with quercetin, a flavonoid with an inhibition role in cell proliferation, by the miR-217/KRAS axis (Zhang, Guo, Chen, & Chen, 2015). Further studies are certainly required but RNA therapeutics may be considered a promising area for sarcomas.

Concluding remarks

miRNAs play an important role in the pathogenesis and progression of sarcoma, as documented by numerous studies. In particular, the findings suggest the possibility that circulating miRNAs could serve as novel, noninvasive biomarkers whose signatures may prove invaluable in the early diagnosis of sarcoma while offering more accurate prognosis and even prediction of recurrence. However, no consensus has emerged regarding which miRNAs are most suitable for clinical application and most of the data are inconsistent. Prospective studies using standard methods for the preparation and detection of miRNAs on multicenter clinical trials are needed to provide high-quality evidence. In addition, the presence of oncogenic fusion genes in a variety of sarcomas represents a valuable opportunity to model the miR-regulated pathways that may have a fundamental role in the formation of these cancers.

Acknowledgments

This work is supported by grants received from the Italian Ministry of Health (5 × 1000 contributions to the IRCCS, Istituto Ortopedico Rizzoli) and from AIRC (IG 2019 – ID. 22805 project – P.I. Scotlandi Katia). We thank Cristina Ghinelli for the graphic support. The authors would like to apologize to investigators whose work was not cited in this revision due to space limitations.

References

Andersen, G. B., & Tost, J. (2020). Circulating miRNAs as biomarker in cancer. *Recent Results in Cancer Research*, *215*, 277–298.

Asano, N., Matsuzaki, J., Ichikawa, M., Kawauchi, J., Takizawa, S., Aoki, Y., … Ochiya, T. (2019). A serum microRNA classifier for the diagnosis of sarcomas of various histological subtypes. *Nature Communications*, *10*(1), 1299.

Ban, J., Jug, G., Mestdagh, P., Schwentner, R., Kauer, M., Aryee, D. N., … Kovar, H. (2011). Hsa-mir-145 is the top EWS-FLI1-repressed microRNA involved in a positive feedback loop in Ewing's sarcoma. *Oncogene*, *30*(18), 2173–2180.

Beg, M. S., Brenner, A. J., Sachdev, J., Borad, M., Kang, Y. K., Stoudemire, J., … Hong, D. S. (2017). Phase I study of MRX34, a liposomal miR-34a mimic, administered twice weekly in patients with advanced solid tumors. *Investigational New Drugs*, *35*(2), 180–188.

Bilbao-Aldaturriga, N., Gutierrez-Camino, A., Martin-guerrero, I., Pombar-Gomez, M., Zalacain-Diez, M., Patino-Garcia, A., … Garcia-Orad, A. (2015). Polymorphisms in miRNA processing genes and their role in osteosarcoma risk. *Pediatric Blood & Cancer*, *62*(5), 766–769.

Borjigin, N., Ohno, S., Wu, W., Tanaka, M., Suzuki, R., Fujita, K., … Kuroda, M. (2012). TLS-CHOP represses miR-486 expression, inducing upregulation of a metastasis regulator PAI-1 in human myxoid liposarcoma. *Biochemical and Biophysical Research Communications*, *427*(2), 355–360.

Boro, A., Bauer, D., Born, W., & Fuchs, B. (2016). Plasma levels of miRNA-155 as a powerful diagnostic marker for dedifferentiated liposarcoma. *American Journal of Cancer Research*, *6*(2), 544–552.

Brohl, A. S., Solomon, D. A., Chang, W., Wang, J., Song, Y., Sindiri, S., … Khan, J. (2014). The genomic landscape of the Ewing sarcoma family of tumors reveals recurrent STAG2 mutation. *PLoS Genetics*, *10*(7), e1004475.

Cai, C. K., Zhao, G. Y., Tian, L. Y., Liu, L., Yan, K., Ma, Y. L., … Ma, B. A. (2012). miR-15a and miR-16-1 downregulate CCND1 and induce apoptosis and cell cycle arrest in osteosarcoma. *Oncology Reports*, *28*(5), 1764–1770.

Cao, Z. Q., Shen, Z., & Huang, W. Y. (2013). MicroRNA-802 promotes osteosarcoma cell proliferation by targeting p27. *Asian Pacific Journal of Cancer Prevention*, *14*(12), 7081–7084.

Cao, Y., Wu, B., Wang, D., & Bi, Z. (2018). miR-196 acts as a tumor suppressor in osteosarcoma by targeting HOXA9. *International Journal of Clinical and Experimental Pathology*, *11*(9), 4579–4584.

Casadei, L., & Pollock, R. E. (2020). Extracellular vesicle cross-talk in the liposarcoma microenvironment. *Cancer Letters*, *487*, 27–33.

Casadei, L., Calore, F., Creighton, C. J., Guescini, M., Batte, K., Iwenofu, O. H., … Pollock, R. E. (2017). Exosome-derived miR-25-3p and miR-92a-3p stimulate liposarcoma progression. *Cancer Research*, *77*(14), 3846–3856.

Chen, S. M., Chou, W. C., Hu, L. Y., Hsiung, C. N., Chu, H. W., Huang, Y. L., … Shen, C. Y. (2015). The effect of microRNA-124 overexpression on anti-tumor drug sensitivity. *PLoS ONE*, *10*(6), e0128472.

Chen, R., Wang, G., Zheng, Y., Hua, Y., & Cai, Z. (2019). Drug resistance-related microRNAs in osteosarcoma: Translating basic evidence into therapeutic strategies. *Journal of Cellular and Molecular Medicine*, *23*(4), 2280–2292.

Chen, J., Yan, C., Yu, H., Zhen, S., & Yuan, Q. (2019). miR-548d-3p inhibits osteosarcoma by downregulating *KRAS*. *Aging (Albany NY)*, *11*(14), 5058 5069.

Chen, S. S., Tang, C. H., Chie, M. J., Tsai, C. H., Fong, Y. C., Lu, Y. C., … Wang, S. W. (2019). Resistin facilitates VEGF-A-dependent angiogenesis by inhibiting miR-16-5p in human chondrosarcoma cells. *Cell Death & Disease*, *10*(1), 31.

Colletti, M., Paolini, A., Galardi, A., Di Paolo, V., Pascucci, L., Russo, I., … Di Giannatale, A. (2019). Expression profiles of exosomal miRNAs isolated from plasma of patients with desmoplastic small round cell tumor. *Epigenomics*, *11*(5), 489–500.

Cortés-Cirano, I., Lee, J. J. K., Xi, R., Jain, D., Jung, Y. L., Yang, L., … PCAWG Consortium. (2020). Comprehensive analysis of chromothripsis in 2,658 human cancers using whole-genome sequencing. *Nature Genetics*, *52*(3), 331–341.

Crompton, B. D., Stewart, C., Taylor-Weiner, A., Alexe, G., Kurek, K. C., Calicchio, M. L., … Stegmaier, K. (2014). The genomic landscape of pediatric Ewing sarcoma. *Cancer Discovery*, *4*(11), 1326–1341.

De Vito, C., Riggi, N., Suvà, M. L., Janiszewska, M., Horlbeck, J., Baumer, K., … Stamenkovic, I. (2011). Let-7a is a direct EWS-FLI-1 target implicated in Ewing's sarcoma development. *PLoS ONE*, *6*(8), e23592.

Delattre, O., Zucman, J., Plougastel, B., Desmaze, C., Melot, T., Peter, M., … et al. (1992). Gene fusion with an ETS DNA-binding domain caused by chromosome translocation in human tumours. *Nature*, *359*(6391), 162–165.

Duan, Z., Choy, E., Harmon, D., Liu, X., Susa, M., Mankin, H., & Hornicek, F. (2011). MicroRNA-199a-3p is downregulated in human osteosarcoma and regulates cell proliferation and migration. *Molecular Cancer Therapeutics*, *10*, 1337–1345.

Duan, Z., Gao, Y., Shen, J., Choy, E., Cote, G., Harmon, D., … Hornicek, F. J. (2017). miR-15b modulates multidrug resistance in human osteosarcoma in vitro and in vivo. *Molecular Oncology*, *11*(2), 151–166.

Dylla, L., Moore, C., & Jedlicka, P. (2013). MicroRNAs in Ewing sarcoma. *Frontiers in Oncology*, *3*, 65.

Ebert, M. S., Neilson, J. R., & Sharp, P. A. (2007). MicroRNA sponges: Competitive inhibitors of small RNAs in mammalian cells. *Nature Methods*, *4*(9), 721–726.

Fellenberg, J., Lehner, B., Saehr, H., Schenker, A., & Kunz, P. (2019). Tumor suppressor function of miR-127-3p and miR-376a-3p in osteosarcoma. *Cells Cancers (Basel)*, *11*(12), 2019.

Franzetti, G. A., Laud-Duval, K., Bellanger, D., Stern, M. H., Sastre-Garau, X., & Delattre, O. (2013). MiR-30a-5p connects EWS-FLI1 and CD99, two major therapeutic targets in Ewing tumor. *Oncogene*, *32*(33), 3915–3921.

Fricke, A., Ullrich, P. V., Heinz, J., Pfeifer, D., Scholber, J., Herget, G. W., … Braig, D. (2015). Identification of a blood-borne miRNA signature of synovial sarcoma. *Molecular Cancer*, *14*, 151.

Gao, Y., Feng, Y., Shen, J. K., Lin, M., Choy, E., Cote, G. M., … Duan, Z. (2015). CD44 is a direct target of miR-199a-3p and contributes to aggressive progression in osteosarcoma. *Scientific Reports*, *5*, 11365.

Gao, G., Tian, Z., Zhu, H. Y., & Ouyang, X. Y. (2018). miRNA-133b targets FGFR1 and presents multiple tumor suppressor activities in osteosarcoma. *Cancer Cell International*, *18*, 210.

Gao, S. S., Wang, Y. J., Zhang, G. X., & Zhang, W. T. (2020). Potential diagnostic value of miRNAs in peripheral blood for osteosarcoma: A meta-analysis. *Journal of Bone Oncology*, *23*, 100307.

Gatta, G., Van der Zwan, J. M., Casali, P. G., Siesling, S., Dei Tos, A. P., Kunkler, I., … RARECARE working group. (2011). Rare cancers are not so rare: the rare cancer burden in Europe. *European Journal of Cancer*, *47*, 2493–2511.

Grünewald, T. G. P., Cidre-Aranaz, F., Surdez, D., Tomazou, E. M., de Álava, E., Kovar, H., … Dirksen, U. (2018). Ewing sarcoma. *Nature Reviews. Disease Primers*, *4*(1), 5.

Guled, M., Pazzaglia, L., Borze, I., Mosakhani, N., Novello, C., Benassi, M. S., & Knuutila, S. (2014). Differentiating soft tissue leiomyosarcoma and undifferentiated pleomorphic sarcoma: A miRNA analysis. *Genes, Chromosomes & Cancer*, *53*(8), 693–702.

He, M., Shen, P., Qiu, C., & Wang, J. (2019). miR-627-3p inhibits osteosarcoma cell proliferation and metastasis by targeting PTN. *Aging (Albany NY)*, *11*(15), 5744–5756.

Hill, K. E., Kelly, A. D., Kuijjer, M. L., Barry, W., Rattani, A., Garbutt, C. C., … Spentzos, D. (2017). An imprinted non-coding genomic cluster at 14q32 defines clinically relevant molecular subtypes in osteosarcoma across multiple independent datasets. *Journal of Hematology & Oncology*, *10*(1), 107.

Hisaoka, M., Matsuyama, A., Nagao, Y., Luan, L., Kuroda, T., Akiyama, H., … Hashimoto, H. (2011). Identification of altered microRNA expression patterns in synovial sarcoma. *Genes, Chromosomes & Cancer*, *50*(3), 137–145.

Hong, D. S., Kang, Y. K., Borad, M., Sachdev, J., Ejadi, S., Lim, H. Y., … Beg, M. S. (2020). Phase 1 study of MRX34, a liposomal miR-34a mimic, in patients with advanced solid tumours. *British Journal of Cancer*, *122*(11), 1630–1637.

Horng, C. T., Shieh, P. C., Tan, T. W., Yang, W. H., & Tang, C. H. (2014). Paeonol suppresses chondrosarcoma metastasis through up-regulation of miR-141 by modulating PKCδ and c-Src signaling pathway. *International Journal of Molecular Sciences*, *15*(7), 11760–11772.

Huang, K., Chen, J., Yang, M. S., Tang, Y. J., & Pan, F. (2017). Inhibition of Src by microRNA-23b increases the cisplatin sensitivity of chondrosarcoma cell. *Cancer Biomarkers*, *18*(3), 231–239.

Huang, W. T., Liu, A. G., Cai, K. T., He, R. Q., Li, Z., Wei, Q. J., … Ma, J. (2019). Exploration and validation of downregulated microRNA-199a-3p, downstream messenger RNA targets and transcriptional regulation in osteosarcoma. *American Journal of Translational Research*, *11*(12), 7538–7554.

Ji, Q., Xu, X., Song, Q., Xu, Y., Tai, Y., Goodman, S. B., … Wang, Y. (2018). miR-223-3p inhibits human osteosarcoma metastasis and progression by directly targeting CDH6. *Molecular Therapy*, *26*(5), 1299–1312.

Jones, K. B., Salah, Z., Del Mare, S., Galasso, M., Gaudio, E., Nuovo, G. J., … Aqeilan, R. I. (2012). miRNA signatures associate with pathogenesis and progression of osteosarcoma. *Cancer Research*, *72*(7), 1865–1877.

Kansara, M., Teng, M. W., Smyth, M., & Thomas, D. M. (2014). Translational biology of osteosarcoma. *Nature Reviews. Cancer*, *14*(11), 722–735.

Kapodistrias, N., Mavridis, K., Batistatou, A., Gogou, P., Karavasilis, V., Sainis, I., … Scorilas, A. (2017). Assessing the clinical value of microRNAs in formalin-fixed paraffin-embedded liposarcoma tissues: Overexpressed miR-155 is an indicator of poor prognosis. *Oncotarget*, *8*(4), 6896–6913.

Kato, H., Tjernberg, A., Zhang, W., Krutchinsky, A. N., An, W., Takeuchi, T., … Roeder, R. G. (2002). SYT associates with human SNF/SWI complexes and the C-terminal region of its fusion partner SSX1 targets histones. *The Journal of Biological Chemistry*, *277*(7), 5498–5505.

Kawano, M., Tanaka, K., Itonaga, I., Iwasaki, T., & Tsumura, H. (2016). MicroRNA-301a promotes cell proliferation via PTEN targeting in Ewing's sarcoma cells. *International Journal of Oncology*, *48*(4), 1531–1540.

Kawano, M., Tanaka, K., Itonaga, I., Iwasaki, T., & Tsumura, H. (2017). MicroRNA-20b promotes cell proliferation via targeting of TGF-β receptor II and upregulates MYC expression in Ewing's sarcoma cells. *International Journal of Oncology*, *51*(6), 1842–1850.

Kawano, M., Tanaka, K., Itonaga, I., Iwasaki, T., & Tsumura, H. (2018). MicroRNA-181c prevents apoptosis by targeting of FAS receptor in Ewing's sarcoma cells. *Cancer Cell International*, *18*, 37.

Kelly, A. D., Haibe-Kains, B., Janeway, K. A., Hill, K. E., Howe, E., Goldsmith, J., … Spentzos, D. (2013). MicroRNA paraffin-based studies in osteosarcoma reveal reproducible independent prognostic profiles at 14q32. *Genome Medicine*, *5*(1), 2.

Lee, D. H., Amanat, S., Goff, C., Weiss, L. M., Said, J. W., Doan, N. B., … Koeffler, H. P. (2013). Overexpression of miR-26a-2 in human liposarcoma is correlated with poor patient survival. *Oncogene*, *2*(5), e47.

Lei, Y., Junxin, C., Yongcan, H., Xiaoguang, L., & Binsheng, Y. (2020). Role of microRNAs in the crosstalk between osteosarcoma cells and the tumour microenvironment. *Journal of Bone Oncology*, *25*, 100322.

Li, W., Li, Y., Guo, J., Pan, H., Zhang, Y., & Wang, X. (2015). Overexpression of miR199b5p inhibits Ewing's sarcoma cell lines by targeting CCNL1. *Molecular Medicine Reports*, *12*(3), 3359–3364.

Li, J., Wang, L., Liu, Z., Zu, C., Xing, F., Yang, P., … Wang, K. (2015). MicroRNA-494 inhibits cell proliferation and invasion of chondrosarcoma cells in vivo and in vitro by directly targeting SOX9. *Oncotarget*, *6*(28), 26216–26229.

Li, Y., Shao, G., Zhang, M., Zhu, F., Zhao, B., He, C., & Zhang, Z. (2017). miR-124 represses the mesenchymal features and suppresses metastasis in Ewing sarcoma. *Oncotarget*, *8*(6), 10274–10286.

Li, Y., Song, X., Liu, Z., Li, Q., Huang, M., Su, B., … Chen, H. (2019). Upregulation of miR-214 induced radioresistance of osteosarcoma by targeting PHLDA2 via PI3K/Akt signaling. *Frontiers in Oncology*, *9*, 29.

Li, D. F., Yuan, Y., Tu, M. J., Hu, X., Li, Y. Z., Yi, W. R., … Yu, A. X. (2020). The optimal outcome of suppressing Ewing sarcoma growth *in vivo* with biocompatible bioengineered miR-34a-5p prodrug. *Frontiers in Oncology*, *10*, 222.

Li, G., Yang, Y., Xu, S., He, M., & Zhang, Z. (2021). mir-21-5p inhibits the progression of human chondrosarcoma by regulating CCR7/STAT3/NF-κB pathway. *Connective Tissue Research*, *62*(3), 313–324.

Liao, Y. Y., Tsai, H. C., Chou, P. Y., Wang, S. W., Chen, H. T., Lin, Y. M., … Fong, Y. C. (2016). CCL3 promotes angiogenesis by dysregulation of miR-374b/ VEGF-A axis in human osteosarcoma cells. *Oncotarget*, *7*(4), 4310–4325.

Lima, J. F., Cerqueira, L., Figueiredo, C., Oliveira, C., & Azevedo, N. F. (2018). Anti-miRNA oligonucleotides: A comprehensive guide for design. *RNA Biology*, *15*(3), 338–352.

Lin, C. Y., Wang, S. W., Chen, Y. L., Chou, W. Y., Lin, T. Y., Chen, W. C., … Tang, C. H. (2017). Brain-derived neurotrophic factor promotes VEGF-C-dependent lymphangiogenesis by suppressing miR-624-3p in human chondrosarcoma cells. *Cell Death & Disease*, *8*(8), e2964.

Liu, G. T., Chen, H. T., Tsou, H. K., Tan, T. W., Fong, Y. C., Chen, P. C., … Tang, C. H. (2014). CCL5 promotes VEGF-dependent angiogenesis by down-regulating miR-200b through PI3K/Akt signaling pathway in human chondrosarcoma cells. *Oncotarget*, *5*(21), 10718–10731.

Liu, G. T., Huang, Y. L., Tzeng, H. E., Tsai, C. H., Wang, S. W., & Tang, C. H. (2015). CCL5 promotes vascular endothelial growth factor expression and induces angiogenesis by down-regulating miR-199a in human chondrosarcoma cells. *Cancer Letters*, *357*(2), 476–487.

Liu, Y., Zhu, S. T. T., Wang, X., Deng, J., Li, W. H., Zhang, P., & Liu, B. S. (2017). MiR-200c regulates tumor growth and chemosensitivity to cisplatin in osteosarcoma by targeting AKT2. *Scientific Reports*, *7*(1), 13598.

Liu, B., Song, X., Yan, Z., Yang, H., Shi, Y., & Wu, J. (2019). MicroRNA-525 enhances chondrosarcoma malignancy by targeting F-spondin 1. *Oncology Letters*, *17*(1), 781–788.

Liu, W., Jiang, D., Gong, F., Huang, Y., Luo, Y., Rong, Y., … Cai, W. (2020). miR-210-5p promotes epithelial-mesenchymal transition by inhibiting PIK3R5 thereby activating oncogenic autophagy in osteosarcoma cells. *Cell Death and Disease*, *11*(2), 93.

Lu, Y., Li, F., Xu, T., & Sun, J. (2016). miRNA-497 negatively regulates the growth and motility of chondrosarcoma cells by targeting Cdc25A. *Oncology Research*, *23*(4), 155–163.

Luo, D. J., Li, L. J., Huo, H. F., Liu, X. Q., Cui, H. W., & Jiang, D. M. (2019). MicroRNA-29b sensitizes osteosarcoma cells to doxorubicin by targeting matrix metalloproteinase 9 (MMP-9) in osteosarcoma. *European Review for Medical and Pharmacological Sciences*, *23*(4), 1434–1442.

Luo, Y., Liu, W., Tang, P., Jiang, D., Gu, C., Huang, Y., … Cai, W. (2019). miR-624-5p promoted tumorigenesis and metastasis by suppressing hippo signaling through targeting PTPRB in osteosarcoma cells. *Journal of Experimental & Clinical Cancer Research*, *38*(1), 488.

Manara, M. C., Pasello, M., & Scotlandi, K. (2018). CD99: A cell surface protein with an oncojanus role in tumors. *Genes (Basel)*, *9*(3), 159.

Mancarella, C., & Scotlandi, K. (2018). IGF system in sarcomas: A crucial pathway with many unknowns to exploit for therapy. *Journal of Molecular Endocrinology*, *61*(1), T45–T60.

Marino, M. T., Grilli, A., Baricordi, C., Manara, M. C., Ventura, S., Pinca, R. S., … Scotlandi, K. (2014). Prognostic significance of miR-34a in Ewing sarcoma is associated with cyclin D1 and ki-67 expression. *Annals of Oncology*, *25*(10), 2080–2086.

Martin-guerrero, I., Bilbao-Aldaturriga, N., Gutierrez-Camino, A., Santos-Zorrozua, B., Dolzan, V., Patino-Garcia, A., & Garcia-Orad, A. (2018). Variants in the 14q32 miRNA cluster are associated with osteosarcoma risk in the Spanish population. *Scientific Reports*, *8*(1), 15414.

Mazzu, Y. Z., Hu, Y., Soni, R. K., Mojica, K. M., Qin, L. X., Agius, P., … Singer, S. (2017). miR-193b-regulated signaling networks serve as tumor suppressors in liposarcoma and promote adipogenesis in adipose-derived stem cells. *Cancer Research*, *77*(21), 5728–5740.

Mazzu, Y. Z., Hu, Y., Shen, Y., Tuschl, T., & Singer, S. (2019). miR-193b regulates tumorigenesis in liposarcoma cells via PDGFR, TGFβ, and Wnt signaling. *Scientific Reports*, *9*(1), 3197.

Mazzu, Y. Z., Yoshikawa, Y., Nandakumar, S., Chakraborty, G., Armenia, J., Jehane, L. E., … Kantoff, P. W. (2019). Methylation-associated miR-193b silencing activates master drivers of aggressive prostate cancer. *Molecular Oncology*, *13*(9), 1944–1958.

McKinsey, E. L., Parrish, J. K., Irwin, A. E., Niemeyer, B. F., Kern, H. B., Birks, D. K., & Jedlicka, P. (2011). A novel oncogenic mechanism in Ewing sarcoma involving IGF pathway targeting by EWS/Fli1-regulated microRNAs. *Oncogene*, *30*(49), 4910–4920.

Minami, Y., Kohsaka, S., Tsuda, M., Yachi, K., Hatori, N., Tanino, M., … Tanaka, S. (2014a). SS18-SSX-regulated miR-17 promotes tumor growth of synovial sarcoma by inhibiting p21WAF1/CIP1. *Cancer Science*, *105*(9), 1152–1159.

Montanini, L., Lasagna, L., Barili, V., Jonstrup, S. P., Murgia, A., Pazzaglia, L., … Benassi, M. S. (2012). MicroRNA cloning and sequencing in osteosarcoma cell lines: differential role of miR-93. *Cellular Oncology (Dordrecht)*, *35*(1), 29–41.

Monterde-Cruz, L., Ramírez-Salazar, E. G., Rico-Martínez, G., Linares-González, L. M., Guzmán-González, R., Delgado-Cedillo, E., … Hidalgo-Bravo, A. (2018). Circulating miR-215-5p and miR-642a-5p as potential biomarker for diagnosis of osteosarcoma in Mexican population. *Human Cell*, *31*(4), 292–299.

Moore, C., Parrish, J. K., & Jedlicka, P. (2017). MiR-193b, downregulated in Ewing sarcoma, targets the ErbB4 oncogene to inhibit anchorage-independent growth. *PLoS ONE*, *12*(5), e0178028.

Morlando, M., Dini, M. S., Torrelli, G., Rosa, A., Di Carlo, V., Caffarelli, E., & Bozzoni, I. (2012). FUS stimulates microRNA biogenesis by facilitating co-transcriptional Drosha recruitment. *The EMBO Journal*, *31*(24), 4502–4510.

Nakatani, F., Ferracin, M., Manara, M. C., Ventura, S., Del Monaco, V., Ferrari, S., … Scotlandi, K. (2012). miR-34a predicts survival of Ewing's sarcoma patients and directly influences cell chemo-sensitivity and malignancy. *The Journal of Pathology*, *226*(5), 796–805.

Nakka, M., Allen-Rhoades, W., Li, Y., Kelly, A. J., Shen, J., Taylor, A. M., … Man, T. K. (2017). TARGET osteosarcoma consortium. Biomarker significance of plasma and tumor miR-21, miR-221, and miR-106a in osteosarcoma. *Oncotarget*, *8*(57), 96738–96752.

Namløs, H. M., Meza-Zepeda, L. A., Barøy, T., Østensen, I. H., Kresse, S. H., Kuijjer, M. L., … Myklebost, O. (2012). Modulation of the osteosarcoma expression phenotype by microRNAs. *PLoS ONE*, *7*(10), e48086.

Nezu, Y., Hagiwara, K., Yamamoto, Y., Fujiwara, T., Matsuo, K., Yoshida, A., … Ochiya, T. (2016). miR-135b, a key regulator of malignancy, is linked to poor prognosis in human myxoid liposarcoma. *Oncogene*, *35*(48), 6177–6188.

Nicolle, R., Ayadi, M., Gomez-Brouchet, A., Armenoult, L., Banneau, G., Elarouci, N., … de Pinieux, G. (2019). Integrated molecular characterization of chondrosarcoma reveals critical determinants of disease progression. *Nature Communications*, *10*(1), 4622.

Nie, C. L., Ren, W. H., Ma, Y., Xi, J. S., & Han, B. (2015). Circulating miR-125b as a biomarker of Ewing's sarcoma in Chinese children. *Genetics and Molecular Research*, *14*(4), 19049–19056.

Niu, G., Li, B., Sun, J., & Sun, L. (2015). miR-454 is down-regulated in osteosarcomas and suppresses cell proliferation and invasion by directly targeting c-met. *Cell Proliferation*, *48*(3), 348–355.

Novello, C., Pazzaglia, L., Conti, A., Quattrini, I., Pollino, S., Perego, P., … Benassi, M. S. (2014). p53-dependent activation of microRNA-34a in response to etoposide-induced DNA damage in osteosarcoma cell lines not impaired by dominant negative p53 expression. *PLoS ONE*, *9*(12), e114757.

Palmini, G., Marini, F., & Brandi, M. L. (2017). What is new in the miRNA world regarding osteosarcoma and chondrosarcoma? *Molecules*, *22*(3), 417.

Parafioriti, A., Cifola, I., Gissi, C., Pinatel, E., Vilardo, L., Armiraglio, E., … Berardi, A. C. (2020). Expression profiling of microRNAs and isomiRs in conventional central chondrosarcoma. *Cell Death Discovery*, *6*(1), 46.

Pasello, M., Manara, M. C., & Scotlandi, K. (2018). CD99 at the crossroads of physiology and pathology. *Journal of Cell Communication and Signaling*, *12*(1), 55–68.

Pazzaglia, L., Novello, C., Conti, A., Pollino, S., Picci, P., & Benassi, M. S. (2017). miR-152 down-regulation is associated with MET up-regulation in leiomyosarcoma and undifferentiated pleomorphic sarcoma. *Cellular Oncology (Dordrecht)*, *40*(1), 77–88.

Pazzaglia, L., Pollino, S., Vitale, M., Bientinesi, E., Benini, S., Ferrari, C., … Benassi, M. S. (2019). miR494.3p expression in synovial sarcoma: Role of CXCR4 as a potential target gene. *International Journal of Oncology*, *54*(1), 361–369.

Pei, Y., Yao, Q., Li, Y., Zhang, X., & Xie, B. (2019). microRNA-211 regulates cell proliferation, apoptosis and migration/invasion in human osteosarcoma via targeting EZRIN. *Cellular & Molecular Biology Letters*, *24*, 48.

Perry, J. A., Kiezun, A., Tonzi, P., Van Allen, E. M., Carter, S. L., Baca, S. C., … Janeway, K. A. (2014). Complementary genomic approaches highlight the PI3K/mTOR pathway as a common vulnerability in osteosarcoma. *Proceedings of the National Academy of Sciences of the United States of America*, *111*(51), E5564–E5573.

Picci, P. (2020). Osteosarcomas (OS). In P. Picci, M. Manfrini, D. M. Donati, M. Gambarotti, A. Righi, D. Vanel, & A. P. Dei Tos (Eds.), *Diagnosis of musculoskeletal tumors and tumor-like conditions* (2nd ed., pp. 185–212). Cham, Switzerland: Springer.

Picci, P., Manfrini, M., Fabbri, N., Gambarotti, M., & Vanel, D. (2016). *Atlas Muscoloskeletal Sarcomas and tumor likes lesion*. Lesions Sprigers.

Qi, X. J., Wang, J. F., Wang, G. D., Xu, Q., & Sun, H. L. (2016). Pivotal role of microRNA-9 in osteosarcoma tumorigenesis and tumor progression. *Genetics and Molecular Research*, *15*(1).

Qiu, J., Zhang, Y., Chen, H., & Guo, Z. (2018). MicroRNA-488 inhibits proliferation, invasion and EMT in osteosarcoma cell lines by targeting aquaporin 3. *International Journal of Oncology*, *53*(4), 1493–1504.

Quambo, J., Xu, X., Li, L., Goodman, S., Bi, W., Xu, M., … Wang, Y. (2017). miR-216a inhibits osteosarcoma cell proliferation, invasion and metastasis by targeting CDK14. *Cell Death and Disease*, *8*(10), e3103.

Ren, J., Yang, M., Xu, F., & Chen, J. (2019). microRNA-758 inhibits the malignant phenotype of osteosarcoma cells by directly targeting *HMGA1* and deactivating the Wnt/β-catenin pathway. *American Journal of Cancer Research*, *9*(1), 36–52.

Renner, M., Czwan, E., Hartmann, W., Penzel, R., Brors, B., Eils, R., … Mechtersheimer, G. (2012). MicroRNA profiling of primary high-grade soft tissue sarcomas. *Genes, Chromosomes & Cancer*, *51*(11), 982–996.

Riggi, N., Suvà, M. L., De Vito, C., Provero, P., Stehle, J. C., Baumer, K., … Stamenkovic, I. (2010). EWS-FLI-1 modulates miRNA145 and SOX2 expression to initiate mesenchymal stem cell reprogramming toward Ewing sarcoma cancer stem cells. *Genes & Development*, *24*, 916–932.

Roberto, G. M., Vieira, G. M., Delsin, L. E. A., Silva, M. O., Hakime, R. G., Engel, E. E., … Brassesco, M. S. (2019). MiR-708-5p is inversely associated with EWS/FLI1 Ewing sarcoma but does not represent a prognostic predictor. *Cancer Genetics*, *230*, 21–27.

Robin, T. P., Smith, A., McKinsey, E., Reaves, L., Jedlicka, P., & Ford, H. L. (2012). EWS/FLI1 regulates EYA3 in Ewing sarcoma via modulation of miRNA-708, resulting in increased cell survival and chemoresistance. *Molecular Cancer Research*, *10*(8), 1098–1108.

Sachdeva, M., Zhu, S., Wu, F., Wu, H., Walia, V., Kumar, S., … Mo, Y. Y. (2009). p53 represses c-Myc through induction of the tumor suppressor miR-145. *Proceedings of the National Academy of Sciences of the United States of America*, *106*(9), 3207–3212.

Sandberg, A., & Bridge, J. (2002). Updates on the cytogenetics and molecular genetics of bone and soft tissue tumors: Alveolar soft part sarcoma. *Cancer Genetics and Cytogenetics*, *136*(1), 1–9.

Sarver, A. L., Li, L., & Subramanian, S. (2010). MicroRNA miR-183 functions as an oncogene by targeting the transcription factor EGR1 and promoting tumor cell migration. *Cancer Research*, *70*(23), 9570–9580.

Sarver, A. L., Thayanithy, V., Scott, M. C., Cleton-Jansen, A. M., Hogendoorn, P. C., Modiano, J. F., & Subramanian, S. (2013). MicroRNAs at the human 14q32 locus have prognostic significance in osteosarcoma. *Orphanet Journal of Rare Diseases*, *8*, 7.

Sasaki, R., Osaki, M., & Okada, F. (2019). MicroRNA-based diagnosis and treatment of metastatic human osteosarcoma. *Cancers (Basel)*, *11*(4), 553.

Satterfield, L., Shuck, R., Kurenbekova, L., Allen-Rhoades, W., Edwards, D., Huang, S., … Yustein, J. T. (2017). miR-130b directly targets ARHGAP1 to drive activation of a metastatic CDC42-PAK1-AP1 positive feedback loop in Ewing sarcoma. *International Journal of Cancer*, *141*(10), 2062–2075.

Sciandra, M., De Feo, A., Parra, A., Landuzzi, L., Lollini, P. L., Manara, M. C., … Scotlandi, K. (2020). Circulating miR34a levels as a potential biomarker in the follow-up of Ewing sarcoma. *Journal of Cell Communication and Signaling*, *14*(3), 335–347.

Scotlandi, K., Hattinger, C. M., Pellegrini, E., Gambarotti, M., & Serra, M. (2020). Genomics and therapeutic vulnerabilities of primary bone tumors. *Cell*, *9*(4), 968.

Shi, J., An, G., Guan, Y., Wei, T., Peng, Z., Liang, M., & Wang, Y. (2019). miR-328-3p mediates the anti-tumor effect in osteosarcoma via directly targeting MMP-16. *Cancer Cell International*, *19*, 104.

Smolle, M. A., Leithner, A., Posch, F., Szkandera, J., Liegl-Atzwanger, B., & Pichler, M. (2017). MicroRNAs in different histologies of soft tissue sarcoma: A comprehensive review. *International Journal of Molecular Sciences*, *18*(9), 1960.

Sohn, E. J., Park, J., Kang, S. I., & Wu, Y. P. (2012). Accumulation of pre-let-7g and downregulation of mature let-7g with the depletion of EWS. *Biochemical and Biophysical Research Communications*, *426*(1), 89–93.

Song, B., Wang, Y., Titmus, M. A., Botchkina, G., Formentini, A., Kornmann, M., & Ju, J. (2010). Molecular mechanism of chemoresistance by miR-215 in osteosarcoma and colon cancer cells. *Molecular Cancer*, *9*, 96.

Subramanian, S., Lui, W. O., Lee, C. H., Espinosa, I., Nielsen, T. O., Heinrich, M. C., … van de Rijn, M. (2008). MicroRNA expression signature of human sarcomas. *Oncogene*, *27*(14), 2015–2026.

Sun, X., Chen, Y., Yu, H., Machan, J. T., Alladin, A., Ramirez, J., … Terek, R. M. (2019). Anti-miRNA oligonucleotide therapy for chondrosarcoma. *Molecular Cancer Therapeutics*, *18*(11), 2021–2029.

Tang, X. Y., Zheng, W., Ding, M., Guo, K. J., Yuan, F., Feng, H., … Gao, L. (2016). miR-125b acts as a tumor suppressor in chondrosarcoma cells by the sensitization to doxorubicin through direct targeting the ErbB2-regulated glucose metabolism. *Drug Design, Development and Therapy*, *10*, 571–583.

Taylor, B. S., Barretina, J., Maki, R. G., Antonescu, C. R., Singer, S., & Ladany, M. (2011). Advances in sarcoma genomics and new therapeutic targets. *Nature Reviews. Cancer*, *11*(8), 541–557.

Thayanithy, V., Sarver, A. L., Kartha, R. V., Li, L., Angstadt, A. Y., Breen, M., … Subramanian, S. (2012). Perturbation of 14q32 miRNAs-cMYC gene network in osteosarcoma. *Bone*, *50*(1), 171–181.

Tirode, F., Laud-Duval, K., Prieur, A., Delorme, B., Charbord, P., & Delattre, O. (2007). Mesenchymal stem cell features of Ewing tumors. *Cancer Cell*, *11*(5), 421–429.

Tirode, F., Surdez, D., Ma, X., Parker, M., Le Deley, M. C., Bahrami, A., … Delattre, O. (2014). Genomic landscape of Ewing sarcoma defines an aggressive subtype with co-association of STAG2 and TP53 mutations. *Cancer Discovery*, *4*(11), 1342–1353.

Tombolan, L., Millino, C., Pacchioni, B., Cattelan, M., Zin, A., Bonvini, P., & Bisogno, G. (2020). Circulating miR-26a as potential prognostic biomarkers in pediatric rhabdomyosarcoma. *Frontiers in Genetics*, *11*, 606274.

Tsai, C. H., Yang, D. Y., Lin, C. Y., Chen, T. M., Tang, C. H., & Huang, Y. L. (2017). Sphingosine-1-phosphate suppresses chondrosarcoma metastasis by upregulation of tissue inhibitor of metalloproteinase 3 through suppressing miR-101 expression. *Molecular Oncology*, *11*(10), 1380–1398.

Ugras, S., Brill, E., Jacobsen, A., Hafner, M., Socci, N. D., Decarolis, P. L., … Singer, S. (2011). Small RNA sequencing and functional characterization reveals microRNA-143 tumor suppressor activity in liposarcoma. *Cancer Research*, *71*(17), 5659–5669.

Uotani, K., Fujiwara, T., Yoshida, A., Iwata, S., Morita, T., Kiyono, M., … Ozaki, T. (2017). Circulating microRNA-92b-3p as a novel biomarker for monitoring of synovial sarcoma. *Scientific Reports*, *7*(1), 14634.

Urdinez, J., Boro, A., Mazumdar, A., Arlt, M. J., Muff, R., Botter, S. M., … Gvozdenovic, A. (2020). The miR-143/145 cluster, a novel diagnostic biomarker in chondrosarcoma, acts as a tumor suppressor and directly inhibits Fascin-1. *Journal of Bone and Mineral Research*, *35*(6), 1077–1091.

Vallabhaneni, K. C., Hassler, M. Y., Abraham, A., Whitt, J., Mo, Y. Y., Atfi, A., & Pochampally, R. (2016). Mesenchymal stem/stromal cells under stress increase osteosarcoma migration and apoptosis resistance via extracellular vesicle mediated communication. *PLoS ONE*, *11*(11), e0166027.

Vares, G., Ahire, V., Sunada, S., Ho, K. E., Sai, S., Chevalier, F., … Saintigny, Y. (2020). A multimodal treatment of carbon ions irradiation, miRNA-34 and mTOR inhibitor specifically control high-grade chondrosarcoma cancer stem cells. *Radiotherapy and Oncology*, *150*, 253–261.

Wang, N. G., Wang, D. C., Tan, B. Y., Wang, F., & Yuan, Z. N. (2015). Down-regulation of microRNA152 is associated with the diagnosis and prognosis of patients with osteosarcoma. *International Journal of Clinical and Experimental Pathology*, *8*(8), 9314–9319.

Wang, T., Ji, F., Dai, Z., Xie, Y., & Yuan, D. (2015). Increased expression of microRNA-191 as a potential serum biomarker for diagnosis and prognosis in human osteosarcoma. *Cancer Biomarkers*, *15*(5), 543–550.

Wang, Y., Jia, L. S., Yuan, W., Wu, Z., Wang, H. B., Xu, T., … Shi, J. G. (2015). Low miR-34a and miR-192 are associated with unfavorable prognosis in patients suffering from osteosarcoma. *American Journal of Translational Research*, *7*, 111–119.

Wang, Y., Zhang, L., Zheng, X., Zhong, W., Tian, X., Yin, B., … Zhang, W. (2016). Long non-coding RNA LINC00161 sensitises osteosarcoma cells to cisplatin-induced apoptosis by regulating the miR-645-IFIT2 axis. *Cancer Letters*, *382*(2), 137–146.

Wang, H., Tang, M., Ou, L., Hou, M., Feng, T., Huang, Y. E., … Zuo, G. (2017). Biological analysis of cancer specific microRNAs on function modeling in osteosarcoma. *Sci Rep*, *7*(1), 5382.

Wang, J. W., Wu, X. F., Gu, X. J., & Jiang, X. H. (2019). Exosomal miR-1228 from cancer-associated fibroblasts promotes cell migration and invasion of osteosarcoma by directly targeting SCAI. *Oncology Research*, *27*(9), 979–986.

Wang, Y., Yang, J., Chen, P., Song, Y., An, W., Zhang, H., … Yan, J. (2020). MicroRNA-320a inhibits invasion and metastasis in osteosarcoma by targeting cytoplasmic polyadenylation element-binding protein 1. *Cancer Medicine*, *9*(8), 2833–2845.

Wang, Y., Shang, G., Wang, W., Qiu, E., Pei, Y., & Zhang, X. (2020). Magnoflorine inhibits the malignant phenotypes and increases cisplatin sensitivity of osteosarcoma cells via regulating miR-410-3p/HMGB1/NF-κB pathway. *Life Sciences*, *256*, 117967.

WHO Classification of Tumours Editorial Board. (2020). *WHO classification of tumours of soft tissue and bone* (5th ed.). Lyon, France: IARC Press.

Wong, P., Hui, A., Su, J., Yue, S., Haibe-Kains, B., Gokgoz, N., … Liu, F. F. (2015). Prognostic microRNAs modulate the RHO adhesion pathway: A potential therapeutic target in undifferentiated pleomorphic sarcomas. *Oncotarget*, *6*(36), 39127–39139.

Wu, M. H., Huang, P. H., Hsieh, M., Tsai, C. H., Chen, H. T., & Tang, C. H. (2016). Endothelin-1 promotes epithelial-mesenchymal transition in human chondrosarcoma cells by repressing miR-300. *Oncotarget*, *7*(43), 70232–70246.

Xie, Y., Sun, W., Deng, Z., Xie, Z., Hu, C., & Cai, L. (2017). MiR-302b suppresses osteosarcoma cell migration and invasion by targeting Runx2. *Scientific Reports*, *7*(1), 13388.

Xie, W., Xiao, J., Wang, T., Zhang, D., & Li, Z. (2019). MicroRNA-876-5p inhibits cell proliferation, migration and invasion by targeting c-met in osteosarcoma. *Journal of Cellular and Molecular Medicine*, *23*(5), 3293–3301.

Xu, N., Papagiannakopoulos, T., Pan, G., Thomson, J. A., & Kosik, K. S. (2009). MicroRNA-145 regulates OCT4, SOX2, and KLF4 and represses pluripotency in human embryonic stem cells. *Cell*, *137*(4), 647–658.

Xu, W., Li, Z., Zhu, X., Xu, R., & Xu, Y. (2018). miR-29 family inhibits resistance to methotrexate and promotes cell apoptosis by targeting COL3A1 and MCL1 in osteosarcoma. *Medical Science Monitor*, *24*, 8812–8821.

Xu, X. Z., Li, X. A., Luo, Y., Liu, J. F., Wu, H. W., & Huang, G. (2019). MiR-9 promotes synovial sarcoma cell migration and invasion by directly targeting CDH1. *The International Journal of Biochemistry & Cell Biology*, *112*, 61–71.

Yoshitaka, T., Kawai, A., Miyaki, S., Numoto, K., Kikuta, K., Ozaki, T., … Asahara, H. (2013). Analysis of microRNAs expressions in chondrosarcoma. *Journal of Orthopaedic Research*, *31*(12), 1992–1998.

Yu, P. Y., Lopez, G., Braggio, D., Koller, D., Bill, K. L. J., Prudner, B. C., … Guttridge, D. C. (2018). miR-133a function in the pathogenesis of dedifferentiated liposarcoma. *Cancer Cell International*, *18*, 89.

Zhang, P., Bill, K., Liu, J., Young, E., Peng, T., Bolshakov, S., … Lev, D. (2012). MiR-155 is a liposarcoma oncogene that targets casein kinase-1α and enhances β-catenin signaling. *Cancer Research*, *72*(7), 1751–1762.

Zhang, X., Guo, Q., Chen, J., & Chen, Z. (2015). Quercetin enhances cisplatin sensitivity of human osteosarcoma cells by modulating microRNA-217-KRAS axis. *Molecules and Cells*, *38*(7), 638–642.

Zhang, L., Lyer, A. K., Yang, X., Kobayashi, E., Guo, Y., Mankin, H., … Duan, Z. (2015). Polymeric nanoparticle-based delivery of microRNA-199a-3p inhibits proliferation and growth of osteosarcoma cells. *International Journal of Nanomedicine*, *10*, 2913–2924.

Zhang, H., Guo, X., Feng, X., Wang, T., Hu, Z., Que, X., … Li, X. (2017). MiRNA-543 promotes osteosarcoma cell proliferation and glycolysis by partially suppressing PRMT9 and stabilizing HIF-1α protein. *Oncotarget*, *8*(2), 2342–2355.

Zhang, P., Li, J., Song, Y., & Wang, X. (2017). MiR-129-5p inhibits proliferation and invasion of chondrosarcoma cells by regulating SOX4/Wnt/β-catenin signaling pathway. *Cellular Physiology and Biochemistry*, *42*(1), 242–253.

Zhang, L., Lv, Z., Xu, J., Chen, C., Ge, Q., Li, P., … Sun, X. (2018). MicroRNA-134 inhibits osteosarcoma angiogenesis and proliferation by targeting the VEGFA/VEGFR1 pathway. *The FEBS Journal*, *285*(7), 1359–1371.

Zhang, S., Li, D., Jiao, G. J., Wang, H. L., & Yan, T. B. (2018). miR-185 suppresses progression of Ewing's sarcoma via inhibiting the PI3K/AKT and Wnt/β-catenin pathways. *Oncotargets and Therapy*, *11*, 7967–7977.

Zhang, L., Liao, Y., & Tang, L. (2019). MicroRNA-34 family: A potential tumor suppressor and therapeutic candidate in cancer. *Journal of Experimental & Clinical Cancer Research*, *38*(1), 53.

Zhang, K., Dong, C., Chen, M., Yang, T., Wang, X., Gao, Y., … Zhou, Y. (2020). Extracellular vesicle-mediated delivery of miR-101 inhibits lung metastasis in osteosarcoma. *Theranostics*, *10*(1), 411–425.

Zhao, Y., Tu, M. J., Wang, W. P., Qiu, J. X., Yu, A. X., & Yu, A. M. (2016). Genetically engineered pre-microRNA-34a prodrug suppresses orthotopic osteosarcoma xenograft tumor growth via the induction of apoptosis and cell cycle arrest. *Scientific Reports*, *6*, 26611.

Zhao, X., Li, J., & Yu, D. (2019). MicroRNA-939-5p directly targets IGF-1R to inhibit the aggressive phenotypes of osteosarcoma through deactivating the PI3K/Akt pathway. *International Journal of Molecular Medicine*, *44*(5), 1833–1843.

Zhou, Y., Huang, Z., Wu, S., Zang, X., Liu, M., & Shi, J. (2014). miR-33a is up-regulated in chemoresistant osteosarcoma and promotes osteosarcoma cell resistance to cisplatin by down-regulating TWIST. *Journal of Experimental & Clinical Cancer Research*, *33*(1), 12.

Zhou, G., Lu, M., Chen, J., Li, C., Zhang, J., Chen, J., … Wu, S. (2015). Identification of miR-199a-5p in serum as noninvasive biomarkers for detecting and monitoring osteosarcoma. *Tumour Biology*, *36*(11), 8845–8852.

Zhou, Y., Zhao, R. H., Tseng, K. F., Li, K. P., Lu, Z. G., Liu, Y., … Min, D. L. (2016). Sirolimus induces apoptosis and reverses multidrug resistance in human osteosarcoma cells in vitro via increasing microRNA-34b expressio. *Acta Pharmacologica Sinica*, *37*, 519–529.

Zhou, X., Chen, J., Xiao, Q., Wang, T., Yu, Y., Li, B., … Zhang, Z. (2018). MicroRNA-638 inhibits cell growth and tubule formation by suppressing VEGFA expression in human Ewing sarcoma cells. *Bioscience Reports*, *38*(1), BSR20171017.

Zhu, Z., Wang, C. P., Zhang, Y. F., & Nie, L. (2014). MicroRNA-100 resensitizes resistant chondrosarcoma cells to cisplatin through direct targeting of mTOR. *Asian Pacific Journal of Cancer Prevention*, *15*(2), 917–923.

Zhu, K., Liu, L., Zhang, J., Wang, Y., Liang, H., Fan, G., … Zhou, G. (2016). MiR-29b suppresses the proliferation and migration of osteosarcoma cells by targeting CDK6. *Protein & Cell*, *7*(6), 434–444.

Zhu, Y., Tang, L., Zhao, S., Sun, B., Cheng, L., Tang, Y., … Long, H. (2018). CXCR4-mediated osteosarcoma growth and pulmonary metastasis is suppressed by MicroRNA-613. *Cancer Science*, *109*(8), 2412–2422.

Zhu, R., Li, X., & Ma, Y. (2019). miR-23b-3p suppressing PGC1a promotes through reprogramming metabolism in osteosarcoma. *Cell Death and Disease*, *10*(6), 381.

Zhuang, M., Qiu, X., Cheng, D., Zhu, C., & Chen, L. (2018). MicroRNA-524 promotes cell proliferation by down-regulating PTEN expression in osteosarcoma. *Cancer Cell International*, *18*, 114.

Index

Note: Page numbers followed by *f* indicate figures and *t* indicate tables.

D

N

9780128222874